THE FAMILIES
AND GENERA
OF VASCULAR PLANTS

Edited by K. Kubitzki

Springer

Berlin
Heidelberg
New York
Hong Kong
London
Milan
Paris
Tokyo

The Families
and Genera
of Vascular Plants

Edited by K. Kubitzki

VI *Flowering Plants · Dicotyledons*
Celastrales, Oxalidales, Rosales, Cornales, Ericales

Volume Editor:
K. Kubitzki

With 137 Figures

 Springer

Professor Dr. Klaus Kubitzki
Institut für Allgemeine Botanik
Ohnhorststraße 18
22609 Hamburg
Germany

ISBN 3-540-06512-1 Springer-Verlag Berlin Heidelberg New York

Library of Congress Cataloging-in-Publication Data
Flowering plants, Dicotyledons : Celastrales, Oxalidales, Rosales, Cornales, Ericales / volume editor, K. Kubitzki.
    p. cm. – (Families and genera of vascular plants; 6)
    Includes bibliographical references (p.).
    ISBN 3-540-06512-1 (alk. paper)
    1. Dicotyledons – Classification.   I. Kubitzki, Klaus, 1933–   II. Series.
 QK495.A12F56 2004
 583′.01′2 – dc22   2003058977

Springer-Verlag is a part of Springer Science+Business Media

springeronline.com

© Springer-Verlag Berlin Heidelberg 2004
Printed in Germany

Production: PRO EDIT GmbH, Heidelberg
Cover design: design & production GmbH, Heidelberg
Typesetting: SNP Best-set Typesetter Ltd., Hong Kong
Printed on acid free paper   5 4 3 2 1 0 – 31/315Di

Preface

During recent years the findings of molecular systematics for the first time have provided a reliable lead for a higher order classification of angiosperms, as reflected in the classification of the Angiosperm Phylogeny Group. From Volume III of this series onward, these new concepts have been used as a guideline for treating angiosperm families in a meaningful context within individual volumes of this series. Among the five orders treated in this volume, Celastrales and Oxalidales represent novel concepts of which at least the latter makes sense also in morphological terms. The concept of Ericales followed here has largely been expanded beyond its conventional limits to include parts of former Ebenales, Lecythidales, Primulales and other orders. Also, the circumscription of many families has been revised on the basis of recent molecular studies. Among the recircumscribed families, Ericacae, Celastraceae, Cunoniaceae and Elaeocarpaceae appear now in revised and expanded circumscriptions, and the former Primulales families have been profoundly reorganised. I am most grateful to my authors who have incorporated into their treatments the most recent results of molecular and cladistic studies. At the same time I, as the editor, also feel the dilemma of prejudicing those authors who responded at an early stage of the preparation of these volumes and whose contributions had to wait for publication until late authors were able to comply. Lamentably, R.C. Carolin's treatments of Stackhousiaceae and Tremandraceae, submitted in the early 1990s, have been superseded by later developments, and I am most grateful to Roger Carolin for his understanding and generous permission to use his treatments in a changed family context.

I would like to direct the attention of the users of this series to the remarkable treatment of Ericaceae in this volume. Its author, P.F. Stevens, is introducing here an interesting kind of "hierarchical" description, which emphasises apomorphic characters and leads to great clearness and conciseness, the constantly aspired aims of this series.

It would hardly have been possible to complete this volume without the help so generously offered by many colleagues. Most substantial support has been provided by C. Bayer and P.F. Stevens who reviewed many contributions and suggested most helpful improvements. A.S. George is thanked for building up a team that contributed Styphelioideae and thus made possible completion of Ericaceae in the new circumscription; I am also grateful for his kind assistance in various other matters. Deep thanks for reading and commenting on particular contributions are due to R. Archer (Celastraceae), I.K. Ferguson (Cornaceae), J.W. Kadereit (Introduction), S.R. Manchester (Cornaceae), M. Matthews (Cunoniaceae), B. Ståhl (Symplocaceae), S. Renner (Sarraceniaceae), R. Rutishauser (Cunoniaceae) and J. Thompson (tremandraceous Elaeocarpaceae). D. Albach and L. Hufford are thanked for their comments on problems concerning Hydrostachyaceae, and M. Matthews and P.K. Endress for showing me papers prior to publication.

My thanks go also to the copyright holders of the illustrations who so generously authorised the use of the material included in this volume.

Finally, it is a pleasure to acknowledge the agreeable collaboration with the staff of Springer-Verlag and ProEdit, Heidelberg, who kindly responded to all requests I had in connection with the completion of this volume.

Hamburg, October 2003   K. KUBITZKI

Contents

List of Contributors

ANDERBERG, A.A.   Department of Phanerogamic Botany, Swedish Museum of Natural History, PO Box 50007, 104 05 Stockholm, Sweden

APPEL, O.   Bredkamp 36e, 22589 Hamburg, Germany

BARNES, R.   Department of Primary Industries, Water and Environment, GPO Box 44, Hobart, Tasmania 7001, Australia

BARTISH, I.V.   Institute of Botany, Academy of Sciences, 25243 Pruhonice, Czech Republic

BAYER, C.   Palmengarten, Siesmayerstr. 61, 60323 Frankfurt/M., Germany

BELL, T.L.   Department of Botany, University of Western Australia, Nedlands, Western Australia 6987, Australia

BRADFORD, J.C.   Environmental Science and Policy, University of California at Davis, 2132 Wickson Hall, Davis, CA 95616, USA

BRETELER, F.J.   Herbarium Vadense, Landbouwuniversiteit Wageningen, Postbus 80120, 6700 ED Wageningen, The Netherlands

BROWN, E.A.   National Herbarium of New South Wales, Royal Botanic Gardens, Mrs Macquarie's Road, Sydney, New South Wales 2000, Australia

COCUCCI, A.A.   Instituto Multidisciplinario de Biología Vegetal, C.C. 495, 5000 Córdoba, Argentina

CONRAN, J.G.   Department of Environmental Biology, University of Adelaide, SA 5005, Australia

COODE, M.J.E.   Royal Botanic Gardens Kew, Richmond, Surrey TW9 3AS, UK

CROWDEN, R.K.   Plant Science, University of Tasmania, GPO Box 252-55, Hobart, Tasmania 7001

DRESSLER, S.   Forschungsinstitut Senckenberg, Senckenberganlage 25, 60325 Frankfurt/Main, Germany

ERBAR, C.   Biodiversität und Pflanzensystematik, Heidelberger Institut für Pflanzenwissenschaften (HIP), Universität Heidelberg, Im Neuenheimer Feld 345, 69120 Heidelberg, Germany

FISCHER, E.   Institut für Biologie, Universität Koblenz-Landau, Universitätsstr. 1, 56070 Koblenz, Germany

FORTUNE-HOPKINS, H.   Royal Botanic Gardens Kew, Richmond, Surrey TW9 3AS, UK

FRITSCH, P.W.   Department of Botany, California Academy of Sciences, Golden Gate Park, San Francisco, CA 94118, USA

GEORGE, A.S.   'Four Gables', 18 Barclay Road, Kardinya, Western Australia 6163, Australia

HUFFORD, L.   School of Biological Sciences, Washington State University,
                           Pullman, WA 99164-4236, USA

JONGKIND, C.C.H.   Herbarium Vadense, Landbouwuniversiteit Wageningen,
                           Postbus 80120, 6700 ED Wageningen, The Netherlands

JORDAN, G.J.   Plant Science, University of Tasmania, GPO Box 252-55,
                           Hobart, Tasmania 7001

KALKMAN, C.   Deceased, formerly National Herbarium Netherlands,
                           Leiden branch

KUBITZKI, K.   Institut für Allgemeine Botanik, Universität Hamburg,
                           Ohnhorststr. 18, 22609 Hamburg, Germany

LADD, P.   Department of Environmental Science, Murdoch University,
                           Murdoch, Western Australia 6150, Australia

LEINS, P.   Biodiversität und Pflanzensystematik, Universität Heidelberg,
                           Im Neuenheimer Feld 345, 69120 Heidelberg, Germany

LEMMENS, R.H.M.J.   Herbarium Vadense, Landbouwuniversiteit Wageningen,
                           Postbus 80120, 6700 ED Wageningen, The Netherlands

LEMSON, K.   School of Natural Sciences, Edith Cowan University,
                           100 Joondalup Drive, Western Australia 6027, Australia

LUTEYN, J.   The New York Botanical Garden, Bronx, New York 10458, USA

McLEAN, C.B.   ICFR, 500 Yarra Boulevard, Richmond, VIC 3121, Australia

MEDAN, D.   Cátedra de Botánica, Facultad de Agronomía de la
                           Universidad de Buenos Aires, Avenida San Martín 4453,
                           C1417DSQ Buenos Aires, Argentina

MENADUE, Y.   Plant Science, University of Tasmania, GPO Box 252-55,
                           Hobart, Tasmania 7001

MORI, S.A.   The New York Botanical Garden, Bronx, New York 10458, USA

NOOTEBOOM, H.P.   National Herbarium of the Netherlands, Leiden branch,
                           Einsteinweg 2, 2333 CC Leiden, The Netherlands

OLIVER, E.G.H.   Compton Herbarium, National Botanical Institute,
                           Kirstenbosch, P. Bag x7, Claremont 7735, South Africa

PATE, J.S.   Department of Botany, University of Western Australia,
                           Nedlands, Western Australia 6987, Australia

PENNINGTON, T.D.   Royal Botanic Gardens Kew, Richmond,
                           Surrey TW9 3AS, UK

PRANCE, G.T.   The Old Vicarage, Silver Street, Lyme Regis,
                           Dorset DT7 3HS, UK

SCHIRAREND, C.   Botanischer Garten der Universität Hamburg,
                           Hesten 10, 22609 Hamburg, Germany

SCHNEIDER, J.V.   Forschungsinstitut Senckenberg, Senckenberganlage 25,
                           60325 Frankfurt/Main, Germany

SCOTT, P.J.   Memorial University of Newfoundland, St. John's,
                           Newfoundland A1 C 5S7, Canada

SIMMONS, M.P.   Department of Biology, Colorado State University
                           Fort Collins, Colorado 80523, USA

STACE, H.M.   Department of Botany, University of Western Australia,
                      Nedlands, Western Australia 6987, Australia

STÅHL, B.   Högskolan på Gotland, 621 57 Visby, Sweden

STEVENS, P.F.   Missouri Botanical Garden, PO Box 299,
                      St. Louis Missouri 64507, USA

SWENSON, U.   Department of Phanerogamic Botany, Swedish Museum of
                      Natural History, PO Box 50007, 104 05 Stockholm, Sweden

WALLNÖFER, B.   Naturhistorisches Museum Wien, Botanische Abteilung,
                      Burgring 7, Postfach 417, 1014 Wien, Austria

WEIGEND, M.   Institut für Biologie/Systematische Botanik, Freie Universität
                      Berlin, Altensteinstr. 6, 14195 Berlin, Germany

WEILLER, C.M.   Research School for Biological Sciences, Australian National
                      University, Australian Capital Territory 0200, Australia

WEITZMAN, A.L.   Department of Botany, Smithsonian Institution, Washington,
                      DC, 20560, USA

WILKEN, D.H.   Santa Barbara Botanic Garden, 1212 Mission Canyon Road,
                      Santa Barbara, CA 93105, USA

Introduction to Families Treated in This Volume

K. Kubitzki

1. Celastrales

1. Herbs with conspicuous tannin sacs in epidermis; leaves estipulate; flowers hermaphrodite; disk 0; stigmas commissural; [ovules tenuinucellate, numerous per loculus, bitegmic or unitegmic; capsules loculicidal]. 2/71, Northern Hemisphere and South America **Parnassiaceae**
– Usually woody without conspicuous tannin sacs in epidermis; stipules small and caducous or 0; flowers hermaphrodite to unisexual; plants (andro)monoecious, (gyno)dioecious, or polygamous; disk +; stigmas carinal
2
2. Leaves unifoliolate with articulated petioles and solitary stipels; flowers in racemes or spikes; ovules 2 per locule, collateral; stylodia 2 or 3, nearly free, or very short; [capsules septicidal; endocarp and exocarp separating; seed arillate; endosperm 0]. 2/2, Africa and Mesoamerica
Lepidobotryaceae
– Leaves simple; flowers in thyrsoids or inflorescences derived there from but rarely racemes or spikes; ovules 1–numerous/locule (alternating if >1); style simple or rarely branched or cleft into stylodia; [fruit loculicidally and/or septicidally capsular, schizocarpic, baccate, samaroid, or drupaceous; seed arillate or not, winged or not; endosperm + or 0]. 98/c. 1200, of worldwide distribution **Celastraceae**

In traditional classifications Celastrales comprised families of usually woody plants with simple leaves, haplostemonous flowers with a disk and apotropous ovules which, in view of this character combination, formed an utterly heterogeneous assemblage. In the classification of Engler and Gilg (1912), for instance, families as divergent as Buxaceae, Rhamnaceae, Aquifoliaceae and Balsaminaceae were dumped into their Sapindales (= Celastrales). In later classifications, most authors gave up this broad circumscription but the Celastrales of Takhtajan (1987), to give just one example, comprised 12 families of which nine, according to our present knowledge, would have to be excluded from this order. In the absence of convincing morphological evidence, only sequence-based phylogenetic studies have led to the recognition of a monophyletic order Celastrales which represents a clade of eurosids I that is sister to Oxalidales + Malpighiales, these three being sister to all remaining eurosids I (Fagales, Rosales, Zygophyllales, Curcurbitales, Fabales; Savolainen, Fay et al. 2000; Soltis et al. 2000; APG II 2003). The Celastrales clade comprises only three families, Lepidobotryaceae, Celastraceae s.l. and Parnassiaceae. The *rbcL* analysis by Savolainen, Fay et al. (2000) provided strong support for a position of Lepidobotryaceae as sister to the other two families. Parnassiaceae, comprising *Parnassia* and *Lepuropetalum*, have often been included in Saxifragaceae. There exists, however, strong morphological (see Simmons on Parnassiaceae, this volume) and molecular evidence for their exclusion from Saxifragaceae. Various analyses of plastid and nuclear genes (Savolainen, Chase et al. 2000; Soltis et al. 1997, 2000) have resolved Parnassiaceae as sister to Celastraceae. In a multigene analysis of Celastraceae (Simmons et al. 2001b), *Parnassia* and *Lepuropetalum* have been resolved as members of an early branching but weakly supported lineage of that family, in which they are sister to *Perrottetia* and *Mortonia*. The latter two genera, as well as the early-derived *Quetzalia*, are somewhat anomalous among Celastraceae in lacking an aril in favour of a sarcotesta, and partly in possessing scalariform vessel perforations. For the time being it seems therefore justified to retain family status for Parnassiaceae, as suggested by Simmons in his contribution to this volume. As a result of Simmons' (2001a, 2001b) analysis, Celastraceae are now re-circumscribed to comprise the genera *Brexia*, *Canotia*, *Plagiopteron*, *Siphonodon*, Stackhousiaceae and Hippocrateaceae, all of which at one time or another had been related to Celastraceae, and all of which have now been shown to be nested within that family.

As discussed above, Celastrales and also Celastraceae are weakly characterised morphologically, as is obvious from the four morphological synapomorphies Simmons et al. (2001b) adduce for the latter, all being subject to reversals: stamen plus staminode number equals petal number; filaments inserted at the outer margin of the disk; styles undivided; and presence of 2–4 ovules per locule. None of these characters are found in all three families, but it may be added that an indumentum is weakly developed in all of them.

2. Oxalidales

1. Hetero-(tri- and di-)-stylous; rapanone present; ellagic acid 0 [petals postgenitally fused into a tube above the free petal bases; ovules almost orthotropous] 2
   - Homostylous; rapanone 0; ellagic acid present or not 3
2. Herbaceous or woody; carpels 5(4), ovarial portions ± united; fruit capsular or baccate; ovules bitegmic/tenuinucellate. 6/880, widely distributed in temperate and tropical regions **Oxalidaceae**
   - Woody throughout; carpels 5–1, always free; fruit follicular; ovules bitegmic/crassinucellate. 12/110–200, pantropical **Connaraceae**
3. Petals 0; apocarpous 4
   - Petals +; syncarpous, if apocarpous then leaves 3–4 per node or flowers unisexual; [pollen grains exceptionally small] 5
4. Herbaceous rosulate perennial with insect-trapping pitchers; flowers always 6-merous, carpels completely plicate; stylodia with punctate stigmas; seeds exarillate. 1/1, Western Australia **Cephalotaceae**
   - Woody; leaves simple or imparipinnate, stipulate; flowers 4–6(–8)-merous; carpels ascidiate; stylodia with decurrent stigmas; seeds arillate. 1/61, Mexico to Bolivia, Greater Antilles **Brunelliaceae**
5. Leaves opposite or ternate (spiral in *Davisonia*), simple or compound; stipules often interpetiolar; filaments exceeding the petals; anthers opening longitudinally (apically in *Bauera*), dorsifixed; disk inter- or intrastaminal; rarely apocarpous; stylodia free. 27/c. 300, Meso- and South America, S Africa, Madagascar, Australia, Malesia, and islands of W Indian Ocean and S Pacific **Cunoniaceae**
   - Leaves usually alternate, simple, stipules lateral or 0; petals often fringed but all but one of the numerous New World species of *Sloanea* apetalous; anthers dehiscing with terminal pores or slits, basifixed; disk extrastaminal; not apocarpous; styles nearly always undivided. 12/c. 550, widely distributed in tropical and temperate regions, not in Africa **Elaeocarpaceae**

On morphological grounds, Oxalidaceae have often been included in Geraniales, but molecular studies have defined Oxalidales as a clade of eurosids I that receives strong statistic support in independent large-scale analyses (e.g. Savolainen, Chase et al. 2000; Savolainen, Fay et al. 2000; Soltis et al. 2000), and always groups with Malpighiales and Celastrales in various constellations. Within Oxalidales, relationships are still unclear but Connaraceae and Oxalidaceae appear closely related and may be sister to the remaining families. Morphologically, Connaraceae and Oxalidaceae are very close; apart from the tristyly and (near) apocarpy, they have similar seed coats (Corner 1976) and agree in sieve element plastids with proteinaceous inclusions (Behnke 1982). The molecular data also suggest a close relationship among Cunoniaceae/Elaeocarpaceae, Cunoniaceae/Brunoniaceae, and Brunelliaceae/Cephalotaceae, but statistic support is not satisfactory in any of these cases (Bradford and Barnes 2001). Tremandraceae are nested within Elaeocarpaceae but details of how the three tremandraceous genera are linked to that family are still unknown. Morphologically, both Elaeocarpaceae and Cunoniaceae are difficult to characterise but, interestingly, racemisation of thyrso-paniculate inflorescences seems to have taken place in both.

Striking similarities in floral structure between Cunoniaceae and Anisophyllaceae have been emphasised by Matthews and Endress (2002); these are interpreted as convergence by molecular workers (A. Schwarzbach, in litt., Dec. 2002).

3. Rosales

1. Androecium of several whorls, basically 10 + 5 + 5, but sometimes reduced to a single whorl; [woody or herbaceous; condensed tannins generally present, ellagitannins and cyanogenic glycosides often present; nodes trilacunar; stipules free or adnate to petiole, rarely 0; hypanthium usually well-developed; epicalyx (when present), sepals, petals and stamens inserted on its rim; nectariferous disk sometimes present, intrastaminal; carpels 1–many, free or variously connate; seed coat mostly mesotestal-sclerotic]. 85/c. 2000, almost cosmopolitan **Rosaceae**
   - Androecium of a single 3–6-merous whorl (in *Shepherdia* 2 whorls) 2
2. Flowers with sepals and petals; syncarpous 3
   - Flowers apetalous 4
3. Sepals not keeled, not hooded; nectary disk 0; flowers solitary. 1/2, Socotra and Somalia **Dirachmaceae**
   - Sepals often keeled; petals mostly hooded; nectary disk present; flowers in cymes or fascicles; [woody or herbaceous; calyx valvate; stamens alternisepalous]. 52/925, almost cosmopolitan **Rhamnaceae**
4. Gynoecium apparently 1-carpellate or apocarpous; stipules 0; cystoliths 0; nodes unilacunar; ellagic acid present 5
   - Gynoecium syncarpous/2-carpellate; stipules present; cystoliths often present; nodes trilacunar; ellagic acid 0 6
5. Indumentum of simple hairs; nectary disk 0; gynoecium 1(–3)-carpellate/apocarpous; ovules pendulous. 1/1, arid regions of NE Africa and adjacent Arabian Peninsula **Barbeyaceae** (see Vol. II of this series)
   - Indumentum of stellate-peltate hairs; nectary disk present; gynoecium apparently 1-carpellate (pseudomonomerous); ovules ascending. 3/30–50, temperate regions of Northern Hemisphere, extending to SE Asia and E Australia **Elaeagnaceae**
6. Flowers bisexual or unisexual; embryo straight; hypanthium present; cystoliths rare; pollen 4–6-porate; secondary leaf veins usually ending in teeth; [laticifers 0]. 7/c. 48, Northern Hemisphere and extending to Africa and South America **Ulmaceae** s.str. (see Vol. II of this series)
   - Flowers unisexual throughout; embryo curved; hypanthium 0; cystoliths common; pollen 2–3-porate; secondary leaf veins ending subterminally 7
7. Laticifers 0 (*Humulus* and *Cannabis* with secretory canals in the phloem containing a watery liquid); fruits drupaceous, solitary; plants woody. 8/130–140, temperate and tropical regions worldwide
   Celtidaceae (incl. Cannabaceae) (see under Ulmaceae and Cannabaceae, Vol. II of this series)

– Laticifers present; fruits drupes or achenes, usually densely clustered; plants woody or herbaceous; [condensed tannins ± 0]   8

8. Woody to herbaceous with laticifers throughout the plant and containing milky latex; stamens straight; ovules (sub)apically attached; stigmatic branch(es) 2 (1). 37/1100, worldwide but mainly tropical

Moraceae (see Vol. II of this series)

– Herbs, herbaceous vines, or trees; laticifers restricted to cortex and containing mucilaginous latex; stamens straight or inflexed; ovules (sub-)basally attached; stigmatic branch 1. 51/1120, widely distributed in tropical and temperate regions of the world

Urticaceae (incl. Cecropiaceae) (see Vol. II of this series)

In the past, Rosaceae had often been linked to Chrysobalanaceae but, starting with the work of Chase et al. (1993), many molecular phylogenetic analyses have contributed to shaping Rosales in the present circumscription, as reflected in the ordinal classification of the Angiosperm Phylogeny Group (1998, 2003). Rosales are part of the nitrogen-fixing clade of eurosids I, which otherwise includes Fabales, Cucurbitales and Fagales (Soltis et al. 1995). Both the nitrogen-fixing clade and its subclades are all strongly supported as monophyletic in various broad-based gene sequence analyses (e.g. Soltis et al. 2000; Savolainen, Chase et al. 2000).

Rosaceae are resolved as the first branch of Rosales. The family is well characterised by its recently discovered, very specific androecial structure; the gynoecium is largely apocarpous, the carpels often contain more than one ovule, and the seeds are (nearly) exalbuminous. The androecium of Rosaceae consists of an outer whorl of paired antepetalous stamens and two simple antesepalous and antepetalous whorls (10 + 5 + 5; Lindenhofer and Weber 1999a, 1999b, 2000). Androecia with antepetalous stamen pairs followed by simple stamen whorls are otherwise known from families such as Alismataceae and Aristolochiaceae (Leins and Erbar 1985), and in Rosaceae may be plesiomorphic. The remaining Rosales families nearly always have only a single stamen whorl. Among these families, two clades, the Urticalean and the Rhamnalean clades, may be distinguished. In both clades the carpels are uniovulate and the corolla has been lost in part of them, and the intron of the mitochondrial *nad1* gene is *trans*-spliced (in contrast to the *cis*-spliced intron in Rosaceae and most other dicots, Qiu et al. 1998).

Support for the Urticalean clade is strong both in terms of morphological and molecular evidence (Sytsma et al. 2002; Stevens 2003), and may call for a revision of family limits in comparison with the treatment of these families ten years ago in Vol. II of this series. Thus, Celtidaceae may merit family status independent from Ulmaceae; Cannabaceae may be a derived lineage of Celtidaceae; and Cecropiaceae appear nested within Urticaceae, as indicated in the conspectus.

The Rhamnalean clade is less well characterised morphologically and its monophyly appears still problematic, partly due to contradictions between morphological and molecular evidence, and partly due to poor statistic support (Thulin et al. 1998), yet the inclusion of Barbeyaceae and Dirachmaceae in this lineage seems appropriate.

4. Cornales and Ericales

Families of Cornales

1. Submerged herbs with perianthless flowers; primary root lacking; stylodia 2; endosperm cellular, in mature seed scanty or none; stipule intrapetiolar; [male flowers of 1 (or 2?) stamens; female flowers bicarpellate]. 1/22, C and South Africa, Madagascar   **Hydrostachyaceae**

– Terrestrial plants with primary root; flowers usually with a perianth (except *Davidia*); style simple or with style branches; stipules none   2

2. Ovule 1 per carpel; ovary inferior, with epigynous disk (except *Davidia*); fruits drupaceous; trichomes not tuberculate   3

– Ovules more than 1 per carpel; ovary superior to inferior; fruits capsular, rarely baccate; trichomes tuberculate with basal cell pedestals   5

3. Leaves strongly serrate; fruits 4-seeded; [disk barbate; gynoecium with axile bundle supply to ovules; inflorescence terminal, thyrso-paniculate]. 1/1, southern Africa

Curtisiaceae

– Leaves entire (*Davidia, Camptotheca* serrate/dentate); fruits 1(–3–5)-seeded   4

4. Inflorescences capitate or cone-like, compound of dichasia; ovaries of flowers of one partial inflorescence coherent or connate; anthers 2-sporangiate, dehiscing with valves; disk barbate; ovary 2-carpellate; fruit without germination valve. 1/3, South Africa   **Grubbiaceae**

– Inflorescences thyrso-paniculate; ovaries not coherent or connate; anthers 4-sporangiate, dehiscing by longitudinal slits; disk glabrous; ovary 1–2(–9)-carpellate; fruit often with germination valve. 7/110, temperate and tropical regions mostly of Northern Hemisphere   **Cornaceae**

5. Tanniniferous shrubs, vines, rarely herbs; vessel elements with scalariform, rarely simple perforation; stylodia free, or style solitary, sometimes branched. 17/220, North and South America, Eurasia, Pacific islands   **Hydrangeaceae**

– Non-tanniniferous coarse herbs, rarely shrubs, vines, treelets, ferociously armed with scabrid, glochidiate and/or stinging hairs; vessel elements usually with simple perforation; style unbranched; [ovules tenuinucellate, some crassinucellate?, some with terminal haustoria]. 20/330, mostly American, some in Africa and Polynesia

Loasaceae

Families of Ericales

1. Placenta free, central; [vessel element perforation simple; flowers haplostemonous with sympetalous corolla; stamens antepetalous; ovules bitegmic, unitegmic in *Aegiceras* and *Cyclamen*, tenuinucellate; endosperm development nuclear]   2
 - Placenta axile or parietal   6
2. Flowers with staminodes alternating with stamens   3
 - Staminodes lacking   4
3. Woody; upper and lower part of anther thecae containing meal of calcium oxalate crystals; ovules not immersed in placentae. 6/90; neotropical   **Theophrastaceae**
 - Herbaceous; anthers not containing meal of calcium oxalate; ovules immersed in placentae. 1/12, almost cosmopolitan   **Samolaceae**
4. Pedicels provided with prophylls; trees with mixed uniseriate and multiseriate rays. 1/150, tropical regions of Old World   **Maesaceae**
 - Prophylls lacking; woody or herbaceous, rays only multiseriate, or rayless   5
5. Herbs, shrubs, or trees with secretory cavities in vegetative and reproductive organs; flowers with short tube; ovules immersed in placentae (*Ardisiandra* excepted). 48/1500, almost cosmopolitan   **Myrsinaceae**
 - Herbs; secretory cavities lacking; flowers with long or short tube; ovules not immersed in placentae (*Dionysia* excepted). 14/600, predominantly north-temperate   **Primulaceae**
6. Raphide cells present; [endosperm cellular; vessel elements usually with simple perforation]   7
 - Raphide cells lacking   11
7. Ovules unitegmic; iridoids present; [shrubs or lianas with spiral, serrate leaves and often rather flat, setose hairs; polystemonous or obdiplostemonous; anthers sometimes poricidal; gynoecium with simple style or stylodia]. 3/360, America, Asia   **Actinidiaceae**
 - Ovules bitegmic (rarely unitegmic in Balsaminaceae; unknown in Pellicieraceae); iridoids unknown; [with integumentary tapetum]   8
8. Fleshy herbs; branched sclereids absent; [flowers vertically monosymmetric, inverting during growth; sepals 5(3), the functional abaxial sepal with prominent nectariferous spur; fruit an explosive capsule]. 2/850, mostly Old World, Africa, esp. Madagascar to mountainous SE Asia   **Balsaminaceae**
 - Woody; with branched sclereids (lacking in Tetrameristaceae)   9
9. Inflorescences with abaxially ascidiate, nectar-secreting bracts; lianas, epiphytes, often heterophyllous; petals lacking glandular pits; ovules many per carpel; [fruits with numerous small seeds borne on fleshy, coloured placenta]. 7/130, Central and South America   **Marcgraviaceae**
 - Inflorescence without ascidiate bracts; erect shrubs and trees; petals with adaxial glandular pits; ovule 1/carpel 10
10. Mangrove tree with fluted trunk bases; prophylls and sepals petaloid, surpassing the petals; ovary imperfectly 2-locular; fruit 1-seeded, indehiscent; seeds large, exalbuminous. 1/1, Central and N South America   **Pellicieraceae**
 - Trees and shrubs lacking sclereids; trunks not fluted; prophylls and sepals not petaloid; ovary 4–5-locular; fruit a 4–5-seeded berry; endosperm copious. 2/2–4, disjunct between N South America and SE Asia/Malesia   **Tetrameristaceae**

11. Seed coat reduced to 1 cell layer or lacking altogether; style hollow, fluted; [vessel element perforation scalariform; endosperm cellular]   12
 - Seed coat usually well-developed, several cells thick; style usually not hollow (hollow in many Styracaceae)   14
12. Corolla sympetalous, cylindrical-urceolate/choripetalous; anthers inverting late/early in floral ontogenesis, with/without 2(4) appendages; pollen in monads/tetrads; [anthers tetrasporangiate/bisporangiate, dehiscing with longitudinal slits/pores; endothecium present/reduced; ovary 5(-1 or –12)-carpellate]. 124/4100, warm-temperate and montane-tropical   **Ericaceae**
 - Corolla choripetalous, or petals fused at most up to 1/3; anthers inverting at most in bud, not appendaged; pollen in monads   13
13. Indumentum of simple and fasciculate hairs; pedicels without prophylls; anthers ventrifixed/versatile, inverting at anthesis, dehiscing by apical, pore-like slits; ovary 3-locular. 2/95, North and tropical America, Asia   **Clethraceae**
 - Glabrous; pedicels with prophylls; anthers dorsifixed, not inverting at anthesis, dehiscing by longitudinal slits; ovary 2–5-locular, each locule with 1–3 ovules and only 1 seed; [sieve element plastids Pcf (the only example within an otherwise Ss type order)]. 2/2, North, Central and N South America   **Cyrillaceae**
14. Insect-trapping herbs or shrubs; [vessel element perforation scalariform; inflorescence a botryoid]   15
 - Not insect trapping   16
15. Shrubs; leaves bearing stalked insect-trapping glands; stamens 5; anthers curved in bud and swinging upward when touched at anthesis to become erect; endothecium lacking fibrous thickenings. 1/2, Cape Province of South Africa   **Roridulaceae**
 - Perennial herbs; leaves ascidiate traps partly filled with digestive liquid; stamens 10–numerous; anthers basifixed or versatile; endothecium well-developed. 3/15, North and N South America   **Sarraceniaceae**
16. Low-growing mycotrophic herbs or shrublets, nearly rayless; [functional stamens 5, alternipetalous, inserted in floral tube; vessel elements commonly with simple perforations]. 5/13, circumboreal   **Diapensiaceae**
 - Herbs, vines, shrubs or trees; rays usually well-developed (not so in some Polemoniaceae)   17
17. Pollen 3–4-porate or 4–60-aperturate   18
 - Pollen 3(4, very rarely [Sapotaceae] 5)-colp(or)ate; woody   19
18. Trees; pollen 3-porate; flowers 4-merous throughout, sympetalous, with an 8-lobed corona inserted distally in floral tube; nectary disk absent; style simple; ovary inferior; ovules 2 per carpel. 1/c. 5, tropical South America   **Lissocarpaceae**
 - Herbs, rarely vines, shrubs; pollen strikingly sculptured, (4-)6–60-porate or colp(or)ate with deeply immersed pores or colpi; flowers usually 5-merous with 3-carpellate gynoecium, corona lacking; intrastaminal nectary disk usually present; style 3-branched; ovary superior; ovules 1–many per carpel; [vessel element perforation simple]. 18/380, America, some in Eurasia   **Polemoniaceae**
19. Bizarre spinose shrubs or small trees with woody or succulent trunks; [vessel elements with simple perforation; nodes unilacunar; prophylls present; flowers sympetalous; ovules bitegmic]. 1/11, S USA and Mexico   **Fouquieriaceae**
 - Woody, not succulent, not spiny (except Chinese *Sinojackia*)   20

20. Young axis with cortical vascular bundles; stipules often present 21
   - Young axis without cortical vascular bundles; stipules usually (Sapotaceae excepted) absent 23
21. Seeds albuminous; [vessel element perforation scalariform; petals replaced by pseudocorolla of 6–28 fused staminodes; nectary disk absent; seeds often ruminate]. 6/21, West Africa, tropical South America **Scytopetalaceae**
   - Seeds exalbuminous; [vessel element perforation simple and scalariform] 22
22. Petals free, 3–6(–18) (lacking in *Foetidia*); stamens usually connate at base and often forming a strap-like structure (sometimes nectariferous) arching over the ovary; nectary disk absent; fruit sometimes operculate; seeds arillate or not; embryos undifferentiated or with fleshy cotyledons. 17/282, pantropical **Lecythidaceae**
   - Pseudocorolla of 30–35 fused staminodes; flowers fully actinomorphic; nectary disk annular; fruit drupaceous; seeds exarillate; embryos well-differentiated. 2/10, West Africa **Napoleonaeaceae**
23. Pubescence of stellate or stellate-peltate trichomes; style usually hollow; [cork deep-seated; vessel element perforation scalariform; prophylls absent; sympetalous, usually campanulate; anthers basifixed; nectary disk absent; fruit a capsule or drupe; ovules bitegmic or unitegmic; testa vascularized]. 11/160, America, Europe, E and SE Asia, Malesia **Styracaceae**
   - Pubescence, if present, different; style not hollow 24
24. Ovules bitegmic 25
   - Ovules unitegmic 28
25. Vessel element perforations simple; bark and heartwood black; [secretory cells common; cork surficial; prophylls lacking; sympetalous; anthers basifixed; ovary 2–8-locular, superior; ovules 2 per loculus; style usually very short but divided in long style branches; fruit a berry; x = 15]. 2/500–600, pantropical **Ebenaceae**
   - Vessel element perforations scalariform; bark and heartwood not black 26
26. Inflorescence cymose; sclereids absent; anthers dehiscing by apical pores or slits, [basifixed; cork surficial; fruit a schizocarp, with persistent columella]. 2/3, SE Asia, East Africa **Sladeniaceae**
   - Inflorescence racemose, or flowers solitary; sclereids widely present; anther dehiscence by slits; [leaves serrate, elongating while still rolled up; accumulating aluminium, and condensed and hydrolysable tannins] 27
27. Anthers versatile; cork usually deep-seated; pseudopollen produced from connective; capsule with persistent column; embryo straight. 7/125–420, SE Asia, Indomalesia, America **Theaceae**
   - Anthers basifixed; cork usually surficial; pseudopollen absent; fruit a berry or drupe; embryo often curved. 12/340, worldwide, tropical and subtropical regions, few in Africa **Ternstroemiaceae**
28. Vessel element perforation simple; latex present; leaves usually entire; hairs often T-shaped, unicellular; inflorescence fasciculate, rarely a raceme; prophylls lacking; stamens antepetalous; ovules 1 per locule; testa shiny, smooth, with rough scar. 53/1100, pantropical **Sapotaceae**
   - Vessel element perforation scalariform; no latex; leaves usually serrate; plants glabrous or with multicellular hairs and stalked glands; inflorescence thyrso-paniculate; prophylls present; ovules 2 per locule; seed enclosed in hard endocarp, testa thin. 1/c. 300, tropical and subtropical America, S and E/SE Asia, Australia to Fiji **Symplocaceae**

The recognition of the orders Cornales and Ericales in the circumscription of the Angiosperm Phylogeny Group II (2003), as followed in this volume, is the result of numerous studies in molecular systematics, starting with Olmstead et al. (1992) and Chase et al. (1993), based on sequence analyses of the *rbc*L gene, and many subsequent investigations, which often employed several plastid, mitochondrial and/or nuclear genes. More recently, Xiang et al. (2002) have presented detailed molecular analyses of Cornales, and Källersjö et al. (2000) and Anderberg et al. (2002) have analysed the phylogenetic relationships of the Ericales. There is now ample molecular evidence for the monophyly both of Cornales and Ericales at the base of a larger monophyletic group, Asteridae, with Cornales and Ericales subsequently being sister to all remaining asterid taxa (Albach et al. 2001c and references therein). Since the decisive criterion for these ordinal concepts is the statistical support in parsimony and/or maximum likelihood analyses, there is little wonder that their circumscription is a far cry from that of traditional orders, and particularly the Ericales are an inclusive group embodying also the Primulales, parts of the traditional Theales and Ebenales, and fragments of other traditional orders as well. These components of the Ericales often seem morphologically to have little, if anything, in common, and a cladistic analysis based on morphological and chemical characters by Anderberg (1992) failed to discover synapomorphies for members of these two orders.

In the pre-molecular era, Cornales and Ericales had been part of what today is considered the rosid alliance and were not included in the erstwhile Sympetalae which, by and large, coincide with the "higher" asterids or euasterids. Whereas in the euasterids character expressions such as sympetalous and haplostemonous flowers, 2-carpellate ovaries, unitegmic and tenuinucellate ovules, cellular endosperm development, simple vessel perforations and the lack of tannins prevail, in the rosids choripetalous and polystemonous or (ob)diplostemonous flowers, ovaries with more than two carpels, bitegmic and crassinucellate ovules, cellular and/or nuclear endosperm development, scalariform vessel element perforations and presence of condensed and/or hydrolysable tannins are the rule. In Cornales and Ericales, a mixture of character expressions of the two major groups is to be found which highlights the intermediate position of these orders. Although traditionally Cornales and Ericales were placed distantly from one another in the system, some authors recognised similarities between them. In a broad-based comparative study, and guided mainly by the Cornus-type wood anatomy, Huber (1963) grouped many families from different traditional orders together, among which Cornaceae and Hydrangeaceae and also Styracaceae and Symplocaceae were deemed to be particularly close to one another. He also emphasised the closeness of Cornales and Ericales, which he thought to be perhaps only separable on the absence vs. presence of terminal endosperm haustoria. A few years later the significance of the monoterpene-derived iridoids was recognised and, building on the pioneering work of, amongst others, Hegnauer (1966), Bate-Smith and Swain (1966) and Kooiman (1969), Dahlgren (1975, 1983) shaped his superorder Cornanae based on the correlation between sympetaly, unitegmic and tenuinucellate ovules, cellular endosperm formation, terminal endosperm haustoria and the occurrence of iridoids. He also stated (Dahlgren et al. 1981) that "[a]n independent origin of several (or even numerous) groups of plants with this combination of characters is highly unlikely". Later molecular work (e.g. Xiang et al. 1998) led to the exclusion from Cornales of several smaller families now included in Garryales, Solanales and Apiales. Although amply showing the iridoid theme, they lack condensed tannins and ellagitannins and also polyandric flowers, but often contain caffeic (or chlorogenic) acid, and thus also morphologically/chemically fit the asterid pattern.

The most detailed molecular analyses of the **Cornales** (Soltis et al. 2000; Albach et al. 2001c; Xiang et al. 2002) recover a monophyletic group comprising several well supported clades, yet leave us uncertain as to their interrelationships. The following clades can be recognised: 1. Grubbiaceae + Curtisiaceae, 2. *Cornus* + *Alangium*, 3. mastixioids + nyssoids (2 and 3 here combined in the family Cornaceae, see "Phylogeny" under Cornaceae, this volume), 4. Hydrangeaceae + Loasaceae, and 5. *Hydrostachys*, the latter placed either at the very basis of the Cornales clade or on a long branch in different positions nested in Hydrangeaceae or more rarely Loasaceae. The resolution of basal relationships among these clades remains unknown.

From the morphological point of view, one would see Hydrangeaceae in a basal position within Cornales, at least according to their floral and fruit morphology: they have often hemiapocarpous gynoecia with free stylodia and many-seeded capsular fruits, all perhaps plesiomorphic; wood anatomically, they are somewhat more derived than Cornaceae (Huber 1963). The position of Loasaceae has been rather dubious for a long time, but their close relationship with Hydrangeaceae is now well settled. On the basis of morphological characters, Takhtajan (1959) and Hufford (1992) suggested a relationship between these families; subsequent molecular work has confirmed their sister group status. The many similarities between the basal subfamily of Hydrangeaceae, Jamesioideae, and Loasaceae (see discussion under "Affinities" in Loasaceae, this volume) are quite convincing in this respect but Loasaceae, in their floral specialisations and prevailing herbaceousness, significantly accompanied by the lack of condensed tannins, have strongly diverged from Hydrangeaceae.

In contrast, Cornaceae, Curtisiaceae and Grubbiaceae have mainly haplostemonous, often 4-merous epigynous flowers with inferior ovaries containing solitary ovules, drupaceous fruits and much-reduced seed coats. Eyde (1988) considered the transition to epigyny with the concomitant evolution of a single-seeded fruit chamber and the fruit stone as the key innovation of the *Cornus*-nyssoid-mastixioid alliance. These traits must have been introduced into this lineage before the branching-off of the *Grubbia/Curtisia* alliance, because *Curtisia* has retained a 4-seeded fruit with an axile bundle supply, whereas in Cornaceae the 1-seeded condition has gradually evolved during the Tertiary (see "Palaeobotany" in Cornaceae, this volume). Yet wood anatomically, *Cornus* and allies appear to be most archaic; among Cornaceae and Hydrangeaceae, both fossil and extant, a link as direct as between Hydrangeaceae and Loasaceae can not be recognised.

In molecular studies *Hydrostachys* has been found nearly consistently a member of Cornales, albeit in unstable positions, moving around in Hydrangeaceae, more rarely in Loasaceae, and sometimes even outside the Cornales, obviously due to long-branch attraction. In the analyses of Xiang et al. (2002), measures were taken to minimise the influence of long branches, and *Hydrostachys* switched into a position at the base of the cornalean clade.

The original ordinal concept of **Ericales** comprised little more than Ericaceae and their satellite families today included in that family. Later, Diapensiaceae, Clethraceae, Cyrillaceae, Actinidiaceae and Grubbiaceae were added, but the strong expansion of the group, as it stands today, became necessary through insights from molecular studies. These included the work of Morton et al. (1996, 1997), Källersjö et al. (2000) and Albach et

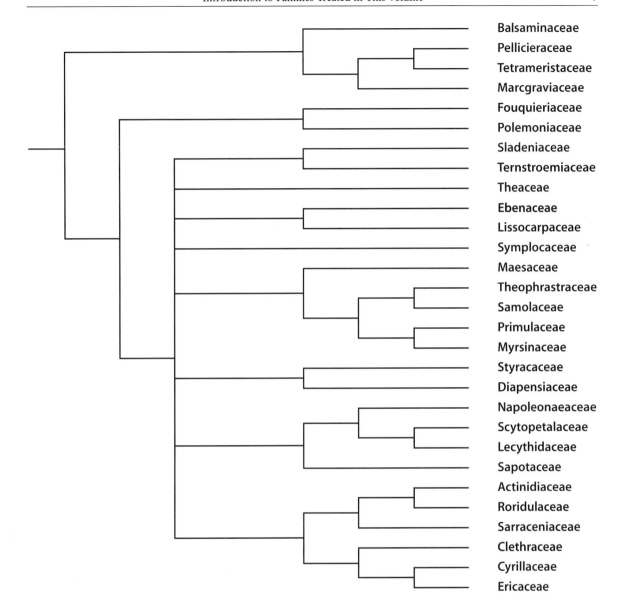

Fig. 1. Parsimony jackknife tree of Ericales families, based on an analysis of sequences from the five genes *atb*P, *ndh*F, *rbc*L, *atp*1, and *mat*K. Redrawn, with permission, from Anderberg et al. (2002); Scytopetalaceae added after Morton et al. (1997)

al. (2001c), and finally the broad-based five-gene analysis of Anderberg et al. (2002), whose lead I am following here. The tree in Fig. 1 shows two clades: Balsaminaceae, Pellicieraceae, Tetrameristaceae with Marcgraviaceae; and Fouquieriaceae with Polemoniaceae subsequently sister to the remaining families, which form a polytomy of eight clades, six of which comprise only one or two families.

The relationship among the four families of the basal clade has first been recognised by Morton et al. (1996). Although strongly supported, they are dissimilar yet have in common the possession of raphides, simple vessel perforations, bitegmic ovules, cellular endosperm formation, and micropylar endosperm haustoria (embryological data are known only for Marcgraviaceae and Balsaminaceae). Marcgraviaceae and Pellicieraceae are Neotropical, Tetrameristaceae are disjunct between northern South America and South-East Asia, and Balsaminaceae are widespread but mainly in the Old World. The families of the following clades have bitegmic or unitegmic ovules, the bitegmic having mostly nuclear endosperm, the unitegmic usually with cellular, just as would be expected (see Dahlgren 1991; Albach et al. 2001a).

The second clade, consisting of Fouquieriaceae and Polemoniaceae, is basal to all remaining taxa of the Ericales and has been confirmed by several molecular studies. Nevertheless, bootstrap support for this clade is only moderate (Anderberg et al. 2002), and Fouquieriaceae and Polemoniaceae differ in the number of integuments, the mode of endosperm development, and the presence of a chalazal endosperm haustorium and iridioids. Nash (1903) had already pointed to a certain similarity between *Gilia* and *Fouquieria*, which both inhabit the same region, and Hufford (1992) focused on the spiny long-shoots of *Fouquieria* and the polemoniaceous *Acanthogilia*; these are, however, morphologically quite distinct and the relationship between the two may not be as close. For a possible relationship between Polemoniaceae and the Primulales, which is not supported by the five-gene analysis, see Albach et al. (2001c: 181).

Sladeniaceae and Ternstroemiaceae (the ex-Theaceae-Ternstroemioideae) form a further clade, again modestly supported. As far as is known, they have bitegmic ovules and nuclear endosperm but differ in various traits. Theaceae s. str. are on a separate clade of the molecular tree; this is reflected by their morphological characters in which they differ from Ternstroemiaceae, which include versatile anthers with the stamens sometimes in five antepetalous groups, branched styles and fruit structure.

In the past Ebenaceae and Lissocarpaceae have always been grouped together; they form another clade in the tree of the five-gene-analysis, which has high bootstrap support. However, the suggestion by Berry et al. (2001) to combine both families into one is not followed here, as both differ in several important characters (see Conspectus).

Symplocaceae are well characterised by reproductive and vegetative traits and are unique in having antesepalous stamen fascicles; increase in stamen number in other Ericales families (Sapotaceae, Styracaceae, Ebenaceae, Theaceae, Actinidiaceae) is from antepetalous stamens. The reputedly primitive wood characters of *Symplocos*, such as solitary vessels with scalariform perforation plates and mainly scalariform to opposite wall pitting, diffuse parenchyma, fibre tracheids and heterogeneous rays of two size classes, are also found in Ternstroemiaceae, together with the accumulation of aluminium. van den Oever et al. (1981) therefore raise the question whether Symplocaceae could be closely related to Ternstroemiaceae, which would merit further consideration. Previously suggested phylogenetic links between Symplocaceae and Sapotaceae (Morton et al. 1996) are not supported by more recent analyses.

As expected, the five families of (ex-)Primulales are firmly grouped together, with Maesaceae sister to the other families, and Samolaceae and Theophrastaceae both basal to Primulaceae and Myrsinaceae (Anderberg and Ståhl 1995). Morphological and molecular analyses have shown Primulaceae and Myrsinaceae in their traditional circumscription to be profoundly interdigitated, and a complete reorganisation has become inevitable, with *Maesa* and *Samolus* elevated to family rank (see Maesaceae, Samolaceae, this volume).

In molecular analyses, Diapensiaceae have shifted from their traditional position close to Ericales into a position sister to Styracaceae, for which strong support is available. This alignment was first recognised by Morton et al. (1996), who discussed the morphological traits common to both families; diplostemony may be basic in them. However, profound differences between the two should not be overlooked and include gynoecium position, seed coat anatomy and, of course, the suite of traits in Diapensiaceae that are related to their pronounced mycotrophy. Also, the seeds differ strongly: exotestal-theoid in Diapensiaceae, exotestal-mesotestal, not theoid in Styracaceae. The unitegmic ovules are parallelisms, as *Styrax* is bitegmic.

The phylogenetic relationships among Lecythidaceae and their relatives have been explored on the basis of morphological and molecular data by Morton et al. (1996). They found a weak association of these families with Sapotaceae, which has been confirmed, although with moderate support, by the five-gene analysis of Anderberg et al. (2002). This is surprising, because the closest relatives of Sapotaceae had been expected to share their antepetalous stamens, but this sapotaceaous trait has no counterpart in the invariably polyandric lecythidaceous families. Sapotaceae, which are highly autapomorphic (see Conspectus), agree with Lecythidaceae and relatives in having mostly trilacunar nodes and sometimes stipulate leaves, rare character states in the Ericales.

Scytopetalaceae, differing from Lecythidaceae, i.a. in albuminous seeds, are treated here as a separate family; the combined morphological/molecular topology of Morton et al. (1996) there-

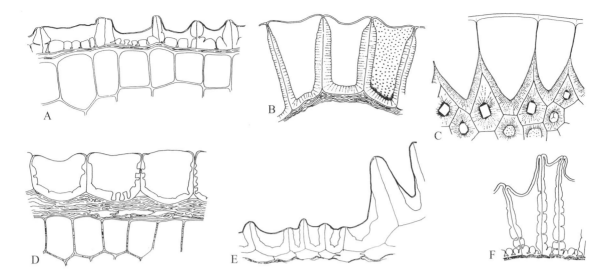

Fig. 2. Ericoid seed coats. **A** *Philadelphus coronarius* (Hydrangeaceae). **B** *Souroubea sympetala* (Marcgraviaceae). **C** *Eurya emarginata* (Ternstroemiaceae). **D** *Symplocos rigidissima* (Symplocaceae). **E** *Roridula gorgonias* (Roridulaceae). **F** *Corema album* (Ericaceae). (Huber 1991)

fore requires treating *Napoleonaea* (with *Crateranthus*) at the same rank. It is interesting – and also somewhat disturbing – that in the basal branches of their lecythidalean clade (see Figs. 4 and 5 in Morton et al. 1997) petals have been lost and replaced by a staminodial corolla substitute (Napoleonaeaceae and Scytopetalaceae), whereas in *Foetidia* within Lecythidaceae, this loss has occurred again without having been repaired for.

The largest clade within the Ericales is formed by the core Ericales, comprising Roridulaceae, Sarraceniaceae, and Actinidiaceae, and Clethraceae, Cyrillaceae, and Ericaceae. All have unitegmic ovules, cellular endosperm, and terminal endosperm haustoria; all except for Actinidiaceae have scalariform vessel perforation, and iridoids are present, Cyrillaceae and Clethraceae excepted. Actinidiaceae, Roridulaceae and Sarraceniaceae share the possession of a hypostase, which otherwise is rare in the Ericales (Anderberg et al. 2002). The absence of a fibrous endothecium in most Ericaceae, *Roridula* and *Heliamphora* is obviously due to the peculiarities of pollen release and may be autapomorphic in each group. Ericaceae, Cyrillaceae, and Clethraceae are held together by traits such as the reduced seed coat and the ovary cavity extending into the "hollow" style, and (except for Cyrillaceae) inversion of the anthers. This is a particularly interesting character, as in its ontogenetically late form (occurring in bud or at anthesis) it is found in Actinidiaceae (Fig. 4D, p. 17), Roridulaceae (Fig. 108E, F, p. 340), Clethraceae (Fig. 22B, p. 70) and basal Ericaceae (see Fig. 55D, p. 166) [but not in Cyrillaceae!], whereas its ontogenetically early expression (Hermann and Palser 2000) characterises the higher Ericaceae. The ex-

otestal "theoid" (Nandi et al. 1998) or "ericoid" seed coat (Huber 1991) is well-developed in core Ericales (Stevens 1971), but also known from other (expanded) Ericales (Fig. 2), and even Hydrangeaceae and Loasaceae (Netolitzky 1926; Huber 1991).

This brief survey of the distribution of some anatomical, embryological and chemical characters in Cornales and Ericales calls for a comparative evaluation. This has been attempted in a cladistic context by several earlier authors such as Hufford (1992) and Albach et al. (2001a) but it is hampered by the lack of resolution of the basal branches in the analyses available for these groups. Thus, at present only some general comments are possible.

In Cornaceae, the reproductive specialisation (inferior ovaries with one-seeded drupes) contrasts with the "primitive" traits of the xylem, whereas in Hydrangeaceae, the situation is inverse. This reminds us that wood anatomists currently emphasise that xylem structure is not a mere measure of "phylogenetic advancement" but reflects how a plant responds to the special requirements of its environment. Even so, it is difficult to understand why Theaceae, Ternstroemiaceae, Symplocaceae, Styracaceae, and Scytopetalaceae have scalariform vessel perforation, whereas the perforation is simple in families such as Sapotaceae and Ebenaceae.

In Ericales the presence of bitegmic ovules is usually correlated with nuclear endosperm, although in Balsaminaceae, Marcgraviaceae, and Ebenaceae bitegmicity correlates with the "advanced" (Dahlgren 1991; Albach et al. 2001a) cellular state. Unitegmic ovules and cellular endosperm are the more common combination; unitegmic ovules and nuclear endosperm are found in *Nyssa*, *Alangium* (here with crassinucellar ovules, which also are found in some *Cornus*), Polemoniaceae, and Sapotaceae. Thus, is appears most likely that the transition from bitegmic to unitegmic ovules and from nuclear to cellular endosperm has taken place independently in several lineages. Endosperm haustoria (Dahlgren 1991) appear scattered in Ericales just like iridoids. The distribution of both of them is best interpreted in terms of a loss of ancestral traits (see also discussion in Albach et al. 2001a).

References

Albach, D.C., Soltis, P.S., Soltis, D.E. 2001a. Patterns of embryological and biochemical evolution in the asterids. Syst. Bot. 26: 242–262.

Albach, D.C., Soltis, D.E., Chase, M.W., Soltis, P.S. 2001b. Phylogenetic placement of the enigmatic angiosperm *Hydrostachys*. Taxon 50: 781–805.

Albach, D.C., Soltis, D.E., Soltis, P.S. 2001c. Phylogenetic analysis of the asterids based on sequences of four genes. Ann. Missouri Bot. Gard. 88: 163–210.

Anderberg, A.A. 1992. The circumscription of the Ericales and their cladistic relationships to other families of "higher" dicotyledons. Syst. Bot. 17: 660–675.

Anderberg, A.A., Ståhl, B. 1995. Phylogenetic interrelationships in the order Primulales, with special emphasis on the family circumscription. Can. J. Bot. 73: 1699–1730.

Anderberg, A.A., Rydin, C., Källersjö, M. 2002. Phylogenetic relationships in the order Ericales s.l.: analysis from molecular data from five genes from the plastid and mitochondrial genome. Am. J. Bot. 89: 677–687.

APG (Angiosperm Phylogeny Group) 1998, 2003. See general references.

Artopoeus, A. 1903. Über den Bau und die Öffnungsweise der Antheren und die Entwickelung der Samen der Erikaceen. Flora 92: 309–345.

Bate-Smith, E.-C., Swain, T. 1966. The asperulosides and the aucubins. In: Swain, T. (ed.) Comparative phytochemistry. London: Academic Press, pp. 159–174.

Behnke, H.-D. 1982. Sieve element plastids of Connaraceae and Oxalidaceae. Bot. Jahrb. Syst. 103: 1–8.

Berry, P.E., Savolainen, V., Sytsma, K.J., Hall, J.C., Chase, M.W. 2001. *Lissocarpus* is sister to *Diospyros* (Ebenaceae). Kew Bull. 56: 725–729.

Bradford, J.C., Barnes, R.W. 2001. Phylogenetics and classification of Cunoniaceae (Oxalidales) using chloroplast DNA sequences and morphology. Syst. Bot. 26: 354–385.

Chase, M.W. et al. 1993. See general references.

Corner, E.J.H. 1976. See general references.

Dahlgren, R. 1975. A system of classification of the angiosperms to be used to demonstrate the distribution of characters. Bot. Notiser 128: 119–147.

Dahlgren, R. 1983. General aspects of angiosperm evolution and macrosystematics. Nord. J. Bot. 3: 119–149.

Dahlgren, G. 1991. Steps toward a natural system of the dicotyledons: embryological characters. Aliso 13: 107–165.

Dahlgren, R.M.T., Rosendahl-Jensen, S., Nielsen, B.J. 1981. A revised classification of the angiosperms with comments on correlation between chemical and other characters. In: Young, D.A., Seigler, D.S. (eds.) Phytochemistry and angiosperm phylogeny. New York: Praeger, pp.149–204.

Engler, A., Gilg, E. 1912. Syllabus der Pflanzenfamilien, 7th edn. Berlin: Borntraeger.

Eyde, R.H. 1988. Comprehending *Cornus*: puzzles and progress in the systematics of the dogwoods. Bot. Rev. 54: 233–351.

Hegnauer, R. 1966. Aucubinartige Glycoside. Über die Verbreitung und Bedeutung als systematisches Merkmal. Pharm. Acta Helv. 41: 577–587.

Hermann, P.M., Palser, B.F. 2000. Stamen development in Ericaceae. I. Anther wall, microsporogenesis, inversion, and appendices. Am. J. Bot. 87: 934–957.

Huber, H. Die Verwandtschaftsverhältnisse der Rosifloren. Mitt. Bot. Staatssamml. München 5: 1–48.

Huber, H. 1991. Angiospermen. Leitfaden durch die Ordnungen und Familien der Bedecktsamer. Stuttgart: Fischer.

Hufford, L. 1992. Rosidae and their relationships to other nonmagnoliid dicotyledons: a phylogenetic analysis using morphological and chemical data. Ann. Missouri Bot. Gard. 79: 218–248.

Källersjö, M., Bergqvist, G., Anderberg, A.A. 2000. Generic realignments in primuloid families of the Ericales s.l.: a phylogenetic analysis based on DNA sequences from three chloroplast genes and morphology. Am. J. Bot. 87: 1325–1341.

Kooiman, P. 1969. The occurrence of asperulosidic glycosides in the Rubiaceae. Acta Bot. Neerl. 18: 124–137.

Leins, P., Erbar, C. 1985. Ein Beitrag zur Blütenentwicklung der Aristolochiaceen, einer Vermittlergruppe zu den Monokotylen. Bot. Jahr. Syst. 107: 343–368.

Lindenhofer, A., Weber, A. 1999a. Polyandry in Rosaceae: evidence for a spiral origin of the androecium in Spiraeoideae. Bot. Jahrb. Syst. 121: 553–582.

Lindenhofer, A., Weber, A. 1999b. The spiraeoid androecium of Pyroideae and Amygdaloideae (Rosaceae). Bot. Jahrb. Syst. 121: 583–605.

Lindenhofer, A., Weber, A. 2000. Structural and developmental diversity of the androecium of Rosoideae (Rosaceae). Bot. Jahrb. Syst. 122: 63–91.

Matthews, M.L., Endress, P.K. 2002. Comparative floral structure and systematics in Oxalidales (Oxalidaceae, Connaraceae, Brunelliaceae, Cephalotaceae, Cunoniaceae, Elaeocarpaceae, Tremandraceae). Bot. J. Linn. Soc. 140: 321–381.

Matthews, J.R., Taylor, G. 1926. The structure and development of the stamen in *Erica hirtiflora*. Trans. Bot. Soc. Edinburgh 29: 235–242.

Morton, C.M. et al. 1996. See general references.

Morton, C.M., Mori, S.A., Prance, G.T., Karol, K.G., Chase, M.W. 1997. Phylogenetic relationships of Lecythidaceae: a cladistic analysis using *rbcL* sequence and morphological data. Am. J. Bot. 84: 530–540.

Nandi, O.I. 1998. See general references.

Nash, G.V. 1903. A revision of the family Fouquieriaceae. Bull. Torrey Bot. Club 30: 449–459.

Netolitzky, F. 1926. Anatomie der Angiospermen-Samen. Handbuch der Pflanzenanatomie, Band 10. Berlin: Borntraeger.

Oever, L. van den, Baas, P., Zandee, M. et al. 1981. Comparative wood anatomy of *Symplocos* and latitude and altitude. IAWA Bull. n.s. 2: 3–24.

Olmstead, R.G., Michaelis, H.J., Scott, K.M., Palmer, J.D. 1992. Monophyly of the Asteridae and identification of their major lineages inferred from DNA sequences of *rbc*L. Ann. Missouri Bot. Gard. 79: 249–265.

Qiu, Y.-L., Chase, M.W., Hoot, S.B., Conti, E., Crane, P.R., Sytsma, K.J., Parks, C.R. 1998. Phylogenetics of the Hamamelidae and their allies: parsimony analyses of nucleotide sequences of the plastid gene *rbc*L. Int. J. Pl. Sci. 159: 891–905.

Savolainen, V., Chase, M.W. et al. 2000. See general references.

Savolainen, V., Fay, M.F. et al. 2000. See general references.

Simmons, M.P., Clevinger, C.C., Savolainen, V., Archer, R.H., Mathews, S., Doyle, J.J. 2001a. Phylogeny of Celastraceae inferred from phytochrome B and morphology. Am. J. Bot. 88: 313–325.

Simmons, M.P., Savolainen, V., Clevinger, C.C., Archer, R.H., Davis, J.I. 2001b. Phylogeny of the Celastraceae inferred from 26 SnrDNA, phytochrome B, *atp*B, *rbc*L, and morphology. Molec. Phylo. Evol. 19: 353–366.

Soltis, D.E., Soltis, P.S., Morgan, D.R., Swensen, B.C., Mullin, B.C., Dowd, J.M., Martin, P.G. 1995. Chloroplast gene sequence data suggests a single origin of the predisposition for symbiotic nitrogen fixation in angiosperms. Proc. Natl. Acad. Sci. USA 92: 2647–2651.

Soltis, D.E. et al. 1997, 2000. See general references.

Stevens, P.F. 1971. A classification of the Ericaceae: subfamilies and tribes. Bot. J. Linn. Soc. 64: 1–53.

Stevens, P.F. 2003. Angiosperm phylogeny website, v. Jan 2003. http://www.mobot.org/mobot/research APweb/.

Sytsma, K.J., Morawetz, J., Pires, J.C., Nepokroeff, M., Conti, E., Zihra, M., Hall, J.C., Chase, M.W. 2002. Urticalean rosids: circumscription, rosid ancestry, and phylogenetics based on *rbc*L, *trn*L-F, and *ndh*F sequences. Am. J. Bot. 89: 1531–1546.

Takhtajan, A. 1959. Die Evolution der Angiospermen. Jena: G. Fischer.

Takhtajan, A. 1987. Systema Magnoliophytorum. Leningrad: Nauka (in Russian).

Thulin, M., Bremer, B., Richardson, J., Niklasson, J., Fay, M.F., Chase, M.W. 1998. Family relationships of the enigmatic genera *Barbeya* and *Dirachma* from the Horn of Africa region. Plant Syst. Evol. 213: 103–119.

Xiang, Q.-Y. 1999. Systematic affinities of Grubbiaceae and Hydrostachyaceae within Cornales – insights from *rbc*L sequences. Harvard Pap. Bot. 4: 527–542.

Xiang, Q.-Y., Soltis, S.E., Soltis, P.S. 1998. Phylogenetic relationships of Cornaceae and close relatives inferred from *mat*K and *rbc*L sequences. Am. J. Bot. 85: 285–297.

Xiang, Q.-Y., Moody, M.L., Soltis, D.E., Fan, C.-Z., Soltis, P.S. 2002. Relationships within Cornales and circumscription of Cornaceae – *mat*K and *rbc*L sequence data and effects of outgroups and long branches. Molec. Phylog. Evol. 24: 35–57.

General References

Morphology, Anatomy, Embryology, Chromosomes, and Palynology

Behnke, H.-D. 1991. Distribution and evolution of forms and types of sieve-element plastids in the dicotyledons. Aliso 13: 167–182.

Corner, E.J.H. 1976. The seeds of dicotyledons., 2 vols. Cambridge: Cambridge University Press.

Davis, G.L. 1966. Systematic embryology of the angiosperms. New York: Wiley.

Eichler, A.W. 1875–1878. Blüthendiagramme. 2 vols. Leipzig: W. Engelmann.

Erdtman, G. 1952. Pollen morphology and plant taxonomy. Stockholm: Almquist & Wiksell.

Fedorov, Al.A. (ed.) 1969. Chromosome numbers of flowering plants. Leningrad: Nauka (in Russian).

Johri, B.M., Ambegoakar, K.B., Srivastava, P.S. 1992. Comparative embryology of angiosperms. 2 vols. Berlin Heidelberg New York: Springer.

Metcalfe, R.C., Chalk, L. 1950. Anatomy of dicotyledons. 2 vols. Oxford: Clarendon Press (2nd edn 1979).

Netolitzky, F. 1926. Anatomie der Angiospermen-Samen. In Linsbauer, K. (ed.) Handbuch der Pflanzenanatomie, 2. Abt., 2. Teil, vol. 10. Berlin: Borntraeger.

Takhtajan, A. (ed.) 1991–2000. Anatomia seminum comparative. Leningrad: Nauka. (Vol. 3: Caryophyllidae-Dilleniidae; Vol. 4. Dicotyledones Dilleniidae; Vol. 5. Rosidae I; Vol. 6. Rosidae II) (in Russian).

Systematics and Classification

APG (Angiosperm Phylogeny Group) 1998. An ordinal classification for the families of flowering plants. Ann. Missouri Bot. Gard. 85: 531–553.

APG II (Angiosperm Phylogeny Group) 2003. An update of the Angiosperm Phylogeny Group Classification for the orders and families of flowering plants: APG II. Bot. J. Linn. Soc. 141: 399–436.

Cronquist, A. 1981. An integrated system of classification of flowering plants. New York: Columbia University Press.

Cronquist, A. 1988. The evolution and classification of flowering plants, 2nd edn. Bronx, N.Y.: The New York Botanical Garden.

Stevens, P.F. 2002. Angiosperm phylogeny website. Version 3. http://www.mobot.org/MOBOT/research/APweb/.

Takhtajan, A. (ed.) 1981. Plant life, vol. 5 (2). Leningrad: Nauka.

Takhtajan, A. 1997. Diversity and classification of flowering plants. New York: Columbia University Press.

Thorne, R.F. 2001. The classification and geography of the flowering plants: Dicotyledons of the class Angiospermae. Bot. Rev. 66: 441–647.

Phytochemistry

Bate-Smith, E.C. 1962. The phenolic constituents of plants and their taxonomic significance. I. Dicotyledons. J. Linn. Soc. Bot. 58: 95–173.

Gibbs, R.D. 1974. Chemotaxonomy of flowering plants, 4 vols. Montreal: McGill-Queen's University Press.

Hegnauer, R. 1962–1992. Chemotaxonomie der Pflanzen. Basel: Birkhaeuser. (Vol. 1: 1962; vol. 2: 1963; Vol. 3: 1964; Vol. 4: 1966; Vol. 5: 1969; Vol. 6: 1973; Vol. 7: 1986; Vol. 8: 1989; Vol. 9: 1990; Vol. 10: 1992).

Nandi, O.I., Chase, M.W., Endress, P.K. 1998. A combined cladistic analysis of angiosperms using *rbc*L and non-molecular data. Ann. Missouri Bot. Gard. 85: 137–212.

Palaeobotany

Knobloch, E., Mai, D. 1986. Monographie der Früchte und Samen in der Kreide Mitteleuropas. Rozpravy ústredního ústavu geologického svazek 47. Praha: Czechoslovakian Academy.

Krutzsch, W. 1989. Paleogeography and historical phytogeography (paleochorology) in the Neophyticum. Plant Syst. Evol. 162: 5–61.

Muller, J. 1981. Fossil pollen records of extant angiosperms. Bot. Rev. 47: 1–142.

Molecular Systematics

Albach, D.C., Soltis, D.E., Chase, M.W., Soltis, P.S. 2001a. Phylogenetic placement of the enigmatic angiosperm *Hydrostachys*. Taxon 50: 781–805.

Albach, D.C., Soltis, P.S., Soltis, D.E., Olmstead, R.G. 2001b. Phylogenetic analysis of asterids based on sequences of four genes. Ann. Missouri Bot. Gard. 88: 163–212.

Anderberg, A.A., Rydin, C., Källersjö, M. 2002. Phylogenetic relationships in the order Ericales s.l.: analyses of molecular data from five genes from the plastid and mitochondrial genomes. Am. J. Bot. 89: 677–687.

Chase, M.W., Soltis, D.E., Olmstead, R.G., Morgan, D., Les, D.H. and 37 further authors. 1993. Phylogenetics of seed plants: an analysis of nucleotide sequences from the plastid gene *rbc*L. Ann. Missouri Bot. Gard. 80: 528–580.

Källersjö, M., Bergqvist, G., Anderberg, A.A. 2000. Generic realignement in primuloid families of the Ericales s.l.: a phylogenetic analysis based on DNA sequences from three chloroplast genes and morphology. Am. J. Bot. 87: 1325–1341.

Morton, C.M., Chase, M.W., Kron, K.A., Swensen, S.M. 1996. A molecular evaluation of the monophyly of the order Ebenales based upon *rbc*L sequence data. Syst. Bot. 21: 567–586.

Savolainen, V., Fay, M.F., Albach, D.C., Backlund, A., van der Bank, M., Cameron, K.M., Johnson, S.A., Lledó, M.D., Pintaud, J.C., Powell, M., Sheahan, M.C., Soltis, D.E., Soltis, P.S., Weston, P., Whitten, W.M., Wurdack, K.J., Chase, M.W. 2000. Phylogeny of the eudicots: a nearly complete familial analysis based on *rbc*L gene sequences. Kew Bull. 55: 257–309.

Savolainen, V., Chase, M.W., Hoot, S.B., Morton, C.M., Soltis, D.E., Bayer, C., Fay, M.F., de Bruijn, A.Y., Sullivan, S., Qiu, Y.-L. 2000. Phylogenetics of flowering plants based on combined analysis of plastid *atp*B gene sequences. Syst. Biol. 49: 306–362.

Soltis, D.E., Soltis P.S., Nickrent, D.L., Johnson, L.A., Hahn, W.J., Hoot, S.B., Sweere, J.A., Kuzoff, R.K., Kron, K.A., Chase, M.W., Swensen, S.M., Zimmer, E.A., Chaw, S.M., Gillespie, L.J., Kress, W.J., Sytsma, K.J. 1997. Angiosperm phylogeny inferred from 18S ribosomal DNA sequences. Ann. Missouri Bot. Gard. 84: 1–49.

Soltis, D.E., Soltis, P.S., Chase, M.W, Mort, M.E., Albach, D.C., Zanis, M., Savolainen, V., Hahn, W.H., Hoot, S.B., Fay, M.F., Axtell, M., Swensen, S.M., Prince, L.M., Kress, W.J., Nixon, K.C., Farris, J.S. 2000. Angiosperm phylogeny inferred from 18S rDNA, *rbc*L, and *atp*B sequences. Bot. J. Linn. Soc. 133: 381–461.

Xiang, Q.-Y., Moody, M.L., Soltis, D.E., Fan, C.Z., Soltis, P.S. 2002. Relationships within Cornales and circumscription of Cornaceae – *mat*K and *rbc*L sequence data and effects of outgroups and long branches. Molec. Phylogen. Evol. 24: 35–57.

Wikström, N., Savolainen, V., Chase, M.W. 2001. Evolution of the angiosperms: calibrating the family tree. Proc. Roy. Soc. Lond. B, 268: 2211–2220.

Actinidiaceae

S. Dressler and C. Bayer

Actinidiaceae Gilg & Werderm. in Engler & Prantl, Nat. Pflanzenfam., ed. 2, 21: 36 (1925), nom. cons.
Saurauiaceae J. Agardh (1858).

Trees, shrubs or climbers, with simple or variously branched trichomes. Leaves alternate, simple, usually serr(ul)ate or dentate, pinnatinerved, petiolate; stipules absent or minute. Flowers in axillary cymes or thyrso-paniculate inflorescences, sometimes solitary, pedicellate, actinomorphic, hermaphroditic or unisexual; sepals (3–)5(–8), imbricate-quincuncial in bud, usually persistent; petals (3–)5(–9), longer than sepals, distinct or ± fused at base, imbricate in bud; stamens (10–)20–240, filaments distinct, sometimes adnate to petals and falling with these, anthers dithecal, dorsifixed, versatile, extrorsely dehiscent by longitudinal slits or subapical pores; ovary superior, syncarpous, (3–)5- to many-locular, sometimes incompletely septate, pubescent or glabrous, sometimes with apical depression; placentation axile, placentae sometimes split; ovules numerous, anatropous, unitegmic; stylodia distinct, as many as locules, sometimes persistent, or style simple, stigma sometimes capitate; fruit usually a berry, sometimes dehiscent; seeds numerous, exarillate, enclosed by pulp, albuminous, embryo large, usually straight.

A family of c. 360 spp. in three genera native to tropical Asia and America, a few spp. of *Actinidia* in temperate E Asia.

VEGETATIVE ANATOMY. Young organs are covered with simple or multicellular, sometimes glandular hairs, which can be of diagnostic value (see Hunter 1966). Raphides contained in elongated idioblasts occur in most tissues. Crystal sand is reported for *Clematoclethra* (Lechner 1914).

The leaves are dorsiventral and sometimes have an arm palisade mesophyll. Some species form a multilayered hypodermis. Stomata of the ranunculaceous type are restricted to the abaxial side (Lechner 1914; Metcalfe and Chalk 1950). Venation is pinnate and of the camptodromous type. The ultimate marginal veins are looped, but some branches extend into the teeth (Yu and Chen 1991).

Vessels of the primary xylem are mostly solitary, often large, and often arranged in radial rows. They can have annular, helical, reticulate, or scalariform thickenings. Perforation plates are scalariform, more rarely simple. The pericycle includes a continuous ring of sclerenchyma. Young stems have a solid pith, which may later disintegrate and/or become lamellate in some *Actinidia* species. The lamellae and a cylinder of outer pith cells become sclerenchymatous with age.

The wood of *Actinidia* has dimorphic vessel elements: few very large, moderately long vessel elements with simple perforations, and numerous smaller, usually scattered or solitary, with oblique scalariform perforation plates. *Saurauia* has exclusively solitary vessels with many-barred scalariform perforation plates. Parenchyma is diffuse apotracheal. Non-septate fibre tracheids have numerous bordered pits. Rays are uniseriate and multiseriate, heterogeneous (Lechner 1914; Metcalfe and Chalk 1950).

Wang et al. (1994) studied the root anatomy of five *Actinidia* species; cortex and endodermis persist during secondary thickening.

REPRODUCTIVE STRUCTURES. Cauliflory is reported for some species. *Saurauia callithrix* is special by having leafless inflorescences at the base of the trunk which spread in the soil and leaf litter, elevating the flowers just above the forest floor (Gilg and Werdermann 1925). The flowers are often arranged in cymes, which are subtended by foliage leaves or bracts. Some *Saurauia* have axillary thyrso-paniculate inflorescences that include monochasia. Bracteose prophylls are frequently found.

The flowers are mostly pentamerous but a few *Saurauia* and *Actinidia* species have tetramerous flowers, and other exceptions occur also. In pentamerous flowers, quincuncial aestivation of sepals is evident even in open flowers, since the abaxial face is often pubescent in the two outermost sepals, half pubescent in one other, and glabrous in the two innermost sepals. For anatomical features see Dickison (1972) and Schmid (1978a).

Petals and stamens may be fused to various degrees and often fall together. The microsporangia of each theca merge. Pollen is released through longitudinal slits or pores that become extrorse by inversion of the anthers.

Brown (1935) reported nectar-secreting tissue at the base of the petals of *Saurauia subspinosa*. According to his study, the androecium is basically diplostemonous. The outer whorl is formed by single stamens in front of the sepals, whereas the majority of stamens develop in centrifugal succession by splitting of antepetalous primordia. For *Actinidia deliciosa*, Brundell (1975) described two or three whorls of staminal initials, depending on the cultivar. In a more detailed study of *A. melanandra* and *A. deliciosa*, van Heel (1987) found a single whorl of staminal primordia in the former, and centrifugal multiplication of such primary primordia in the latter. The numerous carpel primordia arise in a single whorl.

EMBRYOLOGY. The anther wall consists of epidermis, fibrous endothecium, 2–3 ephemeral middle layers, and a secretory tapetum of multinucleate cells the nuclei of which may fuse and become polyploid. Raphides are present in the anther wall and connective. Meiotic division of microspore mother cells is simultaneous. Pollen is shed at the two-celled stage.

The ovules are anatropous, unitegmic and tenuinucellar. A hypostase is present. The nucellus is thin and ephemeral. A hypodermal archesporial cell forms the chalazal megaspore mother cell which develops into a linear tetrad. Embryo sac development is of the Polygonum type (Vijayaraghavan 1965; An et al. 1983). Inverted embryo sac polarity is reported of *Saurauia nepalensis* (Rao 1953). Endosperm formation is cellular; embryogeny corresponds to the Solanad type (Crété 1944b; Vijayaraghavan 1965). Polyembryony due to the proliferation of suspensor cells was observed in *Actinidia deliciosa* (Crété 1944a).

POLLEN MORPHOLOGY. Pollen of Actinidiaceae is remarkably uniform and rather unspecialised. Pollen is usually shed in monads; *Saurauia elegans* has tetrads. The grains are usually 3(4)-colporate, oblate-spheroidal to prolate, the longest axis 13–26(–33) μm. The colpi are long, crassimarginate and usually exhibit an equatorial bridge of ektexine over the endoaperture. Exine is (1–)1.5–2 μm thick. Sexine is as thick as nexine. The complete tectum is psilate, (micro-)granulate or rugulate; columellae are reduced. Exine stratification and sexine pattern are obscure in light microscopy

(Erdtman 1952; Dickison et al. 1982; Zhang 1987; Li et al. 1989; Kang et al. 1993). Similar pollen occurs in Theaceae, Ochnaceae and Clethraceae (Erdtman 1952; Zhang 1987), but there are some differences to Dilleniaceae (Dickison et al. 1982).

Functionally female flowers of *Actinidia deliciosa* shed nonviable pollen, which is usually enucleate and shrivelled but otherwise similar to viable pollen (Schmid 1978a; White 1990).

KARYOLOGY. Chromosome counts are known for many *Actinidia* species (see Yan et al. 1997). The basic chromosome number is x = 29; diploids, tetraploids, hexaploids and octoploids occur. Intraspecific variation appears to be common (mainly diploid and tetraploid cytotypes, 4× and 6× in *A. valvata*, and 4×, 6× and 8× in *A. arguta*). For *Saurauia*, counts of *n* = 30 (South American species: Soejarto 1969, 1970) and *n* = 20 (Asian species: Mehra 1976) were reported.

REPRODUCTIVE BIOLOGY. *Actinidia* is usually dioecious, *Clematoclethra* and *Saurauia* have mostly bisexual flowers. American *Saurauia*, however, is described as functionally dioecious with dimorphic flowers: long-styled, functionally female flowers with malformed, sterile pollen, and short-styled ones with fertile pollen (Soejarto 1969). Brown (1935) reported protandry.

In *Actinidia deliciosa*, staminate flowers open about seven days before pistillate ones. Plants of the pistillate cultivar 'Hayward' in New Zealand bloom 10–18 days, staminate clones usually flower 3–5 days longer (Hopping 1990). Because of this limited overlap in flowering and the linear relationship between seed number and fruit weight, fruit set is improved by artificial pollination. Pollen tubes from the distinct stylar branches are evenly distributed to the numerous carpels and ovules by a compitum ("pollen tube distributor cup": Howpage et al. 1998). For pollen–pistil interaction, pollen tube growth and fertilisation of *Actinidia*, see Hopping and Jerram (1979), Harvey et al. (1987) and González et al. (1996).

Actinidia deliciosa is said to be bee- or wind-pollinated. Despite the abundance of literature on the floral biology of cultivated kiwifruit (e.g. Schmid 1978a; Harvey et al. 1987; Harvey and Fraser 1988; Hopping 1990; Howpage et al. 1998), there is only little information on pollination in the wild. Gilg and Werdermann (1925) suspect insect pollination for *Actinidia* and *Clematoclethra*. For *Saurauia*, observations on insect visits (e.g. Hymenoptera) are reported (Soejarto 1969). The flowers are showy, often fragrant and at least sometimes nectariferous (Brown 1935 for *S. subspinosa*).

FRUIT AND SEED. Fruits are usually berries with massive placentae, and the seeds embedded in a mucilaginous pulp which is mostly greenish, sticky, sweet and edible. Since this pulp originates from the placenta, it does not correspond to an aril (Schmid 1978a). Berries of *Actinidia* may include more than 40 locules and 1500 seeds. In some *Saurauia* species, the pericarp is dehiscent and more or less dry (Soejarto 1969; Dickison 1972). The fruits of *Clematoclethra* have been described as capsular (Gilg and Werdermann 1925), indehiscent (Ying et al. 1993) or drupaceous (Lechner 1914). Generally, endozoochory seems most likely. In *Saurauia*, dispersal is by rain and animals such as birds (Soejarto 1969).

The seed coat is thin and finely reticulate. As in other Ericales, the seeds of Actinidiaceae are exotestal, with thickened inner walls of the outer epidermis (Crété 1944b; Corner 1976; Huber 1991; Takhtajan 1991).

PHYTOCHEMISTRY. Iridoid compounds were found in *Actinidia* and *Roridula* (Jensen et al. 1975). Webby et al. (1994) studied leaf flavonoids of several *Actinidia*, which are based on common flavonols including myricetin. Procyanidin and prodelphinidin point to the presence of condensed tannins (Hegnauer 1964). Actinidin, a proteinase similar to papain, was detected in kiwifruit (McDowall 1970), which makes it a potential cause of contact dermatitis. The same name was applied to a terpenoid pseudoalkaloid found in *Actinidia polygama* (Hegnauer 1964). The mucilage of *Actinidia* contains acidic polysaccharides (Redgwell 1983).

AFFINITIES. Actinidiaceae have been variously placed; especially Theaceae, Dilleniaceae, Cornales and Ericales were considered as closest relatives (see, for example, Schmid 1978a). Ericalean affinities are now mostly accepted, which is supported by agreement in embryological characters (Crété 1944b), floral features (Dickison 1972) and results of molecular analyses (Kron and Chase 1993; Chase et al. 1993 and subsequent studies). According to recent analyses, Actinidiaceae are sister to *Roridula*, both being sister to Sarraceniaceae (Anderberg et al. 2002). In all analyses, nonetheless, statistical support for this topology is not strong, and some studies have found different topologies (e.g. Savolainen, Fay et al. 2000). A potential synapomorphy of a clade comprising

Actinidiaceae, Roridulaceae and Sarraceniaceae is the presence of a hypostase. Other features considered by Anderberg et al. (2002), such as the presence of stylar branches and a fibrous endothecium, are not found in all members of this clade.

DISTRIBUTION AND HABITATS. *Actinidia* is centred in hilly S and E China (between 25 and 30 °N), but some species occur in the cold-temperate and arctic forests of Siberia, Korea and Japan, and the tropics. They inhabit the lower forest storey. *Clematoclethra* occupies similar habitats up to 3100 m but is confined to China. In contrast, *Saurauia* species are trees and shrubs mainly of humid montane forests of tropical Asia and America, and occur in altitudes up to 3600 m. Some species are rheophytic (van Steenis 1981).

PALAEOBOTANY. Seeds resembling those of extant *Saurauia* were recorded from the Maastrichtian onwards in Europe (Knobloch and Mai 1986), and *Actinidia*-like seeds are known from the Upper Eocene in Europe, sometimes in abundance (Mai and Gregor 1982; Friis 1985; Mai 2001). Leaves similar to those of *Saurauia* were found in the Middle Eocene of North America (Taylor 1990), and *Actinidiophyllum* was described from the Tertiary of Japan (Nathorst 1888). A Pliocene wood from the German Westerwald was described as *Actinidioxylon* (Müller-Stoll and Mädel-Angeliewa 1969). Flowers from the early Campanian of Georgia, North America, were described as *Parasaurauia*, mainly differing from modern *Saurauia* in having only 10 stamens (Keller et al. 1996).

ECONOMIC IMPORTANCE. Kiwifruit, produced by *Actinidia deliciosa* cultivars, are an economically important crop, especially in New Zealand but also in Italy, Spain, China and other countries. Since this species was formerly treated mostly as a variety of *A. chinensis*, most literature on cultivated *A. chinensis* refers to what is now *A. deliciosa* (Liang and Ferguson 1986; Ferguson 1990). All commercial plantations in New Zealand can be traced back to a single introduction of seed from China in 1904 (Ferguson and Bollard 1990). There are indications that the hexaploid *A. deliciosa* originated from diploid *A. chinensis* (Crowhurst et al. 1990; Atkinson et al. 1997). Breeding of *A. chinensis*, up to 1997 collected only in the wild and industrially processed in China, led to the recent introduction of the Kiwi Gold (cv. 'Hort 16 A') from New Zealand. Relatively hardy species such as *A. arguta* and *A. kolomikta* are cultivated as ornamentals in Europe and North America. In the former Soviet Union, attempts were made to use their aromatic fruits, which contain even more vitamin C than do kiwifruit. Berries of some *Saurauia* species are sold at local markets in South America. Wood is occasionally used for construction, fire wood and charcoal, but is of no commercial importance. Some uses in folk medicine are reported (Soejarto 1980).

KEY TO THE GENERA

1. Trees or shrubs; stamens basally connate with petals; Asia and America **3. Saurauia**
- Woody climbers; stamens free; Asia 2
2. Flowers usually functionally unisexual; stamens numerous; stylar branches distinct **1. Actinidia**
- Flowers usually hermaphroditic; stamens 10(–30); style simple **2. Clematoclethra**

GENERA OF ACTINIDIACEAE

1. *Actinidia* Lindl.   Fig. 3

Actinidia Lindl., Intr. Nat. Syst., ed. 2: 439 (1836); Li, J. Arnold Arbor. 33: 1–61 (1952), rev.; Liang, Fl. Reip. Pop. Sin. 49(2): 196–268 (1984), reg. rev.; Wei, Higher plants of China 4: 657–672 (2000), reg. rev.

Woody climbers, dioecious or polygamous, ± pubescent; sepals distinct or somewhat fused at base; petals white or yellow to reddish; stamens numerous, anthers longitudinally dehiscent; ovary pubescent or glabrous, many-locular; stylar branches distinct, persistent; fruit a globose to oblong berry, sometimes pubescent; seeds numerous, embedded in pulp. $n = 29$ or multiples. About 60 spp.; E Asia, mostly W to E China, north to Sakhalin and Kuril Is., south to Taiwan, Himalayas, NE India, Indochina, Malaysia.

2. *Clematoclethra* (Franch.) Maxim.   Fig. 4

Clematoclethra (Franch.) Maxim., Trudy Imp. S.-Peterburgsk. Bot. Sada 11: 36 (1890); Tang & Xiang, Acta Phytotax. Sin. 27: 81–95 (1989), rev.; Wei, Higher plants of China 4: 672–674 (2000), rev.

Woody climbers, deciduous, ± pubescent; flowers in up to 12-flowered cymes or solitary, hermaphroditic or unisexual; sepals united at base; petals white or reddish; stamens c. 10–30, filaments dilated at base; anthers longitudinally dehiscent; ovary globose, 5-angled, (4)5-locular; style simple, persistent, stigma small, swollen; ovules numerous; fruit berry-like, blackish. One species, *C. scandens* (Franch.) Maxim. (or 5), in montane forests above 1000 m in W and C China.

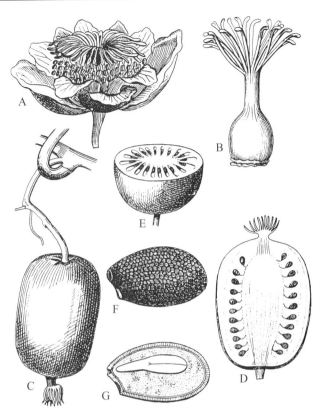

Fig. 3. Actinidiaceae. A, F, G *Actinidia strigosa*. B–E *A. polygama*. **A** Flower. **B** Pistil. **C** Fruit. **D, E** Same, vertical and transverse section. **F** Seed. **G** Same, longitudinal section. (Schneider 1912)

Fig. 4. Actinidiaceae. *Clematoclethra lasioclada*. **A** Flowering twig. **B** Flower. **C** Androecium and gynoecium of young flower, longitudinal section. **D** Androecium and gynoecium at anthesis, note inversion of anthers. **E** Young anthers. **F** Mature anther. **G** Ovary in cross section. (Gilg and Werdermann 1925)

3. *Saurauia* Willd.

Saurauia Willd., Ges. Naturf. Freunde Berlin Neue Schriften 3: 407 (1801), nom. cons.; Hunter, Ann. Missouri Bot. Gard. 53: 47–89 (1966), reg. rev.; Soejarto, Fieldiana Bot. N.S. 2: 1–141 (1980), reg. rev.; Wei, Higher plants of China 4: 674–677 (2000), reg. rev.

Trees or shrubs, usually pubescent; flowers hermaphroditic or sometimes functionally unisexual; sepals fused at very base, petals white or pink; stamens 15-numerous, filament bases fused with petals, pubescent, anthers dehiscing by pores; ovary (3–)5(–8)-carpellate, usually glabrous, stylar branches entirely or distally distinct, stigmas simple to capitate or discoid; fruit a berry, rarely a leathery capsule; seeds numerous, minute, embedded in pulp. $n = 30$ (20 in Asian spp.). About 300 spp., Asia (Himalayas to E and SE Asia) to Fiji and tropical America (C Mexico to Chile but absent from Antilles, Guianas, Brazil); 1 sp. in Queensland.

Selected Bibliography

An, H.-X., Cai, D.-R., Wang, J.-R., Qian, N.-F. 1983. Investigations on early embryogenesis of *Actinidia chinensis* Planch. var. *chinensis*. Acta Bot. Sin. 25: 99–104, 2 pl.

Anderberg, A.A. et al. 2002. See general references.

Atkinson, R.G., Cipriani, G., Whittaker, D.J., Gardner, R.C. 1997. The allopolyploid origin of kiwifruit, *Actinidia deliciosa* (Actinidiaceae). Pl. Syst. Evol. 205: 111–124.

Brown, E.G.S. 1935. The floral mechanism of *Saurauia subspinosa* Anth. Trans. Proc. Bot. Soc. Edinburgh 31: 485–497.

Brundell, D.J. 1975. Flower development of the Chinese gooseberry (*Actinidia chinensis* Planch.) II. Development of the flower bud. New Zeal. J. Bot. 13: 485–496.

Chase, M.W. et al. (1993). See general references.

Corner, E.J.H. 1976. See general references.

Crété, P. 1944a. Polyembryonie chez l'*Actinidia chinensis* Planch. Bull. Soc. Bot. France 91: 89–92.

Crété, P. 1944b. Recherches anatomiques sur la séminogenèse de l'*Actinidia chinensis* Planch. Affinités des Actinidiacées. Bull. Soc. Bot. France 91: 153–160.

Crowhurst, R.N., Lints, R., Atkinson, R.G., Gardner, R.C. 1990. Restriction fragment length polymorphisms in the genus *Actinidia* (Actinidiaceae). Pl. Syst. Evol. 172: 193–203.

Dickison, W.C. 1972. Observations on the floral morphology of some species of *Saurauia*, *Actinidia* and *Clematoclethra*. J. Elisha Mitchell Sci. Soc. 88: 43–54.

Dickison, W.C., Nowicke, J.W., Skvarla, J.J. 1982. Pollen morphology of the Dilleniaceae and Actinidiaceae. Am. J. Bot. 69: 1055–1073.

Erdtman, G. 1952. See general references.

Ferguson, A.R. 1990. Botanical nomenclature: *Actinidia chinensis*, *Actinidia deliciosa* and *Actinidia setosa*. In: Warrington I.J., Weston G.C. (eds.) Kiwifruit: science and management. Auckland: Ray Richards, pp. 36–56.

Ferguson, A.R., Bollard, E.G. 1990. Domestication of the Kiwifruit. In: Warrington I.J., Weston G.C. (eds.) Kiwifruit: science and management. Auckland: Ray Richards, pp. 165–246.

Friis, E.M. 1985. Angiosperm fruits and seeds from the Middle Miocene of Jutland, Denmark. Biol. Skr. 24: 1–165.

Gilg, E., Werdermann, E. 1925. Actinidiaceae. In: Engler & Prantl, Die natürlichen Pflanzenfamilien, ed. 2, vol. 21. Leipzig: W. Engelmann, pp. 36–47.

González, M.V., Coque, M., Herrero, M. 1996. Pollen-pistil interaction in kiwifruit (*Actinidia deliciosa*; Actinidiaceae). Am. J. Bot. 83: 148–154.

Guédès, M., Schmid, R. 1978. The peltate (ascidiate) carpel theory and carpel peltation in *Actinidia chinensis* (Actinidiaceae). Flora 167: 525–543.

Harvey, C.F., Fraser, L.G. 1988. Floral biology of two species of *Actinidia* (Actinidiaceae). II. Early embryology. Bot. Gaz. 149: 37–44.

Harvey, C.F., Fraser, L.G., Pavis, S.E., Considine, J.A. 1987. Floral biology of two species of *Actinidia* (Actinidiaceae). I. The stigma, pollination, and fertilization. Bot. Gaz. 148: 426–432.

Heel, W.A. van 1987. Androecium development in *Actinidia chinensis* and *A. melanandra* (Actinidiaceae). Bot. Jahrb. Syst. 109: 17–23.

Hegnauer, R. 1964. See general references.

Hopping, M.E. 1976. Structure and development of fruit and seeds in Chinese gooseberry (*Actinidia chinensis* Planch.). New Zeal. J. Bot. 14: 63–68.

Hopping, M.E. 1990. Floral biology, pollination, and fruit set. In: Warrington, I.J., Weston, G.C. (eds.) Kiwifruit: science and management. Auckland: Ray Richards, pp. 71–96.

Hopping, M.E., Jerram, E.M. 1979. Pollination of kiwifruit (*Actinidia chinensis* Planch.): stigma-style structure and pollen tube growth. New Zeal. J. Bot. 17: 233–240.

Howpage, D., Vithanage, V., Spooner-Hart, R. 1998. Pollen tube distribution in the kiwifruit (*Actinidia deliciosa* A. Chev. C.F. Liang) pistil in relation to its reproductive process. Ann. Bot. II, 81: 697–703.

Huber, H. 1991. Angiospermen. Leitfaden durch die Ordnungen der Familien der Bedecktsamer. Stuttgart, New York: G. Fischer.

Hunter, G.E. 1966. Revision of Mexican and Central American *Saurauia* (Dilleniaceae). Ann. Missouri Bot. Gard. 53: 47–89.

Jensen, S.R., Nielsen, B.J., Dahlgren, R. 1975. Iridoid compounds, their occurrence and systematic importance in the angiosperms. Bot. Notiser 128: 148–180.

Johri, B.M. et al. 1992. See general references.

Kang, N., Wang, S., Huang, R., Wu, X. 1993. Studies on the pollen morphology of nine species of genus *Actinidia*. J. Wuhan Bot. Res. 11: 111–116, 5 pl.

Keller, J.A., Herendeen, P.S., Crane, P.R. 1996. Fossil flowers and fruits of the Actinidiaceae from the Campanian (Late Cretaceous) of Georgia. Am. J. Bot. 83: 528–541.

Knobloch, E., Mai, D.H. 1986. Monographie der Früchte und Samen in der Kreide von Mitteleuropa. Rozpr. Ustr. Ustsr. Geol. 47: 1–219.

Kron, K.A., Chase, M.W. 1993. Systematics of the Ericaceae, Empetraceae, Epacridaceae and related taxa based upon *rbc*L sequence data. Ann. Missouri Bot. Gard. 80: 735–741.

Lechner, S. 1914. Anatomische Untersuchungen über die Gattungen *Actinidia*, *Saurauia*, *Clethra* und *Clematoclethra* mit besonderer Berücksichtigung ihrer Stellung im System. Beih. Bot. Zentralbl. 32: 431–467.

Li, J.W., Rui, G., Liang, M.Y., Pang, C. 1989. Studies on the pollen morphology of the *Actinidia*. Guihaia 9: 335–339, 1 pl.

Liang, C.-F. 1983. On the distribution of actinidias. Guihaia 3: 229–248.

Liang, C.-F., Ferguson, A.R. 1986. The botanical nomenclature of the kiwifruit and related taxa. New Zeal. J. Bot. 24: 183–184.

Mai, D.H. 2001. Die mittelmiozänen und obermiozänen Floren aus der Meuroer und Raunoer Folge in der Lausitz, Teil 2: Dicotyledonen. Palaeontographica Abt. B, 257: 35–174.

Mai, D.H., Gregor, H.J. 1982. Neue und interessante Arten aus dem Miozän von Salzhausen im Vogelsberg. Feddes Repert. 93: 405–435, pl. XVII–XXIII.

McDowall, M.A. 1970. Anionic proteinase from *Actinidia chinensis*. Preparation and properties of the crystalline enzyme. Eur. J. Biochem. 14: 214–221.

Mehra, P.N. 1976. Cytology of Himalayan hardwoods. Calcutta: Sree Saraswaty Press.

Metcalfe, C.R., Chalk, L. 1950. See general references.

Müller-Stoll, W.R., Mädel-Angeliewa, E. 1969. *Actinidioxylon princeps* (Ludwig) n. comb., ein Lianenholz aus dem Pliozän von Dernbach im Westerwald. Senckenb. Leth. 50: 103–115.

Nathorst, A.G. 1888. Zur fossilen Flora Japans. Palaeont. Abh. 4(3): 197–250, 14 pl.

Rao, A.N. 1953. Inverted polarity in the embryo-sac of *Saurauia napaulensis* DC. Curr. Sci. 22: 282.

Redgwell, R.J. 1983. Composition of *Actinidia* mucilage. Phytochemistry 22: 951–956.

Savolainen, V., Fay, M.F. et al. 2000. See general references.

Schmid, R. 1978a. Reproductive anatomy of *Actinidia chinensis* (Actinidiaceae). Bot. Jahrb. Syst. 100: 149–195.

Schmid, R. 1978b. Actinidiaceae, Davidiaceae and Paracryphiaceae: systematic considerations. Bot. Jahrb. Syst. 100: 196–204.

Schneider, C.K. 1912. Illustriertes Handbuch der Laubholzkunde, Vol. 2. Jena: Fischer.

Soejarto, D.D. 1969. Aspects of reproduction in *Saurauia*. J. Arnold Arbor. 50: 180–196.

Soejarto, D.D. 1970. *Saurauia* species and their chromosomes. Rhodora 72(789): 81–93.

Soejarto, D.D. 1980. Revision of South American *Saurauia* (Actinidiaceae). Fieldiana Bot. N.S. 2: 1–141.

Steenis, C.G.G.J. van 1981. Rheophytes of the world. Alphen a/d Rijn & Rockville: Sijthoff & Nordhoff.

Takhtajan, A.L. (ed.) 1991. Anatomia seminum comparativa. Vol. 3. Dicotyledones Caryophyllidae–Dilleniidae. Leningrad: Nauka.

Taylor, D.W. 1990. Paleobiogeographic relationships of angiosperms from the Cretaceous and early Tertiary of the North American area. Bot. Rev. 56: 279–416.

Vijayaraghavan, M.R. 1965. Morphology and embryology of *Actinidia polygama* Franch. & Sav. and systematic position of the family Actinidiaceae. Phytomorphology 15: 224–235.

Walton, E.F., Fowke, P.J., Weis, K., McLeay, P.L. 1997. Shoot axillary bud morphogenesis in Kiwifruit (*Actinidia deliciosa*). Ann. Bot. (London) 80: 13–21.

Wang, Z.Y., Gould, K.S., Patterson, K.J. 1994. Comparative root anatomy of five *Actinidia* species in relation to rootstock effects on kiwifruit flowering. Ann. Bot. (London) 73: 403–413.

Warrington, I.J., Weston, G.C. (eds.) 1990. Kiwifruit: science and management. Auckland: Ray Richards.

Webby, R.F., Wilson, R.D., Ferguson, A.R. 1994. Leaf flavonoids of *Actinidia*. Biochem. Syst. Ecol. 22: 277–286.

Wei, Y. 2000. Actinidiaceae. In: Fu, L., Chen, T., Lang, K., Hong, T., Lin, Q. (eds.) Higher plants of China (in Chinese), vol. 4. Quingdao: Quingdao Publishing House, pp. 656–677.

White, J. 1990. Pollen development in *Actinidia deliciosa* var. *deliciosa*: histochemistry of the microspore mother cell walls. Ann. Bot. (London) 65: 231–239.

Xiong, Z.-T., Huang, R.-H. 1988. Chromosome numbers of 10 species and 3 varieties in *Actinidia* Lindl. Acta Phytotax. Sin. 26: 245–247, 2 pl.

Yan, G., Yao, J., Ferguson, A.R., McNeilage, M.A., Seal, A.G., Murray, B.G. 1997. New reports of chromosome numbers in *Actinidia* (Actinidiaceae). New Zeal. J. Bot. 35: 181–186.

Ying, T.-S., Zhang, Y.-L., Boufford, D.E. 1993. The endemic genera of seed plants of China. Peking: Science Press.

Yu, C.H., Chen, Z.L. 1991. Leaf architecture of the woody dicotyledons from tropical and subtropical China. Oxford: Pergamon.

Zhang, Z. 1987. A study on the pollen morphology of Actinidiaceae and its systematic position. Acta Phytotax. Sin. 25: 9–23.

Balsaminaceae

E. Fischer

Balsaminaceae A. Rich. in Bory, Dict. Class. Hist. Nat. 2: 173 (1822), nom. cons.

Annual or perennial herbs, sometimes with tubers or rhizomes, occasionally subshrubs; stems erect or procumbent, succulent, rarely woody below. Leaves spirally arranged, rarely decussate or verticillate, simple, petiolate or sessile, pinnately veined, margins crenate, dentate or serrate, teeth or crenations apiculate, the lowermost often gland-tipped, petiole often with short capitate glands or fimbriae, rarely with extra-floral nectaries. Flowering shoots truncate, proliferating; inflorescences axillary racemes or pseudoumbels, often epedunculate and fascicled in leaf axils; flowers zygomorphic, usually resupinate through 180°, not resupinate in some Chinese *Impatiens*; sepals 3 or 5, free, the lower one (by resupination) larger, navicular to saccate, usually tapering or abruptly constricted into a nectariferous spur; petals 5, dorsal petal free, flat or cucullate, often crested dorsally, lower 4 petals free or united into lateral pairs; stamens 5, connate into a ring surrounding ovary and stigma, ripening and usually falling off in one piece before the maturity of the stigma; ovary superior, syncarpous, 5-locular with axile placentation; ovules 5–numerous, anatropous, bitegmic or unitegmic, tenuinucellate; style 1, very short or ± absent; stigmas 1–5. Fruit a berry or a loculicidal fleshy explosive capsule; seeds exalbuminous, seed coat smooth, warted or with simple hairs.

Two genera (one of which monotypic) and c. 1000 species, Europe, Africa, Asia, North and Central America.

VEGETATIVE MORPHOLOGY. Most species of *Impatiens* are perennial herbs with thin rhizomes, adapted to constantly humid conditions in moist forests. Relatively few species are annuals with thin erect stems. Plants growing under temporarily dry conditions have fusiform, fleshy subterranean tubers. Large globose tubers are found in the Madagascan *I. tuberosa*, which inhabits calcareous rocks. *Impatiens etindensis* is an epiphyte forming large globose tubers (Cheek and Fischer 1999). Suffruticose species are present in Africa and India; they produce robust and woody stems and can reach up to 4 m in height. Leaf arrangement in most species is spiral; verticillate leaves with up to 10 leaves per whorl are found in several species, whereas decussate leaves occur more rarely.

VEGETATIVE AND FLORAL ANATOMY. Calcium-oxalate raphides occur in bundles within special cells in the cortex. They are also present in the anthers walls (raphid pollen). Idioblasts with possible mucilage content are found in stems and leaves. The stems of most *Impatiens* species have a pericycle, which is devoid of sclerenchyma. The rigidity of the stem is maintained by the turgescence of the parenchyma (Metcalfe and Chalk 1950). Only few species like *I. sodenii*, *I. niamniamensis* or the epiphytic *I. paucidentata* have basally woody stems and show a special type of primary thickening.

INFLORESCENCE STRUCTURE. The inflorescence consists of a frondose main stem, which proliferates vegetatively, and lateral racemes. Due to the suppression of apical growth, as in *I. palpebrata*, these racemes sometimes appear to be terminal. The racemes may be elongate, as in *Impatiens teitensis*, or subumbellate as in *I. stuhlmannii*. Reduction of flowers may result in single-flowered lateral racemes with the appearance of single axillary flowers (e.g. *I. pseudoviola*). By reduction of the peduncle, axillary flower clusters result (*I. niamniamensis*, *I. keilii*, *I. clavicalcar*). In *I. paucidentata*, the corresponding part is uniflorous. In *I. acaulis*, the main stem is congested and reduced, forming a basal rosette with a pseudoterminal raceme.

FLORAL MORPHOLOGY. The flower is zygomorphic and usually resupinate through twisting of the pedicel. Few species (e.g. *Hydrocera triflora*, *Impatiens tinctoria*, *I. teitensis*, *I. quadrisepala*) have five sepals. Usually, each flower has only two reduced lateral sepals and one petal-like lower sepal, which is modified into a nectary-tipped spur; both upper sepals are lacking. There is a wide range of variation in form and size of the spurs, from shallowly navicular to bucciniform or deeply saccate, short or long filiform, straight, twisted or curved, obviously associated with different pollinators. In a group of Madagascan taxa, the spur is lacking. There are five petals, one of which, the upper dorsal one, is usually hood-like, whereas the others are united into two lateral pairs, each with two unequal lobes. The assumption of Rama Devi (1991) that the dorsal petal is the product of fusion of two sepals and one petal is questionable. The united lateral petals are also very variable in shape and size, again in adaptation to different pollinators, as they provide a suitable landing platform and entrance guide to the spur and the nectar. The five stamens are united by the upper part of the filaments and completely cap the gynoecium.

POLLEN MORPHOLOGY. The pollen is 4-colpate and rectangular in the North American and Eurasian taxa (Fig. 5), 4-colpate rectangular, 4-colpate square or 3-colpate triangular in the taxa from tropical Africa and Asia (Huynh 1968; Lu 1991).

KARYOLOGY. The chromosome number of *Hydrocera triflora* was determined as $2n = 16$ (Govindaran and Subramanian 1986; Rao et al. 1986). Chromosomal variation in *Impatiens* is extensive; the over 170 species studied so far range from $n = 3$ to $n = 33$ (Jones and Smith 1966; Larsen 1981; Zinoveva-Stahevitch and Grant 1984;

Govindaran and Subramanian 1986; Rao et al. 1986; Oginuma and Tobe 1991; Akiyama et al. 1992; Sugawara et al. 1995, 1997). Four basic numbers, x = 7, 8, 9, and 10, are more common than others. African species have mostly the basic numbers x = 7 and 8. In southern India, x = 7, 8, and 10 are predominant. In the western Himalaya x = 7 and 8 are common, whereas in the eastern Himalaya and adjacent SE Asia basic numbers of x = 9 and 10 are prevalent. In Japan, northern Asia, Europe and North America, x = 10 predominates. Jones and Smith (1966) and Akiyama et al. (1992) considered x = 7 as the ancestral condition whereas Rao et al. (1986) proposed x = 8 as the original base number.

POLLINATION AND REPRODUCTIVE SYSTEMS. As far as floral diversity is concerned, *Impatiens* is the dicot counterpart of orchids. As in Orchidaceae, the evolutionary trend towards specialized pollination has implied abundant speciation and often accounts for high rates of endemism in different regions. Due to the strong expression of protandry, most *Impatiens* are functionally monoecious. During the male phase, when the flower just opens, the stamens are fused by their anther walls to form a "brush" which completely covers the pistil. The upper filaments are longer than the lower ones, and the anthers project downwards and can "brush" the pollinator's back or head and transmit the pollen upon the next pollinator visiting the flower. During this stage, the stigmas are non-receptive and not exposed to come into contact with the pollinator. At the end of the male phase, the stamens are shed as a single unit and the stigmas become receptive. Extra-floral nectaries, usually in the form of stipitate glands, are often present and may reward "safety guards" (usually ants) protecting the flowers from damage by non-pollinating animals.

The pollination biology and breeding systems of the majority of species have not yet been studied, and most available studies have been conducted with the temperate species *I. capensis* and *I. pallida*. Published reports and personal field observations suggest the following animals as possible pollinators: honeybees (e.g. *I. austrotanzanica*, *I. gomphophylla*), bumblebees, solitary bees, butterflies (e.g. *I. walleriana*, *I. hoehneliana*), moths, hawkmoths (e.g. *I. teitensis*), flys (e.g. *I. elatostemoides*), and sunbirds (e.g. *I. niamniamensis*, *I. keilii*, *I. paucidentata*) (Arisumi 1974, 1980; Beck et al. 1974; Rust 1977, 1979; Kato 1988; Randall and Hilu 1990; Wilson 1995; Hurlbert et al. 1996).

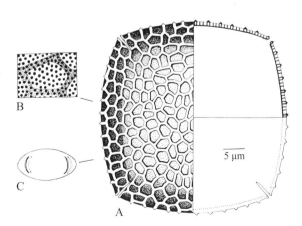

Fig. 5. Balsaminaceae. *Impatiens jurpia*, pollen grain. **A** Polar view (×1600). **B** A mesh of the reticulum. **C** Equatorial view. (Huynh 1968)

Whereas most *Impatiens* have mechanisms to ensure outcrossing, various forms of self-compatibility and cleistogamy have evolved. Grey-Wilson (1980a) suspected self-pollination to be limited to temperate species which, under unfavourable conditions or the absence of appropriate pollinators, occasionally switch to cleistogamy. Meanwhile, self-compatibility and facultative cleistogamy have been established for species such as *I. pallida*, *I. capensis*, and *I. hypophylla*. In these species, cleistogamy is environmentally controlled by factors such as light intensity, drought, flooding, and herbivore damage. Even on the same individual, the upper flowers have been observed to be chasmogamous, whereas the flowers close to ground often are cleistogamous (Schemske 1978, 1984; Antlfinger 1986; Gross et al. 1998; Paoletti and Holsinger 1999). This observation is also confirmed for the Madagascan *I. baroni* (pers. obs.). An extreme case is the Madagascan *I. inaperta* (formerly placed in the monotypic genus *Impatientella*), which produces exclusively cleistogamous flowers.

EMBRYOLOGY AND SEEDS. Mature pollen is 2-nucleate, with the generative nucleus remaining in an arrested metaphase (Huynh 1970).

The ovules are anatropous and bitegmic, unitegmic or intermediate and tenuinucellate; they have a characteristic endothelium. Mature ovules show an epistasis in the micropylar region (Huynh 1970). The outer integument contains raphid bundles. The growth processes involved in the changeover from the bitegmic to the unitegmic condition within *Impatiens* has been analysed by Boesewinkel and Bouman (1991).

In most species, embryo sac development conforms to the Allium type or Polygonum type. The straight embryo fills the seed almost completely. Endosperm formation is *ab initio* cellular with terminal haustoria. The micropylar haustorium is aggressive, and branches of it have been described to reach and enter the raphe, funicle, or even the placenta (Narayana 1965). In the mature seed, the endosperm is restricted to one or only a few layers; it is rich in lipids and poor in starch. The embryo is large and in *I. glandulifera* is provided by well-developed lateral roots which may promote the rooting of the seedling (Boesewinkel and Bouman 1991). The seed coat of *Impatiens* has a lignified exotesta and more or less crushed middle layers. The testa cells have straight or undulating walls. Single cells are modified to papillae or hairs. Grey-Wilson (1980a) distinguished five main seed types: smooth, warted, short-haired, long-haired

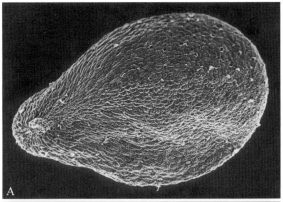
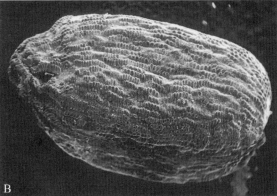
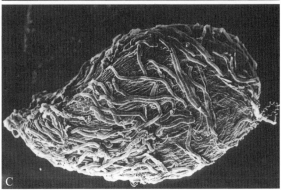

Fig. 6. Balsaminaceae. Seeds with smooth (**A**), warted (**B**) and long-haired (**C**) exotesta. **A** *Impatiens stuhlmannii* (×45). **B** *I. noli-tangere* (×50). **C** *I. niamniamensis* (×40). Photo E. Fischer

and seeds with club-shaped appendages (Fig. 6). The long-haired seeds bear long, helical, hair-like structures that have spiral thickenings (Fig. 6C). As this seed type occurs in groups of *Impatiens* with facultative and obligate epiphytes (e.g. *I. niamniamensis* aggregate), the hairs may play a role as adhesive structures. The seed coat of *Hydrocera* consists of 5–6 thickened layers but lacks a thickened exotesta (Venkateswarlu and Lakhshminarayana 1957).

FRUIT AND DISPERSAL. The mature fruit of *Impatiens* is fleshy and fusiform to cylindric in shape. Dehiscence is caused by a tension in the convex fleshy walls of the ovary. This tension is released by rupture of the walls along the septa, thus throwing the seeds for several meters by the elastic force of the dehiscing fruit wall. At maturity, the slightest touch will cause the explosion of the fruit, hence the folk names "touch-me-not" or "busy lizzy". The fruits of *Hydrocera* are fleshy pseudoberries with a hard endocarp that falls into five distinct units (partial endocarps, each with one seed and two air sacs). These units are dispersed by water and, due to their air sacs, are able to float.

PHYTOCHEMISTRY. *Impatiens* spp. regularly contain raphides of calcium oxalate and large amounts of leucanthocyanins including prodelphinidin. These compounds and raphides are rarely encountered in tender herbs (Hegnauer 1964, 1989). Apart from the occurrence of naphtoquinones such as lawsone, seed oils with acetate and parinarate glycerides are highly characteristic.

DISTRIBUTION AND HABITATS. *Hydrocera triflora* is a semi-aquatic herb native to tropical and subtropical Indo-China and the Indian subcontinent. *Impatiens* is essentially a montane genus centred in tropical Africa (c. 120 spp.), Madagascar (c. 160 spp.), southern India and Sri Lanka (c. 150 spp.), the eastern Himalaya (c. 120 spp.), Indo-China (c. 150 spp.) and adjacent SE Asia. Only a few species of *Impatiens* are known from the temperate montane regions of Europe, North Asia, and Central and North America, and no species are native to South America. A remarkable feature is the high degree of endemism in the centres of distribution. For example, over 90% of the Indian species are endemic and most of them are limited to the Western Ghats (Rao et al. 1986). All native species of Madagascar are endemic, concentrated in the eastern humid high-altitude regions. A similar pattern is observed in SW China and tropical Africa. The rapid radiation and possible subsequent parallel evolution and hybridization (Grey-Wilson 1980b; Merlin and Grant 1986) may obscure the phylogenetic relationships among the species. Recent rapid naturalizations of some species offer further examples of the adaptability of the genus. For example, *I. glandulifera*, a native of the Himalaya, is now widely naturalized in Europe and Japan. It probably escaped from gardens and, within less than 100 years, successfully spread in Europe (Kurtto 1996). *Impatiens*

walleriana, common as a weed throughout tropical South America, may represent another example of such recent naturalization and rapid radiation.

The preferred habitat of *Impatiens* is the herb layer of montane rain forests. Epiphytes are found in Africa (e.g. *I. keilii, I. epiphytica, I. irangiensis, I. paucidentata*), Madagascar (e.g. *I. purroi*) and India (e.g. *I. parasitica*). Seasonally dry habitats are usually avoided with the exception of some specialists with particular morphological adaptations (e.g. tubers in *I. tuberosa* and *I. cinnabarina*, Fig. 7).

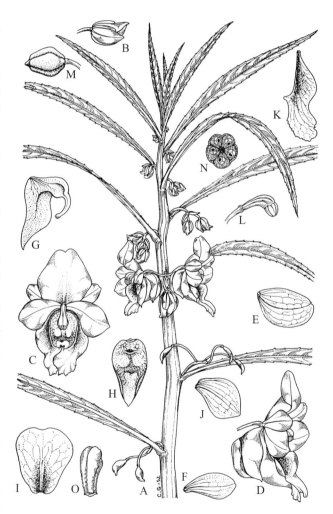

Fig. 7. Balsaminaceae. *Hydrocera triflora*. **A** Habit. **B** Flower bud. **C** Flower, front view. **D** Flower, lateral view. **E** Lower lateral sepal. **F** Upper lateral sepal. **G** Lower sepal and spur, lateral view. **H** Lower sepal, inside. **I** Dorsal petal. **J** Upper lateral petal. **K** Lower lateral petal. **L** Androecium. **M** Fruit. **N** Fruit, cross section. **O** Single partial endocarp. Drawn by C. Grey-Wilson. (Grey-Wilson 1980c)

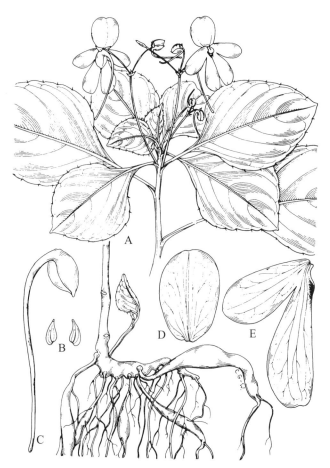

Fig. 8. Balsaminaceae. *Impatiens cinnabarina*. **A** Rootstock, stem base and flowering shoot. **B** Lateral sepals. **C** Lower sepal and spur. **D** Dorsal petal. **E** Lateral united petals. Drawn by C. Grey-Wilson. (Grey-Wilson 1980a)

PHYLOGENY. Morphologically based classifications placed Balsaminaceae in Geraniales and suggested close relationships to Tropaeolaceae and Geraniaceae (e.g. Warburg and Reiche 1895; Cronquist 1981). The tenuinucellate ovule, endothelium, haustoria and exotestal seed coat of *Impatiens* led Boesewinkel and Bouman (1991) to assume a link with sympetalous orders. Recent molecular phylogenetic studies suggest that this family is a member of the newly circumscribed order Ericales (Morton et al. 1997; APG 1998; Soltis et al. 2000). Within this broader alliance, plastid (*rbc*L), nuclear (18SrDNA) and combined sequence data suggest a relationship between Balsaminaceae and Marcgraviaceae, Pellicieraceae, and Tetrameristaceae. It is not easy to understand how these woody families shared a common ancestry with Balsaminaceae.

KEY TO THE GENERA

1. Petals all free; lateral sepals 4, large and coloured like the petals; fruit a pentagonal pseudoberry **1. *Hydrocera***
- Petals not all free, the lower 4 united laterally into pairs; lateral sepals 2–4, small and rather inconspicuous, generally white or greenish; fruit a dehiscent capsule
2. *Impatiens*

1. *Hydrocera* Blume — Fig. 7

Hydrocera Blume, Bijdr.: 241 (1825); Grey-Wilson, Kew Bull. 35: 213–219 (1980).

Semi-aquatic perennial herbs, rather succulent; stems erect, with fleshy and fibrous roots. Leaves alternate, petiolate, lamina linear-lanceolate to elliptic, at base with pair of sessile glands. Flowers pink with carmine to purplish, with 4 lateral sepals (2 pairs), almost as large as petals, lower sepal with short, curved spur; petals 5, free, dorsal petal semi-cucullate. Fruit a fleshy pseudoberry, with hard endocarp which divides into 5 separate units (partial endocarps, each with one seed and two air sacs). One species, *H. triflora* (L.) Wight & Arn., Sri Lanka, S India, S China to Thailand, Cambodia, Laos, Malay Peninsula, Celebes and Java.

2. *Impatiens* L. — Fig. 8

Impatiens L., Sp. Pl.: 937 (1753); Perrier de la Bâthie, Arch. Bot. 7, Mém. 1: 1–124 (1934); Perrier de la Bâthie, Mém. Acad. Sci. Paris II, 67, 2: 1–16 (1948); Shimizu, Acta Phytotax. Geobot. 24: 43–51 (1969); Grey-Wilson, *Impatiens* of Africa (1980); Grey-Wilson, Kew Bull. 34: 661–668 (1980); 35: 203–211 (1980); 44: 61–66 (1989); 44: 67–106 (1989); 44: 115–122 (1989); 44: 711–716 (1989).
Trimorphopetalum Bak. (1887).
Impatientella H. Perr. (1927).
Petalonema Peter (1928).

Terrestrial or epiphytic annual to perennial herbs; stems erect, ascending or decumbent. Leaves alternate, verticillate or opposite, petiolate, lamina linear-lanceolate to orbicular. Flowers white, yellow, orange to purplish, red, greenish or dark brown, lateral petals 2, rarely 4, much smaller than petals, lower sepal with short or long, broad or filiform spur, or spur reduced and lacking (in subg. *Trimorphopetalum*); petals 5, the 4 lateral united into pairs. Fruit an elastically dehiscent capsule, with few to numerous seeds. About 1000 spp. in Europe, North and Central America, Northern Asia, Himalaya, India to China, New Guinea, tropical Africa and Madagascar.

Selected Bibliography

Akiyama, S., Wakabayashi, M., Ohba, H. 1992. Chromosome evolution in Himalayan *Impatiens* (Balsaminaceae). Bot. J. Linn. Soc. 109: 247–257.

Antlfinger, A.E. 1986. Field germination and seedling growth of chasmogamous and cleistogamous progeny of *Impatiens capensis* (Balsaminaceae). Am. J. Bot. 73: 1267–1273.

APG (Angiosperm Phylogeny Group) 1998. See general references.

Arisumi, T. 1974. Chromosome numbers and breeding behavior of hybrids among Celebes, Java and New Guinea species of *Impatiens* L. Hort. Sci. 9: 478–479.

Arisumi, T. 1980. Chromosome numbers and comparative breeding behavior of certain *Impatiens* from Africa, India, and New Guinea. J. Am. Soc. Hort. Sci. 105: 99–102.

Beck, A.R., Weigle, J.L., Kruger, E.W. 1974. Breeding behavior and chromosome numbers among New Guinea and Java *Impatiens* species, cultivated varieties, and their interspecific hybrids. Can. J. Bot. 52: 923–925.

Boesewinkel, F.D., Bouman, F. 1991. The development of bi- and unitegmic ovules and seeds in *Impatiens* (Balsaminaceae). Bot. Jahrb. Syst. 113: 87–104.

Cheek, M., Fischer, E. 1999. A tuberous and epiphytic new species of *Impatiens* (Balsaminaceae) from Southwest Cameroon. Kew Bull. 54: 471–475.

Cronquist, A. 1981. See general references.

Govindarajan, T., Subramanian, D. 1986. Karyotaxonomy of south Indian Balsaminaceae. Cytologia 51: 107–116.

Grey-Wilson, C. 1980a. *Impatiens* of Africa. Rotterdam: A.A. Balkema.

Grey-Wilson, C. 1980b. Hybridization in African *Impatiens*. Studies in Balsaminaceae 2. Kew Bull. 34: 689–722.

Grey-Wilson, C. 1980c. *Hydrocera triflora*, its floral morphology and relationship with *Impatiens*. Kew Bull. 35: 213–219.

Gross, J., Husband, B.C., Stewart, S.C. 1998. Phenotypic selection in a natural population of *Impatiens pallida* Nutt. (Balsaminaceae). J. Evol. Biol. 11: 589–609.

Hegnauer, R. 1964, 1989. See general references.

Hurlbert, A.H., Hosoi, S.A., Temeles, E.J., Ewald, P.W. 1996. Mobility of *Impatiens capensis* flowers: effect on pollen deposition and hummingbird foraging. Oecologia 105: 243–246.

Huynh, K.-L. 1968. Morphologie du pollen des Tropaeolacées et des Balsaminacées I, II. Grana Palynol. 8: 88–184, 277–516.

Huynh, K.-L. 1970. Quelques caractères cytologiques, anatomiques et embryologiques distinctifs du genre *Tropaeolum* et du genre *Impatiens*, et position taxonomique de la famille des Balsaminacées. Bull. Soc. Neuchâtel. Sci. Nat. 93: 165–177.

Jones, K., Smith, J.B. 1966. The cytogeography of *Impatiens* L. (Balsaminaceae). Kew Bull. 20: 63–72.

Kato, M. 1988. Bumblebee visits to *Impatiens* spp. – pattern and efficiency. Oecologia 76: 364–370.

Kurtto, A. 1996. *Impatiens glandulifera* (Balsaminaceae) as an ornamental and escape in Finland, with notes on the other Nordic countries. Symb. Bot. Upsal 31: 221–228.

Larsen, K. 1981. Chromosome numbers in *Impatiens* from Thailand. Nord. J. Bot. 1: 43–44.

Lu, Y.-Q. 1991. Pollen morphology of *Impatiens* L. (Balsaminaceae) and its taxonomic implications. Acta Phytotax. Sin. 29: 352–357.

Merlin, C.M., Grant ,W.F. 1986. Hybridization studies in the genus *Impatiens*. Can. J. Bot. 64: 1069–1074.

Metcalfe, C.R., Chalk, L. 1950. See general references.

Morton, C.M., Mori, S.A., Prance, G.T., Karol, K.G., Chase M.W. 1997. Phylogenetic relationships of Lecythidaceae: a cladistic analysis using *rbc*L sequence and morphological data. Am. J. Bot. 84: 530–540.

Narayana, L.L. 1965. Contributions to the embryology of Balsaminaceae, part 2. J. Jap. Bot. 40: 104–116.

Oginuma, K., Tobe, H. 1991. Karyomorphology of two species of *Impatiens* from Kenya. Acta Phytotax. Geobot. 42: 67–71.

Paoletti, C., Holsinger, K.E. 1999. Spatial patterns of polygenic variation in *Impatiens capensis*, a species with an environmentally controlled mixed mating system. J. Evol. Biol. 12: 689–696.

Rama Devi, D. 1991. Floral anatomy of six species of *Impatiens*. Feddes Repert. 102: 395–398.

Randall, J.L., Hilu, K.W. 1990. Interference through improper pollen transfer in mixed stands of *Impatiens capensis* and *I. pallida* (Balsaminaceae). Am. J. Bot. 77: 939–944.

Rao, R.V.S., Ayyangar, K.R., Sampathkumar, R. 1986. On the karyological characteristics of some members of Balsaminaceae. Cytologia 51: 251–260.

Rust, R.W. 1977. Pollination in *Impatiens capensis* and *Impatiens pallida* (Balsaminaceae). Bull. Torrey Bot. Club 104: 361–367.

Rust, R.W. 1979. Pollination of *Impatiens capensis*: pollinators and nectar robbers. J. Kansas Entomol. Soc. 52(2): 297–308.

Schemske, D.W. 1978. Evolution of reproductive characteristics in *Impatiens* (Balsaminaceae): the significance of cleistogamy and chasmogamy. Ecology 59: 596–613.

Schemske, D.W. 1984. Population structure and local selection in *Impatiens pallida* (Balsaminaceae), a selfing annual. Evolution 38: 817–832

Soltis, D.E. et al. 2000. See general references.

Sugawara, T., Akiyama, S., Murata, J., Yang, Y.-P. 1995. Karyology of ten species of *Impatiens* (Balsaminaceae) from SW Yunnan, China. Acta Phytotax. Geobot. 45: 119–125.

Sugawara, T., Akiyama, S., Yang, Y.-P., Murata, J. 1997. Karyological characteristics of *Impatiens* (Balsaminaceae) in Yunnan, China. Acta Phytotax. Geobot. 48: 7–14.

Venkateswarlu, J., Lakshminarayana, L. 1957. A contribution to the embryology of *Hydrocera triflora* W. et A. Phytomorphology 7: 194–203.

Warburg, O., Reiche, K. 1895. Balsaminaceae. In: Engler & Prantl, Nat. Pflanzenfam. III, 5: 383–392. Leipzig: Engelmann.

Wilson, P. 1995. Selection for pollination success and the mechanical fit of *Impatiens* flowers around bumblebee bodies. Biol. J. Linn. Soc. 55: 355–383.

Zinoveva-Stahevitch, A.E., Grant W.F. 1984. Chromosome numbers in *Impatiens* (Balsaminaceae). Can. J. Bot. 62: 2630–2635.

Brunelliaceae

K. Kubitzki

Brunelliaceae Engler (1897) in Engler & Prantl, Nat. Pflanzen-fam., Nachtr. und Register zu II–IV: 226 (1897), nom. cons.

Evergreen trees, unarmed; twigs with angular internodes alternating with prominent nodes. Leaves simple or pinnate (when pinnate with stipels on the rachis), opposite or ternate, mostly dentate; stipules lateral, sometimes fragmented, free; indumentum of unicellular hairs, ovary and fruit usually bristly. Inflorescences axillary, thyrso-paniculate, provided with small, usually caducous prophylls. Flowers hermaphrodite or through abortion mostly unisexual and dioecious, 4, 5 or 6(–8)-merous; sepals valvate, persistent in fruit; petals 0; nectary disk intrastaminal, adnate to calyx, cupular, 8–10-lobed; stamens free, twice as many as sepals, in 2 whorls, rarely more; filaments inserted in the notches of the nectary disk, those of outer whorl alternate with, of inner opposite to sepals; anthers bithecate, introrse, dehiscing longi-tudinally, the connective with a small protrusion; disk intrastaminal, thick, flat, concave, with as many indentations as stamens or staminodes, tomentose or hispidulous; carpels free, basally immersed in disk, as many as sepals or fewer and alternating with them; ovaries ovoid or ellipsoid, hairy and mostly hispid, biovulate; stylodia verti-cal in flower, apically hooked or curled; stigmas linear, sutural-decurrent; ovules 2 per carpel, col-lateral, bitegmic, anatropous, epitropous, with micropyle directed upwards; obturator +; female flowers with sterile staminodia, male with rudi-mentary pistil. Fruit polyfollicular; follicles tomentose and mostly additionally hispid; stylo-dia diverging horizontally; the hard endocarp detaching from softer exocarp at maturity and expelling seeds which remain attached to a pla-centary stalk continuing the funicle; seeds with hard, shiny testa and raised raphe; embryo large, straight, embedded in carnose, mealy, white endosperm. $n = 14$.

A monogeneric family comprising some 61 species distributed from Mexico through Meso-America and the Greater Antilles south to 18 °S in Bolivia.

VEGETATIVE STRUCTURES. All *Brunellia* species are trees with a straight, cylindrical bole which often reaches the canopy and can attain a height of 40 m. Branching is copious and in dense stands is confined to the upper third of the trunk. Young shoots have a ferrugineous, ochraceous or rufes-cent indumentum which may be lanate, appressed-pubescent, or tomentose and may fall off at an early stage or persist over a long time.

The leaves are simple or pinnate, the former con-dition probably being derived. In pinnate leaves, the insertion of the leaflets on the rachis is accom-panied by stipels. Leaf shape – pinnate vs. simple – was the character on which Cuatrecasas (1970) based his two sections of *Brunellia*, but later (1985) he recognised that the unifoliate species group was not monophyletic. (His criterion for distinguishing between unifoliate and unifoliolate leaves was the presence of stipels in the latter.) Venation of the leaves and leaflets is craspedodro-mous; the margins are serrate, crenate or biserrate to even multi-indentate. The stipules are small, subulate or lanceolate, falling off very early but leaving notable elliptic or circular scars. Often there seem to be more than two stipules which Cuatrecasas (1970) described as "paired" stipules but this may rather be due to a splitting of the stipule primordia.

Nodes are tri(penta)-lacunar. The leaves have a hypodermis of one or two cell layers and conspic-uous bundle sheath extensions which isolate the palisade parenchyma cells of the meshes of the leaf venation. Stomata seem to be anomocytic. In the wood, growth rings are faintly visible or absent. Vessel elements are solitary or arranged in radial multiples; both simple and scalariform perforation plates occur. Vessel members are quite long (720–1480 µm). Pitting between them and between vessels and rays varies from scalariform through transitional to opposite. Libriform fibres are often septate and pitted on radial walls. Rays are heterogeneous and up to 6 cells wide, and usually have long uniseriate wings. Axial parenchyma and crystals are absent (Eyde in Cuatrecasas 1970).

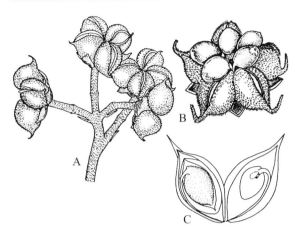

Fig. 9. Brunelliaceae. **A** *Brunellia standleyana*, part of flores-
cence. **B** *B. velutina*, polyfollicle with exposed seeds. **C** *B. latifolia*, follicle structure with exocarp, endocarp, placental cord
and seeds. (Cuatrecasas 1970)

REPRODUCTIVE STRUCTURES. *Brunellia* has
thyrso-paniculate inflorescences (Fig. 10), differing mainly in the complexity of branching (Orozco
and Weberling 1999). The flowers have a cupuliform nectariferous disk interpreted as androgynophore by Matthews and Endress (2002). Notable
features of the flowers include the apocarpous
gynoecium partially adnate to the surrounding
disk, the partially open ventral carpel, the two collateral ovules and the woody endocarp, and the
extended, sutural stigmas (also found in some
Rosaceae but not in Cunoniaceae).

In most *Brunellia*, during fruit development the
ventral side of the carpels grows faster than the
dorsal, pushing the stylodia into a horizontal position and leading to the stellate arrangement
attained in most species. To the reddish or ochraceous indumentum of the fruits are added nearly
always hard, hispid, lignified trichomes. When the
endocarp dehisces through transverse contraction
of the endocarp, the seeds are expelled but remain
attached to the fruit by a narrow ribbon of placental and marginal exocarp tissue continuing the
funicle. This flexuose stalk holds the seeds upright
above the pericarp (Fig. 9), displaying the shining
seeds on the stellate polycarpic fruit.

Details of seed coat structure were given by
López Naranjo and Huber (1971). Both testa and
tegmen are initially 2-layered but eventually the
two layers of the tegmen fuse. The endosperm is
carnose, subhyaline; the embryo is nearly as long
as the endosperm, a feature that distinguishes
Brunelliaceae and Cunoniaceae from families such
as Saxifragaceae, Escalloniaceae and Cornaceae,

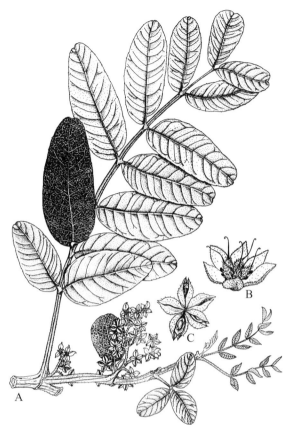

Fig. 10. Brunelliaceae. *Brunellia integrifolia*. **A** Habit.
B Flower. **C** Fruit. (Cuatrecasas 1970)

all formerly considered as possible relatives of
Brunelliaceae but which have small embryos. An
aril, ascribed to *Brunellia* by Cuatrecasas (1970)
and others, has not been seen; the raphe is somewhat raised but relatively dry and hard.

POLLEN MORPHOLOGY. Pollen grains are tricol-
porate, prolate to oblate, $11 - 32.5 \times 10.5 - 28 \mu m$.
The exine is incompletely to almost completely
tectate, and varies from coarsely reticulate to
rugulate and punctate (Orozco 2001; see also
Marticorena in Cuatrecasas 1970).

AFFINITIES. Engler's (1897) main reason for
recognising *Bunellia* as an independent family was
the position of the ovules, which are epitropous
with a ventral raphe, whereas at his time Cunoniaceae were considered to be apotropous throughout. (Cuatrecasas 1970: 45 observed, however, that
ovules in the basal Spiraeanthemoideae [Cunoni-
aceae] are also epitropous.) Perhaps more characteristic may be the peculiar development of the
carpels, including their abaxial deformation, the

sutural stigmas, the dissociation of exocarp and endocarp, and the presentation of seeds on the dehisced carpels.

The abovementioned similarities with Rosaceae are probably parallelisms: the androecium in *Brunellia* lacks the 10 + 5 + 5 pattern of Rosaceae, and the large embryo is oxalidalean rather than rosalean. Incidentally, also Engler (1897) referred to the diagrammatical similarity of *Brunellia* and *Cephalotus*, both now included in Oxalidales. This position is also confirmed by molecular analyses (e.g. Savolainen, Fay et al. 2000; Bradford and Barnes 2001).

DISTRIBUTION AND HABITATS. *Brunellia* is exclusively American, with only six of the 61 known species occurring north of Panama, of which a single, *B. comocladiifolia*, is native to the Greater Antilles, where it has given rise to several well-defined subspecies. The South American species are mostly Andean, with exception of four species that extend to the París Peninsula, Venezuela, and the Roraima Massif in eastern Venezuela. The great majority of the species are narrow endemics, and the majority of them (37) occurs in Colombia; Ecuador has six species, Peru nine, and Bolivia five. The presence of *Brunellia* in the Greater Antilles may indicate that the genus may have been represented north of Panama before the closing of the central American land bridge.

Most species grow in montane forests, either at lower elevations such as from 600 to 2000 m, or much higher, usually between 2800 and 3800 m. A single species, *B. hygrothermica*, grows in the superhumid sea-level region of Colombia.

A single genus:

Brunellia Ruiz & Pavón   Fig. 10

Brunellia Ruiz & Pavón, Prodr. Fl. Peruv. Chil.: 71, pl. 12 (1794); Cuatrecasas, Fl. Neotropica 2 (1970), and Suppl. (1985).

Description as for family. A single genus of 61 described species.

Selected Bibliography

Bradford, J.C., Barnes, R.W. 2001. Phylogenetics and classification of Cunoniaceae (Oxalidales) using chloroplast DNA sequences and morphology. Syst. Bot. 26: 354–385.

Cuatrecasas, J. 1970. Brunelliaceae. Fl. Neotropica Monogr. 2. Darien: Hafner.

Cuatrecasas, J. 1985. Brunelliaceae. Fl. Neotropica Monogr. 2, Suppl. New York: The New York Botanical Garden.

Ehrendorfer, F., Morawetz, W., Dawe, J. 1984. The neotropical angiosperm families Brunelliaceae and Caryocaraceae: first karyosystematical data and affinities. Pl. Syst. Evol. 145: 183–191.

Engler, A. 1897. Brunelliaceae, pp. 182–184. In: Engler & Prantl, Nat. Pflanzenfam., Nachtr. und Register zu II–IV. Leipzig: W. Engelmann.

Engler, A. 1930. Brunelliaceae, pp. 226–229. In: Engler & Prantl, Nat. Pflanzenfam. ed. 2, 18a. Leipzig: W. Engelmann.

López Naranjo, H., Huber, H. 1971. Anatomia comparada de las semillas de *Brunellia* y *Weinmannia* con respecto a su posición sistemática. Pittiera 3: 19–28.

Matthews, M.L., Endress, P.K. 2002. Comparative floral structure and systematics in Oxalidales (Oxalidaceae, Connaraceae, Brunelliaceae, Cephalotaceae, Cunoniaceae, Elaeocarpaceae, Tremandraceae). Bot. J. Linn. Soc. 140: 321–381. With 104 figs.

Orozco, C.I. 1997. Sobre la posición sistemática de *Brunellia* Ruiz & Pavon. Caldasia 19: 145–164.

Orozco, C.I. 2001. Pollen morphology of *Brunellia* (Brunelliaceae) and related taxa in the Cunoniaceae. Grana 40: 245–255.

Orozco, C.I., Weberling, F. 1999. A comparative study of inflorescences in *Brunellia* Ruiz & Pav. (Brunelliaceae) and related taxa. Beitr. Biol. Pfl. 71: 261–279.

Savolainen, V., Fay, M.F. et al. 2000. See general references.

Celastraceae

M.P. SIMMONS

Celastraceae R. Br., in Flinders, Voy. Terra Austral. 2: 554
(1814), nom. cons.
Hippocrateaceae Juss. (1811), nom. cons.
Stackhousiaceae R. Br. (1814), nom. cons.
Brexiaceae Loudon (1830).
Siphonodontaceae (Croizat) Gagnep. & Tardieu (1951).
Canotiaceae Airy Shaw (1965).
Plagiopteraceae Airy Shaw (1965).
Pottingeriaceae? (Engl.) Takht. (1987).

Monoecious, andromonoecious, dioecious, gynodioecious, or polygamous, erect or scandent trees, shrubs, lianas with quickly-deciduous scales, without tendrils, or annual or perennial herbs with erect or prostrate stems, rarely suffrutices, rhizomatous shrubs, ericoid subshrubs, or epiphytic shrubs, glabrous or glabrescent, rarely puberulent, densely pilose, or hirsute and stellate pubescent, often with elastic threads in soft tissues that are evident when broken, unarmed or with thorns, rarely with stems terminating in sharp points, rarely with glandular stems, rarely with buttressed trunks. Leaves simple, alternate, opposite, or subopposite, rarely whorled, subverticillate, or irregularly scattered, fasciculate on short shoots, or opposite on mature branches and alternate on juvenile branches, petiolate or rarely sessile, rarely geniculate, blade laminar, rarely needle- or scale-like, venation pinnate, rarely acrodromous, secondary veins reticulate or rarely distinct crossbars, rarely with abaxial domatia in axils or larger veins, margins entire, crenate, serrate, dentate, spinose, glandular-toothed, spinosedentate, rarely notched; stipules small and caducous, rarely 0. Inflorescences axillary or terminal, rarely epiphyllous or cauliflorous, thyrsoid, cymose, fasciculate, or flowers solitary, rarely paniculate, umbellate-cymose, umbellate, racemose, or in spikes. Flowers actinomorphic, rarely ± zygomorphic with 4 of 5 petals arched, bisexual or unisexual, rarely perigynous with short or cupular hypanthium, perianth (3)4–5(6)-merous, sepals and petals free, rarely petals medially connate; disk intrastaminal, stamens on disk, or extrastaminal, annular, margins upturned, pulvinate, or cupular, fleshy or membranous, sometimes indistinct or 0, continuous, rarely discontinuous, entire, lobed, or angular, rarely lacerate or irregularly lobed; stamens (2)3–5, rarely numerous, rarely 3 long and 2 short, rarely alternating with staminodes, staminodial or 0 in female flowers, alternate with petals when stamen number equals petal number, anthers (1)2-celled, basifixed to dorsifixed, sometimes versatile, dehiscing longitudinally, obliquely, or transversely, introrse, latrorse, or extrorse, rarely apical, connective sometimes apiculate, rarely pustular, with bilobed extension, or tipped by white gland, androgynophore occasionally ±; ovary superior to half-inferior, often partially immersed in disk, present as pistillode or essentially absent in male flowers, completely or incompletely (1)2–5(10)-locular, rarely each locule horizontally divided into 1-ovulate locelli, placentation axile, rarely intruded parietal or basal, ovules erect, axile, or pendulous, 1–12(–numerous) per locule; style terminal, simple, short to 0, stigma simple or lobed, rarely ovary with stigmatic lines on each carpel on margin of an apical hollow with style-like central column arising from base of hollow. Fruit a loculicidally and/or septicidally dehiscent capsule, rarely beaked, schizocarp of 2–5 indehiscent mericarps, drupe, berry, or samara with a single apical wing, 3–5 lateral wings, or a single surrounding wing, rarely an indehiscent capsule or nut with lateral style, pericarp woody, bony, coriaceous, fleshy, or chartaceous, rarely fibrous, capsules smooth, angular, deeply lobed, transversely flattened and lobed to base, or rarely lobed ± halfway to base or entirely connate, rarely echinate, laterally winged, with lateral or oblique horn-like outgrowths, or flattened along each locule. Seeds 1–numerous, smooth or occasionally furrowed, with unbranched or occasionally branched raphe, albuminous or exalbuminous, sometimes winged, wing membranous, basal, sometimes reduced to narrow stipe, apical, or surrounding seed, exarillate or aril basal to completely enveloping seed, aril membranous, fleshy, rarely with basal or apical filamentous extensions, or mucilaginous.

A subcosmopolitan family of 98 genera and about 1211 species that is most diverse in the tropics and subtropics, with fewer temperate species. One genus (*Pottingeria*) is tentatively included, another genus (*Nicobariodendron*) is insufficiently known.

VEGETATIVE MORPHOLOGY. Most members of Celastraceae are shrubs to small trees, although members of some genera reach up to 50 m tall and have buttressed trunks (*Bhesa*, *Kokoona*, and *Lophopetalum* spp.). Most Celastroideae are erect, but some are scandent (*Allocassine*, *Celastrus*, *Euonymus* spp., *Maytenus* spp., *Monimopetalum*, *Ptelidium scandens*, *Tripterygium*), whereas most Hippocrateoideae and Salacioideae are lianas. Scandent taxa may climb with the aid of persistent, sharp, downward-pointed bud scales (e.g., *Celastrus*; Velenovský 1910; Ding Hou 1955), others use adventitious roots (*Euonymus* spp.; Troll 1943). Members of Hippocrateoideae and Salacioideae that grow as lianas do so by thigmotropism and quickly-deciduous scales on dimorphic branches (Troll 1937; Loesener 1942b; Hallé 1962). Scandent *Celastrus* act as stranglers (Troll 1943). Atypical growth forms for the family include: annual or perennial herbs with erect or prostrate stems (Stackhousioideae), suffrutices (*Gymnosporia* spp.), rhizomatous shrubs (*Paxistima*), ericoid subshrubs (*Empleuridium*), and epiphytic shrubs (*Quetzalia standleyi*).

Most genera are unarmed. However, stems terminating in sharp points occur in *Acanthothamnus* and *Canotia*, and thorns occur in *Gloveria*, *Gymnosporia*, *Moya*, and *Putterlickia*. In the same four genera with thorns (and in two genera without thorns: *Schaefferia* spp., *Wimmeria* spp.), leaves are often fasciculate on short shoots, frequently on the thorns themselves. Leaves (or fascicles of leaves) and inflorescences are borne on the thorns (except in *Moya*). These leaves on the thorns are often opposite, even when the rest of the plant has alternate phyllotaxy. A similar pattern is found in *Catha edulis* and *Lydenburgia cassinoides*, which are closely related to *Gymnosporia* and *Putterlickia* (Simmons et al. 2001b). In these two species that lack thorns, leaves are opposite on mature (flowering) branches and alternate on juvenile branches. Based on the similar pattern and close relationship, the thorns of *Gymnosporia* and *Putterlickia* may be homologous to the mature, unarmed branches of *Catha edulis* and *Lydenburgia cassinoides*.

Leaves are simple, and generally alternate, opposite, or subopposite. Leaves may also be whorled (*Brexiella ilicifolia*, *Crossopetalum* spp., *Euonymus* spp., *Euonymopsis* spp.), subverticillate (*Salaciopsis*), or scattered (*Empleuridium*). Leaves are generally petiolate, though they may be sessile (*Apatophyllum*, *Empleuridium*). *Bhesa* is unique in the family with geniculate petioles (at the laminar end; Fig. 11A). Although leaves are generally laminar, they may also be needle-like (*Apatophyllum*, *Empleuridium*), or even reduced to scales (*Canotia*, *Psammomoya*). The scale-shaped leaves of *Canotia* and *Psammomoya* do not appear similar and the two genera are not closely related (Simmons et al. 2001b), indicating that laminar leaves have been reduced to scales independently in the two genera. Leaf dimorphism between juvenile and mature leaves occurs in *Elaeodendron* spp., *Pleurostylia pachyphloea*, *P. putamen*, and *Brexia*. Leaf dimorphism is particularly dramatic in *Elaeodendron orientale* with entire, narrowly-lanceolate juvenile leaves and serrulate, elliptic mature leaves.

VEGETATIVE ANATOMY. Solereder (1908) and Metcalfe and Chalk (1950) have provided overviews of the vegetative anatomy of Celastraceae, and Stant (1951) has examined the anatomy of Stackhousioideae. Wood anatomy of Celastraceae has been examined by Record (1943), Mennega (1972, 1994, 1997), Carlquist (1987; Stackhousioideae) and Archer and van Wyk (1993a). Growth rings are inconspicuous to distinct. Growth ring borders are marked by the presence of unlignified ray cells in some Hippocrateoideae (*Anthodon*, *Apodostigma*, *Cuervea*, *Pristimera*, *Reissantia*, *Simicratea*, *Simirestis*). Vessels are solitary or in radial multiples of 2–3(4). Perforation plates are generally simple, rarely scalariform (*Bhesa*, *Brexia*, *Elaeodendron*, *Perrottetia*). Rays are often strictly uni–biseriate, sometimes of two distinct sizes of 1–2 cells wide and 3–6 cells wide (*Celastrus*, *Elaeodendron*, *Microtropis*, *Schaefferia*, *Siphonodon*), or uniseriate and 4–40 cells wide (Hippocrateoideae). Fibers are thin- to thick-walled, with numerous bordered pits. In some taxa, parenchyma-like bands of thin-walled, septate wood fibers occur (e.g., *Elaeodendron*). Included phloem, generally as conspicuous concentric bands, occasionally as isolated strands, occurs in Salacioideae.

In stems, the cork generally develops in the subepidermis. However, the cork develops in the epidermis of *Euonymus* and *Microtropis*. Cork wings occur in *Euonymus alatus* and *E. phellomanus*. Phloem and xylem are in closed bundles separated by narrow rays. Laticifers containing gutta-percha occur at the interface of the cortex and phloem (Drennan et al. 1987), but perhaps not in Stackhousioideae (Stant 1951). Pith is homogeneous or heterogeneous, absent in *Tripterygium wilfordii*.

Leaf epidermal characters have been examined by Pant and Kidwai (1966), den Hartog and Baas (1978), and Archer (1990). Leaf anatomy has been examined by Jansen and Baas (1973) and Müller (1995). Stomatal types are very diverse (den Hartog and Baas 1978; Archer 1990). Stomata are often strictly abaxial, sometimes also on the midrib of the adaxial surface, rarely adaxial and abaxial. In some genera, crystals occur in small epidermal cells as druses (some *Salacia* spp., *Siphonodon*, some Hippocrateoideae) or solitary, rhomboidal crystals (*Elaeodendron*, some *Lophopetalum* spp., some Hippocrateoideae). *Perrottetia alpestris* and *P. sandwicensis* have mucilaginous epidermal cells (Solereder 1908). Laticifers containing gutta-percha occur in a crescent-shaped layer on the abaxial sides of vascular bundles (Drennan et al. 1987), and sometimes in the mesophyll (some *Wimmeria* spp., many Hippocrateoideae and Salacioideae). The petiole generally has one arc-shaped or circular vascular bundle, rarely three (*Bhesa*, *Brexia*) or two to several vascular bundles (*Lophopetalum*, and some Hippocrateoideae and Salacioideae).

FLORAL MORPHOLOGY. Flowers are actinomorphic, but ± zygomorphic with 4 of 5 arched petals in *Apodostigma*. The perianth is generally 4–5-merous, but rarely 3-merous (*Elaeodendron* spp., *Plagiopteron*) or 6-merous (rarely in *Gymnosporia*).

The nectariferous disk may be intrastaminal (most Celastroideae, Figs. 12–14), extrastaminal (*Brassiantha*, *Dicarpellum*, *Hypsophila*, *Kokoona*, *Sarawakodendron*, Salacioideae, most Hippocrateoideae), or may have stamens on the disk (e.g., *Glyptopetalum*, *Lophopetalum*). In Celastroideae, the disk is generally annular or has the margins upturned, whereas in Hippocrateoideae and Salacioideae the disk is generally pulvinate or cupular. The disk is generally fleshy or membranous, but sometimes indistinct (*Arnicratea*, *Plagiopteron*, *Schaefferia* spp., *Simirestis*) or absent (*Bequaertia*, *Campylostemon*, *Microtropis*, *Schaefferia* spp., *Tristemonanthus*). The disk is generally continuous, but is discontinuous in *Apodostigma* and *Cheiloclinium*. The disk margin may be entire, lobed, or angular, rarely lacerate (*Brexia*) or irregularly lobed (Fig. 18, *Helictonema*). In Stackhousioideae, a thin disk lines the hypanthium.

There are 4–5 stamens in Celastroideae (except 3 stamens in *Dicarpellum*) and 3 stamens in Hippocrateoideae (except *Campylostemon* with 5 stamens) and Salacioideae (except some *Cheiloclinium* spp. with 5 stamens). Exceptions are 2 stamens in *Nicobariodendron* and *Salacia* spp., and numerous stamens in *Plagiopteron*. Loesener (1942a, 1942b) used stamen number as the only character to differentiate Celastraceae s.str. (4–5 stamens) from Hippocrateaceae (2–3 stamens). However, there appear to have been multiple changes in stamen number from 5 to 3 (independently in *Dicarpellum*, and one or more origins in Hippocrateoideae and Salacioideae), and from 3 to 5 (independently in *Campylostemon* and *Cheiloclinium*; Simmons et al. 2001b). In *Siphonodon*, stamens alternate with staminodes. In *Stackhousia* and *Tripterococcus*, the stamens are of unequal lengths with 3 long stamens and 2 short stamens (Fig. 17c).

Anthers are generally 2-celled, but can be 1-celled in some *Euonymus* spp. Anthers may dehisce longitudinally (most Celastroideae), obliquely (e.g., *Glyptopetalum*, *Plenckia*), or transversely (*Hypsophila*, most Hippocrateoideae and Salacioideae). The direction of anther dehiscence may be introrse (most Celastroideae), latrorse (e.g., *Hedraianthera*, *Kokoona*), or extrorse (most Hippocrateoideae and Salacioideae), or rarely apical (*Plagiopteron* [Fig. 19E], *Salacia* spp.). Connectives are generally apiculate or not extended (often variable in a single inflorescence), are pustular in some *Kokoona* spp., have bilobed extensions in some *Peritassa* spp., or are tipped by a white gland in *Macgregoria racemigera*. Androgynophores are generally absent, but they are present in some Hippocrateoideae (Fig. 18c) and Salacioideae.

Ovaries are 2–5-locular (3-locular in Hippocrateoideae and Salacioideae except some *Cheiloclinium* and *Salacia* spp.), but may be 1-locular (*Empleuridium*, *Pottingeria*) or 10-locular (*Siphonodon*). Placentation is axile, rarely intruded parietal (*Pottingeria*) or basal (*Empleuridium*). Ovules may be erect (most Celastroideae), axile (most Hippocrateoideae and Salacioideae), or pendulous (*Euonymus* spp., *Glyptopetalum*, *Gyminda*, *Maurocenia*, *Tetrasiphon*). Styles are simple and range from short to obsolete (e.g., *Brassiantha*), and the stigmas simple or 2–5-lobed.

Flowers of *Plagiopteron*, *Pottingeria*, and *Siphonodon* are unique within the family: *Plagiopteron* (Fig. 19) with its numerous stamens that have apically-dehiscent anthers; *Pottingeria* with its 1-locular, 3-carpellate, paracarpous ovaries with numerous ovules per carpel and septicidally dehiscent capsules; *Siphonodon* with its 10-locular ovaries in which each locule is horizontally divided into 1-ovulate locelli. In *Siphonodon*, a stigmatic line is located at the upper inner margin of each carpel around a central, apical hollow within the ovary; a sterile, style-like central column arises from the base of the hollow (Hooker 1857). Although *Plagiopteron* and *Siphonodon* are well-supported members of Celastraceae, *Pottingeria* has not been incorporated in a phylogenetic analysis, and its inclusion within the family is uncertain (Airy Shaw et al. 1973).

FLORAL ANATOMY. In his study of the vascular traces of *Celastrus*, *Euonymus*, and *Paxistima* flowers, Berkeley (1953) concluded that Celastraceae have a floral tube, and the nectariferous disk is homologous to modified stamens. In *Celastrus*, the vascular traces of the disk diverge from stamen and petal traces. In *Euonymus* and *Paxistima*, no vascular traces lead to the disk.

EMBRYOLOGY. The embryology of Celastraceae has been examined in several species by Mauritzon (1936a, 1936b), Adatia and Gavde (1962), Narang (1965), and Tobe and Raven (1993). Anthers are tetrasporangiate, have a fibrous endothecium, crushed middle layers, and a glandular tapetum. Tapetal cells are two- or more-nucleate. Cytokinesis in meiosis is simultaneous. Microspore tetrads are tetrahedral or decussate. Mature pollen is 2- or 3-celled (Brewbaker 1967).

Ovules are anatropous, bitegmic, crassinucellate or tenuinucellate, and have the Polygonum-type of embryo sac formation. Endosperm formation is nuclear. Endosperm may be present or absent in mature seeds. In mature seeds, the embryos are straight and symmetrical. Polyembryony has been found to be common in several species of *Euonymus* and in *Celastrus paniculatus* (Adatia and Gavde 1962; Brizicky 1964b).

POLLEN MORPHOLOGY. The most thorough survey of pollen in Celastraceae was completed by Lobreau-Callen (1977). Within Celastroideae, pollen is isopolar and shed in monads, whereas in Hippocrateoideae, Salacioideae, and *Lophopetalum*, pollen may also be shed as tetrahedral tetrads, or polyads of eight or 16 grains (Ding Hou 1969; Lobreau-Callen 1977). The apertures of grains shed in tetrads or polyads may be irregularly positioned. Grains are generally tricolporate, but bicolporate and tetracolporate grains have been reported (Ikuse 1956; Ding Hou 1969). Some genera in which grains are shed as tetrads or polyads have triporate grains. Individual grains are suboblate to subprolate, small to medium sized (longest axis 11–35 µm), with reticulate (rarely crotonoid) sculpturing (Lobreau-Callen 1977). An endexine fold (also described as a conduplication or replication) in the aperture is characteristic of Celastraceae s.l., but is absent in *Hippocratea* and *Lauridia* (Lobreau-Callen 1977; Archer and van Wyk 1992). An "annulus" (thickening around the internal margin of the pore) occurs in *Kokoona*, *Lophopetalum*, and Hippocrateoideae (Hallé 1960; Lobreau-Callen 1977).

KARYOLOGY. Relatively few chromosome numbers have been published, mostly in *Celastrus* and *Euonymus*. Base chromosome numbers in Celastraceae include x = 8, 9, 10, 12, 14, 15, 17, and 23. Polyploidy appears to be relatively common in Celastraceae. For example, in *Euonymus*, gametophytic chromosome numbers range from eight (*E. echinatus*; Mehra 1976), 16 (e.g., *E. radicans*; Bowden 1940), and 24 (*E. bullatus*; Mehra 1976) to 32 (e.g., *E. europaeus*; Wulff 1937). Although base chromosome numbers fluctuate in Celastroideae, the base chromosome number has stabilized in Hippocrateoideae and Salacioideae at x = 14 (except *n* = 30 in *Semialarium mexicanum*; Bawa 1973), with *n* = 14 and 28.

POLLINATION. Few taxa have been investigated to determine their specific pollinators. Nectariferous disks, present in most genera, attract bees, beetles, flies, wasps, and even ants (Knuth 1908; Ding Hou 1962; Brizicky 1964a; Sebsebe 1985). Insects are attracted to the thin layer of nectar secreted by the disks of flowers of *Euonymus europaeus*, generally resulting in cross-pollination (Darwin 1877; Müller 1883; Knuth 1908). In addition to the nectariferous disks, the carrion odor produced by flowers of *Gymnosporia buxifolia* attracts insects, especially bluebottle flies (Palmer and Pitman 1972). Flowers of Stackhousioideae produce a sweet smell that attracts insects (Mattfeld 1942).

REPRODUCTIVE SYSTEMS. Most Celastraceae have bisexual flowers. Among those with unisexual flowers, plants may be monoecious (*Elaeodendron* spp., *Gymnosporia* spp.), andromonoecious (*Peripterygia*), gynodioecious (some *Euonymus*, *Maurocenia* [though perhaps dioecious; Archer and van Wyk 1998a]), polygamous (*Elaeodendron* spp., *Gymnosporia*, *Maytenus* spp., *Orthosphenia*, *Plenckia*, *Tricerma*, *Tripterygium*), or dioecious

(*Celastrus*, *Dicarpellum*, *Elaeodendron* spp., *Euonymus* spp., *Gyminda*, *Gymnosporia* spp., *Maytenus* spp., *Menepetalum*, *Microtropis*, *Nicobariodendron*, *Perrottetia*, *Quetzalia*, *Salaciopsis*, *Schaefferia*, *Tetrasiphon*).

Within Celastraceae, the reproductive system of *Euonymus europaeus*, a gynodioecious species, has been most-intensively studied (Darwin 1877; Webb 1979). Bisexual flowers are protandrous, making self-fertilization less likely (Müller 1883; Knuth 1908). Hermaphrodites are significantly more common than females, although the number of flowers per inflorescence is about the same. Females produce significantly more fruits and seeds per inflorescence than hermaphrodites. Furthermore, the females produce larger seeds.

Hybrid individuals and putative hybrid species have been reported in *Euonymus* (Staszkiewicz 1997) and *Celastrus* (Ding Hou 1955; Dreyer et al. 1987). However, artificial hybridization between *Celastrus orbiculatus* and *C. scandens* has been relatively unsuccessful (White and Bowden 1947).

FRUIT AND SEED. Fruit types in Celastroideae include: capsules with loculicidal dehiscence (e.g., *Maytenus*), loculicidal and septicidal dehiscence (*Canotia*), or septicidal dehiscence (*Pottingeria*), schizocarps of indehiscent mericarps (Stackhousioideae), drupes (e.g., *Elaeodendron*), berries (e.g., *Cassine*), nuts (*Pleurostylia*), or samaras with a single apical wing (*Plenckia*, *Rzedowskia*, *Zinowiewia*), 3–5 lateral wings (*Platypterocarpus*, *Tripterygium*, *Wimmeria*), or a single surrounding wing (*Ptelidium*).

The unusual, capsular fruits of Hippocrateoideae are transversely flattened and trilobed, with the lobes generally extending all the way to their base (Fig. 18H). Each of these three segments of the fruit has been referred to as separate capsules (Miers 1872; Smith 1940) or mericarps (Hallé 1962, 1983; Görts-van Rijn and Mennega 1994). However, these segments are not separate capsules because they all develop from a single ovary, and they are not mericarps because they do not split into separate seed-containing parts upon dehiscence (Brizicky 1964a). The seeds of Hippocrateoideae generally have membranous basal wings that may be reduced to narrow stipes (Hallé 1983). The raphe generally traverses the wing near its center (e.g., *Hippocratea*) or rarely along the margin of the wing opposite the thickened margin (*Helictonema velutinaum*, Fig. 18K; Hallé 1962,

1983). The thickened margin of the wing has also been described as the raphe (Miers 1872; Smith 1940). However, the figures by Hallé (1983) conclusively demonstrate that the raphe generally traverses the wing near its center.

The fruits of Salacioideae have been described as baccate (e.g., Lindley 1853; Bentham and Hooker 1862; Hallé 1986), drupaceous (e.g., Miers 1872; Smith 1940; Ding Hou 1964), or drupaceous or baccate (Loesener 1942b). I contend these fruits are properly described as baccate because I have neither seen nor read of a stony endocarp in these fruits. The seeds are surrounded by mucilaginous pulp that is derived from the funiculus and is homologous to an aril (Miers 1872; Baillon 1880; Simmons and Hedin 1999).

Planchon (1845) described the putative arils of *Celastrus scandens* and *Euonymus latifolius* as arillodes – false arils. Planchon cited these false arils as derived from the exostome of the outer integument rather than the funiculus. However, Miers (1856) disputed Planchon's conclusion and, based on his investigation of *Euonymus europaeus*, he concluded that the aril is derived from the funiculus and is therefore a true aril. Pfeiffer (1891) described the arils of *Celastrus*, *Euonymus*, and *Gymnosporia cassinoides* as derived from the exostome and the hilum. Corner (1976) described the aril of *Euonymus glandulosus* as derived entirely from the funiculus, and the aril of other species as derived from the exostome and the funiculus (*Catha edulis*, *Celastrus paniculatus*, *Sarawakodendron filamentosum*). Van der Pijl (1972) concluded that *Euonymus* has an arillode, not an aril.

Alate seeds may have a basal wing with medial attachment of the funiculus vasculature (*Canotia holacantha*, *Catha edulis*), a wing surrounding the seed with medial attachment of the funiculus vasculature (*Kokoona* [appearing primarily as an apical wing], *Lophopetalum*, *Peripterygia*), and basal wings with basal attachment of the funiculus vasculature (Hippocrateoideae). The wing of *Catha edulis* has been described as an aril derived from the funicle and side of the exostome (Loesener 1942a; Corner 1976). The winged seeds of other Celastraceae may also be interpreted as homologous to arils or arillodes (van der Pijl 1972; Simmons and Hedin 1999; but see Miers 1872, and Hallé 1983). Seed wings have been interpreted as being derived from arils in other families (e.g., Forman 1965; Tobe and Raven 1984), and Corner (1954: 156) concluded that "all winged seeds are *prima facie* indications of arillate ancestry". In phylogenetic analyses of Celastraceae, genera with winged seeds have been consistently resolved as nested within clades of obviously-arillate taxa (Simmons and Hedin 1999; Simmons et al. 2001a, 2001b).

Seed anatomy of Celastraceae has been described by Corner (1976). The testa is 5–12 cells thick, generally lignified, and unspecialized, often with a palisade exotesta. The tegmen is 3–10 cells thick, generally with ribbon-like fibers for the outer layer, making the seeds exotegmic. However, the tegmen is crushed and absorbed in Salacioideae. In *Microtropis*, *Perrottetia*, and *Quetzalia*, the pulpy sarcotesta may be mistaken for an aril. In the testa of *Perrottetia*, prismatic Malpighian cells project to different lengths causing the furrowed appearance of the seeds. The embryo has a short to long radicle, and flat, generally green cotyledons that are free to united. In Salacioideae, the cotyledons are generally described as massive (e.g., Miers 1872; Loesener 1942b; Hallé 1962), and may be free or fused (Ding Hou 1964; de Vogel 1980). However, Corner (1976: 94) described the embryo of *Salacia* sp. as "composed entirely of swollen hypocotyl ... without trace of cotyledons". Corner presumably interpreted the fused cotyledons as the swollen hypocotyl. Endosperm, when present (it is absent in *Goniodiscus*, *Kokoona*, *Lophopetalum*, *Polycardia*, Hippocrateoideae and Salacioideae), is oily, generally thin-walled, and ranges from minimal to abundant.

Seed germination of Celastraceae has been studied by Lubbock (1892), Hallé (1962), Rudolf (1974), Wendel (1974), and de Vogel (1980). Seeds of Salacioideae have hypogeal germination. Seeds of Hippocrateoideae have hypogeal (*Cuervea macrophylla*, *Elachyptera* sp., *Loeseneriella africana*, *Simirestis unguiculata*) or epigeal (*Apodostigma pallens*, *Reissantia astericantha*) germination. Other Celastraceae (*Bhesa robusta*, *Celastrus scandens*, *C. paniculatus*, *Euonymus* spp., *Lophopetalum javanicum*, and *Siphonodon celastrineus*) have epigeal germination. In all but *Lophopetalum* and *Salacia*, the seedlings become free from their seedcoats.

DISPERSAL. Several genera are adapted for wind dispersal with samaras (*Platypterocarpus*, *Plenckia*, *Ptelidium*, *Rzedowskia*, *Tripterygium*, *Wimmeria*, and *Zinowiewia*), winged mericarps (*Stackhousia* spp., *Tripterococcus* [Fig. 17E]), or winged seeds (*Canotia holacantha*, *Catha edulis* [Fig. 12G], *Kokoona*, *Lophopetalum* [Ridley 1930; Sinha and Davidar 1992], *Peripterygia*, Hippocra-

teoideae). The winged capsular fruits of some *Euonymus* spp. are unlikely to serve for wind dispersal, but may serve to make the fruits more conspicuous to animals (Blakelock 1951). Birds and/or mammals disperse berries and drupes, as has been reported for drupes of *Elaeodendron australis* (Benson and McDougall 1995), *E. croceum* (Phillips 1927), *E. glaucum*, *E. viburnifolium* (Ridley 1930), *Mystroxylon aethiopicum*, and *M. burkeanum*, and berries of *Cassine peragua* and *Lauridia tetragona* (Palmer and Pitman 1972). Drupes of *Elaeodendron croceum* have been reported to be dispersed by elephants grazing on the foliage (Ridley 1930). The berries of Salacioideae, with their seeds surrounded by mucilaginous arils, have been reported to be dispersed by monkeys (Garber 1986).

Capsular fruits with seeds that have bright-colored arils that contrast with the color of the capsule and seed occur in many members of the family. These seeds are bird dispersed (reported for *Celastrus*, *Euonymus*, and *Maytenus*; Pfeiffer 1891; Ridley 1930; Martin et al. 1951; Ding Hou 1962; Brizicky 1964a; Palmer and Pitman 1972; Dreyer et al. 1987; Benson and McDougall 1995), and both small and large mammal dispersed (reported for *Celastrus*, *Euonymus*, and *Maytenus*; Martin et al. 1951; Dreyer et al. 1987; Rodrigues et al. 1993; Kollmann et al. 1998). Additionally, the seeds of *Maytenus silvestris* have been reported to be ant dispersed (Benson and McDougall 1995).

PHYTOCHEMISTRY. The phytochemistry of Celastraceae has been reviewed by Hegnauer (in Ding Hou 1962), Hegnauer (1964, 1966, 1989), Gibbs (1974), and Brüning and Wagner (1978). Dulcitol, accumulated in leaves and bark, is thought to occur in all members of the family. Dulcitol may be "a highly characteristic biochemical feature" of Celastraceae (Hegnauer in Ding Hou 1962: 229). Two alkaloids, (L(S)-(-)-alpha-aminopropiophenone (cathinone) and *d*-norpseudoephedrine (cathine), concentrated in young leaves and twigs of *Catha edulis*, act as stimulants of the central nervous system, similar to amphetamine (Krikorian and Getahun 1973; Szendrei 1981; Zelger et al. 1981). Maytansine and mayteine, isolated from *Gymnosporia*, *Maytenus*, and *Putterlickia*, have shown tumor-inhibitory activity and have been investigated in clinical trials (summarized by Sebsebe 1985, and Raintree Nutrition, Inc. 2000b). Many taxa have conspicuous, yellow pigments in their bark (of stems and/or roots) that are triterpene derivatives (celastrol, pristemerin or tingenin) and may be ubiquitous within the family (Brüning and Wagner 1978; R. Archer, pers. comm. 2000).

Gutta-percha (E-1,4-polyisoprene) has been isolated from many species in Celastraceae and is apparently universal in the family. Gutta-percha is the trans-configuration of 1,4-polyisoprene, whereas rubber is the cis-configuration. The presence of gutta-percha is generally evident as elastic threads when soft tissues are broken and pulled apart (but not, for example, in *Allocassine*; Archer and van Wyk 1998b). Gutta-percha is produced in laticifers and occurs in the cytoplasm of most types of cells (Drennan et al. 1987).

Subdivision and Relationships Within the Family. The most-recent comprehensive taxonomic treatments of Celastraceae s.str. and Hippocrateaceae were conducted by Loesener (1942a, 1942b) and Hallé (1962, 1986, 1990), respectively. Loesener (1942a) recognized five subfamilies and five tribes of Celastraceae s.str. However, Loesener's subfamilies and tribes have been found to be heterogeneous in wood anatomy (Metcalfe and Chalk 1950), pollen structure (Lobreau-Callen 1977), and leaf anatomy (Den Hartog and Baas 1978). Furthermore, Loesener's subfamilies and tribes have been consistently resolved as unnatural groups in phylogenetic analyses of the family, using morphological and molecular characters (Simmons and Hedin 1999; Simmons et al. 2001a, 2001b).

Loesener (1942b) and Hallé (1962) recognized Hippocrateaceae as a family, separate from Celastraceae. Loesener (1942b) did not formally subdivide Hippocrateaceae. Hallé (1962) described two subfamilies (Hippocrateoideae, Salacioideae) and two tribes of the subfamily Hippocrateoideae: Campylostemonae [sic] (properly Campylostemoneae) and Hippocrateae [sic] (properly Hippocrateeae). Hallé (1986) added a third tribe, Helictonemae [sic] (properly Helictonemateae). Hallé later recognized Hippocrateaceae as a tribe (e.g., Hallé 1978) or as a subfamily (Hallé 1986, 1990) of Celastraceae.

The Hippocrateaceae have been consistently supported as nested within Celastraceae s.str., using morphological and molecular characters (Savolainen et al. 1997; Simmons and Hedin 1999; Savolainen, Chase et al. 2000; Savolainen, Fay et al. 2000; Soltis et al. 2000; Simmons et al. 2001a, 2001b). With the inclusion of *Lophopetalum* and *Plagiopteron*, and the exclusion of *Dicarpellum*, the family Hippocrateaceae and the subfamilies Hippocrateoideae and Salacioideae have been

supported as monophyletic groups, although the tribes Campylostemoneae, Helictonemateae, and Hippocrateeae have not been supported (Simmons and Hedin 1999; Savolainen, Fay et al. 2000; Simmons et al. 2001a, 2001b).

Stackhousia and *Tripterococcus* (Stackhousiaceae) have been supported as nested within Celastraceae, using *rbc*L (Savolainen, Fay et al. 2000), a simultaneous analysis of *atp*B, *rbc*L, and 18S nrDNA (Soltis et al. 2000), and 26S nrDNA and a simultaneous analysis of morphological and molecular characters (Simmons et al. 2001b). However, Stackhousiaceae have been resolved as excluded from Celastraceae using the 5′ flanking region of *rbc*L (Savolainen et al. 1997) and morphology (Simmons and Hedin 1999). Note that these two latter studies were poorly sampled for taxa or characters, respectively, and that the best-supported hypothesis is that Stackhousiaceae should be included within Celastraceae.

The most-comprehensive phylogenetic analysis of Celastraceae sampled 71 species representing 53 genera that have been assigned to Celastraceae for 26S nrDNA, phytochrome B, *rbc*L, *atp*B, and morphology (Simmons et al. 2001b). Although this analysis has been useful to test earlier taxonomic assertions regarding generic circumscriptions and intergeneric relationships, it is insufficiently sampled to use as a basis to propose a new classification of Celastraceae to replace the classifications by Loesener (1942a) and Hallé (1962, 1986, 1990).

The four subfamilies in the classification used here are the three monophyletic subfamilies Stackhousioideae, Hippocrateoideae, and Salacioideae that are independently derived from the paraphyletic Celastroideae. Each of the three monophyletic subfamilies are readily diagnosable using morphological characters. This is an informal classification because of the obvious paraphyly of the subfamily Celastroideae.

Affinities. In broad-scale phylogenetic analyses that have been undersampled for Celastraceae, *Lepuropetalon* and/or *Parnassia* (Parnassiaceae) have been resolved as the sister group of Celastraceae using *rbc*L (Chase et al. 1993; Morgan and Soltis 1993; Savolainen, Chase et al. 2000; Savolainen, Fay et al. 2000), the 5′ flanking region of *rbc*L (Savolainen et al. 1997), and 18S nrDNA (Soltis et al. 1997). In a comparatively well-sampled phylogenetic analysis of Celastraceae, using 26S nrDNA and a simultaneous analysis of morphological and molecular characters, *Lepuropetalon* and *Parnassia* were resolved, in a weakly-supported clade, as part of an early-derived lineage within Celastraceae (Simmons et al. 2001b). However, because of their distinctive morphology, and because the clade from Simmons et al. (2001b) was weakly-supported and based only on characters from 26S nrDNA, *Lepuropetalon* and *Parnassia* are here treated as a separate family, Parnassiaceae.

Afrostyrax (Huaceae) has been resolved as the sister group of the clade of Celastraceae and Parnassiaceae [but Parnassiaceae has not been sampled in a simultaneous analysis of *atp*B and *rbc*L (Savolainen, Chase et al. 2000), and a simultaneous analysis of *atp*B, *rbc*L, and 18S nrDNA (Soltis et al. 2000)], but not in a comparatively well-sampled analysis of *rbc*L (Savolainen, Fay et al. 2000). In the latter study, *Lepidobotrys* and *Ruptiliocarpon* (Lepidobotryaceae), which had not been previously sampled in broad-scale molecular analyses, were resolved as the sister group of the clade of Celastraceae and Parnassiaceae.

DISTRIBUTION AND HABITATS. The family has a subcosmopolitan distribution (from tropical to temperate regions; absent in arctic regions) and is most diverse in the tropics and subtropics. Although numerous species of Celastroideae are native to temperate regions, species of Hippocrateoideae and Salacioideae are restricted to the tropics and subtropics. Major centers of diversity for the family are Australia (for Stackhousioideae), tropical America and Africa (for Hippocrateoideae and Salacioideae), South Africa (for Celastroideae), Madagascar, Macronesia, and Queensland, Australia (for Celastroideae).

Species of Celastraceae generally grow in woodlands, from dry, gallery forests to swamps, from sea level to 3350 m elevation. Wooded habitats range from thickets and bush to primary rain forests. Other members of the family grow in grasslands (*Gymnosporia* spp., *Salacia* spp.), deserts (*Acanthothamnus, Canotia, Mortonia*), and coastal dunes (*Crossopetalum* spp., *Lauridia, Robsonodendron*, and *Salvadoropsis*). At least one species, *Elaeodendron viburnifolium*, grows in tidal mangrove swamps (Ding Hou 1962) and another species, *Quetzalia standleyi*, may grow as epiphytic shrubs (Lundell 1939).

ECONOMIC IMPORTANCE. The leaves and twigs of *Catha edulis* ("kat" or "khat") are chewed or drunk as a tea for their stimulant properties (Krikorian and Getahun 1973), similar to amphetamine (Zelger et al. 1981). The two active components of kat, L(S)-(-)-alpha-aminopropiophenone (cathinone) and *d*-norpseudoephedrine (cathine), are concentrated in young leaves and twigs. Kat, cathinone, and cathine are now controlled drugs in the U.S.A. (Federal Sentencing Guidelines Manual 1998).

Celastrus paniculatus, Kokoona zeylanica, Mystroxylon aethiopicum, and species of *Elaeodendron, Euonymus, Gymnosporia, Hippocratea, Maytenus*, and *Salacia* have been used for making traditional medicinal extracts (Blakelock 1951; Ding Hou 1962; Palmer and Pitman 1972; Burkill 1985; Sebsebe 1985; Chant 1993; Raintree Nutrition, Inc. 2000a, 2000b). *Euonymus atropurpureus, E. europaeus, E. latifolius, Elaeodendron buchananii* and *E. croceum* are poisonous (Blakelock 1951; Palmer and Pitman 1972; Burkill 1985). *Pristimera celastroides* and *Tripterygium wilfordii* have been used as insecticides (Smith 1940; Loesener 1942a). Oil has been extracted from the seeds of *Euonymus europaeus, Goniodiscus elaeospermus*, and *Kokoona zeylanica* (Kuhlmann 1933; Chant 1993). Human food sources have included the arils of *Bhesa*, members of Salacioideae (Smith 1940; Ding Hou 1962), the seeds of *Arnicratea grahamii, Hippocratea volubilis*, and *Hylenaea comosa* (Smith 1940; Loesener 1942b), the drupes of *Mystroxylon aethiopicum*, and the berries of *Lauridia tetragona* and *Maurocenia frangula* (Palmer and Pitman 1972). Various species of *Catha, Celastrus, Elaeodendron, Euonymus, Maytenus, Paxistima, Putterlickia*, and *Tripterygium* are cultivated as ornamentals (Huxley et al. 1992). Wood of *Bhesa paniculata, Euonymus javanicus, Kokoona littoralis, K. reflexa, Lophopetalum wightianum, L. multinervium, Maytenus* spp., and *Siphonodon celastrineus* has been used for timber (Record and Hess 1943; Ding Hou 1962; Palmer and Pitman 1972). Wood of *Euonymus europaeus* and *E. hians* has been used for turnery (Chant 1993). Several lianas, such as *Loeseneriella apocynoides*, are used for binding and basket weaving (Muhwezi 1999). Gutta-percha has been extracted from the bark of *E. verrucosus* (Blakelock 1951). Bark of *Kokoona* and *Lophopetalum*, which contain a thin oily layer, has been used to start fires, even in wet conditions (Ding Hou 1962). Bark extracts of *Lophopetalum javanicum* and *L. pallidum* have been used in dart and arrow poisons (Loesener 1942a; Ding Hou 1962).

CONSERVATION. Although no species of Celastraceae is listed in CITES (Convention on International Trade in Endangered Species), several species are considered threatened. In Australia,

Apatophyllum constablei is endangered and *Apatophyllum olsenii*, *Denhamia parvifolia*, *Hexaspora pubescens*, and *Stackhousia annua* are vulnerable (Briggs and Leigh 1995). In China, *Bhesa sinica* is known from only a single individual (Sheng-ye 1992). In India, *Euonymus angulatus*, *E. assamicus*, *E. serratifolius*, *Salacia jenkinsii*, and *Salacia malabarica* are endangered (Nayar and Sastry 1987). In South Africa, *Empleuridium juniperinum*, *Gymnosporia bachmannii*, *Lydenburgia abbottii*, *Maytenus abbottii*, *M. oleosa*, and *Pseudosalacia streyi* are vulnerable (Hilton-Taylor 1996; Oldfield et al. 1998; R. Archer, pers. comm. 2000). Other endangered and vulnerable species include *Bhesa ceylanica*, *B. nitidissima*, *Gyminda orbicularis*, *Euonymus lanceifolia*, *E. morrisonensis*, *E. pallidifolius*, *E. paniculatus*, *E. serratifolius*, *E. thwaitesii*, *E. walkeri*, *Gymnosporia curtisii*, *Kokoona coriacea*, *K. leucoclada*, *K. sabahana* and *K. sessilis*, *Lophopetalum sessilifolium*, *Maytenus cymosa*, *M. harenensis*, *M. harrisii*, *M. jefeana*, *M. matudai*, *M. microcarpa*, *M. ponceana*, *M. stipitata*, *M. williamsii*, *Microtropis argentea*, *M. borneensis*, *M. densiflora*, *M. fascicularis*, *M. grandifolia*, *M. keningauensis*, *M. rigida*, *M. sabahensis*, *M. sarawakensis*, *M. tenuis*, *Peritassa killipii*, *Perrottetia excelsa*, *Salacia miegei*, *Wimmeria acuminata*, and *W. chiapensis* (Oldfield et al. 1998).

KEY TO THE GENERA

1. Annuals or perennials with woody rootstocks; petals partially connate or free; stamens 5, of unequal or equal length. Australasia   2
   - Trees, shrubs, or lianas; petals free; stamens 2–numerous, of equal length. Old and New World, Australia   4
2. Petals free; stamens of equal length. Australia
     69. Macgregoria
   - Petals partially connate; stamens of unequal length. Australasia   3
3. Style attached between cocci in fruit. Australasia
     70. Stackhousia
   - Style attached terminally in fruit. SW Australia
     71. Tripterococcus
4. Stamens numerous. China, Myanmar   **83. Plagiopteron**
   - Functional stamens 2–5. Old and New World, Australia   5
5. Functional stamens 2; disk intrastaminal. Andaman-Nicobar Islands, India   **97. Nicobariodendron**
   - Functional stamens 3–5, or if 2, disk extrastaminal   6
6. Disk intrastaminal or stamens on disk; flowers unisexual or bisexual; fruit drupaceous, baccate, samaroid, or capsular; seeds with or without surrounding wings, arillate or exarillate. Tropics, subtropics, and temperate   7
   - Disk extrastaminal; flowers bisexual (unisexual in New Caledonia); fruit baccate with seeds surrounded by mucilaginous arils, transversely-flattened, trilobed capsules with seeds with membranous basal wings or narrow stipes, or capsules that are circular or trigonous in cross section with seeds with membranous apical wings or

arils basal to partially enveloping seeds. Tropics and subtropics   88
7. Plants with thorns or stems terminating in sharp points   8
   - Plants unarmed   13
8. Thorns 0; stems terminating in sharp points, glandular. SW North America   9
   - Thorns +; stems not terminating in sharp points, not glandular. Old World, South America   10
9. Fruit drupaceous; leaves laminar, few
     1. Acanthothamnus
   - Fruit capsular, beaked; leaves reduced to scales
     8. Canotia
10. Leaves not borne on thorns; inflorescence fasciculate. South America   **39. Moya**
    - Leaves often borne on thorns; inflorescence generally cymose. Old World   11
11. Ovules 2 per locule; flowers unisexual rarely bisexual. Old World   **24. Gymnosporia**
    - Ovules 3–12 per locule; flowers bisexual. South Africa, S Mozambique   12
12. Nodes >1 per thorn; ovules 3–6 per locule   **20. Gloveria**
    - Nodes 1 per thorn; ovules (4–)6(–12) per locule
      53. Putterlickia
13. Flowers 3–4-merous   14
    - Flowers 5-merous   37
14. Ovary 4-locular and ovules 1 or ± 10 per locule   15
    - Ovary 1–5-locular and ovules 1–12 per locule   20
15. Ovules ± 10 per locule; fruit capsular, woody, oblong. New Guinea   **67. Xylonymus**
    - Ovules 1 per locule; fruit drupaceous or capsular, fleshy or coriaceous, spheroid, obovoid, or deeply lobed. Asia, Macronesia, New World   16
16. Stamens on disk. Asia, Macronesia   17
    - Disk intrastaminal. West Indies, tropical America   18
17. Ovules pendulous; capsule spheroid; seeds with branched raphe. Asia, Macronesia   **21. Glyptopetalum**
    - Ovules erect; capsule deeply lobed; seeds with unbranched raphe. China   **37. Monimopetalum**
18. Ovules pendulous; leaves strictly opposite, entire. Jamaica
      62. Tetrasiphon
    - Ovules erect; leaves alternate, opposite, or whorled, entire or toothed. West Indies, tropical America   19
19. Fruit drupaceous, obovoid; seeds with branched raphe. West Indies, tropical America   **12. Crossopetalum**
    - Fruit capsular, deeply lobed; seeds with unbranched raphe. Cuba, Hispaniola, Puerto Rico   **63. Torralbasia**
20. Leaves alternate or irregularly scattered   21
    - Leaves opposite   27
21. Leaves needle-like, irregularly scattered; ericoid subshrub. South Africa   **16. Empleuridium**
    - Leaves with laminar blade, alternate; trees or shrubs. Old and New World, Australia   22
22. Leaves generally with small abaxial domatia in axils of larger veins; fruit baccate; seeds furrowed
      44. Perrottetia
    - Leaves without domatia; fruit samaroid, capsular or drupaceous; seeds smooth   23
23. Disk indistinct or 0; flowers unisexual, rarely dioecious; fruit drupaceous. Americas, West Indies   **60. Schaefferia**
    - Disk fleshy; flowers bisexual or unisexual; fruit capsular, samaroid, or drupaceous. Old and New World   24
24. Fruit samaroid; ovules 4–9 per locule. C America, Mexico   **66. Wimmeria**
    - Fruit capsular or drupaceous; ovules 1–2(–12) per locule. Old and New World, Australia   25

25. Fruit drupaceous **15. _Elaeodendron_**
  – Fruit capsular 26
26. Capsule with lateral appendages or unornamented; ovules erect and/or pendulous, 2(–12) per locule
  18. _Euonymus_
  – Capsule unornamented; ovules erect, 1–2 per locule
  34. _Maytenus_
27. Low, rhizomatous shrubs. North America **42. _Paxistima_**
  – Trees or erect, rarely scandent shrubs. Old and New World, Australia 25
28. Ovules 1 per locule, pendant. West Indies, tropical America **23. _Gyminda_**
  – Ovules 2–12 per locule, erect, rarely pendant. Old and New World, Australia 29
29. Fruit samaroid. Madagascar, Mexico 30
  – Fruit capsular, drupaceous, or nuts. Old and New World, Australia 31
30. Samara with 1 surrounding wing. Madagascar
  51. _Ptelidium_
  – Samara with 1 apical wing. Mexico **56. _Rzedowskia_**
31. Fruit capsular 32
  – Fruit indehiscent 34
32. Capsules 3–5-locular, loculicidally dehiscent; seeds 1–several. Old and New World, Australia
  18. _Euonymus_
  – Capsules 1-locular, laterally split along 1 side; seed 1. SE Asia, Macronesia, C America, Mexico 33
33. Disk fleshy. C America, Mexico **54. _Quetzalia_**
  – Disk 0 or thin. SE Asia, Macronesia **36. _Microtropis_**
34. Fruit nuts with lateral styles. Old World, Australia
  47. _Pleurostylia_
  – Fruit drupaceous or baccate with styles terminal or 0. Old and New World, Australia 35
35. Fruit drupaceous. Old and New World, Australia
  15. _Elaeodendron_
  – Fruit baccate. South Africa 36
36. Flowers sessile; inflorescence compact thyrsoid or irregularly cymose **30. _Lauridia_**
  – Flowers pedicellate; inflorescence cymose **9. _Cassine_**
37. Leaves opposite or whorled (and subopposite) on mature stems 38
  – Leaves alternate (and subverticillate in New Caledonia) on mature stems 58
38. Leaves reduced to scales. Australia **49. _Psammomoya_**
  – Leaves laminar. Old and New World, Australia 39
39. Ovary 5-locular. Madagascar 40
  – Ovary 2–5-locular. Old and New World, Australia 42
40. Fruit capsular, coriaceous **18. _Euonymus_**
  – Fruit baccate, ± fleshy 41
41. Disk intrastaminal; seeds 1–2, exarillate
  58. _Salvadoropsis_
  – Stamens on disk; seeds 5–10, arillate **17. _Euonymopsis_**
42. Fruit indehiscent 43
  – Fruit dehiscent 52
43. Fruit samaroid. Madagascar, Mexico, C America, South America 44
  – Fruits baccate, drupaceous, or nuts. Old and New World, Australia 45
44. Samara with 1 apical wing, oblanceolate or obovate. Mexico, C America, South America **68. _Zinowiewia_**
  – Samara with 1 surrounding wing, ovate or ovate-lanceolate. Madagascar **51. _Ptelidium_**
45. Fruits as nuts with lateral styles. Old World, Australia
  47. _Pleurostylia_
  – Fruits baccate or drupaceous, with styles terminal or 0. Old and New World 46

46. Fruit drupaceous. Old and New World, Australia 47
  – Fruit baccate. Africa, Madagascar 48
47. Drupes dry, trigonous; seeds 2–3. Madagascar
  25. _Hartogiopsis_
  – Drupes fleshy or coriaceous, not trigonous; seeds 1–2(3). Old and New World, Australia **15. _Elaeodendron_**
48. Ovules pendulous. S Africa **33. _Maurocenia_**
  – Ovules erect. Africa, Madagascar 49
49. Seeds with branched raphes, arillate. Madagascar
  7. _Brexiella_
  – Seeds with unbranched raphes, exarillate. Africa 50
50. Flowers pedicellate; inflorescence cymose **9. _Cassine_**
  – Flowers sessile; inflorescence thyrsoid or cymose 51
51. Inflorescence thyrsoid; berries spheroid **30. _Lauridia_**
  – Inflorescence cymose; berries ellipsoid **2. _Allocassine_**
52. Capsule 1-locular, laterally split along 1 side; seed 1. SE Asia, Macronesia, C America, Mexico 53
  – Capsule 3–5-locular, loculicidally dehiscent; seeds 1–several. Old and New World, Australia 54
53. Disk fleshy. C America, Mexico **54. _Quetzalia_**
  – Disk 0 or thin. SE Asia, Macronesia **36. _Microtropis_**
54. Capsule woody; seeds with wing surrounding seed; stamens on disk; inflorescence thyrsoid. Asia, Macronesia, Australia **31. _Lophopetalum_**
  – Capsule bony to coriaceous; seeds without wing or wing not surrounding seed; disk intrastaminal or stamens on disk; inflorescence cymose or fasciculate. Old and New World, Australia 55
55. Flowers unisexual (dioecious); disk intrastaminal; capsule without wings or protuberances. New Caledonia
  35. _Menepetalum_
  – Flowers bisexual, rarely unisexual; monoecious or dioecious; disk intrastaminal or stamens on disk; capsules with or without wings or protuberances. Old and New World, Australia 56
56. Leaves strictly opposite; ovary 3–5-locular; capsules with or without wings or protuberances; seeds without wings; disk intrastaminal or stamens on disk. Old and New World, Australia **18. _Euonymus_**
  – Leaves opposite on mature stems, alternate on juvenile stems or strictly opposite; ovary 3(4)-locular; capsules without wings or protuberances; seeds with or without wings; disk intrastaminal. Africa 57
57. Seeds without wings; leaves strictly opposite or opposite on mature branches and alternate on juvenile branches. South Africa **32. _Lydenburgia_**
  – Seeds with basal wings; leaves opposite on mature branches and alternate on juvenile branches. Africa, Arabia **10. _Catha_**
58. Leaves needle-like, sessile. Australia **3. _Apatophyllum_**
  – Leaves laminar, petiolate. Old and New World, Australia 59
59. Leaves with distinct crossbar secondary venation, petiole distally geniculate. Asia, Macronesia **4. _Bhesa_**
  – Leaves with reticulate secondary venation, petiole not geniculate. Old and New World, Australia 60
60. Leaves generally with small abaxial domatia in axils of larger veins; ovary 2-locular, seemingly 4-locular at base; seeds furrowed **44. _Perrottetia_**
  – Leaves without domatia; ovary locule number not apparently different at apex and base; seeds smooth 61
61. Leaves conspicuously glandular-serrate and small (<10 mm long). Mexico **41. _Orthosphenia_**
  – Leaves without glandular-toothed margins, or if glandular-toothed, large (>15 mm long). Old and New World, Australia 62

62. Leaf venation acrodromous; ovary 1-locular, 3-carpellate, with intruded parietal placentation; capsule septicidally dehiscent. Myanmar, Thailand **98. *Pottingeria***
- Leaf venation pinnate; ovary 2–7-locular, with axile placentation; fruit indehiscent or capsular, if capsular, loculicidally dehiscent. Old and New World, Australia 63
63. Plants pubescent; inflorescence racemose; fruit linear-oblong; seed 1. Brazil **19. *Fraunhofera***
- Plants glabrous or pubescent; inflorescence thyrsoid, cymose, fasciculate, or flowers solitary; fruit spheroid to ovoid; seeds 1–numerous. Old and New World, Australia 64
64. Inflorescence epiphyllous, rarely axillary; ovary (4)5-locular. Madagascar **48. *Polycardia***
- Inflorescence axillary and/or terminal; ovary 2–10-locular. Old and New World, Australia 65
65. Ovary with apical hollow and style-like central column arising from base of hollow; stamens alternating with small staminodes; drupe with numerous stones. Asia, Macronesia, Australia **61. *Siphonodon***
- Ovary with terminal style; stamens not alternating with staminodes (although disk may be lacerate); fruit dehiscent or indehiscent, not drupe with numerous stones. Old and New World, Australia 66
66. Disk lacerate, alternating with stamens; fruit baccate, woody, becoming pulpy; inflorescence umbellate-cymose; ovary 5–7-locular. Africa, Madagascar **6. *Brexia***
- Disk entire or lobed; fruit dehiscent or indehiscent, chartaceous to fleshy; inflorescence thyrsoid, cymose, fasciculate, or flowers solitary; ovary 1–5-locular. Old and New World, Australia 67
67. Ovary 5-locular 68
- Ovary 2–4-locular 72
68. Inflorescence terminal, thyrsoid; fruits drupaceous, dry. U.S.A., Mexico **38. *Mortonia***
- Inflorescence axillary or terminal, cymose, fasciculate, or condensed thyrsoid; fruits capsular or samaroid, chartaceous to bony. Old and New World, Australia 69
69. Fruit samaroid with 4–5 lateral wings; ovules 2 per locule. Tanzania **45. *Platypterocarpus***
- Fruit capsular with or without lateral wings; ovules 2–12 per locule. Old and New World, Australia 70
70. Inflorescence fasciculate or condensed thyrsoid; anthers obliquely dehiscent, latrorse; disk intrastaminal; leaves entire; ovules 4 per locule; aril at base and one side of seed. E Australia **26. *Hedraianthera***
- Inflorescence cymose (fasciculate); anthers longitudinally or obliquely dehiscent, introrse or latrorse; stamens on disk or disk intrastaminal; leaves entire or toothed; ovules 2–12 per locule; aril basal to enveloping seed. Old and New World, Australia 71
71. Inflorescence terminal or axillary; disk intrastaminal; anthers longitudinally dehiscent, introrse; capsule without wings or protuberances. Australia **13. *Denhamia***
- Inflorescence axillary; stamens on disk or disk intrastaminal; anthers obliquely or longitudinally dehiscent, latrorse or introrse; capsule with or without lateral wings or protuberances. Old and New World, Australia **18. *Euonymus***
72. Plant densely pilose; anthers obliquely dehiscent; ovary 3-locular; erect trees. Queensland, Australia **27. *Hexaspora***
- Plant glabrous or pubescent; anthers longitudinally or obliquely dehiscent; ovary 2–4-locular; scandent or erect shrubs or trees. Old and New World, Australia 73

73. Fruit drupaceous or samaroid; ovules 1–2 per locule 74
- Fruit capsular; ovules 1–12 per locule 79
74. Fruit samaroid 75
- Fruit drupaceous 76
75. Samara with 1 apical wing; erect trees or shrubs. South America **46. *Plenckia***
- Samara with 3 lateral wings; scandent shrubs. Asia **65. *Tripterygium***
76. Disk intrastaminal; ovules 2 per locule; drupe fibrous; leaves entire. Brazil **22. *Goniodiscus***
- Stamens on disk; ovules 1 per locule; drupe fleshy or coriaceous; leaves entire or toothed. Old and New World, Australia 77
77. Inflorescence fasciculate; ovary 2-locular; plant glabrous. South Africa **55. *Robsonodendron***
- Inflorescence cymose or cymose-umbellate; ovary 2–4-locular; plant pubescent or glabrous. Old and New World, Australia 78
78. Inflorescence cymose-umbellate; ovary 3–4-locular; plant pubescent or glabrous. Africa, Madagascar **40. *Mystroxylon***
- Inflorescence cymose; ovary 2–4-locular; plant glabrous. Old and New World, Australia **15. *Elaeodendron***
79. Capsule flattened along each locule, without wings or protuberances; seeds discoid with annular wing. New Caledonia **43. *Peripterygia***
- Capsule not flattened along each locule, with or without wings or protuberances; seeds not winged. Old and New World, Australia 80
80. Stamens on disk 81
- Disk intrastaminal 82
81. Ovary 3-locular; ovules erect, 2 per locule; capsule without wings or protuberances; leaves entire. South Africa **50. *Pseudosalacia***
- Ovary 3–4-locular; ovules pendulous to erect, 2–12 per locule; capsule with or without lateral wings or protuberances; leaves toothed or entire. Old and New World, Australia **18. *Euonymus***
82. Leaves alternate and subverticillate; inflorescence fasciculate; ovary 3-locular; ovules 2–6 per locule. New Caledonia **57. *Salaciopsis***
- Leaves strictly alternate; inflorescence thyrsoid, cymose, fasciculate, or flowers solitary; ovary 2–4-locular; ovules 1–12 per locule. Old and New World, Australia 83
83. Capsule with lateral or oblique horn to wing-like outgrowths; arils nearly enveloping seeds; inflorescence cymose; ovules 2 per locule. SE Africa **52. *Pterocelastrus***
- Capsule with or without lateral wings or protuberances; arils basal to enveloping seeds; inflorescence thyrsoid, cymose, fasciculate, or flowers solitary; ovules 1–12 per locule. Old and New World, Australia 84
84. Ovules 1 per locule; plant puberulent to glabrescent; inflorescence axillary; arils enveloping seeds. Americas **64. *Tricerma***
- Ovules 1–12 per locule; plant glabrous to pubescent; inflorescence terminal or axillary; arils basal to enveloping seeds. Old and New World, Australia 85
85. Scandent shrubs; inflorescence terminal or axillary, thyrsoid, cymose, or flowers solitary; flowers unisexual, rarely bisexual, dioecious; ovules 1–2 per locule; arils nearly enveloping seeds **11. *Celastrus***
- Erect or scandent trees or shrubs; inflorescence axillary, rarely terminal, cymose, fasciculate, or flowers solitary; flowers bisexual, rarely unisexual; monoecious or dioecious; ovules 1–12 per locule; arils basal to enveloping seeds 86

86. Stamens on disk or disk intrastaminal; capsule with or without lateral wings or protuberances; ovary 3–4-locular; ovules pendulous to erect   **18. Euonymus**
  – Disk intrastaminal; capsule without lateral wings or protuberances; ovary 2–4-locular; ovules erect   87
87. Inflorescence cymose; ovules 2–10 per locule. Australia
                                                 13. Denhamia
  – Inflorescence cymose, fasciculate, or flowers solitary; ovules 1–2 per locule. Old and New World, Australia
                                                 34. Maytenus
88. Fruit baccate with seeds surrounded by mucilaginous arils   89
  – Fruit transversely-flattened, trilobed capsules with seeds with membranous basal wings or narrow stipes, or capsules that are circular or trigonous in cross section   96
89. New World   90
  – Old World, Australia   93
90. Disk discontinuous   **91. Cheiloclinium**
  – Disk continuous   91
91. Anthers obliquely or longitudinally dehiscent
                                                 92. Peritassa
  – Anthers transversely dehiscent   92
92. Disk pulvinate, fleshy   **93. Salacia**
  – Disk flattened or cupular, membranous or ± fleshy
                                                 96. Tontelea
93. Anthers obliquely or longitudinally dehiscent   94
  – Anthers transversely dehiscent   95
94. Branches opposite; ovules 2 per locule; berry fusiform. Tropical Africa   **95. Thyrosalacia**
  – Branches alternate; ovules 2–9 per locule; berry spheroid, oblong, or fusiform. Old World, Australia   **93. Salacia**
95. Inflorescence cauliflorous on long pendant branches; leaves subopposite or alternate; androgynophore 0; berry spheroid; ovules 2 per locule. Tropical Africa
                                                 94. Salacighia
  – Inflorescence axillary or cauliflorous, thyrsoid, cymose, or fasciculate; leaves opposite or subopposite, rarely alternate, androgynophore + or 0; berry spheroid, oblong, or fusiform; ovules 2–9 per locule. Old World, Australia
                                                 93. Salacia
96. Capsule circular or trigonous in cross section; seeds with membranous apical wings or arils basal to partially enveloping seeds. Asia, Macronesia, Australia, New Caledonia   97
  – Capsule transversely-flattened, trilobed; seeds with membranous basal wings or narrow stipes. Old and New World tropics and subtropics, Australia   101
97. Stamens 3; ovary 2-locular. New Caledonia
                                                 14. Dicarpellum
  – Stamens 5; ovary 3- or 5-locular. SE Asia, Macronesia, Australia   98
98. Ovary 5-locular. New Guinea   **5. Brassiantha**
  – Ovary 3-locular. Asia, Macronesia, Australia   99
99. Capsule oblong, trigonous, woody; seeds with apical wing. Asia, Macronesia   **29. Kokoona**
  – Capsule fusiform to narrow-elipsoid, coriaceous or bony, seeds without wings. Australia, Borneo   100
100. Inflorescence cymose or flowers solitary; arils without basal filamentous extensions. Australia   **28. Hypsophila**
  – Inflorescence racemose; arils with basal filamentous extensions. Borneo   **59. Sarawakodendron**
101. Disk 0; anthers introrse   102
  – Disk +; anthers extrorse   104
102. Stamens 5   **76. Campylostemon**
  – Stamens 3   103

103. Petals thickened, fleshy   **75. Bequaertia**
  – Petals not thickened or fleshy   **90. Tristemonanthus**
104. Ovules 2 per locule   105
  – Ovules 3–22 per locule   109
105. Androgynophore +. Tropical Africa   **88. Simicratea**
  – Androgynophore 0. Old and New World, Australia   106
106. Disk of 3 discontinuous lobes around stamens; 4 of 5 petals arched, ± zygomorphic. Africa, Madagascar
                                                 73. Apodostigma
  – Disk continuous; flowers actinomorphic. Old and New World, Australia   107
107. Nodes on which inflorescence borne without axillary branches   **85. Pristimera**
  – Nodes on which inflorescence borne with axillary branches   108
108. Disk pulvinate or short cupular; seeds with membranous wings. Old World   **86. Reissantia**
  – Disk deeply cupular; seeds with narrow stipes or membranous wings. Americas, Africa, Madagascar
                                                 78. Elachyptera
109. Androgynophore +. Africa   110
  – Androgynophore 0. Old and New World, Australia   112
110. Plant hirsute and stellate pubescent   **79. Helictonema**
  – Plant glabrous or puberulent   111
111. Plant glabrous   **89. Simirestis**
  – Plant puberulent or glabrescent   **82. Loeseneriella**
112. Petals regularly serrate; capsule lobes entirely connate. C and South America   **72. Anthodon**
  – Petals entire, erosulous, or fimbriate; capsule lobes not connate or connate ± half length. Old and New World, Australia   113
113. Capsule lobes connate ± half length; inflorescence thyrsoid; ovules 6–8 per locule. Mexico, C and South America   **87. Semialarium**
  – Capsule lobes not connate; inflorescence cymose or thyrsoid; ovules 4–22 per locule. Old and New World, Australia   114
114. Capsule woody; seeds with narrow stipes or rarely membranous wings. C and South America   **81. Hylenaea**
  – Capsule coriaceous or chartaceous; seeds with membranous wings or rarely narrow stipes). Old and New World, Australia   115
115. Disk lobed and indistinct; ovary papillose; ovules 4–7 per locule. India, SE Asia, Macronesia   **74. Arnicratea**
  – Disk entire or lobed, distinct, rarely indistinct; ovary smooth, rarely papillose; ovules 3–22 per locule. Old and New World, Australia   116
116. Flowers large (8–20 mm diameter)   117
  – Flowers small (<8 mm diameter)   118
117. Disk membranous, cupular; seeds with narrow stipes or membranous wings. Americas, Africa   **77. Cuervea**
  – Disk fleshy, annular-pulvinate; seeds with membranous wings. Old and New World   **84. Prionostemma**
118. Adaxial surface of petals puberulent; ovules 4–8 per locule; seeds with membranous wings. Americas, Africa
                                                 80. Hippocratea
  – Adaxial surface of petals glabrous; ovules 3–33 per locule; seeds with narrow stipes or membranous wings. Old and New World, Australia   119
119. Petals ± valvate in bud, ovate to lanceolate. Old World, Australia   **82. Loeseneriella**
  – Petals imbricate in bud, ovate, oblong, or suborbicular. Old and New World   120
120. Nodes on which inflorescence borne without axillary branches; ovules 3–10 per locule; seeds with membranous wings. Old and New World   **85. Pristimera**

– Nodes on which inflorescence borne with axillary branches; ovules 4 per locule; seeds with narrow stipes or membranous wings. Americas, Africa, Madagascar
78. *Elachyptera*

GENERA OF CELASTRACEAE

I. SUBFAMILY CELASTROIDEAE Burnett (1835).

Erect or scandent trees, shrubs, suffrutices, rhizomatous shrubs, ericoid subshrubs, or epiphytic shrubs. Flowers unisexual or bisexual, (3)4–5-merous; petals free; disk intrastaminal or extrastaminal or 0, rarely stamens on disk; androgynophore 0; stamens 3–5; ovary (1)2–5(10)-locular. Capsule, drupe, berry, samara, or nut; seeds albuminous or exalbuminous. Included phloem 0.

1. *Acanthothamnus* Brandegee

Acanthothamnus Brandegee, Univ. Calif. Publ. Bot. 3: 383 (1909).
Scandivepres Loes. (1910).

Shrubs, glabrous; stems with dark glands, terminating in sharp points. Leaves few, alternate, entire. Inflorescences axillary, fasciculate, or flowers solitary. Flowers bisexual, 5-merous; disk cupular, lobed, intrastaminal; anthers longitudinally dehiscent, introrse; ovary 2-locular; ovules erect, 2 per locule. Drupe ovoid, ± fleshy; seed 1, albuminous. Only one sp., *A. aphyllus* (Schltdl.) Standl., Mexico, deserts and subdeserts.

2. *Allocassine* N. Robson

Allocassine N. Robson, Bol. Soc. Brot. 39: 30 (1965); Archer & van Wyk, S. Afr. J. Bot. 64: 189–191 (1998), rev.

Scandent shrub, glabrous. Leaves opposite on flowering branches and alternate on vegetative branches, entire or ± glandular-crenate. Inflorescences axillary, cymose. Flowers bisexual, 5-merous; disk fleshy, margins upturned, entire, intrastaminal; anthers longitudinally dehiscent, introrse; ovary 2-locular; ovules erect, 2 per locule. Berry ellipsoid, fleshy; seeds 2–3(4), narrowly ellipsoid, albuminous. Only one sp., *A. laurifolia* (Harv.) N. Robson, eastern S Africa and Zimbabwe, Mozambique, dry forests.

3. *Apatophyllum* McGill.

Apatophyllum McGill., Kew Bull. 25: 401 (1971); Bean & Jessup, Austrobaileya 5: 691–697 (2000), rev.

Shrubs, glabrous. Leaves alternate, rarely subopposite or subverticillate, sessile, needle-like, entire. Flowers axillary, solitary. Flowers bisexual, 5-merous; disk annular or margins upturned, intrastaminal; anthers longitudinally dehiscent, introrse; ovary 2(3)-locular; ovules erect, 2 per locule. Capsule compressed, obovoid or pyriform, loculicidally dehiscent, seeds 1(2), oblong or ovoid, albuminous, aril basal. Five spp., Australia, rocky hillsides and open forests.

4. *Bhesa* Buch.-Ham. ex Arn.   Fig. 11

Bhesa Buch.-Ham ex Arn., Edinburgh New Philos. J. 16: 315 (1834); Ding Hou, Blumea 4: 149–153 (1958), rev.
Kurrimia Wall. (1831).
Kurrimia Wall. ex Meisn. (1837).

Trees with buttressed trunks, glabrous. Leaves alternate, with distinct crossbar secondary venation, entire, petiole geniculate. Inflorescences axillary, paniculate or racemose. Flowers bisexual, 5-merous; disk fleshy, cupular, subentire or

Fig. 11. Celastraceae-Celastroideae. *Bhesa archboldiana.* **A** Fruiting branchlet. **B** Flower. **C** Same, vertical section, pistil removed. **D** Pistil showing basal placentation. **E** Ovary, transverse section. **F** Seed with aril. Drawn by R. van Crevel. (Ding Hou 1962)

5-lobed; anthers longitudinally dehiscent, introrse or extrorse; ovary 2-locular, glabrous or apically pubescent; ovules erect, 2 per locule. Capsule fusiform or deeply 2-lobed, coriaceous, loculicidally dehiscent; seeds 1–2, oblong, albuminous, aril basal or enveloping seed. Five spp., India, Sri Lanka, SE Asia, Macronesia, humid forests, 0–2150 m. The genus is properly *Bhesa*, not *Kurrimia* because the original generic description of *Kurrimia* by Meisner (1837) is based on a taxonomical synonym of *Itea* (Ding Hou 1958).

5. *Brassiantha* A.C. Sm.

Brassiantha A.C. Sm., J. Arnold Arbor. 22: 389 (1941); Ding Hou, Fl. Males. I, 6: 227–291 (1962).

Trees, glabrous. Leaves alternate, entire. Inflorescences axillary, cymose. Flowers bisexual, 5-merous; disk fleshy, pulvinate, lobed, extrastaminal; anthers transversally dehiscent, extrorse; ovary 5-locular; ovules erect, 2–5 per locule. Capsule subspheroid, coriaceous, loculicidally dehiscent; seeds (1–)2–4(–5) per locule, ellipsoid-oblong, albuminous, partially enveloped by aril. Only one sp., *B. pentamera* A.C. Sm., New Guinea, lowland forests, to 100 m (1800–1950 m).

6. *Brexia* Noronha ex Thouars

Brexia Noronha ex Thouars, Gen. Nov. Madag. 20 (1806); Perrier de la Bâthie, Bull. Soc. Bot. France 89: 219–221 (1942); Verdcourt, Fl. Trop. E. Afr. 108 A: 1–3 (1968).

Trees or shrubs, glabrous. Leaves alternate, entire or spinose-dentate. Inflorescences axillary, umbellate-cymose. Flowers bisexual, 5-merous; disk fleshy, lacerate, alternating with stamens; anthers longitudinally dehiscent, introrse; ovary 5–7-locular; ovules axile, numerous. Berry, ovoid or oblong-fusiform, woody, becoming pulpy; seeds numerous, irregularly compressed-ellipsoid, albuminous. Only one sp., *B. madagascariensis* Thouars ex Ker Gawl., Madagascar, Seychelles, coastal east Africa, coastal bush and swamp forests. Eight species of *Brexia* have been recognized by Perrier de la Bâthie (1942), but these are tentatively lumped into one species (Verdcourt 1968).

Brexia has been assigned to Escalloniaceae (Hutchinson 1967), Brexiaceae (Verdcourt 1968), and Grossulariaceae (Cronquist 1981). However, phylogenetic analyses consistently support *Brexia* as a member of Celastraceae (e.g., Savolainen et al. 1997; Simmons et al. 2001b).

7. *Brexiella* H. Perrier

Brexiella H. Perrier, Bull. Soc. Bot. France 80: 204 (1933); Perrier de la Bâthie, Fl. Madag. 116: 1–76 (1946).

Trees or shrubs, glabrous. Leaves opposite, rarely whorled, entire, serrulate, or spinose-dentate. Inflorescences axillary, fasciculate or cymose. Flowers bisexual, 5-merous; disk fleshy, annular, 5-angled or lobed, intrastaminal; anthers longitudinally dehiscent, introrse; ovary 2–3-locular; ovules 2 per locule. Berry spheroid, ± fleshy; seeds 3–4, subspheroid, with branched raphe, albuminous, enveloped by aril. Two spp., Madagascar, humid and littoral forests.

8. *Canotia* Torr.

Canotia Torr., Pacific Railroad Reports 4: 68 (1857); Johnston, Brittonia 27: 119–122 (1975), rev.

Trees or shrubs, glabrous; stems with dark glands, terminating in sharp points. Leaves alternate, reduced to scales. Inflorescences axillary, cymose or thyrsoid. Flowers bisexual, 5-merous; disk fleshy, annular, intrastaminal; anthers longitudinally dehiscent, introrse; ovary 5-locular; ovules axile, 2–6 per locule. Capsule oblong, beaked, woody, septicidally and loculicidally dehiscent; seeds 1(2) per locule, oblong, flattened, with or without membranous basal wing, albuminous. Two spp., Mexico, SE U.S.A., deserts.

Canotia has been assigned to Rutaceae (Gray 1877), Koeberliniaceae (Barnhart 1910), Canotiaceae (Cronquist 1981), and Celastraceae (Hutchinson 1969) as an anomalous genus (Loesener 1942a), or as closely related to *Acanthothamnus* (Johnston 1975). Phylogenetic analyses support the sister-group relationship of *Acanthothamnus* and *Canotia* (Simmons and Hedin 1999; Simmons et al. 2001a, 2001b).

9. *Cassine* L.

Cassine L., Sp. Pl. ed. 1: 268 (1753); Archer & van Wyk, S. Afr. J. Bot. 63: 146–157 (1997), rev.
Hartogiella Codd (1983).

Trees or shrubs, glabrous. Leaves opposite, rarely subopposite, entire or glandular-crenate. Inflorescences axillary, cymose. Flowers bisexual, rarely unisexual, 4–5-merous; disk fleshy, lobed or entire, intrastaminal; anthers longitudinally dehiscent, introrse, rarely extrorse; ovary 2–3-locular; ovules erect, 2 per locule. Berry spheroid, fleshy or coriaceous; seeds 1–2(–6), spheroid or ellipsoid, albuminous. Three spp., S Africa, forests, woodlands, fynbos, and bush.

Cassine is narrowly defined following Archer and van Wyk (1997). *Cassine* is distinct from *Elaeodendron*, based on morphology and anatomy (Archer and van Wyk 1992, 1993a, 1993b, 1997), and a cladistic analysis (Simmons et al. 2001b).

10. *Catha* Forssk. ex Scop.   Fig. 12

Catha Forssk. ex Scop., Intr. Hist. Nat. 228 (1777).

Trees or shrubs, glabrous. Leaves opposite on mature branches and alternate on juvenile branches, glandular-crenulate-denticulate. Inflorescences axillary, cymose. Flowers bisexual, 5-merous; disk thin, margins upturned, 5-lobed, intrastaminal; anthers longitudinally dehiscent, introrse; ovary 3-locular; ovules erect, 2 per locule.

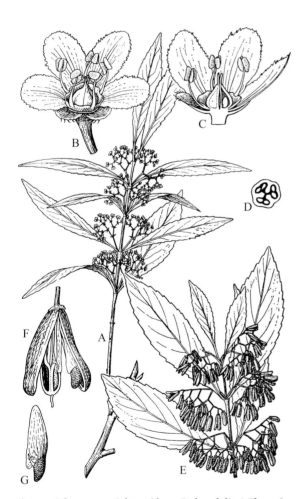

Fig. 12. Celastraceae-Celastroideae. *Catha edulis.* **A** Flowering branchlet. **B** Flower. **C** Same, vertical section. **D** Ovary, transverse section. **E** Infructescence. **F** Dehiscing fruit. **G** Seed. (Robson et al. 1994)

Capsule narrowly oblong, ± bony, loculicidally dehiscent; seeds 1–3, with membranous, basal wing, albuminous. Only one sp., *C. edulis* (Vahl) Forssk. ex Endl., Africa, Arabia, evergreen medium-altitude forest. *Catha abbottii* A.E. van Wyk & M. Prins will be transferred to *Lydenburgia* (van Wyk, in press).

Catha edulis ("kat" or "khat") is intensely cultivated as a cash crop in Ethiopia (Getahun and Krikorian 1973). Two alkaloids, L(S)-(-)-alpha-aminopropiophenone (cathinone) and *d*-norpseudoephedrine (cathine), concentrated in the young leaves and twigs of kat act as stimulants of the central nervous system (Krikorian and Getahun 1973; Szendrei 1981; Zelger et al. 1981). The leaves and twigs of kat are chewed or drunk as a tea in northeastern Africa, the Arabian Peninsula, and Madagascar.

11. *Celastrus* L.   Fig. 13A–D

Celastrus L., Sp. Pl. ed. 1: 196 (1753); Ding Hou, Ann. Missouri Bot. Gard. 42: 215–302 (1955), rev.

Scandent shrubs, glabrous, rarely pubescent. Leaves alternate, subentire or serrate. Inflorescences axillary or terminal, thyrsoid, cymose, or flowers solitary. Flowers rarely bisexual, unisexual, rarely dioecious, 5-merous; disk membranous or fleshy, annular to cupular, entire or 5-lobed, intrastaminal; anthers longitudinally dehiscent, introrse; ovary 3-locular; ovules erect, 1–2 per locule. Capsule subspheroid, rarely oblong, coriaceous, loculicidally dehiscent; seeds 1–6, albuminous, nearly enveloped by aril. $2n = 46$. Thirty-one spp., Americas, Madagascar, India, SE Asia, Macronesia, Queensland, Australia, humid forests to dry thickets. Two species, *C. orbiculatus* Thunb. and *C. scandens* L., are widely cultivated as ornamentals for the winter color of their fruits and arillate seeds (Ding Hou 1955).

12. *Crossopetalum* P. Browne

Crossopetalum P. Browne, Civ. Nat. Hist. Jamaica ed. 1: 145 (1756); Brizicky, J. Arnold Arbor. 45: 206–234 (1964), part. rev.
Rhacoma L. (1759).
Myginda Jacq. (1760).

Trees or shrubs, glabrous. Leaves alternate, opposite, or whorled, entire, serrulate or spinose-dentate. Inflorescences axillary, cymose. Flowers bisexual, 4-merous; disk annular to cupular, 4-lobed, intrastaminal; anthers longitudinally dehiscent, introrse; ovary 4-locular; ovules erect, 1 per

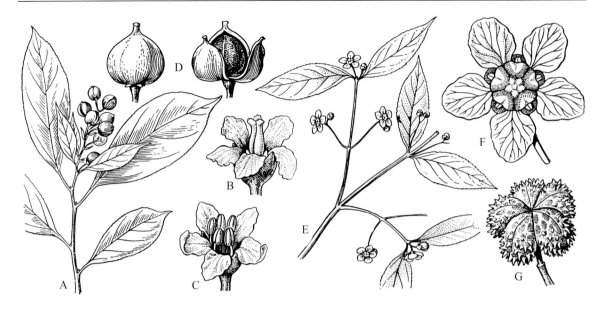

Fig. 13. Celastraceae-Celastroideae. **A–D** *Celastrus scandens.*
A Fruiting branch. **B** Female flower. **C** Male flower. **D** Capsules.
E–G *Euonymus americana.* **E** Flowering branch. **F** Flower.
G Capsule. (Takhtajan 1981)

locule. Drupe obovoid, ± fleshy; seeds 1(2),
obovoid, with branched raphe, albuminous. About
26 spp., West Indies, tropical America, coastal sand
dunes and scrub, hummocks, and pinelands.

13. *Denhamia* Meisn.

Denhamia Meisn., Pl. Vasc. Gen. 18 (1837); Jessup, Fl. Austr. 22:
150–180 (1984).

Trees or shrubs, glabrous. Leaves alternate, entire
or serrate. Inflorescences axillary or terminal,
cymose. Flowers bisexual, 5-merous; disk fleshy,
annular, 5-lobed, intrastaminal; anthers longitudinally dehiscent, introrse; ovary 2–5-locular; ovules
erect, 2–10 per locule. Capsule ovoid or spheroid,
bony, loculicidally dehiscent; seeds 1–several,
ellipsoid or ovoid, albuminous, aril basal to
enveloping seed. Seven spp., eastern and northern
Australia, humid forests to dry thickets.

14. *Dicarpellum* (Loes.) A.C. Sm.

Dicarpellum (Loes.) A.C. Sm., Am. J. Bot. 28: 443 (1941);
Simmons, Fl. Nouv.-Caléd. (in press).
Salacia L. subgenus *Dicarpellum* Loes. (1907).

Trees or shrubs, glabrous. Leaves alternate, entire.
Inflorescences axillary, condensed thyrsoid.

Flowers unisexual, rarely dioecious, (2–)5-merous;
disk fleshy, cupular, entire, extrastaminal; stamens
3; anthers obliquely or longitudinally dehiscent,
introrse; ovary 2-locular; ovules erect, 2 per locule.
Capsule obovoid or ellipsoid, bony, loculicidally
dehiscent; seeds 1–2 per locule, ellipsoid, albuminous, aril basal. Four spp., New Caledonia,
humid forests and infrequently in maquis, 75–
1250 m.

Dicarpellum is recognized as distinct from
Salacia (Smith 1941; Simmons, in press).

15. *Elaeodendron* Jacq. Fig. 14

Elaeodendron Jacq., Icon. Pl. Rar. t. 48 (1782); Archer & van
Wyk, S. Afr. J. Bot. 64: 93–109 (1998), reg. rev.; Kostermans,
Gard. Bull. Sing. 39: 177–191 (1986), reg. rev.
Crocoxylon Eckl. & Zeyh. (1834/5).
Telemachia Urb. (1916).

Shrubs or trees, glabrous. Leaves opposite, subopposite, or alternate, entire, crenate, dentate, or
spinose. Inflorescences axillary, cymose. Flowers
bisexual or unisexual (dioecious, polygamous, or
monoecious), 3–5-merous; disk fleshy, annular,
entire, 4–5-angled, or lobed, intrastaminal or
stamens on disk; anthers longitudinally dehiscent,
introrse or extrorse; ovary 2–4-locular; ovules
erect, 2 per locule. Drupe spheroid or ellipsoid,
fleshy or coriaceous; seeds 1–2(–3), ellipsoid or
ovoid, flattened or triangular, albuminous. $2n = 34$.
About 40 spp. (R. Archer, pers. comm. 2000), West
Indies, Africa, Madagascar, India, Macronesia,
Australia, forests, woodlands, scrub.

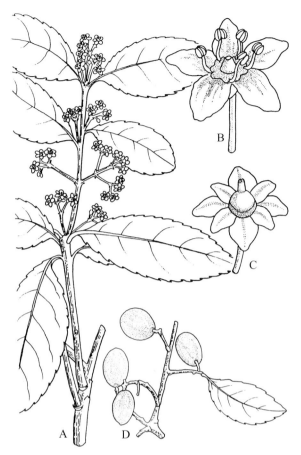

Fig. 14. Celastraceae-Celastroideae. *Elaeodendron buchananii*. **A** Flowering branchlet. **B** Male flower. **C** Female flower. **D** Fruiting branch. Drawn by Jill Lowett. (Robson et al. 1994)

Many species of *Elaeodendron* have been described as *Cassine*. Although most species have been described as both *Elaeodendron* and *Cassine*, some Austral-Asian species have only been described as *Cassine*, and new combinations are required for them (R. Archer, pers. comm. 2000).

16. *Empleuridium* Sond. & Harv.

Empleuridium Sond. & Harv. in Harvey, Thes. cap. 1: 49 (1859); Goldblatt et al., Ann. Missouri Bot. Gard. 72: 167–183 (1985), rev.

Ericoid subshrub, glabrous. Leaves irregularly scattered, sessile, needle-like, entire. Flowers axillary, solitary. Flowers bisexual, 4-merous; disk annular, 4-angled or lobed, intrastaminal; anthers longitudinally dehiscent, introrse; ovary 1-locular; ovules erect, 4. Capsule linear-oblong, laterally split along one side, seed 1, oblong-fusiform, albuminous, aril basal with apical, filamentous exten-

sions. Only one sp., *E. juniperinum* Sond. & Harv., S Africa, rocky sandstone slopes, above 600 m.

When first described, *Empleuridium* was doubtfully assigned to Rutaceae (Sonder 1860). However, based on its embryology, morphology, and pollen, the genus has been transferred to Celastraceae (Goldblatt et al. 1985).

17. *Euonymopsis* H. Perrier

Evonymopsis H. Perrier, Notul. Syst. (Paris) 10: 202 (1942).

Trees or shrubs, glabrous. Leaves opposite or whorled, entire, serrulate, spinose-dentate, or notched. Inflorescences axillary, rarely cauliflorous, fasciculate or cymose. Flowers bisexual, 5-merous; disk fleshy, annular, 5-lobed, stamens inserted on disk; anthers longitudinally dehiscent, introrse; ovary 5-locular; ovules axile, 4–12 per locule. Berry ovoid, ellipsoid, or oblong, fleshy; seeds 5–10, spheroid or ovoid, albuminous, enveloped by aril. Four spp., Madagascar, tropophilous forest and xerophytic scrub, to 400 m.

18. *Euonymus* L.   Fig. 13E–G

Euonymus L., Sp. Pl.: 197 (1753); Ma, Thaiszia 11: 1–264 (2001), rev.
Quadripterygium Tardieu (1948).

Trees or shrubs, rarely scandent, glabrous, rarely pubescent. Leaves opposite, rarely alternate or whorled, entire, serrulate, or spinose-denticulate. Inflorescences axillary, cymose, rarely fasciculate. Flowers bisexual, rarely unisexual and dioecious or gynodioecious, 4–5-merous; disk fleshy, annular, 4–5-lobed, intrastaminal or stamens on disk; anthers longitudinally or obliquely dehiscent, introrse or latrorse; ovary 3–5-locular; ovules erect to pendulous, 2(–12) per locule. Capsule subspheroid or obovoid, smooth or echinate, round, angular, deeply lobed, or laterally winged, coriaceous, loculicidally dehiscent; seeds 1–several, ellipsoid, albuminous, aril basal to enveloping seed. $2n = 16, 32, 48, 64$. About 129 species, Old and New World, Australia, humid forests to mesic thickets.

Several species (especially *E. alatus* [Thunb.] Siebold, *E. europaeus* L., and *E. fortunei* [Turcz.] Hand.-Mazz.) are cultivated as ornamentals for the winter color of their fruits and arillate seeds and/or autumn color, as groundcovers or hedges. This large, variable genus seems unlikely to be a monophyletic group, and some genera, such as

Glyptopetalum, *Monimopetalum*, and *Torralbasia*, may be nested within it.

19. *Fraunhofera* Mart.

Fraunhofera Mart., Nov. Gen. Sp. Pl. 3: 85, t. 235 (1831).

Trees or shrubs, pubescent. Leaves alternate, serrate. Inflorescences axillary and terminal, racemose. Flowers bisexual, 5-merous; disk thin, cupular, 5-lobed, intrastaminal; anthers longitudinally dehiscent, introrse; ovary 2-locular; ovules erect, 2 per locule. Drupe? (only immature fruit known) coriaceous, linear-oblong; seeds 1, linear-oblong. Only one sp., *F. multiflora* Mart., Brazil, deciduous forests.

20. *Gloveria* M. Jordaan

Gloveria M. Jordaan, S. Afr. J. Bot. 64: 299 (1998), rev.

Shrubs with thorns, glabrous. Leaves alternate and fasciculate on short shoots, entire. Inflorescences axillary, cymose. Flowers bisexual, 5-merous; disk fleshy, annular, undulate, intrastaminal; anthers longitudinally dehiscent; ovary 3-locular; ovules erect, 3–6 per locule. Capsule obovoid-trigonous, chartaceous, loculicidally dehiscent; seeds ellipsoid, enveloped by aril. Only one sp., *G. integrifolia* (L.f.) M. Jordaan, S Africa, karroid broken veld.

21. *Glyptopetalum* Thwaites

Glyptopetalum Thwaites, Hooker's J. Bot. Kew Gard. Misc. 8: 267, t. 7B (1856); Ding Hou, Fl. Males. I, 6: 227–291 (1962), reg. rev.

Trees or shrubs, glabrous. Leaves opposite, rarely subopposite, entire or crenulate. Inflorescences axillary, cymose. Flowers bisexual, 4-merous; disk fleshy, annular, 4-angled or lobed, stamens on disk; anthers obliquely dehiscent, latrorse; ovary 4-locular; ovules pendulous, 1 per locule. Capsule spheroid, coriaceous, loculicidally dehiscent; seeds 1–4, oblong, pendulous, with branched raphe, aril basal. About 20 spp., India, Sri Lanka, SE Asia, Macronesia, forests and thickets, 0–1400 m.

22. *Goniodiscus* Kuhlm.

Goniodiscus Kuhlm., Arch. Jard. Bot. Rio de Janeiro 6: 109 (1933).

Trees, glabrous. Leaves alternate, entire. Inflorescences axillary, cymose. Flowers bisexual, 5-merous; disk fleshy, 5-angled, stamens on disk; anthers longitudinally dehiscent, introrse; ovary 3–4-locular; ovules erect, 1 per locule. Drupe obovoid or ovoid, fibrous; seed 1, oblong, exalbuminous. Only one sp., *G. elaeospermus* Kuhlm., Brazil, Amazon forest.

23. *Gyminda* Sarg.

Gyminda Sarg., Gard. & Forest 4: 4 (1891); Brizicky, J. Arnold Arbor. 45: 206–234 (1964), part. rev.

Trees or shrubs, glabrous. Leaves opposite, entire or crenate. Inflorescences axillary, cymose. Flowers unisexual (dioecious), 4-merous; disk fleshy, annular, 4-angled or lobed, intrastaminal; anthers longitudinally dehiscent, introrse; ovary 2(3)-locular; ovules pendulous, 1 per locule. Drupe obovoid or ellipsoid, ± fleshy; seeds 1–2(3), oblong-ellipsoid, albuminous. Four spp., West Indies, tropical America, hummocks, rocky thickets, and woodlands.

24. *Gymnosporia* (Wight & Arn.) Hook. f.

Gymnosporia (Wight & Arn.) Hook f. in Bentham & Hooker, Gen. Pl. 1: 365 (1862); Sebsebe, Symb. Bot. Upsal. 25: 1–98 (1985); Jordaan and van Wyk, S. Afr. J. Bot. 65: 177–181 (1999), part. rev.
Celastrus L. sect. *Gymnosporia* Wight & Arn. (1834).

Trees, shrubs, or suffrutices with thorns, glabrous or pubescent. Leaves alternate (opposite if on thorns) and fasciculate on short shoots, entire or serrulate. Inflorescences axillary, cymose, fasciculate, or flowers solitary. Flowers rarely bisexual, unisexual (dioecious, polygamous, or monoecious), 5(6)-merous; disk fleshy, annular, 5-lobed, intrastaminal; anthers longitudinally dehiscent, introrse; ovary (2)3(4)-locular; ovules erect, 2(3) per locule. Capsule ovoid, spheroid, or obovoid, chartaceous, coriaceous, or woody, loculicidally dehiscent; seeds 1–4, ellipsoid, albuminous, aril basal to enveloping seed. $2n = 20, 24, 36, 56$. About 80 sp., Africa, Madagascar, India, Sri Lanka, SE Asia, Macronesia, Queensland, Australia, forests, woodlands, thickets, scrub, and grasslands.

Gymnosporia is recognized as distinct from *Maytenus* s.s. (Jordaan and van Wyk 1999; Simmons and Hedin 1999; Simmons et al. 2001a, 2001b). The affinity of the three American species (Lundell 1985; Hammel 1997) needs to be critically examined (R. Archer, pers. comm.).

25. *Hartogiopsis* H. Perrier

Hartogiopsis H. Perrier, Notul. Syst. (Paris) 10: 194 (1942), rev.

Tree, glabrous. Leaves opposite, serrulate. Inflorescences axillary, cymose. Flowers bisexual, 5-merous; disk fleshy, pulvinate, lobed, intrastaminal; anthers longitudinally dehiscent, introrse; ovary 3-locular; ovules erect, 2 per locule. Drupe ellipsoid, trigonous, dry; seeds 2–3, oblong, albuminous. Only one sp., *H. trilobocarpa* (Baker) H. Perrier, Madagascar, humid forests, 500–2000 m.

26. *Hedraianthera* F. Muell.

Hedraianthera F. Muell., Fragm. 5: 58 (1865); Jessup, Fl. Austr. 22: 150–180 (1984).

Trees or shrubs, glabrous. Leaves alternate, entire. Inflorescences axillary, condensed thyrsoid or fasciculate. Flowers bisexual, 5-merous; disk ± fleshy, annular, 5-lobed, intrastaminal; anthers obliquely dehiscent, latrorse; ovary 5-locular; ovules erect, 4 per locule. Capsule ovoid or spheroid, bony, loculicidally dehiscent; seeds 1–4 per locule, ellipsoid or ovoid, triangular, albuminous, aril at base and one side. Only one sp., *H. porphyropetala* F. Muell., eastern Australia, humid forests.

27. *Hexaspora* C.T. White

Hexaspora C.T. White, Contrib. Arnold Arbor. 4: 58 (1933); Jessup, Fl. Austr. 22: 150–180 (1984).

Trees, densely pilose. Leaves alternate, serrulate. Inflorescences axillary, cymose. Flowers bisexual, 5-merous; disk fleshy, annular, 5-angled, intrastaminal; anthers obliquely dehiscent; ovary 3-locular; ovules erect, 2–4 per locule. Fruit unknown. Only one sp., *H. pubescens* C.T. White, Queensland, Australia, humid forest.

28. *Hypsophila* F. Muell.

Hypsophila F. Muell., Victorian Nat. 3: 168 (1887); Jessup, Fl. Austr. 22: 150–180 (1984).

Trees or shrubs, glabrous. Leaves alternate or opposite, entire. Inflorescences axillary, cymose or flowers solitary. Flowers bisexual, 5-merous; disk fleshy, pulvinate, lobed, extrastaminal; anthers transversely dehiscent; ovary 3-locular; ovules erect, 6–10 per locule. Capsule fusiform, bony, loculicidally dehiscent; seeds several, oblong, albuminous, aril basal. Two spp., Queensland, Australia, mesic forests and thickets.

29. *Kokoona* Thwaites

Kokoona Thwaites, Hooker's J. Bot. Kew Gard. Misc. 5: 379 (1853); Ding Hou, Fl. Males. I, 6: 227–291 (1962).

Trees, glabrous. Leaves opposite, rarely subopposite or alternate, entire or crenulate. Inflorescences axillary, paniculate. Flowers bisexual, 5-merous; disk fleshy, cupular, corrugated, extrastaminal; anthers longitudinally dehiscent, latrorse, generally with triangular, pustular connective; ovary 3-locular; ovules axile, 6–16 per locule. Capsule oblong, trigonous, woody, loculicidally dehiscent; seeds 6–10(–16?) per locule, flattened, with membranous, apical wing, exalbuminous. Eight spp., Sri Lanka, southern India, Myanmar, Malaysia, Borneo, Philippines, \ to dry forests, 0–1500 m.

30. *Lauridia* Eckl. & Zeyh.

Lauridia Eckl. & Zeyh., Enum. Pl. Afr. Austral. 1: 124 (1834/5); Archer & van Wyk, S. Afr. J. Bot. 63: 227–232 (1997), rev.

Scandent shrubs or rarely trees, branchlets often sharply deflexed, glabrous. Leaves opposite, entire or glandular-crenate. Inflorescences axillary, compact thyrsoid or irregularly cymose. Flowers bisexual, 4(5)-merous; disk fleshy, entire, intrastaminal; anthers longitudinally dehiscent, introrse; ovary 2-locular; ovules erect, 2 per locule. Berry spheroid, fleshy; seeds 1–2, spheroid, albuminous. Two spp., S Africa, bush, woodlands, and dune scrub.

31. *Lophopetalum* Wight ex Arn.

Lophopetalum Wight ex Arn., Ann. Mag. Nat. Hist. I, 3: 150 (1839); Ding Hou, Fl. Males. I, 6: 227–291 (1962).

Trees, glabrous. Leaves opposite or subopposite, entire. Inflorescences axillary, thyrsoid. Flowers bisexual, 5-merous; disk fleshy, annular, 5-angled or lobed, stamens on disk; anthers longitudinally dehiscent, introrse; ovary 3-locular; ovules axile, 4–18 per locule. Capsule oblong or fusiform, trigonous or flattened along each locule, woody, loculicidally dehiscent; seeds many, oblong, flattened, medially attached, membranous wing surrounding seed, exalbuminous or scanty. $2n = 40$. About 20 spp., India, SE Asia, Macronesia, Northern Territory, Australia, humid to dry forests, swamp forests, 0–1500 m.

32. *Lydenburgia* N. Robson

Lydenburgia N. Robson, Bol. Soc. Brot. 39: 34 (1965).

Trees or shrubs, glabrous. Leaves strictly opposite or opposite on mature branches and alternate on juvenile branches, crenulate-denticulate or glandular-crenate. Inflorescences axillary, cymose. Flowers bisexual, 5-merous; disk fleshy, annular, 5-angled, intrastaminal; anthers longitudinally dehiscent, introrse; ovary 3(4)-locular; ovules erect, 2 per locule. Capsule oblong, loculicidally dehiscent; seeds 1–3(4), trigonous, albuminous. Two spp., S Africa, bush or woodland on rocky hills.

Lydenburgia is recognized as distinct from *Catha* following Robson (1965) and Robson et al. (1994). *Catha abbottii* A.E. van Wyk & M. Prins will be transferred to *Lydenburgia* (R. Archer and Y. Steencamp, pers. comm. 2000; van Wyk, in press).

33. *Maurocenia* Mill.

Maurocenia Mill., Gard. Dict. ed. 4: 859 (1754); Archer & van Wyk, Bothalia 28: 7–10 (1998), rev.

Shrub, glabrous. Leaves opposite, entire. Inflorescences axillary, fasciculate or cymose. Flowers bisexual or unisexual, rarely gynodioecious (or dioecious?), 5-merous; disk fleshy, undulate, intrastaminal; anthers longitudinally dehiscent, introrse; ovary 2(3)-locular; ovules pendulous, 2 per locule. Berry spheroid, fleshy; seeds 1–2 per locule, spheroid, albuminous. Only one sp., *M. frangula* Mill., S Africa, mountain kloofs and coastal bush.

34. *Maytenus* Molina   Fig. 15

Maytenus Molina, Sagg. Stor. Nat. Chili ed. 1: 177 (1782); Carvalho-Okano, Ph.D. Thesis, Instituto de Biologia, Universidade de Campinas, Campinas, Brazil (1992), reg. rev.

Trees or shrubs, erect, rarely scandent, glabrous, rarely pubescent. Leaves alternate, entire or serrate. Inflorescences axillary, fasciculate, cymose, rarely racemose, or flowers solitary. Flowers bisexual or unisexual, rarely polygamous or dioecious, (4)5-merous; disk fleshy, annular, (4)5-angled or lobed, intrastaminal; anthers longitudinally dehiscent, introrse; ovary 2–3-locular; ovules erect, 1–2 per locule. Capsule spheroid or obovoid, coriaceous, loculicidally dehiscent; seeds 1–6, ellipsoid or ovoid, albuminous, aril basal to enveloping seed. About 200 spp., tropics and sub-

Fig. 15. Celastraceae-Celastroideae. *Maytenus magellanica.* **A** Fruiting branch. **B** Part of male inflorescence, showing flower buds with prophylls. **C** Female flower. **D** Petal of female flower. **E** Female flower, vertical section. **F** Male flower bud. **G** Male flower. **H** Dehiscing fruit. (Dimitri 1972)

tropics of Old and New Worlds, Australia, humid forests to dry thickets.

Maytenus is narrowly circumscribed here and does not include *Gymnosporia, Moya,* or *Tricerma. Maytenus* s.s. is a heterogeneous genus that is, perhaps, best divided into segregate genera (Jordaan and van Wyk 1999). *Maytenus* s.s. has been resolved as polyphyletic in phylogenetic analyses (Simmons et al. 2001a, 2001b). Additional work is needed to re-circumscribe this genus.

35. *Menepetalum* Loes.

Menepetalum Loes. in R. Schlechter, Bot. Jahrb. Syst. 39: 163 (1906); Müller, Fl. Nouv.-Caléd. 20: 3–74 (1996).

Trees or shrubs, glabrous. Leaves opposite, crenate or serrate. Inflorescences axillary, rarely cauliflorous, cymose. Flowers unisexual, rarely bisexual, dioecious, 5-merous; disk fleshy, annular, 5-angled or lobed, intrastaminal; anthers longitudinally dehiscent, introrse; ovary 3-locular; ovules erect, 2–6 per locule. Capsule subspheroid, bony, loculicidally dehiscent; seeds 1–5 per locule, ovoid, albuminous, aril basal to enveloping seed. Four spp., New Caledonia, humid forests and maquis, 500–1600 m.

36. *Microtropis* Wall. ex Meisn.

Microtropis Wall. ex Meisn., Numer. List 152, n. 4337–40 (1831); Merrill & Freeman, Proc. Am. Acad. Arts 73: 271–310 (1940), rev.
Otherodendron Makino (1909).

Trees or shrubs, glabrous. Leaves opposite, entire. Inflorescences axillary, fasciculate or cymose. Flowers bisexual, rarely unisexual, dioecious, (4)5-merous; disk 0 or annular, (4)5-angled, intrastaminal; anthers longitudinally dehiscent, introrse, rarely extrorse; ovary 2–3-locular; ovules erect, 2 per locule. Capsule ellipsoid or oblong, coriaceous, 1-locular, laterally split along one side; seed 1, ovoid, furrowed, albuminous, sarcotestal. About 66 spp., SE Asia, Macronesia, forests, 0–2700 m. New World *Microtropis* are recognized as *Quetzalia* following Lundell (1970), *contra* Lundell (1939).

37. *Monimopetalum* Rehder

Monimopetalum Rehder, J. Arnold Arbor. 7: 233 (1926).

Scandent shrubs, glabrous (except inflorescence minutely pubescent). Leaves alternate, serrulate. Inflorescences axillary, cymose. Flowers bisexual, 4-merous; disk annular, 4-lobed, stamens on disk; anthers?; ovary 4-locular; ovules erect, 1 per locule. Capsule deeply 1–2(–4) lobed, loculicidally dehiscent; seeds 1–2(–4), oblong, albuminous, aril basal. Only one sp., *M. chinense* Rehder, China.

38. *Mortonia* A. Gray

Mortonia A. Gray, Pl. Wright. 1: 34, t. 4 (1852); Prigge, Ph.D Thesis, Claremont Graduate School: 1–136 (1983), rev.

Shrubs, glabrous. Leaves alternate, entire. Inflorescences terminal, thyrsoid. Flowers bisexual, 5-merous; disk fleshy, margins upturned, 5-lobed, intrastaminal; anthers longitudinally dehiscent, introrse; ovary 5-locular; ovules erect, 2(4) per locule. Drupe oblong, dry; seed 1, oblong, albuminous. Five spp., southwestern U.S.A., Mexico, deserts.

39. *Moya* Griseb.

Moya Griseb., Pl. Lorentz.: 13, 63 (1874); Lourteig & O'Donell, Natura 1: 181–233 (1955), reg. rev.

Shrubs with thorns, glabrescent or pubescent. Leaves alternate and fasciculate on short shoots, entire or denticulate. Inflorescences axillary, fasciculate or flowers solitary. Flowers unisexual (or bisexual?), 5-merous; disk fleshy, annular, entire or 5-lobed or angled, intrastaminal; anthers longitudinally dehiscent, introrse; ovary 2-locular; ovules erect, 2 per locule. Capsule ellipsoid or subspheroid, loculicidally dehiscent; seeds 1–2, ellipsoid, albuminous, nearly enveloped by aril. Three spp., Argentina, Bolivia, Paraguay. *Moya* has been reduced to *Maytenus* by Lourteig and O'Donell (1955), but has been recognized as distinct by Jordaan and van Wyk (1999).

40. *Mystroxylon* Eckl. & Zeyh.

Mystroxylon Eckl. & Zeyh., Enum. Pl. Afr. Austral.: 125 (1834/5); Robson et al., Fl. Trop. E. Afr. 108: 1–78 (1994); Archer et al., Fl. Pl. Afri. 55: 76–80 (1997), part. rev.

Trees or shrubs, glabrous or pubescent. Leaves alternate, entire or crenulate or glandular-denticulate. Inflorescences axillary, cymose-umbellate. Flowers bisexual, 5-merous; disk thin, annular or margins upturned, 5-angled or lobed, intrastaminal; anthers longitudinally dehiscent, introrse; ovary 3–4-locular; ovules erect, 2 per locule. Drupe spheroid or ellipsoid, ± fleshy; seed 1, spherical or ovoid, albuminous. Only one spp. (R. Archer, pers. comm. 2000), *M. aethiopicum* (Thunb.) Loes., Africa, Madagascar, forests, open woodlands, and scrub, 0–2550 m.

41. *Orthosphenia* Standl.

Orthosphenia Standl., Contrib. U.S. Natl. Herb. 23: 684 (1923); Rzedowski, Ciencia (Mexico) 16: 139–142 (1956), rev.

Shrubs, glabrous. Leaves alternate, glandular-serrate. Inflorescences axillary, fasciculate, or flowers solitary. Flowers bisexual or unisexual (polygamous), 5-merous; disk small, intrastaminal; anthers longitudinally dehiscent, introrse; ovary 2-locular; ovules erect, 2 per locule. Berry ovoid-spheroid, dry; seeds 1–2, ellipsoid, albuminous. Only one sp., *O. mexicana* Standl., Mexico, deserts.

42. *Paxistima* Raf.

Paxistima Raf., Sylva Tellur.: 42 (1838); Navaro & Blackwell, Sida 14: 231–249 (1990), rev.
Pachistima Raf. (1818).
Pachystima 'Raf.' ex Endl. (1841).

Low, rhizomatous shrubs, glabrous. Leaves opposite, entire or serrulate. Inflorescences axillary, cymose or flowers solitary. Flowers bisexual, 4-merous; disk fleshy, annular, 4-angled, intrastaminal; anthers longitudinally dehiscent, introrse; ovary 2-locular; ovules erect, 2 per locule. Capsule oblong, chartaceous, loculicidally dehiscent; seeds 1–2, oblong, albuminous, partially enveloped by lacerate aril. $2n = 32$. Two spp., N America, forests, open hillsides, and bluffs, 600–3350 m. This genus is properly spelled "Paxistima", not "Pachistima" or "Pachystima" (Wheeler 1943; Uttall 1986). Both species are cultivated as groundcovers.

43. *Peripterygia* (Baill.) Loes.

Peripterygia (Baill.) Loes. in R. Schlechter, Bot. Jahrb. Syst. 39: 168 (1906); Müller, Fl. Nouv.-Caléd. 20: 3–74 (1996).
Pterocelastrus Meisn. sect. *Peripterygia* Baill. (1874).

Shrubs, glabrous. Leaves alternate, serrate. Inflorescences axillary or terminal, cymose. Flowers bisexual or unisexual, rarely andromonoecious, 5-merous; disk fleshy, annular, 5-angled or lobed, intrastaminal; anthers longitudinally dehiscent, introrse; ovary 3-locular; ovules erect, 1–2(3) per locule. Capsule flattened along each locule, bony, loculicidally dehiscent; seeds discoid with annular, membranous wing, albuminous. Only one sp., *P. marginata* (Baill.) Loes., New Caledonia, maquis, 0–1050 m.

44. *Perrottetia* Kunth

Perrottetia Kunth in Humboldt, Bonpland and Kunth, Nov. Gen. Sp. 7: 73, t. 622 (1824); Ding Hou, Fl. Males. I, 6: 227–291 (1962); Lundell, Phytologia 57: 231–238 (1985), reg. rev.

Trees or shrubs, puberulent or glabrescent. Leaves alternate, entire or serrulate, with generally small abaxial domatia in axils of larger veins. Inflorescences axillary, cymose or thyrsoid. Flowers bisexual or unisexual (dioecious), (4)5-merous; disk annular or margins upturned, entire or lobed, intrastaminal; anthers longitudinally dehiscent, introrse; ovary 2-locular (seemingly 4-locular at base); ovules erect, 2 per locule. Berry spheroid, ± fleshy; seeds 2–4, subspheroid, furrowed, albuminous, sarcotestal. About 17 spp., northern S America, C America, SE Asia, Macronesia, Queensland, Australia, Hawaii, humid forests and thickets.

Cuatrescasas (1948) noted that the species differ by only minute characteristics, and Ding Hou (1962) suggested that there may be fewer species than are recognized.

45. *Platypterocarpus* Dunkley & Brenan

Platypterocarpus Dunkley & Brenan, Kew Bull. 1948: 47 (1948); Robson et al., Fl. Trop. E. Afr. 108: 1–78 (1994).

Trees, glabrous. Leaves alternate, crenate-dentate. Inflorescences axillary, cymose. Flowers bisexual, 5-merous; disk thin, annular, 5-lobed, intrastaminal; anthers longitudinally dehiscent, introrse; ovary (4?)5-locular; ovules erect, 2 per locule. Samara with 4–5 lateral wings, chartaceous or subcoriaceous; seed 1, flattened. Only one sp., *P. tanganyikensis* Dunkley & Brenan, Tanzania, transition between wet and dry forest, 1500–1900 m.

46. *Plenckia* Reissek

Plenckia Reissek in Martius, Fl. Bras. 11: 30 (1861).
Viposia Lundell (1939).

Trees or shrubs, glabrous. Leaves alternate, entire. Inflorescences axillary, cymose. Flowers bisexual or unisexual (polygamous), 5-merous; disk fleshy, pulvinate, 5-angled, intrastaminal; anthers obliquely dehiscent, latrorse; ovary 2-locular; ovules erect, 2 per locule. Samara with 1 apical wing, obovate or oblanceolate, chartaceous; seeds 1–2, oblong, albuminous. Four spp., S America.

47. *Pleurostylia* Wight & Arn.

Pleurostylia Wight & Arn., Prod. Fl. Ind. Orient.: 157 (1834).
Herya Cordem. (1895).

Trees or shrubs, glabrous. Leaves opposite, entire. Inflorescences axillary, cymose. Flowers bisexual, (4)5-merous; disk fleshy, annular or margins upturned, (4)5-lobed, intrastaminal; anthers longitudinally dehiscent, introrse; ovary 2-locular, rarely becoming 1-locular by abortion; ovules erect, 2–8 per locule. Nut obovoid or ellipsoid, bony, with lateral style; seeds 1(–2), ovoid, albuminous. About 8 spp., Africa, Madagascar, India, Sri Lanka, SE Asia, Macronesia, Queensland, Australia, New Caledonia, forests and woodlands, 0–1600 m.

48. *Polycardia* Juss.

Polycardia Juss., Gen. Pl.: 377 (1789); Perrier de la Bâthie, Fl. Madag. 116: 1–76 (1946), rev.

Trees or shrubs, glabrous or pubescent. Leaves alternate, entire, denticulate, or spinose-dentate. Inflorescences epiphyllous, rarely axillary, fasciculate. Flowers bisexual, 5-merous; disk fleshy, annular, entire or ± 5-lobed, intrastaminal; anthers longitudinally dehiscent, introrse; ovary 5(4)-locular; ovules axile, numerous. Capsule ovoid, coriaceous-woody, loculicidally dehiscent; seeds several per locule, oblong, albuminous or exalbuminous, partially enveloped by lacerate aril. Four spp., Madagascar, littoral and dry forests, 0–1200 m.

49. *Psammomoya* Diels & Loes.

Psammomoya Diels & Loes., Bot. Jahrb. Syst. 35: 339 (1904); Jessup, Fl. Austr. 22: 150–180 (1984), rev.

Shrubs, glabrous. Leaves opposite, reduced to scales. Inflorescences axillary, fasciculate. Flowers bisexual, 5-merous; disk thin or ± fleshy, annular, 5-angled, intrastaminal; anthers longitudinally dehiscent, introrse; ovary 2–3-locular; ovules erect, 2 per locule. Capsule obovoid or ellipsoid, bony, loculicidally dehiscent; seeds 1–2, ovoid or spheroid, albuminous, aril basal. Two spp., western Australia, heath.

50. *Pseudosalacia* Codd

Pseudosalacia Codd, Bothalia 10: 565 (1972).

Trees, glabrous. Leaves alternate, entire. Inflorescences axillary, fasciculate or cymose. Flowers bisexual, 5-merous; disk fleshy, annular, 5-angled, stamens on disk; anthers apically dehiscent, extrorse; ovary 3-locular; ovules erect, 2 per locule. Capsule spheroid, coriaceous, loculicidally dehiscent; seeds 2–5, trigonous, albuminous. Only one sp., *P. streyi* Codd, S Africa, coastal, rocky river banks.

51. *Ptelidium* Thouars

Ptelidium Thouars, Hist. Vég. Isles Austr. Afr. 1: 11, 29 (1805); Perrier de la Bâthie, Fl. Madag. 116: 1–76 (1946), rev.

Scandent or erect shrubs, glabrous. Leaves opposite, entire or crenulate-dentate Inflorescences axillary, cymose. Flowers bisexual, 4–5-merous; disk margins upturned, 4–5-lobed; anthers longitudinally dehiscent, introrse or extrorse; ovary

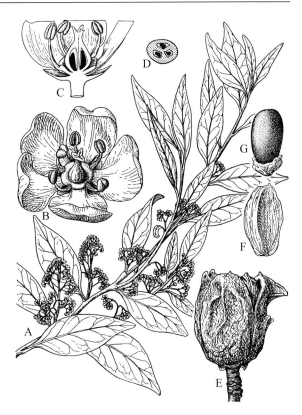

Fig. 16. Celastraceae-Celastroideae. *Pterocelastrus echinatus.* A Flowering branchlet. B Flower. C Same, vertical section. D Ovary, transverse section. E Fruit. F Seed. G Seed with aril removed. (Robson 1966)

2-locular; ovules erect, 2(1?) per locule. Samara with 1 surrounding wing, widely ovate or ovate-lanceolate, membranous; seeds 1–2, oblong, albuminous. Two spp., Madagascar, tropophilous forest, 250–400 m.

52. *Pterocelastrus* Meisn.   Fig. 16

Pterocelastrus Meisn., Pl. Vasc. Gen.: 68 (1837); Robson, Fl. Zambesiaca 2: 355–418 (1966).

Trees or shrubs, glabrous. Leaves alternate, entire. Inflorescences axillary, cymose. Flowers bisexual, 5-merous; disk thin, margins upturned, 5-lobed, intrastaminal; anthers longitudinally dehiscent, introrse; ovary (2)3-locular; ovules erect, 2 per locule. Capsule ellipsoid with lateral or oblique horn to wing-like outgrowths, loculicidally dehiscent; seeds 1–3, ovoid, albuminous, nearly enveloped by aril. Four spp., SE Africa, evergreen forests and rocky slopes.

Pterocelastrus needs to be conserved against *Asterocarpus* Eckl. & Zeyh. (Enum. Pl. Afr. Austral. 1: 122 (1834/5); R. Archer, pers. comm. 2000).

53. *Putterlickia* Endl.

Putterlickia Endl., Gen. Pl. 1086 (1840); Jordaan & van Wyk, S. Afr. J. Bot. 64: 322–329 (1998), rev.

Shrubs or lianas with thorns, glabrous. Leaves alternate (opposite if on thorns) and fasciculate on short shoots, entire or serrulate. Inflorescences axillary, cymose. Flowers bisexual, 5-merous; disk fleshy, annular, 5-lobed, intrastaminal; anthers longitudinally dehiscent, introrse; ovary 3-locular; ovules (4–)6(–12) per locule. Capsule obovoid-trigonous, bony, loculicidally dehiscent; seeds 6–18, enveloped by aril. Four spp., S Africa, southern Mozambique, forests to scrub.

54. *Quetzalia* Lundell

Quetzalia Lundell, Wrightia 4: 137 (1970); Lundell, Contrib. Univ. Michigan Herb. 3: 5–46 (1939), rev. (under *Microtropis*).

Trees or shrubs, rarely lianas, glabrous. Leaves opposite, entire. Inflorescences axillary, cymose. Flowers unisexual (dioecious), (4)5-merous; disk fleshy, annular or margins upturned, (4)5-angled, intrastaminal; anthers longitudinally dehiscent, introrse; ovary 2-locular; ovules erect, 2 per locule. Capsule ellipsoid or oblong, coriaceous, 1-locular, laterally split along one side; seed 1(2), ovoid, furrowed, albuminous, sarcotestal. Eleven spp., C America, Mexico, humid forests, to 3100 m. *Quetzalia* is recognized as distinct from *Microtropis* following Lundell (1970), *contra* Lundell (1939).

55. *Robsonodendron* R.H. Archer

Robsonodendron R.H. Archer in Archer & van Wyk, S. Afr. J. Bot. 63: 116 (1997).

Trees or shrubs, glabrous. Leaves alternate, entire or glandular-crenate. Inflorescences axillary, fasciculate. Flowers bisexual, 5-merous; disk fleshy, margins upturned, entire, intrastaminal; anthers longitudinally dehiscent, introrse; ovary 2-locular; ovules erect, 2 per locule. Drupe spheroid, fleshy; seeds 1–2, spheroid, albuminous. Two spp., S Africa, coastal forest and forest margins, coastal dunes.

56. *Rzedowskia* Medrano

Rzedowskia Medrano, Bol. Soc. Bot. México 41: 41 (1981).

Shrubs, glabrous. Leaves opposite, glandular-denticulate. Inflorescences axillary, cymose.

Flowers bisexual, 4-merous; disk annular, intrastaminal; anthers longitudinally dehiscent; ovary 2-locular; ovules erect, 2 per locule. Samara with 1 apical wing, elliptical, chartaceous; seed 1, furrowed, albuminous. Only one sp., *R. tolantonguensis* Medrano, Mexico, montane scrub, 1400–2000 m.

57. *Salaciopsis* Baker f.

Salaciopsis Baker f., J. Linn. Soc. Bot. 45: 287 (1921); Müller, Fl. Nouv.-Caléd. 20: 3–74 (1996).
Lecardia J. Poiss. ex Guillaumin (1927).

Trees or shrubs, glabrous. Leaves alternate and subverticillate, entire. Inflorescences axillary, rarely cauliflorous, fasciculate. Flowers unisexual (dioecious), 5-merous; disk fleshy, annular, 5-lobed, intrastaminal; anthers longitudinally dehiscent, introrse; ovary 3-locular; ovules erect, 2–6 per locule. Capsule ellipsoid or subspheroid, bony, loculicidally dehiscent; seeds 1–6, ovoid, albuminous, aril basal to enveloping seed. Six spp., New Caledonia, humid and gallery forests, 100–1400 m.

58. *Salvadoropsis* H. Perrier

Salvadoropsis H. Perrier, Bull. Soc. Bot. France 91: 96 (1944).

Trees or shrubs, glabrous. Leaves opposite, entire. Inflorescences axillary, cymose. Flowers bisexual, 5-merous; disk 5-angled, intrastaminal; anthers longitudinally dehiscent, introrse; ovary 5-locular; ovules axile, 12 per locule. Berry spheroid, ± fleshy; seeds 1–2 per locule, oblanceolate, albuminous. Only one sp., *S. arenicola* H. Perrier, Madagascar, coastal dunes.

59. *Sarawakodendron* Ding Hou

Sarawakodendron Ding Hou, Blumea 15: 141 (1967).

Trees, glabrous. Leaves alternate, entire. Inflorescences axillary, racemose. Flowers bisexual, 5-merous; disk fleshy, annular, 5-angled, extrastaminal; anthers transversally dehiscent, extrorse; ovary 3-locular; ovules horizontal, 8 per locule. Capsule narrow-ellipsoid, coriaceous, loculicidally dehiscent; seeds 6–8 per locule, narrow-lanceolate, descending; aril basal, cushion-shaped, with basal, filamentous extensions. Only one sp., *S. filamentosum* Ding Hou, Borneo, humid forests.

60. *Schaefferia* Jacq.

Schaefferia Jacq., Enum. Syst. Pl.: 10 (1760); Brizicky, J. Arnold
  Arbor. 45: 206–234 (1964), part. rev.

Trees or shrubs, glabrous, rarely pubescent. Leaves
alternate or fasciculate on short shoots, entire,
rarely serrulate. Inflorescences axillary, fasciculate
or flowers solitary. Flowers unisexual (dioecious),
4-merous; disk indistinct or 0; anthers longitudinally dehiscent, introrse; ovary 2-locular; ovules
erect, 1(2) per locule. Drupe subspheroid, dry;
seeds 1–2, ellipsoid or ovoid, albuminous. About
23 spp., Americas, West Indies, dry, rocky woodlands, hummocks, thickets.

61. *Siphonodon* Griff.

Siphonodon Griff., Calcutta J. Nat. Hist. 4: 246, t. 14 (1844); Ding
  Hou, Fl. Males. I, 6: 227–291 (1962); Jessup, Fl. Austr. 22:
  150–180 (1984).

Trees, glabrous. Leaves alternate, entire or serrulate. Inflorescences axillary, cymose. Flowers
bisexual, 5-merous; 5 stamens often alternating
with 5 staminodes, anthers obliquely dehiscent,
latrorse; ovary with apical hollow with style-like
central column arising from base of hollow, 10-
locular with each locule horizontally divided into
1-ovulate locelli; ovules in upper locelli erect, in
lower locelli pendulous, 1 per locule. Drupe spheroid or obovoid, hard to fleshy; seeds few to many,
flattened, in 1-seeded, bony pyrenes. Seven spp., SE
Asia, Macronesia, Queensland, Australia, forests
and thickets, 0–1650 m.
  Siphonodon has been considered unusual
within Celastraceae based on the structure of
its gynoecium (Croizat 1947), wood anatomy
(Metcalfe and Chalk 1950), and pollen morphology (Erdtman 1952). However, *Siphonodon*
has been supported as a derived member of
Celastraceae (e.g., Savolainen, Fay et al. 2000;
Simmons et al. 2001b).

62. *Tetrasiphon* Urb.

Tetrasiphon Urb., Symb. Antill. 5: 83 (1904); Fawcett & Rendle,
  Fl. Jamaica 5: 24–35 (1926).

Trees or shrubs, glabrous. Leaves opposite,
entire. Inflorescences axillary, cymose. Flowers
unisexual (dioecious), 4-merous; disk pulvinate, 4-
lobed, intrastaminal; anthers longitudinally dehiscent, introrse; ovary 4-locular; ovules pendulous, 1
per locule. Drupe spheroid, ± fleshy; seeds 2–4,
oblong-ellipsoid, ± flattened, albuminous. Only

one sp., *T. jamaicensis* Urb., Jamaica, thickets on
limestone.

63. *Torralbasia* Krug & Urb.

Torralbasia Krug & Urb. in Seguí, Flora Méd. Tóx. Cuba: 60
  (1900); Liogier, Flora Hispaniola 1: 12–25 (1981).

Trees or shrubs, glabrous. Leaves alternate,
opposite, or subopposite, entire or crenulate. Inflorescences axillary, cymose. Flowers bisexual,
4-merous; disk cupular, 4-lobed, intrastaminal;
anthers longitudinally dehiscent, introrse; ovary 4-
locular; ovules erect, 1 per locule. Capsule deeply
1–4-lobed, coriaceous, longitudinally dehiscent;
seeds 1–4, ellipsoid, albuminous, aril basal. Only
one sp., *T. cuneifolia* (Wright) Krug & Urb., Cuba,
Hispaniola, Puerto Rico, dwarf forests, 800–1500 m.

64. *Tricerma* Liebm.

Tricerma Liebm., Vidensk. Meddel. Dansk Naturhist. Foren.
  Kjöbenhavn: 97 (1853); Lundell, Wrightia 4: 153–172 (1971),
  rev.

Trees or shrubs, puberulent or glabrescent. Leaves
alternate, entire or serrulate. Inflorescences axillary, fasciculate, condensed thyrsoid, or flowers
solitary. Flowers bisexual or unisexual (polygamous), 5-merous; disk fleshy, annular, 5-angled or
lobed, intrastaminal; anthers longitudinally dehiscent, introrse; ovary 3(4)-locular; ovules erect, 1
per locule. Capsule obovoid or fusiform, coriaceous, loculicidally dehiscent; seeds 1–3, ellipsoid
or obovoid, albuminous, enveloped by aril. $2n = 80$.
Seven spp., southern North America, South
America, dry thickets.
  Tricerma is recognized as distinct from
Maytenus (Lundell 1971; Simmons et al. 2001b).

65. *Tripterygium* Hook. f.

Tripterygium Hook f. in Bentham & Hooker, Gen. Pl. 1: 368
  (1862); Ma, Edinburgh J. Bot. 56: 35–46 (1999), rev.

Scandent shrubs, glabrous or tomentose. Leaves
alternate, serrate. Inflorescences axillary or terminal, thyrsoid. Flowers bisexual or unisexual
(polygamous), 5-merous; disk fleshy, margins
upturned, 5-lobed, intrastaminal; anthers longitudinally dehiscent, introrse; ovary 3-locular; ovules
erect, 2 per locule. Samara with 3 lateral wings,
chartaceous; seed 1, trigonous, albuminous. $2n = 24$. One sp., China, Taiwan, Korea, Japan, thickets.
Cultivated as ornamentals for their winged fruits.

66. *Wimmeria* Schltdl. & Cham.

Wimmeria Schltdl. & Cham., Linnaea 6: 427 (1831); Lundell,
Contrib. Univ. Michigan Herb. 3: 5–46 (1939), rev.

Trees or shrubs, glabrous or pubescent. Leaves alternate, some fasciculate on short shoots, serrulate. Inflorescences axillary, cymose. Flowers bisexual, 4-merous; disk fleshy, annular or margins upturned, 5-angled, intrastaminal; anthers longitudinally dehiscent, introrse; ovary (2)3-locular; ovules axile, 4–9 per locule. Samara with (2)3(4) lateral wings, chartaceous; seed 1(2), linear, albuminous. Twelve spp. (C. Clevinger, pers. comm. 2000), C America, Mexico, humid to dry forests.

67. *Xylonymus* Kalkman ex Ding Hou

Xylonymus Kalkman ex Ding Hou, Fl. Males. I, 6: 243 (1962).

Shrub, glabrous. Leaves alternate, entire. Inflorescences axillary, cymose. Flowers bisexual, 4-merous; disk fleshy, annular, 4-angled, intrastaminal; anthers obliquely dehiscent, latrorse; ovary 4-locular; ovules axile, ± 10 per locule. Capsule, oblong, woody, loculicidally dehiscent; seeds several per locule, ellipsoid, albuminous, partially enveloped by aril. Only one sp., *X. versteeghii* Kalkman, New Guinea, humid lowland forests.

68. *Zinowiewia* Turcz.

Zinowiewia Turcz., Bull. Soc. Imp. Nat. Moscou 32: 275 (1859);
Lundell, Contrib. Univ. Michigan Herb. 3: 5–46 (1939), rev.

Trees or shrubs, glabrous. Leaves opposite, entire. Inflorescences axillary, cymose. Flowers bisexual, 5-merous; disk fleshy, annular, 5-angled or lobed, intrastaminal; anthers longitudinally dehiscent, introrse; ovary 2-locular; ovules erect, 2 celled. Samara with 1 apical wing, oblanceolate or obovate, chartaceous; seeds 1(2), linear-oblong, albuminous. Seventeen spp., Mexico, C America, northern S America, humid forests to mesic thickets, to 3150 m.

II. SUBFAMILY STACKHOUSIOIDEAE
Burnett (1835).

Stackhousiaceae R. Br. (1814), nom. cons.

Annual or perennial herbs with erect or prostrate stems. Leaves alternate, entire, sometimes reduced to scales. Inflorescences terminal, racemose, in spikes, umbellate, or flowers solitary. Flowers actinomorphic or ± zygomorphic, bisexual, perigynous with short or cupular hypanthium; 5-merous; petals free or medially connate; disk thin, lining hypanthium; stamens 5, sometimes 3 long and 2 short; anthers longitudinally dehiscent, introrse; ovary 2–5-locular, with basal placentation; ovules 1 per locule, basal, erect. Schizocarp of 2–5 indehiscent mericarps; seeds albuminous.

69. *Macgregoria* F. Muell.

Macgregoria F. Muell., Fragm. 8: 160 (1874).

Erect to prostrate annual herbs, glabrous. Leaves linear. Inflorescences racemose. Flowers subtended by one bract each; hypanthium shallow; petals free; stames equal in length; anthers longer than filaments, tipped by white glands; ovary 5-locular; style with membranous cup at base. Schizocarp of 5 mericarps, mericarps densely papillose with hooked hairs. Only one sp., *M. racemigera* F. Muell., central Australia, arid grasslands and woodlands.

70. *Stackhousia* Sm.

Stackhousia Sm., Trans. Linn. Soc. Lond. 4: 218 (1798); Barker,
Fl. Austr. 22: 186–199 (1984).

Perennial or annual herbs, sometimes rhizomatous. Leaves obovate to linear, sometimes reduced to scales. Inflorescences racemose, in spikes, umbellate or flowers solitary. Flowers bracteolate; hypanthium cupular; petals medially connate; stamens 3 long and 2 short; anthers as long or shorter than filaments; ovary 3–5-locular. Schizocarp of (1–)3–5 mericarps; style inserted between mericarps; mericarps attached only at base, sometimes each with 3 wings. $2n = 18, 20, 30, 36$. About 16 spp., Australasia, dunes, grasslands to forests.

Barker (1984) suggested a number of subgeneric groups but did not formalize them. In particular, Barker noted the group of species related to *S. viminea* with clustered flowers subtended by numerous bracts on the inflorescence, and the group of species related to *S. monogyna* with separate tri-bracteate flowers on the inflorescence.

71. *Tripterococcus* Endl. Fig. 17

Tripterococcus Endl., Enum. Pl. Hueg.: 17 (1837).

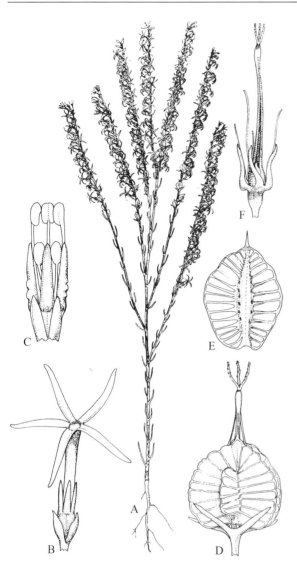

Fig. 17. Celastraceae-Stackhausioideae. *Tripterococcus brunonis.* **A** Plant. **B** Flower with two of the three subtending bracts visible. **C** Flower with corolla removed. **D** Fruit. **E** Mericarp. **F** Gynoecial column after shedding mericarps. (Barker 1984)

Perennial herbs, glabrous. Leaves linear to ovate or sometimes reduced to scales. Inflorescences in one-sided spikes. Flowers subtended by three bracts each; hypanthium cupular; petals medially connate; stamens 3 long and 2 short; anthers shorter than filaments; ovary 3-locular. Schizocarp of (1–)3 mericarps; style terminal, persistent; mericarps each with 3 prominently-veined wings, glabrous. $2n =$ c. 30, c. 32. Two spp., SW Australia, heath, sclerophylous forests, woodlands.

Barker (1984) cited two species in the genus, one undescribed. Only one sp., *T. brunonis* Endl., has been published. Recognition of *Tripterococcus* probably makes *Stackhousia* paraphyletic.

III. Subfamily Hippocrateoideae
(Juss.) Lindl. (1836).

Hippocrateaceae Juss. (1811), nom. cons.

Lianas or scandent, rarely completely erect trees or shrubs. Inflorescences axillary or terminal. Flowers bisexual, (3–4)5-merous; petals free; disk extrastaminal or 0, rarely stamens on disk; androgynophore wide, narrow, or 0; stamens 3, rarely 5 or numerous; anthers transversely dehiscent, introrse or extrorse (apical); ovary 3-locular; ovules axile, 2–16(–22) per locule. Capsule transversely flattened, lobed to base, rarely lobed ± halfway to base or locules entirely connate, coriaceous, rarely woody or chartaceous, loculicidally dehiscent; seeds 2–numerous, with membranous basal wing or narrow stipe, exalbuminous. Included phloem 0.

72. *Anthodon* Ruiz & Pav.

Anthodon Ruiz & Pav., Fl. Peruv. 1: 45, t. 74 (1798); Görts-van Rijn & Mennega, Fl. Guianas, 16: 3–81 (1994).

Lianas, glabrous. Leaves opposite or subopposite, crenulate or serrulate. Inflorescences axillary, cymose. Flowers bisexual, 5-merous; petals regularly serrate; disk fleshy, cupular, entire, extrastaminal; stamens 3; anthers transversely dehiscent, extrorse; ovary 3-locular; ovules axile, 8–14 per locule. Capsule transversely flattened, locules entirely connate, coriaceous, loculicidally dehiscent; seeds 8–14 per locule, with membranous basal wing, exalbuminous. Two spp., C and S America, humid forests, 100–900 m.

73. *Apodostigma* R. Wilczek

Apodostigma R. Wilczek, Bull. Jard. Bot. État 26: 402 (1956); Robson et al., Fl. Trop. E. Afr. 108: 1–78 (1994).

Lianas, glabrous. Leaves opposite or subopposite, entire or denticulate. Inflorescences axillary, cymose. Flowers bisexual, 5-merous; 4 of 5 petals arched, ± zygomorphic; disk of 3 discontinuous lobes around stamens, extrastaminal; stamens 3; anthers transversely dehiscent, extrorse; ovary 3-locular; ovules axile, 2 per locule. Capsule transversely flattened, lobed to base, coriaceous, loculicidally dehiscent; seeds 2 per locule, with membranous basal wing, exalbuminous. Only one sp., *A. pallens* (Oliv.) R. Wilczek, Africa, Madagascar, humid forests to dry thickets.

74. *Arnicratea* N. Hallé

Arnicratea N. Hallé, Bull. Mus. Natl. Hist. Nat., B, Adansonia 6: 12 (1984).

Lianas, glabrous or pubescent. Leaves opposite, rarely subopposite, entire, rarely serrulate. Inflorescences axillary, thyrsoid. Flowers bisexual, 5-merous; disk indistinct, lobed, extrastaminal; anthers transversely dehiscent, extrorse; ovary 3-locular; ovules axile, (4–)6(7) per locule. Capsule transversely flattened, lobed to base, coriaceous, loculicidally dehiscent; seeds numerous, with membranous basal wing, exalbuminous. Three spp., India, SE Asia, Macronesia, lowland forests, to 700 m.

75. *Bequaertia* R. Wilczek

Bequaertia R. Wilczek, Bull. Jard. Bot. État 26: 399 (1956); Robson et al., Fl. Trop. E. Afr. 108: 1–78 (1994).

Lianas, glabrous. Leaves opposite, crenulate-dentate. Inflorescences axillary, cymose. Flowers bisexual, 5-merous; petals thick and fleshy; disk 0; stamens 3; anthers transversely dehiscent, extrorse; ovary 3-locular; ovules axile, 6–8(–12) per locule. Capsule transversely flattened, lobed to base, coriaceous, loculicidally dehiscent; seeds numerous, with membranous basal wing, exalbuminous. Only one sp., *B. mucronata* (Exell) R. Wilczek, tropical Africa, humid forests and river banks, 900–1000 m.

76. *Campylostemon* Welw.

Campylostemon Welw. in Bentham & Hooker, Gen. Pl. 1: 998 (1862); N. Hallé, Fl. Cameroun 32: 3–243 (1990).

Lianas, glabrous. Leaves opposite, entire or serrate. Inflorescences axillary, cymose, rarely thyrsoid. Flowers bisexual, 5-merous; disk 0; stamens 5; anthers transversely dehiscent, introrse; ovary 3-locular; ovules axile, 4–16 per locule. Capsule transversely flattened, lobed to base, coriaceous, loculicidally dehiscent; seeds numerous, with membranous basal wing, exalbuminous. $2n = 56$. Eight or more species, Africa, forests and thickets.

77. *Cuervea* Triana ex Miers

Cuervea Triana ex Miers, Trans. Linn. Soc. Lond. 28: 370 (1872); Smith, Brittonia 3: 341–555 (1940), reg. rev.; N. Hallé, Fl. Cameroun 32: 3–243 (1990).

Trees, shrubs, or lianas, glabrous. Leaves opposite or subopposite, entire, crenulate, or denticulate. Inflorescences axillary, rarely terminal, cymose or thyrsoid. Flowers bisexual, 5-merous; disk membranous, cupular, entire or lobed, extrastaminal; stamens 3; anthers transversely dehiscent, extrorse; ovary 3-locular; ovules axile, 4–9 per locule. Capsule transversely flattened, lobed to base, coriaceous, loculicidally dehiscent; seeds 2–numerous, with membranous basal wing or narrow stipe, exalbuminous. Five spp., C and S America, West Indies, Africa, humid to gallery forests.

78. *Elachyptera* A.C. Sm.

Elachyptera A.C. Sm., Brittonia 3: 383 (1940); N. Hallé, Fl. Cameroun 32: 3–243 (1990).

Lianas or scandent trees or shrubs, glabrous (inflorescences sometimes puberulent). Leaves opposite or subopposite, entire, crenulate, or serrate. Inflorescences axillary or terminal, cymose or thyrsoid. Flowers bisexual, 5-merous; disk ± fleshy, cupular, lobed, extrastaminal; stamens 3; anthers transversely dehiscent, extrorse; ovary 3-locular; ovules axile, 2 or 4 per locule. Capsule transversely flattened, lobed to base, coriaceous, loculicidally dehiscent; seeds 2–4 per locule, with membranous basal wing or narrow stipe, exalbuminous. Seven spp., C and S America, Africa, Madagascar, humid forests to dry thickets, mangrove swamps.

79. *Helictonema* Pierre   Fig. 18

Helictonema Pierre, Bull. Mens. Soc. Linn. Paris N.S. 9: 73 (1898); Robson et al., Fl. Trop. E. Afr. 108: 1–78 (1994).

Lianas, hirsute and stellate pubescent. Leaves opposite, entire. Inflorescences axillary, cymose. Flowers bisexual, 5-merous; disk fleshy, annular, irregularly lobed, extrastaminal; androgynophore wide; stamens 3; anthers transversely dehiscent, extrorse; ovary 3-locular; ovules axile, numerous. Capsule transversely flattened, lobed to base, coriaceous, loculicidally dehiscent; seeds numerous, with membranous basal wing, exalbuminous. Only one sp., *H. velutinum* (Afzel.) N. Hallé, tropical Africa, forests.

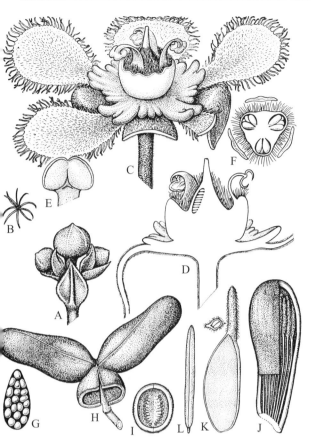

Fig. 18. Celastraceae-Hippocrateoideae. *Helictonema velutinum.* **A** Cymule of inflorescence. **B** Stellate leaf hair. **C** Flower with artificially expanded petals, note disk and androgynophore. **D** Flower, vertical section. **E** Anther. **F** Ovary, transverse section, with stamen filaments. **G** Ovary locule opened, showing the ovules. **H** Trilobed capsule. **I** Basal part of capsule lobe showing two placentas. **J** Fruit lobe, opened. **K** Winged seed, the cotyledons transversely sectioned. **L** Embryo. Drawn by N. Hallé. (Hallé 1990)

80. *Hippocratea* L.

Hippocratea L., Gen. Pl. ed. 1: 363 (1737); Sp. Pl. ed. 1: 1191 (1753); Smith, Brittonia 3: 341–555 (1940), reg. rev.

Lianas, puberulent or glabrescent. Leaves opposite, entire, crenate, or serrate. Inflorescences axillary, cymose or thyrsoid. Flowers bisexual, 5-merous; disk fleshy, pulvinate or cupular, entire, extrastaminal; stamens 3; anthers transversely dehiscent, extrorse; ovary 3-locular; ovules axile, 4–8 per locule. Capsule transversely flattened, lobed to base, chartaceous or coriaceous, loculicidally dehiscent; seeds numerous, with membranous basal wing, exalbuminous. $2n = 28$. Three spp., Americas, West Indies, tropical Africa, humid to gallery forests, 0–1800 m.

81. *Hylenaea* Miers

Hylenaea Miers, Trans. Linn. Soc. Lond. 28: 366 (1872); Görts-van Rijn & Mennega, Fl. Guianas, 16: 3–81 (1994).

Trees or lianas, glabrous. Leaves opposite or subopposite, entire. Inflorescences axillary or terminal, thyrsoid. Flowers bisexual, 5-merous; disk membranous or ± fleshy, cupular, entire; stamens 3; anthers transversely dehiscent, extrorse; ovary 3-locular; ovules axile, 4–8 per locule. Capsule transversely flattened, lobed to base, woody, loculicidally dehiscent; seeds 4–8 per locule, enlarged embryoniferous portion with narrow stipe (or membranous wing), exalbuminous. Three spp., C and S America, forests.

82. *Loeseneriella* A.C. Sm.

Loeseneriella A.C. Sm., Am. J. Bot. 28: 438 (1941); N. Hallé, Fl. Cameroun 32: 3–243 (1990).

Lianas or scandent, rarely erect shrubs, puberulent or glabrescent. Leaves opposite, entire or crenulate. Inflorescences axillary, cymose. Flowers bisexual, 5-merous; disk fleshy, pulvinate or cupular, extrastaminal; androgynophore 0 or wide; stamens 3; anthers transversely dehiscent, extrorse; ovary 3-locular; ovules 4–15(–22) per locule. Capsule transversely flattened, lobed to base, coriaceous, loculicidally dehiscent; seeds numerous, with membranous basal wing, exalbuminous. $2n = 56$. Sixteen spp., Africa, Madagascar, India, SE Asia, Macronesia, Queensland, Australia, humid forests to dry thickets.

83. *Plagiopteron* Griff.   Fig. 19

Plagiopteron Griff., Calcutta J. Nat. Hist. 4: 244, t. 13 (1843); Tang, Wu & Li, Acta Bot. Yunnan. 12: 126–128 (1990), rev.

Scandent shrubs or lianas, stellate pubescent. Leaves opposite, entire. Inflorescence axillary, thyrsoid. Flowers bisexual, (3)4(5) merous; disk indistinct, stamens on disk; stamens numerous; anthers transversely, apically dehiscent; ovary 3-locular; ovules erect, 2 per locule. Capsule transversely flattened, lobed to base, coriaceous; seeds unknown. Only one sp., *P. suaveolens* Griff., China, Myanmar, forests.

Plagiopteron has been assigned to Tiliaceae (Bentham and Hooker 1862), Flacourtiaceae (Warburg 1893), and Plagiopteraceae (Airy Shaw 1965). However, *Plagiopteron* has been consistently supported as related to Hippocrateoideae in phylogenetic analyses (e.g., Simmons and Hedin

1999; Soltis et al. 2000; Simmons et al. 2001b), and the anatomy of *Plagiopteron* is consistent with Celastraceae (Baas et al. 1979).

84. *Prionostemma* Miers

Prionostemma Miers, Trans. Linn. Soc. Lond. 28: 354 (1872); N. Hallé, Bull. Mus. Natl. Hist. Nat., B, Adansonia 3: 5–14 (1981), rev.

Lianas, glabrous or glabrescent. Leaves opposite, entire, rarely dentate. Inflorescences axillary, rarely terminal, thyrsoid. Flowers bisexual, 5-merous; disk fleshy, annular-pulvinate, entire, extrastaminal; stamens 3; ovary 3-locular; ovules axile, 6–20 per locule. Capsule transversely flattened, lobed to base, coriaceous, loculicidally dehiscent; seeds numerous, with membranous basal wing, exalbuminous. Five spp., C and South America, tropical Africa, India, forests and savannas.

85. *Pristimera* Miers

Pristimera Miers, Trans. Linn. Soc. Lond. 28: 360 (1872); N. Hallé, Bull. Mus. Natl. Hist. Nat., B, Adansonia 3: 5–14 (1981), rev.

Lianas or scandent shrubs, glabrous or puberulent. Leaves opposite or subopposite, entire or serrate. Inflorescences axillary, rarely terminal, cymose, rarely thyrsoid. Flowers bisexual, 5-merous; disk fleshy, indistinct, annular or cupular, entire or lobed; stamens 3; anthers transversely dehiscent, extrorse; ovary 3-locular; ovules axile, 2–10 per locule. Capsule transversely flattened, lobed to base, chartaceous or ± coriaceous, loculicidally dehiscent; seeds 2–10 per locule, with membranous basal wing, exalbuminous. Twenty-four spp., Old and New World tropics, humid forests to dry thickets. This genus, which has been resolved as polyphyletic in a 26 S gene tree and a simultaneous analysis (Simmons et al. 2001b), will need further attention.

86. *Reissantia* N. Hallé

Reissantia N. Hallé, Bull. Mus. Natl. Hist. Nat. 30: 466 (1958); Robson et al., Fl. Trop. E. Afr. 108: 1–78 (1994).

Lianas or scandent shrubs, glabrous or puberulent. Leaves opposite or subopposite, entire, crenulate, or serrate. Inflorescences axillary, rarely terminal, cymose, rarely thyrsoid. Flowers bisexual, 5-merous; disk ± fleshy, pulvinate or cupular, lobed, extrastaminal; stamens 3; anthers transversely dehiscent, extrorse; ovary 3-locular; ovules axile, 2

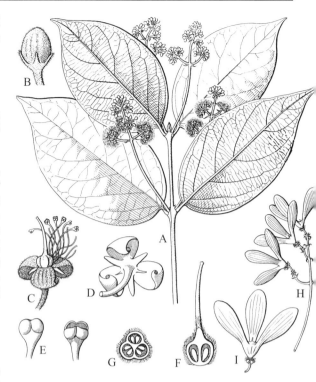

Fig. 19. Celastraceae-Hippocrateoideae. *Plagiopteron suaveolens.* **A** Flowering branchlet. **B** Flower bud. **C** Flower with most stamens removed. **D** Flower from underside, showing unequal sepals. **E** Anthers showing transverse dehiscence. **F** Pistil, ovary vertically sectioned, showing basal placentation. **G** Ovary, transverse section. **H** Infructescence. **I** Fruit. Drawn by R. van Crevel. (Baas et al. 1979)

per locule. Capsule transversely flattened, lobed to base, coriaceous, loculicidally dehiscent; seeds 2 per locule, with membranous basal wing, exalbuminous. $2n = 28$. Six spp., Africa, Madagascar, India, SE Asia, Macronesia, humid forests to dry thickets.

87. *Semialarium* N. Hallé

Semialarium N. Hallé, Bull. Mus. Natl. Hist. Nat., B, Adansonia 5: 22 (1983).

Tree, shrub, or liana, puberulent or glabrescent. Leaves opposite, rarely subopposite, crenulate or serrulate. Inflorescences axillary, rarely terminal, thyrsoid. Flowers bisexual, 5-merous; fleshy, annular-pulvinate, entire, extrastaminal; stamens 3; anthers transversely dehiscent, extrorse; ovary 3-locular; ovules axile, 6–8. Capsule transversely flattened, lobed ± halfway to base, coriaceous, loculicidally dehiscent; seeds 6–8 per locule, with membranous basal wing, exalbuminous. $2n = 30$.

Two spp., Mexico, C and South America, forests, to 1300 m.

88. *Simicratea* N. Hallé

Simicratea N. Hallé, Bull. Mus. Natl. Hist. Nat., B, Adansonia 1: 18 (1983); Robson et al., Fl. Trop. E. Afr. 108: 1–78 (1994).

Lianas, glabrous. Leaves opposite, entire or crenulate. Inflorescences axillary, cymose. Flowers bisexual, 5-merous; disk fleshy, annular, 5-angled, extrastaminal; androgynophore narrow; stamens 3; anthers transversely dehiscent, extrorse; ovary 3-locular; ovules axile, 2 per locule. Capsule transversely flattened, lobed to base, coriaceous, loculicidally dehiscent; seeds 2 per locule, with membranous basal wing, exalbuminous. Only one sp., *S. welwitschii* (Oliv.) N. Hallé, tropical Africa, forest, scrubland.

89. *Simirestis* N. Hallé

Simirestis N. Hallé, Bull. Mus. Natl. Hist. Nat. 30: 464 (1958); N. Hallé, Bull. Mus. Natl. Hist. Nat., B, Adansonia 6: 3–18 (1984), rev.

Lianas, glabrous. Leaves opposite or subopposite, entire or dentate. Inflorescences axillary, cymose, rarely terminal and paniculate. Flowers bisexual, 5-merous; disk indistinct, lobed, extrastaminal; androgynophore wide; stamens 3; anthers transversely dehiscent, extrorse; ovary 3-locular; ovules axile, (4–)6–16 per locule. Capsule transversely flattened, lobed to base, coriaceous, loculicidally dehiscent; seeds numerous, with membranous basal wing, exalbuminous. Eight spp., Africa, forests.

90. *Tristemonanthus* Loes.

Tristemonanthus Loes., Wiss. Ergebn. Deutsch. Zentr.-Afr.-Exped. 1910–11, 2: 77 (1922) sine descr.; Loes., Feddes Repert. 49: 226 (1940); N. Hallé, Fl. Cameroun 32: 3–243 (1990).

Lianas, glabrous. Leaves opposite, entire or denticulate. Inflorescences axillary, cymose. Flowers bisexual, 5-merous; disk 0; stamens 3; anthers transversely dehiscent, introrse; ovary 3-locular; ovules axile, 4–8 per locule. Capsule transversely flattened, lobed to base, coriaceous, loculicidally dehiscent; seeds numerous, with membranous basal wing, exalbuminous. Two spp., tropical Africa, swamp forests.

IV. Subfamily Salacioideae N. Hallé (1962)
[published without Latin diagnosis].

Lianas or scandent, rarely completely erect trees or shrubs. Inflorescences axillary, rarely terminal or cauliflorous. Flowers bisexual, 5-merous; petals free; disk extrastaminal; androgynophore wide, narrow, or 0; stamens (2)3(5); anthers longitudinally, obliquely, or transversely dehiscent, extrorse, rarely apical; ovary (2)3(5)-locular; ovules axile, 2–9 per locule. Berry spheroid, oblong, or fusiform, coriaceous, rarely chartaceous; seeds 1–many, oblong, angled, exalbuminous, surrounded by mucilaginous arils. Included phloem generally present.

91. *Cheiloclinium* Miers

Cheiloclinium Miers, Trans. Linn. Soc. Lond. 28: 420 (1872); Görts-van Rijn & Mennega, Fl. Guianas, 16: 3–81 (1994).

Lianas or scandent trees or shrubs, glabrous or glabrescent. Leaves opposite, entire, crenate, or serrate. Inflorescences axillary, thyrsoid or cymose. Flowers bisexual, 5-merous; disk of 3–5 discontinuous lobes around stamens, extrastaminal; stamens 3 or 5; anthers transversely dehiscent, extrorse; ovary 3 or 5-locular; ovules axile, 2 or 4 per locule. Berry subspheroid, coriaceous; seeds 2–6, oblong, angled, exalbuminous, surrounded by mucilaginous arils. Eleven spp., C and South America, humid forests, 0–1300 m.

92. *Peritassa* Miers

Peritassa Miers, Trans. Linn. Soc. Lond. 28: 402 (1872); Smith, Brittonia 3: 341–555 (1940), rev.

Lianas or scandent trees or shrubs, glabrous (some inflorescences pilose). Leaves opposite, subopposite, or alternate, entire, crenate, or serrate. Inflorescences axillary, thyrsoid or cymose. Flowers bisexual, 5-merous; disk ± membranous, cupular, lobed, extrastaminal; stamens 3; anthers obliquely or longitudinally dehiscent, extrorse, with or without bilobed extended connective; ovary 3-locular; ovules axile, 2 or 4 per locule. Berry spheroid or oblong-ellipsoid, chartaceous or coriaceous; seeds 2–6, oblong, angled, exalbuminous, surrounded by mucilaginous arils. Thirteen spp., C and South America, humid forests.

93. *Salacia* L.

Salacia L., Mant. Pl. 2: 159, 293 (1771); Loesener, Nat. Pflanzenfam. ed. 2, 20b: 198–231 (1942), part. rev.
Salacicratea Loes. (1910).

Lianas or scandent, rarely completely erect trees or
shrubs, glabrous or glabrescent. Leaves opposite
or subopposite, rarely alternate, entire, crenate, or
serrate. Inflorescences axillary or cauliflorous,
thyrsoid, cymose, or fasciculate. Flowers bisexual,
5-merous; disk fleshy, annular, margins upturned,
or cupular, entire or lobed, extrastaminal; androgynophore wide, narrow, or 0; stamens (2)3; anthers
longitudinally, obliquely, or transversely dehiscent, extrorse or apical; ovary (2)3-locular; ovules
axile, 2–9 per locule. Berry spheroid, oblong, or
fusiform, coriaceous; seeds 1–many, oblong,
angled, exalbuminous, surrounded by mucilaginous arils. $2n = 28$. About 200 spp., Old and New
World tropics, Australia, humid forests to dry
scrub and grasslands.

This large, heterogeneous genus is unlikely to
be monophyletic and has been resolved as paraphyletic in phylogenetic analyses (Simmons et al.
2001a, 2001b).

94. *Salacighia* Loes.

Salacighia Loes., Wiss. Ergebn. Deutsch. Zentr.-Afr.-Exped.
1910–11 2: 77 (1922) sine descr.; Loes., Feddes Repert. 49:
228 (1940); N. Hallé, Fl. Cameroun 32: 3–243 (1990).

Lianas, glabrous. Leaves subopposite or alternate,
entire. Inflorescences cauliflorous, fasciculate
on long pendant branches. Flowers bisexual, 5-
merous; disk fleshy, pulvinate, 5-lobed, extrastaminal; stamens 3; anthers transversely dehiscent,
extrorse; ovary 3-locular; ovules axile, 2 per locule.
Berry spheroid; seeds 6, subspheroid, exalbuminous, surrounded by mucilaginous arils. $2n =$
28. Two spp., tropical Africa, dry to humid forests,
10–1000 m.

95. *Thyrosalacia* Loes.   Fig. 20

Thyrosalacia Loes., Feddes Repert. 49: 229 (1940); N. Hallé, Fl.
Cameroun 32: 3–243 (1990).

Scandent shrubs or lianas, glabrous. Leaves opposite, entire. Inflorescences axillary or terminal,
thyrsoid or racemose. Flowers bisexual, 5-merous;
disk fleshy, annular, entire or 5-lobed, extrastaminal; stamens 3; anthers obliquely or longitudinally
dehiscent, extrorse; ovary 3-locular; ovules axile,
2 per locule. Berry fusiform; seeds 1–3, ellipsoid,

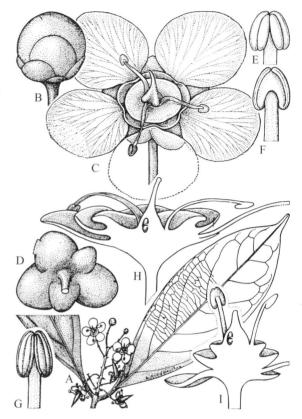

Fig. 20. Celastraceae-Salacioideae. *Thyrsosalacia pararace-
mosa.* **A** Flowering branchlet. **B** Flower bud. **C** Flower. **D** Calyx,
with quincuncial aestivation. **E** Anther, dorsal view. **F** Anther,
ventral view. **G** Dehiscing anther. **H** Flower, vertical section.
I Young flower, same. Drawn by N. Hallé. (Hallé 1990)

exalbuminous, surrounded by mucilaginous arils.
Four spp., tropical Africa, humid forests or rocky
slopes, to 1000 m.

96. *Tontelea* Aubl.

Tontelea Aubl., Hist. Pl. Guiane 1: 31, t. 10 (1775); Smith, Brittonia 3: 341–555 (1940), rev.

Lianas or scandent trees or shrubs, glabrous or
glabrescent. Leaves opposite or subopposite,
entire, crenate or serrate. Inflorescences axillary,
thyrsoid, rarely cymose. Flowers bisexual, 5-
merous; disk membranous or ± fleshy, cupular,
entire or lobed, extrastaminal; stamens 3; anthers
transversely dehiscent, extrorse; ovary 3-locular;
ovules 2–4(–8) per locule. Berry subspheroid or
ellipsoid, coriaceous or woody; seeds few, rarely
many, oblong, angled, exalbuminous, surrounded
by mucilaginous arils. Thirty-one spp., C and
South America, humid forests, to 1800 m.

Genera Doubtfully Included in Celastraceae

97. *Nicobariodendron* Vasudeva Rao & Chakrab.

Nicobariodendron Vasudeva Rao & Chakrab., J. Econ. Taxon. Bot. 7: 513 (1985).

Trees, glabrous. Leaves opposite, entire. Inflorescences axillary, racemose. Flowers unisexual (dioecious), 4–5-merous; disk ± fleshy, annular, intrastaminal; stamens 2; anthers longitudinally dehiscent; ovary and ovules unknown. Drupe oblong-ellipsoid or narrowly obovoid; seed 1, basal. Only one sp., *N. sleumeri* Vasudeva Rao & Chakrab., Andaman-Nicobar Islands, India, forests, to 100 m.

This incompletely known genus is morphologically unusual relative to other Celastraceae, with its racemose inflorescences and two stamens (although these characteristics are not unique within the family), and is questionably included in Celastraceae (Vasudeva Rao and Chakrabarty 1985).

98. *Pottingeria* Prain

Pottingeria Prain, J. Asiat. Soc. Bengal 67: 291 (1898); Airy Shaw, Cutler & Nilsson, Kew Bull. 28: 97–104 (1973), rev.

Trees or shrubs, glabrous. Leaves alternate, entire. Inflorescences axillary, thyrsoid. Flowers bisexual, 5-merous; disk fleshy, annular, 5-lobed, intrastaminal; anthers longitudinally dehiscent, introrse; ovary 1-locular, 3-carpellate, with intruded parietal placentation; ovules numerous. Capsule ovoid, septicidally dehiscent; seeds numerous, narrowly fusiform, albuminous. Only one sp., *P. acuminata* Prain, Myanmar, Thailand, forests, thickets, and scrub, 900–2400 m.

Pottingeria, originally described as a member of Saxifragaceae tribe Escallonieae (King and Prain 1898), is questionably included in Celastraceae (Airy Shaw et al. 1973). Based on pollen morphology, *Pottingeria* appears unrelated to Celastraceae (Hideux and Ferguson 1976; Lobreau-Callen 1977). *Pottingeria* is unusual relative to other members of Celastraceae, with its basal acrodromous leaf venation, intruded parietal placentation, and septicidally dehiscent capsules.

Selected Bibliography

Adatia, R.D., Gavde, S.G. 1962. Embryology of the Celastraceae. In: Plant embryology: a symposium. New Delhi: Council of Scientific & Industrial Research, pp. 1–11.

Airy Shaw, H.K. 1965. Diagnosis of new families, new names, etc., for the seventh edition of Willis's 'Dictionary'. Kew Bull. 18: 249–273.

Airy Shaw, H.K., Cutler, D.F., Nilsson, S. 1973. *Pottingeria*, its taxonomic position, anatomy and palynology. Kew Bull. 28: 97–104.

Archer, R.H. 1990. The taxonomic status of *Cassine* L. *s.l.* (Celastraceae) in southern Africa. M.Sc. Thesis. Pretoria: University of Pretoria.

Archer, R.H., van Wyk, A.E. 1992. Palynology and intergeneric relationships in some southern African species of subfamily Cassinoideae (Celastraceae). Grana 31: 241–252.

Archer, R.H., van Wyk, A.E. 1993a. Bark structure and intergeneric relationships of some Southern African Cassinoideae (Celastraceae). IAWA J. 14: 35–53.

Archer, R.H., van Wyk, A.E. 1993b. Wood structure and generic status of some southern African Cassinoideae (Celastraceae). IAWA J. 14: 373–389.

Archer, R.H., van Wyk, A.E. 1997. A taxonomic revision of *Cassine* L. *s.str.* (Cassinoideae: Celastraceae). S. Afr. J. Bot. 63: 146–157.

Archer, R.H., van Wyk, A.E. 1998a. A taxonomic revision of *Maurocenia* (Celastraceae), a Western Cape monotypic endemic. Bothalia 28: 7–10.

Archer, R.H., van Wyk, A.E. 1998b. A taxonomic revision of *Allocassine* N. Robson (Celastraceae). S. Afr. J. Bot. 64: 189–191.

Baas, P., Geesink, R., van Heel, W.A., Muller, J. 1979. The affinities of *Plagiopteron suaveolens* Griff. (Plagiopteraceae). Grana 18: 69–89.

Baillon, H.E. 1880. The natural history of plants. London: L. Reeve & Co.

Barker, W.R. 1984. Stackhousiaceae. In George, A.S. (ed.) Flora of Australia 22: 186–199. Canberra: Australian Government Publishing Service.

Barnhart, J.H. 1910. Koeberliniaceae. North Am. Flora 25: 101–102.

Bawa, K.S. 1973. Chromosome numbers of tree species of a lowland tropical community. J. Arnold Arbor. 54: 422–434.

Benson, D., McDougall, L. 1995. Ecology of Sydney plant species part 3: dicotyledon families Cabombaceae to Euphorbiaceae. Cunninghamia 4: 217–424.

Bentham, G., Hooker, J.D. 1862. Genera plantarum. London: A. Black.

Berkeley, E. 1953. Morphological studies in the Celastraceae. J. Elisha Mitchell Sci. Soc. 69: 185–208.

Blakelock, R.A. 1951. A synopsis of the genus *Euonymus* L. Kew Bull. 2: 210–290.

Bowden, W.M. 1940. Diploidy, polyploidy, and winter hardiness relationships in the flowering plants. Am. J. Bot. 27: 357–371.

Brewbaker, J.L. 1967. The distribution and phylogenetic significance of binucleate and trinucleate pollen grains in the angiosperms. Am. J. Bot. 54: 1069–1083.

Briggs, J.D., Leigh, J.H. 1995. Rare or threatened Australian plants. Canberra: CSIRO Australia.

Brizicky, G.K. 1964a. The genera of Celastrales in the southeastern United States. J. Arnold Arbor. 45: 206–234.

Brizicky, G.K. 1964b. Polyembryony in *Euonymus* (Celastraceae). J. Arnold Arbor. 45: 251–259.

Brüning, R., Wagner, H. 1978. Übersicht über die Celastraceen-Inhaltsstoffe: Chemie, Chemotaxonomie, Biosynthese, Pharmakologie. Phytochemistry 17: 1821–1858.

Burkill, H.M. 1985. The useful plants of west tropical Africa. Kew: Royal Botanic Gardens.

Carlquist, S. 1987. Wood anatomy and relationships of Stackhousiaceae. Bot. Jahrb. Syst. 108: 473–480.

Chant, S.R. 1993. Celastraceae. In: Heywood, V.H. (ed.) Flowering plants of the world. New York: Oxford University Press.

Chase, M.W. et al. 1993. See general references.

Corner, E.J.H. 1954. The durian theory extended – II. The arillate fruit and the compound leaf. Phytomorphology 4: 152–165.

Corner, E.J.H. 1976. See general references.

Croizat, L. 1947. A study in the Celastraceae: Siphonodonoideae *subf. nov.* Lilloa 13: 31–43.

Cronquist, A. 1981. See general references.

Cuatrecasas, J. 1948. Studies in South American plants, I. Lloydia 11: 185–225.

Darwin, C. 1877. The different forms of flowers on plants of the same species. New York: D. Appleton and Company.

den Hartog née Van Ter Tholen, R.M., Baas, P. 1978. Epidermal characters of the Celastraceae sensu lato. Acta Bot. Neerl. 27: 355–388.

de Vogel, E.F. 1980. Seedlings of Dicotyledons: structure, development, types. Descriptions of 150 woody Malesian taxa. Wageningen: Centre for Agricultural Publishing and Documentation.

Dimitri, M.J. 1972. La region de los bosques andino-patagonicos. Buenos Aires: INTA.

Ding Hou, 1955. A revision of the genus *Celastrus*. Ann. Mo. Bot. Gard. 42: 215–302.

Ding Hou, 1958. A conspectus of the genus *Bhesa* (Celastraceae). Blumea 4: 149–153.

Ding Hou, 1962. Celastraceae – I. In: van Steenis, C.G.G.J. (ed.) Flora Malesiana I, 6: 227–291. Leyden: Flora Malesiana Foundation.

Ding Hou, 1964. Celastraceae – II. In: van Steenis, C.G.G.J. (ed.) Flora Malesiana I, 6: 389–421. Leyden: Flora Malesiana Foundation.

Ding Hou, 1969. Pollen of *Sarawakodendron* (Celastraceae) and some related genera, with notes on techniques. Blumea 17: 97–120.

Drennan, P.M., Drewes, S.E., van Dtaden, J., MacRae, S., Dickens, C.W.S. 1987. An anatomical, phytochemical and ultrastructural characterization of the elastic threads of *Maytenus acuminata*. S. Afr. J. Bot. 53: 17–24.

Dreyer, G.D., Baird, L.M., Fickler, C. 1987. *Celastrus scandens* and *Celastrus orbiculatus*: comparisons of reproductive potential between a native and an introduced woody vine. Bull. Torrey Bot. Club 114: 260–264.

Erdtman, G. 1952. See general references.

Federal Sentencing Guidelines Manual, 1998. Chapter 2, parts D–K. Retrieved 13 June 2000 from the World Wide Web http://www.ussc.gov/1998guid/98chap~1.htm.

Forman, L.L. 1965. A new genus of Ixonanthaceae with notes on the family. Kew Bull. 19: 517–526.

Garber, P.A. 1986. The ecology of seed dispersal in two species of callitrichid primates (*Saguinus mystax* and *Saguinus fuscicollis*). Am. J. Primatol. 10: 155–170.

Getahun, A., Krikorian, A.D. 1973. Chat: coffee's rival from Harar, Ethiopia. I. Botany, cultivation and use. Econ. Bot. 27: 353–377.

Gibbs, R.D. 1974. Chemotaxonomy of flowering plants. Montreal: McGill-Queen's University Press.

Goldblatt, P., Tobe, H., Carlquist, S., Patel, V.C. 1985. Familial position of the cape genus *Empleuridium*. Ann. Mo. Bot. Gard. 72: 167–183.

Görts-van Rijn, A.R.A., Mennega, A.M.W. 1994. 110. Hippocrateaceae. In: Görts-van Rijn, A.R.A. (ed.) Flora of the Guianas 16: 3–81. Königstein: Koeltz Scientific Books.

Gray, A. 1877. Characters of some little known or new genera of plants. Proc. Am. Acad. Arts 12: 159–165.

Hallé, N. 1960. Essai de clé pour la détermination des pollens des Hippocratéacées Ouest-Africaines. Pollen Spores 2: 5–12.

Hallé, N. 1962. Monographie des Hippocratéacées d'Afrique occidentale. Mém. Inst. Franç. Afri. Noire 64: 1–245.

Hallé, N. 1978. Révision monographique des Hippocrateæ (Celastr.). 1. Les espèces de Madagascar. Adansonia 17: 397–414.

Hallé, N. 1983. Révision des *Hippocrateæ (Celastreæ)*. 3. Fruits, graines et structures placentaires. Bull. Mus. Natl. Hist. Nat. Paris Adansonia IV, 5B: 11–26.

Hallé, N. 1986. Celastraceae Hippocrateoideae. In: Morat, P. (ed.) Flore du Gabon (avec compléments pour d'autres pays d'Afrique et Madagascar) 29: 1–287. Paris: Muséum National d'Histoire Naturelle, Laboratoire de Phanérogamie.

Hallé, N. 1990. Celastracées (Hippocrateoidées). In: Satabié, B., Morat, P. (eds.) Flore du Cameroun 32: 3–243. Yaoundé: Ministère de l'Enseignement Supérieur de l'Informatique et de la Recherche Scientifique Mesires.

Hammel, B.E. 1997. Three new species of Celastraceae from Costa Rica, one disjunct from Mexico. Novon 7: 147–155.

Hegnauer, R. 1964, 1966, 1989. See general references.

Hideux, M.J., Ferguson, I.K. 1976. The stereostructure of the exine and its evolutionary significance in Saxifragaceae sensu lato. In: Ferguson, I.K., Muller, J. (eds.) The evolutionary significance of the exine. London: Academic Press, pp. 327–378.

Hilton-Taylor, C. 1996. Red data list of southern African plants. Pretoria: National Botanical Institute.

Hooker, J.D. 1857. On the growth and composition of the ovarium of *Siphonodon celastrineus* Griffith, especially with reference to the subject of its placentation. Trans. Linn. Soc. Lond. 22: 133–139.

Hutchinson, J. 1967. The genera of flowering plants: (Angiospermae) based principally on the Genera Plantarum of G. Bentham and J. D. Hooker. Oxford: Clarendon Press.

Hutchinson, J. 1969. Evolution and phylogeny of flowering plants: Dicotyledons: facts and theory with over 550 illustrations and maps by the author. London: Academic Press.

Huxley, A., Griffiths M., Levy, M. 1992. The new royal horticultural society dictionary of gardening. London: MacMillan Press.

Ikuse, M. 1956. Pollen grains of Japan. Tokyo: Hirokawa Publishing Co.

Jansen, W.T., Baas, P. 1973. Comparative leaf anatomy of *Kokoona* and *Lophopetalum* (Celastraceae). Blumea 21: 153–178.

Johnston, M.C. 1975. Synopsis of *Canotia* (Celastraceae) including a new species from the Chihuahuan Desert. Brittonia 27: 119–122.

Jordaan, M., van Wyk, A.E. 1999. Systematic studies in subfamily Celastroideae (Celastraceae) in southern Africa: reinstatement of the genus *Gymnosporia*. S. Afr. J. Bot. 65: 177–181.

King, G., Prain, D. 1898. Descriptions of some new plants from the north-eastern frontiers of India. J. Asiat. Soc. Bengal 67: 284–305.

Knuth, P. 1908. Handbook of flower pollination. Oxford: Clarendon Press.

Kollmann, J., Coomes, D.A., White, S.M. 1998. Consistencies in post-dispersal seed predation of temperate fleshy-fruited species among seasons, years and sites. Funct. Ecol. 12: 683–690.

Krikorian, A.D., Getahun, A. 1973. Chat: coffee's rival from Harar, Ethiopia. II. Chemical composition. Econ. Bot. 27: 378–389.

Kuhlmann, J.G. 1933. Novo genero de Celastraceas da flora amazonica. Arch. Jard. Bot. Rio de Janeiro 6: 109–110.

Lindley, J. 1853. The vegetable kingdom, or, the structure, classification, and uses of plants, illustrated upon the natural system. London: Bradbury & Evans.

Lobreau-Callen, D. 1977. Les pollens des Celastrales (illustrations, commentaires). Mém. Trav. Inst. Montpellier École Prat. Hautes Études 3: 1–116.

Loesener, T. 1942a. Celastraceae. In: Engler, A., Harms, H., Mattfeld, J. (eds.) Die natürlichen Pflanzenfamilien 20b. Berlin: Duncker & Humblot, pp. 87–197.

Loesener, T. 1942b. Hippocrateaceae. In: Engler, A., Harms, H., Mattfeld, J. (eds.) Die natürlichen Pflanzenfamilien 20b. Berlin: Duncker & Humblot, pp. 198–231.

Lourteig, A., O'Donell, C.A. 1955. Las Celastraceas de Argentina y Chile. Natura 1: 181–233, pl. 1–12.

Lubbock, J. 1892. A contribution to our knowledge of seedlings. London: Kegan Paul, Trench, Trübner & Co.

Lundell, C.L. 1939. Revision of the American Celastraceae I. *Wimmeria*, *Microtropis*, and *Zinowiewia*. Contrib. Univ. Michigan Herb. 3: 5–46, pl. 1–10.

Lundell, C.L. 1970. Studies of American plants – II. Wrightia 4: 129–152.

Lundell, C.L. 1971. Studies of American plants – III. Wrightia 4: 153–172.

Lundell, C.L. 1985. Two species of the genus *Gymnosporia* (Celastraceae) in South America. Phytologia 57: 313–314.

Martin, A.C., Zim H.S., Nelson, A.L. 1951. American wildlife & plants. New York: Dover Publications.

Mattfeld, J. 1942. Stackhousiaceae. In: Engler, A., Harms, H., Mattfeld, J. (eds.) Die natürlichen Pflanzenfamilien 20b. Berlin: Duncker & Humblot, pp. 240–254.

Mauritzon, J. 1936a. Zur Embryologie und systematischen Abgrenzung der Reihen Terebinthales und Celastrales. Bot. Not. 1936: 161–212.

Mauritzon, J. 1936b. Embryologische angaben über Stackhousiaceae, Hippocrateaceae und Icacinaceae. Svensk Bot. Tidskr. 30: 541–550.

Mehra, P.N. 1976. Cytology of Himalayan hardwoods. Calcutta: Sree Saraswaty Press.

Meisner, C.F. 1837. Plantarum vascularium genera secundum ordines naturales digesta eorumque differentiae et affinitates tabulis diagnosticis expositae. Lipsiae: Libraria Weidmannia.

Mennega, A.M.W. 1972. A survey of the wood anatomy of the New World Hippocrateaceae. In: Ghouse, A.K.M. (ed.) Research trends in plant anatomy – K.A. Chowdhury commemoration volume. New Delhi: Tata McGraw-Hill, pp. 61–72.

Mennega, A.M.W. 1994. Wood and timber: Hippocrateaceae. In: Görts-van Rijn, A.R.A. (ed.) Flora of the Guianas 16: 110–140. Königstein: Koeltz Scientific Books.

Mennega, A.M.W. 1997. Wood anatomy of the Hippocrateoideae (Celastraceae). IAWA J. 18: 331–368.

Metcalfe, C.R., Chalk, L. 1950. See general references.

Miers, J. 1856. Remarks on the nature of the outer fleshy covering of the seed in the Clusiaceæ, Magnoliaceæ, etc., and on the development of the raphe in general, under its various circumstances. Trans. Linn. Soc. Lond. 22: 81–96.

Miers, J. 1872. On the Hippocrateaceae of South America. Trans. Linn. Soc. Lond. 28: 319–432, pl. 16–32.

Morgan, D.R., Soltis, D.E. 1993. Phylogenetic relationships among members of Saxifragaceae sensu lato based on *rbc*L sequence data. Ann. Mo. Bot. Gard. 80: 631–660.

Muhwezi, O. 1999. The use of *Loeseneriella apocynoides* around Bwindi Impenetrable National Park, southwest Uganda. In: Timberlake, J., Kativu, S. (eds.) African plants: biodiversity, taxonomy and uses. Kew: Royal Botanic Gardens, pp. 523–527.

Müller, H. 1883. The fertilisation of flowers. London: Macmillan and Co.

Müller, I.H. 1995. Systematics and leaf anatomy of the Celastraceae *sensu stricto* of New Caledonia. Doctoral dissertation. Zürich: Universität Zürich, 97, pl. 1–14.

Narang, N. 1965. The life-history of *Stackhousia linariaefolia* A. Cunn. with a discussion on its systematic position. Phytomorphology 3: 485–493.

Nayar, M.P., Sastry, A.R.K. 1987. Red data book of Indian plants. Calcutta: Botanical Survey of India.

Oldfield, S., Lusty, C., MacKinven, A. 1998. The world list of threatened trees. Cambridge: World Conservation Press.

Palmer, E., Pitman, N. 1972. Trees of southern Africa. Cape Town: A.A. Balkema.

Pant, D.D., Kidwai, P.F. 1966. Epidermal structure and stomatal ontogeny in some Celastraceae. New Phytol. 65: 288–295.

Perrier de la Bâthie, H. 1942. Au sujet des affinités des *Brexia* et des Célastracées et de deux *Brexia* nouveaux de Madagascar. Bull. Soc. Bot. France 89: 219–221.

Pfeiffer, A. 1891. Die Arillargebilde der Pflanzensamen. Bot. Jahrb. Syst. 13: 492–540.

Phillips, J.F.V. 1927. The role of the "bushdove" *Columba arquatrix* T. & K., in fruit-dispersal in the Knysna forests. S. Afr. J. Sci. 24: 435–440.

Planchon, J.-E. 1845. Développements et caractères des vrais et des faux arilles. Ann. Sci. Nat. III, 3: 275–312.

Raintree Nutrition Inc., 2000a. Chuchuhuasi. Retrieved 13 June 2000 from the World Wide Web http://www.raintree.com/chuchuhuasi.htm.

Raintree Nutrition Inc., 2000b. Espinheira santa. Retrieved 13 June 2000 from the World Wide Web http://www.rain-tree.com/chuchuhuasi.htm.

Record, S.J. 1943. The American woods of the orders Celastrales, Olacales, and Santalales. Trop. Woods 53: 11–38.

Record, S.J., Hess, R.W. 1943. Timbers of the New World. New Haven: Yale University Press.

Ridley, H.N. 1930. The dispersal of plants throughout the world. Ashford, Kent: L. Reeve & Co.

Robson, N. 1965. New and little known species from the Flora Zambesiaca area XVI: taxonomic and nomenclatural notes on Celastraceae. Bol. Soc. Brot. 39: 5–55.

Robson, N.K.B. 1966. Celastraceae (incl. Hippocretaceae). In: Exell, A.W., Fernandes, A., Wild, H. (eds.) Flora Zambesiaca 2, pt. 2. London: Crown Agents, pp. 355–418.

Robson, N.K.B., Hallé, N., Mathew, B., Blakelock, R. 1994. Celastraceae. In: Polhill, R.M. (ed.) Flora of tropical east Africa 108. Rotterdam: A.A. Balkema, pp. 1–78.

Rodrigues, M., Olmos, F., Galetti, M. 1993. Seed dispersal by tapir in southeastern Brazil. Mammalia 57: 460–461.

Rudolf, P.O. 1974. *Euonymus* L. In: Schopmeyer, C.S. (ed.) Seeds of woody plants in the United States. Washington, D.C.: Forest Service, U.S. Department of Agriculture, pp. 393–397.

Savolainen, V., Spichiger, R., Manen, J.F. 1997. Polyphyletism of Celastrales deduced from a chloroplast noncoding DNA region. Molec. Phylo. Evol. 7: 145–157.

Savolainen, V., Chase, M.W. et al. 2000. See general references.

Savolainen, V., Fay, M.F. et al. 2000. See general references.

Sebsebe, D. 1985. The genus *Maytenus* (Celastraceae) in NE tropical Africa and tropical Arabia. Acta Univ. Upsal Symb. Bot. 25: 1–98.

Sheng-ye, L. 1992. *Bhesa sinica*. In: Li-kuo, F., Jian-ming, J. (eds.) China plant red data book – rare and endangered plants 1. Beijing: Science Press, pp. 206–207.

Simmons, M.P. in press. Hippocrateaceae. In: Morat, P. (ed.) Flore de la Nouvelle-Calédonie. Paris: Muséum National d'Histoire Naturelle.

Simmons, M.P., Hedin, J.P. 1999. Relationships and morphological character change among genera of Celastraceae sensu lato (including Hippocrateaceae). Ann. Mo. Bot. Gard. 86: 723–757.

Simmons, M.P., Clevinger, C.C., Savolainen, V., Archer, R.H., Mathews, S., Doyle, J.J. 2001a. Phylogeny of the Celastraceae inferred from phytochrome B and morphology. Am. J. Bot. 88: 313–325.

Simmons, M.P., Savolainen, V., Clevinger, C.C., Archer, R.H., Davis, J.I. 2001b. Phylogeny of the Celastraceae inferred from 26 S nrDNA, phytochrome B, *atpB, rbcL,* and morphology. Molec. Phylo. Evol. 19: 353–366.

Sinha, A., Davidar, P. 1992. Seed dispersal ecology of a wind dispersed rain forest tree in the western Ghats, India. Biotropica 24: 519–526.

Smith, A.C. 1940. The American species of Hippocrateaceae. Brittonia 3: 341–555.

Smith, A.C. 1941. Notes on old world Hippocrateaceae. Am. J. Bot. 28: 438–443.

Solereder, H. 1908. Systematic anatomy of the dicotyledons: a handbook for laboratories of pure and applied biology. Oxford: Clarendon Press.

Soltis, D.E., Soltis, P.S., Nickrent, D.L., Johnson, L.A., Hahn, W.J., Hoot, S.B., Sweere, J.A., Kuzoff, R.K., Kron, K.A., Chase, M.W., Swensen, S.M., Zimmer, E.A., Chaw, S.M., Gillespie, L.J., Kress, W.J., Sytsma, K.J. 1997. Angiosperm phylogeny inferred from 18 S ribosomal DNA sequences. Ann. Mo. Bot. Gard. 84: 1–49.

Soltis, D.E. et al. 2000. See general references.

Sonder, O.W. 1859–1860. Rutaceae. In: Harvey, W.H., Sonder, O.W. (eds.) Flora capensis: being a systematic description of the plants of the Cape colony, Caffraria, & Port Natal, vol. 1. Dublin: Hodges, Smith and Co., pp. 369–442.

Stant, M.Y. 1951. Notes on the systematic anatomy of *Stackhousia*. Kew Bull. 1951: 309–318.

Staszkiewicz, J. 1997. The variability of seeds of *Euonymus europaeus* and *E. verrucosus* (Celastraceae). Fragm. Florist. Geobot. Series Polon. 10: 151–159.

Szendrei, K. 1981. The chemistry of khat. Bull. Narcotics 32: 5–36.

Takhtajan, A. 1981. See general references.

Tobe, H., Raven, P.H. 1984. The embryology and relationships of *Alzatea* Ruiz & Pav. (Alzateaceae, Myrtales). Ann. Mo. Bot. Gard. 71: 844–852.

Tobe, H., Raven P.H. 1993. Embryology of *Acanthothamnus, Brexia* and *Canotia* (Celastrales): a comparison. Bot. J. Linn. Soc. 112: 17–32.

Troll, W. 1937. Vergleichende Morphologie der höheren Pflanzen 1. Berlin: Borntraeger.

Troll, W. 1943. Vergleichende Morphologie der höheren Pflanzen 3. Berlin: Borntraeger.

Uttal, L.J. 1986. Once and for all it is *Paxistima*. Castanea 51: 67–68.

van der Pijl, L. 1972. Principles of dispersal in higher plants. Berlin Heidelberg New York: Springer.

van Wyk, A.E. 2002. Celastraceae. In: Coates Palgrave, K. (ed.) Trees of southern Africa.

Vasudeva Rao, M. K., Chakrabarty, T. 1985. *Nicobariodendron* Vasud. & T. Chakrab. (Celastraceae): a new genus from the Nicobar Islands, India. J. Econ. Taxon. Bot. 7: 513–516.

Velenovský, J. 1910. Vergleichende Morphologie der Pflanzen. Prag: Fr. Řivnáč.

Verdcourt, B. 1968. Brexiaceae. In: Milne-Redhead, E., Polhill, R.M. (eds.) Flora of tropical east Africa, vol. 108A. London: Crown Agents for Oversea Governments and Administrations, pp. 1–3.

Warburg, O. 1894. Flacourtiaceae. In: Engler, A., Prantl, K. Die natürlichen Pflanzenfamilien III, 6a. Leipzig: Engelmann, pp. 1–56.

Webb, C.J. 1979. Breeding system and seed set in *Euonymus europaeus* (Celastraceae). Plant Syst. Evol. 132: 299–303.

Wendel, G.W. 1974. *Celastrus scandens* L. In: Schopmeyer, C.S. (ed.) Seeds of woody plants in the United States. Washington, D.C.: Forest Service, U.S. Department of Agriculture, pp. 295–297.

Wheeler, L.C. 1943. History and orthography of the Celastraceous genus "*Pachystima*" Rafinesque. Am. Midl. Nat. 29: 792–795.

White, O.E., Bowden, W.M. 1947. Oriental and American bittersweet hybrids. J. Heredity 38: 125–127.

Wulff, H.D. 1937. Chromosomenstudien an der schleswig-holsteinischen Angiopermen-Flora I. Ber. Deutsch. Bot. Ges. 55: 262–269.

Zelger, J.L., Schorno, H.X., Carlini, E.A. 1981. Behavioural effects of cathinone, an amine obtained from *Catha edulis* Forsk.: comparisons with amphetamine, norpseudoephedrine, apomorphine and nomifensine. Bull. Narcotics 32: 67–82.

Cephalotaceae

J.G. CONRAN

Cephalotaceae Dumort., Anal. Fam. Pl.: 59, 61 (1829), nom. cons.

Perennial carnivorous evergreen herbs. Rootstock a thick, knotty rhizome, roots fibrous. Stems 0. Leaves borne in rosettes at rhizome apices, alternate, exstipular, petiolate, fine-hairy, alternately spathulate-obovate or pitcher-shaped and insect-trapping. Inflorescence a thyrse with scorpoid cymes. Flowers hermaphroditic, small, calyx well developed, basally connate, rotate, purplish-white; limb 6-lobed; lobes imbricate in bud; corolla 0; stamens 12 in two whorls of six; petals 0; filaments free, divergent; anthers dorsifixed, versatile, with a dorsal-connective appendage, 2-thecate, 4-sporangiate, introrse, opening by lateral slits; disk intrastaminal, trichomatous-papillose; gynoecium of 6 free carpels alternating with the inner stamens and sepals; ovaries superior, 1-locular with 1-2 erect anatropous ovules; stylodia terminal, straight, ventrally papillate near the apex. Fruit 1-2-seeded, hairy indehiscent leathery follicles; seeds small, ovoid, brown with a membranous testa; endosperm copious, granulose; embryo minute, linear.

A monotypic family with the single species *Cephalotus follicularis* Labill. confined to far SW Western Australia. The plant is grown widely as an ornamental curiosity.

VEGETATIVE MORPHOLOGY. Roots are fibrous and consist of a short-lived taproot and numerous adventitious roots arising from the rhizome. The rhizome is spreading, much-branched and sheathed with numerous scale leaves and bears apical rosettes of photosynthetic and trapping leaves. The flowering stems are erect, scapose, bracteate; the lower bracts are leaf-like, the scape bracts linear and deciduous, and the upper bracts and those subtending the flowers are small, clustered and incurved.

Leaves are of three types: scale leaves on the rhizomes, aerial rosette leaves with non-sheathing, petiolate and spathulate-obovate photosynthetic blades, and highly modified pitchers with the lamina modified into a pitcher-shaped, hooded trap, these leaves alternating with the non-trap leaves in the shoot apex rosette. All leaves bear unicellular hairs and numerous sessile glands. Dickson (1883) also noted various intermediate forms between normal and trapping leaves.

The trapping leaves are highly modified, with the lamina replaced by a lidded pitcher (Dickson 1883). The pitcher is apical on the petiole and ovoid with two prominent anterior ribs and an ovate, sub-apical opening covered by a protective lid with conspicuous, areolate achlorophyllous patterns. The outer pitcher surface is glandular, with fluid (considered to be nectar) secreted over its surface (Hamilton 1904). The rim of the pitcher is strongly ribbed with a series of large, inward-pointing teeth, and the upper third to half of the inner pitcher surface is covered with small glands, whereas the lower pitcher has two, generally red-coloured patches of larger glands (Lloyd 1942). Inside the rim, past the large teeth, there is a further region, the cornice, consisting of down-wardly-pointed, tooth-like epidermal cells and a smooth transition zone above the glands (Parkes and Hallam 1984).

The pitcher is not a simple epiascidium, and ontogenetic studies have revealed that it is a specifically modified rhachis of a bijugal pinnate leaf (Froebe and Baur 1988).

VEGETATIVE ANATOMY. Raphides are absent. Vessels occur in the roots and rhizome and are heavily pitted with scalariform perforations (Macfarlane 1911). Sieve-tube plastids are of the dicotyledonous Ss-type (Behnke 1988, 1991). Laticifers are absent.

The roots are triarch with a 2-3-layered cortex and thin-walled endodermis. A cork cambium is formed, resulting in 3-4 clear cork cell layers, but the primary cortical tissues are not shed, instead turning brown with age. There is also a starch storage region of uncertain origin within the cork of old roots (Macfarlane 1911).

The rhizome epidermis is initially hairy, these hairs becoming tubercles with age. There is a broad, starchy primary cortex and a partially incomplete ring of vascular tissue with a broad phloem and xylem and a parenchymatous pith. Secondary development results in a 3–5-cell cork layer, starchy secondary cortex and starch deposition in the pith, followed by eventual accumulation of tannins in the parenchymatous tissues (Macfarlane 1911).

Paracytic stomata occur on both leaf surfaces, as do so-called large and small glands and both simple and complex hairs. The mesophyll consists of an adaxial and abaxial palisade layer, with a loose, central spongy mesophyll. There is a larger midrib vein, two smaller lateral veins and a series of smaller, sub-marginal veinlets (Macfarlane 1911).

The pitchers are covered externally with numerous stomata and large glands, the latter consisting of cells with dense cytoplasma arranged like a tetracytic stoma, with a central inner cell pair surrounded by four symmetrically arranged cells, and another pair of internally associated cells (Dickson 1878; Solereder 1908). In section, the pitchers have several layers of chlorenchyma subtended by achlorophyllous mesophyll. The translucent areas are parenchymatous and non-vascularised (Parkes and Hallam 1994). The pitcher interiors have numerous non-functional stomata, referred to by Goebel (1891) as water stomata. Small glands in the upper pitcher consist of 6–12 concentrically arranged cells, whereas the lateral glandular patches in the lower pitcher have numerous large, vascularised, multicellular glands (Macfarlane 1911). These large glands are flask-shaped, with a ventral region of irregularly polygonal cells and an apical region of palisade-like cells (Solereder 1908), and they are bounded internally by transversely thickened, endodermal cells (Parkes and Hallam 1984).

INFLORESCENCE STRUCTURE. The inflorescence is a thyrse. The flowers are borne in short, condensed, lateral scorpoid cymes.

FLORAL MORPHOLOGY. The sepals have three to seven anastomosing vascular traces (Chrtek et al. 1989). The anthers are introrse and cruciform, dorsifixed medially, dehiscing laterally by slits. There is a thin connective and prominent dorsal connective of large cells. The endothecium is 1-layered, discontinuous over the connective or the inner thecal surface. There is a septum between the pollen sacs (Endress and Stumpf 1991). Vogel (1998) reported that the floral nectary in *Cephalotus* forms a contiguous perigynous disk derived from a series of pillars, each with an apical stomatal pore.

EMBRYOLOGY. The ovules are anatropous, bitegmic and crassinucellate. The embryology of the family is the subject of ongoing research (Conran and Macfarlane, in prep.).

POLLEN MORPHOLOGY. Pollen grains are spheroidal, tricolpate (angulaperturate) with a reticulate sexine and slightly thinner nexine and c. 18 μm diameter (Erdtman 1952). The tectum is perforate with dense pitting and the apertures are complex (Hideux and Ferguson 1976).

KARYOLOGY. Kondo (1969), Kress (1970) and Peng and Goldblatt (1983) all determined the chromosome number in Cephalotaceae to be $2n = 20$.

Pollination. Cephalotaceae flowers are pollinated by small insects.

Fruit and Seed Morphology. The fruit is a densely hairy follicle, the hairs derived from the carpel papillae which elongate after fertilisation. The small, single seed is enclosed inside the follicle and is thin-walled and ovoid. The copious endosperm cells are granular (Macfarlane 1911).

Germination is epigeal and phanerocotylar. The seedlings have a well-developed but ephemeral taproot and the hypocotyl is glabrous. The cotyledons are ovoid and glabrous, the first leaves are alternate, resemble the later ones and are glandular-hairy. The seedlings are unusual, with an expanded non-vascularised outgrowth of the hypocotyl growing into the follicle, filling the loculus and possibly acting as a non-starchy food reserve for the seedling. In addition, the hairy follicle coat may assist in keeping the seedling hydrated (Conran and Denton 1996).

Ecology. The Cephalotaceae are considered to be carnivorous, the modified pitcher leaves acting to trap the prey (Lloyd 1942). The trapping leaves have ultraviolet absorption patterns which are thought to attract prey in several families of carnivorous plants (Joel et al. 1985). Dakin (1919) found no direct evidence of digestive enzymes, suggesting indirect digestion through putrefaction, although Joel (in Juniper et al. 1989) reported protease activity in the intercellular plugs of the small glands within the pitchers. Brown (1866) reported the presence of live insect larvae and coccoid unicellular green algae living inside the pitchers.

Cephalotus grows in the peat-soil swamps formed over granite substrata, often in seepage

Fig. 21. Cephalotaceae. *Cephalotus follicularis*. **A** Flowering plant. **B** Flower. **C** Flower, vertical section. (Baillon 1872)

areas, along creeks and under grass tussocks (Luffitz 1966). DeBuhr (1976) reported that the plants re-sprout from the rhizome following fire, but germination is not dependent on fire (Piliciauskas 1989). Plants propagate readily from rhizome pieces, as well as from single leaves or pitchers grown in *Sphagnum*.

PHYTOCHEMISTRY. Tannin cells are present. Myricetin, ellagic acid, quercetin and gallic acid are present (Bate Smith 1962) but iridoids are absent (Jensen et al. 1975).

AFFINITIES. *Cephalotus* has traditionally been associated with the Saxifragaceae but with some features linking it to the Crassulaceae (Macfarlane 1911). More recent molecular studies of its affinities (e.g. Albert et al. 1992; Chase et al. 1993; APG

1998; Savolainen, Fay et al. 2000) place it instead in the Oxalidales near the Cunoniaceae. Its floral diagram is identical with that of Brunelliaceae.

DISTRIBUTION AND HABITATS. The family grows only in the extreme SW of Western Australia where the plant is widespread and locally abundant. The plants are generally found growing in or near *Sphagnum* mounds in well-drained areas of swamps relatively close to the coast, in an area extending over a distance of some 400 km.

ECONOMIC IMPORTANCE. *Cephalotus* is popular with carnivorous plant enthusiasts and widely cultivated. Although listed previously on CITES Appendix II, the plant was recently removed from CITES, given its relative abundance in protected areas of SW Western Australia and its ease of vegetative propagation.

Only one genus:

Cephalotus Labill.   Fig. 21

Cephalotus Labill., Nov. Holl. Pl. Sp. 2: 6, t. 145 (1806).

Description as for family.

Selected Bibliography

Albert, V.A., Williams, S.E., Chase, M.W. 1992. Carnivorous plants: phylogeny and structural evolution. Science. 257: 1491–1495.

APG (Angiosperm Phylogeny Group) 1998. See general references.

Bate Smith, E.C. 1962. See general references.

Baillon, H. 1872. Histoire des plantes, vol. 3. Paris: Hachette.

Behnke, H.-D. 1988. Sieve-element plastids and systematic relationships of Rhizophoraceae, Anisophyllaceae and allied groups. Ann. Missouri Bot. Gard. 75: 1387–1409.

Behnke, H.D. 1991. Distribution and evolution of forms and types of sieve-element plastids in the dicotyledons. Aliso 13: 167–182.

Brown, R. 1866. General remarks on the botany of *Terra Australis*. Misc. Bot. Works. 1: 76–78.

Chase et al. 1993. See general references.

Chrtek, J., Slavíková, Z., Studnička, M. 1989. Beitrag zur Leitbündelanordnung in den Kronblättern von ausgewählten Arten der fleischfressenden Pflanzen. Preslia 61: 107–124.

Conran, J.G., Denton, M.D. 1996. Germination in the Western Australian Pitcher Plant *Cephalotus follicularis* and its unusual early seedling development. W.A. Nat. 21: 37–42.

Corner, E.J.H. 1976. See general references.

Dakin, W.J. 1919. The West Australian pitcher plant (*Cephalotus follicularis*), and its physiology. J. Roy. Soc. W.A. 4: 37–53.

DeBuhr, L.E. 1976. Field notes on *Cephalotus follicularis* in Western Australia. Carn. Pl. Newslett. 5: 8–9.

Dickson, A. 1878. The structure of the pitcher of *Cephalotus follicularis*. J. Bot. 16: 1–5.

Dickson, A. 1883. On the morphology of the pitcher of *Cephalotus follicularis*. Trans. Proc. Edinburgh Bot. Soc. 14: 172–181.

Diels, L. 1928. Cephalotaceae. In: Engler, A., Prantl, K. Die natürlichen Pflanzenfamilen, ed. 2, 18a. Leipzig: W. Engelmann, pp. 71–74.

Endress, P.K., Stumpf, S. 1991. The diversity of stamen structures in 'Lower' Rosidae (Rosales, Fabales, Proteales, Sapindales). Bot. J. Linn. Soc. 107: 217–293. With 294 figures.

Erdtman, G. 1952. See general references.

Froebe, H.A., Baur, N. 1988. Die Morphogenese der Kannenblätter von Cephaltus follicularis Labill. Akad. Wiss. Lit. Mainz, Abh. Math.-Naturwiss. Kl. Jg. 1988, 3, 19 pp.

Goebel, K. 1891. Pflanzenbiologische Schilderungen, part 2. Marburg: Elwert.

Hamilton, A.G. 1904. Notes on the West Australian pitcher plant (*Cephalotus follicularis* Labill.). Proc. Linn. Soc. N.S.W. 29: 36–53.

Hideux, M., Ferguson, I.K. 1976. The stereostructure of the exine and its evolutionary significance in Saxifragaceae *sensu lato*. In: Ferguson, I.K., Muller, I. (eds.) The evolutionary significance of the exine. Linnean Society Symposium Series no. 1. London: Academic Press, pp. 327–377.

Jay, M., Lebreton, P. 1972. Chemotaxonomic research on vascular plants. XXVI. The flavinoids of the Sarraceniaceae, Nepenthaceae, Droseraceae and Cephalotaceae; a critical study of the Sarraceniales. Nat. Can. (Québec) 99: 607–613.

Jensen, S.R., Nielsen, B.J., Dahlgren, R. 1975. Iridoid compounds, their occurrence and systematic importance in the angiosperms. Bot. Notiser 128: 148–180.

Joel, D.M., Juniper, B.E., Dafni, A. 1985. UV patterns in the traps of carnivorous plants. New Phytol. 101: 585–594.

Juniper, B.E., Robins, R.J., Joel, D.M. 1989. Carnivorous plants. London: Academic Press.

Kondo, K. 1969. Chromosome numbers of carnivorous plants. Bull. Torrey Bot. Club 96: 322–328.

Kress, A. 1970. Zytotaxonomie Untersuchungen an einigen Insektenfängern (Droseraceae, Byblidaceae, Cephalotaceae, Roridulaceae, Sarraceniaceae). Ber. Deutsch. Bot. Ges. 83: 55–62.

Lloyd, F.M. 1942. The carnivorous plants, 2nd edn. Waltham, Mass.: Chronica Botanica Co.

Luffitz, F. 1966. The West Australian pitcher plant (*Cephalotus follicularis* Labill.). Austral. Pl. 12: 34–35.

Macfarlane, J.M. 1911. Cephalotaceae. In: Engler, A. (ed.) Das Pflanzenreich IV, 116. Leipzig: W. Engelmann, pp. 1–15.

Nicholls, K.W., Bohm, B.A., Ornduff, R. 1985. Flavonoids and affinities of the Cephalotaceae. Biochem. Syst. Ecol. 13: 261–264.

Parkes, D.M., Hallam, N.D. 1984. Adaptation for carnivory in the West Australian pitcher plant (*Cephalotus follicularis*). Austral. J. Bot. 32: 595–604.

Peng, C.-I., Goldblatt, P. 1983. Confirmation of the chromosome number in Cephalotaceae and Roridulaceae. Ann. Missouri Bot. Gard. 70: 197–198.

Piliciauskas, E. 1989. *Cephalotus follicularis* and how to grow them from seed. Vic. C. P. Soc. Newslett. 6: 12–15.

Savolainen, V., Fay, M.F. et al. 2000. See general references.

Schulze, W., Schulze, E.D., Pate, J.S., Gillison, A.N. 1997. The nitrogen supply from soils and insects during growth of the pitcher plants *Nepenthes mirabilis*, *Cephalotus follicularis* and *Darlingtonia californica*. Oecologia. 112: 464–471.

Solereder, H. 1908. Systematic anatomy of the Dicotyledons, vol. 1. Introduction, Polypetalae and Gamopetalae. Oxford: Clarendon Press.

Vogel, S. 1998. Remarkable nectaries: structure, ecology, organophyletic perspectives. II. Nectarioles. Flora (Jena) 193: 1–29.

Clethraceae

J.V. Schneider and C. Bayer

Clethraceae Klotzsch, Linnaea 24: 12 (1851), nom. cons.

Trees or shrubs, evergreen or deciduous; indumentum of simple, fasciculate and/or stellate trichomes. Leaves alternate, simple, entire or (glandular-)serrate to dentate, petiolate or sessile, exstipulate. Inflorescences terminal or axillary racemes, simple or compound in panicle-, fascicle-, or umbel-like aggregates; bracts present, often caducous, prophylls absent (sometimes rudimentary?); pedicels articulated, sometimes very short. Flowers actinomorphic or slightly zygomorphic, hermaphroditic or rarely functionally female with reduced anthers; sepals 5(6), quincuncial-imbricate, distinct or fused up to 3/4, equal in size or the outer conspicuously larger than the inner; petals 5(6), distinct or rarely fused up to 1/3, imbricate; stamens 10(12), in 2 whorls, free or adnate to petals at their very base; anthers ventrifixed, bilobed-sagittate to caudate, versatile, tetrasporangiate, swinging at anthesis from the extrorse bud position into an introrse, inverted position, dehiscing by eventually apical pore-like slits; disk hypogynous or absent; ovary superior, sometimes with nectary at base, 3–5-locular; placentation axile on upper portion of ovary; ovules numerous or 1 per locule, anatropous to pendent-orthotropous, unitegmic, tenuinucellate; style entire or apically branched into 3(4) stylar branches. Fruit indehiscent, 1–5-seeded, 3–5-ribbed or indistinctly lobed, or a 3-loculicidal, many-seeded capsule, usually pubescent, sometimes ± enclosed by the persistent calyx. Seeds (in *Clethra*) flattened or subtrigonal, ellipsoid to ovoid, irregularly angled to winged, seed coat absent (*Purdiaea*) or with foveolate-reticulate impressions or inconspicuously prominent cells; endosperm cellular, copious, fleshy; embryo short, straight, cylindrical.

Two genera with about 95 species, mostly from tropical montane America and SE Asia (to India), some in temperate North America and Asia, one species on Madeira.

VEGETATIVE MORPHOLOGY. Tropical Clethraceae are generally evergreen, temperate species are deciduous. The Chinese-Indochinese *C. faberi* is usually evergreen but appears to be deciduous in the southeastern part of its area. Leaves often appear apically crowded on vegetative shoots. In *Purdiaea*, the branches often exhibit a zigzag pattern due to sympodial growth (Thomas 1960).

Leaf venation in *Clethra* corresponds to the eucamptodromous, brochidodromous (Yu and Chen 1990), or semi-craspedodromous patterns, i.e. the secondaries are connected by a series of arching veins, sometimes forming prominent arches, or they branch near the margin with one branch running into the margin, the other one towards the superadjacent secondary. *Purdiaea* shows an acrodromous to palinactinodromous pattern (Rodríguez and Berazaín 1991). Domatia formed by hair tufts are found in several Japanese and Chinese *Clethra* species.

Most Clethraceae are pubescent, particularly on buds, leaves, inflorescences and flowers. Single and fascicled trichomes are found. The latter are longer, more robust, free or basally connate, in some tropical American species forming a kind of a stipe. True stellate trichomes occur as well.

VEGETATIVE ANATOMY. The nodes are unilacunar (Thomas 1960). Sections of the bifacial leaves show about two layers of palisade parenchyma. Stomata are paracytic, anomocytic or anisocytic, confined to the abaxial surface and, in *Clethra*, slightly raised above the surface. Secretory cells are found in the petiole and in the phloem of the midrib. Crystals, particularly clustered ones, are abundant (Metcalfe and Chalk 1950; Thomas 1960; Vales et al. 1988).

The wood exhibits a comparatively primitive structure. Vessel elements are predominantly solitary, angular, narrow (diameter mostly 25–50 μm), and in *Clethra* twice as long as in *Purdiaea* (Record 1932), sometimes with fine, spiral thickening in the tips. Perforation plates are simple or scalariform with many bars (Thomas 1960; Giebel and Dickison 1975). As in most Ericales, vestured pits are absent (Jansen et al. 2001). Parenchyma is paratracheal or apotracheal and diffuse. Rays are uni- to multiserial, up to 6 cells wide (Record 1932; Vales et al. 1988). Fibers and moderately long to long fiber-tracheids are present.

Kubota et al. (2001) reported the formation of arbuscular mycorrhiza in *Clethra*.

INFLORESCENCES. In mature evergreen species of *Clethra*, each branchlet usually produces an inflorescence, whereas in deciduous species only the vigorous shoots develop inflorescences. Inflorescences can be terminal and axillary, even on the same individual. The basic type appears to be an open raceme, which can be variously aggregated to form panicle-, fascicle-, or umbel-like inflorescences. Some *Purdiaea* have very short pedicels, which may occasionally lead to dense spiciform inflorescences. Each flower is subtended by a bract, which is sometimes broad and showy or even almost foliose.

FLOWER STRUCTURE. The flowers are basically pentamerous and perfect, but some *Clethra* exhibit functionally unisexual flowers and gynoedioecy caused by reduced anthers and pollen production. In *Clethra*, the petals arise in the same quincuncial succession as the sepals (Nishino 1983). In some species the petals are basally coherent or connate up to the lower third, although easily separable along a marked line (Sleumer 1967). A surface analysis revealed the petals to be irregularly striate, the adaxial side being more distinctly differentiated (Christensen and Jensen 1998).

In *Purdiaea*, the outer staminal whorl is alternipetalous, the androecium being diplostemonous throughout the entire development (Thomas 1960). According to Leins (1964), both staminal whorls of *Clethra* originate from a single circle. The alternipetalous stamens are initiated before the alternisepalous ones but become pushed towards the centre by the petal margins. Thus, the androecium is diplostemonous in origin and only secondarily appears obdiplostemonous. In both genera, the anthers are extrorse in bud but become inverted at anthesis. As a consequence, the original base with the pore-like clefts becomes the top, resulting in an introrse position (*Clethra*: Kavaljian 1952; *Purdiaea*: Ståhl 1992, Fig. 22). The two pollen sacs of each theca merge into a single cavity before the pollen is shed, which often occurs before anthesis.

Brown (1938) reported nectar secretion from the lower part of the ovary and speculated that the secretory region of *Clethra* can be compared to the disk of Ericaceae.

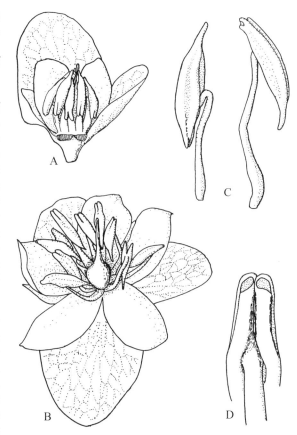

Fig. 22. Clethraceae. *Purdiaea nutans.* **A** Young flower, part of perianth removed. **B** Flower at anthesis with inverted anthers. **C** Stamens, from bud (left) and open flower (right). **D** Apical part of inverted anther, showing pores and filament attachment. (Gustafsson 1992)

EMBRYOLOGY. The data refer to *Clethra*. The anthers are tetrasporangiate, with a fibrous endothecium, two ephemeral layers and a secretory tapetum. Simultaneous divisions of the microspore mother cells follow meiosis; the tetrads are tetrahedral. Ovules are unitegmic with a single layer of integumentary cells. The megagametophyte develops according to the Polygonum type (Kavaljian 1952). Endosperm is cellular throughout its development; chalazal and micropylar haustoria are weakly developed (Johri et al. 1992). Embryogeny corresponds to the Asterad type (Anderberg 1993).

POLLEN MORPHOLOGY. The pollen monads are tricolporate and oblate to spheroidal. The exine is psilate, tectate with obscure stratification. The sexine is thinner than the nexine, the tectum is psilate or rugulate, and the infratectum granular (Erdtman 1952; Alvarado and Ludlow-Wiechers

1982; Barros and Barth 1994; Zhang and Anderberg 2002).

KARYOLOGY. The few data available suggest a basic chromosome number of $n = 8$ for *Clethra*. A polyploid series with diploid (*Clethra arborea*, *C. lanata*), tetraploid (*C. alnifolia*), and decaploid (*C. barbinervis*) species is discerned (Hagerup 1928; Kyhos 1965; Tanaka and Oginuma 1980).

FRUIT AND SEED. The fruit is indehiscent or a three-valved loculicidal capsule that may be more or less enclosed by the persistent calyx. Only few of the numerous ± anatropous ovules per locule develop into seeds in *Clethra*, whereas in *Purdiaea* there is a single, pendent, orthotropous ovule per locule; most of these solitary ovules grow to a mature seed. The seeds are exotestal-theoid. Mature seeds of *Clethra* are subovoid or irregularly angular to subtrigonous and show a foveolate-reticulate seed coat or are flat and more or less winged. In *Purdiaea*, due to vacuolization during ripening, the seed coat is lacking in mature seeds. The cotyledons are short and embedded in the copious cellular (*Purdiaea*) or in the oily and proteinaceous endosperm (*Clethra*). The small, foveolate or winged seeds may be wind-dispersed. In *Purdiaea* the naked seeds are most likely dispersed by wind with the entire fruit, the sepals functioning as wings (Berazaín and Rodríguez 1992).

PHYTOCHEMISTRY. Hegnauer (1964, 1989) reported high contents of galli- and ellagitannins as well as proanthocyanidins in twigs, leaves and inflorescences of *Clethra*. Triterpenes and waxes were found in the cortex, and urolic acid in the fruits of *C. barbinervis*.

SUBDIVISION. Until recently, Clethraceae were recognized as a monogeneric family. The recent inclusion of *Purdiaea*, formerly assigned to Cyrillaceae, is well supported by morphological and molecular data (Anderberg and Zhang 2002).

A subdivision of *Clethra* was based on the length of stamens and style, distribution and deciduousness (De Candolle 1839). Sleumer (1967) found these criteria not useful and retained only sections *Clethra* and *Cuellaria*, based on seed shape and structure of the seed coat. The first comprises the Asian and temperate American, the second the tropical American and Macaronesian species.

AFFINITIES. A placement in Ericales is generally accepted. Relationships with Ericaceae-Arbutoideae, especially the tribe Arbuteae, were considered mainly based on the (analogous) agreement in anther inversion at anthesis. On the other hand, the trimerous ovary led to speculations about relationships with Rhododendroideae (see Sleumer 1967).

Based on the pentacyclic, hermaphroditic, choripetalous, hypogynous flowers and the well-developed endothecium, *Clethra* was considered to represent a basal clade within Ericales (Sleumer 1967). In a cladistic analysis based on morphology and anatomy (Anderberg 1993), *Clethra* was found sister to Ericaceae s.l. and distinguished by its 3-carpellate gynoecium and Asterad-type embryogeny. Sequence analyses of various genes yielded different topologies, with *Clethra* being either sister to Styracaceae or Theaceae (Olmstead et al. 1993; Savolainen, Chase et al. 2000; Savolainen, Fay et al. 2000) or sister to Cyrillaceae/Ericaceae (Kron 1996; Savolainen, Fay et al. 2000; Soltis et al. 2000). This last placement is supported by a molecular study that includes *Purdiaea* (Anderberg and Zhang 2002).

DISTRIBUTION AND HABITATS. *Clethra* has a rare, disjunct distribution. One centre is tropical SE Asia, with some species extending into temperate China, Japan and India. Another centre is tropical Central and South America, with a southernmost limit at about 29° in SE Brazil. Two species are known from SE North America, and one from Madeira. Such a distribution pattern is known for few other plant genera, such as *Persea*. Sleumer (1967) postulated an earlier development from a tropical mountain flora, although it remained unclear whether this was situated in Asia or in America. The Macaronesian species seems to be most closely related to the tropical American ones, indicating an old transatlantic connection.

Purdiaea extends from Belize to Peru with the centre in Cuba.

Clethra prefers acid soils. Some species grow in swamps; there are also records from volcanic soils and limestone. The tropical species are predominantly found in humid montane habitats, reaching their upper altitudinal limit at about 3800 m in China as well as in the Andes. The few species reported from tropical lowlands may all be in secondary vegetation. Conversely, the temperate species are generally found in lowlands or lower montane vegetation. *Clethra* species are light-demanding and thus well-adapted invaders of cleared areas and secondary forests.

PALAEOBOTANY. Fossil records assigned to *Clethra* and the extinct genus *Disoclethra* would point to a widespread occurrence of the family in Europe, with the first records dating back to the Middle Eocene (Friis 1985; Knobloch and Mai 1986; Mai 2001). In the Japanese flora, records of *Clethra* are from the Late Miocene and Pliocene (Ozaki 1991). Other fossils from the Tertiary were reported for East Asia and North America, but some of them are doubtful as to their identity (Sleumer 1967).

ECONOMIC IMPORTANCE. Economic importance is very minor. Several temperate *Clethra* species are cultivated as ornamental plants; especially *C. alnifolia* is popular for its fragrant flowers. In China, the leaves of *Clethra barbinervis* are eaten. The timber of some species is used for furniture and constructions, mostly in Central and South America (Sleumer 1967).

KEY TO THE GENERA

Sepals ± equal in size; ovary 3(4)-locular; disk lacking; petals white, rarely pinkish to cream; fruit capsular **1. Clethra**
- Sepals distinctly unequal, the outer much larger than the inner; ovary (3–)5-locular, disk hypogynous; petals pink to violet; fruit indehiscent **2. Purdiaea**

1. *Clethra* L.

Clethra L., Sp. Pl. 396 (1753): Sleumer, Bot. Jahrb. Syst. 87: 36–175 (1967), rev.

Trees and shrubs with entire to serrate leaves. Racemes single or in panicle-, fascicle-, or umbel-like aggregates. Flowers actinomorphic, 5(6)-merous, hermaphroditic, rarely functionally androdioecious; outer stamens alternisepalous; disk absent; ovary 3-locular; ovules many per locule; style entire to branched. Fruit a 3-loculicidal, many-seeded capsule; seeds irregularly angled to winged, seed coat with foveolate-reticulate impressions or inconspicuously prominent cells; endosperm fleshy, embryo short, straight, cylindrical. Eighty-three spp., tropical America and SE Asia (to India), temperate North America and Asia, one sp. on Madeira.

2. *Purdiaea* Planch.   Fig. 22

Purdiaea Planch., Lond. J. Bot. 5: 250 (1846); Thomas, Contrib. Gray Herb. 186: 1–114 (1960), rev.; Berazaín & Rodríguez, Rev. Jard. Bot. Nac. 13: 21–25 (1992), Cuban spp.

Small trees or shrubs with sessile entire leaves. Racemes simple. Flowers slightly zygomorphic, hermaphroditic, 5-merous, the outer sepals conspicuously larger than the inner; stamens 10, the outer whorl alternipetalous; disk hypogynous; ovary (3–)5-locular; ovules 1 per locule, pendent, orthotropous; style unbranched. Fruit indehiscent, 1–5-seeded, 3–5-ribbed or indistinctly lobed; seed coat absent, endosperm copious, embryo straight, cylindrical. About 12 spp., Belize to Peru, most diverse in Cuba.

Selected Bibliography

Alvarado, J.L, Ludlow-Wiechers, B. 1982. Catálogo palinológico para la flora de Veracruz, no. 10. Familia Clethraceae. Biotica 8: 619–629.

Anderberg, A.A. 1993. Cladistic interrelationships and major clades of Ericales. Pl. Syst. Evol. 184: 207–231.

Anderberg, A.A., Zhang, X. 2002. Phylogenetic relationships of Cyrillaceae and Clethraceae (Ericales) with special emphasis on the genus *Purdiaea* Planch. Org. Divers. Evol. 2: 127–136.

Barros, M.A., Barth, O.M. 1994. Catálogo sistemático do pólen das plantas arbóreas do Brasil meridional, vol. 28. Burseraceae e Clethraceae. Revista Brasil. Biol. 54: 317–322.

Berazaín, R., Rodríguez, S. 1992. Novedades taxonómicas en el género *Purdiaea* Planchon (Cyrillaceae) en Cuba. Revista Jard. Bot. Nac. Univ. Habana 13: 21–25.

Brown, W.H. 1938. The bearing of nectaries on the phylogeny of flowering plants. Proc. Am. Philos. Soc. 79: 549–595.

Candolle, A.P. de 1839. Prodromus systematis naturalis, Vol. 7 (2). Paris: Treuttel & Würtz, , pp. 588–590.

Christensen, K.I., Hansen, H.V. 1998. SEM-studies of epidermal patterns of petals in the angiosperms. Opera Bot. 135: 1–91.

Drude, O. 1889. Clethraceae. In: Engler & Prantl, Die natürlichen Pflanzenfamilien IV, 1. Leipzig: Engelmann, pp. 1–2.

Erdtman, G. 1952. See general references.

Friis, E.M. 1985. Angiosperm fruits and seeds from the Middle Miocene of Jutland, Denmark. Biol. Skr. Dansk Vid. Selsk. 24(3): 1–165.

Giebel, K.P., Dickison, W.C. 1975. Wood anatomy of Clethraceae. J. Elisha Mitchell Sci. Soc. 91: 17–26.

Gustafsson, C. 1992. Cyrillaceae. In: Harling, G., Andersson, L. (eds.) Flora of Ecuador no. 45, pp. 29–34.

Hagerup, O. 1928. Morphological and cytological studies of Bicornes. Dansk Bot. Ark. 6(1): 1–27.

Hegnauer, R. 1964, 1989. See general references.

Jansen, S., Baas, P., Smets, E. 2001. Vestured pits: their occurrence and systematic importance in eudicots. Taxon 50: 135–167.

Johri, B.M., Ambegaokar, K.B., Srivastava, P.S. 1992. See general references.

Kavaljian, L.G. 1952. The floral morphology of *Clethra alnifolia* with some notes on *C. acuminata* and *C. arborea*. Bot. Gaz. 113: 392–413.

Knobloch, E., Mai, H.D. 1986. Monographie der Früchte und Samen in der Kreide von Mitteleuropa. Rozpr. ústred. úst. Geol. 47: 1–219, 56 pl.

Kron, K.A. 1996. Phylogenetic relationships of Empetraceae, Epacridaceae, Ericaceae, Monotropaceae, and Pyrolaceae: evidence from nuclear ribosomal 18S sequence data. Ann. Bot. Lond. 77: 293–303.

Kron, K.A., Chase, M.W. 1993. Systematics of Ericaceae, Empetraceae, Epacridaceae, and related taxa based upon *rbc*L sequence data. Ann. Missouri Bot. Gard. 80: 735–741.

Kubota, M., McGonigle, T.P., Hyakumachi, M. 2001. *Clethra barbinervis*, a member of the order Ericales, forms arbuscular mycorrhizae. Can. J. Bot. 79: 300–306.

Kyhos, D.W. 1965. Documented chromosome numbers of plants. Madroño 18: 122–126.

Lechner, S. 1914. Anatomische Untersuchungen über die Gattungen *Actinidia*, *Saurauia*, *Clethra* und *Clematoclethra* mit besonderer Berücksichtigung ihrer Stellung im System. Beih. Bot. Zentralbl. 32: 431–467.

Leins, P. 1964. Entwicklungsgeschichtliche Studien an Ericales-Blüten. Bot. Jahrb. Syst. 83: 57–88.

Mai, D.H. 2001. Die mittelmiozänen und obermiozänen Floren aus der Meuroer und Raunoer Folge in der Lausitz, Teil 2: Dicotyledonen. Palaeontographica, Abt. B., 257(1–6): 35–176.

Metcalfe, C.R., Chalk, L. 1950. See general references.

Nishino, E. 1983. Corolla tube formation in the Primulaceae and Ericales. Bot. Mag. Tokyo 96: 319–342.

Olmstead, R.G., Bremer, B., Scott, K.M., Palmer, J.D. 1993. A parsimony analysis of the Asteridae sensu lato based on *rbcL* sequences. Ann. Missouri Bot. Gard. 80: 700–722.

Ozaki, K. 1991. Late Miocene and Pliocene floras in Central Honshu. Bull. Kanagawa Pref. Mus. Nat. Sci. Spec. Issue, 244 pp.

Record, S.J. 1932. Woods of the Ericales, with particular reference to *Schizocardia*. Trop. Woods 32: 11–14.

Rodríguez, S., Berazaín, R. 1991. Caracterización de la nervadura foliar en el género *Purdiaea* Planch. (Cyrillaceae). Revista Jard. Bot. Nac. Univ. Habana 12: 69–73.

Savolainen, V., Chase, M.W. et al. 2000. See general references.

Savolainen, V., Fay, M.F. et al. 2000. See general references.

Sleumer, H. 1967. Monographia Clethracearum. Bot. Jahrb. Syst. 87: 36–175.

Soltis, D.E. et al. 2000. See general references.

Ståhl, B. 1992. Cyrillaceae. In: Harling, G., Andersson, L. (eds.) Flora of Ecuador, vol. 45, pp. 29–34.

Tanaka, R., Oginuma, K. 1980. Karyomorphological studies on *Clethra barbinervis* and two allied species. J. Jap. Bot. 55: 65–72.

Thomas, J.L. 1960. A monographic study of the Cyrillaceae. Contrib. Gray Herb. 186: 1–114.

Vales, M.A., Moncada, M., Machado, S. 1988. Anatomía comparada de Clethraceae en Cuba. Revista Jard. Bot. Nac. Univ. Habana 9: 69–73.

Yu, C.H., Chen, Z.L. 1990. Leaf architecture of the woody dicotyledons from tropical and subtropical China. Oxford: Pergamon Press.

Zhang, X.-P., Anderberg, A.A. 2002. Pollen morphology in the ericoid clade of the order Ericales, with special emphasis on Cyrillaceae. Grana 41: 201–215.

Connaraceae

R.H.M.J. Lemmens, F.J. Breteler and C.C.H. Jongkind

Connaraceae R. Brown in Tuckey, Narr. Exped. Congo: 431 (1818).

Small trees, shrubs, or lianas, evergreen or sometimes deciduous. Leaves alternate, imparipinnate, trifoliolate or unifoliolate, exstipulate, exstipellate; petiole pulvinate at base; petiolules entirely pulvinate; leaflets entire, opposite or not. Inflorescences axillary, sometimes clustered near the tip of the branches, occasionally cauliflorous, usually paniculate, sometimes less branched and resembling a raceme, or condensed. Flowers actinomorphic, 5(4)-merous, bisexual (unisexual), heterodistylous or heterotristylous (homostylous); pedicels usually jointed, prophylls 0; sepals imbricate or valvate, free or united at base (in *Rourea solanderi* almost entirely fused in bud), caducous to persistent and sometimes accrescent in fruit; petals imbricate, free to connivent near the base; stamens in 2 whorls, free or united at base, the outer antesepalous, longer than the inner antepetalous ones, the latter sometimes staminodial, anthers dorsifixed, dehiscing lengthwise, introrse; gynoecium 1- or 5-carpellate, superior; carpels free, each carpel with a terminal stylodium and capitate stigma; ovules 2 per carpel, collateral, nearly basal to nearly apical, anatropous to hemitropous, micropyle always directed upwards. Fruit consisting of 1–5 follicles, dry or more or less fleshy, usually dehiscing by a ventral suture, sometimes along the dorsal side as well, rarely circumscissile at the base (some *Rourea*), or indehiscent (*Hemandradenia, Jollydora*). Seed 1 per follicle, sometimes 2 (*Jollydora*); testa partly to entirely fleshy (sarcotesta), yellow to red; endosperm present or not; hilum lateral to basal; cotyledons thin and flat to planoconvex; radicle situated at the margin of the cotyledons and then apical, ventral or dorsal, or central and covered by the cotyledons. $x = 14, 16$.

A family comprising 12 genera and 110–200 species (depending on the species concept used), almost exclusively found in the tropics. Only a few species surpass the Tropic of Capricorn (e.g. *Cnestis polyphylla* in southern Africa) and the Tropic of Cancer (e.g. some *Connarus* species on the Asiatic mainland). The family is largely restricted to lowland rain forest, but some species occur in mountains and others in thickets in savannas.

Vegetative Morphology. Connaraceae are all woody plants, mostly lianas climbing by means of the winding ends of young branches, sometimes additionally provided with leaves transformed into woody hooks. Several species are present as small shrubs in the undergrowth of the forest and may hardly show any growth for many years but, when the forest canopy opens, they rapidly produce long shoots and become lianas. However, all species of the genera *Burttia, Ellipanthus, Hemandradenia, Jollydora* and *Vismianthus* do not show any lianescent tendencies and are shrubs or small trees, rarely medium-sized trees up to 30 m tall (*Ellipanthus*). A few *Cnestis, Connarus* and *Rourea* always have a shrub- or tree-like habit.

The main stem of lianescent species is usually (sub)cylindrical in cross section, but some species (e.g. *Rourea minor, Agelaea pentagyna*) have a strongly ridged stem showing a number of concentric rings. These latter species have included phloem. The bark is usually smooth, more rarely fissured, and in a few species strongly suberized (*Connarus suberosus, Rourea coccinea* and *R. orientalis*). Lenticels are usually present on the branchlets. Some species produce a red sticky (*Connarus, Agelaea gabonensis*) or colourless, slimy (*Rourea myriantha, R. solanderi*) exudate.

The leaves are pinnate, trifoliolate or unifoliolate, with the petiole pulvinate at base (Fig. 24B) and the petiolules entirely so; this implies that unifoliolate species have a pulvinus at the base and at the apex of the petiole. In *Connarus, Hemandradenia* and *Vismianthus*, glands can be found on the surface of the leaves. The tertiary venation is open except in *Manotes* where it consists of a closed pattern of very fine, parallel veinlets (Fig. 25C).

Vegetative Anatomy. The indumentum consists of various hair types: multicellular (*Cnestis, Connarus, Jollydora, Manotes* and *Rourea*) or unicellular glandular hairs (*Manotes*), unicellular,

1-armed non-glandular hairs (all genera except *Jollydora*), unicellular 2-branched hairs (*Burttia*, *Connarus*, *Hemandradenia*, some species of *Rourea*, *Vismianthus*) and chained, 2-armed unicellular hairs (*Connarus*). For more details, see Jongkind (in Breteler 1989). Stomata are paracytic in *Cnestis* and in some species of *Rourea*, anisocytic including helicocytic in *Agelaea*, *Cnestidium* and some species of *Rourea*, anomo-cyclocytic in *Burttia*, *Connarus*, *Ellipanthus*, *Hemandradenia*, *Manotes*, *Vismianthus* and some species of *Rourea*, and bicyclic in *Jollydora*. A papillose lower epidermis is present in many species of *Cnestis*, *Pseudoconnarus* and *Rourea*. Mucous cells are sometimes present in the upper epidermis. Sometimes a pattern of more or less parallel lines is present in the lower epidermis (in *Rourea calophylla*, *R. camptoneura* and *Manotes macrantha*). The secondary xylem is anatomically rather uniform and moderately highly specialized. The wood is diffuse-porous, the perforations are simple and slightly oblique to transverse, the ground tissue usually consists of libriform fibres together with fibre-tracheids and/or vasicentric tracheids or only libriform fibres, parenchyma is mostly absent or scanty, rays are usually exclusively uniseriate or biseriate, crystals are often present (usually in fibres), latex tubes generally absent (but present in *Connarus*), vertical intercellular canals often present in libriform tissue. For more details, see Dickison (1972b) and den Outer and van Veenendaal (in Breteler 1989).

INFLORESCENCE STRUCTURE. The inflorescences are basically axillary panicles. They often arise from the axils of reduced leaves near the apex of young branches, giving the impression of a compound terminal inflorescence, particularly in *Agelaea*, *Connarus*, *Manotes*, *Pseudoconnarus* and some *Cnestis* and *Rourea* species. Such flowering branches always end in a vegetative bud which sometimes continues the branch after flowering but usually shrivels. The panicles are often reduced to raceme-like or strongly condensed inflorescences (e.g. in *Hemandradenia mannii* and *Rourea obliquifoliolata*). Cauliflory with large nodose proliferations on the stem occurs in *Jollydora*, some *Cnestis* spp., *Rourea calophylloides* and *Manotes macrantha*.

FLOWER STRUCTURE. The flowers have 5 sepals, 5 petals, 10 stamens in 2 whorls and either 1 or 5 carpels; they usually have a jointed pedicel. The sepals are distinctly valvate (*Manotes*) to distinctly imbricate (most species), or in between. The petals are usually imbricate and may have glandular hairs or glands on their surface. *Manotes* flowers have a distinct androgynophore (Fig. 25E). The inner, antepetalous whorl of stamens is shorter or subequal to the outer antesepalous whorl, and is staminodial in *Ellipanthus*, *Hemandradenia* and in some *Connarus* spp.

A wide diversity in forms of heterostyly is known to occur in Connaraceae (Figs. 23–25). Heterotristyly is found in most *Agelaea*, *Jollydora* and *Manotes* spp., heterodistyly (either with 10 or 5 fertile stamens) in many genera and species, and homostyly in *Cnestis ferruginea*. The flowers of several species show a form of heterostyly transitional between tristyly and distyly. *Ellipanthus beccarii* and possibly some American *Connarus* spp. are dioecious; male plants have rudimentary pistils, female plants sterile stamens (lacking pollen).

EMBRYOLOGY. Ovules are hemi-anatropous, bitegmic and crassinucellate, and have a pronounced preraphe; the micropyle is formed by the outer integument, and a funicle is absent. A primary parietal cell divides at least once. The embryo sac develops according to the Polygonum type, endosperm development is nuclear and the tissue becomes cellular only after the formation of numerous free nuclei (Mauritzson 1939; Corner 1976).

POLLEN MORPHOLOGY. The pollen is rather uniform, with the exception of that of *Jollydora*. Pollen in the 5-carpellate tribes Cnestideae and Manoteae is spheroidal or subspheroidal and small (20–30 μm), with a thin wall and a finely reticulate ornamentation. In Connareae it is suboblate to subprolate and larger (30–40 μm), with a thicker wall and more coarsely reticulate. *Jollydora* (Jollydoreae) shows a completely different pollen type; its pollen is large (40–50 μm), oblate and tetracolpate, with a thick wall and a coarse reticulum, a most unusual type in Dicotyledons, otherwise only known from *Impatiens*. For details, see Dickison (1979) and van den Berg (in Breteler 1989).

KARYOLOGY. There is karyological information on African taxa only. The chromosome numbers known indicate that there are two groups, one comprising taxa with $2n = 28$ and the other taxa with $2n = 32$. The first group comprises species from the tribes Cnestideae and Manoteae, the second group species from the tribe Connareae. The somatic chromosomes of the Connareae have lengths varying in the range 0.5–1 μm, whereas those in Cnestideae and Manoteae are 1–2 μm long. It appears that the karyotype in Connareae differs from the other tribes investigated in both number and length of the chromosomes. For more details, see Arends (in Breteler 1989).

POLLINATION. Observations on pollination are very scarce. Many species are sweet scented and probably pollinated by insects. Bees have been observed visiting the flowers.

FRUIT AND SEED. In 5-carpellate species, usually some carpels fail to produce seeds; often only a single follicle is formed. The follicles are mostly dehiscent, but indehiscent in *Hemandradenia* and *Jollydora*, and usually orange to red. They are glabrous to hairy outside, sometimes with long, stinging hairs (some *Cnestis* spp.), and usually glabrous inside except in *Cnestis* and several *Connarus* spp.

Usually only one of the two ovules per carpel develops into a seed, but in *Jollydora* often both ovules mature. The seeds are often two-coloured with a black, glossy testa strongly contrasting with the yellow to orange sarcotesta, as well as with the follicle. In most species the larger part of the seed is squeezed out of the follicle that retains a firm grip on the part that remains inside (Fig. 24D), but in *Manotes* and *Vismianthus* the seed is almost completely released from the follicle, remaining attached by means of a threadlike appendix of the sarcotesta, which remains wedged in the narrowly funnel-shaped, bottom part of the follicle. In these latter two genera, endocarp and exocarp separate from each other at maturity.

In the seed-coat the testa is unlignified but the outer epidermis is developed as a thickened palisade. It becomes sarcotestal in the raphe-chalaza region or more or less all over the seed (*Hemandradenia*, *Jollydora*, *Manotes macrantha* and some *Rourea* spp.); the sarcotesta is yellow to red, pulpy, and in *Connarus* often has a crenulate border or short limb. The tegmen usually is crushed or disintegrates early, except for an outer layer of lignified fibres.

The structure of the seeds concerning endosperm and cotyledons is very variable. The cotyledons can be flat and thin and embedded in abundant endosperm (*Manotes*, some *Cnestis* spp.) to planoconvex without endosperm (*Agelaea*, *Connarus*, *Jollydora*, most *Rourea*), but intermediate conditions are common. The radicle is not always located near the micropyle. In some *Connarus* and *Rourea* spp. it is situated more or less ventrally or dorsally (relative to the follicle) and, in some African *Connarus* (*C. congolanus*, *C. staudtii*), the radicle is found in the centre of the seed and quite hidden between the peltate cotyledons.

DISPERSAL. The strikingly coloured, dehiscing follicles showing the often two-coloured seed-coat seem to advertise the seeds to birds such as fruit pigeons and hornbills, which are probably the main dispersers. Most Connaraceae are lianas of the canopy or savannah shrubs, and birds will readily detect the fruits. The fruits of *Hemandradenia* and *Jollydora*, which are both understorey treelets in the rain forest, are indehiscent; they may escape the attention of birds and are more likely to be dispersed by small mammals.

PHYTOCHEMISTRY. There are indications that many species of the family contain chemically interesting substances. The seeds and sometimes also other parts of the plants (particularly roots) are reported as poisonous. They are often used as dog poison, and they seem to be poisonous for other animals like sheep, goats and rats as well. All parts of *Cnestis polyphylla* are reported to contain a neurotoxical compound which has been named glabrin (Jeannoda et al. 1985); this compound is methionine sulphoximine (S-(3-amino-3-carboxypropyl)-S-methyl sulphoximine), and also has ichthyotoxic properties. Methionine sulphoximine is probably also the toxic compound in several other species (e.g. of *Agelaea* and *Rourea*). However, in most species the toxic compounds have not yet been identified.

Roots of some *Connarus* spp. were demonstrated to contain rapanon, embelin and bergenin, and roots of *Rourea minor* β-sitosterol, its glucoside, hentriacontan and mesoinosit.

Tannins are reported as common, and both hydrolysable and condensed tannins may be present. Gallic acid has been isolated from fruits and leaves of South American *Connarus* spp. In leaf extracts from the Asiatic *Connarus semidecandrus*, myricetin, quercetin, kaempferol, delphinidin and cyanidin have been identified.

SUBDIVISION AND RELATIONSHIPS WITHIN THE FAMILY. Planchon (1850) divided the family into two tribes, the Connareae and the Cnestideae, based on differences in the aestivation of the sepals and the presence or absence of endosperm. Gilg (1891) maintained these tribes and, in 1897, after the description of *Jollydora*, he divided the Connaraceae in two subfamilies, Connaroideae and Jollydoroideae. Schellenberg (1910) proposed a new subdivision of the family, distinguishing the subfamilies Connaroideae (including *Jollydora*) and Cnestoideae; he divided the former subfamily into two tribes, Connareae and Roureeae, the latter of which was again divided into two subtribes. In 1938 Schellenberg reverted to Gilg's subdivision into two subfamilies, and the number of genera reached 24. Subsequently the number of genera was reduced by Leenhouts (1958), and even further by Breteler (1989) to 12. Lemmens (in Breteler 1989) proposed a tribal classification which is followed in this treatment.

AFFINITIES. Relationships of Connaraceae with other families have been ambiguous for a long time. Most often a close relationship with Leguminosae and Rosaceae has been postulated, but some authors considered Oxalidaceae, Anacardiaceae, Meliaceae and Dilleniaceae as related families. The pinnate leaves and apocarpous, pod-like fruits are reminiscent of Leguminosae, and the exstipulate leaves and the presence of a sarcotesta would emphasize affinities with Sapindaceae. However, any preference expressed as to the position of Connaraceae in any of the orders Rosales, Sapindales and Dilleniales left them with part of their characters in disagreement with that position.

Phylogenies based on plastid gene sequences place Connaraceae as sister of Oxalidaceae in an order Oxalidales (=Cunoniales) (APG 1998). The two families share the presence of heterostyly, the benzoquinone rapanone, exotegmic seeds and some other morphological characters that have been listed by Nandi et al. (1998).

DISTRIBUTION AND HABITATS. The Connaraceae are a pantropical family of which only a few species are found beyond the tropics. The largest number of genera (10), representing all four tribes, is found in Africa, followed by Asia (6) and then America (4), each with representatives of two tribes. The main centre of distribution is Central Africa, particularly Cameroon and Gabon. Two genera, *Cnestidium* and *Pseudoconnarus*, are endemic to tropical America, where *Connarus* and *Rourea* (which are also well represented in tropical Asia) are very rich in species. Some species have a very large area of distribution; for example, *Rourea minor* is found in Africa, Madagascar, islands of the Indian Ocean and the Pacific, and in tropical Asia. In genera like *Agelaea*, *Connarus*, *Ellipanthus*, *Rourea* and *Vismianthus*, groups of closely related species are common to Africa and Asia. This must be an old disjunction, as their large seeds are not likely to be dispersed over long distances and have hardly been dispersed by man.

Connaraceae are largely restricted to the lowland rain forest. More rarely they are found in mountain vegetation; some species grow in thickets or in remnants of forests in savannas.

ECONOMIC IMPORTANCE. Representatives of Connaraceae are little used by man. The seeds and roots of several species are toxic and sometimes used to poison dogs and rats (particularly *Cnestis* and *Rourea* spp.). Diluted in small doses, they are occasionally used as medicine, mainly against stomach-ache and dysentery (*Agelaea macrophylla*, *Cnestis ferruginea* and several *Rourea* species), and as anthelmintic (*Connarus africanus*). *Ellipanthus* species that reach the size of medium-sized trees are sometimes cut in South-East Asia for their timber. The fatty oil present in seeds of some *Connarus* spp. has been used in India in soap production.

KEY TO THE GENERA

1. Flowers with 5 carpels 2
 – Flowers with 1 carpel 8
2. Follicles (nearly) glabrous 3
 – Follicles covered with a continuous indumentum of (sometimes glandular) hairs 4
3. Seeds with abundant endosperm; leaves always 3-foliolate; America **9. Pseudoconnarus**
 – Seeds with little or no endosperm; leaves pinnate (but branches with unifoliolate or 3-foliolate leaves may occur); circumtropical **12. Rourea**
4. Flowers with a distinct androgynophore; Africa **7. Manotes**
 – Flowers without a distinct androgynophore 5
5. Follicles hairy inside; Africa, Asia **8. Cnestis**
 – Follicles glabrous inside 6
6. Sepals in fruit strikingly accrescent; circumtropical **12. Rourea**
 – Sepals in fruit inconspicuous or caducous 7
7. Leaves 3- or 5-foliolate; Africa, Asia **10. Agelaea**
 – Leaves pinnate with variable number of leaflets within one plant; America **11. Cnestidium**
8. Leaves unifoliolate 9
 – Leaves with more than one leaflet (except sometimes uppermost leaves on flowering branches) 12
9. Follicles glabrous; Africa, Asia **5. Vismianthus**
 – Follicles velutinous 10
10. Petals glabrous; follicles dehiscent; seeds with abundant endosperm, cotyledons thin and flat; Africa **4. Burttia**
 – Petals hairy; follicles dehiscent or indehiscent; seeds with or without endosperm, cotyledons flat to planoconvex 11
11. Follicles dehiscent; testa for about 1/4 fleshy; Africa, Asia **2. Ellipanthus**
 – Follicles indehiscent; testa almost entirely fleshy; Africa **3. Hemandradenia**
12. Understorey treelet; petals and filaments glabrous; follicles indehiscent, 1–2-seeded; Africa **6. Jollydora**

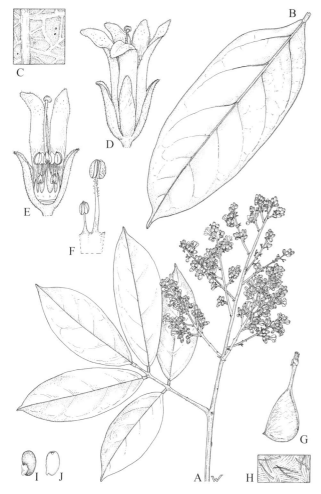

Fig. 23. Connaraceae-Connareae. *Connarus longistipitatus.*
A Flowering branchlet. B Leaflet. C Detail leaf underside. D
Flower. E Flower, sepals and petals partly removed. F Short and
long stamen. G Follicle. H Detail inner side of follicle. I Seed.
J Seed, vertical section. Drawn by W. Wessel-Brand (Orig.)

– Treelet, shrub or liana; petals and/or filaments with many
  (sometimes glandular) hairs; follicles dehiscent, 1-seeded;
  circumtropical   **1.** *Connarus*

Tribes and Genera of Connaraceae

I. Tribe Connareae Planchon (1850).

Lianas, shrubs or small trees; leaves unifoliolate to
pinnate; carpel 1; follicle dehiscent or (*Hemandradenia*) indehiscent with 1 ventrally attached
seed; pollen tricolporate.

1. *Connarus* L.   Fig. 23

Connarus L., Sp. Pl.: 675 (1753); Leenhouts, Flora Males. I, 6:
  525 (1958); Forero, Flora Neotropica 36: 36 (1983); Lemmens
  in Breteler (ed.) The Connaraceae: 239 (1989).

Small trees, shrubs or lianas. Leaves trifoliolate or
imparipinnate (upper leaves unifoliolate); leaflets
often glandular punctate. Inflorescence paniculate, often pseudoterminal. Flowers bisexual (unisexual), 5-merous, heterodistylous; sepals, petals,
filaments and style usually glandular, carpel 1.
Follicle often stipitate, often glabrescent outside,
glabrous or pilose and often glandular inside; seed
solitary, attached to the ventral side of the follicle,
below the hilum with sarcotesta, endosperm 0,
cotyledons thick, planoconvex, radicle apical,
dorsal or almost in the centre of the seed between
the cotyledons. About 75 spp., circumtropical.

2. *Ellipanthus* Hook.f.

Ellipanthus Hook.f. in Benth. & Hook.f., Gen. Pl. 1, 1: 434
  (1862); Leenhouts, Flora Males. I, 6: 520 (1958); Lemmens in
  Breteler (ed.) The Connaraceae: 268 (1989); Lemmens, Bull.
  Mus. Natl. Hist. Nat. Adansonia 14: 99 (1992).

Shrubs or small trees. Leaves unifoliolate. Inflorescence a small panicle. Flowers bisexual or unisexual, 4–5-merous, heterodistylous, petals pilose
outside and usually also inside, filaments and
style pilose, carpel 1. Follicle more or less stipitate,
densely tomentose outside, glabrous inside, dehiscent; seed 1(2) per follicle, attached to the ventral
side of the follicle, basal part with sarcotesta,
endosperm thin, cotyledons thick, radicle apical.
About six spp., East Africa, Madagascar and tropical Asia.

3. *Hemandradenia* Stapf

Hemandradenia Stapf, Bull. Misc. Inform. (Kew Bull.): 288
  (1908); Eimunjeze in Breteler (ed.) The Connaraceae: 275
  (1989).

Shrubs or small trees. Leaves unifoliolate. Inflorescence a glomerule or panicle. Flowers bisexual,
(4–)5(–7)-merous, heterodistylous, petals pilose
outside and usually also inside, filaments and style
pilose, carpel 1. Follicle not stipitate, densely
tomentose outside, glabrous inside, indehiscent.
Seed solitary, attached to the ventral side of the
follicle, sarcotesta completely covering the seed,
endosperm thin to copious, cotyledons thin or
thick, radicle apical. Two spp., West and Central
Africa.

4. *Burttia* Baker f. & Exell   Fig. 24

Burttia Baker f. & Exell, J. Bot. 69: 249 (1931); Breteler &
  Brouwer in Breteler (ed.) The Connaraceae: 169 (1989).

Shrubs or small trees. Leaves unifoliolate. Inflorescence 1–3-flowered. Flowers bisexual, (4–)5-merous, heterodistylous, petals, filaments and style glabrous, carpel 1. Follicle not stipitate, densely tomentose outside, glabrous inside, dehiscent. Seed solitary, attached near the top of the ventral side of the follicle, sarcotesta covering whole length of the seed on one side, endosperm copious, cotyledons thin and narrow, radicle apical. One sp., *B. prunoides* Baker f. & Exell, East Africa.

5. *Vismianthus* Mildbr.

Vismianthus Mildbr., Notizbl. Bot. Gart. Mus. Berlin 12 (115): 706 (1935); Breteler & Brouwer in Breteler (ed.) The Connaraceae: 369 (1989).
Schellenbergia Parkinson (1936).

Shrubs or small trees. Leaves unifoliolate, glandular punctate. Inflorescence racemose. Flowers bisexual, 5-merous, heterodistylous, petals glabrous but glandular, filaments glabrous, style hairy, carpel 1. Follicle more or less stipitate, (sub)glabrous outside, glabrous inside, dehiscent, inner and outer pericarp separating. Seed solitary, attached to the ventral side of the follicle, sarcotesta fimbriate or undulate with a long appendage and covering the base of the seed, endosperm very thin, cotyledons thin and folded or thick and plano-convex, radicle apical. Two spp., one in Tanzania, the other in Myanmar.

II. Tribe Jollydoreae (Gilg) Lemmens (1989).

Treelets, usually unbranched; leaves large, pinnate; carpel 1; follicle indehiscent, seeds often 2, ventrally attached; pollen tetracolpate.

6. *Jollydora* Pierre

Jollydora Pierre, unprinted drawing by Delpy (1895); Breteler & van Ziel in Breteler (ed.) The Connaraceae: 284 (1989).

Usually unbranched treelets. Leaves imparipinnate. Inflorescence racemose, in cauliflorous clusters or in axils of leaves. Flowers bisexual, 5-merous, heterotristylous, petals and filaments glabrous, style sometimes with a few hairs, carpel 1. Follicle stipitate or not, glabrous or glabrescent outside, glabrous inside, indehiscent; seeds 1 or 2, attached to the ventral side of the follicle, sarcotesta almost completely covering the seed, endosperm 0, cotyledons thick and almost horny, radicle apical. Three spp., Central Africa.

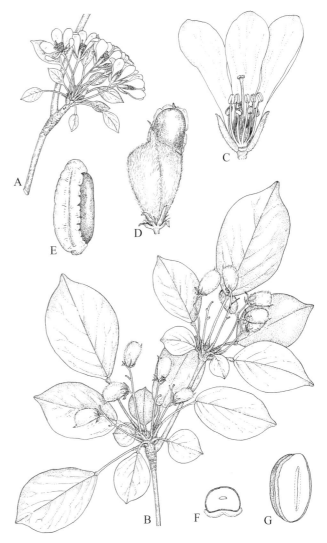

Fig. 24. Connaraceae-Connareae. *Burttia prunoides*. **A** Flowering branchlet. **B** Fruiting branchlet. **C** Flower, sepals and petals partly removed. **D** Dehiscing fruit with seed. **E** Seed with sarcotesta. **F** Transverse section of seed. **G** Section of seed lengthwise, sarcotesta removed. Drawn by W. Wessel-Brand (Orig.)

III. Tribe Manoteae Lemmens (1989).

Lianas; leaves pinnate, leaflets striately veined; androgynophore +; carpels 5; follicles dehiscent, seed 1, ventrally attached; pollen tricolporate.

7. *Manotes* Planchon   Fig. 25

Manotes Planchon, Linnaea 23: 438 (1850); Jongkind in Breteler (ed.) The Connaraceae: 294 (1989).

Lianas or scandent shrubs. Leaves imparipinnate, leaflets with a dense pattern of very fine parallel

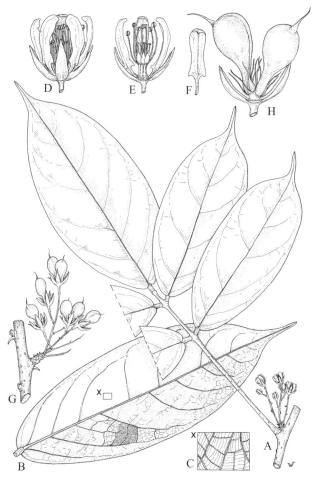

Fig. 25. Connaraceae-Manoteae. *Manotes macrantha.* **A** Flowering branchlet. **B** Leaflet. **C** Detail of leaflet. **D** Flower. **E** Flower, sepals and petals partly removed. **F** Petal. **G** Branchlet with fruits. **H** Follicles. Drawn by Wessel-Brand (Orig.)

veinlets. Inflorescence racemose or paniculate. Flowers bisexual, 5-merous, heterotristylous or heterodistylous, with distinct androgynophore; petals, filaments and style hairy, often also with glandular hairs; carpels 5. Follicle more or less stipitate, minutely velutinous to long-hairy outside, glabrous inside, dehiscent, inner and outer pericarp splitting apart at maturity; seed solitary, attached to the ventral side of the follicle, sarcotesta (almost) completely covering the seed and with a thread-like appendix attaching the seed to the base of the follicle, endosperm copious, cotyledons thin, radicle apical. Four or five spp., humid tropical Africa.

IV. TRIBE CNESTIDEAE Planchon (1850).

Shrubs or lianas; leaves pinnate or trifoliolate; carpels 5; follicles dehiscent follicles with 1 seed basally attached; pollen tricolporate.

8. *Cnestis* A.L. Juss.

Cnestis A.L. Juss., Gen. Pl.: 374 (1789); Leenhouts, Flora Malesiana I, 6: 497 (1958); Lemmens in Breteler (ed.) The Connaraceae: 174 (1989).

Lianas, climbing shrubs (small trees). Leaves imparipinnate. Inflorescence a panicle, raceme or fascicle, axillary or cauliflorous. Flowers bisexual, 5-merous, heterodistylous or sometimes more or less homostylous, petals glabrous or with some hairs outside, filaments glabrous, styles often hairy near base, carpels 5. Follicle not or indistinctly stipitate, densely tomentose, sometimes also with long, rigid hairs outside, velutinous, sometimes with rigid hairs inside, dehiscent; seed solitary, attached to the base of the follicle, basal part with sarcotesta, endosperm copious, cotyledons thin, radicle apical. Thirteen spp., 12 in tropical Africa and one in tropical Asia.

9. *Pseudoconnarus* Radlk.

Pseudoconnarus Radlk., Sitzb. Math.-Phys. Cl. Akad. Münch. 16: 356 (1886); Forero, Flora Neotropica 36: 130 (1983).

Lianas. Leaves trifoliolate. Inflorescence paniculate. Flowers bisexual, 5-merous, more or less distinctly heterodistylous, petals glabrous but glandular, filaments glabrous, styles glabrous or hairy, carpels 5. Follicle not stipitate, glabrous outside and inside, dehiscent; seed solitary, attached to the base of the follicle, endosperm copious, cotyledons thin, radicle apical. Five spp., tropical South America.

10. *Agelaea* Planchon

Agelaea Planchon, Linnaea 23: 437 (1850); Leenhouts, Flora Males. I, 6: 500 (1958); Jongkind in Breteler (ed.) The Connaraceae: 136 (1989).

Lianas. Leaves trifoliolate, sometimes 5-foliolate and imparipinnate to almost palmately compound. Inflorescence paniculate. Flowers bisexual, 5-merous, heterotristylous or heterodistylous, petals and filaments glabrous, styles often hairy, carpels 5. Follicle not or indistinctly stipitate, velutinous outside, glabrous inside, dehiscent; seed solitary, attached to the base of the follicle, basal part with sarcotesta, endosperm 0, cotyledons thick, planoconvex, radicle apical. Eight spp., tropical Africa and tropical Asia.

11. *Cnestidium* Planchon

Cnestidium Planchon, Linnaea 23: 439 (1850); Forero, Flora Neotropica 36: 29 (1983).

Lianas. Leaves imparipinnate. Inflorescence paniculate. Flowers bisexual, 5-merous, ± distinctly heterodistylous, petals and filaments glabrous, styles often hairy, carpels 5. Follicle not stipitate, pubescent outside, glabrous inside, dehiscent; seed solitary, attached to the base of the follicle, basal part with sarcotesta, endosperm scarce, cotyledons thick, radicle apical. Two spp., Central and South America.

12. *Rourea* Aublet

Rourea Aublet, Pl. Guiane 1: 467 (1775); Leenhouts, Flora Males. I, 6: 510 (1958); Forero, Flora Neotropica 36: 138 (1983); Jongkind in Breteler (ed.) The Connaraceae: 310 (1989).
Bernardinia Planchon (1850); Forero, Flora Neotropica 36: 25 (1983).
Roureopsis Planchon (1850); Leenhouts, Flora Males. I, 6: 505 (1958).

Lianas, shrubs or small trees. Leaves trifoliolate or imparipinnate, occasionally unifoliolate. Inflorescence racemose or paniculate, sometimes nearly globose. Flowers bisexual, 5-merous, heterodistylous, petals glabrous or with some hairs apically, filaments glabrous, styles often hairy near base, carpels 5. Follicle not or indistinctly stipitate, glabrous or velutinous outside, glabrous inside, dehiscent; seed 1(2) per follicle, attached to the base of the follicle, with a sarcotesta around the basal part to almost completely covering the seed, endosperm 0, cotyledons thick and plano-convex, radicle apical to ventral. About 40–70 spp., circumtropical.

Selected bibliography

APG (Angiosperm Phylogeny Group) 1998. See general references.

Breteler, F.J. (ed.) 1989. The Connaraceae. A taxonomic study with emphasis on Africa. Agric. Univ. Wageningen Papers 89-6, 403 pp.

Corner, E.J.H. 1976. See general references.

Dickison, W.C. 1971. Anatomical studies in the Connaraceae 1: carpels. J. Elisha Mitchell Sci. Soc. 87: 77–86.

Dickison, W.C. 1972. Anatomical studies in the Connaraceae 2: wood anatomy. J. Elisha Mitchell Sci. Soc. 88: 120–136.

Dickison, W.C. 1973a. Anatomical studies in the Connaraceae 3: leaf anatomy. J. Elisha Mitchell Sci. Soc. 89: 121–138.

Dickison, W.C. 1973b. Anatomical studies in the Connaraceae 4: the bark and young stem. J. Elisha Mitchell Sci. Soc. 89: 166–171.

Dickison, W.C. 1979. A survey of pollen morphology of the Connaraceae. Pollen Spores 21: 31–79.

Forero, E. 1976. Connaraceae. Flora Neotropica Monogr 36. New York: New York Botanical Garden.

Gilg, E. 1891. Connaraceae. In: Engler & Prantl, Nat. Pflanzenfam. III, 3: 61–70.

Gilg, E. 1897. Connaraceae. In: Engler & Prantl, Nat. Pflanzenfam., Nachträge 1: 189–190.

Hegnauer, R. 1964, 1989. See general references.

Jeannoda, V.L.R., Valisolalao, J., Creppy, E.E., Dirheimer, G. 1985. Identification of the toxic principle of *Cnestis glabra* as methionine sulphoximine. Phytochemistry 24: 854–855.

Jongkind, C.C.H. 1991. A new section and a new species in *Agelaea* Sol. ex Planch. (Connaraceae). Bull. Jard. Bot. Nat. Belgique 61: 71–75.

Leenhouts, P.W. 1958. Connaraceae. In: van Steenis, C.G.G.J. (ed.) Flora Malesiana I, 5: 495–591.

Lemmens, R.H.M.J. 1992. A reconsideration of *Ellipanthus* (Connaraceae) in Madagascar and continental Africa, and a comparison with the species in Asia. Bull. Mus. Natl. Hist. Nat. B Adansonia 14: 99–108.

Mauritzson, J. 1939. Contributions to the embryology of the orders Rosales and Myrtales. Acta Univ. Lund 35: 1–121.

Nandi, O.I. et al. 1998. See general references.

Planchon, J.E. 1850. Prodromus monographiae ordinis Connaracearum. Linnaea 23: 411–442.

Schellenberg, G. 1910. Beiträge zur vergleichenden Anatomie und zur Systematik der Connaraceen. Mitt. Bot. Mus. Univ. Zürich 50, 158 pp.

Schellenberg, G. 1938. Connaraceae. In: Engler, A. (ed.) Das Pflanzenreich IV: 127.

Takhtajan, A. 1969. Flowering plants, origin and dispersal. Edinburgh: Oliver and Boyd.

Cornaceae

K. Kubitzki

Cornaceae (Dumort.) Dumort., Anal. Fam. Pl.: 33, 34 (1829),
    nom. cons.
Alangiaceae DC., Prodr. 3: 203 (1828).
Nyssaceae Juss. ex Dumort. (1829).
Mastixiaceae Calest. (1905).
Davidiaceae (Harms) H.L. Li (1955).

Trees or shrubs (rhizomatous halfshrubs). Hairs
simple, unicellular. Leaves alternate or opposite,
simple, entire (dentate), exstipulate. Inflorescences
many-flowered thyrso-panicles, heads, or few-
flowered cymes; flowers hermaphrodite to unisexual, actinomorphic, epigynous, mostly 4- or
5-merous; calyx adnate to ovary with 4–5(–10)
lobes or teeth or obsolete; petals free (basally
connate), imbricate or valvate, reduced or lacking
in pistillate flowers; stamens as many as and alternating with petals, or in 2 isomerous whorls (in 1
whorl of up to 40 stamens), mostly attached to or
around the edge of an epigynous nectary disk
(lacking in *Davidia*); filaments free; anthers bithecate, dehiscing longitudinally; pistil 1–2(–9)-
locular; style with lobed or capitate stigma, or
with 2–3 style branches; ovary inferior, with 1
pendulous ovule per locule; ovules anatropous
(hemitropous), unitegmic and crassinucellate
(*Nyssa* and red-fruited *Cornus* tenuinucellate).
Fruits drupes, often with a germination valve,
1(2–6)-seeded (syncarps); seeds small to medium-
sized, exotestal, with straight embryo and copious
endosperm.

A family of 7 genera and ca. 110 species, in
warm-temperate and humid tropical regions predominantly of the northern hemisphere, most
genera in East and Southeast Asia.

VEGETATIVE MORPHOLOGY. *Cornus canadensis*
and *C. suecica* are the only herbaceous (or more
precisely halfshrubby) representatives of the
family, all other being woody. These range from
shrubs to trees, sometimes vast, but rarely buttressed; some *Alangium* are scandent. Tupelos
(*Nyssa sylvatica* subsp. *biflora* and *N. aquatica*),
growing in swamps, develop protruding under-
water lenticels and roots proliferating from older
roots or from the stem bases (Hook et al. 1970). In
N. aquatica intercellular spaces are present in the
cambium. Through these spaces, oxygen can reach

the xylem of the stem and thence the roots. It is
likely that the swollen bases of aquatic tupelos also
provide intercellular space.

The complex branching pattern in *Cornus* was
analysed in a series of papers by Hatta (see Hatta
et al. 1999).

The leaves are alternate or opposite, mostly
constantly so for whole genera, but among the
otherwise opposite-leaved *Cornus* two species
have alternate leaves, and in *Mastixia* phyllotaxis
changes from alternate to (sub)opposite, often in
the same individual.

VEGETATIVE ANATOMY. Sertorius (1893) and
Adams (1949) gave detailed descriptions of the
anatomy of the family, and Noshiro and Baas
(1998) studied the wood anatomy. The following
features are of particular interest. Stomates are of
the ranunculid type. Non-glandular and glandular
hairs are always 1-celled. Two-armed 1-celled hairs
are found in *Mastixia*, *Diplopanax*, *Cornus* and in
a slightly different form (with asymmetric arms)
in one *Alangium*; in *Cornus* they are calcified. Ca-
oxalate is deposited in druses. The wood is of a
primitive type, and markedly heterogeneous rays
and relatively long vessel members with scalariform perforations are general. *Davidia* has the
largest number of bars in its perforation plates,
followed by *Mastixia*, *Nyssa*, and *Cornus*; in most
Alangium perforations are simple. In cross section, vessels appear solitary or in radial groups.

In *Mastixia* and *Diplopanax*, ducts with a secretory epithelium containing resin are found in the
periphery of the pith, along the vascular bundles
of the leaves, and in the flowers and fruits. Judging
from fruit remains, the earliest mastixioids such as
Beckettia and *Mastixiopsis* lacked secretory ducts
(Eyde 1988).

Alangium has articulated laticifers which
develop from parenchyma cells, the transverse
walls of which disintegrate or are ruptured. They
are usually arranged in an arc or ring around the
phloem of the vascular bundles. They can occur in
any part of the *Alangium* flower except the ovule
and style, but terminate at the base of the flower
or extend only a short distance into the pedicel.
In the leaves, they accompany the major vascular

bundles. They are also plentiful in fruits of sect. *Alangium* but are hard to find there in flowers, whereas in sects. *Marlea* and *Rhytididandra* they are conspicuous at all stages of flower and fruit development. In sect. *Conostigma* they seem to appear as the fruits develop (Eyde 1968a).

INFLORESCENCES. Their structure has been studied in detail by Jahnke (1986). Murrell (1993), who used inflorescence characters in a phylogenetic analysis, made some additions. In most Cornaceae the inflorescences are terminal, which sometimes leads to the formation of sympodia. The basic structure is a thyrsoid. Especially in taxa with alternate leaves, the inflorescences appear (thyrso-)paniculate. Variations of this basic scheme include the formation of condensed capitula, sometimes in combination with showy involucra; the suppression of bracts; metatopies (both concaulescence and recaulescence occur); suppression to loss of terminal flowers on inflorescences and partial inflorescences; and the formation of open racemes.

FLORAL MORPHOLOGY. Cornaceae have a peculiar gynoecial bundle supply, which has been analysed by Eyde (1968b). Central bundles, usually present in syncarpous gynoecia of other plant groups and corresponding to the ventral bundles of the individual carpels, are lacking. In *Davidia* the ovules are supplied by ramifying bundles that extend longitudinally through the septa from the base to the apex of the locules. In *Nyssa*, *Alangium* and *Cornus*, the pendent ovules are supplied by ovular bundles that pass in an arc above each septum. These bundles connect the ovules with paired or fused longitudinal bundles that raise through the ovary wall opposite the septa. They can be interpreted as ventral carpel bundles. Recent *Mastixia* is usually pseudomonomerous but, according to Eyde (1968b), fossil fruits related to it reveal a vascular pattern similar to that of *Cornus*.

KARYOLOGY. In *Cornus* and *Alangium*, the gametophytic chromosome number $n = 11$ is widespread; some *Cornus* have $n = 10$ or 9; *C. canadensis* is tetraploid ($2n = 44$); for one *Alangium*, $n = 11, 9$ and 8 have been reported (Goldblatt 1978; Eyde 1988). *Nyssa* and *Camptotheca* have $n = 22$, *Davidia* 21. In *Mastixia*, *M. arborea* has $n = 13$ and *M. trichotoma* has $n = 11$. This may point to an ancient base number $x = 11$, which was retained in the cornoids and doubled in the nyssoids, and has undergone some aneuploid change.

POLLEN MORPHOLOGY. Pollen of *Mastixia*, *Cornus* and *Curtisia* has been described by Ferguson (1977) and Ferguson and Hideux (1978) as 3-colporate, oblate (*Mastixia*) or spheroidal (to prolate, some *Cornus*), with a (rounded-)triangular amb, short columellae, and usually a perforate tectum (complete in *Curtisia*). Although habitually looking quite different, these pollen grains agree in having a characteristic, complex H-shaped endoaperture with lamellated thinnings of the endexine (the lamellations are lacking in *Mastixia*). An H-shaped endoaperture may also be present in *Camptotheca* and *Nyssa*, although the analysis of their pollen morphology by Sohma (1963, 1967) and Eramijan (1971) is not fully conclusive in this respect, and hardly so in *Alangium* (Reitsma 1970). Pollen of Recent *Alangium* species is mostly 3-colporate and its ornamentation ranges from reticulate to rugulate, striate and verrucate or gemmate (Reitsma 1970; for fossil pollen see Eyde et al. 1969; Eyde 1972; Morley 1982). A fossil *Alangium* pollen named *Lanagiopollis regularis* Morley exhibits great similarity to pollen of Recent *Mastixia*.

EMBRYOLOGY. The pollen of *Davidia*, *Nyssa*, *Alangium*, *Cornus* is 2-celled when shed (Eyde 1988).

The ovules are anatropous or rarely (*Cornus*) hemitropous, unitegmic, and crassinucellate or rarely (*Nyssa* and red-fruited *Cornus*) tenuinucellate (Tandon and Herr 1971; Eyde 1988). An endothelium is reported from *Davidia*. In the past, much stress has been laid on the orientation of the ovules. The micropyle is always directed upward and, as far as I could ascertain, in all genera except *Alangium* and *Cornus* the raphe is ventral. In *Alangium* and also in *Cornus*, at least initially, the raphe is dorsal. A developing *Cornus* ovule turns from the first apotropous position so that the raphe is to one side of the locule and the micropyle to the other (Eyde 1988).

The development of the embryo sac usually follows the Polygonum type, although in *Cornus* the Fritillaria type and some aberrant modes have been observed. Endosperm development is Cellular in *Davidia* and *Cornus*, Nuclear in *Nyssa* and *Alangium* (Johri et al. 1992).

FRUIT AND SEED. Germination valves are a characteristic trait of all cornaceous fruit stones except for *Alangium* (Fig. 26; Hill 1933; Eyde 1988). In *Mastixia*, *Diplopanax*, and *Davidia* the germination valves extend full length of the locules (Fig. 26K–N), whereas in *Cornus*, *Nyssa*, *Camptotheca*,

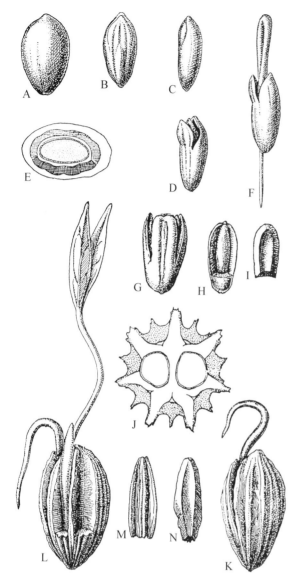

Fig. 26. Cornaceae. A–F *Nyssa sylvatica*. **A** Fruit. **B** Fruit stone, showing the valve in face view and side view (**C**), and in **D** the valve pushed away by the emerging radicle. **E** Fruit stone in transverse section, the valve is shown by hatching. **F** Germinating seed, the valve pushed aside. **G–I** *Cornus officinalis*. **G** Bilocular fruit stone showing the two valves commencing to split away. **H** Side view of fruit stone with a valve removed exposing the seed. **I** Detached valve. **J–N** *Davidia involucrata*. **J** Fruit stone in transverse section, showing 7 locules, only 2 in this case fertile; the valves shaded. **K** Fruit stone with one germinating seed. **L** Later stage in germination, showing three seedlings and the cavities in which the seeds lie. **M, N** Detached valves seen from outside and inside. (Hill 1933)

excepted) is 2-locular (Fig. 26G) but *C. quadrilocularis* Chandler from the Eocene was 4-locular; Palaeocene *Nyssa* fruits were 3–4-locular, Recent are 1(rarely 2)-chambered (Fig. 26A–E; Eyde and Barghoorn 1963), except for the recently discovered Central American *N. talamancana*, which is a 2–3-chambered leftover. Fossil mastixioids had fruits with up to four chambers, each provided with a sulcus surrounded by a germination valve, and a protrusion of the fruit stone into the locule. In Recent mastixioids, *Davidia* excepting, the plurilocular condition has vanished altogether. Recent *Mastixia* and *Diplopanax* have a deeply indented stone, with a C- or U-shaped seed chamber, which is not found in *Nyssa* or *Cornus*.

Fruit wall histology is very similar in *Nyssa*, *Mastixia*, *Curtisia*, *Cornus* and *Alangium* (Reidt 1997); a striking feature is the "giant" cells of the mesocarp (amounting to 800 μm long) of *Mastixia*, *Alangium* and some *Cornus*.

The seeds have copious endosperm, which in *Alangium* is oily and contains hemicellulose. The embryo is small (*Mastixia*) to large and, in *Alangium*, has foliaceous cotyledons and an elongated hypocotyl.

POLLINATION. *Cornus* seems to have a generalized pollination system; beetles, bees, butterflies, and flies having been observed to be attracted by their flowers (Eyde 1988).

SEED DISPERSAL. Both blue and red-fruited *Cornus* attract many kinds of bird, especially thrushes (Eyde 1988). *Nyssa aquatica* and the dry-site ecotypes of *N. sylvatica* have juicy fruits and are dispersed by many kinds of bird, including white-throated sparrow, catbird, brown thrasher, thrushes, and scarlet tanager. Also many tetrapods play an important role in fruit dispersal; bears, raccoons, foxes, and opossums are known to eat the fruits. Rodents gnaw the stones, and toothmarks on fossils show that they did so in the past (Eyde 1997). The swamp dwellers, *N. sylvatica* subsp. *biflora*, *N. aquatica*, and *N. ogeche*, seem to be efficiently water-dispersed, *N. aquatica* having been observed floating for a period of 85 days. Less is known about dispersal of the Asian species, in which birds certainly play an important role (Eyde 1997).

It is obvious that Cretaceous mastixioid and *Cornus* fruits were considerably smaller than their Tertiary relatives (Eyde 1988). This supports the notion that the shift from abiotic to mammal-mediated dispersal in the Palaeogene may have been causal for an increase of the fruits in fleshiness and mechanical and chemical protection, the

and fossil *Amersinia* they are confined to the apical portion of the locule. Fruit stone traits of the extant genera are well differentiated, but lessen when they are traced back in time. Cretaceous and Palaeogene Cornaceae had fruits with more chambers than presently; Recent *Cornus* (the 3-locular *C. oblonga* latter including resin, latex, iridoids and the alkaloids based on them.

PHYTOCHEMISTRY. Among the phenolic compounds, ellagic acid, gallic acid, and the tannins based upon them are reported for *Davidia, Nyssa, Camptotheca, Cornus* and *Mastixia* but not for *Alangium*. Iridoids are strongly diversified, comprising normal iridoids and secoiridoids of route I sensu Jensen (1991), and are known from *Davidia, Mastixia, Nyssa, Alangium*, and *Cornus*, cornin being the most widespread compound. The alkaloids of *Camptotheca* (camptothecin) and *Alangium* (alangiside and tubulosin) are based on secoiridoids (Hegnauer 1964, 1989; Bate-Smith et al. 1975). In the species of the blue-fruited line of *Cornus* including the two alternate-leaved species, iridoids are lacking and obviously are replaced by salidroside and its oxidised congener, whereas red-fruited cornelian cherries and showy-bracted dogwoods contain iridoids and secoiridoids (Jensen et al. 1975a).

PHYLOGENY. Relationships and circumscription of this group have always been problematic; progress in this field is largely due to the efforts of Eyde (1968b, 1988) and the results of molecular studies. On the basis of morphological evidence, three groups of genera have been recognised: the nyssoids (*Davidia, Nyssa*, and *Camptotheca*); the mastixioids (*Mastixia* and *Diplopanax*); and the cornels (*Cornus*); *Alangium*, which possesses many traits in common with *Cornus*, has been accepted as cornaceous since the time of Hooker (1867) and Harms (1897), but not so by Eyde (1988).

With increasing knowledge of their floral, pollen and general morphology, genera formerly included in Cornaceae, such as *Helwingia, Kaliphora, Melanophylla, Corokia, Garrya, Aucuba* and *Griselinia*, have been excluded from them upon morphological arguments, as exposed by Eyde (1988: 309 seq.). Since most of these genera contain iridoids (Jensen et al. 1975b), but none ellagic acid or the tannins based upon it (Bate-Smith et al. 1975), and since otherwise they are similar to Cornaceae in floral and embryological traits, they had been retained in Cornales or their surroundings (Dahlgren 1980, 1989; Takhtajan 1987). Only with the application of molecular techniques (e.g. Xiang et al. 1993, 1997, 1998) has it been possible to determine their true affinity, and now they are placed in two orders, Garryales and Apiales, forming part of two different, major clades (Euasterids I and II; see APG 1998).

Several detailed molecular studies, including the most recent by Xiang et al. (2002), show Cornaceae as the sister-group of Hydrangeaceae + Loasaceae within a re-modelled order Cornales. In the circumscription adopted in this volume, the family Cornaceae comprises two well-supported clades, a *Cornus-Alangium* clade, and a nyssoid-mastixioid clade with *Nyssa, Davidia, Camptotheca, Mastixia* and *Diplopanax*. In some molecular analyses, the latter grouping dissociates into the pairs *Nyssa/Camptotheca* and *Mastixia/Diplopanax*, with *Davidia* in a vacillating position. In the desire for basing families only on clades that have broad statistic support, Xiang et al. (2002) suggest three digeneric families, Cornaceae, Nyssaceae, and Mastixiaceae, and the monogeneric Davidiaceae. Accordingly, they put into question the phylogenetic value of the peculiar gynoecial vasculature, the germination valves in the fruit stones, and the H-shaped endoapertures.

Here I prefer a broader circumscription of the family that is based on the combination of epigyny, fruit stones with one seed per chamber dehiscing with valves, and the chromosome base number x = 11. Transseptal bundles, which are an attribute of many inferior ovaries (Eyde 1988), may be less decisive, and H-shaped endoapertures, also found in *Curtisia* but perhaps not in nyssoids, may be a cornalean plesiomorphy. That *Diplopanax* would go with *Mastixia* in the molecular analysis could be expected from the outset, whereas the close association of *Cornus* with *Alangium* expressed in all sequence analyses is the clarification of a contentious issue of long standing. Now it becomes clear why the flowers, fruits and ovules of *Alangium* are much like those of *Cornus*, in spite of the substantial differences between the two genera: Hooker (1867) and Harms (1897) had the right instinct in including the former in their Cornaceae.

It is remarkable that this circumscription of the Cornaceae coincides exactly with Takhtajan's (1997) morphologically based concept of his order Cornales, except for his inclusion of *Curtisia*.

FOSSILS AND DISTRIBUTIONAL HISTORY. The oldest cornaceous fossils are mastixioid fruit stones from the Upper Cretaceous and Palaeogene of Europe, where the "Mastixioidean Flora" seems to have originated. Most of these fruit stones are traversed by resin ducts and each locule opens with a separate valve. Various form genera have been based on these fruit remains; some of the earliest were *Eomastixia* Chandl. and *Beckettia* Reid & Chandl., both of which were 2–4-locular but still lacked secretory ducts. Later forms included the 1-locular *Mastixiopsis* Kirchh., *Retinomastixia* Kirchh., *Mastixicarpum* Chandl.

and *Tectocarya* Kirchh. and, although Mai (1993) suggested they could easily be included into a broadly construed genus *Mastixia*, they may exhibit different affinities to extant cornaceous genera. Thus, the fruit stones of extinct *Mastixiopsis* had the diagnostic infold of mastixioids but on the surface had vertical ribs and intervening bundles of modern *Nyssa sinensis* and *N. sylvatica*. *Mastixicarpum* is considered as congeneric with the Recent *Diplopanax*.

In Europe the record of the Mastixioidean Flora extends from the Upper Cretaceous into the Pliocene; in North America these fossils are found in the Palaeocene and Eocene (Mai 1993). From Asia, where the two extant genera live today, no fossils are known to me.

Among the Nyssoids, a fossil species of the genus *Davidia*, which is still alive in eastern Asia, is known from the Palaeocene of mid-latitude North America and eastern Russia (Manchester 2002); the fruits of the fossil species (*D. antiqua*) were definitely smaller than those of the extant species. The extinct genus *Amersinia* Manchester, Crane & Golovneva (Manchester et al. 1999) from the Palaeocene of North America and East Asia had fruit characters reminiscent of extant *Camptotheca*, whereas associated leaves described as *Beringiaphyllum* Manchester, Crane and Golovnea resemble those of extant *Davidia*. The erstwhile and present geographic restriction of *Davidia* and *Amersinia* to East Asia and North America may indicate that these plants dispersed across Beringia in the late Cretaceous or early Tertiary.

Nyssa has an extensive fossil record which can be attributed to the preference of its species for swampy and lakeside habitats, where chances for fossilisation were high. Although today disjunct between eastern Asia, eastern North America and Central America, during the Tertiary the genus was widespread in the Northern Hemisphere. The oldest fossils of *Nyssa* appeared in the lower Eocene of Europe, but the genus disappeared from this continent during the Pliocene. The first records in North America are from the middle Eocene, whereas in Asia *Nyssa* appeared only in the Miocene (see also Wen and Stuessy 1993). Three types of *Nyssa* endocarp sculpture can be distinguished in the fossil material from the Eocene of Europe and North America, and indicate that considerable diversification of the genus had been attained at that time. These types include ridged fruits with sunken bundles, ridged fruits with raised bundles (today represented by *N. aquatica*), and smooth fruits (today exemplified by *N. javanica*) (Eyde 1997).

In *Cornus*, the blue-fruited line is known since the Upper Cretaceous, whereas the red-fruited line appeared only in the Tertiary. Early *Cornus* species had more than 2-locular fruits, such as *C. quadrilocularis* from the Eocene. The big-bracted dogwoods with separate fruits, which today are purely American, were present in Europe during the mid-Oligocene. Their compound-fruited descendents (Fig. 29), surviving in East Asia, are known from Pliocene beds from western Europe and from the Pleistocene of Japan (Eyde 1985).

Alangium has a fossil record based on fruits and pollen (Eyde et al. 1969; Reitsma 1970; Morley 1982). In the Upper Cretaceous restricted to North America, by the Eocene *Alangium* had extended until East Asia along the northern shore of the Sea of Tethys. From there it seems to have dispersed to Africa/Madagascar. During the Neogene, its main diversification took place in India and Southeast Asia; by the Pliocene, *Alangium* had retreated to its present range.

KEY TO THE GENERA

1. Leaf margin serrate or dentate; perianth 0 or indistinct; ovary 6–10-locular   **3. *Davidia***
- Leaf margin entire (dentate in capitate-flowered *Camptotheca*); ovary 1–3-locular   2
2. Two-armed 1-cellular hairs + on leaves and flowers; petals ± valvate   3
- Two-armed 1-cellular hairs 0 (but see *Alangium grisolleoides*); petals ± imbricate   5
3. Two-armed hairs calcified; petals not inflexed at apex; resin ducts 0; style with capitate or truncate stigma; fruit 2(3)-locular   **6. *Cornus***
- Two-armed hairs not calcified; petals inflexed at apex; resin ducts + in vegetative parts, flowers and fruits; style with punctiform or rarely bifid or lobed stigma; fruit 1-locular   4
4. Inflorescence thyrso-paniculate; epicarp soft; fruit stone sulcate   **4. *Mastixia***
- Inflorescence ± spicate; epicarp hard; fruit stone not sulcate   **5. *Diplopanax***
5. Petals pronouncedly strap-shaped; stamens 4–40, in a single whorl; articulated laticifers present in various tissues   **7. *Alangium***
- Petals not strap-shaped; stamens 8–15, in 2 isomerous whorls; laticifers lacking   6
6. Pistillate and staminate inflorescences in globular heads of ca. 50 flowers, these compound in racemes; fruits dry, flattened, grouped in heads   **2. *Camptotheca***
- Pistillate flowers solitary, staminate in racemes or heads of up to 20 flowers; fruits drupaceous with fleshy epicarp   **1. *Nyssa***

GENERA OF CORNACEAE

1. *Nyssa* L.   Fig. 26A–F

Nyssa L., Sp. Pl.: 1058 (1753); Burckhalter, Sida 15: 323–342 (1992) (rev. N Am. spp.).

Polygamodioecious deciduous trees. Leaves alternate. Inflorescences axillary, pedunculate; staminate flowers in short racemes or heads, pistillate flowers single or in few-flowered racemes; staminate flowers: calyx lobes minute; petals 5–10, imbricate; stamens 8–15, usually in 2 distinct whorls; disk pulvinate; gynoecium abortive; hermaphroditic or pistillate flowers: floral tube adnate to ovary; stamens fewer than in staminate flowers, or abortive, or 0; disk pulvinate or conical; ovary 1(–3)-locular; style simple or with 2–3 stylar branches; stigma reflexed or revolute. Drupe ovoid or ellipsoid, 1(–3)-chambered, blue-black, or more rarely purple or red; endocarp with longitudinal ridges or wings, opening in apical part by a triangular abaxial valve; embryo straight; endosperm copious. $n = 22$. About 8 species, 3 in SE North America, 1 in Costa Rica, and 4 in East and Southeast Asia; preferably on flooded ground.

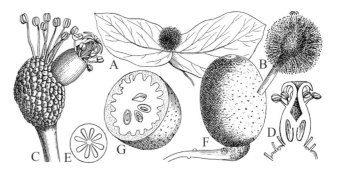

Fig. 27. Cornaceae. *Davidia involucrata.* **A** Inflorescence subtended by showy bracts. **B** Same, without bracts. **C** Same, with most male flowers removed, the single female flower to the right. **D** Female flower, vertical section. **E** Ovary, transverse section. **F** Fruit. **G** Same, transverse section. (Schneider 1912)

2. *Camptotheca* Dcne.

Camptotheca Dcne., Bull. Soc. Bot. France 20: 157 (1873).

Monoecious deciduous tree. Leaves alternate, usually entire but occasionally dentate. Inflorescences of globular heads grouped in a raceme, the heads composed of up to 50 flowers; flowers in dyads, 5-merous, in terminal head hermaphroditic, in lateral heads staminate; petals overlapping little if at all; stamens 5 + 5, inserted on disk; ovary (0 in staminate flowers) 1-locular; style immersed in disk, halfway split into 2 style branches. Fruits grouped in heads, flattened-spatulate, crowned by persistent disk and style, epicarp fibrous, endocarp leathery, 1-seeded; embryo elongated, endosperm fleshy. $n = 22$. A single sp., *C. acuminata* Dcne., C, S, and SE China, in lowland warm-temperate mixed mesophytic forest.

3. *Davidia* Baillon   Fig. 27

Davidia Baillon, Adansonia 10: 114 (1871).

Deciduous, andromonoecious tree. Leaves alternate, serrate. Inflorescences on short lateral shoots, in solitary globose heads subtended by 2(3) large, creamy-white subopposite spreading bracts; staminate flowers apetalous, numerous, densely packed on globose axis, 1–7-staminate; hermaphroditic or pistillate flowers inserted at top of head, with calyx adnate to ovary; stamens, if present, 15–26 arranged around the base of style; style short and conical, with 6–10 spreading style branches; ovary 6–10-locular; nectary disk 0.

Drupe 3–5-seeded, ellipsoid, mesocarp fleshy, endocarp bony, longitudinally ca. 12-ridged and 6-valved; seeds with fleshy endosperm and straight embryo subequal to endosperm. $n = 21$. A single sp., *D. involucrata* Baill., C China, in montane mixed mesophytic forest.

4. *Mastixia* Bl.   Fig. 28

Mastixia Bl., Bijdr.: 654 (1826); Matthew, Blumea 23: 51–93 (1976), rev.

Resinous evergreen trees with 2-armed hairs. Leaves alternate or (sub)opposite. Inflorescences thyrso-paniculate, many-flowered; flowers hermaphroditic, 4–5(–7)-merous; calyx toothed or lobed; petals usually valvate in bud (contorted), thick, inflexed at apex and bidentate or fimbriate; stamens 1 or 2 whorls inserted at base of nectary disk; ovary 1-locular; ovule 1, pendulous; style short, thick, with punctiform or rarely 2-lobed stigma. Drupes with thin or thick epicarp, blue to dark purple at maturity; endocarp woody, sulcate with an intrusive, incomplete, wedge-shaped (subgen. *Mastixia*) or swollen (subgen. *Manglesia*) septum on the germination valve; seeds with small embryo in copious endosperm. $n = 11, 13$. About 19 spp. or more (Matthew 1976); extending from India through warm-temperate Southeast Asia and Malesia to New Guinea and the Solomon Is. The status of subgen. *Manglesia* and its relationship to *Diplopanax* requires further study.

5. *Diplopanax* Hand.-Mazz.

Diplopanax Hand.-Mazz., Sinensia 3: 197 (1933); Ying et al., The endemic genera of seed plants of China: 142, fig. 36.1 (1993); Averyanov & Nguyen, Novon 12: 433–436 (2002).

Fig. 28. Cornaceae. *Mastixia pentandra.* **A** Flowering twig. **B** Part of inflorescence. **C** Flower, vertical section. **D** Stamens. **E** Fruit. Drawn by R. van Crevel. (Matthew 1976)

Fig. 29. Cornaceae. *Cornus japonica,* flowering branch and coenocarp. (Big-bracted dogwoods with coenocarps in the Tertiary extended to Europe, but today are restricted to E Asia.) Drawn by T. Yamada. (Nakai 1927)

Resinous deciduous tree with 2-armed hairs. Leaves subverticillate. Inflorescence spicate(-paniculate); flowers sessile, hermaphroditic, 5-merous; sepals short; petals valvate in bud, inflexed at apex; stamens 5 + 5; ovary 1-locular, 1-ovulate; style conical. Drupe (oblong-)ovoid; exocarp hard; endocarp woody, with an intrusive, swollen, incomplete septum on the germination valve; seed 1, compressed. Two species in mountain forests of S China and Vietnam.

6. *Cornus* L.   Figs. 26G–I, 29

Cornus L., Sp. Pl.: 117 (1753).
Chamaepericlymenum Hill (1756).
Swida Opiz (1838).
Benthamidia Spach (1839).
Afrocrania (Harms) Hutch (1942).
Dendrobenthamia Hutch. (1942).
Bothrocaryum (Koehne) Pojarkova (1950).
Yinquania Zhu (1984).

Evergreen or deciduous trees or shrubs, or rarely rhizomatous herbs, with 2-armed, calcified trichomes. Leaves usually opposite (alternate in *C. alternifolia* and *C. controversa*). Inflorescences terminal or axillary and thyrso-paniculate, or in heads surrounded by petal-like involucral bracts; flowers hermaphroditic, rarely dioecious, tetramerous; calyx distinct or obsolete; petals valvate; stamens alternipetalous; disk fleshy; anthers dorsifixed; style with capitate or truncate stigma; ovary (1)2(3)-locular. Drupe areolate at apex; pericarp fleshy, blue, white, or red; endocarp bony or crustaceous, 2(1)-locular, opening with valves. Seeds oblong, compressed, with elongate embryo and fleshy endosperm. $n = 9$–11. About 60 spp., mainly of temperate regions of the northern hemisphere, 1 or 2 spp. in South America, 1 in Africa.

The infrageneric classification is complex and controversial; no formal subdivision is given here. However, the molecular data available support the distinction of five groups in two major branches.

The first major branch comprises the blue- or white-fruited dogwoods with 1) alternate-leaved and 2) opposite-leaved *Cornus*. The second major branch represents the red-fruited dogwoods with 3) the cornelian cherries, 4) the dwarf dogwoods, and 5) the big-bracted dogwoods.

7. *Alangium* Lamk.   Fig. 30

Alangium Lamk., Encycl. Méth., Bot. 1: 174 (1783), nom. cons.;
   Bloembergen, Bull. Jard. Bot. Buitenzorg III, 16: 139–235
   (1939), monogr.

Evergreen (*A. platanifolium* summergreen) trees, shrubs, or lianas, sometimes thorny, with stellate and glandular, rarely 2-armed (*A. grisolleoides*) hairs with unequal arms; articulated laticifers in various tissues of the shoot. Leaves alternate, undivided or lobed. Inflorescences axillary cymes; flowers with articulated pedicels, hermaphroditic or rarely (*A. grisolleoides*) dioeciously distributed; calyx with 4–10 lobes or teeth, or truncate; petals 4–10, strap-shaped, valvate or rarely contorted, sometimes connate at the base; stamens 4–40 in a single cycle around a conspicuous nectary disk; filaments barbate or villose at inner side; anthers linear, basifixed or dorsifixed and versatile and widely sagittate; style elongate, filiform, with clavate, 2–4-lobed or capitate stigma or rarely style short and with 2 elongate stylodia; ovary 1(2)-locular. Drupe crowned with calyx and disk, with crustaceous or woody endocarp, 1(2)-locular, each locule with 1 seed or one locule empty; seeds with large, straight embryo; endosperm copious, oily, also containing hemicellulose. $n = 8, 11$. About 21 spp., Africa, Madagascar, tropical and warm-temperate S and SE Asia from India to Japan, and extending further through Malesia to Queensland and Fiji. Four sections distinguished by Bloembergen (1939) and Eyde (1968a); sect. *Conostigma* probably basal, position of *A. grisolleoides* Capuron from Madagascar still problematic (Eyde 1988). Flowering in the dry season, fruiting at the end of the dry season and often in the following wet season.

Selected Bibliography

Adams, J.E. 1949. Studies in the comparative anatomy of the Cornaceae. J. Elisha Mitchell Soc. 65: 218–244.
APG (Angiosperm Phylogeny Group) 1998. See general references.
Bate-Smith, E.C., Ferguson, I.K., Hutson, K., Jensen, S.R., Nielsen, B.J., Swain, T. 1975. Phytochemical interrelationships in the Cornaceae. Biochem. Syst. Ecol. 3: 79–89.

Fig. 30. Cornaceae. *Alangium chinense*. **A** Flowering branchlet. **B** Flower. **C** Corolla and stamens. **D** Gynoecium with disk. **E** Fruit. **F** Transverse section through fruit. **G** Fruit, longitudinal section. (Cannon 1978)

Bloembergen, S. 1939. A revision of the genus *Alangium*. Bull. Jard. Bot. Buitenzorg III, 6: 139–235.
Cannon, J.F.M. 1978. Alangiaceae. In: Launert, E. (ed.) Flora Zambesiaca vol. 4. London: Flora Zambesiaca Managing Committee, pp. 633–635.
Dahlgren, R.M.T. 1980. A revised system of classification of the angiosperms. Bot. J. Linn. Soc. 80: 91–124.
Dahlgren, G. 1989. The last Dahlgrenogram. System of classification of the dicotyledons. In: Kit Tan (ed.) Plant taxonomy. The Davis and Hedge Festschrift. Edinburgh: Edinburgh University Press, pp. 249–260.
Eramijan, E.M. 1971. Palynological data on the systematics and phylogeny of Cornaceae Dumort. and related families. In: Kuprianova, L.A., Yakovlev, M.S. (eds.) Pollen morphology of Cucurbitaceae, Thymelaeaceae, Cornaceae. Leningrad: Nauka, pp. 235–273.
Eyde, R.H. 1966. The Nyssaceae of the southeastern United States. J. Arnold Arbor. 47: 117–125.
Eyde, R.H. 1968a. Flowers, fruits, and phylogeny of Alangiaceae. J. Arnold Arbor. 49: 167–192.

Eyde, R.H. 1968b ('1967'). The peculiar gynoecial vasculature of Cornaceae and its systematic significance. Phytomorphology 17: 172–182.

Eyde, R.H. 1972. Pollen of *Alangium*: toward a more satisfactory synthesis. Taxon 21: 471–477.

Eyde, R.H. 1985. The case for monkey-mediated evolution by big-bracted dogwoods. Arnoldia 45: 2–9.

Eyde, R.H. 1988. Comprehending *Cornus*: puzzles and progress in the systematics of the dogwoods. Bot. Rev. 54: 233–351.

Eyde, R.H. 1997. Fossil record and ecology of *Nyssa* (Cornaceae). Bot. Rev. 63: 97–123.

Eyde, R.H., Barghoorn, E.S. 1963. Morphological and paleobotanical studies of the Nyssaceae. II. The fossil record. J. Arnold Arbor. 44: 328–376.

Eyde, R.H., Xiang, Q.-Y. 1990. Fossil mastixioid (Cornaceae) alive in eastern Asia. Am. J. Bot. 77: 689–692.

Eyde, R.H., Bartlett, A., Barghoorn, E.S. 1969. Fossil record of *Alangium*. Bull. Torrey Bot. Club 96: 288–314.

Fan, C., Xiang, Q.-Y. 2001. Phylogenetic relationships within *Cornus* (Cornaceae) based on 26S rDNA sequences. Am. J. Bot. 88: 1131–1138.

Ferguson, I.K. 1966. The Cornaceae in the southeastern United States. J. Arnold Arbor. 47: 106–116.

Ferguson, I.K. 1977. Cornaceae. World Pollen and Spore Flora 6. Stockholm: Almqvist & Wiksell.

Ferguson, I.K., Hideux, M.J. 1978. Some aspects of the pollen morphology and its taxonomic significance in Cornaceae sens. lat. In: 4th Int. Palyn. Conf., Lucknow, 1976–1977, 1, pp. 240–249.

Goldblatt, P. 1978. A contribution to cytology in Cornales. Ann. Missouri Bot. Gard. 65: 650–655.

Harms, H. 1897. Cornaceae. In: Engler, A., Prantl, K. Die natürlichen Pflanzenfamilien III, 8. Leipzig: W. Engelmann, pp. 250–270.

Hatta, H., Honda, H., Fisher, J.B. 1999. Branching principles governing the architecture of *Cornus kousa* (Cornaceae). Ann. Bot. 84: 183–193.

Hegnauer, R. 1964, 1989. See general references.

Hill, A.W. 1933. The method of germination of seeds enclosed in a stony endocarp. Ann. Bot. 47: 873–887.

Hook, D.D., Brown, C.L., Kormanik, P.P. 1970. Lenticel and water root development of swamp tupelo under various flooding conditions. Bot. Gaz. 131: 217–224.

Hooker, J.D., 1867. Cornaceae. In: Bentham, G., Hooker, J.D., Genera Plantarum, vol. 1 (3). London: Reeve & Co, pp. 947–952.

Horne, A.S. 1909. The structure and affinities of *Davidia incolucrata*, Baill. Trans. Linn. Soc. Lond. II, 7: 303–326, pl. 31–33.

Jahnke, C. 1986. Der Infloreszenzbau der Cornaceen sensu lato und seine systematischen Konsequenzen. Trop. subtrop. Pflanzenwelt 57. Akad. Wiss. Lit. Mainz. Stuttgart: Steiner.

Jensen, S.R. 1991. Plant iridoids, their biosynthesis and distribution in angiosperms. In: Harborne, J.B., Tomas-Barberan, F.A. (eds.) Ecological chemistry and biochemistry of plant terpenoids. Oxford: Clarendon Press, pp. 133–158.

Jensen, S.R., Kjaer, A., Nielsen, B.J. 1975a. The genus *Cornus*: non-flavonoid glucosides as taxonomic markers. Biochem. Syst. Ecol. 3: 75–78.

Jensen, S.R., Nielsen, B.J., Dahlgren, R. 1975b. Iridoid compounds, their occurrence and systematic importance in the angiosperms. Bot. Not. 128: 148–180.

Johri, B.M. et al. 1992. See general references.

Kostermans, A.J.G.H. 1982. The genus Mastixia Bl. (Cornaceae) in Ceylon. Reinwardtia 10: 81–92.

Mai, D. 1993. On the extinct Mastixiaceae (Cornales) in Europe. Geophytology 23: 53–63.

Manchester, S.R. 2002. Leaves and fruits of *Davidia* (Cornales) from the Paleocene of North America. Syst. Bot. 27: 368–382.

Manchester, S.R., Crane, P.R., Golovnea, L.B. 1999. An extinct genus with affinities to extant *Davidia* and *Camptotheca* (Cornales) from the Paleocene of North America and eastern Asia. Int. J. Plant Sci. 160: 188–207.

Matthew, K.M. 1976. A revision of the genus *Mastixia* (Cornaceae). Blumea 23: 51–93.

Morley, R.J. 1982. Fossil pollen attributable to *Alangium* Lamarck (Alangiaceae) from the Tertiary of Malesia. Rev. Palaeobot. Palynol. 36: 65–94.

Murrell, Z.E. 1993. Phylogenetic relationships in *Cornus* (Cornaceae). Syst. Bot. 18: 469–495.

Nakai, T. 1927. Flora Sylvatica Koreana, part XVI, Araliaceae & Cornaceae. Seoul: Forestal Exp. Station.

Noshiro, S., Baas, P. 1998. Systematic wood anatomy of Cornaceae and allies. IAWA J. 19: 43–97.

Reidt, G. 1997. Zur Anatomie der Früchte ausgewählter Cornaceen und einiger möglicher verwandter Gattungen. Inaugural-Dissertation, Nat.-Math. Fakultät, Ruprecht-Karls-Universität Heidelberg.

Reitsma, T. 1970. Pollen morphology of the Alangiaceae. Rev. Palaeobot. Palynol. 10: 249–332.

Schneider, C.K. 1912. Illustriertes Handbuch der Laubholzkunde, Vol. 2. Jena: G. Fischer.

Sertorius, A. 1893. Beiträge zur Kenntnis der Anatomie der Cornaceae. Bull. Herb. Boisier 1: 469–484, 496–512, 551–570, 614–639.

Sohma, K. 1963. Pollen morphology of the Nyssaceae, I. *Nyssa* and *Camptotheca*. Sci. Rep. Tohoku Univ. IV (Biol.) 29: 389–392.

Sohma, K. 1967. Pollen morphology of the Nyssaceae, II. *Nyssa* and *Davidia*. Sci. Rep. Tohoku Univ. IV (Biol.) 33: 527–532.

Takhtajan, A. 1980. Systema Magnoliophytorum. Leningrad: Nauka (in Russian).

Takhtajan, A. 1997. See general references.

Tandon, S.R., Herr, J.M. Jr. 1971. Embryological features of taxonomic significance in the genus *Nyssa*. Can. J. Bot. 49: 505–514.

Wen, J., Stuessy, T.F. 1993. The phylogeny and biogeography of *Nyssa* (Cornaceae). Syst. Bot. 18: 68–79.

Xiang, Q.-Y. 1999. Systematic affinities of Grubbiaceae and Hydrostachyaceae within Cornales – insights from *rbc*L sequences. Harvard Pap. Bot. 4: 527–542.

Xiang, Q.Y., Soltis, D.E., Morgan, D.R., Soltis, P.S. 1993. Phylogenetic relationships of *Cornus* L. sensu lato and putative relatives inferred from *rbc*L sequence data. Ann. Mo. Bot. Gard. 80: 723–734.

Xiang, Q.-Y., Brunsfeld, S.J., Soltis, D.E., Soltis, P.S. 1997. Phylogenetic relationships in *Cornus* based on chloroplast DNA restriction sites: implications for biogeography and character evolution. Syst. Bot. 21: 515–534.

Xiang, Q.-Y., Soltis, D.E., Soltis, P.S. 1998. Phylogenetic relationships of Cornaceae and close relatives inferred from *mat*K and *rbc*L sequences. Am. J. Bot. 85: 285–297.

Xiang, Q.-Y., Moody, M.L., Soltis, D.E., Fan, C.-Z., Soltis, P.S. 2002. Relationships within Cornales and circumscription of Cornaceae – *mat*K and *rbc*L sequence data and effects of outgroups and long branches. Molec. Phylog. Evol. 24: 35–57.

Cunoniaceae

J.C. Bradford, H.C. Fortune Hopkins and R.W. Barnes

Cunoniaceae R.Br. in Flinders, Voy. Terra Austral. 2: 548 (1814), nom. cons.
Baueraceae Lindl. (1830).
Eucryphiaceae Endl. (1841), nom. cons.
Davidsoniaceae. (1952).

Trees, shrubs, occasionally hemiepiphytic or strangling (some *Weinmannia*); hairs generally simple, sometimes trichomes stellate and peltate, long, stiff and irritating in *Davidsonia*; bark usually grey to light brown; fissures longitudinal; stems commonly with lenticels. Leaves evergreen, winter deciduous in *Eucryphia glutinosa*; opposite-decussate in pairs or sometimes 3 or 4 leaves per node (spiral in *Davidsonia*); imparipinnate, trifoliolate, palmately compound, or unifoliolate, pinnately veined, stipellate or not, firm, often coriaceous; leaf margins toothed, often glandular-serrate, sometimes entire; venation craspedodromous, semicraspedodromous, occasionally brochidodromous or reticulodromous (some *Eucryphia*); small tuft or pocket domatia along midvein sometimes present; stipules often conspicuous, often interpetiolar. Inflorescences terminal, axillary, sometimes cauliflorous, paniculate, thrysoid or cymose and with straight peduncle, or capitate, rarely flowers solitary and axillary; bracts usually stipulate; flowers often with prophylls; pedicels present or sometimes 0. Flowers actinomorphic, occasionally protandrous, commonly hermaphroditic, plants dioecious or polygamodioecious in *Pancheria*, *Vesselowskya*, and some *Weinmannia*, andromonoecious in some *Schizomeria* and some *Cunonia*; sepals (3)4 or 5(–10), valvate or imbricate, separate or basally connate; petals as many as sepals and alternate with them, sometimes 0, more numerous than the sepals in *Bauera*; androecium usually diplostemonous, sometimes uniseriate or multiseriate, rarely polyandric; filaments slender, exceeding petals; anthers dithecal, opening longitudinally, somewhat apically in *Bauera*; disc free, annular or composed of segments, less often adnate to the ovary or 0; ovary usually superior or partially to mostly inferior, usually 2- or 3–5-carpellate, 4–14-carpellate in *Eucryphia*, effectively 1-carpellate in *Hooglandia* with second carpel possibly vestigial, syncarpous or less often the carpels ± distinct; stylodia free, often diverging, with terminal, rarely decurrent stigmas; ovules (1)2–many in each locule, often in two rows, anatropous or hemianatropous; apotropous or epitropous, bitegmic; placentation axile to pendulous. Fruit dehiscent, follicular or capsular; seeds usually exposed from open carpel sutures, the carpels separating distally or (*Cunonia*) from the base, or fruit indehiscent, drupe-like, or dry with wing-like expanded sepals or carpel-walls, or carpels hairy, or carpels swollen and bladder-like (*Platylophus*). Seeds small, winged or not, glabrous or hairy (*Weinmannia* and *Ackama*); endosperm starchy, 0 in *Davidsonia*.

A family of 27 genera and c. 300 species, occurring in South and Meso America, Australia, Malesia, Madagascar, and islands of the Caribbean, South Pacific, and western Indian Ocean, only two species in southern Africa and no species in mainland Asia north of the Malay Peninsula.

Vegetative Morphology. All known species are potentially long-lived plants that flower repeatedly over many years. Nearly all species require moderate to high light intensity, and as a consequence most are canopy species, but life form varies depending on the habitat. In lowland or premontane rainforest, Cunoniaceae may be trees over 30 m tall and 1 m in diameter. In montane forest, subalpine zones, or on poor soils, plants are commonly small trees to rounded shrubs. Extremely windswept, cold or rocky conditions produce ground-creeping shrubs. In neotropical cloud forests, some *Weinmannia* grow as hemiepiphytes and may even develop into large stranglers (J. Bradford, pers. obs.; A. Gentry, pers. comm.). Many Cunoniaceae coppice readily, especially *Anodopetalum biglandulosum*, an endemic of Tasmania that forms dense tangles of stems called 'horizontal scrub' by sending vertical stems from bent and fallen layers of once vertical stems (Barker and Brown 1994). Branch nodes in contact with the ground can root and sprout shoots, enabling an individual genet to disperse and dominate an area. The growth form of *Schizomeria ovata*, from eastern Australia, differs depending on its habitat (Stanley and Ross 1983). In tall *Eucalyptus* woodlands, which generally have a light, open understorey, it grows as a spreading shrub, whereas in shorter-statured rainforests, which generally have little understorey light penetration, it grows as a canopy tree. On ultramafic substrates in New Caledonia, several species have tall, spindly growth forms with branching rare above the base.

In general, members of the family are easy to recognize based on vegetative characters such as dentate, decussately arranged leaves (sometimes whorled, or alternate in *Davidsonia*) and prominent stipules. Stipules may arise as a pair of primordia that remain separate throughout development, as pairs of primordia that fuse during growth, or as a single primordium (Rutishauser and Dickison 1989; Dickison and Rutishauser 1990). Mature stipular form depends upon the number of stipular primordia, their location, and degree of fusion during development. Most often stipules are interpetiolar or called lateral if in unfused pairs (Figs. 35, 37) but axillary stipules occur in South Pacific *Geissois* species. Leaves are imparipinnate, unifoli(ol)ate, trifoli(ol)ate, or palmately compound. Stipels are sometimes present. Tuft or pocket domatia may occur on the abaxial leaf surface near the midvein. Hairs are unbranched, variable in length, density and distribution on the plant, and may be straight or curled to various degrees. Stellate hairs are present on stems and sepals in *Gillbeea*. *Spiraeopsis* has stellate hairs and orbicular, gland-like trichomes on leaves, stems and sepals, and *Ackama* has orbicular, gland-like trichomes on leaves. *Davidsonia* has long, glandular, urticating hairs. Extra pairs of axillary buds occur in some taxa.

VEGETATIVE ANATOMY. (Refer to Ingle and Dadswell 1956; Eyde 1970; Dickison 1975b, 1978, 1980a, 1980b; Rao and Dickison 1985a, 1985b; Rutishauser and Dickison 1989; Dickison and Rutishauser 1990). Leaf venation is craspedodromous, semicraspedodromous, or less commonly brochidodromous and reticulodromous (some *Eucryphia*). Areolation ranges from incompletely closed to completely closed and well-developed. Major veins are often ensheathed with fibres. Ultimate veinlets often have specialized thick-walled, solitary or clustered, terminal elements. Elongated, tubular cells are a characteristic feature of the vein endings of *Codia*. Mature stomata can be anomocytic, brachyparacytic, encyclocytic or anisocytic, and are confined to the abaxial lamina surface. Extracuticular wax occurs in the form of flakes, platelets, or coarse rods or rodlets. Very prominent cuticular, papilla-like protrusions occur on the abaxial surface cells of *Eucryphia* (Hill 1991; Barnes and Jordan 2000). Epidermal cells often have periclinal divisions in addition to being enlarged and mucilaginous. Abaxial epidermal cells vary from smooth to papillose (e.g. *Acrophyllum*). An adaxial and sometimes abaxial hypodermis composed of thick-walled cells is present in some taxa. The mesophyll is tanniniferous and contains large solitary crystals and druses.

Young stem anatomy is characterized by a continuous ring of xylem. The secondary phloem includes groups of fibres. The cortex is sometimes provided with sclereids and often a ring of perivascular fibres. Nodal anatomy is trilacunar, 3-trace, multilacunar, multi-trace, or unilacunar, one-trace (*Bauera*). Species with whorled leaves typically possess 'split-lateral' traces, including the very uncommon condition of multilacunar, multi-trace nodes with split-lateral traces in *Pancheria confusa*, *P. hirsuta* and *P. robusta*.

Wood anatomy exhibits a wide range of variation. Vessel distribution is exclusively or predominantly solitary, or commonly a combination of solitaries, radial pore multiples, and clusters. Vessel element perforation plate type is either (1) exclusively scalariform (less than ten to over 30 bars), (2) predominantly simple but with variable frequencies of few-barred scalariform plates, or (3) exclusively simple (*Pseudoweinmannia*). Intervessel pitting ranges between scalariform and transitional to opposite and alternate. Spiral thickenings are present in the vessels of some *Eucryphia* species. Imperforate tracheary elements are tracheids or libriform fibres. Rays are heterocellular or heterocellular and homocellular, with both uniseriates and multiseriates. Axial parenchyma is variable, being absent, diffuse and diffuse-in-aggregates, or distributed in continuous or discontinuous bands (1–5 cells wide), sometimes terminal. Prismatic crystals are common.

INFLORESCENCE STRUCTURE. Inflorescence variation occurs in (1) the number and position of flower-bearing axes within the stem system, (2) the timing of floral maturation within a flower-bearing axis, and (3) the pattern and orders of branching within the flower-bearing axis.

Flower-bearing axes (e.g. racemes, panicles) may be borne in leaf axils, terminal on shoots, or both; cauliflory is occasional (Pacific *Geissois*, *Davidsonia*, some *Acsmithia*). Within a species or higher-level clade, the position and number of flower-bearing axes may be highly predictable or very plastic. For example, nearly all c. 75 species in *Weinmannia* sect. *Weinmannia* have racemes restricted to axils of distal leaves. In contrast, *Davidsonia* species may bear panicles from near the base of the trunk to the crown of the tree.

Bradford and Barnes (2001) found a significant variation in the timing of floral maturation within a flower-bearing axis among tribes and genera of Cunoniaceae. Floral maturation occurs either in terminal flowers first (i.e. basipetal), or more or less at the same time to slightly acropetally (i.e. synchronous/acropetal). Tribe Spiraeanthemeae appears to have the derived form of synchronous timing, whereas most outgroups and a basal grade of ingroup taxa, including tribe Schizomerieae, *Davidsonia*, *Hooglandia*, *Aistopetalum* and *Gillbeea*, have basipetalous flower sequence. A possible clade of four tribes, Geissoieae, Caldcluvieae, Codieae and Cunonieae, and the genus *Acrophyllum* have the synapomorphy of synchronous flower sequence.

In branched flower-bearing axes (e.g. paniculiform or thrysoid to cymiform axes), phyllotaxis is usually basally decussate and distally spiral but, depending upon the taxa, this transition may occur near the stem, not until the end of the axis, or not at all. The flower-bearing axis often consists of a dominant, primary shoot that bears secondary lateral axes that are longer and more branched basally, and shorter and less branched distally. This typically leads to a conical outline, which may be called paniculiform, and typifies tribe Spiraeanthemeae and the genera *Davidsonia*, *Hooglandia*, *Aistopetalum*, *Gillbeea*, *Ackama* and *Spiraeopsis*. In some genera, no dominant inflorescence axis or only a weakly dominant one exists. For example, in tribe Schizomerieae and the genera *Caldcluvia* and *Opocunonia* in tribe Caldcluvieae, there is no elongate main axis and, therefore, the inflorescence has a rounded shape. The degree of branching varies from as little as one order in taxa with small inflorescences, such as in *Acrophyllum*, to very large and profuse branching systems with up to five or six orders, as in *Aistopetalum*, *Ceratopetalum*, *Schizomeria* and *Spiraeopsis*. Within a tree, the branching order or profusion of inflorescence units may also differ among flower-bearing axes in relation to their position along the stem (Bradford 1998). For example, *Ackama* has more prolific panicles at terminal nodes than at subterminal ones (i.e. acrotonic total inflorescence) whereas in some species of *Weinmannia* sect. *Inspersae,* inflorescence axes are more branched at subterminal nodes (i.e. basitonic total inflorescence).

Tribes Geissoieae and Cunonieae have flowers borne along unbranched axes, mainly racemes. Flowers may arise either solitarily or in fascicles, and may be pedicellate or not. These racemose structures are borne directly from leaf axils or, more often, in groups from a lateral or central peduncle (i.e. Inflorescence Module, sensu Bradford 1998). There are several different patterns of raceme organization along Inflorescence Modules. For example, in some groups the Inflorescence Module only bears racemes laterally and produces a vegetative bud in the terminal position, whereas in other groups the Inflorescence Module bears a terminal and lateral racemes.

Three genera, *Pancheria* (Fig. 36), *Codia* and *Callicoma*, have flowers arranged in capitula. These are borne at the ends of slender stalks (peduncles), either directly from leaf axils, or peduncles are borne in pairs or series from an axillary shoot (i.e. within an Inflorescence Module). *Codia* and *Callicoma* are closely related to *Pullea*, which has paniculate flower-bearing axes and often clustered flowers (Bradford and Barnes 2001), whereas *Pancheria* is more closely related to *Cunonia* and *Weinmannia* with racemose flower-bearing axes (Bradford 2002).

The inflorescence in *Acrophyllum* is unusual, with the main axis alternating between producing a sequence of vegetative nodes with whorled leaves and a sequence of several reproductive nodes with axillary inflorescence structures (Fig. 34). At each reproductive node is a whorl of small, once-branched panicles bearing about a dozen flowers (Hoogland 1960).

Solitary flowers are found in *Eucryphia*, *Anodopetalum* and *Bauera*. In the former two genera, the flower is provided with a pedicel, bracts, and a peduncle suggesting that solitary flowers are derived from reduced cymes. In fact, three-flowered cymes rarely occur in *Anodopetalum* (Barnes and Rozefelds 2000). *Bauera* is the only genus in which pedicels arise directly from leaf axils.

FLORAL MORPHOLOGY AND ANATOMY. Floral morphology has been best described in the various treatments of genera (see references in taxonomic part), in the context of floral biology by Hopkins (see discussion re Pollination below) and in a comparative phylogenetic context by Bradford and Barnes (2001). Flowers typically have two perianth whorls, twice the number of stamens as petals (usually diplostemonous, sometimes obdiplostemonous; see Endress and Stumpf 1991), and a nectary surrounding a bicarpellate, superior ovary with terminal stylodia. Variations include the presence of only one perianth whorl, stamen number equal to petal number, numerous stamens, absence of nectary, a few to several carpels, a semi-inferior to inferior ovary, and decurrent stigmas. Perianth parts are distinct or only weakly fused. Petal or, when absent, sepal colour is usually white, yellow or pink, less often red to purple. Filaments usually exceed the perianth and are slender, cylindrical or strap-like. Anthers are dithecal, with or without an apically protruding connective, and open longitudinally or, rarely, somewhat apically (*Bauera*). The floral disc is most often distinct from the ovary but may be adnate to it. When distinct, it may be entire or composed of inter- and intrastaminal segments. The carpels are usually fused into a single ovary but always have free stylodia (only at the tip in *Gillbeea adenopetala*), locule number is equal to carpel number, except in the bicarpellate, unilocular *Hooglandia*; in some groups the carpels are more or less distinct.

Aspects of floral anatomy, including details of floral vascularization, were presented by Eyde (1970), Bensel and Palser (1975), Dickison (1975a, 1975c, 1978), and Matthews et al. (2001). Tannin and druses are common throughout all floral parts, with mucilage-producing cells localized in some tissues. Sepals usually have three-trace vasularization with fusion of lateral traces of adjacent sepals. Petals (when present) and stamens are vascularized by one trace, except for *Eucryphia* which has prominent stamen fascicle (trunk) bundles. Gynoecia are predominantly vascularized by three major bundles, a dorsal and a pair of ventral vascular bundles. The ventral bundles of adjacent (usually opposing) carpels may be fused in the ovary and separate into the stylodia of adjacent carpels. *Pullea* is the only genus studied in which ventral carpel vasculature terminates in the ovules and does not enter the stylodia. Ventral carpel sutures are open at, or slightly above, the level of placentation in most genera, or are closed throughout (*Aistopetalum*). Carpels of the essentially apocarpous genera *Acsmithia* and *Spiraeanthemum* (Fig. 31) are vascularized by a dorsal bundle and either three or four bundles constituting the ovular and wing vasculation in the ventral position. Ovules are (1)2–many in each locule along the axile to pendulous placenta. Ovules are anatropous or hemianatropous, and the micropyle is either apotropous (derived) or epitropous (plesiomorphic) (Bradford and Barnes 2001).

EMBRYOLOGY. Prakash and McAlister (1977) give a detailed account of the embryology of *Bauera*

capitata and compare it to previous studies in Cunoniaceae and Saxifragaceae in general (e.g. Davis 1966; Johri et al. 1992). Cunoniaceae anthers are tetrasporangiate with a glandular tapetum and tanniniferous endothecium. Pollen is 2-celled. Ovules are anatropous, crassinucellate, bitegmic, and the micropyle is zigzag in shape. The embryo sac contains abundant starch at maturity. Seeds have nuclear endosperm, lack haustoria, and have a coat formed by both integuments. Embryos are straight and usually embedded in abundant endosperm, but endosperm is lacking in *Davidsonia* (Doweld 1998).

POLLEN MORPHOLOGY. Pollen grains are small (10–13 μm), generally tricolporate (*Cunonia, Opocunonia, Spiraeanthemum, Weinmannia*) or dicolp(or)ate (*Ceratopetalum, Eucryphia, Geissois, Pullea, Schizomeria*), but syncolpate in *Gillbeea*. Grains vary in shape from suboblate to subprolate, generally with a circular equatorial outline, but sometimes dicolp(or)ate grains are bilaterally flattened (*Ceratopetalum, Schizomeria*). Only *Cunonia capensis* has oblate, angulaperturate grains which differ markedly from the normal suboblate grains with a circular outline found in the New Caledonian species of *Cunonia*.

Ectoapertures are colpate (except *Gillbeea*) or indistinct, as in *Pullea*. Endoapertures vary from indistinct or simply constructed (*Pullea, Ceratopetalum, Schizomeria*) to simple-complex/complex (*Weinmannia, Aistopetalum, Opocunonia, Spiaeopsis, Spiraeanthemum*). Tectum sculpture also varies from uniformly perforate (homogeneous) or finely reticulate (*Caldcluvia, Spiraeopsis, Opocunonia, Pullea*, some *Weinmannia*) to partial and coarsely reticulate (heterogeneous) in the centre of the mesocolpia (*Gillbeea, Ceratopetalum*, some *Schizomeria*) (Hideux and Ferguson 1976). In some genera the sculpture in the centre of the mesocolpia tends towards rugulate (*Schizomeria*, some *Eucryphia*).

KARYOLOGY. Published chromosome counts are summarised in Fortune Hopkins and Hoogland (2002) and include species from 9 genera. Most taxa have $2n = 32$ (*Ackama, Bauera, Ceratopetalum, Cunonia, Geissois* and *Lamanonia*). New Zealand species of *Weinmannia* have $2n = 30$, as does *Eucryphia lucida* (Labill.) Baill. *Pancheria sebertii* Guillaumin has $2n = 24$. More chromosome counts of species in tribe Cunonieae (*Vesselowskya, Weinmannia, Pancheria* and *Cunonia*) may be warranted since they appear variable and relationships in this tribe are not well understood (Bradford 2002).

FRUIT AND SEED. Dickison (1984) provides a review and description of most Cunoniaceae fruits and seeds. Corrections are made to previous descriptions of *Anodopetalum* fruits by Barnes and Rozefelds (2000). Bradford and Barnes (2001) discuss evolution of fruit morphology. For variation in carpel number, see discussion of flowers. Usually, all carpels develop into pericarps of mature fruits, although locule and seed abortion may occur in some indehiscent fruits.

Seventeen genera have capsular or multi-carpellate follicular fruits. Free carpels are probably the plesiomorphic condition in Cunoniaceae, being present in related families (Brunelliaceae and Cephalotaceae) and in the most basal tribe Spiraeanthemeae (see Bradford and Barnes 2001, their Fig. 5). Most other Cunoniaceae have a syncarpous gynoecium. Reversal to pseudo-follicular fruits has occurred in *Pancheria* (Cunonieae), but the carpels remain basally fused and their ventral margins are inrolled. Capsular fruits are similar to follicular ones, in that they open through their septa (usually dehiscing basipetally) to expose adaxial (ventral) surfaces of separate carpels that release seeds from open sutures. *Cunonia* has a unique circumbasal, often acropetal dehiscence, but is also septicidal. Each carpel may contain one to numerous small (c. 0.5–2.0 mm long) seeds. Dispersal is frequently facilitated by appendages of wings or hairs, but a few genera have ovoid seeds lacking prominent appendages.

Indehiscent fruits occur in ten genera, including members of tribes Schizomerieae, Geissoieae, Codieae, as well as *Davidsonia*, *Hooglandia*, *Aistopetalum* and *Gillbeea* (Bradford and Barnes 2001; Sweeney and Bradford, in prep.). Several genera have dry, indehiscent fruits, but each is different from another or, if similar, this is due to convergence. For example, *Ceratopetalum* (Schizomerieae) and *Pullea* (Codieae) have an enlarged calyx that persists as wings. In *Codia* (Codieae) and *Pseudoweinmannia* (Geissoieae) the exocarp is covered with hairs, but these are wavy and stiff in *Pseudoweinmannia* and wooly in *Codia*. In *Gillbeea* the carpel walls form wings as lateral phalanges. *Platylophus* (Schizomerieae) fruits lack appendages and develop into swollen bladders (utricles). Drupaceous fruits, with a range of exocarp toughness and mesocarp fleshiness, are found in *Davidsonia*, *Schizomeria*, *Aistopetalum* and *Hooglandia*. Doweld (1998) reports that *Davidsonia* fruits are dry and split into two carpels at maturity, but field observations indicate otherwise. Ripe, syncarpous *Davidsonia* fruits fall from the parent when still fleshy (not dry), and are eaten whole by vertebrates (see Dis-

tribution and Ecology). Indehiscent Cunoniaceae fruits typically have only one or two seeds per carpel. In *Hooglandia*, one seed develops in a unilocular ovary, and the fruit is elongate and flattened rather than orbicular as in other genera. *Davidsonia* and *Hooglandia* seeds are the largest in the family, being about 2 cm long.

Dickison (1984) describes seed morphology, seed coat development and mature seed anatomy of most Cunoniaceae, and Doweld (1998) describes *Davidsonia* seeds. Wing-like outgrowths are formed primarily by two epidermal layers and have variable sizes and shapes. Wings form opposite the hilum, or both chalazally and micropylarly with possible lateral extension. Seed hairs, found in *Weinmannia* and *Ackama*, are unicellular, thin-walled and often wavy. Seed surfaces may be warty, striate, smooth, papillate or a combination of these features. At early stages of development, the epidermis of the outer integument forms a tanniniferous exotesta in mature seeds. In most taxa, the outer epidermis of the inner integument gradually becomes sclerotized to produce exotegmic mature seeds. However, seed coats in *Pullea* and *Codia* are composed of only three cell layers and lack a mechanical tissue. A sclerotic layer is also absent in *Platylophus*, *Anodopetalum* and *Gillbeea*. Most taxa have a visible hilar scar, an elongate, raised raphe, and a raphal bundle that is usually unbranched or, in drupaceous fruits, well-branched near the chalaza. The large seeds of *Davidsonia* have an expanded pachychalaza layer.

FRUIT AND SEED DISPERSAL. Capsular or follicular fruits generally remain attached to the parent, and seeds disperse individually. Plants are rarely found with open fruits in which seeds remain. (Open fruits with seeds on herbarium specimens were likely collected when closed, and only opened during drying). In 1995, Bradford and Hopkins (pers. obs.) found a tree of *Weinmannia richii* A. Gray in Fiji that had open capsules with seeds. The seeds were found blowing out of the fruit and being carried high by a light breeze of hot, dry air. Seeds of *Weinmannia* and *Ackama* are covered with light hairs, whereas in most genera they have wing-like appendages and a few are simply ovoid. In contrast to the dispersal potential of *Weinmannia* seeds, *Acrophyllum*, which is a narrow endemic on specialized substrates, has unappendaged seeds that may inhibit dispersal.

Most indehiscent fruits fall from the parent, but some smaller, fleshy fruits may persist while ripe. Wind or water may disperse the fruits of *Pseudoweinmannia* and *Codia*, which are covered by hairs. Fruits with wing-like appendages, including laterally expanded carpel walls in *Gillbeea* and enlarged calyces on fruits in *Pullea* and *Ceratopetalum*, may also be wind dispersed. Dry bladders in *Platylophus* may be water dispersed, and consumption by wild pigs is reported (Palmer and Pitman 1961). Drupaceous fruits in *Aistopetalum*, *Davidsonia*, *Hooglandia* and *Schizomeria* are probably vertebrate dispersed. In SE Australia, *Schizomeria* fruits are eaten by pigeons (Waterhouse 2001), in New Guinea by fruit bats, arboreal marsupials and birds, including flightless cassowaries (Pratt 1983). *Davidsonia* fruits are eaten by cassowaries (Cooper and Cooper 1994). Most species probably require moderate to high light levels for successful establishment and are long-lived in the canopy (e.g. Lusk 1999).

PHYTOCHEMISTRY. Large amounts of aluminium and mucilage are typically found in the leaves, and the family is characterized by a richly developed phenyl propanoid metabolism. In the leaves and bark, both condensed and hydrolyzable tannins are prominent, the latter sometimes based on mono- and di-O-methylellagic acid. Among flavonol glycosides, those based on kaempferol and quercetin are widespread, whereas myricetin glycosides are found in *Ceratopetalum*, *Bauera*, *Davidsonia* and some *Weinmannia*, often in combination with prodelphinidin (Hegnauer 1964, 1989; Bate-Smith 1977). Wollenweber et al. (2000) found numerous flavonoids in the leaf and bud excretes of *Eucryphia*, some species of which also share the possession of azaleatin with other Cunoniaceae.

SUBDIVISION AND RELATIONSHIPS WITHIN THE FAMILY. Cladistic analyses of Cunoniaceae by Hufford and Dickison (1992) and Orozco (1997) used only morphology and could not resolve or provide strong support for many clades. With outgroups from other members of Oxalidales, improved morphological data, and especially DNA sequences, strong support for some intrafamilial relationships was provided by Bradford and Barnes (2001). Based on their phylogenetic analyses, six tribes were circumscribed. Six genera could not be placed in any tribe, but were placed informally into either a 'Basal Grade' (*Davidsonia* and *Bauera*) or a 'Core Cunon' clade (*Eucryphia*, *Gillbeea* and *Acrophyllum*). A slightly expanded analysis to include *Hooglandia* showed that it is not a member of any tribe but falls within the 'Basal Grade' and appears closely related to *Aistopetalum* (Sweeney and Bradford 2003).

The genera *Acsmithia* and *Spiraeanthemum* (tribe Spiraeanthemeae) from New Guinea and the South Pacific form a small sister clade to the rest of the family. Within this larger clade, *Bauera* and a group of genera with mostly thyrsoid to cymiform inflorescences and drupaceous or indehiscent dry fruits (including tribe Schizomerieae and *Davidsonia*, *Aistopetalum* and *Hooglandia*) form a basal grade. The largest, i.e. the 'Core Cunon', clade includes four tribes representing genera such as *Caldcluvia*, *Callicoma*, *Geissois* and *Weinmannia*, with paniculiform, racemose, or capitate inflorescences and usually capsular fruits. About two-thirds of Cunoniaceae species richness is in a clade of the three largest genera, *Weinmannia*, *Cunonia*, and *Pancheria* (tribe Cunonieae).

Most recognized genera have clear apomorphies (Bradford and Barnes 2001). *Weinmannia* was weakly supported as monophyletic, based on the presence of hairs on seeds (Bradford 1998). In addition, most species studied have a multicellular hair base unique in the family (Barnes, unpubl. data). However, molecular data is thus far inconclusive as to the monophyly of *Weinmannia* with respect to *Cunonia* (Bradford 1999, 2002; Bradford and Barnes 2001), although the five sections of *Weinmannia* are clearly monophyletic. Although no known morphological character unites the species of *Geissois*, cladistic analyses of DNA sequences suggest that *Geissois* is monophyletic. However, greater sampling of species and loci is warranted (Bradford and Barnes 2001).

FAMILY CIRCUMSCRIPTION AND AFFINITIES. At one time or another, *Bauera*, *Eucryphia*, *Davidsonia* and *Brunellia* have either been retained as distinct families, or included within Cunoniaceae s.l. *Bauera* is sometimes placed in Saxifragaceae, where Cunoniaceae in general once resided. Until recently, little has been known about how this group of families relates to other angiosperms. Large-scale cladistic analyses of DNA sequences have now placed Cunoniaceae s.l. in the order Oxalidales, along with Elaeocarpaceae (including Tremandraceae), Cephalotaceae, Connaraceae and Oxalidaceae (APG 1998). On the basis of cladistic analyses of plastid DNA sequences and morphology (Bradford and Barnes 2001), Cunoniaceae are here circumscribed to include *Bauera*, *Eucryphia* and *Davidsonia* within the family, but *Brunellia* outside of Cunoniaceae and possibly near *Cephalotus*. *Aphanopetalum*, traditionally placed in Cunoniaceae, belongs in Saxifragales (Dickison et al. 1994; Bradford and Barnes 2001; Fishbein et al. 2001). Relationships within Oxalidales need further study, but Tremandraceae are nested within Elaeocarpaceae; Brunelliaceae, Cephalotaceae and Elaeocarpaceae are probably more closely related to Cunoniaceae than to Oxalidaceae and Connaraceae.

Only by examining a combination of morphological characters can one distinguish Cunoniaceae from the related families Brunelliaceae and Elaeocarpaceae. A single pair of interpetiolar stipules is synapomorphic for a Cunoniaceae clade, and is unique among Oxalidales families. However, while Brunelliaceae and Elaeocarpaceae have lateral stipules, which may be one way to differentiate among these families, lateral stipules are also derived within Cunoniaceae (Geissoieae, *Bauera*, *Caldcluvia* and *Gillbeea*). Character combinations typical for these three families are:

- **Cunoniaceae**: leaves usually opposite, simple or compound, toothed; stipules often interpetiolar; filaments exceeding the petals; floral disc inter- or intra-staminal.
- **Brunelliaceae**: leaves usually opposite, simple or compound, toothed; stipules lateral; brown hairs on stems and leaves; dioecious; floral disc intra-staminal; large decurrent stigmas; shiny, arillate seeds exposed from follicles.
- **Elaeocarpaceae**: leaves usually alternate, simple, toothed; stipules lateral; petals often fringed; anthers open by terminal pores or slits; floral disc extra-staminal.

DISTRIBUTION AND HABITATS. Cunoniaceae are predominantly found in tropical (especially montane) and wet temperate regions of the Southern Hemisphere, with most generic diversity in Australia and Tasmania (16 genera), New Guinea (9 genera), and New Caledonia (7 genera). Some groups are widespread in Malesia, Melanesia and Polynesia (especially *Weinmannia*). Two genera are in South Africa (*Cunonia* and *Platylophus*). About 40 species of *Weinmannia* are located in Madagascar, the Comores and the Mascarenes. In the Americas only 4 genera occur, but the region is species rich because of the diversity of *Weinmannia*. The American distribution is from Mexico and the West Indies to Central America (*Weinmannia*), and through tropical South America (*Weinmannia* and *Lamanonia*) to temperate South America (*Caldcluvia*, *Eucryphia*, *Lamanonia* and *Weinmannia*).

The Cunoniaceae lineage was probably widespread and diversified prior to the breakup of Gondwana based on phylogenetic reconstructions, continental disjunctions of some genera and

tribes, and the fossil record. *Caldcluvia paniculata* from South America is the sister taxon to *Opocunonia*, *Ackama* and *Spiraeopsis* from Austro-Malesia and New Zealand (tribe Caldcluvieae); *Lamanonia* from South America is the sister group to *Geissois* and *Pseudoweinmannia* from Australia and the South Pacific (tribe Geissoieae); and *Platylophus* from South Africa is the sister taxon to *Anodopetalum* and *Ceratopetalum* from Australia and New Guinea (tribe Schizomerieae). *Cunonia* has one species endemic to South Africa and c. 25 species endemic to New Caledonia. Two species in section *Weinmannia* from the Mascarene Islands (Indian Ocean) are most closely related to about 70 species from the Americas (Bradford 1998, 2002; Bradford and Barnes 2001).

Although the majority of species inhabit tropical montane rainforests, there are many lowland rainforest species in Madagascar, eastern Malesia, western Melanesia and Australia, and some in cool-temperate rainforests in Chile, Tasmania and New Zealand. A few species occur in moderately dry habitats, such as *Lamanonia* and *Weinmannia* in South American cerrado vegetation, or *Codia* and *Pancheria* in seasonally dry maquis in New Caledonia. In dry Australian heathlands, *Bauera* may be found in riparian habitats.

Species occur on a diversity of soils, from lowland clays to upland humus. Many species are opportunists on poorly developed, shallow soils derived from substrates such as sandstone, quartz or lava, suggesting a good primary colonizing ability. *Acrophyllum australe* is narrowly endemic on soils derived from exposed layers of shale along moist sandstone cliffs in Australia's Blue Mountains (Cooper 1986). Cunoniaceae often successfully occupy ultramafic (e.g. serpentine) substrates. Some species occur both on and off ultramafics, but many are substrate endemics, which is especially common in New Caledonia in the genera *Cunonia*, *Pancheria*, *Codia*, *Acsmithia* and *Geissois*. *Weinmannia clemensiae* Steenis and *W. devogelii* H.C. Hopkins are serpentine endemics from Mt. Kinabalu, Borneo and Lake Matano, Sulawesi respectively (Hopkins 1998a). Some plants living on ultramafics tend to hyperaccumulate heavy metals in their leaves (Jaffré 1980).

Because many Cunoniaceae species tend to be locally common and their pollen has a short disperal radius (McGlone 1982), palaeoecologists have used pollen records to track the movement of Cunoniaceae species, especially *Weinmannia*, and thereby infer the historic locations of vegetation associations (Pocknall 1980; McGlone 1983; Helmens and Kuhry 1986; Hooghiemstra 1989; Villagran 1990; Heusser 1993). This indicates, for example, that many species characteristic of upper elevation Andean vegetation occurred in the lower Amazon basin during glacial periods (Bush et al. 1990).

FLORAL AND POLLINATION BIOLOGY. Cunoniaceae flowers are generally small to medium sized and, when not individually showy, they are frequently massed into highly visible inflorescences. Exceptions to this include *Eucryphia*, with large, white, solitary flowers, up to 6 cm in diameter in *E. glutinosa*, and *Anodopetalum* which has inconspicuous, solitary, green flowers (Barnes and Rozefelds 2000).

Floral colour is generally white or a pale pastel shade, including cream-yellow, greenish or pink, but magenta (some *Bauera*), purplish (some *Spiraeopsis* and *Cunonia*) or bright scarlet (Pacific *Geissois*, some *Cunonia*) also occur. A range of floral structures contributes to the colour, including the sepals, especially when the petals are small, fugaceous or absent. Filaments are usually white, sometimes with contrasting yellow or pink anthers, and the disc and gynoecium sometimes contrast sharply with the pale tepals and stamens. The numerous, often exserted stamens in *Eucryphia*, *Bauera*, *Lamanonia* and *Geissois* may also enhance visibility. Included stamens occur in *Davidsonia*.

Where present, the floral disc appears nectariferous but the abundance and composition of the nectar have not been assessed. Large nectar droplets are produced in red-flowered *Geissois*, but in some genera the volumes per flower may be very small indeed. For instance, honey bees at capitula of *Pancheria gatopensis* Guillaumin (Plateau de Tango) inserted their tongues into flowers, suggesting foraging for nectar, but none was visible to the human eye and it was scarcely detectable to the human tongue (Donovan, Bradford and Hopkins, pers. obs. 2000). The small, rounded, shiny structures at each tip of the crescent-shaped petals in *Gillbeea* are also apparently secretory (Endress 1994). Where information is available, flowers are generally sweetly scented and sometimes cloyingly so (e.g. *Codia*), with notable exceptions in Pacific *Geissois* and several species of *Cunonia*.

Cunoniaceae flowers are generally cup-, dish- or bowl-shaped and most species are probably entomophilous, although data are few and largely anecdotal. A wide range of visitors has been noted, including wasps, flies, Lepidoptera and beetles, although bees of various sorts are likely to be the main pollinators in most cases (see Economic Importance) and have been reported from genera with a variety of inflorescence types, including *Codia*, *Cunonia*, *Eucryphia*, *Pancheria*, *Platylophus* and *Weinmannia*.

Some flower-visiting vertebrates are probably part of a suite of generalist potential pollinators; for example, various birds have been reported from the flowers *Opocunonia*, *Weinmannia*, *Platylophus* and *Cunonia capensis*, and geckos have been seen licking the white, candle-like racemes of *C. balansae* Brongn. & Gris in New Caledonia (Hopkins, Bradford and Donovan, pers. obs. 2000). A few *Cunon* species appear to be specialised for bird-pollination, such as the Pacific species of *Geissois* whose red, generally cauliflorous racemes attract numerous Meliphagidae, Nectariniidae and Zosteropidae, as well as some insects, geckos (Bavary 1869; S. Zona, pers. comm.) and perhaps pteropodid bats (MacKee, specimen labels). In New Caledonia, *Cunonia macrophylla* Brongn. & Gris has bright green, scentless flowers in one-sided, bottle-brush racemes that attract Meliphagidae. Anemophily is possible in the dioecious *Vesselowskya* from eastern Australia, which has catkin-like racemes, decurrent stigmas and which grows gregariously along streams and rivers (Bradford, pers. obs.; Rozefelds et al. 2001).

As inferred from floral morphology, most species have bisexual, homogamous flowers. However, dioecy occurs in *Spiraeanthemum*, *Hooglandia*, *Pancheria*, *Vesselowskya* and some groups of *Weinmannia*, where it may have evolved three times (Bradford 1998), being present in sections *Weinmannia*, *Fasciculatae* and *Leiospermum*. In the latter genus at least, dioecy is frequently incomplete ('leaky'), with geographical variation in sexual expression occurring in *W. fraxinea*, and temporal switches, usually from male to bisexual or vice versa, are seen in several Pacific species (Hopkins 1998a, 1998b; Hopkins and Florence 1998). Protandry is marked in *Opocunonia* and *Spiraeopsis*, and perhaps occurs in *Ackama* (all tribe Caldcluvieae), and andromonoecy has been reported in some species of both *Schizomeria* and *Cunonia*.

PALAEOBOTANY. The fossil record of Cunoniaceae has been reviewed by Barnes and colleagues (1999, 2001). Confirmed macrofossils of Cunoniaceae represent 11 genera from a broad spectrum of morphological diversity in the family. Most of these records are leaf compressions or impressions or preserved cuticles and, with the exception of a single, fossil infructescence from New Zealand (?*Weinmannia* sp. indet.), all accepted leaf and reproductive macrofossils of Cunoniaceae are from Australian Cainozoic deposits. The oldest of these records are from the Late Palaeocene (*Eucryphia*), Eocene (*Acsmithia/Spiraeanthemum*, *Ceratopetalum*, *Codia*, *Weinmannia*) and Oligocene (*Acsmithia*, *Callicoma*, *Ceratopetalum*, *Schizomeria*, *Vesselowskya*), suggesting that the family had diversified by the mid to late Cretaceous or Early Palaeocene.

A Late Cretaceous fossil flower from Sweden has been assigned to a fossil genus within Cunoniaceae, *Platydiscus* Schönenberger & Friis (Schönenberger et al. 2001), although strong similarities to Anisophylleaceae (Matthews et al. 2001) make this placement less certain.

Fossil wood identifications include *Weinmannioxylon eucryphioides* from Eocene sediments at King George Island (Poole et al. 2001), which has strong affinities to *Eucryphia*, and two *Weinmannioxylon* species from Late Cretaceous sediments in Antarctica (Poole et al. 2000; Poole and Cantrill 2001). The fossil *Cunonioxylon parenchymatosum* from the Eocene of Lower Saxony in Germany was directly compared to wood of extant *Cunonia* (Gottwald 1992). There is significant taxonomic confusion generated by Cunoniaceae fossil wood identifications, as a single fossil genus may contain species with affinities to numerous extant genera. On this basis, Cunoniaceae fossil wood species should not be considered to represent extant genera. However, Antarctic specimens suggest that the family has been a common element of southern temperate rainforest for over 100 million years.

Leaf and reproductive macrofossils indicate that the extant geographic range of some genera has been reduced or is different to that in the past. *Acsmithia*, *Callicoma*, *Schizomeria*, *Vesselowskya* and *Weinmanniaphyllum* (form genus encompassing leaf macrofossils indistinguishable from *Weinmannia* and *Cunonia*) were present in Tasmania during the Early Oligocene but are no longer locally extant (Carpenter and Buchanan 1993; Barnes 1999; Barnes and Hill 1999b; Barnes et al. 2001). *Ceratopetalum*, *Codia* and *Spiraeanthemum/Acsmithia* are known from Eocene-Oligocene deposits in southern and south-western Australia (Carpenter and Pole 1995; Barnes 1999; Barnes and Hill 1999a, 1999b; Barnes et al. 2001), which are regions presently too dry for extant species to occur, based on the climate envelope of their modern habitat.

Elaeocarpaceae and Cunoniaceae leaf macrofossils may be difficult to distinguish, especially when specimens lack preserved cuticle, or are incomplete. This is the case with many historical records. It is also difficult to distinguish between Elaeocarpaceae and Cunoniaceae pollen, which are both very small and not easily identifiable to the generic level.

Given the above caveats, the oldest pollen records of Cunoniaceae date from the Late Palaeocene, Eocene and Oligocene. These ancient deposits are evidence of regional extinction of some taxa, such as Cunoniaceae-dominated forests in the Late Palaeocene of South Australia (Sluiter 1991), *Weinmannia* from Australia (Hill and Macphail 1983), and *Gillbeea* (*Concolpites leptos* fossil pollen type) from South America (Romero and Castro 1986) and southern Australia (Stover and Partridge 1973).

ECONOMIC IMPORTANCE. References to human use of Cunoniaceae species come mainly from local or regional economic botany texts or floras, and industry sources (e.g. timber, apiculture). The wood of Cunoniaceae species is used in general construction as poles, wall panelling, flooring, plywood, furniture, cabinetry, joinery, tools, boat keels, canoes, packing boxes, musical instruments and, historically, to make wagons and coaches (*Ceratopetalum apetalum*, *Cunonia capensis*). The most important commercial timber regions for the family are Malesia, Australia, and the western Pacific, principally of the genera *Ackama*, *Callicoma*, *Ceratopetalum*, *Eucryphia*, *Geissois*, *Pseudoweinmannia*, *Schizomeria* and *Weinmannia*. In the Americas, *Weinmannia* species are mostly of local construction and cabinetry use. The Chilean *Eucryphia cordifolia* is valued for the high quality of its wood, which has diverse uses. Mascarene island and South African species are used mainly for furniture making. Where the human population relies on wood for fuel, especially in the Andes and Madagascar, *Weinmannia* trees are often chopped down or whole forests are burned to make charcoal. Several species have been used in local tanning industries (e.g. *Weinmannia tinctoria*).

Bee keepers use *Weinmannia* forests for commercial production of honey in Madagascar and New Zealand. *Eucryphia*-dominated forests are honey resources in Tasmania and Chile. *Platylopus trifoliatus* honey is popular in South Africa. *Davidsonia* is a highly regarded 'bush fruit' in Australia where it is used for jams and wine, and *Schizomeria* fruits may also be used in jams and pies.

Horticultural use is greatest in Australia, including *Ceratopetalum gummiferum* (the New South Wales Christmas Bush). Because a variety of ethnomedicinal sources indicate that Cunoniaceae extracts may treat infections, a research program in New Caledonia has begun on the pharmacology of Cunoniaceae (Fogliani et al. 2002).

CONSERVATION. Little has been done generally to identify species of Cunoniaceae at risk of extinction, although 28 taxa are listed as threatened in the most recent global assessment (Walter and Gillett 1998). Unfortunately, many species occur in areas with extremely high deforestation rates, such as portions of the Andes and Madagascar. These are also the areas where systematic understanding is poorest. The greatest threats are the conversion of forested land to subsistence agricultural land, habitat degradation from logging, grazing, fire, mining and invasion of exotic species, and expanding suburban development. In the following, some regions are selected where high diversity and/or endemism of Cunoniaceae coincide with elevated rates of environmental interference.

The tropical rainforest in the Atherton Tableland of NE Queensland harbours 13 endemic species in eight genera (Hyland and Whiffin 1993) and therefore it is especially critical for Cunoniaceae conservation. Small areas of upland habitat in north-eastern and eastern Australia support two narrow endemic *Eucryphia* and *Ceratopetalum* species (Forster and Hyland 1997; Rozefelds and Barnes 2002). *Bauera sessiliflora* and *Acrophyllum australe* are lowland forest/heath endemics, the latter especially being at risk of extinction from frequent fires and encroaching human disturbance. *Davidsonia johnsonii* and *D. jerseyana* are restricted to a small region of north-eastern New South Wales and south-eastern Queensland heavily impacted by small-scale agriculture, and the former species may not reproduce sexually (Harden and Williams 2000).

New Caledonia, with seven genera and over 90 endemic species occurring in a relatively small area, is a major conservation priority for Cunoniaceae. Many species are restricted to a single or a few populations, often on top of mountains. Species occurring at the highest elevations are somewhat protected from the anthropogenic fires that frequently overrun vegetation at lower elevations (Lowry 1998), but may be more susceptible to vanishing climatic envelopes due to global warming (Still et al. 1999). Currently, the most endangered species have narrow ranges at low to mid elevations and do not establish well in fire-

disturbed habitats. Outside of New Caledonia, the most imperilled, cunon-rich Pacific flora is Fiji, which has five genera and 13 species with 12 endemic (Smith 1985). The spectacular *Geissois* species are most endangered because they live at relatively low elevations where there is the greatest loss of habitat from agriculture, grazing, and the invasion of exotics.

Madagascar is home to about 40 endemic species of *Weinmannia* that are members of two endemic and diverse sections. Species occur primarily in the northern and eastern regions of the island, from sea level (which is unusual in *Weinmannia*) to near the summit of the highest peaks. Burning and clearing forest to grow hill rice is the major cause of habitat loss for *Weinmannia*, especially in the lowlands and central plateau. Comprehensive systematic information is lacking and several species are undescribed (Bradford and Miller 2001). Very little habitat remains for two endemic Mascarene island species of *Weinmannia* (Cadet 1984), and this is being maintained by manual weeding of non-native, invasive plants (W. Strahm, pers comm.).

In tropical America, *Weinmannia* is highly diversified in the Andes where natural vegetation is being destroyed rapidly in some regions, especially Ecuador and Colombia. Presently, montane forest is being eliminated by fires being set to clear farmland below, and to improve grazing and create more grasslands above. *Lamanonia* occurs in moist, evergreen, semideciduous forests and savannah-like vegetation of southern Brazil, northern Argentina and northern Paraguay, in habitats that are some of the most endangered in South America due to expanding soybean cultivation and cattle grazing. *Eucryphia glutinosa* occurs in riparian habitats in the Mediterranean climate region of Chile and is threatened by grazing and water management (Muñoz Pizarro 1973).

KEY TO THE GENERA

1. Leaves alternate   **5. *Davidsonia***
 – Leaves decussate or whorled   2
2. Flowers solitary in leaf axils   3
 – Flowers borne on an inflorescence axis   5
3. Flowers subtended by a pedicel only   **6. *Bauera***
 – Flowers subtended by a pedicel, pair of bracteoles and a
   peduncle   4
4. Polyandric; carpels usually >3   **11. *Eucryphia***
 – Diplostemonous; bicarpellate   **9. *Anodopetalum***
5. Inflorescences condensed spherical heads (capitula)   6
 – Inflorescences not capitate, e.g. racemose, paniculiform or
   cymiform   9
6. Flowers unisexual; fruits capsular; leaves (2)3–4 per
   node   **25. *Pancheria***

– Flowers bisexual; fruits indehiscent or capsular; leaves 2 per node, rarely more 7

7. Fruits capsular; leaves deeply serrate **23. Callicoma**
– Fruits indehiscent; leaves entire or weakly serrate 8
8. Leaves entire, rarely toothed; ovary inferior; calyx small **22. Codia**
– Leaves toothed; ovary half inferior; calyx prominent **21. Pullea**
9. Flowers maturing synchronously to acropetally in an inflorescence 10
– Flowers maturing at different times in basipetalous sequence 23
10. Polyandric (more than twice as many stamens as perianth parts); leaves palmately compound or trifoliate; inflorescences racemes 11
– Diplostemonous, rarely haplostemonous, leaves imparipinnate, unifoliolate or rarely palmately compound; inflorescences racemes, capitula, panicles or cymes 13
11. Leaves 3-foliolate; fruit indehiscent, dry, densely covered in wavy hairs **15. Pseudoweinmannia**
– Leaves 3- or more-foliolate; fruit capsular, glabrous or hairy 12
12. Stipules lateral and unfused; flowers never red **14. Lamanonia**
– Stipules axillary or lateral and fused; flowers red or not **16. Geissois**
13. Inflorescence axes unbranched (racemose) 14
– Inflorescence axes branched (e.g. cymes and panicles) 16
14. Leaves palmately compound or trifoliolate; flowers 3-merous; stigma decurrent **24. Vesselowskya**
– Leaves simple or imparipinnate; flowers 4–5-merous; stigma terminal 15
15. Seeds with hairs; fruits dehisce basipetally; nectary disc distinct from ovary **27. Weinmannia**
– Seeds without hairs; fruits dehisce circumbasally; nectary disc adnate to ovary **26. Cunonia**
16. Carpels fused; petals + or 0 17
– Carpels unfused or only basally so; petals 0 22
17. Petals 0; ovary half-inferior; fruit indehiscent; leaves unifoliolate **21. Pullea**
– Petals +; ovary superior; fruit capsular; leaves unifoliolate or imparipinnate 18
18. Leaves 3 per node, unifoliolate **13. Acrophyllum**
– Leaves 2 per node, unifoliolate or imparipinnate 19
19. Stipules lateral; leaves unifoliolate; flowers not protandrous **17. Caldcluvia**
– Stipules interpetiolar; leaves imparipinnate, rarely unifoliolate; flowers often protandrous 20
20. Stellate hairs + on leaves **20. Spiraeopsis**
– Stellate hairs 0 on leaves 21
21. Inflorescence conical in outline (paniculate); seeds with hairs **19. Ackama**
– Inflorescence rounded in outline (cymiform); seeds without hairs **18. Opocunonia**
22. Leaves 2 per node; flowers unisexual **2. Spiraeanthemum**
– Leaves 3–4 per node; flowers bisexual **1. Acsmithia**
23 Inflorescence a 3-flowered cyme **9. Anodopetalum**
– Inflorescence of numerous flowers 24
24. Stellate hairs + on floral parts; fruit dry with laterally expanded carpel walls **12. Gillbeea**
– Stellate hairs 0 on floral parts; fruit fleshy or dry, not as above 25
25. Fruit drupaceous 26
– Fruit dry and indehiscent 28
26. Leaves simple; petals + **7. Schizomeria**

– Leaves pinnately compound; petals 0 27
27. Flowers hermaphroditic; gynoecium of 4–6 locules **3. Aistopetalum**
– Flowers unisexual (plants dioecious); gynoecium unilocular **4. Hooglandia**
28. Ovary half inferior; sepals enlarged and often colourful in fruit; petiole swollen and articulated at junction of blade **8. Ceratopetalum**
– Ovary superior; sepals inconspicuous in fruit; petiole not swollen or articulated at junction of blade **10. Platylophus**

GENERA OF CUNONIACEAE

Tribes and genera are mainly organized according to the branching sequence of cladograms in Bradford and Barnes (2001 with Figs. 4 and 5), who also give descriptions of tribes, and with some modification based on Sweeney and Bradford (2003), notably the inclusion of *Hooglandia* and the new placement of *Aistopetalum*. Several genera with uncertain relationships are not placed in a tribe, but informally included in either a 'Basal Grade' or 'Core Cunon Clade'.

Characters applying to all genera unless stated otherwise

Leaves decussate; non-stipellate; margin toothed; hairs simple; stipules interpetiolar; one pair of stipules borne per node. Flowers bisexual, pedicellate; perianth 4–5-merous, of sepals and petals, calyx valvate; androecium diplostemonous; ovary bicarpellate, syncarpous, superior, stylodia free, stigmas terminal.

I. TRIBE SPIRAEANTHEMEAE Engl. (1928).

1. *Acsmithia* Hoogl.

Acsmithia Hoogl., Blumea 25: 492–501 (1979), rev.; Hoogland, Bull. Mus. Natl. Hist. Nat., B Adansonia 9: 393–397 (1987).
Spiraeanthemum A.Gray (1854) p.p.

Trees and shrubs; nodes of stems knobby. Leaves whorled, unifoliolate, margin entire or toothed, pocket domatia along the midrib often present. Inflorescence axillary, axillary and terminal, or sometimes cauliflorous, paniculate; floral maturation synchronous; perianth of one whorl; nectary of small segments; carpels 2–5, free almost to base. Fruits follicular; seeds winged. Fourteen species: 4 from E. Malesia, 1 from Australia, 8 from New Caledonia, and 1 from Fiji.

2. *Spiraeanthemum* A. Gray Fig. 31

Spiraeanthemum A. Gray, Proc. Am. Acad. Arts 3: 128 (1854); Hoogland, Blumea 25: 501–505 (1979), rev.

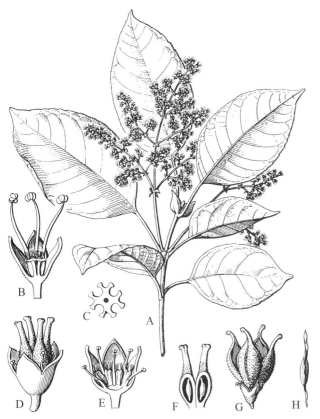

Fig. 32. Cunoniaceae. *Aistopetalum viticoides*. **A** Flowering branch. **B** Flower. **C** Same, longitudinally sectioned. **D** Fruit. Drawn by R. van Crevel. (Hoogland 1960)

Fig. 31. Cunoniaceae. *Spiraeanthemum katakata*. **A** Flowering branch. **B** Male flower, vertical section. **C** The same, transverse section of disk. **D** Female flower. **E** Same, pistil removed. **F** Same, vertical section. **G** Fruit. **H** Seed. Drawn by R. van Crevel. (Hoogland 1960)

Trees and shrubs. Leaves unifoliolate, often with pocket domatia along the midvein. Inflorescence axillary; paniculate; floral maturation synchronous. Flowers unisexual, dioecious; perianth of one whorl, nectary of small segments; gynoecium of (2–)4–5(6) free carpels. Fruit follicular; seeds winged, 2 per carpel. Six species from the Bismarck Archipelago, Bougainville, Solomons, Vanuatu, Fiji, and Samoa.

II. Unplaced 'Basal Grade' Genera

3. *Aistopetalum* Schltr.   Fig. 32

Aistopetalum Schltr., Bot. Jahrb. Syst. 52: 142–144 (1914); Perry, J. Arnold Arb. 30: 158–159 (1949); Hoogland, Austral. J. Bot. 8: 333–337 (1960), rev.

Trees. Leaves imparipinnate. Inflorescence axillary; paniculiform, large, with decussate branching throughout; floral maturation centrifugal; perianth of one 4–6-merous whorl; floral disc large, adnate to the ovary; gynoecium of 4–6 carpels. Fruit drupaceous, one seed per carpel. Two species (*A. multiflorum* Schltr. and *A. viticoides* Schltr.), endemic to New Guinea.

4. *Hooglandia* McPherson & Lowry

Hooglandia McPherson & Lowry, Ann. Missouri Bot. Gard. (in prep.); Sweeney and Bradford, Ann. Missouri Bot. Gard. (in prep.).

Trees. Leaves imparipinnate; stipules interpetiolar or occasionally divided partially or completely to the base. Inflorescence axillary; paniculate; floral maturation subsynchronous with terminal flowers opening first. Flowers unisexual, dioecious; perianth of one whorl, 4-merous, imbricate; nectary adnate to the ovary; gynoecium bilaterally symmetrical, with a single locule and a single bent style, but with a small bump opposite the style suggesting the fusion of two carpels, two ovules. Fruit drupaceous, flattened and elongate, with a thin fleshy exocarp, hard endocarp, and a single large seed. One species, *Hooglandia ignambiensis* McPherson and Lowry, from upper elevation montane forest in the Panie range of north-eastern New Caledonia.

5. *Davidsonia* F. Muell.

Davidsonia F. Muell., Fragm. 4: 4 (1867); Bange, Blumea 7: 293–296 (1952); Harden & Williams, Telopea 8: 413–428 (2000), rev.

Trees. Leaves alternate, imparipinnate with prominent teeth and a jagged rachis; hairs rigid, caducous, mildly urticating. Inflorescence axillary, terminally, and adventitiously along branches and the trunk, an elongate panicle; inflorescence very large in *D. pruriens* F.Muell. but smaller in other taxa and sometimes nearly racemose in *D. jerseyana* (F.M. Bailey) Harden & Williams and *D. johnsonii* Harden & Williams; floral maturation basipetalous. Flowers with a single perianth whorl, the lobes fused half their length; nectary of small segments. Fruit drupaceous, with fleshy, edible pulp containing 4 seeds. Three species from Eastern Australia. Used as an ornamental and fruit tree to make jams and wine, with the common name 'Davidson's Plum'.

6. *Bauera* Banks ex Andrews

Bauera Banks ex Andrews, Bot. Repos. 3: t. 198 (1801); Bailey, The Queensland Flora, 2: 543 (1900); Stanley & Ross, Flora of South-eastern Queensland 1: 219 (1983); Everett in Harden, Flora of New South Wales 1: 521 (1990).

Scrambling shrub with narrow stems. Leaves unifoliolate (appearing 3-foliolate, but lateral 'leaflets' considered to be modified stipules, see Dickison and Rutishauser 1990), sessile or nearly so; stipules leaf-like, borne beside each leaf. Flowers solitary in leaf axils; perianth parts variable in number, corolla of one or sometimes two whorls; stamens numerous; floral nectary 0; ovary of 2–3 carpels. Fruit capsular; seeds elliptical, appendages 0. Four species in eastern and southern Australia and Tasmania.

III. TRIBE SCHIZOMERIEAE J.C. Bradford & R.W. Barnes (2001).

7. *Schizomeria* D.Don

Schizomeria D. Don, Edinburgh New Philos. J. 9: 94– 95 (1830); Perry, J. Arnold Arb. 30: 151–158 (1949); Whitmore, Guide to the Forest of the British Solomon Islands (1966); Stanley & Ross, Flora of South-eastern Queensland 1: 225 (1983); Harden, Flora of New South Wales 1: 520 (1990); Fortune Hopkins & Hoogland, Fl. Males. I, 16: 53–165 (2002).
Cremnobates Ridl. (1916).

Trees and shrubs. Leaves unifoliolate. Inflorescence terminal or less often axillary; thrysoid;

floral maturation basipetalous. Flowers bisexual or male and plants hermaphrodite or andromonecious; perianth (4)5(6)-merous; petals tridentate; floral nectary annular, often with large lobes. Fruit drupaceous. Ten species distributed in the Moluccas, New Guinea, the Solomons, and eastern Australia.

8. *Ceratopetalum* Sm.

Ceratopetalum Sm., Specim. Bot. New Holl. 1: 9, t. 3 (1793); Hoogland, Austral. J. Bot. 8: 318–328 (1960), rev.; Hoogland, Brunonia 4: 213–215 (1981); Stanley & Ross, Flora of South-eastern Queensland 1: 225 (1983); Harden, Flora of New South Wales 1: 519 (1990); Rozefelds & Barnes, Int. J. Plant Sci. 163: 651–673 (2002).

Trees and shrubs, resinous. Leaves unifoliolate or trifoliolate. Inflorescence terminal or axillary; thrysoid to cymiform, often retaining decussate branching, or somewhat helical distally; floral maturation basipetalous; petals 0 or small and deeply incised (*C. gummiferum* Sm.); nectary large, adnate to the ovary; ovary half inferior. Fruit dry, indehiscent, with enlarged, often reddish calyx lobes. About 9 species occurring in Eastern Australia and New Guinea, the 'Christmas Bush' (*C. gummiferum* Sm.) used ornamentally, and 'Coachwood' (*C. apetalum* D.Don) a valued timber species.

9. *Anodopetalum* A. Cunn. ex Endl.

Anodopetalum A. Cunn. ex Endl., Gen. Pl. 11: 818 (1839); Curtis, The Student's Flora of Tasmania 1: 176 (1956); Barnes and Rozefelds, Austral. Syst. Bot. 13: 267–282 (2000).

Trees and shrubs. Leaves unifoliolate or trifoliolate. Inflorescence axillary; reduced to a single flower, (a 3-flowered cyme); floral maturation basipetalous; calyx large, petals slender and apically notched; anthers with prominent horn; nectary annular. Fruit capsular. One species, *A. biglandulosum* (A. Cunn ex. Hook.) Hook. f., endemic to Tasmania. The plant's unusual growth form can produce dense, overlaying thickets of branches (Barker and Brown 1994).

10. *Platylophus* D. Don

Platylophus D. Don, Edinburgh New Philos. J. 9: 92 (1830); Coates Palgrave, Trees of Southern Africa (1983).

Trees. Leaves 3-foliolate. Inflorescence axillary; thrysoid; floral maturation basipetalous; petals small and tridentate; nectary annular and lobed.

Fruits indehiscent, swelling into a dry bladder; seeds few small, wingless. One species, *P. trifoliatus* (L.) D.Don, in South Africa.

IV. UNPLACED 'CORE CUNON' CLADE GENERA

11. *Eucryphia* Cav.   Fig. 33

Eucryphia Cav., Icon. 4: 48, t. 372 (1797); J. Bausch, Bull. Misc. Inform. 8: 317–349 (1938); Curtis, The Student's Flora of Tasmania 1: 179 (1956); Harden, Flora of New South Wales 1: 523 (1990); P.I. Forst. & B. Hyland, Austrobaileya 4: 589–596 (1997).

Trees and shrubs; resinous. Leaves evergreen and peltate with white undersurface, or deciduous and lacking peltations in *E. glutinosa* (P. & E.) Baill.; unifoliolate or imparipinnate; margin entire or toothed. Inflorescence an axillary peduncle with a single, pedicellate flower. Flowers large, enclosed by imbricate, apically coherent sepals that are shed at maturity; petals 4, large and white; stamens numerous, fasciculate, at the margins of a broad receptacle; ovary of 4–14 fused carpels. Fruit capsular; seeds winged, numerous. Seven species from E Australia (3 spp.), Tasmania (2 spp.), and southern S America (2 spp.). Cultivated as an ornamental.

12. *Gillbeea* F. Muell.

Gillbeea F. Muell., Fragm. 5: 17 (1865); Hoogland, Austral. J. Bot. 8: 328–333 (1960), rev; Rozefelds & Pellow, Nordic J. Bot. 20: 435–441 (2000).

Trees. Leaves imparipinnate; stipules in lateral pairs; stellate hairs restricted to inflorescence axes and flowers. Inflorescence axillary and terminal; paniculiform; floral maturation basipetalous; petals bifid with apical glands; nectary annular, large; gynoecium of three fused carpels. Fruit indehiscent, dry, with carpel walls expanding to form lateral wings. Three species, two from northeastern Australia (*G. adenopetala* F. Muell. and *G. whypallana* Rozefelds & Pellow) and one from New Guinea (*G. papuana* Schltr.).

13. *Acrophyllum* Benth.   Fig. 34

Acrophyllum Benth. in Maund, Botanist 2: t. 95 (1838); Hoogland, Aust. J. Bot. 8: 318–341 (1960); Hoogland, Brunonia 4: 216 (1981).
Calycomis D. Don (1830).

Long-stemmed shrub. Leaves in whorls of 3, unifoliolate, margin deeply serrate, indumentum on undersurface white, peltate. Inflorescence at several successive nodes along the main stem, producing small, axillary panicles, the first node sometimes branching to form a lateral inflorescence, and the main axis resuming vegetative growth; floral maturation synchronous; perianth 5-merous; nectary small, annular at base of ovary, or 0. Fruit capsular; seeds tiny, ellipsoid, unwinged, with a granular surface. One species, *A. australe* (A. Cunn.) Hoogland, restricted to a small area of moist sandstone and shale cliffs between 520 and 660 m above sea level in the lower Blue Mountains of New South Wales, Australia (Cooper 1986).

Fig. 33. Cunoniaceae. *Eucryphia cordifolia.* **A** Branch with fruits and flower buds. **B** Flower. **C** Pistil. **D** Fragment of dehisced fruit. **E** Seed. Drawn by E. Gasipi. (Dimitri 1972)

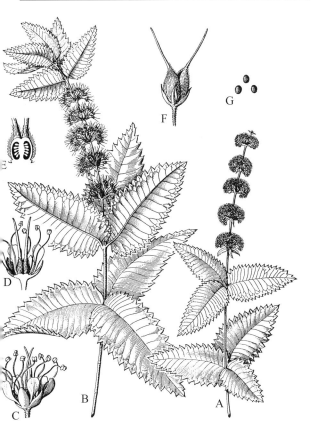

Fig. 34. Cunoniaceae. *Acrophyllum australe.* **A** Flowering branch. **B** Fruiting branch. **C** Flower. **D** Flower, vertically sectioned outside the ovary. **E** Ovary, vertical section. **F** Fruit. **G** Seeds. Drawn by R. van Crevel. (Hoogland 1960)

Fig. 35. Cunoniaceae. *Lamanonia speciosa.* **A** Flowering branch. **B** Flower, vertical section. **C** Stamens. **D** Pistil. **E** Ovary, transverse section. **F** Ovary, vertical section. Drawn by P.J.F. Turpin. (St.-Hilaire 1828)

V. Tribe Geissoieae Endl. ex Meisn. (1838).

14. *Lamanonia* Vell.   Fig. 35

Lamanonia Vell., Fl. Flum. 228 (1825); Zickel & Leitão, Revista Brasil. Bot. 16: 73–91 (1993), rev.

Small trees. Leaves palmately compound with 3–5 leaflets, stipellate, sometimes with weakly-formed tuft domatia along midribs; stipules in lateral pairs. Inflorescence axillary; racemose; floral maturation synchronous; perianth of a single, 5–7-merous whorl; androecium polystemonous, c. 20–60 stamens; disc annular, adnate to ovary. Fruits capsular; seeds winged, numerous. Five species in seasonally dry forest and woodland in southern Brazil, Argentina and Paraguay.

15. *Pseudoweinmannia* Engl.

Pseudoweinmannia Engl., Nat. Pflanzenfam. ed. 2, 18a: 249 (1928); Bailey, The Queensland Flora 2: 542 (1900); Stanley & Ross, Flora of South-eastern Queensland 1: 226 (1983); Harden, Flora of New South Wales 1: 518 (1990).

Trees. Leaves 3-foliolate; stipels minute, caducous. Inflorescence axillary, often proximal to current leaves; racemose, racemes solitary, sometimes branching into 2–3 axes, or in pairs from axillary short-shoots; floral maturation synchronous; perianth of one, usually 6-merous whorl; androecium of numerous (c. 20) stamens, strap-like basally and imbricate; nectary annular and adnate to ovary; gynoecium of 2(3) fused carpels, with false septa to appear four-locular. Fruits indehiscent, ovary covered with long, dense, wavy hairs, easily dispersed by wind; seeds small, unwinged. Two species, *P. apetala* F.M.Bailey and *P. lachnocarpa* (F. Muell.) Engl. in E Australia.

16. *Geissois* Labill.

Geissois Labill., Sert. Austro-Caledon.: 50, t. 50 (1825); Bailey, The Queensland Flora 2: 541 (1900); Guillaumin, Bull. Soc. Bot. France 87: 242–245 (1940); Stanley & Ross, Flora of South-eastern Queensland 1: 226 (1983); Harden, Flora of

New South Wales 1: 518 (1990); Schimaski & Rozefelds, Aust. Syst. Bot. 15: 221–236 (2002), reg. rev.

Trees or shrubs. Leaves palmately compound, 3–9 leaflets; stipellate; tuft domatia occasional along midrib; stipules axillary, rarely connate into a cup-like phytotelm, or lateral, foliaceous, and often fused across the leaf axil and node. Inflorescence axillary and coeval to current leaves, or cauliflorous, or medial and distal to current leaves; racemose, racemes solitary or grouped into a compound structure; floral maturation synchronous; perianth a single whorl, large, 4-merous and red in Pacific species, and 5–6-merous and cream-coloured in Australian ones; stamens several to numerous (c. 8–32), curled in bud in Pacific species, erect in bud in Australian species; nectary annular, adnate to base of ovary. Fruit capsular; seeds winged, numerous. About 18 species distributed in eastern Australia, Fiji, New Caledonia, Vanuatu, and Santa Cruz (Solomon Islands).

VI. Tribe Caldcluvieae J.C. Bradford & R.W. Barnes (2001).

17. *Caldcluvia* D.Don

Caldcluvia D. Don, Edinburgh New Philos. J. 9: 92 (1830); R. Rodríguez et al., Flora Arbórea de Chile (1983).

Trees. Leaves unifoliolate; small, tuft domatia along the midvein; stipules in lateral pairs. Inflorescence axillary; cymiform; floral maturation synchronous; perianth 5-merous; floral nectary segmented. Fruit capsular; seeds small with minute wings and a papillate surface. One species, *C. paniculata* (Cav.) D. Don, temperate rain forests of Chile and Argentina.

18. *Opocunonia* Schltr.

Opocunonia Schltr, Bot. Jahrb. Syst. 52: 154–155 (1914); Perry, J. Arnold Arb. 30: 143–145 (1949); Hoogland, Blumea 25: 486–487 (1979); Fortune Hopkins & Hoogland, Fl. Males. I, 16: 53–165 (2002).

Trees. Leaves imparipinnate; tuft domatia often present along midrib. Inflorescence axillary or terminal; cymiform; floral maturation synchronous. Flowers protandrous; perianth (5)6-merous; nectary annular. Fruit capsular; seeds winged, numerous (c. 40). According to Hoogland (1979), there is one species, *O. nymanii* (K. Schum.) Schltr., endemic to New Guinea but common and widespread there. The species is apparently so

variable in many characters that Schlechter (Bot. Jahrb. Syst. 52, 1914) considered it to be several species and even a separate genus.

19. *Ackama* A. Cunn.

Ackama A. Cunn., Ann. Nat. Hist. 2: 358 (1839); Bailey, The Queensland Flora 2: 540 (1900); Allan, Flora of New Zealand v. 1 (1961); Salmon, The Native Trees of New Zealand 188 (1986); Stanley & Ross, Flora of South-eastern Queensland 1: 226 (1983); Harden, Flora of New South Wales 1: 519 (1990).

Tree. Leaves, imparipinnate, stipellate or not, tuft domatia along midvein, orbicular trichomes present. Inflorescence axillary, often in a series; paniculate; floral maturation synchronous. Flowers sometimes protandrous, sessile or nearly so; nectary segmented; ovary 2- or 3–4-carpellate. Fruit capsular, with hairy seeds that lack wings. Four species, 2 in New Zealand and 2 in Australia.

20. *Spiraeopsis* Miq.

Spiraeopsis Miq., Fl. Ind. Bat. 1, 1: 719 (1856); Hoogland, Blumea 25: 481–490 (1979); Fortune Hopkins & Hoogland, Fl. Males. I, 16: 53–165 (2002).

Trees. Leaves imparipinnate; indumentum stellate and of small, orbicular-glandular trichomes. Inflorescence axillary; paniculate, often large; floral maturation synchronous; disc segmented; ovary of 2–5 fused carpels. Fruit capsular; seeds winged, small. Six species, E Malesia and Solomon Islands.

VII. Tribe Codieae G. Don (1834).

21. *Pullea* Schltr.

Pullea Schltr., Bot. Jahrb. Syst. 52: 164–166 (1914); Bailey, The Queensland Flora 2: 536 (1900); Perry, J. Arnold Arb. 30: 163–165 (1949); Hoogland, Blumea 25: 490–492 (1979), rev. *Stutzeria* F. Muell. (1865).

Trees. Leaves decussate or in whorls of 3; unifoliolate. Inflorescence axillary, terminal, or cauliflorous; paniculate, often in a series from a common peduncle, flowers well dispersed or grouped in glomerules; floral maturation synchronous. Flowers sessile or with a short pedicel; perianth a single, 5-merous whorl, imbricate; filaments strap-like; nectary of 5 segments; gynoecium half-inferior. Fruit dry, indehiscent, with enlarged, persistent sepals. There are at least three species: *P. stutzeri* (F. Muell.) Gibbs in NE Australia; *P. mollis*

Schltr. in New Guinea, and the highly variable *P. glabra* Schltr. from eastern Malesia and Fiji.

22. *Codia* J.R. Forst. & G. Forst.

Codia J.R. Forst. & G. Forst., Char. Gen. Pl.: 59, t. 30 (1776); Brongn. & Gris, Bull. Soc. Bot. France 9: 76–77 (1862); Brongn. & Gris, Ann. Sci. Nat. Bot. V, 1: 377–378 (1864); Guillaumin, Bull. Soc. Bot. France 87: 254–256 (1940).

Shrubs and trees. Leaves decussate in pairs (whorled in *C. albifrons* Vieill. ex Guillaumin), unifoliolate, margin usually entire. Inflorescence axillary, spherical heads (i.e. capitula), on slender peduncles, or peduncles borne from short-shoots; floral maturation synchronous; petals slender or 0; ovary inferior. Fruits indehiscent with an outer covering of lanate hairs; seeds with a thin, undifferentiated seed coat. About 12 species endemic to New Caledonia.

23. *Callicoma* Andrews

Callicoma Andrews, Bot. Repos. 9: t. 566 (1809); Kennedy, Aust. J. Bot. 29: 721–731 (1981).

Trees. Leaves unifoliolate, deeply toothed; glossy green upper surface, lower surface with minute, white trichomes. Inflorescence axillary; spherical heads (i.e. capitula), on slender peduncles from leaf axils, or peduncles borne from axillary short-shoots, often in a series; floral maturation synchronous; perianth of one 4–6-merous whorl, valvate or imbricate; stamens 11–15; nectary 0; ovary of 2–3 carpels. Fruit capsular; seeds unwinged with short papillae. One species, *C. serratifolia* Andrews, common in Eastern Australia.

VIII. Tribe Cunonieae (R.Br.) Schrank & Mart. (1829).

24. *Vesselowskya* Pamp.

Vesselowskya Pamp., Ann. Bot. (Rome) 2: 93 (1905); Rozefelds et al., Aust. Syst. Bot. 14: 175–192 (2001), rev.

Small trees. Leaves palmately compound, 3–5-foliolate. Inflorescence axillary; racemose, 3–5 racemes borne on a peduncle; floral maturation synchronous. Flowers unisexual, dioecious; perianth 3-merous, calyx valvate, petals present in male flowers, 0 in female flowers; nectary segmented; stigmas decurrent. Fruit capsular; seeds winged. Two species from eastern Australia,

growing gregariously in cool-temperate rainforests along rocky streams.

25. *Pancheria* Brongn. & Gris   Fig. 36

Pancheria Brongn. & Gris, Bull. Soc. Bot. France 9: 74–76 (1862); Brongn. & Gris, Ann. Sci. Nat. Bot. V, 1: 367–368, 374–377 (1864); Guillaumin, Bull. Soc. Bot. France 87: 249–254 (1940).

Trees and shrubs. Leaves whorled or decussate (*P. confusa* Guillaumin), unifoliolate or imparipinnate. Inflorescence axillary; spherical heads (i.e. capitula), capitula on slender peduncles from leaf axils, or peduncles borne from axillary short-

Fig. 36. Cunoniaceae. *Pancheria elegans.* **A** Flowering branch. **B** Fruiting branch. **C** Male flower. **D** Same, opened to show pistillode. **E** Female flower. **F** Dehiscing fruit. **G** Seed. Drawn by Riocreux. (Brongniart and Gris 1868)

shoots; floral maturation synchronous. Flowers unisexual; perianth 3- or 4-merous, calyx imbricate; nectary annular or segmented; gynoecium of two, mostly free carpels. Fruits capsular/pseudofollicular; seeds winged, small. About 30 species, endemic to New Caledonia.

26. *Cunonia* L.

Cunonia L., Syst. Nat., ed. 10, 2: 1013, 1025, 1368 (1759); Guillaumin, Bull. Soc. Bot. France 87: 246–248 (1940); Coates Palgrave, Trees of Southern Africa (1983); Hoogland et al., Adansonia III, 19: 7–20 (1997).

Trees and shrubs. Leaves unifoliolate or imparipinnate. Inflorescence axillary or terminal; racemose; racemes borne in pairs or threes from a peduncle, occasionally arising directly from leaf axils; floral maturation synchronous. Flowers bisexual, or male and bisexual on a raceme; perianth 5-merous, calyx imbricate; diplostemonous (haplostemonous); floral nectary adnate to ovary. Fruits capsular, opening circumbasally and septicidally, stylodia often remaining fused so that dehiscence is acropetal; seeds small, winged. A genus of 25 species, with 24 in New Caledonia and one in South Africa. The South African species, *C. capensis* L., is morphologically similar to two distinctive species in New Caledonia, *C. macrophylla* Brong. & Gris and *C. schinziana* Däniker (Bradford 1998).

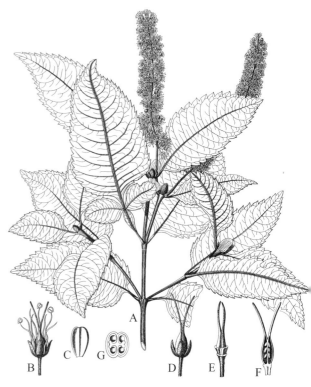

Fig. 37. Cunoniaceae. *Weinmannia heterophylla.* **A** Flowering branch; note stipules on lateral branch to the right. **B** Flower. **C** Petal. **D** Pistil with petals and stamens removed. **E** Young pistil with disk. **F** Ovary, vertical section. **G** Ovary, transverse section. Drawn by P.J.F. Turpin. (Kunth in Humboldt et al. 1823, tab. 522)

27. *Weinmannia* L.   Fig. 37

Weinmannia L., Syst. Nat. ed. 10: 1005 (1759), nom. cons.; Bernardi, Candollea 17: 123–189 (1961), reg. rev.; Bernardi, Candollea 18: 285–334 (1963), reg. rev.; Bernardi, Bot. Jahrb. Syst. 83: 126–184 (1964), reg. rev.; Bernardi, Flore de Madagascar et des Comores (1965), reg. rev.; Hopkins et al., Adansonia III, 20: 5–130 (1998), reg. rev.; J.C. Bradford, Ann. Missouri Bot. Gard. 85: 565–593 (1998); G. Harling, Flora of Ecuador. Cunoniaceae (1999), reg. rev.; Bradford & Miller, Adansonia III, 23: 219–236 (2001), reg. rev.

Arnoldia Blume (1826) = section *Fasciculatae* Bernardi ex Hoogland & H.C. Hopkins.

Leiospermum D.Don (1830) = section *Leiospermum.*

Ornithrophus Bojer ex Engl. (1870) = section *Spicatae* Bernardi ex J.C.Bradford

Pterophylla D.Don (1830) = section *Fasciculatae* Bernardi ex Hoogland & H.C. Hopkins.

Windmannia P.Browne (1756) = section *Weinmannia*

Trees and shrubs, sometimes hemiepiphytic and strangling. Leaves usually decussate in pairs but whorled in *W. commersonii* Bernardi, unifoliolate or imparipinnate, rachis winged or not, margins usually toothed (entire). Inflorescence axillary or terminal, racemose, borne solitary in leaf axils or in compound units of various forms; floral maturation synchronous. Flowers bisexual or unisexual, dioecious or polygamodioecious; solitary or fasciculate, pedicellate or sessile; calyx imbricate, petals persistent or caducous; nectary annular or segmented. Fruits capsular; seeds with hairs, lacking wings. This large genus has five sections: sect. *Weinmannia* (Americas and Mascarene Islands; includes sect. *Simplicifoliae*), sect. *Fasciculatae* (Malesia and western South Pacific), sect. *Spicatae* Bernardi ex J.C. Bradford (Madagascar and Comores), sect. *Inspersae* Bernardi ex J.C. Bradford (Madagascar), and sect. *Leiospermum* (South Pacific). About 150–160 species with the following distribution: c. 75 Americas, c. 40 W Indian Ocean (Madagascar, Comores and Mascarenes), 40 Malesia and South Pacific.

Excluded Genera

Adenilema Bl. is now commonly included in *Neillia* D. Don (Rosaceae).

Aphanopetalum Endl. will probably be recognized as its own family in Saxifragales (Fishbein et al. 2001.; Bradford and Barnes 2001).

Gummellia Ruíz & Pav. is now part of Simaroubaceae (Hoogland in Gunn et al., U.S.D.A. Tech. Bull. no. 1796 [1992]).

Kaernbachia Schltr. is synonymous with *Turpinia* Vent. (Staphyleaceae; Linden in Van Stennis, Fl. Males. I, 6: 49 [1960]).

Paracryphia Baker f. was originally included in Eucryphiaceae, but most subsequent workers have placed it elsewhere. Dickison and Baas (1977) suggested affinities with Sphenostemonaceae, Theaceae and Actinidiaceae. It is now considered a monotypic family. Molecular data place it in the Euasterid II clade and in a subclade with Quintiniaceae and Sphenostemonaceae (Savolainen, Fay et al. 2000; Lundberg 2001).

Selected Bibliography

APG (Angiosperm Phylogeny Group) 1998. See general references.

Barker, P.C.J., M.J. Brown. 1994. *Anodopetalum biglandulosum:* growth form and abundance in Tasmanian rainforest. Aust. J. Ecol. 19: 435–443.

Barnes, R.W. 1999. Palaeobiogeography, extinctions and evolutionary trends in the Cunoniaceae. A synthesis of the fossil record. Ph.D. Thesis. School of Plant Science. Hobart: University of Tasmania.

Barnes, R.W., Hill, R.S. 1999a. *Ceratopetalum* fruits from Australian Cainozoic sediments and their significance for petal evolution in the genus. Aust. Syst. Bot. 12: 635–645.

Barnes, R.W., Hill, R.S. 1999b. Macrofossils of *Callicoma* and *Codia* (Cunoniaceae) from Australian Cainozoic Sediments. Aust. Syst. Bot. 12: 647–670.

Barnes, R.W., Jordan, G.J. 2000. *Eucryphia* (Cunoniaceae) reproductive and leaf macrofossils from Australian Cainozoic sediments. Aust. Syst. Bot. 13: 373–394.

Barnes, R.W., Rozefelds, A.C. 2000. Comparative morphology of *Anodopetalum* (Cunoniaceae). Aust. Syst. Bot. 13: 267–282.

Barnes, R.W., Hill, R.S., Bradford, J.C. 2001. The history of Cunoniaceae in Australia from macrofossil evidence. Aust. J. Bot. 49: 301–320.

Bate-Smith, E.C. 1977. Chemistry and taxonomy of the Cunoniaceae. Biochem. Syst. Ecol. 5: 95–105.

Bavary, A. 1869. Catalogue des reptiles de la Nouvelle-Calédonie et descriptions d'espèces nouvelles. Mém. Soc. Linn. Normandie 15: 1–37.

Bensel, C.R., Palser, B.F. 1975. Floral anatomy in the Saxifragaceae sensu lato. IV. Baueroideae and conclusions. Am. J. Bot. 62: 688–694.

Bradford, J.C. 1998. A cladistic analysis of species-groups in *Weinmannia* (Cunoniaceae) based on morphology and inflorescence architecture. Ann. Missouri Bot. Gard. 85: 565–593.

Bradford, J.C. 1999. Cunoniaceae. In: Jørgensen, P.M., Léon-Yañez, S. Catalogue of the vascular plants of Ecuador. St. Louis: Missouri Bot. Gard. Press, pp. 419–420.

Bradford, J.C. 2002. Molecular phylogenetics and morphological evolution in Cunonieae (Cunoniaceae). Ann. Missouri Bot. Gard. 89: 491–503.

Bradford, J.C., Barnes, R.W. 2001. Phylogenetics and classification of Cunoniaceae (Oxalidales) using chloroplast DNA sequences and morphology. Syst. Bot. 26: 354–385.

Bradford, J.C., Miller, J.S. 2001. New taxa and nomenclatural notes on the flora of the Marojejy massif, Madagascar. V. Cunoniaceae: *Weinmannia*. Adansonia III, 23: 219–236.

Brongniart, A., Gris, A. 1868. Description de quelques plantes remarquables de la Nouvelle-Calédonie. Nouv. Arch. Mus. Hist. Nat. Paris 4: 1–48, pl. 1–15.

Bush, M.B., Colinvaux, P.A., Wiemann, M.C., Piperno, D.R., Liu, K. 1990. Late Pleistocene temperature depression and vegetation change in Ecuadorian Amazonia. Quat. Res. 34: 330–345.

Cadet, T. 1984. Plantes rares ou remarquables des Mascareignes. Paris: Agence de Coopération Culturelle et Technique.

Carpenter, R.J., Buchanan, A.M. 1993. Oligocene leaves, fruits and flowers of the Cunoniaceae from Cethana, Tasmania. Aust. Syst. Bot. 6: 91–109.

Carpenter, R.J., Pole, M. 1995. Eocene plant fossils from the Lefroy and Cowan paleodrainages, Western Australia. Aust. Syst. Bot. 6: 91–109.

Cooper, M.G. 1986. A pilot survey of six rare plants in New South Wales. Sydney, NSW: Parks and Wildlife Service.

Cooper, W., Cooper, W.T. 1994. Fruits of the Rain Forest. Chatswood: Geo Productions.

Davis, G.L. 1966. See general references.

Dickison, W.C. 1975a. Floral morphology and anatomy of *Bauera*. Phytomorphology 25: 69–76.

Dickison, W.C. 1975b. Leaf anatomy of Cunoniaceae. Bot. J. Linn. Soc. 71: 275–294.

Dickison, W.C. 1975c. Studies on the floral anatomy of the Cunoniaceae. Am. J. Bot. 62: 433–447.

Dickison, W.C. 1978. Comparative anatomy of Eucryphiaceae. Am. J. Bot. 65: 722–735.

Dickison, W.C. 1980a. Comparative wood anatomy and evolution of the Cunoniaceae. Allertonia 2: 281–321.

Dickison, W.C. 1980b. Diverse nodal anatomy of the Cunoniaceae. Am. J. Bot. 67: 975–981.

Dickison, W.C. 1984. Fruits and seeds of the Cunoniaceae. J. Arnold Arb. 65: 149–190.

Dickison, W.C., Baas, P. 1977. The morphology and relationships of *Paracryphia* (Paracryphiaceae). Blumea 23: 417–438.

Dickison, W.C., Rutishauser, R. 1990. Developmental morphology of stipules and systematics of the Cunoniaceae and presumed allies. II. Taxa without interpetiolar stipules and conclusions. Bot. Helvetica 100: 75–95.

Dickison, W.C., Hils, M.H., Lucansky, T.W., Stern, W.L. 1994. Comparative anatomy and systematics of woody Saxifragaceae. *Aphanopetalum* Endl. Bot. J. Linn. Soc. 114: 167–182.

Dimitri, M.J. 1972. La región de los bosques andino-patagónicos. Buenos Aires: INTA.

Doweld, A.B. 1998. The carpology and taxonomic relationships of *Davidsonia* (Davidsoniaceae). Edinburgh J. Bot. 55: 13–25.

Endress, P.K. 1994. Diversity and evolution of tropical flowers. Cambridge: Cambridge University Press.

Endress, P.K., Stumpf, S. 1991. The diversity of stamen structures in 'Lower' Rosidae (Rosales, Fabales, Proteales, Sapindales). Bot. J. Linn. Soc. 107: 217–293.

Eyde, R.H. 1970. Anatomy. In: Cuatrecasas, J. (ed.) Flora Neotropica Monogr. 2. Brunelliaceae. Darien, Conn.: Organization for Flora Neotropica, pp. 32–43.

Fishbein, M., Hibsch-Jetter, C., Soltis, D.E., Hufford, L. 2001. Phylogeny of Saxifragales (Angiosperms, Eudicots): analysis of a rapid, ancient radiation. Syst. Biol. 50: 817–847.

Fogliani, B., Bouraima-Madjebi, S., Medevielle, V., Pineau, R. 2002. Screening of 50 Cunoniaceae species from New Caledonia for antimicrobial properties. New Zeal. J. Bot. 40: 511–520.

Forster, P.I., Hyland, B.P.M. 1997. Two new species of *Eucryphia* (Cunoniaceae) from Queensland. Austrobaileya 4: 589–596.

Fortune Hopkins, H.C.F., Hoogland, R.D. 2002. Cunoniaceae. Flora Males. I, 16: 53–165. Leyden: Foundation Flora Malesiana.

Gottwald, H. 1992. Hölzer aus marinen Sanden des oberen Eozän von Helmstedt (Niedersachsen). Palaeontographica B. 225: 27–103, 20 t.

Govil, C.M., Saxena, N.P. 1976. Anatomy and embryology of *Weinmannia fraxinea* Sm. (Cunoniaceae). J. Indian Bot. Soc. 55: 219–226.

Harden, G.J., Williams, J.B. 2000. A revision of *Davidsonia* (Cunoniaceae). Telopea 8: 413–428.

Hegnauer, R. 1964, 1989. See general references.

Helmens, K.F., Kuhry, P. 1986. Middle and Late Quaternary vegetational and climatic history of the Paramo de Agua Blanca (Eastern Cordillera, Colombia). Palaeogeogr. Palaeoclimatol. Palaeoecol. 56: 291–335.

Heusser, C.J. 1993. Late Glacial of Southern South America. Quat. Sci. Rev. 12: 345–350.

Hideux, M., Ferguson, I.K. 1976. The stereostructure of the exine and its evolutionary significance in Saxifragaceae *sensu lato*. In: Ferguson, I.K., Muller, J. (eds.) The Evolutionary significance of the exine. Linn. Soc. Symp. Ser. 1, pp. 327–377.

Hill, R.S. 1991. Leaves of *Eucryphia* (Eucryphiaceae) from Tertiary sediments in south-eastern Australia. Aust. Syst. Bot. 4: 481–497.

Hill, R.S., Macphail, M.K. 1983. Reconstruction of the Oligocene vegetation at Pioneer, northeast Tasmania. Alcheringa 7: 281–299.

Hooghiemstra, H. 1989. Quaternary and Upper Pliocene glaciations and forest development in the tropical Andes: evidence from a long high resolution pollen record from the sedimentary basin of Bogotá, Colombia. Palaeogeogr. Palaeoclimatol. Palaeoecol. 72: 11–26.

Hoogland, R.D. 1960. Studies in the Cunoniaceae. I. The genera *Ceratopetalum*, *Gillbeea*, *Aistopetalum*, and *Calycomis*. Aust. J. Bot. 8: 318–341.

Hopkins, H.C.F. 1998a. A revision of *Weinmannia* (Cunoniaceae) in Malesia and the Pacific 1. Introduction and an account of the species of Western Malesia, the Lesser Sundas and the Moluccas. Adansonia III, 20: 5–41.

Hopkins, H.C.F. 1998b. A revision of *Weinmannia* (Cunoniaceae) in Malesia and the Pacific 3. New Guinea, Solomon Islands, Vanuatu and Fiji, with notes on the species of Samoa, Rarotonga, New Caledonia and New Zealand. Adansonia III, 20: 67–106.

Hopkins, H.C.F., Florence, J. 1998. A revision of *Weinmannia* (Cunoniaceae) in Malesia and the Pacific 4. The Society, Marquesas and Austral Islands. Adansonia III, 20: 107–130.

Hufford, L., Dickison, W.C. 1992. A phylogenetic analysis of Cunoniaceae. Syst. Bot. 17: 181–200.

Humboldt, F.H.A. von, Bonpland, A.J., Kunth, C.S. 1823. Nova genera et species plantarum, vol. 6, t. 522. Paris.

Hyland, B.P.M., Whiffin, T. 1993. Australian tropical rain forest trees: an interactive identification system, vol. 2. Canberra: CSIRO.

Ingle, H.D., Dadswell, H.E. 1956. The anatomy of the timbers of the south-west Pacific area IV. Cunoniaceae, Davidsoniaceae, and Eucryphiaceae. Aust. J. Bot. 4: 125–151.

Jaffré, T. 1980. Étude écologique de peuplement végétal des sols dérivés de roches ultrabasiques en Nouvelle-Calédonie. Paris: O.R.S.T.O.M.

Johri, B.M. et al. 1992. See general references.

Jordan, G.J., Carpenter, J.R., Hill, R.S. 1991. Late Pleistocene vegetation and climate near Melaleuca Inlet, south-west Tasmania, as inferred from fossil evidence. Aust. J. Bot. 39: 315–333.

Jordan, G.J., Macphail, M.K., Barnes, R.W., Hill, R.S. 1995. An Early to Middle Pleistocene flora of subalpine affinities in lowland Western Tasmania. Aust. J. Bot. 43: 231–242.

Lowry, P.P.I. 1998. Diversity, endemism and extinction in the flora of New Caledonia: a review. In: Peng, C.-I., Lowry II, P.P. (eds.) Rare, threatened, and endangered floras of Asia and the Pacific Rim. Taipei: Institute of Botany, pp. 181–206.

Lundberg, J. 2001. A well resolved and supported phylogeny of Euasterids II based on a Bayesian inference, with special emphasis on Escalloniaceae and other incertae sedis. In: Lundberg, J. (ed.) Phylogenetic studies in the Euasterids II with particular reference to Asterales and Escalloniaceae, chap. V. Uppsala, Acta Universitatis Upsaliensis.

Lusk, C.H. 1999. Long-lived light-demanding emergents in southern temperate forests: the case of *Weinmannia trichosperma* (Cunoniaceae) in Chile. Plant Ecol. 140: 111–115.

Matthews, M.L., Endress, P.K., Schönenberger, J., Friis, E.M. 2001. A comparison of floral structures of Anisophylleaceae and Cunoniaceae and the problem of their systematic position. Ann. Bot. II, 88: 439–455.

McGlone, M.S. 1982. Modern pollen rain, Egmont National Park, New Zealand. New Zeal. J. Bot. 20: 253–262.

McGlone, M.S. 1983. Holocene pollen diagrams, Lake Rotorua, North Island, New Zealand. J. Roy. Soc. New Zeal. 13: 53–65.

Muñoz Pizarro, C. 1973. Chile: plantas en extinción. Santiago: Editorial Universitaria.

Orozco, C.I. 1997. Sobre la posicion sistematica de *Brunellia* Ruiz & Pavon. Caldasia 19: 145–164.

Palmer, E., Pitman, E. 1961. Trees of South Africa. Cape Town: A.A. Balkema.

Pocknall, D.T. 1980. Modern pollen rain and Aranuian vegetation from Lady Lake, north Westland, New Zealand. New Zeal. J. Bot. 18: 275–284.

Poole, I., Cantrill, D.J. 2001. Fossil woods from Williams Point Beds, Livingston Island, Antarctica: a Late Cretaceous southern high latitude flora. Palaeontology.

Poole, I., Cantrill, D.J., Hayes, P., Francis, F. 2000. The fossil record of Cunoniaceae: new evidence from Late Cretaceous wood of Antarctica? Rev. Palaeobot. Palynol. 111: 127–144.

Poole, I., Hunt, J.R., Cantrill, D.J. 2001. A fossil wood flora from King George Island: ecological implications for an Antarctic Eocene vegetation. Ann. Bot. II, 88: 33–54.

Prakash, N., McAlister, E.J. 1977. An embryological study of *Bauera capitata* with comments on the systematic position of *Bauera*. Aust. J. Bot. 25: 615–622.

Pratt, T.K. 1983. Diet of the Dwarf Cassowary *Casuarius bennettii picticollis* at Wau, Papua New Guinea. The Emu 82 Suppl.: 283–285.

Rao, T.A., Dickison, W.C. 1985a. The veinsheath syndrome in Cunoniaceae I *Pancheria* Brongn. & Gris. Proc. Ind. Acad. Sci. (Plant Sci.) 95: 87–94.

Rao, T.A., Dickison, W.C. 1985b. The veinsheath syndrome in Cunoniaceae. II. The genera *Acsmithia*, *Codia*, *Cunonia*, *Geissois*, *Pullea* and *Weinmannia*. Proc. Ind. Acad. Sci. (Plant Sci.) 95: 247–261.

Romero, E.J., Castro, M.T. 1986. Material fungio y granos de polen de angiospermas de la formación Río Turbio (Eoceno), Provincia de Santa Cruz, Republica Argentina. Ameghiniana 23: 101–118.

Rozefelds, A.C., Barnes, R.W. 2002. The systematic and bio-geographical relatonships of *Ceratopetalum* (Cunoniaceae) in Australia and New Guinea. Int. J. Plant Sci. 163: 651–673.

Rozefelds, A.C., Barnes, R.W., Pellow, B. 2001. A new species and comparative morphology of *Vesselowskya* (Cunoniaceae). Aust. Syst. Bot. 14: 175–192.

Rutishauser, R., Dickison, W.C. 1989. Developmental morphology of stipules and systematics of the Cunoniaceae and presumed allies. I. Taxa with interpetiolar stipules. Bot. Helvetica 99: 147–169.

Saint-Hilaire, A.F.C.P. de 1828. Flora brasiliae meridionales, vol. 2, t. 117. Paris.

Savolainen, V., Fay, M.F. et al. 2000. See general references.

Schönenberger, J., Friis, E.M., Matthews, M.L., Endress, P.K. 2001. Cunoniaceae in the Cretaceous of Europe: evidence from fossil flowers. Ann. Bot. 88: 423–437.

Sluiter, I.R.K. 1991. Early Tertiary vegetation and climates, Lake Eyre region, northeastern South Australia. In: Williams, M.A.J., Deckker, P.D., Kershaw, A.P. (eds.) The Cainozoic in Australia: a reappraisal of the evidence.Melbourne: Geological Society of Australia, Spec. Publ. 18, pp. 99–118.

Smith, A.C. 1985. Flora Vitiensis Nova: a new flora of Fiji. Volume 3. Lawai, Hawaii: Pacific Tropical Botanical Garden.

Stanley, T.D., Ross, E.M. 1983. Flora of south-eastern Queensland. Brisbane: Queensland Department of Primary Industries.

Still, C.J., Foster, P.N., Schneider, S.H. 1999. Simulating the effects of climate change on tropical montane cloud forests. Nature 398: 608–610.

Stover, L.E., Partridge, A.D. 1973. Tertiary and Late Cretaceous spores and pollen from the Gippsland Basin, Southeastern Australia. Proc. Roy. Soc. Victoria 85: 237–286.

Sweeney, P., Bradford, J.C. 2003. The phylogenetic position and comparative morphology of *Hooglandia* within Cunoniaceae. Ann. Missouri Bot. Gard. 90.

Villagran, C. 1990. Glacial climates and their effects on the history of the vegetation of Chile: a synthesis based on the palynological evidence from Isla de Chiloé. Rev. Palaeobot. Palynol. 65: 17–24.

Walter, K.S., Gillett, H.J. (eds.) 1998. 1997 IUCN Red List of Threatened Plants. Gland (Switzerland), Cambridge (UK): IUCN – World Conservation Union.

Waterhouse, R.D. 2001. Observations on the diet of the Topknot Pigeon *Lopholaimus antarcticus* in the Illawarra Rainforest, New South Wales. Corella 25: 32–38.

Wollenweber, E., Dörr, M., Rozefelds, A.C., Minchin, P., Forster, P.I. 2000. Variation in flavonoid exudates in *Eucryphia* species from Australia and South America. Biochem. Syst. Ecol. 28: 111–118.

Curtisiaceae

K. Kubitzki

Curtisiaceae (Harms) Takht., Sist. Magnoliof.: 214 (1987).

Evergreen tree. Leaves opposite, coarsely dentate, rusty velvety beneath; hairs unicellular. Inflorescences terminal, tricussately thyrso-paniculate, many-flowered, prophylls present; flowers minute, 4-merous, hermaphroditic, or rarely plants dioecious; calyx tube turbinate with open, triangular lobes; petals (sub)valvate, triangular-ovate; stamens 4, alternipetalous; filaments subulate; anthers cordate at base, introrse, longitudinally dehiscing; disk epigynous, broad, 4-angled, densely barbate; ovary inferior, 4-locular; ovules 1 per locule, with ventral raphe and micropyle directed upwards; style short, subconical, with 4-lobed stigma. Fruits drupaceous, small, subglobose, 4-locular, 4-seeded, areolate at apex, crowned by the calyx, eventually dehiscing; seeds elongate-oblong, subcylindric, with elongate embryo in copious endosperm. $n = 13$.

A single sp., *C. dentata* (N.L. Burm.) C.A. Smith, southern and eastern S Africa, in forest and bushland.

FLORAL STRUCTURE. Most Cornaceae, in which *Curtisia* has always been included, have a peculiar gynoecial bundle supply that was studied by Eyde (1968). In *Davidia, Cornus, Alangium* and *Nyssa*, ovular bundles pass through the septa or opposite to them and enter the placentae arching above each septum. The syncarpous, 4-carpellate inferior ovary of *Curtisia* differs from them in having an axile bundle supply to the ovules, just as would be expected for a syncarpous ovary, and which is probably plesiomorphic in relation to the cornaceous condition.

POLLEN MORPHOLOGY. Pollen is sphaeroidal and relatively small (12–20×12–$17\,\mu$m, i.e. half the size of most *Cornus*), tricolporate, provided with colpi costae, a complete tectum and a lamellated ektexine in the areas of the endoapertures; the tectum has impressions that appear as perforations but do not connect with the intercolumellar space (Ferguson and Hideux 1978). Similar pollen is found in *Cornus* and *Mastixia* and possibly also other genera of Cornaceae.

KARYOLOGY. *Curtisia* and *Mastixia arborea* share the gametophytic chromosome number $n = 13$ (Goldblatt 1978).

FRUITS. *Curtisia* fruits are red at maturity and are eagerly sought after by birds, wild pigs, monkeys and baboons. Fruit opening seems to be with valves, just as in Cornaceae, as Eyde (1988: 296 [in legend to *Curtisia* fruit]) writes: "If boiled and dried repeatedly, this stone would open like a *Cornus* stone".

PHYTOCHEMISTRY. The red wood of *Curtisia* quickly darkens in fresh air and the bark contains 3–21.5% tannin (Watt and Breyer-Brandwijk 1962). Proanthocyanins, ellagitannins and two unidentified iridoids have been reported from leaf tissue by Bate-Smith et al. (1975).

AFFINITIES. Most plant systematists have included *Curtisia* in Cornaceae, which made sense in view of its pollen morphology and the valvate fruit stone, apart from the general agreement with Cornaceae in floral and vegetative structures. Eyde (1988) excluded *Curtisia* because its gynoecium lacks the apomorphic transseptal bundles of the cornaceous genera. In contrast, I would not object against including the plesiomorphic condition in the family, and would like to follow taxonomists such as Hooker (1867) and Harms (1897) in the placement of *Curtisia* in Cornaceae, had not the molecular analyses of Xiang et al. (2002) resolved *Grubbia* and *Curtisia* in a well-supported association and sharing a deletion, the latter making an artifactial long branch attraction of the two less probable (see also under Grubbiaceae, this volume).

DISTRIBUTION AND HABITATS. *Curtisia* extends from the Cape Peninsula eastward to Natal and from there to the north through the Transvaal into the mountains of eastern Zimbabwe. In closed

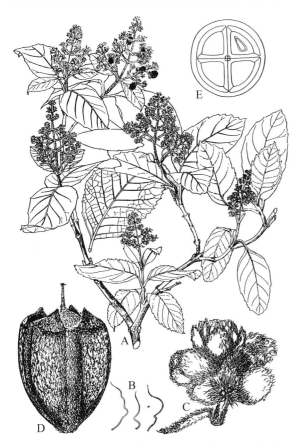

Fig. 38. Curtisiaceae. *Curtisia dentata.* **A** Flowering/fruiting twig. **B** Hairs. **C** Group of flowers. **D** Fruit. **E** Transverse section of fruit. (Cannon 1978)

vegetation it is a tree up to 20 m high and in open vegetation it has a more shrubby habit; it grows up to an altitude of 1800 m.

One genus:

Curtisia Aiton   Fig. 38

Curtisia Aiton, Hort. Kew. 1: 162 (1789), nom. cons.

Description as for family.

Selected Bibliography

Bate-Smith, E.C., Ferguson, I.K., Hutson, K., Jensen, S.R., Nielsen, B.J., Swain, T. 1975. Phytochemical interrelationships in the Cornaceae. Biochem. Syst. Ecol. 3: 79–89.

Cannon, J.F.M. 1978. Cornaceae. In: Launert, E. (ed.) Flora Zambesiaca vol. 4. London: Flora Zambesiaca Managing Committee, pp. 635–638.

Eyde, R.H. 1968. The peculiar gynoecial vasculature of Cornaceae and its systematic significance. Phytomorphology 17: 172–182.

Eyde, R.H. 1988. Comprehending *Cornus*: puzzles and progress in the systematics of the dogwoods. Bot. Rev. 54: 233–351.

Ferguson, I.K. 1977. Cornaceae. World pollen and spore flora 6. Stockholm: Almqvist and Wiksell, 34 pp.

Ferguson, I.K., Hideux, M.J. 1978. Some aspects of the pollen morphology and its taxonomic significance in Cornaceae s.l. In: 4th Int. Palynol. Conf., Lucknow, 1976–77, 1: 240–249.

Goldblatt, P. 1978. A contribution to cytology in Cornales. Ann. Missouri Bot. Gard. 65: 650–655.

Harms, H. 1897. Cornaceae. In: Engler, A., Prantl, K. Die natürlichen Pflanzenfamilien III, 8. Leipzig: W. Engelmann, pp. 250–270.

Hooker, J.D., 1867. Cornaceae. In: Bentham, G., Hooker, J.D. Genera Plantarum, vol. 1 (3). London: Reeve & Co., pp. 947–952.

Marais, W. 1985. *Curtisia*. Kew Mag. 2: 368.

Watt, J.M., Breyer-Brandwijk, M.G. 1962. The medicinal and poisonous plants of southern and eastern Africa. Edinburg: E. & S. Livingstone.

Xiang, Q.-Y. et al. 2002. See general references.

Cyrillaceae

K. Kubitzki

Cyrillaceae Endl., Ench. Bot.: 578 (1841), nom. cons.

Deciduous or evergreen and glabrous small trees or shrubs. Leaves alternate, simple, entire, short-petioled or sessile, exstipulate. Inflorescences terminal or axillary racemes; pedicels with paired prophylls. Flowers perfect, regular, hypogynous; sepals 5(–7), imbricate, sometimes quincuncially arranged, persistent, the outer larger than the inner, often accrescent in fruit; petals as many as sepals, distinct, imbricate; androecium diplostemonous or (*Cyrilla*) haplostemonous; anthers dorsifixed, tetrasporangiate, not inverted at anthesis, dehiscing by longitudinal slits; disk intrastaminal, surrounding the base of the ovary; gynoecium syncarpous; ovary (2–)5-locular with axile placentas; style very short or nearly suppressed with lobed stigma or with 2–4 very short stylar branches with apical stigmas; ovules 1–3 per locule (pendulous from near the top of the locule or from a pendulous, stalk-like placenta), anatropous, unitegmic, tenuinucellar. Fruit indehiscent, 1-seeded, dry and drupaceous or samaroid; seed coat completely reduced; embryo slender, straight, with small, slightly expanded cotyledons, embedded in a copious, fleshy endosperm. $x = 10$.

A family of two probably monotypic genera, extending from the coastal plains of SE North America to Central and N South America, including the West Indies.

VEGETATIVE STRUCTURES. Young stems are provided with ridges decurrent from the leaf bases; these vanish at the end of the second year. Periderm develops early in the inner layers of the cortex. The secondary xylem is composed of vessel elements with scalariform perforation plates with numerous bars, tracheids and fiber-tracheids, uniseriate and multiseriate rays, and diffuse axial parenchyma; nodes are unilacunar throughout (Thomas 1961).

Leaf venation is pinnate. Stomata are anomocytic. Solitary crystals are found in the leaves and druses are abundant. Ligulate, glandular structures on each side of the buds and of bracts are of unknown homology but, according to Thomas (1960), may be vestigial stipules.

The root system is shallow. Along the horizontally running roots numerous adventitious roots arise, by means of which *Cyrilla* and *Cliftonia* are able to build up clones of vast extension. Thomas (1961) excavated a group of *Cyrilla* bushes covering an area 15 feet wide and 30 feet long, comprising over 50 bushes. These plants were interconnected with each other by their root system and remained so for several years.

Sieve element plastids have protein filaments and crystals but no starch (Pcf type; Behnke 1982, 1991), and these are probably autapomorphic.

EMBRYOLOGY. In the anthers the epidermis, endothecium and middle layer contain tannin; the tapetum is of the secretory type. Ovules are unitegmic and tenuinucellate; both normal Polygonum type and aposporous embryo sacs develop in the same ovule (*Cliftonia*), the latter remaining functionless. Haustoria develop at both ends of the embryo sac. Degeneration of the egg and zygote is frequent. During embryo development, the integumentary cells become vacuolised and degenerate; they have vanished in the mature seed. The relatively large embryo is straight and nearly as long as the seed. The pericarp is tanniniferous and contains druses (Copeland 1953; Vijayaraghavan 1969; Vijayaraghavan and Dhar 1978).

POLLEN MORPHOLOGY. Pollen is tri-(to hexa-)colporate, spheroidal to suboblate with wide colpi, the tectum psilate and microperforate, the infratectum granular, the equatorial diameter 18–29 μm (Zhang and Anderberg 2002).

REPRODUCTIVE BIOLOGY. The vegetative reproduction by root sprouting of *Cyrilla* and *Cliftonia* is often observed in cut-over or otherwise disturbed areas, but disturbance or injury is not a necessary prerequisite for the formation of clones by root sprouting. Within these clones, the abundant fruits annually produced rarely contain seeds, and parthenocarpoy has been observed to be more frequent towards the centre of large clones. This may indicate a pronounced self-sterility; Thomas (1961) concluded that self-fertilization may provide sufficient stimulus to initiate fruit development, whereas seed development seems to require pollination from a different individual.

In North America Uphof (1942) observed that *Cyrilla* and *Cliftonia* are protogynous. The flowers are showy and produce scent and nectar; they are visited by several classes of insects.

FRUIT AND SEED. The complete lack of a seed coat has been noted by several students of the family and there is not the slightest rudiment of a testa (Gilg 1892; Thomas 1960).

PHYTOCHEMISTRY. Bate-Smith (1962) found myricetin, ellagic acid and procyanidin; i.e. both condensed and hydrolysable tannnins are likely to be present.

FAMILY CIRCUMSCRIPTION. The South and Central American genus *Purdiaea*, usually included in Cyrillaceae, has on the basis of morphological data and DNA sequence analyses been transferred to Clethraceae, with which it agrees in slight sympetaly and versatile, ventrifixed, inverted and poricidal anthers (Anderberg and Zhang 2002).

AFFINITIES. Formerly associated with Aquifoliaceae or Celastraceae, DNA sequence analyses place Cyrillaceae in the broadly circumscribed Ericales sensu APG (1998), in which they are the sister group of Ericaceae, whereas Ericaceae + Cyrillaceae are sister to Clethraceae (Anderberg et al. 2002; Anderberg and Zhang 2002).

DISTRIBUTION AND HABITATS. Cyrillaceae prefer the wet and acidic habitats on the coastal plains of the SE USA; *Cyrilla* further extends into Mexico, Central America, the Caribbean, the Roraima sandstone formation in Venezuela and Guiana, Colombia, and N Brazil, ranging there from the lowland to the mountain summits, and growing in wet heath formations.

PALAEOBOTANICAL RECORD. In the Brandon Lignite of Vermont (USA) of Upper Oligocenic age, the most abundant wood and second most abundant pollen was found to be that of *Cyrilla* (Barghoorn and Spackman 1949). *Cyrilla*-like wood is also known from Tertiary brown coal beds in western Europe (van der Burgh 1973), accompanied by ample pollen. Other pollen records from

the Maestrichtian onwards have been referred to Clethraceae and/or Cyrillaceae, but the distinction between the two remains problematic (Muller 1981). The same may be true of several fossil fruit remains from the European Upper Cretaceous and Tertiary that have been related to Cyrillaceae (see Knobloch and Mai 1986), the taxonomic position of which is contentious (Friis 1985).

KEY TO THE GENERA

1. Stamens 10; petals not glandular; fruit distinctly 3–5-winged   1. *Cliftonia*
– Stamens 5; petals glandular on adaxial surface; fruit (sub)globose, longitudinally bisulcate or trisulcate
                                                                   2. *Cyrilla*

1. *Cliftonia* Banks ex Gaertn.f.

Cliftonia Banks ex Gaertn.f., Fructus 3: 246 (1805); Thomas, Contrib. Gray Herb. 186: 74–76 (1960), rev.

Evergreen. Pedicels with two caducous prophylls; stamens 5 + 5; filaments laterally expanded and petaloid in lower half; ovary 3–5-angled and -locular, each locule containing a single, pendulous ovule; stigma massive, subsessile, 2–5-lobed. Fruit 2–5-winged, up to 5-seeded but usually devoid of seeds. A single sp., *C. monophylla* (Lam.) Britton ex Sarg., on the coastal plains of the SE U.S.A.

2. *Cyrilla* Garden ex L.

Cyrilla Garden ex L., Mant. Pl. 1: 5 (1767); Thomas, Contrib. Gray Herb. 186: 76–104 (1960), rev.

Evergreen or deciduous. Pedicels with two persistent prophylls; petals thickened and glandular on upper surface medially and below the middle; stamens 5, antepetalous; filaments subulate; ovary 2–4-locular; ovules pendulous, 1–3 per locule; style short, persistent, with 2–4 short stylar branches. Fruit longitudinally 2–3-sulcate with no more than 1 seed in each locule, often devoid of seeds. $2n = 40$. A single sp., *C. racemiflora* L., N South America, Central America, Caribbean, and SE U.S.A. Up to eleven spp. have been distinguished, but Thomas (l.c.), who studied their variation pattern, was unable to detect real gaps or discontinuities separating them.

Selected Bibliography

Anderberg, A.A. et al. 2002. See general references.
Anderberg, A.A., Zhang Xiaoping 2002. Phylogenetic relation-

ships of Cyrillaceae and Clethraceae (Ericales) with special emphasis on the genus *Purdiaea* Planch. Org. Div. Evol. 2: 127–137.

APG (Angiosperm Phylogeny Group) 1998. See general references.

Barghoorn, E.S., Spackman, W. 1949. A preliminary study of the flora of the Brandon Lignite. Am. J. Sci. 247: 33–39.

Bate-Smith, E.C. 1962. See general references.

Behnke, H.-D. 1982. Sieve element plastids of Cyrillaceae, Erythroxylaceae and Rhizophoraceae: description and significance of subtype PV plastids. Plant Syst. Evol. 141: 31–39.

Behnke, H.-D. 1991. See general references.

Copeland, H.F. 1953. Observations on the Cyrillaceae, particularly on the reproductive structures of the North American species. Phytomorphology 3: 405–411.

Friis, E.M. 1985. Angiosperm fruits and seeds from the Middle Miocene of Jutland (Denmark). Biol. Skr. Vid. Selsk. København 24 (3): 1–66.

Gilg, E. 1892. Cyrillaceae. In: Engler & Prantl, Nat. Pflanzenfam. III, 5. Leipzig: W. Engelmann, pp. 179–182.

Knobloch, E., Mai, D.H. 1986. Monographie der Früchte und Samen in der Kreide von Mitteleuropa. Rozpr. Ústř. úst. geol., Praha 47: 1–220.

Muller, J. 1981. See general references.

Thomas, J.L. 1960. A monographic study of the Cyrillaceae. Contrib. Gray Herb. 186: 1–114.

Thomas, J.L. 1961. The genera of the Cyrillaceae and Clethraceae of the southeastern United States. J. Arnold Arbor. 42: 96–106.

Uphof, J.C.Th. 1942. Cyrillaceae. In: Die natürlichen Pflanzenfamilien, 2nd edn., 20b. Leipzig: W. Engelmann, pp. 1–12.

van der Burgh, J. 1973. Hölzer der niederrheinischen Braunkohlenformation. 2. Hölzer der Braunkohlengruben "Maria Theresia" zu Herzogenrath, "Zukunft West" zu Eschweiler und "Viktor" (Zülpich Mitte) zu Zülpich. Nebst einer systematisch-anatomischen Bearbeitung der Gattung *Pinus*. Rev. Palaeobot. Palynol. 15: 73–275.

Vijayaraghavan, M.R. 1969. Studies in the family Cyrillaceae. 1. Development of male and female gametophyte in *Cliftonia monophylla* (Lam.) Britton ex Sarg. Bull. Torrey Bot. Club 96: 484–489.

Vijayaraghavan, M.R., Dhar, U. 1978. Embryology of *Cyrilla* and *Cliftonia* (Cyrillaceae). Bot. Not. 131: 127–138.

Zhang, X.-P., Anderberg, A. 2002. Pollen morphology in the ericoid clade of the order Ericales, with special reference on Cyrillaceae. Grana 41: 201–215.

Diapensiaceae

P.J. Scott

Diapensiaceae (Link) Lindl., Nat. Syst. Bot., ed. 2: 233 (1836), nom. cons.

Low-growing, generally mat-forming, evergreen herbs or shrublets with ecto- and endotrophic mycorrhizae. Leaves alternate, simple, linear-lanceolate, exstipulate, usually crowded on stems, or leaves round-cordate, petiolate, and fewer. Flowers solitary or in compact racemes, perfect, actinomorphic, hypogynous, slightly protogynous or homogamous; prophylls usually present; perianth biseriate, pentamerous, imbricate; sepals free or connate into a 5-lobed tube, persistent; petals 5, white, purple, or rarely yellow, caducous, connate into a sympetalous corolla or (*Galax*) almost free; stamens 5, attached to corolla tube and alternate with its lobes, distinct or connate basally in a ring, in all but *Diapensia* and *Pyxidanthera* alternating with 5 staminodia; anthers tetrasporangiate or (*Galax*) disporangiate, dehiscing longitudinally or (*Galax*, *Pyxidanthera*) transversely; nectary disk absent; gynoecium 3-carpellate; ovary syncarpous, 3-locular; ovules few to many on more or less intrusive placentas; ovules anatropous to campylotropous, unitegmic, tenuinucellate; style simple, hollow; stigma 3-lobed (*Diapensia* with 3 style branches). Fruit a loculicidal capsule; seeds thin, with exotestal seed coat, a central straight or slightly curved embryo and abundant fleshy endosperm.

A family of 5 genera and 13 species distributed in North America and eastern Asia, and one cultivar.

VEGETATIVE MORPHOLOGY. Leaf shape varies from reniform to orbicular and petiolate in *Galax* and *Shortia* to linear and petiolate in *Berneuxia* to linear-lanceolate or linear-oblanceolate and (nearly) sessile in *Diapensia* and *Pyxidanthera*. The size decreases in the series given above, and the reduction in size and shape is correlated with adaptation to xeric or arctic-alpine habitats. Trichomes are absent or infrequent except in *Pyxidanthera*. *Diapensia* forms domes or mats which are adapted to windy habitats, and act as solar heat traps, which restrict their distribution in temper-

ate regions. The domes have the surface covered with living leaves and the interior is packed with dead leaves which decompose to duff (Day and Scott 1981).

VEGETATIVE ANATOMY. The leaf venation pattern of *Diapensia* and *Pyxidanthera* is unique among dicotyledons. It is inverted-eucamptodromous, having the secondary veins curve basally and become attenuated in that direction without rejoining the primary vein or other secondaries. *Galax* and *Shortia* have actinodromous or camptodromous patterns, and *Berneuxia* has a brochidodromous pattern (Murphy and Hardin 1976). Leaf epidermal cells have undulating lateral walls and, particularly on the upper surface, a very thick outer wall traversed by numerous pits. On the lower leaf surface, *Berneuxia thibetica* has papillose epidermal cells and *Diapensia lapponica* has a swollen, thick cuticle over the middle of each epidermal cell. Mesophyll is either bifacial or isodiametric in *Shortia* and *Galax*. Chlorophyll is present in the epidermis of *Diapensia* and *Galax*. Stomata are found only on the lower surface in *Berneuxia*, some *Shortia*, and *Diapensia lapponica*, but on both surfaces in other species of *Diapensia*, *Galax*, *Pyxidanthera*, and *Shortia*. Subsidiary cells are absent. The number of layers of palisade cells varies from 3 in *Berneuxia*, 2–3 in *Diapensia lapponica*, and 1 in *Shortia uniflora*. The spongy mesophyll in *Diapensia lapponica* has some thick-walled cells (Solereder 1908; Metcalfe and Chalk 1950).

Calcium oxalate is excreted along the leaf margins as clustered crystals or as solitary crystals in *Galax*. Internal glands are absent (Solereder 1908; Metcalfe and Chalk 1950). A number of members of the family undergo annual changes in the coloration of their leaves. *Diapensia lapponica* leaves are olive green in the summer and dark burgundy in winter due to anthocyanin accumulation (Day and Scott 1981). In the wood, primary medullary rays are observed only immediately above the nodes; secondary rays are absent, and the vessels have small lumina (Solereder 1908). Pith is often heterogeneous, consisting of thin-

and thick-walled cells. Growth rings are poorly defined in *Diapensia lapponica* (Metcalfe and Chalk 1950).

The principal reserve foods in alpine and arctic plants are starches, sugars, and lipids. In evergreen subshrubs like *Diapensia lapponica*, lipids and some carbohydrates are stored in the old stems and old leaves (Petersen 1912).

The roots have ecto- and endotrophic mycorrhizae. In *Diapensia lapponica*, the tap and adventitious roots are generally shallow (<30 cm) and convoluted. Cork is poorly developed and phloem shows collenchymatous thickening in *Diapensia lapponica* (Metcalfe and Chalk 1950).

FLOWER STRUCTURE. Flowers are actinomorphic, hypogynous, and pentamerous except for the ovary which is tricarpellate. The sepals are persistent and free or slightly connate. The petals are free in *Galax* but connate in the other genera. The family has a modified diplostemonous androecium. The outer whorl consists of 5 functional stamens and the inner whorl has 5 staminodia.

The stamens are adnate to the corolla and alternate with the corolla lobes. The staminodia are sometimes reduced, or absent in *Diapensia* and *Pyxidanthera*. When present, the staminodia occur opposite the corolla lobes and are adnate to varying degrees. The anthers vary in length, orientation, number of pollen sacs, and plane of dehiscence. The anthers of *Pyxidanthera* and *Galax* have appendages. The relative size, number of thecae and pollen chambers, orientation of dehiscence, and presence of appendages show considerable variation. There are, however, a number of features in common. Anthers are definitely or obliquely introrse. The vascular strand of the filament extends to the apex of the connective where it tapers off in *Diapensia* and *Shortia*. It ends in a knob of vascular tissue below the horizontal pollen sacs in *Galax* and *Pyxidanthera*.

The ovary has three locules with several to many ovules. The placentae are axile, usually enlarged. There is a split in the central column in the upper part of the ovary which, in some genera, extends completely through the placentae so that the placentation becomes parietal. The ovary tapers into the style which varies in relative length and has a stylar canal. The stigma is 3-lobed. Ovules vary from few to many. In some species, nectar is secreted by a few layers of cells on the outer surface of the ovary base.

For the vascular supply of the flowers, the reader is referred to the study by Palser (1963).

EMBRYOLOGY. The mature anther walls have a well-developed fibrous endothecium. The tapetum is binucleate and glandular. The ovules are generally anatropous, but vary from hemitropous to campylotropous. In *Shortia uniflora* the ovules are unitegmic and the nucellus is single-layered and disappears during megaspore development. The inner layer of the integument constitutes an endotheliaceous tissue. The nucellus consists of an epidermis and a few cells at the base. The epidermis disintegrates so that the gametophyte is in contact with the inner layer of the single integument (Palser 1963). The pollen tube passes down the style through the stylar canal and enters the ovules porogamously. Seed production is a problem for many species. Examination of ovules shows comparatively few with endosperm or embryo development and considerable signs of disintegration, suggesting failure of pollination or fertilization (Palser 1963).

The endosperm is cellular and there are no endosperm haustoria. The embryo is straight, cylindrical, approximately two-thirds the length of the seed. Cotyledons are approximately one-third the total length. The seed coat consists of 2–3 cell layers (Palser 1963; Yamazaki 1966).

POLLEN MORPHOLOGY. Pollen grains are monads, tricolp(oroid)ate, semitectate or intectate, per-reticulate, with at least some brochi more than 1 μm wide. The sexine is as thick as the nexine or thicker; colpi are tenuimarginate with an equatorial constriction (Fig. 39). The pollen grains are not of a very distinctive type; they resemble those of many families (Erdtman 1952; Xi and Tang 1990).

KARYOLOGY. A chromosome number of $2n = 12$ has been determined for the family except in *Galax urceolata* where diploids and tetraploids exist that are indistinguishable morphologically. The tetraploid cytotypes are considered an excellent example of autopolyploidy in natural populations (Nesom 1983).

POLLINATION AND REPRODUCTIVE SYSTEMS. Flowers of *Diapensia lapponica* are visited by a number of members of Hymenoptera and Diptera (Fig. 40; Petersen 1912; Day and Scott 1981). Problems with reproduction have been noted in the literature and few seedlings have been observed in populations. Although some species are abundant locally, their distributions are often limited. There have been a number of reasons suggested: poor seed dispersal, seed being retained in the capsule

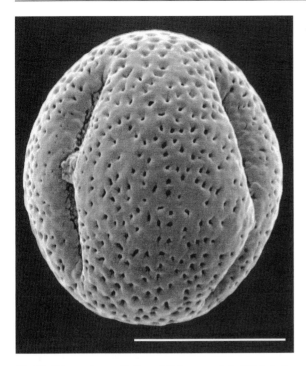

Fig. 39. Diapensiaceae. Pollen of *Galax urceolata*, SEM photo, bar = 10 μm (Palynological Laboratory, Stockholm)

Fig. 40. Diapensiaceae. *Diapensia lapponica* flowers, visited by Diptera. Photo P.J. Scott

ponica (Bliss 1962). Seeds germinate in 21–28 days (in 11 days in *Diapensia lapponica*) after 3 months of chilling. There may be as many as 50 flowers per year on very large plants of *Diapensia lapponica*, with each flower giving rise to a mean production of 121 seeds per capsule, with seed set averages of 70–75% and an average of 240 ovules/capsule (Day and Scott 1981).

DISPERSAL. The upright capsules of many species would suggest dislodgement and limited dispersal by wind. Capsules of *Diapensia lapponica* have been observed projecting through the snow, dehisced and with mature seed in mid-winter. This suggests dispersal across the surface of the snow (Day and Scott 1981).

PHYTOCHEMISTRY. The family is noted for accumulating significant amounts of aluminium. *Shortia* is surpassed by only a couple of other plants in the amounts accumulated (Gibbs 1974).

AFFINITIES AND PHYLOGENY. Diapensiaceae are characterized by the presence of staminodia, single pollen grains, absence of a disk, 3-carpellate gynoecium, 1-layered testa, absence of integumentary tapetum, absence of terminal and chalazal haustoria, and aluminium accumulation (Anderberg 1993). The family has often been associated with the narrowly defined Ericales (Takhtajan 1980, among others) and more recently has been included in the broadly defined Ericales sensu APG (1998). The five-gene sequence analysis by Anderberg et al. (2002) places Diapensiaceae in close association with Styracaceae, a position also supported by embryological traits.

A phylogenetic analysis based on morphology and four genes (Rönblom and Anderberg 2002) showed that *Galax* is sister to all other genera, and *Pyxidanthera* sister to the remaining genera at the next higher node.

DISTRIBUTION AND HABITATS. The present distribution suggests that the family was formerly more widespread. Presently, it is represented by a few relict and well-defined species. Most species are confined to the two centres of diversity for the family, the Appalachian Mountain system of eastern North America and eastern Asia, except for *Diapensia lapponica* which is circumpolar. These plants are probably relicts of the Arctotertiary forest which had a circumboreal distribution before the Pleistocene. The majority of species do not compete well with other plants and are confined to arctic/alpine habitats except *Pyxidanthera*

where it germinates, inviability of seeds, and fungal attack of seeds (Ross 1936; Primack and Wyatt 1975). In localities at the southern limit of the range of *Diapensia lapponica*, some plants in each population bloom in late summer and disperse their seeds through the winter. This appears to be an adaptation to avoid fungal infection (Day and Scott 1981).

SEED. The seeds have a straight embryo with ample endosperm, high in fat in *Diapensia lap-*

which grows in sandy pine barrens and *Galax* which grows in open hardwood forests. *Diapensia lapponica* is described as snow-fearing and is usually found growing in areas blown clear of snow during the winter (Tiffney 1972).

ECONOMIC IMPORTANCE. Members of the family have little importance in commerce. The leaves of *Galax urceolata* are preserved and used by florists. Specialist growers cultivate some of the *Diapensia*, *Pyxidanthera*, and *Shortia*.

KEY TO THE GENERA

1. Staminodia absent; flowers solitary; leaves narrow and imbricate, sessile or very shortly petiolate   2
– Staminodia 5, opposite the petals (absent in some Asian *Shortia*); flowers solitary or not; leaves orbicular to obovate or spatulate, long-petiolate   3
2. Anther loculi dehiscing transversely, awned at base; flowers sessile   **4. Pyxidanthera**
– Anther loculi dehiscing lengthwise, unawned; flowers pedicellate   **3. Diapensia**
3. Corolla lobes toothed or fringed   **2. Shortia**
– Corolla lobes entire   4
4. Anthers 2-locular; style elongated   **1. Berneuxia**
– Anthers 1-locular; style very short   **5. Galax**

1. *Berneuxia* Decne.

Berneuxia Decne., Bull. Soc. Bot. France 20: 159 (1873).

Perennial herb, subacaulescent; caudex thickened and creeping. Leaves long-petiolate. Inflorescence with slender peduncle about 18 cm long (about same length as leaves), 10–15-flowered; anthers 2-locular; stamens and staminodia fused and forming a ring at base of petals; staminodia reduced to scales; style elongated, inserted in depression on top of ovary. A single species, *B. thibetica* Decne., in Himalayan Mts. (Li 1943).

2. *Shortia* Torr. & Gray   Fig. 41A–H

Shortia Torr. & Gray, Am. J. Sci. Arts 42: 48 (1842), nom. cons. *Schizocodon* Sieb. & Zucc. (1843).

Low perennial herbs forming dense clumps or carpets, spreading by horizontal rhizomes. Leaves long-petiolate. Flowers solitary or in racemes, nodding; peduncle bracted; petals united, the lobes undulate-crenate notched; stamens 2-loculed, bent sharply inward; staminodia borne near base of petals; style elongated, filiform. Five species in East Asia and one in E North America.

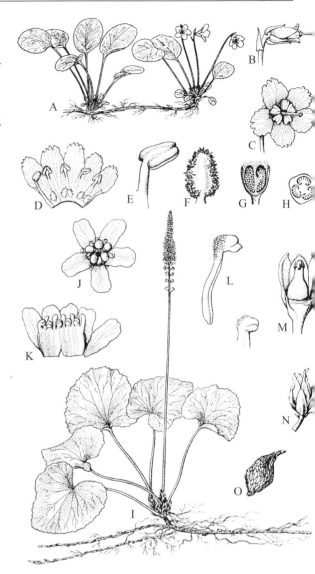

Fig. 41. Diapensiaceae. **A–H** *Shortia galacifolia*. **A** Habit. **B** Lateral view of flower, corolla removed. **C** Flower. **D** Expanded corolla. **E** Stamen. **F** Staminode. **G** Ovary, vertical section. **H** Ovary, transverse section. **I–O** *Galax urceolata*. **I** Habit. **J** Flower. **K** Corolla and staminal tube, expanded. **L** Stamen and anther. **M** Ovary with 3 sepals and prophyll. **N** Fruit. **O** Seed. (From Wood and Channell 1959)

3. *Diapensia* L.   Fig. 40

Diapensia L., Sp. Pl. 1: 141 (1753).

Cushion-like subshrubs. Leaves short-petiolate. Flowers terminal on branchlets, solitary; peduncles elongated and scape-like after anthesis; stamens adnate to base of corolla; anthers 2-loculed, with oblique, divergent loculi; staminodia absent or vestigial. Four species, *D. lapponica* L. circumboreal, three in the Himalayan Mts.

4. *Pyxidanthera* Michx.

Pyxidanthera Michx., Fl. Bor.-Am. 1: 152 (1803).

Prostrate creeping evergreen subshrub. Leaves awl-pointed, sessile. Flowers solitary and sessile on short, densely leafy branches; petals united by broad stamen-filaments to form a tube; anther locules 2, awned at base and dehiscing transversely; pollen only discharged when awned tip pushed downward; staminodia absent; style as long as the corolla tube. A single species, *P. barbulata* Michx., in eastern North America.

5. *Galax* Sims   Fig. 41I–O

Galax Sims, Curtis Bot. Mag. 5: t. 754 (1804).

Herbaceous perennial forming thick matted tufts; rhizomes scaly-bracted, branched. Leaves petiolate. Inflorescence a raceme on a naked peduncle. Petals distinct to base; stamens and staminodia approximately equal, their filaments united and forming a tube adnate to petals at base and falling with petals; anthers 1-loculed, dehiscing horizontally across the top, lower part of anthers tapering to an obtuse point; style short. A single species, *G. urceolata* (Poir.) Brummitt, in eastern N America, with diploid and tetraploid races.

Selected Bibliography

Anderberg, A.A. 1992. The circumscription of the Ericales and their cladistic relationships to other families of "higher" dicotyledons. Syst. Bot. l7: 660–675.

Anderberg, A.A. 1993. Cladistic interrelationships and major clades of the Ericales. Pl. Syst. Evol. 184: 207–231.

Anderberg, A.A. et al. 2002. See general references.

APG (Angiosperm Phylogeny Group) 1998. See general references.

Bliss, L.C. 1962. Caloric and lipid content in alpine tundra plants. Ecology 43: 481–529.

Brummitt, R.K. 1972. Nomenclatural and historical considerations concerning the genus *Galax*. Taxon 21: 303–317.

Darlington, C.D., Wylie, A.P. 1955. Chromosome atlas of flowering plants. London: Allen and Unwin.

Day, R.T., Scott, P.J. 1981. Autecological aspects of *Diapensia lapponica* L. in Newfoundland. Rhodora 83: 101–109.

Diels, L. 1914. Diapensiaceen-Studien. Bot. Jahrb. Syst. Suppl. 50: 304–330.

Erdtman, G. 1952. See general references.

Evans, W.E. 1927. A revision of the genus *Diapensia* with special reference to the Sino-Himalayan species. Notes Roy. Bot. Gard. Edinburgh 15: 209–236.

Gibbs, R.D. 1974. Chemotaxonomy of flowering plants. Montreal: McGill-Queen's University Press.

Li, H.-L. 1943. On the Sino-Himalayan species of *Shortia* and *Berneuxia*. Rhodora 45: 333–337.

Metcalfe, C.R., Chalk, L. 1950. See general references.

Murphy, H.T., Hardin, J.W. 1976. A new and unique venation pattern in the Diapensiaceae. Torreya 103: 177–179.

Nesom, G.L. 1983. *Galax* (Diapensiaceae): geographic variation in chromosome numbers. Syst. Bot. 8: 1–14.

Palser, B.F. 1963. Studies of floral morphology in the Ericales. VI. The Diapensiaceae. Bot. Gaz. 124: 200–219.

Petersen, H.E. 1912. The structure and biology of arctic flowering plants. II. Diapensiaceae. *Diapensia lapponica* L. Meddel. Gronland 36: 141–154.

Primack, R.B., Wyatt, R. 1975. Variation and taxonomy of *Pyxidanthera* (Diapensiaceae). Brittonia 27: 115–118.

Rönblom, K., Anderberg, A.A. 2002. Phylogeny of Diapensiaceae based on molecular data and morphology. Syst. Bot. 27: 383–395.

Ross, M.N. 1936. Seed reproduction of *Shortia galacifolia*. Bot. Gdn. J. 37: 208–211.

Samuelson, G. 1913. Studien über Entwicklungsgeschichte der Blüten einiger Bicornestypen. Svensk Bot. Tidskr. 7: 97–188.

Scott, P.J., Day, R.T. 1983. Diapensiaceae: a review of the taxonomy. Taxon 32: 417–423.

Solereder, H. l908. Systematic anatomy of the dicotyledons. Oxford: Clarendon Press.

Takhtajan, A.L. 1980. Outline of the classification of flowering plants. Bot. Rev. 46: 225–359.

Tiffney, W.N. Jr. 1972. Snow cover and the *Diapensia lapponica* habitat in the White Mountains, New Hampshire. Rhodora 74: 358–377.

Wood, C.E. Jr., Channell, R.B. 1959. The Empetraceae and Diapensiaceae of the southeastern United States. J. Arnold Arbor. 40: 161–171.

Xi, Y.-Z., Tang, Y.-C. 1990. Pollen morphology and phylogenetic relationships in the Diapensiaceae. Cathaya 2: 89–112.

Yamazaki, T. 1966. The embryology of *Shortia uniflora* with brief review of the systematic position of the Diapensiaceae. J. Jap. Bot. 41: 245–251.

Dirachmaceae

C. BAYER

Dirachmaceae Hutch., Fam. Fl. Pl., ed. 2: 248 (1959).

Trees or shrubs; hairs simple and glandular-peltate. Leaves alternate, simple, lobed to serrate, petiolate; venation pinnate; stipules linear-triangular. Flowers pedicellate, single in the axils of foliage leaves; pedicel provided with 4–8 lanceolate or linear-triangular bracts. Flowers actinomorphic, hermaphrodite, tetracyclic, perigynous; sepals 5–6 or 8, with valvate aestivation, pubescent; petals white, free, imbricate, as many as petals, with ventral, basal appendages; stamens 5–6 or 8, opposite the petals and basally fused with them; filament bases deltoid; anthers basifixed, extrorse, tetrasporangiate, opening by longitudinal slits; gynoecium syncarpous; ovary pubescent, 5–6- or 8-locular, radially lobed; style distally glabrous; stigma clavate to cylindrical; ovules 1 per locule, bitegmic, anatropous-apotropous, placentation basal-axile. Fruits capsular, septicidal and septifragal, opening from base to apex, with long hairs on the inner surface; seeds smooth, brown; endosperm scanty.

A monogeneric family with two species, *Dirachma socotrana* Schweinf. ex Balf. f. from Socotra, and *D. somalensis* D.A. Link from Somalia.

VEGETATIVE STRUCTURES. The branching system of *Dirachma socotrana* includes long and short shoots, the latter arising from the axils of foliage leaves on long shoots. Yakovleva (1994) reported the presence of mucilage-containing cells in the leaf epidermis of *Dirachma socotrana*.

REPRODUCTIVE STRUCTURES. Apart from simple hairs, peltate glandular trichomes occur on the pedicels and on the outside of the sepals of *Dirachma somalensis*, and on the ovary of *D. socotrana* (Link 1994). The pedicels are provided with four or up to eight sterile bracts, which are sometimes described as epicalyx or hypocalyx. The flowers are octamerous in *D. socotrana*, and pentamerous or hexamerous in *D. somalensis*. The ovary is enclosed by a cup of receptacular or sepaline origin. Petals and stamens are inserted near the margin of this cup. Between each petal and the stamen in front of it, a pubescent projection of the petal is found. According to Link (1994), nectar is probably produced by epithelial glands underneath these appendages. The cup-shaped cavity of the flower is apically closed by the petal appendages and their indumentum, and may retain the nectar (Fig. 42).

The fruit valves spread when ripe, but their apical portions remain attached to the style. They expose clusters of long unicellular hairs that enclose the seeds. Based on his observations on flower and fruit structure of *Dirachma*, Link (1994) suspected melittophily and anemochory.

Boesewinkel (1988, 1995) and Boesewinkel and Bouman (1997) described the seed as exotestal with one antiraphal integumentary bundle and an endotegmic pigment layer.

POLLEN MORPHOLOGY. According to the short descriptions given by Erdtman (1952) and Link (1991, 1994), the pollen grains are prolate and tricolporate; the exine is finely reticulate.

EMBRYOLOGY. Some embryological features of *Dirachma* have been described by Boesewinkel and Bouman (1997). The tapetum is secretory and includes binuclear cells. Mature pollen contains starch grains. The crassinucellate ovules are laterally flattened and have a zigzag-shaped micropyle. Embryo sac development probably corresponds to the Polygonum type; the antipodals degenerate after fertilisation.

PHYTOCHEMISTRY. Comparing several genera of Geraniaceae, Bate-Smith (1973) found trivial flavonoids in *Dirachma*, similar to those of *Geranium*.

AFFINITIES. Following the original description by Balfour (1888), who found most affinities with Vivianieae and Wendtieae (formerly included in Geraniaceae), many subsequent authors placed *Dirachma* at the level of tribe or family in or close to Geraniaceae. However, Balfour (1888) himself had indicated a certain resemblance with Tili-

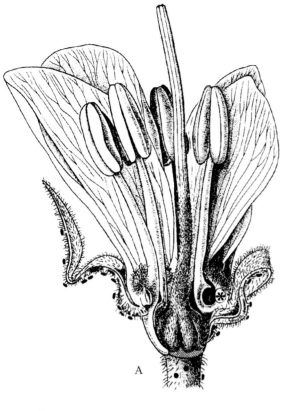

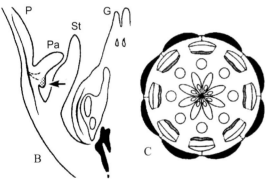

Fig. 42. Dirachmaceae. **A** Flower of *Dirachma somalensis* (perianth and androecium removed in part; asterisk: nectar gland). **B** *D. socotrana*. Flower segment in vertical section (G: gynoecium, St: stamen, P: petal, Pa: petal appendage, arrow: nectary). **C** Floral diagram of *D. somalensis*. (Link 1994)

(Ronse Decraene and Smets 1995). However, it is unlikely that these characters justify the position of *Dirachma* within this alliance, because Malvalean characters such as the presence of stellate hairs, palmate venation, mucilage in tissues other than the epidermis, and exotegmic seed coats, are missing. Several other important characters of *Dirachma*, including the shoot anatomy, are unknown so that the systematic position of *Dirachma* remains uncertain. Recent studies exclude close affinities with both Geraniales and Malvales but suggest a placement in a redefined rhamnalean clade, even if the relationships to *Barbeya*, another isolated genus placed in this alliance, remain unresolved (Boesewinkel and Bouman 1997; Thulin et al. 1998).

DISTRIBUTION, HABITATS AND CONSERVATION. *Dirachma socotrana* is endemic to Socotra, occurring together with *Buxus hildebrandtii* on limestone boulders. *D. somalensis* was collected in Central Somalia on lime as well as on sand. Both species are very rare and highly endangered (Bazara'a et al. 1991; Link 1994).

One bispecific genus:

Dirachma Schweinf. ex Balf. f.   Fig. 42

Dirachma Schweinf. ex Balf. f., Trans. Roy. Soc. Edinburgh 31: 45, t. 8 (1888).

Characters as for family.

Selected Bibliography

Balfour, B. 1888. Botany of Socotra: *Dirachma*. Trans. Roy. Soc. Edinburgh 31: 45–46, pl. 8.

Bate-Smith, E.C. 1973. Chemotaxonomy of *Geranium*. Bot. J. Linn. Soc. 67: 347–359.

Bazara'a, M., Guarino, L., Miller, A., Obadi, N. 1991. *Dirachma socotrana* – back from the brink? Oryx 25: 229–232.

Boesewinkel, F.D. 1988. The seed structure and taxonomic relationships of *Hypseocharis* Remy. Acta Bot. Neerl. 37: 111–120.

Boesewinkel, F.D. 1995. Seed coat structure and the delimitation of the Geraniales. Acta Bot. Neerl. 44: 491 (Abstract).

Boesewinkel, F.D., Bouman, F. 1997. Ovules and seeds of *Dirachma socotrana* (Dirachmaceae). Plant Syst. Evol. 205: 195–204.

Erdtman, G. 1952. See general references.

Hutchinson, J. 1959. The families of flowering plants, ed. 2, vol. 2. Oxford: Clarendon Press.

Knuth, R. 1931. Geraniaceae. In: Engler & Prantl, Die natürlichen Pflanzenfamilien, ed. 2, 19a. Leipzig: Engelmann, pp. 43–66.

aceae. Hutchinson (1959) considered *Dirachma* to represent a distinct family and included it in his Tiliales. Characters of *Dirachma* that have been used to support this placement include the presence of a hypocalyx, a valvate, amucronate calyx, petals with ventral projection and ventral nectary (albeit epithelial instead of trichomatic), hairs on the ventral side of the fruit wall (Link 1991, 1994), special features of the mucilage cells (Yakovleva 1994), and the obhaplostemonous androecium

Link, D.A. 1991. *Dirachma somalensis* D.A. Link sp. nov. A new species of a remarkable and highly endangered mono-generic family. Bull. Jard. Bot. Belg. 61: 3–13.

Link, D.A. 1994. *Dirachma* Schweinf. (Dirachmaceae) a highly remarkable and endangered bispecific genus. In: Seyani, J.H., Chikuni, A.C. (eds.) Proc. 13th Plenary Meet. A.E.T.F.A.T., Zomba, Malawi, 2–11 April 1991. Vol. 2. Plants for the people. Zomba: National Herbarium and Botanic Gardens of Malawi, pp. 1229–1238.

Ronse Decraene, L.P., Smets, E.F. 1995. The distribution and systematic relevance of the androecial character oligomery. Bot. J. Linn. Soc. 118: 193–247.

Thulin, M., Bremer, B., Richardson, J., Niklasson, J., Fay, M.F., Chase, M.W. 1998. Family relationships of the enigmatic rosid genera *Barbeya* and *Dirachma* from the Horn of Africa region. Plant Syst. Evol. 213: 103–119.

Yakovleva, O.V. 1994. The ultrastructure of mucilage cells in the leaf epidermis of *Dirachma socotrana* (Dirachmaceae). Bot. Zhurn. 79: 52–58, pl. 2.

Ebenaceae

B. WALLNÖFER

Ebenaceae Gürke in Engler & Prantl, Nat. Pflanzenfam. IV, 1: 153 (1891), nom. cons.

Mostly evergreen dioecious, rarely monoecious or polygamous trees or shrubs; bark, roots and heartwood often black. Unicellular trichomes simple or two-armed; multicellular glandular or peltate hairs sometimes present. Leaves simple, mostly alternate, spirally or distichously arranged, mostly petiolate, exstipulate; leaf-margins usually entire; extrafloral nectaries usually present on abaxial leaf surface. Inflorescences axillary, cymose, fasciculate, raceme-like or paniculate. Flowers usually unisexual, less frequently hermaphroditic, articulated at base, regular, 3–5(–8)-merous; calyx mostly gamosepalous, valvate or imbricate, spreading or reflexed, persistent and often accrescent in fruit; corolla gamopetalous, isomerous with calyx, shortly to deeply lobed; tube prominent or very short in male flowers; male flowers: stamens (3–)12–20(–100), mostly unequal in size, often setulose, commonly inserted at base of corolla tube, hidden or slightly exserted, often arranged in two hardly discernible whorls; filaments usually short and flattened, free or united in pairs, triads or fascicles, or sometimes into a central cylinder; anthers usually linear, erect, basifixed, distally often apiculate, tetrasporangiate, the outermost introrse, the innermost extrorse, and those in-between latrorse, longitudinally dehiscing; pistillode usually reduced, lacking ovules, style and stylodia, rarely totally absent, or well-formed in the terminal (structurally or rarely functionally hermaphroditic) flower of a male cyme; female flowers: staminodes arranged in a single whorl, usually inserted at base of corolla tube, often markedly reduced in shape, fewer than stamens in male flowers, rarely absent; rudimentary anthers introrse; disk absent or less frequently well-developed; gynoecium syncarpous, 2–8-carpellate; ovary superior, sessile; style short, style branches usually longer than style, same number as carpels; stigmas often variously lobed; carpels biovulate, mostly bilocular due to a false, secondary septum or rarely unilocular; placentation apical; ovules pendulous, anatropous, bitegmic,

tenuinucellate. Fruit a berry usually with a pulpy endocarp; seeds 1–16 per fruit, pendulous; hilum small, apical, inconspicuous; seed coat exotestal, pigmented, thin, parenchymatous, soft to leathery; endosperm hard, abundant, oily, smooth to ruminate; embryo surrounded by endosperm, turned upside-down (radicle superior), with two, oval, well-developed foliaceous cotyledons and a strongly developed radicle.

A family comprising two genera and c. 500–600 species, distributed mainly in the tropical and subtropical regions of both the Old and the New World.

VEGETATIVE MORPHOLOGY. Most species of *Diospyros* are small to medium-sized trees in the forest understorey. Only few species grow up to large canopy trees. *Euclea* species are subshrubs, shrubs or trees up to 12 m high. The growth architecture of many species of *Diospyros* conforms to Massart's model (Hallé et al. 1978). Leaf arrangement is distichous on the branches and spiral on the trunk. In the latter case the leaves are often much reduced in size or only scale-like. Some species of the fire-adapted savannahs in South America and Africa are geoxylic subshrubs. Certain species have been reported to develop root-suckers and to establish often large populations or coppices (Skallerup 1953; Rathore 1971; White 1983; for further details see also Wallnöfer 2001).

VEGETATIVE ANATOMY. The bark of many tropical species is black outside. In cross-section the bark of twigs shows a characteristic thin layer of black tissue. Stone cells are present in young twigs, leaves, and fruits of various species. The indumentum consists of four kinds of grey, brownish or often characteristically rusty-brown hairs: unicellular, simple trichomes; unicellular, two-armed, sessile or rarely short-stalked trichomes; small, club-shaped, grey, multicellular, glandular hairs, and peltate, multicellular hairs (Busch 1913; Solereder 1899, 1908, 1914; Ng 1971). Extrafloral nectaries of the specialized Benincasa type (Fig. 43L) occur on abaxial leaf surfaces near the base of the lamina and often scattered along, but somewhat away from the midvein (Contreras and Lersten 1984; Elsler 1907). Characteristic epidermal papillae are present on the abaxial leaf surfaces of several species of *Diospyros*. Each hairless epidermal cell yields on its surface a more or less centrally placed, coronulate papilla, which is star-like connected to the papillae of the surrounding cells by cuticular folds (Elsler 1907). The stomata are usually confined to the abaxial leaf surfaces and are mostly anomocytic, less frequently actinocytic (Ng 1971; Wilkinson 1979).

The wood is usually hard, heavy, fine-textured, and pale or dark coloured. The heartwood of about 50 species is coloured black or dark brown ("ebonised") due to a dark substance (derivatives and oxidative decomposition products of naphthoquinones) filling the lumina. The heartwood of some species is irregularly composed of coloured and uncoloured parts and has a variegated, banded, or marbled aspect. According to Wright (1904) and Hillis and Soenardi (1994), the black deposits are formed as a response after injury and subsequent infection by fungi (for further details see also Wallnöfer 2001).

The structure of the wood is remarkably constant throughout the family: vessels solitary and in radial multiples of 2, 3, 4 or sporadically more, with simple perforations; pits small (to 8μm diameter) between vessels as well as between vessels and ray cells or parenchyma; parenchyma predominantly apotracheal, scattered and in numerous uniseriate lines; sometimes forming vasicentric sheaths round the vessels; rays 1–2 cells, rarely to 3 or 4 cells wide, less than 1 mm high, heterogeneous; fibers with small pits, and with walls thinner than lumina, rarely thicker (Ng 1971, 1991b; Metcalfe and Chalk 1950).

INFLORESCENCE STRUCTURE. The Ebenaceous inflorescence (Ng 1991b, Franceschi 1993) is conventionally referred to as a cyme. It is always axillary; the pedicels are articulated and, if lateral, subtended by a bract. If there is one flower only (especially in the female condition), it occupies a terminal position on a bracteate peduncle. Quite often, the inflorescence is a three-flowered dichasium. Numerous variations are seen ranging between three-flowered dichasia and multibranched paniculate inflorescences but, in nearly all cases, the terminal units are dichasia. Usually within the same species the male inflorescences are many-flowered, whereas the female ones are only 1(3)-flowered. The inflorescences of

Euclea are mostly unbranched, multiflorous racemes.

FLOWER MORPHOLOGY. Ebenaceae have unisexual or rarely hermaphroditic flowers and are mostly dioecious; less frequently they are monoecious or even more rarely polygamous. In *Euclea* the male flowers are usually larger than the female but similar in shape. In *Diospyros* the male flowers are smaller than the female and different in shape. Hermaphroditic flowers occur generally in *Diospyros* sect. *Royena* (the former genus *Royena*) and, according to Salter (1953) and White and Barnes (1958), are only structurally but not functionally hermaphroditic. Among species of other sections of *Diospyros*, hermaphroditic flowers are sporadically found and occupy the terminal positions in otherwise male inflorescences (Yasui 1915; Namikawa et al. 1932; Franceschi 1993).

The number of floral parts is known to vary within species and cannot be used for species delimitation or to segregate smaller genera, as had been done in the past. The carpels are biovulate and in some species remain unilocular but mostly are divided by a false longitudinal septum that originates from the carpellary wall opposite the placenta. In the first case, each locule is biovulate and ovaries are 2-, 3- or 4-locular. In case of the uniovulate locules, ovaries are 4-, 6-, 8-, 10-, 12-, 14- or 16-locular (Ng 1971, 1991b).

EMBRYOLOGY. The anther wall comprises the epidermis, the fibrous or non-fibrous endothecium, two middle layers, and multinucleate tapetal cells. The ovule is pendulous, oblong, anatropous, bitegmic, and tenuinucellar. The archesporial cell functions as megaspore mother cell, undergoes meiotic divisions, and produces a linear tetrad. The chalazal megaspore develops into a Polygonum type of embryo sac. Endosperm development is cellular. The embryogeny and organogenesis probably correspond to the Polygonad type-Chenopodiad variation or Chenopodiad type (Davis 1966; Johri et al. 1992). The seed coat is formed only by the outer integument.

POLLEN MORPHOLOGY. The pollen grains are tricolporate and prolate-spheroidal to prolate. The sexine is as thick as the nexine or thicker and smooth or very finely warty. The ora are mostly well defined, usually lalongate, sometimes with indistinct lateral edges or rarely indistinguishable. The pollen is remarkably constant throughout the family, and the main variations are in size and shape of the grains and the apertures (Erdtman 1952; Ng 1971; Franceschi 1993).

KARYOLOGY. In both genera the basic chromosome number is x = 15. An euploid series with $2n$ = 30, 60, 90, and 135 has been observed in *Diospyros*, with 36 species reported to be diploid, and a few others tetra- and hexaploid (Franceschi 1993; Yonemori et al. 2000). Eight species of *Euclea* are diploid (White and Vosa 1980).

POLLINATION. The flowers of various species are reported to be visited by bees (Hague 1911; Ayala-Nieto and Ludlow-Wiechers 1983). Beetles (considered to be the main pollinators), flies and wasps (all c. 3–6 mm long) were observed visiting the male and female flowers of *Diospyros pentamera* in Australia (Irvine and Armstrong 1990). *Diospyros hispida*, a species of the Brazilian cerrado, flowers during the night and attracts small nocturnal Lepidoptera (moths) (Silberbauer-Gottsberger and Gottsberger 1988: 657). Also various rainforest-species are reported to flower during the night, but nothing is known about the pollinators.

FRUIT AND SEED. The fruit is a multilocular, usually indehiscent berry subtended by a persistent, 3–8-lobed or rarely unlobed (truncate) calyx, which is usually accrescent in *Diospyros* but not in *Euclea*. The fruits of some African species of *Diospyros* sect. *Royena* become dry and occasionally are tardily dehiscent (Winter 1963; White 1983). In no species, however, is the fruit completely and spontaneously dehiscent. At fruit-maturity the pericarp is either completely soft or possesses a thinner or thicker, leathery or hard outer layer. A hypodermal layer of stone-cells, very rarely of radial fibers or a mixture of fibers and stone-cells is developed. The mesocarp is often exiguous and sometimes two-layered. A layer of pulpy, sometimes fibrous endocarp, which is mostly present, adheres closely to the seeds and resembles a sarcotesta (Ng 1971, 1991a). Idioblasts containing tannins occur generally in the pericarp of the fruits (Corner 1976; Utsunomiya et al. 1998). Especially in the immature fruits, tannins are responsible for the characteristic astringent taste, which diminishes when the fruits ripen but rarely disappears completely as in some cultivars (Yonemori et al. 2000). The fruits of some species are poisonous, presumably when raw, and are used in various parts of the world to catch fish. The diameter of mature fruit varies in the range 5–100 mm (Ng 1971, 1991a). Due to parthenocarpy, the fruits of various cultivars of *D. kaki* are seedless (Yonemori et al. 2000).

Each fruit of *Diospyros* contains 1–16 seeds, which are 8–40 mm long. The seeds of *Euclea* are usually solitary and 3–10 mm in diameter when ripe, if more than one, then forming a sphere together. During the development of the seed of *Euclea*, the other (abortive) ovules and the axis are pushed to one side of the fruit. The solitary seeds are, therefore, encircled by a curved vascular loop (like on a tennis ball), which is apically linked by a shallow, vertical groove which represents the impression of the displaced axis (Ng 1971; Winter 1963).

The rumination of the endosperm in *Diospyros*, if present, is of the Annona type (Periasamy 1966), whereas in *Euclea* it takes the peculiar form of ingrowths from the testa forming a cylinder round the whole radicle (Ng 1971; White 1983). The rumination in *Diospyros* sect. *Royena* forms only a shallow ingrowth round the distal part of the radicle. The embryo is turned upside-down (radicle superior), possessing two oval, well-developed foliaceous cotyledons and a strongly developed radicle.

DISPERSAL. The fruits and seeds are dispersed by various fruit-eating birds and mammals (Ridley 1930; White 1983; Pannell and White 1988). The fruits of coastal and riverine species as well as those of species growing in periodically inundated habitats may perhaps also be water dispersed (for further details see also Wallnöfer 2001).

GERMINATION AND SEEDLING. The germination of *Diospyros* seeds is mostly epigeal, less frequently hypogeal or of the type "durian" (=crypto-epigeal). The cotyledons are emergent and photosynthetic or not, or are non-emergent, and shed together with the seed body. The epicotyledonary leaves are usually alternate, but in a few species the first two are subopposite (Ng 1971, 1991a; for further details see also Wallnöfer 2001).

PHYTOCHEMISTRY. Naphthoquinones, terpenoids, benzopyrones, and condensed tannins (mostly based on prodelphinidin) are widely distributed and are characteristic of Ebenaceae (Mallavadhani et al. 1998; Hegnauer 1966, 1989). Naphthoquinones are frequent and their oxidation products provide the dark colour of the bark, heartwood, leaves and fruits. Idioblasts containing tannins occur in various organs, including fruits (Yonemori et al. 2000).

AFFINITIES. Ebenaceae have traditionally been placed in the order Ebenales, together with all or some of the following families: Sapotaceae, Styracaceae, Symplocaceae and Lissocarpaceae (Morton et al. 1997). Recent molecular studies, based upon *rbc*L sequence data, have demonstrated that Ebenaceae are monophyletic, whereas Ebenales in their traditional circumscription are not. In these studies Ebenaceae appear as the sister-group of the Primulales within the expanded Ericales as defined by APG (1998). *Lissocarpa*, the only genus of Lissocarpaceae, had been included in Ebenaceae by some authors and, according to Ng (1971, 1991b) and Franceschi (1993), is the closest relative of Ebenaceae. This is strongly endorsed by the molecular studies of Berry et al. (2001) and Anderberg et al. (2002).

DISTRIBUTION, HABITAT AND ECOLOGY. Ebenaceae occur mainly in the lowlands of the tropical and, to a lesser extent, subtropical regions of the world. Only a few species are found at higher elevations or in warm-temperate regions. *Euclea* is restricted to Africa and Arabia, whereas *Diospyros* is pantropical. Many species of the latter genus are small to medium-sized trees of the forest understorey and often have a remarkably low population density. Several species grow along rivers; some are rheophytes (White 1998), and others are characteristic elements of swamps, beach communities and periodically flooded habitats. Some species occur in dryer vegetation, in fire-prone savannahs, or in deciduous forests.

PALAEOBOTANY. The fossil record has been compiled by Hiern (1873), Franceschi (1993), Collinson et al. (1993), and Mai (1995). Most finds are from the Tertiary. Fossil leaves, because of their rather nondescript and very variable appearance, can probably not be assigned to the family with absolute certainty. It seems that in the past not much attention was paid to the presence of extrafloral nectaries and the coronulate papillae on the abaxial leaf surface. Probably many fossils which have been assigned to Ebenaceae may have to be reconsidered.

USES AND ECONOMIC IMPORTANCE. Ebenaceae are the source of some economically important products, the most valuable being the persimmon fruits and ebony timbers. *Diospyros kaki* is the most important fruit-yielding species and is cultivated in many varieties on a large scale, especially in temperate east Asia (Yonemori et al. 2000). Other frequently cultivated species are *D. digyna*

(zapote prieto, black sapote), *D. blancoi* (velvet apple or mabolo), and *D. lotus*. The fruits of many other species are of more regional importance. The leaves of *D. melanoxylon* ("tendu") are utilized in India as wrappers for cigarettes called "bidi" (Hunter 1981). Ebony timbers are used mainly for the production of musical instruments, for carved work, cabinet making, and as decorative veneer for the production of furniture. The true ebony timbers are obtained from various species having a completely or partially black heartwood, the economically most important of which are *Diospyros ebenum* and *D. melanoxylon* from India and Sri Lanka.

KEY TO THE GENERA

1. Calyx usually accrescent on fruits; berry usually many-seeded; seeds usually laterally flattened, encircled by a more or less straight vascular loop round the periphery, which is apically not linked by a vertical groove **1. *Diospyros***
- Calyx not accrescent on fruits; berry usually 1(3)-seeded; seeds usually subglobose, encircled by a curved vascular loop, which is apically linked by a shallow, vertical groove **2. *Euclea***

1. *Diospyros* L.   Fig. 43

Diospyros L., Sp. Pl., ed. 1: 1057 (1753); Hiern, Trans. Cambridge Philos. Soc. 12 (1): 27–300 (1873), rev.; Bakhuizen van den Brink, Bull. Jard. Bot. Buitenzorg, III, 15 (1–5): 1–515 (1936–1955), SE Asian spp.; White, Bull. Jard. Bot. Belg. 48: 245–358 (1978), 58: 325–448 (1988), African spp.; Perrier de la Bâthie, Flore de Madagascar et des Comores (1952).
Royena L. (1753).
Maba J.R. Forster & J.G.A. Forster (1775).
Tetraclis Hiern (1873).
Rhaphidanthe Hiern ex Gürke (1891).

Leaves usually alternate and with entire margins. Inflorescences usually cymose or fasciculate if male and uniflorous (less frequently few-flowered) if female. Male flowers usually smaller than the female ones, and differing in shape. Calyx variable, 3–8-lobed, or cup-shaped and entire, persistent and usually accrescent on fruits. Stamens 2 to ca. 100. Ovary 2–8-carpellate. Fruit a 1–16-seeded berry. $2n = 30, 60, 90, 135$. About 500–600 spp., with 200–300 in Asia and the Pacific area, ca. 95 in Madagascar, 94 on the African mainland, ca. 100 in the Americas, and 15 in Australia.

A comprehensive monograph is badly in need since the last one dates back more than a century (Hiern 1873). Besides *Diospyros* (with 15 sections), Hiern (1873) recognized also *Maba* (with 6 sections), *Royena* and *Tetraclis*, which now are all included in *Diospyros*. Bakhuizen van den Brink (1936–1955) grouped the SE Asiatic and Pacific

Fig. 43. Ebenaceae. **A–K** *Diospyros curranii.* **A** Fruiting branch. **B** Calyx with ovary. **C** Ovary. **D** Transverse section of ovary. **E** Transverse section of fruit. **F** Fruit. **G** Calyx of fruit, outside view. **H** Flower. **I** Same, vertical section. **J** Stamens. **K** Staminodes. **L** *Diospyros koeboeensis,* underside of leaf with nectaries. (Bakhuizen van den Brink 1936–1955)

species into 5 subgenera and 36 sections, and White (1980) the African species into 18 sections.

2. *Euclea* Murray

Euclea Murray, Syst. Veg., ed. 13: 747 (1774); Hiern, Trans. Cambridge Philos. Soc. 12 (1): 27–300 (1873), rev.; de Winter, Fl. S. Afr. 26: 54–99 (1963).

Leaves alternate, subopposite, or subverticillate in whorls of three, with entire or rarely finely crenulate margins. Inflorescences usually raceme-like, less frequently compound or occasionally flowers solitary. Male flowers usually larger than female but similar in shape. Calyx 4–5-lobed, shallowly cyathiform or patelliform, persistent but not accrescent in fruit. Stamens 10–30. Ovary 2–3-carpellate. Fruit usually a 1(3)-seeded berry. $2n = 30$. About 12–20 spp. confined to Africa, Arabia, Socotra and the Comoro Islands; many species restricted to the Cape flora.

The species pertaining to subgen. *Rymia* (Endl.) Verdc. occur only in South Africa, Namibia and SW Angola and have an urceolate to subglobose, shallowly 5–8-lobed corolla. The species belonging to subgen. *Euclea* are widespread and have a campanulate, deeply 4–5-lobed corolla.

Selected Bibliography

Anderberg, A.A. et al. 2002. See general references.

APG (Angiosperm Phylogeny Group) 1998. See general references.

Ayala-Nieto, M.L., Ludlow-Wiechers, B. 1983. Catálogo Palinológico para la Flora de Veracruz. No. 13. Familia Ebenaceae. Biótica 8: 215–226.

Bakhuizen van den Brink, R.C. 1936–1955. Revisio Ebenacearum Malayensium. Bull. Jard. Bot. Buitenzorg III, 15 (1–5): 1–515.

Berry, P.E., Savolainen, V., Sytsma, K.J., Hall, J.C., Chase, M.W. 2001. *Lissocarpa* is sister to *Diospyros* (Ebenaceae). Kew Bull. 56: 725–729.

Busch, P. 1913. Anatomisch-systematische Untersuchung der Gattung *Diospyros.* Crefeld: Wilhelm Greven.

Collinson, M.E., Boulter, M.C., Holmes, P.L. 1993. Magnoliophyta ('Angiospermae'). In: Benton, M.J. (ed.) The fossil record 2. London: Chapman & Hall, pp. 809–841.

Contreras, L.S., Lersten, N.R. 1984. Extrafloral nectaries in Ebenaceae: anatomy, morphology, and distribution. Am. J. Bot. 71: 865–872.

Corner, E.J.H. 1976. See general references.

Davis, G.L. 1966. See general references.

Elsler, E. 1907. Das extraflorale Nektarium und die Papillen der Blattunterseite bei *Diospyros discolor* Willd. Sitzungsber. Kaiserl. Akad. Wiss., Math.-Naturwiss. Cl., Abt. 1, 116: 1563–1590.

Erdtman, G. 1952. See general references.

Franceschi, D. de 1993. Phylogénie des Ebénales: analyse de l'ordre et origine biogéographique des espèces indiennes. Publications du Département d'Ecologie, Institut Français de Pondichérry 33: 1–153.

Hague, S.M. 1911. A morphological study of *Diospyros virginiana.* Bot. Gaz. (Crawfordsville) 52: 34–44, pl. 1–3.

Hallé, F., Oldeman, R.A.A., Tomlinson, P.B. 1978. Tropical trees and forests. Berlin Heidelberg New York: Springer.

Hegnauer, R. 1966, 1989. See general references.

Hiern, W.P. 1873. A monograph of Ebenaceae. Trans. Cambridge Philos. Soc. 12: 27–300.

Hillis, W.E., Soenardi, P. 1994. Formation of ebony and streaked woods. IAWA J. 15: 425–437.

Hunter, J.R. 1981. Tendu (*Diospyros melanoxylon*) leaves, bidi cigarettes, and resource management. Econ. Bot. 35: 450–459.

Irvine, A.K., Armstrong, J.E. 1990. Beetle pollination in tropical forests of Australia. In: Bawa, K.S., Hadley, M. (eds.) Reproductive ecology of tropical forest plants. Man Biosph. Ser. 7: 135–148.

Johri, B.M. et al. 1992. See general references.

Mai, D.H. 1995. Tertiäre Vegetationsgeschichte Europas. Jena: Gustav Fischer.

Mallavadhani, U.V., Panda, A.K., Rao, Y.R. 1998. Pharmacology and chemotaxonomy of *Diospyros*. Phytochemistry 49: 901–951.

Metcalfe, C.R., Chalk, L. 1950. See general references.

Morton, C.M., Chase, M.W., Kron, K.A., Swensen, S.M. 1997. A molecular evaluation of the monophyly of the order Ebenales based upon *rbcL* sequence data. Syst. Bot. 21: 567–586.

Namikawa, I., Sisa, M., Asai, K. 1932. On flower types of *Diospyros kaki* L.f. Jap. J. Bot. 6: 139–172.

Ng, F.S.P. 1971. A taxonomic study of the Ebenaceae with special reference to Malesia. Thesis, University of Oxford, 221 pp.

Ng, F.S.P. 1991a. Manual of forest fruits, seeds and seedlings, vol. 1. Kuala Lumpur: Forest Research Institute Malaysia, Malayan Forest Record no. 34: 61–62, 319–327.

Ng, F.S.P. 1991b. The relationships of the Sapotaceae within the Ebenales. In: Pennington, T.D. (ed.) The genera of Sapotaceae, chap. 1. Kew: Royal Botanic Gardens & New York: New York Botanical Garden, pp. 1–13.

Pannell, C.M., White, F. 1988. Patterns of speciation in Africa, Madagascar, and the tropical Far East: regional faunas and cryptic evolution in vertebrate-dispersed plants. Monogr. Syst. Bot. Missouri Bot. Gard. 25: 639–659.

Periasamy, K. 1966. Studies on seeds with ruminate endosperm. VI. Rumination in the Araliaceae, Aristolochiaceae, Caprifoliaceae and Ebenaceae. Proc. Indian Acad. Sci. 60B: 127–134.

Rathore, J.S. 1971. Studies in the root system and regeneration of *Diospyros melanoxylon* Roxb. Indian Forester 97: 379–386.

Ridley, H.N. 1930. The dispersal of plants throughout the world. Ashford: L. Reeve & Co.

Salter, T.M. 1953. A note on sex in *Royena glabra* L. (Ebenaceae). J. S. Afr. Bot. 19: 29–30.

Silberbauer-Gottsberger, I., Gottsberger, G. 1988. A polinização de plantas do cerrado. Revista Brasil. Biol. 48: 651–663.

Skallerup, H.R. 1953. The distribution of *Diospyros virginiana* L. Ann. Missouri Bot. Gard. 40: 211–225.

Solereder, H. 1899. Systematische Anatomie der Dicotyledonen. Stuttgart: Ferdinand Enke.

Solereder, H. 1908. Systematic anatomy of the dicotyledons, vol. 1–2 (English translation by L.A. Boodle & F.E. Fritsch). London: Oxford University Press, 1182 pp.

Solereder, H. 1914. Zwei Beiträge zur systematischen Anatomie. Bot. Jahrb. Syst. 50 (Suppl.): 578–585.

Utsunomiya, N., Subhadrabandhu, S., Yonemori, K., Oshida, M., Kanzaki, S., Nakatsubo, F., Sugiura, A. 1998. *Diospyros* species in Thailand: their distribution, fruit morphology and uses. Econ. Bot. 52: 343–351.

Wallnöfer, B. 2001. The biology and systematics of Ebenaceae: a review. Ann. Naturhist. Mus.Wien B, 103: 485–512.

White, F. 1980. Notes on the Ebenaceae. VIII. The African sections of *Diospyros*. Bull. Jard. Bot. Belg. 50: 445–460.

White, F. 1983. Ebenaceae. In: Launert, E. (ed.) Flora Zambesiaca 7(1): 248–300. London: Flora Zambesiaca Managing Committee.

White, F. 1998. The vegetative structure of African Ebenaceae and the evolution of rheophytes and ring species. In: Hopkins, H.C.F., Huxley, C.R., Pannell, C.M., Prance, G.T., White, F. (eds.) The biological monograph. Kew: Royal Botanic Gardens, pp. 95–113.

White, F., Barnes, R.D. 1958. Generic characters in the Ebenaceae. J. Oxford Univ. Forest Soc. IV, 6: 31–34.

White, F., Vosa, C.G. 1980. The chromosome cytology of African Ebenaceae with special reference to polyploidy. Bol. Soc. Brot. II, 53: 275–297.

Wilkinson, H.P. 1979. The plant surface (mainly leaf). In: Metcalfe, C.R., Chalk, L. (eds.) Anatomy of the dicotyledons, 2nd edn, vol. 1. Oxford: Clarendon Press, pp. 97–165.

Winter, B. de 1963. Ebenaceae. In: Dyer, R.A., Codd, L.E., Rycroft, H.B. (eds.) Flora of Southern Africa 26: 54–99. Pretoria: Government Printer.

Wright, H. 1904. The genus *Diospyros* in Ceylon: its morphology, anatomy, and taxonomy. Ann. Roy. Bot. Gard. (Peradeniya) 2: 1–106, 133–210, 20 pl.

Yasui, K. 1915. Studies of *Diospyros kaki*. I. Bot. Gaz. (Crawfordsville) 60: 362–373.

Yonemori, K., Sugiura, A., Yamada, M. 2000. Persimmon genetics and breeding. In: Janick, J. (ed.) Plant breeding reviews, vol. 19: 191–225. New York: Wiley.

Elaeagnaceae

I.V. Bartish and U. Swenson

Elaeagnaceae Juss., Gen. Pl.: 74 (1789), nom. cons.

Anemophilous or entomophilous small trees, shrubs or rarely woody climbers; shoots often reduced to spines; young branches, leaves, and calyx tube covered with peltate and stellate trichomes; nodules on the roots containing nitrogen-fixing bacteria. Leaves alternate, opposite, or rarely in pseudowhorls, simple, petiolate, without stipules; lamina entire, pinnately veined. Inflorescences axillary, short, fasciculate, spicate or racemose, or rarely flowers solitary. Flowers actinomorphic, apetalous, often fragrant, perfect, or plants monoecious, dioecious, or rarely polygamous, the female flowers without staminodes, the male without pistil; sepals 2 or 4(–6), joined into a hypanthium; hypanthium free, constricted above the gynoecium, white, cream, or yellow; receptacle tubular (bisexual and female flowers) or mostly flat (male flowers); stamens in one (rarely two) whorl(s), (2–)4 or 8(–12), adnate to hypanthium, equal in length, all fertile, alternisepalous and/or oppositisepalous, erect in bud; filaments free, very short; anthers dorsifixed or basifixed, non-versatile, longitudinally dehiscing; gynoecium one-carpellate (probably pseudomonomerous; Takhtajan 1997); ovary superior, 1-celled and 1-ovulate; style elongate; stigma dry, non-papillate; ovule with a short and broad funicle, anatropous, bitegmic, crassinucellate; placentation basal. Fruit drupe- or berry-like with thin, membranous pericarp, enclosed by the persistent calyx tube, which becomes fleshy. Seed solitary; testa hard; embryo straight and achlorophyllous; endosperm scanty or absent; cotyledons fleshy, plano-convex. Germination phanerocotylar.

A family with three genera and 30–50 species, distributed mainly in the temperate regions of the Northern Hemisphere, some in tropical SE Asia and eastern Australia.

VEGETATIVE STRUCTURES. The stems and leaves are covered with peltate or scaly trichomes. The leaves are leathery, simple, entire and without stipules. Stomata are anomocytic. Vessels have simple perforations, and small to vestigial vestured pits are present, and nonvestured pits are also common. Rays are heterogeneous to almost homogeneous in some species of *Elaeagnus*. Axial parenchyma is diffuse, sometimes very scanty. Phloem is usually tangentially stratified into hard and soft layers (Jansen et al. 2000). Sieve-element plastids are of Ss-type.

EMBRYOLOGY. Microsporogenesis is simultaneous, and pollen grains are 2-celled or 3-celled when shed. Embryo sac development is of the Polygonum type or (*Shepherdia*) Allium type. Endosperm formation is nuclear. An endospermal chalazal haustorium develops in *Elaeagnus*.

POLLEN MORPHOLOGY. The reader is referred to Erdtman (1952), Leins (1967), and Sorsa (1971). Pollen is (2)3(4)-colporate, oblate or oblate-spheroidal. In *Elaeagnus*, three areas with a comparatively thin sexine are separated by three longitudinal, aperturiferous strands of incrassate sexine. Pollen grains of *Shepherdia argentea* and *S. rotundifolia* are angulaperturate or oblate. In contrast, grains of *S. canadensis* are longicolpate-parvorate and, as to their shape, distinctly different from other elaeagnaceous pollen grains. The pollen of *Hippophae* is suboblate or oblate-spheroidal.

KARYOLOGY. The basic chromosome number in *Hippophae* is x = 12 (Rousi 1965) and in *Elaeagnus* x = 14 (Arohonka and Rousi 1980). *Shepherdia canadensis* has x = 11 and *S. argentea* x = 13 (Arohonka and Rousi 1980, and references therein).

POLLINATION. *Hippophae* is fully anemophilous, while *Elaeagnus* and *Shepherdia* often have fragrant flowers and are entomophilous.

SEED DISPERSAL. The fleshy fruits of all three genera are eaten by birds and may be carried for long distances. Fruits of *Hippophae* dried during winter are known to float and disperse by spring floods along rivers.

REPRODUCTIVE SYSTEMS. *Hippophae* and *Shepherdia* are dioecious, *Elaeagnus* is monoecious or polygamous.

PHYTOCHEMISTRY. Elaeagnaceae are strongly tanniniferous; both ellagitannins and condensed tannins are present. Further compounds reported by Hegnauer (1966, 1989) include indol alkaloids, sinapinic acid, flavonols, pentacyclic triterpenes and L-quebrachit.

SYSTEMATICS. Phylogenetic studies of morphological, karyological, and molecular characters have identified three clades, corresponding to the genera *Elaeagnus*, *Hippophae* and *Shepherdia* (Hyvönen 1996; Bartish et al. 2002). Sequence analysis of the chloroplast DNA *rbc*L fragment indicates a close relationship between *Elaeagnus* and *Shepherdia*, *Hippophae* being sister to that clade (Richardson et al. 2000).

AFFINITIES. Elaeagnaceae have been placed either into Proteales (Cronquist 1988) or Rhamnales (Thorne 1992), or in a monotypic order, Elaeagnales, close to the two former (Takhtajan 1997). More recent molecular and morphological analyses (Thulin et al. 1998; Richardson et al. 2000; Soltis et al. 2000) place Elaeagnaceae together with Barbeyaceae, Cannabaceae, Cecropiaceae, Celtidaceae, Dirachmaceae, Moraceae, Rhamnaceae, Rosaceae, Ulmaceae, and Urticaceae in a group of families forming the expanded order Rosales sensu APG (1998). Barbeyaceae attach as sister to Elaeagnaceae with a moderate support; *Barbeya*, a small dioecious tree with opposite leaves and wind-pollinated flowers, is in many respects similar to *Hippophae*. It is unclear, however, whether Rhamnaceae or Ulmaceae are most closely related to Elaeagnaceae and Barbeyaceae.

DISTRIBUTION AND HABITATS. The family is mostly deciduous and has a typical Northern Hemisphere range; in SE Asia it extends to Malesia and NE Queensland. *Elaeagnus* occurs throughout this range, whereas *Hippophae* is found throughout Eurasia (Lian et al. 1998). *Shepherdia* is restricted to North America. Elaeagnaceae occur chiefly in steppes, river deltas and mountain terraces, valley slopes and along coasts at altitudes from 0 to 5000 m (Rousi 1971; Lian et al. 1998) but, in SE and E Asia, Malesia and NE Queensland, *Elaeagnus* is a lianaceous rain forest element (Veldkamp 1986).

PALAEOBOTANY. The oldest fossil pollen attributed to Elaeagnaceae is recorded from the upper Eocene of Central Asia (Pulatova 1973). In Europe, the oldest, well-documented fossil record for *Elaeagnus* or a related genus appears to be *Slowakipollis hippophaeoides* from the Oligocene (Krutzsch 1962). *Shepherdia* pollen has been reported from the Miocene of Oregon (Graham 1963), and from the Miocene of Idaho (Smiley et al. 1975). *Boehlensipollis hohli* from the Oligocene of Europe may represent an extinct genus of Elaeagnaceae (Muller 1981).

MYCOTROPHY. Mycorhiza fungi are present in the cortex of the rootlets of *Hippophae* and *Shepherdia*. All genera possess root nodules containing nitrogen-fixing actinomycetes (*Frankia*) with nitrogenase activity (Gardner 1958; Vanstraten et al. 1977). *Frankia* strains isolated from *Elaeagnus* and *Hippophae* are able to cross the inoculation barriers and infect *Alnus* species (Bosco et al. 1992).

Fig. 44. Elaeagnaceae. *Shepherdia canadensis*. **A** Branch of female plant with young fruits. **B** Fascicle of male flowers. **C** Fascicle of female flowers. Drawn by Pollyanna von Knorring (Orig.)

Fig. 45. Elaeagnaceae. *Elaeagnus angustifolia*. **A** Flowering branch armed with thorns. **B** Hermaphrodite flower, dissected, one stamen fallen off. **C** Fruiting branch. Drawn by Pollyanna von Knorring (Orig.)

Fig. 46. Elaeagnaceae. *Hippophae rhamnoides*. **A** Fruiting branch armed with thorns. **B** Short shoot of male plant with flowers. **C** Male flower with two sepals and four stamens (one of which removed). **D** Female flowers. Drawn by Pollyanna von Knorring (Orig.)

ECONOMIC IMPORTANCE. A number of *Elaeagnus* species are grown as ornamental shrubs. *Elaeagnus angustifolia*, *E. macrophylla*, *E. pungens*, and *E. umbellata* are mainly grown as deciduous or evergreen shrubs for their attractive foliage. The fruits of a number of species are edible, for example, those of *Hippophae rhamnoides* (sea buckthorn), *Elaeagnus angustifolia* (Russian olive, or oleaster), and *Shepherdia argentea* (silver buffalo berry). Fruits of sea buckthorn are rich in vitamins A, C, E, and have been used for centuries in Europe and Asia for the production of juices, jams, and beverages. Seed oil of sea buckthorn was formally listed in the Pharmacopoeia in 1977 and clinically tested in Russia and China (Xu 1994). The most important pharmacological functions attributed to sea buckthorn oil are: anti-inflammatory, bactericidal, pain relief, and promotion of tissue regeneration. Currently, *Hippophae* plantations are also used in China to prevent erosion of soil (Lu 1992).

KEY TO THE GENERA

1. Leaves exclusively opposite; stamens 8 (dioecious)
   1. *Shepherdia*
- Leaves alternate, rarely subopposite or in pseudowhorls; stamens 4   2

2. Polygamous or monoecious; flowers often scented, sepals 4
   2. *Elaeagnus*
- Dioecious; flowers odourless; sepals 2 (male flowers)
   3. *Hippophae*

1. *Shepherdia* Nutt.   Fig. 44

Shepherdia Nutt., Gen. N. Am. Pl. 2: 240 (1818); Servettaz, Beih. Bot. Centralbl. 25: 1–420 (1909).

Dioecious. Leaves opposite. Flowers scented, entomophilous; in male flowers stamens 8, alternate to and opposite with the 4 sepals. Three species in temperate North America.

2. *Elaeagnus* L.   Fig. 45

Elaeagnus L., Sp. Pl.: 121 (1753); Servettaz, Beih. Bot. Centralbl. 25: 1–420 (1909).

Polygamous or monoecious. Leaves alternate. Flowers 4-merous, often scented and entomophilous; perianth quadrangular; stamens equal in number and alternating with the sepals. Fruit usually with eight longitudinal ribs. About 20–45 species, most in Eurasia and from there to North America (only one species, *Elaeagnus commutata*), and through tropical SE Asia and Malaysia to N Queensland. The most widespread species is *Elaeagnus angustifolia* with discontinuous range from China and C Asia to Europe. The genus is in need of revision.

3. *Hippophae* L. Fig. 46

Hippophae L., Sp. Pl.: 1023 (1753); Rousi, Ann. Bot. Fennici 8: 177–227 (1971); Swenson and Bartish, Nord. J. Bot. 22: 369–374 (2002).

Dioecious. Leaves usually alternate, rarely in pseudowhorls or subopposite. Flowers odourless, anemophilous; calyx with only 2 lobes, valvate; male flowers with 4 stamens, alternate to and opposite with the sepals. Fruit with longitudinal ribs in several species. Seven species, four of them recognized recently, in temperate areas of Eurasia. *Hippophae rhamnoides* is the most widespread species with fragmented range from C China to the Atlantic coast of Europe.

Selected Bibliography

APG (Angiosperm Phylogeny Group) 1998. See general references.

Arohonka, T., Rousi, A. 1980. Karyotypes and C-bands in *Shepherdia* and *Elaeagnus*. Ann. Bot. Fennici 17: 258–263.

Bartish, I.V., Jeppsson, N., Nybom, H., Swenson, U. 2002. Phylogeny of *Hippophae* (Elaeagnaceae) inferred from parsimony analysis of chloroplast DNA and morphology. Syst. Bot. 27: 41–54.

Bosco M., Fernandez, M.P., Simonet P., Materassi R., Normand, P. 1992. Evidence that some *Frankia* sp. strains are able to cross boundaries between *Alnus* and *Elaeagnus* host specificity groups. Appl. Env. Microbiol. 58: 1569–1576.

Cronquist, A. 1988. The evolution and classification of flowering plants, 2nd edn. New York: New York Bot. Garden.

Erdtman, G. 1952. See general references.

Gardner, I.C. 1958. Nitrogen fixation in *Elaeagnus* root nodules. Nature 181: 717–718.

Graham, A. 1963. Systematic revision of the Sucker Creek and Trout Creek Miocene floras of southeastern Oregon. Am. J. Bot. 50: 921–936.

Hegnauer, R. 1966, 1989. See general references.

Hyvönen, J. 1996. On phylogeny of *Hippophae* (Elaeagnaceae). Nordic J. Bot. 16: 51–62.

Jansen, S., Piesschaert, F., Smets, E. 2000. Wood anatomy of Elaeagnaceae, with comments on vestured pits, helical thickenings, and systematic relationships. Am. J. Bot. 87: 20–28.

Krutzsch, W. 1962. Stratigraphisch bzw. botanisch wichtige neue Sporen- und Pollenformen aus dem deutschen Tertiär. Geologie 11: 265–307.

Leins, P. 1967. Morphologische Untersuchungen an Elaeagnaceen-Pollenkörnern. Grana Palynologica 7: 390–399.

Lian, Y.S., Chen, X., Lian, H. 1998. Systematic classification of the genus *Hippophae* L. Seabuckthorn Res. 1: 13–23.

Lu, R. 1992. Seabuckthorn: A multipurpose plant species for fragile mountains. Katmandu: ICIMOD Publication Unit.

Muller, J. 1981. See general references.

Pulatova, M.Z. 1973. The upper Eocene flora of the Tadjik depression by palynological data. Palinologiya kaynofita. Moscow: Nauka, pp. 114–121 (in Russian)

Richardson, J.E., Fay, M.F., Cronk, Q.C.B., Bowman, D., Chase, M.W. 2000. A phylogenetic analysis of Rhamnaceae using *rbcL* and *trnL-F* plastid DNA sequences. Am. J. Bot. 87: 1309–1324.

Rousi, A. 1965. Observations on the cytology and variation of European and Asiatic populations of *Hippophae rhamnoides*. Ann. Bot. Fennici 2: 1–18.

Rousi, A. 1971. The genus *Hippophae* L.: a taxonomic study. Ann. Bot. Fennici 8: 177–227.

Smiley, C.J., Gray, J., Huggins, L.M. 1975. Preservation of Miocene fossils in unoxidized lake deposits, Clarkia, Idaho. J. Paleontol. 49: 833–844.

Soltis, D.E. et al. 2000. See general references.

Sorsa, P. 1971. Pollen morphological study of the genus *Hippophae* L., including the new taxa recognized by A. Rousi. Ann. Bot. Fennici 8: 228–236.

Swenson, U., Bartish, I.V. 2002. Taxonomic synopsis of *Hippophae* (Elaeagnaceae). Nord. J. Bot. 22: 369–374.

Takhtajan, A. 1997. See general references.

Thorne, R.F. 1992. Classification and geography of flowering plants. Bot. Rev. 58: 225–348.

Thulin, M., Bremer, B., Richardson, J., Niklasson, J., Fay, M.F, and Chase, M.W. 1998. Family relationships of the enigmatic rosid genera *Barbeya* and *Dirachma* from the Horn of Africa region. Pl. Syst. Evol. 213: 103–119.

Vanstraten J., Akkermans, A.D.L., Roelofsen, W. 1977. Nitrogenase activity of endophyte suspensions derived from root-nodules of *Alnus*, *Hippophae*, *Shepherdia* and *Myrica* spp. Nature 266: 257–258.

Veldkamp, J.F. 1986. Elaeagnaceae. In: Flora Malesiana I, 10: 151–156.

Xu, M. 1994. The medical research and exploitation of sea buckthorn. *Hippophae* 7: 32–34.

Elaeocarpaceae

M.J.E. COODE

Elaeocarpaceae Juss. ex DC., Essai Propr. Méd. Pl., ed. 2: 87
(1816), 'Elaeocarpeae', nom. cons.
Tremandraceae R. Br. ex DC. (1824), nom. cons.

Trees to suffrutescent, sometimes ericoid; sometimes buttressed; bark without exudate; indumentum of simple hairs, sometimes gland-tipped or
stellate. Leaves spirally arranged, distichous, opposite or rarely whorled, simple (rarely pinnatisect
or pinnate in juveniles), entire to serrate, venation
usually pinnate, ± trinerved at base in *Vallea* and
in some *Aristotelia* and *Sloanea*; stipules present,
or 0 when colleters may occur. Flowers regular,
bisexual (unisexual in some *Aristotelia* and
Elaeocarpus), solitary and axillary, in fascicles or
in terminal or axillary simple or compound
inflorescences, either cymose or racemose; sepals
usually free, valvate, rarely imbricate at tip; petals
present (0 in some *Sloanea*), free, valvate at insertion and usually (induplicate-)valvate further up
but sometimes overlapping at apex (in some
Sloanea variously fused even into a tube), usually
expanded and variously toothed or lobed, sometimes entire and sepaloid; disk a toothed ring or
separate lobes or pulvinate or flat-topped, or 0;
stamens 4–300, free, inserted on or above the disk
or around free disk-lobes, filaments shorter to
longer than the anthers, anthers basifixed,
tetrasporangiate, opening at apex by 1–2 pores
(sometimes extended into a tube) or short slits
(sometimes extending to base), connective sometimes extended ('awned'); ovary superior, syncarpous (internal fusion sometimes incomplete at
apex); loculi 2–8(9); ovules 1–30/loculus, pendulous, anatropous, attached to axis in 1–2 series;
style simple or apically branched, rarely stylodia
free. Fruit indehiscent (a drupe or berry) or dehiscent (walls thin, horny or woody, initially fleshy in
Vallea), splitting loculicidally or loculicidally and
septicidally, sometimes spiny or bristly (*Sloanea*
p.p.). Seeds 1–many, up to c. 15/loculus, variously
'arillate' or with watery or fleshy sarcotesta or
seed coat dry when sometimes with fine or rarely
stout (*Platytheca*) hairs; embryo embedded in
endosperm, usually straight with broad (sometimes narrow) cotyledons.

A family comprising 12 genera and c. 550
species, widely distributed in tropical and warm-
temperate southern regions but lacking in continental Africa.

VEGETATIVE MORPHOLOGY. The family has some
suffrutescent subshrubs, sometimes ericoid, and a
few shrubs; most species are small to large trees.
There are occasional reports of epiphytes or
scrambling shrubs (*Elaeocarpus bilobatus*). Many
can have buttresses or prop-roots, sometimes
(*Sloanea* and *Elaeocarpus*) very large. Terminalia-
branching ('modèle d'Aubréville': Hallé et al. 1978)
occurs sporadically in *Sloanea* and particularly in
Elaeocarpus.

VEGETATIVE ANATOMY. The reader is referred
to Metcalfe and Chalk (1950), Carlquist (1977),
and Gasson (1996). The leaf epidermis is often
mucilaginous; the indumentum consists of unicellular and glandular hairs; stellate hairs are found
in the tremandraceous genera. Stomata are paracytic or encyclocytic. Crystals are mostly solitary.
Cork arises superficially. Nodes are trilacunar.
Sieve element plastids are of the Ss type. In the
wood, growth rings are usually discernable;
vessels are often in radial multiples. Perforation
plates are mainly simple but sometimes are
accompanied by some scalariform perforations.
Vessel ray pitting has reduced borders and varies
from circular to elongated (vertical, oblique, or
scalariform). Fibre-tracheids are absent, but libriform fibres are generally present; they are sometimes septate. Axial parenchyma is sparse and
tends to be marginal. Rays are heterocellular,
mixed uniseriate and pluriseriate, the latter up to
7 cells wide, and varying strongly in height. Their
margins are composed of several rows of upright
cells, and uniseriate rays consist entirely of
uprights. Prismatic crystals are found in all genera
except *Aristotelia* and *Vallea*, located in chambered or non-chambered ray cells or chambered
axial parenchyma cells.

INFLORESCENCE STRUCTURE. Inflorescences are
terminal or usually axillary panicles, cymes,

thyrses or racemes, but are also solitary flowers or fascicles (all can be found in *Sloanea*). The most floriferous are probably the panicles of *Elaeocarpus stipularis* var. *brevipes* with up to 90 flowers. Bracts are usually early caducous; a few species of *Elaeocarpus* have prophylls on the pedicels as well.

FLOWER STRUCTURE. Flowers are regular, mostly bisexual (some dioecious species in *Aristotelia* and monoecious species in *Elaeocarpus*). The calyx is nearly always of free valvate sepals, but is fused in *Crinodendron* and occasionally ± imbricate in *Sloanea*. Petals vary greatly; absent in two Australian and all but one New World species of *Sloanea*. Disk usually present, variable, absent in *Tetratheca* and *Platytheca*. Stamens are sometimes twice the petals ('Tremandraceae'), more usually not an exact number, from 4 in *Aristotelia fruticosa* of New Zealand and 5 in *Elaeocarpus bilongvinas* of New Guinea to 300 in some *Sloanea*; filaments various; anthers basifixed and opening initially by 1–2 apical pores or short slits, sometimes the connective extended into an awn (*Sloanea* p.p., *Elaeocarpus* p.p.) or tube. The ovary is superior, carpels 2–9-fused, style usually single, although sometimes there are style branches as many as carpels and even ± free stylodia (*Aristotelia* p.p., *Sloanea* p.p.); stylodia in *Peripentadenia mearsii* are free at base. For floral anatomy, see Payer (1857), van Heel (1966) and Matthews and Endress (2002).

EMBRYOLOGY. Pollen grains are 2-celled when shed. The ovule is anatropous, bitegmic and crassinucellate with a zigzag micropyle formed by both integuments. An endothelium and hypostase have been found in an *Elaeocarpus* species. The embryo sac is of the Polygonum type, endosperm formation is nuclear (Johri et al. 1992).

POLLEN MORPHOLOGY. Pollen grains are unusually small, prolate to suboblate (longest axis 8–25 μm) and tricolopor(oid)ate, sometimes syncolpate; the exine is often distinctly columellate; the sexine is crassitectate and the tectum is obscurely to faintly psilate, rugulate or finely reticulate and microperforate (Erdtman 1952; Sharma 1969; Barth and Barbosa 1973; Lobreau-Callen in Tirel 1985; Tang and Wu 1990).

KARYOLOGY. Several *Aristotelia* have been counted with $2n = 28$, two *Elaeocarpus* with $2n = 24$ and 30 respectively (Fedorov 1969); *Crinodendron hookerianum* has $n = 21$, and *Dubouzetia elegans* $n =$ ca. 90.

POLLINATION. Very few observations have been made on pollination of Elaeocarpaceae. The red-flowered *Crinodendron hookerianum* has all the features of a bird-pollinated flower and seems to be pollinated by a humming-bird (Bricker 1992). The rest appear to be entomophilous; thrips have been found in flowers. It is currently assumed that the majority at least are adapted to 'buzz-pollination' (Matthews and Endress 2002) which tends to be associated with pendant flowers, short stamen filaments, basifixed anthers with apical dehiscence and punctiform stigmas; the assumption is also made that the flowers are generally not nectariferous, which is not supported by field experience (M.J.E. Coode) that flowers of most species of *Elaeocarpus* at least are scented and some obviously contain nectar.

REPRODUCTIVE SYSTEMS. Dioecy is known in the New Zealand species of *Aristotelia*, and perhaps also in *A. chilensis* of South America. Unisexual flowers are known in some species of the *Acronodia* and *Polystachyus* groups of *Elaeocarpus* (W and C Malesia, Borneo: Coode 1996a, 1996b) and in Madagascar (Tirel 1985). Tirel reports that some individuals have bisexual flowers while others have male flowers only. Corner (1939: 309) said of the 'subgenus *Acronodia*' that there were 'male and monoecious individuals' while in the *Polystachyus* group there are species with separate male and female inflorescences, but whether always on separate trees is not established yet. Otherwise, the flowers are bisexual, but see Coode (1985) for possible unisexual flowers in *Vallea*. In some species of *Elaeocarpus*, fruits which are apparently perfect may contain no seeds. Whether there are barriers to self-pollination seems to be unknown.

FRUIT AND SEED. There is wide variation in fruit type and consequently seed type. Of those with dehiscent fruits, the three 'tremandraceous' genera have small fruits with thin, flimsy walls; *Crinodendron*, *Peripentadenia* and *Dubouzetia* have much larger fruits with tough, horny walls often separating into two layers and splitting loculicidally at first, sometimes septicidally subsequently – the resultant open fruits can resemble fruits of Euphorbiaceae at a similar stage; fruit of *Sloanea* have woody walls generally much thicker than those of the three preceding genera and never separating into layers (Fig. 47E), whereas *Vallea* fruits open while the walls are still fleshy. The four genera with indehiscent fruits are few-seeded and have either berries (*Sericolea*, *Aristotelia*) or drupes (*Elaeocarpus*, Fig. 48E; *Aceratium*). Seeds may have dry coats or sarcotestas (fleshy in *Sloanea*, thinner and ± watery in *Crinodendron*, *Sericolea* and some *Aristotelia*, thick and spongy in *Dubouzetia galorei*). In *Aristotelia*, *Dubouzetia*, *Tetratheca* and *Tremandra* there may be ± waxy, apical outgrowths (strophioles) of the chalaza, often spirally coiled. 'Arils' (formed from the chalaza or raphe or even seed-coat or all of them) may partly or wholly cover seeds in some *Sloanea* and *Peripentadenia*, while in *Vallea* there is a lop-sided development from the raphe. Seeds of *Sericolea* are usually curved and in *Elaeocarpus* a range of shapes (Weibel 1968; Coode 1984), from straight through curved to hooked, is useful in defining infrageneric groupings. Testa and tegmen are multiplicative; the exotesta is thick-walled, and in Sloanea and possibly in *Aristotelia* as well the tegmen has many outer layers of fibres which, according to Corner (1976), are relictual and deny a close affinity with Malvales.

DISPERSAL. Drupes of some *Elaeocarpus* and *Aceratium* are known to be dispersed by birds, such as fruit-pigeons and cassowaries (Crome 1975a, 1975b). *Elaeocarpus* fruits are often blue; this is due not to pigment but to diffraction (Lee 1991). *Aceratium* fruits are usually red. Presumably the other indehiscent (fleshy) fruits are also bird-dispersed. The colour of seeds (whether dry seed coat, sarcotestas or arils) contrasts with the colour of the loculus interior in many species of *Sloanea* and would suggest attraction to birds such as hornbills; the role of broad-based or bristle spines on the fruit exterior is not documented. Strophioles (elaiosomes) are attractive to ants; *Tetratheca* is discussed in Berg (1976) in this respect; whether the strophioles of the tree-forming genera are also indicative of ant-dispersal is not known. Clifford and Drake (1985) report on finding endocarps in kangaroo dung. The seeds of *Dubouzetia galorei* float and are bright red; within the spongy outer layers is a dense, hard sclerotesta which sinks. Cassowaries are known to feed on the fallen seeds of this species (label of *Balgooy* 6596 from Aru), but perhaps fish also play some part along the flood-plains of southern New Guinea.

PHYTOCHEMISTRY. Both condensed and hydrolysable tannins are present, particularly in the bark, the former based on procyanidin, the latter containing gallic acid, ellagic acid and methylellagic acid. Common triterpenes and saponins also occur; one *Sloanea* is known to be cyanogenic. Many different alkaloids have been isolated from *Aristotelia* species, which are tryptamin-monoterpene derived. *Elaeocarpus* and *Peripentadenia* species have furnished alkaloids of the elaeocarpidin group, which are derived from tryptamin or 1,3 diaminopropan or putrescin and a polyketid moiety. A bitter principle, elaeocarpid, has been isolated from *Elaeocarpus* species. Halphen-positive seed oils and large amounts of mucilage, both characteristic of Malvales families, have not been found in Elaeocarpaceae (Hegnauer 1973, 1990).

SUBDIVISION AND RELATIONSHIPS. Previous attempts at subdividing the family have been only partly successful (e.g. Schlechter 1916). However, it does seem possible to group the genera as follows: *Aristotelia*+*Vallea*; *Sloanea*; *Crinodendron*+*Peripentadenia*+*Dubouzetia*; *Elaeocarpus*+*Aceratium*+*Sericolea*. *Sloanea* is isolated; it would seem to be near the base if not basal. The three Tremandraceous genera (*Tremandra*, *Tetratheca* and *Platytheca*) have been found, on molecular evidence, to be nested in Elaeocarpaceae (Savolainen, Fay et al. 2000; Bradford and Barnes 2001). That morphology which appears comparable (especially the 'anthers' opening by a single pore) supports this placement, but in other respects it is difficult to evaluate their true position. Thompson (1976) provides a table of distinctions, and it is clear that while *Platytheca* and *Tremandra* are very distinct, *Tetratheca* is less easy to define.

AFFINITIES. Elaeocarpaceae were generally thought to be part of, or close to, Tiliaceae and placed in Malvales; more recently other positions were proposed (Dahlgren 1988; Takhtajan 1997; Thorne 2000). *Muntingia* was often included and, more understandably, *Petenaea* (both currently treated in Malvales, see vol. V of this series). Recently, Elaeocarpaceae have been placed on molecular evidence in Oxalidales. The family, as enlarged recently by the inclusion of Tremandraceae also on molecular grounds, is difficult to delimit on visible characteristics. Surveys with less patchy sampling are desirable to confirm the conclusions already being widely followed.

DISTRIBUTION AND HABITATS. Elaeocarpaceae are an essentially southern Gondwana family, of which only *Elaeocarpus* and *Sloanea* reach the northern hemisphere in any numbers; they are not found in continental Africa. *Sloanea* is found in Central and South America, Madagascar, NE India and Nepal through to S and C China, throughout Malesia to the Solomon Islands, New Caledonia and N and E Australia; it is the only genus of the family to be reported from the London Clay, as seeds and abraded fruits. *Elaeocarpus* has a wider distribution in the Old World but is missing from the New: Madagascar, Mauritius, Ceylon, E India through to C China and S Japan, Hawai'i, throughout Malesia and the western Pacific, Australia and New Zealand; fossils are known from Australia. The rest have a more limited distribution. *Vallea* is known from northern South America to Peru and Bolivia; *Aristotelia* from Chile and Argentina, New Zealand, Tasmania and Australia. *Crinodendron* is known from the southern part of South America, *Peripentadenia* is endemic to NE Queensland, while *Dubouzetia* is known from the Moluccas, New Guinea, New Caledonia and Australia (Northern Territory and NE Queensland). *Tremandra* and *Platytheca* are endemic to SW West Australia; *Tetratheca* extends from there across the southern part of the continent to E Australia. *Aceratium* is known from the Moluccas through New Guinea and the Bismarck Archipelago, Solomon Islands, Vanuatu with different species from NE Queensland. *Sericolea* is endemic to highland New Guinea, once found on Fergusson Island. Mostly they are species of high, or disturbed, forest; at mid-altitudes especially, forests can be found with several species co-existing, with large numbers of individuals. Some reach altitudes of 3000 m or more, in Malesia at least (*Sericolea calophylla* up to 3500 m), where they tend to be shrubby. Some are limited to seral habitats while a few are pioneers, e.g. *Aristotelia serrata* and *Elaeocarpus murukkai*, the latter forming even-age stands with a species of *Macaranga* and virtually nothing else on landslips at mid-latitude in Papua New Guinea. *Aristotelia chilensis*, after having been introduced to the Island of Juan Fernandez off the Chilean coast, has become a threat to the endemic flora at lower and middle elevations. The two Australian *Dubouzetia* are shrubby species which inhabit rocky gorges at low altitudes. The three 'tremandraceous' genera are found in open 'bushland' vegetation.

PALAEOBOTANY. Rozefelds (1990) and Rozefelds and Christophel (1996a, 1996b) report on endocarps of *Elaeocarpus* from the Oligo/Miocene in Australian deposits. *Sloanea* is recorded from the London Clay; recent material in better condition would be welcome.

ECONOMIC IMPORTANCE. *Elaeocarpus angustifolius* (syn. *E. sphaericus* or *Ganitrus sphaerica*) is the source of endocarps used as beads (*rudraksha*)

that are strung together to make *malas* or rosaries, important in eastern religious practice. *Aceratium oppositifolium* fruits are edible. A few species of *Elaeocarpus* have edible mesocarp and are sometimes pickled (*E. robustus*) in SE Asia. The seeds of *Elaeocarpus* appear to be edible but only worth extracting in large-fruited species (e.g. *E. schlechterianus*). Some large Australian *Sloanea* and *Elaeocarpus* yield useful timber. Species of *Crinodendron*, *Vallea* and *Aristotelia* are grown in north-temperate horticulture, while species of other genera are found in tropical gardens. Several local uses for *Sloanea* in the New World are listed by Earle Smith (1954).

CONSERVATION. Many species of most, if not all genera probably need listing as under threat of some sort, e.g. the 1–2 individuals of *Elaeocarpus bojeri* and hardly more numerous *E. integrifolius*, both of Mauritius (Coode 1987b; see also 1997 IUCN Red List of Threatened Plants, Walter and Gilletts 1998: pp. 242 and 584). Many well-characterised species are known from the type collection only.

KEY TO THE GENERA

1. Suffrutescent; indumentum with some gland-tipped hairs, or with stellate hairs; stipules absent, sometimes replaced by colleters; ovary flattened or narrowed in flower, ovules in a single row or 1; fruits dehiscent, fragile, ± flattened, less than 1.5 cm 2
 - Shrubs to trees; indumentum of simple non-glandular hairs only; stipules usually present though sometimes early caducous; ovary terete in flower, rarely somewhat angular, ovules in pairs or biseriate; fruits dehiscent with tough walls, mostly exceeding 1.5 cm, or indehiscent 4
2. Indumentum mostly of stellate hairs; disk of minute, free lobes distant from each other; stamens with filaments about as long as anthers ..... **10. Tremandra**
 - Indumentum of simple (some gland-tipped) hairs, none stellate; disk absent; stamens with filaments shorter than anthers 3
3. Petals with a central vein stronger than the rest and with pinnate side-branches; stamens with the 4 pollen sacs arranged in 1 series; fruit septum present; seed surface with thick hairs, seeds without appendages ..... **11. Platytheca**
 - Petals with net-venation; stamens with the four pollen sacs in 2 series; fruit septum 0; seed surface glabrous or with fine hairs, seeds with an appendage at apex ..... **12. Tetratheca**
4. Stamens dehiscing at least initially by 2 separate apical pores 5
 - Stamens dehiscing at least initially by 1 apical pore or slit 7
5. Ovules 8–16/loculus (seeds usually many); petals variously much-divided, or entire and sepaloid, or entire and expanded at apex, or absent ..... **3. Sloanea**
5. Ovules 2/loculus (seeds few); petals 3-lobed, not further divided, or merely 3-notched 6

6. Leaves opposite or in whorls of 3; stipules not foliaceous; stamen filaments terete; fruits indehiscent, seeds with or without a short terminal appendage and sometimes with a thin sarcotesta **2. Aristotelia**
  - Leaves spirally arranged; stipules when present foliaceous; stamen filaments flattened, broad-based and contiguous; fruits dehiscent, initially ± fleshy, seeds with an obvious asymmetric outgrowth of raphe **1. Vallea**
7. Fruits indehiscent 8
  - Fruits dehiscent 10
8. Leaves alternate to spirally arranged or very rarely opposite; fruit a drupe with ± woody endocarp usually separable from mesocarp (rarely with persistent radial fibres) **9. Elaeocarpus**
  - Leaves opposite, sometimes alternate in young shoots; fruit a drupe with weakly woody 'core' and fibrous mesocarp, or a berry 9
9. Petals variously much-divided; stamens with filaments longer than anthers and setae arranged around the apical pore or slit of anther; ovules 6–14/loculus; fruit a drupe, seed surface often with fine hairs, embryos straight **8. Aceratium**
  - Petals merely notched or ± entire; stamens with filaments as long as anthers or less, no setae at anther apex; ovules 2/loculus, seeds glabrous with thin sarcotesta, embryos curved **7. Sericolea**
10. Sepals fused into a cup, finally splitting along 1–2 lines; seeds without appendages but sometimes with a watery or thin sarcotesta **4. Crinodendron**
  - Sepals free; seeds with strophiole (sometimes spiral) at apex, or with complete arillode (an outgrowth of chalaza) or with thick spongy sarcotesta 11
11. Disk of large free lobes, the stamens inserted around the lobes; petals without pockets at base; stamens with filaments as long as anthers or less, apical pores extending to lateral slits; seeds with complete arillode, an outgrowth of chalaza **6. Peripentadenia**
  - Disk pulviniform to annular, stamens inserted mostly between disk and ovary base; petals with pockets at base corresponding with disk-lobes; stamens with filaments longer than anthers, apical pores not extending laterally into slits; seeds with strophiole at apex or with thick sarcotesta **5. Dubouzetia**

I. CORE ELAEOCARPS (genera 1–9)

Shrubs to trees; hairs simple, not gland-tipped; leaves spirally arranged or opposite, simple when mature, venation usually pinnate; pedicels continuous with calyx; sepals usually free; petal veins, when visible, netted, petals with or without basal pockets; stamen filaments usually terete; anthers with 2 bilobed contiguous thecae; ovary terete to angular; ovules bi-seriate; fruits indehiscent or dehiscent with tough walls; septum 0; seeds usually glabrous; embryo usually straight with broad cotyledons.

1. ARISTOTELIA ALLIANCE (see Coode 1985).

Stipules foliaceous or reduced; petals free, expanded; stamen filaments ± equalling anthers; anthers initially dehiscing by 2 apical pores, often splitting to base; ovules 2/loculus, collateral but overlapping; styles branched or not; fruit septum 0; seeds 1/loculus at most, glabrous. Two genera, New World and southern hemisphere.

1. *Vallea* L. Fil.

Vallea Mutis ex L. Fil., Suppl. Pl. Syst. Veg.: 42 (1782).

Shrubs; leaves with ternate basal venation; stipules foliaceous, often early caducous; inflorescence cymose, rarely flowers solitary; petals imbricate at tip, 3-lobed; disk mostly replaced by spread ovary-base; stamens 15–60, inserted in a ± single ring; filaments, sigmoid, bases broad, contiguous; ovary 3–5-locular; styles branched at apex; fruit fleshy at first, loculicidally part-dehiscing, valves persistent; seeds with a lopsided outgrowth of raphe, testa dry. 1–2 species from Colombia and Venezuela to Peru and Bolivia, in the lower slopes of the Andes.

2. *Aristotelia* L'Hér.

Aristotelia L'Hér., Stirp. Nov. 31 t. 16 (1786) nom. cons. *Friesia* DC.

Shrubs to medium trees; leaves opposite, rarely in 3s or alternate, juveniles sometimes pinnatisect, basal venation sometimes ternate; stipules reduced (to 0?); inflorescence a panicule or thyrse, rarely a 3-flowered cyme or flowers ± solitary; flowers bisexual or plants dioecious; petals imbricate at tip, 3-lobed, -notched or ± entire; disk cushion-shaped to annular, rarely minutely lobed; stamens 4–15; ovary 2–4-locular; styles notched at apex or with as many branches as loculi, sometimes to base; fruit a berry; seeds sometimes with a thin sarcotesta, with or without a short terminal appendage. 5 species from Chile, New Zealand and Australia (Tasmania, New South Wales).

2. SLOANEA ALLIANCE

3. *Sloanea* L. Fig. 47

Sloanea L., Sp. Pl. :512 (1753); Earle Smith, Contrib. Gray Herb. Harv. Univ. 175: 1–114 (1954), New World spp.; Coode, Kew Bull. 38: 347–427 (1983), conspectus of Old World spp. *Echinocarpus* Blume *Antholoma* Labill.

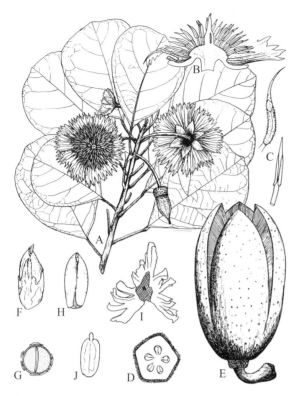

Fig. 47. Elaeocarpaceae. *Sloanea rhodantha.* **A** Flowering branchlet. **B** Flower, vertical section. **C** Stamens, inside and outside view. **D** Ovary, cross section. **E** Dehisced fruit. **F** Arillate seed. **G** Seed, cross section, with adhering fringes of aril. **H** Seed, aril removed, the area of adherence of aril cross-hatched. **I** Aril, expanded. **J** Embryo. (Capuron around 1965)

Small to large trees, often buttressed; occasionally shrubby; leaves sometimes subopposite, juveniles sometimes pinnate, basal venation sometimes ternate; stipules linear to foliaceous, often caducous, rarely 0?; inflorescence cymose, thyrsoid, a fascicle or raceme, or flowers solitary; sepals sometimes imbricate at tip; petals expanded and divided or toothed (rarely entire) and imbricate at tip or fused into a partial or complete corona or sepaloid and acute or 0 (most New World, some Australia); disk flat-topped to pulviniform; stamens (8–)20–300; filaments shorter than to exceeding anthers, sometimes ± 0; anthers often awned, apical setae 0, dehiscing by 2 apical pores to short apical slits sometimes extending down the sides; ovary 2–5-locular; ovules 8–30/loculus in 2 rows; styles unbranched or with as many branches as loculi, sometimes to base; fruit woody, loculicidally dehiscent, valves generally massive, persistent (often brightly coloured inside), often with spines (broad based to bristly;

seeds 1–many, glabrous, with a coloured sarcotesta or variable cover of hard seed-coat by often coloured arillode. About 100 species from C & tropical South America, and over 50 from the Old World including Madagascar (3), NE India & Nepal (4), Burma (7), China (12–15), Vietnam (6), Borneo (2), New Guinea (18), New Caledonia (9), Australia (4).

3. CRINODENDRON ALLIANCE (see Coode 1987).

Stipules minute, often early caducous; petals free, expanded; stamen filaments straight to weakly sinuous, at least equalling anthers; anthers without connective extension and apical setae, dehiscing by a single apical pore sometimes developing into slits; ovules biseriate; styles unbranched above; fruit without spines, walls horny often of 2 layers, dehiscent loculicidally at first, sometimes also finally septicidally; seeds few to several per fruit, 1–4 per loculus, usually glabrous. Three genera, southern hemisphere.

4. *Crinodendron* Molina

Crinodendron Molina, Sagg. Chil. ed. 1: 179 (1782); Bricker, Syst. Bot. 16: 77–88 (1991), rev.
Tricuspidaria Ruiz & Pavon

Shrubs to medium trees; flowers solitary with articulated pedicels; sepals fused at first, later splitting along 1–2 lines; petals 3-lobed to 3(–5)-toothed, 2-pocketed at base; disk tall, resembling a cogwheel; stamens 13–18, inserted between disk and ovary; ovary 3–5-locular; ovules 4–15/loculus; styles unbranched; fruit-valves not detaching, with columella sometimes persistent; seeds glabrous with complete watery sarcotesta and without appendage. About 5 species from S South America; one very distinct with red, bird-pollinated flowers (*C. hookerianum*).

5. *Dubouzetia* Brongn. & Gris

Dubouzetia Pancher ex Brongn. & Gris, Bull. Soc. Bot. Fr. 8: 199 (1861).

Inflorescence short-racemose to fasciculate, rarely flowers solitary with articulated pedicels; petals rounded, ± entire to minutely serrulate, 2-pocketed at base; disk pulviniform to slightly 5-lobed; stamens 15–45, inserted on disk surface and/or at ovary base; filaments longer than anthers; ovary 3–5-locular; ovules 4–12/loculus; fruit valves not detaching; seeds few to several, 1–3 per loculus, glabrous or with fine spreading hairs, seed-coat hard with an apical spiral waxy strophiole but *D. galorei* with thick, spongy sarcotesta surrounding a small sclerotesta. 6 species from New Caledonia, of which 1 is also in Moluccas to New Guinea, 3 more from New Guinea and 2 endemic to small areas of northern Australia.

6. *Peripentadenia* L.S. Sm.

Peripentadenia L.S. Sm., Proc. Roy. Soc. Queensl. 68: 45 (1957).

Juvenile leaves sometimes pinnatisect; flowers grouped to fasciculate, perhaps condensed-solitary, pedicels not articulated; petals 3-lobed with lobes further few-divided, incurved but without pockets at base; disk with 5 separate masses; stamens 50–75, inserted around the disk-lobes; filaments sometimes exceeding anthers; anthers eventually splitting down sides to base; ovary 3-locular; ovules 4–9/loculus; styles fused and unbranched above (in *P. mearsii* free at base); fruit with valves detaching, columella persistent; seeds 1(–2) per fruit, glabrous, with a hard seed-coat with complete aril attached only at chalaza. 2 species endemic to NE Queensland.

4. Elaeocarpus Alliance

Stipules minute to foliaceous, often early caducous; petals free, usually expanded; stamen filaments various; anthers dehiscing by a single apical pore or slit, with or without connective extension and/or apical setae; ovules biseriate; styles unbranched above; fruit a berry or drupe; seeds few to several per fruit, 1–2 per loculus, usually glabrous. Three genera, (sub)tropical Asia and Pacific Islands to Australia and Madagascar.

7. *Sericolea* Schltr.

Sericolea Schltr., Bot. Jahrb. 54: 95 (1916); van Balgooy, Blumea 28: 103–141 (1982), rev.

Shrubs to small trees; leaves opposite; stipules linear, minute; inflorescence paniculate or racemose, branches and flowers opposite; petals small, minutely lobed or notched, without pockets at base; disk weakly to distinctly lobed; stamens (8–)10–15(–18), inserted between disk and ovary base; filaments straight to sinuous, shorter than to equalling anthers; anthers without connective extension or apical setae, dehiscing by a single

apical confluent slit; ovary 2-locular; ovules 2/loculus; styles unbranched but notched at tip; fruit a berry; seeds 1–4 per fruit, glabrous, with thin complete sarcotesta (not obvious when dried), without appendages. Embryo weakly curved; cotyledons fairly broad. About 16 species from submontane to subalpine New Guinea.

8. *Aceratium* DC.

Aceratium DC., Prodr. 1: 519 (1824).

Shrubs to small trees; leaves opposite, sometimes alternate on young shoots; stipules minute or ?0; inflorescence short-racemose to sub-umbellate; petals variously divided to laciniate, without pockets at base; disk tall, resembling a cogwheel; stamens 12–20(–30), inserted between disk and ovary base; filaments straight in young bud, later usually sigmoid-inrolled, longer than anthers; anthers without connective extension, apical setae present and obvious, surrounding the apical pore or short slit not extending down sides; ovary 3–5-locular; ovules (5)6–14/loculus; styles unbranched; fruit ± drupoid, the mesocarp with radiating fibres persistent on thin-walled or ± woody endocarp; seeds 1–2 per fruit, glabrous or with fine spreading hairs, lacking sarcotesta or other appendage. About 20 species, 5 endemic to NE Queensland, the rest from Moluccas to New Guinea, Solomon Islands and Vanuatu.

9. *Elaeocarpus* L. Fig. 48

Elaeocarpus L., Sp. Pl.: 515 (1753).
Many synonyms, listed in e.g. Tirel 1985: 5.

Shrubs to large trees; leaves spirally arranged to distichous, very rarely opposite (Mauritius, New Guinea); stipules minute to large, linear to foliaceous, sometimes early caducous; inflorescence a raceme, rarely short-racemose or paniculate; flowers usually bisexual, sometimes unisexual (on monoecious plants?); petals expanded or sepaloid, sometimes small, variously lobed and/or laciniate, or entire and ± sepaloid, rarely entire and expanded, with or without pockets at base; disk variously tall or low, sometimes resembling a cogwheel, or lobed, sometimes merely annular with minute teeth, rarely pulviniform; stamens 5–200, inserted on disk surface and/or at ovary base; filaments straight to sigmoid, shorter than, equalling or exceeding anthers, sometimes ± 0; anther connective extension often present as awn or 0, apical setae sometimes present (but never surrounding

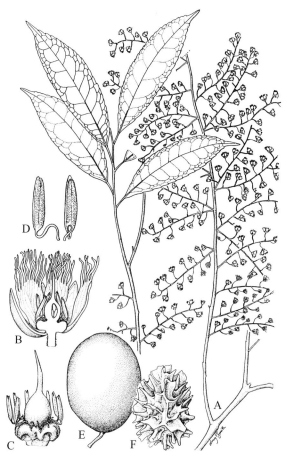

Fig. 48. Elaeocarpaceae. *Elaeocarpus sericoloides.* **A** Flowering branchlets. **B** Flower, longitudinal section. **C** Pistil and disk, some stamens removed. **D** Stamens. **E** Fruit. **F** Fruit stone. Drawn by Lucy T. Smith. (Coode 2002)

apical pore), dehiscing by a single apical confluent slit often extending down sides sometimes to base; ovary 2–9-locular; ovules 2–16/loculus; styles unbranched; fruit a drupe with woody often sculptured endocarp usually separable from the mesocarp, sometimes with ± persistent fibres as in *Aceratium*; seeds 1–5(–7)/fruit, at most 1/loculus, glabrous and without sarcotesta or appendages. Embryo straight with broad cotyledons to curved with broad or narrow cotyledons. The largest genus, with about 350 species altogether: Madagascar 8, Ceylon & India c. 30, Thailand c. 18, China c. 20, Japan 2, Peninsular Malaysia c. 30, Java c. 15, Borneo c. 70, Sulawesi c. 40, Philippines c. 50, New Guinea c. 85, New Caledonia 29, Australia c. 25, Lord Howe Is. 1, New Zealand 2, Fiji 22, Hawai'i 1.

II. TREMANDRACEOUS GENERA (genera 10–12; Thompson, Telopea 1: 139–215 (1976), and pers. comm. March 2003.

Suffruticose to sometimes ericoid shrublets; hairs simple (with some gland-tipped) or stellate; leaves usually opposite to verticillate, simple, where visible often penninerved; stipules absent or replaced by colleters; flowers solitary; pedicels discontinuous with calyx; sepals free, valvate; petals free, expanded or linguiform, without basal pockets, veins netted or pinnate; anthers with 4 thecae, without connective extension and apical setae, dehiscing by a single apical pore; disk usually 0; ovary somewhat to clearly flattened or narrowed, 2-locular (rarely 3); ovules 1–2, rarely more, uniseriate or single on a central axis; styles unbranched; fruits somewhat to clearly compressed, dehiscent, walls flimsy, septum present or 0.

10. *Tremandra* R. Br. Fig. 49

Tremandra R. Br. in Flinders, Voy. Terra Austral. 2: app. 3: 544 (1814).

Subshrubs rarely to 2 m; hairs mostly stellate (simple on anthers and petals); leaves opposite, simple, sometimes ± lobed and/or ± trinerved at base; flowers scattered; petals entire and rounded, expanded, ± net-veined; disk with small distinct lobes; stamens 10, inserted at ovary base; alternate stamens with thickened disk-like structure, filaments ± straight, about equalling anthers, thecae in 2 series; ovary flattened; fruit regularly dehiscent, valves detaching, septum 0; seeds 1–2 per fruit, with fine slender-spreading to ± adpressed hairs, seedcoat hard with apical appendage. 2 species endemic to SW West Australia.

11. *Tetratheca* Sm.

Tetratheca Sm., Specimen. Bot. New Holland: 5, t. 2 (1793); Thompson, Telopea 1: 139–215 (1976), rev.

Subshrubs, sometimes ericoid or trailing, usually to 1 m, rarely more; indumentum often mixed of fine hairs, thick-based bristles and gland-tipped hairs; colleters sometimes present in Western Australia; leaves alternate to opposite, sometimes in verticels, often reduced or caducous; flowers scattered or sometimes grouped; petals free, ligulate to expanded, entire; disk 0; stamens 8–10, inserted at ovary base, sometimes fused at base to varying extents into pairs, filaments straight but

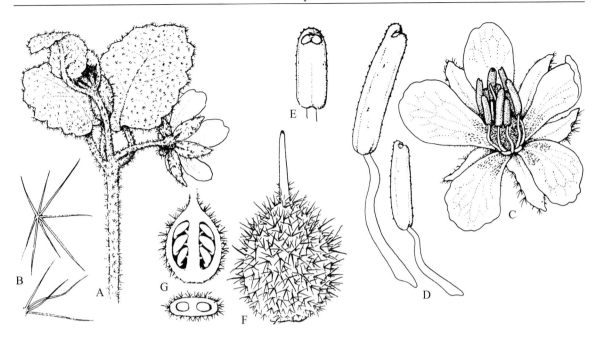

often much expanded, shorter than anthers, thecae in 2 series, apical pore usually extended into a tube, ovary often aberrantly 3-locular; ovules 1–2/loculus (1 sp. with 8–10 ovules), rarely more; fruit dehiscent loculicidally (occasionally later along septa), valves persistent, septum 0; seeds 1–4 per fruit, so 1–2 per loculus, usually with fine (?retrorse) hairs, seed-coat hard with apical appendage. About 40 species from across southern Australia.

12. *Platytheca* Steetz

Platytheca Steetz in Lehmann, Pl. Preiss. 1: 220 (1845).

Subshrubs. Hairs simple, often short and stiff, sometimes gland-tipped; thick-based bristles perhaps also present; colleters sometimes present; leaves in verticels, reduced, venation obscured; flowers scattered; petals with a central penninerved vein to tip stronger than the rest, expanded, short-mucronate; disk 0; stamens 10, inserted at ovary base in 2 series, filaments straight, stout, ± flat, shorter than anthers, thecae in 1 series, apical pore extended into a long tube; ovules 1/loculus; fruit dehiscent loculicidally, then septicidally, valves detaching, septum present; seeds 1–2 per fruit, with adpressed hairs narrow at base and thick above, without appendage. 2 species from SW West Australia.

Fig. 49. Elaeocarpaceae. *Tremandra stelligera*. **A** Flowering branch. **B** Stellate hairs. **C** Flower. **D** Outer and inner stamen. **E** Anther, inside view. **F** Pistil. **G** Ovary, vertical and transverse section. Drawn by B. Johnsen. (Thorne 1983)

Selected Bibliography

Balgooy, M.M.J. van 1982. A revision of *Sericolea* Schlechter (Elaeocarpaceae). Blumea 28: 103–141.

Barth, O.M., Barbosa, A.F. 1973. Catálogo sistemático das pólens das plantas arbóreas do Brasil meridional. XVII. Elaeocarpaceae e Tiliaceae. Mem. Inst. Oswaldo Cruz 71: 203–217.

Berg, R.Y. 1976. Myrmecochorous plants in Australia and their dispersal by ants. Austr. J. Bot. 23: 475–508.

Bradford, J.C., Barnes, R.W. 2001. Phylogenetics and classification of Cunoniaceae (Oxalidales) using chloroplast DNA sequences and morphology. Syst. Bot. 26: 354–385.

Bricker, J.S. 1991. A revision of the genus *Crinodendron* (Elaeocarpaceae). Syst. Bot. 16: 77–88.

Bricker, J.S. 1992. Pollination biology of the genus *Crinodendron* (Elaeocarpaceae). J. Arizona-Nevada Acad. Sci. 24–25: 51–54.

Capuron, R. (without year, about 1965). Études sur les essences forestières de Madagascar: Voanama, *Sloanea rhodantha* (Baker) R. Cap. – Elaeocarpacées. Centre technique forestier tropical (Section de Madagascar). Mimeographed.

Carlquist, S. 1977. Wood anatomy of Tremandraceae: phylogenetic and ecolgical implications. Am. J. Bot. 64: 704–713.

Clifford H.T., Drake, W.E. 1985. Seed dispersal by kangaroos and their relatives. J. Trop. Ecol. 1: 373–374.

Coode, M.J.E. 1978. A conspectus of Elaeocarpaceae in Papuasia. Brunonia 1: 131–302.

Coode, M.J.E. 1981. Elaeocarpaceae. In: Henty, E.E. (ed.) Handbooks of the flora of Papua New Guinea, vol. II. Melbourne: Melbourne University Press.

Coode, M.J.E. 1983. A conspectus of *Sloanea* (*Elaeocarpaceae*) in the Old World. Kew Bull. 38: 347–427.

Coode, M.J.E. 1984. *Elaeocarpus* in Australia and New Zealand. Kew Bull. 39: 509–586.

Coode, M.J.E. 1985. *Aristotelia* and *Vallea*, closely related in Elaeocarpaceae. Kew Bull. 40: 479–507.

Coode, M.J.E. 1987a. *Crinodendron, Dubouzetia* and *Peripentadenia*, closely related in Elaeocarpaceae. Kew Bull. 42: 777–814.

Coode, M.J.E. 1987b. 55. Eléocarpacées: In: Bosser et al. (eds.) Flore des Mascareignes, fams. 51. Malvacées – 62. Oxalidacées. Mauritius: MSIRI; Paris: ORSTOM; London: RBG Kew.

Coode, M.J.E. 1995. *Elaeocarpus* in the Flora Malesiana area – *E. kraengensis* and ten new species from Sulawesi. Kew Bull. 50: 267–294.

Coode, M.J.E. 1996a. *Elaeocarpus* for Flora Malesiana – notes, new taxa and combinations in the *Acronodia* group. Kew Bull. 51: 267–300.

Coode, M.J.E. 1996b. *Elaeocarpus* for Flora Malesiana: the *Polystachyus* group. Kew Bull. 51: 649–666.

Coode, M.J.E. 2001. *Elaeocarpus* for Flora Malesiana – the *E. stipularis* complex, *E. nitidus* group & *E. barbulatus*. Kew Bull. 56: 513–565.

Coode, M.J.E. 2002. *Elaeocarpus* in New Guinea: *E. sericoloides* (*Fissipetalum* group), *E. royenii* & *E. multisectus* (sect. *Elaeocarpus*). Contributions to the flora of Mt. Jaya, IX. Kew Bull. 57: 925–935.

Corner, E.J.H. 1939. *Elaeocarpus*. Gard. Bull. Str. Settlem. 10: 308–329.

Corner, E.J.H. 1976. See general references.

Crome, F.H.J. 1975a. Some observations on the biology of the cassowary in northern Queensland. Emu 76: 8–14.

Crome, F.H.J. 1975b. The ecology of fruit pigeons in tropical northern Queensland. Austr. Wildlife Res. 2: 155–185.

Dahlgren, R.M.T. 1988. Rhizophoraceae and Anisophyllaceae: summary statement, relationships. Ann. Miss. Bot. Gard. 75: 1259–1277.

Earle Smith, C. 1954. The New World species of *Sloanea* (Elaeocarpaceae). Contrib. Gray Herb. 175: 1–114.

Erdtman, G. 1952. See general references.

Federov, A.A. 1969. See general references.

Frith, H.J., Crome, F.H.J., Wolfe, T.O. 1976. Food of fruit pigeons in New Guinea. Emu 76: 49–58.

Gasson, P. 1996. Wood anatomy of the Elaeocarpaceae. In: Donaldson, L.A., Singh, A.P., Butterfield, B.G., Whitehouse, L.J. (eds.) Recent advances in wood anatomy. Rotorua: New Zealand Forest Research Institute Ltd, pp. 47–71.

Hallé, F., Oldeman R.A.A., Tomlinson P.B. 1978. Tropical trees and forests. Berlin Heidelberg New York: Springer.

Heel, W.A. van 1966. Morphology of the androecium in Malvales. Blumea 13: 177–394.

Hegnauer, R. 1973, 1990. See general references.

Johri, B.M. et al. 1992. See general references.

Lee, D.W. 1991. Ultrastructural basis and function of iridescent blue colour of fruits in *Elaeocarpus*. Nature 349: 260–263.

Matthews M.L., Endress P.K. 2002. Comparative floral structure and systematics in Oxalidales (Oxalidaceae, Connaraceae, Brunelliaceae, Cephalotaceae, Cunoniaceae, Elaeocarpaceae, Tremandraceae). Bot. J. Linn. Soc. 140: 321–381.

Metcalfe, R.C., Chalk, L. 1950. See general references.

Payer, J.-B. 1857. Traité d'organogénie comparée de la fleur. Paris: Masson.

Rozefelds, A.C. 1990. A mid Tertiary rainforest flora from Capella, central Queensland. In: Proc. 3rd IOP Conf., Melbourne, pp. 123–136.

Rozefelds A.C., Christophel, C. 1996a. *Elaeocarpus* (Elaeocarpaceae) from the Early to Middle Miocene Yallourn Formation of Eastern Australia. Muelleria 9: 229–237.

Rozefelds A.C., Christophel, C. 1996b. *Elaeocarpus* (Elaeocarpaceae) endocarps from the Oligo-Miocene of Eastern Australia. Papers & Proc. Roy. Soc. Tasmania 130: 41–48.

Savolainen, V., Fay, M.F. et al. 2000. See general references.

Schlechter, R. 1916. Die Elaeocarpaceen Papuasiens. Bot. Jahrb. Syst. 54: 92–155.

Sharma, B.D. 1969. Pollen morphology of Tiliaceae in relation to plant taxonomy. J. Palynol. 5: 7–29.

Straka, H., Friedrich, B. 1983. Palynologia madagassica et mascarenica. Fam. 121–127. Microscopie électronique à balayage et addenda. Pollen Spores 25: 49–73.

Takhtajan, A. 1997. See general references.

Tang,Y., Wu, Z.Y. 1990. Study on the pollen morphology of Chinese Elaeocarpaceae. Acta Bot. Yunnan. 12: 397–403, 4 pl. (in Chinese with English summary).

Thompson, J. 1976. A revision of the genus *Tetratheca* (Tremandraceae). Telopia 1: 139–215.

Thorne, R.F. 1983. Proposed new realignements in the angiosperms. Nord. J. Bot. 3: 85–117.

Thorne, R.F. 2000. The classification and geography of the flowering plants: dicotyledons of the class Angiospermae (subclasses Magnoliidae, Ranunculidae, Caryophyllidae, Dilleniidae, Rosidae, Asteridae and Lamiidae). Bot. Rev. 66: 441–647.

Tirel, C. 1982. Eléocarpacées. In: Hallé, N. (ed.) Flore de la Nouvelle Calédonie et dépendances 11: 3–124.

Tirel, C. 1985. *Eléocarpacées*. In: Badré, F. (ed.) Flore de Madagascar et des Comores, Famille 125: 3–54.

Tirel, C., Raynal, J. 1980. Recherches bibliographiques sur trois espèces d'*Elaeocarpus* (Elaeocarpaceae). Adansonia II, 20: 169–177.

Walter, K.S., Gilletts, H.J. (eds.) 1998. 1997 IUCN Red List of Threatened Plants. Gland: IUCN – World Conservation Union.

Weibel, R. 1968. Morphologie de l'embryon et de la graine des *Elaeocarpus*. Candollea 23: 101–108.

Ericaceae

P.F. STEVENS

with J. LUTEYN (most Vaccinieae); E.G.H. OLIVER (*Erica*); T.L. BELL, E.A. BROWN, R.K. CROWDEN, A.S. GEORGE, G.J. JORDAN, P. LADD, K. LEMSON, C.B. MCLEAN, Y. MENADUE, J.S. PATE, H.M. STACE, C.M. WEILLER (Styphelioideae)

Ericaceae Juss., Gen. Pl.: 159 (1789) ("Ericae"), nom. cons.
Epacridaceae R. Brown (1810).
Monotropaceae Nuttall (1818).
Empetraceae S.F. Gray (1821).
Vacciniaceae DC. ex Gray (1821).
Pyrolaceae Dumort. (1829).

Evergreen or deciduous shrubs, rarely scandent, lianes, or trees, epiphytic or not, or herbs, rarely achlorophyllous and/or rhizomatous; hair roots present, with investing mycorrhizal fungal hyphae forming a loose covering over hair roots and penetrating only the outer cortical cells; indumentum unicellular and multicellular hairs, or unicellular hairs only, or rarely none; terminal bud scaly, rarely naked, or aborting; leaves spiral, opposite or whorled, entire or serrate, rarely margins strongly revolute and leaves needle-like ("ericoid"), exstipulate; inflorescences terminal or axillary, usually racemose; prophylls paired, rarely 0; flowers rarely single, rarely multibracteolate, usually conspicuous, hermaphroditic, rarely unisexual, polysymmetrical, rarely monosymmetrical, sepals (2–)4–5(–7), fused at the very base, petals (3)4–5(–7), fused, rarely free or fused as a cap; stamens (2–)5(–8), 10(–16), free from the corolla, rarely adnate; anthers tetrasporangiate, rarely bisporangiate, inverting during development, with 2(4) apparently terminal or dorsal appendages or not, dehiscence introrse or terminal, rarely latrorse or extrorse, with pores or short, rarely long slits or a slit; endothecium lacking, rarely present; pollen in tetrahedral tetrads or rarely in monads; nectary present, rarely absent; ovary superior to inferior, (1–)4–5(–12)-carpellate, placentation axile to intruded parietal, rarely apical or basal; ovules (1–)numerous/carpel, anatropous to subcampylotropous, unitegmic, tenuinucellate; style usually about as long as corolla, hollow, rarely expanded at the apex; stigma punctate to lobed; fruit a berry, drupe, or capsule, rarely calyx fleshy; seeds small to minute, testa usually single-layered, variously winged or not; embryo straight, fusiform, rarely embryo minute, undifferentiated; endosperm cellular, fleshy, well developed, with haustoria at both ends; germination epigeal.

A mainly temperate, warm temperate and montane tropical family of some 124 genera and 4100 species.

CHARACTERS OF RARE OCCURRENCE. Cushion plants in some species of *Dracophyllum*, and in *Disterigma empetrifolium* and *Rhododendron saxifragoides*. Leaves distichous in *Vaccinium distichum* and perhaps some species of *Disterigma*. Leaf like the finger of a glove, or peltate, in some *Cassiope*; *Calluna* with tails from the abaxially recurved portions of the lamina. Linear leaves relatively common in Styphelioideae but otherwise sporadic, e.g. *Diplycosia*, *Rhododendron*, *Rusbya*, and *Agarista* (where they are V-shaped in transverse section). Leaves 60 cm to 1 m long in arborescent *Dracophyllum* and *Richea pandanifolia*. Whole inflorescence concolourous in *Oligarrhena* and some Vaccinieae. Calyx much longer than the corolla in a few *Erica*. Corolla in *Vaccinium* section *Polycodium* open during development, fused into a cap that falls at anthesis in *Richea*, scarcely opening in *Conostephium* and *Cosmelia*, blue in a number of species of *Andersonia*. Most Styphelioideae lack glandular hairs, but corolla hairs glandular in *Melichrus*. Corolla tube of *Utleya* with solid spurs; that of most *Kalmia* with pouches in which anthers are held under tension. Antesepaline whorl of stamens in *Diplarche* adnate to the corolla, antepetaline whorl largely free; calyx free, corolla adnate halfway up ovary in *Erica serrata*. Flowers 4-merous, with 2 fertile stamens, in *Oligarrhena*. Filaments reflexed and anthers lying along corolla lobes in *Acrotriche*; anthers cohering at margins in *Sprengelia incarnata*, which has bulbous hairs on the anther, *S. monticola* and some Vaccinieae, connate in *Coleanthera*; filaments in particular connate sporadically in Vaccinieae. The mesocarp of *Pentachondra dehiscens* splits to reveal the pyrenes inside.

VEGETATIVE MORPHOLOGY. Ericaceae are evergreen, sometimes deciduous, often rather xeromorphic shrubs, rarely root climbers, lianes, small trees or herbs. Roots developing from the stem come from immediately above axillary buds. Pyroleae are herbs or slightly woody; *Pyrola picta* may have almost leafless stems (*P. aphylla*!; Haber 1987). Pterosporeae and Monotropeae are herbaceous and completely lack chlorophyll (Henderson 1919); shoots develop from roots (as also in *Moneses*). The young shoots of many tropical Vaccinieae are dark red, and the whole plant is red in some Monotropeae, the plant drying black. Many Ericaceae are rhizomatous, or have lignotubers from which they sprout after fire, while some epiphytic and epilithic tropical Vaccinieae in particular have root swellings.

Temple (1975) found infraspecific variation in architectural models sensu Hallé et al. (1978), and it is difficult to place many taxa in models (Stevens 1982; Temple in Oliver 1991). In many Vaccinioideae the apical bud aborts, in many Rhododendreae it produces an inflorescence, but in Ericeae it may remain vegetative. Stems are commonly orthotropic, but in *Vaccinium* section *Cyanococcus* and in *Oxydendrum* they are mixed, being orthotropic initially and more or less plagiotropic by the end of the season. Some tropical Vaccinieae have orthotropic axes that sprawl under their own weight unless supported, or the axes may be pendulous. Buds are generally perulate, excepting many Ericeae and a few Rhodoreae; the outer pair of perulae is prominent and acicular in several neotropical Vaccinieae (incorrectly called stipules). The relationship of vegetative branches to inflorescences varies; vegetative branches in *Enkianthus* and *Rhododendron* sect. *Sciadorhodion* arise within the terminal bud from axils of perulae; more normally such branches arise from buds in leaf axils.

The leaves are usually spirally arranged – pseudoverticillate in Pyroleae and sometimes elsewhere – or whorled to opposite, as in Ericeae; almost all conditions occur even in small genera like *Kalmia* and *Pieris*. Empetreae, particularly *Empetrum*, are labile, with spiral and whorled leaves on the one plant. Distichy is very rare. Leaves may be convolute or revolute in bud. Ericoid leaves, in which the lamina is more or less linear, the margins being sharply recurved and concealing the lower surface and its stomata (Hagerup 1953; cf. Hara 1958), are common. Here the serrulate "margin" with its multicellular hairs probably represents the point of reflexion of the leaf surface, not the true margin. Such leaves are almost restricted to Ericoideae, especially Ericeae, and intergrade with leaves that have strongly revolute margins, as in *Erica* itself (Oliver 2000). Some *Cassiope* have peltate leaves in which the abaxial

surface seems to be obliterated developmentally (Stevens 1970a). Styphelioideae have xeromorphic, flat to recurved leaf blades that are often pungent at the apex. Variation in leaf type within genera such as *Kalmia*, *Agarista*, *Diplycosia* and *Cassiope* is high (Stevens 1970a), and variation in vegetative features provides much systematic information; Philipson (1985) summarises variation within *Rhododendron*.

The leaf margin is usually entire to only slightly serrate; the serrations are often (but not always in Arbutoideae and never in most Styphelioideae) associated with multicellular hairs. Some tropical Vaccinieae have large glands at the base of the margin or on the petiole. Secondary veins are pinnate, but most tropical Vaccinieae are strongly pli-nerved and with a clear midrib while most Styphelioideae are pli- or parallel-nerved and lack an obvious midrib. Fine details of the venation have been little studied, but taxa like *Agarista* are notable in having all veins almost equally prominent, so forming a dense, even reticulum (e.g. Lems 1962b; Judd 1984).

Ericaceae have many hair types, variation within *Rhododendron* and *Lyonia* being particularly notable (see Seithe 1980; Judd 1981); unicellular and multicellular hairs usually occur together. Multicellular hairs are eglandular and dendroid to stellate, lepidote, or flattened-setose, or variously glandular. In glandular hairs secretion may accumulate on the outside of the hair (*Gaylussacia*), or within the hair, separating individual cells (*Rhododendron*). Multicellular hairs, when present, are likely to be found at least on the leaf margin or base of the shoot. Most Styphelioideae lack multicellular hairs.

VEGETATIVE ANATOMY. The anatomy of the mycorrhizal associations in the root provides interesting systematic information (Cullings 1994). In some Monotropoideae lateral roots arise exogenously (e.g. Copeland 1935).

The vascular anatomy of the family needs study, general surveys such as those by Cox (1948a, 1948b) being dated and recent work limited in scope (e.g. Noshiro et al. 1995 and references therein). Clemants (1995) provided a detailed survey of the wood of several species of *Bejaria*; Carlquist (1989) described woods of Empetreae. Vessels are numerous, scattered, and have scalariform to simple perforation plates, tracheids occur, axial parenchyma is sparsely vasicentric, and rays are often uniseriate, sometimes wider, especially in lignotubers (see below), and are comprised of upright cells. Pith can be homogeneous and more or less lignified, heterogeneous (small cells with thick walls and large cells with thin walls intermingled), or *Calluna*-type (thick-walled cells restricted to the periphery of the pith; Watson 1964a). Homogeneous pith is widespread. There are intermediates between these types, and infrageneric variation, as in *Gaultheria* and *Leucothoë*, but the character is valuable; variation is similar, but less extensive, in the cortex. Position of initiation of the cork cambium provides important information in Vaccinieae, where it can be superficial, or deep-seated, being initiated inside the ring of pericyclic fibres (the basal condition for the family), or develop in a canker-like fashion (Stevens 1974; Odell et al. 1989). Monotropoideae lack a cork cambium (Henderson 1919).

Whether fibres in the secondary phloem are in bands or scattered separates groups of genera in Vaccinioideae (Stevens 1970b). Nodes in Ericaceae are predominantly unilacunar, but *Rhododendron* may have several traces in the cortex arising from one or more gaps (Philipson and Philipson 1968), as also *Therorhodion* (inflorescence only), *Bejaria* and *Lebetanthus* have three traces coming from a single gap, while Richieae are tri- or multilacunar (Watson 1967b). Styphelioideae often have several arcuate vascular bundles in the petiole; in other Ericaceae the single bundle varies from arcuate to annular. There are sclereids of various types in the cortex, especially in the petiole, but their variation and distribution are poorly understood. Fibres occur in the pericyclic area in the stem (and leaf); they are less well developed in the basal clades. In Ericeae in particular the petiole bundle has broad flanges of lignified tissue on either side, a feature that needs more study.

The leaves are usually bifacial, but in some *Arctostaphylos* they are held vertically; only rarely are there stomata on the adaxial surface of the lamina, as in *Cassiope* and Cosmelieae. The two main stomatal types are anomocytic and paracytic, rarely parallelocytic (Richeeae) or stephanocytic. Stomatal type tends to be constant in groups of genera, but it varies within small genera like *Elliottia* and *Enkianthus*. Stomata are unoriented or oriented parallel to the long axis of the lamina, but in Ericeae, *Cassiope*, etc., they may be oriented transversely. The adaxial epidermis is lignified in Styphelioideae, some genera of the *Lyonia*-group (Stevens 1970b; Judd 1979) and *Notopora*. The adaxial epidermis and/or hypodermis, rarely the abaxial epidermis as well, may have "mucilaginous" cells which bulge into the palisade mesophyll (e.g. Lavier-George 1936; Copeland 1943); this is known only in some Ericoideae, Arbutoideae and perhaps Pyroleae (Pyykkö 1968), although it is rarely constant within genera. A 1-, 2- or more layered hypodermis is common, especially in tropical Vaccinieae.

The palisade mesophyll of some thick-leaved tropical epiphytic Vaccinieae is especially tall with transverse thickened bands on the walls (e.g. *Disterigma*), it (and the hypodermis) perhaps being involved in water retention during dry periods. In several Vaccinieae and also in large-leaved Richeeae, the spongy mesophyll adjacent to the abaxial epidermis is thick-walled and lignified. The thickening of the cellulose walls of the spongy mesophyll varies considerably within *Vaccinium*. In *Diplycosia* and some species of *Gaultheria*, fibres wander through the mesophyll, and the dry leaf is notably tough. The mesophyll is dorsiventral or centric in symmetry, inverse dorsiventral in *Cassiope* and also in the bases of leaves of Cosmelieae, where symmetry becomes isobilateral then centric along the leaf from base to tip. Veins are deeply embedded within the mesophyll, sometimes offset to the abaxial side. In Styphelioideae they are usually all about equal in size and are associated with abaxial bands of sclerenchyma. These reach the abaxial epidermis in Styphelieae and some Epacrideae; fibres form transcurrent bands in Richeeae and one species of *Leucopogon* (e.g. Watson 1967b). The fibres link with the pericyclic fibres of the stem. Young leaves especially of Styphelioideae are bordered abaxially by a layer of unthickened cells that produce crystals within their walls. These cells disintegrate with maturity, leaving free crystals in mature leaves. Druses and sometimes crystals of calcium oxalate commonly occur in the petiole, lamina and elsewhere in the plant.

INFLORESCENCE STRUCTURE. Flowers with paired prophylls (bracteoles) are borne in the axils of bracts and arranged in a raceme, or spike in several Styphelioideae. Racemes range from being much elongated and with large, leafy bracts to short and with small bracts. They are often borne on wood of the previous season or flush, or on older wood (Lems 1962a), while in some *Lyonia* (Judd 1981) and Styphelioideae all leaf axils of the uppermost part of the previous season's innovation may be occupied by inflorescences. Many taxa have axillary fascicles, probably reduced racemes. In Ericeae inflorescences may be reduced to single flowers, which in turn are aggregated and localised on the stem, so suggesting a raceme (Hansen 1950; Oliver 1991, 2000). The terminal, corymbose or capitate inflorescences of many Rhodoreae and the axillary panicles along defoliate branches in some *Thibaudia* are all indeterminate. Single flowers with paired prophylls occur in the axils of leaves at the beginning of the current season's growth in *Vaccinium* section *Myrtillus* and at the end in *Chamaedaphne*, while the panicles in *Oxydendrum* terminate the season's growth. The single, axillary flowers of some *Gaultheria* and Styphelioideae are here called multibracteolate; the "bracteoles" are probably reduced inflorescence bracts. In genera like *Cyathodes* with such flowers, the basal bracteoles (which have been described as bracts elsewhere) are strongly keeled, and the number of bracteoles varies considerably (see also Virot 1975). In *Styphelia*, *Coleanthera* and some *Astroloma*, the terminal branch rudiment is a flattened, expanded bract-like structure, and what appear to be single, axillary flowers are reduced inflorescences (Watson 1964b; see also Lemson 1996; Cherry et al. 2001). In the generic descriptions, it can be assumed that flowers in racemes or spikes are subtended by a bract and have two prophylls; multibracteolate flowers are probably incorrectly designated, but pending more detailed comparative studies they describe the single flowers mentioned above. Truly terminal flowers are uncommon, but are known from *Craibiodendron* (Judd 1986), *Elliottia* (Bohm et al. 1978), *Monotropa*, *Enkianthus*, and perhaps *Calluna* (Oliver 2000) and Epacrideae. Inflorescences in Richeeae are particularly complex.

Inflorescences of many Rhodoreae have large and usually deciduous perulae surrounding them (Fig. 56, p. 171). Similar perulae in *Cavendishia*, as well as bracts, are brightly coloured and (sub)persistent (Luteyn 1983); they are known from a few other Vaccinieae. Leaves near the flowers are sometimes brightly coloured in *Erica* (Oliver 1991).

The position of the prophylls on the pedicel varies greatly, although they are usually towards or at the base. In *Diplycosia* and *Tepuia* they are characteristically immediately under the calyx (see Fig. 61B, p. 182) and in *Disterigma* they envelop the inferior ovary. In many Ericeae both bracts and prophylls are recaulescent and become part of the calycine whorl, perhaps even "replacing" the sepals (Oliver 2000). *Monotoca elliptica* may lack bracts; prophylls are absent in Pyroleae and some Monotropeae and Ericeae.

FLORAL STRUCTURE. Ericaceae show extensive variation in meristicity and in the degree of fusion of the corolla. The androecium is especially diverse, "Bicornes", the name early given to part of the family, referring to the distinctive appendages on the stamens. Genera of Vaccinieae in particular have been characterised by staminal features (but see below).

The flowers range from 1–60 mm long and are typically bisexual, polysymmetrical and 5-merous. In *Rhododendron* and some related genera the median sepal is abaxial, rather than the normal adaxial condition. Ericeae are 4-merous, apart from *Erica* subgenus *Pentapera*; some *Erica* are 3-merous. The ovary is superior, except in Vaccinieae, however, a few *Gaultheria* are semi-inferior, while in some *Vaccinium* the ovary is not completely inferior; ovary condition was the main feature used to justify the segregation of Vacciniaceae.

Calyx and corolla are usually distinct from each other and from bracts or prophylls. The calyx consists of free or almost free, imbricate (quincuncial) sepals. In inferior ovaries, "calyx tube" describes the inferior portion, "calyx limb" the free part borne on top of the ovary. Bracts and calyx are coloured in some taxa. In Empetreae calyx and corolla are almost indistinguishable, in Monotropeae both may be similar to the bracts, and in some Styphelioideae prophylls and calyx may be similar (Fig. 58, p. 174). In *Rhododendron* the calyx is often obsolete. Both calyx and corolla are dry and marcescent in Ericeae and the calyx is scarious in many Styphelioideae. Elsewhere they are more or less fleshy and wither after flowering, although the calyx becomes fleshy in most *Gaultheria* and *Diplycosia*. The sepals are largely free, although joined at the very base, and are over 10 mm long in some *Ceratostema*; in some tropical Vaccinieae they form a tube 12 mm long with calyx lobes barely evident. Winging of the inferior ovary and of the calyx tube has been used as a generic character in Vaccinieae.

The corolla is usually sympetalous, forming a cylindrical, campanulate or urceolate tube with free lobes; the lobes are valvate or imbricate in bud. Empetreae, most Monotropoideae and scattered genera elsewhere are polypetalous. Polypetalous taxa have often been called primitive, but polypetaly is derived from sympetaly (Copeland 1943, 1947; Kron and Judd 1990; Judd and Kron 1993); the only exception could be Monotropoideae. The shape of the corolla is broadly correlated with taxonomy – Rhodoreae often have a broadly open corolla; Arbutoideae are overwhelmingly urceolate; Vaccinioideae and Styphelioideae are often urceolate to tubular; Pyroleae are rotate. However, in *Phyllodoce* (sympetalous) the corolla is campanulate or urceolate, in *Bejaria* (polypetalous), rotate to tubular. The corolla of *Anthopterus* in particular is strongly winged. Polysymmetry is pervasive, but many *Rhododendron* have monosymmetrical flowers often associated with colour patterning on the adaxial and adjacent corolla lobes, while in *Elliottia* and in several Pyroleae bisymmetry is evident mainly in the curvature of the style and, in the latter, in the positioning of the anthers as well. Some *Paphia*, *Thibaudia* and *Erica* have curved corollas. Indumentum on the outside of the corolla is like that elsewhere on the plant. On the inside there are often unicellular hairs (uniseriate in Prionoteae) that are especially conspicuous in Styphelioideae. They may be in distinct tufts, as in *Brachyloma* and *Astroloma*, sometimes with beautiful surface sculpturing (e.g. Weiller 1996b).

Most Ericaceae have twice as many stamens as petals, but there are numerous exceptions. Styphelioideae have only as many stamens as corolla lobes. These are borne opposite the sepals and are more or less strongly epipetalous, being inserted at the base or throat of the corolla tube; there are no staminodes. Epipetaly is rare elsewhere (but see *Diplarche*). Filaments are free to fused, variously shaped, sometimes geniculate, and often hairy (but almost never in Styphelioideae). Anther dehiscence is predominantly by two, terminal subintrorse pores (Figs 50, 51; but see below); these may be confluent in some Vaccinieae and rarely latrorse or extrorse. There are slits almost the length of the anther in the basal *Enkianthus* (Fig. 55, p. 166), scattered in Monotropoideae and Ericoideae (including the wind-pollinated Empetreae, see Fig. 57, p. 172) and throughout Styphelioideae. In Vaccinieae in particular the anther thecae often have very long, apical tubules through which the pollen exits by pores or short slits (Fig. 51). In one section of *Leucopogon*, the anthers are sterile above the thecae, while in *Melichrus* there is a sterile basal portion (Paterson 1961). Anthers are sometimes connate, most notably in *Coleanthera* and some Vaccinieae. The pollen in some Styphelioideae is red or purple; the usual colour is white or cream.

There has been much argument over the nature of the stamen appendages. They may be either paired, varying in position from the tips of the anthers to the filaments, or four, when they are always on the tips of the anthers. Appendages on the "backs" of anthers (see below), including the tubules, or on the filaments are called spurs; they are flattened to terete, and where they are borne often varies infragenerically, as in *Erica* and *Pieris*. Appendages on the tips of the anthers are called awns (e.g. *Enkianthus*, Arbutoideae, *Cassiope*, etc.), and may be morphologically comparable to the

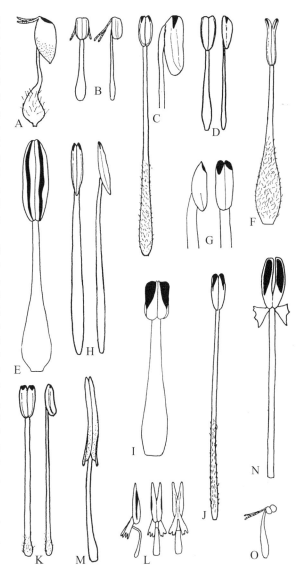

Fig. 50. Ericaceae. Stamens in Arbutoideae, Cassiopoideae, Ericoideae and Harrimanelloideae. **A** *Arbutus andrachne*. **B** *Cassiope lycopodioides*. **C** *Bejaria racemosa*. **D** *Bryanthus gmelinii*. **E** *Ledothamnus sessiliflorus*. **F** *Menziesia lasiophylla*. **G** *Rhododendron fictolacteum*. **H** *Rhododendron tanakae*. **I** *Elliottia pyroliflora*. **J** *Epigaea asiatica*. **K** *Rhodothamnus chamaecistus*. **L** *Calluna vulgaris*. **M** *Daboecia cantabrica*. **N** *Erica australis*. **O** *Harrimanella stelleriana*. (Stevens 1971)

tubules of many Vaccinieae; the apex of the anther may bifurcate, hence the paired awns of some *Gaultheria* (Hermann and Palser 2000). The presence of spurs is loosely correlated with the stamens being included (Matthews and Knox 1926; Hermann and Palser 2000).

There is usually a nectary at the base of the ovary (or on top, as in Vaccinieae); in some Monotropeae (Wallace 1977) and Ericeae it has prominent

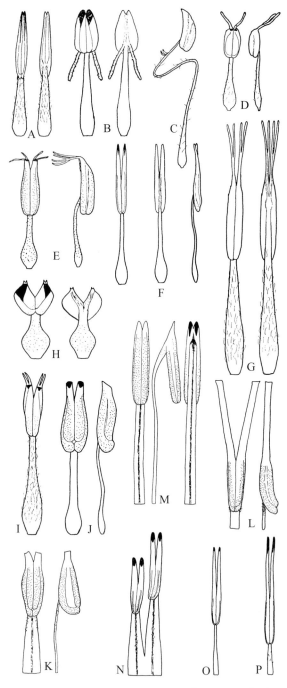

Fig. 51. Ericaceae. Stamens in Vaccinioideae. **A** *Oxydendron arboreum*. **B** *Pieris nana*. **C** *Lyonia foliosa*. **D** *Andromeda polifolia*. **E** *Zenobia pulverulenta*. **F** *Chamaedaphne calyculata*. **G** *Leucothoë keiskii*. **H** *Gaultheria suborbicularis*. **I** *Gaultheria procumbens*. **J** *Diplycosia malayana*. **K** *Vaccinium poasanum*. **L** *Symphysia racemosa*. **M** *Notopora schomburgkii*. **N** *Orthaea apophysata*. **O** *Themistoclesia compacta*. **P** *Gaylussacia braziliensis*. (Stevens 1971)

projections. Palser (e.g. 1961) provides extensive details on the morphology of the nectary.

The gynoecium is usually about as long as the corolla. The ovary has 1–12 locules with apical to basal and axile to parietal placentation; there are one to hundreds of ovules per carpel. The ovary is glabrous to variously hairy; in some Arbutoideae it is covered with multicellular papillae that are fleshy in fruit. The style is long, slender, and usually hollow, but it is relatively short and stout, especially in some Ericeae and Monotropoideae. It may join the apex of the ovary smoothly (style continuous) or be more or less deeply impressed. When it is very deeply impressed (e.g. *Agarista*, *Prionotes*), the placentae, which of course always arise below where the style and ovary join, appear to be basal. The stigma is punctate to variously lobed, the apex of the style sometimes also being expanded, and is at least sometimes wet and papillate.

FLORAL ANATOMY. Palser and collaborators and Copeland have carried out extensive studies of floral anatomy; for Styphelioideae information may also be found in Étienne (1919), Copeland (1954), Jackes (1968) and Lemson (unpublished). Palser (1961) found that variation in floral anatomy was not very highly correlated with that of other features, and thus species with inferior ovaries vary greatly in how much the whole vascular supply is fused. Details of floral vasculature are largely ignored below.

There is extensive variation in the distribution of stomata on both surfaces of the calyx and corolla (Watson 1962, 1967a; Stevens 1971). The stomata can be difficult to detect, and more extensive surveys are needed, but there is a correlation with phylogeny.

When there are twice as many stamens as petals, the androecium is generally obdiplostemonous, but the traces of the sepaline stamens may diverge from the central stele before those of the petaline stamens (e.g. Leins 1964; Palser and Murty 1967). If there is only one whorl of stamens, it is always the antesepaline whorl. The anthers are tetrasporangiate. In Styphelioideae, however, they are bisporangiate and dehisce by two (Prionoteae, a few other genera) or one (the rest) slits; the latter condition is caused by the breakdown of the tissue separating the two sporangia. The anthers usually invert during development and the vascular bundle forms an apical hook as it turns 180°. Hence, the "apical" pores of many Ericaceae are really basal and the "introrse" slits of, for example, Styphelioideae (see above) are extrorse; we refer to the apparent position in the discussion here. Inversion in *Enkianthus*, Monotropoideae and Arbutoideae is late or may not occur at all (e.g. Copeland 1941, 1947); note that anthers in *Cyrilla* (Cyrillaceae, sister to Ericaceae) do not invert. The development of the anther wall is unexpectedly variable (Hermann and Palser 2000). There is much variation in how the anthers dehisce. There is a well-developed endothecium in *Enkianthus* and *Bejaria* (Copeland 1943), although it is otherwise absent, or in Monotropeae and perhaps a few Rhodoreae only poorly developed. Elsewhere an exothecium only is found. Anthers in Ericoideae open either by collapse tissue, or by resorbtion tissue, the cells of the latter containing much oxalic acid (D'Arcy et al. 1996); resorbtion tissue is widespread, occurring in i.a. Styphelioideae (Paterson 1961). Vaccinioideae may have white disintegration tissue on the backs of the anthers that sometimes extends to the apices of the filaments.

In some Ericeae and Styphelioideae the carpels are opposite the petals, rather than opposite the sepals, as is common (Palser 1961; Palser and Murty 1967). Numerous species of *Vaccinium* in particular have inpushings of the outer walls of the carpel, and are described as being pseudo-10-locular. The ten carpels of *Gaylussacia* alternate with both sepals and petals (Palser 1961); it is unclear if they are derived from a pseudo-10-locular gynoecium or not. In Arbutoideae extra carpels appear to develop in the plane of the septae (Palser 1954). *Epigaea* has carpellary bundles in the plane of the septae (Palser 1952). The style is hollow and the stylar canal is confluent with the ovary chamber.

EMBRYOLOGY. The tapetum is secretory, and microsporogenesis is simultaneous. Tapetal cells may be uninucleate (Epacridoideae), binucleate (Monotropoideae, *Vaccinium*), or multinucleate (Ericoideae, ?some Vaccinioideae; see Johri et al. 1992). Pollen is binucleate.

The ovules are usually anatropous, unitegmic, tenuinucellate, and have an endothelium. The megagametophyte is 8-nuclear and develops from a single archesporial cell; in *Styphelia* the three megaspores that do not form the embryo sac may remain functional (Brough 1924). The megagametophyte has sharp, ear-like processes in *Enkianthus* and *Epigaea* (Palser 1952).

The embryo in *Monotropa* may consist of only two cells (Olson 1980); all Monotropeae have much reduced, undifferentiated embryos. There are usually well-developed endosperm haustoria at both ends of the embryo, although they may be absent in Monotropeae (Copeland 1941), or the micropylar haustorium alone may be present, as in *Elliottia* (Copeland 1943).

POLLEN MORPHOLOGY. Most woody Ericaceae show little variation in pollen morphology (see Oliver 1987, 1991; various contributors in Luteyn 1995; Warner and Chinnappa 1986). The grains are usually in isopolar, radiosymmetric, triangular to globular tetrahedral tetrads 23–82 μm in diameter. They are usually tricolporate, sometimes tricolpate or tricolporoidate, and with definite transverse furrows. Aperture number varies from 3 to 5 at the species level in *Enkianthus* (Anderberg 1994b). Some Monotropeae have porate pollen (Wallace 1995). The exine is scabrate, verrucate, microrugulate or psilate (the latter especially by aperture margins). *Calluna vulgaris* has distinctive irregular tetrads that have only faint colpi and distinct ora (Foss and Doyle 1988). Viscin threads occur in most Rhodoreae, Phyllodoceae, and Bejarieae (Fig. 52). *Enkianthus*, some *Erica* (Oliver 1991), and many Monotropeae have pollen in monads, as does *Andersonia macranthera*. *Chimaphila* has polyads, tetrads and monads, and may even have very reduced colpi (Takahashi 1986, 1987).

In Styphelioideae, pollen is in tetrads 23 μm (*Sprengelia*) to 82 μm (*Epacris*) across, or less commonly in reduced tetrads or monads. Monads are rare outside Styphelieae, but occur in some species of *Richea*. In *Astroloma* and some species of genera that otherwise have normal tetrads, irreg-

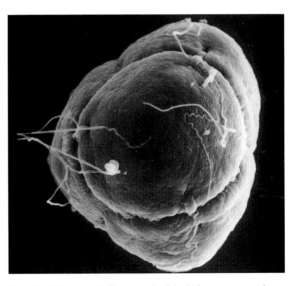

Fig. 52. Ericaceae. Pollen tetrad of *Ledothamnus atroadenus* with adherent viscin threads (×2350). Photo J. Luteyn

ular tetrads with 0–4 viable grains are produced. About half the genera of Styphelieae have pollen grains that appear to be monads, and these range from 10 μm (*Oligarrhena*) to 100 μm (*Astroloma*) across. They have been described as "pseudomonads" (Smith-White 1948a, 1955), but in most cases no evidence of aborted grains remains at maturity. In *Monotoca* the pollen is a reduced tetrad (T-monad) with three aborted grains attached to the pole of the fertile grain.

KARYOLOGY. All chromosome numbers in the accounts are given as haploid (*n*) numbers. North temperate Ericaceae and some Styphelioideae are moderately well-known cytologically, but the rest of the family is very poorly known. Janaki Ammal et al. (1950) surveyed *Rhododendron*, finding numbers as high as $2n = 156$ (= 12x) in *R. manipurense*. Intrageneric polyploid series are quite common, and include some Styphelioideae (Smith-White 1955; Weiller 1996a, 1996c; Dawson 2000 for numbers in that subfamily). For chromosome numbers in *Gaultheria* and its relatives, see Middleton and Wilcock (1990) and Luteyn (in Luteyn 1995); Atkinson et al. (1995) give counts for tropical Vaccinieae which suggest that a more extended survey will provide valuable systematic information. In general, base numbers are constant within genera or groups of genera, but *Gaultheria* in particular varies from $n = 11$, 12, or 13. The base number for the family is unknown, but in Enkianthoideae $n = 11$.

Many Styphelioideae are $n = 13$ or $n = 12$; no counts are known from Prionoteae. It is likely that either $n = 12$ or $n = 13$ is the plesiomorphic character state (Stace et al. 1997). Probable chromosome number reduction series (including $n = 13$, 12, 11, 10, 9, 8, 7, 6, 4) are especially marked in Styphelieae (Smith-White 1948a, 1955); possible infrageneric reduction series are *Leucopogon* sens. str. ($n = 12$, 11, 10), *Cyathodes* ($n = 12$, 10), *Brachyloma* ($n = 9$, 7), and *Sphenotoma* ($n = 7$, 6). Derived chromosome numbers may provide synapomorphies to mark certain problematic lineages in Styphelieae (Stace et al. 1997). A remarkable case of stabilised triploidy was reported in "*Leucopogon*" *juniperinus*, in which haploid chromosome numbers ($n = 4$) are transmitted to the pollen and diploid numbers ($2n = 8$) are transmitted in the megaspore. The maternal and paternal chromosomes may be responding differentially to a cytoplasmic gradient that leads to the formation of derived monad pollen by apoptosis of cells of the ericoid tetrad pollen grain (Smith-White 1948b).

POLLINATION AND REPRODUCTIVE SYSTEMS. Little is known about pollination in Ericaceae, especially its tropical members, despite its popularity among horticulturalists and its often large or conspicuous flowers (although the corolla is only some 0.7 mm long in *Erica petricola*). Most records of "pollinators" are simply of flower visitors, and how effective these visitors may be in pollination is unknown.

Meiosis in many temperate Ericaceae occurs in the late summer of the year preceding anthesis. Inflorescences overwinter either enclosed in a perulate bud (*Rhododendron*) or completely exposed (*Pieris*); the flowers open in the early spring.

Flowers are usually hermaphroditic. Autogamy is quite common (e.g. Reader 1977; Oliver 1991; Kraemer 2001; see also Luteyn 2002). Preliminary studies of *Cyathodes glauca* indicate that it is an obligate oucrosser (pers. comm. C.M. Weiller), as is *Pentachondra pumila* (Godley 1966). Gynodioecism is known in *Gaultheria* in particular (e.g. Middleton 1991; Hermann and Cambi 1992) and in Styphelieae. *Leptecophylla* is functionally dioecious with the flowers effectively unisexual: all species produce larger and apparently hermaphrodite but actually male flowers with apparently functional gynoecia and male-sterile flowers. In the latter, anther development is aborted at an early stage, resulting in reduced anther size and no pollen formation, but fruit is set. The functionally dioecious *L. juniperina* has a prezygotic incompatability mechanism in the "male" flower, causing the arrest of pollen tube growth and associated pollen tube abnormalities at various stages from the stigma to the base of the style (pers. comm. C.M. Weiller). All species are protandrous and the stigmas are normally coated with pollen at anthesis (as in *Planocarpa* and *Cyathodes*). In *Planocarpa* two species (*P. petiolaris* and *P. sulcata*) are assumed to be gynodioecious, with both hermaphrodite and male-sterile flowers setting fruit. *Monotoca*, *Pentachondra* and *Trochocarpa* all have some species with unisexual flowers, while dioecy occurs in the wind-pollinated Empetreae. *Monotoca* exhibits a range in states of reduction of the anthers. Some species of the genus have bisexual flowers, some are unisexual with reduced and barren anthers, while in others the anthers are completely aborted with only bare filaments visible.

Flower colour is an important attractant, although flowers of some *Acrotriche*, *Andersonia*, *Erica*, *Vaccinium*, and *Rhododendron*, etc., smell, whether sweetly or unpleasantly (see also Knudsen and Olesen 1993). A common colour is bright white or cream, often with a pinkish tinge, while many tropical Vaccinieae have red flowers. Orange, pink, yellow, blue and mauve/purple are less common. In species with cryptic flowers, the colours may be dull cream/brownish (*Monotoca tamariscina*) or greenish, in which case they may have a strong odour (*Acrotriche* spp.). The corolla is usually single coloured. Simple bicolours are known in *Rhododendron*, *Erica*, tropical Vaccinieae, *Epacris*, and large-flowered species of *Agapetes* (where there is often horizontal banding); many temperate species of *Rhododendron* have spots on their adaxial petal(s), presumably honey guides. In *Epacris impressa*, there are four races with short white corollas, long pink corollas, and broad white or pink corollas (Stace and Fripp 1977). Many plants occur as pure-colour stands in different habitats and show some differences in flowering times, but some populations are panmictic with pink- and white-flowered plants, and at Wilson's Promontory there are heterogeneous populations with red and white flowers.

Flowers are usually animal-pollinated. Bees are common visitors and effective pollinators of many Ericaceae with small, scented urceolate flowers (e.g. Reader 1977; contributors to Luteyn 1995), and they also visit many larger-flowered species, e.g. *Rhododendron*, as well as many Monotropeae (Wallace 1977). In Ericoideae that have viscin threads mixed with the pollen, strands of tetrads become attached to the flower visitors. In some *Kalmia* pollination is explosive; bees release the anthers, held under tension in pockets in the corolla, and get dusted with pollen. Pollination by hovering flies with long probosces is common in southern Africa *Erica* (Rebelo et al. 1985), while pollination by thrips is suspected both there (Rebelo et al. 1985; cf. Oliver 1991) and in northwestern Europe (Hagerup and Hagerup 1953). Species pollinated by sphingids have a very sweet and heavy scent, as have possibly bat-pollinated taxa (in *Rhododendron*: the flowers are very large and robust, and are often 6-merous or more); flowers pollinated by bees and butterflies are also scented.

Perhaps 90% of Western Australian Styphelioideae are insect-pollinated (Keighery 1996). The corolla tube in *Leucopogon* is variable in length but in most species is relatively short (1–3 mm). Visitors include *Tarsipes* (seen on several species), but more generally bees, including feral *Apis mellifera*, moths, butterflies and dipterans (Clifford and Drake 1981; Keighery 1996; Brown et al. 1997). *Sphenotoma* and *Lysinema* have a relatively long

corolla tube with a narrow orifice and attract moths, butterflies and long-tongued bees such as *Amegilla* species. *Andersonia micrantha* has a foetid odour and attracts flies. Some *Monotoca* species have a very short corolla tube forming an open bowl (*M. tamariscina*) and are visited by small dipterans (Keighery 1996).

The corolla tube of *Conostephium* has a narrow entrance, and the stigma, rather than blocking the entrance, is exserted one to several millimetres from the tube. Insect visitors are relatively large and quite specialised (*Leioproctus* bees and *Amegilla pulchra*), but no nectar is produced and the flower is buzz-pollinated (T. Houston, pers. comm., cf. Keighery 1996). Buzz pollination occurs elsewhere, as in *Vaccinium stamineum* (Cane et al. 1985; Knudsen and Oleson 1993).

Hummingbirds visit the small, urceolate flowers of a number of Arbutoideae in eastern North America (Diggs 1995), although confirmation of pollination is needed. In general, species of Ericaceae known or suspected to be bird-pollinated lack scent and many, but by no means all, have long, tubular, red flowers (in Styphelioideae mostly *Astroloma*, *Cosmelia*, a few *Andersonia*, several *Styphelia*); the ovary may be superior or inferior. Gullet-type flowers are almost restricted to *Rhododendron* within Ericaceae. Hummingbirds are pollinators in the New World (Luteyn and Sylva 1999; Luteyn 2002 and references therein) and sunbirds, honeyeaters (Meliphagidae) and other groups in the Old World. Flower mites are dispersed by birds in both hemispheres (Colwell 1973; Stevens 1976; Naskrecki and Colwell 1998). Nectar content in neotropical Vaccinieae averaged 20.4(7.1–32)% sucrose, largely appropriate for hummingbirds (Luteyn and Sylva 1999). About 60 species of *Erica* may be bird pollinated, but few species of birds are involved, and only one, the sunbird *Nectarinia violacea*, is largely restricted to the genus (Rebelo et al. 1985; Oliver 1991). In Papuasia bird pollination seems to predominate in species of *Rhododendron* living at high altitudes (over 3000 m), species probably pollinated by butterflies, sphingids and even bats being found more commonly below 3000 m (Stevens 1976). *Cosmelia* and *Astroloma*, Styphelioideae visited by honeyeaters and spinebills, have a relatively long corolla in which the entrance to the tube is occluded by either adpression of the very short lobes or stigma and corolla hairs blocking the entrance; access to nectar is restricted to species that are reasonably forceful in probing the flower. Of particular interest is the pollination of the functionally dioecious *Leptecophylla divaricata* by several species of birds including honeyeaters (Higham and McQuillan 2000).

The Honey Possum (*Tarsipes rostratus*) utilises a number of Styphelioideae (Turner 1982) although, given their floral morphology, it is unclear how it would pollinate many species that it visits. *Acrotriche* species with small, cauliflorous flowers appear, on the basis of flower size, poorly adapted to pollination by a comparatively large animal (McConchie et al. 1986) but, on the New England plateau, flowers of *A. aggregata* form an important part of the diet of the marsupial mouse (*Antechinus stuartii*) and it may also be the pollinator (Fletcher, in McConchie et al. 1986). In Western Australia, the only recorded visitor to *Acrotriche cordata* is a muscid fly (Keighery 1996), so it is unclear if mammals are always involved in the pollination of the genus.

Secondary pollen presentation in clumps at the ends of the petals has been suggested for *Elliottia* (Copeland 1943). Many Ericaceae have various forms of hairs associated with the corolla, and in many cases these are related, either directly or indirectly, to pollination. In *Acrotriche* the anthers dehisce in bud and shed the sticky pollen onto the hair tufts at the tips of the corolla lobes. At anthesis the lobes recurve, taking the pollen with them – again, secondary pollen presentation (McConchie et al. 1986; Ladd 1994). In *Leucopogon* and *Styphelia* hairs extend from the corolla lobes into the mouth of the tube. The situation is similar in *Astroloma*. In many cases the dehiscing anthers shed pollen among the hairs, which may assist in retaining pollen within the corolla tube where it more effectively contacts pollinators. (In *Astroloma*, *Brachyloma*, *Lissanthe* and *Melichrus* the corolla hairs seem to assist in retaining nectar at the base of the corolla tube.)

Empetreae are likely to be wind pollinated. Their inconspicuous perianth has free parts, the anthers are exserted, and the stigmatic surface relatively much enlarged; as is common in wind-pollinated taxa, they are largely dioecious and ovule number is reduced. Anderberg (1994a) examined the correlation between ploidy level, breeding system and fruit colour in *Empetrum*. Some 100 species of *Erica*, especially species previously placed in *Philippia* and other minor genera, are also probably wind pollinated (Rebelo et al. 1985; Oliver 1991); there is no association with dioecy here. *Richea sprengelioides* and *R. procera* are candidates for wind-pollination.

Anthers are commonly enclosed in the corolla tube, but they may form a funnel-shaped structure that presents pollen at the mouth of the corolla tube (*Leucopogon* spp., *Vaccinium* spp., *Brachyloma preissii*). In many Rhodoreae, *Styphelia* and *Cyathodes* the anthers are well exserted, as they are in *Richea sprengelioides* and the "brush blossoms" of *Andersonia setifolia* (Keighery 1996). The stickiness of the pollen also varies, as in Styphelioideae, from very dry and loose in the buzz-pollinated *Coleanthera* (Keighery 1996), *Sprengelia* and *Conostephium* to very sticky in *Acrotriche* and species of *Astroloma* and *Cosmelia* visited by birds. In *Leucopogon*, species with pendulous flowers tend to have enclosed anthers that contain dry, loose pollen, while species with erect flowers have anthers in the corolla throat that have relatively sticky pollen. In buds of *Brachyloma preissii* a globule of sticky fluid is exuded by the stigma onto the unopened anthers. The globule is then spread across the anthers and pollen as the flower opens and the anthers dehisce, ensuring that the pollen is well coated in sticky material.

In many species the stigma is turgid and appears very wet from the mature bud stage to anthesis (see also Heslop-Harrison and Shivanna 1977). In other species, especially a number of Styphelioideae, particularly those with an exserted style, the stigma is not enlarged and appears virtually dry when mature.

Barriers to hybridization are often poorly developed. In *Rhododendron* hybridization between species in secondary (often anthropogenic) habitats is common (for *Erica*, see Oliver 1991), and introgression may occur (Kobayashi et al. 2000). Hybrids between several "genera" have been reported. A hybrid between *Gaultheria* and *Pernettya* (×*Gaulnettya* Marchant, Luteyn 1995 for references), was one of the first synthesized. However, in *Erica* × *Simocheilus* (Oliver 1991), *Rhododendron* × *Ledum*, and *Gaultheria* × *Pernettya*, the latter members of the pairs have been synonymized, and this is a likely course with *Phyllodoce*, *Rhodothamnus* and *Kalmiopsis*, given the hybridization reported to occur between the genera (e.g. Dome 1999) and their underlying similarity. Purported hybrids between *Kalmia* and *Rhododendron* are misidentified.

FRUIT AND SEED. The flowers are usually horizontal to pendulous and the fruits are erect (Fig. 53). *Rhododendron* subsect. *Ledum* has erect flowers and pendulous capsules opening from the base first; some *Vaccinium* have pendulous flowers and fruits. Capsule dehiscence is either septicidal or loculicidal, rarely septifragal, the two common types characterising large groups of genera. Details of how the wall is thickened or the margins

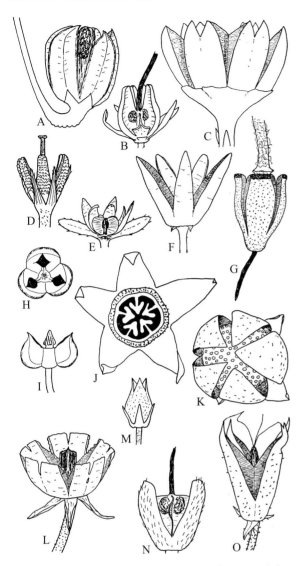

Fig. 53. Ericaceae. Fruit. **A** *Enkianthus pauciflorus.* **B** *Cassiope mertensiana.* **C** *Bejaria racemosa.* **D** *Ledothamnus sessiliflorus.* **E** *Diplarche multiflora.* **F** *Rhododendron lapponicum.* **G** *Rhododendron tomentosum.* **H, I** *Elliottia (Tripetaleia) bracteata,* in I one valve removed. **J** *Epigaea repens,* ovary with top removed. **K** *Epigaea asiatica,* note seeds on placentae. **L** *Phyllodoce aleutica.* **M** *Kalmia (Leiophyllum) buxifolia.* **N** *Calluna vulgaris.* **O** *Daboecia azorica.* (Stevens 1971)

separate helps delimit individual genera (e.g. *Lyonia, Ledothamnus, Pyrola*). Capsule shape (elongated versus spherical) is of systematic significance in Rhodoreae and *Erica*.

Some fruits are baccate, developing from an inferior (Vaccinieae) or, rarely, a superior (some *Diplycosia* and *Gaultheria*) ovary; in both the latter genera, the calyx alone is usually fleshy. The endocarp surrounding the loculi often has one or two layers of fibres. These layers are much developed and form the stone surrounding the seeds

individually, or all seeds together in the drupes that occur in Empetreae, Stphelieae, many Arbutoideae, and *Gaylussacia*.

The testa is usually only a single cell layer thick, but several cells thick in the region of the raphe, although only the walls of the outer layer are much thickened; in *Enkianthus* vascular tissue persists in this region (Netolitzky 1926). The testa of *Vaccinium* sect. *Oxycoccus* and Andromedeae is thicker. All cells may be papillate (*Daboecia*, rarely in *Erica*), or a few cells may be more or less balloon-like and in aggregate form a wing (*Enkianthus*, many *Leucothoë*). Seeds also have elongated "tails" (*Rhododendron*), outgrowths from one end (*Pterospora*) or one side (*Andromeda*, where it is many cells thick), *Craibiodendron*), or they may be generally flattened (*Elliottia racemosa*). The cells of the testa are more or less isodiametric in surface view, or strongly elongated, at least 2× longer than wide; intermediates and infrageneric variation occur in *Kalmia, Vaccinium, Erica* and *Pieris*. The anticlinal walls are usually straight, but sinuous in some *Erica* (Oliver 1991). Thickening is best developed on the anticlinal and inner periclinal walls, and varies from less than 3 μm (especially in taxa with elongated cells) to over 50 μm thick (in some *Costera*). The outer periclinal walls are rarely much thickened, although *Gaultheria* and its relatives in particular are exceptions; in *Erica* these walls may have surface ornamentations (Oliver 1991). The other walls have fine plasmodesmata 2 μm or less in diameter, but pits are larger in some Pyroleae (Takahashi 1993), *Calluna* and *Elliottia*. Some tropical Vaccinieae have broad, scalariform bands of thickening on the testal cells; the testa often becomes mucilaginous on wetting. The walls of the testa cells are particularly thin when the endocarp is thick, as in Arbutoideae, while in some *Erica* (Oliver 2000) the ultrathin testa is transparent.

The straight, white and more or less terete, rarely spatulate embryo is embedded in endosperm; the cotyledons are relatively short. Some tropical Vaccinieae have green embryos; in these taxa the testa is mucilaginous.

DISPERSAL. There are two main dispersal syndromes in the family, by wind and animals. Capsular fruits have small and quite often variously winged or tailed seeds which are presumably dispersed by wind; the walls of the testa are usually thin. The capsule in *Erica* is xerochastic, and seeds of some *Erica* have an elaiosome, presumably for ant dispersal (E.G.H. Oliver, pers. comm.). The flowers of *Richea scoparia* are visited by the snow skink, *Niveoscincus microlepidotus*, which removes the corolla to get at the nectar and so later facilitates seed release (Olsson et al. 2000).

Baccate fruits are often dark purple or red when ripe. The testa cells are rather thick-walled, and the fruits are presumably eaten by birds, dispersal being endozoochorous. However, in *Vaccinium* section *Oxycoccus* the fruits float, and the seeds can also (they have a multilayered testa); water dispersal seems an option. In some tropical Vaccinieae the testa is mucilaginous and the embryo is green; exozoochorous dispersal is a possibility. Old World Vaccinieae with such seeds often have more or less translucent white, yellow or orange fruits; in the New World white or lilac fruits are common. Seeds with this syndrome of characters germinate better on moss than do seeds with a thick-walled testa, which germinate better on soil (S. Vander Kloet, pers. comm.); taxa with a mucilaginous testa are often epiphytic or epilithic, places where moss or humus accumulations are high. Seeds in taxa which have capsules and fleshy calyces have thick-walled testa cells and a white embryo; again, endozoochory is likely, although wind dispersal is also possible (Luteyn 1995, *Gaultheria*). In those taxa in which single seeds individually or all the seeds together are surrounded by endocarp, the unit of dispersal is the pyrene (aggregate). Mammals eat the fruits of a number of taxa, as the name bearberry (*Arctostaphylos uva-ursi*) suggests; bears are also very fond of the fruits of *Vaccinium* spp. (McCloskey 1948). The more or less indehiscent fruits of some Ericeae are barely fleshy; the development of pyrenes varies considerably, and dispersal mechanisms are unknown (Oliver 1991). Drupes of Styphelieae are dispersed by animals, usually after ingestion and passing through in the faeces. Birds (in Australia, especially the emu, *Dromaius novaehollandiae*) are the usual agents, but several marsupials have been recorded eating the fruit. In *Epigaea* the placentae alone are fleshy and seeds, which have thick walls, may be dispersed by terrestrial animals.

PHYTOCHEMISTRY. Harborne and Williams (1973) surveyed the flavonoids and simple phenols of Ericaceae, while Moore et al. (1970) provided details for Empetreae; anthocyanins (Jarman and Crowden 1974) and flavones (Jarman 1975, Jarman and Crowden 1977) have been surveyed for Styphelioideae. The group has a fairly distinctive chemical spectrum.

The flavonol gossypetin 3-galactoside links Empetreae with other Ericaceae, being especially common in Ericoideae. In a survey of flavonoids in Monotropoideae, *Pterospora* and *Hypopitys* alone were found to lack flavonol glycosides (Bohm and Averett 1989). Ellagic acid is known mainly from arbutoids (Stevens 1971; Harborne and Williams 1973) and methyl salicylate from *Gaultheria* and its relatives (cf. Luteyn et al. 1980).

In Styphelioideae, eight types of compounds occur in leaf epicuticular waxes, viz. alkanes, aldehydes, triterpenes, primary alcohols, esters, β-diketones, flavonoids and diterpenes. Most are widespread, but β-diketones are confined to Richeeae, flavonoid waxes are restricted to *Prionotes*, and diterpenes to Richeeae and Epacrideae (Salasoo 1983, 1985; Mihaich 1989). Both the morphology of the wax layer, which includes plates, ribbons and tubes (Mihaich 1989; Weiller et al. 1994), and wax composition show more variation in Styphelieae, both at the level of the tribe and of individual species.

RELATIONSHIPS WITHIN THE FAMILY. Ericaceae are sister to the rather poorly-known Cyrillaceae, and in turn these are sister to Clethraceae (Anderberg et al. 2002). Detailed studies of these groups, particularly their stamen development and the nature of any mycorrhizal associations, are integral to understanding the evolution of Ericaceae.

Ericaceae were the subject of important phenetic studies (e.g. Watson et al. 1967). Stevens (1995) summarized some earlier classifications, while Copeland (1941) divided the Monotropoideae into tribes. A phenetic analysis of pollen characters by Warner and Chinnappa (1986) resulted in the break-up of most groups studied, but their taxonomic conclusions have mostly not been confirmed. Recent morphological and molecular work (e.g. Anderberg 1993; Judd and Kron 1993; Crayn et al. 1996; Kron 1996; Kron et al. 2002a, 2000b) has changed our ideas of higher-level relationships, and Kron et al. (2002a) provide a detailed review of the family and a new classification (Fig. 54). Earlier conventional classifications (Stevens 1971; Watson 1976) are decidedly unsatisfactory, but the informal classification of Watson (1967b) for Epacridaceae has many good points.

Enkianthus (Fig. 55, p. 166) is sister to the rest of the family. Within Monotropoideae, the relationships of Monotropeae and Pterosporeae have been controversial. Furman and Trappe (1971) thought that they were derived from Pyroleae, Anderberg (1993) the reverse, and Copeland (1941) that they were of polyphyletic origin from Arbutoideae; see also Wallace (1975). Bidartondo and Bruns (2001) have recently begun to clarify relationships within

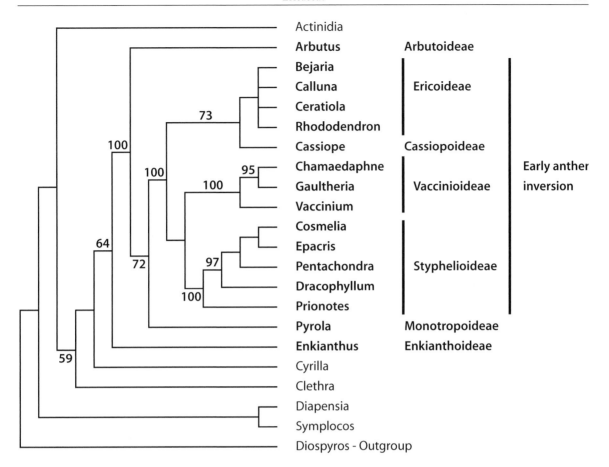

Fig. 54. Ericaceae. Strict consensus of three trees found in the combined sequence analysis of 18s, *rbc*L, and *mat*K genes for 22 taxa of Ericaceae and their closest relatives. Bootstrap values greater than 50% are given above/below branches. Taxa in boldface type represent Ericaceae. Redrawn from Kron et al. (2002a)

the group (cf. also Kron et al. 2002a); we follow their work in recognising *Hypopitys*, but the number of monotypic genera is disconcerting. Variation in fruit type, inflorescence size, and anther morphology do not correlate well (Wallace 1987). Freudenstein (1999) looked at relationships and character transformation in Pyroleae. Within Arbutoideae, Hileman et al. (2001) suggest that *Arbutus*, plesiomorphic at least as regards some aspects of fruit anatomy, is paraphyletic.

A major lineage has Cassiopoideae as sister to the diverse and speciose Ericoideae that includes *Bejaria*, *Erica*, *Rhododendron* and *Empetrum* and their relatives. The inclusion of the Empetraceae in Ericaceae is supported by chemical (Moore et al. 1970) and molecular (e.g. Kron 1996) data, and by the distribution of hosts of the fungi *Chrysomyxa* and *Exobasidium* (Savile 1979). Anderberg (1994a) discussed relationships in Empetreae.

Harrimanella, with *Vaccinium* and *Epacris* and their numerous relatives, form another clade. Epacridaceae had traditionally been divided into two subfamilies, Styphelioideae and Epacridoideae, based on features of the ovary and

fruit, although Drude (1889), Watson (1967b) and Powell (1983) separated *Richea*, *Dracophyllum* and *Sphenotoma* as a separate subfamily, Richeoideae, based on the presence of annular leaf scars. Styphelioideae were then divided into tribes. Styphelioideae s.l. form a monophyletic group within Ericaceae (Anderberg 1992; Crayn et al. 1996, 1998; Kron 1996; Kron et al. 1999b), and we recognize seven monophyletic tribes within them. Prionoteae (*Prionotes* and *Lebetanthus*) are basal, and possess plesiomorphic features such as multicellular hairs and filaments free from the corolla. The monotypic Archerieae are probably sister to the rest of the subfamily, but relationships between the remaining five tribes are unclear (Kron et al. 2002b; for morphological studies, see Powell et al. 1996, 1997). Styphelieae, the largest tribe, are in some respects most derived, but generic limits here and in Epacrideae in particular need attention (e.g. Cherry et al. 2001).

Evolutionary studies had suggested two groupings of genera within the erstwhile Andromedeae, one including *Lyonia* and its relatives, the other *Gaultheria* and its relatives (Stevens 1970b). Judd (1979) clarified cladistic relationships within Lyonieae. Recent studies only partly confirm other tribal groupings in Vaccinioideae (Kron et al. 1999a, 2002b), and relationships around *Gaultheria* need much study (Powell and Kron 2001).

Characters involved in pollination syndromes and/or variation in the numbers of flower parts have previously been considered very important in generic delimitation. Larger or smaller groups of species distinguished by such features were removed from *Rhododendron*, *Erica* and *Vaccinium* which have, hardly surprisingly, turned out to be highly paraphyletic. Realignments in Rhodoreae were suggested by Kron and Judd (1990), while both molecular (Kron and King 1996) and morphological (N. Gift, pers. comm.) data suggest major changes in generic limits in the Phyllodoceae. We adopt a broad circumscription of *Kalmia* below (see Kron et al. 2002a for the formal transfers), but more changes may be needed. Oliver (2000, for a summary) showed that meristic and fruit differences used to segregate genera from *Erica* are unsatisfactory; thus, genera like *Salaxia* and *Simocheilus* (= *Thoracospermum*), previously placed in the Salaxidae, are probably related to separate sections of *Erica*, hence the broad limits of that genus below.

Luteyn (e.g. 1997) suggested some changes to generic limits in neotropical Vaccinieae but, as Vander Kloet (1985) noted, the limits of *Vaccinium* itself were unsatisfactory. Indeed, Kron et al. (2002b, see also 1999a) indicate the magnitude of the problem. Major clades they note include Asian *Agapetes*, perhaps with many other SE Asian species of *Vaccinium* (see the description of *Agapetes* below), Oceanic *Agapetes* (= *Paphia*) + *Dimorphanthera*, and a large, only weakly supported New World group that in turn includes well-supported Andean and Meso-American/Caribbean clades. Many genera described below are para- or polyphyletic. Other smaller and perhaps independent lineages include mainly species of *Vaccinium*. Unfortunately, it is too early to perceive the limits of the major groupings; cytological data (Atkinson et al. 1995) may help clarify the limits of the clades. Formal intergeneric transfers should be made only with great caution.

HABITATS AND SYMBIONTIC RELATIONSHIPS. Ericaceae are particularly prominent in arctic and temperate regions, the montane tropics of South East Asia-Malesia and Central and South America, where Vaccinieae in particular are frequently lianes and epiphytes. They are also common in Mediterranean climates, including California and especially in heathland in South Africa (fynbos) and Australia (kwongan of SW Australia, wallum of SE Queensland). Ericaceae are heliophilous, most members preferring acid conditions, notable exceptions being *Rhododendron hirsutum* in the Alps and some Ericeae in the Cape region (Oliver 1991). Such habitats are characteristically very low in available nutrients, contain mostly low levels of organic matter, and are often seasonally dry, as in the Mediterranean climates. Some epiphytic and epilithic tropical Vaccinieae in particular, and rarely *Rhododendron*, have root tubers apparently involved in water storage (Luteyn 2002).

Ericaceae also grow successfully in acid bogs and as epiphytes, both habitats often being high in organic matter and the latter also liable to dry out intermittently. They are a major component of vegetation in north temperate barrens and other rocky outcrops. Some *Vaccinium* are epilithic in Malesia, while Monotropeae and Pterosporeae are mycotrophic herbs of the forest floor.

Ericaceae are noted for being mycorrhizal. These fungal associations develop on specialised, very fine (40–70 μm diameter) "hair roots" that effectively replace ordinary root hairs which may even be absent (Wood 1961). Hair roots are ephemeral, particularly in species growing in habitats subjected to summer drought (Bell and Pate 1996a). The cortex of the hair root is reduced to two single-celled layers, an endodermis and an exodermis, both with Casparian strips on their radial walls (Allaway and Ashford 1996); the exodermis in some Styphelioideae may have thickened walls and a suberised lamella which constitutes the outermost layer in older hair roots once the epidermis collapses or is sloughed off (Briggs and Ashford 2001). The vasculature of a hair root is minimal – a single xylem tracheid and a single sieve element plus companion cell (McLennan 1935).

In ericoid mycorrhizae (see Read 1996) the outer cortical cells of hair roots are almost filled by branching fungal hyphae. The hyphae first develop a loose network around each hair root, from the region behind the root apex where epidermal cells are differentiating to the older regions of the same root where epidermal cells have enlarged and matured (Duddridge and Read 1982). Hyphae positioned alongside the outermost epidermal cells branch at right angles and eventually penetrate the wall, usually at a single specific point per cell (Hutton et al. 1994), presumably using enzymes. Despite earlier reports, there is no lateral or internal spread of infection laterally via adjacent epidermal cells (see Duddridge and Read 1982; Hutton et al. 1994). The hyphae in the epidermal cells branch, forming prolific intracellular coils which retain their structural integrity after the infected root cell senesces (Duddridge and Read 1982). In *Lysinema ciliatum* certain thick-walled epidermal cells containing hyphal coils separate at the middle lamella and may then be released in the soil, whereas others remain attached to their parent roots (Ashford et al. 1996). Either may be a source for infection of other hair roots formed in the same or subsequent seasons.

Arbutoid mycorrhizae, which also occur in the Pyroleae, show the same penetration of the cells, but the root is invested by a Hartig net, the hyphae following the cell walls or even forming a complete sheath. In monotropoid mycorrhizae, the root is invested in the same way, but the hyphae form peg-like intrusions of the cell wall. This categorization may be something of an oversimplification (Largent et al. 1980), and the mycorrhizal status of *Enkianthus* (and Cyrillaceae) appears to be unknown.

Several species of ascomycete fungi form ericoid mycorrhizae and some of these fungi associate with ectomycorrhizal roots as well (Bergero et al. 2000). Eight species of fungus have been recorded from *Gaultheria shallon* alone (Xiao and Berch 1995). There is little evidence of host specificity at both the genus and species level (Pearson and Read 1973; Hutton 1997). Morphologically and culturally distinct groups of fungi have been isolated from hair root material of Styphelioideae across southern Australia, but most had little in common with *Hymenoscyphus ericae* and *Oidiodendron* spp. that infect other Ericaceae (Hutton 1997; McLean 1999). Basidiomycetes have been implicated in the formation of arbutoid and monotropoid mycorrhizae (Read 1983; Leake 1994; Cullings et al. 1996), and Bidartondo and Bruns (2001; see also Kretzer et al. 2000) found strong specificity in the relationship between fungus and host in Monotropeae and Pleuricosporeae, at least. One (or more, but similar) species of fungus is associated with the host throughout its range, different hosts being associated with different basidiomycetes. When different hosts grow together, they maintain their association with their "own" species of fungus.

The fungal association may (a) improve the mobilization of nutrients from the soil rhizosphere, (b) increase the exploitation of a larger volume of soil and (c) make otherwise intractable sources of nutrients available. Mobilization of nutrients is the key issue in improving competitive effectiveness of ericaceous hosts whose preferred habitats are often high in organic matter and/or low in available nutrients, especially nitrogen. Facilitating the uptake of nitrogen from substrates such as proteins, peptides and other high-molecular weight insoluble forms is an important function of the fungal component, not making phosphorus available, as was originally thought (Read 1983; Perotto et al. 1995). Inorganic nitrogen in the form of ammonium and nitrate is readily taken up by non-infected and mycorrhizally-infected roots, and the absorbed N is incorporated into the insoluble fraction of the plant shoot (Read 1996; Bell and Pate 1996a). Nitrate is the greatest component of xylem sap when inorganic N is in good supply in the soil, as it is after fire or during major nitrification events following initial winter rains. Ammonium is dominant when N is limited, as during the greater part of the growing season (Bell and Pate 1996a). Ammonium is likely to be the form in which N passes from fungus to host root, and it moves to the shoot in organic form, principally as glutamine, but in some species arginine predominates (Bell and Pate 1996a; Smith and Read 1997). The endophyte receives carbohydrate from the host. Finally, the achlorophyllous *Monotropa*, at least, apparently receives both ^{14}C- and ^{32}P-labelled compounds from trees by way of its mycorrhizal associate (Björkman 1960; Leake 1994).

Fire is a major formative influence on vegetation of the Mediterranean-type ecosystems in Australia and South Africa in which many Styphelioideae and *Erica* occur. Growth, morphology, carbohydrate storage and accumulation, and partitioning of biomass differ between fire-sensitive (killed by fire) and fire-tolerant (surviving fire) plants (Bell and Pate 1996b; Bell and Ojeda 1999). Although most species are fire-sensitive or "seeders", the remainder, the "resprouters", survive the fire. A seeder typically has a single main stem associated with a compact canopy and limited lateral root extension, while a resprouter generally shows a spreading, multi-stemmed canopy with large-diameter roots or lignotuberous root stocks (Bell and Ojeda 1999). However, there are intermediate forms of fire response types (facultative seeder sprouters) and a great deal of infraspecific variation, particularly in southern African ericas.

Seeder species have much smaller starch reserves in roots than resprouters. This starch is mostly confined to rays of xylem parenchyma and scattered groups of inter-ray parenchyma, and the usually much greater storage capacity of roots of resprouters is associated with their broader rays (Bell and Pate 1996b; Bell and Ojeda 1999). In Stphelioideae, amounts of starch in shoots are similarly low in both seeders and resprouters, but seeders tend to begin flowering earlier (within one to three years after germination) and achieve at least a three-fold or greater shoot: root dry weight ratio (Bell and Pate 1996b). Other Ericaceae have lignotubers that also sprout after fire or in response to damage to the above-ground parts (e.g. Judd 1981).

South-western Australian Stphelioideae resprout in several ways. The shoot bud bank can be exhausted while not fully depleting the starch reserves held in the root stock – "bud-limited" sprouting, as in *Leucopogon verticillatus*. Alternatively, root starch may be totally depleted before the shoot bud bank is exhausted – "energy reserve-limited" sprouting, as in *Conostephium pendulum* (Bell and Pate 1996b). *Leucopogon striatus*, commonly found in banksia woodlands of Western Australia, shows a third resprouting pattern, the potential for limited resprouting during the first few years after seedling establishment, followed by a loss of this ability as the plant matures (Bell 1995).

Ericaceae often tolerate heavy metals. In California they are notable components of serpentine vegetation, while in Papua New Guinea *Rhododendron beyerinckianum* can grow directly on ultramafic rock. In New Caledonia, *Dracophyllum* and *Styphelia* (*Leucopogon*) are prominent on maquis on cuirasse derived from serpentine rocks. Ericaceous shrubs are often found in active volcanic areas in both Central America and Malesia. Lateritic soils are the habitat of some *Lyonia* in the Caribbean (Judd 1981) and of Stphelioideae in Australia.

DISTRIBUTION. Ericaceae show a diversity of distributional patterns. These include an amphi-south Atlantic disjunction within *Agarista* (Stevens 1970b; Judd 1984), an amphi-north Atlantic disjunction within *Corema*, a disjunction between North and South America (*Gaylussacia*), and numerous north-temperate disjunctions within *Rhododendron*, *Lyonia*, *Elliottia*, and *Epigaea*. Within Stphelioideae, there are several vicariant relationships between taxa growing in the SW and SE of Australia (Smith-White 1948a). *Empetrum* is widely disjunct, growing in the northern hemisphere and the Falkland Islands,

Tristan da Cunha and the south of South America. A few Ericaceae (*Vaccinium* spp., related to the widespread north temperate/Arctic section *Myrtillus*, some Stphelioideae) grow in Hawaii, and several Stphelioideae and two Vaccinieae in New Caledonia. There are few dates for the events causing these patterns, but relationships within Arbutoideae suggest a vicariant event between western North America and the Mediterranean at the Palaeogene/Neogene boundary; western North American species of *Arbutus* were not involved (Hileman et al. 2001).

Within Vaccinieae, the clade of *Paphia* and *Dimorphanthera*, with some 100 species, is restricted to Papuasia and the west Pacific (*Dimorphanthera* is not immediately related to the New World *Satyria*, cf. Stevens 1974), while another probable clade including many species currently placed in *Vaccinium* and also *Agapetes* s. str. (= *Agapetes* below), some 400+ species, grows from China and the Himalayas to eastern Malesia. Many tropical American Vaccinieae are related with Andean and Central America-Caribbean sub-groups (Kron et al. 2002b).

Massings of rather closely related species are common. These include the ca. 400 species of *Agapetes* s.l. from Southeast Asia and Malesia, and the ca. 285 species of *Rhododendron* sect. *Vireya* (monophyletic) in Malesia, with only a few elsewhere. Stphelioideae are especially diverse in Australia, where nine genera containing about 60 species are endemic in the south-west; Luteyn (2002) discusses diversification and diversity in neotropical Vaccinieae. An even more remarkable aggregation is the 207 species of *Erica* found in less than 625 km^2 of the south-western Cape (Oliver et al. 1983), with over 450 species in the south-western Cape as a whole (Rebelo et al. 1985). Species of *Agarista* and *Gaylussacia* sect. *Gaylussacia* form smaller aggregations in eastern Brasil, while *Lyonia* is diverse in the Antilles.

In general, the greatest density and presumably also diversification of Ericaceae is in Mediterranean climates or in high-rainfall temperate to tropical climates in geologically young and active areas with high relief. The great majority of species in the family is found in these areas.

Heads (2003) has produced a lengthy panbiogeographic analysis of Malesian Ericaceae which appeared too late for incorporation into the account.

FOSSIL RECORD. Much of the fossil record of Ericaceae is of leaf impressions and of doubtful utility (e.g. Stevens 1970b). However, reports of large-leaved species of *Kalmia* like *K. cuneata* from Europe (Fischer 1992) are of great biogeographical significance, such species currently being restricted to North America. There are seeds and/or fruits of several other taxa no longer growing there, e.g. *Zenobia*, *Lyonia*, *Eubotrys*, ?*Enkianthus*; *Epacricarpidium* from the Middle Miocene from Denmark (Friis 1985); *Leucothoë*-like seeds from the Maastrichtian from central Europe (Knobloch and Mai 1986); and seeds of *Rhododendron* from Palaeocene deposits in England (Collinson and Crane 1978). Styphelioideae have a sparse fossil record (see Jordan and Hill 1996). The record extends back to the Early Tertiary and possibly the Late Cretaceous and is concentrated in south-eastern Australia, particularly Tasmania. Generalised ericalean pollen from the Late Cretaceous may indicate the first appearance of Styphelioideae, and more distinctive epacrid pollen appears in the Middle Eocene. Fossil leaves of Richeeae and Epacrideae are known from the early Oligocene, and of Styphelieae from the latest Oligocene–early Miocene. *Trochocarpa* and Cosmelieae (probably *Sprengelia*) are at least as old as the Early Pleistocene, the former possibly the Late Oligocene/Early Miocene. *Monotoca* pollen is known from the mid-Miocene. Endocarps identified as Epacridaceae from the Eocene in England should be studied further. Pollen probably of Ericeae is found between 71 and 64 Ma in Namaqualand, although similar pollen is known elsewhere as early as the middle Cretaceous (Scholtz 1985).

The discovery of the minute (<2 mm long) flower buds of *Paleoenkianthus* from the Turonian (90 Ma B.P.) is more tantalizing than anything else. The plant has anthers suggestive of those of *Enkianthus*, but it also has viscin threads; features of the pollen surface and the stigma-style complex are not known from any extant Ericaceae (Nixon and Crepet 1993). The Cretaceous Ericalean flora of the Atlantic Coastal Plain is very diverse (Crepet et al. 2001), and we may need to change our ideas on the evolution of the family when we understand this flora better.

ECONOMIC IMPORTANCE. *Rhododendron* is one of the more popular ornamental shrubs in north temperate gardens. Many cultivars have been developed and hybridization is common; kurume azealeas (*R. kaempferi* × *R. kiusianum*) have been developed in Japan. Tropical species of *Rhododendron* sect. *Vireya* also have their devotees. However, naturalised *R. ponticum* forms almost impenetrable thickets over substantial areas of the western British Isles. Genera like *Pieris*, *Kalmia* and *Erica* are grown for their flowers and foliage.

A few Styphelioideae are grown as container plants. However, there are propagation problems (McLean 1999); seeds are difficult to germinate and cutting success is variable (Thompson 1986; Williams 1986), although the latter is improved by mycorrhizas (McLean 1995).

Blueberries (*Vaccinium* sect. *Cyanococcus*), bilberries (*Vaccinium* sect. *Myrtillus*), cranberries (*Vaccinium* sect. *Oxycoccus*, see Eck 1990) and, to a lesser extent, cowberries or lingonberries (*V. vitis-idaea*) are cultivated or gathered in the wild and used for jams and jellies, and even Christmas ornaments (cranberries). *Astroloma humifusum* has potential for the "bush food" industry; its fruits have been used for jam. The fruits of the Andean *Gaultheria myrsinoides* are reported to be variously hallucinogenic, poisonous, or harmless (Luteyn 1995, under *Pernettya*), and those of *Comarostaphylos discolor* (Diggs 1995) may also be poisonous. Andromedotoxins can cause serious poisoning or even death (Ewan-Nyambi et al. 1993 and references therein), while the honey of *Rhododendron* can be fatal, as Xenophon found in his retreat across Asia Minor.

There are more minor uses of the family, and there are many records of its use in local pharmacopeias. For instance, tea has been made from the leaves of *Rhododendron* sect. *Ledum* and from *Vaccinium arctostaphylos*. *Gaultheria procumbens* was an important source of salicylic acid before it was made by synthesis; *G. shallon* and *Kalmia latifolia* are much used as "greens" by florists in North America; the root burls of *Erica arborea* and *E. scoparia* are used for briar (from "bruyère", heather) pipes (Oliver 1991). *Calluna vulgaris* ("Scotch heather"), grazed by sheep and grouse alike, is an important element in the economic and social fabric of much of highland England and Scotland.

CONSERVATION. In Australia, more than 100 species of Styphelioideae are considered under threat from various sources (Briggs and Leigh 1989). Most occur in the south-west of Western Australia (Keighery 1996 lists five species as Declared Rare Flora). Pathogenic *Phytophthora* spp. are now widespread in southern Australia and are a threat to Styphelioideae and many other species. In Western Australia, *Coleanthera virgata* and *Leucopogon cryptanthus* have not been seen since the types were collected in the 1840s and are considered extinct. In Victoria *Choristemon humilis*, now assigned to *Leucopogon*, has not been collected since 1923 and is also thought to be extinct. Several *Erica* spp. in South Africa are very localized and are in danger of extinction.

Conspectus of Ericaceae

Key to the Genera (to those of Styphelioideae partly based on Powell 1983)

1. Plants without chlorophyll   2
– Plants with chlorophyll   12
2. Sepals very different from petals, the latter fused   3
– Two perianth whorls not clearly distinguishable, inner
   members usually free   5

3. Plant lacking multicellular hairs; axes nodding
                                              12. Monotropopsis
– Plant with multicellular hairs; axes erect   4
4. Axis slender, bracts not overlapping; corolla urceolate;
   anthers with long spurs; seeds winged   **6. Pterospora**
– Axis stout, bracts overlapping; corolla campanulate;
   anthers with very short spurs; seeds not winged
                                                  7. Sarcodes
5. Axes nodding   6
– Axes erect   8
6. Style continuous; fruit baccate   **11. Monotropastrum**
– Style impressed; fruit a capsule   7
7. Flowers single, white   **10. Monotropa**
– Flowers 2–4, purplish   **14. Hypopitys**
8. Style impressed   **8. Allotropa**
– Style continuous   9
9. Petals fused   **15. Hemitomes**
– Petals free   10
10. Flowers 3-merous; filaments flattened **9. Cheilotheca**
– Flowers 4–5-merous; filaments terete   11
11. Anthers elongated, not inverting, with two long slits
                                           16. Pleuricospora
– Anthers hippocrepiform, inverting, with a single apical
   slit   **13. Pityopus**
12. Plants ± herbaceous; petals free   13
– Plants woody; petals usually fused   16
13. Plant with erect, leafy stems; inflorescence umbellate
                                              4. Chimaphila
– Plant with rosette of leaves ± at ground level; inflorescence not umbellate   14
14. Inflorescence with >6 flowers, racemose   15
– Inflorescence of a single flower   **3. Moneses**
15. Lamina finely serrate; inflorescence secund   **5. Orthilia**
– Lamina at most obscurely crenate; inflorescence with
   flowers all around the axis   **2. Pyrola**
16. Ovary superior (if inferior, then anthers with awns) 17
– Ovary inferior   91
17. Flowers nearly always 4-merous; corolla ± scarious,
   persistent; leaves opposite or whorled, nearly always
   ericoid   18
– Flowers rarely 4-merous; corolla not scarious; leaves
   usually spiral, rarely ericoid   19
18. Leaves sessile, tailed   **38. Calluna**
– Leaves petiolate, not tailed   **40. Erica**
19. Fruit septicidal; pollen usually with viscin threads 20
– Fruit loculicidal, or indehiscent; pollen lacking viscin
   threads   34
20. Leaves opposite or whorled; anthers usually with slits
   their entire length   21
– Leaves scattered to pseudo-verticillate; anthers rarely
   with slits their entire length   22
21. Lamina ericoid   **24. Ledothamnus**
– Lamina with recurved margins, but not ericoid
                                           35. Kalmia (part)
22. Inflorescence bud with large, usually brown perulae;
   bracts and bracteoles usually deciduous   23
– Inflorescence bud lacking such perulae; bracts and
   bracteoles usually persistent   25
23. Inflorescence axis elongated, with leaves below the
   flowers; bracts and prophylls leafy   **25. Therorhodion**
– Inflorescence axis not elongated, lacking leaves below the
   flowers; bracts and prophylls not leafy   24
24. Capsule subspherical   **27. Menziesia**
– Capsule longer than wide   **28. Rhododendron**
25. Petals free   26
– Petals fused   28

26. Leaves ericoid; vegetative plant prostrate **23. Bryanthus**
  – Leaves rarely ericoid; plant erect   27
27. Filaments ± terete, anthers with terminal pores
                                22. Bejaria
  – Filaments flattened, anthers with elongated slits
                                29. Elliottia
28. Anthers with slits at least half their length   29
  – Anthers with terminal pores or short slits   30
29. Lamina <2 cm long; antesepalous stamens strongly epipetalous   **26. Diplarche**
  – Lamina >5 cm long; all stamens free from petals
                                30. Epigaea
30. Corolla with 10 pouches   **34. Kalmia** (part)
  – Corolla lacking pouches   31
31. Lamina ericoid, or with strongly recurved margins   32
  – Lamina plane   33
32. Lamina ericoid; style impressed   **31. Phyllodoce**
  – Lamina with strongly recurved margins; style and ovary continuous   **39. Daboecia**
33. Lamina with ciliate margins; plant lacking sessile glandular hairs   **32. Rhodothamnus**
  – Lamina lacking ciliate margins; plants with sessile glandular hairs   **33. Kalmiopsis**
34. Lamina rarely xeromorphic, margin often with gland-tipped teeth, venation pinnate, midrib usually evident; plant nearly always with multicellular hairs   67
  – Lamina xeromorphic, with entire margins, if serrulate, not gland-tipped, venation closely parallel- or pli-nerved, midrib not evident; plant lacking multicellular hairs   35
35. Stems with annular scars after leaf fall   36
  – Stems without annular scars after leaf fall   38
36. Corolla ovoid or conical, forming a cap that falls at anthesis, the lobes not opening   **56. Richea**
  – Corolla cylindrical or campanulate, not shed at anthesis, the lobes opening   37
37. Flowers in short spikes; bracts persistent; corolla tube narrow   **57. Sphenotoma**
  – Flowers in compound racemes or panicles, rarely solitary; bracts falling; corolla broadly cylindrical or campanulate   **55. Dracophyllum**
38. Leaves sheathing   39
  – Leaves not sheathing   41
39. Filaments adnate to corolla tube   **45. Cosmelia**
  – Filaments free   40
40. Corolla glabrous; tube very short; lobes imbricate; anthers connivent or cohering   **46. Sprengelia**
  – Corolla hairy inside; tube cylindrical; lobes imbricate; anthers free   **47. Andersonia**
41. Flowers 4-merous   42
  – Flowers 5-merous   43
42. Stamens 4   **66. Cyathopsis**
  – Stamens 2   **54. Oligarrhena**
43. Fruit drupaceous, indehiscent, rarely splitting irregularly and transversely   44
  – Fruit a loculicidal capsule   62
44. Corolla lobes incurved at the apex, style impressed into ovary   **53. Needhamiella**
  – Corolla lobes rarely incurved; style and ovary continuous   45
45. Anthers exserted above corolla tube   46
  – Anthers included within corolla tube, or lying between spreading lobes   49
46. Anthers connate in a cone around style
                              62. Coleanthera
  – Anthers free   47
47. Corolla <1.6 mm long, the lobes spreading
                              59. Androstoma

  – Corolla >5 mm long, with distinctly revolute lobes   48
48. Leaves scattered along stem   **75. Styphelia**
  – Leaves distinctly clustered in false whorls
                              65. Cyathodes
49. Corolla tube conical in upper part, lobes very small, erect; anthers 2-lobed   **63. Conostephium**
  – Corolla tube cylindrical, campanulate or rotate, lobes spreading or recurved; anthers rounded   50
50. Corolla tube with tufted hairs or scales below middle inside; filaments usually flat   51
  – Corolla tube glabrous below middle inside, rarely puberulent or pubescent; filaments terete, usually thin   52
51. Corolla tube short, broad, 5 tufts of hairs or ring of glands at base, lobes spreading   **71. Melichrus**
  – Corolla tube elongated, cylindrical, 5 tufts of hairs or fringed scales in tube, lobes erect around anthers, spreading or recurved at top   **60. Astroloma**
52. Locules often more than 5; fruit with separate stones
                              53
  – Locules usually 5 or fewer; fruit with a single stone   55
53. Corolla valvate in bud; filaments thin, terete
                            73. Pentachondra
  – Corolla ± imbricate in bud; filaments short, thick and tapered   54
54. Adaxial surfaces of corolla lobes with hairs towards the base   **74. Trochocarpa**
  – Adaxial surfaces of corolla lobes with widespread hairs
                            67. Decatoca
55. Flowers pedicellate, prophylls basal   **70. Lissanthe**
  – Flowers sessile, prophylls inserted immediately below calyx   56
56. Ovary and style with hairs   **64. Croninia**
  – Ovary and style glabrous   57
57. Corolla lobes imbricate in bud   **61. Brachyloma**
  – Corolla lobes valvate in bud   58
58. Corolla tube elongate   59
  – Corolla tube short; filaments terete or filiform
                            68. Leptecophylla
59. Corolla lobes glabrous or papillose; ovary 1–2-locular
                            72. Monotoca
  – Corolla lobes hairy, rarely glabrous; ovary (2–)5(–10)-locular   60
60. Corolla lobes with erect tufts of hairs near apex and fine hairs at throat; flowers often yellow-green to green
                            58. Acrotriche
  – Corolla lobes sparsely to densely bearded, rarely glabrous; flowers white, cream or red   61
61. Drupe ± spherical; mesocarp fleshy or thin
                            69. Leucopogon
  – Drupe depressed; mesocarp thick, pulpy
                            74. Planocarpa
62. Flowers pedicellate, bracteoles two, basal, quite different from sepals   **44. Archeria**
  – Flowers sessile, bracteoles several, imbricate, passing into sepals   63
63. Corolla lobes contorted in bud   64
  – Corolla lobes imbricate in bud   65
64. Lamina ovate, acuminate, pungent, erect above base then spreading widely   **52. Woollsia**
  – Lamina linear to ovate-lanceolate, erect, appressed to stem   **50. Lysinema**
65. Filaments long, inserted at base of corolla tube; anthers adnate, connivent around style   **51. Rupicola**
  – Filaments short, inserted in corolla throat; anthers attached above middle   66
66. Filaments terete, shorter than anthers; anthers dorsally attached to filament, included   **49. Epacris**

– Filaments compressed, about as long as anthers; anthers adnate to filament for most of their length, exserted
48. Budawangia

67. Lamina ericoid and <2 cm long, fruit a drupe   68
– Lamina rarely ericoid-like, if so, >3 cm long, or fruit a capsule   70
68. Inflorescence capitulate   **36. Corema**
– Inflorescence not capitulate   69
69. Lamina with wooly hairs in abaxial groove; style longer than stigmatic lobes   **37. Ceratiola**
– Lamina lacking wooly hairs in abaxial groove; style shorter than stigmatic lobes   **35. Empetrum**
70. Inflorescence terminal; ovary superior; fruit fleshy, usually indehiscent   71
– Inflorescence usually axillary; fruit proper (i.e. not including calyx) fleshy only if ovary is inferior   74
71. Leaves in whorls of three   **19. Ornithostaphylis**
– Leaves scattered   72
72. Ovary smooth   **20. Arctostaphylos**
– Ovary papillate   73
73. Ovary with several ovules/loculus; fruit a berry; lamina usually <3 times longer than broad   **17. Arbutus**
– Ovary with 1 ovule/loculus; fruit a drupe; lamina usually >3 times longer than broad   **18. Comarostaphylis**
74. Lamina <1 cm long   75
– Lamina >1 cm long   76
75. Leaves decussate; flowers axillary   **21. Cassiope**
– Leaves spiral; flowers terminal   **41. Harrimanella**
76. Stamens with geniculate filaments and/or with spurs   77
– Stamens with ± straight filaments, awned or not   80
77. Corolla urceolate to campanulate; seeds with wing on one side   **78. Craibiodendron**
– Corolla urceolate to tubular; seeds unwinged, spindle-shaped or rounded   78
78. Capsule with much thickened, often whitish ribs, or filaments with delicate spurs that have disintegration tissue   **81. Lyonia**
– Capsule lacking thickened ribs; stamen spurs lacking disintegration tissue   79
79. Lamina with veins on the abaxial surface graded in prominence; stamens spurred   **79. Pieris**
– Lamina with veins on the abaxial surface about equal in prominence; stamens without spurs   **80. Agarista**
80. Inflorescence terminal; anthers with a single pair of awns   81
– Inflorescence rarely terminal; if so, then anthers not with awns   84
81. Inflorescence perulate; anthers with slits for much of their length   **1. Enkianthus**
– Inflorescence not perulate; anthers with short, terminal slits or pores   82
82. Indumentum of scales; inflorescence a foliaceous raceme   **84. Chamaedaphne**
– Indumentum never of scales; inflorescence racemose, not foliaceous   83
83. Inflorescence >10 cm long, branched   **77. Oxydendrum**
– Inflorescence <3 cm long, unbranched   **82. Andromeda**
84. Anthers with slits their entire length   85
– Anthers with pores or short slits   86
85. Corolla <5 mm long   **42. Lebetanthus**
– Corolla >10 mm long   **43. Prionotes**
86. Anthers with terminal tubules, lacking awns   87
– Anthers usually lacking terminal tubules, frequently awned   88
87. Inflorescence racemose, rarely paniculate; lamina lacking fibres in mesophyll, with one or more pairs of marginal glands near base   **89. Tepuia**

– Inflorescence fasciculate; lamina with fibres free in mesophyll (break lamina!), lacking marginal glands
88. Diplycosia
88. Fruit fleshy, or fruit dry but calyx fleshy
87. Gaultheria
– Fruit and calyx dry   89
89. Corolla >1 cm long, campanulate   **83. Zenobia**
– Corolla <1 cm long, ± urceolate   90
90. Prophylls basal   **85. Leucothoë**
– Prophylls apical   **86. Eubotrys**
91. Filaments and/or anthers of alternating stamens strongly unequal   92
– Filaments and anthers of alternating stamens similar to somewhat unequal (but spurs may differ considerably)   95
92. Filaments unequal; anthers not notably robust   93
– Filaments ± equal; anthers rather woody, the tubules widening distally   94
93. Anthers equal; stamens <1/2 the length of the corolla; bracts small or early deciduous   **102. Orthaea**
– Anthers unequal; stamens as long as corolla (–1/2 as long); bracts usually large and persistent
101. Cavendishia
94. Filaments connate their entire length (New World)
105. Satyria
– Filaments free (Old World)   **98. Dimorphanthera**
95. Fruit pyrenoid   **124. Gaylussacia**
– Fruit baccate   96
96. Anthers with slits the length of the thecae, tubules minute
94. Lateropora
– Anthers with pores or short slits, tubules usually well developed   97
97. Anther tubules with other than terminal or introrse pores or short clefts, much shorter than the thecae   98
– Tubules with terminal or introrse pores or short clefts (if lateral, then tubules longer than thecae)   99
98. Anther tubules with short, extrorse clefts; pedicels stout
95. Notopora
– Anther tubules with short, latrorse clefts; pedicels slender to stout   **93. Didonica**
99. Prophylls at apex of pedicel enveloping ovary (often with inflorescence bracts); flowers (3)4(5)-merous
114. Disterigma
– Prophylls rarely at apex of pedicel, not enveloping ovary; flowers usually 5-merous   100
100. Ovary 5-locular, with 5 additional inpushings; cork-cambium superficial (surface of leafy stems lacking longitudinal cracks)   **92. Agapetes**
– Ovary usually simply 5-locular; cork cambium deep-seated, or more superficial, but localised (surface of leafy stems often developing longitudinal cracks)   101
101. Stamens 5–10; calyx lobes free, 5–9 mm long
111. Oreanthes
– Stamens (4, 5, 8) 10; calyx lobes usually smaller, connate or not   102
102. Calyx continuous with pedicel   103
– Calyx articulated with pedicel   117
103. Anther tubules elongate, half as wide (or less) than thecae   104
– Anther tubules at most slightly longer than thecae, rarely thin, if elongate, more than half as wide as thecae   105
104. Calyx lobes >2 mm long; filaments with tuft of retrorse hairs on back   **113. Pellegrinia**
– Calyx lobes <1 mm long; filaments without tuft of retrorse hairs on back   **110. Semiramisia**

105. Calyx and corolla conspicuously winged or rarely angled
     106
   – Calyx terete or angled, rarely winged; corolla terete 108
106. Leaves usually opposite or pseudoverticillate; corolla
     wings quite broad distally   **122. *Anthopterus*** (part)
   – Leaves scattered; corolla wings at most narrow distally
     107
107. Corolla terete   **118. *Themistoclesia*** (part)
   – Corolla narrowly winged, wings broadest at the base
     122. *Anthopterus* (part)
108. Calyx angled, rarely winged   109
   – Calyx terete   112
109. Calyx angles opposite the lobes   **107. *Polyclita***
   – Calyx angles alternating with the lobes   110
110. Calyx limb 1<cm long   **100. *Agapetes scortechinii***
   – Calyx limb <4 mm long   111
111. Lamina linear, with midrib alone evident **117. *Rusbya***
   – Lamina not linear, with more complex venation
     118. *Themistoclesia* (part)
112. Corolla (10–)14 mm long, carnose to coriaceous; embryo
     white   113
   – Corolla <10 mm long (if longer, then filaments
     longer than anthers), thinly fleshy; embryo green (white)
     114
113. Lamina 8 cm long; inflorescence racemose, flowers
     numerous; anther tubules at most 2 × length of thecae
     120. *Thibaudia* (part)
   – Lamina <3(–6) cm long; inflorescence fasciculate, flowers
     1–3 (racemose, flowers numerous); anther tubules 2–5 ×
     longer than thecae   **121. *Demosthenesia***
114. Pedicels usually long and filiform, cernuous; corolla urce-
     olate; filaments usually longer than anthers
     115. *Sphyrospermum*
   – Pedicels at most thin, not cernuous; corolla various;
     filaments usually shorter than anthers   115
115. Lamina pli-nerved, with prominent marginal glands near
     base, lacking surface glandular hairs; inflorescence fasci-
     culate   **123. *Costera***
   – Lamina pli- or penni-nerved, lacking marginal glands
     near base, usually with surface glandular hairs; inflores-
     cence various   116
116. Lamina ± pli-nerved; corolla tubular, rarely urceolate;
     embryo green   **116. *Diogenesia***
   – Lamina ± penni-nerved; corolla urceolate, rarely cam-
     panulate; embryo white   **90. *Vaccinium*** (part)
117. Anther tubules elongate, half as wide (or less) than thecae
     118
   – Anther tubules at most slightly longer than thecae, rarely
     thin, if elongate, more than half as wide as thecae 120
118. Corolla with lobes at least 4 mm long, the base often
     ventricose   **109. *Ceratostema***
   – Corolla with lobes <2 mm long, the base never ventricose
     119
119. Filaments connate; anther pores terminal
     112. *Siphonandra*
   – Filaments free or connate; anther pores oblique, or
     tubules with short slits   **97. *Gonocalyx***
120. Thecae conspicuously papillate; tubules rigid, connate or
     not; stamens often <2/5 the length of corolla   121
   – Thecae smooth or papillate; tubules flexible; stamens
     often as long as corolla   123
121. Calyx turbinate, with wings or angles opposite the lobes;
     connective extending dorsally up each anther tubule
     106. *Mycerinus*
   – Calyx with wings (if present) alternating with the lobes;
     connective terminating at the bases of anther tubules
     122

122. Anthers spurred, if not, connective thickened; tubules dis-
     tinct; corolla never angled, often abruptly constricted
     104. *Psammisia*
   – Anthers not spurred; tubules often connate; corolla
     angled or not, gradually narrowed distally
     103. *Macleania*
123. Corolla ca. 4 mm long, tube with solid spurs opposite the
     lobes; calyx with wings alternating with the lobes
     96. *Utleya*
   – Corolla often longer, tube lacking spurs; calyx usually not
     winged   124
124. Corolla urceolate to campanulate, <12 mm long (–22 mm
     long, tubular), the lobes often imbricate, thin and drying
     membranaceous, or flowers polypetalous; filaments often
     longer than the anthers   **90. *Vaccinium*** (part)
   – Corolla ± tubular, (8–)12 mm long, the lobes thick, drying
     coriaceous, valvate; filaments shorter than the anthers
     125
125. Anther tubules usually 2–5 times longer than thecae;
     calyx limb often 2–4 times longer than tube
     119. *Plutarchia*
   – Tubules up to 2 times longer than thecae; calyx limb as
     long as or shorter than tube   126
126. Calyx with wings longer than lobes; anthers alternately
     spurred; stamens about half as long as corolla
     108. *Anthopteropsis*
   – Calyx with at most obscure wings; anthers lacking spurs;
     stamens about as long as corolla   127
127. Flowers 6–8-merous; corolla flaring   **91. *Symphysia***
   – Flowers 5-merous; corolla urceolate to tubular   128
128. Calyx limb to as long as tube; filaments much shorter to
     longer than anthers; tubules to 2.5× longer than thecae
     99. *Paphia*
   – Calyx limb often longer than tube; filaments much
     shorter than anthers; tubules about as long as thecae
     120. *Thibaudia* (part)

GENERA OF ERICACEAE
Basic features

Plant woody; hair roots invested by fungal hyphae that form a Hartig net; pericyclic fibres at most poorly developed; node unilacunar; pith homogeneous; bud scales present; leaves spiral, convolute in bud, scattered, with midrib and reticulate venation; inflorescence terminal; flowers 5-merous, pedicels present, not articulated with calyx, prophylls basal; corolla sympetalous; anthers with slits, endothecium present; nectary at base of ovary; placentation axile, numerous ovules/locule; capsule loculicidal; raphe with vascular bundle; embryo white, terete.

Note: these features can be understood to be applicable to all genera of the family, unless modified in the subfamilial or tribal characterizations, and the characters in these behave similarly.

I. SUBFAMILY ENKIANTHOIDEAE Kron, Judd & Anderberg (2002).

Enkiantheae P.F. Stevens (1971).

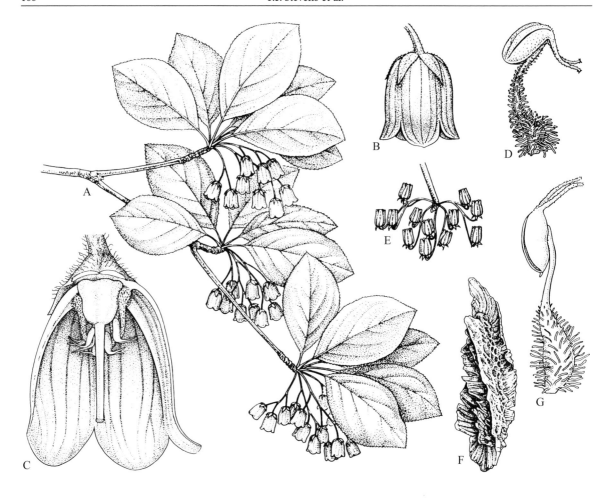

Fig. 55. Ericaceae. **A–F** *Enkianthus campanulatus.* **A** Flowering branchlet. **B** Flower. **C** Same, opened, showing awned anthers. **D** Stamen with inverted, awned anther; one of the thecae dehiscing with longitudinal slit. **E** Capsules. **F** Seed. **G** *Enkianthus perulatus*, stamen, the anther with protruding connective. Drawing by Polyanna von Knörring. (Anderberg 1994b)

Pith heterogeneous; inflorescence perulate; calyx valvate; anthers with paired awns, megagametophyte with ears; testa cells not elongated, rather tall.

1. *Enkianthus* Loureiro   Fig. 55

Enkianthus Loureiro, Fl. Coch. 1: 339 (1793); Anderberg, Nordic J. Bot. 14: 385–401 (1994).

(Sub)deciduous shrubs; leaves pseudoverticillate; inflorescences perulate, corymbose to umbellate, rarely paniculate, with terminal flower, bracts and prophylls poorly developed; corolla campanulate to urceolate, lobed at most ca. 1/3; anthers smooth apart from awns; style impressed or not; sutures of capsule more or less thickened; seeds winged or not; $n = 11$. 16 spp., four sections, E Asia; low to mid altitude.

All other Ericaceae

Raphe lacking vascular bundle; exothecium present.

II. Subfamily Monotropoideae Arnott (1832).

Pyroloideae Kostel. (1834).

Herbs; multicellular hairs absent; inflorescences racemes, eperulate.

1. Tribe Pyroleae Dumortier (1829).

Rarely subshrubs; fungal hyphae with complex, coiled intrusions into exodermal cells of hair roots; multicellular hairs absent; leaves pseudoverticillate, ± serrate; flowers weakly monosymmetrical or not, prophylls absent; petals free; filaments thin, anthers with short terminal tubules (rarely absent) and pores; testa cells moderately elongated, little thickened; embryo small, hardly differentiated.

2. *Pyrola* L.

Pyrola L., Sp. Plant. 1: 396 (1753); Krísa, Bot. Jahrb. Syst. 90: 476–508; Haber, Syst. Bot. 12: 324–335 (1987), Can. J. Bot. 66: 1993–2000 (1988).

Rhizomatous herb; leaves in rosette, thick; pollen in tetrads; no disc; style with collar at the apex; capsule valves with cobwebby edges; testa cells with moderate-sized to no plasmodesmata; $n = 23, 46$. Ca. 30 spp., north temperate, Sumatra, Guatemala. Infrageneric classification and species limits difficult; the anthers of *P. minor* L. lack tubules, the buds of *P. chlorantha* lack scales.

3. *Moneses* S.F. Gray

Moneses S.F. Gray, Nat. Arr. Brit. Pl. 2: 396, 403 (1821).

Herb, shoots arising from roots; leaves in rosette, thick; flower single, on long stalk, bract somewhat below flower; pollen in tetrads; no disc; style with collar at the apex; testa cells with thin scalariform thickenings; $n = 13$. 2 spp., north temperate.

4. *Chimaphila* Pursh

Chimaphila Pursh, Fl. Am. Sept. 1: 279 (1814).

Rhizomatous subshrub; leaves coriaceous; inflorescence a corymb, bracts ± borne up pedicel; pollen in tetrads to monads; style short, with collar at the apex; testa cells with small or no plasmodesmata; $n = 13$. Five spp., boreal and north-temperate, S to Guatemala and Hispaniola. Pollen in tetrads, monads or polyads.

5. *Orthilia* Rafinesque

Orthilia Rafinesque, Aut. bot.: 103 (1840).

Rhizomatous herb; leaves thin; inflorescence secund; anthers lacking tubules, pollen in monads; disc lobed; style long, no collar at the apex; capsule valves with cobwebby edges; cells of testa with moderate to small plasmodesmata; $n = 19$. One sp., *O. secunda* (L.) House, north-temperate.

2. TRIBE PTEROSPOREAE Baillon (1891).

Echlorophyllous; glandular multicellular hairs present; leaves entire; axes pink to reddish, many-flowered; petals fused; anthers with 2 spurs; style impressed, rather short.

6. *Pterospora* Nuttall

Pterospora Nuttall, Gen. N. Am. Pl. 1: 269 (1818); Wallace, Wasmann J. Biol. 33: 63–68 (1975).

Rootballs tight; axes pink to reddish; corolla urceolate; stamens 1/2 as long as corolla, anthers long-spurred, inverting late?, with apical and lateral slits; nectary not obvious; fruit dehiscing from the base first; seeds winged, cells somewhat elongated, thin-walled; $n = 8$. One sp., *P. andromedea* Nutt., W North America including Mexico, the Great Lakes region; 60–3675 m alt.

7. *Sarcodes* Torrey

Sarcodes Torrey, Smithson. Contrib. Know. 6: 18, pl. 10 (1853); Wallace, Wasmann J. Biol. 33: 68–72 (1975).

Roots fibrous; axes red to orange; corolla campanulate, slightly saccate at base; stamens ca. 2/3 length corolla, anthers elongated, with very small spurs, with terminal extrorse slits, nectaries with low lobes; fruit with irregular dehiscence; seeds subovoid, cells not elongated, thick-walled; $n = 32$. One sp., *S. sanguinea* Torrey, USA (California and adjacent Oregon) and Mexico (Baja California); 1070–3050 m alt.

3. TRIBE MONOTROPEAE Dumortier (1829).

Echlorophyllous; shoots arising from roots; fungal hyphae forming peg-like inpushings into exodermal cells of hair roots; roots rarely endogenous; leaves entire; bracts (sub)persistent, as large or larger than and as conspicuous as sepals; prophylls usually absent; sepals usually as large as petals/corolla; style rather short, stigma expanded; embryo minute, undifferentiated.

8. *Allotropa* A. Gray

Allotropa A. Gray in Newberry, Rep. Expl. Surv. Railroad 6(3): 80 (1858); Wallace, Wasmann J. Biol. 33: 13–18 (1975).

Roots fibrous; axes erect, red and white striped, many-flowered; perianth a single whorl, 5, free,

campanulate; stamens (9)10(11), as long as perianth, anthers inverting late, with pore-like slits
ca. 1/3 the length of anther; style impressed; seeds
linear, cells elongated, thin-walled; $n = 13$. One sp.,
A. virgata A. Gray, W USA; 80–3000 m alt.

9. *Cheilotheca* J.D. Hooker

Cheilotheca Bentham & J.D. Hooker, Gen. Plant. 2: 607 (1876);
   Wallace, Wasmann J. Biol. 33: 18–23 (1975).
Andresia Sleumer, Fl. Males. I, 6: 669 (1967).

Root mass loose; axes erect, white to purple, 1-few-
flowered; sepals 2–5; corolla tubular-campanulate,
petals 3, free, concave; stamens 6, 1/2–2/3 length of
corolla; anthers linear, thecae free, with long slits,
or fused, hippocrepiform, with a single terminal
slit; ovary 6-carpellate, placentation parietal, style
continuous; fruit baccate; seeds? Two spp., Assam
(India), the Malay Peninsula (Malaysia) and
Sumatra (Indonesia); 600–1500 m alt.

10. *Monotropa* L.

Monotropa L., Sp. Plant. 1: 387 (1753); Wallace, Wasmann J.
   Biol. 33: 29–42 (1975).

Rootballs ± tight; axes nodding, white to reddish,
1-few-flowered; flowers 3–8-merous; sepals absent
or not; corolla tubular, petals free, saccate at base;
stamens as long as corolla, anthers horizontally
reniform, with terminal slit or slits; nectary with
paired lobes between antepetalous stamens; ovary
with axile placentation, style impressed, stigma
umbilicate to funneliform; fruit capsular; seeds
spindle-shaped, cells elongated, thin-walled;
$n = 16, 24$. 1 sp., *M. uniflora* L., north temperate;
0–2000 m alt.

11. *Monotropastrum* H. Andres

Monotropastrum H. Andres, Notizbl. Berl.-Dahl. 12: 696, fig. 8
   (1935); Wallace, Wasmann J. Biol. 33: 42–48 (1975).
Eremotropa H. Andres, Bot. Jahrb. Syst.76: 103, pl. 6 (1953);
   Wallace, Taxon 36: 128–130 (1987).

Root mass dense; axes nodding, white, 1-few-flow-
ered; flowers 3–5-merous; sepals absent or not,
more or less distinct or not; corolla tubular, petals
free, more or less saccate at base; stamens as long
as corolla, anthers horizontally reniform, with terminal slit or slits; nectary with paired lobes
between antepetalous stamens; ovary with parietal
placentation, style continuous, stout to slender,
stigma funneliform; fruit baccate; seeds ovoid,
cells reticulate. Two spp., Punjab (India) to Korea

Laos and Vietman, Sumatra (Indonesia); 1000–
2700 m alt.

12. *Monotropsis* Schweinitz

Monotropsis Schweinitz in Elliott, Sketch Bot. S. Carol. Georgia
   1: 478 (1817); Wallace, Wasmann J. Biol. 33: 49–53 (1975).

Roots coralloid; axes nodding, violet to purple,
several-flowered; flowers (4)5(6)-merous; sepals
present, different in texture; corolla subtubular,
petals fused, slightly saccate at the base; stamens
ca. 3/5 the length of the corolla, with extrorse
pores; nectary crenate; ovary with largely parietal
placentation, style continuous; fruit baccate; seeds
ovoid, cells little elongated, thick-walled. One sp.,
M. odorata Schweinitz, SE USA, 600–1350 m alt.

13. *Pityopus* Small

Pityopus Small, N. Am. Fl. 29: 161 (1914); Wallace, Wasmann J.
   Biol. 33: 53–58 (1975).

Rootballs tight; axes erect, white to yellowish,
(1-)several-flowered, branched or not; flowers 4-
(5-)merous; sepals present; corolla tubular, petals
free; stamens 2/3 the length of the anthers, thecae
horizontal, connate, hippocrepiform, with a single
apical slit; ovary with parietal placentation, style
continuous, stigma umbiculate; fruit baccate;
seeds ellipsoid, cells not elongated, thick-walled.
One sp., *P. californicus* (Eastwood) H.F. Copeland:
W USA; 30–2000 m alt.

14. *Hypopitys* Crantz

Hypopitys Crantz, Inst. Rei Herb. 2: 467 (1766).

Rootballs ± tight; axes nodding, cream to red or
purple, (1)2- to several-flowered; flowers 4–5-
merous, subcampanulate; sepals (0–)4–5; corolla
free, saccate at the base; stamens almost as long as
the sepals, anthers horizontal, thecae connate,
hippocrepiform, with single terrminal slit;
nectary with paired lobes between opposi-
tipetalous stamens; style impressed, stigma umbilicate; seeds spindle-shaped, cells elongated,
thin-walled; $n = 8, 16, 24$. 1 sp., *H. monotropa*
Crantz, north temperate region, Central America;
to 400 m alt. Possibly to be divided (see Bidartondo
and Bruns 2001).

15. *Hemitomes* A. Gray

Hemitomes A. Gray in Newberry, Rep. Expl. Surv. Railroad 6(3):
   80, pl. 12 (1858); Wallace, Wasmann J. Biol. 33: 24–29 (1975).

Roots fibrous; axes erect, cream to reddish, 1- to several-flowered, branched or not; flowers 4(–6)-merous; sepals and prophylls apparently present; corolla tubular-campanulate, petals connate; stamens ca. 2/3 the length of the corolla, anthers not inverting, with long slits; ovary 8-carpellate, with parietal placentation, style continuous; fruit baccate; seeds ellipsoid, testa cells little elongated, ?thick-walled. One sp., *H. congesta* A. Gray, W USA; 30–2700 m alt.

16. *Pleuricospora* A. Gray

Pleuricospora A. Gray, Proc. Am. Acad. Arts Sci. 7: 369 (1868); Wallace, Wasmann J. Biol. 33: 58–63 (1975).

Roots coralloid; axes erect, cream to yellowish, (1) to several-flowered; flowers usually 4-merous, tubular; sepals present; petals free; stamens as long as corolla, anthers not inverting, with long slits; nectary not seen; ovary with parietal placentation, style continuous, stigma barely expanded; fruit indehiscent; seeds ovoid, testa cells not elongated, thick-walled; $n = 26$. One sp., *P. fimbriolata* A. Gray; W North America; 150–2750 m alt. As with several other Montropeae, the flower can also be described as lacking a corolla, but with large and closely-associated prophylls.

All other Ericaceae

Evergreen; fungal hyphae with complex, coiled intrusions into exodermal cells of hair roots; anthers without endothecium, pollen in tetrads.

III. Subfamily Arbutoideae (Meisner) Niedenzu (1890).

Arbuteae Meisner (1839).

Bark usually flaking, the surface smooth; leaves convolute in bud, often serrulate; inflorescences eperulate; corolla urceolate, lobes short; filaments swollen at the base, anthers with paired, reflexed awns, almost smooth, with terminal pores or short slits; placentae apical, style continuous, stigma slightly lobed; fruit fleshy; testa cells moderately elongated, thick-walled; $n = 13$. Ellagic acid is common.

17. *Arbutus* L.

Arbutus L., Sp. Plant. 1: 395 (1753); P.D. Sorenson, Fl. Neotrop. 66: 194–221 (1995).

Trees or large shrubs; lamina often serrate, smooth abaxially; anthers with apical projection; ovary papillate, ovules 2–several/carpel; fruit baccate, papillate; seeds few; embryo spatulate. 10 spp., British Columbia S to Nicaragua, circum-Mediterranean; seasonal climates, s.l.–3100 m alt. Possibly paraphyletic (Hileman et al. 2001).

18. *Comarostaphylos* Zuccarini

Comarostaphylos Zuccarini, Abh. Math.-Phys. Cl. Königl. Bayer. Acad. Wiss. 2: 331 (1837); G.M. Diggs Jr., Fl. Neotrop. 66: 146–193 (1995).

Shrubs to trees; lamina often serrate, papillate abaxially; anthers lacking apical projection; ovary papillate, carpels 4–6, 1 ovule/carpel; fruit drupaceous, papillate, pyrenes not separating; embryo spatulate. 10 spp., S California to Panama; montane.

19. *Ornithostaphylos* Small

Ornithostaphylos Small, N. Am. Fl. 24: 101 (1914).

Shrub; leaves whorled, lamina entire, smooth but densely hairy abaxially; anthers with apical projection; ovary smooth, 2 ovules/carpel; fruit drupaceous; embryo spatulate; $n = ?$ One sp., *O. oppositifolia* (C. Parry) Small, S California, Baja California.

20. *Arctostaphylos* Adanson

Arctostaphylos Adanson, Fam. Pl. 2: 165, 520 (1763), nom. cons.
Xylococcus Nuttall (1889).
Arctoüs Niedenzu (1890).

Evergreen, rarely deciduous shrubs to small trees; lamina entire to serrate, smooth abaxially; anthers without apical projections; ovary smooth, carpels 5–10, ovules 1(2)/carpel; fruit drupaceous, 4–10 seeded, segments separating or not; embryo allantoid, rarely spatulate. To 60 spp., mostly W USA S to Guatemala, two spp. circumboreal, s.l. to alpine. Species limits difficult; *Arctoüs* (perulate inflorescence, deciduous) and *Xylococcus* (spatulate embryo, barely swollen filaments) often segregated (see also Hileman et al. 2001). The leaves may be held edge on.

All other Ericaceae

Hartig net absent; pericyclic fibres in stem and fibres associated with leaf bundles; anthers with pores, or slits, inverting early; style impressed.

Cassiopoideae + Ericoideae

Leaves entire(!).

IV. SUBFAMILY CASSIOPOIDEAE Kron & Judd (2002).

Cassiopeae P.F. Stevens (1971).

Leaves a variant of ericoid; flowers single, axillary; corolla campanulate, 1/3 lobed; anthers with pores and paired, reflexed awns; stigma truncate; testa cells elongated, thin-walled.

21. *Cassiope* D. Don

Cassiope D. Don, Edinburgh New Phil. J. 17: 157 (1834).

Small shrubs, pith *Calluna*-type; leaves opposite, acicular, tubular or channelled and with auricles, sessile; flowers single, axillary, with 4–6 basal bracteoles; $n = 13$. Ca. 12 spp., circumboreal, most spp. Himalayas-East Asia; s.l. to alpine.

V. SUBFAMILY ERICOIDEAE Link (1829) ("Ericeae")

Subfamily Rhododendroideae (Juss.) Sweet (1828).

Anthers lacking appendages, pollen with viscin threads; fruit a septicidal capsule.

1. TRIBE BEJARIEAE Copeland (1943).

Leaves revolute in bud; petals free, corolla rotate; anthers with resorbtion tissue; testa cells ca. 2× longer than broad, thin-walled.

22. *Bejaria* L.

Bejaria L., Mant. Plant.: 152, 242 (1771) ("*Befaria*"); S. Clemants, Fl. Neotrop. 66: 54–106 (1995).

Shrubs to trees; stomata tetracytic; flowers 5–7-merous; corolla also tubular; anthers with short apical slits, with endothecium; fruit partly loculicidal; seeds elongated; $n = ?$ 15 spp., sect. *Racemosae* Fedtsch. & Basil. one sp. in SE USA, sect. *Bejaria* Cuba, central Mexico to N Peru and the Guianas; s.l.–3900 m alt.

23. *Bryanthus* J.G. Gmelin

Bryanthus J.G. Gmelin, Fl. Sibir. 4: 132 (1769).

Prostrate shrub; leaves ericoid, inflorescences racemes, pedicels articulated; anthers with short terminal slits, resorbtion tissue?; style continuous, stigma subtruncate. One sp., *B. gmelinii* D. Don, NE Asia. The petals may be connate basally.

24. *Ledothamnus* Meisner

Ledothamnus Meisner, Fl. Bras. 7: 171 (1863); Luteyn, Fl. Neotrop. 66: 107–122 (1995), in P.E. Berry et al. (eds) Fl. Venez. Guyana 4: 745–748 (1998).

Shrubs; leaves ericoid, decussate to whorled; flowers (5)6(–9)-merous; stamens as many as petals, rarely twice as many, filaments flattened, anthers with slits their entire length; ovary strongly verrucose, 5(–7)-carpellate; stigma subtruncate; seeds irregular, testa cells little thickened. Seven spp., Guayana Highlands of S Venezuela and adjacent Brasil; 1000–2800 m alt.

2. TRIBE RHODOREAE DC. ex Duby (1828).

Diplarcheae Airy Shaw (1964).

Inflorescences perulate; median petal adaxial; anthers with collapse tissue; testa cells elongated, thin-walled. Gossypetin is present.

25. *Therorhodion* Small

Therorhodion Small, N. Am Fl. 29: 45 (1914).

Prostrate deciduous shrub; leaves revolute in bud; inflorescence elongated, borne above leaves of current flush, bracts and prophylls leafy; flowers monosymmetrical; corolla subrotate; anthers with pores; seeds?; $n = 12$. Two spp., W Alaska, NE Asia.

26. *Diplarche* J.D. Hooker & Thompson

Diplarche J.D. Hooker & Thompson, J. Bot. Kew Misc. 6: 380 (1854).

Small evergreen shrubs; pith?; bud scales?; leaves convolute in bud; inflorescences eperulate, corymbs, elongating greatly in fruit; corolla hypocrateriform; antesepalous stamens epipetalous, antepetalous stamens less so, anthers with slits their entire length, resorbtion tissue?; capsule septifragal, the outer walls separating into two layers periclinally. Two spp., E Himalayas; 3300–4500 m alt.

27. *Menziesia* J.E. Smith

Menziesia J.E. Smith, Plant. Icon. Hact. Ined. 3: t. 56 (1791).

Deciduous shrubs; leaves revolute in bud; indumentum often of flattened hairs; flowers 4- or 5-merous, tubular or urceolate, lobes short; stamens 5, 8 or 10, anthers with short slits; seeds spindle-shaped. 7 spp., temperate East Asia and North America, low alt. Viscin threads absent, but cf. Copeland (1943).

28. *Rhododendron* L.   Fig. 56

Rhododendron L., Sp. Plant. 1: 392 (1753); Sleumer, Fl. Males. I, 6: 474–668 (1967); W.R. Philipson, Notes Roy. Bot. Gard. Edinburgh 34: 1–71 (1975); Luteyn & O'Brien, Contributions towards a classification of *Rhododendron* (1980); Cullen, Notes Roy. Bot Gard. Edinburgh 39: 1–207 (1980); Chamberlain, Notes Roy. Bot. Gard. Edinburgh 39: 209–486 (1982); Davidian, The *Rhododendron* species, vol. 1 (1982), vol. 2 (1989), vol. 3 (1992); W.R. Philipson & M.N. Philipson, Notes Roy. Bot. Gard. Edinburgh 44: 1–23 (1986); Kron, Edinburgh J. Bot. 50: 249–364 (1993); Judd & Kron, Edinburgh J. Bot. 52: 1–54 (1995).
Azalea L. (1753).
Ledum L. (1753).
Tsusiophyllum Maxim. (1870).

Evergreen, less often deciduous shrubs or more rarely trees; indumentum diverse, often glandular or eglandular scales, or branched eglandular hairs; leaves convolute or revolute in bud; inflorescences rarely axillary; corolla variable in shape, shallowly lobed to polypetalous, monosymmetrical or not; stamens 5–20, anthers with pores or rarely slits; ovary 5–12-locular; capsule ca. 2× longer than broad; seeds spindle-shaped to ellipsoid or flattened, variously winged, testa cells rarely polygonal; $n = 13(-12x)$. 8 subgenera, ca. 850 spp., north temperate, esp. SE Asia, Malesia (very large section *Vireya*), two spp. in NE Australia. *R. tanakae* (Maxim.) has anthers that dehisce by slits and a 3-carpellate ovary.

3. TRIBE PHYLLODOCEAE Drude (1897).

Epigaeae Britton & Brown (1913).
Cladothamneae Copeland (1943).

Pith heterogeneous; inflorescences eperulate; pedicels ± articulated; median sepal abaxial; anthers with resorbtion tissue; style impressed; testa cells not elongated, little thickened.

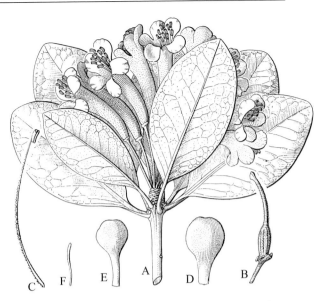

Fig. 56. Ericaceae. *Rhododendron scabribracteatum.* **A** Flowering branchlet. **B** Pistil. **C** Stamen. **D, E** Two inner perulae from outside, their scars at tip of branchlet. **F** Prophyll. Drawn by C. van Crevel. (Sleumer 1967)

29. *Elliottia* Muhlenberg

Elliottia Muhlenberg in Elliott, Bot. S. Carol. Georgia 1: 448 (1817); Bohm et al., J. Arnold Arbor. 59: 311–341 (1978).
Cladothamnus Bongard (1833).
Tripetaleia Siebold & Zuccarini (1873).

Deciduous to evergreen shrubs; multicellular hairs absent; rarely pith almost *Calluna*-type; leaves convolute in bud, thin; inflorescences panicles to racemes or cymes; flowers 3–5-merous; petals free; anthers with long slits; style curved, stigma much expanded; seeds ovoid to flattened or winged; testa cells with large pitted areas; embryo with small chalazal haustorium; $n = 11$. Four spp., SE and W North America, Japan; s.l. to tree line.

30. *Epigaea* L.

Epigaea L., Sp. Plant. 1: 395 (1753).

Procumbent shrublets; stomata tetracytic; leaves revolute in bud; calyx lobes large, green; flowers with uniseriate hairs; corolla hypocrateriform to widely campanulate; anthers with slits at least half their length; stigma expanded; capsule spherical, thin-walled, with fleshy placentae; testa cells not elongated, rather thick-walled. Three spp, SE Black Sea region, Japan, E North America; s.l. to montane.

31. *Phyllodoce* Salisbury

Phyllodoce Salisbury, Parad. Lond.: t. 36 (1805).

Small shrubs; bud scales?; leaves revolute in bud, ericoid; inflorescences corymbs; corolla urceolate or campanulate, lobed less than 1/3; anthers with short slits; stigma subcapitate; testa cells at most slightly elongated, not thickened; $n = 12$. 7 spp., north temperate; s.l. to alpine.

32. *Rhodothamnus* Reichenb.

Rhodothamnus Reichenb., Fl. Germ. Exc. 1: 417 (1830).

Small shrubs; corolla subrotate, lobed ca. 2/3; anthers with short slits or pores; stigma capitate; capsule often slightly loculicidal; testa cells not elongated; $n = 12$. Two spp., Europe, Turkey; mid elevation. Hybridizes with *Phyllodoce* (x*Phyllothamnus* Schneider).

33. *Kalmiopsis* Rehder

Kalmiopsis Rehder, J. Arnold Arbor. 13: 30 (1934).

Small shrubs; inflorescences corymbs; corolla broadly campanulate, lobed 1/3–1/2; anthers with short flaring slits; stigma slightly capitate; testa cells little elongated. One sp., *K. leachiana* (L.F. Henderson) Rehder, W USA (Oregon); montane. Doubtfully distinct from *Phyllodoce*, with which it hybridizes.

34. *Kalmia* L.

Kalmia L., Sp. Plant. 1: 391 (1753); Jaynes, Kalmia: The Laurel Book II (1988).
Leiophyllum Hedwig f., Gen. Plant.: 313 (1806).
Loiseleuria Desvaux, J. Bot. 1: 35 (1813).

Small to large, rarely deciduous shrubs; pith rarely homogeneous; leaves alternate, opposite or pseudoverticillate, revolute in bud; inflorescences often corymbs, rarely axillary; corolla broadly campanulate, with 10 pouches, or polypetalous, rotate; stamens 10(5), anthers with short to long slits; ovary 2–5-locular; style rarely continuous, stigma more or less capitate; capsule often with short loculicidal slits; testa cells moderately elongated or not, fairly thick to thin-walled; $n = 12$, 24. 11 spp., *Kalmia* (= *Loiseleuria*) *procumbens* L. ± circumboreal, the rest North America, Cuba; s.l. to above timberline.

4. TRIBE EMPETREAE Pax (1891).

Shrubs; leaves basically whorled, ericoid; plants dioecious, or rarely monoecious, or rarely flowers hermaphroditic; perianth parts free; anthers without appendages, with elongated slits, pollen lacking viscin threads; disc absent; 1 basal ovule/loculus; fruit a drupe, stones separate; $n = 13$.

35. *Empetrum* L.   Fig. 57G–N

Empetrum L., Sp. Plant. 2: 1022 (1753); Vasilyev, Rod [the genus] *Empetrum* (1961).

Prostrate shrubs; leaves also spiral, without woolly hairs in abaxial channel; flowers axillary, solitary; perianth segments 6, 2-seriate; stamens 3; ovary 6–9-locular; fruit red or black. Two spp., north

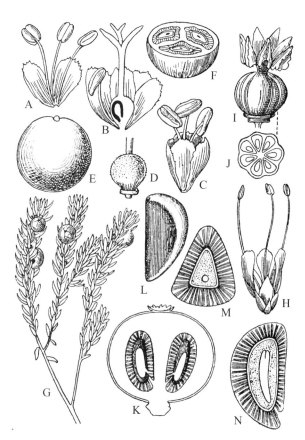

Fig. 57. Ericaceae. A–F *Corema conradi*. **A** Male flower, two perianth segments removed. **B** Female flower. **C** Male flower. **D** Ovary. **E** Fruit. **F** Same, transverse section showing pyrenes. **G–N** *Empetrum nigrum*. **G** Habit. **H** Male flower. **I** Pistil. **J** Ovary, transverse section. **K** Fruit, vertical section. **L** Pyrene. **M, N** Same, transverse and vertical section. (Schneider 1912)

temperate, extreme S South America, Falkland Islands, Tristan da Cunha; s.l. to alpine. Species limits uncertain.

36. *Corema* D. Don — Fig. 57A–F

Corema D. Don, Edinburgh New Phil. J. 2: 63 (1826).

Erect shrubs; leaf without woolly hairs; inflorescence terminal, subcapitate; perianth segments 3–4, 1-seriate; stamens 3(4); ovary 3-locular; fruit white. Two spp., E USA, SW Europe, Azores; low alt.

37. *Ceratiola* Michaux

Ceratiola Michaux, Fl. Bor.-Am. 2: 221 (1803).

Erect shrubs, leaves with woolly hairs; inflorescences axillary, fasciculate; perianth segments 4(–6), 2-seriate; stamens 2; ovary 2-carpellate; fruit red. One sp., *C. ericoides* Michaux, SE USA, low alt.

5. TRIBE ERICEAE DC. ex Duby (1829).

Salaxidae Drude (1889).
Calluneae P.F. Stevens (1971).
Daboecieae P.F. Stevens (1971).

Evergreen shrubs; leaves ericoid or very strongly revolute, rarely recurved, whorled, rarely opposite or scattered; flowers nearly always 4-merous; calyx and corolla marcescent; anthers often spurred, pores more or less elongated, pollen lacking viscin threads.

38. *Calluna* R.A. Salisbury

Calluna R.A. Salisbury, Trans. Linn. Soc. Lond. 6: 317 (1802).

Pith *Calluna*-type; leaves sessile, tailed; flowers single in leaf axils, with 3–5(+) pairs of bracteoles; sepals large, coloured; corolla tubular-capanulate, lobed halfway; capsule septifragal; testa cells barely elongated, thick-walled, with large pits; $n = 8$. One sp., *C. vulgaris* (L.) Hull, Europe, Turkey; s.l. to montane.

39. *Daboecia* D. Don

Daboecia D. Don, Edinburgh New Phil. J. 17: 150 (1834).

Pith very heterogeneous; leaves strongly revolute; inflorescences racemes terminating season's growth; prophylls absent; flowers 4–5(6)-merous; corolla urceolate; anthers lacking appendages, papillate, with short, terminal slit, resorbtion tissue present; ovary continuous; testa cells thick-walled, papillate. One or two spp., SW Europe, Azores; s.l.–moderate alt.

40. *Erica* L.

Erica L., Sp. Pl. 1: 352 (1753); Schumann & Kirsten, Ericas of South Africa (1992); E.G.H. Oliver, Contrib. Bolus Herb. 19: 1–483 (2000).
Blaeria L. (1753).
Salaxis Salisbury (1802).
Sympieza Roem. & Schultes (1818).
Scyphogyne Brongn. in Duperry (1829).
Bruckenthalia H.G.L. Reichenbach (1831).
Eremia D. Don (1834).
Thamnus Klotzsch (1834) ("1835").
Thoracosperma Klotzsch (1834) ("1835").
Philippia Klotzsch, Linnaea (1834) ("1835").
Coccosperma Klotzsch (1838).
Ericinella Klotzsch (1838).
Grisebachia Klotzsch (1838).
Acrostemon Klotzsch (1838).
Coilostigma Klotzsch, Linnaea 12: 234 (1838); E.G.H. Oliver, Bothalia 17: 163–170 (1987).
Simocheilus Klotzsch (1838).
Anomalanthus Klotzsch (1838).
Syndesmanthus Klotzsch (1838).
Platycalyx N.E. Brown (1905).
Arachnocalyx Compton (1934).
Eremiella R.H. Compton (1953).
Nagelocarpus Bullock (1954).
Stokeanthus E.G.H. Oliver (1976).

Prostrate shrubs to trees to 10(–30) m tall; pith homogeneous; bud scales sometimes absent; bracts and prophylls rarely recaulescent; sepals (1–3)4(5), small and bract-like to large and coloured; stamens (4–6)8(–10), filaments straight to S-shaped, spurs on the backs of anthers or apex of filaments, flattened to terete; pollen rarely in monads; carpels (1–)4(–8), rarely 1 ovule/carpel; style much longer than ovary or rarely almost absent, rarely continuous, slightly expanded or rarely much expanded, cup-shaped, or reflexed; stigmatic lobes none to well developed; fruit loculicidal or indehiscent, ± drupaceous; cells of testa variable, not to much elongated, thin to thick, pits, if present, small; $n = 12$, rarely 18. 860 spp., southern Africa (90%), also Europe, Middle East, south-west Arabian peninsula, Africa, Madagascar; s.l. to afroalpine. Infrageneric classification is unresolved; *Bruckenthalia* has monads, $n = 18$, but cannot otherwise be separated.

VI. Subfamily Harrimanelloideae
Kron & Judd (2002).

41. *Harrimanella* Coville

Harrimanella Coville, Proc. Wash. Acad. Sci. 3: 570 (1901).

Small shrubs; bud scales? absent; leaves acicular, petiolate; flower single, terminal, prophylls absent; corolla subcampanulate; anthers with long spurs; $n = 8$. 2 spp., interruptedly circumboreal.

Styphelioideae + Vaccinioideae

Leaves with gland-hairs on the margins; inflorescences axillary.

VII. Subfamily Styphelioideae Sweet (1828).

Subfamily Epacridoideae Link (1829).

Epidermis lignified; adaxial fibre caps of leaf veins absent, abaxial fibre caps massive, not contacting the abaxial epidermis; leaves xeromorphic, pungent, usually discolorous, secondary veins palmate; inflorescence axillary; sepals free, imbricate, persistent; stamens equal to and alternate with corolla lobes, filaments glabrous, anthers bisporangiate, with two long introrse slits.

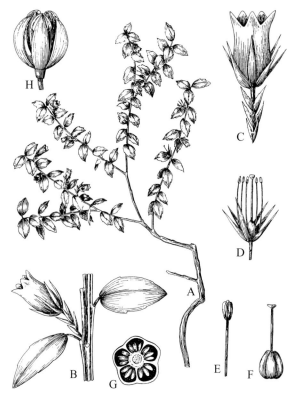

Fig. 58. Ericaceae. *Lebetanthus americanus*. **A** Flowering branch. **B** Same, detail. **C** Flower. **D** Same, petals removed. **E** Stamen. **F** Pistil. **G** Ovary, transverse section. **H** Dehisced fruit. Drawn by N. Raspini. (Dimitri 1972)

1. Tribe Prionoteae Drude (1889).

Scrambling or climbing; leaves spreading, subsessile; flowers solitary, pendulous; bracteoles several; corolla glabrous, lobes imbricate in bud, later ± spreading; anthers included.

42. *Lebetanthus* Endl.

Lebetanthus Endl., Gen. Pl. 10: 749 (1839).

Scrambling shrubs; pedicel short; corolla suburceolate; nectary scales 5; ovules ca. 10 per locule, stigma discoid. One sp., *Lebetanthus americanus* Endl., Patagonia and Tierra del Fuego.

43. *Prionotes* R. Br.

Prionotes R. Br., Prodr.: 552 (1810).

Climbing or epiphytic shrubs; leaves not crowded; pedicel long, slender; corolla cylindrical, somewhat constricted at throat; nectary scales very small; ovules many per locule, style strongly

impressed, stigma small. One sp., *P. cerinthoides* R. Br., Australia (Tas.), endemic.

All other Styphelioideae

Multicellular hairs absent; leaf teeth not glandular, midrib not evident, veins commonly evident on abaxial surface only; anthers with a single slit.

2. Tribe Archerieae Crayn & Quinn (1998).

Ovary deeply 5-lobed, style inserted between the lobes.

44. *Archeria* J.D. Hooker

Archeria J.D. Hooker, Fl. Tasmania 1: 262, t. 80, 81 (1857).

Shrubs, erect or spreading; leaves distichous or imbricate, shortly stalked or sessile; veins parallel; inflorescences short terminal racemes, or flowers solitary, axillary towards stem apex, bracteoles several, basal, caducous; corolla cylindrical or campanulate, glabrous, lobes valvate, imbricate or contorted in bud, later ± recurved, pilose or glabrous; stamens inserted towards top of tube, filaments short, anthers at throat; nectary annular disk or scales free; ovules several per locule, style included or shortly exserted, stigma capitate. Seven spp., five endemic in Australia (Tasmania), two in New Zealand.

Remaining Styphelioideae

Pedicels short.

3. TRIBE COSMELIEAE Crayn & Quinn (2002).

Stomata often cyclocytic; leaves amphistomatic, spirally twisted, crowded, imbricate, sessile, sheathing at base, scars absent, veins campylodromus; flowers single, terminal; bracteoles several, imbricate, foliose, grading upwards in size to sepals, persistent.

45. *Cosmelia* R. Br.

Cosmelia R. Br., Prodr.: 553 (1810).

Stiff, erect shrubs; lamina erect to spreading; flowers on short branches; sepals not petaloid, ciliolate, otherwise glabrous; corolla much longer than sepals, 5-lobed, red, tube cylindrical, glabrous; lobes elliptic, short, imbricate in bud, later suberect; stamens inserted near throat, filaments flat, hirsute, anthers included, linear, upper half dorsally adnate to filament; nectary annular; ovules 20–30 per locule, style long, not exserted, stigma capitate, lobed; seeds numerous, reticulate; $n = 13$. One sp., *C. rubra* R. Br., SW Western Australia.

46. *Sprengelia* J.E. Sm.

Sprengelia J.E. Sm., Kongl. Vetensk. Acad. Nya Handl. 15: 260 (1794).

Slender glabrous shrubs; flowers also crowded in "heads"; sepals with wholly hyaline margins, glabrous or ciliate; corolla tube short, or petals free or shortly cohering, glabrous, lobes imbricate to almost valvate in bud, later spreading widely, glabrous; stamens free, anthers with two slits, cohering around style, rarely hairy on outer surface, exposed when corolla lobes spread; nectary 0; ovules several per locule, style shortly exserted, stigma small, lobed or truncate; $n = 12$. Four spp., Australia, all States except W.A., N.T.; one sp. in New Zealand but possibly introduced.

47. *Andersonia* R. Br.

Andersonia R. Br., Prodr.: 553 (1810); L. Watson, Kew Bull. 16: 85–128 (1962).

Shrubs, usually slender, glabrous or hairy; flowers also solitary in axils of bracts; sepals linear to narrowly ovate, glabrous to hairy and usually ciliate, if hyaline then only below the middle, usually equal to corolla; corolla tubular to campanulate, hairy inside, tube at least half as long as lobes, lobes valvate in bud, later recurved, rarely incurved; stamens free, anthers included, or partially or fully exserted; nectary scales free, or partially to fully fused; placentation basal, ovules 1–25 per locule, style exserted or not, stigma clavate or capitate, 5-lobed; seeds small, reticulate; $n = 13$. About 35 spp., two sections, endemic in SW Australia.

4. TRIBE EPACRIDEAE Dumortier (1829) ("Epacreae").

Leaves crowded; bracts many, persistent, imbricate, grading to sepals; flowers single in upper axils; corolla lobes later spreading; ovules at least several per locule.

48. *Budawangia* I. Telford   Fig. 59

Budawangia I. Telford, Telopea 5: 231 (1992).

Decumbent, rhizomatous shrubs; leaves shortly petiolate, with 3 sub-parallel veins below; flowers "pedunculate"; bracts ciliate; corolla broad-campanulate, short, glabrous, lobes imbricate in bud, mainly glabrous; stamens inserted in throat, exserted, spreading; filaments filiform, anthers dorsally adnate to filament, oblong; nectary 0; ovules many per locule, style well-exserted, stigma small, truncate; seeds numerous, reticulate. One sp., *B. gnidioides* (Summerh.) Telford, endemic in Australia (N.S.W.).

49. *Epacris* Cav.

Epacris Cav., Icon. 4: 25, t. 344 (1797).

Shrubs, usually erect; leaves sometimes stem-clasping, also scattered, parallel-veined, commonly concolorous; sepals usually ciliate; corolla cylindical or campanulate, glabrous, lobes imbricate in bud, glabrous; stamens inserted in throat, filaments short, anthers included or shortly

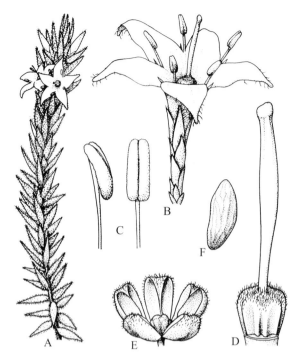

Fig. 59. Ericaceae-Styphelioideae. *Budawangia gnidioides*. **A** Flowering stem. **B** Flower. **C** Stamens. **D** Pistil. **E** Fruit. **F** Seed. Drawing by I. Telford. (Telford 1992)

exserted; nectary scales 5, occasionally united; style short to exserted, stigma small or capitate; *n* = 13. 40–50 spp., mainly Australia (all states except W.A. and N.T.), several in New Zealand and one, probably introduced, in New Caledonia.

50. *Lysinema* R. Br.

Lysinema R. Br., Prodr.: 552 (1810).

Slender shrubs; leaves also scattered, 1 vein abaxially; inflorescences spike-like; corolla slender, tubular, lobes contorted in bud, glabrous; stamens free, anthers versatile, included or barely exserted; nectary scales 5; ovules many, stigma capitate; *n* = 13. Five spp., endemic in SW Australia.

51. *Rupicola* Maiden & Betche

Rupicola Maiden & Betche, Proc. Linn. Soc. New South Wales 23: 774 (1898); Telford, Telopea 5: 234–239 (1992).

Erect or decumbent shrubs; leaves subsessile, with 3–5 sub-parallel veins; flowers "pedunculate"; bracteoles usually glabrous except ciliate margin; corolla broad campanulate, glabrous, lobes imbricate in bud, glabrous; stamens inserted at base of corolla-tube, filaments thick, anthers with one slit,

oblong, exserted, fused to filament from midpoint or below, connivent, slit short, apical; nectary absent; style well exserted, stigma small, truncate; seeds many, reticulate. Four spp., endemic in Australia (N.S.W.).

52. *Woollsia* F. Muell.

Woollsia F. Muell., Fragm. 8: 52, 55 (1873).

Erect shrubs; leaves crowded; finely parallel-veined abaxially; sepals ciliate; corolla cylindrical, lobes contorted in bud, later spreading widely, glabrous; stamens free, anthers included; nectary scales distinct, acuminate; style exserted, stigma truncate; *n* = 13. One sp., *W. pungens* (Cav.) F. Muell., endemic in Australia (Qld and N.S.W.).

5. TRIBE OLIGARRHENEAE Crayn & Quinn (2002).

Very small shrubs without lignotuber; leaves crowded, appressed, veins parallel, obscure; prophylls 2; corolla glabrous; stamens episepalous, anthers included; ovary 2-locular, 1 apical ovule per locule, style short; fruit a drupe.

53. *Needhamiella* L. Watson

Needhamiella L. Watson, Kew Bull. 18: 272 (1965).
Needhamia R. Br. (1810) non Scop. (1777).

Leaves opposite and spiral; flowers in upper axils; prophylls densely ciliate; sepals ciliate; corolla tube cylindrical, sometimes sparsely hairy outside, lobes induplicate-valvate in bud, later widely spreading, apex incurved, hairy around throat; stamens inserted in tube; nectary scales united in a disc; style short, stigma capitate, 5-lobed, below anthers; fruit a dry drupe. One sp., *N. pumilio* (R. Br.) L. Watson, SW Australia.

54. *Oligarrhena* R. Br.

Oligarrhena R. Br., Prodr.: 549 (1810).

Inflorescence a spike; prophylls thin, ciliate; flowers 4-merous; sepals thin, ciliate; corolla campanulate, glabrous, lobes valvate in bud, later spreading; stamens 2, inserted near apex of tube, 2 staminodes sometimes present; nectary scales 4, distinct; stigma minute; fruit falling with calyx and corolla as a unit. One sp., *O. micrantha* R. Br., SW Australia.

6. Tribe Richeeae Crayn & Quinn (2002).

Nodes tri- or multilacunar; pith heterogeneous; stomata paracytic; leaves crowded, bases sheathing stem, scars annular; veins parallel, with adaxial and abaxial fibre caps extending to the epidermis; inflorescences terminal.

55. *Dracophyllum* Labill.

Dracophyllum Labill., Voy. Rech. Pérouse 2: 211, t. 40 (1800).

Erect or prostrate shrubs, rarely small trees; leaves minutely serrate; inflorescences also axillary, panicles, racemes or spikes, or flowers solitary; pedicels often present, bracts usually caducous; corolla cylindrical or campanulate, lobes imbricate in bud, later spreading to recurved; stamens free or inserted on tube, included, anthers dorsifixed; nectary scales free; ovules many per locule, stigma 5-lobed; $n = 13$. About 50 spp., E Australia (Qld, N.S.W., Tas.), Lord Howe Is., New Caledonia, New Zealand.

56. *Richea* R. Br.

Richea R. Br., Prodr.: 555 (1810).

Shrubs or small trees; flowers crowded in a spike or panicle; bracts and bracteoles up to 6; sepals glabrous; corolla ovoid or conical, glabrous, almost closed at apex, splitting transversely near base and falling early; stamens 4 or 5, free, exserted, filaments persistent, anthers versatile; nectary scales large, obscure or absent; ovules many per locule; stigma capitate or small, lobed; $n = 13$ (Smith-White 1955). 11 spp., endemic in Australia (N.S.W., Vic., Tas.).

57. *Sphenotoma* (R. Br.) Sweet

Sphenotoma (R. Br.) Sweet, Fl. Australas. t. 44 (1828).

Slender erect shrubs; inflorescences dense ovate to oblong spikes; bracts persistent; sepals glabrous or hairy, ciliolate; corolla hypocrateriform, glabrous, tube shorter or longer than sepals, lobes valvate in bud, later spreading, with prominent longitudinal ridges at base constricting corolla throat; stamens inserted on tube, anthers included, dorsally attached at apex or midpoint to filament; nectary annular or of distinct scales; ovules several per locule, style included in tube, stigma 5-lobed; $n = 6, 7$. Six spp., endemic in Western Australia.

7. Tribe Stphelieae Bartling (1830).

Veins parallel; stamens epipetalous; one apical ovule per locule, style continuous; fruit a drupe.

58. *Acrotriche* R. Br.

Acrotriche R. Br., Prodr.: 547 (1810).

Erect to spreading, rarely procumbent shrubs; leaves crowded, sessile or shortly petiolate, entire or toothed, mucronate or aristate, glabrous or hirsute, veins also radiating, glaucous abaxially; inflorescences spikes, rarely cauliflorous; sepals obtuse, ciliolate; corolla tube cylindrical or campanulate, with tufts of hair spreading across throat, glabrous below, lobes valvate in bud, later spreading, with a tuft of long hairs at the thickened and commonly recurved apex; stamens inserted in throat, filaments short, anthers exserted, oblong or orbicular, versatile, lying between the corolla lobes; nectary annular; ovary 2–10-locular, style conical or cylindrical, not exserted, stigma small, lobed; mesocarp dry or pulpy; endocarp sometimes splitting into nutlets; $n = 9$. 15 spp., endemic in Australia, all states except N.T. The pollen is presented on the tuft of hairs on the corolla lobes.

59. *Androstoma* J.D. Hooker

Androstoma J.D. Hooker, Fl. Antarctica: 44 (1844).

Straggling shrubs; leaves crowded, linear, coriaceous, with callus tip, slightly discolorous, with 3 veins below, shortly petiolate; inflorescences terminal and axillary spikes of 1–3 flowers; corolla campanulate, glabrous, lobes valvate in bud, later spreading; stamens inserted in throat, filaments filiform, anthers exserted; ovary 3–4-locular, style equal to or longer than tube; mesocarp pulpy; $2n = 24$. One sp., *A. empetrifolia* J.D. Hooker, endemic in New Zealand.

60. *Astroloma* R. Br.

Astroloma R. Br., Prodr.: 538 (1810).

Low shrubs, commonly mat-like; leaves crowded, sessile to shortly petiolate, veins also radiating; flowers solitary with several bracteoles or 2 or 3 on a short peduncle; bracts ciliolate; sepals ciliate; corolla cylindrical, with tufts of hairs in throat or near base, lobes valvate in bud, later revolute, commonly bearded inside; stamens inserted in throat, filaments flat or terete, anthers included or exserted; nectary annular; style not or shortly exserted, stigma capitate, 5-lobed; n = 4, 7, 8, 16. 28 spp., endemic in Australia, in all states except N.T.

61. *Brachyloma* Sonder

Brachyloma Sonder, in J.G.C. Lehmann, Pl. Preiss. 1: 304 (1845).

Shrubs; leaves crowded, veins also radiating; flowers solitary or in small spikes; corolla tubular, with 5 tufts of reflexed hairs in throat, lobes ± imbricate in bud, later spreading, papillate adaxially; stamens inserted near top of tube, anthers not exserted; nectary annular; ovary 4–10-locular, style short, stigma lobed. Seven spp., endemic in Australia, all states except N.T.

62. *Coleanthera* Stschegl.

Coleanthera Stschegl., Bull. Soc. Imp. Nat. Moscou 32 (1): 4 (1859).

Shrubs; leaves crowded; inflorescences 1–3-flowered spikes; sepals ciliate; corolla tube short, hairy in throat, lobes valvate in bud, later strongly revolute, hairy inside; stamens inserted in throat, anthers exserted and forming a cone around the style, with two slits; nectary scales none or united in an obscure disc; stigma small. Three spp., endemic in SW Australia.

63. *Conostephium* Benth.

Conostephium Benth. in Endl., Enum. Pl. Huegel.: 76 (1837).
Conostephiopsis Stschegl., Bull. Soc. Imp. Nat. Moscou 32: 5 (1859).

Small shrubs; leaves crowded; flowers solitary, pedicellate, usually pendulous; bracteoles several; corolla tubular, conical towards apex, hairy in throat; lobes very small, valvate in bud, later not spreading; stamens inserted at base of cone of corolla, anthers included, with two slits; nectary scales 5, distinct, or absent; style exserted; stigma small; ovary 5-locular; n = 8. Six spp., endemic in south-western Australia.

64. *Croninia* J. Powell

Croninia J. Powell, Nuytsia 9: 125 (1993).

Small shrubs; leaves strongly veined; flowers solitary; bracteoles several, grading into sepals, scarious, ciliolate; sepals scarious, ciliolate; corolla tube cylindrical, pubescent inside, lobes triangu-

lar, thick, fleshy, valvate in bud, later spreading, shortly bearded inside; stamens inserted just below throat, filaments terete, anthers attached close to apex, bifurcate; nectary annular; ovary villous, style equal to or exceeding tube, hirsute, stigma truncate; mesocarp thin. One sp., *C. kingiana* (F. Muell.) J. Powell, endemic in SW Western Australia.

65. *Cyathodes* Labill.

Cyathodes Labill., Nov. Holl. Pl. Spec. 1: 57, t. 81 (1805); C.M. Weiller, Aust. Syst. Bot. 9: 491–507 (1996).

Low, diffuse to erect compact shrubs; leaves pseudoverticillate, spreading; lower surface glaucous; flowers solitary, terminal or axillary, erect, subsessile; bracteoles many, closely imbricate; corolla tube suburceolate or cupular, exceeding calyx, sparsely pubescent inside; lobes valvate in bud, later revolute, shorter than tube, sparsely pubescent; stamens inserted near top of tube, filaments thick and tapered, anthers exserted; nectary annular, truncate; ovary (5)6–10-locular, stigma small, lobed; mesocarp pulpy; n = 12 or 18. Three spp.: endemic in Australia (Tasmania).

Cyathodes dealbata R. Br. is an anomalous species from Tasmania and New Zealand. It is an alpine mat plant with small, silver-backed leaves, a narrow, cylindrical corolla with short, densely-bearded lobes, and a succulent orange drupe.

66. *Cyathopsis* Brongn. & Gris

Cyathopsis Brongn. & Gris, Bull. Soc. Bot. France 11: 66 (1864); Virot, Fl. Nouv. Caled. 6: 24–29 (1975), as *Styphelia*.

Compact shrubs; leaves crowded, subsessile, veins radiating; inflorescence a spike, crowded in upper axils, basal bracts 8–12; sepals 4; corolla tube short, glabrous inside and out, lobes twice as long as tube, valvate in bud, later spreading, hirsute inside; stamens inserted between lobes and almost as long, the anthers exserted; nectary scarcely lobed; style short, stigma small. One sp., *C. floribunda* Brongn. & Gris, New Caledonia.

67. *Decatoca* F. Muell.

Decatoca F. Muell., Trans Roy. Soc. Victoria n.s. 1: 25 (1889).

Shrubs or small trees; leaves petiolate, veins radiating; inflorescence a terminal or axillary spike; sepals 4 or 5; corolla tube cylindrical, 8–10-ribbed in lower half, hairy inside in upper two-thirds, lobes ± imbricate in bud, later spreading, glabrous inside above hairy base; stamens attached below lobes, filaments short, thick and tapered, anthers scarcely exserted; nectary 5-lobed; ovary 10-locular, style thick, tapering, stigma small; mesocarp thick, fruit with multiple pyrenes. One sp., *D. spenceri* F. Muell., Papua-New Guinea.

68. *Leptecophylla* C.M. Weiller

Leptecophylla C.M. Weiller, Muelleria 12: 196 (1999).

Erect shrubs, rarely prostrate or a tree to 6 m tall; leaves spreading or suberect, usually pungent, lower surface glaucous, striate; flowers solitary, terminal and axillary; bracteoles many, usually closely imbricate; flowers effectively unisexual, the plants dioecious; corolla campanulate or suburceolate, exceeding or about equalling calyx, glabrous or pubescent inside, lobes valvate in bud, later spreading, glabrous to densely bearded inside; stamens inserted in throat, filaments terete or filiform, anthers wholly or partly included; nectary annular, truncate, or lobed and toothed; ovary 5–7-locular, style usually continuous, short, stigma at or below level of anthers, or long and with a conspicuous bend and the stigma exserted; mesocarp pulpy; $n = 10, 12$ (Smith-White 1955, as *Cyathodes*). 13 spp., Australia, New Zealand, Papua New Guinea, and several Pacific islands.

69. *Leucopogon* R. Br.

Leucopogon R. Br., Prodr.: 541 (1810).

Shrubs or small trees; leaves rarely opposite or pseudoverticillate, petiolate or sessile, commonly crowded, entire, ciliate or denticulate; flowers solitary, multibracteolate, or in spikes; sepals commonly ciliolate; corolla tube commonly cylindrical, glabrous inside or hairy in throat, lobes valvate in bud, later spreading, commonly bearded inside; stamens inserted at top of tube, anthers enclosed or very shortly exserted, commonly with sterile tips; nectary scales 5 or united as a disc; ovary 2–5-locular, style slender, stigma capitate to truncate, strongly to slightly lobed. About 230 spp., most in Australia (especially temperate regions), a few in New Zealand, New Caledonia, SE Asia and some Pacific islands. Probably to be split; includes taxa often placed in *Styphelia* in the past.

70. *Lissanthe* R. Br.

Lissanthe R. Br., Prodr.: 540 (1810).

Erect to spreading or bushy shrubs; leaves petiolate, entire, tip mucronate or aristate; inflorescences racemes, terminal or in upper axils, flowers pedicellate; sepals ciliolate; corolla tube cylindrical, hairy inside above middle, lobes valvate in bud, later recurved, glabrous to villous inside; stamens inserted at throat, filaments short, filiform, anthers half exserted or obvious at throat; nectary annular; ovary 3–9-locular, style cylindrical, not exserted, stigma lobed; $n = 7, 14$. Six spp., endemic in Australia, all states except W.A. and N.T.

71. *Melichrus* R. Br.

Melichrus R. Br., Prodr.: 539 (1810).

Prostrate to erect shrubs; leaves crowded towards stem apex; flowers solitary, ± sessile, commonly below current season's growth; bracteoles several, ciliolate; sepals ciliate; corolla urceolate or almost rotate, 5 tufts of hairs or a ring of glandular hairs near base, lobes valvate in bud, later spreading, sparsely hairy inside; stamens inserted near base, filaments short, flat, anthers usually included; nectary annular; ovary 4–6-locular, style short, thick, stigma small; $n = 8$. Four spp., endemic in Australia (Qld, N.S.W., Victoria).

72. *Monotoca* R. Br.

Monotoca R. Br., Prodr.: 546 (1810).

Shrubs or small trees; leaves petiolate, commonly glaucous abaxially; inflorescences (terminal) spikes or racemes; flowers bisexual or unisexual; sepals 4 or 5, ciliate; corolla campanulate; lobes valvate in bud, later recurved, glabrous or papillose; stamens inserted at top of corolla tube, filaments short, anthers enclosed or half-exserted; nectary scales distinct or united in a disk; ovary 1- or 2-locular, style short, stigma small, lobed; $n = 12$. 17 spp., endemic in Australia, all states except N.T.

73. *Pentachondra* R. Br.

Pentachondra R. Br., Prodr.: 549 (1810).

Small shrubs, sometimes prostrate; leaves sessile or shortly petiolate, crowded; spike (2)3–5(–7)-flowered, or flowers solitary, terminal and in upper axils, multibracteolate, bisexual or pistillate; corolla cylindrical, sparsely hairy inside, lobes valvate in bud, later recurved, bearded adaxially; stamens inserted in throat, anthers almost sessile, enclosed or ± exserted; nectary scales distinct or united in a disk; ovary 5–11-locular, style included or exserted, stigma lobed; fruit with multiple pyrenes; $n = 14$. Four or five spp., Australia (N.S.W., Vic., Tas.) and New Zealand.

74. *Planocarpa* C.M. Weiller

Planocarpa C.M. Weiller, Austral. Syst. Bot. 9: 510 (1996).

Small shrubs; leaves crowded towards tips of innovations, erect to spreading, oblong to ovate, flat or slightly convex, the margin sometimes recurved; plants hermaphrodite or gynodioecious; flowers erect, solitary and multibracteolate, or in 1–3-flowered spikes near branch tips, on short, erect peduncle; corolla tube campanulate or cylindrical, glabrous or puberulent inside, lobes valvate in bud, later spreading, glabrous or densely bearded, shorter than tube; stamens inserted in throat, anthers included; nectary annular, glabrous; style short, stigma small; ovary 5–8-locular; fruit depressed globose, mesocarp thick, pulpy; $n = 9$. Three spp., endemic in Australia (Tasmania). The perulae of leaf buds persist, becoming grey.

75. *Styphelia* J.E. Sm.

Styphelia J.E. Sm., Spec. Bot. New Holland: 45 (1795).

Shrubs, occasionally prostrate, with or without lignotuber; leaves crowded; flowers pedicellate, solitary, multibracteolate, or rarely a 1–3-flowered raceme; corolla tubular, hairy in throat and at base, lobes valvate in bud, later revolute, hairy inside; stamens inserted at base, anthers exserted, versatile; nectary scales distinct or united in a shallow cup; stigma small, exserted; $n = 4$. 15 spp., endemic in southern Australia.

76. *Trochocarpa* R. Br.

Trochocarpa R. Br., Prodr. 548 (1810).

Shrubs or small trees; leaves crowded, shortly petiolate; inflorescences terminal and axillary spikes; flowers unisexual or bisexual; sepals ciliate; corolla cylindrical or campanulate, with reflexed hairs in throat, lobes ± imbricate in bud, later recurved, glabrous or sparsely hairy adaxially; stamens inserted in throat, filaments short, thick

and tapered, anthers shortly exserted; nectary annular; ovary 8–11-locular, style long, rarely very short, stigma lobed; fruit with multiple pyrenes; $n = 10$. About 12 spp., Malesia (Papua-New Guinea, Borneo, Celebes) and Australia (all states except N.T.).

VIII. Subfamily Vaccinioideae Arnott (1832).

Apical bud of vegetative shoot aborting; stigma truncate; $n = 12$.

1. Tribe Oxydendreae Cox (1948).

Pith *Calluna*-type; stomata paracytic; anthers lacking appendages, with slits half their length.

77. *Oxydendrum* DC.

Oxydendrum DC., Prodr. 7: 601 (1839).

Deciduous trees; hairs not glandular; inflorescences terminating the flush, panicles; corolla subtubular; filaments stout, anthers ?with disintegration tissue dorsally; seeds elongated, testa cells elongated, moderately thick walled. One sp., *O. arboreum* (L.) DC., SE USA; low alt.

2. Tribe Lyonieae Kron & Judd (1999).

Epidermis lignified; filaments geniculate.

78. *Craibiodendron* W.W. Smith

Craibiodendron W.W. Smith, Rec. Bot. Surv. India 4: 276 (1911); Judd, J. Arnold Arbor. 67: 441–469 (1986).

Trees; leaves entire; inflorescences racemes or panicles; corolla urceolate to campanulate, thick; anthers lacking appendages; seeds with wing on one side, testa cells much elongated, thin-walled; $n = ?$ Five spp., SE Asia, 200–3200 m alt.

79. *Pieris* D. Don.

Pieris D. Don, Edinburgh New Phil. J. 17: 159 (1834); Judd, J. Arnold Arbor. 63: 103–144 (1982).
Arcterica Coville (1901).

Shrubs, rarely vines; pith heterogeneous, rarely homogeneous; leaves rarely whorled, serrate, rarely entire; inflorescence terminal or axillary panicles or racemes; filaments expanded at the base, rarely straight, anther spurs at the filament junction; disintegration tissue rarely absent; seeds angular, rarely spindle-shaped, testa cells not to much elongated, thin- to moderately thick-walled. 7 spp., subgenus *Arcterica* (Coville) Judd monotypic; subgenus *Pieris* with two sections: East Asia, E USA, Cuba; low to moderate alt.

80. *Agarista* G. Don

Agarista G. Don, Gen. Syst. 3: 837 (1834); Judd, J. Arnold Arbor. 65: 255–352 (1984), Fl. Neotrop. 66: 295–344 (1995).
Agauria (DC.) Bentham & J.D. Hooker, Gen. Plant. 2: 586 (1876); Sleumer, Bot. Jahrb. 69: 374–394 (1938).

Shrubs; pith *Calluna*-type (± heterogeneous); leaves revolute, rarely convolute in bud, entire, rarely serrate; inflorescences racemes, rarely terminal, rarely panicles; corolla urceolate to tubular; anthers lacking appendages; style slightly (rarely not) expanded near apex; seeds spindle-shaped, testa cells much elongated, thin-walled; $n = 12$. 30 spp., SE USA (one sp.) to South America, esp. E Brasil, Africa (scattered), Madagascar, and Réunion; s.l.–3400 m alt. Only one sp. with several variants reported (Sleumer) in the Old World, but variation in Madagascar is very great.

81. *Lyonia* Nuttall   Fig. 60

Lyonia Nuttall, Gen. N. Am. Pl. 1: 266 (1818), nom cons.; Judd, J. Arnold Arbor. 62: 63–209, 315–436 (1981), Fl. Neotrop. 66: 222–294 (1995).

Evergreen to deciduous shrubs, rarely trees; pith heterogeneous or homogeneous; leaves entire to serrate; inflorescences racemes to fascicles, rarely panicles; flower 4–7(8)-merous; anthers with spurs on filament, rarely none; fruit with pale, thickened sutures down lines of dehiscence; seeds spindle-shaped, testa cells much elongated, thin-walled; $n = 12$. 35 spp., four sections: SE Asia (sect. *Pieridopsis* (Rehder) Airy Shaw) and SE USA (sects. *Arsenococcus* (Small) Judd, *Lyonia* and *Maria* (DC.) C.E. Wood) to Mexico and the Greater Antilles; low alt. to montane.

3. Tribe Andromedeae Klotzsch (1838).

No multicellular hairs except on leaf margin; pith heterogeneous; anthers with awns; testa several-layered, cells not elongated, all walls moderately thickened.

Fig. 60. Ericaceae. *Lyonia ovalifolia.* **A** Habit. **B** Flower. **C** Pistil. **D** Ovary. **E** vertical section of ovary. **F** Transverse section of ovary. **G** Stamens. Drawn by R. van Crevel. (Sleumer 1697)

82. *Andromeda* L.

Andromeda L., Sp. Plant 1: 393 (1753).

Shrub; leaves revolute in bud, entire; inflorescences terminating the flush, subcorymbose; anthers with one pair of suberect awns, disintegration tissue absent; seeds angular. One to two spp., north temperate.

83. *Zenobia* D. Don

Zenobia D. Don, Edinburgh New Phil. J. 17: 158 (1834).

Deciduous shrub; leaves serrulate; inflorescences fasciculate; anthers with two pairs of erect awns; $n = 33$. One sp., *Z. pulverulenta* (Willd.) Pollard, SE USA; low alt.

4. Tribe Gaultherieae Niedenzu (1889).

Stomata usually paracytic; corolla often urceolate; filaments swollen, anthers with short slits, with disintegration tissue; $n = ?$

84. *Chamaedaphne* Moench

Chamaedaphne Moench, Meth. 457 (1794).

Subevergreen shrubs; pith heterogeneous; indumentum lepidote; leaves serrate?; inflorescences terminating the flush, racemes or panicles, bracts foliaceous; filaments flattened, anthers lacking appendages, disintegration tissue?, tubules ca. 1/2 their length; testa cells ca. 2× longer than broad, rather thick-walled; $n = ?$. One sp., *C. calyculata* (L.) Moench, circumboreal, low to medium alt.

85. *Leucothoë* D. Don

Leucothoë D. Don, Edinburgh New Phil. J. 17: 159 (1834).

Evergreen, rarely deciduous shrubs, rarely small trees; pith heterogeneous or almost *Calluna*-type; leaves ± serrulate; inflorescences racemes; anthers with 1 or 2 pairs of erect awns, or muticous; seeds rounded or winged by fringe of enlarged cells, testa cells little elongated or thickened; $n = 11$. 6 spp., rather diverse (five sections!), Southeast Asia, SE USA; low alt.

86. *Eubotrys* Nutt.

Eubotrys Nutt., Trans. Am. Phil. Soc. 8: 269 (1842).

Deciduous shrubs; pith ± homogeneous; leaves serrulate; inflorescences racemes; prophylls apical; anthers with 1 or 2 pairs of erect awns; seeds rounded or winged by fringe of enlarged cells, testa cells little elongated or thickened; $n = 11$. 2 spp., SE USA; low alt.

87. *Gaultheria* L.

Gaultheria Kalm ex L., Gen. Pl. ed. 5, 187 (1754); Sleumer, Fl. Males. I, 6: 677–696 (1967); Middleton & Wilcock, Edinburgh J. Bot. 47: 291–301 (1990); Middleton, Bot. J. Linn. Soc. 106: 229–258 (1991); Luteyn, Fl. Neotrop. 66: 384–488 (1995).
Pernettya Gaudichaud, Ann. Sci. Nat. 5: 102 (1825) (as *Pernettia*); Luteyn, Fl. Neotrop. 66: 365–383 (1995).
Chiogenes Salisbury (1817).

Shrubs, sometimes procumbent; pith homogeneous to heterogeneous; leaves serrate or not; flowers in racemes, rarely terminal, or rarely panicles, or rarely flowers solitary and multibracteolate; corolla rarely tubular or campanulate; anthers with (1)2 pairs of awns, these sometimes muticous, tubules rarely present, rarely with slits almost the length of the anther; capsule surrounded by fleshy calyx, or rarely a berry with persistent calyx, or not fleshy; seeds angular, testa cells not elongated, all walls moderately thick to thick; $n = 11$ (most common), 12, 13, polyploidy on base

11 and 12. 130 spp., circum-Pacific, extending to the Himalayas, E USA and E Brasil; s.l. to subalpine. Sometimes smelling of wintergreen.

88. *Diplycosia* Bl.   Fig. 61

Diplycosia Bl., Bijd.: 857 (1856); Sleumer, Fl. Males. I, 6: 696–740 (1967).
Pernettyopsis King & Gamble, J. Asiat. Soc. Bengal 74: 79 (1905); Sleumer, Fl. Males. I, 6: 675–677 (1967).

Shrubs, sometimes the roots or stem bases swollen, often with ± setose indumentum; leaf margins entire, rarely serrulate; inflorescences fascicles, rarely 1-flowered; flowers rarely 4-

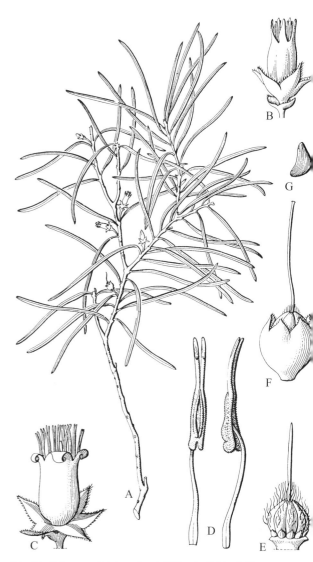

Fig. 61. Ericaceae. *Diplycosia pinifolia.* **A** Habit. **B** Flower at beginning of anthesis. **C** Flower at full anthesis. **D** Stamens in front and side view. **E** Pistil and disk. **F** Young fruit. **G** Seed. Drawn by R. van Crevel. (Sleumer 1967)

merous; anthers lacking appendages, with tubules ca. 1/2 their length, rarely with pores; capsule surrounded by fleshy calyx, rarely a berry; seeds angular, testa cells elongated, somewhat thickened; $n = 18$. Ca. 100 spp., but poorly known, Malesia; often at moderate altitudes. Sometimes smelling of wintergreen; mesophyll with fibres.

89. *Tepuia* Camp

Tepuia Camp in Gleason & Killipp, Brittonia 3: 178, figs 4–5 (1939); Luteyn, Fl. Neotrop. 66: 351–364 (1995), in Fl. Venez. Guyana 4: 760–763 (1998).

Shrubs; lamina often strongly recurved, entire, with marginal glands at or near base; inflorescences racemes, rarely panicles; anthers lacking appendages, tubules ca. 1/2 their length; fruit baccate, rarely with persistent calyx; seeds angular, testa cells isodiametric. 7 spp., Guayana Highlands of SE Venezuela; 1500–2500 m alt. Sometimes smelling of wintergreen.

5. TRIBE VACCINIEAE Reichenbach (1831).

Thibaudieae Bentham & J.D. Hooker (1876).

Stomata paracytic; leaves usually coriaceous, pli-nerved; inflorescences axillary; pedicel articulated; calyx and corolla usually valvate, calyx limb as long or shorter than tube; stamens about equalling corolla, isomorphic or alternately slightly dimorphic, anthers with tubules, rarely with spurs; ovary inferior; fruit fleshy, many-seeded; testa hard; embryo white. Generic limits in Vaccinieae are particularly unsatisfactory.

90. *Vaccinium* L.　　　　　　Fig. 62

Vaccinium L., Sp. Pl.: 349 (1753); Sleumer, Bot. Jahrb. Syst. 71: 423–493 (1941); Vander Kloet, The genus *Vaccinium* in North America (1988), Pacific Sci. 47: 76–85 (1993); Vander Kloet & Dickinson, Bot. Mag. Tokyo 105: 601–614 (1992).
Oxycoccus J. Hill (1756).

Shrubs, rarely trees, lianes, or rhizomatous; leaves rarely thin, deciduous, penni-nerved; inflorescences racemes, rarely terminal, or rarely fascicles, or flowers single, axillary; flowers (4)5(6)-merous; rarely pedicel continuous; calyx limb erect, rarely spreading, often shorter than tube, lobes variable; corolla <1.5 cm long, cylindric, urceolate, or campanulate, lobes imbricate, rarely valvate, rarely parted to base; filaments short to long, anthers with or without spurs, disintegration tissue rarely

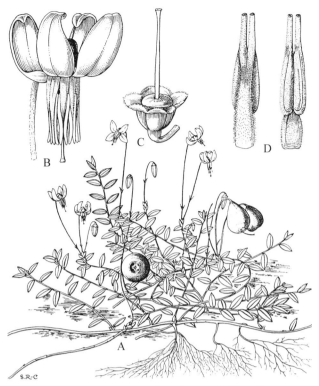

Fig. 62. *Vaccinium oxycoccus.* **A** Flowering and fruiting plants. **B** Flower. **C** Calyx and gynoecium. **D** Stamens, left back view, right front view. (Ross-Craig 1979)

present, thecae smooth or papillate, tubules ± long and slender, with terminal pores, rarely with slits; ovary 4–5(rarely falsely 10)-locular; fruit purple to black, rarely red or white; $n = 12, 24$. Ca. 140 spp., arctic and north temperate (few spp.), Central and South America, SE Asia, Pacific, Africa and Madagascar. Rarely epiphytic. Wildly paraphyletic, limits unclear.

91. *Symphysia* K.B. Presl

Symphysia K.B. Presl, Epist. Symphysia (1827); Vander Kloet, Taxon 34: 440–447 (1985).
Hornemannia Vahl, Skrift. Naturh.-Selsk. Kiøben. 6: 120 (1827), non Willd. (1809).

Flowers (5)6–8-merous; inflorescence racemes; pedicels expanded at the apex; calyx limb shorter than the massive tube, erect, lobes minute; corolla <1 cm long, flaring tubular, thick, sub-bistratose; stamens shorter than corolla, filaments long, anther thecae papillate, tubules shorter, narrower, with oblique pores; fruit black. $n = 24?$. One sp., *S. racemosa* (Vahl) Stearn, N Caribbean, Jamaica and E Cuba to Martinique.

92. *Agapetes* G. Don

Agapetes G. Don, Gen. Hist. 3: 862 (1834); Sleumer, Bot. Jahrb. Syst. 71: 423–493 (1941) (*Vaccinium*, in part), Fl. Males. I, 6: 746–878 (1967) (*Vaccinium*, excl. sect. *Pachyantha*)

Shrubs, rarely trees, trailing vines, sometimes with lignotubers; phellogen superficial; leaves rarely penni-nerved; inflorescences racemes, rarely fascicles; flowers (4)5-merous, pedicels articulated, rarely continuous, calyx limb rarely longer than tube, erect, rarely spreading, lobes variable; corolla <5 cm long, cylindric or urceolate, rarely campanulate, lobes imbricate (to 1/3 tube); filaments short to long, anthers with or without spurs, rarely disintegration tissue present, thecae papillate, tubules short and broad to long and slender, with terminal pores or slits; ovary falsely 10-locular; fruit purple to black, orange; testa hard or mucilaginous; embryo white or green; $n = 12, 24$. Ca. 400 spp., SE Asia and Malesia, montane, perhaps to New Caledonia. Sometimes epiphytic or even epilithic. Most Indo-Malesian species of *Vaccinium* are provisionally included here (see Kron et al. 2002b: the *Agapetes* and *Bracteata-Oarianthe* clades).

93. *Didonica* Luteyn & Wilbur

Didonica Luteyn & Wilbur, Brittonia 29: 255 (1977); Luteyn, Syst. Bot. 16: 587–597 (1991).

Epiphytic shrubs, leaves flat to involute; inflorescences fascicles or racemes; flowers 5–6-merous; calyx limb usually campanulate; corolla 0.7–2.2 cm long, thin or succulent, cylindric-campanulate; filaments short, anthers finely papillate, incurved at the base, tubules slightly shorter, as wide, rigid, somewhat spreading, with short, oval, latrorse slits which do not extend to the actual tip; fruit ?white; seed? Four spp., Costa Rica and Panama; 500–1400 m. alt.

94. *Lateropora* A.C. Smith

Lateropora A.C. Smith, Contrib. U.S. Natl. Herb. 28: 333 (1932); Wilbur & Luteyn, Ann. Miss. Bot. Gard. 68: 162–163 (1981).

Terrestrial or epiphytic shrubs; inflorescences racemes, often umbelliform; pedicel short, stout; calyx limb erect; corolla <1 cm long, thick, broadly urceolate or squatly campanulate; filaments short, weakly adherent to the base of corolla, thecae papillate, strongly incurved at base with the lower 1/3 protruding inwardly or even turning upward, tubules short, with lateral slits almost the entire length of the anther; fruit blue-black; seed? Three spp., Costa Rica and Panama; medium alt.

95. *Notopora* J.D. Hooker

Notopora J.D. Hooker, Icon. Pl. 12: 53, pl. 1159 (1876); Steyermark, Acta Bot. Venez. 2: 288–298 (1967); Luteyn, Fl. Venez. Guyana 4: 750–752 (1998).

Terrestrial shrubs; leaves congested, pinniveined, margins often strongly recurved; inflorescences fascicles or racemes; calyx limb slightly longer than tube, erect, lobed halfway; corolla 2–4 cm long, carnose, cylindric to urceolate-subcylindric; filaments longer than anthers, weakly adherent to base of corolla, thecae papillate, tubules slightly narrower and shorter, with wide, extrorse clefts; fruit unknown. Five spp: Guayana highland region of Venezuela and adjacent Guyana; 400–2500 m alt.

96. *Utleya* Wilbur & Luteyn

Utleya Wilbur & Luteyn, Brittonia 29: 267 (1977).

Epiphytic shrub; leaves pinniveined; inflorescences short racemes; calyx tube strongly 5-winged to sinuses, limb erect; corolla <1 cm long, thick, urceolate, conspicuously 5-spurred opposite lobes; filaments short, thecae smooth, curved inwards at the base, tubules as long and wide, flexible, with elongate, oval slits; fruit unknown. One sp., *U. costaricensis* Wilbur & Luteyn, endemic to Costa Rica; medium alt.

97. *Gonocalyx* Planchon & Lindley

Gonocalyx Planchon & Lindley, Gard. Chron. 1856: 152 (1856); Luteyn, Syst. Bot. 15: 747 (1990), key.

Small epiphytic, rarely terrestrial shrubs; leaf margin flat to revolute; inflorescences one- to few-flowered fascicles or loose racemes; flowers (4)5-merous; calyx limb spreading, the tube rarely narrowly winged; corolla (0.6–)1 cm long, tubular or cylindric-campanulate to globose-urceolate, bistratose at the apex; filaments short, free or connate, thecae papillate, tubules ca. 3× as long, narrower, with small, subterminal, oblique pores or short slits; fruit (red) purple to black; $n = 23/24$. 9 spp., Costa Rica to N Colombia, Caribbean; low to medium alt.

98. *Dimorphanthera* F. Muell.   Fig. 63

Dimorphanthera F. Muell., Wing's South. Sci. Rec. n.s. 2 (1886); Sleumer, Fl. Males. I, 6: 747–753 (as *Vaccinium* sect. *Pachyanthum*), & 885–914 (1967); P.F. Stevens, Contrib. Herb. Austral. 8: 1–34 (1974).

Trees to shrubs or lianes, rarely epiphytes; stem base rarely swollen; inflorescences racemes to fascicles; pedicel rarely continuous; calyx tube rarely winged, limb to as long, more or less erect, rarely deeply lobed; corolla (0.5–)1.2–4.5 cm long, thick, tubular or campanulate, ± bistratose lobes rarely to 1/3; stamens ca. 1/2 the corolla length, filaments short, anthers spurred or not, usually woody, dimorphic, thecae papillate or not, tubules as wide, ca. as long, broadened and flaring in larger anthers, erect in smaller anthers, with elongate pores, or rarely smaller anthers with connate tubules and single pore; fruit purplish black; $n = 24, 36$. 80 spp., most New Guinea (also Philippines, Moluccas, New Britain and New Ireland); (s.l.–)750–4000 m alt. Section *Pachyantha* (6 spp.) probably in the *Paphia* lineage.

99. *Paphia* Seemann

Paphia Seemann, J. Bot. 2: 77 (1864); Sleumer, Fl. Males. I, 8: 878–885 (1967), as *Agapetes*; P.F. Stevens, Edinburgh J. Bot. 60 (2003), in press.

Small tree to rhizomatous shrublet, lianes, rarely epiphytes; leaves rarely penni-nerved; inflorescences fascicles; pedicel rarely continuous; calyx tube rarely shortly winged, limb to as long, erect to subspreading; corolla 1.3 cm long, thick, tubular, terete to angled; filaments much shorter to longer than anthers, thecae slightly papillate, tubules shorter to 2.5× as long, broad to narrow, with elongated slits; fruit blackish; $n = 36$. 20 spp., E New Guinea (16 spp.), Bougainville, Australia (Queensland), New Caledonia and Fiji; 1000–4000 m alt.

100. "*Agapetes scortechinii*"

Climbing (epiphytic) shrub; inflorescences fascicles; pedicel continuous; calyx tube winged, limb narrower, longer, erect, lobes long; corolla ca. 2 cm long, more or less tubular; stamens rather shorter than corolla, filaments as long as anthers, thecae papillate, tubules as long, wide, with much elongated slits; fruit colour?; $n = 12$. *A. scortechinii* (King & Gamble) Sleumer, Malay Peninsula; 1200–2100 m alt.

Fig. 63. Ericaceae. *Dimorphanthera amblyornidis*. **A** Habit. **B** Inflorescence. **C** Ovary, disk and style. **D** Major stamens, front and back view. **E** Minor stamens, front and back view. Drawn by R. van Crevel. (Sleumer 1967)

101. *Cavendishia* Lindley, nom. cons.

Cavendishia Lindley, Bot. Reg. 21: pl. 1791 (1835); Luteyn, Fl. Neotrop. Monogr. 35: 1–290 (1983).

Epiphytic or terrestrial shrubs; leaves penni- or pli-nerved; inflorescences racemes or fascicles, rarely flowers solitary; bracts usually large, showy, persistent; calyx limb erect or spreading, lobed to halfway; corolla (0.6–)1–4.6 cm long, usually thick, tubular; stamens rarely shorter than corolla, dimorphic, filaments short, thecae smooth or slightly papillate, tubules twice as long, ca. as wide, with elongate slits; fruit dark blue-black; $n = 24/$ca. 26. About 130 spp., Mexico (Oaxaca) to Bolivia, E to Brazil (Amapá); lowland to high alt.

102. *Orthaea* Klotzsch

Orthaea Klotzsch, Linnaea 24: 23 (1851); Luteyn, Nordic J. Bot. 7: 31–37 (1987).
Lysiclesia A.C. Smith, Contrib. U.S. Natl. Herb. 28: 517 (1932).
Empedoclesia Sleumer, Notizbl. Bot. Gart. Mus. Berlin-Dahlem 12: 124 (1934).

Epiphytic shrubs; leaves rarely subcoriaceous, penni-nerved; inflorescences racemes to fascicles, rarely flowers solitary; bracts rarely large, showy and caducous; calyx tube 3-winged, limb ca. 2× as long, erect to suberect, lobes rarely greatly enlarged; corolla >1 cm long, tubular to subcylindric; stamens usually ca. 1/3 the corolla length, filaments free or connate, long, dimorphic, anthers equal, rarely slightly dimorphic; disintegration tissue rarely present, thecae smooth, tubules shorter, ca. as wide, with terminal or slightly oblique pores; fruit blue-black. 35 spp., subgen. *Orthaea* and *Lysiclesia* (A.C. Smith) Luteyn, Mexico to Bolivia, E to Guyana and Trinidad. *O. crinita* A.C. Smith may have minute spurs on the filaments.

103. *Macleania* W.J. Hooker

Macleania W.J. Hooker, Icon. Pl. 2: pl. 109 (1837); Yeo, Baileya 15: 45–59 (1967); Luteyn, BioLlania 6: 455–465 (1997).

Epiphytic or terrestrial shrubs; leaves pinni- or pli-nerved; inflorescences racemes, rarely terminal, or fascicles, rarely flowers solitary; pedicel rarely continuous, often angled to conspicuously winged, wings rarely protruding beyond the calyx limb as a spur; calyx erect to spreading, lobes (3–4)5; corolla >1 cm long, rarely shorter, subcylindric or elongate-urceolate, bistratose at apex; stamens ca. 1/4 to nearly equalling corolla, filaments free or connate, anther thecae strongly papillate, tubules ca. as long, somewhat narrower, completely connate or not, rigid, broadly conical, with elongate, distinct or fused clefts; fruit blue-black or white; seeds often mucilaginous; embryo sometimes green; $n = 48$. 40 spp., S Mexico to Peru; most diverse in W Ecuador; s.l. to high alt.

104. *Psammisia* Klotzsch

Psammisia Klotzsch, Linnaea 24: 42 (1851); Luteyn, Opera Bot. 92: 120–121 (1987), spp. with globose corollas.

Epiphytic or terrestrial shrubs; leaves pinni- or pli-nerved; inflorescences, ± fascicles to racemes, rarely terminal, rarely flowers solitary; calyx limb equal or ca. 2× as long as tube, erect or spreading, rarely winged, lobes (2–4)5; corolla 0.5–4 cm long, thick, subcylindric, elongate-urceolate or subglobose, bistratose at apex; stamens (8)10(–12), from ca. 1/3 to often nearly equalling corolla, filaments free or connate, thecae all or alternately 2-spurred, rarely spurs lacking, strongly papillate, tubules 1/4 to as long, narrower, with elongate clefts; fruit usually green at maturity; testa mucilaginous; embryo green. 60 spp., Costa Rica to Bolivia, E to French Guiana and Trinidad; lowland rainforest to high alt.

105. *Satyria* Klotzsch

Satyria Klotzsch, Linnaea 24: 14 (1851).

Epiphytic or terrestrial shrubs; inflorescences racemes to fascicles, often ramiflorous; calyx limb flaring or spreading, rarely longer then tube, lobes (3–4)5; corolla (0.7–)1 cm long, cylindric to slightly flaring, or rarely globose-urceolate, rarely lobed ca. 1/3; stamens up to 1/3 corolla length, filaments short, connate, anthers dimorphic, thecae ± woody, slightly papillate to smooth, tubules ca. as long, as wide or wider, longer anthers: tubules often flaring distally and with incurved and ornately decorated tips, with latrorse pores, shorter anthers: tubules usually not flaring distally, with introrse pores; fruit dark blue-black; $n = 24$. 25 spp., S Mexico to Bolivia, E to French Guiana; lowland rainforest to cloud forest.

106. *Mycerinus* A.C. Smith

Mycerinus A.C. Smith, Bull. Torrey Bot. Club 58: 441 (1931); Luteyn, Fl. Venez. Guyana 4: 749–750 (1998).

Terrestrial or epiphytic shrub; leaves thick-coriaceous, brittle, the margin flat or strongly recurved and with large glands at lamina base; inflorescences ± fascicles; prophylls apical; calyx limb dilated, narrowly winged to the tips of lobes, lobes long; corolla 1–2 cm long, carnose, cylindric to campanulate-subinfundibuliform; stamens shorter than corolla, filaments short, stout, anther thecae finely papillate, connective continuing to the apex of tubules, tubules equaling thecae, membranous, with oval clefts more than half their length, style stout; fruit reddish purple; seed? 3 spp., Guayana Highland region of Venezuela; 1300–2700 m alt. Related to *Macleania* and *Psammisia*.

107. *Polyclita* A.C. Smith

Polyclita A.C. Smith, Bull Torrey Bot. Club 63: 314 (1936).

Terrestrial or epiphytic shrub; leaves penni-nerved; inflorescence short racemes; pedicel continuous; calyx tube angled or narrowly winged opposite the lobes, limb erect, lobed ca. halfway; corolla ca. 1.5 cm long, thick, tubular, bistratose at the apex; filaments short, anthers shortly spurred or not, thecae slightly papillate, rigid, tubules slightly shorter, ca. as wide, flexible, with elongated slits; fruit blackish; seed? One sp., *P. turbinata* (O. Kuntze) A.C. Smith, Bolivia; mid alt.

108. *Anthopteropsis* A.C. Smith

Anthopteropsis A.C. Smith, Ann. Miss. Bot. Gard. 28: 441 (1941).

Epiphytic shrub; leaves subcoriaceous; inflorescences terminal, rarely axillary racemes; calyx tube conspicuously and broadly 5-winged to the sinuses, limb rather longer, erect, apiculate; corolla 1.5–2 cm long, thick, cylindric, inconspicuously 5-angled; stamens ca. 1/2 corolla length, filaments short, anthers alternately and inconspicuously spurred, thecae slightly to coarsely papillate, basally shortly mucronate, tubules shorter, little narrower, flexible, with slits their entire length; fruit ?red; seed? One sp., *A. insignis* A.C. Smith, central Panama; medium alt. Close to *Macleania* and *Psammisia*.

109. *Ceratostema* Jussieu

Ceratostema Jussieu, Gen. Pl.: 163 (1789); Luteyn, J. Arnold Arbor. 67: 485–492 (1986); Luteyn, Fl. Ecuad. 54: 210–253 (1998).
Periclesia A.C. Smith, Contrib. U.S. Natl. Herb. 28: 357 (1932).

Terrestrial or epiphytic, rarely scandent shrubs, lignotubers common; bud scales rarely subulate; leaves penni- or pli-nerved; inflorescences fascicles to racemes, rarely flowers solitary; flowers (4)5(6)-merous, bracts small to large and showy, pedicel rarely continuous; calyx tube rarely winged, limb to 4× as long, suberect or spreading, lobes short to long, often distinctly veined; corolla thick, >1 cm long, rarely less, subcylindric, often ventricose at base, bluntly angled, rarely terete, lobed to halfway; filaments short, distinct or connate, slightly dimorphic or not, anthers often slightly dimorphic, thecae usually strongly papillate, tubules much narrower, 2–5× longer or rarely shorter, flexible, separate to connate nearly to apex, with short, oblique clefts or nearly terminal,

sometimes flaring pores; fruit dark blue-black, rarely translucent white; testa mucilaginous; embryo green. 32 spp., Venezuela to N Peru, 26 spp. endemic to eastern (Oriente) Ecuador; (600–)800–3700(–1950) m alt.

110. *Semiramisia* Klotzsch

Semiramisia Klotzsch, Linnaea 24: 25 (1851); Luteyn, Syst. Bot. 9: 359–367 (1984).

Terrestrial or epiphytic shrubs, rarely lianoid; leaves coricaeous, pli-nerved; inflorescences short racemes or 1-flowered; pedicel continuous; calyx tube angled or winged, limb to 2× longer, erect to spreading, lobed to halfway; corolla >1 cm long, thick, tubular, bistratose; stamens slightly shorter than corolla, filaments short, connate or free, anther thecae strongly papillate, tubules ca. 1/2 as broad, 5–7× longer, with terminal or subterminal, oblique pores, sometimes flaring; fruit unknown. Four spp., Venezuela to Peru; mid alt.

111. *Oreanthes* Bentham

Oreanthes Bentham, Pl. Hartweg.: 140 (1844); Luteyn, Brittonia 29: 173–176 (1977); Fl. Ecuad. 54: 329–337 (1996).

Terrestrial or epiphytic shrubs; leaves thick-coriaceous, obscurely pli-nerved, usually succulent; inflorescences 1-flowered to fascicles; pedicel continuous; calyx tube rarely bluntly 5-angled, limb to 3× longer, lobed to base; corolla >1 cm long, cylindric, terete to angled (lobes ca. 1/4); stamens 5 or 10, filaments free or connate, anther thecae essentially smooth, tubules ca. 1/2 as broad, slightly to up to 6× longer, with terminal or subterminal and oblique, usually flaring pores or rarely elongate clefts; fruit unknown. 7 spp., Ecuador and adjacent Peru; 1340–3290 m alt.

112. *Siphonandra* Klotzsch

Siphonandra Klotzsch, Linnaea 24: 24 (1851).

Terrestrial or epiphytic; leaves penni-nerved, rarely crenulate; inflorescence racemes; calyx limb spreading; corolla ca. 2.5 cm long, cylindrical; filaments short, connate, thecae papillate, tubules very narrow, 4–5× longer, flexible, with flaring terminal pores; fruit black, testa ?mucilaginous; embryo green. 1–2 spp., S Peru to N Bolivia; high alt.

113. *Pellegrinia* Sleumer

Pellegrinia Sleumer, Notizbl. Bot. Gart. Mus. Berlin-Dahlem 12: 287 (1935).

Usually epiphytic; bud scales rarely subulate; leaves pli- or penni-nerved; inflorescences fascicles or 1-flowered; pedicel continuous; calyx limb suberect or spreading, lobes long; corolla 2 cm long, subcylindrical, subinflated; filaments shorter than anthers, with dense retrorse hairs abaxially, anther thecae smooth, tubules narrow, 2–4× longer, flexible, with small flaring or oblique pores; fruit blue-black; seed? Five spp., Peru; high alt.

114. *Disterigma* (Klotzsch) Niedenzu

Disterigma (Klotzsch) Niedenzu, Bot. Jahrb. Syst. 11: 160, 209 (1889); A.C. Smith, Brittonia 1: 203–232 (1933); Wilbur, Bull. Torrey Bot. Club 119: 280–288 (1992); Luteyn, Fl. Ecuad. 54: 253–286 (1996).
Killipiella A.C. Smith, J. Wash. Acad. Sci. 33: 242 (1943).

Terrestrial or epiphytic shrubs; leaves often congested, usually less than 3 cm long and obscurely pli-nerved, rarely crenate; inflorescences fascicles, flowers rarely solitary; flowers (3)4(5)-merous, pedicel with apical prophylls investing ovary, very short, obscurely articulated; calyx limb somewhat longer than the tube, erect, lobed to base, rarely imbricate; corolla <1 cm long, subcylindric or campanulate-cylindric; stamens rarely same number as corolla lobes, ca. 2/3 corolla length, filaments longer or shorter than anthers, thecae smooth, tubules to 2× longer, little narrower, rarely narrower, rarely fused, with elongate, elliptical, introrse clefts or subterminal pores; fruit blue-black or translucent white; testa mucilaginous (or not?); embryo green; n = >40. 35 spp., Guatemala S to Bolivia and E to Guyana, 2/3 in Ecuador; low to high alt.

115. *Sphyrospermum* Poeppig & Endlicher

Sphyrospermum Poeppig & Endlicher, Nov. Gen. Sp. Pl. 1: 4 (1835); A.C. Smith, Brittonia 1: 203–232 (1933); Luteyn, Fl. Ecuad. 54: 341–363 (1996).

Epiphytic or terrestrial shrubs with pendent branches; leaves coriaceous, sometimes very thickly, small, obscurely pli-nerved; inflorescences fascicles, rarely flowers solitary; flowers 4–5-merous, pedicels slender, cernuous, distally swollen; calyx limb ± spreading, lobes short to long; corolla to 1.2(–2.0) cm long, campanulate or cylindrical to infundibuliform; stamens 4–5(–10), ca. 3/4 corolla length, slightly dimorphic or not, fil-

aments usually longer than anthers, somewhat sigmoid at the base, anthers membranous, thecae smooth, rarely papillate, tubules to 3× longer, usually little narrowed, with oval clefts; fruit translucent white to lavender or purplish; testa mucilaginous; embryo green; n = 24, 48, ca. 75. 22 spp., S Mexico to Bolivia, esp. Ecuador, E to French Guiana, and Haiti to Trinidad;

116. *Diogenesia* Sleumer

Diogenesia Sleumer, Notizbl. Bot. Gart. Berlin-Dahlem 12: 121 (1934); Notes Roy. Bot. Gard. Edinburgh 36: 251–258 (1978).
Eleutherostemon Herzog (1915), non Klotzsch (1838).

Usually epiphytic shrubs, branches slender, often pendent, rarely climbing; bud scales subulate; inflorescences fascicles or racemes, rarely flowers solitary; flowers 4–5-merous, pedicel slender, continuous; calyx suberect; corolla usually <1 cm long, cylindric, urceolate or campanulate (lobed 1/2 way); stamens 4–6, 8 or 10, filaments longer or shorter than anthers, anther thecae smooth or faintly papillate, tubules equal or longer and slightly narrowed, with oval pores or clefts; fruit purple-black or rarely white; testa mucilaginous; embryo green; n = 12. 13 spp., Venezuela to N Bolivia; mid alt. Close to *Sphyrospermum*.

117. *Rusbya* Britton

Rusbya Britton, Bull. Torrey Bot. Club 20: 68 (1893).

Epiphytic shrub; bud scales subulate; leaves linear, one-nerved; inflorescences 1-flowered; pedicel continuous; calyx tube narrowly angled, limb suberect; corolla <1 cm long, cylindrical-urceolate; filaments dimorphic, anthers dimorphic, almost smooth, tubules not narrowed and slightly longer, with elongated slits; fruit not known. One sp., *R. taxifolia* Britton, N Bolivia; mid alt.

118. *Themistoclesia* Klotzsch

Themistoclesia Klotzsch, Linnaea 24: 41 (1851); Sleumer, Notizbl. Bot. Gart. Mus. Berlin-Dahlem 13: 108–111(1936).

Often epiphytic shrubs; leaves subcoriaceous to coriaceous, rarely thick-fleshy, obscurely pli-nerved, rarely pinnate; inflorescences fascicles or racemes, rarely flowers solitary; flowers (4)5-merous; pedicel continuous (articulated?); calyx tube bluntly angled, rarely strongly winged, limb erect to spreading, lobes rarely long; corolla ca. 1 cm long, ovoid to cylindric, terete to angled; stamens 4, 5, 8 or 10, more than 1/2 to as long as corolla, filaments to as long as the anthers, anther thecae smooth, tubules nearly equal to 4× as long, rarely fused, little to moderately narrowed, with short clefts; fruit purple, black or cream; testa mucilaginous; embryo green. 25 spp., Costa Rica to Venezuela and Peru; mid to high alt.

119. *Plutarchia* A.C. Smith

Plutarchia A.C. Smith, Bull. Torrey Bot. Cl. 63: 311 (1936).

Terrestrial shrubs; leaves also opposite; inflorescences fascicles; flowers 4–5-merous, bracts often conspicuous, pedicel articulated; calyx tube rarely winged, limb erect, to 2–4 times longer, usually lobed to base; corolla >1 cm long, cylindric to urceolate-cylindric, terete or angled; stamens (5–)8–10, filaments short, slightly dimorphic or not, thecae slightly papillate, tubules 2–4 times longer, rather narrower, flexible, with elongate clefts; fruit blue-black, rarely white; testa mucilaginous; embryo green. 11 spp., most in Colombia and Ecuador; high-elevation montane forest and páramo.

120. *Thibaudia* Jaume Saint-Hilaire

Thibaudia Jaume Saint-Hilaire, Expos. Fam. 1: 362 (1805).
Calopteryx A.C. Smith, J. Arnold Arbor. 27: 100 (1946).

Epiphytic or terrestrial shrubs; bud scales rarely subulate; leaves rarely penni-nerved, crenate; inflorescences axillary or terminal, rarely ramiflorous fascicles, racemes or panicles; pedicel rarely continuous; calyx tube rarely narrowly 5-angled, or rarely strongly winged to sinuses, limb often longer, erect to slightly spreading, lobed to halfway; corolla >1 cm long, cylindric to subcylindric, rarely 5-angled to strongly winged; filaments short, separate or connate, anthers firm or membranous, thecae smooth or slightly papillate, tubules ca. as long and wide, with elongate clefts; fruit purple to blackish; testa mucilaginous or hard; embryo white, rarely green. 60 spp., Panama and Costa Rica, mostly from Colombia to Bolivia and E to Guayana and Brazil; low to high alt.

121. *Demosthenesia* A.C. Smith

Demosthenesia A.C. Smith, Bull. Torrey Bot. Club 63: 310 (1936).

Usually epiphytic shrubs, rarely lianes, lignotubers frequent; bud scales often subulate, leaves rarely penni-nerved; inflorescences fasciculate or racemose; pedicel continuous; calyx tube rarely slightly winged, limb to 2× as long, suberect, often deeply lobed; corolla >1 cm long, cylindrical, rarely urceolate; filaments very short, often slightly dimorphic, thecae submembranous, slightly papillate, tubules flexible, not narrowed, to 6× longer, with elongate slits or oblique flaring pores; fruit blue-black; seed? 11 spp., C Peru to N Bolivia; high alt.

122. *Anthopterus* Hooker

Anthopterus Hooker, Icon. Pl. 3: pl. 243 (1840); Luteyn, Opera Bot. 92: 109–113 (1987); Brittonia 48: 605–610 (1997).

Shrubs or small trees, rarely epiphytes; bud scales rarely subulate; leaves rarely subopposite or pseudoverticillate, penni- or pli-nerved; inflorescences racemes; bracts sometimes conspicuous; pedicel continuous; calyx tube strongly winged, limb also longer, suberect to spreading, moderately lobed; corolla ca. 1(–2) cm long, subcylindric to subglobose, narrowly to broadly 5-winged, lobes rarely long; filaments shorter than anthers (rarely same length), often connate, anthers membranous, thecae smooth, tubules slightly narrowed, 1–3× longer, with short to elongate clefts; fruit blue; seed? 11 spp., SE Panama to NE Peru; s.l.–2500 m alt. Most species poorly known; close to *Thibaudia* and *Themistoclesia*.

123. *Costera* J.J. Smith

Costera J.J. Smith, Icon. Bog. 4: 77, tab. 324 (1910); Sleumer, Fl. Males. I, 6: 740–746 (1967).

Shrubs, climbing or epiphytic; inflorescences fascicles; flowers 4–5-merous; pedicel continuous; calyx limb erect to spreading, lobed; corolla <1 cm long, urceolate to campanulate, lobed to halfway or less; stamens 5, 8, or 10, filaments to as long as anthers, anther thecae papillate, tubules much to slightly narrower and shorter to 1.5× longer, with oblique pores or short slits; fruit white to dark purple; testa hard or mucilaginous; embryo white or green; $n = 12$. 9 spp., W Malesia; mostly 500–1300 m alt. Circumscription and relationships uncertain.

124. *Gaylussacia* Kunth, nom. cons.

Gaylussacia Kunth, Nov. Gen. Sp. Pl. 3, ed. fol.: 215, ed. qu.: 275, t. 275 (1819); Sleumer, Bot. Jahrb. Syst. 86: 309–384 (1967).

Shrubs or subshrubs; indumentum sometimes glandular; leaves coriaceous to membranous, evergreen to deciduous, pinni-nerved, apex glandular; inflorescences racemes, flowers rarely solitary; flowers 4–5-merous; calyx limb deeply lobed; corolla <1 cm long, cylindric, urceolate, or campanulate, often angled; filaments short to much longer than anthers, thecae smooth to papillate, tubules ca. 2× longer, narrowed or not, with narrow lateral to subterminal pores; ovary 10-locular; fruit drupaceous, with 10 pyrenes; $n = 24$. 48 spp., North America (sections *Decamerium* (Nuttall) J.D. Hooker and *Vitis-Idaea* J.D. Hooker) to N Argentina, esp. SE Brazil (sect. *Gaylussacia*); s.l.–3450 m alt. The walls of the testa cells have large pits.

Selected Bibliography

Allaway, W.G., Ashford, A.E. 1996. Structure of the hair roots in *Lysinema ciliatum* R. Br. and its implications for their water relations. Ann. Bot. II, 77: 383–388.

Anderberg, A.A. 1992. The circumscription of the Ericales, and their cladistic relationships to the families of "higher" dicotyledons. Syst. Bot. 17: 660–675.

Anderberg, A.A. 1993. Cladistic interrelationships and major clades of the Ericales. Pl. Syst. Evol. 184: 207–231.

Anderberg, A.A. 1994a. Phylogeny of the Empetraceae, with special emphasis on character evolution in the genus *Empetrum*. Syst. Bot. 19: 35–46.

Anderberg, A.A. 1994b. Cladistic analysis of *Enkianthus* with notes on the early diversification of the Ericaceae. Nordic J. Bot. 14: 385–401.

Anderberg, A.A. et al. 2002. See general references.

Ashford, A.E., Allaway, W.G., Reed, M.L. 1996. A possible role for thick-walled epidermal cells in the mycorrhizal hair roots of *Lysinema ciliatum* R. Br. and other Epacridaceae. Ann. Bot. II, 77: 375–381.

Atkinson, R., Jong, K., Argent, G. 1995. Cytotaxonomic observations in tropical Vaccinieae (Ericaceae). Bot. J. Linn. Soc. 117: 135–145.

Bell, T.L. 1995. Biology of Australian Epacridaceae: with special reference to growth, fire response and mycorrhizal nutrition. Ph.D. thesis, Department of Botany, University of Western Australia.

Bell, T.L., Ojeda, F. 1999. Underground starch storage in *Erica* species of the Cape floristic region – differences between seeders and resprouters. New Phytol. 144: 143–152.

Bell, T.L., Pate, J.S. 1996a. Nitrogen and phosphorus nutrition in mycorrhizal Epacridaceae of South-west Australia. Ann. Bot. II, 77: 389–397.

Bell, T.L., Pate, J.S. 1996b. Growth and fire response of selected Epacridaceae of south-western Australia. Austr. J. Bot. 44: 509–526.

Bell, T.L., Pate, J.S., Dixon, K.W. 1994. Response of mycorrhiza seedlings of SW Australian sandplain Epacridaceae to added nitrogen and phosphorus. J. Exp. Bot. 45: 779–790.

Bell, T.L., Pate, J.S., Dixon, KW. 1996. Relationships between fire response, morphology, root anatomy and starch distribution in south-west Australian Epacridaceae. Ann. Bot. II, 77: 357–364.

Bergero, R., Perotto, S., Girlanda, M., Vidano, G., Luppi, A.M. 2000. Ericoid mycorrhizal fungi are common root associates of a Mediterranean ectomycorrhizal plant (*Quercus ilex*). Molec. Ecol. 9: 1639–1649.

Bidartondo, M.I., Bruns, T.D. 2001. Extreme specificity in epiparasitic Montropoideae (Ericaceae): widespread phylogenetic and geographic structure. Molec. Ecol. 10: 2285–2295.

Björkman, E. 1960. *Monotropa hypopitys* L. – an epiparasite on tree roots. Physiol. Plant. (Copenhagen) 13: 308–327.

Bohm, B.A., Averett, J.E. 1989. Flavonoids in some Monotropoideae. Biochem. Syst. Ecol. 17: 399–401.

Bohm, B.A., Brim, S.W., Hebda, R.J., Stevens, P.F. 1978. Generic limits in the tribe Cladothamneae (Ericaceae) and its position in the Rhododendroideae. J. Arnold Arbor. 59: 311–341.

Briggs, C.L., Ashford, A.E. 2001. Structure and composition of the thick wall in hair root epidermal cells of *Woolsia pungens*. New Phytol. 149: 219–232.

Briggs, J.D., Leigh, J.H. 1989. Rare or threatened Australian plants: 1988 revised edition. Canberra: Australian National Parks and Wildlife Service.

Brough, P. 1924. Studies in the Epacridaceae. I. The life history of *Styphelia longifolia* (R. Br.). Proc. Linn. Soc. New S. Wales 49: 162–178.

Brown, E.M., Buridge, J.D., Dell, J., Edinger, D., Hopper, S.D., Wills, R.T. 1997. Pollination in Western Australia. A database of animals visiting flowers. Handbook no. 15, Western Australian Naturalists' Club.

Cane, J.H., Eickwort, G.C., Wesley, F.R., Spielholz, J. 1985. Pollination ecology of *Vaccinium stamineum* (Ericaceae: Vaccinioideae). Am. J. Bot. 72: 135–142.

Carlquist, S. 1989. Wood and bark anatomy of Empetraceae; comments on paedomorphosis in woods of certain small shrubs. Aliso 12: 497–515.

Cherry, W., Gadek, P.A., Brown, E.A., Heslewood, M.M., Quinn, C.J. 2001. *Pentachondra dehiscens* sp. nov. – an aberrant new member of Stephelieae. Austr. Syst. Bot. 14: 513–533.

Clemants, S.E. 1995. *Bejaria* Mutis ex Linnaeus. In: Luteyn, J.L. (ed.) Flora neotropica. Monograph 66. Ericaceae Part II, the superior-ovaried genera. New York: New York Botanical Garden, pp. 54–106.

Clifford, H.T., Drake, W.E. 1981. Pollination and dispersal in eastern Australian heathlands. In: Specht, R.L. (ed.) Ecosystems of the World 9B. Heathlands and related shrublands. Amsterdam: Elsevier, pp. 39–49.

Collinson, M.E., Crane, P.R. 1978. *Rhododendron* seeds from the Palaeocene of England. Bot. J. Linn. Soc. 76: 195–205.

Colwell, R.K. 1973. Competition and coexistence in a simple tropical community. Am. Nat. 107: 737–760.

Copeland, H.F. 1935. On the genus *Pityopus*. Madroño 3: 154–168.

Copeland, H.F. 1941. Further studies on Monotropoideae. Madroño 6: 97–119.

Copeland, H.F. 1943. A study, anatomic and taxonomic, of the genera of Rhododendroideae. Am. Midl. Nat. 30: 533–625.

Copeland, H.F. 1947. Observations on the structure and classification of the Pyroleae. Madroño 9: 65–102.

Copeland, H.F. 1954. Observations on certain Epacridaceae. Am. J. Bot. 41: 215–222.

Cox, H.T. 1948a. Studies in the comparative anatomy of the Ericales. I. Ericaceae – subfamily Rhododendroideae. Am. Midl. Nat. 39: 220–245.

Cox, H.T. 1948b. Studies in the comparative anatomy of the Ericales. II. Ericaeae subfamily Arbutoideae. Am. Midl. Nat. 40: 493–516.

Crayn, D.M., Kron, K.A., Gadek, P.A., Quinn, C.J. 1996. Delimitation of Epacridaceae: preliminary molecular evidence. Ann. Bot. II, 77: 317–321.

Crayn, D.M., Kron, K.A., Gadek, P.A., Quinn, C.J. 1998. Phylogenetics and evolution of epacrids: a molecular analysis using the plastid gene *rbc*L with a reappraisal of the position of *Lebetanthus*. Austr. J. Bot. 46: 187–200.

Crepet, W.L., Nixon, K.C., Gandolfo, M.A. 2001. A Cretaceous Atlantic Coastal Plain "ericoid" complex. In: Botany 2001: plants and people, p. 62.

Cullings, K.W. 1994. Molecular phylogeny of the Monotropoideae (Ericaceae) with a note on the placement of the Pyroloideae. J. Evol. Biol. 7: 501–516.

Cullings, K.W. 1996. Single phylogenetic origin of ericoid mycorrhizads within the Ericaceae. Can. J. Bot. 74: 1896–1909.

Cullings, K.W., Szaro, T.M., Bruns, T.D. 1996. Evolution of extreme specialization within a lineage of ectomycorrhizal epiparasites. Nature 379: 63–66.

D'Arcy, W.G., Keating, R.C., Buchmann, S.L. 1996. The calcium oxalate package or so-called resorbtion tissue in some angiosperm anthers. In: D'Arcy, W.G., Keating, R.C. (eds.) The anther: form, function and phylogeny. Cambridge: Cambridge University Press, pp. 158–191.

Dawson, M.I. 2000. Index of chromosome numbers in the Epacridaceae. Proc. Linn. Soc. New S. Wales 73: 37–56.

Diggs, G.M. Jr. 1995. *Comarostaphylis* Zuccarini. In: Luteyn, J.L. (ed.) Flora neotropica. Monograph 66. Ericaceae Part II, the superior-ovaried genera. New York: New York Botanical Garden, pp. 146–193.

Dimitri, M.J. 1972. La región de los bosques andino-patagónicos. Buenos Aires: INTA.

Dome, A.P. 1999. × *Phylliopsis* – *Phyllodoce* spp. × *Kalmiopsis* – ericaceous aristocrats. Rock Gard. Quart. 57: 35–45.

Drude, O. 1889. Epacridaceae. In: Engler, A., Prantl, K. Die natürlichen Pflanzenfamilien, vol. IV, 1. Leipzig: W. Engelmann, pp. 66–79.

Duddridge, J., Read, D.J. 1982. An ultrastructural analysis of the development of mycorrhizas in *Rhododendron ponticum*. Can. J. Bot. 60: 2345–2356.

Eck, P. 1990. The American cranberry. New Brunswick: Rutgers University Press.

Étienne, P. 1919. Étude anatomique de la famille des Épacridacées. Thèse Doctorale. Toulouse: Faculté Pharmacologique, Université de Toulouse.

Ewane-Nyambi, G., Bois, P., Raymond, G. 1993. The effects of *Agauria salicifolia* leaf extract on calcium current and excitation-contraction coupling of isolated frog muscle cells. J. Ethnopharmacol. 38: 55–61.

Fischer, O. 1992. New knowledges about the distribution of the genus *Kalmia* L. in the European Tertiary. Cour. Forsch.-Inst. Senck. 147: 383–391.

Foss, P.J., Doyle, G.J. 1988. A palynological study of the Irish Ericaceae and *Empetrum*. Pollen Spores 30: 151–178.

Freudenstein, J.V. 1999. Relationships and character transformation in Pyroloideae (Ericaeae) based on ITS sequences, morphology and development. Syst. Bot. 24: 398–408.

Friis, E.M. 1985. Angiosperm fruits and seeds from the Middle Miocene of Jutland (Denmark). Kongel. Danske Vidensk. Selsk. Biolog. Skr. 24(3): 1–165.

Furman, T.E., Trappe, J.M. 1971. Phylogeny and ecology of mycotropic echlorophyllous angiosperms. Quart. Rev. Biol. 46: 219–225.

Godley, E.J. 1966. Breeding systems in New Zealand plants. 4. Self sterility in *Pentachondra pumila*. New Zeal. J. Bot. 53: 324–355.

Haber, E. 1987. Variability, distribution, and systematics of *Pyrola picta* s.l. (Ericaceae) in western North America. Syst. Bot. 12: 324–335.

Hagerup, O. 1953. The morphology and systematics of the leaves in Ericales. Phytomorphology 3: 459–464.

Hagerup, E., Hagerup, O. 1953. Thrips pollination of *Erica tetralix*. New Phytol. 52: 1–7.

Hallé, F., Oldeman, R.A.A., Tomlinson, P.B. 1978. Tropical trees and forests. Berlin Heidelberg New York: Springer.

Hansen, I. 1950. Die europäischen Arten der Gattung *Erica* L. Bot. Jahrb. Syst. 75: 1–81.

Hara, N. 1958. Structure of vegetative shoot apex and development of the leaf in the Ericaceae and their allies. J. Fac. Sci. Univ. Tokyo 7: 367–450.

Harborne, J.B., Williams, C.A. 1973. A chemotaxonomic survey of flavonoids and simple phenols in leaves of the Ericaceae. Bot. J. Linn. Soc. 66: 37–54.

Henderson, M.W. 1919. A comparative study of the structure and saprophytism of the Pyrolaceae and Monotropaceae, with reference to their derivation from the Ericaceae. Contrib. Bot. Lab. Morris Lab. Univ. Penn. 5: 42–109.

Hermann, P.M., Cambi, V.N. 1992. Nuevas datos sobre la sexualidad de *Gaultheria caespitosa* P. & E. (Ericaceae). Parodiana 7: 83–90.

Hermann, P.M., Palser, B.F. 2000. Stamen development in the Ericaceae. I. Anther wall, microsporogenesis, inversion and appendages. Am. J. Bot. 87: 934–957.

Heslop Harrison, Y., Shivanna, K.R. 1977. The receptive surface of the angiosperm stigma. Ann. Bot. II, 41: 1233–1258.

Higham, R.K., McQuillan, P.B. 2000. *Cyathodes divaricata* (Epacridaceae) – the first record of a bird-pollinated dioecious plant in the Australian flora. Austr. J. Bot. 48: 93–99.

Hileman, L.C., Vasey, M.C., Parker, V.T. 2001. Phylogeny and biogeography of the Arbutoideae (Ericaceae): implications for the Madrean-Tethyan hypothesis. Syst. Bot. 26: 131–143.

Hutton, B.J. 1997. Biology and ecology of endophytes of Australian native heaths (Epacridaceae). Ph.D Thesis, Department of Botany, University of Western Australia.

Hutton, B.J., Dixon, K.W., Sivasithamparam, K. 1994. Ericoid endophytes of Western Australian heaths (Epacridaceae). New Phytol. 127: 557–566.

Jackes, B.R. 1968. Floral anatomy of the genus *Oligarrhena* R. Br. (Epacridaceae). Austr. J. Bot. 16: 451–454.

Janaki Ammal, E.K., Enoch, I.C., Bridgwater, M. 1950. Chromosome numbers in species of *Rhododendron*. Rhodod. Year Book 5: 78–95.

Jarman, S.J. 1975. Experimental taxonomy in the family Epacridaceae. PhD Thesis, Department of Botany, University of Tasmania.

Jarman, S.J., Crowden, R.K. 1974. Anthocyanins in the Epacridaceae. Phytochemistry 13: 743–750.

Jarman, S.J., Crowden, R.K. 1977. The occurrence of flavonol arabinosides in the Epacridaceae. Phytochemistry 16: 929–930.

Johri, B.M. et al. 1992. See general references.

Jordan, G.J., Hill, R.S. 1996. The fossil record of the Epacridaceae. Ann. Bot. II, 77: 341–346.

Judd, W.S. 1979. Generic relationships in the Andromeae (Ericaceae). J. Arnold Arbor. 60: 477–503.

Judd, W.S. 1981. A monograph of *Lyonia* (Ericaceae). J. Arnold Arbor. 62: 63–128, 129–209, 315–436.

Judd, W.S. 1984. A taxonomic revision of the American species of *Agarista* (Ericaceae). J. Arnold Arbor. 65: 255–342.

Judd, W.S. 1986. A taxonomic revision of *Craibiodendron* (Ericaceae). J. Arnold Arbor. 67: 441–469.

Judd, W.S., Kron, K.A. 1993. Circumscription of Ericaceae (Ericales) as determined by preliminary cladistic analyses based on morphological, anatomical and embryological features. Brittonia 45: 99–114.

Keighery, G.J. 1996. Phytogeography, biology and conservation of Western Australian Epacridaceae. Ann. Bot. II, 77: 347–355.

Knobloch, E., Mai, D.H. 1986. Monographie der Früchte und Samen in der Kreide von Mitteleuropa. Rosp. Ústřed. Ústav Geol. 47: 1–215.

Knudsen, J.T., Oleson, M.J. 1993. Buzz-pollination and patterns of variation in several traits in north European Pyrolaceae. Am. J. Bot. 80: 900–193.

Knudsen, J.T., Tollesten, L. 1991. Floral scent and intrafloral variation in *Moneses* and *Pyrola* (Pyrolaceae). Plant Syst. Evol. 177: 81–91.

Kobayashi, N., Handa, T., Yoshimyura, K., Tsumura, Y., Arisumi, K., Takayanagi, K. 2000. Evidence for introgressive hybridization based on chloroplast DNA polymorphisms and morphological variation in wild evergreen azalea populations of the Kirishima mountains, Japan. Edinburgh J. Bot. 57: 209–219.

Kraemer, M. 2001. On the pollination of *Bejaria racemosa* Mutis ex Linné f. (Ericaceae), an ornithophilous Andean páramo shrub. Flora 196: 59–62.

Kretzer, A.M., Bidartondo, M.I., Grubisha, L.C., Spatafora, J.W., Szaro, T.M., Bruns, T.D. 2000. Regional specialization of *Sarcodes sanguinea* on a single fungal symbiont from the *Rhizopogon ellenae* (Rhizopogonaceae) complex. Am. J. Bot. 87: 1778–1782.

Kron, K.A. 1996. Phylogenetic relationships of Empetraceae, Epacridaceae, Ericaceae, Monotropaceae, and Pyrolaceae: evidence from nuclear ribosomal 18s sequence data. Ann. Bot. II, 77: 293–303.

Kron, K.A., Chase, M.W. 1993. Systematics of the Ericaceae, Empetraceae, Epacridaceae and related taxa based on *rbcL* sequence data. Ann. Miss. Bot. Gard. 80: 735–741.

Kron, K.A., Judd, W.S. 1990. Phylogenetic relationships within the Rhodoreae (Ericaceae) with specific comments on the placement of *Ledum*. Syst. Bot. 15: 57–68.

Kron, K.A., King, J.M. 1996. Cladistic relationships of *Kalmia*, *Leiophyllum*, and *Loiseleuria* (Phyllodoceae, Ericaceae) based on nucleotide sequences from *rbcL* and nuclear ribosomal internal transcribed spacer regions (ITS). Syst. Bot. 21: 17–29.

Kron, K.A., Fuller, R., Crayn, D.M., Gadek, P.A., Quinn, C.J. 1999a. Phylogenetic relationships of Styphelioideae and vaccinioids (Ericaceae s.l.) based on *mat*K sequence data. Plant Syst. Evol. 218: 55–65.

Kron, K.A., Judd, W.S., Crayn, D.M. 1999b. Phylogenetic analyses of Andromedeae (Ericaceae subfam. Vaccinioideae). Am. J. Bot. 86: 1290–1300.

Kron, K.A., Judd, W.S., Stevens, P.F., Crayn, D.M., Anderberg, A.A., Gadek, P.A., Quinn, C.J., Luteyn, J.L. 2002a. Phylogenetic classification of Ericaceae: molecular and morphological evidence. Bot. Rev. 68: 335–423.

Kron, K.A., Powell, E.A., Luteyn, J.L. 2002b. Phylogenetic relationships within the blueberry tribe (Vaccinieae, Ericaceae) based on sequence data from *mat*K and nuclear ribosomal ITS regions, with comments on the placement of *Satyria*. Am. J. Bot. 89: 327–336.

Ladd, P.G. 1994. Pollen presenters in the flowering plants – form and function. Bot. J. Linn. Soc. 115: 163–193.

Largent, D.L., Sugihara, N., Wishner, C. 1980. Occurrence of mycorrhizae on ericaceous and pyrolaceous plants in northern California. Can. J. Bot. 58: 2274–2279.

Lavier-George, L. 1936. Recherches sur les épidermes foliaires des *Philippia* de Madagascar, utilisation de leurs caractères comme bases d'une classification. Bull. Mus. Hist. Nat. Paris 8(2): 173–199.

Leake, J.R. 1994. Tansley Review no. 69. The biology of myco-heterotrophic ('saprophytic') plants. New Phytol. 127: 171–216.

Leins, P. 1964. Entwicklungsgeschichtliche Studien an Ericales-Blüten. Bot. Jahrb. Syst. 83: 57–88.

Lems, K. 1962a. Adaptive radiation in the Ericaceae. 1. Shoot development in the Andromedeae. Ecology 43: 524–528.

Lems, K. 1962b. Evolutionary studies in the Ericaceae. 2. Leaf anatomy as a phylogenetic index in the Andromedeae. Bot. Gaz. 125: 178–186.

Lemson, K.L. 1996. Current problems in the taxonomy of *Andersonia* R. Br. Ann. Bot. II, 77: 323–326.

Lemson, K.L. 2001. The phylogeny and taxonomy of *Andersonia* R. Br (Ericaceae/Epacridaceae). Ph.D. Thesis, The University of Western Australia.

Lens, F., Gasson, P., Smets, E., Jansen, S. 2003. Comparative wood anatomy of epacroids (Styphelioideae, Ericaceae s.l.). Ann. Bot. II, 91: 835–856.

Luteyn, J.L. 1983. *Cavendishia*. Flora Neotropica Monograph 35. New York: New York Botanical Garden.

Luteyn, J.L. 1995 (ed.). Flora Neotropica. Monograph 66. Ericaceae Part II, the superior-ovaried genera. New York: New York Botanical Garden.

Luteyn, J.L. 1996 [1997]. Redefinition of the neotropical genus *Anthopterus* (Ericaceae: Vaccinieae). Brittonia 48: 605–614.

Luteyn, J.L. 2002. Diversity, adaptation and endemism in neotropical Ericaceae: biogeographical patterns in the Vaccinieae. Bot. Rev. 68: 55–87.

Luteyn, J.L., Harborne, J.B., Williams, C.A. 1980. Survey of the flavonoids and simple phenols in the leaves of *Cavendishia* (Ericaceae). Brittonia 32: 1–16.

Luteyn, J.L., Sylva S., D.S. 1999. "Murrí" (Antioquia Department, Colombia): hotspot for neotropical blueberries (Ericaceae: Vaccinieae). Brittonia 51: 280–302.

Matthews, J.R., Knox, E.M. 1926. The comparative morphology of the stamen in the Ericaceae. Trans. Proc. Bot. Soc. Edinburgh 29: 243–281.

McCloskey, R. 1948. Blueberries for Sal. New York: Viking.

McConchie, C.A., Hough, T., Singh, M.B., Knox, R.B. 1986. Pollen presentation on petal combs in the geoflorous heath *Acrotriche serrulata* (Epacridaceae). Ann. Bot. II, 57: 155–164.

McLean, C. 1995. Mycorrhizae of the Epacridaceae and its use in propagation. Combined Proc. Int. Plant Propagators 45: 108–111.

McLean, C.B. 1999. Investigation of mycorrhizas of the Epacridacaeae. Ph.D. Thesis, Department of Applied Biology and Biotechnology, RMIT University.

McLennan, E. 1935. Non-symbiotic development of seedlings of *Epacris impressa* Labill. New Phytol. 34: 55–63.

Middleton, D.J. 1991. Ecology, reproductive biology and hybridization in *Gaultheria* L. Edinburgh J. Bot. 48: 81–89.

Middleton, D.J., Wilcock, C.C. 1990. Chromosome counts in *Gaultheria* and related genera. Edinburgh J. Bot. 47: 303–313.

Mihaich, C.M. 1989. Leaf epicuticular waxes in the taxonomy of the Epacridaceae. Ph.D. Thesis, Department of Plant Sciences, University of Tasmania.

Moore, D.M., Harborne, J.B., Williams, C.A. 1970. Chemotaxonomy, variation and geographical distribution of the Empetraceae. Bot. J. Linn. Soc. 63: 277–293.

Naskrecki, P., Colwell, R.K. 1998. Systematics and host plant affiliations of hummingbird flower mites of the genera *Tropicoseius* Baker & Yunker and *Rhinoseius* Baker & Yunker (Acari: Mesostigmata: Ascidae). Lanham, Maryland: Entomological Society of America.

Netolitzky, F. 1926. Handbuch der Pflanzenanatomie. Band 10. Anatomie der Angiospermen-Samen. Berlin: Borntraeger.

Nixon, K.C., Crepet, W.L. 1993. Late Cretaceous fossil flowers of Ericalean affinity. Am. J. Bot. 80: 616–623.

Noshiro, S., Suzuki, M., Ohba, H. 1995. Ecological wood anatomy of Nepalese *Rhododendron*. 1. Interspecific variation. J. Plant Res. 108: 1–9.

Odell, A.E., Vander Kloet, S.P., Newell, R.E. 1989. Stem anatomy of *Vaccinium* section *Cyanococcus* and related taxa. Can. J. Bot. 67: 2328–2334.

Oliver, E.G.H. 1987. Studies in the Ericoideae (Ericaceae). V. The genus *Coilostigma*. Bothalia 17: 163–170.

Oliver, E.G.H. 1991. The Ericoideae (Ericaceae) – a review. Contrib. Bolus Herb. 13: 158–208.

Oliver, E.G.H. 2000. Systematics of Ericeae (Ericaceae: Ericoideae) species with indehiscent and partially dehiscent fruits. Contrib. Bolus Herb. 19: 1–483.

Oliver, E.G.H., Linder, H.P., Rourke, J.P. 1983. Geographical distribution of present-day Cape taxa and their phytogeographical significance. Bothalia 14: 427–440.

Olson, A.R. 1980. Seed morphology of *Monotropa uniflora* L. (Ericaceae). Am. J. Bot. 67: 968–974.

Olsson, M., Shine, R., Ba'k-Olsson, E. 2000. Lizards as a plant's 'hired help': letting pollinators in and seeds out. Biol. J. Linn. Soc. 71: 191–202.

Palser, B.F. 1952. Studies of floral morphology in the Ericales. II. Megasporogenesis and megagametophyte development in the Andromedeae. Bot. Gaz. 114: 33–52.

Palser, B.F. 1954. Studies of floral morphology in the Ericales. III. Organography and vascular anatomy in several species of the Arbuteae. Phytomorphology 4: 335–354.

Palser, B.F. 1961. Studies of floral morphology in the Ericales. V. Organography and vascular anatomy in several United States species of the Vaccinieae. Bot. Gaz. 123: 79–111.

Palser, B.F., Murty, Y.S. 1967. Studies in the floral morphology of the Ericales 8. Organography and vascular anatomy in *Erica*. Bull. Torrey Bot. Club 94: 243–320.

Paterson, B.R. 1961. Studies of floral morphology in the Epacridaceae. Bot. Gaz. 122: 259–279.

Pearson, V., Read, D.J. 1973. The biology of mycorrhiza in the Ericaceae. I. The isolation of the endophyte and synthesis of mycorrhizas in aseptic culture. New Phytol. 72: 371–379.

Perotto, S., Perotto, R., Faccio, A., Schubert, A., Varma, A., Bonfante, P. 1995. Ericoid mycorrhizal fungi: cellular and molecular bases of their interactions with the host plant. Can. J. Bot. 73 (suppl.): S557–S568.

Philipson, W.R. 1985. Shoot morphology in *Rhododendron*. Notes Roy. Bot. Gard. Edinburgh 43: 161–171.

Philipson, W.R., Philipson, M.N. 1968. Diverse nodal types in *Rhododendron*. J. Arnold Arbor. 49: 193–225.

Powell, E.A., Kron, K.A. 2001. An analysis of the phylogenetic relationships in the wintergreen group (*Diplycosia*, *Gaultheria*, *Pernettya*, *Tepuia*; Ericaceae). Syst. Bot. 26: 808–817.

Powell, J.M. 1983. Epacridaceae. In: Morley, B.D., Toelken, H.R. (eds.) Flowering plants in Australia. Adelaide: Rigby Publishers, pp. 111–114.

Powell, E.A., Kron, K.A. 2002. Hawaiian blueberries and their relatives – a phylogenetic analysis of *Vaccinium* sections *Macropelma*, *Myrtillus*, and *Hemimyrtillus* (Ericaceae). Syst. Bot. 27: 768–779.

Powell, J.M., Chapman, A.R., Doust, A.N.L. 1987. Classification and generic status in the Epacridaceae – a preliminary analysis. Austr. Syst. Bot. Soc. Newslett. 53: 70–78.

Powell, J.M., Crayn, D.M., Gadek, P.A., Quinn, C.J., Morrison, D.A., Chapman, A.R. 1996. Re-assessment of relationships within Epacridaceae. Ann. Bot. II, 77: 305–315.

Powell, J.M., Morrison, D.A., Gadek, P.A., Crayn, D.M., Quinn, C.J. 1997. Relationships and generic concepts within Styphelieae (Epacridaceae). Austr. Syst. Bot. 10: 15–29.

Pyykkö, M. 1968. Embryological and anatomical studies on Finnish species of the Pyrolaceae. Ann. Bot. Fenn. 5: 153–165.

Read, D.J. 1983. The biology of mycorrhiza in the Ericales. Can. J. Bot. 61: 985–1004.

Read, D.J. 1996. The structure and function of the ericoid mycorrhizal root. Ann. Bot. N.S. 77: 365–374.

Reader, R.J. 1977. Bog ericad flowers: self compatability and relative attractiveness to bees. Can. J. Bot. 55: 2279–2287.

Rebelo, A.G., Siegfried, W.R., Oliver, E.G.H. 1985. Pollination syndromes of *Erica* species in the south-western Cape. S. Afr. J. Bot. 51: 270–280.

Salasoo, I. 1983. Alkane distribution in epicuticular waxes of Epacridaceae. Phytochemistry 22: 937–942.

Salasoo, I. 1985. Rimuene in Epacridaceae and Ericaceae: existence of chemotypes? Austr. J. Bot. 33: 239–43.

Savile, D.B.O. 1979. Fungi as aids in higher plant classification. Bot. Rev. 45: 377–503.

Schneider, C.K. 1912. Illustriertes Handbuch der Laubholzkunde, Bd. 2. Jena: G. Fischer.

Scholtz, A. 1985. The palynology of the upper lacustrine sediments of the Arnot Pipe, Banke, Namaqualand. Ann. S. Afr. Mus. 95: 1–109.

Seithe, A. 1980. *Rhododendron* hairs and taxonomy. In: Luteyn, J.L., O'Brien, M. (eds.) Contributions toward a classification of *Rhododendron*. New York: New York Botanical Garden, pp. 89–115.

Sleumer, H. 1967. Ericaceae. In: Flora Malesiana I, 6: 469–914. Leiden: Noordhoff.

Smith, S.E., Read, D.J. 1997. Mycorrhizal symbioses. San Diego: Academic Press.

Smith-White, S. 1948a. A survey of chromosome numbers in the Epacridaceae. Proc. Linn. Soc. New S. Wales 73: 37–56.

Smith-White, S. 1948b. Polarised segregation in a stable triploid. Heredity 2: 119–129.

Smith-White, S. 1955. Chromosome numbers and pollen types in the Epacridaceae. Austr. J. Bot. 3: 48–67.

Stace, H.M., Fripp, Y.J. 1977. Raciation in *Epacris impressa*. III. Polymorphic populations. Austr. J. Bot. 25: 325–336.

Stace, H.M., Chapman, A.R., Lemson, K.L, Powell, J.M. 1997. Cytoevolution, phylogeny and taxonomy in Epacridaceae. Ann. Bot. II, 79: 283–290.

Stevens, P.F. 1970a. *Calluna*, *Cassiope*, and *Harrimanella*, a taxonomic and evolutionary problem. New Phytol. 69: 1131–1148.

Stevens, P.F. 1970b. *Agauria* and *Agarista*: an example of tropical translatlantic affinity. Notes Roy. Bot. Gard. Edinburgh 30: 341–359.

Stevens, P.F. 1971. A classification of the Ericaceae: subfamilies and tribes. Bot. J. Linn. Soc. 64: 1–53.

Stevens, P.F. 1974. Circumscription and relationships of *Dimorphanthera* (Ericaceae) and notes on some Papuasian species. Contrib. Herb. Austr. 8: 1–34.

Stevens, P.F. 1976. The altitudinal and geographical distribution of flower types in *Rhododendron* section *Vireya*, especially in the Papuasian species, and their significance. Bot. J. Linn. Soc. 72: 1–33.

Stevens, P.F. 1982. Phytogeography and evolution of the Ericaceae of New Guinea. In: Gressitt, J.L. (ed.) Biogeography and ecology of New Guinea. Monogr. Biol. 42, pp. 331–354.

Stevens, P.F. 1995. Familial and interfamilial relationships. In: Luteyn, J.L. (ed.) Flora Neotropica. Monograph 66. Ericaceae Part II, the superior-ovaried genera. New York: New York Botanical Garden, pp. 1–12.

Takahashi, H. 1986. Pollen polyads and their variation in *Chimaphila* (Pyrolaceae). Grana Palynol. 25: 161–169.

Takahashi, H. 1987. Pollen morphology and its taxonomic significance of the Monotropoideae. Bot. Mag. (Tokyo) 100: 385–405.

Takahashi, H. 1988. Pollen morphology and systematics in two subfamilies of the Ericaceae: Pyroloideae and Monotropoideae. Korean J. Plant Tax. 18: 9–17.

Takahashi, H. 1993. Seed morphology and its systematic implications in Pyroloideae (Ericaceae). Int. J. Plant Sci. 154: 175–186.

Telford, I.R. 1992. *Budawangia* and *Rupicola*, new and revised genera of Epacridaceae. Telopea 5: 229–239.

Temple, A. 1975. Ericaceae: étude architecturale de quelques espèces. Montpellier.

Thompson, W.K. 1986. Effects of origin, time of collection, auxins and planting media on rooting of cuttings of *Epacris impressa* Labill. Scientia Hortic. 30: 127–134.

Turner, V. 1982. Marsupials as pollinators in Australia. In: Armstrong, J.A., Powell, J.M., Richards, A.J. (eds.) Pollination and evolution. Sydney: Royal Botanic Gardens, pp. 55–66.

Vander Kloet, S.P. 1985. On the generic status of *Symphysia*. Taxon 34: 440–447.

Vander Kloet, S.P., Dickinson, T.A., Strickland, W. 2003. From Nepal to Formosa, a much larger foot print for *Vaccinium* sect. *Aëthopus*. Acta Bot. Yunn. 25: 1–24.

Virot, R. 1975. Epacridacées. In: Flore de la Nouvelle Calédonie et Dépendances 6. Paris: Muséum National d'Histoire Naturelle.

Wallace, G.D. 1975. Studies of the Monotropoideae (Ericaceae): taxonomy and distribution. Wasmann J. Biol. 33: 1–88.

Wallace, G.D. 1977. Studies of the Monotropoideae (Ericaceae). Floral nectaries: anatomy and function in pollination ecology. Am. J. Bot. 64: 199–206.

Wallace, G.D. 1987. Transfer of *Eremotropa sciaphila* to *Monotropastrum* (Ericaceae). Taxon 36: 128–130.

Wallace, G.D. 1995. Ericaceae subfamily Monotropoideae. In: Luteyn, J.L. (ed.) Flora Neotropica. Monograph 66. Ericaceae Part II, the superior-ovaried genera. New York: New York Botanical Garden, pp. 13–27.

Warner, B.G., Chinnappa, C.C. 1986. The implications and evolutionary trends in pollen of Canadian Ericales. Can. J. Bot. 64: 3113–3126.

Watson, L. 1962. The taxonomic significance of stomatal distribution and morphology in Epacridaceae. New Phytol. 61: 36–40.

Watson, L. 1964a. The taxonomic significance of certain anatomical observations on Ericaceae – the Ericoideae, *Calluna*, and *Cassiope*. New Phytol. 63: 274–280.

Watson, L. 1964b. Some remarkable inflorescences in the Ericales and their taxonomic significance. Ann. Bot. II, 28: 311–318.

Watson, L. 1967a. The taxonomic significance of certain anatomical variations among Ericaceae. J. Linn. Soc. Bot. 59: 111–125.

Watson, L. 1967b. Taxonomic implications of a comparative study of Epacridaceae. New Phytol. 66: 495–504.

Watson, L. 1976. Ericales revisited. Taxon 25: 269–271.

Watson, L., Williams, W.T., Lance, G.N. 1967. A mixed-data approach to Angiosperm taxonomy: the classification of Ericales. Proc. Linn. Soc. Lond. 178: 25–35.

Weiller, C.M. 1996a. Reassessment of *Cyathodes* (Epacridaceae). Austr. Syst. Bot. 9: 491–507.

Weiller, C.M. 1996b. *Planocarpa* (Epacridaceae), a new generic name. Austr. Syst. Bot. 9: 509–519.

Weiller, C.M. 1996c. Reinstatement of the genus *Androstoma* Hook.f. (Epacridaceae). New Zeal. J. Bot. 34: 179–185.

Weiller, C.M., Crowden, R.K., Powell, J.M. 1994. Morphology and taxonomic significance of leaf epicuticular waxes in the Epacridaceae. Austr. Syst. Bot. 7: 125–152.

Williams, R. 1986. Research into propagation of Australian native plants. Int. Pl. Propag. Soc. 36: 183–187.

Wood, C.E. Jr. 1961. The genera of Ericaceae in the southeastern United States. J. Arnold Arbor. 42: 10–80.

Xiao, G., Berch, S.M. 1995. The ability of known ericoid mycorrhizal fungi to form mycorrhizae with *Gaultheria shallon*. Mycologia 87: 467–470.

Generic concepts in Styphelieae are currently being revised. The following genera do not appear to be monophyletic, and will undergo significant change: *Astroloma* R.Br., *Brachyloma* Sond., *Leucopogon* R. Br. *sensu lato*, *Lissanthe* R.Br., *Monotoca* R.Br. and *Styphelia* Sm. *sensu* Benth.

Fouquieriaceae

K. Kubitzki

Fouquieriaceae DC., Prodr. 3: 349 (1828), nom. cons.

Shrubs to small trees with woody or succulent trunks, bearing simple to branched, outwardly arched to horizontal, spinose, heteroblastic branches; periderm translucent, often exfoliating and associated with the persistent epidermis; stem cortex with peripheral groups of sclereids and assimilatory tissue and inner starch- and anastomosingly arranged water-storage tissue; nodal spines 2–45 mm long, subtended by tapering, continuous decurrent ridges laterally separated by distinct or shallow furrows, or by recurrent ridges widely separated; nodes unilacunar, 1-trace. Leaves alternate, simple, entire, exstipulate, petiolate to nearly sessile, glabrous or slightly pubescent below, those of long-shoots elliptical, long-petiolate; spines rigid, separating abaxially from petioles of long-shoot leaves and continuing into the decurrent ridges of the cortex; short-shoot leaves in axillary fascicles, shortly petiolate to nearly sessile. Inflorescence determinate, terminal or axillary, spicate, racemose to paniculate or corymbosely paniculate; flowers with 2 prophylls, hypogynous, perfect; sepals 5, distinct, strongly imbricate, quincuncial, persistent; petals 5, connate into a tube, lobes imbricate; stamens hypogynous, 10(–23), unevenly exserted; filaments slightly adnate to corolla base, occasionally with basal ligulate spur; anthers tetrasporangiate, introrse, acuminate at apex, 2-lobed at base, longitudinally dehiscent; gynoecium fused of 3 carpels; ovary superior, with septiform parietal but basally axile placentation; style in upper half usually branched into 3 style branches; ovules anatropous, bitegmic, 6–20. Fruit a loculicidal capsule with columnar, axile placentae; seeds 6–15, oblong-elliptical with membranous margins of unicellular trichomes; endosperm scanty, oily and proteinaceous; embryo 3–9 mm long; cotyledons flat. x = 12.

A monogeneric family with 11 xerophytic species from the southern U.S.A. to south Mexico.

VEGETATIVE STRUCTURES. The growth habit of *Fouquieria* species varies from small woody shrubs or small trees to columnar stem succulents. The woody habit is found in six species. The remaining species are stem succulents with a highly parenchymatized, central xylary water-storage tissue. In *Fouquieria fasciculata* and *F. purpusii*, the succulent tissue is restricted to the basal or central portion of the stem, whereas the distal parts are woody. *Fouquieria columnaris* is the tallest species and has conical, tapering trunks, which are succulent from the base to the tip. Mature individuals are commonly 14, exceptionally 25 m high and may consist of a single unbranched or forked central trunk. In these stems, in which a solid xylem forms only a small portion of the stem, mechanical strength is provided by the turgescence of the parenchymatous water-storing tissue, and upon drought the trunks may collapse and bend downward (Fig. 64).

The long-shoots have simple petiolate leaves, which tend to wilt at dry periods. The short-shoots, produced in the axils of long-shoot leaves, bear fascicles of shortly petiolate or sessile leaves, which are retained even after prolonged drought. The spines are associated with long-shoot leaves in a most characteristic fashion (Fig. 65). The abaxial side of the petioles develops into a rigid, conical spine composed of narrow, elongate fibres, which continue from the spine through the underlying stem cortex to form the decurrent ridge characteristic of the stems. Upon maturation of the long-shoot, an abscission layer forms between the leaf and spine tissue and eventually the leaf is shed. The development of the spines suggests that they are of cortical origin (Humphrey 1931; Henrickson 1969).

The decurrent ridges cover young stems and are separated only by deep or shallow furrows. With the enlargement of the stems, the ridges become tangentially separated and eventually detach from the stem. The underlying, translucent periderm consists of alternating fibrous and suberinous layers and usually exfoliates from older stems in thin papery sheets. A rigid periderm forms only under the ridges, whereas the intervening areas are covered by a persistent epiderms. The cortex consists of an outer, thin, chlorenchymatous layer,

Fig. 64. Fouquieriaceae. *Fouquieria splendens*, Baja California. Photo K. Kubitzki

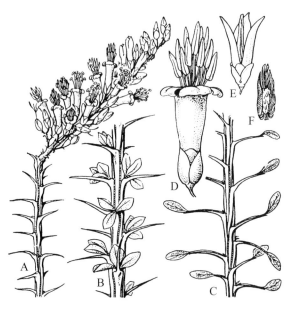

Fig. 65. Fouquieriaceae. *Fouquieria splendens*. **A** Long-shoot ending in an inflorescence. **B** Long-shoot with axillary, leaf-bearing short-shoots, the long shoot leaves fallen off. **C** Long-shoot, the leaves abscising from the spines, which appear as extensions of the cortical ridges of the axis. **D** Flower. **E** Dehiscing capsule. **F** Seed. (Takhtajan 1981, from Henrickson 1969)

which follows the pattern of translucent furrows and series of tightly packed sclereid nests. The inner cortex consists of starch-storing tissue, except for a network of water-storage tissue, which also follows the anastomosing pattern of translucent furrows (Humphrey 1935; Henrickson 1969, 1972).

In the wood, the vessel elements have simple perforations; imperforate tracheary elements have bordered pits; and wood rays are mostly homocellular and 1–8 cells wide; wood parenchyma is diffuse.

A detailed description of the vegetative anatomy of *Fouquieria* has been given, among others, by Henrickson (1969, 1972).

INFLORESCENCES. Inflorescences range from elongate or congested panicles to racemoids or spicoids and seem to be consistently determinate; in some species such as *F. columnaris* the sequence of flowering is basipetalous but acropetalous sequence predominates [the latter, in spite of the presence of terminal flowers in the schemes of Henrickson (1972), interpreted by him as indeterminate].

FLOWER STRUCTURE. The histology and vascularization of the flowers has been described by Henrickson (1972). Stamen number is variable in *Fouquieria* but in all species stamens are borne from the receptacle in a single whorl. Ten is the predominating and apparently basic number.

EMBRYOLOGY. The ovules are anatropous, bitegmic, and tenuinucellar with an integumentary tapetum; the micropyle is formed by the inner integument. The embryo sac is of Polygonum type, and endosperm development is ab initio cellular, with a chalazal haustorium (Johansen 1936; Khan 1943).

POLLEN MORPHOLOGY. (Henrickson 1973). The pollen grains are 3-colporate, more or less oblate-spheroidal and eureticulate; distinguishing features include: (1) lumina diminishing in size towards the poles and colpus margins; (2) a tendency to form variable, often striate patterns; (3) thickening of the nexine towards the poles; and (4) absence of nexine thickening around the apertures. Most distinctive are the grains of *F. columnaris*, which have a crassitectate nexine with large brochi.

KARYOLOGY. Chromosome numbers are known for all but one species, seven of which are diploid ($2n = 24$), one tetraploid, and two hexaploid (Henrickson 1972).

POLLINATION. The family exhibits a great diversity in floral features and modes of pollination (Henrickson 1972). In all species, sweet nectar is produced from nectaries around the ovary base. Whitish flowers with relatively short floral tubes, sometimes producing a sweet smell, are visited by various kinds of insects. Hummingbird visitation has been observed to the red tubular flowers borne in terminal inflorescences of species such as *F. splendens* and *F. formosa*. Some species have been shown to be self-compatible, as isolated cultivated plants set seeds, but this may be irrelevant in the wild because floral structure does not seem to favour self-pollination (Henrickson 1972).

FRUIT AND SEED. The fruit is a loculicidal capsule containing a large, parenchymatous central axis and 5–18 broad, winged seeds; its histology has been described by Henrickson (1972). Placentation changes characteristically during the maturation of the fruit: at anthesis, the ovary has three parietal placentae borne on thin septa extending into the locule but, in the process of growth of the ovary, the three septa are broken and the placentae fuse into a common, central parenchymatous column.

The flattened seeds contain a small embryo and a thin layer of endosperm. The seed wing is formed from unicellular trichomes derived from the epidermis of the outer integument (Henrickson 1972).

PHYTOCHEMISTRY. Presence of ellagic acid, caffeic acid, flavonols, glycosides of cyanidin and (in red-flowered spp.) pelargonidin, dammaran (Hegnauer 1989) and seco-iridoid glucosides (Dahlgren et al. 1976) have been reported. Ocotillo wax from the cortex of *Fouquieria splendens* is a dammaran derivative.

AFFINITIES. Fouquieriaceae are notable for exhibiting several basal traits: their ovules are bitegmic, and their wood has tracheids and diffuse axial parenchyma. The family has often been compared with Styracaceae and Ericaceae; Nash (1903) and Thorne (1968), among others, have suggested a relationship with Polemoniaceae. Molecular studies have resolved all these families as forming part of a major clade, Ericales s.l., although the precise position of *Fouquieria* within this clade remains uncertain in most recent comprehensive studies (for example, Soltis et al. 2000; Albach et al. 2001). The relationship between Fouquieriaceae and Polemoniaceae has been supported in the combined tree of a five-gene study (Anderberg et al. 2002), but not by the combined plastid genes or the mitochondrial gene alone.

DISTRIBUTION AND HABITATS. *Fouquieria* ranges from western Baja California to northern Arizona and from eastern Texas to south-eastern Oaxaca. Six species are narrow endemics. Species of the southern United States and northern Mexico occur mainly in desert habitats receiving 50–350 mm rainfall. Species in southern Mexico grow in deciduous tropical forest and arid tropical scrub averaging mostly 400–700 mm precipitation. Between desert and forest species, no difference in ecological preferences can be recognised.

A single genus:

***Fouquieria* Kunth in HBK.**   Figs. 64, 65

Fouquieria Kunth in HBK., Nov. Gen. Sp. 6: 81 (1821); Henrickson, Aliso 7: 439–537 (1972), rev.
Idria Kellogg (1860).

Description as for family.
Eleven spp. in the arid regions of Mexico and adjacent south-western U.S.A. Three subgenera were distinguished by Henrickson (1972): **subgen.**

Fouquieria, woody throughout; periderm exfoliating in thin sheets; style exserted, 8 spp.; **subgen. Bronnia** (H.B.K.) Henrickson, stems succulent, epidermis persistent; style exserted, 2 spp.; **subgen. Idria** (Kellogg) Henrickson, tall, succulent, periderm not exfoliating, style included, only *F. columnaris* (Kellog) Kellog ex Curran.

Selected Bibliography

Albach, D.C., Soltis, P.S., Soltis, D.E., Olmstead, R.G. 2001. Phylogenetic analysis of asterids based on sequences of four genes. Ann. Missouri Bot. Gard. 88: 163–212.

Anderberg, A.A. et al. 2002. See general references.

Bate-Smith, E.C. 1964. Chemistry and taxonomy of *Fouquieria splendens*: a new member of the asperuloside group. Phytochemistry 3: 623–625.

Behnke, H.-D. 1976. Sieve-element plastids of *Fouquieria*, *Frankenia* (Tamaricales), and *Rhabdodendron* (Rutaceae), taxa sometimes allied with Centrospermae (Caryophyllales). Taxon 25: 265–268.

Dahlgren, R., Jensen, S.R., Nielsen, B.J. 1976. Iridoid compounds in Fouquieriaceae and notes on its possible affinities. Bot. Not. 129: 207–212.

Hegnauer, R. 1989. See general references.

Henrickson, J. 1969. Anatomy of periderm and cortex of Fouquieriaceae. Aliso 7: 97–126.

Henrickson, J. 1971. Anatomy of periderm and cortex of Fouquieriaceae. Aliso 7: 97–126.

Henrickson, J. 1972. A taxonomic revision of the Fouquieriaceae. Aliso 7: 439–537.

Henrickson, J. 1973. Fouquieriaceae DC. World Pollen Spore Flora 1: 1–12.

Humphrey, R.R. 1931. Thorn formation in *Fouqieria splendens* and *Idria columnaris*. Bull. Torrey Bot. Club 58: 263–264.

Humphrey, R.R. 1935. A study of *Idria columnaris* and *Fouquieria splendens*. Am. J. Bot. 22: 184–207.

Johansen, D.A. 1936. Morphology and embryology of *Fouquieria*. Am. J. Bot. 23: 95–99.

Khan, R. 1943. The ovule and embryo sac of *Fouquieria*. Proc. Natl. Inst. Sci. India 9: 253–256.

Nash, G.V. 1903. A revision of the family Fouquieriaceae. Bull. Torrey Bot. Club 30: 449–459.

Soltis, D.E. et al. 2000. Angiosperm phylogeny inferred from 18 S DNA, *rbc*L, and *atp*B sequences. Bot. J. Linn. Soc. 133: 381–461.

Takhtajan, A. (ed.) 1981. See general references.

Thorne, R.F. 1968. Synopsis of a putative phylogenetic classification of the flowering plants. Aliso 6: 57–66.

Grubbiaceae

K. KUBITZKI

Grubbiaceae Endl., Gen. Pl.: 327 (1837), nom. cons.

Shrubs; leaves decussate, simple, entire, petiolate to nearly sessile; stipules wanting, but the leaf bases connected by a transversal ridge. Hairs unicellular. Inflorescences small, axillary, 3(2)-flowered dichasia or many-flowered, cone-like compound dichasia; the flowers of one inflorescence with coherent or connate inferior ovaries. Flowers subtended by bracts and provided with 2 prophylls, tiny, hermaphroditic, actinomorphic, epigynous, monochlamydeous, tetramerous, diplostemonous; tepals distinct, valvate, sepaloid, canescent-hairy on outer surface, pink to red on inner surface; stamens 8, 4 of these longer, opposite the tepals and basally attached to them, and 4 alternating with the tepals; filaments linear, ending as a blunt tip beyond the thecae; anthers dorsifixed, 2-thecate, 2-sporangiate; dehiscence lengthwise extending over the upper and lower shoulders of the anther, with 2 valves opening towards the ventral side; disk epigynous, papillate to shortly hairy; gynoecium 2-carpellate, transversal; style unbranched, with 2 short apical lobes but appearing simple; ovary inferior, initially 2-locular, becoming 1-locular by disintegration of upper part of septum above insertion of the ovules, these solitary per carpel, pendulous from the upper part of the septum, anatropous, unitegmic, tenuinucellate. Fruits 1-seeded drupes, those of one infructescence becoming laterally compressed and fused, forming a coenocarp suggesting a small cupressaceous cone; seeds with thin testa; endosperm copious, oily, its outer surface heavily cutinized; embryo straight, central, the cotyledons shorter than hypocotyl and radicle.

A single genus of 3 spp., restricted to the Cape Province of South Africa.

MORPHOLOGY. The species of *Grubbia* are evergreen shrubs; they branch from a single stem and a taproot, or ascend with numerous shoots from a lignotuber at the surface of the ground (*G. tomentosa*). Young branches are longitudinally ridged and covered with a tomentum. The leaves are leathery and linear or narrowly triangular-lanceolate, and usually have strongly revolute margins (Fig. 66A, B). The leaf bases are connected by a transverse, long-pubescent ridge or wing on each side across the stem.

The anthers are 2-sporangiate (Endress and Stumpff 1990); Hieronymus (1889) and Carlquist (1977a) observed that one microsporangium in each theca is sterile. Cronquist (1983) stated that "the orientation of the anthers of *Grubbia* is very suggestive of the Ericales, although they are not exactly like those of any other family: they are minute, inverted and adnate for their whole length to the distal part of the filament, thus appearing to be extrorse". I have been unable to confirm Cronquist's observation, but an ontogenetic study may clarify the situation; inverted anthers often have a characteristic vascular hook where the vascular bundle from the filament enters the connective (Anderberg 1992).

ANATOMY. The vegetative anatomy, including wood anatomy, has been treated by Carlquist (1977b, 1978). The nodes are trilacunar. The upper leaf epidermis has a thick cuticle or a cutinized thick wall; the lower epidermis bears anomocytic stomata and non-glandular unicellular trichomes. Rhombic crystals and crystal druses of calcium oxalate are found in the leaf tissue, and crystals also occur in the ray cells of the wood. In the secondary xylem the vessels are long and provided with oblique perforation plates with many (16–43) bars. Lateral wall pitting is alternate on vessel-tracheid contacts, and scalariform on vessel-ray contacts. Tracheids are of almost the same length as the vessels. Rays are narrow multiseriate plus uniseriate, with procumbent cells only in central portions of multiseriate rays. Axial parenchyma is sparse and diffuse. The wood of Bruniaceae and Geissolomataceae is very similar.

EMBRYOLOGY. The ovules are provided with an endothelium and a long micropyle. The sporogenous cell functions as megaspore mother cell, and the chalazal megaspore develops into a Polygonum-type embryo sac. Endosperm formation seems to be cellular (Dahlgren and van

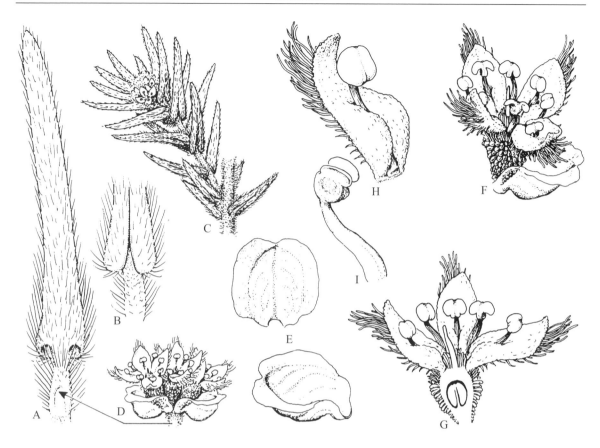

Fig. 66. Grubbiaceae. *Grubbia rosmarinifolia*. **A** Leaf, upper side, with position of floral triad indicated. **B** Leaf, base of blade seen from below. **C** Floriferous branchlet. **D** Floral triad. **E** Prophylls supporting the lateral flowers in the triad. **F** Lateral flower of triad and its bract. **G** Flower, vertical section. **H** Stamen and tepal opposite this. **I** Stamen alternating with tepals, anther locules opened. (Dahlgren and van Wyk 1988)

Wyk 1988). The chalazal end has a conspicuous, globose hypostase. At the cotyledonary stage the massive cellular endosperm shows chalazal and micropylar haustoria (Fagerlind 1947).

POLLEN MORPHOLOGY. Pollen grains of *G. rosmarinifolia* are prolate (17 × 12 μm), tricolporate with narrow colpi and lalongate ora, and tectate-baculate. The tectum is psilate and not or inconspicuously pitted; tectal perforations seem to be lacking (Dahlgren and van Wyk 1988).

FRUIT AND SEED. The individual fruits are drupes with a thin, fleshy exocarp and a thick, bony endocarp. The syncarps of *G. rosmarinifolia* are composed of 2–3 fused ovaries and do not exceed 1 mm in diameter; they may be barochorous or wind-dispersed (sect. *Ophira* Burm. & L.). Many

more (6–20) fruits fuse to form the syncarps of *G. tomentosa* and *G. rourkei*, which are 5 mm or more in length (sect. *Strobilocarpus* Klotzsch); at maturity, they turn reddish or purple and are probably dispersed by birds (Carlquist 1977a).

PHYTOCHEMISTRY. The leaves are tanninniferous, but there are negative reports of flavonoids or iridoids.

PHYLOGENY. At one time or another, families such as Santalaceae, Ericaceae, Hamamelidaceae and Bruniaceae have been considered as the closest relatives of Grubbiaceae (Carlquist 1977b; Dahlgren and van Wyk 1988). (The santalalean affinity of *Grubbia* was based on its alleged ategmic condition, which has proven to be wrong.) Carlquist (1977b) was unable to definitely exclude either of these possibilities but stressed the similarity in wood anatomy between Grubbiaceae, Geissolomataceae and Bruniaceae. Supported by the presence of allegedly inverted anthers, endosperm haustoria and unitegmic, tenuinucellate ovules, Cronquist (1981) and Takhtajan (1987) included Grubbiaceae in their narrowly circumscribed Ericales. This was also supported by various cladistic (Anderberg 1992) and molecular (summarised by APG 1998) analyses, in which Grubbiaceae together with Actinidiaceae appeared among the "lower" Ericales. More recent molecular studies, based on a broader sampling and carefully selected outgroups and employing both chloroplast and nuclear genes (Xiang 1999; Xiang et al. 2002), have resolved *Grubbia* with high statistical support in the Cornales clade sister to *Curtisia*. The close relationship between these two genera is also corroborated by a uniquely shared, short insertion in the *mat*K sequences (Xiang et al. 2002). Xiang et al. (2002) suggest to merge *Curtisia* and *Grubbia* in a single family and adduce about 11 morphological characters common to them, most of which, perhaps with the exception of the ridge connecting the bases of the opposite leaves, appear to be generalised cornalean characters, whereas the fusion of the ovaries and resulting syncarp structure in *Grubbia* and other differences (e.g. fruit structure) to my mind forbid such a shotgun marriage.

DISTRIBUTION AND HABITATS. On the basis of their relatively inefficient water conducting system and the extensive infraspecific differentiation of *G. rosmarinifolia*, Carlquist (1977a, 1977b) regarded the genus as a relictual group that has persisted in mesic pockets within the otherwise xeromorphic Mediterranean-type flora of the Table Mountain sandstone of the Cape Province. However, each of the three species occupies a different mesic habitat.

A single genus:

Grubbia Bergius   Fig. 66

Grubbia Bergius, Kong. Vet. Acad. Handl. 28: 34 (1767); Carlquist, J. S. Afr. Bot. 43: 115–128 (1977), rev.

Description as for family; 3 spp., S Africa.

Selected Bibliography

Anderberg, A. 1992. The circumscription of the Ericales, and their cladistic relationships to other families of "higher" dicotyledons. Syst. Bot. 17: 660–675.

APG (Angiosperm Phylogeny Group) 1998. See general references.

APG II (Angiosperm Phylogeny Group) 2003. See general references.

Carlquist, S. 1977a. A revision of Grubbiaceae. J. S. Afr. Bot. 43: 115–128.

Carlquist, S. 1977b. Wood anatomy of Grubbiaceae. J. S. Afr. Bot. 43: 129–144.

Carlquist, S. 1978. Vegetative anatomy and systematics of Grubbiaceae. Bot. Notiser 131: 117–126.

Cronquist, A. 1981. See general references.

Cronquist, A. 1983. Some realignments in the dicotyledons. Nord. J. Bot. 3: 75–83.

Dahlgren, R., van Wyk, A.E. 1988. Structures and relationships of families endemic to or centered in southern Africa. Ann. Missouri Bot. Gard. 25: 1–94.

Endress, P.K., Stumpff, S. 1990. Non-tetrasporangiate stamens in the angiosperms: structure, systematic distribution and evolutionary aspects. Bot. Jahrb. Syst. 112: 193–240.

Fagerlind, F. 1947. Die systematische Stellung der Familie Grubbiaceae. Svensk Bot. Tidskr. 41: 315–320.

Harms, H. 1935. Grubbiaceae. In: Engler, A., Harms, H. (eds.) Die natürlichen Pflanzenfamilien, ed. 2, 16b. Leipzig: W. Engelmann, pp. 46–51.

Hieronymus, G. 1889. Grubbiaceae. In: Engler, A., Prantl, K. Die natürlichen Pflanzenfamilien III, 1. Leipzig: W. Engelmann, pp. 228–230.

Takhtajan, A.L. 1987. See general references.

van Tieghem, P. 1897. Sur les caractères et les affinitées des Grubbiacées. J. Bot., Paris 11: 127–138.

Xiang, Q.-Y. 1999. Systematic affinities of Grubbiaceae and Hydrostachyaceae within Cornales – insights from *rbc*L sequences. Harvard Pap. Bot. 4: 527–542.

Xiang, Q.-Y., Moody, M.L., Soltis, D.E., Fan, C.-Z., Soltis, P.S. 2002. Relationships within Cornales and circumscription of Cornaceae – *mat*K and *rbc*L sequence data and effects of outgroups and long branches. Molec. Phylog. Evol. 24: 35–57.

Hydrangeaceae

L. Hufford

Hydrangeaceae Dumort., Anal. Fam.: 36, 38 (1829), nom. cons., excluding *Kania* and *Pottingeria*.

Perennial shrubs, vines, or herbs from woody rhizomes; bark exfoliating in strips or sheets. Leaves evergreen or deciduous, generally opposite, less commonly whorled or alternate, exstipulate; petioles prominent to inconspicuous; lamina simple, entire or toothed, uncommonly lobed. Inflorescences cymose, sometimes in corymbs, thyrses, or panicles; flowers few and large to numerous and small; sterile flowers with enlarged, showy calyces in some. Flowers bisexual or unisexual, rarely dioecious; calyx imbricate or valvate; sepals 4–12, free or basally united; corolla imbricate, valvate, or convolute; petals 4–12, basally fused, or petals completely united and calyptrate; androecium haplostemonous, diplostemonous, or polystemonous, stamens 4–numerous, free or basally united; filaments flat and linear, subulate, or filiform, distal forks present or absent; anthers basifixed, tetrasporangiate, with distal connective protrusion in some; gynoecium 2–12-carpellate, syncarpous; ovary partially to completely inferior; placentation usually axile at the base and parietal above, rarely strictly axile or strictly parietal; ovules 2–many, anatropous; stylodia free, or a single style, sometimes with distal style branches; stigmas usually papillate, less commonly smooth. Fruits capsules or berries. Seed length 10 mm or less; testa sculpture reticulate.

A family of 17 genera and about 220 species distributed in temperate and subtropical regions of the Americas, Pacific islands, Asia, and Europe.

VEGETATIVE MORPHOLOGY. Shoots are sympodial (Zhou and Hara 1988, 1989). Inflorescences form at the apices of branches and the distal nodes of branches. Most species are shrubs that have numerous branches at ground level, although these appear also to form subterranean rhizomes that affect asexual reproduction. Ishii and Takeda (1997) suggested that the strong basitonic branching and density-dependent turnover of stems lead to intense intra-clonal competition for resources

among shoot axes. *Cardiandra*, *Deinanthe*, and *Kirengeshoma* have only subterranean perennating organs, and their aerial branches are herbaceous and annual. Populations of these three genera often consist of dense clusters of aerial stems. *Cardiandra* and *Deinanthe* evolved herbaceous aerial stems independent of those in *Kirengeshoma*. Scrambling and vining shoots that climb by shoot-borne roots evolved separately at least twice in Hydrangeae: (1) in *Hydrangea anomala* and *Hydrangea* section *Cornidia*, and (2) in *Schizophragma*, *Pileostegia*, and *Decumaria*. Shoots of *Whipplea* are scandent, and scandent branches occur also in *Platycrater* and *Philadelphus*. Barykina and Kapranova (1983) reported asexual reproduction through the rooting of scandent shoots in *Philadelphus*.

Leaves of most species are opposite, although some are whorled or alternate. Leaves are simple and are lobed only in *Hydrangea quercifolia*, *H. sikokiana*, and *Kirengeshoma*. Most leaves of *Deinanthe bifida* have a bilobed apex. Laminas of most species are largely ovate or elliptical. Their margins range from prominently to inconspicuously toothed or entire. The laminas of *Carpenteria*, *Fendlera*, *Fendlerella*, and various *Philadelphus* from the arid North American West have entire margins. Entire margins evolved in parallel in the two clades of climbing species, notably in *Pileostegia* and in *Hydrangea* section *Cornidia*.

ANATOMY. Gregory (1998) provided a comprehensive review of vegetative anatomy. All Hydrangeaceae, except *Hydrangea* section *Cornidia*, *Kirengeshoma*, and *Pileostegia*, have erect unicellular hairs that typically have a tuberculate surface and usually a pointed apex, and that are raised on multicellular bases (Hardin and Pilatowski 1981; Gregory 1998). Similar trichomes are present also in the sister family Loasaceae. Two-armed, unicellular trichomes are present in *Kirengeshoma*, whereas two-armed, multicellular trichomes are present in *Deinanthe*. Stellate, unicellular trichomes characterize *Deutzia* (Fig. 68F), whereas stellate, multicellular trichomes are found in both *Pileostegia* and *Hydrangea* section *Cornidia*. Trichomes that have a multicellular stalk and a single apical cell occur in some species of *Philadelphus*.

Druse crystals have been reported in *Carpenteria*, *Fendlera*, *Jamesia*, *Philadelphus*, and *Whipplea*. Many species of *Deutzia* and *Kirengeshoma* share small, spherical or lenticular crystals. Crystalliferous idioblasts that bear raphides embedded in mucilage are characteristic of tribe Hydrangeae (Umemoto 1974; Gregory 1998). Excluding the wood, crystals have not been observed in *Fendlerella* but have been reported for all other genera.

Nodes are generally trilacunar and three-trace, although additional traces have been reported in *Broussaisia*, *Kirengeshoma*, *Schizophragma* (Dhillon 1975; Kapranova 1976; Gregory 1998). Some *Hydrangea* are 5- or 7-lacunar and 5- or 7-trace, respectively.

Wood anatomy has been described by Burkett (1932), Quibell (1972), Stern (1978), Styer and Stern (1979a, 1979b), and Snezhkova (1986, 1990). Growth rings are generally present. The wood ranges from diffuse-porous to ring-porous. Vessels are usually solitary, in radial multiples, and clusters of a few cells, but *Broussaisia* is reported to have strictly solitary vessels. Perforation plates are oblique and exclusively scalariform in most species. *Whipplea* and some species of *Dichroa* and *Philadelphus* have both scalariform and simple perforation plates. Perforation plates are generally simple in *Fendlerella*. Imperforate elements consist of tracheids and fiber tracheids in *Broussaisia*, *Carpenteria*, *Deutzia*, *Dichroa*, *Fendlerella*, some species of *Philadelphus*, and *Hydrangea arborescens*; only tracheids in *Fendlera*, *Fendlerella*, *Jamesia*, some species of *Philadelphus*, *H. heteromalla*, *H. paniculata*, and *H. quercifolia*; only fiber tracheids in *Decumaria*, *Schizophragma*, most species of *Hydrangea*, and in the late wood of *Whipplea*; and fiber tracheids and septate fibers in *Platycrater*. Axial parenchyma is present in most genera, although absent in *Platycrater*, some species of *Deutzia*, and most *Hydrangea*. Rays are both uniseriate and multiseriate in most genera, except bi- and multiseriate in *Schizophragma*, uni- and biseriate in *Fendlerella*, and only uniseriate in *Whipplea*. Uniseriate rays are generally heterocellular, but consist only of upright cells in *Deutzia*, *Philadelphus*, and *Whipplea*. Multiseriate rays are heterocellular (also the biseriate rays of *Fendlerella*), except in *Platycrater* and some species of *Hydrangea* in which cells are upright and square.

Bark that exfoliates in strips or sheets is characteristic of the family (absent only in a few species of *Hydrangea*) and is possibly a derived state shared with the Loasaceae, in which the condition is also common.

Stomates are usually anomocytic, varying to paracytic in *Deutzia*, but are strictly paracytic in *Dichroa* and some species of *Hydrangea* (Gregory 1998).

Leaf venation is acrodromous in *Fendlera*, *Fendlerella*, *Philadelphus*, and *Whipplea*, but pinnate in all other genera (Gregory 1998). Secondary veins in the pinnately veined taxa vary slightly: *Carpenteria*, *Deutzia*, and *Jamesia* have basal secondary veins that are eucamptodromous, but most others are brochidodromous (or semicraspedodromous in *Deutzia*). *Kirengeshoma* is craspedodromous, with transitions to actinodromous. Among Hydrangeae, *Cardiandra*, *Deinanthe*, and *Platycrater* are eucamptodromous; *Decumaria* and *Pileostegia* are brochidodromous; *Schizophragma* is cladodromous, whereas *Hydrangea* s.s. is brochidodromous, although tending to semicraspedodromous in section *Cornidia* (Watari 1939; Stern 1978; Hao and Hu 1996a; Gregory 1998).

FLOWER STRUCTURE. The floral dimorphism centered on perfect fertile flowers and sterile display flowers that have an enlarged calyx (Fig. 70) is a derived feature characteristic of Hydrangeae (but lost in *Broussaisia*, *Dichroa*, *Pileostegia*, *Decumaria*, and some species of *Hydrangea* section *Cornidia*) (Hufford 1997). Floral dimorphism associated with dioecy is characteristic of the Hawaiian island endemic *Broussaisia*.

The fertile flowers are actinomorphic and epigynous. Ovaries are largely completely inferior in genera such as *Deutzia*, *Decumaria*, some *Dichroa* and most *Hydrangea*; closer to half inferior in *Carpenteria*, most *Philadelphus*, *Broussaisia* (female flowers), some *Hydrangea* and *Dichroa*; and less than half inferior for instance in *Jamesia*, *Kirengeshoma* and *Broussaisia* (male flowers).

Among most Hydrangeae, perianth merosity varies between tetramerous and pentamerous. *Jamesia*, *Fendlerella*, and *Whipplea* typically have a pentamerous perianth, whereas that of *Fendlera* and *Philadelphus* is typically tetramerous. In *Carpenteria* and *Deinanthe*, the perianth is pentamerous to octamerous. The *Decumaria* perianth is most commonly octamerous (Fig. 71B), although it ranges from hexamerous to dodecamerous. Sepals are usually basally connate (most prominent in *Kirengeshoma*), but they appear more-or-less free in *Carpenteria*, *Deinanthe*, and *Philadelphus*. The petals are free or tenuously connate at the base in all except *Pileostegia* and *Hydrangea anomala*, in which the petals can be entirely united and abscise as a unit at anthesis. The corolla is funnel- to bowl-shaped, rotate, erect, or reflexed.

Both diplostemonous and polystemonous androecia are common (Wettstein 1893; Gelius 1967; Hufford 1998). Haplostemony is present only in *Dichroa pentandra* and *D. platyphylla*. Stamen filaments of *Fendlera* and various species of *Deutzia* are distally flanged, and these form a corona in the center of the flower. Filaments are usually bifacial, at least basally, and strongly so in *Fendlerella*, *Jamesia*, *Whipplea*, and especially in *Fendlera* and some *Deutzia*, although they are unifacial over most of their length in *Carpenteria*, some *Deutzia*, *Kirengeshoma*, *Philadelphus*, and Hydrangeae. Anthers are basifixed, bithecate and tetrasporangiate. The connectives have distal protrusions that project beyond the thecal junction in *Fendlera*, *Fendlerella*, *Jamesia*, *Whipplea* and some species of *Deutzia*. *Fendlerella*, *Jamesia*, *Whipplea*, and *Decumaria sinensis* have a fleshy protrusion on the abaxial side of the connective. The endothecium consists of a single cell layer, except at the ends adjacent to the connective on the abaxial side of the anther at which 2–4 layers are present for a short distance. Endothecium is limited to the thecal regions. Two to three layers of thin, elongate cells are positioned between the endothecium and tapetum. The tapetum may be present or absent at dehiscence (Saxena 1971). Anther dehiscence is slightly introrse in *Jamesia*, *Fendlera*, *Fendlerella*, *Whipplea*, *Carpenteria*, *Philadelphus*, *Kirengeshoma*, and some *Deutzia*, whereas Hydrangeae have evolved latrorse dehiscence, but *Broussaisia* and *Decumaria sinensis* have a reversal to introrse dehiscence.

Ovaries are connate over their entire length (Morf 1950). They are synascidiate at the base (characterized by continuous septa), but symplicate in the mid and upper regions (characterized by discontinuous septa) (Klopfer 1971, 1973). The genera differ in the extent of stylar connation: *Cardiandra*, *Deutzia*, *Fendlera*, *Hydrangea*, *Kirengeshoma*, *Platycrater*, and some *Dichroa* have completely separate stylodia; genera such as *Fendlerella* and *Philadelphus* have styles with free style branches; and others (e.g., *Broussaisia* and *Schizophragma*) have simple styles. *Deinanthe* also has a simple style, but this arises through the postgenital fusion of initially free stylodia.

Placentae of most genera are axile at the base and parietal distally. Some species of *Deutzia* have

only parietal placentation. Examined *Fendlerella* and *Whipplea* have axile placentation.

Ovules are pendant in *Decumaria*, *Fendlera*, *Fendlerella*, *Kirengeshoma*, *Pileostegia*, *Philadelphus*, *Schizophragma*, and *Whipplea*; oriented upward in examined *Deutzia*, and *H. heteromalla*; and horizontal in *Broussaisia*, *Carpenteria*, *Dichroa*, *H. anomala* subsp. *petiolaris*, *H. hirta*, and *H. macrophylla*. In *Cardiandra*, *Deinanthe*, *Jamesia*, and *Platycrater* ovule orientation varies with location.

Taxa that have style branches or stylodia have a stigma at the apex of each; the stigmatic zone extends down the furrow on the ventral side in *Fendlera*, *Fendlerella*, and *Whipplea*. The completely synstylous genera, including *Deinanthe*, have stigmas positioned along furrows on the outer surface of the stylar apex. Stigmas are papillate, except in *Deinanthe*. Stigmas of *Deinanthe* have a smooth surface that becomes coated by secretions.

The functionally dioecious *Broussaisia* has male flowers that have a sterile pistillode, and the female flowers lack stamens or have rudimentary staminodes.

EMBRYOLOGY. Pollen is released at the binucleate stage. The unitegmic, tenuinucellate ovules are anatropous. Embryo sac development is Polygonum type. The lower half of the embryo sac is enclosed by an endothelium. The embryo sac extends into or beyond the micropyle in various genera. These genera have micropylar haustoria. The egg apparatus of examined *Deutzia* and *Philadelphus* is positioned outside of the ovule. The antipodal cells of *Kirengeshoma* become haustorial. The polar nuclei fuse before fertilization. Endosperm formation is ab initio cellular, except in *Fendlera* (Mauritzon 1933, 1939; Davis 1966; Saxena 1971).

POLLEN MORPHOLOGY. Pollen is relatively uniform among the genera (Agababyan 1961; Wakabayashi 1970; Pastre and Pons 1973; Hideux and Ferguson 1976; Hao and Hu 1996b). The tricolporate grains are spheroidal to slightly prolate or oblate. The exine is tectate to semitectate. The tectum is perforate to reticulate. The exine can have a secondary sculpture of minute granulations. Pollen grains are small (11.2–22.4 × 9.8–21.0 μm).

KARYOLOGY. Base chromosome numbers among the genera include x = 11, 13, 14, 15, 16, 17, and 18. Sax (1931) reported the regular formation of polyploid series among species of *Deutzia*. Tetraploids have also been reported in *Hydrangea* (Sax 1931), although they are uncommon. Chromosome numbers do not vary among species of *Philadelphus*, but morphological differences among the chromosomes of different species interfere with their pairing in interspecific crosses and limit natural hybridizations (Ammal 1951). Hamel (1951) demonstrated size differentiation among chromosomes in *Deinanthe*, *Kirengeshoma*, and *Schizophragma*. Chromosome complements of *Deutzia* are reported to be little differentiated in either size or shape (Ohba and Akiyama 1992).

REPRODUCTIVE SYSTEMS. The reproductive biology of Hydrangeaceae remains largely unexamined. Flower size and number as well as ovule size and number vary among the clades. Trade-offs in these features, which are associated with changes in stamen number and pollen:ovule ratios, appear to be central to reproductive strategies (Hufford, unpublished data).

In *Hydrangea*, Pilatowski (1981) reported that the densely flowered inflorescences serve as a platform on which various unspecialized pollinators land and forage. Robertson (1892) found *Hydrangea arborescens* was visited primarily by many kinds of bees (including various species of *Halictus*) and flies as well as by a few coleopterans and lepidoterans.

The tropical *Broussaisia* is the only member of the family known to be dioecious. Degener (1945) reported that the flowers were visited by beetles and wasps.

FRUIT. Fruits are capsular, except in the tropical *Broussaisia* and *Dichroa*. Capsular fruits open variously. Septicidal dehiscence that proceeds basipetally from the base of the stylodia is plesiomorphic and characteristic of *Fendlera*, *Jamesia*, *Fendlerella*, *Whipplea*, and *Carpenteria*. The basipetal splits extend to mid fruit in *Fendlera* and *Jamesia*. Dehiscence proceeds to near the base of the fruit in *Fendlerella* and *Carpenteria*, although the style of *Carpenteria* restricts the apical separation of the fruit, creating window-like openings maintained until the styles break. In *Whipplea*, the individual, single-seeded carpels separate entirely. *Philadelphus* has loculicidal capsules that dehisce basipetally from the base of the style to the lower half of the fruit. *Kirengeshoma* has both septicidal and loculicidal dehiscence that progresses basipetally but extends little from the apex of the ovary, mimicking the dehiscence of Hydrangeae (especially *Deinanthe*).

Deutzia fruits have two regions of dehiscence. They are septicidal but, unlike the above taxa, split acropetally along ovarian septa from the base of the fruit. The individual carpels spread slightly at the base, creating a broad gap through which most seeds appear to be dispersed. In addition, the fruits also dehisce apically, creating an opening between the stylodia.

Most Hydrangeae, including *Cardiandra*, *Deinanthe*, *Platycrater*, and *Hydrangea*, in which seed sizes are relatively small (Hufford 1995), have dehiscence that creates a gap between the stylodia. This has been lost in association with an increase in seed length in *Schizophragma*, *Pileostegia*, and *Decumaria*, in which intercostal portions of the lateral walls of the ovary separate from the costal ribs. After the intercostal pieces have fallen from the fruits, they have a cage-like appearance. Among Hydrangeae, the tropical genera *Broussaisia* and *Dichroa* shifted from dry, dehiscent fruits to berries. Those of *Dichroa* are bright pink to bluish, and of *Broussaisia* purplish, red, or pinkish. Seeds of *Broussaisia* are surrounded by finger-like ingrowths of the fruit wall.

The inner wall layers surrounding the locules differentiate during late ovary development as a mechanical region that has lignified cell walls. The cells of this region are oriented perpendicular to the long axis of ovary. These layers of lignified cells are lacking in *Dichroa*.

SEEDS. Seeds are 5 mm or less in length, except in *Kirengeshoma* in which seeds are up to 10 mm. The embryo is straight and embedded in endosperm (Corner 1976). Seeds are mostly ellipsoidal, ovate, or spindle-shaped, but funnel-form in *Philadelphus*, urceolate in *Broussaisia*, *Dichroa*, and their close relatives *Hydrangea hirta* and *H. scandens*, obovate, disc-like in *H. anomala*, horn-shaped in *Decumaria barbara*, and bottle-form in *Pileostegia* (Hufford 1995). All have a reticulate testa sculpture, although secondary sculptures vary, including papillae, granulations, rugae, striations, and smooth. Exotesta cell walls are lignified (Nemirovich-Danchenko and Lobova 1998). Seeds are typically winged, and the wing is usually restricted to the chalazal end but extends entirely around the seed in *Kirengeshoma* and *Hydrangea anomala*. *Schizophragma*, most *Philadelphus*, most *Deutzia*, and *Decumaria sinensis* have a micropylar flange that may function as wing. *Fendlerella* and *Whipplea* have a funicular appendage.

DISPERSAL. The small, winged seeds of most Hydrangeaceae are undoubtedly wind dispersed.

They are readily thrown from the dehisced fruits when branches move in wind. A seed wing is lacking in *Broussaisia, Dichroa*, and their close relatives *H. hirta* and *H. scandens*, which have urceolate seeds. This is associated with a shift to indehiscent, fleshy fruits and presumably animal dispersal in *Broussaisia* and *Dichroa*.

PHYTOCHEMISTRY. Most genera of Hydrangeaceae contain secoiridoids and loganin (Jensen et al. 1975). *Deutzia* is unique among examined members of the family in having iridoids that lack C-10 (Frederiksen et al. 1999). Kaempferol and quercetin are widespread; myricetin was found in *Decumaria* and *Jamesia* (Bohm et al. 1985). Procyanidin and prodelphinidin have repeatedly been recorded (Bate-Smith 1978), and some *Hydrangea* were found to contain ellagic and gallic acids (Hegnauer 1973). The alkaloids febrifugin and isofebrifugin (see Economic Importance below) have been isolated from *Dichroa* and *Hydrangea* (Hegnauer 1973). *Hydrangea* has fatty acid compounds dominated by linoleic acid, as is common among most other Cornales (Breuer et al. 1987).

SUBDIVISION AND RELATIONSHIPS IN THE FAMILY. Phylogenetic analyses (Soltis et al. 1995; Hufford et al. 2001) indicate that *Jamesia* and *Fendlera* form a monophyletic group (subfam. Jamesioideae) that is the sister clade to the rest of the family (subfam. Hydrangeoideae). Hydrangeoideae consist of the sister tribes Philadelpheae and Hydrangeae (sensu Hufford et al. 2001). In the Philadelpheae, current phylogenetic sampling indicates that *Carpenteria* and *Philadelphus* are sister clades, *Deutzia* and *Kirengeshoma* are sister clades, and *Fendlerella* and *Whipplea* are sister clades. The sister of the *Fendlerella-Whipplea* clade within Philadelpheae remains unclear. Relationships are less well resolved in Hydrangeae (Hufford 1997; Hufford et al. 2001). *Cardiandra* and *Deinanthe* are monophyletic, and there is some support for their placement as the sister of the rest of the Hydrangeae (=*Hydrangea* s.l.). *Hydrangea* s.str. is paraphyletic (Soltis et al. 1995; Hufford 1997; Hufford et al. 2001). *Broussaisia, Decumaria, Dichroa, Pileostegia, Platycrater*, and *Schizophragma* evolved among the clades of *Hydrangea* s.str. *Broussaisia* and *Dichroa* form a monophyletic group with *H. hirta, H. macrophylla*, and *H. scandens*. The vines *Decumaria, Pileostegia*, and *Schizophragma* form a monophyletic group in *Hydrangea* s.l. (Soltis et al. 1995; Hufford 1997; Hufford et al. 2001).

AFFINITIES. Hydrangeaceae are the sister clade of Loasaceae (Downie and Palmer 1992; Hempel et al. 1995; Soltis et al. 1995; Xiang et al. 1998; Hufford et al. 2001). They are well supported as a clade of Cornales (sensu APG 1998).

Hydrostachys has been affiliated with Hydrangeaceae in various molecular systematic studies (Hempel et al. 1995; Albach et al. 2001); however, its exceptional branch length makes this placement dubious (Hufford 1997; Hufford et al. 2001).

DISTRIBUTION AND HABITATS. Soltis et al. (1995) suggested that Hydrangeaceae had a New World origin. Hydrangeaceae and their sister clade Loasaceae may have originated in mesophytic or xerophytic environments of Central America or southwestern North America. The three largest genera, *Deutzia, Hydrangea* (also the monophyletic *Hydrangea* s.l.) and *Philadelphus*, are disjunct between the Old and New Worlds. The two species of *Decumaria*, which are part of *Hydrangea* s.l., are disjunct between eastern Asia and the southeastern United States. Hu (1954) suggested that the most primitive species of *Philadelphus* are in mesophytic areas of Central America, and that major clades of the genus continue to have extant representatives in mountainous areas of Mexico, in which the genus radiated. *Deutzia* is disjunct between Mexico and Central and eastern Asia. It is unclear whether the Mexican species of *Deutzia* (sect. *Neodeutzia*), which are uniquely polystemonous in the genus, are a relictual group of an initial radiation of the genus in Central America or a derived clade. *Hydrangea* sect. *Cornidia*, which is limited to Central and South America, except for the disjunct *H. integrifolia* Hayata of the Philippine Islands and Taiwan, is nested as a derived clade of *Hydrangea* s.l. and, thus, may represent a secondary expansion into Central America. As discussed below (Paleobotany), clades of Hydrangeoideae were widespread in the Northern Hemisphere during the Tertiary, and the fragmentation and migration of Tertiary forest elements may account for the current distribution of *Hydrangea* sect. *Cornidia*.

Fendlera, Fendlerella, Jamesia, and *Whipplea* are restricted to northern Mexico and western North America. Axelrod and Raven (1985) suggested *Jamesia* originated in the Southern Rocky Mountains (U.S.A) in the Eocene. Holmgren and Holmgren (1989) suggested *Jamesia* spread to the west after Miocene uplifts.

Various smaller genera of Hydrangeae are limited in distribution to Asia. The vining *Pileoste-*

gia and *Schizophragma* are found on Taiwan and in eastern Asia and the Himalayan region. *Deinanthe* and *Platycrater* are distributed in southern Japan and China in moist, temperate forests. *Cardiandra*, the sister clade of *Deinanthe*, is also found in Japan and China, and also on Taiwan. *Dichroa*, which is distributed through southern China, the Himalayas, and Southeast Asia (including the Malay Archipelago, Philippine Islands, and New Guinea), is nested within *Hydrangea* s.l. as part of the "macrophylla" clade (Hufford et al. 2001). The "macrophylla" clade also includes *H. macrophylla*, *H. scandens*, and *H. hirta*, and currently extends from the Himalayas through China to Taiwan and Japan. The Hawaiian endemic *Broussaisia* is also associated with the "macrophylla" clade, and its present distribution may represent a long distance disperal from Asia.

Jamesia and *Fendlera*, which compose subfam. Jamesioideae, grow in rocky, often steep habitats that are exposed and relatively arid. In tribe Philadelpheae (subfam. Hydrangeoideae), *Fendlerella* and some species of *Philadelphus* grow in environments very similar to those of *Jamesia* and *Fendlera*. *Carpenteria*, *Whipplea*, and many other species of *Philadelphus* are found in only slightly less exposed, arid habitats. *Kirengeshoma* and members of tribe Hydrangeae (subfam. Hydrangeoideae) are found in more mesophytic habitats. Many Hydrangeae are found in shaded to open forests or forest margins, especially on slopes, cliffs, riparian banks. *Broussiasia*, *Dichroa*, and *Deinanthe* are found in more moist, shaded habitats. The shifts of *Broussaisia* and *Dichroa* into wet tropical forests are clearly derived in the family.

PALEOBOTANY. Hydrangeaceae have a rich fossil record that includes seeds, leaves, sterile flowers, and fruits. Most fossil Hydrangeaceae are from the Tertiary when members of the family appear to have been common elements of mesophytic forests in the northern hemisphere (Hu and Chaney 1940; Ozaki 1991; Manchester 1994; Mai 1998).

Tribe Hydrangeae was more widespread in the Northern Hemisphere during the Tertiary than currently. *Hydrangea* fossils are known variously from the Oligocene to Pliocene in Europe and western North American where they are not members of the extant flora (Chaney and Sanborn 1933; MacGinitie 1941; Gorbunov 1970; Lancucka-Srodoniowa 1975; Mai 1985; Manchester 1994). Similarly, *Schizophragma* is known in extant floras only from eastern Asia, but Mai (1985) has described Pliocene fossils of *Schizophragma*

polonica from Europe. *Dichroa bornensis*, described from the Middle Oligocene of Europe by Mai (1985), lacks character states sufficient to include it in the extant genus *Dichroa* (Hufford 1995) but can be reasonably assigned to the "macrophylla" clade of *Hydrangea* (Hufford et al. 2001).

Philadelphus fossils similar to the extant *P. lewisii* have been found from the Pliocene of California, U.S.A. (Condit 1944) and Upper Cretaceous of Vancouver Island, Canada (Bell 1957), which are close to the current range of the extant species. Subfam. Jamesioideae is represented by fossil leaves of *Jamesia caplanii* Axelrod from the Oligocene of southwestern Colorado, U.S.A. (Axelrod 1987). These fossil leaves are very similar to those of the extant *J. americana*, which is currently part of the flora of Colorado and the southwestern U.S.A.

ECONOMIC IMPORTANCE. Roots of *Dichroa febrifuga* have been used as a traditional source of antimalarial drugs in Asia (Fairbairn and Lou 1950; Steinmetz 1972; Murata et al. 1998; Kim et al. 2000). Compounds from these roots are a potential source of anti-inflammatory drugs (Kim et al. 2000). Dried roots of *Hydrangea arborescens* have been used as a diuretic and diaphoretic (Spongberg 1972). *Deutzia* has been used in China to treat enuresis, malaria, and scabies (He 1990).

The Japanese prepare the sweet beverage *amacha* from steamed and dried leaves of *Hydrangea macrophylla* (Spongberg 1972).

Many genera are important ornamentals, especially *Deutzia* (Zaïkonnikova 1966), *Hydrangea* (McClintock 1956), *Kirengeshoma* (Cannon 1981), *Philadelphus* (Wyman 1965), and *Schizophragma* (Nevling 1964).

CONSERVATION. *Carpenteria* is of conservation concern in California, U.S.A. because of habitat destruction. *Kirengeshoma* is rare in Japan.

SUBFAMILIAL CLASSIFICATION

I. Subfam. Jamesioideae Hufford (2001).
   Genera 1–2.

II. Subfam. Hydrangeoideae A. Br. (1864).
1. Tribe Philadelpheae Reichenbach (1828).
   Genera 3–8.
2. Tribe Hydrangeae DC. (1830).
   Genera 9–17.

KEY TO THE GENERA

1. Aerial shoots strictly herbaceous and annual ... 2
- Aerial shoots woody and perennial ... 4

2. All flowers fertile (sterile display flowers lacking); sepals united for over half of their length; petals largely erect, curving outward only at their tips 　　**6. Kirengeshoma**
– Both fertile and sterile flowers (the latter with enlarged calyces); sepals only slightly united at the base; petals reflexed from the base 　　3
3. Flowers less than 1 cm wide; stamens 12–20; stylodia separate, stigmas terminal and papillate
　　10. Cardiandra
– Flowers more than 1.5 cm wide; stamens >100; style bottle-brush-like because of protrusive stigmatic zone, stigmas smooth 　　**9. Deinanthe**
4. Style simple; stigma either capitate or with bottle-brush form 　　5
– Stylodia free or style branched into style branches, lacking capitate or bottle-brush stigmas 　　9
5. Shrubs with numerous basal branches; fruits septicidal capsules or indehiscent berries 　　6
– Vines; fruits dehisce through loss of lateral intercostal walls 　　7
6. Leaves deciduous; flowers bisexual; flowers more than 3 cm wide; stamens >200; fruits septicidally dehiscent; seeds elliptical with a thin wing 　　**8. Carpenteria**
– Leaves evergreen; flowers unisexual, plants dioecious; flowers less than 1.5 cm wide, male flowers with 8–12 stamens; fruits fleshy, indehiscent (berry-like); seeds urceolate, lacking a wing 　　**13. Broussaisia**
7. Leaves deciduous; petals separate or tenuously connate at their apices; seeds ellipsoidal with micropylar flange and chalazal wing or horn-shaped with broadened micropylar pole and pointed chalazal wing 　　8
– Leaves evergreen; petal margins connate and corolla calyptrate (abscising at anthesis); seeds bottle-shaped with neck-like micropylar pole and rounded chalazal pole
　　15. Pileostegia
8. Inflorescences of both sterile and fertile flowers; petals 4–5; stamens 8–10; seeds ellipsoidal 　　**14. Schizophragma**
– Inflorescences of only fertile flowers; petals usually 6–12; stamens >12; seeds horn-shaped 　　**16. Decumaria**
9. Ovules 1 per locule; seeds with funicular appendage (easily detached in *Fendlerella*) 　　10
– Ovules at least 4 per locule; seeds without funicular appendage 　　11
10. Erect shrubs; leaves entire; capsules ovoid to ellipsoidal; seeds straight 　　**3. Fendlerella**
– Trailing shrubs; leaves toothed; capsules globose; seeds slightly crescent-shaped 　　**4. Whipplea**
11. Inflorescences generally with both sterile and fertile flowers; petals reflexed or absent; fruits dehiscing only apically, forming an opening between the stylodia 　　12
– Inflorescences only with fertile flowers; petals funnel-form to rotate or reflexed; fruits apically or basally septicidal (if the latter then also with apical interstylar opening), or loculicidal, or berries 　　13
12. Flowers less than 1 cm wide; stamens fewer than 30
　　11. Hydrangea
– Flowers greater than 1 cm wide; more than 200 stamens
　　17. Platycrater
13. Corolla valvate; petals reflexed; fruits berries; seeds urceolate 　　**12. Dichroa**
– Corolla imbricate or valvate; petals funnel-form to rotate; fruits capsular; seeds ellipsoidal to funnel-form 　　14
14. Stellate trichomes on vegetative and reproductive structures; stylodia completely separate; fruits dehiscing both apically between the stylodia and septicidally from base acropetally 　　**5. Deutzia**

– Stellate trichomes lacking; stylodia free, or style simple or branched; fruits either loculicidal or septicidal but not both septicidal and apical between stylodia 　　15
15. Stamens >15; fruits loculicidal 　　**7. Philadelphus**
– Stamens <15; fruits septicidal 　　16
16. Leaves entire; stamen filaments have two distal flanges; seeds >4 mm long 　　**2. Fendlera**
– Leaves toothed; stamen filaments lack distal flanges; seeds <2 mm long 　　**1. Jamesia**

Genera of Hydrangeaceae

I. Subfam. Jamesioideae Hufford (2001).

Shrubs. Sterile flowers absent. Perianth 4–5-merous, petals imbricate, androecium diplostemonous, carpels 3–5. Fruit capsular, seeds <6 per locule.

1. *Jamesia* Torr. & Gray 　　Fig. 67

Jamesia Torr. & Gray, Fl. N. Am. 1: 593 (1840); Holmgren and Holmgren, Brittonia 41: 335–350 (1989), rev.

Shrubs. Leaves deciduous, petiolate, lamina toothed. Inflorescence cymose panicle. Sterile flowers absent. Sepals 4–5, valvate. Petals 4–5, imbricate, spathulate, white to pinkish. Stamens 8–10, filaments subulate, anthers with abaxial

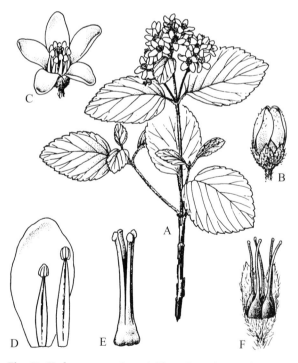

Fig. 67. Hydrangeaceae-Jamesioideae. *Jamesia americana*. **A** Flowering branch. **B** Floral bud. **C** Flower. **D** Petal and stamens. **E** Pistil with free stylodia. **F** Fruit, after dehiscence. (Holmgren and Holmgren 1989)

and distal connective protrusion. Ovary 3–5-carpellate, partly inferior, style short with long style branches, or stylodia free; ovules oriented variously. Fruit septicidal capsule. Seeds ~20, 0.5–1 mm long, winged, without micropylar flange. $n = 16$ (Sax 1931; Hamel 1953; Löve 1969). Two species. Most common in the Southern Rocky Mountains, U.S.A., but scattered in various mountain ranges of the American Great Basin and Southwest and northern Mexico.

2. *Fendlera* Engelm. & Gray

Fendlera Engelm. & Gray, Smithson. Contrib. 3: 77 (1852).

Shrubs, bark of twigs longitudinally ridged. Leaves deciduous, petiole inconspicuous, lamina entire. Inflorescence dichasial cyme. Sterile flowers absent. Sepals 4, valvate. Petals 4, imbricate, spathulate, white. Stamens 8, filaments broad, linear, dorsiventrally flattened with forked apex, anther with distal connective protrusion. Ovary 4-carpellate, partly inferior, stylodia free, ovules pendant. Fruit septicidal capsule. Seeds 15–25, 4–5.5 mm long, wing and micropylar flange lacking. $n = 11$ (Sax 1931; Hamel 1953). Two to four species, in the American Southwest and northern Mexico in shrub or pinyon pine-juniper communities on steep slopes and canyon walls of high deserts. Revision needed.

II. Subfam. Hydrangeoideae A. Br. (1864).

Shrubs or herbaceous perennials. Sterile flowers absent or present. Perianth 4–12-merous, petals imbricate or valvate, androecium diplostemonous, haplostemonous, or polystemonous, carpels 2 or more. Fruit capsule or berry, seeds 1–many per locule.

1. Tribe Philadelpheae Reichenbach (1828).

Shrubs or herbaceous perennials. Sterile flowers absent. Capsular fruits either loculicidal or septicidal or both.

3. *Fendlerella* A.A. Heller

Fendlerella A.A. Heller, Bull. Torrey Bot. Club 1828: 626 (1828).

Shrubs, bark of twigs longitudinally ridged. Leaves deciduous, petiole inconspicuous, lamina entire. Inflorescence cymose panicle. Sterile flowers

absent. Sepals (4)5, apert. Petals (4)5, apert, spathulate, white. Stamens (8)10, filaments dorsiventrally flattened and subulate, anthers have abaxial and distal connective protrusion. Ovary (2)3(4–5)-carpellate, partly inferior; short simple style with style branches, ovules pendant. Fruit septicidal capsule. One seed per carpel, 1–2 mm long, wing and micropylar flange lacking, funicular appendage present. Four species, in northern Mexico and American Southwest in shrub or pinyon pine-juniper communities on steep slopes and canyon walls. Revision needed.

4. *Whipplea* Torr.

Whipplea Torr., Pacif. Rail. Rep. 4: 90 (1856).

Shrubs, trailing or low growing, bark of twigs longitudinally ridged. Leaves deciduous, petiolate, lamina toothed. Cymose panicle. Sterile flowers absent. Sepals (4)5(6), apert. Petals (4)5(6), apert, spathulate, white. Stamens (8)10(12), filaments dorsiventrally flattened and subulate, anthers with abaxial and distal connective protrusion. Ovary (2-3)4–5-carpellate, partly inferior, style short with style branches; ovules pendant. Fruit septicidal capsule. One seed per locule, 1–1.5 mm long, wing and micropylar flange lacking, funicular appendage present. One species, *Whipplea modesta* Torr. In California, Oregon, and Washington, U.S.A., in dry, rocky areas of open forests of coast ranges.

5. *Deutzia* Thunb.   Fig. 68

Deutzia Thunb., Nov. Gen.: 19 (1781); Zaikonnikova, Deutzias – ornamental shrubs – a monograph of the genus *Deutzia* Thunb. (1966), rev.; Zaikonnikova, Baileya 19: 133–144 (1975), key; He, Phytologia 69: 332–339 (1990), rev. Sichuan, China spp.

Shrubs. Leaves deciduous, petiolate, lamina toothed. Infloresence cymose panicle, dichasial cymes. Sterile flowers absent. Sepals 5, valvate. Petals 5, imbricate or valvate, spathulate to ovate, white to pinkish. Stamens 10–15, stamen filaments subulate to linear with forked apex. Ovary 3–5-carpellate, completely inferior, stylodia free, ovules horizontal. Fruit capsular, dehisces both apically between styles and acropetally septicidal from the base of the fruit. Seeds numerous, 0.5–3 mm long, winged. $n = 13, 26, 39, 52, 65$ (Sax 1931; Löve 1969; Ohba and Akiyama 1992). Approximately 60 species. Zaikonnikova (1966) recognized the sections *Neodeutzia* Engl., *Mesodeutzia* C.K. Schn., and *Deutzia* as well as six subsections and 16

Fig. 68. Hydrangeaceae-Hydrangioideae. *Deutzia paniculata.* **A** Flowering branch. **B** Flower. **C** Flower, seen from beneath. **D** Ovary, transverse section. **E** Fruiting twig. **F** Underside of leaf. Drawn by T. Yamada. (Nakai 1926)

series. The genus is disjunct between Mexico (only the approximately four species of sect. *Neodeutzia*) and Asia. Species of sects. *Mesodeutzia* and *Deutzia* are distributed from Korea, Japan, Taiwan, and the Philippine Islands throughout most of China and the Himalayan region. Most species are Asian (Zaikonnikova 1975; He 1990).

6. *Kirengeshoma* Yatabe

Kirengeshoma Yatabe, Bot. Mag. (Tokyo): 5: 1 (1890).

Herbaceous perennial from rhizome. Leaves opposite-subopposite, petiolate, palmately lobed, lamina toothed. Inflorescence cymose panicle. Sterile flowers absent. Sepals 5, valvate, prominent calyx tube. Petals 5, imbricate, spathulate, yellow to creamy white. Stamens ~15, stamen filaments subulate. Ovary 3–4-carpellate, partly inferior, stylodia free, ovules pendant. Fruit basipetally septicidal and loculicidal capsule. Seeds numerous, 4–10 mm long, strongly flattened laterally,

Fig. 69. Hydrangeaceae-Hydrangeoideae. **A–D** *Philadelphus inodorus.* **A** Flowering branch with dehisced capsules from previous season. **B** Flower and bud with valvate calyx lobes. **C** Style with style branches. **D** Transverse section of ovary. **E–H** *P. hirsutus.* **E** Ovary in vertical section showing auriculate placentae and pendulous ovules. **F** Young fruit soon after disarticulation of corolla and androecium showing the ovarian disk. **G** Mature, dehisced capsule. **H** Seed. (Spongberg 1972)

winged. $2n = 52, 54$ (Hamel 1951; Funamoto and Nakamura 1989). Two species. Japan and Korea. In mountain forests.

7. *Philadelphus* L.   Fig. 69

Philadelphus L., Sp. Pl.: 1: 470 (1753); Hu S.-Y., J. Arnold Arbor. 35: 275–333 (1954), 36: 52–109 (1955), 37: 15–90 (1956), rev.; Hwang, Acta Bot. Austro Sinica 7: 1–12 (1991), synopsis of Chinese species.

Erect or semi-scandent shrubs. Leaves deciduous or persistent, petioles often inconspicuous, lamina entire or toothed. Flowers solitary, in dichasial cymes, or cymes aggregated in panicles. Sterile flowers absent. Flowers often strongly fragrant. Sepals 4(5), valvate. Petals 4(5), imbricate, spathulate to ovate, white. Stamens >15, filaments subulate to filiform. Ovary (3)4(5)-carpellate, partly to completely inferior, style with style branches; ovules pendant. Fruit loculicidal capsule. Seeds numerous, 1–4 mm long, chalazal wing, micropylar flange (rarely absent). $n = 13$ (Bangham 1929; Sax 1931). Approximately 80 species. Asia, Europe,

North America and Central America. Tropical, temperate, and arid environments in open forests, scrublands, rocky slopes, and riparian banks. Hu (1954) recognized four subgenera: *Gemmatus* (Koehne) Hu, *Macrothyrsus* Hu, *Deutzioides* Hu, and *Philadelphus*. Hu (1954) suggested that the genus originated in the New World, and pointed to tropical and subtropical species of subgenus *Gemmatus* in Central America as the most primitive extant species.

8. *Carpenteria* Torr.

Carpenteria Torr., Smithson. Contrib. 6: 12 (1854).

Shrubs. Leaves deciduous, entire, petioles inconspicuous. Inflorescence terminal and axillary dichasial cymes. Sterile flowers absent. Flowers often strongly fragrant. Sepals 5–7, valvate. Petals 5–7, imbricate, largely ovate with very short claw, white. Stamens >200, filaments filiform. Ovary 5–7-carpellate, partly inferior, style simple; ovules pendant. Fruit septicidal capsule. Seeds numerous, 0.5–1 mm long, winged, no micropylar flange. One species, *Carpenteria californica* Torr. Restricted to dry ridges and slopes of the Sierra Nevada foothills in California, U.S.A, between the San Joaquin and King rivers. A conservation concern because of habitat loss.

2. TRIBE HYDRANGEAE DC. (1830).

Shrubs or herbaceous perennials. Sterile flowers usually present, rarely absent. Fruits either berries or capsules with apical dehiscence or fragmenting lateral walls.

9. *Deinanthe* Maxim.

Deinanthe Maxim. Mem. Acad. Petersb. VII, 16: 2 (1867).

Herbaceous perennial from rhizome. Leaves petiolate, lamina apex acute or bifid, margin toothed. Inflorescence cymose panicle. Sterile flowers present. Sepals 5–6, imbricate. Petals 5–8, imbricate, ovate, white, pinkish purple to blue, reflex at anthesis. Stamens >200, stamen filaments filamentous. Ovary 4–6-carpellate, partly inferior, stylodia connate postgenitally, ovules horizontal and pendant. Fruit capsule with apical dehiscence pulling apart the connate stylodia. Seeds numerous, 0.5–1.5 mm long, wing and micropylar flange present. $2n = 34$ (Hamel 1951). Two species: *D. bifida* from Japan (southwest Honshu, Shikoku,

and Kyushu) and *D. caerulea* from China. Both in moist forests of mountains (Lancaster 1996).

10. *Cardiandra* Sieb. & Zucc.

Cardiandra Sieb. & Zucc. Fl. Jap.: 119 (1835).

Herbaceous perennial from rhizome. Leaves alternate, petiolate, lamina toothed. Inflorescence cymose panicle. Sterile flowers present. Sepals 4–5(6), valvate. Petals 4–5(6), imbricate, ovate, white, reflex at anthesis. Stamens 12–25, filaments filiform. Ovary 2–3(4–5)-carpellate, partly inferior, stylodia free, most ovules erect. Fruit capsular with apical dehiscence. Seeds numerous, 0.5–1.0 mm long, wing and micropylar flange present. About six species. Eastern Asia. Moist forests in mountains. Revision needed.

11. *Hydrangea* L.   Fig. 70

Hydrangea L. Sp. Pl.: 397 (1754); McClintock, Proc. Calif. Acad. Sci. 29: 147–256 (1957), rev.; Wei, Guihaia 14: 101–121 (1994), rev. Chinese species.

Fig. 70. Hydrangeaceae-Hydrangioideae. *Hydrangea serrata*. **A** Flowering branchlet. **B** Flower bud. **C** Flower. **D** Fruiting branch. **E** Fruit. Drawn by T. Yamada. (Nakai 1926)

Erect or scandent shrubs or vines, climbing by shoot-borne roots. Leaves deciduous or persistent, petiolate, margins rarely lobed, toothed or entire. Inflorescence cymose panicle, often flat-topped or pyramidal. Sterile flowers present usually or absent. Sepals 4–5, valvate. Petals 4–5, valvate, persisting until fruiting and reflexed, abscising after anthesis, or calyptrate and abscising at anthesis, usually ovate, rarely spathulate, white, pink, blue, or yellow. Stamens 8–20, stamen filaments filiform or subulate. Ovary 2–4(5)-carpellate, partly to completely inferior, stylodia free, ovules pendant, erect, or horizontal. Fruit capsular with apical interstylar dehiscence. Seeds numerous, 0.5–3 mm long; ellipsoidal, spindle-shaped, obovoid, or urceolate, winged and some with micropylar flange or both wing and micropylar flange lacking. $n = 18$, 36 (Sax 1931; Schoenagel 1931; Löve 1966, 1969). About 29 species. Eastern Asia (extending westward through the Himalayan region), including Sumatra, Java, Philippine Islands, Ryukyu Islands and Taiwan, southeastern North America, Central America, and western South America. In mesophytic or moist forests and often on slopes, cliffs, or riparian banks. Five sections recognized, but their monophyly, except *Cornidia*, is dubious. Various temperate deciduous species that have sterile flowers are important ornamentals, especially cultivars of *H. macrophylla* and *H. paniculata*.

12. *Dichroa* Lour.

Dichroa Lour., Fl. Cochinch.: 301 (1790); Hwang, Acta Phytotax. Sinica 25: 384–389 (1987), synopsis and key.

Shrubs. Leaves persistent, petiolate, lamina toothed. Inflorescence cymose panicle. Sterile flowers absent. Sepals (4)5(6), valvate. Petals (4)5(6), valvate, ovate, blue or pinkish; reflex at anthesis. Stamens 4–5 or 8–12, filaments filiform. Ovary (4)5(6)-carpellate, partly to completely inferior, stylodia free or style with style branches; ovules horizontal. Fruit a berry. Seeds numerous, about 0.5 mm long; urceolate, lacking both wing and micropylar flange. $n = 18$ (Löve 1969). Twelve species in two sections (sect. *Dichroa*, 10 species; sect. *Silvicola* S.M. Hwang, 2 species). Distributed through mainland SE Asia southward to Pacific islands. Three species endemic to New Guinea, two species endemic to Philippine Islands. The plants have been poorly collected, patterns of character variation are not known, and a revision is needed.

13. *Broussaisia* Gaudich.

Broussaisia Gaudich., Voy. Bot.: 479 (1826).

Shrub, stems hirsute (rarely glabrous), especially when young; leaf scars prominent (internodes short). Leaves persistent at ends of branches, opposite or whorled, petiolate, lamina obovate to ovate or elliptical, margins toothed. Inflorescence cymose panicle. Sterile flowers absent. Dioecious.

Male flowers: sepals (4)5(6), valvate; petals (4)5(6), free, valvate, ovate, greenish blue, pink, or greenish white, reflex at anthesis; stamens (8–)10(–12), stamen filaments taper to apex; Ovaries (4)5(6)-carpellate, shallowly inferior, style simple; ovules horizontal, sterile. Female flowers: pink to dark blue; sepals (4)5–6, valvate; corolla rudimentary or lacking; stamens lacking; carpels (4)5(6)-carpellate, completely inferior, single style, stigma capitate, ovules horizontal. Fruit a fleshy berry. Seeds numerous and surrounded by projections from inner wall of ovary, about 0.5 mm long, urceolate. $2n = 32$ (Wagner et al. 1990). One species (*B. arguta* Gaud.), although various varieties and forma have been described to account for polymorphism (Fosberg 1939). Endemic to Hawaiian Islands, where located on all main islands, except Ni'ihau and Kaho'olawe, in dense, wet forests (Degener 1945).

14. *Schizophragma* Sieb. & Zucc.

Schizophragma Sieb. & Zucc., Fl. Jap.: 58: t. 26 (1835).

Vines, climbing by shoot-borne roots. Leaves deciduous, petiolate, margins toothed or entire. Inflorescence cymose panicle. Sterile flowers present, one-lobed. Sepals (4)5, valvate. Petals (4)5, valvate, abscising after anthesis, ovate, white. Stamens 8–10, stamen filaments filiform. Ovaries (4)5-carpellate, deeply inferior, style simple; stigmas capitate, ovules pendant. Fruits capsules with intercostal portions of the lateral walls separating from costal ribs. Seeds numerous, 2.5–4.5 mm long, spindle-shaped, winged and with micropylar flange. $n = 14$, 36 (Sax 1931; Hamel 1951, Nevling 1964). About 10 species. Eastern Asia. In mesophytic forests. Revision needed.

15. *Pileostegia* Hook. f. & Thoms.

Pileostegia Hook. f. & Thoms., J. Linn. Soc. 2: 57 (1858).

Perennial vines, climbing by shoot-borne roots. Leaves persistent, margins entire. Inflorescence cymose panicle. Sterile flowers present, one-lobed. Sepals (4)5, valvate, persist as rim on fruit. Petals (4)5, valvate, triangular, white, calyptrate, abscising at anthesis. Stamens (8)10, stamen filaments filiform. Ovaries (4)5-carpellate, deeply inferior; style simple; stigmas capitate, ovules pendant. Fruits capsules with intercostal portions of the lateral walls separating from costal ribs. Seeds numerous, 1–2 mm long, bottle-shaped, winged. $n = 18$. Four species. Eastern Asia (China, Taiwan, Ryukyu Islands, India, and Himalayan region). Mesophytic forests at mid elevations in mountains. Revision needed.

16. *Decumaria* L.   Fig. 71

Decumaria L., Sp. Pl. ed. 2, 2: 1663 (1763).

Perennial vines, climbing by shoot-borne roots. Leaves deciduous, margins entire. Inflorescence cymose panicle. Sterile flowers absent. Sepals 6–12, valvate. Petals 6–12, valvate, abscising at

Fig. 71. Hydrangeaceae-Hydrangioideae. *Decumaria barbara.* **A** Shoot with terminal corymb. **B** Flowers – note capitate stigmas with radiating lines. **C** Stamen. **D** Vertical section of ovary (petals and stamens removed) showing stylar column with stigma and axile placentation. **E** Transverse section of ovary. **F** Immature capsule with persistent style and stigma. **G** Dehisced capsule. **H** Seed. (Spongberg 1972)

fruiting, ovate, white. Stamens >15, stamen filaments filiform. Ovaries 6–12-carpellate, deeply inferior; style simple; stigmas capitate, ovules pendant. Fruits capsules with intercostal portions of the lateral walls separating from costal ribs. Seeds numerous, 1.5–3 mm long, horn-shaped, winged and with micropylar flange. $n = 14$ (Sax 1931). Two species: *D. barbara* L., southeastern North America; *D. sinensis*, China. In mesophytic forests.

17. *Platycrater* Sieb. et Zucc.

Platycrater Sieb. et Zucc., Fl. Jap. 62: t. 27 (1835).

Shrubs. Leaves deciduous, petiolate, toothed. Inflorescence dichasial cymes or cymose panicle. Sterile flowers present, with (2)3–4(5) lobes. Sepals 4, valvate. Petals 4, valvate, ovate, white, abscising at fruiting. Stamens >200, filaments filiform. Ovaries 2–4-carpellate, deeply inferior, stylodia free, stigmas terminal, ovule variable. Fruit capsular with apical interstylar dehiscence. Seeds numerous, 1.0–1.5 mm long; spindle-shaped, winged, micropylar flange lacking. One species: *Platycrater arguta* Sieb. et Zucc. Warm temperate areas of Japan and China.

Selected Bibliography

Agababyan, V.S. 1961. A contribution to the palynomorphology of the family Hydrangeaceae Dum. Izv. Akad. Nauk SSSR, Ser. Biol. 14 (11): 17–26.

Albach, D.C. et al. 2001. See general references.

Ammal, J. 1951. Chromosomes and evolution of garden *Philadelphus.* J. Roy. Hort. Soc. 76: 269–275.

APG (Angiosperm Phylogeny Group) 1998. See general references.

Axelrod, D.I. 1987. The late Oligocene Creede flora, Colorado. Univ. Calif. Publ. Geol. Sci. 130: 1–235.

Axelrod, D.I., Raven, P.H. 1985. Origins of the cordilleran flora. J. Biogeogr. 12: 21–47.

Bangham, W. 1929. The chromosomes of some species of the genus *Philadelphus.* J. Arnold Arbor. 10: 167–169.

Barykina, R.P., Kapranova, N.N. 1983. Ontomorphogenesis of some *Philadelphus* species. Biol. Nauki 9: 71–76.

Bate-Smith, E.C. 1978. Astringent tannins of *Viburnum* and *Hydrangea* species. Phytochemistry 17: 267–270.

Bell, W.A. 1957. Flora of the Upper Cretaceous Nanaimo Group, Vancouver Island, British Columbia. Mem. Geol. Surv. Can. 292: 1–89.

Bensel, C.R., Palser, B.F. 1975a. Floral anatomy in the Saxifragaceae sensu lato. III. Kirengeshomoideae, Hydrangeoideae and Escallonioideae. Am. J. Bot. 62: 676–687.

Bensel, C.R., Palser, B.F. 1975b. Floral anatomy in the Saxifragaceae sensu lato. IV. Baueroideae and conclusions. Am. J. Bot. 62: 688–694.

Bohm, B.A., Nicholls, K.W., Bhat, U.G. 1985. Flavonoids of the Hydrangeaceae Dumortier. Biochem. Syst. Ecol. 13: 441–445.

Breuer, B., Stuhlfauth, T., Fock, H., Huber, H. 1987. Fatty acids of some Cornaceae, Hydrangeaceae, Aquifoliaceae, Hamamelidaceae and Styracaceae. Phytochemistry 26: 1441–1445.

Burkett, G.W. 1932. Anatomical studies within the genus *Hydrangea*. Proc. Indiana Acad. Sci. 41: 83–95.

Cannon, R. 1981. *Kirengeshoma palmata* Saxifragaceae. Am. Hort. 60: 32–33.

Chaney, R.W., Sanborn, E.I. 1933. The Goshen Flora of west central Oregon. Publ. Carnegie Inst. Wash. 439: 1–103.

Condit, C. 1944. The Table Mountain Flora. In: Chaney, R.W. (ed.) Pliocene floras of California. Publ. Carnegie Inst. Wash. 553: 57–90.

Corner, E.J.H. 1976. See general references.

Davis, G.L. 1966. See general references.

Degener, O. 1945. Plants of Hawaii National Park. Ann Arbor: Edwards Brothers.

Dhillon, M. 1975. Morphology and vascular anatomy of the node and flower of *Deutzia staminea* R. Brown (Saxifragaceae). J. Res. Punjab Agric. Univ. 12: 156–160.

Downie, S.R., Palmer, J.D., 1992. Restriction site mapping of the chloroplast DNA inverted repeat: a molecular phylogeny of the Asteridae. Ann. Missouri Bot. Gard. 79: 266–283.

Engler, A. 1930. Saxifragaceae. In: Engler, A., Prantl, K. Die natürlichen Pflanzenfamilien, ed. 2, 18a. Engelmann: Leipzig, pp. 74–226.

Fairbairn, J.W., Lou, T.C. 1950. A pharmacognostical study of *Dichroa febrifuga* Lour., a Chinese antimalarial plant. J. Pharm. Pharmacol. 2: 162–177.

Flora of Taiwan, 1993. Taipei, Taiwan: Editorial Committee, Flora of Taiwan, vol. 3.

Fosberg, F.R. 1939. Taxonomy of the Hawaiian genus *Broussaisia* (Saxifragaceae). Occ. Pap. Bishop Mus. 15: 49–60.

Frederiksen, L.B., Damtoft, S., Jensen, S.R. 1999. Biosynthesis of iridoids lacking C-10 and the chemotaxonomic implications of their distribution. Phytochemistry 52: 1409–1420.

Funamoto, T., Nakamura, T. 1989. Karyomorphological study on *Kirengeshoma palmata* Yatabe in Japan (Saxifragaceae). Chromosome Inf. Serv. 47: 22–23.

Gelius, L. 1967. Studien zur Entwicklungsgeschichte an Blüten der Saxifragales sensu lato mit besonderer Berücksichtigung des Androeceums. Bot. Jahrb. Syst. 87: 253–303.

Gorbunov, M.G. 1970. On the fossil remains of the representatives of the genus *Hydrangea* in the flora of the locality Compasskiy Bor on the River Tym (west Siberia). Bot. Zhurn. 55: 795–806.

Gregory, M. 1998. Hydrangeaceae. In: Cutler, D.F., Gregory, M. (eds.) Anatomy of dicotyledons, ed. 2, vol. 4. Saxifragales. Oxford: Oxford University Press, pp. 87–108.

Hamel, J.L. 1951. Notes préliminaires à l'étude caryologique des Saxifragacées. VI. Les chromosomes somatiques des *Kirengeshoma palmata* Yatabe, *Deinanthe coerulea* Stapf, et *Schizophragma integrifolia* (Franch) Oliv. Bull. Mus. Hist. Nat. (Paris) II 23: 651–654.

Hamel, J.L. 1953. Contribution à l'étude cytotaxinomique des Saxifragacées. Rev. Cytol. Biol. Vég. 14: 113–313.

Hao, G., Hu, C. 1996a. A study of leaf venation of Hydrangeoideae (Hydrangeaceae). Guihaia 16: 155–160.

Hao, G., Hu, C. 1996b. A study of pollen morphology of Hydrangeoideae (Hydrangeaceae). J. Trop. Subtrop. Bot. 4: 26–31.

Hardin, J.W., Pilatowski, R.E. 1981. Atlas of foliar surface features in woody plants. III. *Hydrangea* (Saxifragaceae) of the United States. J. Elisha Mitchell Sci. Soc. 97: 29–36.

He, P. 1990. Taxonomy of *Deutzia* (Hydrangeaceae) from Sichuan, China. Phytologia 69: 332–339.

Hegnauer, R. 1973. See general references.

Hempel, A.L., Reeves, P.A., Olmstead, R.G., Jansen, R.K. 1995. Implications of *rbc*L sequence data for higher order relationships of the Loasaceae and the anomalous aquatic plant *Hydrostachys* (Hydrostachyaceae). Pl. Syst. Evol. 194: 25–37.

Hideux, M.J., Ferguson, I.K. 1976. The stereostructure of the exine and its evolutionary significance in Saxifragaceae sensu lato. In: Ferguson, I.K., Muller, J. (eds.) The evolutionary significance of the exine. Linn. Soc. Symp. Series 1, pp. 327–377.

Holmgren, N.H., Holmgren, P.K. 1989. A taxonomic study of *Jamesia* (Hydrangeaceae). Brittonia 41: 335–350.

Hsu, C.-C. 1968. Preliminary chromosome studies on the vascular plants of Taiwan (II). Taiwania 14: 11–27.

Hu, S.-Y. 1954. A monograph of the genus *Philadelphus*. J. Arnold Arbor. 35: 275–333.

Hu, S.-Y. 1955. A monograph of the genus *Philadelphus*. J. Arnold Arbor. 36: 52–109.

Hu, S.-Y. 1956. A monograph of the genus *Philadelphus*. J. Arnold Arbor. 37: 15–90.

Hu, H.-H., Chaney, R.W. 1940. A Miocene flora from Shantung Province, China. Publ. Carnegie Inst. Wash. 507: 1–147.

Hufford, L. 1995. Seed morphology of Hydrangeaceae and its phylogenetic implications. Int. J. Plant Sci. 156: 555–580.

Hufford, L. 1997. A phylogenetic analysis of Hydrangeaceae using morphological data. Int. J. Plant Sci. 158: 652–672.

Hufford, L. 1998. Early development of androecia in polystemonous Hydrangeaceae. Am. J. Bot. 85: 1057–1067.

Hufford, L., Moody, M., Soltis, D.E. 2001. A phylogenetic analysis of Hydrangeaceae based on sequences of the plastid gene *mat*K and their combination with *rbc*L and morphological data. Int. J. Plant Sci. 162: 835–846.

Hutchinson, J. 1927. Contributions towards a phylogenetic classification of flowering plants. VI. A. The genera of Hydrangeaceae. Kew Bull. 1927: 100–107.

Ishii, H., Takeda, H. 1997. Effects of the spatial arrangement of aerial stems and current-year shoots on the demography and growth of *Hydrangea hirta* in a light-limited environment. New Phytol. 136: 443–453.

Jensen, S.R., Nielsen, B.J., Dahlgren, R. 1975. Iridoid compounds, their occurrence and systematic importance in the angiosperms. Bot. Not. 128: 148–180.

Kapranova, N.N. 1976. Ontogeny, morphology and anatomy of the annual branch of *Philadelphus schrenkii* Rupr. Nauch. Dokl. vyssh. Shk., biol. Nauki SSSR 19(4): 83–89.

Kim, Y.H., Ko, W.S., Ha, M.S., Lee, C.H., Choi, B.T., Kang, H.S., Kim, C.H. 2000. The production of nitric oxide and TNF-α in peritoneal macrophages is inhibited by *Dichroa febrifuga* Lour. J. Ethnopharmacol. 69: 35–43.

Klink, V.P. 1995. Flower ontogeny in the dioecious Hawaiian endemic *Broussaisia arguta* Gaud. (Hydrangeaceae). MS thesis. Durham: University of New Hampshire.

Klopfer, K. 1971. Beiträge zur floralen Morphogenese und Histogenese der Saxifragaceae 6. Die Hydrangeoideen. Wiss. Z. Pädagog. Hochsch. Potsdam, Math.-Naturwiss. 15: 77–95.

Klopfer, K. 1973. Florale Morphogenese und Taxonomie der Saxifragaceae sensu lato. Feddes Rep. 84: 475–516.

Lancaster, R. 1996. *Deinanthe bifida* and *D. caerulea*. Garden (London) 121: 364–365.

Lancucka-Srodoniowa, M. 1975. *Hydrangea* L (Saxifragaceae) and *Schefflera* Forst. (Araliaceae) in the Tertiary of Poland. Acta Palaeobot. 16: 103–112.

Löve, A. (ed.) 1966. IOPB chromosome number reports. VI. Taxon 15: 117–128.

Löve, A. (ed.) 1969. IOPB chromosome number reports. XX. Taxon 18: 213–221.

MacGinitie, H.D. 1941. A Middle Eocene flora from the central Sierra Nevada. Publ. Carnegie Inst. Wash. 534: 1–178.

Mai, D.H. 1985. Beiträge zur Geschichte einiger holziger Saxifragales-Gattungen. Gleditschia 13: 75–88.

Mai, D.H. 1998. Contribution to the flora of the middle Oligocene Calau Beds in Brandenburg, Germany. Rev. Palaeobot. Palynol. 101: 43–70.

Manchester, S.R. 1994. Fruits and seeds of the Middle Eocene Nut Beds flora, Clarno Formation, Oregon. Palaeontogr. Am. 58: 1–205.

Mauritzon, J. 1933. Studien über die Embryologie der Familien Crassulaceae und Saxifragaceae. Ph.D. diss. Lund University.

Mauritzon, J. 1939. Contributions to the embryology of the orders Rosales and Myrtales. Lunds Univ. Årsskr., N.F. Adv. 2 35: 1–121.

McClintock, E. 1956. The cultivated hydrangeas. Baileya 4: 165–175.

Morf, E. 1950. Vergleichend-morphologische Untersuchungen am Gynoeceum der Saxifragaceen. Ber. Schweiz. Bot. Gesellsch. 60: 516–590.

Murata, K., Takano, F., Fushiya, S., Oshima, Y. 1998. Enhancement of NO production in activated macrophage in vivo by an antimalarial crude drug, Dichroa febrifuga. J. Nat. Prod. (Lloydia) 61: 729–733.

Nakai, T. 1926. Flora sylvatica Koreana, Pars XV, Saxifragaceae. Seoul: Forest Exp. Station.

Nemirovich-Danchenko, E.N., Lobova, T.A. 1998. The seed coat structure in some representatives of the order Hydrangeales. Bot. Zhurn. 83: 1–9.

Nevling, L.I. 1964. Climbing hydrangeas and their relatives. Arnoldia 24: 17–39.

Ohba, H., Akiyama, S. 1992. A taxonomic revision of Deutzia (Saxifragaceae, s. l.) in the Ryukyu Islands, S. Japan. J. Jap. Bot. 67: 154–165.

Ozaki, K. 1991. Late Miocene and Pliocene floras in central Honshu, Japan. Spec. Issue Bull. Kanagawa Pref. Mus.

Pastre, A., Pons, A. 1973. Quelques aspects de la systématique des Saxifragacées à la lumière des données de la palynologie. Pollen Spores 15: 117–133.

Pilatowksi, R.E. 1981. A taxonomic study of the Hydrangea arborescens complex. Castanea 47: 84–98.

Quibell, C.F. 1972. Comparative and systematic anatomy of Carpentetrieae (Philadelphaceae). PhD diss. Berkeley: University of California.

Robertson, C. 1892. Flowers and insects. IX. Bot. Gaz. 17: 269–276.

Roels, P., Ronse Decraene, L.-P., Smets, E.F. 1997. A floral ontogenetic investigation of the Hydrangeaceae. Nord. J. Bot. 17: 235–254.

Sax, K. 1931. Chromosome numbers in the ligneous Saxifragaceae. J. Arnold Arbor. 12: 198–205.

Saxena, N.P. 1971. Studies in the family Saxifragaceae VI. Structure and development of gametophyte in Deutzia corymbosa R. Br. and Dichroa febrifuga Lour. Ber. Schweiz. Bot. Gesellsch. 81: 91–96.

Schoenagel, E. 1931. Chromosomenzahl und Phylogenie der Saxifragaceen. Bot. Jahrb. Syst. 64: 266–288.

Snezhkova, S.A. 1986. Wood structure in some lianes of the Far East. Bot. Zhurn. 71: 768–773.

Snezhkova, S.A. 1990. Structure of woods of some representatives of Hydrangeaceae and Grossulariaceae. Byull. Glav. Bot. Sada 158: 78–80.

Soltis, D.E., Xiang, Q.-Y., Hufford, L. 1995. Relationships and evolution of Hydrangeaceae based on rbcL sequence data. Am. J. Bot. 82: 504–514.

Spongberg, S.A. 1972. The genera of Saxifragaceae in the southeastern United States. J. Arnold Arbor. 53: 409–498.

Steinmetz, E.F. 1972. Dichroae febrifugae cortex radicis. Acta Phytotherap. 19: 3–5.

Stern, W.L. 1978. Comparative anatomy and systematics of woody Saxifragaceae. Hydrangea. Bot. J. Linn. Soc. 76: 83–113.

Styer, C.H., Stern, W.L. 1979a. Comparative anatomy and systematics of woody Saxifragaceae. Deutzia. Bot. J. Linn. Soc. 79: 291–319.

Styer, C.H., Stern, W.L. 1979b. Comparative anatomy and systematics of woody Saxifragaceae. Philadelphus. Bot. J. Linn. Soc. 79: 267–289.

Umemoto, K. 1974. Pattern of calcium oxalate crystals in the leaves of Hydrangea genus plants. Yakugaku Zasshi 94: 110–115.

Wagner, W.L., Herbst, D.R., Sohmer, S.H. 1990. Manual of the flowering plants of Hawai'i. Honolulu: University of Hawaii Press.

Wakabayashi, M. 1970. On the affinity in Saxifragaceae s. lat. with special reference to the pollen morphology. Acta Phytotax. Geobot. 24: 128–145.

Watari, S. 1939. Anatomical studies on leaves of some saxifragaceous plants, with special reference to the vascular system. J. Fac. Sci. Univ. Tokyo III 5: 195–316.

Wettstein, R. v. 1893. Über das Androeceum von Philadelphus. Ber. Deutsch. Bot. Gesellsch. 11: 480–484.

Wyman, D. 1965. The mock-oranges. Arnoldia 25: 29–36.

Xiang, Q.-Y., Soltis, D.E., Soltis, P.S. 1998. Phylogenetic relationships of Cornaceae and close relatives inferred from matK and rbcL sequences. Am. J. Bot. 85: 285–297.

Zaïkonnikova, T.I. 1966. Deutzias – ornamental shrubs – a monograph of the genus Deutzia Thunb. Moscow: Nauka.

Zaïkonnikova, T.I. 1975. A key to the species of the genus Deutzia Thunberg (Saxifragaceae). Baileya 19: 133–144.

Zhou, T.S., Hara, N. 1988. Development of shoot in Hydrangea macrophylla I. Terminal and axillary buds. Bot. Mag. Tokyo 101: 281–291.

Zhou, T.S., Hara, N. 1989. Development of shoot in Hydrangea macrophylla II. Sequence and timing. Bot. Mag. Tokyo 102: 193–206.

Hydrostachyaceae

C. Erbar and P. Leins

Hydrostachyaceae Engler, Syllabus, ed. 2: 215 (1898), nom. cons.

Submerged aquatic perennial rosette plants with a short disk-like stem, attached to rocks in fast-flowing water; vascular system reduced. Leaves alternate, entire to 2–3 times pinnatifid, leaf base enlarged with intrapetiolar stipule; stomata wanting; stem, leaves, and adventitious roots provided with numerous small, rounded, scale-like or fringed emergences. Plants dioecious, very seldom monoecious; unisexual flowers small, without perianth, sessile in bracts of dense spikes; staminate flower interpreted as consisting of a single stamen which is deeply divided into two thecae or of two independent bisporangiate stamens; pistillate flower consisting of a single superior, bicarpellate, unilocular ovary with two elongate, persistent, filiform stylodia; ovules numerous on two parietal placentas, anatropous, unitegmic, tenuinucellar. Fruit a septicidal capsule with numerous tiny seeds.

A monogeneric family of 22 species native to Madagascar and tropical and southern Africa.

VEGETATIVE MORPHOLOGY. The perennial hapaxanthous rosette plants start growing in fresh, fast-flowing water. The shoot of the water plants is flat and disk-like and attached to rocks and stones. Its diameter is usually a few centimetres. A primary root is not developed but numerous adventitious roots fasten the plants to the rocks. The leaves are arranged spirally in tufts (Fig. 73A). The floating leaves are 30–50(–100) cm long. They have an enlarged leaf base with a small, membranous, median stipule (intrapetiolar stipule), a stout petiole and a flexible lamina. The leaves are simple or more frequently 2–3 times pinnatifid. The rhachis, the pinnae as well as the petiole are densely covered with numerous rounded, tuberculate, scale-like, thread-like or fringed emergences. Their size varies between 1.5 and 5 mm. Emergences may be condensed at the top of the leaflets.

The plants remain unbranched during the vegetative phase. The shoot system becomes sympodial only at the initiation of inflorescences. This happens already during the submerged stage whereas flowering starts only when the plants fall dry.

At the top of the main axis, reduced leaves (scale-like hypsophylls) are formed and the axis terminates with a spike. Additional inflorescences are formed in the axils of the uppermost pinnate leaves (polytelic synflorescence). At the onset of flowering the plants have a *Plantago*-like appearance (Warming 1891a, 1891b; Rauh and Jäger-Zürn 1966; Cusset 1973; Jäger-Zürn 1998).

VEGETATIVE ANATOMY. The vascular system is reduced. In the disk-like stem, procambial tissue with annular- and spiral vessels is present. In the peduncle of the spike, proto- and metaxylem and phloem groups are diffusely scattered in the vascular tissue. The single bundle is closed. Some vessels with annular or spiral thickenings are close together and surrounded by thin-walled cells in younger stages. Later these cells between the vessels become sclerified at the base and collenchymatous in the upper part of the inflorescence axis. The single vascular bundles differ from the usual mode in that the proto-xylem and protophloem are not compressed in older stages but continue functioning. Due to the later-on strong elongation of the peduncle of the spike, the vessels get disrupted. In older stages no entire vessels are found. Adult plants have only schizogenic vascular spaces with remnants of vessels.

The leaves have neither stomata nor special aerenchyma. The petiole has a central main bundle and unusually numerous, small bundles supplying the emergences. The main bundle is surrounded by thin-walled parenchymatous cells which later become collenchymatous. Annular and spiral vessels can be found only in young leaves whereas in older ones schizogenous vascular spaces occur. Sieve tubes with simple sieve plates have been observed only sporadically, e.g. in *Hydrostachys maxima* (Warming 1891a; Rauh and Jäger-Zürn 1966; Jäger-Zürn 1998).

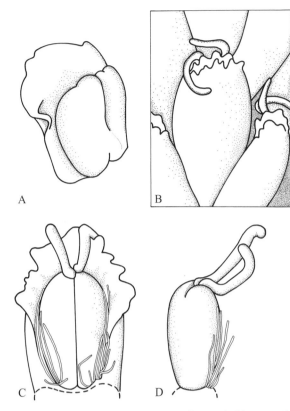

Fig. 72. Hydrostachyaceae. *Hydrostachys multifida*. **A** Male flower in the axil of its bract, adaxial view. **B** Part of female inflorescence; the flowers covered by bracts, only stylodia are visible. **C** Adaxial view of adult female flower with subtending bract. **D** Side view of female flower with silky hairs on lateral flank. (Orig.)

FLOWER STRUCTURE. Nearly all species of *Hydrostachys* are dioecious. Only *Hydrostachys monoica* is monoecious and *Hydrostachys stolonifera* is facultatively monoecious (Cusset 1973; Jäger-Zürn 1998). Long, dense spikes of unisexual flowers (Fig. 73B, D) terminate the main axis as well as the axillary branches arising in the axils of the uppermost pinnate leaves. All spikes are open (no terminal flower developed) and the uppermost bracts are sterile.

The simple male flower lies in the deepening of its subtending, shell-like bract (Fig. 72A). The male flower is either interpreted as consisting of a single tetrasporangiate stamen which is deeply divided into two theca (e.g. Rauh and Jäger-Zürn 1966; Cronquist 1981; Jäger-Zürn 1998) or as consisting of two independent, bisporangiate stamens (Warming 1891a, 1891b; Leins and Erbar 1988, 1990; Takhtajan 1997). The microsporangia, which face the bracts, open longitudinally by slits. In many members of *Hydrostachys*, the androecium

is laterally flanked by a group of silky hairs each (Cusset 1973).

Floral developmental studies support the existence of two androecial organs. In the earliest developmental stage of the male flower, two androecial primordia arise separately on the lateral edges of a transversally elongated, low apex. The common base of the two bisporangiate organs, which is formed by the floral apex, remains rather short (Leins and Erbar 1988, 1990). Because of their fully separate initiation, we can interpret the male flower as consisting of two bisporangiate stamens. These may derive from originally tetrasporangiate stamens in which the sporangia of the neighbouring thecae have fused apically. In a second phylogenetic process, the synthecal sporangia shifted into a lateral position.

The female flower consists only of a dimerous and unilocular gynoecium with parietal placentation and the carpels arranged in a transversal position. Silky hairs are formed on the lateral flanks of the floral apex. Abaxially the ovary is wholly enclosed by its subtending bract so that only the two subulate stylodia with long stigmatic areas of short papillae are visible in flowering inflorescences (Fig. 72B, C). Arrangement and shape of the emergences on the dorsal side of the female bracts are of taxonomical value, distinguishing the species (Cusset 1973).

The early development of the gynoecium is remarkable. When the stylodia become visible at the top of the two carpel primordia, plication takes place in that the carpels start forming margins. It is very unusual that the stylodia are unaffected by this process. Below the insertion of the stylodia, the broad margins are inversely U-shaped. In early stages the ovary becomes zygomorphic: coenocarpy is more pronounced on the abaxial than on the adaxial side. Later on, a suture extends to the base of the ovary on its adaxial side (along this suture the ripe fruit opens septicidally). Only the base of the ovary bears a low septum. Most of the ovary is unilocular, with many ovules on parietal placentae. The base of the stylodia becomes overtopped by growth of the ovary on the dorsal side of its upper choricarpous region (Fig. 72D; Leins and Erbar 1988, 1990).

EMBRYOLOGY. The anther can be interpreted as tetrasporangiate (Rauh and Jäger-Zürn 1966; Jäger-Zürn 1998) or bisporangiate (Warming 1891a, 1891b; Leins and Erbar 1988, 1990). The tapetum is secretory and its cells become 2-nucleate. Cell wall formation after meiosis is simultaneous and the pollen grains are 2-celled when shed as permanent tetrads.

The ovule is anatropous, unitegmic and tenuinucellate. The nucellus is weakly developed and short, whereas the chalaza is conspicuous. One subepidermal cell of the apex becomes the archesporial cell which functions directly as the megaspore mother cell. The development of the chalazal megaspore follows the Polygonum type. Endosperm formation is ab initio cellular and a micropylar haustorium is formed. Embryogeny follows the Onagrad type (Rauh and Jäger-Zürn 1966; Jäger-Zürn 1998).

POLLEN MORPHOLOGY. The pollen grains are arranged in permanent tetrahedral or hexahedral to rhomboidal tetrads. The individual pollen grains are heteropolar trisymmetrical and inaperturate. The sexine is one-layered with a thick tectum and as thick as the nexine. The tectum partly exhibits microfurrows and suprategillar microspinules. The exine has irregularly formed, thinner areas (Erdtman 1952; Straka 1988). These are reduced and unspecific palynological features.

KARYOLOGY. The only count is $2n = 20–24$ for *Hydrostachys imbricata* (Palm 1915).

POLLINATION. Flowering sets in when the water level drops. The flowers elevated above the water surface are wind-pollinated (flowers apetalous and stylodia with a long, papillate stigmatic area). Although the yellow anthers in the male spikes and the gay red or (seldom) yellow stylodia in the female spikes are somewhat striking, visit by insects was never observed (Rauh and Jäger-Zürn 1966).

REPRODUCTIVE SYSTEMS. The plants grow totally submerged during the rainy season. Only in the dry season, when the water level in the rivers sinks, do the vegetative organs emerge from the water, the peduncles of the spikes enlarge, and flowering and seed ripening take place. The inflorescences have already been initiated during the submerged phase. Only a few species, such as *Hydrostachys distichophylla* and *H. verruculosa*, grow permanently submerged in some places. They never come in bloom and propagate vegetatively by sprouts of the adventitious roots. Also deep-water species (e.g., *Hydrostachys verruculosa*) may propagate vegetatively during some years if conditions for flowering are not given (Rauh and Jäger-Zürn 1966).

FRUIT AND SEED. The fruit of *Hydrostachys* is a septicidal capsule. Adaxially it opens in full length along a preformed suture, abaxially only in the short choricarpous region. The capsule releases numerous, tiny seeds (0.25 × 0.5 mm) lacking endosperm (only negligible remains of the endosperm are present). The outer epidermis of the single integument develops into the testa whose outer cell walls are thickened and contain pectin. Getting wet, they swell up and the seed surface becomes mucilaginous (Rauh and Jäger-Zürn 1966; Jäger-Zürn 1998).

DISPERSAL. Since the fruit ripens and opens above the water level (Rauh and Jäger-Zürn 1966), the tiny seeds may be released from the septicidal capsule by wind. A further transport by water also seems possible, since seeds may easily be transported by currents. Contact with water, however, causes swelling of the epidermis which thus becomes mucilaginous. The mucilage fastens the seed to the rocks during germination (myxospermy; Rauh and Jäger-Zürn 1966; Jäger-Zürn 1998).

PHYTOCHEMISTRY. Apart from kaempferol glycosides (Scogin 1992), caffeic acid has been recorded from *Hydrostachys* (Rønsted et al., 2002). Iridoids or alkaloids have not been found in *Hydrostachys* up to the present.

AFFINITIES. The placement of the aquatic family in angiosperm classifications is difficult due to the reductions in vegetative and floral morphology. In addition, in this taxon pollen characters and chemical compounds are not relevant for systematics. The family has been associated with the Podostemaceae by many authors for a long time, although Warming (1891b, 1891c) separated Hydrostachyaceae as a different family and suggested a position near Saxifragaceae (Warming 1891a; but see Les et al. 1997). Podostemaceae are quite different in having bisexual flowers with centrifugal stamen inception on a ring primordium (Rutishauser and Grubert 1999, 2000). On the basis of their floral morphological and embryological studies, Rauh and Jäger-Zürn (1966) assigned the family a position within Asteridae near Scrophulariaceae-Plantaginaceae. Subsequently, most authors accepted a close relationship to the Scrophulariales (e.g., Cronquist 1981; Leins and Erbar 1988, 1990; Wagenitz 1992; Takhtajan 1997). Comparative DNA (*rbc*L) sequence analyses provided the suggestion that Hydrostachyaceae are allied with the Hydrangeaceae in the Cornales

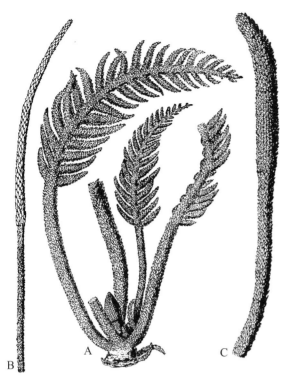

Fig. 73. Hydrostachyaceae. *Hydrostachys imbricata.* **A** Habit. **B** Male inflorescence. **C** Female inflorescence. (Warming 1891c)

(Hempel et al. 1995; APG 1998). Further molecular studies of *rbc*L sequences (Xiang 1999; Savolainen, Fay et al. 2000) and of combined sequences from the *rbc*L, 18 S rDNA, and *atp*B genes (Albach et al. 2001b; Soltis et al. 2000) confirm a relationship of Hydrostachyaceae to Hydrangeaceae, although the remarkably long branch leading to *Hydrostachys* is problematic (Xiang 1999). In the analysis of the *ndh*F gene, *Hydrostachys* is sister to Escalloniaceae (Albach et al. 2001b).

DISTRIBUTION AND HABITATS. The habitats of Hydrostachyaceae are rapids and waterfalls of tropical and subtropical rivers. Most of the species (15) are endemic to Madagascar, a few (7) are found in southern Africa, hardly transgressing the equator. In Madagascar the species are found in large populations in rocky rapids of the central highlands (between 800 and 2000 m altitude). In Africa Hydrostachyaceae are widely spread in rivers throughout the mountain regions of Angola, Congo, Zaire, Tanzania, Malawi, Zimbabwe, Mozambique and South Africa (Jäger-Zürn 1998). They are submerged during the rainy season. In the dry season at low water level, the plants become emerged and flowering and seed ripening take place.

One genus:

Hydrostachys Thouars   Fig. 73

Hydrostachys Thouars, Nova Madag. Gen. 2 (1806); Cusset, Adansonia II, 13: 75–119 (1973). About 22 spp., Madagascar, tropical and southern Africa.

Characters of the family.

Selected Bibliography

Albach, D.C. et al. 2001b. See general references.

APG (Angiosperm Phylogeny Group) 1998. See general references.

Cronquist, A. 1981. See general references.

Cusset, C. 1973. Révision des Hydrostachyaceae. Adansonia II, 13: 75–119.

Engler, A. 1895. Hydrostachydaceae africanae. Bot. Jahrb. Syst. 20: 136–137.

Engler, A. 1898. Syllabus der Pflanzenfamilien. 2. ed. Berlin: Borntraeger.

Erdtman, G. 1952. See general references.

Hempel, A.L., Reeves, P.A., Olmstead, R.G., Jansen, R.K. 1995. Implications of *rbc*L sequence data for higher order relationships of the Loasaceae and the anomalous aquatic plant *Hydrostachys* (Hydrostachyaceae). Pl. Syst. Evol. 194: 25–37.

Jäger-Zürn, I. 1998. Anatomy of the Hydrostachyaceae. In: Handbuch der Pflanzenanatomie 13, 4. Landolt, E. (ed.) Extreme adaptations in angiospermous hydrophytes. Berlin: Borntraeger, pp. 129–196.

Leins, P., Erbar, C. 1988. Einige Bemerkungen zur Blütenentwicklung und systematischen Stellung der Wasserpflanzen *Callitriche, Hippuris* und *Hydrostachys*. Beitr. Biol. Pflanzen 63: 157–178.

Leins, P., Erbar, C. 1990. The possible relationship of Hydrostachyaceae based on comparative ontogenetical flower studies. Mitt. Inst. Allg. Bot. Hamburg 23b: 723–729.

Les, D.H., Philbrick, C.T., Novelo R.A. 1997. The phylogenetic position of river-weeds (Podostemaceae): insights from *rbc*L sequence data. Aquat. Bot. 57: 5–27.

Palm, B. 1915. Studien über Konstruktionstypen und Entwicklungswege des Embryosacks der Angiospermen. Akad. Afhandl. Stockholm 1915: 1–259.

Rauh, W., Jäger-Zürn, I. 1966. Zur Kenntnis der Hydrostachyaceae. I. Blütenmorphologische und embryologische Untersuchungen an Hydrostachyaceen unter besonderer Berücksichtigung ihrer systematischen Stellung. Sitzungsber. Heidelb. Akad. Wiss.-Math.-Naturwiss. Kl. 1966: 1–177.

Rønsted, N., Strandgaard, H., Jensen, S.R., Mølgaard, P. 2002. Chlorogenic acid from three species of *Hydrostachys*. Biochem. Syst. Ecol. 30: 1105–1108.

Rutishauser, R., Grubert, M. 1999. The architecture of *Mourera fluviatilis* (Podostemaceae): developmental morphology of inflorescences, flowers, and seedlings. Am. J. Bot. 86: 907–922.

Rutishauser, R., Grubert, M. 2000. Developmental morphology of *Apinagia multibranchiata* (Podostemaceae) from the Venezuelan Guyanas. Bot. J. Linn. Soc. 132: 299–323.

Savolainen, V., Fay, M.F. et al. 2000. See general references.

Scogin, R. 1992. Phytochemical profile of *Hydrostachys insignis* (Hydrostachyaceae). Aliso 13: 471–474.

Soltis, D.E. et al. 2000. See general references.

Straka, H. 1988. Die Pollenmorphologie von *Hydrostachys*, Hydrostachyaceae, und ihre Bedeutung für die systematische Einreihung der Familie. Beitr. Biol. Pflanzen 63: 413–419.

Takhtajan, A. 1997. See general references.

Wagenitz, G. 1992. The Asteridae: evolution of a concept and its present status. Ann. Missouri Bot. Gard. 79: 209–217.

Warming, E. 1891a. Podostemaceae. In: Engler, A., Prantl, K. Die natürlichen Pflanzenfamilien III, 2a. Leipzig: Engelmann, pp. 1–22.

Warming, E. 1891b. Note sur le genre *Hydrostachys*. Overs. Kongel. Danske Vidensk. Selsk. Forh. Medlemmers Arbeider 1891: 37–43.

Warming, E. 1891c. Familien Podostemaceae IV. Kongel. Danske Vidensk. Selsk. Naturvidensk. Math. Afh. Ser. 6, VII 4: 135–179.

Xiang, Q.-Y. 1999. Systematic affinities of Grubbiaceae and Hydrostachyaceae within Cornales – insights from *rbc*L sequences. Harvard Pap. Bot. 4: 527–542.

Lecythidaceae

G.T. Prance and S.A. Mori

Lecythidaceae Poit., Mém. Mus. Hist. Nat. Paris 13: 141–165
    (1925).
Barringtoniaceae F. Rudolphi (1830).
Foetidiaceae (Nied.) Airy Shaw (1956).

Small to large trees; rarely shrubs; axis with cortical vascular bundles. Leaves alternate, simple, margins usually entire (opposite and clustered in one species of *Abdulmajidia*), pinnately nerved; stipules absent or minute and caducous. Inflorescences terminal, axillary or cauline, simple racemes, panicles with 2 or 3 orders of racemose or spicate branches or fascicles. Flowers actinomorphic or zygomorphic, hermaphrodite; sepals 2–6 or rarely unlobed; petals 3–6(8), infrequently 12 or 18, imbricate, free (absent in *Foetidia*); stamens numerous, 10–1210, connate at base into a short or long staminal ring (free to base in *Foetidia*), the ring actinomorphic or prolonged on one side into a strap-like structure which arches over the summit of the ovary; anthers bilocular, latrorse, introrse or rarely poricidal; ovary inferior or semi-inferior, usually 2-, 3-, 4- or 6-locular, with 2–115 anatropous ovules in each loculus, the axile placenta at the apex, base or throughout the length of the locule; ovules bitegmic, tenuinucellate; style short or more rarely long, undivided. Fruits indehiscent, then dry, fleshy or woody or dehiscent by a circumscissile operculum, then woody, sometimes very large; seeds exotestal/mesotestal, winged (*Cariniana* and *Couroupita*) or without wings in remaining genera; endosperm lacking or very scanty; arils present or absent; embryos undifferentiated or with fleshy plano-convex or foliaceous cotyledons.

Seventeen genera in three subfamilies, the Lecythidoideae exclusively neotropical with 210 species, the Planchonioideae with 55 species in Africa, Madagascar and tropical Asia to N Australia, and the unigeneric Foetidioideae with 17 species in Madagascar, Mauritius and E Africa.

VEGETATIVE MORPHOLOGY. Almost all species of Lecythidaceae are trees with only three species growing as shrubs. Some species, such as *Bertholletia excelsa*, are extremely large, emergent trees

up to 55 m tall. The bark of Neotropical Lecythidaceae (Mori et al. 1987a) is easily recognized in the field because of its fibrous nature and the presence of files of unilaterally thickened, crystalbearing cells in the phloem. Crystalliferous phloem can be observed at low magnifications by making a tangential peel of the bark with a sharp knife. The bark can often be peeled from the trunk in long strips, and then the inner bark can be beaten into thin sheets. It has been used for caulking boats, cigarette paper, native clothing, and cordage. An excellent field feature is that the bark can be peeled from the twigs in long strips; this is shared by species of Annonaceae and therefore sterile collections of these two families are sometimes confused. External bark morphology of Neotropical Lecythidaceae is useful in differentiating species. Some species possess essentially smooth bark (e.g. *Eschweilera collina*), the bark of others is scalloped (e.g. *E. apiculata*, *E. laevicarpa*, and *E. micrantha*) and still others have very deeply furrowed bark (e.g. *Corythophora rimosa* and *Lecythis zabucajo*). The presence of cortical bundles helps characterize the Lecythidaceae, and the orientation of these bundles (inversely oriented in the Planchonioideae and Foetidioideae vs. normally oriented in the Lecythidoideae) has been used to distinguish subfamilies within the family (Prance and Mori 1979). However, the validity of cortical bundle orientation needs further investigation. The stem-node-leaf continuum of Lecythidaceae has been accurately described by Lignier (1890). Its principal features, at least in the New World Lecythidaceae, are (1) numerous cortical bundles that enter the petiole and are indirectly connected to the stele via fusion with stelar traces; and (2) three traces which depart from the stele at different levels, the median trace usually departs at the same node as the leaf it enters whereas the lateral traces depart from below that node (Mori and Black 1987). The leaves of most species of Planchonioideae and of the more primitive genera of Lecythidoideae (*Gustavia*, *Grias*, *Couroupita*) tend to be large and clustered towards the ends of the branches. These comply to Corner's model of architecture (e.g. *Gustavia grandibracteata*, *G. monocaulis*, and *Barringtonia calyptrocalyx*), Schoute's model (all species of *Grias*, *Gustavia grandibracteata*, and *G. superba*), and Rauh's model (the cauliflorous *Couroupita guianensis*). Most other species have smaller leaves and a much branched crown with typically leptocaul form. In all, eight types of architecture have been reported in Lecythidaceae, the others of which are: – Leeuwenberg's model: *Barringtonia edulis*; – Koriba's model: *Petersianthus africanus*; – Massart's model: *Couratari stellata*; – Roux's model: *Bertholletia excelsa*; and – Troll's model: *Lecythis pisonis*, *Eschweilera* sp.

Stipules are usually absent but, in a few species, minute caducous stipules occur. Several species of Lecythidoideae and *Planchonia* have buttressed trunks (e.g. most species of *Couratari*). The considerable variation in bark characteristics is discussed in Roth (1969) and Prance and Mori (1979).

WOOD ANATOMY. The wood anatomy of New World Lecythidaceae has been studied by the late Carl de Zeeuw (1987, 1990) who summarized its structure (1990) as follows. "Pores diffuse, moderately small to moderately large, numerous to very few, elliptical in outline, solitary or in radial multiples of 2 to 3 with a small component of long radial strings or clusters. Vessel elements medium length, with ends at a large angle to cell axis; perforations simple in all genera, with the addition of some reticulate or scalariform plates in *Grias*; intervessel pitting alternate (except in *Grias* and *Gustavia* which have a mixture of alternate and opposite to irregular pitting), the pits usually medium size (except very small in *Gustavia*); ray-vessel pitting in part similar to intervessel pitting and in part half-bordered to simple linear-irregular pits of larger size; tyloses present or absent, usually thin walled. Rays entirely heterocellular Type II or III in *Grias* and *Gustavia*, mixed heterocellular and homocellular in other genera, mostly less than one millimeter in height, 2- to 3-seriate, the tallest in *Grias* and *Gustavia*, the widest up to 6-seriate in *Grias* and *Gustavia*; crystals present in *Gustavia* and in some samples of *Grias*; silica sparse to abundant in all but *Bertholletia*, *Couroupita*, *Grias*, and *Gustavia*; dark gum often present. Axial parenchyma apotracheal banded in all genera except *Grias* and *Gustavia* where it is reticulate to diffuse-in-aggregates, and with scanty paratracheal in addition in all genera, mostly in 1- to 2-seriate bands but uniseriate in *Grias* and *Gustavia* and up to 6-seriate in part of *Lecythis*; strands average 4–7 cells long; strands of crystal cells associated with parenchyma bands in all genera, crystal cell walls unilaterally or uniformly thickened, the latter always occurring within uniseriate parenchyma bands; crystal cell wall thickening either exclusively one or the other type, or in mixture in part of the genera; gum often present. Fibers medium to moderately long, pitting small to very small, clearly bordered in *Allantoma*, *Bertholletia*, *Cariniana*, *Couratari*, and *Couroupita*; gelatinous fibers present in most genera".

Before his death, de Zeeuw had completed a series of manuscripts on the wood anatomy of *Foetidia* (de Zeeuw, manuscript), *Careya* and *Chydenanthus* (de Zeeuw, manuscript), *Petersianthus* (de Zeeuw, manuscript), and *Planchonia* (de Zeeuw, manuscript). Information from these studies is summarized below.

Foetidia possesses uniformly encapsulated crystals in strands in the axial parenchyma and silica bodies in the ray cells. Ray crystals, however, are not present. De Zeeuw's studies of *Careya*, *Chydenanthus*, and *Planchonia* support a close relationship among these genera. He reports septations in the fibers of all three genera and crystals in the axial parenchyma of *Careya* for the first time. De Zeeuw also found septate fibers in *Barringtonia samoensis* and suggests that their presence supports a close relationship between *Barringtonia* and other genera of Planchonioideae. He reports the presence of prismatic crystals in the axial parenchyma and rays of *Planchonia* for the first time. Although the two species of *Petersianthus* are geographically isolated from one another, de Zeeuw found them to be similar in their general wood anatomy but different enough to be distinguished by their wood alone. *Petersianthus macrocarpus* emits a strong, disagreeable odour from the green wood, and silica is virtually absent in both the rays and the axial parenchyma. In contrast, the wood of *P. quadrialatus* lacks a strong odour but possesses rough, rounded silica bodies in the rays and axial parenchyma.

FLOWER STRUCTURE. Flowers are rather large and have an extremely variable morphology. They are actinomorphic in the Planchonioideae, Foetidioideae and basal members of Lecythidoideae, but highly zygomorphic in the more advanced Lecythidoideae. The calyx is usually imbricate in bud and 4–6-lobed (2-lobed in *Bertholletia*) but is valvate in *Foetidia* and unlobed in *Barringtonia* section *Barringtonia*. Petals vary from 3 to 18 and are free. There are numerous stamens in most genera, up to 1200 in *Gustavia* and as few as 10 in

Cariniana. In all actinomorphic androecia, except *Foetidia*, the stamens are united at their base. Zygomorphy is caused by variations of the androecium which, in some genera of Lecythidoideae, forms a strap-like structure to one side which arches over the summit of the ovary to form a hood over a ring of fertile stamens at its base (Fig. 74). The hood appendages may bear anthers (*Lecythis*, *Corythophora*) or may be sterile and produce nectar at their base. In zygomorphic androecia of the Planchonioideae, the inner whorl of stamens is often reduced and staminodial. An annular nectar-secreting disk occurs in the Planchonioideae and the Foetidioideae but not in the Lecythidoideae. The ovaries are inferior or semi-inferior. Short styles are characteristic of most Lecythidoideae except *Bertholletia* and a few species of *Lecythis*, whereas the styles are long and filamentous in Planchonioideae and Foetidioideae and usually equal the filaments in length. There are few to numerous ovules attached in various positions to the axile placenta.

EMBRYOLOGY. The embryology has been studied in detail by Tsou (1994a). The ovules are anatropous, bitegmic, tenuinucellate with multi-layered outer and inner integuments, the inner integument forms the micropyle, and one or more vascular bundles lead to the outer integument. Anther wall formation is of the basic type and the tapetum is glandular.

POLLEN MORPHOLOGY. The pollen of Lecythidaceae has been studied by Muller (1972, 1979) and by Tsou (1994a, 1994b). Two distinct types of pollen occur in the family, syntricolpate in Planchonioideae and tricolpate or tricolporate in Lecythidoideae and Foetidioideae (Fig. 75). The latter type also occurs in the related Napoleonaeaceae. The *Planchonia* pollen type is characterized by grains which are usually larger than 40 μm, possess a thick ectexine and syntricolpate colpi, and generally lack clearly defined endoapertures. Specialized structures in the marginal zones of the ectoaperture and in the polar area are found in some species. The *Lecythis* type is less specialized, smaller (the polar axis usually less than 35 μm), tricolporoidate or tricolporate, and the exine is generally simpler, composed of a thin endexine, a layer of more or less distinct columellae, and a rather thin tectum which is smooth, finely reticulate, foveolate or scabrate-verrucate.

Dimorphic pollen occurs in *Couroupita* and some species of *Lecythis* (Jacques 1965; Mori et al. 1980; Mori and Boeke 1987). In species with

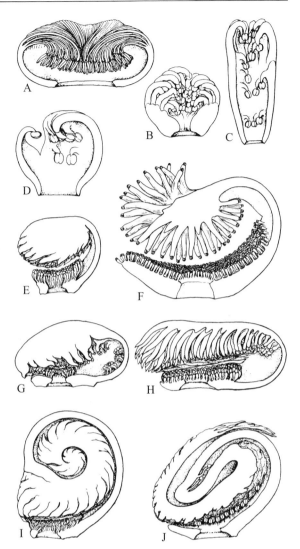

Fig. 74. Lecythidaceae. Androecial structures in Lecythidoideae, showing increasing zygomorphy in connection with formation of a strap-like hood over the ovary and the ring of fertile stamens. A *Gustavia hexapetala*, stamens often >1000, offering pollen reward. B *Grias cauliflora*, androecium a symmetric stamen ring; stamens all fertile, offering pollen reward; pollination probably by beetles. C *Allantoma lineata*, androecium fused into an actinomorphic tube; stamens all fertile; pollination probably by beetles. D *Cariniana pauciramosa*, antheriferous tube slightly zygomorphic; stamens all fertile. E *Lecythis corrugata*, the androecium zygomorphic, the hood without anthers but nectariferous. F *Couroupita guianensis*, fertile stamens around gynoecium with monad pollen, sterile stamens on hood with tetrad fodder pollen. G *Corythophora alta*, the closed antheriferous hood with sterile fodder pollen. H *Lecythis pisonis*, hood flat, its proximate appendages antheriferous with sterile fodder pollen. I *Eschweilera longipes*, androecium zygomorphic, the hood closed with a triple coil, offering nectar. J *Couratari stellata*, androecium zygomorphic, hood with double coil inwards and an external flap outwards, offering nectar from closed hood. (Modified from Mori and Prance 1987)

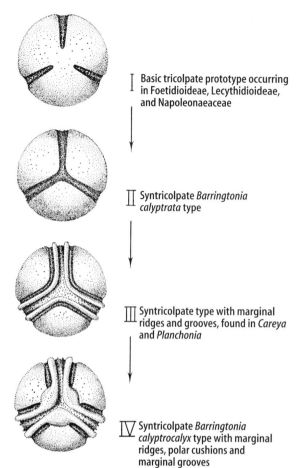

I Basic tricolpate prototype occurring in Foetidioideae, Lecythidioideae, and Napoleonaeaceae

II Syntricolpate *Barringtonia calyptrata* type

III Syntricolpate type with marginal ridges and grooves, found in *Careya* and *Planchonia*

IV Syntricolpate *Barringtonia calyptrocalyx* type with marginal ridges, polar cushions and marginal grooves

Fig. 75. Lecythidaceae. Main pollen types of Lecythidaceae; pollen grains drawn in polar view. (Muller 1973)

dimorphic pollen, the staminal ring pollen is fertile and the hood pollen does not germinate. In *Couroupita guianensis* the ring pollen is in monads and the sterile hood pollen remains in tetrads. In *Lecythis pisonis* the hood pollen turns black 24 hours after anthesis while that of the ring remains yellow. In species of *Corythophora*, the sterile pollen is yellow whereas the fertile pollen is white. In this genus, the sterile pollen may be situated in the staminal ring (*C. amapaensis*) or it may be located on the hood (*C. rimosa*). In the few species that have been studied, the sterile or fodder pollen serves as the pollinator reward. For example, in *C. amapaensis*, euglossine bees enter the flower, and extract yellow, fodder pollen which they subsequently place in their corbiculae. At the same time, fertile, white pollen is deposited on the head and back of the bee from where it is transferred to the stigma of subsequent flowers visited (Mori and Boeke 1987).

KARYOLOGY. The most detailed study of chromosome number is that of Kowal et al. (1977). The Lecythidoideae has a basic number of x = 17 and the Planchonioideae has x = 13. Tetraploid counts of n = 34 have been made for *Gustavia superba* and n = 26 for *Petersianthus macrocarpus* (Mangenot and Mangenot 1962) and *Barringtonia racemosa* (Morawetz 1986). The genera separated from Lecythidaceae have different numbers – *Napoleonaea* n = 16 and *Asteranthos* n = 21 (Kowal 1989).

POLLINATION. Details of the pollination of New World Lecythidaceae can be found in Prance (1976), Mori and Kallunki (1976), Mori et al. (1978), Prance and Mori (1979), Mori and Boeke (1987), Nelson et al. (1987), Mori and Prance (1990b), and Knudsen and Mori (1996). Observed pollinators are principally bees, but bats and beetles have also been observed visiting the flowers.

Species of *Gustavia*, all with actinomorphic flowers, non-differentiated pollen, and open flowers, offer pollen as a reward to their bee pollinators (Mori and Boeke 1987). Species of another New World, actinomorphic-flowered genus, *Grias*, possess floral scents with fatty acid derivatives typical of beetle-pollinated flowers (Knudsen and Mori 1996). The discovery of beetles in the flowers of an Ecuadorian species of *Grias* suggests that this genus is indeed beetle-pollinated (J. Knudsen, personal communication to S.A. Mori 1995). Nothing is known about the pollination of *Allantoma* but, based on the similarity of the floral structure of this genus with that of *Grias*, beetle pollination is also suspected.

Pollination of zygomorphic-flowered species of Lecythidaceae is more complicated. Species with this floral type vary in the type of pollinator reward, the degree of openness of the androecial hood, and in the complexity of the androecial hood (Mori 1989). Some species possess fodder pollen on an open hood (e.g. *Lecythis zabucajo*), others have fodder pollen on a closed hood (e.g. *Corythophora rimosa*), others seem to offer a combination of fodder pollen and nectar on a closed hood (e.g. *Lecythis corrugata*; Mori and Boeke 1987), and the most advanced species florally offer only nectar from a closed hood (e.g. species of *Eschweilera* and *Couratari*) as rewards to the floral visitors. The floral visitors are mostly bees, but bats have also been reported to visit at least two species (Mori et al. 1978; Mori and Prance 1990b).

A large number of different types of bees have been reported to visit the flowers of zygomorphic-flowered Lecythidaceae. Detailed studies of individual species of Lecythidaceae have not shown a one-to-one relationship between species of plant and species of bee. The most important trait of the bee pollinator of zygomorphic-flowered Lecythidaceae with closed hoods is that the pollinator be large enough to force open the hood, and possess a tongue long enough to reach the nectar. For example, in the Brazil nut (*Bertholletia excelsa*), species of *Bombus*, *Centris*, *Epicharis*, *Eulaema*, and *Xylocopa* have been observed entering the flowers (Mori and Prance 1990a; Mori 1992). Some species with more complicated androecial hoods, e.g. *Eschweilera pedicellata*, may be exclusively visited by euglossine bees because they are the only New World bees with tongues long enough to easily reach the nectar. It is noteworthy that the flowers of *E. pedicellata* are pollinated by trap-lining euglossine bees and that the large and showy flowers do not have any scent (Knudsen and Mori 1996).

Two species, *Lecythis poiteaui* and *L. barnebyi*, are known to be pollinated by bats. Both of these species produce flowers on racemes that project from the ends of the branches, have very large flowers with more than 1,000 stamens, emit a musty scent from the flowers, and produce copious, relatively dilute nectar (Mori et al. 1978; Mori and Prance 1990a). Many species of the actinomorphic-flowered, Old World Planchonioideae, especially species of *Barringtonia* and *Abdulmajidia*, are also known to be bat-pollinated.

FRUIT AND SEED. The fruits of Lecythidaceae are extremely variable. Those of *Foetidia* and Planchonioideae are indehiscent drupes with 1 to many seeds. *Petersianthus* is remarkable for the large wings on the exterior of the fruit and *Barringtonia* for their rectangular 4-ridged or 4-winged fruit. The fruits of several species of *Barringtonia* are buoyant and are dispersed by ocean currents which accounts for their widespread distribution, e.g. *B. racemosa* and *B. acutangula*. The varied structure of the fruit of Lecythidoideae is described and illustrated in Prance and Mori (1978, 1979) and Mori and Prance (1990b). The basal genera *Gustavia*, *Grias* and *Couroupita* have indehiscent fruits with the seeds embedded in pulp. These are dispersed by a variety of forest-floor mammals such as agouti, arboreal mammals such as monkeys and, for riverine species, by fish. The remaining genera of Lecythidoideae have a great variety of dehiscent fruits. *Cariniana* has unilaterally winged seeds and *Couratari* has wings that surround the seed. Both genera are tall, emergent trees, or savanna or

riverine species and are wind-dispersed. The seeds of *Lecythis* sections *Pisonis* and *Lecythis* have a swollen funicular aril (see Fig. 77G) and remain in the fruit after dehiscence of the operculum and are bat-dispersed. The large, globose, woody pyxidium of *Bertholletia* is secondarily indehiscent and has only a small operculum that falls inwards (see Fig. 76E). This species drops the pyxidium to the ground and is dispersed by agoutis. *Lecythis lurida* is also secondarily indehiscent. The seeds of *Allantoma lineata* are water-dispersed. The seeds of *Bertholletia excelsa*, *Allantoma lineata* and *Abdulmajidia* spp. all have an extremely hard, woody testa.

Three embryo types occur in the seeds of Lecythidaceae:

1) macropodial or undifferentiated in *Grias*, *Abdulmajidia*, *Allantoma*, *Chydenanthus*, *Corythophora*, *Eschweilera*, *Lecythis*, *Barringtonia* and *Careya*;
2) *Gustavia* type with fleshy, plano-convex cotyledons;
3) *Couroupita* type with leaf-like cotyledons and a long, well-differentiated radicle in *Couroupita*, *Cariniana*, *Foetidia*, *Planchonia* and *Petersianthus*.

The germination of all species observed so far is phanerocotylar.

PHYTOCHEMISTRY. Some species are accumulators of selenium. The seeds of two species of *Lecythis*, *L. ollaria* and *L. minor* are toxic because of the presence of the selenium-containing analog of the sulphur amino acid cystathionine: $HOOC-CH(NH_2)-CH\ Se-CH_2CH(NH_2)COOH$ (Kerdel-Vargas 1966). The bark of *Planchonia careya* (F. Muell.) Kunth contains saponins and is used by natives of New Guinea to stun fish.

AFFINITIES. The family has traditionally been placed in the Myrtales. Thorne (1968) suggested a Thealean relationship, and Cronquist (1981) and Takhtajan (1987) both established the order Lecythidales somewhere between the Theales and Malvales. A Myrtalean relationship is no longer acceptable on the basis of differences in wood anatomy, pollen and floral morphology such as the basal staminal ring. Tsou (1994a) was the first to suggest a relationship with Scytopetalaceae, which seems highly likely. *Asteranthos*, which was formerly in the Lecythidaceae, is better placed in the Scytopetalaceae. The Lecythidaceae, Napoleonaeaceae and Scytopetalaceae all seem to be closely related. This is supported by DNA sequence analyses (Morton et al. 1997); however, if *Asteranthos* is removed from Lecythidaceae, it is also necessary to remove *Napoleonaea* to obtain a monophyletic group of Lecythidoideae, Planchonioideae and Foetidioideae. The multigene sequence analysis by Anderberg et al. (2002) places these families sister to Sapotaceae within the broadly defined Ericales sensu APG (1998).

PALAEOBOTANY. There are few, reliable fossil records of Lecythidaceae. *Lecythidoanthus kugleri* described by Berry (1924a, 1924b) from the Miocene of Trinidad almost certainly represents a lecythidaceous flower. Milanez (1935) described in detail a Cretaceous wood from Piauí, Brazil which he named *Lecythioxylon brasiliense*. Huertas (1969) described a fossil fruit of an *Eschweilera* from Cundinamarca, Colombia which he named *Lecythidopyxion girardotanum*. *Barringtonia*-like fossils have frequently been described, for example, *Barringtonioxylon deccanense* from the upper Cretaceous or Early Eocene of Muhurzari, India and *B. eopterocarpum* (Prakash and Dayal 1965).

Fossil pollen that is similar to some extant *Barringtonia* and *Chydenanthus* species is known from the lower Eocene of Borneo and India; a slightly different type, referable to recent *Abdulmajidia*, *Barringtonia* and *Planchonia*, first appeared in the lower Eocene of Cameroon (see Tsou 1994b for references).

ECONOMIC IMPORTANCE. The most well-known economic product from the family is the Brazil nut from the seeds of *Bertholletia excelsa* (see Mori and Prance 1990a; Mori 1992). Various other species of *Lecythis* also have edible seeds but are only used locally. Two species of *Grias* have avocado-like fruit, the pulp of which is relished in Amazonian Peru, and the yellowish pulp around the seeds of *Gustavia speciosa* subsp. *speciosa* is eaten in Colombia. The fruit and seeds of *Planchonia careya* are eaten in New Guinea. The wood of many species of Lecythidoideae is used locally for general construction, but the quantity of silica in some species restricts their use. *Cariniana pyriformis* has been overexploited for its timber which has been exported from Colombia as Colombian mahogany or albarco. The wood of *Planchonia valida* is much used in house construction in Indonesia. The seeds and bark of various species of *Barringtonia* are used locally as a fish poison.

CONSERVATION. Many species of Lecythidaceae are of extremely restricted distribution and grow in highly threatened areas such as western Colombia and Ecuador, eastern Brazil, and Sumatra and Java, and are therefore potentially threatened with extinction. The 15 Madagascan species of *Foetidia* are in particular danger.

INFRAFAMILIAL CLASSIFICATION
Flowers actinomorphic or zygomorphic; calyx usually imbricate in bud; petals present; styles usually short or of medium length; stamens united at base, sometimes prolonged into a hood or tube; pollen tricolpate or tricolporate. Fruit often dehiscent, less frequently indehiscent. Cortical bundles with normal orientation; secondary xylem with crystal chains.

Subfam. Lecythidoideae (genera 1–10)
Flowers actinomorphic; calyx imbricate in bud; petals present; styles long and linear; stamens united at base; pollen syntricolpate. Fruit indehiscent. Cortical bundles with reversed orientation (xylem outside; phloem inside); secondary xylem without crystal chains.

Subfam. Planchonioideae (genera 11–16)
Flowers actinomorphic; calyx valvate in bud; petals absent; styles long and linear; stamens free to base; pollen syntricolpate. Fruit indehiscent. Cortical bundles with reversed orientation; secondary xylem without crystal chains.

Subfam. Foetidioideae (genus 17)

KEY TO THE GENERA

1. Petals present; calyx usually imbricate in bud; stamens united at base or into tube or hood   2
   – Petals absent; calyx valvate in bud; stamens free to base
                                                        17. Foetidia
2. Floral nectary disc absent; styles usually short; flowers actinomorphic or zygomorphic; fruit dehiscent or indehiscent; secondary xylem with crystal chains   3
   – Floral nectary disc present; styles long and linear; flowers actinomorphic; fruit indehiscent; secondary xylem without crystal chains   14
3. Androecium actinomorphic   4
   – Androecium zygomorphic   7
4. Flowers 2.5–20 cm diam. at anthesis; stamens united only at base; ovules inserted on upper 1/2 of septum; fruits fleshy, globose, to ovoid indehiscent   5
   – Flowers less than 2.5 cm diam. at anthesis; stamens united into a tube with some anthers inserted on interior; ovules inserted on lower 1/2 of septum; fruits woody, campanulate or cylindrical, dehiscent   6
5. Petals 6–8(–18); stamens 500–1210, the anthers linear, 2–5 mm long, dehiscing by apical pores; placentae expanded, the ovules 7–93 per locule, horizontal or slightly descending; fruits usually with 2 or more seeds; embryo with plano-convex, fleshy cotyledons   **1. Gustavia**
   – Petals 4; stamens 85–210, the anthers globose, less than 1 mm long, dehiscing by longitudinal slits; placentae not

expanded, the ovules 2–4, pendulous; fruits 1-seeded; embryo undifferentiated **2. Grias**

6. Buds oblong; calyx rim-like or with 5 inconspicuous, broadly triangular lobes at anthesis; petals 5, the apices pointed; ovary usually 4-locular; seeds not winged; embryo undifferentiated **3. Allantoma**
 – Buds globose; calyx with 6 triangular lobes at anthesis; petals 6, the apices rounded; ovary usually 3-locular; seeds unilaterally winged; embryo with 2 foliaceous cotyledons **4. Cariniana**

7. All or at least three-fourths of hood appendages bearing anthers **8**
 – Usually with all of hood appendages sterile, or rarely with less than half of hood appendages bearing anthers **10**

8. Ovary 6-locular; ovules 30–115 per locule, on bilamellar placentae throughout length of locule; fruit globose, indehiscent; seed as long as broad; embryo with 2 foliaceous, highly convoluted cotyledons **5. Couroupita**
 – Ovary 2–3(–5)-locular; ovules 5–8 per locule, the placentae not bilamellar, attached towards base of locule; fruit campanulate or cylindric, dehiscent; seeds longer than broad; embryo undifferentiated or with 2 foliaceous cotyledons, these are highly convoluted **9**

9. Androecium elongated on one side into a strap-like structure which curves over summit of ovary; ovary 2(–5)-locular; seeds not winged; embryo undifferentiated **6. Corythophora**
 – Androecium elongated on one side but not forming a strap-like structure over summit of ovary; ovary 3-locular; seeds with unilateral wings; embryo with foliaceous cotyledons **4. Cariniana**

10. Buds enclosed by calyx except for horizontal slit at apex; calyx with 2 lobes at anthesis; style greater than 10 mm long; fruit functionally indehiscent, with small, inwardly falling operculum only; seeds with thick, boney testa, remaining inside fruit at maturity **7. Bertholletia**
 – Buds not enclosed by calyx; calyx with 6 lobes at anthesis; style usually less than 10 mm long; fruit usually dehiscent, with freely falling operculum; seeds without thick, boney testa, usually falling from fruit at maturity **11**

11. Androecial hood coiled inwards, with outwardly extended flap at apex of coil; fruit cylindric or campanulate; ovary 3-locular; seeds with wing around circumference; embryo with 2 foliaceous cotyledons **8. Couratari**
 – Androecial hood flat or if coiled inward without outwardly extended flap at apex of coil; fruit usually globose; ovary not 3-locular; seeds without wings; embryo undifferentiated **12**

12. Androecial hood usually forming complete coil inwards, with blunt tipped appendages at apex of coil, these differentiated from more abundant, echinate hood appendages; ovary usually 2-, infrequently 4-locular; seeds with lateral arils or, less frequently, aril completely surrounding testa **9. Eschweilera**
 – Androecial hood flat or expanded at apex but not forming complete coil inwards, the hood appendages not differentiated; ovary usually 4-, less frequently 2-locular; seeds usually with basal aril, less frequently without aril **13**

13. Ovary 2-locular, the style not differentiated from summit of ovary, the summit umbonate; fruit campanulate **6. Corythophora**
 – Ovary 4-locular, the style usually differentiated from summit of ovary, the summit usually truncate; fruit usually globose **10. Lecythis**

14. Fruit broadly winged; anthers introrse **11. Petersianthus**
 – Fruit not winged; anthers extrorse **15**

15. Ovules 4 per loculus, inserted at apex of axis, pendulous; fruit one-seeded **12. Barringtonia**
 – Ovules 8–20 per loculus and usually inserted all along axis or towards base; or ovules 1–2 per loculus, inserted at base of axis and ascending; fruit usually 2–many-seeded **16**

16. Pedicel articulate; ovules 1–2 per loculus, inserted at base of axis and ascending; fruit 1–2-seeded; carpels 2 **13. Chydenanthus**
 – Pedicel not articulate; ovules 8–20 per loculus, inserted all along axis, at top or middle of axis; fruit usually 5–many-seeded, rarely 2–4-seeded; carpels 3 or 4 **17**

17. Petals 3; ovules inserted on upper part of axis; outer integument of seed thin; carpels 3 **14. Abdulmajidia**
 – Petals 4; ovules inserted along or on mid part of axis; outer integument of seed arilloid; carpels 4 **18**

18. Embryo undivided, undifferentiated **15. Careya**
 – Embryo divided, with two foliaceous plicate cotyledons **16. Planchonia**

GENERA OF LECYTHIDACEAE

Users wanting to make species determinations are referred to the Flora Neotropica monograph 21 (1, 2), in which all Neotropical genera have been revised by the authors. The Pflanzenreich treatments of Knuth (1939a, 1939b) are largely outdated.

1. *Gustavia* L.

Gustavia L., Pl. Surinam. 12: 17–18 (1775).

Small to large trees, leaves often clustered at end of branches. Inflorescences suprafoliar, axillary or cauline, solitary or racemose. Flowers actinomorphic. Calyx nearly entire to 4–6-lobed. Petals 6, 8, 12 or 18. Androecium a symmetrical ring of 500–1210 stamens, the filaments fused at base, all fertile. Ovary 4–6(–10)-locular, 7–93 ovules per loculus. Fruit berry-like, indehiscent, with seeds embedded in pulp; seeds with yellow, expanded, contorted funicles or without well-developed funicles. Embryo with large, fleshy, plano-convex cotyledons and minute hypocotyl and plumule. $2n$ = 34, 68. 40 species, from Costa Rica to Amazonia and coastal NE Brazil.

2. *Grias* L.

Grias L., Syst. Nat. ed. 10: 1075 (1759).

Trees to 30 m tall, leaves clustered at end of branches. Inflorescences on trunk or branches, racemes or fascicles, rarely in axils of lower leaves. Flowers actinomorphic. Calyx 4-lobed. Petals 4. Androecium a symmetrical ring of 85–210 stamens, the filaments fused at base, all fertile, curved inwards. Ovary 4-locular, 2–4 pendent ovules per loculus. Fruit a drupe, indehiscent, with

1 seed embedded in a fleshy pulp; seeds fusiform, lacking cotyledons or developed funicle. $2n = 34$. Six species, Central America, Jamaica, NW South America.

3. *Allantoma* Miers

Allantoma Miers, Trans. Linn. Soc. Lond. 30: 170, 294 (1874), pro parte, androphoro excl.

Small to large trees. Leaves not clustered at end of branches. Inflorescences terminal or subterminal racemes or once-branched panicles. Flowers actinomorphic. Calyx 5–6-lobed. Petals 5–6. Androecium a slightly asymmetrical urceolate tube with apex divided in 8–10, inwardly reflexed lacinae bearing anthers and additional stamens inserted sparsely over interior; stamens ca. 30, all fertile. Ovary (3)4–5 locular, with ca. 20 ovules inserted all along septum of each loculus. Fruit an elongate, cylindrical, circumscissile capsule, with tack-shaped operculum, dehiscent from base; seeds numerous, narrowly linear longate, with hard, woody testa, and caducous flattened stipe-like funicle, not embedded in pulp. One species, Neotropical, Amazonian Venezuela and Brazil along upper Orinoco, Rio Negro and lower Amazon rivers.

4. *Cariniana* Casar.

Cariniana Casaretto., Nov. Stirp. Bras. Dec. 4: 34–37 (1842).

Small to large trees to 55 m tall. Leaves not clustered at end of branches. Inflorescences terminal or axillary racemes or panicles. Flowers nearly actinomorphic to zygomorphic. Calyx 5–6-lobed. Petals 5–6. Androecium obliquely hood-shaped, almost symmetrical to markedly asymmetric, with stamens inserted in a complete circle all over interior or at apex only; stamens 10–150, all fertile. Ovary 3-locular with ovules inserted all along septum. Fruit a cylindrical or campanulate circumscissile capsule, with tack-shaped operculum; seeds numerous, unilaterally winged, not embedded in pulp. 15 species, Neotropical, from northern Colombia and Venezuela to Central Brazil, predominantly Amazonian.

5. *Couroupita* Aubl.

Couroupita Aubl., Hist. Pl. Guiane 2: 708–711, t. 282 (1775).

Medium- to large-sized trees. Leaves clustered at end of branches. Inflorescences cauliflorous or ramiflorous racemes or panicles with continuous growth at apex. Flowers zygomorphic. Calyx 6-lobed. Petals 6. Androecium with strap-like hood arching over summit of ovary, with anthers bearing sterile pollen in hood and fertile pollen in basal ring, with open gap between ring and hood stamens, stamens numerous. Ovary 6-locular with numerous ovules attached to septum. Fruit round, large and indehiscent, exocarp woody, numerous seeds embedded in a fleshy pulp; seeds with funicle 10–15 mm long, cotyledons foliaceous. $2n = 34$. 3 species, Neotropical, Nicaragua to Colombia, Venezuela, Amazonia and the Guianas.

6. *Corythophora* R. Knuth

Corythophora R. Knuth in Engl. Pflanzenr. IV. 219a: 50–51 (1939).

Large trees. Leaves not clustered at end of branches. Inflorescences terminal or subterminal racemes or panicles. Flowers zygomorphic. Calyx 6-lobed. Petals 6. Androecium with strap-like hood arching over summit of ovary, the hood appendages with or without anthers, 40–230 stamens in staminal ring, hood closed over staminal ring. Ovary 2–5-locular, with 4–26 ovules in each loculus, ovules attached to base or lower part of septum. Fruit a woody, campanulate or cylindrical circumscissile capsule; seeds elongate, not embedded in pulp, with basal aril and macropodial embryo. $2n = 34$. Four species, Neotropical, E Amazonian Brazil and Guianas.

7. *Bertholletia* Humb. & Bonpl.   Fig. 76

Bertholletia Humb. & Bonpl., Pl. Aequinoct. 1: 122–127, t. 36 (1807).

Large trees to 55 m tall. Leaves not clustered at ends of branches. Inflorescences terminal and axillary spikes or panicules of spikes. Flowers zygomorphic. Calyx 2-lobed. Petals 6. Androecium with strap-like hood arching over summit of ovary and tightly pressed against staminal ring; hood appendages all sterile, ring stamens 80–135. Ovary (3)4(–6)-locular, with 16–25 ovules inserted at base of septum of each loculus. Fruit a large, globose, woody circumscissile capsule with small, internally detaching operculum, thus secondarily indehiscent; seeds 10–25, triangular in cross section, testa hard and woody, not embedded in pulp, the embryo undifferentiated, cotyledons represented by two small scales only. $2n = 34$.

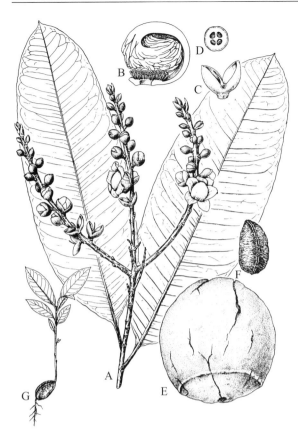

Fig. 76. Lecythidaceae. *Bertholletia excelsa*. **A** Flowering branch. **B** Androecium, medial section. **C** Calyx and pistil. **D** Ovary, vertical section. **E** Pyxidium. **F** Seed. **G** Seedling. Drawn by B. Angell. (Mori and Prance 1990b)

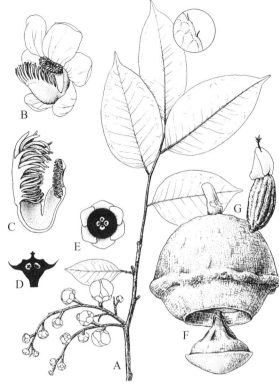

Fig. 77. Lecythidaceae. *Lecythis pisonis*. **A** Flowering branch. **B** Flower. **C** Androecium, medial section. **D** Ovary, vertical section. **E** Ovary, transverse section. **F** Pyxidium with detached operculum. **G** Seed with fleshy aril and cord-like funicle at base. Drawn by B. Angell. (Mori and Prance 1990b)

One species, *B. excelsa* Humb. & Bonpl., confined to Amazonia.

8. *Couratari* Aubl.

Couratari Aubl., Hist. Pl. Guiane 2: 723, t. 290 (1775).

Medium to large trees. Leaves not clustered at ends of branches. Inflorescences terminal or axillary racemes or panicles. Flowers zygomorphic. Calyx 6-lobed. Petals 6. Androecium with strap-like hood arching over summit of ovary, the hood apically coiled inwards, the apical part abruptly folded back over hood to form outwardly folded flap; hood appendages all sterile; ring stamens 10–75. Ovary 3-locular, with numerous ovules inserted along septum or at base of loculus. Fruit an elongate, cylindrical or campanulate circumscissile capsule; operculum tack-shaped, seeds numerous, not embedded in pulp, with symmetrical wings, cotyledons foliaceous. 20 species, Neotropical from Costa Rica to E Brazil.

9. *Eschweilera* Mart. ex A.P. DC.

Eschweilera Martius ex A.P. DC., Prodr. 3: 393 (1828).

Small to large trees. Leaves not clustered at end of branches. Inflorescences terminal, axillary, ramiflorous or cauliflorous racemes, spikes or panicles. Flowers zygomorphic. Calyx 4- or 6-lobed. Petals 4 or 6. Androecium with strap-like hood arching over summit of ovary, the appendages sterile, coiled inwards; ring stamens 82–500. Ovary usually 2-locular, rarely 4-locular, the ovules usually basal. Fruit small to medium-sized, circumscissile capsules always dehiscent; seeds 1–35, not embedded in pulp, often with a lateral appressed aril, rarely without aril. $2n = 34$. About 85 species, Neotropical, Mexico, Central America, Colombia, Venezuela, Amazonia to E and C Brazil.

10. *Lecythis* Loefl.   Fig. 77

Lecythis Loefl., Iter Hispan.: 189 (1758).

Small to very large trees. Leaves not clustered at end of branches. Inflorescences terminal or axillary racemes, spikes or once-branched panicles, rarely on young stems below leaves. Flowers zygomorphic. Calyx 6-lobed, the lobes often with mucilage ducts. Petals 6. Androecium with strap-like hood arching over summit of ovary, the hood appendages sterile or with anthers; ring stamens 70–1000. Ovary (3)4(–6)-locular, ovules inserted on lower part of septum. Fruits small to very large, woody circumscissile capsules, sometimes indehiscent and falling to the ground with seeds inside, generally campanulate or urceolate, the base usually remaining on tree after dehiscence; seeds, often with a basal swollen aril projecting beyond micropylar end, rarely with lateral aril or exarillate. $2n = 34$. 26 species. Neotropical, Central America to São Paulo, Brazil.

11. *Petersianthus* Merrill

Petersianthus Merrill, Philipp J. Sci. C. Bot. 11: 200 (1916).
Combretodendron A. Chev.

Large trees. Leaves not clustered at ends of branches. Inflorescences terminal panicles. Flowers actinomorphic. Calyx 4-lobed. Petals 4, less than 10 mm long. Disc well developed, lining area above receptacle. Androecium a symmetrical ring of numerous stamens, fused at base to 4 mm, all fertile. Ovary 4-locular, 3–8 ovules per loculus, placentation central axile. Fruits dry, indehiscent, fibrous, fusiform, broadly 3–4-winged. Seeds 4–5; embryo with 2 plicate cotyledons and a long radicle. 2 species, tropical W Africa and Philippines.

12. *Barringtonia* Forst.   Fig. 78

Barringtonia Forst., Char. Gen.: 75, t. 38 (1776); Payens, Blumea
    15: 157–263 (1967), monogr.

Small to large trees. Leaves clustered at ends of branches. Inflorescences axillary or less frequently terminal or cauline racemes or spikes, usually pendulous. Flowers actinomorphic. Calyx distinctly 4-lobed or completely fused or with only an apical pore in early development. Petals (3)4(5), exceeding 10 mm long. Disc a well-developed ring. Androecium a symmetrical ring of numerous stamens in 3–8 whorls, fused for 1–10 mm, inner whorl staminodial, second inner whorl often staminodial. Ovary (2–)4-locular, 4 pendulous ovules per loculus, placentation axile. Fruits 1-seeded drupes, often rectangular, 4-ridged or 4-winged;

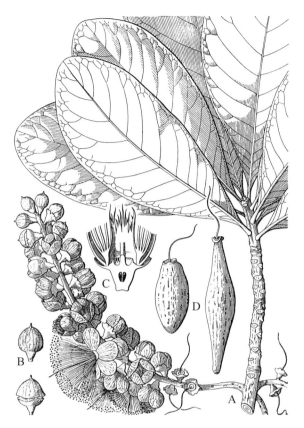

Fig. 78. Lecythidaceae. *Barringtonia calyptrata*. **A** Habit. **B** Flower buds, beneath with circumscissile calyx. **C** Flower in vertical section, note nectary disk and ramification of androecium. **D** Fruits with persistent style. Drawn by R. van Crevel. (Payens 1967)

embryo undifferentiated. $2n = 26$. 41 species, E Africa, Madagascar and tropical Asia.

13. *Chydenanthus* Miers

Chydenanthus Miers, Trans. Linn. Soc. II, 1: 111 (1875).

Medium to large trees. Leaves not clustered at ends of branches. Inflorescences terminal panicles. Flowers actinomorphic. Calyx cupuliform, with 4 small lobes. Petals 4, exceeding 10 mm long. Disc a well-developed ring. Androecium a symmetrical ring of numerous stamens in 3–4 whorls, the innermost of which is staminodial, distinctly fused at base. Ovary cylindrical, 2-locular, 1–2 erect ovules per loculus, placentation basal, axile. Fruits long, elliptic, unridged 1-seeded drupes; embryo undifferentiated. $2n = 26$. 2 species, Java and Sumatra.

14. *Abdulmajidia* Whitmore

Abdulmajidia Whitmore, Kew Bull. 29: 207–211 (1974).

Trees. Leaves clustered at ends of branches. Inflorescences terminal, axillary or cauline racemes or spikes. Flowers actinomorphic. Calyx 3–4-lobed, coloured deep red. Petals 3, ca. 28 mm long. Disc a well-developed ring. Androecium a symmetrical ring of numerous stamens in 7–8 whorls, the innermost of which is staminodial, distinctly fused at base. Ovary 3–4-locular, 8–20 ovules per loculus, inserted on upper portion of placenta. Fruits 2–5-seeded drupes, rounded not ridged; seeds with hard, woody testa. 2 species, Malay Peninsula.

15. *Careya* Roxburgh

Careya Roxburgh, Hort. Bengal.: 52 (1814).

Shrubs to large trees. Leaves clustered at ends of branches. Inflorescences terminal or axillary spikes. Flowers actinomorphic. Calyx 4-lobed. Petals 4, 15–40 mm long. Disc a well-developed ring. Androecium a symmetrical ring of numerous stamens in 5–8 whorls, the innermost of which is staminodial, outermost also staminodial, distinctly fused at base. Ovary 4(5)-locular, 20–40 ovules per loculus inserted in two or four rows on placenta; placentation central axile. Fruits drupes, 4-locular, with several seeds in each loculus, crowned by 4 persistent sepals; seeds with undifferentiated embryos. $2n = 26$. 4 species, tropical Asia.

16. *Planchonia* Blume

Planchonia Blume in Van Houtte, Fl. Sevres 7: 24 (1851).

Small to large trees. Leaves clustered at ends of branches. Inflorescences terminal spikes or racemes or solitary flowers. Flowers actinomorphic. Calyx 4-lobed. Petals 4, 30–55 mm long. Disc a well-developed ring. Androecium a symmetrical ring of numerous stamens in many whorls, the innermost of which is staminodial, the filaments distinctly fused at base. Ovary 3–4-locular, ca. 20 ovules per loculus, inserted in two rows on placenta, placentation central axile. Fruits 1–many-seeded drupes; embryo with two foliaceous, plicate cotyledons and coiled or curved radicle. $2n = 26$. 8 species, Andaman Is., tropical Asia, N Australia.

17. *Foetidia* Commerson

Foetidia Commerson ex Lam. Encyl. 2: 457 (1786).

Small to large trees. Leaves clustered at ends of branches. Inflorescences terminal or axillary racemes or flowers solitary. Flowers actinomorphic. Calyx 3–5-lobed, valvate, lobes thick, hard, purple on both surfaces. Petals absent. Disc a slightly elevated area on surface of ovary adjacent to androecium, purple coloured. Stamens numerous, free to base, all fertile. Ovary 3–4-locular, 10–15 ovules per loculus, inserted in two rows on central placenta. Fruits turbinate, 1–4-locular, few-seeded drupes; embryo slightly curved, with long radicle. 17 species, Madagascar, Mauritius and E Africa.

Selected Bibliography

Anderberg, A.A. 2002. See general references.
APG (Angiosperm Phylogeny Group) 1998. See general references.
Berry, E.W. 1924a. A fossil flower from the miocene of Trinidad. Am. J. Sci. 7(38): 103–108.
Berry, E.W. 1924b. The Tertiary Flora of the Island of Trinidad. B.W.I. Johns Hopkins Studies in Geology 6: 120–123.
Cronquist, A. 1981. See general references.
Huertas, G.G. 1969. Un nuevo genero y especie fosiles de las lecitidaceas. Caldasia 10(48): 365–369.
Jacques, F. 1965. Morphologie de pollen et des ovules de *Couroupita guianensis* Aubl. (Lécythidacées). Pollen Spores 7: 175–180.
Kartawinata, E.K. 1965. The genus *Planchonia* Blume (Lecythidaceae). Bull. Bot. Surv. India 7: 162–187.
Kerdell-Vargas, F. 1966. The depilatory and cytotoxic action of Coco de Mono (*Lecythis ollaria*) and its relationship to chronic seloniosis. Econ. Bot. 20: 187–195.
Knudsen, J., Mori, S.A. 1996. Floral scents and pollination in Lecythidaceae. Biotropica 28: 42–60.
Knuth, R. 1939a. Barringtoniaceae. In: Engler, A. (ed.) Das Pflanzenreich IV, 219, pp. 1–82.
Knuth, R. 1939b. Lecythidaceae. In: Engler, A. (ed.) Das Pflanzenreich IV, 219a, pp. 1–146.
Kowal, R.R. 1989. Chromosome numbers of *Asteranthos* and the putatatively related Lecythidaceae. Brittonia 41: 131–135.
Kowal, R.R., Mori, S.A., Kallunki, J.A. 1977. Chromosome numbers of Panamanian Lecythidaceae and their use in subfamilial classification. Brittonia 29: 399–410.
Lignier, O. 1890. Recherches sur l'anatomie des organes végétatifs des Lécythidées, des Napoléonées et des Barringtoniées (Lécythidacées). Bull. Sci. France Belgique 21: 291–420.
Mangenot, S., Mangenot, G. 1962. Enquête sur les nombres chromosomiques dans une collection d'espèces tropicales. Rev. Cytol. Biol. Vég. 25: 411–447.
Miers, J. 1874. On the Lecythidaceae. Trans. Linn. Soc. 30: 157–318.
Miers, J. 1875. On the Barringtoniaceae. Trans. Linn. Soc. II 1: 47–118.

Milanez, F.R. 1935. Estudo de um Dicotyledoneo fossil do cretáceo. Rodriguesia 1(2): 83–89.

Monteiro-Scanavacca, W.R. 1974. Vascularização do gineceu em Lecythidaceae. Bol. Bot. Univ. São Paulo 2: 53–69.

Monteiro-Scanavacca, W.R. 1975a. Vascularização e natureza de estruturas do androceu em Lecythidaceae. Bol. Bot. Univ. São Paulo 3: 61–74.

Monteiro-Scanavacca, W.R. 1975b. Estudo da placentação em Lecythidaceae. Bol. Bot. Univ. S. Paulo 3: 75–86.

Morawetz, W. 1986. Remarks on karyological differentiation patterns in tropical woody plants. Pl. Syst. Evol. 152: 49–100.

Mori, S.A. 1989. Diversity of Lecythidaceae in the Guianas. In: Holm-Nielsen, L.B., Nielsen, I.C., Balslev, H. (eds.) Tropical forests: botanical dynamics, speciation, and diversity. New York: Academic Press, pp. 319–331.

Mori, S.A. 1992. The Brazil nut industry – past, present, and future. In: Plotkin, M., Famolare, L. (eds.) Sustainable harvest and marketing of rain forest products. Washington, D.C.: Island Press, pp. 241–251.

Mori, S.A., Black, D. 1987. Stem and leaf. In: Mori, S.A. et al. The Lecythidaceae of a lowland Neotropical forest: La Fumée Mountain, French Guiana, chap. VII. Mem. New York Bot. Gard. 44: 72–85.

Mori, S.A., Boeke, J.D. 1987. Pollination. In: Mori, S.A. (ed.) The Lecythidaceae of a lowland neotropical forest: La Fumée Mountain, French Guiana. Mem. New York Bot. Gard. 44: 137–155.

Mori, S., Kallunki, J. 1976. Phenology and floral biology of *Gustavia superba* (Lecythidaceae) in Central Panama. Biotropica 8: 184–192.

Mori, S.A., Prance, G.T. 1981. Relações entre a classificação genérica de Lecythidaceae do Novo Mundo seus polinizadores e dispersadores. Rev. Bras. Bot. 4: 31–37.

Mori, S.A., Prance, G.T. 1987. A guide to collecting Lecythidaceae. Ann. Missouri Bot. Gard. 74: 321–330.

Mori, S.A., Prance, G.T. 1990a. Taxonomy, ecology and economic botany of the Brazil nut (*Bertholletia excelsa* Humb. & Bonpl.: Lecythidaceae). Adv. Econ. Bot. 8: 130–150.

Mori, S.A., Prance, G.T. 1990b. Lecythidaceae – part II. The zygomorphic-flowered New World genera. Fl. Neotrop. Monogr. 21(2): 1–376.

Mori, S.A., Prance, G.T., Bolten, A.B. 1978. Additional notes on the floral biology of neotropical Lecythidaceae. Brittonia 30: 113–130.

Mori, S.A., Orchard, J.E., Prance, G.T. 1980. Intrafloral pollen differentiation in the New World Lecythidaceae subfamily Lecythidoideae. Science 209: 400–403.

Mori, S.A., Black, D., de Zeeuw, C. 1987a. Habit and bark. In: Mori, S.A. et al. The Lecythidaceae of a lowland Neotropical forest: La Fumée Mountain, French Guiana, chap. VIII. Mem. New York Bot. Gard. 44: 86–99.

Mori, S.A. et al. 1987b. The Lecythidaceae of a lowland neotropical forest. La Fumée Mountain, French Guiana. Mem. New York Bot. Gard. 44: 1–190.

Moritz, A. 1984. Estudos biológicos da floração e da frutificação da Castanha-do-brasil (*Bertholletia excelsa* H.B.K.). Bélem: EMBRAPA-CPATU, Documentos 29: 1–82.

Morton, C.M., Mori, S.A., Prance, G.T., Karol, K.G., Chase, M.W. 1997. Phylogenetic relationships of Lecythidaceae: a cladistic analysis using *rbcL* sequence and morphological data. Am. J. Bot. 84: 530–540.

Muller, J. 1972. Pollen morphological evidence for subdivision and affinities of Lecythidaceae. Blumea 20: 351–355.

Muller, J. 1973. Pollen morpholohy of *Barringtonia calyptrocalyx* K. Sch. (Lecythidaceae). Grana 13: 29–44.

Muller, J. 1979. Pollen. In: Prance, G.T., Mori, S.A. (eds.) Lecythidaceae – part I. The actinomorphic-flowered New World Lecythidaceae (*Asteranthos, Gustavia, Grias, Allantoma & Cariniana*). Fl. Neotrop. Monogr. 21: 72–76.

Nelson, B.W., Absy, M.L., Barbosa, E.M., Prance, G.T. 1987. Observations on flower visitors to *Bertholletia excelsa* H.B.K. and *Couratari tenuicarpa* A.C. Sm. (Lecythidaceae). Acta Amazonica 15 (1/2) Suppl.: 225–234.

Payens, J.P.D.W. 1967. A monograph of the genus *Barringtonia* (Lecythidaceae). Blumea 15: 157–263.

Prakash, V., Dayal, R. 1965. *Barringtonioxylon eopterocarpum* sp. nov. A fossil wood of Lecythidaceae from Deccan intertrappean beds of Mahurzari. Paleobotanist 13: 25–29.

Prance, G.T. 1976. The pollination and androphore structure of some Amazonian Lecythidaceae. Biotropica 8: 235–241.

Prance, G.T., Mori, S.A. 1978. Observations on the fruits and seeds of neotropical Lecythidaceae. Brittonia 30: 21–33.

Prance, G.T., Mori, S.A. 1979. Lecythidaceae – part I. The Actinomorphic-flowered New World Lecythidaceae. Fl. Neotrop. Monogr. 21: 1–270.

Roth, I. 1969. Estructura anatómica de la corteza de algunas especies arbóreas venezolanas de Lecythidaceae. Acta. Bot. Venez. 4: 89–117.

Takhtajan, A. 1987. See general references.

Thorne, R.F. 1968. Synopsis of a putatively phylogenetic classification of flowering plants. Aliso 6: 57–66.

Tsou, C.-H. 1994a. The embryology, reproductive morphology and systematics of Lecythidaceae. Mem. New York Bot. Gard. 71: 1–110.

Tsou, C.-H. 1994b. The classification and evolution of pollen types of Planchonioideae (Lecythidaceae). Plant Syst. Evol. 189: 15–27.

Whitmore, T.C. 1974. *Abdulmajidia*, a new genus of Lecythidaceae from Malaysia. Kew Bull. 29: 207–211.

Zeeuw, C.H. de. 1987. Wood anatomy. In: Mori, S.A. et al. The Lecythidaceae of a lowland neotropical forest: La Fumée Mountain, French Guiana, chap. IX. Mem. New York Bot. Gard. 44: 100–112.

Zeeuw, C.H. de. 1990. Secondary xylem of Neotropical Lecythidaceae. In: Mori, S.A., Prance, G.T. (eds.) Lecythidaceae – part II. The zygomorphic-flowered New World genera, chap. II. Fl. Neotrop. Monogr. 21 (II): 4–59.

Zeeuw, C.H. de. Manuscript. Wood anatomy of the Old World Lecythidaceae. Part I – *Foetidia* Commerson ex Lamarck, Foetidioideae.

Zeeuw, C.H. de. Manuscript. Wood anatomy of the Old World Lecythidaceae. Part II – *Crateranthus* E.G. Baker and *Napoleonaea* P. Beauv., Napoleonaeoideae.

Zeeuw, C.H. de. Manuscript. Wood anatomy of the Old World Lecythidaceae. Part III. *Careya* Roxb. and *Chydenanthus* (Bl.) Miers, Planchonioideae.

Zeeuw, C.H. de. Manuscript. Wood anatomy of the Old World Lecythidaceae. Part IV – *Petersianthus* Merrill, Planchonioideae.

Zeeuw, C.H. de. Manuscript. Wood anatomy of the Old World Lecythidaceae. Part V – *Planchonia* King, Planchonioideae.

Zeeuw, C.H. de. Manuscript. Wood anatomy of the Old World Lecythidaceae. Part VI – *Barringtonia* J.R. & G. Forster, Planchonioideae.

Lepidobotryaceae

K. KUBITZKI

Lepidobotryaceae J. Léonard in Bull. Jard. Bot. Etat 20: 38 (1950), nom. cons.

Dioecious trees with very sparse indumentum of unicellular hairs. Leaves unifoliolate, rachis with a stipel and disarticulation at base of pulvinate petiolule; stipules and stipels fugaceous. Inflorescences racemes or spikes, sometimes compound; flowers unisexual; sepals and petals 5, free, imbricate (calyx with quincuncial prefloration); stamens 5 + 5, the antesepalous longer than the antepetalous; filaments fused at base into a nectariferous disk or tube; anthers tetrasporangiate, dehiscing lengthwise; gynoecium 2–3-carpellate, ovary syncarpous, with apical-axile placentation; ovules 2 per locule, collateral, anatropous, with placental obturator above the micropyle; stylodia 2 or 3, nearly free or very short. Fruit a septicidal capsule, exocarp and endocarp separating; seeds black, usually 1 per fruit, with orange aril attached to apex of locule; embryo oily, endosperm 0.

A family comprising two monotypic genera; West Africa, C and Andean South America south to Peru.

VEGETATIVE STRUCTURES. Both genera agree in the characteristic leaf structure. The petiole of the unifoliolate leaf is pulvinate at its base (Fig. 79H). The petiolulus is pulvinate for its entire length, articulated at the juncture with the petiole and is provided with an adaxial stipel. In *Lepidobotrys* this stipel has been found to be 2-nerved and is probably the fusion product of two stipels (Léonard 1950).

Leaf anatomy of *Lepidobotrys* is noteworthy for paracytic and laterocytic stomata, vertical bundle sheath extensions of some of the veins, and cristarque cells (van Welzen and Baas 1984). In the wood of *Ruptiliocarpon* growth rings are discernable; the vessels are solitary or grouped and have simple perforations with oblique plates; intervascular pits are vestured; fibres are thin-walled, non-septate; rays are uniseriate homogeneous or nearly so, with mostly procumbent cells; parenchyma is scattered, in bands, paratracheal, and terminal (Mennega 1993). The wood of *Lepidobotrys* is similar but has no vestured pits.

Germination of *Ruptiliocarpon* is epigeal (Hammel and Zamora 1993).

FLORAL MORPHOLOGY. The nectary disk of Lepidobotryaceae has attracted considerable attention; it is shallowly bowl-shaped in *Lepidobotrys* and tubular in *Ruptiliocarpon*. The disk is found in male and female flowers but in the latter is usually smaller. Its nectar-secreting function, obvious from the rich development of glandular tissue, has been observed in the greenhouse (Link 1991). In *Lepidobotrys* the stamens are inserted with long filaments at the margin of the disk which protrudes between the filament bases (Fig. 79D). The disk is comprised of glandular tissue which extends deeply into the floral axis but is never vascularized itself but closely associated with the sepal and carpel strands; it is covered by an epidermis on which nectarial stomata are sparsely dispersed (Link 1991). In *Ruptiliocarpon* the nectary disk is tubular and the stamens are inserted on its rim. The antesepalous stamens have short free filaments, whereas the anthers of the antepetalous stamens are sessile on the margin of the tube. In the male flowers the adaxial side of the staminal tube has profusely branched vascular tissue and densely stained cells (Tobe and Hammel 1993). The disk of *Lepidobotrys* is interpreted by Link (1991) as receptacular, whereas Tobe and Hammel (1993) contend that the tubular structure in *Rutiliocarpon* is the product of fusion of the stamen filaments.

EMBRYOLOGY. Ovules of *Ruptiliocarpon* are anatropous, bitegmic, and crassinucellar. The inner and outer integument are multiplicative; both take part in the formation of the micropyle. The ovules and young seeds are pachychalazal. The embryo sac is 8-celled. An obturator is formed from funicular tissue near the micropyle (Tobe and Hammel 1993).

POLLEN MORPHOLOGY. The pollen of *Lepidobotrys* is tricolporoidate, subprolate,

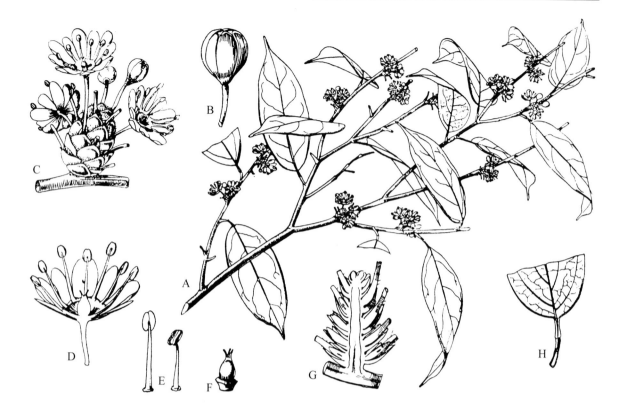

Fig. 79. Lepidobotryaceae. *Lepidobotrys staudtii*. **A** Inflorescence. **B** Flower bud. **C** Inflorescence. **D** Flower, vertical section. **E** Stamens. **F** Pistil. **G** Inflorescence, vertical section. **H** Base of leaf and petiole showing joint. (Hutchinson and Dalziel 1927)

crassitegillate-intectate, with intraluminal columellae and excessivley wide ectoapertures; the nexine has a granular inner surface with which it is linked together with the intine (Erdtman 1952; Oltmann 1971; Huynh 1969). The exine of *Rupotiliocarpon* is described as verrucate to fossulate/foveolate (obviously from SEM analysis) (Hammel and Zamora 1993). Both Oltmann (1971) and Huynh (1969) stressed the palynological similarity between *Lepidobotrys*, *Sarcotheca* and *Dapania*.

FRUIT AND SEED. The capsule of *Lepidobotrys* dehisces with 2 or 3 valves, while that of *Ruptiliocarpon* ruptures irregularly. Both integuments take part in the formation of the seed coat but mechanical strength is provided by the fibrous exotegmen (only present in *Ruptioliocarpon*, Tobe and Hammel 1993).

AFFINITIES. *Lepidobotryum* and *Ruptiliocarpon* have been compared with Linaceae, Oxalidaceae,

Erythroxylaceae, Meliaceae, and several other families (see Hallier 1923; Léonard 1950; Hamel and Zamora 1993). Léonard (1950), who proposed family rank for *Lepidobotrys*, described various characters distinguishing it from Oxalidaceae:

– Lepidobotryaceae: dioecious; stipules and stipels +; disk +; ovary 3–2-merous; ovules collateral; capsule septicidal.
– Oxalidaceae: hermaphrodite or androdioecious; stipules and stipels 0; disk 0; ovary 5-merous; ovules superposed.

In a molecular analysis (*rbc*L), Savolainen, Fay et al. (2000) found strong support for a Celastrales clade, in which *Lepidobotryaceae* are basal to Parnassiaceae and Celastraceae. Since in this analysis Celastrales appear in relatively close proximity to Oxalidaceae, the matter may require further attention.

The relationship between *Lepidobotrys* and *Ruptiliocarpon*, doubted by Takhtajan (1997), is so close that, to my mind, the generic rank of *Ruptiliocarpon* is only weakly justified.

KEY TO THE GENERA

1. Flowers pedicellate; ovary 3-merous   **1. *Lepidobotrys***
– Flowers sessile; ovary 2-merous   **2. *Ruptiliocarpon***

1. *Lepidobotrys* Engl.

Fig. 79

Lepidobotrys Engl., Bot. Jahrb. Syst. 32: 108 (1902); Léonard,
Bull. Jard. Bot. Etat 20: 33 (1950).

Flowers pedicellate, male in many-flowered stro-
biliform racemes, female in few-flowered, short
fasciculate racemes, unisexual with rudimentary
organs of opposite sex; stamens with versatile
anthers on long filaments inserted at margin of
fleshy disk surrounding the ovary; ovary 3-locular;
stylodia 3. Exocarp leathery, endocarp pergamen-
taceous. A single species, *L. staudtii* Engl., Africa,
Guineo-Congo region.

2. *Ruptiliocarpon* Hammel & N. Zamora

Ruptiliocarpon Hammel & N. Zamora, Novon 3: 408 (1993).

Flowers sessile, in leaf-opposed compound spikes,
cryptically unisexual (with rudimentary ovules or
anthers without pollen); stamens with basifixed
anthers and filaments fused into a nectariferous
tube; stamens inserted on margin of tube, the
antesepalous on short filaments, the antepetalous
with sessile anthers; connectives produced to form
a small pubescent appendage; ovary 2-locular; stylodia very short. Exocarp coriaceous to woody,
irregularly rupturing and exposing 2 horny endocarps, one nearly completely surrounding the
seed, the other usually empty and smaller. Only
one species, *R. caracolito* Hammel & N. Zamora, in
humid lowland forests of Costa Rica, Colombia,
Peru, and Suriname.

Selected Bibliography

Erdtman, G. 1952. See general references.
Hallier, H. 1923. *Lepidobotrys* Engl.: Die Oxalidaceen und die
    Geraniaceen. Beih. Bot. Centralbl. 39, II: 163.
Hammel, B.E., Zamora, N.A. 1993. *Ruptiliocarpon* (Lepi-
    dobotryaceae): a new arborescent genus and tropical
    American link to Africa, with a reconsideration of the
    family. Novon 3: 408–417.
Hutchinson, J., Dalziel, J.M. 1927. Flora of west tropical Africa,
    vol. 1. London: Crown Agents.
Huynh, K.-L. 1969. Etude du pollen des Oxalidaceae I. Mor-
    phologie générale – palynotaxonomie des *Oxalis* améri-
    cains. Bot. Jahrb. Syst. 89: 271–303.
Léonard, J. 1950. *Lepidobotrys* Engl., type d'une famille nou-
    velle des Spermatophytes: les Lepidobotryaceae. Bull. Jard.
    Bot. Nat. Belg. 20: 31–40.
Link, D.A. 1991. The floral nectaries of *Geraniales*. III. *Lepi-
    dobotryaceae* J. Léonard. Bull. Jard. Bot. Natl. Belge 61:
    347–354.
Mennega, A.M.W. 1993. Comparative wood anatomy of
    Ruptiliocarpon caracolito (Lepidobotryaceae). Novon 3:
    418–422.
Oltmann, O. 1971. Pollenmorphologisch-systematische Unter-
    suchungen innerhalb der Geraniales. Diss. bot. 11, 163 + XI
    pp., 30 t. Lehre: Cramer.
Savolainen, V., Fay, M.F. et al. 2000. See general references.
Takhtajan, A. 1997. See general references.
Tobe, H., Hammel, B. 1993. Floral morphology, embryology,
    and seed anatomy of *Ruptiliocarpon caracolito* (Lepi-
    dobotryaceae). Novon 3: 423–428.
Welzen, P. van, Baas, P. 1984. A leaf anatomical contribution to
    the classification of the Linaceae complex. Blumea 29:
    453–479.

Lissocarpaceae

B. Wallnöfer

Lissocarpaceae Gilg in Engler, Syllabus, ed. 9 & 10: 324 (1924), nom. cons.

Small, glabrous trees without latex. Leaves alternate, simple, exstipulate, finely and pinnately veined, with entire margins; extrafloral nectaries often present on abaxial leaf surface. Inflorescences axillary, racemose. Flowers with prophylls, articulated at base, sessile or shortly pedicellate, actinomorphic, 4-merous, unisexual (plants probably dioecious); calyx campanulate with 4 imbricate, retuse lobes, persistent but not enlarging in fruit; corolla sympetalous, isomerous with calyx, with lobes contorted sinistrorsely; tube prominent, bearing distally a corona of 8 lobes which, however, is missing in some species; male flowers with 8 stamens in one whorl, the filaments shortly connate and attached to the corolla tube below the middle; anthers linear, erect, basifixed, with an apiculate-prolonged connective, 4-sporangiate, dehiscing by longitudinal slits; locules and ovules not developed; female flowers: staminodes resembling stamens, but anthers collapsed and without pollen; gynoecium syncarpous, 4-carpellate; ovary inferior, 4-locular; style terminal; stigma shallowly 4-lobed; ovules 2 per locule, anatropous; placentation apical. Fruit an ovoid berry with a persistent calyx; seeds 1–2 by abortion, pendulous; hilum relatively small, apical; testa smooth, thin, coriaceous; endosperm horny, abundant, smooth; embryo straight, with two small foliaceous cotyledons and a strongly developed radicle.

A monogeneric family with 8 species from tropical South America.

VEGETATIVE ANATOMY. Extrafloral nectaries occur on abaxial leaf surfaces scattered along the midvein, but are lacking on some leaves (Busch 1913; Gentry 1996). The stomata are anomocytic and are restricted to the abaxial leaf surface. Diffuse vesiculose sclereids are present in the petiole and mesophyll of the leaves. The vasculature of the petioles consists of an arc with outwardly curving ends. The marginal ultimate venation is fimbriate.

For further details see Schadel and Dickison (1979); compare also the description of the leaf anatomy of *L. tetramera* by Busch (1913).

In the wood the vessels are solitary and in radial multiples of 2–9 cells; perforation plates are simple or less frequently scalariform; pits among vessels and ray cells or parenchyma are small (mostly 4–8 μm in diameter); axial parenchyma is predominantly apotracheal; imperforate tracheary elements are of the libriform fiber type; rays are heterocellular, 1–2 cells wide, predominantly more than 1 mm high; fibers are without spiral thickenings (Ng 1971, 1991; Dickison and Phend 1985). The wood of Ebenaceae is very similar except that the rays are mostly over 1 mm high.

INFLORESCENCE AND FLOWER STRUCTURE. The inflorescence is of racemose nature (for further details, see Wallnöfer, in press). Erroneously flowers were often considered to be hermaphroditic, but Ng (1971, 1991) pointed out that they are unisexual and the plants probably dioecious. Gynoecium structure is quite similar to Ebenaceae except that the ovary is inferior and false septa are not developed (Ng 1971, 1991). Each of the four locules bears a pair of anatropous ovules at the apex of the ovary. The four locules all open into a common stylar passage. As in Ebenaceae, the radicle of the embryo is superior.

EMBRYOLOGY AND SEED. The ovules are pendulous, oblong, anatropous, with the raphe descending on the outer side, but otherwise unknown. The vascular system of the seed consists of one bundle descending from the placenta to the chalaza which then sends 6–12 branches back up to the apex of the locule. These branches are visible as prominently raised, longitudinal ridges on the seed surface (Ng 1971; White 1981).

POLLEN MORPHOLOGY. The pollen is triporate and suboblate to oblate-spheroidal or somewhat flattened; the pores are tenuimarginate and about 7 μm wide. The sexine is thinner than the nexine

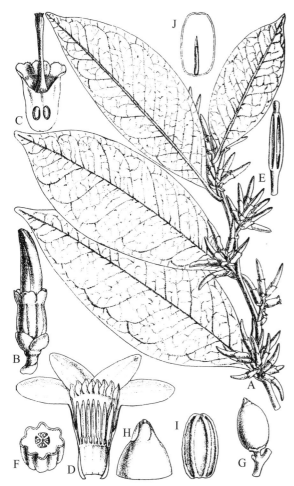

Fig. 80. Lissocarpaceae. *Lissocarpa benthamii.* **A** Flowering branch. **B** Flower with prophylls. **C** Flower with calyx and ovary in vertical section. **D** Corolla laid open. **E** Stamen. **F** Ovary in transverse section. **G** Fruit. **H** Apex of fruit with persistent calyx. **I** Seed. **J** Same in vertical section. (Oliver 1895)

and columellate (Erdtman 1952). According to Ng (1971) and White (1981), the surface shows a prominent reticulate sculpturing. Later workers, however, found the tectum fossulate to somewhat rugulate (Morton and Dickison 1992) or granulated and undulate. Size ranges from 46 × 53 to 65 × 70 μm (Ng 1971).

AFFINITIES. When Bentham (1876) described *Lissocarpa* he placed it in Styracaceae. Oliver (1895) considered that it better belonged in Ebenaceae. Perkins (1907) excluded it from Styracaceae, and Gilg (1924) placed it in a family of its own. Ng (1971, 1991) and Franceschi (1993) found

that *Lissocarpa* is the closest relative of Ebenaceae. This was strongly endorsed by the molecular studies of Berry et al. (2001) and Anderberg et al. (2002).

DISTRIBUTION AND HABITATS. *Lissocarpa* occurs in Bolivia, Peru, Colombia, Venezuela, northern Brazil, and Guyana (White 1981; Berry 1999, 2001; Wallnöfer, in press). Four species are characteristic elements of the specialized vegetation on white sand and along blackwater rivers, four are found in montane forests.

One genus:

Lissocarpa Benth.   Fig. 80

Lissocarpa Benth. in Benth. & Hook. f., Gen. Pl. 2: 671 (1876); Berry, Brittonia 51: 214–216 (1999), synop.

Description as for the family.

Selected Bibliography

Anderberg, A.A. 2002. See general references.
Bentham, G. 1876. Styraceae. In: Bentham, G., Hooker, J.D. Genera plantarum 2: 666–671. London: Reeve & Co.
Berry, P.E. 1999. A synopsis of the family Lissocarpaceae. Brittonia 51: 214–216.
Berry, P.E. 2001. Lissocarpaceae. In: Berry, P.E., Yatskievych, K., Holst, B.K. (eds.) Flora of the Venezuelan Guayana. 6: 19–20. St. Louis: Missouri Botanical Garden Press.
Berry, P.E., Savolainen, V., Sytsma, K.J., Hall, J.C., Chase, M.W. 2001. *Lissocarpa* is sister to *Diospyros* (Ebenaceae). Kew Bull. 56: 725–729.
Busch, P. 1913. Anatomisch-systematische Untersuchung der Gattung *Diospyros*. Crefeld: Wilhelm Greven, 95 pp.
Dickison, W.C., Phend, K.D. 1985. Wood anatomy of the Styracaceae: evolutionary and ecological considerations. I.A.W.A. Bull., N.S., 6: 3–22.
Erdtman, G. 1952. See general references.
Franceschi, D. de 1993. Phylogénie des Ebénales: analyse de l'ordre et origine biogéographique des espèces indiennes. Publ. Dépt. Ecol. Institut Français de Pondichérry 33: 1–153 (+ annexe A-E, + 61 planches).
Gentry, A.H. 1996. A field guide to the families and genera of woody plants of northwest South America. Chicago: University of Chicago Press.
Gilg, E. 1924. Lissocarpaceae. In: Engler, A. Syllabus der Pflanzenfamilien. Ed. 9 & 10: 324.
Morton, C.M., Dickison, W.C. 1992. Comparative pollen morphology of the Styracaceae. Grana 31: 1–15.
Ng, F.S.P. 1971. A taxonomic study of the Ebenaceae with special reference to Malesia. Doctoral Thesis, University of Oxford, 221 pp.
Ng, F.S.P. 1991. The relationships of the Sapotaceae within the Ebenales. In: Pennington, T.D. The genera of Sapotaceae, chap. 1. Kew: Royal Botanic Gardens & New York: New York Botanical Garden, pp. 1–13.

Oliver, D. 1895. *Lissocarpa benthamii* Gürke. Hooker's Icon. Pl. 25: pl. 2413.

Perkins, J. 1907. Styracaceae. In: Engler, A. (ed.) Das Pflanzenreich IV 241: 1–111. Leipzig: Engelmann.

Schadel, W.E., Dickison, W.C. 1979. Leaf anatomy and venation patterns of the Styracaceae. J. Arnold Arbor. 60: 8–37.

Wallnöfer, B., in press. A revision of *Lissocarpa* Benth (Lissocarpaceae). Ann. Naturhist. Mus. Wien, B, 105.

White, F. 1981. Lissocarpaceae. In: Maguire, B. et al. The botany of the Guayana Highland – part XI. Mem. New York Bot. Gard. 32: 329–330.

Loasaceae

M. WEIGEND

Loasaceae Juss., Ann. Mus. Natl. Hist. Nat. 5: 19–27 (1804), "Loaseae".

Annual or perennial, erect, decumbent, or winding herbs, rarely rosette plants, vines, woody lianas, subshrubs, shrubs or small trees, from 5 cm to 10 m tall; stem with white pith, rarely solid, terete, rectangular or grooved, often with lenticels and/or tuberculate; rhizome, xylopodium or runners sometimes present; roots fibrous, primary root dominant or evanescent, occasionally primary and/or secondary roots tuberous, or only adventitious roots present in mature plant; indumentum typically with scabrid/glochidiate trichomes and urticant setae, sometimes with uniseriate and/or glandular trichomes. Leaves opposite or some alternate, petiolate or sessile, exstipulate or pseudostipulate; lamina linear, ovate, or circular, membranaceous to coriaceous, 2–50 cm long, base cuneate, cordate or peltate, usually divided, rarely pinnatisect, trifoliate, bipinnatifid, or palmate; margin serrate or mucronate, rarely entire, with hydathode teeth. Inflorescences often terminal thyrsoids, sometimes reduced to dichasia, monochasia or monads, rarely racemes; bracts present or rarely absent, mostly green. Flowers perfect, proterandric, usually actinomorphic, chasmogamous, rarely cleistogamous; perianth heterochlamydeous, (4)5(–8)-merous; calyx tube conical to globose, densely covered with trichomes; calyx lobes usually persistent and accrescent, rarely caducous; aestivation apert; corolla choripetalous, sympetalous or pseudosympetalous, aestivation apert, contort or imbricate; petals erect, spreading or reflexed, linear, spatulate, ovate or circular, planar or deeply boat-shaped, margin entire, irregularly serrate, or laciniate, sometimes with filiform apical appendages or longitudinal lamelliform flaps, often with short claw and one lateral tooth on each side, membranaceous to carnose, green, white, yellow, orange or red; androecium haplostemonous, obdiplostemonous or polystemonous (150 or more), all stamens fertile or some staminodial; initiation centripetal; filaments filiform, inserted basally or epipetalous, rarely very short, or petaloid; anthers basifixed, with 1–4 microsporangia; connective usually undifferentiated; staminodia highly diversified, free and filiform or petaloid or united to form elaborate staminodial complexes (nectar scales); nectaries antepetalous glands, or ring- or cup-shaped disc, or absent; ovary completely inferior to 4/5 superior; style filiform, included or exserted; stigma punctiform or with 2–5 papillose, parallel or divergent lobes; ovules anatropous, either single, pendent, or numerous on simple or divided parietal placentae, in some *Petalonyx* with obturator, in Gronovioideae and Petalonychoideae possibly crassinucellar, in Mentzelioideae and Loasoideae unitegmic and tenuinucellar with well-developed chalazal and micropylar haustoria. Fruit a capsule or cypsela, typically dehiscent with 3–5 apical valves, septicidal or rarely septifragous, rarely with 3–5 apical valves and 1 longitudinal suture, or only with 3–5 longitudinal sutures, rarely indehiscent; seeds 1–many, globose, ovoidal or angular, sometimes winged; testa usually pale or dark brown to black, reticulate with or without central papillae, sometimes with fenestrate anticlinal walls, or striate; embryo straight, embedded in copious and oily endosperm. Seedling with 2 ovate to obcordate, densely trichomatose, apically emarginate cotyledons, midvein ending in hydathode tooth.

Comprising 20 genera in four subfamilies with ca. 330 species, mostly American, outliers in Africa and on the Marquesas Islands (Polynesia).

VEGETATIVE MORPHOLOGY. The vast majority of taxa are annual to perennial herbs and this appears to be the plesiomorphic condition for most groups. Subshrubs (frequently with ligneous underground structures – *Cevallia*, some *Petalonyx*, *Mentzelia*) and shrubs (e.g. in *Mentzelia*, *Aosa*, *Presliophytum*, *Nasa*, *Loasa*, *Kissenia*) evolved several times independently in desert habitats from herbaceous ancestors. *Gronovia* and *Fuertesia* are the only winding genera in Loasaceae with alternate leaves. *Fuertesia* is the only woody liana in the family. Winding herbs with opposite leaves are restricted to one monophyletic assemblage of primarily South Andean genera: *Loasa, Caiophora, Blumenbachia, Scyphanthus*. Irrespective of habit, the basic structure of the plants follows one pattern: at least the first two foliage leaves are always opposite (usually there are additional opposite leaf pairs). These are mostly followed by alternate leaves and the shoot nearly always terminates in a distal dichasium or an inflorescence with opposite leaves (and paracladia) at least in its terminal element(s). Typically all internodes are elongated, but there are rosulate taxa with shortened basal internodes (e.g. in *Loasa* and *Caiophora*).

Underground structures are diverse. *Nasa* has an evanescent primary root which is quickly replaced by adventitious roots from the decumbent basal portion of the stem. In some species of *Nasa* (e.g. *N. cymbopetala*) this basal, decumbent region turns into a thick, horizontal rhizome. *Caiophora, Blumenbachia, Loasa* and *Scyphanthus* have a strong primary root which develops either into a more or less thickened tap-root (*C. pentlandii, B. prietea*) or into a spindle-shaped root tuber (*L. asterias*); sometimes the primary root is fibrous and only the secondary roots are tuberous (*C. contorta*), sometimes both are tuberous (*C. pterosperma, B. dissecta*). Additionally, runners with cataphylls are widespread but not universally present in *Loasa* (*L. gayana, L. nana*) and *Caiophora* (*C. nivalis, C. contorta*). A long-lived (>7 years) xylopodium is found in *Schismocarpus* and *Xylopodia*, with annual and perennial shoots, respectively, arising from them.

Foliage in Loasaceae is usually evergreen, but some taxa are deciduous during the dry season (*Xylopodia*, some *Mentzelia*) or during winter (south temperate taxa, many species of *Loasa, Blumenbachia*). Some of these deciduous taxa produce narrowly spaced, spirally inserted leaves in a rapid flush at the beginning of the moist season (e.g. some *Nasa* and *Aosa*).

VEGETATIVE ANATOMY. The stem is typically covered with numerous trichomes and terete, rarely quadrangular (*Blumenbachia* sect. *Blumenbachia*), or deeply grooved (some species of the *Nasa triphylla* group). The stem is usually filled with a white pith and may be externally covered with dark green to black lenticels (many *Nasa*) or cork layers (some *Petalonyx*). Frequently the bark starts exfoliating in the second year (e.g. *Petalonyx, Presliophytum*, many *Mentzelia*).

The numerous different trichome types can be technically divided into three basic types. Scabridglochidiate trichomes are unicellular, silicified or calcified with one to numerous processes. These trichomes are called scabrid if the processes are shorter than the diameter of the shaft, and glochidiate if they are longer and typically hook-shaped (Fig. 81). There are numerous derivations from these two basic types (e.g. Dostert and Weigend 1999). Stinging hairs or setae are also unicellular, but they have a smooth surface, a slightly bulbous base, a sharp tip, and they are filled with a clear to brownish fluid. A more or less violent skin rash develops upon contact with these stinging hairs. The chemical identity of the irritants is unknown. Flexible (non-silicified) uniseri-

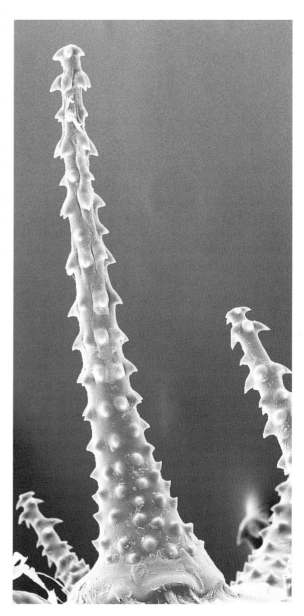

Fig. 81. Loasaceae. *Klaprothia fasciculata*, glochidiate trichome from fruit surface, SEM (×500). Photo M. Weigend

ate hairs represent a third trichome type. They are two to over 20 cells long and usually gland-tipped. Setae (stinging hairs) and scabrid-glochidiate trichomes are widespread in Loasaceae, but uniseriate trichomes are largely restricted to the genus *Nasa*. Subspecies, species, species groups and sometimes genera can be defined by trichome cover: slender glochidiate trichomes with only two apical hooks are found in *Gronovia* and *Fuertesia*, glochidiate hairs with porrect instead of deflexed branches in *Petalonyx*. Long (>10 cells), brown or white uniseriate trichomes are typical of the *Nasa grandiflora* group, and T-shaped (medifixed) trichomes are restricted to the *Nasa triphylla* group; symmetrically T-shaped trichomes occur only in *Nasa humboldtiana*.

The brittle wood of Loasaceae is always soft and sometimes slightly succulent. The data here follow Carlquist (1984, 1987) unless otherwise indicated. Growth rings are often visible, e.g. in *Xylopodia* (pers. obs.) and some *Eucnide*, *Mentzelia* and *Petalonyx*, but sometimes absent (*Plakothira frutescens*). Vessels have simple (most) or scalariform (rare: *Nasa picta*) perforation plates. Lateral wall pitting consists typically of circular to oval (6×6–$8\,\mu m$) pits. Fibre tracheids are present; they have bordered pits or the borders are vestigial. Tall and wide, homogeneous or heterogeneous vascular rays (multi- and uniseriate) are usually present, rarely missing (*Petalonyx*). The ray cells have thin but lignified walls. Axial parenchyma is diffuse and abundant (most), or absent (*Presliophytum*), or vasicentric (*Plakothira frutescens*). Sieve-elements are of the S-type.

Cystoliths (calcium carbonate, silica) and crystal druses/raphides (calcium oxalate) are frequently present, e.g. in the wood parenchyma, in the leaf lamina or the ovary roof.

INFLORESCENCE STRUCTURE. Urban (1892, 1910) provided the first detailed analyses of some taxa, and a comprehensive survey was carried out by Weigend (1997). Most Loasaceae have acrotonic, terminal thyrsoids with a distal dichasium, or structures derived from it (see Figs. 84D, 85A). Re- and concaulescence is widespread in Loasaceae inflorescences, and only relatively few taxa (some *Mentzelia* and *Loasa*, all *Blumenbachia*, *Scyphanthus* and *Caiophora*) lack metatopy. Reductions of dichasia to monochasia have occurred repeatedly. In *Caiophora* and *Scyphanthus* the inflorescences are winding monochasia derived from the distal dichasium. By a shortening of the internodes, rosulate species with apparently axillary flowers have evolved from these monochasial taxa in *Caio-*

phora. Many species in *Loasa* and some species in *Blumenbachia* have lateral inflorescences which probably evolved from a terminal thyrsoid via early proliferation. Other derivations concern the number and size of bracts: *Aosa* lacks bracts; in *Nasa* only one instead of two bracts is present on the monochasial branches. *Mentzelia* sect. *Mentzelia* has terminal thyrsoids with dichasial branches and long internodes. All other sections in *Mentzelia* have strongly basitonic inflorescences with shortened distal internodes and often single terminal flowers (distal dichasium reduced to terminal flower); the primary flower is often preceded by 1–12 sterile bracts. *Petalonyx* is aberrant with racemose inflorescences with one recaulescent bract and two prophylls per flower, departing strongly from the thyrsoid/cymose patterns of the rest of the family.

FLOWER STRUCTURE. Gronovioideae and Petalonychoideae have petals with a single principal vein, a cup-shaped nectary, a single pendent ovule and only five stamens. In Gronovioideae the calyx lobes are usually larger than the petals and persist after anthesis, petals are free and the anthers have introrse dehiscence and are densely trichomatose. In Petalonychoideae the calyx lobes are smaller than the petals, the petals are free or united, the anthers have lateral dehiscence and are glabrous, 2–3 of the stamens can be staminodial and filiform (Davis and Thompson 1967). Mentzelioideae and Loasoideae share the presence of petals with numerous veins from the base, an annular nectary or distinct antesepalous nectar glands, and numerous ovules and stamens. Mentzelioideae have free, flat staminodia, if any, petals are planar and membranaceous. Loasoideae have antesepalous staminodial complexes of 3–12 staminodia in an inner and an outer tier, the outer ones are usually united into a nectar scale. The typical staminodial complex in Loasoideae consists of three outer, united staminodia forming a nectar scale and two more or less enclosed, free, inner staminodia. The nectar scales are extremely diverse morphologically and can be decorated with dorsal calli, arches or filaments, apical wings, and colour patterns. Petals are usually shortly clawed and deeply boat-shaped (cymbiform), white or yellow, reflexed or spreading, and the immature stamens are reflexed into the petals. Stamens move autonomously into an upright position when mature, but this process can also be triggered by manipulation of the nectar scale in many species (Wittmann and Schlindwein 1995; Schlindwein and Wittmann 1997). More or less campanulate corollas with red or orange, carnose, unclawed petals and erect stamens have evolved in hummingbird-pollinated groups. The staminodial complexes are here robust and provide a mechanical guide to the copiously produced nectar. *Schismocarpus* and some species of *Mentzelia* (*M*. sect. *Mentzelia*) show distinct heteranthery.

Ovary position is completely inferior to largely superior, both conditions are sometimes found in very closely related taxa. The parietal placentae in Loasoideae provide good characters for generic delimitation: Most taxa have placentae which are either flat or globose in transverse section but in *Caiophora* they are Y-shaped with ovules sitting only on the morphologically adaxial side of the placenta, and in *Presliophytum* and *Huidobria* they are cross-shaped in section and have ovules all over their surface.

EMBRYOLOGY. The only detailed studies on embryology available are from *Loasa* and *Blumenbachia* (Garcia 1962): the ovules are unitegmic and tenuinucellate, with very distinctive micropylar and chalazal haustoria and cellular endosperm. This type is widespread in Loasoideae (*Loasa, Nasa, Blumenbachia, Presliophytum, Caiophora*). *Mentzelia* has basically similar ovules (Vijayaraghavan and Prabhakar 1984). The embryology of *Schismocarpus* and *Eucnide* is still unknown. *Gronovia* and *Petalonyx* have surficially very distinctive ovules which are probably crassinucellar and certainly lack well-developed micropylar and chalazal haustoria. *Petalonyx thurberi* (and maybe other *Petalonyx*) ovules have an obturator. More data and a critical anatomical study of the embryology of Loasaceae are evidently required.

POLLEN MORPHOLOGY. Pollen morphology is comparatively uniform in Loasaceae. The pollen grains are usually spherical to prolate and tricolpate or tricolporate. The exine can be punctate and spinose, echinate or rugulose (Gronovioideae) or striate with thin lirae (*Petalonyx*, Poston and Nowicke 1993). In *Mentzelia, Schismocarpus* and *Eucnide* the pollen is striate to striate-reticulate with exclusively longitudinal striations. The pollen in Loasoideae is usually reticulate, but sometimes equatorially striate (*Loasa* ser. *Floribundae* and *Deserticolae*). The colpus margin may be differentiated or undifferentiated. *Kissenia* has syncolpate pollen grains with microreticulate exine. The rainforest genus *Chichicaste* is the only taxon with tricolporoidate pollen with conspicuous stoppers, a feature that has evolved convergently in many plant species from perhumid habitats.

KARYOLOGY. Chromosome counts are available for the majority of genera and some of these are strongly supported by karyology. Gronovioideae have $2n = 26$ (*Cevallia*) or $2n = 74$ (*Gronovia*). Petalonychoideae (*Petalonyx*) have $2n = 44, 46$ (Davis and Thompson 1967). Base numbers of these two groups are unknown and all of the genera may represent palaeopolyploids. Mentzelioideae seem to have a base number of $x = 7$. From *Mentzelia* the numbers $2n = 18, 20, 22, 28, 36, 54, 72$ have been reported (Hill 1976), the apparent tetraploid $2n = 28$ being the most frequent one. Numbers over $n = 14$ are exclusively from weedy, annual *Mentzelia* sect. *Trachyphytum* (Hill 1976). *Eucnide* is apparently strictly hexaploid with $2n = 42$ (Thompson and Ernst 1967). Loasoideae seem to have the base number $x = 6$. *Loasa malesherbioides* has the lowest chromosome number so far known from the family, $2n = 12$ (Grau 1988), and it is possibly the only diploid, all other taxa being polyploids. The tetraploid chromosome number $2n = 24$ is widely found (*Blumenbachia, Aosa, Presliophytum, Kissenia, Klaprothia, Xylopodia*; Coleman and Smith 1969; Poston and Thompson 1977; pers. obs.). Higher ploidy levels are encountered in *Klaprothia* (Poston and Thompson 1977) and *Plakothira* (pers. obs., octaploid with $2n = 48$), *Nasa* (octaploid with $2n = 56$, *N. dyeri*, pers. obs.), *Loasa* ($2n = 36$, *L. triloba*, Grau 1988) and *Huidobria* (hexaploid with $2n = 36$, Grau 1997). Most species of *Loasa* have $2n = 24, 26$ (Grau 1988) and most of the genus *Nasa* $2n = 28$ (pers. obs.). In all groups of Loasaceae, polyploidization and dysploid changes thus seem to have been fairly common events.

POLLINATION. Loasaceae are animal- or rarely self-pollinated. The reward in polystemonous Loasaceae is usually pollen and nectar, but some taxa (e.g. *Mentzelia*) lack nectaries and offer pollen as the only reward (Thompson and Ernst 1967). A floral odour is usually not perceptible but has been reported from the nectariferous flowers of *Mentzelia decapetala* (Keeler 1981) and *Nasa ferruginea* (pers. obs.). Facultative autogamy is widespread in annuals and has been reported from *Mentzelia, Eucnide* (Thompson and Ernst 1967), *Petalonyx* (Davis and Thompson 1967), *Gronovia, Klaprothia, Nasa, Blumenbachia* and *Loasa*. Cleistogamy is known from *Loasa triloba* (Gilg 1925) and has now also been observed in *Nasa chenopodiifolia*. In *Eucnide* hawkmoths and bees have been reported as pollinators, and pollination by colibris has been suggested for some taxa (Thompson and Ernst 1967). *Klaprothia* is only visited by Empid flies, whereas *Presliophytum* is visited by a very wide range of pollinators including Lepidoptera and Hymenoptera (Euglossine and Colletid bees, *Xylocopa*, *Apis*, *Bombus*), all of which are likely to effect pollination. Many taxa of Loasoideae share pendent flowers with spreading petals and strongly contrasting nectar scales and thigmonasty (stamen movement triggered by manipulation of the nectar scale in *Caiophora*, *Loasa*, *Blumenbachia*, *Aosa*, *Nasa*, *Xylopodia*). These are primarily pollinated by specialized Colletid bees (Schlindwein and Wittmann 1997). Colletids are the only regular visitors of these flowers. Many high Andean members of *Nasa* and *Caiophora* have large, orange corollas with large nectar scales and copious nectar production. These are visited and pollinated by humming birds and bumble bees. *Caiophora coronata* has bowl-shaped flowers close to the ground and is pollinated by small rodents (Cocucci and Sérsic 1997).

FRUIT, SEED AND DISPERSAL. Loasaceae have capsular fruits, with the only exception of Gronovioideae and Petalonychoideae which have cypselas. The capsules are typically many-seeded and dehiscent and the seeds have a dark brown to nearly black testa.

Most Loasoideae have a reticulate testa; a group of closely allied genera from southern South America (*Caiophora*, *Loasa*, *Scyphanthus*) shares very high, fenestrate anticlinal walls (Fig. 82). *Nasa* has low, apically thickened anticlinal walls. The seeds of *Eucnide* are striate whereas the seeds of *Mentzelia* are reticulate or polyhedral with central

papillae – one of the very few morphological characters separating these two genera (Hill 1976; Hufford 1988).

Corner (1976) interpreted the seeds as neotenic enlarged ovules with an exotesta often characteristic of the genera. For example, in *Loasa* he found the exotesta as a palisade of cells with strongly thickened outer walls, and in *Eucnide* as tabular polygonal cells with thick cuticle.

Seeds are usually barochoric, i.e. they have no obvious means of dispersal. In *Caiophora* the seeds remain trapped in hanging fruits with longitudinal slits which only release them during dry and windy conditions. Anomochory is the dispersal mechanism of *Caiophora*, *Scyphanthus* and many *Loasa* (seeds very light with a deeply pitted surface) and *Presliophytum* (seeds very small). Epizoochory is found in *Blumenbachia*, *Klaprothia mentzelioides* and *Mentzelia aspera*: fruits are tardily deshiscent and densely covered with hook-shaped trichomes. Developing fruits of some *Eucnide* are negatively phototropic and grow into rock crevices where they dehisce and shed their seeds (Thompson and Ernst 1967). The seeds of *Blumenbachia* sect. *Gripidea* have two air-filled wings which considerably slow down the speed of the falling seed in the air but they also lend buoyancy to the seed when it falls into water and may thus represent an adaptation to hydrochory or/and anemochory. The seeds of *Loasa* ser. *Macrospermae* are large, round and heavy. They roll down between the rock debris (scree slope habitat), thus reaching the moister lower layers. There they germinate and have enough stored assimilates to produce a long hypocotyl that carries them back to the light. *Nasa* lacks obvious dispersal mechanisms, apart from a pronounced elongation of the pedicel (in taxa with erect, cylindrical capsules) so that the seeds are sprinkled out by strong wind in dry weather. Other species (many *Nasa* ser. *Grandiflorae*) have horizontal capsules opening with apical valves and a longitudinal slit, thus forming an open platform on which the seeds are exposed to rain and wind. Anemochory of the entire fruit is found in Gronovioideae (accrescent calyx) and *Petalonyx* (fruit remains attached to bract and prophylls).

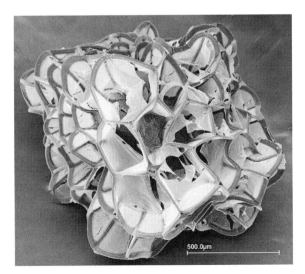

Fig. 82. Loasaceae. *Caiophora madrequisa*, deeply pitted seed with thin anticlinal cell walls, SEM (×40). Photo M. Weigend

PHYTOCHEMISTRY. The phytochemistry of Loasaceae is poorly known, with the only exception of iridoids, on which extensive data are available (Hegnauer 1989; Müller et al. 1999; Weigend et al. 2000; Rodríguez et al. 2002). Iridoids and secoiridoids are nearly universally present in Loasaceae and only very few taxa have no or very low concentrations of these compounds. Some monomeric iridoids are widespread throughout the group, e.g. sweroside, 8-epi-kingiside and loganin, loganic acid. Others are restricted to certain groups: tricoloroside and similar hetero-oligomeric compounds are found only in some taxa of *Loasa*, 10-hydroxyoleoside dimethylester is widespread in *Caiophora* and has not been reported from any other taxon in the family. Loasaceae contain relatively small amounts of phenolic compounds such as quercetin and caffeic acid (Hegnauer 1966); cyanogenic compounds and alkaloids have not been reported from the family. The chemical composition of the seeds has been investigated for only very few taxa (three species of *Mentzelia*) and these contained ca. 20% storage protein and 34–43% fat (oleic, linolic and palmitic acids, Hegnauer 1966). Many other interesting phytochemical questions have not yet been addressed, e.g. the identity of the irritant substances in the stinging hairs or the type of floral pigments.

DISTRIBUTION AND HABITATS. Loasaceae are largely restricted to the Americas, ranging form the northern temperate zone to the southern temperate zone. They are absent from northern temperate forests, but are found in the Mediterranean regions of both hemispheres, in open vegetation such as prairie, pampa, semi-desert, desert, loma, and in forest vegetation such as rain-green forests, rainforests, cloud forests, and southern temperate forests. The family extends from sea level up to the subparamo (ca. 3500 m) and puna (ca. 4800 m) of the Andes. The highest concentration of taxa and the most narrowly endemic taxa are found in the largely cloud forest-dwelling genus *Nasa* between elevations of 1500 and 3500 m in the northern and central Andes. Peru is the most important centre of diversity for the family, with at least 80 species in seven genera (mostly subfamily Loasoideae), two genera and most species being endemics. Chile and Mexico, each with approximately 40 species and seven genera, are secondary centres of diversity. Loasoideae are mainly South American and extend north into southern Mexico. This subfamily has outliers in Africa (*Kissenia*) and Polynesia (*Plakothira*). Mentzelioideae, Gronovioideae and Petalonychoideae have their centre of diversity in Mexico and the south-western U.S.A., and only the former two reach South America with one genus each (*Mentzelia* and *Gronovia* respectively).

SUBDIVISION. The relationships within the family are now well understood and the molecular data presented by Xiang et al. (1998) recognize four lineages, representing the four subfamilies Gronovioideae, Petalonychoideae, Mentzelioideae and Loasoideae (Weigend 1997), but indicating a basal position of Loasoideae rather than Petalonychoideae. This is congruent with morphological data: Loasoideae show the closest similarity to the sister group of Loasaceae (Hydrangeaceae). Both groups share predominantly opposite phyllotaxy, reticulate testa, parallel stigmatic lobes, pluriovulate ovaries, polystemonous androecia, petals with numerous veins from the base, and annular nectaries. Loasoideae contain approximately 200 species in 13 genera. Mentzelioideae are recognized as a natural group of genera (comprising *Mentzelia*, *Schismocarpus* and *Eucnide*) which is difficult to define morphologically, apart from the alternate phyllotaxy and lack of specialized staminodial complexes. A basal position of *Eucnide* in Loasaceae has recently been proposed (Hufford 1998), but morphological support for this is weak and a more detailed molecular analysis will be required to sort out the exact relationships between the subfamilies of Loasaceae. Both Mentzelioideae and Loasoideae have retained various plesiomorphic characters such as pluriovulate ovaries, polystemonous androecia, petals with numerous veins from the base, dilated filaments and annular nectaries. However, these plesiomorphic characters can not be used to define a core family Loasaceae, from which the haplostemonous subfamilies Gronovioideae and Petalonychoideae are excluded, or united in a different subfamily (Takhtajan 1997). Mentzelioideae and Loasoideae share only plesiomorphic characters, and the haplostemonous subfamilies seem to have arisen from a common ancestor with Mentzelioideae (Xiang 1999; Moody et al. 2001). Apart from haplostemonous androecia, they share uniovulate ovaries, petals with a single vein from the base, and cup-shaped nectaries but, in anther and fruit morphology, perianth structure, and pollen surface, they are very distinct from each other.

Within subfamily Loasoideae only two tribes are accepted here – tetramerous Klaprothieae and penta- to octamerous Loaseae. Kissenieae Urb. & Gilg (1900)) can not reasonably be treated as a group distinct from Loaseae, and are here reduced under that tribe. Klaprothieae (6 species) are very clearly defined by a variety of characters, such as the lack of urticant setae, strictly opposite phyllotaxy, and longitudinal lamellae on the petals. Loaseae can not be similarly defined by derived characters. They comprise ca. 196 species and are ecologically and morphologically the most diverse group within the family. Loaseae can be informally divided into "Lower Loaseae" with erect flowers with white (rarely yellowish or greenish white) petals and pale (white, greenish or yellowish white) nectar scales and without thigmonasty (eight species in the genera *Huidobra, Presliophytum, Chichicaste, Kissenia*). These "Lower Loaseae" include all taxa without specialized pollination by Colletid bees or ornithophily. They are a grade, not a clade. The "Higher Loaseae" typically have pendent flowers, brightly coloured petals, variously elaborated and coloured nectar scales and show thigmonasty (ca. 188 species in the genera *Aosa, Blumenbachia, Caiophora, Loasa, Scyphanthus, Nasa*) and are pollinated by specialized Colletid bees or hummingbirds. The "Higher Loaseae" are probably monophyletic and fall into three, probably natural groups: the eastern Brazilian/Caribbean genus *Aosa* (6 spp.), the large North/Central Andean genus *Nasa* (predominantly tropical montane in mesic habitats), and the South Andean group comprising *Blumenbachia, Caiophora, Loasa* and *Scyphanthus* (high Andean, and south temperate, in Mediterranean and desert habitats).

Within subfamily Mentzelioideae two taxa were recognized by Gilg (1925): Eucnideae with *Eucnide*, and Mentzelieae with *Mentzelia* and *Schismocarpus*. The relationships between these tribes are still unclear and no subdivision of subfamily Mentzelioideae is proposed here.

AFFINITIES. Until very recently the affinities of Loasaceae were highly controversial. Traditionally the family had been placed near Turneraceae and Passifloraceae (Cronquist 1981; Weigend 1997) or even united with Turneraceae in one family (Humboldt, Bonpland and Kunth 1823). The superficial similarity between the Turneraceae and some Loasaceae is considerable, but refers only to characters which are quite variable within Loasaceae themselves (Gilg 1925). Phytochemistry and embryology indicated that the affinities of Loasaceae lie elsewhere, and Takhtajan (1980) was the first to propose a placement close to Corniflorae as a separate superorder Loasiflorae. Recent molecular studies confirmed a relationship to Cornales (Hempel et al. 1995, Xiang et al. 1998, Xiang 1999), and the Loasaceae are now known to be sister to the Hydrangeaceae (s.l., incl. *Philadelphus, Jamesia, Fendlera, Fendlerella, Deutzia* and allies). A taxonomically isolated position of Loasaceae (order or superorder), as repeatedly proposed, is not justified. The genera *Jamesia* and *Fendlera* are the most basal branch of the Hydrangeaceae clade in Xiang (1999), and a close

examination of their morphology reveals that they also display the closest morphological similarity to Loasaceae: characteristic "Loasaceae" trichomes (Behnke and Barthlott 1983) are found in *Jamesia* (scabrid) and *Fendlera* (glochidiate). In Loasaceae and *Fendlera* they cover even petals, style and anthers. Filaments (esp. the first, antesepalous one) are dilated (*Jamesia*) to forked (*Fendlera*): this is a character which is widespread in, and highly characteristic of Loasaceae, and it gave rise to both the simple staminodes of Mentzelioideae and the staminodial complexes of Loasoideae (Weigend 1997). The seeds of *Jamesia* and *Fendlera* and all primitive members of Loasoideae have a straight embryo surrounded by fleshy endosperm (Takhtajan 1997) and a dark brown, reticulate testa with slightly elevated anticlinal walls. The basic patterns of centripetal stamen initiation are also identical between Loasaceae and Hydrangeaceae (Payer 1853; Leins and Winhard 1973; Hufford 1998). Other characters uniting at least some Hydrangeaceae with Loasaceae are fruit morphology (semi-superior, septicidal capsules opening with apical valves in both *Jamesia/Fendlera* and most Loasaceae), petal morphology (clawed, boat-shaped petals), style and stigma morphology (style short, parallel stigmatic branches in *Fendlera* and Loasaceae), stem filled with white pith, bark exfoliating, flower typically 4–5-merous, aestivation contort or imbricate, androecium mostly polystemonous in both groups, ovary unilocular, typically of 3–5 carpels, with intruding, parietal placentae, embryology (unitegmic, tenuinucellate ovules, cellular endosperm formation, endosperm haustoria; Takhtajan 1980), pollen typically prolate to spherical, tricolpate to tricolporate, inflorescence morphology (thyrsoids with a distal dichasium in *Fendlera* and *Fendlerella* and most Loasaceae), presence of hydathode teeth on the leaves (e.g. in *Philadelphus, Jamesia* and in all Loasaceae), phytochemistry (very similar iridoid spectra; Takhtajan 1980; Hegnauer 1989), and the occurrence of raphides or oxalate druses. A close relationship between Cornales and Loasaceae is also congruent with data from wood anatomy (Carlquist 1992).

On the basis of the phylogenetic hypothesis adopted here, subfamily Loasoideae appear to have retained a proportionately larger number of plesiomorphic character states than the other three subfamilies. A basal position of *Eucnide* (in Mentzelioideae; Hufford 1990) is currently supported by neither published molecular data (Xiang 1999) nor morphology. Gronovioideae and Petalonychoideae seem to represent the most derived clade in the family (haplostemonous flowers, one-seeded, inferior cypsela, exclusively alternate phyllotaxy, testa reduced, petals one-veined) and are clearly sister groups (Weigend 1997; Moody et al. 2001). The family Loasaceae as such is strongly supported as monophyletic by its general morphological coherence (Weigend 1997) and by molecular markers (Hempel et al. 1995; Xiang 1999), but it can not be defined by autapomorphic characters, since the "typical" trichomes are now known to be present also in hydrangeaceous genera such as *Fendlera*. Characters present in Loasaceae and altogether absent from Hydrangeaceae are the stinging hairs and pronounced recaulescence and concaulescence in the inflorescence, but there are some Loasaceae (e.g. *Mentzelia* sect. *Mentzelia*) which lack both.

ECONOMIC USES. No Loasaceae are currently cultivated for economic purposes, but many taxa are collected from the wild and play some role in folk medicine: species of *Caiophora* and *Nasa* are used to treat a variety of disorders (e.g. allergies, bronchial diseases, and liver complaints) in Andean South America. Species of *Mentzelia* are used against stomach complaints in South America and exported to Europe for that purpose. The seeds of *Mentzelia* used to be an important food source for Native Americans in western North America, and seeds and other parts of the plants were used in medicinal preparations (Moerman 1998).

CONSPECTUS OF LOASACEAE

 I. Subfamily Loasoideae Urb. & Gilg (1900).
 1. Tribe Loaseae Urb. & Gilg (1900).
    Genera 1–10.
 2. Tribe Klaprothieae Urb. & Gilg (1900).
    Genera 11–13.
 II. Subfamily Mentzelioideae (Rchb.) Urb. & Gilg (1900).
    Genera 14–16.
 III. Subfamily Gronovioideae (Rchb.) Link (1833).
    Genera 17–19.
 IV. Subfamily Petalonychoideae Weigend (1997)
    Genus 20.

KEY TO THE GENERA

1. Flowers haplostemonous; petals with 1 principal vein, membranaceous; ovary with 1 pendent ovule; seed without dark testa 2
 – Flowers polystemonous or obdiplostemonous; petals with 3–5 principal veins; ovary with 1–many ovules on parietal placentae; testa dark 5
2. Flowers in racemes; calyx lobes caducous; bracts and prophylls remaining attached to mature fruit. Petalonychoideae **20. Petalonyx**

 – Flowers in cymoids or thyrsoids; calyx lobes persistent; mature fruit free of bracts and prophylls. Gronovioideae 3
3. Erect subshrub with narrow, sinuate-lobate leaves; calyx lobes and petals linear, isomorphic; connective protracted into long appendage **19. Cevallia**
 – Liana or climber with cordate to subpalmately lobed leaves; calyx lobes spatulate; petals lanceolate; connective not differentiated 4
4. Leaves cordate, margin entire; inflorescence branches dichasial; petals laciniate **18. Fuertesia**
 – Leaves subpalmately lobed; inflorescence branches monochasial; petals entire **17. Gronovia**
5. Petals flat and membranaceous; staminodia, if present, not in antesepalous groups. Mentzelioideae 6
 – Petals deeply boat-shaped or carnose if flat; staminodia always present, in antesepalous groups. Loasoideae 8
6. Flowers obdiplostemonous with 5 large and 5 small anthers; xylopodium present **16. Schismocarpus**
 – Flowers polystemonous, if stamens few then all the same size; xylopodium absent 7
7. Testa tuberculate; stinging hairs absent; petals always free **14. Mentzelia**
 – Testa striate; stinging hairs usually present; petals free or united **15. Eucnide**
8. Flowers tetramerous; petals with longitudinal lamellae; urticant hairs always absent. Klaprothieae 9
 – Flowers penta- to octamerous; petals without longitudinal lamellae; urticant hairs often present. Loaseae 11
9. Lamina with two lobes on each side, margin serrate; petals green, narrowly obovate; erect shrub from underground xylopodium **13. Xylopodia**
 – Lamina entire, margin serrate; petals usually white, widely ovate if pale green; erect annual or perennial herbs or shrubs without xylopodium 10
10. Capsule straight, dehiscent, with >20 seeds; erect perennial herbs or shrubs **12. Plakothira**
 – Capsule straight or twisted, if twisted seeds <15; erect annual or decumbent perennial herbs **11. Klaprothia**
11. Flowers erect; petals white, cream or greenish white (never with serrate margin); nectar scales pale yellow, white or cream coloured, flat, dorsal threads filiform if present 12
 – Flowers usually pendent, very rarely erect (petals always with deeply serrate margin if flowers erect); petals and/or nectar scales brightly coloured, scales usually red and yellow or green and with various morphological elaborations such as dorsal calli, wings, or a double arch, dorsal threads often dilated if present 15
12. Nectar scales of more than 3 staminodia, 4–5 free staminodia present; plant without urticant setae **3. Huidobria**
 – Nectar scales of 3 staminodia, 2 free staminodia; plant with or without urticant setae 13
13. Calyx lobes much larger than petals; nectar scales >5× as long as wide; desert shrub without urticant setae **2. Kissenia**
 – Calyx lobes smaller than petals, nectar scales as long as wide or up to 2× as long as wide; plants with urticant setae 14
14. Erect desert shrubs with leaves up to 5 cm in diam.; petals with long claw, narrowly ovate; nectar scales with dorsal threads much exceeding the scale neck in length **4. Presliophytum**
 – Tall rainforest herb (to 4 m) with leaves up to 40 cm long; petals with short claw, subcircular to widely ovate; nectar scales without threads **1. Chichicaste**

15. Inflorescence ebracteose; testa tuberculate   **5.** *Aosa*
 – Inflorescence bracteose or frondose; testa deeply pitted, irregularly fibrous, reticulate or rugulate, never tuberculate   16
16. Every flower on the branches of the inflorescence with one bract only (primary flower usually with two); nectar scales with conspicuous dorsal sacs and apical wings, often with dorsal callus   **10.** *Nasa*
 – Every flower of the inflorescence with two bracts; nectar scales rectangular or with double arch and/or flags but never with conspicuous dorsal sacs or apical wings or dorsal calli   17
17. Fruits always twisted anticlockwise; seeds winged or with fibrous testa or angular, but never deeply pitted with fenestrate anticlinal walls nor rugulose; nectar scales rectangular, equalling the free staminodia in size
                                                       6. *Blumenbachia*
 – Fruits straight or twisted, if twisted fruits twisted clockwise and anticlockwise alternating in the inflorescence; testa deeply pitted with fenestrate anticlinal walls, very rarely with irregularly rugulose testa; nectar scales with double arch and often with flags, or nectar scales reduced in size (<< than free staminodia)   18
18. Fruits completely inferior, narrowly cylindrical, straight, more than 10× as long as wide   **9.** *Scyphanthus*
 – Fruits partially superior to completely inferior, cylindrical, ovoidal, clavate or globose, never more than 7× as long as wide, often twisted   19
19. Fruits straight, opening with apical valves   **7.** *Loasa*
 – Fruits straight or twisted, opening with longitudinal slits (very rarely: rosette herb with oblong, cream-coloured petals, and flowers singly on ebracteose peduncles if capsule opening with apical valves)   **8.** *Caiophora*

I. Subfamily Loasoideae Urb. & Gilg (1900).

Leaves opposite below and alternate above or opposite throughout; stinging hairs sometimes present. Inflorescences thyrsoids, often with numerous paracladia and strong metatopies, or dichasia, rarely reduced to few-flowered monochasia; flowers pendent or erect, 4–8-merous; petals free, mostly clawed, either deeply boat-shaped or carnose (never flat and membranaceous), or sometimes with longitudinal lamelliform flaps, margin entire or serrate, with >3 principal veins from base; polystemonous, fertile stamens usually in antepetalous position, filaments filiform; anthers latrorse; staminodia in antesepalous groups of 3–11, free or united with outer staminodia united to form a nectar scale (typically outer 3 united and inner 2 free), scales variously decorated with calli, filiform appendages and/or saccate thickenings, sometimes winged apically, white, red, yellow, usually with colour patterns; free staminodia usually S-shaped, papillose; parietal placentae simple or divided, usually with numerous ovules. Capsule terete, often with 10 prominent veins on the outside, calyx persistent, opening with 3–7 apical valves and/or 1–7 longitudinal sutures, very rarely indehiscent, straight or narrowly twisted; seeds usually numerous; testa mostly reticulate, pale brown to black.

1. Tribe Loaseae Urb. & Gilg (1900).

Leaves alternate or opposite; stinging hairs mostly present. Flowers 5(–8)-merous, mostly pendent; petals lacking longitudinal lamellae, yellow, orange, red, white, very rarely green; staminodial complexes always present, outer staminodes united to form a nectar scale. Capsule usually with 10 prominent veins on the outside.

1. *Chichicaste* Weigend

Chichicaste Weigend, *Nasa* and the Conquest of South America: 215 (1997).

Erect, sparsely branched, short-lived herb up to 4 m tall, with stinging hairs. Leaves opposite below, alternate above; lamina ovate with shallowly lobed and serrate margin, up to 40 cm long. Inflorescences terminal thyrsoids with monochasial or rarely dichasial paracladia; each flower with 2 small prophylls; flower erect; petals cream white to green; nectar scale white, apex forming 3–4 triangular lobes. Capsule 1/3 superior, subglobose, opening with 5 apical valves; testa reticulate. Only one species, *Chichicaste grandis* (Standl.) Weigend from the lowland rainforests of NW Colombia to Costa Rica.

2. *Kissenia* Endl.

Kissenia Endl., Gen. Pl. suppl.: 76 (1842), (sphalm. *Fissenia* Endl.).

Erect, densely branched shrubs without stinging hairs. Dominant, carnose tap-root present. Leaves opposite below, alternate above; lamina ovate to reniform with shallowly lobed and crenate margin. Inflorescences terminal dichasia or thyrsoids with 1–3 paracladia; each flower with 2 small, entire prophylls; flowers erect, numerous; petals cream coloured, much shorter than calyx lobes; nectar scale oblong, yellowish; ovary largely inferior, placentae reduced with few ovules. Capsule inferior, indehiscent, crowned with conspicuously accrescent calyx lobes; seeds 1–2, irregularly ovoidal; testa poorly developed. $2n = 24$. Two species from arid SW and NE Africa and the Arabian Peninsula.

3. *Huidobria* Gay

Huidobria Gay, Fl. Chil. 2: 440 (1847); Grau, Sendtnera 4: 77–93 (1997), rev.

Erect, densely branched annual herbs or shrubs without stinging hairs. Dominant, carnose tap-root present. Leaves opposite below, alternate above; lamina linear with entire margin or ovate and with shallowly lobed and crenate margin. Inflorescences complex asymmetrical dichasia (flowers apparently irregularly alternating with foliage leaves); each flower with 2 frondose prophylls; flowers erect; petals white to cream; nectar scale formed from 5–7 outer staminodia, white, with 5–7 long dorsal appendages, inner 3–5 staminodia free, white; placentae deeply divided into 3 lamellae. Capsule largely inferior, subglobose, opening with 4–5 apical valves; testa striate or reticulate. $2n = 36$. Two spp., deserticolous shrubs from Chile.

4. *Presliophytum* (Urb. & Gilg) Weigend

Presliophytum (Urb. & Gilg) Weigend, *Nasa* and the Conquest of South America: 215 (1997).

Erect, densely branched shrubs with stinging hairs. Dominant, carnose tap-root present. Leaves opposite below, often alternate above, lamina ovate to reniform with shallowly lobed and crenate margin. Inflorescences complex, asymmetrical dichasia; each flower with 2 frondose prophylls (flowers apparently irregularly alternating with foliage leaves); erect; petals white; nectar scale white, with 3 long dorsal appendages; placentae deeply divided into 3 lamellae. Capsule inferior, subglobose, opening with 4–5 apical valves; testa foveate-reticulate. $2n = 24$. Three species from the desert of western Peru (below 2000 m), on rocky slopes. *Loasa longiseta* Phil. from N Chile probably also belongs into this genus, differing form the other members only in fruit and seed morphology.

5. *Aosa* Weigend

Aosa Weigend, *Nasa* and the Conquest of South America: 214 (1997).

Annual or perennial herbs or shrubs, with stinging hairs. Leaves opposite below, often spirally inserted above; lamina oblong, ovate to suborbicular, lobed or entire, margin crenate or serrate. Inflorescence complex terminal thyrsoids with monochasial or rarely dichasial paracladia; prophylls absent; petals cream to green; nectar scale

green and brown, or red and yellow, with 3 dorsal filaments, scale apex forming 3 lobules, or entire and recurved. Capsule inferior to 3/4 superior, subglobose to clavate, sometimes curved, opening with 5 apical valves; testa reticulate, epidermis cells longer than wide ($210 \times 60\,\mu m$), at their chalazal pole with crescent-shaped, rugulose bulge. $2n = 24$. Seven spp. from E Brazil (6 spp.) and Hispaniola (1 sp.).

6. *Blumenbachia* Schrad.

Blumenbachia Schrad., Goett. Gel. Anz. 3/171: 1706 (1805); Weigend, Sendtnera 4: 202–220 (1997), rev.

Scandent or ascending, annual or perennial herbs, with stinging hairs. Root system fibrous, rarely primary root developing into root tuber or thickened tap-root. Leaves opposite; lamina usually widely ovate, subpalmately lobed, sometimes bipinnatisect. Inflorescences terminal thyrsoids with dichasial or monochasial branches (sect. *Angulatae*), or dichasia (sect. *Gripidea*) or borne singly (sect. *Blumenbachia*); each flowers with 2 frondose or bracteose prophylls; petals white; nectar scale with 3 long, dorsal, filiform appendages, white, red and yellow; placentae simple. Capsule inferior, cylindrical to globose, twisted anticlockwise only, septicidal or septifragous, apex coherent; seeds angular; testa polyhedral with irregular lateral wings (sect. *Angulatae*), or ovoidal and testa fibrous with narrow epidermis cells (sect. *Blumenbachia*), or widely reticulate, with 1–2 terminal wings (sect. *Gripidea*). $2n = 24, 26$. Twelve spp., Brazil, Uruguay, Paraguay, Argentina, Chile.

7. *Loasa* Adans.

Loasa Adans., Fam. Pl. 2: 501 (1763); Urban & Gilg, Nova Acta Acad. Caes. Leop.-Carol. German. Nat. Cur. 76 (1900), rev.; Sleumer, Bot. Jahrb. Syst. 76: 411–462 (1955), rev.

Rosulate, erect or scandent annual or perennial herbs, rarely small shrubs, stinging hairs present. Primary root dominant, thin, or forming a tap-root or thickened to a root tuber, sometimes rhizomatose. Leaves opposite throughout or opposite below and alternate above; lamina most often ovate, pinnatifid or bipinnatisect, sometimes widely ovate or subcircular and subpalmately lobed or ternate. Inflorescences terminal thyrsoids or dichasia, rarely flowers single; each flower with 2 frondose prophylls; flowers pendent, or erect in some rosulate species; petals white, yellow or red; nectar scale with (rarely without) 3 long, often flag-shaped dorsal appendages and decorated with a double arch on back, usually white, red and yellow. Capsule cylindrical to subglobose, opening with 3–5 apical valves; seeds ovoidal; testa deeply pitted, anticlinal walls fenestrate, rarely seeds large, globose and testa rugose. $2n = 12, 24, 26, 36$. Ca. 36 species, mostly Chile and adjacent Argentina, one species ranging into coastal Peru. *Loasa* is very closely related to *Caiophora*, *Scyphanthus* and *Blumenbachia* and is likely to be paraphyletic. On the other hand, *Loasa malesherbioides* Phil. from Chile and Argentina is aberrant in this generic group and should probably be segregated from *Loasa*.

8. *Caiophora* K. Presl

Caiophora K. Presl., Reliq. Haenk. 2: 41, tab. 42 (1836) (sphalm. *Cajophora*); Weigend, Sendtnera 4: 221–242 (1997), rev.

Rosulate, erect or scandent perennial herbs with stinging hairs. Primary root dominant, sometimes thickened, or primary and/or secondary roots tuberous, often rhizomatose. Leaves opposite; lamina mostly ovate, pinnatifid or bipinnatisect, sometimes ternate. Inflorescences terminal symmetrical or highly asymmetrical thyrsoids or dichasia, rarely flowers borne singly from basal rosette; each flower with 2 frondose prophylls; flowers mostly pendent, 5–8-merous; petals green, white, yellow or red; nectar scale usually with 3 long, often flag-shaped dorsal appendages and decorated with a double arch on back (occasionally dorsal appendages lost, arches poorly differentiated), usually white, red, yellow, or green; free staminodia sometimes with basal appendages; placentae Y-shaped, ovules only the morphologically adaxial side. Capsule mostly inferior, subglobose to narrowly cylindrical, frequently antidromously twisted, usually opening with 3–5 longitudinal slits, rarely opening with 3 apical valves; seeds angular; testa deeply pitted with fenestrate anticlinal walls. $2n = 14, 16$. Ca. 56 spp., Peru, Bolivia, Argentina, Chile, one species each in Uruguay, Brazil and Ecuador.

9. *Scyphanthus* D. Don

Scyphanthus D. Don in Sweet, Brit. Fl. Gard. I, 3: tab. 238 (1828).

Scandent annual herbs, without stinging hairs. Leaves opposite; lamina widely ovate, bipinnatisect. Root system fibrous. Inflorescences terminal, highly asymmetrical dichasia; each flower with 2 frondose prophylls; flowers yellow; nectar scale

with 3 flag-shaped dorsal appendages and double arch, white, red and yellow. Capsule inferior, narrowly cylindrical, opening with longitudinal slits; seeds angular; testa deeply pitted with fenestrate anticlinal walls. One or two spp., Mediterranean scrub lands in Chile, very close to *Caiophora* and to *Loasa* ser. *Pinnatae*.

10. *Nasa* Weigend   Fig. 83

Nasa Weigend, Nasa and the Conquest of South America: 214 (1997); Dostert and Weigend, Harvard Papers Bot. 4: 439–468 (1999), rev.; Weigend, Arnaldoa 5: 159–170 (1998), rev.; Weigend, Nordic J. Bot. 20: 15–31 (1999), rev.; Weigend, Flora of Ecuador 64: 1–90 (2000), rev.

Herbs, subshrubs or shrubs, 5–400 cm, with stinging hairs. Mature plant only with adventitious from the decumbent basal stem portion (primary root evanescent). Leaves opposite or alternate; lamina ovate to orbicular, lobulate, palmate or pinnatifid to bipinnate, sometimes peltate. Inflorescences terminal thyrsoids, dichasia or monochasia, each flower with one recaulescent bract; petals white, yellow, orange or red, often bicoloured; nectar scales with 0–3 calli on back and/or apical wings and/or 2 dorsal nectar sacs; staminodia L-shaped, sometimes with basal appendages; placentae simple. Capsule cylindrical to globose, usually opening with 3–5 apical valves; seeds ovoidal, globose or slightly angular; testa shallowly reticulate with wide anticlinal walls. $2n = 28, 56$. Ca. 100 spp., most in Colombia, Ecuador and Peru, few in Chile, Bolivia, Venezuela and C America (to S Mexico). Most diverse in cloud forests, some species weeds in cultivated lands, a few found in semi-deserts, rain-green coastal forests, rainforests, puna, subparamo or paramo.

2. TRIBE KLAPROTHIEAE Urb. & Gilg (1900).

Leaves opposite; stinging hairs always absent. Flowers 4-merous, mostly erect; petals each with 2 longitudinal lamellae, white or green; staminodial complex with outer staminodia free or rarely united to form a nectar scale; placentae simple. Capsule without prominent veins on the outside.

11. *Klaprothia* Kunth.   Fig. 81

Klaprothia Kunth in Humb., Bonpl. & Kunth, Nov. Gen. Sp. 6: 96 (1823); Poston & Nowicke, Syst. Bot. 15: 671–677 (1990), rev.

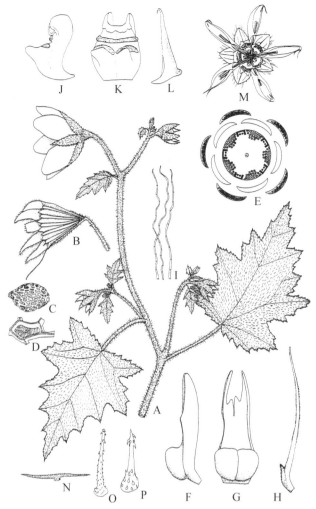

Fig. 83. A–Q Loasaceae-Loasoideae. **A–I** *Nasa argemonoides*. **A** Habit. **B** Mature Fruit. **C** Seed. **D** Detail of testa epidermis. **E** Floral diagram, staminodial complexes black, U-shaped structure: nectar scale of three united staminodes. **F–G** Nectar scale. **H** Free staminode. **I** Uniseriate trichomes (stem). **J–M** *Nasa rudis* subsp. *australis*. **J–K** Nectar scale. **L** Free staminode. **M** Flower. **N** *Nasa humboldtiana*, T-shaped, scabrid trichome (stem). **O** *Nasa triphylla*, glochidiate trichome. **P** *Nasa aequatoriana*, glochidiate trichome. (Drawn by M. Weigend)

Erect or ascending annual or perennial herbs. Lamina undivided. Inflorescences terminal, dichasia (*K. mentzelioides* Kunth) or complex thyrsoids (*K. fasciculata* (K. Presl) Poston); each flower with 2 minute prophylls; flowers erect; petals white; staminodial groups with 2–6 staminodia each, all free or (*K. mentzelioides*) 3–4 outer ones only slightly connate basally, very densely papillose-hairy, apex often irregularly club-shaped and lobed, yellow.

Capsule inferior, ovoidal and straight to slightly twisted and tardily dehiscent with apical valves, or clavate and twisted and opening with longitudinal slits; seeds narrowly ovoidal; testa reticulate with aequatorial striae. $2n = 24, 48$. Two widespread and weedy species from S Mexico to Brazil and Bolivia, also Caribbean, Galapagos.

12. *Plakothira* Florence

Plakothira Florence, Adansonia II, 7: 240 (1985); Florence, Allertonia 7: 238–253 (1997).

Shrub or perennial herb with fibrous roots. Lamina undivided. Inflorescences terminal dichasia or complex thyrsoids; each flower with 2 minute prophylls; flowers erect; petals green or white; staminodial groups antesepalous with 7–9 staminodia each, 4–6 outer ones fused to form a green and yellow nectar scale or free to base, often papillose-hairy. Capsule semisuperior, subglobose, straight; seeds narrowly ovoidal; testa reticulate with aequatorial striae. $2n = 48$. Three spp., Marquesas Is. (Polynesia), probably congeneric with *Klaprothia* (*Plakothira* is morphologically and karyologically much closer to *K. mentzelioides* than *K. mentzelioides* is to *K. fasciculata*).

13. *Xylopodia* Weigend   Fig. 84C–I

Xylopodia Weigend, *Nasa* and the Conquest of South America: 215 (1997).

Shrub with erect branches from horizontal xylopodium. Lamina with 2–3 lobes on each side, margin serrate. Inflorescences terminal dichasia; each flower with 2 minute prophylls; flowers pendent; petals green; staminodial groups with 7–9 staminodia, outer 4–6 staminodes fused to form a green and yellow scale, densely papillose-hairy, inner ones free, greenish yellow. Capsule semisuperior, subglobose, straight, opening with 4 apical valves; seeds narrowly ovoidal; testa reticulate. $2n = 24$. Only one species, *X. klaprothioides* Weigend, rocky slopes in Cajamarca, Peru.

II. Subfamily Mentzelioideae (Rchb.) Urb. & Gilg (1900).

Leaves usually opposite below and alternate above, with scabrid-glochidiate trichomes, sometimes with stinging hairs. Inflorescences terminal thyrsoids, often without metatopy; flowers erect, merous, sessile or pedicellate; prophylls frondose or bracteose; petals free or rarely united, planar or shallowly boat-shaped, never clawed, margin entire, membranaceous, with >3 principal veins from base; polystemonous, rarely obdiplostemonous; filaments filiform or sometimes widened or apically forked; anthers latrorse; staminodia, if present, petaloid and free to base; parietal placentae simple, with numerous ovules. Capsule terete, without prominent veins on the outside, calyx persistent, opening with 3–5 apical valves, straight; seeds ovoidal or angular or laterally compressed; testa mostly polyhedral or striate, grey or brown.

14. *Mentzelia* L.

Fig. 84A, B

Mentzelia L., Sp. Pl. 1: 516 (1753); Darlington, Ann. Missouri
  Bot. Gard. 21: 103–226 (1934), rev.; Hill, Brittonia 28: 86–112
  (1976), seed morph.
Bartonia Pursh ex Sims (1812).
Acrolasia K. Presl (1835).
Trachyphytum Nutt. ex Torrey & Gray (1840).
Bicuspidaria Rydb. (1903).

Annual or perennial herbs, subshrubs, shrubs or small trees with scabrid-glochidiate trichomes, without stinging hairs. Dominant and persistent tap-root present. Leaves opposite below and alternate above, rarely opposite throughout (*M. arborescens* Urb. & Gilg); lamina ovate to triangular-ovate, occasionally lyrate, usually lobed, margin serrate or mucronate. Inflorescences complex, terminal, basitonic thyrsoids; each flower with 2 frondose prophylls; petals white, yellow, or orange; stamens numerous (rarely 10); filaments equal to very unequal in size, sometimes outer filaments dilated or apically forked; sometimes with petaloid staminodes (*M. decapetala* (Pursh) Urb. & Gilg). Stigma punctiform. Capsule inferior, sessile or pedicellate, opening with 3–7 apical valves; seeds angular or laterally compressed; testa reticulate or polyhedral, sometimes with central papillae, sometimes narrowly winged. $2n = 9, 10, 11, 14, 18, 27, 36$. Ca. 80 spp., from Argentina to Canada incl. Caribbean and Galapagos Is., most in SW USA and Mexico. The majority of species in desert and semi-desert habitats, some penetrate into grasslands and rain-green forests.

15. *Eucnide* Zucc.

Eucnide Zucc., Del. Hort. Monac. 1844/4: 28 (1844); Thompson
  & Ernst, J. Arnold Arbor. 48: 56–88 (1967), rev.; Thompson
  & Powell, Phytologia 49: 16–32 (1981), rev.

Annual or perennial herbs; stinging hairs usually present. Leaves petiolate; lamina widely ovate to subcircular, margin shallowly lobed and deeply

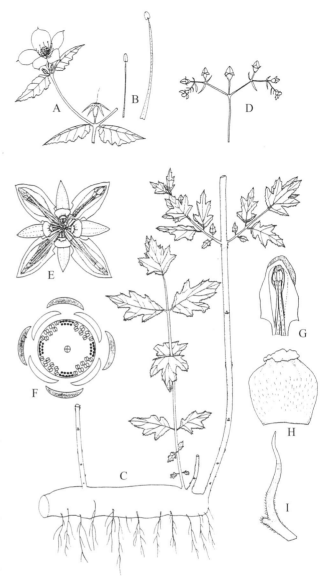

Fig. 84. A, B Loasaceae-Mentzelioideae, *Mentzelia scabra*. A Inflorescence branch. B Inner and outer stamen showing distinct heteranthery. C–I Loasaceae-Loasoideae, *Xylopodia klaprothioides*. C Habit. D Infructescence. E Flower. F Floral diagram, staminodial complexes black. G Petal with reflexed anthers held by longitudinal lamellae. H Nectar scale. I Free staminode. Drawn by M. Weigend

serrate or mucronate. Inflorescences terminal thyrsoids (sometimes with long, terminal monochasia) or simple dichasia; each flower with 2 frondose or bracteose prophylls; petals narrowly ovate or ovate, sometimes post- or congenitally united (sympetalous), yellowish white, yellow, or orange; filaments free to base or stamens epipetalous, all equal, staminodia absent; stigma with 3–5 divergent lobes. Capsule inferior, cylindrical to conical, pedicellate, pedicel sometimes strongly elongating after anthesis (chasmocarpous), opening with 3–5 apical valves; seeds ovoidal; testa striate. $2n = 42$. Thirteen spp., Guatemala to SW USA, most species in Mexico.

16. *Schismocarpus* Blake

Schismocarpus Blake, Contrib. Gray Herb. II, 53: 61 (1918).

Subshrub without stinging hairs. Annual shoots arising from an underground xylopodium. Leaves alternate, petiolate; lamina widely ovate to subcircular, entire, margin serrate. Inflorescences terminal thyrsoids, typically with one distal dichasium and one additional monochasial paraclade; each flower with 2 bracteose prophylls; petals narrowly ovate or ovate, membranaceous, planar, apex acute, green, white; stamens 10, 5 shorter, 5 longer, obdiplostemonous; staminodia absent; stigma with 5 divergent lobes. Capsule semisuperior, ovoidal, pedicellate, opening with 5 apical valves; seeds ovoidal; testa striate. Only one species, *S. pachypus* Blake, from Oaxacá, Mexico.

III. Subfamily Gronovioideae (Rchb.) Link (1833).

Leaves (upper) always alternate (first pair in seedling opposite), always with glochidiate trichomes, sometimes with stinging hairs. Inflorescences terminal thyrsoids with a terminal dichasium and 1–2 additional paracladia, each flower with 2 filiform prophylls; flowers erect, 5-merous; petals free, spatulate, planar, not clawed, margin entire or laciniate, thinly membranaceous, with one principal vein from base; haplostemonous, staminodia absent; filaments filiform; anthers introrse; ovary inferior, with 1 pendent ovule. Fruit a cypsela, pentagonous in cross section, alate on ribs, calyx persistent and accrescent, fruit free from bracts and prophylls; testa white to beige.

17. *Gronovia* L.

Gronovia L., Sp. Pl. 1: 202 (1753).

Scandent, annual herbs; stinging hairs and glochidiate hairs with 2 hooks at the tip always present. Lamina reniform, deeply sinuate-lobate with 3–4 lobes on each side, lobes ovate-lanceolate, long acuminate, membranaceous, margins entire.

Inflorescences terminal thyrsoids, typically with distal dichasium and one additional, monochasial paraclade; calyx lobes free or united nearly to apex, yellow or yellowish green; petals entire, included or long and exserted from the calyx, thinly membranaceous, pale yellow; stamens without differentiated connective. $n = 37$. Two spp., fast-growing annual herbs in rain-green forests, NW Peru to Mexico.

18. *Fuertesia* Urb.   Fig. 85A–L

Fuertesia Urb., Ber. Deutsch. Bot. Ges. 28: 515 (1910).

Liana with splinter hairs and glochidiate hairs with 2 hooks at the tip. Lamina cordate, coriaceous, margin entire. Inflorescences terminal thyrsoids, typically with distal dichasium and 1–3 additional dichasial paracladia; calyx lobes free, yellowish green; petals laciniate, included in the calyx, thinly membranaceous, pale yellow; stamens without differentiated connective. One species, *F. dominguensis* Urb. from Hispaniola.

19. *Cevallia* Lagasc.

Cevallia Lagasc., Gen. Sp. Pl.: 35 (1805); Davis & Thompson, Madroño 19: 1–18 (1967).

Subshrub with glochidiate hairs with spreading (not reflexed) branches, without stinging hairs. Lamina narrowly ovate, margin sinuate. Inflorescences terminal thyrsoids, typically with distal dichasium and 1 additional, mono- or dichasial paraclade, strongly condensed; calyx lobes and petals identical, free to base, whitish, membranaceous; stamens with connective long protracted into club-shaped appendage. $n = 13$ (Davis and Thompson 1967). One species, *C. sinuata* Lagasc., in the deserts of SW USA and N Mexico.

IV. Subfamily Petalonychoideae Weigend (1997).

Leaves alternate with glochidiate hairs with short, spreading (not reflexed) branches, without stinging hairs. Inflorescences terminal racemes; each flower with 2 prophylls, in the axil of 1 ovate pherophyll; flowers erect, 5-merous; petals free to base or postgenitally fused, often clawed, with one principal vein from base; haplostemonous, stamens antesepalous, sometimes displaced outside the pseudosympetalous corolla or filaments locked between the petal margins; filaments

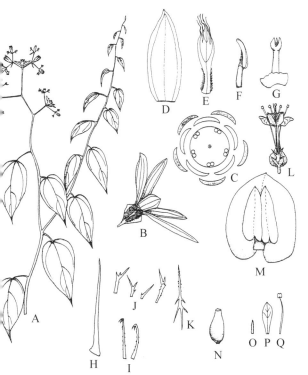

Fig. 85. A–L Loasaceae-Gronovioideae, *Fuertesia domingensis.* **A** Habit. **B** Fruit. **C** Floral diagram. **D** Calyx lobe. **E** Petal. **F** Stamen. **G** Style and stigma in cup-shaped nectary. **H** Stinging hair (seta). **I** Hook-shaped, glochidiate hairs. **J** Modified glochidiate trichomes. **K** Splinter hair. **L–Q** Loasaceae-Petalonychoideae, *Petalonyx* **L** *P. parryi*, flower. **M–Q** *P. linearis.* **M** Fruit, enclosed in bract and bracetoles. **N** Fruit, bracts and bracteoles removed. **O** Sepal. **P** Petal. **Q** Stamen. (Original; L Davis and Thompson 1967)

Selected Bibliography

Behnke, H.-D., Barthlott, W. 1983. New evidence from the ultrastructural and micromorphological fields in angiosperm classification. Nordic J. Bot. 3: 43–66.

Carlquist, S. 1984. Wood anatomy of Loasaceae with relation to systematics, habit and ecology. Aliso 10: 583–602.

Carlquist, S. 1987. Wood anatomy of *Plakothira* (Loasaceae). Aliso 11: 563–569.

Carlquist, S. 1992. Wood anatomy of sympetalous dicotyledon families: a summary, with comments on systematical relationships and evolution of woody habit. Ann. Missouri Bot. Gard. 79: 303–332.

Cocucci, A.A., Sérsic, A.N. 1997. Evidence of rodent pollination in *Cajophora coronata* (Loasaceae). Plant Syst. Evol. 211: 113–128.

Coleman, J.R., Smith, L.B. 1969. Chromosome numbers of some Brazilian Angiosperms. Rhodora 71: 548–551.

Corner, E.J.H. 1976. See general references.

Cronquist, A. 1981. See general references.

Darlington, J. 1934. A monograph of *Mentzelia.* Ann. Missouri Bot. Gard. 21: 103–226.

Davis, W.S., Thompson, H.J. 1967. A revision of *Petalonyx* (Loasaceae) with a consideration of affinities in subfamily Gronovioideae. Madroño 19: 1–18.

Dostert, N., Weigend, M. 1999. A synopsis of the *Nasa triphylla* complex (Loasaceae), including some new species and subspecies. Harvard Pap. Bot. 4: 439–467.

Garcia, V. 1962. Embryological Studies in the Loasaceae: development of endosperm in *Blumenbachia hieronymi* Urb. Phytomorphology 12: 307–312.

Gilg, W. 1925. Loasaceae. In: Engler, A., Prantl, K. Die natürlichen Pflanzenfamilien ed. 2., 21. Leipzig: W. Engelmann, pp. 522–543.

Grau, J. 1988. Chromosomenzahlen chilenischer Loasaceae. Mitt. Bot. Staatssamml. München 27: 7–14.

Grau, J. 1997. *Huidobria*, eine isolierte Gattung der Loasaceae aus Chile. Sendtnera 4: 77–93.

Hegnauer, R. 1966, 1989. See general references.

Hempel, A.L., Reeves, P.A., Olmstead, R.G., Jansen, R.K. 1995. Implications of *rbcL* sequence data for higher order relationships of the Loasaceae and the anomalous aquatic plant *Hydrostachys* (Hydrostachyaceae). Plant Syst. Evol. 194: 25–37.

Hill, R.J. 1976. Taxonomic and phylogenetic significance of seed coat microsculpturing in *Mentzelia* (Loasaceae) in Wyoming and adjacent western States. Brittonia 28: 86–112.

Hufford, L. 1988. Seed morphology of *Eucnide* and other Loasaceae. Syst. Bot. 13: 154–167.

Hufford, L. 1990. Androecial ontogeny and the problem of monophyly of Loasaceae. Can. J. Bot. 68: 402–419.

Hufford, L. 1998. Early development of androecia in polystemonous Hydrangeaceae. Am. J. Bot. 85: 1057–1067.

Humboldt, F.H.A. v., Bonpland, A.J., Kunth, C.S. 1815–1823 [1825] Nova genera et species plantarum. Paris.

Keeler, K.H. 1981. The nectaries of *Mentzelia nuda*: from pollinator attraction to seed protection. Am. J. Bot. 68: 295–299.

Leins, P., Winhard, W. 1973. Entwicklungsgeschichtliche Studien an Loasaceen-Blüten. Österr. Bot. Zeitschr. 121/122: 145–165.

Moerman, D.E. 1998. Native American ethnobotany. Portland: Timber Press.

Moody, M.L., Hufford, L., Douglas E. Soltis, D.E., Soltis P.S. 2001. Phylogenetic relationships of Loasaceae subfamily Gronovioideae inferred from *mat*K and ITS sequence data. Am. J. Bot. 88: 326–336.

filiform; anthers laterorse; filiform staminodia sometimes present; ovary with 1 pendent ovule. Cypsela remaining attached to bract and prophylls, irregularly pentagonous, not alate on ribs; calyx caducous; testa white to beige. $2n = 46$.

Only one genus:

20. *Petalonyx* A. Gray   Fig. 85M–R

Petalonyx A. Gray, Pl. Nov. Thurb.: 319 (1854); Davis & Thompson, Madroño 19: 1–18 (1967), rev.

Characters as for subfamily.

Five spp. of deserticolous shrubs and subshrubs in N Mexico and SW USA.

Müller, A.A., Kufer, J.K., Dietl, K.G., Weigend, M. 1999. Iridoid glucosides – chemotaxonomic markers in Loasoideae. Phytochemistry 52: 67–78.

Payer, M. 1853. Organgénie des familles Myrtacées, Punicées, Philadelphacées, Loasées et Ombelliféres. Ann. Sci. Nat., Paris, III, 20: 100–135.

Poston, M.E., Nowicke, J.W. 1993. Pollen morphology, trichome types, and relationships of the Gronovioideae (Loasaceae). Am. J. Bot. 80: 689–704.

Poston, M.E., Thompson, H.J. 1977. Cytotaxonomic observations in Loasaceae subfamily Loasoideae. Syst. Bot. 2: 28–35.

Rodríguez, V., Schripsema, J., Jensen, S.R. 2002. An iridoid glucoside from *Gronovia scandens* (Loasaceae). Biochem. Syst. Ecol. 30: 243–247.

Schlindwein, C., Wittmann, D. 1997. Microforaging routes of *Bicolletes pampeana* (Colletidae) and bee-induced pollen presentation in *Cajophora arechavaletae* (Loasasceae). Bot. Acta 109: 1–7.

Takhtajan, A.L. 1980. Outline of the classification of flowering plants (Magnoliophyta). Bot. Rev. 46: 225–259.

Takhtajan, A.L. 1997. See general references.

Thompson, H.J., Ernst, W.R. 1967. Floral biology and systematis of *Eucnide* (Loasaceae). J. Arnold Arbor. 48: 56–88.

Urban, I. 1892. Die Blütenstände der Loasaceen. Ber. Deutsch. Bot. Ges. 10: 220–225.

Urban, I. 1910. Zwei neue Loasaceen von Sto Domingo. Ber. Deutsch. Bot. Ges. 28: 515–520.

Vijayaraghavan, M.R., Prabhakar, K. 1984. The endosperm. In: Johri, B.M. (ed.) Embryology of angiosperms, chap. 7. Berlin Heidelberg New York: Springer, pp. 319–376.

Weigend, M. 1997. *Nasa* and the conquest of South America. München: Eigenverlag M. Weigend.

Weigend, M. 2000. 132. Loasaceae. In: Harling, G., Andersson, L. (eds.) Flora of Ecuador 64: 1–92.

Weigend, M. 2001. Loasaceae. In: Bernal, R., Forero, E. (eds.) Flora de Colombia, vol. 22. Sta. Fé de Bogotá: Instituto de Ciencias Naturales, pp. 1–100.

Weigend, M., Kufer J., Mueller, A.A. 2000. Phytochemistry and the systematics and ecology of Loasaceae and Gronoviaceae (Loasales). Am. J. Bot. 87: 1202–1210.

Wittmann, D., Schlindwein C. 1995. Melittophilous plants, their pollen and flower visiting bees in southern Brazil. 1. Loasaceae. Biociências (Porto Alegre) 3: 19–34.

Xiang, Q.-Y. 1999. Systematic affinities of Grubbiaceae and Hydrostachyaceae within Cornales – insights from *rbc*L sequences. Harvard Pap. Bot. 4: 527–542.

Xiang, Q.-Y., Soltis D.E., Soltis P.S. 1998. Phylogenetic relationships of Cornaceae and close relatives inferred from *mat*K and *rbc*L sequences. Am. J. Bot. 85: 285–297.

Maesaceae

B. Ståhl and A.A. Anderberg

Maesaceae (A. DC.) Anderberg, Ståhl & Källersjö, Taxon 49: 185 (2000).

Small to medium-sized trees, shrubs or sometimes lianas with schizogenous cavities; indumentum of non-glandular, uniseriate trichomes and scales. Leaves simple, petiolate, exstipulate, distinctly alternate (generally not clustered), with prismatic crystal druses; margins entire or variously serrate or crenate. Inflorescences axillary racemes or compound racemes; pedicels subtended by a bract and bearing prophylls just below the calyx. Flowers (6)5–4-merous, bisexual or functionally female, if unisexual then plants polygamous or functionally dioecious; calyx lobes mostly broadly ovate, persistent; corolla white, yellow or pinkish, campanulate or sometimes urceolate, lobes broadly ovate to subrotund, quincuncial in bud; stamens homomerous, antepetalous, included or slightly exserted; filaments inserted on the corolla tube; anthers almost square, or somewhat shorter than wide, apically rounded or retuse, dorsifixed, introrsely dehiscent by longitudinal slits; ovary semi-inferior; style well demarcated; stigma truncate or capitate, entire or 2–5-lobed; placentation free central; ovules many, arranged in 1–5 whorls, not or only slightly immersed into the placenta. Fruit berry-like with a somewhat woody mesocarp, globular or ovoid, indehiscent. Seeds numerous, small, angular, dark brown; seed coat reticulate, two-layered, with rhomboid crystals; endosperm abundant with evenly thickened cell walls; embryo with short hypocotyl and short, narrow cotyledons. $x = 10$.

A monogeneric family with about 150 species, distributed in tropical and subtropical regions of the Old World.

VEGETATIVE STRUCTURES. Most species are shrubs, but some are small or medium-sized trees, scandent shrubs or lianas.

Schizogeneous secretory cavities are common and probably occur in the leaves of all species; they are often visible as brownish-orange lines on vegetative and reproductive parts (Smith 1973). Calcium oxalate druses are found in the mesophyll. The wood in Maesaceae is unique among primulalean families in having a mixture of uni- and multiseriate rays (Janssonius 1920).

INFLORESCENCE AND FLORAL STRUCTURE. Inflorescences are axillary, usually appearing along a major part of the young shoots. Most species have compound racemes (Fig. 86), but species with simple racemes are not uncommon. The flowers are perigynous. The calyx lobes are usually well demarcated, being broadly obtuse-ovate to triangular in shape; the corolla is campanulate or sometimes urceolate with a well-developed tube which, depending on species, varies from being somewhat shorter to more than twice the length of the lobes. There is a single whorl of antepetalous stamens, which are usually completely included in the corolla. The filaments are basally fused to the corolla tube. The anthers are basifixed, shortly oblong-ovoid and obtuse or retuse at apex; they open introrsely by means of longitudinal slits. The ovary is conical and tapers into a short but well-demarcated style; the stigma is capitate or truncate, or sometimes 3–5-lobed. There is no nectariferous disk. The ovules are spirally inserted in one or 2–8 series on the central placenta.

Floral development in the family was studied by Caris et al. (2000).

EMBRYOLOGY. According to Johri et al. (1992), the anthers of *Maesa* have a fibrous endothecium and a secretory tapetum and, like in other primulalean families, the ovules are bitegmic, tenuinucellate and have a Polygonum-type embryo sac and nuclear endosperm formation. An integumentary tapetum is present and the embryogeny is of the onagrad type.

POLLEN MORPHOLOGY. Pollen of Philippine *Maesa cumingii* is 3-colporoidate, ± spheroidal ($11 \times 9\,\mu m$) and subreticulate (Erdtman 1952), and similar to pollen of most other members of primulalean families.

KARYOLOGY. A basic chromosome number of $x = 10$ has so far been recorded in about five species of *Maesa*.

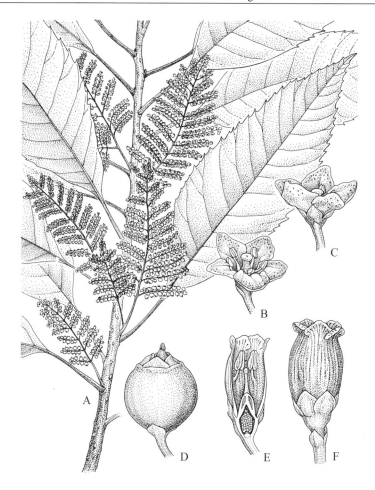

Fig. 86. Maesaceae. **A–C** *Maesa lanceolata*. **D–F** *M. japonica*. **A** Habit, flowering branch. **B, C, E, F** Flowers (**E** in vertical section showing placentation and ovules). **D** Fruit. (Anderberg and Ståhl 1995)

REPRODUCTIVE SYSTEMS AND POLLINATION. Morphologically, the flowers of all species appear to be hermaphrodite. However, Mez (1902) and Sleumer (1987) described *Maesa* as dioecious, but there is little support of this in other treatments of the genus. Although admitting that it is difficult to ascertain whether the anthers are functional or not, Sleumer (1987) found the anthers of female flowers reduced in size, thus suggesting cryptic or functional dioecy. Flowers within a given individual are sometimes unisexual and the plants functionally polygamous. Flower morphology suggests pollination by insects, but anemophily may be expected in some species.

FRUIT AND SEED. Being many-seeded, the fruits are best described as berries, although the mesocarp is often thin and the endocarp rather hard and brittle. The seeds are small, angular, and dark brown to black; the testa is two-layered and contains rhomboid crystals; endosperm is abundant and has smooth cells walls. The embryo is short with cotyledons of about the same length and width as the hypocotyl.

PHYTOCHEMISTRY. A special kind of quinone, maesaquinone, occurs in fruits of *Maesa lanceolata* and *M. japonica*, and other quinones are present in vegetative parts of the former species (Hegnauer 1969). Saponins have been reported from several species of the family, both in vegetative parts and fruits. In addition, ubiquitous flavonols have been isolated from leaves of *M. chisa*, and alkaloids have been detected in leaves of *M. ramentacea* and *M. perlarius*. The endosperm contains amyloid and fatty oil, although the former appears to be sparse in the family (Seegeler 1983).

SUBDIVISION AND RELATIONSHIPS WITHIN THE FAMILY. The single genus, *Maesa*, has been divided into two subgenera (Mez 1902). Thus, a few African species with ovules arranged in a single series on the placenta are referred to subg. *Monotaxis* Mez and the remaining species (also some African), with ovules in 2 or several series, are placed in subg. *Maesa*.

AFFINITIES. Maesaceae have mostly been treated as a subfamily of Myrsinaceae, both agreeing in woody habit, sympetalous, 5-merous flowers, free central placentation, and schizogenous secretory cavities. However, the differences in reproductive structures and wood anatomy were noted by earlier workers (de Candolle 1841; Janssonius 1920). Recent phylogenetic studies, including both morphological and DNA sequence data (Anderberg and Ståhl 1995; Anderberg et al. 1998; Källersjö et al. 2000; Savolainen, Fay et al. 2000), strongly support the position of Maesaceae distinct from Myrsinaceae and Primulaceae. According to Källersjö et al. (2000), Maesaceae are sister to Theophrastaceae-Primulaceae-Myrsinaceae.

DISTRIBUTION AND HABITATS. The family is distributed throughout the Old World tropics, extending into subtropical regions in Asia. The largest concentration of species is found in SE Asia, but many species are also present in India, China, and the Pacific. About six species occur in Africa, one of which has seemingly been introduced in Madagascar (Perrier de la Bâthie 1953), and three species occur in tropical Australia (Queensland). In China, *Maesa* grows in evergreen broad-leaved or mixed forests at 600–2700 m altitude, often in damp areas such as riverbanks. Towards the Equator, the family continues as primary lowland or mid-altitude element with few species reaching subalpine elevations. Several species occur in disturbed and secondary forests.

ECONOMIC IMPORTANCE. In Ethiopia, the oily seeds of *Maesa lanceolata* are used for frying, and crushed fruits are used to treat infections of intestinal parasites (Seegeler 1983). The fruits of some Indian species are sometimes used as a fish-poison, and the leaf of one species, *M. indicia* Wall., is a curry ingredient (Ambasta 1986). The wood is sometimes used for construction works. Extracts of leaves, twigs and bark have insecticidal effects (Hegnauer 1969).

One genus:

Maesa Forssk.   Fig. 86

Maesa Forssk., Fl. Aegypt. Arab. 66 (1775); Mez, Pflanzenreich IV 236: 15–54 (1902); Ridley, Fl. Malaya Peninsula 2: 227–229 (1923), reg. rev.; Walker, Philippine J. Sci. 73: 12–48 (1940), reg. rev.; Smith, J. Arnold Arb. 54: 3–36 (1973), Fijian spp.; Fosberg & Sachet, Phytologia 44: 362–369 (1979), Pacific spp.; Taton, Fl. d'Afrique Centrale (no vol.) 3–12 (1980), reg. rev.; Halliday, Fl. Trop. E. Afr. (no vol.) 2–5 (1984), reg. rev.; Sleumer, Blumea 32: 39–65 (1987), New Guinean spp.

Description as for the family. About 150 spp., throughout the Old World tropics and subtropics; two subgenera: subg. *Monotaxis* Mez with ovules in a single row, and subg. *Maesa* in 2 or more series.

Selected Bibliography

Ambasta, S.P. (ed.) 1986. The useful plants of India. New Delhi: Publication and Information Directorate.

Anderberg, A.A., Ståhl, B. 1995. Phylogenetic interrelationships in the order Primulales, with special emphasis on the family circumscriptions. Can. J. Bot. 73: 1699–1730.

Anderberg, A.A., Ståhl, B., Källersjö, M. 1998. Phylogenetic relationships in the Primulales inferred from *rbc*L sequence data. Plant Syst. Evol. 211: 93–102.

Anderberg, A.A., Ståhl, B., Källersjö, M. 2000. Maesaceae, a new primuloid family in the order Ericales s.l. Taxon 49: 183–187.

Candolle, A. de 1841. Troisième mémoire sur la famille de Myrsinacées. Ann. Sci. Nat. Bot. Biol. Vég. 16: 129–176.

Caris, P., Ronse Decraene, L.P., Smets, E., Clinckemaillie, D. 2000. Floral development of three *Maesa* species, with special emphasis on the position of the genus within Primulales. Ann. Bot. II, 86: 87–97.

Erdtman, G. 1952. See general references.

Hegnauer, R. 1969. See general references.

Janssonius, H.H. 1920. Mikrographie des Holzes der auf Java vorkommenden Baumarten 4. Leiden: Brill.

Johri, B.M. et al. 1992. See general references.

Källersjö, M. et al. 2000. See general references.

Mez, C. 1902. Myrsinaceae. In: Engler, A. (ed.) Das Pflanzenreich IV, 236. Leipzig: W. Engelmann.

Perrier de la Bâthie, H. 1953. Myrsinacées. In: Humbert, H. (ed.) Flore de Madagascar et des Comores. Paris: Mus. Natl. Hist. Nat.

Savolainen, V., Fay, F. et al. 2000. See general references.

Seegeler, C.J.P. 1983. Oil plants in Ethiopia, their taxonomy and agricultural significance. Wageningen: Pudoc.

Sleumer, H. 1987. A revision of the genus *Maesa* Forssk. (Myrsinaceae) in New Guinea, the Moluccas, and the Solomon Islands. Blumea 32: 39–65.

Smith, A.C. 1973. Studies of Pacific island plants, XXV. The Myrsinaceae of the Fijian region. J. Arnold Arbor. 54: 1–41, 228–292.

Marcgraviaceae

S. Dressler

Marcgraviaceae Choisy in DC., Prodr. 1: 565 (1824), nom. cons.

Terrestrial, hemiepiphytic or epiphytic lianas or shrubs, rarely small trees. Hypophyllous glands, raphide cells and variously shaped sclereids frequently present. Leaves simple, alternate, exstipulate, glabrous, margins entire or (some *Marcgravia*) minutely crenate. Inflorescences terminal, racemose, sometimes resembling umbels or spikes, erect or pendulous; bracts transformed into variously shaped nectaries; pedicels mostly with two sepaloid prophylls. Flowers hermaphroditic, actinomorphic, hypogynous; sepals (4)5, unequal, free or nearly so, imbricate, persistent; petals 3–5, imbricate, free or connate; stamens 3–many; filaments free or basally connate, uni- to biseriate; anthers basifixed or nearly so, tetrasporangiate, introrse, longitudinally dehiscent. Ovary superior, completely or incompletely 2–20-locular; ovules few to numerous, anatropous. Fruit subglobose, apiculate with persistent style and stigma, capsular, loculicidally and septifragously dehiscent from the base or berry-like and irregularly dehiscent, pulpy inside. Seeds exotestal-theoid, hemispherical to reniform, few to numerous with a shiny reticulate testa; endosperm scanty or lacking; embryo straight.

A family comprising 7 genera and about 130 species, distributed in tropical Central and South America including the West Indies.

VEGETATIVE MORPHOLOGY. The developing leaves convolutely enclose the shoot-tip, which is rather characteristic of the family and similar to some Theaceae. All Marcgraviaceae develop an apical mucro that is often caducous at the fully expanded leaf.

In *Marcgravia*, the leaves are distichously arranged and often provided with a drip-tip, whereas phyllotaxis of the other genera is spiral. *Marcgravia* has dimorphic branches: sterile, creeping or root-climbing branches that are angled and bear two rows of small, juvenile, cordate leaves; and fertile and pendulous branches that are rootless and terete. The latter bear the typical adult leaves and may develop terminal inflorescences. Both forms are connected by intermediate leaf-forms.

A noteworthy feature of the family is the presence of solid or poriform "hypophyllous glands", which occur often in specific patterns on the abaxial face of the leaf. In *Marcgravia*, they can intergrade with "marginal glands". They have been regarded as domatia, hydathodes, extrafloral nectaries, or resin-secreting organs, which certainly is not appropriate. A nectar-secreting activity has been observed in early developmental stages of the leaf. There is still no certainty about their ecological significance. They may attract ants as protection against predators while the leaf tissue is still soft (Dressler 1997).

Leaf venation is relatively uniform, pinnate-brochidodromous with ascending secondary arches. Leaf venation on juvenile branches in *Marcgravia* displays an earlier ontogenetic stage.

VEGETATIVE ANATOMY. The leaves of Marcgraviaceae are dorsiventral and hypostomatic. Stomata (at least of *Marcgravia*) belong to the staurocytic type (Fryns-Claessens and van Cotthem 1973). In the adult leaves of *Marcgravia* a hypodermis and a thick cuticle are usually present, and the palisade parenchyma includes one to three layers. Leaves of the juvenile branches lack a hypodermis, the upper epidermis is often papillate and stomata occur also on the upper side. The other genera lack a hypodermis as well as marginal glands, and the palisade parenchyma usually includes only one layer.

Different kinds of idioblasts occur in the family. Very characteristic is the presence of various types of sclereids in the leaves (de Roon 1967), nectaries, perianth, pistil and pericarp as well as in the cortex and medulla of the shoots. Raphide cells are found in parenchymatous tissues. Styloids and oil-containing or mucilaginous cells occur rarely (de Roon 1975).

An unilacunar node with a single trace in the petiole closed to a siphonostele is common throughout the family (Schofield 1968).

The main features of the wood are as follows: vessels mostly solitary or in radial multiples of

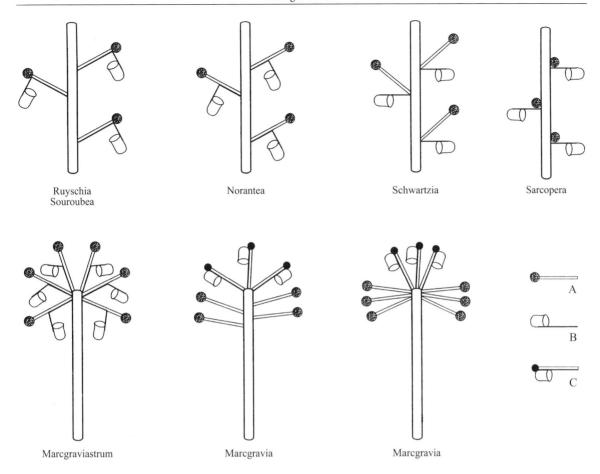

Ruyschia
Souroubea

Norantea

Schwartzia

Sarcopera

Marcgraviastrum

Marcgravia

Marcgravia

A

B

C

2–4, medium-sized (*Marcgravia*) to rather long (*Norantea* s.l., *Souroubea*), perforations simple, rarely scalariform (e.g. in *Norantea* s.l., *Souroubea*); fibres septate, c. 800 µm long in *Marcgravia*, 1200–1800 µm in *Norantea* s.l. and *Souroubea*; parenchyma scanty paratracheal to vasicentric; rays homogeneous, uni- to multiseriate, cells mostly upright (de Roon 1975). *Norantea* s.l. has been reported being a caulinar myrmecophyte with a non-disintegrating pith (Jolivet 1996).

The sieve-element plastids belong to the Ss-type (Behnke 1991). The epicuticular wax platelets are unspecific (Ditsch and Barthlott 1994). An indumentum of multicellular hairs is rarely present on young shoots or the inflorescence.

INFLORESCENCE STRUCTURE. All inflorescences are racemose. In *Sarcopera* they are spike-like, whereas in *Marcgraviastrum* and *Marcgravia* the inflorescence axis is extremely contracted, which leads to umbellate inflorescences. A characteristic, autapomorphic trait of the family is the presence of extrafloral nectaries in the

Fig. 87. Marcgraviaceae. Inflorescence structure with insertion of the nectarial bracts and the degree of fusion of their stalk with the pedicel. **A** Fertile flower. **B** Nectarial bract. **C** Nectary with abortive flower. (after de Roon 1975)

inflorescence that consist of a bract fused with its petiole (Fig. 87). These nectaries are generally developed on all flowers of the inflorescence in all Marcgraviaceae except *Marcgravia*. In this genus the apical (central) flowers are sterile and provided with a nectary that is fused with the abortive pedicel, whereas the fertile flowers lack nectaries.

The nectaries are foliaceous to gibbose in *Ruyschia*; cup- or spur-shaped and often auriculate in *Souroubea*; cup-, sac,- or pitcher-shaped in *Norantea*, *Sarcopera*, *Schwartzia*, and *Marcgraviastrum*; and tubular-clavate to pitcher- or boat-shaped in *Marcgravia*. Often they are conspicuously coloured and, from a pair of glands, secrete a sweetish liquid as an attractant for pollinators. Inflorescence type and insertion of nectaries are highly characteristic for each genus and facilitate identification.

In *Sarcopera*, the apical flowers often have fully developed nectaries, while the nectaries become reduced towards the base of the inflorescence and the lowermost flowers lack the nectaries. This feature as well as the contraction of the axis (*Marcgraviastrum*) are interpreted as advanced states. A combination of both is found in *Marcgravia*.

FLORAL STRUCTURE AND ANATOMY. The flowers are pentamerous (except *Marcgravia*). The aestivation is imbricate and the sepals, if five, are quincuncially arranged. The petals are free or somewhat fused and reflexed at anthesis. In *Marcgravia*, the corolla is tetramerous and completely fused into a calyptra that falls off at anthesis.

The stamens are arranged in one to two whorls; their number is (3–)5 in *Ruyschia* and *Souroubea*, and ranges from (5–)7–100 (e.g. in *Marcgraviastrum* and *Marcgravia*). The anthers are tetrasporangiate, dorsifixed and open introrsely by longitudinal slits. The filaments are filiform to flattened and free, rarely adnate to the base of the petals. Pentandrous flowers always have antesepalous stamens.

The ovaries are superior, (in)completely 2–15-locular and have axile placentation with the placentae proliferating into the locules. In *Marcgravia* and *Marcgraviastrum* ovules are numerous, whereas the other genera have only 5–20. The style is distinct but in *Souroubea*, *Sarcopera*, and *Marcgravia* the stigma is mostly (sub)sessile.

Sclereids are common in the pericarp and all floral tissues except the androecium. For further anatomical details see Juel (1887), Weber (1956), and de Roon (1975).

EMBRYOLOGY. Only *Marcgravia* spp. have been studied. The anther tapetum is secretory and consists of uninucleate cells. Ovules are anatropous, bitegmic and tenuinucellar, with the micropyle formed by the inner integument. The nucellus is highly reduced. The chalazal megaspore develops into a Polygonum-type embryo sac. Endosperm development is cellular and the cell closest to the micropyle develops into a non-aggressive haustorium. For further details see Davis (1966) and Johri et al. (1992).

POLLEN MORPHOLOGY. The pollen grains are mostly spheroidal (suboblate to subprolate), 3(4)-colporate and relatively small (12–31 μm long) (Fig. 88). The colpi are usually long and narrow, only some *Marcgraviastrum* and *Marcgravia* spp. have shorter colpi. *Souroubea* and *Ruyschia* differ

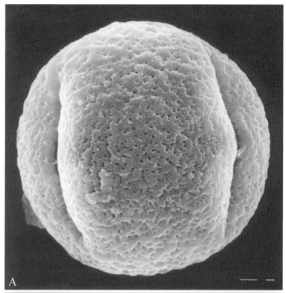

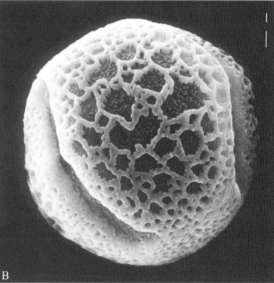

Fig. 88. Marcgraviaceae. Pollen in equatorial view, SEM (×4000). **A** *Marcgravia sintenisi*, sexine tectate-perforate. **B** *Marcgravia umbellata*, sexine coarsely reticulate. Photo S. Dressler

from the rest of the family in having an imperforate or perforate tectum whereas the other genera may also have reticulate pollen. Another palynological trait of *Souroubea* and *Ruyschia* is the comparatively thin (<2 μm) exine. Punt (1971) postulates evolutionary trends: from tectate to coarsely reticulate and tectate-perforate sexine; from long and slit-like exocolpi to shorter and broader ones; and from small, distinct endoapertures to large, lalongate endoapertures with parallel to divergent or diffuse lateral ends.

KARYOLOGY. No data on chromosome numbers are published, but preliminary studies revealed small chromosomes (0.7–2.5 μm) with semi-reticulate interphase nuclei and proximal-anterior chromosome condensing in prophase to prometaphase. For *Marcgravia evenia* ssp. *calcicola* $2n = 36$ has been determined (A.M. Benko-Iseppon, pers. comm.).

POLLINATION. The pollination system in *Ruyschia* is unknown; fly pollination was suggested (Delpino 1869; Vogel 1993). In some species of *Souroubea* the flowers emit a very strong, sweet smell and are psychophilous (Gentry in Putz and Mooney 1991). Sphingophily has also been suggested (Vogel 1993). A detailed study of *S. guianensis* (Machado and Lopes 2000) revealed "non-legitimate" visits of a hummingbird and the occurrence of "tapetal oil". *Norantea*, *Sarcopera*, and *Schwartzia brasiliensis* show large, lively coloured nectaries and bird pollination has been observed in a few species (Sazima et al. 1993). Species of *Marcgraviastrum* and *Marcgravia*, genera that for long have been regarded as excellent examples of bird-pollination, have proved to be rather chiropterophilous (Fig. 89; see also Vogel 1958; Sazima and Sazima 1980; Tschapka and von Helversen 1999; Zusi and Hamas 2001; Dressler and Tschapka 2002). These genera may have independently shifted from entomophily to vertebrate pollination. Ornithophily seems well developed in the *Galeatae* group of *Marcgravia* subgen. *Marcgravia* (syn. *Plagiothalamium*), in which red nectaries, which subtend the pseudo-umbel, ensure nototribic pollination by hummingbirds. In certain spp. of *Marcgravia* subgen. *Orthothala-*

Fig. 89. Marcgraviaceae. *Marcgravia serrae* with flower bat *Hylonycteris underwoodii* approaching for nectar bracts central in the inflorescence. (Tschapka and von Helversen 1999)

mium, erect inflorescences, many light stamens and nocturnal anthesis facilitate sternotribic pollination by phyllostomid bats. Opossum pollination was observed in *Marcgravia nepenthoides* (Tschapka and von Helversen 1999).

Although formerly reported as being generally protandrous, as observed in some *Marcgravia* spp. (Tschapka and von Helversen 1999) and *Souroubea guianensis* (Machado and Lopes 2000), stigma and anthers mature at about the same time in other taxa (e.g. *Schwartzia brasiliensis*, cf. Sazima et al. 1993). Despite their elaborate inflorescences, many Marcgraviaceae are probably autogamous and even cleistogamous (Bailey 1922), since self-pollination takes place in the bud.

FRUIT, SEED AND DISPERSAL. All Marcgraviaceae have depressed globose fruits with a persistent calyx. The pericarp is more or less coriaceous rather than woody, and unripe fruits could be described as baccate. The pericarp is reddish-green to purple or brownish and eventually dehisces irregularly or loculicidally and septifragously from the base and the apex, exposing a pulpy, purple placentar tissue with numerous small seeds. Thus, the fruits are capsular.

The seeds are flattened, hemispherical to reniform, and have a shiny, reddish, reticulate testa. They are irregularly distributed in the pulp, where often also some black aborted seeds are found. Endosperm is starchy but scanty or lacking. The embryo is straight, relatively large and provided with a short radicle and long cotyledons.

At maturity, the pericarp and the pulpy placentae are often reddish. When the fruit dehisces, the seeds, pulp, and pericarp attract birds and mammals. Toucans, aracaries and others are reported to consume the fruits (Davis and Yost 1983). Vegetative propagation by sterile shoots also plays a role.

PHYTOCHEMISTRY. Melchior (1924) reported the occurrence of inulin crystals from the leaves of two *Marcgravia* species, but Weber (1955) was unable to detect this substance in a different *Marcgravia* species and *Norantea guianensis*. Attempts to trace inulin or fructan with fructanase in several taxa were unsuccessful (Albrecht and Dressler, unpubl. data).

Unspecified terpenes, tannins, saponins, alkaloids, and phenolics including leucopelargonidin and myricetin have been reported from the family (Hegnauer 1969; Saleh and Towers 1974; Merz 1991).

RELATIONSHIPS WITHIN THE FAMILY. *Norantea* s.l., which shows a high diversity in inflorescence structure and pollen morphology, was recognised as a heterogeneous assemblage by de Roon (1975) and Bedell (1985). It was separated into four genera by de Roon and Dressler (1997).

The unspecialised inflorescence morphology and pollination system of *Ruyschia* and *Souroubea* may be plesiomorphic within the family. In contrast, characters such as heterophylly, contraction of the inflorescence axis, differentiation of sterile and fertile flowers, increase in number of stamens and ovary locules, oligomerisation of the corolla, pollination system and specialised wood anatomy characterise *Marcgravia* as a highly advanced group.

AFFINITIES. Many traits of Marcgraviaceae such as external morphology, polymorphic sclereids, pollen morphology and epicuticular waxes are also found in Theaceae/Ternstroemiaceae. Seed structure has its next equivalent in tribe Adinandreae of Ternstroemiaceae. The possession of raphide bundles is shared with Tetrameristaceae and Pellicieraceae, both long regarded as close relatives of Theaceae. In both families each locule contains only a solitary ovule and terminal inflorescences are lacking. *Pelliciera* has solitary subsessile axillary flowers with two large, coloured prophylls. Tetrameristaceae have axillary pedunculate umbelliform racemes. Recent molecular analyses place Tetrameristaceae and Pellicieraceae sister to Marcgraviaceae and all these close to Balsaminaceae, somewhat distant from Theaceae and Ternstroemiaceae in the Ericales (Savolainen et al. 2000; Anderberg et al. 2002).

DISTRIBUTION AND HABITATS. The family is confined to the Neotropics, where it extends from S Mexico to N Bolivia including the West Indian islands with higher altitudes. Members of the family are relatively rare and prefer primary vegetation ranging from humid tropical lowland forests to montane rain and cloud forests. Many species prefer higher altitudes and the genus *Ruyschia* is restricted to them. Sometimes altitudinal vicariance between species is observed.

PALAEOBOTANY. Single pollen grains of *Marcgravia* and of *Norantea* s.l. were recorded from the Lower Oligocene (c. 45 million years ago) of San Sebastian, Puerto Rico (Graham and Jarzen 1969). Lacking a detailed description, however, this record was not confirmed by Muller (1981). Fossilized wood was described as *Ruyschioxylon sumatrense* from the Tertiary of Sumatra [sic!] by Hofmann (1884).

ECONOMIC IMPORTANCE. Apart from occasional horticultural use in the tropics, the family has no economic significance. Miscellaneous applications in folk medicine have been reported, which include use against wounds, headaches, sleeplessness, rheumatism, amoebiasis, centipede stings, haemorrhaging during menstruation, diarrhoe and syphilis (Descourtilz 1827; Davis and Yost 1983; Schultes and Raffauf 1990; Putz and Mooney 1991; Merz 1991; herbarium labels). Some local tribes in Peru and Ecuador eat the fruits of *Marcgravia* ("purum hijos"= false figs) or drink the sap; Angely (1977) reports that *Schwartzia adamantia* is believed to indicate the presence of diamonds.

KEY TO THE GENERA

1. Inflorescence umbellate or subumbellate 2
- Inflorescence spicate or racemose 3
2. Inflorescence completely fertile (all flowers developed); sepals and petals 5; petals free or variously connate; leaves spiral **6. Marcgraviastrum**
- Inflorescence partly sterile (central flowers aborted with only the bracteal nectaries developed); sepals and petals 4; petals calyptrately connate; leaves distichous **7. Marcgravia**
3. Inflorescence spicate; nectaries inserted next to the sessile flowers **5. Sarcopera**
- Inflorescence racemose; nectaries variously inserted on the pedicel 4

4. Stamens 3–5; nectariferous bracts inserted at the base of
    the calyx   5
 – Stamens (5–)12 or more; nectariferous bracts inserted at
    various distances from the calyx, but never at its base   6
5. Ovary 2-locular; bracts gibbose or somewhat leaf-like,
    solid or almost so   **1. Ruyschia**
 – Ovary 3–5-locular; bracts spur-like, tubular, hollow, often
    auriculate   **2. Souroubea**
6. Inflorescence an elongated raceme, (20–)35–65 cm long;
    pedicels short, 2–5(–7) mm long; nectariferous bracts
    inserted above the middle of the pedicel, never attached at
    the base   **3. Norantea**
 – Inflorescence a short raceme, 4–25(–35) cm long; pedicels
    elongate, (20–)30–70 mm long; nectariferous bracts
    inserted below the middle of the pedicel, rarely at the
    base   **4. Schwartzia**

1. *Ruyschia* Jacq.

Ruyschia Jacq., Enum. Syst. Pl. 2: 17 (1760); de Roon, Contrib.
   tow. Monogr. Marcgraviaceae: 141–150 (1975), rev.
Caracasia Szyszył. (1893).

Climbing shrubs or lianas. Leaves shortly petiolate. Inflorescences dense multiflorous racemes
with 20–30(–50) flowers. Flowers small, shortly
pedicellate. Nectaries small, sessile, inserted at the
base of flowers or on upper half of pedicel, gibbose
to semi- or subglobose, mostly solid. Petals free or
basally slightly united. Stamens 3 or 5. Filaments ±
fused with the base of petals. Ovary 2-locular,
ovules few to up to 20, stigma capitate. Seven
spp., mostly from higher altitudes of Central
America (3), northern Andes (3), and Lesser
Antilles (1).

2. *Souroubea* Aubl.   Fig. 90

Souroubea Aubl., Hist. Pl. Guiane 1: 244 (1775); de Roon,
   Contrib. tow. Monogr. Marcgraviaceae: 151–187 (1975), rev.

Climbing shrubs or lianas, often epiphytic. Leaves
shortly petiolate. Inflorescences lax or dense
racemes with 15–60(–100) flowers. Nectaries on
the upper part of the pedicel, mostly below calyx,
sessile or sometimes stipitate, hollow, mostly spur-
like and auriculate. Flowers 5-merous (rarely 3- to
6-merous). Petals free or connate up to 2/3 of their
length. Filaments free or basally adnate to corolla.
Ovary (3–)5-locular, often pentagonal, stigma with
(3–)5 radiating lobes. Seeds rather few. Nineteen
spp., Mexico to Bolivia, not in the West Indies.

3. *Norantea* Aubl.

Norantea Aubl., Hist. Pl. Guiane 1: 554 (1775); Bedell, Generic
   rev. Marcgraviac. I. *Norantea* complex: 69–105 (1985), rev.
Norantea subgen. *Sacciophyllum* Delpino (1869).

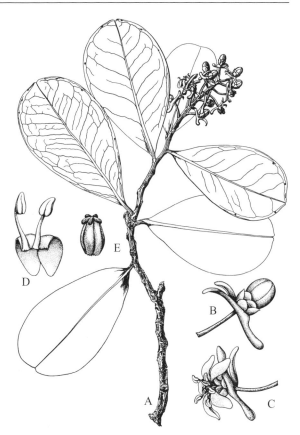

Fig. 90. *Souroubea vallicola.* **A** Habit. **B** Flower with nectary. **C**
Flower at anthesis. **D** Petals with stamens. **E** Ovary with stigma.
(de Roon 1970)

Lianas or sprawling shrubs, often epiphytic. Leaves
petiolate. Inflorescences long, dense racemes with
100–300 flowers. Flowers shortly pedicellate. Nectaries on the upper half of the pedicel, stipitate,
saccate. Sepals 5. Petals 5, free to slightly basally
connate. Stamens 20–35; filaments basally adnate
to corolla. Ovary 5-locular with 10–20 ovules per
locule. Two spp., N South America and S Brazil,
Bolivia.

4. *Schwartzia* Vell.

Schwartzia Vell., Fl. Flumin.: 221 (1829); Bedell, Generic rev.
   Marcgraviac. I. *Norantea* complex: 106–215 (1985), rev.
Norantea subgen. *Cochliophyllum* Delpino (1869).

Sprawling shrubs, occasionally small trees. Leaves
sessile to petiolate. Inflorescences short racemes
with 8–60 flowers (only *S. brasiliensis* with long
racemes of 60–300 flowers). Pedicels elongate.
Nectaries adnate to the lower third of the pedicel,
mostly stipitate, cup-, sac- or boat-shaped. Sepals
5. Petals 5, free to variously connate. Stamens

(5–)12–27 or 50–80 in one or several whorls. Filaments free or basally connate or adnate to corolla. Ovary (in)completely 3–5-locular. Fourteen spp., from Costa Rica along the Andes to Bolivia, Caribbean, E Brazil.

5. *Sarcopera* Bedell

Sarcopera Bedell in de Roon & S. Dressler, Bot. Jahrb. Syst. 119: 328 (1997); Bedell, Generic rev. Marcgraviac. I. *Norantea* complex: 226–295 (1985), rev.
Norantea subgen. *Pseudostachyum* Delpino (1869).

Sprawling shrubs and lianas, rarely small trees, often epiphytic. Leaves sessile to petiolate, occasionally asymmetrical. Inflorescences spicate with (35–)100–450 flowers. Pedicels absent (rarely short). Flowers brilliantly coloured. Nectaries petiolate, inserted to the inflorescence axis at flower base (at least flowers of the upper portion of inflorescence), cup- or sac-shaped. Sepals 5. Petals 5, free or slightly basally connate. Stamens (6–)8–25; filaments free or variously connate or adnate to corolla. Ovary completely 2-, 3- (5-)locular with 4–8(–12) ovules per locule; style lacking. About 10 spp., from Honduras through the Andean Cordillera to N Bolivia, Guayana Highlands.

6. *Marcgraviastrum* (Wittm. ex Szyszył.) de Roon & S. Dressler

Marcgraviastrum (Wittm. ex Szyszył.) de Roon & S. Dressler, Bot. Jahrb. Syst. 119: 332 (1997); Bedell, Generic rev. Marcgraviac. I. *Norantea* complex: 296–420 (1985), rev.
Norantea subgen. *Byrsophyllum* Delpino (1869).
Norantea subsect. *Marcgraviastrum* Wittm. ex Szyszył. (1893).

Sprawling shrubs or lianas, often epiphytic. Leaves sessile to petiolate. Inflorescences umbelliformly contracted racemes with (2–)5–14(–22) flowers. Flowers erect on pedicels, subtended by sessile, rarely stalked, pendulous, saccate to tubular nectaries attached to the lower part of the pedicel. Sepals 5. Petals 5, free to variously connate. Stamens 12–many; filaments mostly free with the outer whorl basally adnate to corolla. Ovary (in)completely 5–9-locular with numerous ovules per locule. Fifteen spp.; Nicaragua along Andean Cordillera to Peru and Surinam, 2 spp. E Brazilian Shield.

7. *Marcgravia* L.

Marcgravia L., Sp. Pl. 1: 503 (1753); de Roon, Ann. Missouri Bot. Gard. 57: 39–50 (1970), reg. rev.

Climbing shrubs or vines with dimorphic branches: sterile, juvenile branches creeping, appressed to the substrate, and attached by roots; fertile ones free, mostly pendulous and provided with wart-like lenticels. Leaves distichous, those of sterile branches small, thin, sessile, mostly asymmetrically cordate; those of fertile ones larger, thicker, often with a drip-tip. Inflorescences umbelliform, the apical (central) flowers abortive with only the bracts (nectaries) well developed, the fertile flowers without bracts, long-pedicellate. Nectaries tubular-saccate or boat-shaped. Sepals 4, decussate. Petals 4, connate into a deciduous cap. Stamens 6–many, free. Ovary 3–20-locular with numerous ovules per locule. Stigma capitate to umbonate. About 60 spp. in Central and South America and the West Indies.

Two subgenera (sections according to Gilg & Werdermann) are recognised: subgen. *Marcgravia* (=subgen. *Plagiothalamium* Wittm.) with flowers deflexed on the pedicels and stalked leaves, and subgen. *Orthothalamium* Delpino with flowers erect on the pedicels and mostly (sub)sessile leaves.

Selected Bibliography

Anderberg, A.A. et al. 2002. See general references.
Angely, J. 1977. Flora descritiva do Paraná, ed. 2, 10 vols. São Paulo: Centro de Pesquisas Básicas e Instituto Paranaense de Botânica.
Bailey, I.W. 1922. The pollination of *Marcgravia*: a classical case of ornithophily? Am. J. Bot. 9: 370–384.
Bedell, H.G. 1985. A generic revision of Marcgraviaceae I. The *Norantea* complex. Ph.D. Dissertation. College Park, MD: University of Maryland.
Behnke, H.-D. 1991. See general references.
Davis, G.L. 1966. See general references.
Davis, W.E., Yost, J.A. 1983. Ethnobotany of the Waorani of Eastern Ecuador. Bot. Mus. Leafl. 29: 159–211.
Delpino, F. 1869. Rivista monografica della famiglia delle Marcgraviaceae. Nuovo Giorn. Bot. Ital. 1: 257–290.
Descourtilz, M.E. 1827. Flore médicale des Antilles. 8 vols. Paris: Pichard.
Ditsch, F., Barthlott, W. 1994. Mikromorphologie der Epicuticularwachse und die Systematik der Dilleniales, Lecythidales, Malvales und Theales. Trop. Subtrop. Pflanzenwelt 88.
Dressler, S. 1997. 321. *Marcgravia umbellata*. [Curtis's] Bot. Mag., ser. 6, 14: 130–136.
Dressler, S., Tschapka, M. 2002. Bird versus bat pollination in the genus *Marcgravia* and the description of a new species. [Curtis's] Bot. Mag. VI, 19: 104–114.
Fryns-Claessens, E., van Cotthem, W. 1973. A new classification of the ontogenetic types of stomata. Bot. Rev. (Lancaster) 39: 71–138.
Gilg, E., Werdermann, E. 1925. Marcgraviaceae. In: Engler & Prantl, Die natürlichen Pflanzenfamilien, ed. 2, vol. 21. Leipzig: W. Engelmann, pp. 94–106.

Graham, A., Jarzen, D.M. 1969. Studies in neotropical paleo-botany. 1. The Oligocene communities of Puerto Rico. Ann. Missouri Bot. Gard. 56: 308–357.

Hegnauer, R. 1969. See general references.

Hofmann, H.L. 1884. Untersuchungen über fossile Kiesel-hölzer. Diss. Univ. Leipzig. Z. Naturwiss. 57: 156–195.

Howard, R.A. 1970. The ecology of an elfin forest in Puerto Rico. X. Notes on two species of *Marcgravia*. J. Arnold Arbor. 51: 41–55.

Johri, B.M. et al. 1992. See general references.

Jolivet, P. 1996. Ants and plants. An example of coevolution. Leiden: Backhuys Publishers.

Juel, H.O. 1887. Beiträge zur Anatomie der Marcgraviaceae. Kongl. Svenska Vetenskapsacad. Handl. 12, Afd. 3, Nr. 5, Bihang: 1–28.

Machado, I.C., Lopes, A.V. 2000. *Souroubea guianensis* Aubl.: quest for its legitimate pollinator and the first record of tapetal oil in the Marcgraviaceae. Ann. Bot. II, 85: 705–711.

Melchior, H. 1924. Über das Vorkommen von Inulin in den Blättern der Marcgraviaceen. Ber. Deutsch. Bot. Ges. 42: 198–204.

Merz, K.S. 1991. Untersuchungen an tropischen epiphytischen Lebensgemeinschaften: Biomasse, Wasserretention, Natur-stoffe. Dissertation. Heidelberg: Universität Heidelberg.

Muller, J. 1981. Fossil pollen records of extant angiosperms. Bot. Rev. (Lancaster) 47: 1–142.

Nandi, O.I., Chase, M.W., Endress, P.K. 1998. A combined cladistic analysis of angiosperms using *rbc*L and non-molecular data sets. Ann. Missouri Bot. Gard. 85: 137–212.

Punt, W. 1971. Pollen morphology of the genera *Norantea*, *Souroubea*, and *Marcgravia* (Marcgraviaceae). Pollen Spores 13: 199–232.

Putz, F.E., Mooney, H.A. (eds.) 1991. The biology of vines. Cambridge: Cambridge University Press.

Roon, A.C. de 1967. Foliar sclereids in the Marcgraviaceae. Acta Bot. Neerl. 15: 585–628.

Roon, A.C. de 1970. Flora of Panama, VI. 121. Marcgraviaceae. Ann. Missouri Bot. Gard. 57: 29–50.

Roon, A.C. de 1975. Contributions towards a monograph of the Marcgraviaceae. Ph.D. Thesis. Utrecht.

Roon, A.C. de, Dressler, S. 1997. New taxa of *Norantea* Aubl. s.l. (Marcgraviaceae) from Central America and adjacent South America. Bot. Jahrb. Syst. 119: 327–335.

Saleh, N.A.M., Towers, G.H.N. 1974. Flavonol glycosides of *Norantea guianensis* flowers. Phytochemistry 13: 2012.

Savolainen, V. et al. 2000. See general references.

Sazima, M., Sazima, I. 1980. Bat visits to *Marcgravia myriostigma* Tr. et Pl. (Marcgraviaceae) in southeastern Brazil. Flora 169: 84–88.

Sazima, I., Buzato, S., Sazima, M. 1993. The bizarre inflores-cence of *Norantea brasiliensis* (Marcgraviaceae): visits of hovering and perching birds. Bot. Acta 106: 507–513.

Schofield, E.K. 1968. Petiole anatomy of the Guttiferae and related families. Mem. New York Bot. Gard. 18: 1–55.

Schultes, R.E., Raffauf, R.F. 1990. The healing forest. Portland: Dioscorides Press.

Tschapka, M., von Helversen, O. 1999. Pollinators of syntopic *Marcgravia* species in Costa Rican lowland rainforest: bats and opossums. Plant Biol. 1: 382–388.

Vogel, S. 1958. Fledermausblumen in Südamerika. Oesterr. Bot. Z. 104: 491–530.

Vogel, S. 1993. Floral adaptive radiation in the Marcgraviaceae (abstr.). Sprengel Symposium Berlin-Spandau 1993: 21.

Weber, H. 1955. Haben die Marcgraviaceen "Inulinblätter"? Ber. Deutsch. Bot. Ges. 68: 408–412.

Weber, H. 1956. Über die Blütenstände und die Hochblätter von *Norantea* Aubl. (Marcgraviaceae). Beitr. Biol. Pflanzen 32: 313–329.

Zusi, R.L., Hamas, M.J. 2001. Bats and birds as potential polli-nators of three species of *Marcgravia* lianas on Dominica. Caribbean J. Sci. 37: 274–278.

Myrsinaceae

B. STÅHL and A.A. ANDERBERG

Myrsinaceae R. Br., Prodr.: 399 (1810).
Aegicerataceae Blume (1833).
Coridaceae J. Agardh, Theoria Syst. Pl.: 332 (1858).

Perennial or sometimes annual herbs, subshrubs, shrubs, trees or lianas; secretory cavities with a red or brownish content often present in vegetative and reproductive parts; indumentum of multicellular, branched or unbranched trichomes, peltate and capitate trichomes common. Leaves simple, exstipulate, alternate, sometimes opposite or whorled, if alternate then sometimes in a basal rosette or condensed into pseudowhorls; margin entire or variously serrate. Inflorescences of terminal or axillary racemes, if axillary then sometimes reduced into fascicles or solitary flowers. Flowers 5-merous, sometimes 3- (*Cybianthus* subg. *Triadophora*, *Pelletiera*), 4- (e.g. *Cybianthus*), or 7–9-merous (*Trientalis*), hermaphrodite or unisexual, if unisexual then plants mostly dioecious; calyx herbaceous, rotate to campanulate, rarely hypocrateriform; corolla rotate, campanulate, urceolate or rarely absent (*Glaux*), if rotate then often divided to near base or with free lobes; stamens in one whorl, antepetalous, filaments partly to entirely fused with the corolla, sometimes united to form a temporary or permanent tube; anthers tetrasporangiate, dithecal, introrse, opening by longitudinal slits or apical pore-like slits, rarely transversely septate, free or rarely fused, if free then often connivent and forming a protruding cone; gynoecium syncarpous, 5-carpellate; ovary superior, unilocular, with a free central placental column; style short to long, sometimes exserted; stigma punctate, truncate to capitate, or rarely discoid; ovules few to many, in one to several series and often immersed into the placenta, anatropous and bitegmic, or campylotropous and unitegmic (*Cyclamen*), tenuinucellate. Fruits fleshy, often berry-like drupes (mainly tropical groups) or capsules opening with valves, a circular lid, or by disintegration (mainly temperate groups), rarely viviparous (*Aegiceras*). Seeds of capsular fruits mostly angular, papillose, reticulate or rugose, those of drupaceous fruits subglobose, with or without immersed hilum area; endosperm abundant, in seeds from drupaceous fruits sometimes ruminate; embryo straight to curved, hypocotyl short or long, cotyledons short and narrow.

An almost cosmopolitan family of 49 genera and about 1500 species.

CHARACTERS OCCURRING IN RELATIVELY FEW GENERA AND SPECIES. Anthers transversely septate in *Aegiceras* and a few species of *Ardisia*. Flowers apetalous in *Glaux*, choripetalous in *Embelia* and *Heberdenia*; zygomorphic in *Coris*; 3-merous in *Cybianthus* subg. *Triadophora* and *Pelletiera*; 7-merous in *Trientalis*. Fruits viviparous in *Aegiceras*; operculate in *Anagallis*. Leaves with subepidermal fibre strands in *Cybianthus* subg. *Triadophora*. Most species of *Embelia* and *Grenacheria* scandent. Tuberous hypocotyl in *Cyclamen*. *Grammadenia* is mostly epiphytic.

VEGETATIVE MORPHOLOGY. Most temperate genera are perennial herbs, although some are annuals (*Asterolinon*, *Pelletiera*, some *Anagallis*) or subshrubs (*Coris* and some *Lysimachia*), whereas the tropical genera are mostly shrubs or small to medium-sized trees. Among the tropical groups, however, some are herbaceous representatives, such as *Labisia* and *Ardisia* subg. *Bladhia*. Some tropical Myrsinaceae are pachycaul (e.g. *Badula*, *Discocalyx*). In *Cyclamen*, the hypocotyl develops into a large tuber.

The leaves are mostly alternate, and in some groups they are condensed into pseudowhorls at shoot apices. Opposite leaves characterise *Asterolinon* and *Pelletiera* as well as several species of *Anagallis* and *Lysimachia*; some species of *Lysimachia* have whorled leaves. In most taxa the leaves are short-petiolate, an exception being *Grammadenia* with sessile leaves and *Ardisiandra* and *Cyclamen* with mostly long-petiolate leaves. In the latter two genera the leaf blade is distinct from the petiole and has a palmate venation. Otherwise, most members of the family have oblanceolate to oblong leaf blades. The leaf margins vary from entire to serrulate or serrate. In *Amblyanthus*, *Amblyanthopsis* and *Ardisia* subg. *Crispardisia*, the leaf margins are crenulate, a feature related to bacterial symbiosis (Lersten and Horner 1976). The leaves of many species of *Cyclamen* have cartilaginous outgrowths marginally, but these are not homologous with the leaf-teeth of other genera. The young, developing leaves are typically folded inwards (involute or conduplicate), except in *Ardisiandra* which has revolute vernation.

VEGETATIVE ANATOMY. Conspicuous secretory cavities, appearing as reddish brown or black dots and streaks, are present in vegetative and reproductive parts of virtually all tropical Myrsinaceae, but occur also in *Coris* and several species of *Lysimachia*. Extraxylary foliar bundles of unbranched fibres like those of Theophrastaceae are lacking in the Myrsinaceae, but sclereids arranged in a similar way adjacent to the leaf epidermis occur in *Cybianthus* subg. *Triadophora* (Grosse 1908). Stone cells occur in the mesophyll of many tropical members of the family, being particularly common in *Tapeinosperma*.

Multicellular trichomes occur in great variety, both glandular and non-glandular (Grosse 1908). Immersed trichomes with a short, one- or few-celled stalk and multicellular head are common in tropical groups, a feature shared among related families only by Theophrastaceae. Likely homologous to these immersed trichomes are the scales present in several tropical genera. In *Cybianthus* subg. *Conomorpha*, these scales often cover major parts of the plant giving it a brownish colour. Small but protrusive, capitate trichomes are also present on the inner surface of the corolla in many woody Myrsinaceae, being particularly abundant on the corollas of *Embelia*, *Heberdenia*, *Loheria*, *Pleiomeris* and many *Cybianthus*, in which the petal lobes often are more or less papillose. Articulated hairs are found in *Ardisiandra* and temperate genera except *Cyclamen*, and non-articulated, unbranched or irregularly branched trichomes in many other genera. *Parathesis* is characterised by stellate trichomes, which occur on both young shoots and floral parts. Short-lived, secretory trichomes in the buds of some *Ardisia* subg. *Crispardisia* evidently function as colleters and are important for the bacterial symbiosis in this group (Lersten 1977).

According to Metcalfe and Chalk (1950), the wood is characterised by rather small vessels arranged in radial multiples of 3 or 4 cells and in irregular groups; perforation plates are simple or scalariform. The parenchyma is paratracheal and usually sparse. The rays are typically multiseriate and although some uniseriate rays may occur, the mixture of uni-and multiseriate rays present in

Maesaceae does not occur in the Myrsinaceae. In lignified Hawaiian species of *Lysimachia* the wood lacks rays, but has septate fibres with small simple pits (Carlquist 1992).

INFLORESCENCE STRUCTURE. The inflorescence in the Myrsinaceae is basically racemose, being developed as a terminal or axillary raceme, spike, compound raceme, pseudo-umbel or fascicle (Mez 1902). In addition, several herbaceous genera and species have solitary flowers in the leaf axils. In contrast to Maesaceae, prophylls are lacking.

FLOWER STRUCTURE. The flowers are hypogynous and obhaplostemonous. The corolla is usually open and rotate or campanulate (see Fig. 91). A few genera deviate with respect to corolla shape: *Coris* with a two-lipped, zygomorphic corolla, and *Emblemantha*, *Sadiria* and *Synardisia* with urceolate corollas. Generally the corolla is sympetalous, but in *Embelia* and *Heberdenia* the lobes are distinct and in many other genera they are united at the very base only. *Asterolinon* and *Pelletiera* have inconspicuous petals which are much shorter than the calyx. In *Glaux*, the corolla is lacking and the synsepalous calyx is pink and showy. Usually, corolla aestivation is either imbricate, sometimes quincuncial, or twisted. However, *Lysimachia* subg. *Seleucia*, *Parathesis*, and many species of *Cybianthus* subg. *Conomorpha* have valvate corollas. *Labisia* has a conduplicate corolla aestivation with the margins of each lobe clasping one of the anthers. Generally the corolla lobes are entire but emarginate lobes occur in flowers of *Amblyanthus*, *Discocalyx*, *Monoporus*, *Oncostemum*, and *Wallenia*. In most genera, the corolla lobes are spreading or erect but in *Aegiceras*, *Cyclamen*, *Geissanthus* and many *Ardisia* the lobes are reflexed at anthesis, and in *Parathesis* they are recurved.

The stamens, which develop from the same primordium as the petals (Sattler 1962), are placed in front of these. In many genera the stamens are more or less free and connate only at the very base near the receptacle, whereas in others the filaments are partly or entirely fused. In addition, the filaments or filament ring are often fused with the corolla for a longer or shorter distance. Several genera have more or less exserted stamens but this depends also on how much the petals are reflexed. In some cases this feature is combined with the occurrence of versatile anthers, e.g. in *Aegiceras* and *Wallenia*. In *Cyclamen*, *Lysimachia*, *Ardisia* and several other genera, the filaments are shorter than the anthers, which are connivent to form a

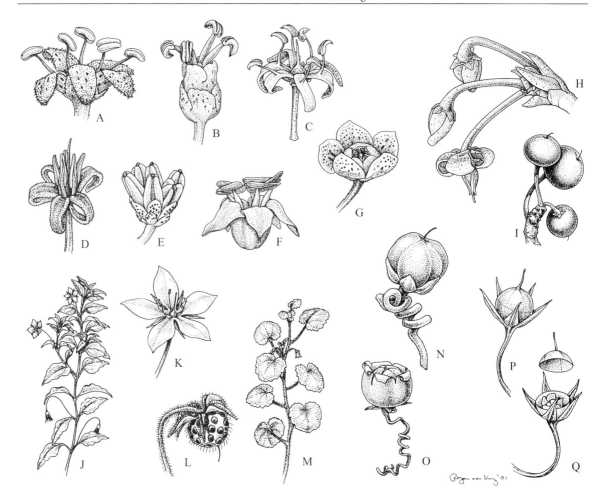

Fig. 91. Myrsinaceae. **A** *Embelia philippinensis*, flower. **B** *Wallenia laurifolia*, flower. **C** *Heberdenia bahamensis*, flower. **D** *Parathesis cubana*, flower. **E** *Myrsine africana*, flower. **F** *Aegiceras corniculatum*, flower. **G** *Oncostemum* sp., flower. **H, I** *Ardisia crenata*. **H** Inflorescence. **I** Fruit. **J, K** *Lysimachia nemorum*. **J** Habit. **K** Flower. **L, M** *Ardisiandra wettsteinii*. **L** Fruit. **M** Habit. **N, O** *Cyclamen hederifolium*, unopened and dehisced fruit. **P, Q** *Anagallis arvensis*, same. Drawn by Pollyanna von Knorring

prominent cone. In *Amblyanthus*, *Conandrium* and *Hymenandra*, the anthers are laterally connate to form a permanent cone surrounding the ovary. In *Oncostemum*, both the filaments and anthers are fused to form a ring-like, tubular structure. The anthers are introrse, have a fibrous endothecium and usually open with longitudinal slits. In some groups (e.g. *Grammadenia*, *Monoporus*, and species of *Lysimachia*) the anthers open by apical pores. *Aegiceras* and some Asiatic *Ardisia* (Stone 1993) have transversely septate, locellate anthers. The anthers vary in shape from short, globose or almost square (e.g. *Asterolinon*, *Cybianthus*) to

narrowly (*Cyclamen*, *Hymenandra* and some species of *Ardisia*) or triangularly sagitatte.

Staminode structures present in related families are lacking in Myrsinaceae. The alleged staminodes of *Lysimachia* subg. *Seleucia* (= *Stieronema*) lack vascular strands and may be protuberances rather than staminodes.

The ovary is unilocular and evidently syncarpous, although it is initiated from a ring-shaped primordium. The ovules vary from few to numerous and are arranged in one or several series on the placenta; they are immersed into the placental tissue, except in *Ardisiandra*. The stigma is punctate, truncate or capitate, rarely discoid. In the short-styled genera *Myrsine* and *Rapanea*, the stigma is often irregularly shaped, e.g. fimbriate or ligulate.

EMBRYOLOGY. Although data from many, particularly tropical groups are lacking, the embryology of the Myrsinaceae seems to agree largely with that of other primulalean families (Davis 1966; Johri et al. 1992). The anther tapetum is secretory with uninucleate cells. Pollen is binucleate when shed. The ovules are anatropous, tenuinucellate and usually bitegmic with the micropyle formed by both integuments. The inner integument forms an endothelium, except in *Coris*, *Cyclamen*, *Aegiceras*, *Ardisia*, and *Myrsine*. The archespore is multicellular and the embryo sac is of the Polygonum type with ephemeral antipodes, but lacks endosperm haustoria. The seeds are albuminous with nuclear endosperm formation, excepting *Aegiceras* which has viviparous fruits and exalbuminous seeds. Embryogeny follows the caryophyllad type except in *Trientalis*, which seems to have an onagrad type of embryogeny. The embryo is short, or long and curved, the latter condition being found in taxa with drupaceous fruits; the cotyledons are generally short and narrow. The embryo is dicotyledonous except in *Cyclamen*, in which the development of one of the cotyledons is suppressed. Polyembryony has been recorded in species of *Ardisia*, *Parathesis*, and *Rapanea* (Braun 1859; Anderberg and Ståhl 1995).

Aegiceras and *Cyclamen* differ from other genera by having unitegmic ovules. However, during development the inner integument is partly resorbed, at least in *Cyclamen*, indicating that this feature is a reduction from the prevailing condition in the family.

POLLEN MORPHOLOGY. The pollen grains of Myrsinaceae are oblate-spheroidal to almost spherical, and 10–45 μm in diameter (Erdtman 1952). Most taxa have 3-colporate pollen but 4-colporate grains have been reported from Hawaiian *Lysimachia* (subg. *Lysimachiopsis*), *Myrsine*, and *Rapanea*, and pantocolporate grains are known from *Lysimachia* subg. *Idiophyton* (Nowicke and Skvarla 1977; Ståhl 1996; Hope 2000).

KARYOLOGY. Particularly the herbaceous members of the Myrsinaceae show a great variability in chromosome number, but numbers based on x = 9, 10, 11, and 12 are common. At least in *Lysimachia* and *Trientalis* polyploidy seems to be frequent. In woody, tropical members of the family $2n = 46$ is by far most commonly reported chromosome number. However, $2n = 96$ reported in some species of *Ardisia* (Faure 1968) suggests some variability in the woody groups as well.

POLLINATION AND REPRODUCTIVE SYSTEMS. Most Myrsinaceae appear to be pollinated by insects, notably bees and flies. However, although species of *Anagallis* and *Elingamita* have nectar-producing trichomes on floral parts, most members seem to have dry, nectarless flowers with only pollen offered as a reward. A structure similar to a nectar disc is present in *Ardisiandra* but does not seem to produce any nectar. A unique feature is the oil-producing glandular hairs on the petals and stamen filaments of certain *Lysimachia* species. These species do not produce nectar, and Vogel (1986) showed that they are pollinated by a specific genus of oil-collecting bees, *Macropis*. Some genera have flowers with a protruding anther cone and these are likely to be buzz-pollinated by pollen-gathering bees, as was shown by Pascarella (1997a) for *Ardisia escallonioides*. Autogamy has been reported for several species of *Ardisia* (Pascarella 1997b), and at least *Cyclamen* shows an increased tendency for selfing if not visited by pollinators. Anemophily has been reported in *Rapanea* (Otegui and Cocucci 1999).

FRUIT AND SEED. With the present circumscription, both single-seeded drupes and many- to few-seeded capsules are present in the Myrsinaceae. Drupaceous fruits characterise all woody tropical and subtropical groups except *Aegiceras*, the fruits of which are viviparous. When ripe, the drupes are red to black and provided with a variously thick and juicy mesocarp and a woody but usually brittle endocarp. They are often somewhat compressed and usually less than 1 cm wide. Somewhat larger fruits occur in *Elingamita* and *Tapeinosperma*, and in the latter genus the mesocarp is more or less fibrous. In *Fittingia*, which also has comparatively large fruits, the soft mesocarp forms a distinct wing in herbarium specimens.

More or less globose capsules characterise the remaining, chiefly herbaceous genera. In most species the capsules open with apical valves or crumble irregularly, but in *Anagallis* the capsules open with a circular lid. The valvate capsule in *Trientalis* has a papery wall that falls off early, leaving the seeds on the placenta. A similar capsule is found also in certain species of *Lysimachia*.

The seeds of all Myrsinaceae except *Aegiceras* have copious endosperm. Seeds in taxa with drupaceous fruits are more or less globose and have an immersed hilum area forming a concavity at the base of the seed; the testa is thin and devoid of crystals and the endosperm is often ruminate with irregularly thickened cells walls (Anderberg and Ståhl 1995). In genera with capsular fruits the seeds are many, except in *Asterolinon* and *Pelletiera* which have few seeds; the seeds are normally angular with smooth, reticulate or papillose surfaces, and the hilum is more or less flush with the surface; the testa is rather thick, usually distinctly two-layered and generally provided with rhomboid crystals. The seeds of *Asterolinon linum-stellatum* and *Pelletiera* deviate in being coarsely rugose, and by having a distinctly immersed, concave hilum area. Seeds of *Trientalis* have a white outer reticulum covering the seeds. Some *Lysimachia*, e.g. *L. vulgaris*, have an outer porous, almost spongy tissue covering the seed surface, and in Neotropical *Lysimachia* subg. *Theopyxis* the seeds are winged (Ståhl 1990). The endosperm is composed of cells with smooth and evenly thickened walls except in *Cyclamen*, which has irregularly thickened endosperm cell walls provided with distinct narrow constrictions.

DISPERSAL. Most groups with drupaceous fruits are probably dispersed by birds and other fruit-eating animals. In groups with capsular fruits the seeds are released ballistically through the opening of the capsule. However, in most *Cyclamen* the fruit is pulled to the ground by coiling of the pedicel, and upon drying the capsule opens near the ground to expose the seeds, the coat of which contains edible substances that attract dispersing insects.

REPRODUCTIVE BIOLOGY. Most Myrsinaceae have perfect, bisexual flowers, but many of the tropical and subtropical woody genera are dioecious. In addition, due to incomplete separation of sexes or incomplete reductions of reproductive organs, many woody genera are often referred to as polygamous or cryptically dioecious. Vegetative propagation is known in many species of *Anagallis*, *Ardisia* subg. *Bladhia* and *Lysimachia*, which are prostrate and root at the nodes. Propagation by means of root suckers has been reported in *Myrsine* (Burrows 1999), and may also occur in other shrubby taxa. *Trientalis* propagates with rhizomes ending in a bulb-like winter bud, and a similar mode is known from *Lysimachia thyrsiflora* and *Glaux*.

PHYTOCHEMISTRY. Similarly to the other primulalean families, most Myrsinaceae produce saponins. Calcium oxalate in the form of druses and simple crystals are present in the leaves of most tropical woody genera. The seeds store oil and amylose, but no starch (Hegnauer 1969).

SUBDIVISION AND RELATIONSHIPS WITHIN THE FAMILY. Myrsinaceae form the largest family of the order Primulales of earlier classifications (e.g. Cronquist 1988). It is sister to Primulaceae, the two being the sister group of Theophrastaceae.

The present circumscription of Myrsinaceae is based on phylogenetic research by Anderberg and Ståhl (1995), Anderberg et al. (1998), Källersjö et al. (2000), and Anderberg et al. (2001) and differs in two important aspects from most previous classifications. Firstly, the genus *Maesa*, which often has been treated as a subfamily of Myrsinaceae, had to be transferred to a family of its own (Anderberg et al. 2000), and secondly, several genera previously treated as members of Primulaceae have been included in Myrsinaceae as circumscribed here. It should also be noted that *Coris* and *Aegiceras*, two genera that in some earlier classifications were treated in separate families, Coridaceae and Aegicerataceae respectively, are now included in Myrsinaceae. In the classification of Primulaceae by Pax and Knuth (1905), the genera now transferred to Myrsinaceae belonged to four different tribes: Androsaceae (*Ardisiandra*, *Stimpsonia*), Lysimachieae (*Anagallis*, *Asterolinon*, *Glaux*, *Lysimachia*, *Pelletiera*, *Trientalis*), Cyclamineae (*Cyclamen*), and Corideae (*Coris*). In the analysis of Källersjö et al. (2000), based on a combination of morphological and cpDNA sequence data, *Coris* is placed as sister to the rest of Myrsinaceae, with *Ardisiandra* being placed as the sister group to a more inclusive group formed by three clades: (1) the genera of the erstwhile Lysimachieae, (2) *Cyclamen*, and (3) the genera of Myrsinaceae s. str. (except *Maesa* but including *Aegiceras*). The analyses by Källersjö et al. (2000) suggest that both *Anagallis* and *Glaux* are nested within *Lysimachia* as presently circumscribed. Among woody tropical Myrsinaceae, generic alignments are also rather unclear, although many genera with few ovules in a single series on the placenta tend to appear in the same general clade (Ståhl 1996).

DISTRIBUTION AND HABITATS. Myrsinaceae have a worldwide distribution. The herbaceous genera are distributed mainly in (cool) temperate regions, whereas the woody genera are mainly tropical. However, several herbaceous or suffrutescent representatives occur in tropical or subtropical areas, and many woody genera show high species diversity in tropical-montane habitats. *Coris* and *Cyclamen* are chiefly distributed in the Mediterranean region and North Africa; *Glaux* is confined to grassy seashores in the North Temperate zone. The largest genera, *Ardisia* and *Rapanea*, are pantropical, occurring in lowland as well as in montane forests. Some genera have more restricted distributions, like *Heberdenia* and *Pleiomeris*, endemic to Macaronesia, *Sadiria* (Assam-Bhutan), *Solonia* (Cuba), and *Vagaea* (Hispaniola). Among the genera with regionally restricted distributions should be mentioned *Ardisiandra* (East Africa), *Wallenia* (West Indies), and *Oncostemum* (Madagascar). The most widely distributed species is found in *Anagallis*, viz. *A. arvensis* and *A. foemina*, two weedy, cosmopolitan species with a European/Mediterranean origin, *A. pumila*, with a pantropical (submontane) distribution, and *A. minima* which is cosmopolitan.

FOSSIL RECORD. Pollen of the *Myrsine* type appears in the Oligocene of Australia and New Zealand (Mildenhall 1980; Martin 1994).

KEY TO THE GENERA

1. Herbs or subshrubs, rarely shrubs; fruit dry, capsular; seeds usually angular   2
   - Shrubs, trees or lianas, sometimes subshrubs; fruit ± fleshy, drupaceous (or rarely viviparous); seeds subglobose   11
2. Flowers with corolla and calyx   3
   - Flowers with calyx only   **8. *Glaux***
3. Flowers regular, actinomorphic   4
   - Flowers two-lipped, distinctly zygomorphic   **1. *Coris***
4. Corolla-lobes conspicuously reflexed   **10. *Cyclamen***
   - Corolla-lobes spreading or erect, never reflexed   5
5. Capsule opening with a circular lid   **9. *Anagallis***
   - Capsule opening with valves or irregularly   6
6. Corolla much smaller than calyx   7
   - Corolla equal to, or larger than calyx   8
7. Petals and capsule five-merous   **6. *Asterolinon***
   - Petals and capsule three-merous   **7. *Pelletiera***
8. Flowers 7–9-merous; flowers white; leaves in a terminal pseudowhorl; seeds with white reticulum, persistent
                                                                                        5. *Trientalis*
   - Flowers usually 5-merous; flowers white, pink or yellow; leaf arrangement variable; seeds not with loose white reticulum   9
9. Leaves coarsely dentate   10
   - Leaf margins entire   **4. *Lysimachia***
10. Flowers in axillary fascicles   **3. *Ardisiandra***
    - Flowers solitary in the shoot apex   **2. *Stimpsonia***
11. Lianas or scandent shrubs   12
    - Plants not scandent   13
12. Corolla lobes free to base   **11. *Embelia***
    - Corolla lobes united to form a distinct tube
                                                                                      12. *Grenacheria*
13. Flowers arranged in fascicles   14
    - Flowers arranged in panicles, racemes or umbel-like inflorescences   17
14. Ovules many, in 2–4 series on the placenta (Macaronesia)
                                                                                                                    15
    - Ovules few, in a single series on the placenta   16
15. Corolla lobes free to base   **13. *Heberdenia***
    - Corolla lobes united to form a distinct tube
                                                                                        14. *Pleiomeris*
16. Style well demarcated; stigma discoid or discoid-fimbriate
                                                                                                15. *Myrsine*
    - Style very short and inconspicuous; stigma variously shaped (ligulate, capitate), but not discoid   **16. *Rapanea***
17. Style two to several times longer than the ovary, often slender; stigma mostly punctate (without obvious stigmatic surface)   18

- Style shorter than or slightly longer than the ovary, not slender; stigma usually truncate, capitate, or discoid   33
18. Filaments (or filament tube) longer than the anthers   19
    - Filaments shorter or as long as the anthers   23
19. Anthers versatile   20
    - Anthers not versatile   22
20. Fruits viviparous, mangrove plants   **32. *Aegiceras***
    - Fruits not viviparous   21
21. Filaments fused into a tube, free towards apex
                                                                                                22. *Solonia*
    - Filaments not fused   **23. *Geissanthus***
22. Flowers bisexual; ovules in two or several series
                                                                                                26. *Gentlea*
    - Flowers unisexual; ovules 1-seriate   **27. *Stylogyne***
23. Corolla tube about as long as the lobes, or longer   24
    - Corolla tube much shorter than the lobes   26
24. Ovules in two or more series; style very slender (Mesoamerica)   **19. *Synardisia***
    - Ovules 1-seriate; style comparatively thick (Asia)   25
25. Subshrubs; leaf blades decurrent   **24. *Emblemantha***
    - Shrubs or small trees; leaf blade not decurrent
                                                                                                25. *Sadiria*
26. Corolla lobes valvate in bud, stellate- or dendroid-pubescent   **31. *Parathesis***
    - Corolla lobes imbricate or contorted in bud, not stellate- or dendroid-pubescent   27
27. Corolla twisted to the left   **30. *Antistrophe***
    - Corolla twisted to the right   28
28. Flowers 4-merous; ovules 1-seriate   **21. *Tetradisia***
    - Flowers typically 5-merous; ovules in one or more series
                                                                                                                    29
29. Ovules 1-seriate, erect at the base of the placenta   30
    - Ovules in two or more series, if 1-seriate then with ovules inserted around the middle of the placenta   31
30. Inflorescences long-pedunculate, ovules 10–15
                                                                                      28. *Ctenardisia*
    - Inflorescences without long peduncles; ovules 6–8
                                                                                        29. *Yunckeria*
31. Anthers connate, forming an anther tube
                                                                                      20. *Hymenandra*
    - Anthers often coherent at the beginning of anthesis but not connate   32
32. Ovules 1-seriate; inflorescences of axillary, bracteate clusters (African)   **18. *Afrardisia***
    - Ovules mostly in two or more series; inflorescences axillary or terminal, usually appearing as panicles
                                                                                                17. *Ardisia*
33. Free parts of filaments as long as or longer than the anthers, at least in male flowers   34
    - Free parts of filaments shorter than the anthers, or anthers sessile or fused   39
34. Leaf margins crenate; petals emarginate   35
    - Leaf margins entire or serrate to serrulate; petals emarginate or not   36
35. Anthers connate, forming a cone   **33. *Amblyanthus***
    - Anthers coherent but not connate   **34. *Amblyanthopsis***
36. Corolla cup-shaped, shorter than the calyx
                                                                                      35. *Elingamita*
    - Corolla with well-developed lobes, longer than the calyx
                                                                                                                    37
37. Corolla usually ± papillose within, tube shorter than the lobes   38
    - Corolla not papillose within (or near base only), tube as long as lobes or longer   **36. *Wallenia***
38. Flowers borne in panicles on short lateral shoots subtended by numerous bracts (Malesia)   **37. *Loheria***

– Inflorescences variable but not as above (Neotropics)
38. *Cybianthus* p.p.
39. Ovules 2- or 3-seriate, anthers connate   **49. *Conandrium***
– Ovules 1- or rarely 2-seriate, and if 2-seriate, then anthers not connate   40
40. Leaves sessile; epihytic shrubs   **39. *Grammadenia***
– Leaves petiolate; terrestrial plants   41
41. Plants suffrutescent or herbaceous; petals conduplicate and largely concealing the anthers   **45. *Labisia***
– Plants woody; petals not conduplicate   42
42. Leaves sclerophyllous, linear (Hispaniola)   **40. *Vegaea***
– Leaves not linear   43
43. Filaments and anthers united into a distinct ring or tube
41. *Oncostemum*
– Stamens not united into a tube   44
44. Flowers bisexual   45
– Flowers unisexual   47
45. Anthers sessile or subsessile   **42. *Badula***
– Stamens with well-developed filaments   46
46. Anthers dorsifixed, style longer than the ovary
43. *Tapeinosperma*
– Anthers broadly basifixed, style shorter than the ovary
44. *Discocalyx*
47. Corolla papillose on inside   **38. *Cybianthus* p.p.**
– Corolla not papillose on inside   48
48. Flowers 4-merous, borne on short, densely bracteate, axillary inflorescences   **46. *Systellantha***
– Flowers 4- or 5-merous, arranged in ± lax panicles or racemes   49
49. Anthers opening by one or two apical pores
47. *Monoporus*
– Anthers opening by longitudinal slits   50
50. Fruit with spongy or fleshy exocarp (fruits "winged" in herbarium specimens); endocarp longitudinally ridged
48. *Fittingia*
– Fruit with thin exocarp; endocarp smooth
44. *Discocalyx*

Genera of Myrsinaceae

1. *Coris* L.

Coris L., Sp. Pl.: 177 (1753).

Perennial herbs, woody at the base. Leaves small, alternate, linear, flat or revolute. Flowers zygomorphic, 5-merous, in a terminal spike-like raceme; calyx campanulate, spinose, with conspicuous dark resin dots; corolla pink or white, two-lipped with a short tube; corolla-lobes bifid, imbricate in bud; stamens adnate to the tube; stigma capitate. Capsule ovoid, opening with apical valves; seeds few, papillose. $2n = 18$. Two spp., Mediterranean region, Somalia.

2. *Stimpsonia* C. Wright ex A. Gray

Stimpsonia C. Wright ex A. Gray, Mem. Amer. Acad. Arts II, 6: 401 (1858); Røsvik, Årb. Univ. Bergen Ser. Mat.-Nat. 7: 1–15 (1969); Anderberg et al., Bot. Jahrb. Syst. 123: 369–376 (2001).

Annual or biennial herb. Leaves in a basal rosette and scattered along the stem, dentate, elliptic to ovate, with articulated hairs. Flowers solitary in the leaf-axils along the stem; calyx campanulate; corolla white, hypocrateriform with a short tube, hairy; corolla-lobes spreading, emarginate; stamens adnate to the tube; stigma capitate. Capsule globose, opening with apical valves; seeds many, reticulate. One sp., *S. chamaedryoides* Wright, Asia.

3. *Ardisiandra* Hook. f.   Fig. 91L, M

Ardisiandra Hook. f., J. Proc. Linn. Soc. Bot. 7: 205 (1864); Rösvik, Årb. Univ. Bergen Ser. Mat.-Nat. 7: 1–15 (1969).

Perennial herbs. Stem prostrate, with copious articulated hairs. Leaves alternate, petiolate, coarsely dentate, revolute when young. Flowers 5-merous, solitary or in few-flowered fascicles in the leaf-axils; calyx campanulate, lobes cordate; corolla campanulate, white or yellow; corolla-lobes erect, entire, imbricate in bud; stamens connivent, filaments basally connate into a ring, anthers sagittate forming a cone; style protruding from the anther cone; stigma subcapitate. Capsule subglobose, often thin-walled; seeds many, papillose. $2n = 34$. Three spp., E African mountains.

4. *Lysimachia* L.   Fig. 91J, K

Lysimachia L., Sp. Pl.: 146 (1753); Handel-Mazzetti, Not. Roy. Bot. Gard. Edinburgh 16: 51–122 (1928), rev.; Ray, Illinois Bot. Monogr. 24: 1–160 (1956), rev. New World spp.; Huynh, Candollea 25: 267–296 (1970), pollen.
Naumburgia Moench (1802).

Perennial, biennial or annual herbs, rarely shrubs. Leaves entire, alternate, opposite, or whorled, sometimes in terminal pseudowhorls. Flowers 5-merous, in terminal racemes, thyrses, or solitary in the leaf-axils; calyx rotate to campanulate; corolla deeply lobed, rotate to campanulate, white, yellow, red, pink, or purple, twisted or valvate in bud; stamens spreading or connivent and forming a cone; stigma truncate or subcapitate. Capsule globose, opening with valves or disintegrating irregularly; seeds reticulate or papillose, sometimes winged, or with outer spongy tissue. $2n = 16$, 20, 24, 28, 30, 32, 34, 36, 40, 42, 56, 60, 68, 72, 84, 92, 96, 98, 100, 102, 108, 112. About 150 spp., worldwide.

5. *Trientalis* L.

Trientalis L., Sp. Pl.: 344 (1753).

Perennial herbs. Leaves entire, elliptic, in a terminal pseudowhorl. Flowers generally 7-merous, solitary in the leaf-axils; calyx rotate; corolla rotate, deeply lobed, white, twisted in bud; stamens spreading; stigma truncate. Capsule globose, valves caducous; seeds with white outer reticulum. $2n = 88, 90, 96, 100, 130, 160$. Two spp., North temperate regions.

6. *Asterolinon* Hoffmannsegg & Link

Asterolinon Hoffmannsegg & Link, Fl. Portug. 1: 332 (1813–1820).

Annual herbs. Leaves opposite, entire, ovate or lanceolate. Flowers inconspicuous, 5-merous, solitary in the leaf-axils; calyx rotate, deeply lobed; corolla rotate, deeply lobed, white; corolla-lobes rotundate, obtuse, twisted in bud; stamens spreading; stigma subcapitate. Capsule globose, disintegrating irregularly, or with caducous valves; seeds few and rugose, or many and papillose. $2n = 40$. Two spp., Europe and N & NE Africa.

7. *Pelletiera* A. St.-Hil.

Pelletiera A. St.-Hil., Mém. Mus. Hist. Nat. 9: 365 (1822).

Annual herbs. Leaves opposite, entire, lanceolate. Flowers inconspicuous, solitary in the leaf-axils; calyx 5-merous, rotate, deeply lobed; corolla 3-merous, rotate, deeply lobed, white; corolla-lobes lanceolate, acute; stamens spreading; stigma subcapitate. Capsule globose, disintegrating irregularly; seeds few, rugose. Two spp., South America and Macaronesia.

8. *Glaux* L.

Glaux L., Sp. Pl.: 207 (1753).

Perennial herb. Leaves small, fleshy, opposite, broadly triangular, entire. Flowers apetalous, 5-merous, solitary in the leaf-axils; calyx campanulate, pink; corolla absent; stamens spreading; stigma truncate. Capsule globose, opening with apical valves; seeds few, reticulate. $2n = 30$. One sp. (*G. maritima* L.), North temperate region.

9. *Anagallis* L.   Fig. 91P, Q

Anagallis L., Sp. Pl.: 148 (1753); Taylor, Kew Bull. 3: 321–350 (1955), African spp.
Centunculus L. (1753).

Annual or perennial herbs. Leaves ovate to linear, opposite or alternate, entire. Flowers 5-merous, solitary in the leaf-axils or in loose terminal racemes; calyx rotate, deeply lobed; corolla rotate or campanulate, deeply lobed, blue, red, pink or white, twisted in bud; stamens free, filaments often hairy; stigma truncate or subcapitate. Capsule globose, operculate; seeds papillose. $2n = 20, 22, 40, 66, 75, 80$. About 20 spp., Africa, S America, Europe.

10. *Cyclamen* L.   Fig. 91N, O

Cyclamen L., Sp. Pl.: 145 (1753); Grey-Wilson, The genus *Cyclamen* (1988); Anderberg, Kew Bull. 49: 455–467 (1994); Anderberg et al., Pl. Syst. Evol. 220: 147–160 (2000).

Perennial herbs with tuberous hypocotyl. Leaves basal, petiolate, cordate, glabrous, rather fleshy, often with pale green or white patterns. Flowers hypogynous, 5-merous, nodding, axillary, solitary; calyx campanulate; corolla campanulate, white to pink often with purple blotches or stripes; corolla-lobes entire and distinctly reflexed, twisted in bud; peduncles coiling at anthesis; stamens connivent, forming a cone; style often exserted; stigma truncate. Capsule globose, opening with apical valves. Seeds almost smooth. $2n = 20, 22, 30, 34, 48, 68, 72, 84, 96$. 19 spp., S & C Europe, Turkey, Lebanon, N Africa, Somalia.

11. *Embelia* Burm. f.   Fig. 91A

Embelia Burm. f., Fl. Ind. 62, tab. 23 (1786), nom. cons.; Mez in Pflanzenreich IV. 236: 295–332 (1902); Taton, Fl. Afr. Central, Myrsinaceae, pp. 30–49 (1980), reg. rev.; Walker, Philippine J. Sci. 73: 155–184 (1940), rev. E Asian spp.; Perrier de la Bâthie, H., Fl. Madagascar 161: 124–137 (1953), reg. rev.; Smith, J. Arnold Arb. 54: 274–278 (1973), rev. Fijian spp.; Halliday, Fl. Trop. Afr. (no vol.) 11–16 (1984), reg. rev.; Sleumer, Blumea 32: 385–394 (1987), rev. New Guinean spp.; Stone in Ng, Tree Fl. Malaya 4: 278 (1989), reg. rev.; Chen & Pipoly, Fl. China 15: 30–34 (1996), reg. rev.

Lianas, sometimes shrubs or small trees. Leaves petiolate or rarely subsessile, sometimes distichous, margins crenate or entire. Inflorescences terminal or axillary panicles, racemes or fascicles. Flowers 4–5-merous, bi- or unisexual and then the plants dioecious; corolla greenish or white, rarely reddish, rotate, the inside densely papillose, lobes free or united at base only, narrowly ovate to elliptic, imbricate in bud, usually spreading and somewhat recurved at anthesis; stamens usually exserted, filaments distally adnate to and decurrent on lower part of corolla lobes; anthers dorsi- or basifixed, broadly ovoid or oblong, sometimes recurved at anthesis, opening by longitudinal slits; ovary subglobose to ovoid, style short or somewhat longer then the ovary, stigma discoid; ovules few, 1-seriate. Fruit subglobose, sometimes depressed; endosperm ruminate or sometimes smooth. About 130 spp., Old World tropics.

12. *Grenacheria* Mez

Grenacheria Mez in Pflanzenreich IV. 236: 292–294 (1902); Sleumer, Blumea 32: 394–396 (1987), rev. New Guinean spp.

Scandent shrubs or lianas. Leaves short-petiolate, margins entire. Inflorescences axillary racemes forming synflorescences at stem apices. Flowers 5-merous, unisexual, the plants dioecious; corolla whitish, rotate or short-campanulate, densely papillose on inside, lobes ovate, imbricate in bud; stamens included, filaments partly fused with and decurrent on the corolla tube; anthers dorsifixed, ovoid, opening by longitudinal slits; ovary ovoid, tapering into a style of ± equal length, stigma discoid; ovules few, 1-seriate. Fruit globose; endosperm smooth. About six spp., Malesia.

13. *Heberdenia* Banks ex A. DC.   Fig. 91C

Heberdenia Banks ex A. DC., Ann. Sci. Nat. II, 16: 79 (1841); de Wit, Bull. Jard. Bot. Bruxelles 27: 233–242 (1957); Ståhl, Bot. J. Linn. Soc. 122: 315–333 (1996).

Small trees. Leaves short-petiolate, margins entire. Inflorescences axillary fascicles, appearing just below the leaves. Flowers 5-merous, bisexual; corolla pale green, rotate, the inner surface papillose, lobes narrowly oblong, free to base, imbricate in bud, recurved at anthesis; stamens exposed at anthesis; filaments long, at base adnate to and decurrent on corolla lobes; anthers dorsifixed, versatile, narrowly sagittate, recurved at anthesis, opening by longitudinal slits; ovary broadly ovoid, style rather long, stigma punctate; ovules many, 2–4-seriate. Fruit subglobose; endosperm ruminate. One sp., *H. bahamensis* (Gaertn.) Sprague, Macaronesia.

14. *Pleiomeris* A. DC.

Pleiomeris A. DC., Ann. Sci. Nat. II, 16: 79 (1841); Mez in Pflanzenreich IV. 236: 337–338 (1902); de Wit, Bull. Jard. Bot. Bruxelles 27: 233–242 (1957).

Small trees or shrubs. Leaves short-petiolate, margins entire. Inflorescences axillary, few-flowered fascicles. Flowers 5-merous, bisexual; corolla greenish, papillose on inside, rotate with well-developed tube, lobes narrowly ovate, subvalvate in bud; stamens enclosed, filaments largely

fused to corolla tube; anthers basifixed, broadly sagittate, opening through longitudinal slits; ovary ovoid, tapering into a rather slender style equal or slightly longer than the ovary, stigma punctate; ovules 15–20, 2- or 3-seriate. Fruit globose; endosperm ruminate. One sp., *P. canariensis* (Willd.) A. DC., Macaronesia.

15. *Myrsine* L.   Fig. 91E

Myrsine L., Sp. Pl. 196 (1753); Mez in Pflanzenreich IV. 236: 338–342 (1902); Taton, Fl. Afr. Centr., Myrsinac. 49–52 (1980), reg. rev.; Halliday, Fl. Trop. E. Afr., Myrsinac. 6–8 (1984), reg. rev.; Larsen & Hu, Fl. Thailand 6(2): 163–164 (1996), reg. rev.

Shrubs or small trees. Leaves petiolate, margins serrate or entire. Inflorescences axillary, few-flowered fascicles. Flowers 4–5-merous, unisexual or (morphologically) bisexual, plants chiefly dioecious; corolla rotate, greenish, white or pink, petals ovate, imbricate in bud; stamens in male flowers exserted, filaments united for most of their length, forming a distinct collar more or less adnate to the corolla tube; anthers dorsifixed, sagittate, opening through longitudinal slits; ovary in female flowers globose–ovoid, style short but well demarcated, stigma discoid, lobed or fimbriate; ovules few, 1-seriate. Fruit globose, endosperm ruminate. $2n = 46$. About four spp., Azores, Africa, tropical and subtropical Asia.

16. *Rapanea* Aubl.

Rapanea Hist. Pl. Guiane 1: 121 (1775); Mez in Pflanzenreich IV. 236: 342–396 (1902); Lundell, Fl. Guatemala, Fieldiana Bot. 24, 8: 190–192 (1966), reg. rev.; Smith, J. Arnold Arb. 54: 278–292 (1971), Fijian spp.; Pipoly, Sida 17: 115–162 (1996), Philippine spp.; Ricketson & Pipoly, Sida 17: 579–589 (1997), Mesoamerican spp.; Ricketson & Pipoly, Sida 18: 1095–1144 (1999), Venezuelan spp.
Suttonia A. Rich. (1832).

Shrubs or trees. Leaves short-petiolate, margins entire, rarely serrate. Inflorescences axillary fascicles, sometimes on short bracteate shoots. Flowers 4–5(6)-merous, unisexual or bisexual, if unisexual then plants dioecious or polygamous; corolla rotate, rarely campanulate, greenish, lobes narrowly ovate to ovate, usually granular on margins and inner surface, valvate in bud; stamens enclosed, or exposed, filaments inconspicuous, largely fused to the corolla; anthers basifixed, sagittate to oblong, opening by longitudinal slits; ovary globose to ovoid, style short and inconspicuous, stigma variable and often irregular, subglobose, conical, or ligulate; ovules few, 1-seriate.

Fruit subglobose; endosperm smooth or ruminate. $2n = 46, 48$. About 300 spp., pantropical. Often included in *Myrsine*.

17. *Ardisia* Sw.   Fig. 91H, I

Ardisia Sw., Nov. Gen. Sp. Pl. 3, 48 (1788), nom. cons.; Mez in Pflanzenreich IV. 236: 57–154 (1902); Walker, Philippine J. Sci. 73: 48–155 (1940), E Asian spp.; Lundell, Fl. Guatemala, Fieldiana Bot. 24, 8: 136–156 (1966), reg. rev.; Stearn, Bull. Brit. Mus. Nat. Hist., Bot. 4: 156–165 (1969), Jamaican spp.; Lundell, Fl. Panama, Ann. Missouri Bot. Gard. 58: 313–352 (1971), reg. rev.; Stone in Ng, Tree Fl. Malaya 4: 268–278 (1989), reg. rev.; Sleumer, Blumea 33: 115–140 (1988), New Guinean spp.; Yang & Dwyer, Taiwania 34: 192–297 (1989), rev. subg. *Bladhia*; Larsen & Hu, Fl. Thailand 6 (2): 82–151 (1996), reg. rev.; Chen & Pipoly, Fl. China 15: 10–30 (1996), reg. rev.; Pipoly & Ricketson, Sida 18: 433–472 (1998), rev. subg. *Graphardisia*.
Icacorea Adans. (1775).
Bladhia Thunb. (1781).
Auriculardisia Lundell (1981).
Ibarraea Lundell (1981).
Parardisia Nayar & Giri (1986).

Shrubs or treelets, sometimes subshrubs or herbs. Leaves petiolate, margins entire, crenate or serrate. Inflorescences terminal or axillary panicles or racemes, often with flowers arranged in corymbs or umbels. Flowers 5(4)-merous, bisexual; corolla rotate, pink or white, petals contorted in bud, often recurved at anthesis, ovate-lanceolate; stamens enclosed to exposed, filaments short; anthers narrowly sagittate, opening by longitudinal slits or subapical pores; ovary ovoid, style long and slender, stigma punctate; ovules few to numerous, one- to several-seriate. Fruit subglobose, endosperm smooth or ruminate. $2n = 46, 48, 96$. About 250 spp. pantropical, mainly tropical Americas and Asia, a few in temperate Japan.

18. *Afrardisia* Mez

Afrardisia Mez in Pflanzenreich IV. 236: 183–187 (1902); de Wit, Blumea Suppl. IV, 242–262 (1958); Taton, Fl. Afr. Centr., Myrsinac. 12–29 (1980), reg. rev.; Halliday, Fl. Trop. E Afr., Myrsinac. 16–19 (1984), reg. rev.

Shrubs, trees or subshrubs. Leaves short-petiolate, margins serrate or entire. Inflorescences axillary clusters, the peduncles short, bracteate. Flowers 5-merous, bisexual, or female and the plants then gynodioecious; corolla rotate pink, red or white, lobes contorted in bud, spreading at anthesis, ovate–lanceolate; stamens enclosed, filaments short; anthers narrowly sagittate, opening by longitudinal slits; ovary subglobose, style long and slender, stigma punctate; ovules 6–8, 1-seriate.

Fruit globose; endosperm smooth. 16 spp., tropical Africa. Often included in *Ardisia*.

19. *Synardisia* (Mez) Lundell

Synardisia (Mez) Lundell, Wrightia 3: 88–90 (1963); Lundell, Fl. Guatemala, Fieldiana Bot. 24, 8: 195–197 (1966).
Ardisia subg. *Synardisia* Mez (1902).

Shrubs or trees. Leaves short-petiolate, margins entire. Inflorescences terminal panicles. Flowers 5-merous, bisexual; corolla pink, campanulate to urceolate, lobes contorted in bud, broadly ovate; stamens enclosed, filaments short; anthers shortly sagittate, shortly produced at apex, opening through longitudinal slits; ovary subglobose, style long and slender, stigma punctate; ovules 8 or more, 2- or 3-seriate. Fruit depressed-globose. One sp., *S. venosa* (Mez) Lundell, Mexico to Honduras.

20. *Hymenandra* A. DC.

Hymenandra (A. DC.) A. DC. ex Spach, Hist. Veg. Phan. 9: 374 (1840); Stone, Gard. Bull. Singapore 43: 1–17 (1991).
Ardisia sect. *Hymenandra* A. DC., Trans. Linn. Soc. London 27: 126 (1834); Ann. Sci. Nat. II, 2: 297 (1834).

Subshrubs, sometimes unbranched. Leaves petiolate, sometimes shortly so, margins entire or sometimes crenate. Inflorescences axillary, or terminal on reduced lateral branches, often pedunculate and subtended by foliaceous bracts, paniculate but often umbelliform. Flowers 5-merous, bisexual; corolla pink, purple or white, rotate, lobes ovate to narrowly ovate, contorted or imbricate in bud; stamens included, filaments very short; anthers basifixed, narrowly sagittate, laterally connate along thecal margins, opening by longitudinal slits; ovary ovoid, style long and slender, stigma punctate; ovules 5–12, 1- or 2-seriate. Fruit subglobose; endosperm smooth. Eight spp., SE Asia (mainly Borneo). Pipoly and Ricketson (1999) argued for the inclusion of several, chiefly Mesoamerican species of *Ardisia* in *Hymenandra*.

21. *Tetrardisia* Mez

Tetrardisia Mez in Pflanzenreich IV. 236: 189 (1902); Stone, Malayan Nat. J. 46:1–11 (1992).

Shrubs or small trees. Leaves petiolate, margins crenate, serrulate or subentire. Inflorescences axillary or terminal panicles. Flowers usually 4-merous, bisexual; corolla purplish or pink, rotate, lobes lanceolate, imbricate in bud; stamens enclosed, filaments very short; anthers narrowly sagittate, opening by longitudinal slits; ovary subglobose, style long and slender, stigma punctate; ovules few, 1-seriate. Fruits globose, endosperm smooth or ruminate. Three of four spp., Java, Borneo, Malaya Peninsula, Thailand. Treated as a subgenus of *Ardisia* by Larsen and Hu (1995).

22. *Solonia* Urb.

Solonia Urb., Fedde Repert. 18: 22–23 (1922); León & Alain, Fl. Cuba 4: 113 (1957).

Shrubs. Leaves short-petiolate, margins dentate. Inflorescences terminal panicles. Flowers 5-merous, bisexual; corolla rotate, lobes ovate to broadly ovate, contorted in bud, reflexed at anthesis; stamens exposed at anthesis, filaments united into a tube adnate to the corolla; anthers dorsifixed, versatile, sagittate, opening through longitudinal slits; ovary broadly ovoid, style thick, somewhat longer than the ovary, stigma punctate; ovules 12–15, 2- or 3-seriate. Fruits globose. One sp., *S. reflexa* Urb., Cuba.

23. *Geissanthus* Hook. f.

Geissanthus Hook. f. in Benth. & Hook. f., Gen. 2: 642 (1876); Mez in Pflanzenreich IV. 236: 232–241 (1902); Pipoly, Novon 3: 463–474 (1993), Colombian spp.

Trees or shrubs. Leaves petiolate, margins entire or crenulate. Inflorescences terminal panicles. Flowers basically 5-merous, unisexual, the plants dioecious or polygamous; calyx split open by (2)3–6(–8) unequal lobes; corolla white, campanulate, lobes imbricate in bud, ovate–oblong, recurved at anthesis; stamens exserted, filaments long; anthers dorsifixed, versatile, sagittate, opening by longitudinal slits; ovary subglobose, style long and slender, stigma truncate-capitate; ovules 3–5, 1-seriate. Fruit globose; endosperm smooth. About 30 spp., Andes, Venezuela to Bolivia.

24. *Emblemantha* B. C. Stone

Emblemantha B. C. Stone, Proc. Acad. Sci. Philadelphia 140: 275–280 (1988).

Low subshrubs. Leaves petiolate with decurrent leaf blade, margins crenate, lower surface greyish, veins red-purple above. Inflorescences axillary panicles with short lateral branches. Flowers 5-merous, bisexual; corolla urceolate-tubular and almost completely closed, lobes very short-trian-

gular; stamens enclosed, filaments short; anthers basifixed, sagittate, apiculate; ovary subglobose, style rather stout, somewhat exserted, stigma punctate; ovules few, 1-seriate. Fruits unknown. One sp. (*E. urnulata* B. C. Stone), Sumatra.

25. *Sadiria* Mez

Sadiria Mez in Pflanzenreich IV. 236: 181–183 (1902).

Shrubs or trees. Leaves petiolate, margins crenate or entire. Inflorescences axillary panicles, usually appearing as few-flowered fascicles or umbels. Flowers 5-merous, bisexual; corolla campanulate or urceolate, lobes ovate, contorted in bud; stamens included, filaments very short, adnate to lower part of corolla; anthers basifixed, narrowly cordate, sometimes apically produced, opening by longitudinal slits; ovary subglobose, style long and thick, stigma punctate; ovules few, 1-seriate. Fruit globose, endosperm ruminate. Four spp., Assam, Bhutan.

26. *Gentlea* Lundell

Gentlea Lundell, Wrightia 3: 100 (1964); Lundell, Fl. Guatemala, Fieldiana Bot. 24, 8: 156–160 (1966), reg. rev.; Ricketson & Pipoly, Sida 17: 697–707 (1997).
Ardisia subg. *Walleniopsis* Mez (1902).

Shrubs or small trees. Leaves petiolate, margins entire or serrulate. Inflorescences terminal panicles or umbels. Flowers mostly 5-merous, bisexual; corolla greenish white or pink, campanulate or rotate, lobes occasionally free to base, imbricate in bud; stamens exserted, filaments long; anthers dorsifixed, shortly obtuse-cordate, opening by longitudinal slits; ovary subglobose or ovoid, style long and slender, stigma punctate; ovules few to many, in 2 or more series. Fruit subglobose; endosperm smooth or ruminate. Nine spp., Mexico to N and NW South America.

27. *Stylogyne* A. DC.

Stylogyne A. DC., Ann. Sci. Nat. Bot. II, 16: 78 (1841); Mez in Pflanzenreich IV. 236: 263–279 (1902); Lundell, Fl. Panama, Ann. Missouri Bot. Gard. 58: 308–312 (1971), reg. rev.

Shrubs or small trees. Leaves petiolate, margins entire or sometimes crenulate. Inflorescences terminal or axillary panicles, often reduced to umbels or fascicles. Flowers 5(4)-merous, mostly unisexual and the plants dioecious; corolla white, green or pinkish, rotate, lobes oblong, contorted in bud; stamens included, filaments long and slender;

anthers dorsifixed or basifixed, narrowly sagittate, opening by longitudinal slits; ovary ovoid, style long and slender, stigma truncate; ovules few, 1- or 2-seriate. Fruit globose; endosperm smooth. About 60 spp., Neotropics.

28. *Ctenardisia* Ducke

Ctenardisia Ducke, Arch. Jard. Bot. Rio de Janeiro 5: 179–180 (1930); Lundell, Wrightia 7: 42–44 (1982).

Shrubs, monocaulous. Leaves psuedoverticillate at stem apices, short-petiolate, margins serrate or entire. Inflorescences terminal panicles, long-pedunculate. Flowers 5-merous, bisexual; corolla greenish, rotate, lobes lanceolate to oblong, reflexed at antheis, contorted in bud; stamens enclosed, filaments rather short; anthers sagittate, dorsifixed, opening by longitudinal slits; ovary ovoid, style long and slender, stigma punctate; ovules 10–14, 1-seriate at base of placenta. Fruit globose. Two spp., Brazil.

29. *Yunckeria* Lundell

Yunckeria Lundell, Wrightia 3: 111–114 (1964); Lundell, Fl. Guatemala, Fieldiana Bot. 24, 8: 197–200 (1966).

Shrubs or small trees. Leaves long-petiolate, margins entire. Inflorescences terminal panicles. Flowers 5(4)-merous, bisexual; corolla rotate, lobes narrowly oblong, imbricate in bud; stamens included but filaments rather short; anthers narrowly sagittate, dorsifixed, opening by apical pores; ovary ovoid, style long and slender, stigma punctate; ovules 6–8, 1-seriate at base of placenta. Fruit globose. Three spp., Mexico to Nicargua. Often included in *Ctenardisia*.

30. *Antistrophe* A. DC.

Antistrophe A. DC., Ann. Sc. Nat. II, 16: 79 (1841); Mez in Pflanzenreich IV. 236: 150–151, 187–189 (1902); Grosse, Bot. Jahrb. Beibl. 96: 22 (1908), anatomy; Stone in Ng (ed.), Tree Fl. Malaya 4: 268 (1989), reg. rev.

Shrubs or subshrubs. Leaves petiolate, margins entire or serrate. Inflorescences axillary, few-flowered panicles, often reduced to umbels or fascicles. Flowers 4–5-merous, bisexual; corolla rotate, petals contorted to the left, lanceolate; stamens enclosed, filaments very short; anthers narrowly sagittate with long apices, dorsifixed, opening by longitudinal slits; ovary subglobose, style long and slender, stigma punctate; ovules 2–4, 1-seriate. Fruit globose, endosperm ruminate.

Four or five spp., Assam, Malaya.

31. *Parathesis* (A. DC.) Hook. f.   Fig. 91D

Parathesis (A. DC.) Hook. f. in Benth. & Hook. f., Gen. 2: 645 (1876); Mez in Pflanzenreich IV. 236: 173–181 (1902); Lundell, Contrib. Texas Res. Found., Bot. Stud. 5: 1–206 (1966); Lundell, Fl. Guatemala, Fieldiana Bot. 24, 8: 160–189 (1966), reg. rev.; Lundell, Fl. Panama, Ann. Missouri Bot. Gard. 58: 292–304 (1971), reg. rev.

Trees or shrubs, often with stellate or dendroid hairs on young shoots and floral parts. Leaves petiolate, margins entire or rarely crenulate. Inflorecences terminal or axillary panicles. Flowers 5-merous, bisexual; corolla pink or white, rotate with short tube, lobes narrowly ovate to lanceolate, recurved at anthesis, valvate in bud; stamens exposed due to recurving petals, filaments short, fused to corolla tube; anthers narrowly saggitate, basifixed, opening through longitudinal slits, sometimes shortly produced at apex; ovary sub-globose to ovoid, style long and slender, stigma punctate; ovules few, 1-seriate. Fruit globose; endosperm ruminate. About 75 spp., Mesoamerica, Greater Antilles, Andes.

32. *Aegiceras* Gaertn.   Fig. 91F

Aegiceras Gaertn., Fruct. 1: 216 (1788); Mez in Pflanzenreich IV. 236: 55–57 (1902); Walker, Philippine J. Sci. 73: 47–48 (1940), E Asian spp.; Stone in Ng, Tree Fl. Malaya 4: 266 (1989), reg. rev.

Shrubs or small trees. Leaves petiolate, coriaceous, margins entire. Inflorescences terminal or axillary umbels or panicles. Flowers 5-merous, bisexual; corolla campanulate, white, petals contorted in bud, reflexed at anthesis; stamens exserted, filaments distally united into a ring adnate to the corolla base; anthers sagittate, dorsifixed, versatile, transversely septate, opening by longitudinal slits; ovary ovoid, gradually tapering into a thick style of more than twice the length of the ovary, stigma punctate; ovules numerous, pluriseriate. Fruit viviparous, curved and elongate. $2n = 46$. Mangrove genus of two spp., SE Asia, Pacific, NE Australia.

33. *Amblyanthus* A. DC.

Amblyanthus A. DC., Ann. Sc. Nat. II, 16: 79 (1841); Mez in Pflanzenreich IV. 236: 208–210 (1902).

Trees or shrubs. Leaves petiolate, margins glandular-crenate. Inflorescences axillary panicles but appearing as long-pedunculate corymbs. Flowers 5-merous, bisexual; corolla short-camapanulate to rotate, petals contorted in bud, very broadly ovate, emarginate; stamens included, filaments free; anthers fused, ovoid, dorsifixed, opening through longitudinal slits; ovary ovoid, tapering into a style of about the same length, stigma capitate; ovules 2–4, 1-seriate. Fruit globose; endosperm ruminate. Three spp., Bengal, NE India.

34. *Amblyanthopsis* Mez

Amblyanthopsis Mez in Pflanzenreich IV. 236: 210–211 (1902).

Shrubs. Leaves petiolate, margins glandular-crenate. Inflorescences axillary, few-flowered panicles appearing as corymbs or umbels. Flowers 5-merous, bisexual; corolla shortly campanulate to rotate, petals contorted in bud, broadly ovate, emarginate; stamens included, filaments rather long; anthers ovoid, dorsifixed, opening by longitudinal slits; ovary subglobose, style thick and long, stigma capitate; ovules c. 3, 1-seriate. Fruit globose; endosperm smooth. Two spp., NE India, Bhutan.

35. *Elingamita* Baylis

Elingamita Baylis, Rec. Auckland Inst. Mus. 4: 99–102 (1951); Heenan, New Zeal. J. Bot. 38: 569–574 (2000).

Trees. Leaves short-petiolate, margins entire. Inflorescences terminal panicles. Flowers 4–6-merous, unisexual, the plants dioecious; corolla yellowish, tubular, shorter than the calyx, obscurely lobed; stamens exserted, filaments adnate to corolla at base only; anthers oblong, dorsifixed, opening by longitudinal slits; ovary ovoid tapering into a short style, stigma punctate and somewhat excavate; ovules 2–4, 1-seriate. Fruit subglobose and rather large, exocarp thick; endosperm smooth. $2n = 46$. One sp., *E. johnsonii* Baylis, New Zealand.

36. *Wallenia* Sw.   Fig. 91B

Wallenia Sw., Nov. Gen. Sp. Pl. 2: 31 (1788); Mez in Pflanzenreich IV. 236: 241–249 (1902); Stearn, Bull. Br. Mus. Nat. Hist., Bot. 4(4): 165–174 (1969), Jamaican spp.

Trees or shrubs. Leaves petiolate or rarely sessile, margins entire, sometimes clustered at branch tips. Inflorescences terminal panicles (subg. *Wallenia*) or axillary racemes (subg. *Homowallenia* Mez). Flowers 4–5-merous, unisexual and the

plants dioecious; corolla greenish or cream, campanulate (tube shorter in female flowers), lobes broadly ovate, sometimes emarginate, imbricate in bud; stamens in male flowers exserted, filaments inserted at base of the corolla; anthers sagittate, usually recurved at anthesis, dorsifixed, versatile, opening by longitudinal slits; ovary in female flowers ovoid, tapering into a style of about equal length or shorter, stigma truncate; ovules 3–5, 1-seriate. Fruit globose, endosperm smooth. About 20 spp. West Indies.

37. *Loheria* Merr.

Loheria Merr., Philippine J. Sci. Bot. 5: 373 (1910); Sleumer, Blumea 33: 100–102 (1988), New Guinean spp.; Stone, Micronesica 24: 65–80 (1991).

Small trees, sparsely branched or unbranched. Leaves short-petiolate or subsessile, often large, clustered at branch tips, margins entire throughout or distally crenate. Inflorescences lateral, lax panicles (subg. *Loheria*) or arranged in globular clusters on pendant peduncles (subg. *Longicorna* B. C. Stone). Flowers 4–5-merous, unisexual, the plants dioecious; corolla red, pink, white or greenish, rotate (subg. *Loheria*) or ± tubular (subg. *Longicorna*), lobes oblong to elliptic, ± papillose on inside, sometimes reflexed at anthesis, imbricate in bud; stamens enclosed or exposed due to recurving petals, filaments adnate to base of corolla tube; anthers obtusely sagittate, basifixed, opening through longitudinal slits; ovary ovoid with a rather stout style of equal length, stigma capitate to discoid; ovules 3–5, 1-seriate. Fruit subglobose; endosperm ruminate. Six spp., Malesia (Philippines, New Guinea).

38. *Cybianthus* Mart.

Cybianthus Mart., Nov. Gen. Sp. 3: 87 (1829), nom. cons.; Mez in Pflanzenreich IV. 236: 215–229, 249–263, 283–292 (1902); Agostini, Acta Biol. Venez. 10: 129–185 (1980); D'Arcy, Ann. Missouri Bot. Gard. 60: 442–448 (1973), rev. subg. *Triadophora*; Pipoly, Brittonia 35: 61–80 (1983), rev. subg. *Laxiflorus*; Pipoly, Ann. Missouri Bot. Gard. 79: 908–957 (1992), reg. rev., subg. *Conomorpha*; Pipoly, Sida 18: 1–160 (1998), rev. Ecuadorean and Peruvian spp.
Conomorpha A. DC. (1834).
Weigeltia A. DC. (1834).
Comomyrsine Hook. f. in Benth. & Hook. (1876).
Corelliana D'Arcy (1973).

Shrubs or small trees, vegetative and floral parts sometimes brownish-lepidote. Leaves petiolate, sometimes pseudoverticillate, margins entire or sometimes serrate. Inflorescences axillary panicles or racemes. Flowers 4–5-merous (3-merous in subg. *Triadophora* (Mez) Agostini), mostly unisexual and the plants dioecious; corolla green, greenish-brown or yellowish, rotate or short-campanulate, the inner surface often granular, lobes imbricate in bud, ovate, sometimes reflexed at anthesis; stamens enclosed or exposed because of the reflexed petals; filaments short to long; anthers subglobose to broadly oblong, dorsifixed, opening by oblique to round apical pores or longitudinal slits; ovary subglobose, style short, stigma capitate or truncate; ovules 2–5, uniseriate. Fruit globose, endosperm smooth. About 150 spp., Neotropics.

39. *Grammadenia* Benth.

Grammadenia Benth., Pl. Hartw. 218 (1846); Mez in Pflanzenreich IV. 236: 228–232 (1902); Pipoly, Mem. New York Bot. Gard. 43: 1–76 (1987).

Shrubs or small trees, often epiphytic. Leaves sessile, basally auriculate, margins entire or denticulate. Inflorescences axillary racemes. Flowers mostly 5-merous, uni- or bisexual, the plants dioecious, monoecious or hermaphroditic; corolla greenish white, rotate, lobes oblong to ovate, imbricate in bud; stamens included; anthers sessile or subsessile (filaments forming a sheath adnate to the corolla tube), almost square, opening by subapical slits; ovary subglobse, usually depressed, style short, stigma truncate; ovules few, 1- or 2-seriate. Fruit ellipsoid or subglobose, endosperm ruminate. Seven spp., C America, Lesser Antilles, tropical Andes. Sometimes treated as a subgenus of *Cybianthus* s. l.

40. *Vegaea* Urb.

Vegaea Urb., Symb. Antill. 7: 535–537 (1913).

Trees. Leaves clustered at branch tips, thick-coriaceous, sessile, linear and spine-tipped, margins entire. Inflorescences axillary, few-flowered racemes. Flowers 5(6)-merous, bisexual; corolla campanulate, white, waxy and thick, lobes broadly ovate, imbricate in bud; stamens included, filaments short; anthers ovoid-oblong, dorsifixed, opening by longitudinal slits; ovary ovoid, style thick, somewhat longer than the ovary, curved below apex, stigma truncate; ovules few, 1-seriate. Fruits not known. One sp., *V. pungens* Urb., Hispaniola.

41. *Oncostemum* A. Juss.   Fig. 91G

Oncostemum A. Juss., Mém. Mus. Paris 19: 133 (1830); Mez in Pflanzenreich IV. 236: 189–208 (1902); Perrier de la Bâthie, Fl. Madagascar 161: 16–120 (1953).

Shrubs or small trees. Leaves petiolate, margins entire or rarely crenulate. Inflorescences axillary panicles or racemes, through reductions often appearing as umbels. Flowers (4)5-merous, bisexual; corolla white, purple or greenish, rotate or short-campanulate, lobes oblong to ovate, contorted in bud; stamens enclosed, united into a tube with upper part free from the corolla; anthers ± connate, obtuse-triangular, opening by longitudinal slits; ovary ovoid, tapering into a style of equal length, stigma discoid, rarely 5-lobed; ovules 3–5, 1-seriate. Fruit globose, endosperm ruminate. About 90 spp., Madagascar. Distinction from *Badula* unclear (Coode 1976).

42. *Badula* Juss.

Badula Juss., Gen. Pl. 420 (1789); Perrier de la Bâthie, Fl. Madagascar 161: 120–124 (1953); Coode, Fl. Mascareignes 115: 6–25 (1981).

Trees or shrubs, often unbranched or sparsely branched. Leaves petiolate, condensed at branch tips, margins entire. Inflorescences axillary panicles. Flowers 5-merous, bisexual; corolla rotate, white or reddish, petals ovate to subrotund, imbricate in bud; stamens enclosed, filaments very short; anthers shortly sagittate, opening by longitudinal slits; ovary ovoid, style of about the same length as the ovary, stigma discoid; ovules 3–5, 1-seriate. Fruit subglobose, endosperm smooth. About 17 spp., Mascarenes, Madagascar.

43. *Tapeinosperma* Hook. f.

Tapeinosperma Hook. f. in Benth. & Hook., f., Gen. Pl. 2: 647 (1876); Mez in Pflanzenreich IV. 236: 162–171 (1902); Smith, J. Arnold Arb. 54: 228–263 (1973), Fijian spp.; Sleumer, Blumea 33: 102–107 (1988), New Guinean spp.

Trees or shrubs, often unbranched. Leaves often large and clustered at branch tips, petiolate or sessile, margins entire. Inflorescences axillary or terminal panicles. Flowers mostly 5-merous, bisexual; corolla red, rotate or campanulate, lobes ovate, contorted in bud; stamens included, filaments connate at base and adnate to the corolla tube, distally free; anthers oblong-ovoid, dorsifixed near base, opening by longitudinal slits starting as subapical pores; ovary globose or ovoid, style rather thick, longer than the ovary, stigma truncate, capitate or somewhat discoid; ovules (2–)6–12, 1-seriate. Fruits large, variable in shape, often depressed-globose, fleshy when fresh; endosperm smooth or ruminate. About 50 spp., Borneo, New Guinea, New Hebrides, Fiji, New Caledonia, trop. Australia.

44. *Discocalyx* (A. DC.) Mez

Discocalyx (A. DC.) Mez in Pflanzenreich IV. 236: 211–214 (1902); Smith, J. Arnold Arb. 54: 236–273 (1973), Fijian spp.; Sleumer, Blumea 33: 83–93 (1988), New Guinean spp.

Shrubs or trees, sparsely branched or unbranched. Leaves clustered at branch tips, petiolate, margins entire. Inflorescences axillary racemes or panicles. Flowers 4–5-merous, unisexual and plants dioecious, or bisexual; corolla urceolate, campanulate or rotate, white or sometimes yellowish, green or red, lobes imbricate in bud, ovate or oblong, sometimes emarginate; stamens enclosed; filaments very short; anthers appearing sessile, basifixed, broadly sagittate or ovoid, truncate at apex, opening by longitudinal slits; ovary ovoid or subglobose, style very short, stigma discoid, sometimes with erose margins; ovules 2–5, 1-seriate. Fruit subglobose; endosperm smooth. About 50 spp., SE Asia (especially the Philippines), the Pacific.

45. *Labisia* Lindl.

Labisia Lindl., Bot. Reg. 31, tab. 48 (1845), nom. cons.; Mez in Pflanzenreich IV. 236: 171–172 (1902); Stone, Malayan Nat. J. 42: 43–51 (1988), synopsis; Stone in Ng, Tree Fl. Malaya 4: 281 (1989), reg. rev.

Perennial herbs or subshrubs, usually unbranched. Leaves long- to short-petiolate, blade base decurrent, the blade with numerous, densely set secondary veins, margins entire or crenulate-serrulate. Inflorescences axillary panicles with reduced lateral branches. Flowers 5-merous, bisexual; corolla white or pink, rotate, lobes broadly ovate with incurved margins largely concealing the anthers, valvate in bud; stamens included; anthers sessile, sagittate, apically produced, opening by longitudinal slits; ovary globose, style of about the same length as the ovary, stigma punctate; ovules few, 1-seriate. Fruit globose; endosperm smooth. $n = 23$–25. Six spp., W Malesia, New Guinea.

46. *Systellantha* B. C. Stone

Systellantha B. C. Stone, Malayan Nat. J. 46: 13–24 (1992).

Shrubs or treelets with compressed and somewhat sinuous branches. Leaves petiolate, margins entire to crenulate. Inflorescences short axillary racemes. Flowers 4-merous, unisexual with flowers of both sexes occurring on the same inflorescence; corolla pink, rotate, lobes ovate, imbricate in bud; stamens included, filaments short, fused at base of corolla; anthers dorsifixed near base, narrowly sagittate, opening by longitudinal slits starting as subapical pores; ovary subglobose, style thick, longer than the ovary, stigma truncate; ovules 1–3, 1-seriate. Fruit globose, seed not known. Two spp., Borneo.

47. *Monoporus* A. DC.

Monoporus A. DC., Ann. Sci. Nat. II, 16: 78 (1841); Perrier de la Bâthie, Fl. Madagascar 161: 6–14 (1953).

Trees or shrubs. Leaves petiolate, often condensed at branch tips, margins entire. Inflorescences axillary panicles. Flowers 5-merous, unisexual, the plants dioecious; corolla pink to orange-red, rotate, the tube very short, the lobes oblong or ovate, spreading at anthesis, imbricate in bud, emarginate; stamens included, filaments very short; anthers sagittate, opening apically by one or two irregular pores; ovary subglobose to cylindrical, style very short or absent, stigma punctate; ovules many, 1-seriate in several parallel rows. Fruit globose; endosperm ruminate. Eight spp., Madagascar.

48. *Fittingia* Mez

Fittingia Mez, Bot. Arch. 1: 105–106 (1922); Sleumer, Blumea 33: 94–100 (1988); Stone, Sida 16: 267–270 (1994).

Small trees or shrubs, mostly unbranched. Leaves clustered at branch tips, petiolate, margins entire. Inflorescences axillary racemes. Flowers 4–5-merous, unisexual, the plants dioecious; corolla pale red white or greenish, campanulate to urceolate, lobes imbricate in bud, ovate to oblong; stamens included, filaments short; anthers obtusely sagittate, basifixed, opening by longitudinal slits; ovary subglobose, style short, stigma capitate to discoid; ovules unknown. Fruit subglobose with a thick and soft exocarp (flattened in herbarium specimens); endosperm smooth. Six spp., New Guinea.

49. *Conandrium* (K. Schum.) Mez

Conandrium Mez in Pflanzenreich IV. 236: 156–157 (1902); Sleumer, Blumea 33: 109–113 (1988).

Shrubs or small trees. Leaves petiolate, sometimes shortly so, margins entire or vaguely crenate towards apex. Inflorescences axillary panicles or racemes with long slender pedicels. Flowers 5-merous, bisexual; corolla rotate, red to pink, lobes contorted in bud, broadly ovate; stamens enclosed, filaments free, short; anthers broadly sagittate, laterally connate over the ovary, opening by longitudinal slits; ovary ovoid to globose, style short, stigma punctate; ovules many, 2- or 3-seriate. Fruit globose. Two spp., New Guinea, Moluccas, Bismarck Archipelago.

Selected Bibliography

Anderberg, A.A., Ståhl, B. 1995. Phylogenetic interrelationships in the order Primulales, with special emphasis on the family circumscriptions. Can. J. Bot. 73: 1699–1730.

Anderberg, A.A., Ståhl, B., Källersjö, M. 1998. Phylogenetic relationships in the Primulales inferred from *rbc*L sequence data. Pl. Syst. Evol. 211: 93–102.

Anderberg, A.A., Ståhl, B., Källersjö, M. 2000. Maesaceae, a new primuloid family in the order Ericales s.l. Taxon 49: 183–187.

Anderberg, A.A., Peng, C.-I., Trift, I., Källersjö, M. 2001. The *Stimpsonia* problem; evidence from DNA sequences of plastid genes *atp*B, *ndh*F and *rbc*L. Bot. Jahrb. Syst. 123: 369–376.

Braun, A. 1859. Über Polyembryonie und Keimung von *Caelebogyne*. Ein Nachtrag zu der Abhandlung über Parthenogenesis bei Pflanzen. Abh. K. Acad. Wiss. Berlin, 1859: 109–263.

Burrows, J. 1999. *Myrsine*. Plant Life (S. Afr.) 20: 25–26.

Carlquist, S. 1992. Wood anatomy of sympetalous dicotyledon families: a summary with comments on systematic relationships and evolution of the woody habit. Ann. Missouri Bot. Gard. 79: 303–332.

Coode, M.J.E. 1976. Notes on Pittosporaceae and Myrsinaceae of the Mascarenes. Kew Bull. 31: 221–225.

Cronquist, A. 1988. See general references.

Davis, G.L. 1966. See general references.

Erdtman, G. 1952. See general references.

Faure, P. 1968. Contribution à l'étude caryo-taxonomique des Myrsinacées et des Théophrastacées. Mem. Mus. Hist. Nat. Sér. B Bot. 18: 37–57.

Grosse, A. 1908. Anatomisch-systematische Untersuchungen der Myrsinaceen. Bot. Jahrb., Beibl. 95: 1–46.

Hegnauer, R. 1969. See general references.

Hope, G. 2000. Australian national university pollen database. http/www.geo.arizona.edu/palynologfy/sem/anu.html

Johri, B.M. et al. 1992. See general references.

Källersjö, M. et al. 2000. See general references.

Larsen, K., Hu, C.M. 1995. Reduction of *Tetradisia* to *Ardisia*. Nord. J. Bot. 15: 161–162

Lersten, N.R. 1977. Trichome forms in *Ardisia* (Myrsinaceae) in relation to the bacterial leaf nodule symbiosis. Bot. J. Linn. Soc. 75: 229–244.

Lersten, N.R., Horner, H.T. 1976. Bacterial leaf nodule symbiosis in angiosperms with emphasis on Rubiaceae and Myrsinaceae. Bot. Rev. 42: 145–214.

Martin, H.A. 1994. Australian Tertiary phytogeography: evidence from palynology. In: Hill, R.S. (ed.) History of the Australian vegetation: Cretaceous to Recent. Cambridge: Cambridge University Press, pp. 104–142.

Metcalfe, C.R., Chalk, L. 1950. See general references.

Mez, C. 1902. Myrsinaceae. In: Engler, A. (ed.) Das Pflanzenreich 9, IV 236: 1–437.

Mildenhall, D.C. 1980. New Zealand Late Cretaceous and Cenozoic plant biogeography: a contribution. Palaeogeogr Palaeoclimatol Palaeoecol 31: 197–233.

Nowicke, J.W., Skvarla, J.J. 1977. Pollen morphology and the relationships of the Plumbaginaceae, Polygonaceae, and Primulaceae to the order Centrospermae. Smithsonian Contrib. Bot. 37: 1–64.

Otegui, M., Cocucci, A. 1999. Flower morphology and biology of *Myrsine laetevirens*, structural and evolutionary implications of anemophily in Myrsinaceae. Nord. J. Bot. 19: 71–85.

Pascarella, J.B. 1997a. Pollination ecology of *Ardisia escallonioides* (Myrsinaceae). Castanea 62: 1–7.

Pascarella, J.B. 1997b. Breeding systems of *Ardisia* Sw. (Myrsinaceae). Brittonia 49: 45–53.

Pax, F., Knuth, R. 1905. Primulaceae. In: Engler, A. (ed.) Das Pflanzenreich 22, IV 237: 1–386.

Pipoly, J.J., Ricketson, J.M. 1999. Discovery of the Indo-Malesian genus *Hymenandra* (Myrsinaceae) in the Neotropics, and its boreotropical implications. Sida 18: 701–746.

Sattler, R. 1962. Zur frühen Infloreszenz- und Blütenentwicklung der Primulales sensu lato mit Berücksichtigung der Stamen-Petalum-Entwicklung. Bot. Jahrb. Syst. 81: 358–396.

Ståhl, B. 1990. Primulaceae. In: Harling, G., Andersson, L. (eds.) Fl. Ecuador 39: 23–35.

Ståhl, B. 1996. The relationships of *Heberdenia bahamensis* and *H. penduliflora* (Myrsinaceae). Bot. J. Linn. Soc. 122: 314–333.

Stone, B.C. 1993. New and noteworthy Malesian Myrsinaceae, VI. *Scherantha*, a new subgenus of *Ardisia*. Pacific Sci. 47: 276–294.

Vogel, S. 1986. Ölblumen und ölsammelnde Bienen. Zweite Folge, *Lysimachia* und *Macropis*. Akad. Wiss. Lt. Mainz, Abh. Math.-Naturwiss. Kl., Trop. Subtrop. Pflanzenwelt 54: 1–168.

Napoleonaeaceae

G.T. PRANCE

Napoleonaeaceae A. Rich. in Bory, Dict. Class. Hist. Nat. 11: 432 (1827).

Trees, shrubs or suffrutices; young axis with cortical vascular bundles. Leaves alternate, simple, entire or serrate, pinnately nerved; stipules absent. Inflorescences usually of solitary, axillary flowers, less frequently in cauline fascicles or axillary panicles; 4–7 bracts subtending each flower, often with paired glands on abaxial surface. Flowers large, actinomorphic, hermaphrodite; sepals (2)3 and imbricate or 5–6 and valvate, often with paired glands towards apex; corolla sympetalous, plicate, with 30–35 ribs, with a wavy margin, probably representing an outer ring of staminal tissue or pseudo-corolla; androecium either 8–10 whorls of numerous, free stamens fused to base of pseudo-corolla, or of 3 concentric whorls, fused at base, the outer and middle antherless, the inner one bearing 10 anthers, and 10 staminodes; anthers 2- or 4-locular, longitudinally dehiscent; ovary inferior or semi-inferior, 3–5-locular, with 2–numerous anatropous ovules in each loculus, the placentation apical-axile with 15–18 pendulous ovules or central-axile with 2–4 ovules per locule; style slender with a pointed stigma or short with a pentagonal or hexagonal stigma; nectary disk well developed, annular. Fruit indehiscent with 1 to several exalbuminous seeds; embryo with 2 plano-convex cotyledons, and a short radicle and large plumule.

Two genera and 10 species in tropical West Africa.

VEGETATIVE MORPHOLOGY. Most species are shrubs or small trees, at least one species reaches 30 m in height.

The wood anatomy of Napoleonaeaceae was described by the late Carl de Zeeuw in an unpublished paper from which this description is condensed (for reference see Bibliography of Lecythidaceae, this volume). The wood is yellow-brown without differentiation between sapwood and heartwood, and possesses a fine texture with the vessels indistinct to the naked eye. Vessels diffuse, uniformly distributed but rarely with a line of vessels at growth boundary, moderately large, mean diam. 76 µm in *Crateranthus*, 88 µm in *Napoleonaea*, over 50% solitary, the rest in groups of 2–6; perforations mostly simple, in part multiple to branched scalariform; intervessel pits alternate to subopposite, closely spaced, elliptic to angular, 4–7 µm; ray-vessel pits linear, elongate, opposite, tyloses often present in heartwood. Rays heterocellular with 2 classes in *Crateranthus*, 60% 3–7-seriate with uniseriate tails, the rest uniseriate or slightly biseriate; rays at broad end up to 14 cells wide in *Napoleonaea*; crystals present in *Napoleonaea* only. Axial parenchyma apotracheal in uniseriate tangential lines or reticulate in areas of closely spaced, wide rays, paratracheal in limited amounts where lines of parenchyma touch tangential faces of vessels, strands of moderate length, crystal strands common in *Crateranthus* and small prismatic crystals only in *Napoleonaea*, silica absent. Fibres moderately long in *Napoleonaea* with mean length 2295 µm, and long in *Crateranthus* with mean length 2779 µm, 16.3–21 µm in diameter, thick walled.

The wood of the two genera is quite similar indicating a close relationship. The partly scalariform perforation and scalariform to linear ray-vessel pits set Napoleonaeaceae apart from Lecythidaceae but not from *Asteranthos* (Scytopetalaceae), which shares those characters.

FLOWER STRUCTURE. The large, usually solitary hermaphrodite flowers are unique among angiosperms because of the structure of the corolla and androecium. Both genera have a showy, many-ribbed, united corolla with a dentate, wavy margin. Masters (1869) and others have considered this structure to be of staminodial origin which seems likely from the ribs. The numerous, long free stamens of *Crateranthus* (Fig. 92) are fused to the base of the corolla and are arranged in 8–10 whorls. In *Napoleonaea* the androecium consists of 3 concentric whorls. The outer whorl consists of 60–70, narrow strap-like appendages which are free or slightly connate at their base; the middle whorl is of 30–40, ribbon-like appendages which are wider and longer than the outer ones;

they are also fused laterally up to 2/3 of their length. These two whorls are delicately folded and hooked together. The inner whorl consists of ± 20 stamens which are fertile and are similar morphologically to the middle whorl. These stamens curve 180° inwards and their apex is inserted into the narrow space below the larger broad stigma, so that the anthers are situated between the annular nectary disk and the short column of the style. They are grouped into five groups of four, each group coinciding with one side of the pentagonal stigma. The outer two stamens of each group develop extremely large, bisporangiate anthers and the inner two bear only vestigial anthers. The anthers remain hidden inside the flowers even at anthesis. The anthers are latrorse in *Crateranthus* and extrorse in *Napoleonaea*. The calyx of the two genera is quite different: *Crateranthus* has a 3-lobed imbricate calyx which persists in fruit, and *Napoleonaea* possesses a 5–6-lobed valvate calyx. In *Napoleonaea* there are two large glands towards the apex of the outer surface. The ovary of *Napoleonaea* is inferior, that of *Crateranthus* is semi-inferior and there is a well-developed, annular nectary disk in both genera. For further details of flower structure and its development see Crüger (1860) and Tsou (1994), and the detailed study of the flower of *Napoleonaea* by Masters (1869).

EMBRYOLOGY. The embryology of the family was studied by Tsou (1994). The anthers have a glandular tapetum and the endothecial wall thickenings are rod-like. The ovules are bitegmic and anatropous or either anatropous or campylotropous in *Napoleonaea*. The nucellus is tenuinucellate and embryo sac formation is of the Polygonum type. The micropyle is formed from the inner integument. *Crateranthus* differs from *Napoleonaea* in the long-sagittate anthers and a completely closed endothecium that extends to the filament, the extremely long micropyle formed by very slender cells of the inner integument, and non-branching ovular vasculature.

POLLEN MORPHOLOGY. The pollen is 3-colporate with a granulate colpus membrane, and is more similar to that of Lecythidoideae than to Planchonioideae (Lecythidaceae) (Muller 1972; Tsou 1994).

KARYOLOGY. The basic chromosome number of two species of *Napoleonaea* has been reported as $x = 16$ (Mangenot and Mangenot 1957, 1962), which differs from that of Lecythidoideae $x = 17$,

Planchonioideae (Lecythidaceae) $x = 13$, and from *Asteranthos* (Scytopetalaceae), $x = 21$.

FRUIT AND SEED. The fruit is a 1- to several-seeded drupe. In *Crateranthus* the calyx is persistent at the proximal end and the style at the distal end. In *Napoleonaea*, the calyx scars or small persistent calyx are at the distal end of the fruit, there are 1 or 2 seeds, and the embryo consists of 2 plano-convex cotyledons and has a short radicle and a large plumule.

AFFINITIES. An Ebenalean affinity for *Napoleonaea* was suggested by Lindley (1830) and Miers (1874). There is no doubt that Lecythidaceae, Napoleonaeaceae and Scytopetalaceae remain a closely related group of families and that their affinities are with Sapotaceae, Styracaceae and Ebenaceae. *Napoleonaea* has frequently been placed in a family of its own (e.g. de Candolle 1837; Airy Shaw 1973), and this was also suggested by Tsou (1994: 101), but usually Napoleonaeaceae have been regarded as a subfamily of Lecythidaceae. They are most closely related to the genus *Asteranthos*, which has a similar pseudo-corolla and has recently been related or transferred to Scytopetalaceae (Tsou 1994; Morton et al. 1997). If *Asteranthos* is placed in Scytopetalaceae, it is then necessary to recognise the Napoleonaeaceae as a distinct family to maintain a monophyletic family Lecythidaceae (see Tsou 1994; Morton et al. 1997).

DISTRIBUTION AND HABITATS. The family is confined to W Tropical Africa where most species occur in the understorey of lowland, tropical moist forests, especially near to the coast and along rivers. *Napoleonaea gossweilera*, the southernmost species, occurs in savannah and open forests in Zambia.

ECONOMIC USES. No general uses of species of the family are recorded. The bark of *Napoleonaea vogelii* and *N. imperialis* is used locally as a cough medicine. The pulp around the seeds of some species of *Napoleonaea* is eaten. The hard wood of *N. imperialis* is used locally to beat mud floors and to make clogs.

KEY TO THE GENERA

1. Sepals 5–6, valvate; androecium of 3 whorls, the outermost and middle of sterile, strap-like appendages, the inner bearing 10 large fertile, 2-locular anthers and 10 small sterile ones; style short, with an expanded pentagonal or hexagonal stigma; ovules 2–4 per loculus, central-axile
   1. *Napoleonaea*

Fig. 92. Napoleonaeaceae. *Crateranthus talbotii.* **A** Flowering branch. **B** Flower bud. **C** Flower, vertical section. **D** Stamen. **E** Young fruit. (Knuth 1939)

– Sepals (2)3, imbricate; androecium of 8–10 whorls of numerous free stamens inserted on base of corolla; style filamentous, with a pointed stigma; ovules 15–18 per loculus, apical-axile and pendulous   **2.** ***Crateranthus***

1. *Napoleonaea* P. Beauv.

Napoleonaea P. Beauv. ex Fr. Fischer, Mém. Soc. Nat. Mosc. 1: 92 (1806); Fl. Oware 2, 29: 7 (1807); Liben, Bull. Jard. Bot. Natl. Belg. 41: 363–382 (1971), rev.

Trees or shrubs. Flowers usually solitary and axillary, less frequently in cauline fascicles or axillary panicles. Calyx 5(6)-lobed, valvate. Corolla plicate, 35-ribbed, fused. Androecium of 3 whorls, the outer two of free or fused, strap-like sterile appendages, the inner of 20 strap-like filaments, 10 with large fertile, extrorse anthers, 10 with small sterile anthers. Stamens and staminodes in 5 groups of 4, folded inwards beneath stigma between style and disk. Style short, with large,

flattened pentagonal or hexagonal stigma. Ovary 5-locular, inferior. Fruit with 1–2 seeds per locule. x = 16. Eight spp. in W Tropical Africa.

2. *Crateranthus* Baker   Fig. 92

Crateranthus Baker in Rendle, Cat. Talbot's Nig. Pl.: 35 (1913); Knuth in Pflanzenreich IV. 219: 65–67 (1939).

Small trees, stem of young branches winged in 2 species. Flowers solitary and axillary. Calyx (2)3-lobed, imbricate. Corolla plicate, ca. 30-ribbed, fused. Stamens inserted on base of corolla, the filaments free, in 8–10 whorls; anthers introrse. Style filamentous, the stigma pointed. Ovary 3–4-locular, semi-inferior. Fruit with persistent calyx at proximal end. Three spp. in W Tropical Africa.

Selected Bibliography

Airy Shaw, H.K. 1973. A dictionary of the flowering plants and ferns of the late J.C. Willis, 8th edn. Cambridge University Press.

Candolle, A.P. de 1837. Prodr. systematis naturalis, Vol. 7. Paris: Treuttel & Würtz, pp. 550–551.

Crüger, H. 1860. Westindische Fragmente: 11. Die Entwicklung der Blume von *Napoleona imperialis* Beauv. Bot. Z. 18: 361–367.

Knuth, R. 1939. Barringtoniaceae. In: Engler, A. (ed.) Pflanzenreich IV: 219. Leipzig: W. Engelmann.

Liben, L. 1971. Révision du genre africain *Napoleonaea* P. Beauv. (Lecythidaceae). Bull. Jard. Bot. Natl. Belg. 41: 363–382.

Lignier, O. 1890. Recherches sur l'anatomie des organes végétatifs des Lecythidées, des Napoléonées et des Barringtoniées. Bull. Sci. France Belgique 21: 291–420.

Lindley, J. 1830. Introduction to a natural system of botany, 44, Lecythideae, 177, Belvisiaceae. London: Longman, Rees, Orme, Brown & Green.

Mangenot, S., Mangenot, G. 1957. Nombres chromosomiques nouveaux chez diverses Dicotylédones et Monocotylédones d'Afrique occidentale. Bull. Jard. Bot. État 27: 639–654.

Mangenot, S., Mangenot, G. 1962. Enquête sur les nombres chromosomiques dans une collection d'espèces tropicales. Rev. Cytol. Biol. Vég. 25: 411–447.

Masters, M.T. 1869. On the structure of the flower in the genus *Napoleona,* etc. J. Linn. Soc. 10: 492–504.

Miers, J. 1874. On *Napoleona, Omphalocarpum,* and *Asteranthos.* Trans. Linn. Soc. II, 1: 17–19.

Morton, C.M., Mori, S.A., Prance, G.T., Karol, K.G., Chase, M.W. 1997. Phylogenetic relationships of Lecythidaceae: a cladistic analysis using *rbc*L sequence and morphological data. Am. J. Bot. 84: 530–540.

Muller, J. 1972. Pollen morphological evidence for subdivision and affinities of Lecythidaceae. Blumea 20: 351–355.

Tsou, C.-H. 1994. The embryology, reproductive morphology and systematics of Lecythidaceae. Mem. New York Bot. Gard. 71: 1–110.

Oxalidaceae

A.A. Cocucci

Oxalidaceae R. Br. in Tuckey, Narr. Exp. Zaire: 433 (1818), nom. cons.

Perennial, rarely annual herbs, sometimes succulent, often with underground storaging bulbs, tubers or rape-like roots, or shrubs, small trees, or sometimes vines; nodes trilacunar. Leaves alternate, often clustered, digitate or pinnate, rarely unifoliolate, the terminal leaflet lacking in *Biophytum*, leaflets articulate; petiole usually well-developed, sometimes expanded into blade-like phyllodium, or woody and persistent, at the base and insertion of the blade articulated; stipules present or not. Inflorescences thyrso-paniculate, with monochasia or dichasia in spike-like, umbellate or capitate arrangement, or racemes. Flowers regular, perfect, usually heterotri(di)stylous, or plants androcioecious (*Dapania*), rarely cleistogamous and apetalous; calyx lobes 5, quincuncially overlapping, the three external ones often much larger, all persistent in fruit; petals 5, mostly contortly overlapping, free or often postgenitally united at the base above the free claws, caducous, white, red, violet to purple or yellow, never blue; stamens obdiplostemonous, antepetalous stamens shorter than antesepalous or sterile (*Averrhoa* p. p.); filaments united at the base in an annulus; anthers introrse with small connective protrusion; nectaries outside the base of antepetalous stamens (actually at the level of a short androgynophore); ovary superior; carpels 5, antepetalous, united to form a 5-locular ovary; placentas axile or rarely parietal (*Oxalis aberrans*); stylodia always distinct, stigmata capitate or punctiform; ovules 2 per locule, rarely more, anatropous, bitegmic, crassinucellate or tenuinucellate, with endothelium, the micropyle directed upwards, slit-like (*Averrhoa* and *Biophytum*) or funnel-shaped (*Oxalis*) with chalazal appendages. Fruit a loculicidal, sometimes fleshy capsule, or a 5-ribbed berry. Seeds with straight embryo and fleshy endosperm, arillate in *Dapania*.

A family comprising five genera and about 880 species, widespread in tropical and temperate zones.

MORPHOLOGY. Oxalidaceae are commonly perennial herbs, sometimes cushion-forming, sometimes succulent, or shrubs, trees (*Averrhoa*, *Sarcotheca* and some *Oxalis* species) or lianas (*Dapania* and a few species of *Oxalis*). The herbaceous species normally form rosettes; some South African species of *Oxalis* may have an evident aerial stem. Bulbs, which are more frequent among the species of *Oxalis* of the Cape and in the South American section *Jonoxalis*, are formed by fleshy leaf bases that may or may not be protected by a tunica of leathery, scale-like leaves. These bulbs may contribute to propagation by forming new bulbs. The bulbs may form at the end of sinking internodes (Fig. 95A, p. 289); exceptionally bulbs arise on aerial shoots.

The leaves are borne alternately and are often clustered. Rarely they are sessile, as in *Oxalis* sect. *Sessilifoliolatae*, and usually have well-developed petioles. These are sometimes expanded into blade-like phyllodia (species of *Oxalis* sect. *Heterophyllum* such as *O. rusciformis*), or woody and persistent (*Oxalis* sect. *Thamnoxys*). At or somewhat above the leaf base and at the insertion of the leaf blade, there are well-developed articulations that are engaged in leaf movements.

The leaf blade is usually digitate or pinnate; the terminal leaflet is lacking in *Biophytum*, *Dapania*, *Sarcotheca*, and a few species of *Oxalis* have simple leaf blades articulated from the petiole, which suggests that the blade is actually unifoliolate by reduction from a compound leaf. In *Oxalis* the leaf blade is usually trifoliolate (*Oxalis* sect. *Thamnoxys*), but species with two (*Oxalis* sect. *Pteropodae*), four (*O. deppei* and *O. tetraphylla*) or 5–12 leaflets (*Oxalis* sect. *Multifoliolatae*, *Palmatifoliatae* and *Ionoxalis*) exist as well.

The digitate leaf blades are considered to be the result of peltation (Troll 1937–1943). This is particularly evident in four-segmented leaf blades, where one leaflet is produced from the transverse zone (Querzone) and is opposite to the terminal leaflet. Leaf blades with many leaflet pairs are considered to be derived from more simple forms (Knuth 1930).

Leaf sensitivity is common in the family. It was early reported for *Biophytum*, the "living herb". Leaves are sensitive in response to light (sleeping position) or touch (evident in *Averrhoa*, *Biophytum*, less so in *Oxalis*) by moving petioles and leaf blades. Leaves are also sensitive in *Sarcotheca*. In *Oxalis* each leaflet folds lengthwise along the middle nerve and sinks to lie against the petiole. In some *Oxalis* and in *Biophytum* even the cotyledons are sensitive. In *Oxalis acetosella* the sleeping position is attained 1–3 min after darkening while return to the horizontal position takes place within 1/2–1 h after the leaves are exposed the light again. Spontaneous leaf blade oscillations have been reported for *Oxalis* and *Averrhoa* (Knuth 1930).

VEGETATIVE ANATOMY. Bladder-like trichomes occur on the leaf underside (Knuth 1930) and uniseriate, slightly moniliform trichomes are found on the stamen filaments (Matthews and Endress 2002). Lysigenous secretory spaces are commonly situated at the leaf or sepal margins in some African *Oxalis*. Calcium oxalate crystals may be found in the parenchyma cells of every plant organ and in the xylem bundles of the bulb scales. Each cell bears normally one large cubic or prismatic crystal. Crystal-bearing cells may be arranged in cell chains beneath the epidermis. Sieve elements contain a unique variant of P-type plastids (Behnke 1982).

FLOWER STRUCTURE. Floral morphology of Oxalidaceae has been studied by Matthews and Endress (2002) within a broader comparative context.

FRUITS, SEEDS AND SEEDLINGS. The fruits are more (*Dapania*) or less (*Oxalis*) fleshy capsules, berries (*Averrhoa*, *Sarcotheca*), or fleshy capsules eventually becoming schizocarpic (*Biophytum*). The seed coat has a crystalliferous endotesta and a well-developed exotegmen; the endosperm is oily (Bouman 1974; Corner 1976; Boesewinkel 1985).

In some species (*O. rubella*) the primary shoot is initially enclosed in a sheathing cotyledonary tube. The growing petiole of the first leaf pushes and presses the epicotyl down into the deeper parts of the cotyledonary tube.

BREEDING SYSTEMS AND POLLINATION BIOLOGY. At least in *Oxalis*, the stigma is of the dry type and formed by multicellular and multiseriate papillae (Heslop-Harrison and Shivanna 1977). Oxali-

daceae are one of the four families where trimorphic heterostyly is known to occur (the others being Connaraceae, Lythraceae and Pontederiaceae). *Oxalis* is mostly tristylous. Tristyly is also found in *Biophytum*. Loss of tristyly to secondary distyly has taken place in some species of *Oxalis* (Orndurff 1972). Distyly is present in *Dapania* and *Sarcotheca* (Weeler 1992; Richards 1997). *Sarcotheca celebica*, although morphologically distylous, is functionally dioecious, with a short-styled, female-sterile morph, and a long-styled, male-sterile morph (Lack and Kevan 1987). Style heteromorphism is linked with an outcrossing breeding mode, although autogamy and even cleistogamy (*Oxalis acetosella*) are also present.

Flowers are short-lived and open only for a few hours on sunny days. Bee pollination prevails. Some *Oxalis* of South Africa (*O. tubiflora* or *O. annae*) and South America (*O. macrostylis*) have comparatively long, salverform corollas and are probably butterfly pollinated (Vogel 1954). Bees are the most important pollinators in *Sarcotheca* (Lack and Kevan 1987).

DISPERSAL. Seeds of *Oxalis* and *Biophytum* are ballistic: the elastic self-everting endotesta ejects the seed explosively (Overbeck 1923). The outer, elastic skin of the testa splits off from the inner portion of the seed by rupturing of the cell walls of the crystalliferous endotestal cell layer. The inner tangential walls, the thickened bases of the radial walls and the crystals remain attached to the sclerotic exotegmic layer of the ejected seed. The berries of *Averrhoa* and *Sarcotheca* are probably zoochorous.

POLLEN MORPHOLOGY. Pollen grains are described as tricolpate or tricolporoidate; some *Oxalis* are 3-colporate and exceptionally pantocolp(or)ate or 4-colp(or)ate; the exine is finely reticulate (Fig. 93). The shape is oblate to spheroidal (Erdtman 1952; Huynh 1969).

EMBRYOLOGY. Pollen is usually shed in the binucleate state; that of *Biophytum* has been reported to be trinucleate. The ovules are crassinucellate in *Averrhoa*, and tenuinucellate in *Oxalis* and *Biophytum*. The ovules have a slit-like (*Averrhoa* and *Biophytum*) or funnel-shaped (*Oxalis*) micropyle and have chalazal appendages; in *Oxalis* the nucellus together with the inner integument is elevated and separated from the outer integument by a stalk (Matthews and Endress 2002). Megagametophyte development is of the Polygonum type

Fig. 93. Oxalidaceae. *Oxalis lasiopetala*, pollen in polar view, REM, bar = 10 μm. (Palynological Laboratory Stockholm)

(Davis 1966) or Allium type (Herr and Dowd 1968). Endosperm development is nuclear.

KARYOLOGY. The numerous chromosome counts available for *Oxalis* reveal virtually every gameto-phytic chromosome number from 5 to 42, with predominance of $n = 7$ (Fedorov 1969). *Biophytum* has $n = 7$–16 with a mode at $n = 9$; *Averrhoa* has $n = 11$ and 12. This pattern has been interpreted as being due to aneuploid reduction from ancient ploidys with $n = 12$ (Lewis 1979) or, according to De Azkue (2000), from primitive diploids with $n = 6$.

AFFINITIES. Oxalidaceae have traditionally been related to Geraniaceae (Engler 1931; Cronquist 1981; Thorne 2000), with which they share pen-tamerous flowers, obdiplostemony, a syncarpous gynoecium and staminal nectaries. However, studies of plastid and nuclear gene sequences place Oxalidaceae close to Connaraceae and families such as Cephalotaceae, Cunoniaceae and Elaeocarpaceae (Chase et al. 1993; Price and Palmer 1993; Savolainen, Fay et al. 2000; Soltis et al. 2000). The affinity between Oxalidaceae and Connaraceae had already been recognized in the pre-molecular era. Recent morphological studies (Mathews and Endress 2002) strongly supported this affinity by a set of morphological and anatom-ical synapomorphies such as imbricate petal aes-tivation, trimorphic heterostyly, postgenital union of the petals which appear to be hooked together with their margins above the claw, multicellular hairs on the petals, sieve-tube plastid type and absence of oxalate druses. In addition, the absence of ellagic acid and the presence of the rare benzo-quinone rapanone are shared by both families (Hegnauer 1969, 1990). Pentamery and obdiplo-stemony are also shared but probably as ple-siomophic conditions within the Rosid clade (Matthews and Endress 2002). Geographical dis-tribution patterns suggest the origin of the family in the southern hemisphere, prior to the separa-tion of South America and Africa (Raven and Axelrod 1974).

Within the family, *Dapania*, *Sarcotheca* and *Averrhoa* are closely related (Veldkamp 1967). The Malagasy species of *Dapania* is intermediate to *Sarcotheca*. This group has retained several possi-ble plesiomorphic character states: a higher chro-mosome base number, multifoliolate leaf blades, and woody habit.

USES. *Oxalis tuberosa* (oca) tubers, a crop from Andean civilizations, are locally important as an alternative to potatoes. Its wild relatives come from Bolivia and Peru (Emswhiller 2002; Emswhiller and Doyle 2002). Fruits of *Averrhoa carambola* (starfruit or bilimbi) are edible and also used as a bleaching agent.

KEY TO THE GENERA

1. Fruit a dry capsule   2
   - Fruit fleshy, sometimes dehiscent   3
2. Capsule valves not much spreading, ± united with the central columella   **5. *Oxalis***
   - Capsule valves, when spreading out, completely detaching from the central columella   **4. *Biophytum***
3. Leaves plurifoliolate (5 or more leaflets); ovules and seeds 3–7 per locule   **3. *Averrhoa***
   - Leaves unifoliolate or trifoliolate; ovules and seeds 0–2 per locule   4
4. Shrubs or trees; inflorescences thyrso-paniculate; fruit indehiscent; seeds exarillate   **1. *Sarcotheca***
   - Lianas (rarely shrubs); inflorescences racemes; fruit dehis-cent into a 5-rayed star; seeds arillate   **2. *Dapania***

1. *Sarcotheca* Blume   Fig. 94

Sarcotheca Blume, Mus. Bot. Ludg. Bat. 1: 241 (1850); Veldkamp, Blumea 15: 527–543 (1967), rev.
Roucheria Miq. (1859).
Connaropsis Planch. ex Hook. f. (1860).

Shrubs or trees. Leaves unifoliolate or trifoliolate, when trifoliolate, lateral leaflets early caducous. Inflorescences thyrso-paniculate, axillary or pseudoterminal; flowers small, heterodistylous,

Fig. 94. Oxalidaceae. *Sarcotheca glauca.* **A** Habit. **B** Flower. **C** Fruit. **D** Androecium and gynoecium of short-styled flower. **E** Same, gynoecium. **F, G** Same of long-styled flower. Drawn by R. van Crevel. (Veldkamp 1967)

red or white; stamens 10; ovules 2 per locule. Fruit a globose berry with 5 episeptal furrows. Eleven species or probably less, each of local range in Thailand, Malay Peninsula, Borneo and Philippines.

2. *Dapania* Korth.

Dapania Korth., Nederl. Kruidk. Arch. 3: 381 (1854); Veldkamp, Blumea 15: 523–527 (1967), rev.

Woody lianas, rarely erect shrubs. Leaves unifoliolate, stipulate. Inflorescences axillary racemes. Flowers red to white, heterodistylous or androdioecious, stamens all fertile or 5 staminodial; ovules in hermaphrodite flowers 1–2 per locule. Fruit fleshy, loculicidal to the base with patent valves, 1–6-seeded. Seeds arillate, aril bright to whitish yellow. Three species, two of which in Malaya, Sumatra and Borneo, and one in Madagascar.

3. *Averrhoa* L.

Averrhoa L., Sp. Pl.: 428 (1753); Lourteig & Cerceau-Larrival, Phytologia 56: 381–412 (1984), rev.
Carambola Adans. (1763).

Trees. Leaves alternate, pinnate, plurijugate, terminal leaflet lacking. Inflorescences thyrsopaniculate. Flowers small, downy, purple streaked; stamens all fertile or 5 staminodial; ovules 3–7 per locule. Fruit baccate. Seeds sometimes arillate. Two species, *A. carambola* L. and *A. bilimbi* L., probably of Indomalaysian origin and spread by man throughout the tropics.

4. *Biophytum* DC.

Biophytum DC., Prodr. 1: 689 (1824).

Herbs up to 1 m, stem sometimes woody. Leaves sometimes rosulate, pinnate, plurijugate, terminal leaflet lacking. Inflorescences umbelliform or ± capituliform. Flowers yellow, heterotristylous (at least in part); stamens united at the base; ovules many per locule. Fruit a loculicidal capsule. About 50 species widespread in the tropical zones of America, Africa and Asia, reaching the Himalayas and Mexico in the North, and the Zambezi River and Madagascar in the South; particularly species-rich in Madagascar; growing in dry and wet open places.

5. *Oxalis* L. Fig. 95

Oxalis L., Sp. Pl.: 433 (1753); Lourteig, Bradea 7(1): 1–99 (1994), subg. *Thamnoxys*, 7(2): 1–629 (2000), subg. *Oxalis* and *Trifidus*.

Annual and perennial herbs, sometimes succulent, including rhizome- or bulb-bearing geophytes, chamaephytes, shrubs or vines (*O. rhombifolia* Jacq. and *O. scandens* H.B.K. from South America) and an aquatic of the Western Cape (*O. natans* Eckl. & Zeyh.) with a long internode holding a cluster of floating leaves. Leaf blades with (1–)3–many leaflets. Flowers solitary, thyrsic or subumbellate, heterotristylous, distylous, or homostylous. Fruit capsular. Seeds with self-everting endostesta. About 500 species mainly in tropical and subtropical zones, extending into arctic (Lappland) and subantartic (Falkland Islands) zones, particularly species-rich in South Africa and Andean South America. *O. magellanica* Forst. is distributed in S South America, New Zealand, Tasmania and Australia. Some species are widely naturalized weeds. The genus is divided into four subgenera [subg. *Oxalis*, 412 spp., subg. *Monoxalis* (Small) Lourteig, 2 spp., subg. *Thamnoxys* (Endl.) Reiche, 71 spp., subg. *Trifidus* Lourteig, 2 spp.] and 34 sections.

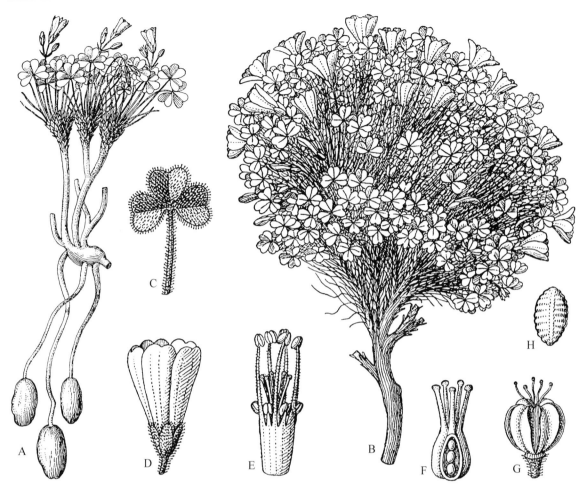

Selected Bibliography

Fig. 95. Oxalidaceae. **A** *Oxalis contracta*, floriferous plant with subterranean tubers. **B–H** *Oxalis leptocaulos*. **B** Habit. **C** Leaf. **D** Flower. **E** Androecium and gynoecium. **F** Pistil. **G** Capsule. **H** Seed. (Knuth 1930)

Behnke, H.-D. 1982. Sieve-element plastids of Connaraceae and Oxalidaceae. A contribution to the knowledge of P-type plastids in dicotyledons and their significance. Bot. Jahrb. Syst. 103: 1–8.

Boesewinkel, F.D. 1985. Development of ovule and seed-coat in *Averrhoa* (Oxalidaceae) with notes on some related genera. Acta Bot. Neerl. 34: 413–424.

Bouman, F. 1974. Developmental studies of the ovule, integuments, and seeds in some angiosperms. Doctoral Thesis, University of Amsterdam. Naarden: LOS.

Chase, M.W. et al. 1993. See general references.

Corner, E.J.H. 1976. See general references.

Cronquist, A. 1981. See general references.

Davis, G.L. 1966. See general references.

De Azkue, D. 2000. Chromosome diversity of South American *Oxalis* (Oxalidaceae) Bot. J. Linn. Soc. 132: 143–152.

Emswhiller, E. 2002. Biogeography of the *Oxalis tuberosa* alliance. Bot. Rev. 68: 128–152.

Emswhiller, E., Doyle, J.J. 2002. Origin of domestication and polyploidy in oca (*Oxalis tuberosa*: Oxalidaceae). 2. Chloroplast expressed glutamin synthetase data. Am. J. Bot. 89: 1042–1056.

Engler, A. 1931. Reihe Geraniales (Historische Entwicklung der Ansichten über die Umgrenzung der Reihe und ihre Zusammensetzung). In: Engler, A., Prantl, E. Die natürlichen Pflanzenfamilien, ed. 2, 19a. Leipzig: Engelmann, pp. 4–6.

Erdtman, G. 1952. See general references.

Federov, A. 1969. See general references.

Hartl, D. 1957. Die Pseudosympetalie von *Correa speciosa* (Rutaceae) und *Oxalis tubiflora* (Oxalidaceae). Abh. Akad. Wiss. Lit. Mainz, Abh. Math. Naturwiss. Kl. 2: 53–63.

Hegnauer, R. 1969, 1990. See general references.

Herr, J.M., Dowd, M.L. 1968. Development of the ovule and megagametophyte in *Oxalis corniculata* L. Phytomorphology 18: 43–53.

Heslop-Harrison, Y., Shivanna, K.R. 1977. The receptive surface of the Angiosperm stigma. Ann. Bot. 41: 1233–1258.

Huynh, K.-L. 1969. Etude du pollen des Oxalidacées. Bot. Jahrb. Syst. 89: 272–303.

Knuth, R. 1930. Oxalidaceae. In: Engler, A. (ed.) Das Pflanzenreich IV. 95. Weinheim: Engelmann, pp. 1–481.

Lack, A.J., Kevan, P.G. 1987. The reproductive biology of a distylous tree, *Sarcotheca celebica* (Oxalidaceae) in Sulawesi, Indonesia. Bot. J. Linn. Soc. 95: 1–8.

Lewis, W. 1979. Polyploidy in Angiosperms: Dicotyledons. In: Lewis, W. et al. (eds.) Polyploidy. New York: Plenum Press, pp. 241–268.

Matthews, M.L., Endress, P.K. 2002. Comparative structure and systematics in Oxalidales (Oxalidaceae, Connaraceae, Brunelliaceae, Cephalotaceae, Cunoniaceae, Elaeocarpaceae, Tremandraceae). Bot. J. Linn. Soc. 140: 321–381. With 104 figs.

Moore, R.J. 1973, 1974, 1977. Index to Plant Chromose Numbers 1967–1971, 1972, 1973–1974. Regnum Vegetabilis 90, 91, 96.

Orndurff, R. 1972. The breakdown of trimorphic incompatibility in *Oxalis* section *Corniculatae*. Evolution 26: 52–65.

Overbeck, F. 1923. Zur Kenntnis des Mechanismus der Samenausschleuderung von Oxalis. Jahrb. wiss. Bot. 62: 258–282.

Price, R.A., Palmer, J.D. 1993. Phylogenetic relationships of Geraniaceae and Geraniales from *rbcL* sequence comparisons. Ann. Missouri Bot. Gard. 80: 661–671.

Raven, P., Axelrod, D.I. 1974. Angiosperm biogeography and past continental movements. Ann. Missouri Bot. Gard. 61: 539–673.

Richards, A.J. 1997. Plant breeding systems. London: Allen & Unwin.

Savolainen V., Fay, M.F. et al. 2000. See general references.

Soltis, D.E. et al. 2000. See general references.

Thorne, R.F. 2000. The classification and geography of the flowering plants: dicotyledons of the class Angiospermae (subclasses Magnoliidae, Ranunculidae, Caryophyllidae, Dilleniidae, Rosidae, Asteridae, and Lamiidae). Bot. Rev. 66: 441–647.

Troll, W. 1937–1943. Vergleichende Morphologie der höheren Pflanzen, vol. 1 (1–3). Berlin: G. Bornträger.

Veldkamp, J.F. 1967. A revision of *Sarcotheca* Bl. and *Dapania* Korth. (Oxalidaceae). Blumea 15: 519–543.

Vogel, S. 1954. Blütenbiologische Typen als Elemente der Sippengliederung. Bot. Stud. 1: X + 338.

Weeler, S.G. 1992. Evolutionary modifications of tristylous breeding systems. In: Barrett, S.C.H. (ed.) Evolution and function of heterostyly. Berlin Heidelberg New York: Springer, pp. 247–272.

Parnassiaceae

M.P. SIMMONS

Parnassiaceae Gray, Nat. arr. Brit. pl. 2: 623, 670 (1821), nom. cons.
Lepuropetalaceae (Engl.) Nakai (1943).

Perennial rosulate herbs or diminutive winter annuals, glabrous or young petiole bases puberulent, with conspicuous tannin sacs in epidermis. Leaves alternate or subopposite, spathulate, ovate, reniform, or orbicular, entire, exstipulate, all petiolate or cauline leaf(s) appearing sessile. Flowers solitary on scapes or on lateral shoots, hermaphrodite, 5-merous, weakly zygomorphic; sepals (4)5(–7), persistent; petals (0, 4)5(–7), minute or showy; stamens 5, antesepalous, anthers longitudinally dehiscent, introrse or extrorse and individually dehiscent above stigmas; staminodes antepetalous, glandular, dilated distally; ovary superior to half-inferior, 3–4(5)-carpellate, placentation axile or parietal; ovules horizontal, numerous, bitegmic or unitegmic; style terminal, short or obsolete; style branches or stylodia distinct; stigmas commissural. Fruit a membranous capsule, loculicidally dehiscent at apex. Seeds numerous, cylindrical or oblong, minute, blackish or testa transparent and embryo opaque, endosperm a single cell layer or 0.

A family of two genera and about 71 species, widely distributed in the Northern Hemisphere, most diverse in China and the Himalayas, and in South America.

VEGETATIVE STRUCTURES. Rhizomes are present only in *Parnassia*. The surface of the rhizome is composed of a 3-layered exoderm with suberised cells. The cortex and solid pith of the rhizome consist of starch-filled parenchyma cells. Adventitious roots originate from the pericambium. The pericycle of *Parnassia* is 6-layered and fibrous, in *Lepuropetalon* 1–2-layered and sclerenchymatous. Three collateral vascular bundles surround a solid pith in *Lepuropetalon*, a hollow pith in *Parnassia* (Korta 1972; Gornall and Al-Shammary 1998). Secretory cells with tanniniferous contents occur in the epidermis of *Lepuropetalon* (Fig. 96F) and *Parnassia* (Metcalfe and Chalk 1950). The lowermost or only, depending on the species, cauline

leaf of *Parnassia* appears sessile. However, the single leaf trace divides from the stem vasculature well below the leaf, and runs parallel to the stem vasculature before entering the lamina. This indicates that the leaf is pseudo-sessile, with the petiole fused to the stem. Other cauline leaves, if present, are truly sessile (Watari 1939). Leaf venation is campylodromous in *Parnassia*, acrodromous in *Lepuropetalon*. Leaf crystals are absent. Stomata are anomocytic, on the abaxial leaf surface in *Parnassia*, on both leaf surfaces in *Lepuropetalon* (Gornall and Al-Shammary 1998). Fimbriate appendages on the young leaves of *Parnassia* secrete mucilage (Solereder 1908; Metcalfe and Chalk 1950).

FLORAL MORPHOLOGY. The flowers of *Parnassia* are weakly zygomorphic (Martens 1936; Hultgård 1987). The calyx is quincuncial, with both right- and left-handed spirals equally frequent. The immature stamens are introrse. Individually, at the rate of about one stamen per day, the stamens elongate and bend inwards, dehiscing extrorsely upwards, directly over the stigmas. After dehiscence, the stamens bend outwards, lying between the petals. The stamens move in order of their age (the stamen opposite the largest sepal first) in one of two zigzag orders (Gris 1868; Martens 1936; Hultgård 1987). Nectar is secreted from the pad of tissue bearing the staminodial rays and not from the glistening globules at the apices of the rays (Fig. 96B, C). A central stalk is evident on each staminode, such that the number of stalks is generally uneven. In addition to visually attracting insects (Daumann 1935), the globules may serve to transfer the released pollen onto the insects (Bennett 1871). The gynoecium of *Parnassia* is syncarpous at the base, but paracarpous and unilocular above, with T-shaped placentae. In contrast, the placentation of *Lepuropetalon* is strictly parietal, without T-shaped placentae.

FLORAL ANATOMY. In *Parnassia*, the vascular traces of the nectaries separate from the petal traces above the point at which the stamen traces separate from the sepal traces. Therefore, the nectaries have been interpreted as an inner androe-cial whorl of staminodes (Arber 1913; Eames 1961; Sharma 1968). The same pattern appears to apply to *Lepuropetalon* (Murbeck 1918). The globiferous filaments of the staminodes of *Parnassia* are each supplied by separate vascular traces. Based on this vasculature pattern, the individual stamens and staminodes have been interpreted as derived from stamen-fascicles that have become connate (Drude 1875; Arber 1913; Eames 1961; Bensel and Palser 1975; but see Klopfer 1972).

EMBRYOLOGY. The embryology of *Parnassia* has been examined by Pace (1912), Saxena (1964), and Sharma (1968). Anthers are tetrasporangiate (bisporangiate at anthesis in *Lepuropetalon*), have a fibrous endothecium, and a secretory tapetum. Tapetal cells are one- or two-nucleate. Microspore tetrads are tetrahedral. Mature pollen is bi-nucleate, with equally-sized vegetative and generative nuclei.

Ovules are anatropous, unitegmic in *Lepuropetalon* (Murbeck 1918), bitegmic in *Parnassia*, tenuinucellate, and have Polygonum type of embryo sac formation. The megaspore mother cell generally forms a linear tetrad of megaspores, although T-shaped tetrads also occur. The second, third, or fourth megaspores may develop, and two megaspores may develop simultaneously. Endosperm formation is nuclear. The endosperm is consumed by the embryo until only a single cell layer remains, or is entirely absent in mature seeds (Arber 1913). The endosperm is composed of a single cell layer in *Lepuropetalon* (Murbeck 1918). Embryos are straight, with short cotyledons.

POLLEN MORPHOLOGY. Pollen grains are shed as monads (Pace 1912). The grains are prolate to spheroidal and tricolporate; tetracolporate and syncolpate grains occur in *Parnassia*. The sexine is as thick or slightly thicker than the nexine, with reticulate sculpturing that is less-developed towards the poles and colpi (Erdtman 1952; Hideux and Ferguson 1976; Hultgård 1987).

KARYOLOGY. Chromosome numbers have been extensively studied in *Parnassia*. *Parnassia* has three base chromosome numbers: x = 7, 8, and 9 (Funamoto et al. 1998). *P. kotzebuei* (2n = 18, 36), *P. laxmannii* (2n = 18, 36), *P. oreophyla* (2n = 18, 36) include both diploid and tetraploid individuals (Funamoto et al. 1996). The tetraploids have been considered to be autopolyploids (*P. oreophyla*; Funamoto et al. 1996; *P. palustris*; Wentworth and Gornall 1996) or intraspecific

hybrid polyploids (*P. palustris*; Hultgård 1987). *P. palustris* includes diploid, triploid, tetraploid, pentaploid, and hexaploid individuals (2n = 17, 18, 27, 36, 32–37, 43–45 and 54; Funamoto et al. 1998 and references cited therein). The reports of 2n = 20 for *P. palustris* are probably erroneous (Gastony and Soltis 1977). The hexaploids of *P. palustris* have been considered to be the product of triploid hybrids that have undergone chromosome doubling (Spongberg 1972). The only differences in gross morphology between diploid and tetraploid populations of *P. palustris* are the generally larger pollen grains and seeds in tetraploids (Hultgård 1987).

POLLINATION. The pollinators of *P. palustris* are non-social wasps (Pompiloidea, Apoidea, and Vespoidea), flies (Syrphidae, Diptera), crane flies (Tipulidae), bottle flies (Calliphoridae), mosquitoes (Culicidae), butterflies and moths (Lepidoptera), and ants (*Myrmica*; Sprengel 1793; Daumann 1935; Spongberg 1972; Hultgård 1987; Proctor et al. 1996). Insects are visually attracted by the globules of the staminodes and the prominent, dark petal venation on the white petals that act as nectar guides to the nectaries at the staminode bases (Sprengel 1793; Daumann 1935; Spongberg 1972). Once in the vicinity of the flowers, the scent from the nectaries induces flies to land on the flowers (Daumann 1935).

REPRODUCTIVE SYSTEMS. The reproductive system of *Lepuropetalon* is essentially unknown. Based on the winter-annual habit and the dehiscence of the anthers directly over the stigmas, *Lepuropetalon* is probably self-pollinated (Spongberg 1972). In contrast, the reproductive system of *Parnassia palustris* has been extensively studied. *P. palustris* has been reported to be protandrous, requiring insect pollination (Sprengel 1793; Gris 1868; Bennett 1871; Hultgård 1987). However, protandry has been shown to be imperfect as the styles are receptive when the anthers dehisce individually above them, and self-fertilization does occur (Martens 1936; Hultgård 1987). Based on seed set and germination, Martens (1936) concluded that autogamy is superior to outcrossing in *P. palustris*, and that the stamen dehiscence above the styles favors autogamy. In contrast, based on seed set, Hultgård (1987) concluded that outcrossing is superior to autogamy. Hultgård suggested that the species primarily reproduces by outcrossing, and interpreted the method of stamen dehiscence to favor outcrossing.

FRUIT AND SEED. The fruit is a loculicidal capsule with a persistent calyx, which is borne erect and opens at the apex, with numerous tiny seeds. From about 100 to 2000 pale- to dark-brown seeds develop in capsules of *Parnassia palustris* (Hultgård 1987).

The testa of *P. nubicola* is 5–6 cells thick. The outer epidermal cells enlarge and accumulate tannin-like deposits (Sharma 1968). In *P. palustris*, these cells are tetragonal or pentagonal, isodiametric to elongated, and have raised borders. An air cavity develops between the testa and the embryo. The testa is transparent, but the embryo is not (Hultgård 1987). Seed germination of *Parnassia* is epigeal (Spongberg 1972). The testa of *Lepuropetalon* is 2 cells thick, membranous, and the cells have raised borders. There are small intercellular spaces between the two layers of cells forming the testa. An air cavity does not develop between the testa and the embryo (Murbeck 1918).

DISPERSAL. The capsules of *Parnassia* are borne erect, so that the very light seeds, which are mainly wind-dispersed (Ridley 1930), only fall out by wind blowing on the capsule (Sprengel 1793). The seeds are buoyant and float for several weeks, allowing dispersal by water (Praeger 1913). The air cavity between the testa and the embryo facilitates dispersal by wind and water (Hultgård 1987).

PHYTOCHEMISTRY. Flavonoids of *Lepuropetalon* and/or *Parnassia* have been investigated by Jay (1971) and Bohm et al. (1986). The flavonoid profiles support the close relationship of the two genera and their separation from Saxifragaceae. Methanolic extracts of *P. nubicola* were shown to have antiproliferative activity against the growth of human keratinocytes (KC and Müller 1997).

AFFINITIES. The close relationship of the two genera is supported by floral and vegetative characters (Murbeck 1918; Spongberg 1972; Gornall and Al-Shammary 1998). Several workers have treated *Lepuropetalon* and *Parnassia* as members of Saxifragaceae (Hooker and Thomson 1858; Engler 1930; Cronquist 1981). However, this putative relationship has been recognized as problematic based on various characters from morphology (Drude 1875; Hallier 1901; Murbeck 1918), stamen vasculature (Eames 1961), embryology (Sharma 1968), flavonoids (Jay 1971), floral development (Klopfer 1972), floral anatomy (Bensel and Palser 1975), and pollen structure (Hideux and Ferguson 1976). *Parnassia* has also been suggested to be closely related or transitional to Droseraceae (based on characters of the gynoecium; Drude 1875; Pace 1912; Arber 1913) and/or Hypericaceae (based on characters of the androecium or flavonoids; Lindley 1846; Drude 1875; Arber 1913; Jay 1971), or even Nymphaeaceae (based primarily on vegetative characters; Hallier 1901). Many workers (Drude 1875; Murbeck 1918; Sharma 1968; Klopfer 1972; Hultgård 1987) have supported recognition of Parnassiaceae as a distinct family (or order, Takhtajan 1997).

In several broad-scale molecular phylogenetic analyses, Parnassiaceae have been resolved as the sister group of Celastraceae (e.g., Chase et al. 1993; Savolainen, Chase et al. 2000; Soltis et al. 2000). In a well-sampled phylogenetic analysis of Celastraceae using 26S nrDNA and a simultaneous analysis of morphological and molecular characters (Simmons et al. 2001), Parnassiaceae were resolved, in a weakly-supported clade, as part of an early-derived lineage within Celastraceae. However, because of their distinctive morphology, and the fact that the clade from Simmons et al. (2001) was weakly-supported and based only on characters from 26S nrDNA, Parnassiaceae are retained as a separate family.

DISTRIBUTION AND HABITATS. The center of diversity of *Parnassia* is in China and the Himalayas, to which most (49) species are endemic (Gu and Hultgård 2001). *Parnassia palustris* is the most widespread species of the genus, with populations in North America, Europe, and Asia, and south to Morocco (Korta 1972). In the New World, *P. townsendii* is the southernmost species and is endemic to the Sierra Madre Occidental of Mexico. *P. townsendii* may represent relictual populations of the *P. fimbriata* complex, the other members of which (*P. fimbriata* and *P. intermedia*) have dispersed northward (Bye and Soltis 1979). *Parnassia* grows in temperate to arctic regions, preferentially in open, moist habitats from shore meadows to mountainsides. *Lepturopetalon* has a disjunct distribution in North and South America, where it grows in seasonally-moist rock outcrops and sandy soils, including disturbed areas. It is probably more widespread than currently known (Ward and Gholson 1987).

ECONOMIC IMPORTANCE. Several species of *Parnassia* are cultivated in rock gardens and bog gardens (Huxley et al. 1992).

CONSERVATION. *Parnassia caroliniana* and *P. cirrata* are rare (Walter and Gillett 1998).

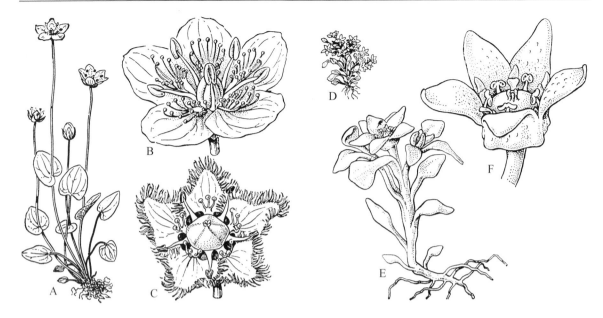

Fig. 96. Parnassiaceae. **A–B** *Parnassia palustris*. **A** Habit. **B** Flower. **C** *Parnassia alpicola*, flower. **D–F** *Lepuropetalon spathulatum*. **D** Habit. **E** Flowering plant. **F** Flower with sepals showing the pattern of epidermal tannin sacs. (Takhtajan 1981)

KEY TO THE GENERA

1. Diminutive winter annuals forming hemispherical clumps, without rhizomes; leaves spathulate; flowers inconspicuous, petals smaller than sepals. Northern and Southern Hemispheres of the New World **2. Lepuropetalon**
- Perennials forming rosettes, usually with rhizomes or rootstocks; leaves ovate, reniform, or orbicular; flowers conspicuous, petals larger than sepals. Northern Hemisphere of the Old and New World **1. Parnassia**

GENERA OF PARNASSIACEAE

1. *Parnassia* L.   Fig. 96A–C

Parnassia L., Sp. Pl.: 273 (1753); Drude, Linnaea 39: 239–324 (1875), rev.; T.C. Ku, Bull Bot. Res., Harbin 7: 1–59 (1987), reg. rev.; Gu and Hultgård, *Parnassia*, pp. 358–379, in: Fl. China vol. 8 (2001).

Perennials, forming rosettes, usually with rhizomes or rootstocks, glabrous or young petiole bases puberulent. Leaves alternate, ovate, reniform, or orbicular; basal leaves petiolate, cauline leaf(s) appear sessile. Flowers borne on scape; sepals (4)5(–7); petals (4)5(–7), showy; staminodes glandular, ornamented with many terminal globular stalks or 2–5-lobed; immature anthers introrse, mature extrorse, individually dehiscent

above stigmas; ovary 3–4(5)-carpellate. Seeds oblong, testa transparent, embryo opaque, endosperm minimal or 0. $2n = 14, 17, 18, 27, 32, 36, 32$–$37, 43$–$45$ and 54. About 70 spp., northern hemisphere–holarctic, south to northern Mexico, Morocco, and India, various open, moist habitats including rock outcrops, bogs, and wet meadows, grasslands, forests, and disturbed areas.

Drude (1875) recognized four sections based on characteristics of the staminodes, ovary position, and carpel number. Engler (1930) followed Drude's classification, and added a fifth section. Building on Drude and Engler's classifications, Ku (1987, 1995) recognized nine sections and nine series for the Chinese species, based on characteristics of the staminodes, petal margins, and cauline leaf number.

2. *Lepuropetalon* Elliott   Fig. 96D–F

Lepuropetalon Elliott, Sketch Bot. S. Carolina 1: 370 (1817); Spongberg, J. Arnold Arbor. 53: 409–498 (1972), part. rev.

Diminutive winter annuals, caulescent or acaulescent, forming hemispherical clumps, glabrous. Leaves alternate or subopposite, spathulate. Flowers minute; petals (0, 4)5, minute; staminodes unornamented; anthers introrse; ovary 3-carpellate. Seeds cylindrical, blackish, endosperm minimal. $2n = 46$. Only one sp., *L. spathulatum* Elliott, with a disjunct distribution in southeastern U.S.A., the Gulf Coast of Mexico, Uruguay, central Chile.

Selected Bibliography

Arber, A. 1913. On the structure of the androecium in *Parnassia* and its bearing on the affinities of the genus. Ann. Bot. 27: 491–510.

Bennett, A.W. 1871. Note on the structure and affinities of *Parnassia palustris*, L.J. Linn. Soc., Bot. 11: 24–31.

Bensel, C.R., Palser, B.F. 1975. Floral anatomy in the Saxifragaceae sensu lato. I. Introduction, Parnassioideae and Brexioideae. Am. J. Bot. 62: 176–185.

Bohm, B.A., Donevan L.S., Bhat, U.G. 1986. Flavonoids of some species of *Bergenia*, *Francoa*, *Parnassia*, and *Lepuropetalon*. Biochem. Syst. Ecol. 14: 75–77.

Bye, R.A., Soltis, D.E. 1979. *Parnassia townsendii* (Saxifragaceae), a Mexican endemic. SouthW. Nat. 24: 209–222.

Chase, M.W. et al. 1993. See general references.

Cronquist, A. 1981. See general references.

Cunnell, G.J. 1959. The arrangement of sepals and petals in *Parnassia palustris* L. Ann. Bot. 23: 441–453.

Daumann, E. 1935. Über die Bestäubungsökologie der *Parnassia*-Blüte. II. Ein weiterer Beitrag zur experimentellen Blütenforschung. Jahrb. Wiss. Bot. 81: 705–717.

Daumann, E. 1960. Über die Bestäubungsökologie der *Parnassia*-Blüte: ein weiterer Beitrag zur experimentellen Blütenökologie. Biol. Pl. 2: 113–125.

Drude, O. 1875. Ueber die Blüthengestaltung und die Verwandtschaftsverhältnisse des Genus *Parnassia*, nebst einer systematischen Revision seiner Arten. Linnaea 39: 239–324.

Eames, A.J. 1961. Morphology of the angiosperms. New York: McGraw-Hill.

Engler, A. 1930. Saxifragaceae. In: Engler & Prantl, Nat. Pflanzenfam., ed. 2, 18a. Leipzig: W. Engelmann, pp. 74–225.

Erdtman, G. 1952. See general references.

Funamoto, T., Kondo, K., Hong, D.-Y., Yang, Q.-E., Ge, S., Hizume, M., Shimada, T. 1996. Karyomorphological studies in Chinese *Parnassia* (II) three species in Qinghai Province. La Kromosomo II 82: 2845–2854.

Funamoto, T., Kondo, K., Hong, D.-Y., Zhou, S.-L., Deguchi, H. 1998. A chromosome study of three *Parnassia* species collected in the Qin Ling Mountains, Shaanxi Province, China. Chromosome Sci. 2: 111–115.

Gastony, G.J., Soltis, D.E. 1977. Chromosome studies of *Parnassia* and *Lepuropetalum* (Saxifragaceae) from the eastern United States. A new base number for *Parnassia*. Rhodora 79: 573–578.

Gornall, R.J., Al-Shammary, K.I.A. 1998. Parnassiaceae. In: Cutler, D.F., Gregory, M. (eds.) Anatomy of the Dicotyledons: Saxifragales, 4: 245–247. Oxford: Clarendon Press.

Gris, M.A. 1868. Sur le mouvement des étamines dans la Parnassie des marais. C.R. Hebd. Séances Acad. Sci. 67: 913–916.

Gu, C., Hultgård, U.-M. 2001. *Parnassia*. In: Wu Z.-Y., Raven, P. (eds.) Flora of China, vol. 8. Beijing, St. Louis: Science Press and Missouri Botanical Garden, pp. 358–379.

Hallier, H. 1901. Über die Verwandtschaftsverhältnisse der Tubifloren und Ebenalen den polyphyletischen Ursprung der Sympetalen und Apetalen und die Anordnung der Angiospermen überhaupt. Abh. Naturwiss. Naturwiss. Verein Hamburg 16: 1–112.

Hideux, M.J., Ferguson, I.K. 1976. The stereostructure of the exine and its evolutionary significance in Saxifragaceae sensu lato. In: Ferguson, I.K., Muller, J. (eds.) The evolutionary significance of the exine. London: Academic Press, pp. 327–378.

Hooker, J.D., Thompson, T. 1858. Præcursores ad floram Indicam. J. Proc. Linn. Soc. 2: 54–103.

Hultgård, U.-M. 1987. *Parnassia palustris* L. in Scandinavia. Acta Univ. Upsal. Symb. Bot. Upsal. 28: 1–128.

Huxley, A., Griffiths M., Levy, M. 1992. The new royal horticultural society dictionary of gardening. London: MacMillan.

Jay, M. 1971. Quelques problèmes taxinomiques et phylogénétiques des Saxifragacées vus à la lumière de la biochimie flavonique. Bull. Mus. Natl. Hist. Nat. 42: 754–775.

KC, S.K., Müller, K. 1997. Antiproliferative activity of selected Nepalese medicinal plants against the growth of human keratinocytes. Pharm. Pharmacol. Lett. 7: 63–65.

Klopfer, K. 1972. Beiträge zur floralen Morphogenese und Histogenese der Saxifragaceae 7: *Parnassia palustris* und *Francoa sonchifolia*. Flora 161: 320–332.

Korta, J. 1972. Anatomical analysis of *Parnassia palustris* L. Acta Biol. Cracov. 15: 31–37.

Ku, T. 1987. A revision of the genus *Parnassia* (Saxifragaceae) in China. Bull. Bot. Res., Harbin 7: 1–59.

Ku, T. 1995. Parnassioideae Engl. In: Lu, L., Hwang, S. (eds.) Flora Reipublica Popularis Sinicae 35: 1–66. Beijing: Science Press.

Lindley, J. 1846. The vegetable kingdom. London: Bradbury and Evans.

Martens, P. 1936. Pollination et biologie florale chez *Parnassia palustris*. Bull. Soc. Roy. Bot. Belgique 68: 183–221.

Metcalfe, C.R., Chalk, L. 1950. See general references.

Murbeck, S. 1918. Über die Organisation und verwandtschaftlichen Beziehungen der Gattung *Lepuropetalon*. Ark. Bot. 15: 1–12.

Pace, L. 1912. *Parnassia* and some allied genera. Bot. Gaz. 54: 306–328.

Praeger, R.L. 1913. On the buoyancy of the seeds of some Britannic plants. Sci. Proc. Roy. Dublin Soc. 14: 13–62.

Proctor, M., Yeo, P., Lack, A. 1996. The natural history of pollination. Portland, Oregon: Timber Press.

Ridley, H.N. 1930. The dispersal of plants throughout the world. Ashford, Kent: L. Reeve & Co.

Savolainen, V., Chase, M.W. et al. 2000. See general references.

Saxena, N.P. 1964. Studies in the family Saxifragaceae. 2. Development of ovule and megagametophyte in *Parnassia nubicola* Wall. Proc. Ind. Acad. Sci. B 60: 196–202.

Sharma, V.K. 1968. Morphology, floral anatomy and embryology of *Parnassia nubicola* Wall. Phytomorphology 18: 193–204.

Simmons, M.P., Savolainen, V., Clevinger, C.C., Archer, R.H., Davis, J.I. 2001. Phylogeny of the Celastraceae inferred from 26 S nuclear DNA, phytochrome B, *rbc*L, and *atp*B, and morphology. Molec. Phylog. Evol. 19: 353–366.

Solereder, H. 1908. Systematic anatomy of the dicotyledons: a handbook for laboratories of pure and applied biology. Oxford: Clarendon Press.

Soltis, D.E., Soltis, P.S., Chase, M.W., Mort, M.E., Albach, D.C., Zanis, M., Savolainen, V., Hahn, W.H., Hoot, S.B., Fay, M.F., Axtell, M., Swensen, S.M., Nixon, K.C., Farris, J.S. 2000. Angiosperm phylogeny inferred from a combined data set of 18 S rDNA, *rbc*L, and *atp*B sequences. Bot. J. Linn. Soc. 133: 381–461.

Spongberg, S.A. 1972. The genera of Saxifragaceae in the southeastern United States. J. Arnold Arbor. 53: 409–498.

Sprengel, C.K. 1793. Das entdeckte Geheimnis der Natur im Bau und in der Befruchtung der Blumen. Berlin: Friedrich Vieweg dem aeltern.

Takhtajan, A. (ed.) 1981. See general references.

Takhtajan, A. 1997. See general references.

Walter, K.S., Gillett, H.J. 1998. 1997 IUCN red list of threatened plants. Gland, Switzerland: IUCN – The World Conservation Union.

Ward, D.B., Gholson, A.K. 1987. The hidden abundance of *Lepuropetalum spathulatum* (Saxifragaceae) and its first reported occurrence in Florida. Castanea 52: 59–67.

Watari, S. 1939. Anatomical studies on the leaves of some Saxifragaceous plants, with special reference to the vascular system. J. Fac. Sci. Univ. Tokyo, Sect. 3, Bot. 5: 195–316.

Wentworth, J.E., Gornall, R.J. 1996. Cytogenetic evidence for autopolyploidy in *Parnassia palustris*. New Phytol. 134: 641–648.

Pellicieraceae

K. KUBITZKI

Pellicieraceae (Triana & Planchon) L. Beauvis. ex Bullock in Taxon 8: 182 (1959).

Usually small (more rarely up to 18 m high) mangrove trees with fluted trunk bases. Leaves spirally arranged, asymmetric, involute in bud, glabrous, sessile, exstipulate, the wider half of the blade initially bearing a series of prominent but ephemeral (salt?) glands rolled innermost in bud but after expansion becoming entire; leaf base narrowed to the insertion with a pair of extrafloral nectaries and decurrent. Flowers solitary in the 1–3 leaf axils below the resting terminal bud, pentamerous, sessile, enclosed by 2 prophylls before anthesis, large (–12 cm wide) at anthesis; calyx lobes short, free, quincuncially arranged, with numerous small glands internally at the base; petals free, lanceate, about 6 cm long, ephemeral and falling with the sepals and stamens; stamens 5, free, lying within the alternate groves of the ovary; anthers narrow, about 2.5 cm long, sagittate, distally pointed, extrorsely dehiscing by longitudinal slits; gynoecium 2-carpellate, syncarpous; ovary superior, tapering into a narrow style with a bifid stigma; ovary cylindrical, woody, about 6 cm long with 10 longitudinal grooves; locules 2, each containing a single anatropous ovule pendulous from the inner angle, only one of them developing, or one locule sterile. Fruit indehiscent, napiform, 8–12 cm long and about as wide, somewhat flattened, the style persisting; fruit wall ridged, leathery externally but spongy within; seed solitary, exalbuminous; seed coat at maturity of ribbon-like fragments; embryo cordate, with elongate hypocotyl pointing into the stylar beak; plumule reddish, long, slender and hooked, enclosed by 2 fleshy cotyledons.

Monotypic, *Pelliciera rhizophorae* Triana & Planchon a mangrove of Central and northern South America.

MORPHOLOGY AND ANATOMY. The evergreen trees have a well-developed, erect trunk and a narrow, elongate crown. They develop terminal "naked" buds (not protected by bud scales) in association with flowering. Branching is discontinuous (Attim's model) and tiers of plagiotropic axes are produced that bear terminal leaf clusters.

Apart from subterranean roots, aboveground aerial roots are formed on the stem base in numerous orthostichies (Fig. 97). These roots do not immediately penetrate the surface of the stem base but grow downwards inside the bark. Thus, the cortical tissue is elevated but the surface remains unbroken, except for numerous prominent lenticels, so that the vertical series of adventitious roots below a covering layer of bark forms the fluted buttresses (Tomlinson 1986). These buttresses form a conical mantle around the stem which at soil level may be up to 1.5 m wide and may extend on the stems up to the high-tide mark (Fuchs 1970).

Anatomically, *Pelliciera* is characterized by the presence of raphides in idioblasts of the parenchymatous tissue; vessel segments with oblique, simple perforations; imperforate tracheary elements with inconspicuously bordered pits; elongate, sinuous fibres in the mesophyll; encyclocytic stomata; branched, sclerenchymatous idioblasts and elongated thick-walled fibres in the cortex and pith; and petioles with an almost cylindrical vascular system (Beauvisage 1920; Metcalfe and Chalk 1950; Baretta-Kuipers 1976).

EMBRYOLOGY AND CHROMOSOMES. Unknown.

POLLEN MORPHOLOGY. Pollen grains are 3-colporate, spheroidal, crassi-exinous, with lalongate ora (Erdtman 1952). Ornamentation of the grains vary both in Recent and fossil material from scabrate to distinctly reticulate, and a size range of 40–90 μm is found likewise in modern and fossil pollen (Graham 1977).

PHYTOCHEMISTRY. Tests for aluminium accumulation were negative, in contrast to the rich occurrence of aluminium in Theaceae and Ternstroemiaceae (Chenery 1948).

AFFINITIES. Traditionally included in Ternstroemiaceae, *Pelliciera* was removed from this family on anatomical grounds by Beauvisage

Fig. 97. Pellicieraceae. Fluted trunk bases of mangrove species *Pelliciera rhizophorae*, Piedra Blancas N.P., Costa Rica. Courtesy W. Huber, A. Weissenhofer, Vienna

(1920). It differs from Ternstroemiaceae and Theaceae in possessing raphides, in having wood with pore multiples and simple vessel perforations, and in lacking aluminium accumulation, features shared with Tetrameristaceae and Marcgraviaceae. Indeed, recent molecular studies employing both plastid and nuclear genes (Savolainen, Chase et al. 2000; Savolainen, Fay et al. 2000; Soltis et al. 2000; Anderberg et al. 2002) provide strong support for a clade consisting of *Pelliciera, Tetramerista*, Marcgraviaceae, and *Impatiens*, whereas Ternstroemiaceae and Theaceae occupy more distant positions within the broadly construed Ericales. *Pelliciera* is closest to *Tetramerista*, and the two share, among other features, pits on the upper surface of the petals.

Distribution, Habitats and Seed Dispersal. *Pelliciera* is distributed on the Pacific coast of Central and northern South America from Costa Rica to northernmost Ecuador, and more sparsely on the Atlantic coast of Colombia. The microfossil record shows that, during the Eocene and Oligocene, *Pelliciera* was a common member of coastal communities throughout the Caribbean region and from southern Mexico to Panama and Guyana, and also was present in Nigeria but, by the end of the Tertiary, it was reduced to its present range (Fuchs 1970; Graham 1977). Its increasing range restriction is attributed, i.a., to changes in climate, sea level, soil salinity (Jímenez 1984) and competition by *Rhizophora*. With *Rhizophora*, *Pelliciera* occupies the intertidal zone, but with a much more restricted ecological range, typically in sheltered sites such as estuarine banks or protected beaches (Tomlinson 1986).

The fallen fruits release the seeds which are dispersed in the water and germinate immediately after becoming stranded. The poor protection of the seed by the fruit wall is of short duration and precludes dispersal of the fruits by ocean currents. Seeds in water swell after a few hours and sink (Collins et al. 1977).

Only one monotypic genus:

Pelliciera Planchon & Triana   Fig. 98

Pelliciera Planchon & Triana in Benth. & Hook., Gen. Pl. 1: 186 (1862); Kobuski, J. Arnold Arbor. 32: 256–262 (1951), rev.

Description as for family.

Fig. 98. Pellicieraceae. *Pelliciera rhizophorae.* **A** Flowering branch. **B** Androecium and gynoecium. **C** Fruit. Drawn by H.J. Cuddy. (Robyns 1967)

Selected Bibliography

Anderberg, A.A. et al. 2002. See general references.
Baretta-Kuipers, T. 1976. Comparative wood anatomy of Bonnetiaceae, Theaceae and Guttiferae. Leiden Bot. Ser. 3: 76–101.

Beauvisage, L. 1920. Contribution à l'étude anatomique de la famille des Ternstroemiacées. Tours.
Chenery, E.M. 1948. Aluminium in the plant world. Pt. I. General survey in dicotyledons. Kew Bull. 1948: 173–183.
Collins, J.P., Berkelhamer, R.C., Mesler, M. 1977. Notes on the natural history of the mangrove *Pelliciera rhizophorae* Tr. & Pl. (Theaceae). Brenesia 10/11: 17–29.
Erdtman, G. 1952. See general references.
Fuchs, H.P. 1970. Ecological and palynological notes on *Pelliciera rhizophorae.* Acta Bot. Neerl. 19: 884–894.
Graham, A. 1977. New records of *Pellicieria* (Theaceae/Pellicieraceae) in the Tertiary of the Caribbean. Biotropica 9: 48–52.
Jímenez, J.A. 1984. A hypothesis to explain the reduced distribution of the mangrove *Pelliciera rhizophorae.* Tr. & Pl. Biotropica 16: 304–308.
Kobuski, C.E. 1951. Studies in the Theaceae XXII. The genus *Pelliciera.* J. Arnold Arbor. 32: 256–262.
Metcalfe, C.E., Chalk, L. 1950. See general references.
Robyns, A. 1967. Flora of Panama, Pt. IV. Fam. 122. Theaceae. Ann. Missouri Bot. Gard. 54: 41–56.
Savolainen, V., Chase, M.W. et al. 2000. See general references.
Savolainen, V., Fay, M.F. et al. 2000. See general references.
Soltis, D.E. et al. 2000. See general references.
Szyszyłowicz, I. von 1893. Theaceae. In: Engler, K., Prantl, K. Die natürlichen Pflanzenfamilien III, 6. Leipzig: W. Engelmann, pp. 175–192.
Tomlinson, P.B. 1986. The botany of mangroves. Cambridge: Cambridge University Press.

Polemoniaceae

D.H. Wilken

Polemoniaceae Juss., Gen. Pl.: 136 (1789), nom. cons.

Annual or perennial herbs, sometimes vines or woody shrubs, rarely small trees; indumentum of multicellular and uniseriate trichomes, eglandular or with terminal unicellular to multicellular glands, rarely plants completely glabrous. Leaves alternate to opposite, pinnately or less often palmately veined, entire to deeply divided, sometimes compound, petiolate or sessile, exstipulate. Inflorescences dichasial, in racemose to paniculate or capitate clusters, rarely flowers solitary. Flowers hermaphroditic, actinomorphic, sometimes zygomorphic, hypogynous; sepals (4)5(6), connate, rarely free, persistent, the tube herbaceous along the midribs, hyaline between the midribs, sometimes herbaceous throughout; petals (4)5(6), the corolla rotate to salverform, funnelform or bilabiate, the lobes mostly convolute in bud; stamens (3–)5(6), alternate with the petals, the filaments equal to unequal in length and attached to the tube, sometimes at differing levels; anthers basifixed to dorsifixed, dithecal, tetrasporangiate, dehiscing by longitudinal slits; ovary inserted on a nectariferous disk, syncarpous, (2)3-locular, placentae axile; style with (2)3 stigmatic branches; ovules 1–many per locule, in 2 rows, (hemi)anatropous, tenuinucellate, with one thick integument. Fruit a dehiscent dry capsule, loculicidal or rarely septicidal (*Acanthogilia*, *Cobaea*), sometimes explosively dehiscent (*Collomia*, *Phlox*), rarely indehiscent; seeds smooth to angled, sometimes winged, the epidermal cells with spiral thickenings and included mucilage, the cells bursting when wetted and forming a mucilaginous coat; mature embryos straight; endosperm oily.

The family consists of 18 genera and about 380 species, distributed primarily in North America, extending into Central and South America, with a few species in Eurasia.

VEGETATIVE MORPHOLOGY. Most members of the Polemoniaceae are annuals or herbaceous perennials. Herbaceous perennials survive dormancy either as rhizomes, basal rosettes, or buds at the tips of vertical, underground axes. Mono-carps living from 2–10 years as vegetative rosettes are known in *Aliciella* and *Ipomopsis*. Vines characterize *Cobaea*, while *Acanthogilia* and *Cantua* are exclusively shrubs. Suffrutescent perennials or subshrubs also occur in several genera with a majority of annuals, including *Aliciella*, *Collomia*, *Eriastrum*, *Gilia*, *Ipomopsis*, and *Linanthus*. *Loeselia*, *Phlox* and *Polemonium* are primarily perennial but each includes annual species.

Leaf arrangement is usually alternate, but opposite in *Phlox* and *Linanthus*. Many annuals develop basal rosettes but in some genera (*Collomia*, *Eriastrum*, *Gilia*, *Navarretia*) the basal leaves are weakly developed or senesce during bolting and efflorescence. The leaves of *Gymnosteris* are reduced to a single whorl of foliaceous bracts subtending the terminal flower cluster. Leaves of Polemoniaceae lack stipules and can be petiolate or sessile. Leaf blades are pinnately veined, varying from entire to compound. Simple leaf blades with entire to toothed or lobed margins predominate in *Cantua*, *Collomia*, *Gymnosteris*, *Loeselia*, and *Phlox*. Although other genera may have entire leaves, the blades of at least the basal and lower cauline leaves are deeply lobed or divided. Palmately divided leaves occur in some species of *Collomia*, *Ipomopsis* and *Linanthus*. *Cobaea* and *Polemonium* have compound leaves, with the terminal leaflet in *Cobaea* modified into a branched tendril.

VEGETATIVE ANATOMY. Primary stems are characterized by a well-developed, often collenchymatous cortex and a well-defined endodermis (Metcalfe and Chalk 1950). Nodes are unilacunar. Phloem elements are arranged either as a continuous cylinder or as separate strands associated with sclerenchyma, but the xylem is in a continuous ring. Woody taxa are generally characterized by nonstoried wood, relatively few vessels per group, vessel-elements with simple perforation plates and alternate circular bordered pits on their lateral walls, axial vasicentric scanty parenchyma, and both multiseriate and uniseriate rays with mostly erect cells (Carlquist et al. 1984). Mean vessel diameters range from over 60 µm in *Cobaea* to less than 30 µm in *Leptodactylon* and *Phlox*. Mean vessel-element lengths range from 116–250 µm in *Leptodactylon* to over 400 µm in *Acanthogilia* and *Cantua buxifolia*. Tracheids are the primary imperforate tracheary elements in such temperate genera as *Eriastrum, Leptodactylon,* and *Phlox*, whereas tropical genera (*Cantua, Cobaea, Huthia,* and *Loeselia*) have fiber-tracheids or libriform fibers. Rayless or near-rayless woods occur in *Cobaea, Eriastrum, Linanthus,* and *Phlox*. Although wide vessels, vessel diameter dimorphism, and abundant ray parenchyma in *Cobaea* are common in dicotyledonous vines, the absence of rays appears to be unique (Carlquist et al. 1984). Overall, ancestors of *Acanthogilia, Cantua* and *Loeselia* appear to have been woody. In contrast, raylessness or predominantly erect ray cells suggest secondary woodiness in *Cobaea, Eriastrum, Linanthus,* and *Phlox* (Carlquist 1992).

Leaves can be either bifacial or isolateral. The cuticle varies from smooth to verrucose. In some species of *Collomia, Gilia,* and *Phlox*, the epidermal cells are mucilaginous. Stomata, sometimes present on both upper and lower surfaces, are anomocytic but approach paracytic in taxa with linear leaves or leaflets (Metcalfe and Chalk 1950). Isolateral leaves are associated with narrowly dissected or linear leaves, especially in species of arid habitats. Large intercellular spaces are notably absent in the meosphyll of bifacial leaves.

Trichomes are either uniseriate eglandular or uniseriate glandular with terminal, globose to ovoid heads that are either unicellular or multicellular and often glandular. Eglandular trichomes predominate in *Eriastrum, Ipomopsis, Langloisia, Linanthus,* and *Phlox*, whereas glandular trichomes predominate in *Aliciella, Allophyllum, Bonplandia, Cantua, Collomia, Gilia, Navarretia,* and *Polemonium*. Although most species of *Gilia* have unicellular glands, those of the related *Aliciella* are multicellular. In some species of *Allophyllum, Collomia, Gilia, Navarretia* and *Polemonium*, glandular trichomes are the source of a distinctive mephitic odor. *Langloisia* is distinctive in its possession of branched trichomes. Densely tangled trichomes in some genera (e.g., *Eriastrum, Ipomopsis, Gilia*) are uniseriate eglandular, repeatedly geniculate, and composed of "rectangular parallelepipeds, joined at right angles to each other" (Grant 1959). Multicellular bristles or setae occur on foliar organs in *Langloisia* and some *Loeselia*.

INFLORESCENCE STRUCTURE. A detailed morphological analysis of the inflorescences of Polemoniaceae was given by Weberling (1998). Most Polemoniaceae have terminal, determinate inflorescences. Many- to few-flowered inflorescences are most common but solitary flowers occur in some species. All inflorescences found in Polemoniaceae can be related to determinate panicles, which are of wide occurrence. Partial inflorescences are often cymose, thus leading to thyrso-paniculate or, if such cymes are reduced to single flowers, to botryoid inflorescences. These cymes are usually dichasia. The occurrence of paired flowers is due to unequal branching of such dichasia (e.g., in *Bonplandia geminiflora*). Other modifications of paniculate inflorescences include condensation to capitate clusters. The arrangement of parts is sometimes obscured by metatopies (e.g., *Langloisia, Gilia*). Both basipetal and acropetal development occurs in various parts of the inflorescence. In some species of hummingbird- or hawkmoth-pollinated *Ipomopsis*, flowers are displaced to one side of thyrso-paniculate inflorescences. Inflorescences can be bracteate throughout or ebractetae distally. Floral bracts are often reduced but can be foliaceous. In *Loeselia* the foliar bracts often have conspicuous "window-like" hyaline or scarious areolae.

FLORAL MORPHOLOGY AND ANATOMY. Polemoniaceae flowers, with few exceptions, are actinomorphic and hypogynous and have a pentamerous perianth composed of fused sepals and fused petals, and 5 alternipetalous stamens. Tetramerous and hexamerous flowers are found in *Linanthus*. The ovary has 3, rarely 2, locules with axile placentation; the style has 3 stigmatic branches and is inserted on an entire to lobed annular nectariferous disk. The sepals of *Cobaea* and some species of *Gilia* are almost completely free, but form a tube in other genera. The tube is herbaceous in *Bonplandia, Cantua, Cobaea, Collomia,* and *Polemonium*. The tube of *Collomia* is notably plicate in late flower and fruit. In all other genera the tube is composed of 5 herbaceous costae that extend into the lobes and are separated by hyaline portions proximal to the sinuses. The hyaline portions can be narrower or wider than the herbaceous costae, and are often ruptured by the developing fruit.

Corollas can be campanulate, salverform, funnelform, or sometimes rotate (Fig. 100). Zygomorphic, often bilabiate, flowers occur in *Bonplandia* and some species of *Ipomopsis, Langloisia* and *Loeselia*. The petal lobes are convolute in bud, although imbricate petals are reported in *Cantua* (Johnson et al. 1999). In most Polemoniaceae, the

5 stamens are inserted at or above the middle of the tube, sometimes between the lobes. In *Bonplandia*, *Cantua*, and *Loeselia*, the stamens are inserted in the lower tube, but those of *Cobaea* and some *Gilia* are nearly basal. Filaments can be equal or unequal in length. Anthers are basifixed in most species, but are dorsifixed in *Cobaea* and *Eriastrum*.

Ten vascular bundles diverge from the central stele of the pedicel (Dawson 1936). Five alternate bundles become the midribs of the sepals. The other five branch several times distally, giving rise sequentially to traces supplying the corolla and stamens, the nectariferous disc, and the ovary. The nectariferous disc in most Polemoniaceae is supplied by 5 bundles, but 10 bundles supply those of *Cantua* and *Cobaea*. In most Polemoniaceae, the central bundle leading to each petal lobe trifurcates at the base of the tube; the 3 traces often anastomose in the throat or distal portion of the lobe but give rise to secondary branches (Day and Moran 1986). Anastomoses are lacking in *Linanthus*, but the 3 primary traces can branch several times. In *Acanthogilia*, *Cantua*, *Cobaea*, and *Phlox*, some of the distal secondary branches can form additional anastomoses. In *Allophyllum*, *Eriastrum*, *Gilia*, *Ipomopsis*, and *Navarretia* the secondary branches often remain free. Small corollas in several genera often have either 1 or 3 unbranched veins.

EMBRYOLOGY. The pollen grains are 2-celled when shed. The ovule is anatropous to hemianatropous, unitegmic, and tenuinucellate. The archesporial cell functions directly as the megaspore mother cell and the chalazal megaspore of a linear tetrad develops into a Polygonum-type embryo sac. Endosperm formation is nuclear, but the tissue later becomes cellular (Johri et al. 1992).

POLLEN MORPHOLOGY. The pollen grains of Polemoniaceae are oblate to spheroidal and mostly pantoporate; zonotreme (Fig. 99A), bizonotreme or anomotreme/porate (Fig. 99B) or colp(or)ate figurations are found in a few taxa. Pollen grains vary in size between 22 and 190 μm, and the number of apertures vary in the range (4–)6–50 (Stuchlik 1967; Taylor and Levin 1975). Pollen size is more closely correlated with style length than with taxonomic affinity (Plitmann and Levin 1983). The sexine can be striate, reticulate, pertectate, or verrucate. Germinal apertures range from circular to oval or elliptic in shape. The lirae in pollen with striate sexine can be vermiform, but most often radiate from the apertures like the

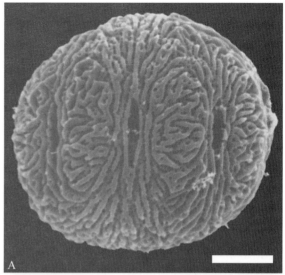
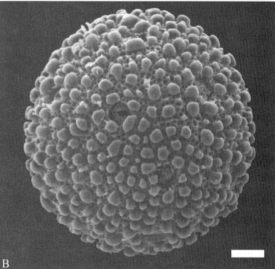

Fig. 99. Polemoniaceae, pollen. **A** *Collomia biflora*, zono(6–8)colporate, REM. **B** *Cantua buxifolia*, pantoporate, sexine insulate, REM. Bars = 10 μm. Photo B.E. Herber

"lines of force in a magnetic field" (Stuchlik 1967). The sexine of pantotreme reticulate pollen varies from homobrochate to heterobrochate.

KARYOLOGY. In Polemoniaceae, chromosome numbers are based on multiples of 6, 7, 8, and 9. Grant (1959) hypothesized x = 9 as ancestral and suggested that taxa with lower numbers were derived, sometimes independently among lineages. Chromosome numbers exclusively with x = 9 are reported in *Acanthogilia*, *Cantua*, *Gilia*, *Loeselia*, *Navarretia*, *Linanthus*, and *Polemonium*. Chromosome numbers with x = 7 occur in *Eriastrum*, *Ipomopsis*, *Langloisia*, and *Phlox*. *Collomia* and *Gymnosteris* have chromosome numbers based on x = 8 and x = 6, respectively. Both *Aliciella* and *Allophyllum* have diploid chromosome numbers based on either x = 8 or x = 9. The most parsimonious origin for the 2n = 30 reported in *Bonplandia* may be tetraploidy involving ancestors with n = 7 and 8. Notably, *Giliastrum* includes species with 2n = 12, 18, 20, 24 and 36, with x = 6 as the most likely base number. The highest numbers are 2n = 54 in *Cantua* (hexaploid based on x = 9) and 2n = 52 in *Cobaea scandens*, which may be derived by an aneuploid reduction from the hexaploid condition. Tetraploid and some hexaploid taxa occur in *Aliciella, Gilia, Giliastrum, Ipomopsis, Linanthus,* and *Phlox*. B-chromosomes have been reported in *Linanthus* and *Phlox* (Smith and Levin 1967; Patterson 1980).

REPRODUCTIVE AND POLLINATION BIOLOGY. A rich diversity of breeding systems is found in Polemoniaceae and has been discussed at length by Grant and Grant (1965). Both self-incompatibility and self-compatibility are found within most genera. Most self-compatible taxa have mixed mating systems, but autogamy has been documented in members of at least 9 genera (Plitmann and Levin 1990). Cleistogamy is known in *Collomia grandiflora* and *Polemonium micranthum*. Distyly, accompanied by a weak self-incompatibility system, occurs in *Gilia heterostyla* and functional gynodioecy has been reported in *Ipomopsis rubra*. The effective pollinators of many Polemoniaceae are Coleoptera, Diptera, Hymenoptera, and Lepidoptera. Variously colored corollas, some with distinctive nectar guides, attract insects that collect either pollen, nectar, or both. The type of insect visitor is often correlated with flower size, corolla shape and color, and inflorescence structure. Taxa visited or pollinated by Coleoptera (e.g., *Ipomopsis congesta, Linanthus parryae*) have broadly campanulate, often light-colored corollas or small flowers disposed in congested or capitate inflorescences. Erect, often fragrant flowers with small corollas with long, narrow tubes are often pollinated by Diptera (Bombyliidae, Cyrtidae) or relatively small Lepidoptera. Hawkmoth pollination is known in at least 7 genera, associated with fragrant, vespertine or nocturnal, white, long-tubed corollas. Spreading or pendent flowers with relatively large, brightly colored, diurnal corollas, and copious nectar attract hummingbirds in at least 8 genera. Large, nocturnal solitary flowers in some taxa of *Cobaea* are bat pollinated. *Polemonium viscosum* displays an unusual polymorphism for floral size and scent, promoting attractiveness to bumblebees while repelling nectar-thieving ants (Galen and Newport 1988).

Most genera include two or more pollination syndromes. *Cobaea* includes taxa that are autogamous or chasmogamous and pollinated by hawkmoths and bats. Although autogamy occurs in some annual taxa of *Ipomopsis*, the genus also includes chasmogamous floral syndromes associated with effective pollination by moths (Sphingidae), bees (Andrenidae, Apidae, Halictidae), butterflies (Hesperiidae), beetles (Melyridae), and hummingbirds (Trochilidae). Diversification also has occurred at the infraspecific level in *Gilia splendens*, which includes geographic races pollinated by hummingbirds or flies, and one predominantly autogamous race.

FRUITS AND SEEDS. Dehiscence is either septicidal in *Cobaea*, or loculicidal in the remaining genera. Except for some species of *Navarretia*, in which fruits dehisce from the base, the fruits of the family are capsules. Mature fruits vary from globose to ovoid. The number of seeds varies from 1 to many per locule. Seeds, which vary from less than 100 μm to over 2 cm, are often ovoid to ellipsoid and can be smooth, angular, or verrucose. Seeds of *Acanthogilia, Cantua,* and *Cobaea* are thin and often prominently winged. The seeds are albuminous and the embryo is straight or highly curved. The seed coats of many Polemoniaceae have cylindrical cells with prominent helical secondary thickenings and contain mucilage (Grant 1959). When wetted, imbibition causes the outer walls to separate from the testa, accompanied by extrusion of the expanded thickenings and mucilage (Grubert and Hambach 1972). *Collomia, Ipomopsis, Phlox,* and *Polemonium* each include species with either copious (annuals) or little, if any, mucilage (perennials). Thus, copious mucilage in the family appears correlated with life form rather than taxonomic affinity.

PHYTOCHEMISTRY. Polemoniaceae are particularly rich in flavonoids, including 6-methoxyflavonols (patuletin, eupalitin, eupatoletin) and C-glycosylfavones (Smith et al. 1977). Floral pigments include delphinidin, cyanidin, and pelargonidin glycosides, seemingly correlated with pollination syndromes (Harborne and Smith 1978), with pelargonidin associated with hummingbird pollination and cyandin or delphinidin associated with insect pollination. The almost exclusive presence of acylated anthocyanins in corollas of *Linanthus, Loeselia,* and *Phlox* is correlated with the presence of C-glycosyl flavones in the vegetative organs. Other constituents of systematic interest include triterpene saponins (Hiller et al. 1979; Jurenitsch et al. 1979) and ketose-isoketose fructose oligosaccharides (Pollard and Amuti 1981).

Subdivisions and Relationships Within the Family. Based on cladistic analyses of molecular and morphological data, Porter and Johnson (2000) subdivided Polemoniaceae into three subfamilies, the Acanthogilioideae, Cobaeoideae and Polemonioideae. The monotypic *Acanthogilia*, an endemic shrub of xeric habitats in Baja California, is the sole member of **Acanthogilioideae**. It has dimorphic leaves, axillary inflorescences, large salverform corollas, and verrucate, zonotreme pollen. The primary leaves are pinnately divided and become spinose, but the axillary short-shoot leaves are simple, entire, and borne in fascicles. **Cobaeoideae** are composed of genera that share a woody or semiwoody habit, herbaceous calyces lacking conspicuous hyaline intervals, pantotreme pollen, and winged seeds. *Cantua* is composed of shrubs and small trees occupying relatively mesic montane and subalpine habitats in the Andes of South America. Leaves in most taxa are simple and the flowers, with salverform to funnelform corollas, are relatively large and disposed in relatively loose, terminal clusters. Calyx tubes are either completely herbaceous or have narrow, relatively thin to hyaline intervals between the herbaceous costae. *Bonplandia* has an herbaceous habit, heterophylly, and paired, zygomorphic, axillary flowers. *Cobaea*, with a distribution restricted to neotropical forests, is the most distinct genus within the family. Its species share a viny habit, deeply compound leaves, terminal tendrils, broadly campanulate corollas, and septicidal capsules.

Polemonioideae are equally diverse but many genera do not have distinct combinations of characters (Mason 1945). *Polemonium* appears unique in the subfamily by having pinnately compound leaves, herbaceous calyces that remain unruptured by the developing fruits, and striate pollen with 30–100 apertures and mostly vermiform lirae.

Most species of *Allophyllum*, *Collomia*, and *Navarretia* have senescent basal rosettes, cauline leaves gradually reduced upwards, congested or somewhat capitate inflorescences, and dark-colored ovoid seeds. Pollen grains may be either pantotreme or zonotreme, with 6–20 apertures. A pertectate or pilate sexine occurs in *Allophyllum* and *Navarretia*. *Collomia* and some *Allophyllum*

have x = 8, but x = 9 occurs throughout *Navarretia* and the remaining species of *Allophyllum*.

Eriastrum, *Langloisia*, and most species of *Ipomopsis* also have senescent basal rosettes, gradually reduced cauline leaves, and terminal, congested or capitate inflorescences. However, all members of this group share x = 7 and have light-colored seeds. With the exception of bilabiate species in *Ipomopsis* and *Langloisia*, most have salverform corollas. The seeds of *Eriastrum* and *Ipomopsis* are often ellipsoid and angled to minutely winged, but those of *Langloisia* are ovoid and smooth or angled. Although some *Ipomopsis* have reticulate pollen, most species in the 3 genera have a striate-reticulate sexine in which the lirae radiate from the apertures. *Ipomopsis minutiflora*, with uniformly glandular, puberulent indumentum, slightly unequal calyx lobes, and bizonotreme pollen, may be better placed elsewhere, perhaps recognized as a distinct genus.

Although Grant (1998) placed *Loeselia* in Cobaeoideae, molecular phylogenies suggest closer relationships to *Eriastrum*, *Gilia*, *Ipomopsis*, and *Langloisia* (Johnson et al. 1996; Porter and Johnson 1998). Most *Loeselia* have a suffrutescent habit, often congested inflorescence, intercostal calyx membranes that are slightly hyaline, pantotreme, pertectate pollen, and minutely winged seeds. Although leaves and bracts are entire to dentate, their margins bear setae similar to those of *Langloisia*. Floral bracts in most *Loeselia* are foliaceous, with hyaline or thinly membranous areolae.

Aliciella, *Gilia*, and *Giliastrum* have flowers disposed in relatively open inflorescences. The sepal lobes often have a well-developed hyaline or scarious margin. *Aliciella* and *Gilia* have paniculate or thyrsoid inflorescences, salverform to funnelform corollas, zonotreme, striate to reticulate pollen, and x = 9. Glands in *Aliciella* are multicellular, but they are unicellular in *Gilia* and *Giliastrum*. *Giliastrum* has either solitary flowers or flowers disposed in open dichasia, rotate to short-funnelform corollas, zonotreme pertectate pollen, and x = 6. Several species in *Gilia* appear sufficiently distinct to merit recognition. They apparently lack basal rosettes and have entire cauline leaves, solitary pedicellate flowers, and short-funnelform to campanulate corollas. *Gilia tenerrima*, with short-funnelform corollas, has anomotreme pollen. *Gilia capillaris*, *G. sinistra*, and *G. leptalea* have funnelform corollas but their pollen is pantotreme. Another group, including *G. filiformis*, *G. inyoensis*, and *G. maculata*, has mostly free sepals, campanulate to rotate corollas, and stamens inserted near the corolla base.

Linanthus and *Phlox* have opposite leaves with entire to palmatifid blades, terminal inflorescences often composed of dichasia or solitary flowers, mostly salverform corollas, pantotreme pollen with reticulate to striate-reticulate sexine, and x = 7. The closely related *Gymnosteris*, with x = 6, salverform corollas, and pantotreme reticulate pollen, apparently lacks cauline leaves, but the terminal capitate inflorescence is subtended by a whorl of photosynthetic bracts.

As an alternative to the above phylogenetic system of classification of the family, another classification based on traditional taxonomy has been proposed by Grant (1998). In a review of these classifications, Grant (2001) reviews and compares both of them.

AFFINITIES. Based on morphology, the Polemoniaceae have been traditionally treated as allied to Convolvulaceae (Takhtajan 1980) or Hydrophyllaceae (Cronquist 1981) within Asteridae. Thorne (1983), among others, placed Polemoniaceae nearest Fouquieriaceae, but with both in Asteridae, based on their pentamerous perianth and androecium, gamopetaly, epipetalous stamens, axile placentation, loculicidal capsule, and well-developed embryos. Fouquieriaceae and some Polemoniaceae also have thin, winged seeds and fasciculate spinescent leaves (*Acanthogilia*).

Several cladistic analyses using diverse molecular data have consistently suggested a closer alliance with Fouquieriaceae and such families as Diapensiaceae, Ericaceae, and Primulaceae within Ericales/Primulales (Downie and Palmer 1992; Olmstead et al. 1993; Johnson et al. 1996, Porter 1997; Porter and Johnson 1998). A tricarpellate gynoecium characterizes Fouquieriaceae, Diapensiaceae, Roridulaceae, and some Ericaceae, but elsewhere is unusual in Asteridae. Ketose and isoketose oligosaccharides also occur in some Ericaceae, but not in Asteridae s.s. (Pollard and Amuti 1981). Accordingly, Polemoniaceae have been included in an expanded Ericales at the base of Asteridae by the Angiosperm Phylogeny Group (APG 1998). In the five-gene sequence analysis by Anderberg et al. (2002), Polemoniaceae and Fouquieriaceae appear in close association, albeit with only moderate support, which is understandable in view of the strong morphological divergence between the two families.

DISTRIBUTION AND HABITATS. *Acanthogilia* is endemic to thorn shrublands of Baja California.

Most members of the Cobaeoideae occur in subtropical to tropical woodlands and forests of Central and western South America. The majority of Polemonioideae occur in western North American woodlands and shrublands, associated primarily with summer-dry Mediterranean, Great Basin, and Sonoran Desert climates. Some genera (*Polemonium*, *Phlox*) include taxa adapted to the relatively moist summer and cool to cold winter climates of both eastern and western North America, including the arctic. *Polemonium* is also represented by a few species in relatively cool moist climates of Eurasia as far south as the northern flank of the Himalayas. Initial diversification may have occurred during the late Cretaceous or early Tertiary in what is now Mexico, Central America, or northern South America. Dispersal to Eurasia may have involved both amphi-Beringian and amphi-Atlantic bridges. However, over 90% of Polemoniaceae occur in western North America, the majority of which are annuals (Grant 1959; Raven and Axelrod 1978). Thus, progressive cooling and drying, successive orogenies, and the development of a Mediterranean climate during the mid to late Tertiary may have played a significant role in radiation, especially within Polemonioideae. Some predominantly western North American genera (*Collomia*, *Gilia*, *Ipomopsis*, *Linanthus*, and *Navarretia*) have 1–2 close relatives in western South America, and two species (*Phlox gracilis*, *Polemonium micranthum*) occur disjunctly. These amphitropical relationships are best attributed to long-distance dispersal during the late Tertiary (Raven 1963).

FOSSIL RECORD. Fossil pollen has only rarely been attributed to the family (see Muller 1981). A megafossil from the Eocene of Utah, USA, *Gilisenium*, which is close to *Gilia*, is important as an early megafossil record of a herbaceous plant (Lott et al. 1998). Pleistocene fruits attributed to *Phlox* have been reported from Alaska (Chaney and Mason 1936).

ECONOMIC IMPORTANCE. Some species of *Cobaea*, *Cantua*, *Collomia*, *Ipomopsis*, *Linanthus*, and *Polemonium* are cultivated for ornament, and cultivars have been developed from several species of *Phlox*. Species of *Ipomopsis*, *Loeselia*, and *Polemonium* are used by native North Americans to produce a mild soap. In Central America, infusions of *Cobaea* flowers have been used as a cough suppressant. Potentially useful anti-inflammatory and cyto-toxic antitumor compounds (triterpene saponins, cucurbitacins) have been isolated from

Ipomopsis and *Polemonium* (Aurada et al. 1982; Arisawa et al. 1984). *Cantua buxifolia*, considered sacred by the Incas, is the national flower of Peru.

Key to the Genera

1. Leaves pinnately compound   2
 - Leaves simple, entire to pinnately or palmately divided 3
2. Vines; terminal leaflet a tendril   **2. Cobaea**
 - Annuals or herbaceous perennials; terminal leaflet foliar
                                              5. Polemonium
3. Axis above cotyledons unbranched and apparently leafless; leaves reduced to a bracteate whorl subtending the terminal inflorescence   **18. Gymnosteris**
3. Cauline nodes and leaves present; cauline leaves
 - alternate or opposite   4
4. Leaves predominantly opposite   5
 - Leaves predominantly alternate   6
5. Stamens inserted at the same level in the tube or throat
                                              16. Linanthus
 - Stamens inserted at different levels in the tube **17. Phlox**
6. Stamens inserted below the middle of the tube   7
 - Stamens inserted at or above the middle of the tube, sometimes between the lobes   10
7. Sepals mostly free; corolla lobes longer than tube
                                                  15. Gilia
 - Sepals mostly fused; corolla lobes shorter than tube   8
8. Shrubs; corolla campanulate or salverform; seeds thin, conspicuously winged   **3. Cantua**
 - Annuals, suffrutescent perennials, or subshrubs; corollas more or less zygomorphic, seeds plump, wingless or inconspicuously winged   9
9. Flowers axillary, in pairs; corolla tube conspicuously bent
                                             4. Bonplandia
 - Flowers terminal, solitary or in clusters, if axillary then corolla actinomorphic; corolla tube straight **12. Loeselia**
10. Plants woody; flowers axillary; seeds winged
                                            1. Acanthogilia
 - Plants herbaceous, if woody then suffrutescent or subshrubs; flowers not axillary; seeds wingless or incompletely and minutely winged   11
11. Inflorescence capitate or corymbose   12
 - Inflorescence paniculate or thrysoid, sometimes axillary
                                                            16
12. Calyx plicate, the intercostal membranes distended distally, not ruptured by the developing fruit; leaf tips acute to rounded   **6. Collomia**
 - Calyx not plicate, the intercostal membranes hyaline, often ruptured by the developing fruit; leaf tips mucronate to spinulose   13
13. Inflorescence bracts spinulose, often pinnatifid; calyx lobes subequal   14
 - Inflorescence bracts acute to mucronate, entire, sometimes pinnatifid to palmatifid; calyx lobes equal   15
14. Indumentum predominantly eglandular   **10. Eriastrum**
 - Indumentum predominantly glandular; eglandular trichomes not arachnoid   **8. Navarretia**
15. Corolla campanulate, funnelform, or rotate, often with a well-developed throat   **15. Gilia**
 - Corolla salverform, throat lacking   **9. Ipomopsis**
16. Basal and cauline leaves mostly entire   **15. Gilia**
 - Basal and cauline leaves pinnately lobed to pinnatifid or palmatifid   17
17. Leaf segments with 1–3 terminal setose bristles
                                             11. Langloisia

 - Leaf segments rounded to mucronate or spinescent, setose bristles lacking   18
18. Upper cauline leaves palmatifid   19
 - Upper cauline leaves entire to pinnately toothed or pinnatifid   20
19. Terminal leaf segment often longer than lateral segments, rounded   **7. Allophyllum**
 - Terminal leaf segment not markedly longer than lateral segments, spinescent   **16. Linanthus**
20. Corolla rotate to short-funnelform, lobes longer than the tube; stamens inserted in the lower tube **13. Giliastrum**
 - Corolla salverform to long-funnelform, lobes equal to or shorter than the tube; stamens inserted in the upper tube or throat   21
21. Indumentum predominantly glandular; trichomes with mostly multicellular glands   **14. Aliciella**
 - Indumentum eglandular to glandular; glandular trichomes with unicellular glands   22
22. Corolla campanulate, funnelform, or rotate, often with a well-developed throat   **15. Gilia**
 - Corolla salverform, throat lacking   **9. Ipomopsis**

I. Subfam. Acanthogilioideae J.M. Porter and L.A. Johnson (2000).

Shrubs; leaves dimorphic, the primary ones alternate, spinescent, and pinnate, the axillary ones simple, in fascicles; corollas actinomorphic, salverform; pollen zonotreme, verrucate; fruit dehiscing both loculicidally and septicidally; seeds flat, winged.

1. *Acanthogilia* A.Day & R.Moran

Acanthogilia A.Day & R.Moran, Proc. Calif. Acad. Sci. 44: 111 (1986).

Spiny shrubs with dimorphic leaves; long-shoot leaves alternate, persistent, pinnate with spinescent divisions; axillary short-shoot leaves in fascicles, linear; pubescence minutely glandular. Inflorescence axillary, 1–3-flowered. Calyx tubular, often with narrow hyaline intervals; corolla actinomorphic, salverform; stamens subequally inserted near the middle of the tube, filaments glabrous; anthers exserted; pollen zonotreme, verrucate; style exserted. Fruit loculicidally and septicidally dehiscent; seeds 1–6 per locule, flat, winged. $2n = 18$. One sp., *A. gloriosa* (Brandegee) Day & Moran, thorn shrublands of N Baja California.

II. Subfam. Cobaeoideae (D. Don) Arnott (1832).

Vines, shrubs, small trees, or suffrutescent perennials; leaves monomorphic, simple to pinnately lobed or compound; corollas campanulate, salverform or somewhat bilabiate; pollen pantotreme, reticulate, striate, or verrucate; fruit dehiscing loculicidally or septicidally; seeds somewhat flat, usually winged.

2. *Cobaea* Cavanilles — Fig. 100B–D

Cobaea Cavanilles, Icon. 1: 11. (1791); Prather, Syst. Bot. Monogr. 57: 1–81 (1999).

Vines with alternate, pinnately compound leaves; lateral leaflets 4–8, terminal leaflet a branched tendril; plants glabrous to variously pubescent. Inflorescence cymose, 1–5-flowered; pedicels often with 2 foliaceous bracts. Calyx with mostly free, herbaceous sepals; corolla actinomorphic, broadly campanulate, constricted near the base; stamens inserted near the base of the tube, filaments basally villous; anthers exserted, dorsifixed; pollen large, pantotreme, reticulate; style exserted. Fruit dehiscing septicidally; seeds 1–many per locule, flat, often broadly winged. $2n = 52$ reported in *C. scandens* Cav. About 20 spp., forests of tropical Central and NW South America.

3. *Cantua* Juss. ex Lam. — Fig. 100A

Cantua Juss. ex Lam., Encycl. 1: 603 (1785); Infantes Vera, Lilloa 31: 75–107 (1962).
Huthia Brand (1908).

Shrubs or small trees, often with dimorphic leaves; principal leaves alternate, sometimes also in fascicles, thin to coriaceous, entire to crenate, rarely lobed; plants glabrous or with eglandular to glandular indumentum. Inflorescence of 3–10 flowers, terminal, raceme-like/botryoid or somewhat capitulate. Calyx tubular to campanulate, herbaceous throughout or thin and hyaline; corolla actinomorphic, salverform to campanulate; stamens inserted at different levels near or below the middle of the tube, filaments glabrous to villous; anthers exserted to included; pollen pantotreme, reticulate to verrucate; style included to exserted. Fruit dehiscing loculicidally; seeds 3–8 per locule, flat, winged. $2n = 54$. Ca. 20 spp., interandine region of Ecuador, Peru and Bolivia.

4. *Bonplandia* Cav. — Fig. 100H

Bonplandia Cav., Anales Hist. Nat. 2: 131 (1800); Rzedowski et al., Acta Bot. Mex. 31: 55–61 (1995).

Suffrutescent perennial, sometimes flowering the first year, with alternate, linear to ovate, simple to

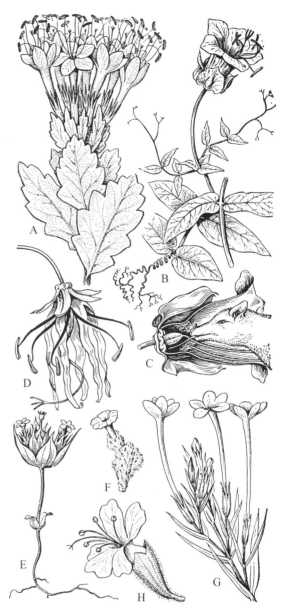

Fig. 100. Polemoniaceae. A *Cantua quercifolia*, flowering shoot. B, C *Cobaea scandens*. B Flowering shoot with tendrils. C Vertical section of flower with pollinating bat. D *Cobaea penduliflora*, flower. E *Gymnosteris parvula*, whole plant. F *Phlox bryoides*. G *Phlox longiflorum*. H *Bonplandia geminiflora*, flower. (Takhtajan 1981)

pinnately lobed leaves, margins entire to serrate; indumentum minutely glandular puberulent, sometimes eglandular. Inflorescence of terminal and axillary sympodia. Flowers usually paired; perianth zygomorphic; calyx herbaceous throughout; corolla somewhat bilabiate, geniculate at the throat; stamens inserted near the mid-tube, filaments basally puberulent; anthers exserted; pollen pantotreme, striate-reticulate; style slightly included to exserted. Fruit dehiscing loculicidally; seeds one per locule, somewhat thick, narrowly winged or wingless. $2n = 30$. One species, *B. geminiflora* Cav., woodlands and forests of Mexico.

III. Subfam. Polemonioideae (Juss.) Arnott (1832).

Annuals, perennials, sometimes shrubs; leaves simple to pinnately or palmately lobed or compound; corollas rotate to salverform, sometimes bilabiate; pollen pantotreme or zonotreme, mostly striate to reticulate, sometimes pertectate; fruit dehiscing loculicidally; seeds ovoid to ellipsoid, wingless.

5. *Polemonium* L.

Polemonium L., Sp. Pl. 1: 162 (1753); Davidson, Univ. Calif. Publ. Bot. 23: 209–282 (1950), rev.
Polemoniella Heller (1904).

Herbaceous perennials, often from rhizomes, rarely annual; cauline leaves alternate, compound; indumentum eglandular or glandular. Inflorescences mostly terminal, paniculate to capitulate. Calyx actinomorphic, herbaceous throughout, lobes equal; corolla actinomorphic, salverform, funnelform, or rotate; stamens inserted at the same level in the lower tube, filaments equal in length, pubescent; anthers included to exserted; pollen pantotreme, mostly vermiform striate; style included to exserted. Fruit dehiscing loculicidally; seeds 1–12 per locule, ovoid to ellipsoid, smooth or angled, strongly to not at all mucilaginous when wetted. $2n = 18$. Ca. 25 species, shrublands and mesic sites in woodlands and forests of North America and Eurasia.

6. *Collomia* Nutt.

Collomia Nutt., Gen. Amer. 1: 126 (1818); Wilken et al., Biochem. Syst. Ecol. 10: 239–243 (1982).

Annual or perennial herbs. Leaves often in basal rosettes; rosette leaves in annuals senescent at bolting, cauline leaves alternate, entire to lobed, rarely palmatifid or pinnatifid; indumentum eglandular to glandular, sometimes with a strong mephitic odor. Inflorescence terminal and capitate, rarely axillary. Calyx actinomorphic, herbaceous or narrowly hyaline between the herbaceous costae, the lobes equal to subequal, the intercostal

membranes distended in fruit; corolla actinomorphic, salverform to funnelform; stamens inserted at the same or different levels in the upper tube, filaments equal to unequal in length, glabrous or basally puberulent; anthers exserted to included; pollen pantotreme or zonotreme, striate to reticulate; style slightly included to exserted. Fruit dehiscing loculicidally; seeds 1–3 per locule, ovoid, smooth or somewhat angled, weakly to strongly mucilaginous when wetted. $2n = 16$. Ca. 15 species, shrublands, woodlands, and forests of W North America, 1–2 in W South America.

7. *Allophyllum* (Nutt.) A. & V. Grant

Allophyllum (Nutt.) A. & V. Grant, Aliso 3: 99 (1955); Grant & Grant, Aliso 3: 93–110 (1955).

Annuals. Leaves sometimes in basal rosettes, basal leaves senescent at bolting, cauline ones alternate, entire to twice pinnatifid, uppermost leaves sometimes palmatifid, the lobes oblong, the terminal ones often broader than the lateral lobes; indumentum predominantly glandular, with a mephitic odor. Inflorescence terminal, mostly 2–3-flowered, somewhat congested. Calyx actinomorphic, hyaline between the herbaceous costae, the lobes equal; corolla actinomorphic to slightly zygomorphic, funnelform; stamens inserted at the same or different levels in the upper tube, filaments equal to unequal in length, glabrous; anthers included to exserted; pollen pantotreme, pertectate; style slightly included to exserted. Fruit dehiscing loculicidally; seeds 1–3 per locule, ovoid, smooth or somewhat angled, strongly mucilaginous when wetted. $2n = 16, 18$. Five species, shrublands and forests of California and Baja California.

8. *Navarretia* Ruiz & Pav.

Navarretia Ruiz & Pav., Fl. Peruv. Prodr.: 20 (1794); Crampton, Madroño 12: 225–256 (1954).

Annuals; basal leaves often in a rosette, senescent at bolting, cauline leaves abruptly to gradually reduced in size. Leaves entire or once to twice pinnatifid, the ultimate lobes entire to lobed; plant glabrous, or with eglandular to glandular trichomes. Inflorescence congested or capitate panicle. Calyx tubular, hyaline between the herbaceous costae, the lobes equal to subequal, mucronate; corolla actinomorphic, funnelform to salverform; stamens 5, equally inserted at the same level in the upper tube or between the lobes; filaments mostly equal, glabrous; anthers included to exserted; pollen pantotreme, mostly pertectate to pilate; style included to exserted. Fruit dehiscing loculicidally; seeds 1–many, strongly mucilaginous when wetted, ovoid to subglobose, mostly smooth, sometimes angled. $2n = 18$. About 30 spp., shrublands and woodlands of western North America.

9. *Ipomopsis* Michx.

Ipomopsis Michx., Fl. Bor. Am. 1: 141 (1803); Grant, Aliso 3: 351–362 (1956).

Annuals, long-lived monocarps, or suffrutescent perennials. Leaves often in persistent basal rosettes, cauline leaves alternate, entire to twice pinnatifid, gradually reduced in size, the ultimate segments often mucronate or spinulose; indumentum glandular to eglandular. Inflorescences mostly terminal, thyrsoid, paniculate, or capitulate, rarely with few flowers. Calyx actinomorphic to slightly zygomorphic, hyaline between the herbaceous costae, the lobes equal to subequal; corolla actinomorphic to slightly zygomorphic, salverform to funnelform; stamens inserted at the same or different levels in the throat or upper tube, rarely lower; filaments equal to unequal in length, glabrous; anthers included to exserted; pollen pantotreme or zonotreme, bizonotreme in one species, striate to reticulate; style included to exserted. Fruit dehiscing loculicidally; seeds 1–12 per locule, ovoid to ellipsoid, angled to minutely winged, mucilaginous when wetted. $2n = 14, 28$. Ca. 30 species, shrublands and forests of W North America, SE North America (1), and W South America (1).

10. *Eriastrum* Woot. & Standl.

Eriastrum Woot. & Standl., Contrib. U.S. Nat. Herb. 16: 160 (1913); Harrison, Brigham Young Univ. Sci. Bull., Biol. Ser. 16: 1–26 (1972).

Annuals or suffrutescent to woody perennials. Leaves in annuals often in basal rosettes, senescent at bolting, alternate, entire to twice pinnatifid, cauline leaves gradually reduced in size, the ultimate segments often mucronate; indumentum predominantly eglandular, often arachnoid. Inflorescences mostly terminal, capitulate. Calyx actinomorphic, hyaline between the herbaceous costae, the lobes slightly unequal; corolla actinomorphic, salverform to funnelform; stamens inserted at the same level in the throat or upper tube, filaments

equal to unequal in length, glabrous; anthers included to exserted, often basifixed; pollen pantotreme or zonotreme, striate to striate-reticulate; style included to exserted. Fruit dehiscing loculicidally; seeds 1–3 per locule, ovoid to ellipsoid, angled to minutely winged, mucilaginous when wetted. $2n = 14$. Fourteen species, shrublands and forests of W North America.

11. *Langloisia* Greene

Langloisia Greene, Pittonia 3: 30 (1896); Timbrook, Madroño 33: 157–174 (1986).
Loeseliastrum (Brand) Timbrook (1986).

Annuals. Leaves sometimes in basal rosettes, senescent at bolting, cauline ones alternate, dentate to once pinnatifid, the teeth or segments with 1–3 terminal bristles; indumentum glandular and eglandular, the trichomes often branched. Inflorescence mostly terminal, capitulate, flowers sometimes axillary and solitary. Calyx actinomorphic, narrowly hyaline between the herbaceous costae, the lobes equal to subequal; corolla actinomorphic and funnelform or strongly zygomorphic and bilabiate; stamens inserted at the same level or slightly different levels in the throat or upper tube, filaments equal to subequal in length, straight to somewhat declinate, glabrous; anthers included to exserted; pollen zonotreme, striate to striate-reticulate; style included to exserted. Fruit dehiscing loculicidally; seeds 1–10 per locule, ovoid, smooth, mucilaginous when wetted. $2n = 14$. Three species, desert shrublands and woodlands of SW North America.

12. *Loeselia* L.

Loeselia L., Gen. Pl. 276 (1754); Turner, Phytologia 77: 318–337 (1994).

Annuals, suffrutescent perennials, or subshrubs. Leaves alternate or opposite, simple, margins entire to aristate; indumentum eglandular, glandular, or glabrous. Inflorescences axillary or terminal, somewhat congested to capitulate. Calyx actinomorphic or slightly zygomorphic, the lobes and tube mostly herbaceous, the tube somewhat hyaline or thinly membranous below the sinuses; corolla actinomorphic to zygomorphic, salverform to somewhat rotate or bilabiate; stamens inserted at the same level below the mid-tube, filaments equal in length, glabrous or pubescent basally; anthers often exserted; pollen pantotreme, pertectate; style included to exserted. Fruit dehiscing loculicidally; seeds 1–10 per locule, globose to ellipsoid, angled to minutely winged, mucilaginous when wetted. $2n = 18$. Ca. 14 species, shrublands and forests of S North America, Central America, and NW South America.

13. *Giliastrum* (Brand) Rydb.

Giliastrum (Brand) Rydb., Fl. Rocky Mts.: 699 (1917); Porter, Aliso 17: 83–85 (1998).

Annuals or suffrutescent perennials, sometimes flowering the first year. Leaves in basal rosettes, cauline ones alternate, entire to pinnatifid, gradually reduced in size; indumentum predominantly glandular puberulent. Inflorescences open, paniculate, sometimes solitary. Calyx actinomorphic, broadly hyaline between narrow herbaceous costae, the lobes equal; corolla actinomorphic, rotate to short-funnelform; stamens equally inserted below the mid-tube, filaments equal in length, glabrous or pubescent basally; anthers often exserted; pollen zonotreme, pertectate; style often exserted. Fruit dehiscing loculicidally; seeds 3–15 per locule, ovoid to globose, smooth to verrucate or angled, mucilaginous when wetted. $2n = 12, 18, 20, 24, 36$. Eight species, shrublands and woodlands of W North America and Central America.

14. *Aliciella* A.Brand

Aliciella A.Brand, Helios 22: 78 (1905); Porter, Aliso 17: 23–46 (1998).

Annuals and perennials, often monocarpic. Leaves in a basal rosette, often senescent at bolting, cauline leaves abruptly (most annuals) to gradually reduced (perennials) in size; pubescence mostly of uniseriate trichomes bearing multicellular glands. Leaves entire to once-pinnatifid. Inflorescence paniculate. Calyx tubular, hyaline between the herbaceous costae; corolla actinomorphic, salverform to funnelform; stamens 5(3), stamens inserted at the same or different levels in the upper tube or in the sinuses of the lobes, sometimes declinate, filaments glabrous, sometimes papillose; anthers included to exserted; pollen zonotreme, striate-reticulate to reticulate; ovary glabrous, style included to exserted. Fruit dehiscing loculicidally; seeds not or only slightly mucilaginous when wetted, generally 1–12 per locule, but 18–28 in *A. latifolia* and *A. ripleyi*, ovoid to subglobose, smooth or angled. $2n = 16, 18, 32, 34, 36, 50$. About 20 spp., shrub-

lands, woodlands, and forests of W North America.

15. *Gilia* Ruiz & Pav.

Gilia Ruiz & Pav., Fl. Peruv. Prodr. 4: 25 (1794); Grant & Grant, Aliso 3: 59–91 (1954), 3: 203–287 (1956).

Annuals or herbaceous to suffrutescent perennials, rarely subshrubs; basal leaves often in a rosette, often senescent at bolting, cauline leaves alternate, abruptly to gradually reduced in size; pubescence eglandular or glandular. Leaves entire or once to thrice-pinnatifid. Inflorescence paniculate, relatively open, sometimes congested and capitate, ultimately of 1–3 subsessile to pedicellate flowers. Calyx tubular, hyaline between the herbaceous costae, the lobes equal; corolla actinomorphic, rotate, campanulate, funnelform, or salverform; stamens 5, equally inserted on the tube or throat, filaments equal to unequal in length, rarely declinate, glabrous; anthers included to exserted; pollen zonotreme, striate to reticulate; ovary glabrous to minutely glandular, style included to exserted. Fruit dehiscing loculicidally; seeds 1–many per locule, mucilaginous when wetted, generally many per locule, ovoid to subglobose, angled to rough. $2n = 18, 36, 72$. About 50 spp., shrublands, woodlands, and forests of W North America.

16. *Linanthus* Benth.

Linanthus Benth., Edward's Bot. Reg. 19: 1622 (1833); Patterson, Madroño 24: 36–48 (1977).
Leptodactylon Hook. & Arn. (1839).
Linanthastrum Ewan (1942).

Annuals, herbaceous or suffrutescent perennials, or shrubs. Leaves opposite, sometimes alternate, simple, entire to palmatifid, sometimes pinnatifid; indumentum eglandular, glandular, or glabrous. Inflorescences axillary or terminal, somewhat congested to capitulate. Calyx actinomorphic, narrowly hyaline between the herbaceous costae, the lobes equal; corolla actinomorphic, campanulate to funnelform or salverform; stamens inserted at the same level in the throat or tube, filaments equal in length, glabrous or minutely puberulent basally; anthers exserted to included; pollen pantotreme, striate-reticulate to reticulate; style included to exserted. Fruit dehiscing loculicidally; seeds 1–10 per locule, ovoid, smooth or angled, either mucilaginous or remaining unchanged when wetted. $2n = 18, 36$. About 55 species, shrublands, woodlands, and forests of W North America and W South America.

17. *Phlox* L.   Fig. 100F, G

Phlox L., Gen. Pl. 1: 52 (1737); Wherry, Morris Arb. Monogr. 3: 1–174 (1955), rev.
Microsteris Greene, Pittonia 3: 301 (1898).

Annuals, herbaceous or suffrutescent perennials, sometimes from rhizomes; indumentum predominantly eglandular. Leaves opposite, simple, entire to toothed, ovate to linear, those of the inflorescence sometimes alternate. Inflorescences mostly terminal, paniculate to capitulate. Calyx actinomorphic, narrowly hyaline between the herbaceous costae, the lobes equal; corolla actinomorphic, salverform; stamens inserted at different levels in the tube, filaments mostly equal in length, glabrous; anthers mostly included; pollen pantotreme, reticulate; style included to exserted. Fruit dehiscing loculicidally, often explosively; seeds 1 per locule, ovoid to ellipsoid, smooth, angled, or minutely winged, either mucilaginous or remaining unchanged when wetted. $2n = 14, 28$. About 70 species, grasslands, shrublands, woodlands, and forests of North America and NE Asia (1).

18. *Gymnosteris* E.Greene   Fig. 100E

Gymnosteris E.Greene, Pittonia 3: 303 (1898); Wherry, Am. Midland Nat. 31: 230–231 (1944).

Diminutive annuals with persistent cotyledons, stems solitary or with 1 branch; indumentum glandular puberulent. Leaves reduced to a whorl of basally connate bracts subtending the terminal inflorescence. Inflorescence of 1–several flowers, terminal, capitate. Calyx actinomorphic, broadly hyaline to scarious between narrow herbaceous costae, the lobes equal; corolla actinomorphic, salverform; stamens inserted at the same level in the upper tube, subsessile, filaments equal in length, glabrous; anthers included; pollen pantotreme, reticulate; style included. Fruit dehiscing loculicidally; seeds 1–3 per locule, ovoid, smooth, strongly mucilaginous when wetted. $2n = 12$. Two species, cold desert shrublands and woodlands of W North America.

Selected Bibliography

Anderberg, A.A. et al. 2002. See general references.

APG (Angiosperm Phylogeny Group) 1998. See general references.

Arisawa, M., Pezzuto, J., Kinghorn, A., Douglas, A., Cordell, G., Farnsworth, N. 1984. Plant anticancer agents. 30. Cucurbitacins from *Ipomopsis aggregata* (Polemoniaceae). J. Pharmaceut. Sci. 73: 411–413.

Aurada, E. Jurenitsch, J., Kubelka, W. 1982. Structure of triterpene sapogenins from *Polemonium caeruleum* L. Sci. Phar. 50: 331–350.

Carlquist, S. 1992. Wood anatomy of sympetalous dicotyledon families: a summary, with comments on systematic relationships and evolution of the woody habit. Ann. Missouri Bot. Gard. 79: 303–332.

Carlquist, S., Eckhart, V., Michener, D. 1984. Wood anatomy of Polemoniaceae. Aliso 10: 547–572.

Chaney, R., Mason, H. 1936. A Pleistocene flora from Fairbanks, Alaska. Am. Mus. Nov.: 887.

Cronquist, A. 1981. See general references.

Dawson, M.L. 1936. The floral morphology of the Polemoniaceae. Am. J. Bot. 23: 501–511.

Day, A., Moran, R. 1986. *Acanthogilia*, a new genus of Polemoniaceae from Baja California, Mexico. Proc. Calif. Acad. Sci. 44: 111–126.

Downie, S., Palmer, J. 1992. Restriction site mapping of the chloroplast DNA inverted repeat: a molecular phylogeny of the Asteridae. Ann. Missouri Bot. Gard. 79: 266–283.

Galen, C., Newport, M. 1988. Pollination quality, seed set, and flower traits in *Polemonium viscosum*: complementary effects of variation in flower scent and size. Am. J. Bot. 75: 900–905.

Grant, V. 1959. Natural history of the phlox family. The Hague, Netherlands: Martinus Nijhoff.

Grant, V. 1998. Primary classification and phylogeny of the Polemoniaceae, with comments on molecular cladistics. Am. J. Bot. 85: 741–752.

Grant, V. 2001. A guide to understanding recent classifications of the family Polemoniaceae. Lundellia 4: 12–24.

Grant, V., Grant, K. 1965. Flower pollination in the phlox family. New York: Columbia University Press.

Grubert, M., Hambach, M. 1972. Untersuchen über die verschleimenden Samen von *Collomia grandiflora* Dougl. (Polemoniaceae). Beitr. Biol. Pflanz. 48: 187–206.

Harborne, J.B., Smith, D. 1978. Correlations between anthocyanin chemistry and pollination ecology in Polemoniaceae. Biochem. Syst. Ecol. 6: 127–130.

Hiller, K., Paulick, A., Doehnert, H., Franke, P. 1979. Saponins of *Polemonium caeruleum* L. Pharmazie 34: 565–566.

Johnson, L., Schultz, J., Soltis, D., Soltis, P. 1996. Monophyly and generic relationships of Polemoniaceae based on *mat*K sequences. Am. J. Bot. 83: 1207–1224.

Johnson, L., Soltis, D., Soltis, P. 1999. Phylogenetic relationships of Polemoniaceae inferred from 18s ribosomal DNA equences. Plant Syst. Evol. 214: 65–89.

Johri, B.M. et al. 1992. See general references.

Jurenitsch, J., Haslinger, E., Kubelka, W. 1979. Structure of sapogenins from *Polemonium reptans*. Pharmazie 34: 445–446.

Kapil, R., Rustagi, P., Venkataraman, R. 1968. A contribution to the embryology of Polemoniaceae. Phytomorphology 17: 403–412.

Lott, T.A., Manchester, S.R., Dilcher, D.L. 1998. A unique and complete polemoniaceous plant from the middle Eocene of Utah, USA. Rev. Palaeobot. Palynol. 104: 39–49.

Mason, H. 1945. The genus *Eriastrum* and the influence of Bentham and Gray upon the problem of generic confusion in Polemoniaceae. Madroño 8: 33–59.

Metcalfe, C., Chalk, L. 1950. See general references.

Muller, J. 1981. See general references.

Olmstead, R., Bremer, B., Scott, K., Palmer, J. 1993. A parsimony analysis of the Asteridae sensu lato based on *rbc*L sequences. Ann. Missouri Bot. Gard. 80: 700–722.

Patterson, R. 1980. The occurrence of B chromosomes in *Linanthus pachyphyllus*. Caryologia 33: 141–149.

Plitmann, U., Levin, D. 1983. Pollen-pistil relationships in the Polemoniaceae. Evolution 37: 957–967.

Plitmann, U., Levin, D. 1990. Breeding systems in the Polemoniaceae. Plant Syst. Evol. 170: 205–214.

Pollard, C., Amuti, K. 1981. Fructose oligosaccharides: possible markers of phylogenetic relationship among dicotyledonous plant families. Biochem. Syst. Ecol. 9: 69–78.

Porter, J. 1997. Phylogeny of Polemoniaceae based on nuclear ribosomal internal transcribed spacer DNA sequences. Aliso 15: 57–77.

Porter, J., Johnson, L. 1998. Phylogenetic relationships of Polemoniaceae: inferences from mitochondrial *nad1b* intron sequences. Aliso 17: 157–188.

Porter, J., Johnson, L. 2000. A phylogenetic classification of Polemoniaceae. Aliso 19: 55–91.

Prather, L.A., Ferguson, C.J., Jansen, R.K. 2000. Polemoniaceae phylogeny and classification: implications of sequence data from the chloroplast gene *ndh*F. Am. J. Bot. 87: 1300–1308.

Raven, P. 1963. Amphitropical relations in the flora of North and South America. Quart. Rev. Biol. 29: 151–171.

Raven, P., Axelrod, D. 1978. Origin and relationships of the California flora. Univ. Calif. Publ. Bot. 72: 1–134.

Smith, D., Levin, D. 1967. Karyotypes of eastern North American *Phlox*. Am. J. Bot. 54: 324–334.

Smith, D., Glennie, C., Harborne, J.B., Williams, C. 1977. Flavonoid diversification in the Polemoniaceae. Biochem. Syst. Ecol. 5: 107–115.

Spencer, S., Porter, J. 1997. Evolutionary diversification and adaptation to novel environments in *Navarretia* (Polemoniaceae). Syst. Bot. 22: 649–668.

Stuchlik, L. 1967. Pollen morphology in the Polemoniaceae. Grana Palynol. 7: 146–240.

Takhtajan, A. 1980. Outline of the classification of flowering plants (Magnoliophyta). Bot. Rev. 46: 225–359.

Takhtajan, A. (ed.) 1981. See general references.

Taylor, T.N., Levin, D.A. 1975. Pollen morphology of Polemoniaceae in relation to systematics and pollination systems: scanning electron microscopy. Grana 15: 91–112.

Thorne, R. 1983. Proposed new realignments in the angiosperms. Nord. J. Bot. 3: 85–117.

Weberling, F. 1998. Die Infloreszenzen. Vol. II/2. Jena: G. Fischer.

Primulaceae

A.A. Anderberg

Primulaceae Vent., Tabl. Règne Vég. 2: 285 (1799), nom. cons.

Perennial or sometimes annual herbs, often cushion-forming, and sometimes woody at the base; stem and leaves with long articulated hairs, short glandular hairs, or sometimes glabrous. Leaves usually alternate, often forming a basal rosette but sometimes scattered on the stem, opposite or whorled, entire, dentate or rarely pinnatifid or pinnate; stipules absent; young leaves involute or conduplicate, seldom revolute. Inflorescence terminal or axillary, racemose, spicate, umbellate, sometimes scapose and with one to many superimposed umbels, or flowers solitary. Flowers perfect, hypogynous, actinomorphic or rarely slightly zygomorphic, with imbricate, sometimes quincuncial corolla aestivation, 5(6–8)-merous, sometimes heterostylous; bracts present or sometimes absent, prophylls lacking; calyx synsepalous, campanulate or cylindrical, with five herbaceous lobes; corolla sympetalous, campanulate to hypocrateriform; corolla-tube usually distinct, short to very long; corolla-lobes entire, emarginate, or fringed (*Soldanella*), sometimes reflexed (*Dodecatheon*); stamens 5(–8), antepetalous, epipetalous; anthers tetrasporangiate, dithecal, introrse, opening with longitudinal slits, completely distinct or connivent and forming a protruding anther cone; gynoecium syncarpous, 5-carpellate; ovary superior, unilocular, with a stipitate, free central placental column, only rarely with rudimentary septa basally; ovules few to many arranged in several series on the surface of or rarely (*Dionysia*) immersed in the placenta, anatropous, bitegmic, tenuinucellate; style sometimes exserted; stigma truncate to capitate. Fruit a capsule opening with apical valves, a circular lid, or by irregular disintegration; seeds more or less angular, reticulate or papillose; endosperm copious.

A predominantly north temperate family with 13 genera and c. 600 species.

VEGETATIVE MORPHOLOGY. All species are herbs but in the cushion-forming species the stem is often more or less lignified. In some *Dodecatheon* the hypocotyl shows an initial swelling but never develops into a tuber. The leaves are typically alternate but are sometimes opposite or whorled. Most species have a basal leaf rosette and a scapose inflorescence terminating the short, often rhizomatous stem. Stolons are formed in some *Androsace*. The leaves are usually short-petiolate or the blades gradually taper towards the base. In some genera such as *Cortusa*, *Soldanella* and part of *Primula*, the petioles are distinctly demarcated from the lamina which has a palmate venation. Most species have dentate or serrate leaf margins and marginal veinlets ending in hydathodes. Ptyxis is important systematically: in most genera it is involute or conduplicate, but revolute in *Dionysia*, *Cortusa*, and most *Primula*.

Immersed capitate hairs, like in Myrsinaceae, are not known from Primulaceae, but long, articulated hairs are typical of virtually all genera, albeit characteristically absent from some (e.g. *Dodecatheon*). Ordinary short glandular hairs are also common. In species of *Androsace*, *Douglasia* and *Vitaliana*, white stellate hairs are frequent but otherwise not present in the family. *Omphalogramma*, *Bryocarpum* and certain *Primula* have blackish, flattened gland-dots on the leaves. Many *Primula* and *Dionysia* have a distinct farinose coating on young leaves, stems and parts of the inflorescence. This can be either powdery or woolly in appearance, and its presence is often species-specific.

VEGETATIVE ANATOMY. Secretory cavities and calcium oxalate crystals in the leaves and flower parts are absent from Primulaceae, and extraxylary bundles of unbranched fibres, as in Threophrastaceae, are lacking. Irregularly shaped foliar sclereids are found in *Dionysia* and some other genera.

INFLORESCENCE STRUCTURE. The inflorescence in Primulaceae is basically racemose and the flowers are arranged in racemes, spikes, corymbs, or umbels. In *Hottonia* and several *Primula*, the flowers are arranged on scapes in superimposed umbels. The bracts which subtend the flowers may sometimes be fused to the pedicel to a greater or lesser extent. In *Kaufmannia* the bract is broad and deeply incised. Only a few taxa are ebracteate. Prophylls are lacking throughout. Flowers are often nodding, particularly in species growing in alpine or montane habitats, such as many *Primula*, *Dodecatheon*, *Soldanella*, and *Bryocarpum*.

FLORAL STRUCTURE. Branching of the vascular system in the corolla of *Soldanella* has been been brought forward as support for a staminodial origin of parts of the corolla, but this needs to be confirmed. In *Androsace*, *Douglasia* and *Vitaliana*, the corolla throat is characteristically constricted and provided with scale-like outgrowths, which are formed by swellings of the corolla tissue.

The heterostylous taxa have short-styled thrum-type flowers with the style included in the corolla-tube, and long-styled pin-type flowers with long styles and the stigma presented at the mouth of the tube. The stigma is usually distinctly capitate.

The gynoecium is probably formed by five carpels, although it is initiated from a ring-shaped primordium (Schaeppi 1937).

EMBRYOLOGY. The anther tapetum is secretory; cells are uninucleate. Pollen is binucleate when shed. The ovules are anatropous, bitegmic, and tenuinucellate with a micropyle formed by both integuments, the inner of which has an integumentary tapetum. In *Soldanella* the inner integument is partly resorbed. The archespore is multicellular, and the embryo sac is of the Polygonum type with ephemeral antipodals but lacks endosperm haustoria. The seeds are albuminous with nuclear endosperm formation. Embryogeny follows the caryophyllad type; the embryo is short and provided with short, narrow cotyledons (Davis 1966; Johri et al. 1992).

POLLEN MORPHOLOGY. The pollen grains of the Primulaceae are basically 3-colporate. Many *Primula* have syncolpate pollen; stephanocolpate pollen is found in *Primula* (incl. *Sredinskya*) and *Dionysia* (Erdtman 1952; Wendelbo 1961; Nowicke and Skvarla 1977; Richards 1993).

KARYOLOGY. Primulaceae show a great variability in chromosome numbers but chromosome numbers based on $n = 9$, 10, 11, and 12 are common. Polyploidy is frequent in *Androsace* and *Primula*.

POLLINATION. Most Primulaceae seem to be pollinated by insects, chiefly bees and flies. Species with showy flowers with well-developed corollas are often visited for nectar and nectaries situated on the gynoecium are found in many genera, e.g. *Primula*, *Androsace*, and *Soldanella*. There are also a number of species with nodding flowers and an anther cone, as in *Dodecatheon*, *Soldanella*, and *Cortusa*. These flowers are likely to be buzz-pollinated.

FRUIT AND SEED. Most Primulaceae have many-seeded capsules, although some genera may have few seeds, e.g. *Douglasia* and *Vitaliana*. The capsules are often globose but in a few genera ovoid, ellipsoidal, or even long cylindrical capsules are found. In most species the capsules dehisce with apical valves; *Pomatosace* has operculate capsules that dehisce with a circular lid. A number of other taxa are often stated to have operculate capsules (*Bryocarpum*, *Soldanella*, *Primula* sect. *Carolinella*) but in these taxa capsules open through a more or less irregular split of the basal portion of the style, leaving short, blunt valves and a part of the style-base as a lid-like structure at the distal end of the mature fruit. It is evident that so-called operculate capsules comprise different fruit types that have arisen several times.

The seeds are normally angular with a smooth, reticulate or papillose surface and a hilum more or less flush with the surface. The testa is rather thick, usually distinctly two-layered and generally provided with rhomboid crystals. The seeds of *Douglasia* and *Vitaliana* are almost semiglobose with a flat or slightly concave ventral surface. The endosperm is copious and composed of cells with smooth and evenly thickened walls except in *Douglasia*, *Vitaliana*, and some *Androsace* which have irregularly thickened endosperm cell walls with distinct, narrow constrictions.

DISPERSAL. The mature fruit is usually borne on a straight pedicel or scape, and the seeds are released ballistically through the openings of the capsule. Some *Primula* are myrmecochorous, with a swollen funicle acting as an elaiosome. In the aquatic *Hottonia* as well as in some wetland *Primula*, the seeds are dispersed by water (van der Pijl 1972).

REPRODUCTIVE BIOLOGY. Heterostyly is known from *Hottonia*, *Dionysia*, *Vitaliana*, and *Primula*, and in the latter genus more than 90% of the species are heterostylous. Wendelbo (1961) considered the genus to be primarily homostylous and stated that other heterostylous genera had evolved within *Primula*. Heterostyly in *Primula* has been studied extensively and it is known that the long-styled pin-type is homozygous, whereas the short-styled thrum-type is heterozygous for the heterostyly-gene. In heterostylous Primulaceae the pin-type flowers also have longer stigmatic papillae and smaller pollen than the thrum-type flowers, except in *Vitaliana* in which the pollen grains are of equal size in pin- and thrum-type flowers (Richards 1993).

PHYTOCHEMISTRY. Primulaceae often contain various saponins but lack compounds like iridoids and ellagic acid. Many genera, including *Primula*, also produce quinoid compounds which are known to be allergenic. The flavonol gossypetin, which has a very restricted distribution in flowering plants, has been found in *Dionysia*, *Douglasia* and several *Primula* species. The seeds store oil and amylose rather than starch (Hegnauer 1969, 1973).

SUBDIVISION AND AFFINITIES. Pax (1889) and Pax and Knuth (1905) subdivided Primulaceae into five tribes. Melchior (1964) hypothesized that the herbaceous growth habit in Primulaceae is derived, and Judd et al. (1994) put forward that Myrsinaceae are a paraphyletic group from which Primulaceae have evolved. Anderberg and Ståhl (1995) performed a preliminary cladistic analysis of family circumscription in Primulales, and the work by Källersjö et al. (2000), which was based on morphological and DNA sequence data from three chloroplast genes, demonstrated that not only Myrsinaceae but also Primulaceae in the conventional circumscription are paraphyletic, since *Samolus* belongs to Theophrastaceae, and genera such as *Coris*, *Ardisiandra*, *Cyclamen*, and the genera of Primulaceae-Lysimachieae belong to Myrsinaceae. Källersjö et al. (2000) therefore reduced Primulaceae to only one of the five primulaceous tribes proposed by Pax (1889), Primuleae, which is characterised by imbricate/quincuncial corolla aestivation. In this circumscription, Primulaceae, with few exceptions, are herbaceous plants with a basal leaf rosette and a scapose inflorescence and dry capsular fruits. A later study by Anderberg et al. (2001) showed that also *Stimpsonia* is part of the herbaceous basal complex of the Myrsinaceae, and that it should be excluded from Primulaceae.

Wendelbo (1961) considered it likely that *Sredinskya*, *Hottonia*, *Dionysia*, *Cortusa* and *Dodecatheon* have their closest relatives within *Primula*, which is supported by the results of Källersjö et al. (2000), Mast et al. (2001), and Trift et al. (2002).

Dodecatheon much resembles *Primula* sect. *Parryi* with their involute leaves, and *Cortusa* comes close to many species of *Primula* sect. *Cortusioides*. *Sredinskya* is closely related to the *Primula veris* clade. *Soldanella*, *Hottonia*, and *Omphalogramma* share many potential synapomorphies with species in *Primula*, but in the studies by Mast et al. (2001) and Trift et al. (2002) the three genera form a clade separate from *Primula*. The *Androsace* complex is the sister group of the other genera in the family, and the genera *Douglasia*, *Vitaliana*, and *Pomatosace* are derived within the *Androsace* complex, in which *Androsace* itself is paraphyletic. Detailed relationships between the *Androsace* complex and the *Primula* complex still need to be clarified.

The studies by Källersjö et al. (2000), Mast et al. (2001), and Trift et al. (2002) have demonstrated that a number of genera have their closest relatives in *Primula* or *Androsace*, and this will lead to a number of new generic circumscriptions. In the following taxonomic account, the traditional generic circumscriptions are maintained, awaiting a new generic classification.

FOSSIL RECORD. Seeds identified as *Androsace* are known from the Siberian Miocene, but other megafossils referred to Primulaceae, and particularly to *Primula*, are less certain.

DISTRIBUTION AND HABITATS. Primulaceae are mainly distributed in temperate and arctic regions of the Northern Hemisphere; only few species grow in the Southern Hemisphere. The Himalayas are particularly rich in species of *Primula* and *Androsace*, and many small genera such as *Pomatosace*, *Bryocarpum* and *Omphalogramma* are also confined to this region. Many species are confined to alpine habitats including meadows, rock crevices, and scree; other species grow in woodlands. *Hottonia*, with its submersed gill-leaves, is adapted to life in freshwater.

ECONOMIC IMPORTANCE. Many species of Primulaceae are appreciated indoor plants, especially cultivars of *Primula vulgaris* and *P. obconica*, which are grown commercially. Species of *Cortusa*, *Androsace*, and *Primula* are common as garden ornamentals and widely used in rock gardens. Many *Dionysia* are very decorative but grown only by specialists.

KEY TO THE GENERA

1. Submerged aquatic plants. Leaves pinnate   **7. *Hottonia***
– Terrestrial plants. Leaves simple or rarely pinnatifid   2

2. Corolla-lobes conspicuously reflexed **13. Dodecatheon**
- Corolla-lobes spreading or erect 3
3. Corolla-lobes 7, flowers solitary yellow; capsule narrowly cylindrical, 3–4 cm long **9. Bryocarpum**
- Corolla-lobes 5(6–8), flowers solitary or many; capsule globose or ellipsoid, not more than 2 cm long 4
4. Throat of corolla constricted; corolla-tube short 5
- Throat of corolla not constricted; corolla-tube long or short 8
5. Capsule opening with an operculate lid; leaves pinnatifid **4. Pomatosace**
- Capsule opening with valves; leaves simple 6
6. Corolla tube generally short **1. Androsace**
- Corolla tube longer 7
7. Homostylous; flowers pink **2. Douglasia**
- Heterostylous; flowers yellow **3. Vitaliana**
8. Corolla six to eight-lobed, slightly zygomorphic **8. Omphalogramma**
- Corolla five-lobed, actinomorphic 9
9. Corolla-lobes fringed; leaves orbicular to reniform **12. Soldanella**
- Corolla-lobes usually entire; leaves of various shapes but if the corolla-lobes are fringed, then the leaves are not entire and orbicular or reniform 10
10. Stamens inserted at the base of the corolla 11
- Stamens inserted in the corolla-tube 12
11. Flowers pink; bracts entire **10. Cortusa**
- Flowers yellow; bracts deeply incised **11. Kaufmannia**
12. Corolla tube at least three times longer than the calyx. Ovules few **6. Dionysia**
- Corolla tube not three times longer than the calyx. Ovules many **5. Primula**

GENERA OF PRIMULACEAE

1. *Androsace* L.

Androsace L., Sp. Pl.: 141 (1753); Smith and Lowe, Androsaces (1977).
Aretia L. (1753).

Suffrutices, perennial, or annual herbs, sometimes densely pulvinate, or prostrate and forming mats. Leaves entire or dentate, alternate, sessile or petiolate, generally in basal rosettes, ovate, spathulate or linear, often with stellate hairs. Inflorescence generally scapose. Flowers homostylous, in terminal umbels or solitary in the leaf-axils; calyx campanulate; corolla hypocrateriform, with a short tube, constricted at the throat and with scale-like outgrowths, white, pink, or red; corolla-lobes entire or emarginate; stamens adnate to the tube; stigma capitate. Capsule globose, opening with valves; seeds reticulate or papillose. $2n = 18$, 20, 22, 30, 36, 38, 40, 48, 78. About 150 spp., north temperate areas, many spp. in China and the Himalayas. Four sections: sect. *Androsace* (= *Andraspis* (Duby) Koch), annual, leaves oblong or linear, entire or dentate, not clearly petiolate, flowers in umbels; sect. *Pseudoprimula* Pax, perennial, leaves long petiolate, large, margin lobate or crenate, flowers in umbels; sect. *Chamaejasme* Koch, caespitose perennials, leaves scarcely petiolate, spathulate to linear, entire or minutely dentate, flowers in umbels; sect. *Aretia* (L.) Duby, caespitose perennials, leaves entire or minutely dentate, flowers solitary.

Androsace is paraphyletic if the three following genera are excluded.

2. *Douglasia* Lindley

Douglasia Lindley, Quart. J. Sci. Lit. Arts 1827: 385 (1827), nom. cons.; Constance, Am. Midl. Nat. 19: 249–259 (1938), rev.; Kelso, Can. J. Bot. 70: 593–596 (1991).

Perennial herbs, sometimes pulvinate. Leaves often in basal rosettes, entire or dentate, often with stellate hairs. Inflorescence scapose. Flowers homostylous, in terminal umbels; calyx campanulate; corolla hypocrateriform, with a short tube, constricted at the throat and with scale-like outgrowths, violet or purple; stamens adnate to the tube; stigma capitate. Capsule globose; seeds few, reticulate or papillose. $2n = 36$, 38, 40. Eight spp., USA, Canada, E Siberia.

Douglasia is a derived part of *Androsace*.

3. *Vitaliana* Sesler

Vitaliana Sesler in Donati, Essai Hist. Nat. Mer Adriat.: 69 (1758); Constance, Am. Midl. Nat. 19: 249–259 (1938); Ferguson, Fl. Europaea 3: 20 (1972).
Douglasia subg. Gregoria (Duby) Knuth (1905).

Perennial herb, branched, prostrate and forming mats. Leaves alternate, linear, with stellate hairs. Flowers heterostylous, subsessile, solitary or paired; calyx tubular; corolla hypocrateriform, with a tube about twice as long as calyx, somewhat constricted at the throat and with scale-like outgrowths, yellow; corolla-lobes linear, longer than wide, imbricate when young; stamens adnate to the tube; stigma capitate. Capsule globose, opening with apical valves, seeds few, almost smooth. $2n = 32$, 40, 80. One sp., *V. primuliflora* Bertol., European Alps.

Vitaliana is a derived part of *Androsace*.

4. *Pomatosace* Maxim.

Pomatosace Maxim., Bull. Acad. Imp. Sci. Saint-Pétersbourg 27: 499 (1882).

Annual herb. Leaves basal, pinnatifid. Inflorescence scapose. Flowers homostylous, in terminal umbels; calyx campanulate with conspicuous triangular, persisting lobes; corolla white, with short tube, constricted at the throat and with scale-like outgrowths; corolla-lobes entire, erect to spreading; stamens adnate to the tube; stigma capitate. Capsule globose, opening with a lid; seeds many, papillose. $2n = 20$. One sp., *P. filicula* Maxim., NW China.

Pomatosace is a derived part of *Androsace*.

5. *Primula* L. Fig. 101D

Primula L., Sp. Pl.: 142 (1753); Wendelbo, Årb. Univ. Bergen, Mat. Nat. Ser. 11: 1–49 (1961), rev.; Wendelbo, Årb. Univ. Bergen. Mat. Nat. Ser. 7: 1–15 (1961), pollen morph.; Smith, Forrest and Fletcher, Plant Monogr. Reprints 11: 1–835 (1977); Richards, Primula (1993); Mast et al. Int. J. Plant Sci. 162: 1381–1400 (2001), phylogeny; Trift et al. Syst. Bot. 27: 396–407 (2002), phylogeny.
Sredinskya (Stein.) Fedorov (1950).

Perennial herbs. Leaves basal, revolute or involute when young, often with a white or yellow farinose coating, entire, dentate or serrate, evenly attenuating or with well demarcated blade and petiole, sometimes pinnatifid, with articulated hairs or glabrous, exceptionally with flattened gland-dots. Inflorescence often scapose. Flowers heterostylous or less often homostylous, in umbels, sometimes in superimposed umbels, spike-like racemes, or solitary; calyx campanulate or cylindrical; corolla campanulate to hypocrateriform, with a short or long tube, white, yellow, pink, red, violet or blue; corolla-lobes spreading or erect, entire, emarginate, bifid, or fringed; stamens adnate to the tube; stigma capitate. Capsule globose or ellipsoid, opening with apical valves, rarely with a lid, or through irregular disintegration; seeds reticulate or papillose. $2n = 18, 20, 22, 23, 24, 29, 32, 36, 40, 44, 48, 54, 62, 66, 72, 126$ ($x = 8, 9, 10, 11, 12$). About 400 spp., mainly N Hemisphere. Six subgenera, often characterized by the following string of features: subg. *Sphondylia* (Duby) Rupr. (1863), generally homostylous, leaves involute, pollen 3-colporate; subg. *Auriculastrum* Schott (1851), heterostylous, leaves involute, pollen 3-colporate or 3-syncolpate; subg. *Primula*, heterostylous, leaves revolute, pollen stephanocolpate; subg. *Auganthus* (Link) Wendelbo (1961), heterostylous or homostylous, leaves revolute, pollen 3-colporate, 3-syncolpate, or stephanocolpate; subg. *Carolinella* (Hemsley) Wendelbo (1961), heterostylous, leaves revolute, capsule operculate; subg. *Aleuritia* (Duby) Wendelbo (1961), heterostylous or homostylous, leaves revolute, pollen 3-colporate, 3-syncolpate, or stephanocolpate.

Primula is paraphyletic in its present circumscription. At least the genera *Dionysia* and *Cortusa*, and *Dodecatheon* have their closest relatives within

Primula. Many of the subgenera are also in need of a new circumscription.

6. *Dionysia* Fenzl

Dionysia Fenzl, Flora 26: 389 (1843); Wendelbo, Årb. Univ. Bergen, Mat. Nat. Ser. 3: 1–83 (1961); Grey-Wilson, The genus *Dionysia* (1989); Al Wadi & Richards, New Phytol. 121: 303–310 (1992).

Perennial herbs, often woody at base and often forming dense tufts. Leaves alternate or whorled, revolute when young, often with articulated hairs and a farinose coating. Inflorescence generally scapose. Flowers heterostylous, in terminal simple, or superimposed umbels, or solitary; calyx tubular; corolla hypocrateriform, with a very long slender tube, violet, pink or yellow; corolla-lobes spreading, entire or emarginate; stamens adnate to the tube; stigma capitate. Capsule subglobose, opening with apical valves; seeds few or many, papillose. $2n = 20$. 40 spp., Mountains of Central Asia.

Dionysia is a derived part of *Primula*, diagnosed by longer corolla tube and fewer ovules.

7. *Hottonia* L.

Hottonia L., Sp. Pl.: 145 (1753).

Perennial aquatic herbs. Leaves submersed, pinnate, alternate or verticillate, glabrous; leaf-lobes linear, involute when young. Inflorescence scapose, held above the water surface. Flowers in several whorls (superimposed umbels), heterostylous; corolla white to pink, longer or shorter than the calyx; corolla-lobes spreading, often emarginate; stamens adnate to the tube; stigma capitate. Capsule opening laterally with apical valve-slits; seeds few, reticulate. $2n = 20, 22$. Two spp., North America, Eurasia.

8. *Omphalogramma* (Franch.) Franch.

Omphalogramma (Franch.) Franch., Bull. Soc. Bot. France 45: 178 (1898); Fletcher, Not. Roy. Bot. Gard. Edinburgh 20: 125–159 (1949).
Primula subg. *Omphalogramma* Franch. (1885).

Perennial herbs. Leaves basal, involute when young, ovate to elliptic, entire to denticulate, with flattened gland-dots. Inflorescence scapose, ebracteate. Flowers (5)6(–8)-merous, homostylous, somewhat zygomorphic, solitary; corolla hypocrateriform, with a rather long tube, blue, rose to dark purple; corolla-lobes spreading,

Fig. 101. Primulaceae. **A** *Soldanella montana*, immature fruit. **B** Same, dehisced, showing stylopodial lid. **C** *Dodecatheon meadia*, flower. **D** *Primula denticulata*, flower, vertical section. **E** *Cortusa matthioli*, habit. **F** Same, inflorescence. Drawn by Pollyanna von Knorring

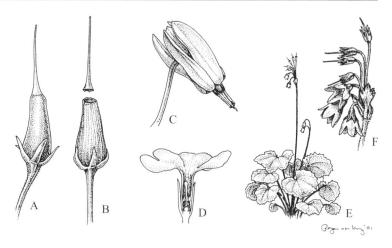

emarginate to dentate; stamens adnate to the tube; stigma capitate. Capsule cylindrical, opening with valves; seeds reticulate with winged margin. $2n = 48, 96$. 15 spp., Himalayan region.

9. *Bryocarpum* Hook. f. & Thoms.

Bryocarpum Hooker f. & Thomson, Hooker's J. Bot. Kew Gard. Misc. 9: 200 (1857).

Perennial herb. Leaves basal, involute when young, ovate with truncated base, with blackish flattened gland-dots. Flowers 7-merous, nodding, homostylous, solitary; calyx campanulate; corolla narrowly campanulate, yellow; corolla-lobes erect, emarginate; stamens adnate to the tube; stigma capitate. Capsule long cylindrical, opening with a lid and with apical valves; seeds many, reticulate. One sp., *B. himalaicum* Hook. f. & Thoms., E Himalayas.

10. *Cortusa* L.   Fig. 101E, F

Cortusa L., Sp. Pl.: 144 (1753).

Perennial herbs. Leaves basal, revolute when young, petiolate, coarsely dentate. Inflorescence scapose; bracts entire. Flowers homostylous, in umbels; calyx campanulate; corolla campanulate, pink with darker blotches in the throat; corolla-lobes erect, entire; stamens connivent; style exserted; stigma capitate. Capsule elliptic, opening with valves; seeds many, papillose. $2n = 24$. Eight spp., mountains of Europe and Asia.

Cortusa is a derived part of *Primula*.

11. *Kaufmannia* Regel

Kaufmannia Regel, Trudy Imp. S.-Peterburgsk Bot. Sada 3: 293 (1875).

Perennial herbs. Leaves basal, coarsely dentate, petiolate. Inflorescence scapose; bracts deeply incised. Flowers homostylous, in umbels; corolla yellow; corolla-lobes entire; stamens connivent; style exserted; stigma capitate. Capsule elliptic, opening with valves; seeds many, papillose. 1–2 spp. in Central Asia.

Kaufmannia is very close to *Cortusa*, and like this probably a derived part of *Primula*.

12. *Soldanella* L.   Fig. 101A, B

Soldanella L., Sp. Pl.: 144 (1753).

Perennial herbs. Leaves ovate to reniform, entire, petiolate, often with articulated hairs. Inflorescence scapose, bracteate. Flowers nodding, homostylous, in terminal few-flowered umbels or solitary; calyx campanulate; corolla blue, campanulate; corolla-lobes erect, laciniate; stamens connivent, forming a cone; style exserted; stigma capitate. Capsule elliptic, opening with a lid and with apical valves; seeds many, reticulate. $2n = 38, 40$. 10 spp., European Alps.

13. *Dodecatheon* L.   Fig. 101C

Dodecatheon L., Sp. Pl.: 144 (1753); Thompson, Contrib. Dudley Herb. 4: 73–154 (1953).

Perennial herbs. Leaves basal, linear to spathulate, glabrous. Inflorescence scapose. Flowers many together in a terminal umbel, nodding, homostylous; calyx campanulate; corolla campanulate, white to pink often with darker blotches; corolla-lobes entire and distinctly reflexed; stamens connivent, forming a protruding cone; style exserted; stigma capitate or truncate. Capsule elliptic, opening with apical valves or with a lid; seeds angular or somewhat winged, reticulate. $2n = 22$, 44, 88. 13 spp. in E North America and Siberia.

Dodecatheon is a derived part of *Primula*, diagnosed e.g. by distinctly reflexed corolla lobes.

Selected Bibliography

Anderberg, A.A, Ståhl, B. 1995. Phylogenetic interrelationships in the order Primulales, with special emphasis on the family circumscriptions. Can. J. Bot. 73: 1699–1730.

Anderberg, A.A., Peng, C.-I., Trift, I., Källersjö, M. 2001. The *Stimpsonia* problem; evidence from DNA sequences of plastid genes *atp*B, *ndh*F and *rbc*L. Bot. Jahrb. Syst. 123: 369–376.

Davis, G. 1966. See general references.

Erdtman, G. 1952. See general references.

Hegnauer, R. 1969, 1973. See general references.

Johri, B.M. et al. 1992. See general references.

Judd, W.S., Sanders, R.W., Donoghue, M.J. 1994. Angiosperm family pairs: preliminary phylogenetic analyses. Harvard Papers Bot. 5: 1–51.

Källersjö, M. et al. 2000. See general references.

Mast, A.R., Kelso, S., Richards, J., Lang, D.J., Feller, D.M., Conti, E. 2001. Phylogenetic relationships in *Primula* L. and related genera (Primulaceae) based on noncoding chloroplast DNA. Int. Y. Plant Sci. 162:1381–1400.

Melchior, H. 1964. Primulales. In: Melchior, H. (ed.) Engler's Syllabus der Pflanzenfamilien 2. Berlin: Bornträger, pp. 389–394.

Nowicke, J.W., Skvarla, J.J. 1977. Pollen morphology and the relationships of the Plumbaginaceae, Polygonaceae, and Primulaceae to the order Centrospermae. Smithsonian Contrib. Bot. 37: 1–64.

Pax, F. 1889. Primulaceae. In: Engler & Prantl, Die natürlichen Planzenfamilien IV. 1. Leipzig: Engelmann, pp. 98–116.

Pax, F., Knuth, R. 1905. Primulaceae. In: Engler, A. (ed.) Das Pflanzenreich IV, 237: 1–386.

Richards, J. 1993. *Primula*. Portland: Timber Press.

Schaeppi, H. 1937. Vergleichend-morphologische Untersuchungen am Gynoecium der Primulaceen. Zeitschr. Gesamte Naturwiss. 3: 239–250.

Trift, I., Källersjö, M., Anderberg, A.A. 2002. The monophyly of *Primula* (Primulaceae) evaluated by analysis of sequences from the chloroplast gene rbcL. Syst. Bot. 27: 396–407.

Van der Pijl, L. 1972. Principles of dispersal in higher plants. Berlin Heidelberg New York: Springer.

Wendelbo, P. 1961. Studies in Primulaceae III. On the genera related to *Primula* with special reference to their pollen morphology. Årb. Univ. Bergen, Mat. Nat. Ser. 7: 1–15.

Zhang, L.-B., Comes, H.P., Kadereit, J.W. 2001. Phylogeny and quaternary history of the European montane/alpine endemic *Soldanella* (Primulaceae) based on ITS and AFLP variation. Amer. J. Bot. 88: 2331–2345.

Rhamnaceae

D. Medan and C. Schirarend

Rhamnaceae Juss., Gen. Pl.: 376 (1789) ('Rhamni'), nom. cons.

Deciduous or evergreen, often thorny trees, shrubs, woody climbers or lianes, rarely herbs. Leaves simple, petiolate, alternate or opposite, with 1 main vein or 3–5 veins, entire to serrate, sometimes much reduced; stipules small, caducous or persistent, sometimes fused intrapetiolarly or interpetiolarly, or transformed into spines, absent in most *Phylica*. Inflorescence basically cymose, cymes mostly axillary, sessile or pedunculate, or reduced to many–few-flowered fascicles. Flowers small, 3–6 mm in diameter, regular, (3)4–5(6)-merous, bisexual or unisexual, plants sometimes dioecious, haplostemonous, hypogynous to epigynous, yellowish to greenish, rarely brightly coloured; hypanthium patelliform or hemispherical to tubular, sometimes absent, at the rim bearing calyx, corolla and stamens; sepals 4 or 5, valvate in bud, triangular, erect to more or less recurved during anthesis, often keeled adaxially; petals 4 or 5, rarely 0, usually smaller than sepals, concave or hooded, rarely almost flat, often shortly clawed, often enfolding the stamens; stamens 4 or 5, antepetalous, filaments thin, adnate to the base of the petals, anthers minute, versatile or not, 2(4)-locular, dehiscing by longitudinal slits, usually introrse; disc intrastaminal, nectariferous, thin to more or less fleshy, entire or lobed, glabrous, rarely pubescent, free from ovary or tightly surrounding it, or adnate to the hypanthium; combined pubescence of hypanthium and style sometimes forming a secondary pollen presenter and pollen-dosing structure; ovary superior to inferior, (1)2–4-locular, with 1(2) ovules in each locule, ovules anatropous, basal and erect. Fruit indehiscent, winged or not, schizocarpic, capsular, rarely explosively dehiscent, or a more or less fleshy drupe with 1–4 indehiscent, rarely dehiscent pyrenes. Seeds with thin, oily albumen, sometimes exalbuminous, embryo large, oily, straight or rarely bent.

An almost cosmopolitan family of 52 genera and about 925 species.

VEGETATIVE MORPHOLOGY. Architecture is intermediate between Attim's and Roux's (*Ziziphus*), or between Koriba's and Roux's (*Ziziphus, Paliurus*) models (Tourn et al. 1992). Dichasial sympodia are common: the apical meristem of orthotropic shoots ceases growth or produces a spine, while buds of two distal nodes continue the growth. Leaves usually subtend a single bud, less commonly serial buds (Colletieae, *Gouania, Ziziphus*, Tourn et al. 1991). The lowermost serial bud can form vegetative long shoots, short-shoots, or flowering shoots, and the distal bud often produces a spine that may ramify (Tortosa et al. 1996). Spinyness is absent only in Gouanieae and Ventilagineae but pervasive only in the Colletieae. Twining shoots occur in *Berchemia, Ziziphus, Ventilago* and *Smythea*, and circinate tendrils in *Gouania, Reissekia* and *Helinus*. Rhizomes were reported for *Colletia*, and shoots adventitious on roots for *Rhamnus, Phylica*, and *Ziziphus*. Some species are able of sprouting after fire (Zedler 1995; Tortosa et al. 1996).

VEGETATIVE ANATOMY. Leaves are generally dorsiventral, the hairs mostly simple (exceptions: two-armed in *Sageretia*, stellate in Pomaderreae). The epidermis is often mucilaginous or includes mucilaginous cells. Stomata are usually anomocytic, less commonly paracytic or anisocytic. Stomata-bearing furrows or cavities are found in *Ceanothus* and *Condalia*. A multiple epidermis is unusual (*Rhamnus*); a hypodermis is reported for several genera. Palisade parenchyma is one to several cells thick, often including isolated, enlarged cells filled with mucilage, tannin, or both. Secretory cavities and secretory channels occur occasionally in Rhamneae and Paliureae. Lysigenous mucilage cavities have been recorded for the leaf vein parenchyma and petioles from all tribes. Druses are common in the mesophyll; acicular crystals are known only from Gouanieae. Small lateral, vertically transcurrent veins are known from several tribes. The petioles have usually a single open or U-shaped bundle. Characters of diagnostic value include type of leaf structure, venation, thickness and striation of the cuticle, presence of a multiple epidermis, type of mucilaginous epidermis, presence of stomata-bearing cavities or furrows, type of stomata, distribution of indumentum, crystals and secretory cells, proportion of palisade tissue, and presence of collenchyma connecting the main vein with the upper epidermis (Gemoll 1902; Herzog 1903; Medan 1986).

The wood is generally diffuse-porous to ring-porous. Growth rings are ± distinct, marked by differences in vessel diameter, latewood fibres or marginal parenchyma bands. Vessels (2–220 per mm²) are solitary or in radial multiples or clusters of 2–5(–15); vessel member length is from 110–700 µm. Perforations are mostly simple, rarely scalariform to reticulate; intervessel pits often coalesce and are rarely vestured. Vessel-ray and vessel-parenchyma pits are exclusively half-bordered. Vascular tracheids intergrading with narrow vessels are rather abundant, often with spiral thickenings. Fibre elements are exclusively libriform, 650–1000 µm long. Wood rays (10–20 per mm) are 1–5 cells wide and 10–30 cells high and distinctly heterogeneous to homogeneous. Axial parenchyma is usually scanty paratracheal to vasicentric, rarely diffuse apotracheal or the parenchyma cells are arranged in uni- to multiseriate, tangential to diagonal bands. Prismatic to diamond-shaped, rarely styloid crystals are very frequent in the axial and ray parenchyma cells or in chambered strands of the axial parenchyma. The heartwood is often filled with brownish, gum-like deposits (Schirarend 1984).

Ultrastructural data are scant. The mucilaginous epidermal cells of *Rhamnus frangula* were studied in detail by Jakovleva (1988); sieve-element plastids are of the S-type in all four genera investigated by Behnke (1974); in *Rhamnus frangula* the mucilage-secreting cavities in shoot apices and flower buds have been found to be of lysigenous origin (Bouchet 1974).

INFLORESCENCE STRUCTURE. The inflorescences are generally described as cymes, or more specifically as umbelliform, corymbiform, racemiform, spiciform cymes (Suessenguth 1953; Brizicky 1964). Solitary terminal flowers have also been reported. Detailed studies are only available for a few species of *Ceanothus, Noltea, Phylica, Paliurus, Pomaderris* and *Rhamnus*, and for Colletieae (Tortosa et al. 1996). Within this tribe, deciduous monotelic synflorescences (either thyrsoids with cymes as paracladia, or botryoids with 1-flowered paracladia) were considered basic, and proliferating synflorescences and perennial short-shoots as derived.

FLOWER STRUCTURE. Claims for an axial respectively appendicular nature of the hypanthium lack adequate support (Medan 1985). Flower size and hypanthium depth are correlated in Colletieae (Medan and Aagesen 1995). The nature of the petal-stamen complex is much disputed among plant morphologists (Bennek 1958; Sattler 1973; Ronse Decraene et al. 1993).

The disc is intrastaminal, either annular (secretory surface a ring surrounding the ovary), adpressed (secretory surface lining the hypanthium, delimited distally by a rim), revolute (a laminar projection of the flower tube above ovary level) or indistinct (as the adpressed type but the boundary with non-secreting area only anatomically discernible) (Medan and Aagesen 1995).

The development of the gynoecium starts with a unilocular primordium; plication and centripetal growth of the fused carpellary flanks build up the septa, which later fuse at the centre, often incompletely, thus leaving a compitum (Medan 1985, 1988). Carpels bear an ovule on one of their margins, the other margin being sterile (van Tieghem 1875; Suessenguth 1953). In 3-carpellate ovaries the sequence is usually fertil-sterile-fertil-sterile-sterile-fertil, i.e. one septum bears two ovules, the second a single ovule, and the third is sterile, but over 15 different septation patterns have been found (Suessenguth 1953; Tortosa 1984; Medan 1985). The stigmata are terminal on the stylar branches; entire styles have various stigma types (Medan and Aagesen 1995). The style is traversed by as many stylar canals as locules (Medan 1985); rarely a single canal is present, or the style is solid (Hanácková and Piñeyro López 1999). Increased functional syncarpy is achieved in genera with a single stylar canal or a solid style, or when a stigmatic compitum is formed (*Colletia*, D'Ambrogio and Medan 1993).

EMBRYOLOGY. Anthers are 2- or 4-locular; the development of anther wall follows the basic type. The endothecium has fibrous thickenings; the two middle layers are ephemeral, and the cells of the glandular tapetum become 2-nucleate. Cytokinesis in the microspore mother cells is simultaneous; tetrads are tetrahedral or isobilateral. Pollen grains are 2-celled when shed. The ovules are anatropous, bitegmic (outer integument thicker), crassinucellate, and the micropyle is usually formed by the outer integument. A hypostase is sometimes present, an epistase is rare (*Ziziphus*).

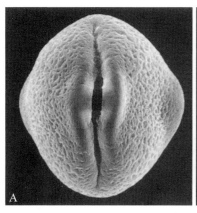 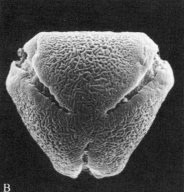 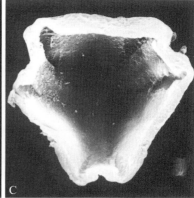

Fig. 102. Rhamnaceae pollen. **A** *Ventilago viminalis*, equatorial view; sexine supra-reticulate. **B** *Karwinskia rocana*, polar view; sexine rugulate. **C** *Auerodendron glaucescens*, equatorial section showing paired exine thinnings parallel to colpi; all SEM (×2000). Photo C. Schirarend

Funicular vascular strands usually almost reach the tip of the outer integument. The archesporium is 1- to several-celled, each cell giving rise to a parietal derivative that produces up to 13 parietal layers. The apical cells of the nucellar epidermis form a nucellar cap 5 cells thick. Sometimes a nucellar beak protrudes through the micropyle. The embryo sac is of the Polygonum type, less commonly of the Allium type. Antipodals are generally ephemeral; in *Ziziphus* they persist or become coenocytic. The polar nuclei fuse at about the time of fertilisation. Multiple embryo sacs are reported for several genera. Endosperm formation is nuclear; the tissue later becomes cellular from the micropylar pole, but cellularization does not always reach the chalazal end. Embryogeny follows the Asterad type, sometimes the Solanad type (*Ziziphus*). Synergid polyembryony is occasionally found (*Ziziphus*; Davis 1966).

POLLEN MORPHOLOGY. Pollen grains are usually isopolar, radially symmetrical, angulaperturate, 3(4)-zono-colporate, suboblate to subprolate, 11–40 µm in polar diameter, 12–34 µm in equatorial diameter, and angular to circular in polar view (Fig. 102). Apertures are composed of comparatively narrow colpi and ± narrow, lalongate to elliptic, rarely lolongate endoapertures. The colpi extend over about 4/5 of the length of the polar axis. The exine is 1–2(–3.5) µm thick, with conspicuous thinnings parallel to the colpi (Fig. 102C). The sexine is mostly pertectate-perforate. The tectum is microreticulate, striate, rugulate, fossulate, verrucate, pilate or ± psilate (LO-

pattern), with ± densely spaced perforations. The boundary between tectum and subtectal elements is rather indistinct; the subtectal layer is very different in thickness and mostly granular. Solid subtectal bacula are rare (*Pomaderris*); the foot layer is of different thickness and has a distinct distal surface. The nexine is always well differentiated, thin at the mesocolpia, and ± thickened at the colpi (costae transversales; Schirarend and Köhler 1993).

KARYOLOGY. The basic chromosome number of Rhamnaceae appears to be x = 12; 2n = 24 is reported for most tribes. Other counts include n = 11, known from *Helinus* and most Colletieae studied, and n = 10, 11, and 13 recorded for *Rhamnus* (Rhamneae). Polyploidy is apparently restricted to cultivated species of *Ziziphus* (e.g. 24, 48, 60, 72, 96 for *Ziziphus jujuba*) and to *Pomaderris* (e.g. 24, 36 and 48; Richardson et al. 2000a).

REPRODUCTIVE BIOLOGY. Relevant data are available for some 14 genera, but detailed studies have been conducted for only 15 species (references in Medan and D'Ambrogio 1998; Medan and Arce 2000). Overall, flowers of Rhamnaceae are small, often inconspicuous, and receive visits of insects that are rewarded with easily accessible nectar and pollen.

Flowering is usually in spring and summer; some taxa start blooming in late winter (*Berchemia, Discaria, Retanilla, Rhamnus*) or continue to flower into the cold season (*Colletia, Phylica*). Individual flowers last 1.5–10 days, being usually protandrous (a shorter male phase generally overlaps with a longer female phase). Homogamy was reported for *Colletia*; a report on protogyny (*Maesopsis*) needs confirmation.

The flowers are generally white, greenish, less commonly yellow, rarely pink or blue (*Ceanothus*) or red (*Colletia*). They are weakly scented (*Rhamnus*) to almost oppresively fragrant (*Discaria*), from foetid or nauseating (*Discaria americana*, *Ziziphus mucronata*) to pleasant (vanilla-, cinnamom-, or honey-like).

Pollen is commonly exploited and may function as a subsidiary reward. Nectar is easily accessible except in species with relatively long hyphanthia; amounts may be minute, but no true nectarless species has been reported. Sugar concentrations are usually lower than 20% but may reach 30% (*Ziziphus*). Both fructose/glucose and saccharose-dominated nectars have been reported.

The cuculate petals cover often the anthers during anthesis, which may prevent pollen robbing and may protect pollen from desiccation (Medan and Aagesen 1995). Centrifugal stamen movements during anthesis, recorded for several genera, possibly prevents pollen-stigma interference (Medan and Hilger 1992). In *Trevoa* and *Retanilla* pollen becomes trapped in the hairs inside the floral tube; this hairy barrier acts as an exposed, secondary pollen presenter because mouthparts of insects must traverse it to reach nectar. Both wet and dry stigmas have been reported.

Generalized, unspecialised entomophily is the rule, with Hymenoptera and Diptera co-dominating nearly always. Increased flower tube length seems associated with a higher proportion of Hymenoptera and Lepidoptera in some assemblages (Colletieae, Medan and Aagesen 1995). Evidence of pollen transport by wind exists for *Rhamnus* (Aronne and Wilcock 1995) and *Ziziphus* (Zietsman 1990).

Monomorphic hermaphroditism combined with protandry and self-incompatibility (SI) is widespread in the family. Incomplete dichogamy, high flower number per individual and sequential flowering favour selfing, which is counterbalanced by gametophytic SI. Pollen-ovule ratios under this system are about 7000:1.

Heterodichogamy occurs in *Ziziphus*. It implies the occurrence of two floral morphs, both being protandrous, one starting flowering several hours after the other. While the flowers of one morph are receptive, those of the other release pollen; later, the situation is reversed, so that each individual is alternately male and female (Galil and Zeroni 1967; Zietsman and Botha 1992). Pollen-ovule ratios vary from 3300:1 to 35000:1.

Self-compatibility occurs in some *Pomaderris* spp. (Harvey and Braggins 1985). Their flowers are monomorphic and apparently protandrous, but SI is largely suppressed (fruit set under self-pollination above 60%).

Protandry and SI are apparently maintained when abortion of the stamens or the pistil give rise to pistillate or staminate flowers. Andromonoecy is reported for *Alphitonia*, *Colubrina*, and *Trevoa*, the latter showing reduced incompatibility. The pollen-ovule ratio is relatively high (*Trevoa quinquenervia*: 14500:1). True dioecy is known from *Rhamnus*, where the pollen-ovule ratio is very high (ca. 32300:1). Less well-known cases of dimorphism occur in *Noltea*, *Alphitonia*, *Phylica*, *Gouania*, *Rhamnus*, and *Discaria*.

Under open pollination, the fruit/flower ratio of SI species is usually below 10%. Fruit set is higher in species with reduced incompatibility (*Trevoa quinquenervia*: 48.8%) and in self-compatible species (*Pomaderris kumeraho*: 43.9%). Crop size varies from 300 to 8000 fruits per individual per year (*Rhamnus*), and from 120 (*Discaria*) to over 17000 viable seeds per plant per year (*Colletia*). Reproductive success is pollen-limited at the flower level in some cases (*Rhamnus*, *Ziziphus*, and *Discaria*).

FRUIT AND SEED. Dry fruits may be indehiscent, schizocarpous, or capsular; dehiscence can be slow or explosive. Fleshy fruits are always drupaceous. All types have a sclerenchymatous endocarp usually developed from the inner carpel epidermis (Vikhireva 1952). Each carpel produces one stone, or the fruit produces a single, compound stone. Parthenocarpic fruits were recorded in *Discaria*, *Rhamnus*, and *Ziziphus*.

The thin-walled, xerochastic, 3–4-carpellate explosive capsules open through sudden aperture of ventricidal, dorsicidal and septifragal dehiscence lines in each of the 3–4 stones (endocarpids), which also separate along septicidal planes. Dehiscence is caused by oblique bending of the endocarpids due to a distinct distribution of endocarpal cells. These occur in parallel layers with crossing micellary structure, which leads to enhancing stress during fruit dessication. A pedestal may persist on the plant after dehiscence (Colletieae, *Ceanothus*: colletioid dehiscence; Medan 1985) or not (*Colubrina* p.p., *Noltea*: colubrinoid dehiscence; Johnston 1971; Medan and Aagesen 1995). The seeds are ejected up to 9 m away.

Slowly dehiscing capsules are similar but the endocarpids open slowly and often tardily and incompletely; the fruit disintegrates into one-seeded endocarpids that dehisce later, or the seeds fall off from the open fruits, or remain on the fruiting pedicel after pericarp breakdown.

Most Gouanieae have schizocarps that split into three indehiscent, winged, one-seeded mericarps which temporarily remain attached to a car-pophore. Mericarps may be solid (*Gouania*, *Crumenaria*) or inflated (*Reissekia*). Fruits transitional between capsules and schizocarps occur in the Gouanieae (*Alvimiantha*, Fig. 105, p. 331), in which the septicidal capsule liberates winged, indehiscent hemimericarps (Medan 1989).

Dry indehiscent winged fruits occur in *Ventilago* (Ventilagineae, with terminal wing) and *Paliurus* (Paliureae, with horizontal wing). A 3-seeded nut with a fleshy, edible pedicel occurs in *Hovenia* (Paliureae). *Trevoa* (Colletieae) has a papery, usually one-seeded, rostrate nut partially covered by the persistent floral tube (Fig. 103E, p. 328).

Typical drupes occur in Rhamneae, with either a compound stone or 2–4 1-seeded individual stones, which may be tardily dehiscent.

Seeds are exotestal, small to medium-sized, rounded and somewhat dorsiventrally compressed (rarely cordate: *Alvimiantha*, or dorsally furrowed: *Rhamnus* p.p.). The testa is generally multiplica-tive, smooth. The outer epidermis is a palisade of thick-walled prismatic cells, often with a linea lucida. The mesophyll is aerenchymatous and ± crushed, rarely with scattered sclerotic cells (*Ventilago*). The inner epidermis is unspecialised. The tegmen is not or slightly multiplicative, soon crushed; the inner epidermis is not lignified. The vascular bundle extends into the raphe or has postchalazal extensions (*Ventilago*, *Ziziphus*). The endosperm is nuclear, oily, generally as a thin layer, rarely well developed and ruminate (*Reynosia*), sometimes absent (Ventilagineae, several Rhamneae, *Ampeloziziphus*). The embryo is straight or rarely bent, invariably chlorophyllous (Yakovlev and Zhukova 1980), oily; cotyledons are comparatively large, usually flat and reaching the seed coat (thus dividing the endosperm in two sectors), rarely curved (some *Rhamnus*). Funicu-lar arils are often present, sometimes conspicuous (*Ceanothus*, *Colubrina*, *Phylica*) and different in colour from the seed coat (*Alphitonia*), perhaps overlooked in some genera (Medan and Aagesen 1995). A tissue of raphal origin serves as an elaio-some (some *Rhamnus*; Aronne and Wilcock 1994). Heat may promote germination (*Rhamnus*, *Phylica*, *Ceanothus*; Keeley 1992; Kilian and Cowling 1992). Embryo dormancy is overcome by stratification in *Discaria*, *Hovenia* and *Retanilla* (Soriano 1960; Frett 1989; Keogh and Bannister 1993). The stone hinders germination in *Ziziphus* (Grice 1996).

DISPERSAL. Entire fruits (samaras) are dispersed in *Paliurus* and *Ventilago*. In *Smythea* the disper-sal by wind precedes dehiscence (Ridley 1930). In the Gouanieae the diaspores are winged mericarps or hemimericarps. Undehisced *Smythea* fruits may float in seawater for months (Guppy 1906; Ridley 1930). The seeds of *Colubrina asiatica* may float in seawater over long time (Carlquist 1966). Transport of whole fruits by rainwater was reported for *Retanilla* (Reiche 1907). All explosive capsules effect ballistic seed dispersal (e.g. *Cean-othus*, Evans et al. 1987).

Birds of over 35 families are known to disperse whole fruits of *Condalia*, *Berchemia*, *Maesopsis*, *Phylica*, *Rhamnus*, *Scutia*, and *Ziziphus*, and apparently also the arillate seeds of *Alphitonia* (Sun and Dickinson 1996). Mammals including some 15 different groups from rodents to ele-phants are involved in dispersal of *Karwinskia*, *Rhamnus*, and *Ziziphus*, and perhaps *Hovenia*. Ants transport arillated seeds of *Phylica* (Kilian and Cowling 1992), several Pomaderreae (Andersen and Ashton 1985), *Ceanothus* (Mills and Kummerow 1989), and (after dehiscence of bird-dispersed fruits) non-arillate seeds of *Rhamnus* (Aronne and Wilcock 1994).

PHYTOCHEMISTRY. The family is generally known for the widespread occurrence of calcium oxalate, tannins and certain alkaloids. Phenolic com-pounds are accumulated in large quantities as flavonols and leucoanthocyanins and the con-densed tannins based on them. Derivatives of anthraquinone are restricted to *Rhamnus*, where they are pharmacologically used as laxatives (e.g. Cortex Rhamni Frangulae, Fructus Rhamni Cathartici). More widely distributed within the family are the derivatives of naphtalin and lupeol and different types of saponins. For many genera the accumulation of alkaloids has been proved, which comprise peptid alkaloids and isochinolin alkaloids (Jossang et al. 1996).

SUBDIVISION AND RELATIONSHIPS WITHIN THE FAMILY. On the basis of a phylogenetic analysis of *rbc*L and *trn*L-F sequences of the plastid genome and morphological data, Richardson et al. (2000a, 2000b) revised the tribal classification of the family. Two of the five tribes recognised by Sues-senguth (1953) were shown to be polyphyletic, whereas three appeared monophyletic. For three of the six, well-supported further groupings, older tribal names were taken up, and three were newly named. The eleven tribes now recognised can be characterised morphologically. At the next higher level, the tribes group together into three clades, for which no morphological characters can be indicated, possibly because they are relics of formerly much more coherent groups (Richardson 2000a).

AFFINITIES. Recent angiosperm classifications placed Rhamnaceae in Rhamnales, either as the sole family of the order (e.g. Takhtajan 1997), or together with Elaeagnaceae (Thorne 1992). Celastrales, Urticales, and Euphorbiales had often been considered as closely related groups. Analyses of DNA sequences, including those by Soltis et al. (1997, 2000), Thulin et al. (1998), Richardson et al. (2000b), and Savolainen, Chase et al. (2000), support the placement of Rhamnaceae together with their closest relatives, Barbeyaceae and Dirachmaceae, in Rosales sensu APG II (2003), as part of the eurosid I assemblage.

DISTRIBUTION AND HABITATS. The family is of worldwide distribution, with habitat preferences ranging from tropical rain forest to moderately arid areas and from the sea level to the treeline. A basic preference for tropical and subtropical regions can be recognized. Many members of the family are genuine forest species, but several taxa occur in extratropical, xerophytic, open vegetation.

SYMBIONTS AND PARASITES. Root nodules forming perennial coralloid masses are found in all genera of Colletieae and in *Ceanothus*. The nodules are interpreted as modified lateral roots and are inhabited by actinomycetes of the genus *Frankia*. Fixation of atmospheric nitrogen takes place in particular structures of the microsymbiont (Baker and Schwintzer 1990; Cruz-Cisneros and Valdes 1991; Swensen 1996; Huss-Danell 1997). Endomycorrhizae have been reported from *Ceanothus*, *Colletia*, *Discaria* and *Trevoa* (Gardner 1986). Mites have been observed in domatia of *Rhamnus* (Lundström 1887). Galls are produced on male flowers of *Rhamnus ludovici-salvatoris* by a cecidomyiid fly (Traveset 1999). Insects associated with European Rhamnaceae have been used in biological control of *Rhamnus cathartica* in Canada (Malicky et al. 1970).

PALAEOBOTANY. One of the oldest fossils attributed to Rhamnaceae are leaf impressions from the Upper Cretaceous (Hollick 1930), but the assignment of such fossils is often problematic, especially before the Oligocene (Johnston 1977). A bisexual, obhaplostemonous flower reported for the mid-Cretaceous of Nebraska (94–96 Ma B.P.; Basinger and Dilcher 1984) has been ascribed to Rhamnaceae. The earliest known pollen records are from the Oligocene (Muller 1981).

ECONOMIC IMPORTANCE. Bark, leaves and fruits of several *Rhamnus* have been used as laxatives (notably *R. frangula* and *R. catharticus*). Diverse Old World species of *Rhamnus* provide yellow and green dyes as well as drugs. Several species of *Karwinskia* are highly toxic (Waksman et al. 1989; Lux et al. 1998) and of potential use in human medicine. *Maesopsis eminii* is widely planted in Africa and Asia for its timber, which is used in house and boat building; timber of *Ziziphus*, *Reynosia*, *Krugiodendron*, *Hovenia*, *Colubrina* species are used for construction, fine furniture, carving, lathework and music instruments. Many *Ziziphus* have edible fruits; among them *Z. jujuba* (Chinese jujube) and *Z. mauritiana* (Indian jujube) are cultivated at a commercial scale. *Hovenia dulcis* is also grown for the edible inflorescence stalks. Species of *Ceanothus*, *Colletia*, *Hovenia*, *Noltea*, *Paliurus*, *Phylica*, *Pomaderris* and *Rhamnus* are cultivated as ornamentals. Introduced species may be invasive: *Ziziphus mauritiana* in Australia (Grice 1996, 1998); *Rhamnus cathartica* in Canada (Malicky et al. 1970); *R. frangula* in the U.S.A. (Possessky et al. 2000); the native *Discaria toumatou* may turn invasive in New Zealand (Bellingham 1998).

CONSERVATION. Studies specifically addressing conservation are scarce (Hall and Parsons 1987, *Discaria*; Godt et al. 1997, *Ziziphus*). Almost 30 genera include endangered or vulnerable species (World Conservation Monitoring Centre 1996). A few Australian species of *Cryptandra*, *Spyridium* and *Trymalium* are considered extinct, and species of *Gouania*, *Lasiodiscus* and *Rhamnus* are also potentially extinct.

CONSPECTUS OF RHAMNACEAE
The following tribal classification follows that of Richardson et al. (2000a) with the addition of genera 25 and 50.

1. Tribe Paliureae Reiss. ex Endl. (1840).
   – Genera 1–3.
2. Tribe Colletieae Reiss. ex Endl. (1840).
   – Genera 4–9.
3. Tribe Phyliceae Reiss. ex Endl. (1840).
   – Genera 10–13.
4. Tribe Gouanieae Reiss. ex Endl. (1840).
   – Genera 14–19.
5. Tribe Pomaderreae Reiss. ex Endl. (1840).
   – Genera 20–26.

6. Tribe Rhamneae Hook. f. (1862).
 – Genera 27–39.
7. Tribe Maesopsideae A. Weberb. (1895).
 – Genus 40.
8. Tribe Ventilagineae Hook. f. (1862).
 – Genera 41–42.
9. Tribe Ampelozizipheae J.E. Richardson (2000).
 – Genus 43.
10. Tribe Doerpfeldieae J.E. Richardson (2000).
 – Genus 44.
11. Tribe Bathiorhamneae J.E. Richardson (2000).
 – Genus 45.

The relationships of seven additional genera (genera 46–52) are uncertain; therefore, these taxa are not included in any of the above tribes. Generic boundaries within tribe *Pomaderreae* are currently under study and new genera will be probably proposed in near future.

Key to the Tribes and Genera

1. Fruit with apical appendages, longitudinal wings, or a transversal membranous ring 2
 – Fruit not as above 9
2. Shrubs; fruit with a transversal membranous ring. **Paliureae p.p.** **1. *Paliurus***
 – Climbers or herbs; fruit not as above 3
3. Ovary 3-locular; fruit with longitudinal wings; tendrils present; endosperm present. **Gouanieae p.p.** 4
 – Ovary 2-locular; fruit with apical appendages; tendrils absent; endosperm absent. **Ventilagineae** 8
4. Perennial or annual herbs **18. *Crumenaria***
 – Erect or climbing shrubs or lianes 5
5. Erect, non-climbing shrubs 6
 – Lianes or climbing shrubs with tendrils 7
6. Fruit a schizocarp, separating into 3–4 indehiscent, winged, inflated mericarps **16. *Reissekia***
 – Fruit longitudinally 2–4-winged, indehiscent **19. *Pleuranthodes***
7. Fruit an explosive capsule **15. *Helinus***
 – Fruit not explosive **14. *Gouania***
8. Fruit a samara **42. *Ventilago***
 – Fruit a capsule **41. *Smythea***
9. Infructescence axis succulent; disc usually hairy. **Paliureae p.p.** **3. *Hovenia***
 – Infructescence axis not succulent; disc not hairy 10
10. Trees or shrubs, usually armed; leaves decussate; roots with actinorhizal nodules. **Colletieae** 11
 – Combination of characters not as above 16
11. Inflorescences with a terminal flower; floral tube pubescent inside; fruit indehiscent or slowly splitting into indehiscent endocarpids; arils remaining enclosed in the endocarpids 12
 – Inflorescences without terminal flowers; floral tube glabrous inside; fruit a explosive capsule; aril detached from endocarpid and seed at dehiscence 13
12. Anthers 4-locular; stigma exserted; fruit a papery rostrate nut **4. *Trevoa***
 – Anthers 2-locular; stigma not exserted; fruit a barely fleshy drupe, indehiscent or slowly splitting into indehiscent endocarpids **5. *Retanilla***
13. Floral tube caducous at fruit maturity; rim of fruit pedestal sinusoid; margin of aril deeply lobed **7. *Kentrothamnus***
 – Rim of fruit pedestal smooth, or floral tube persistent at fruit maturity; margin of aril more or less smooth 14

14. Filaments distally geniculate; nectary inconspicuous, adpressed to lower floral tube; floral tube persistent in fruit **6. *Adolphia***
 – Filaments curved distally; nectary forming a conspicuous disc; floral tube not persistent in fruit 15
15. Bases of opposite leaves united forming a line; stomata anomocytic; disc ring-like, encircling the ovary at bottom of floral tube **9. *Discaria***
 – Bases of opposite leaves not united; stomata paracytic; disc forming a revolute laminar projection of the floral tube located above ovary level **8. *Colletia***
16. Fruit fleshy 17
 – Fruit dry, dehiscent 34
17. Leaf venation pinnate 18
 – Leaf venation palmate 31
18. Ovary 1-locular; stigma mushroom-like. **Maesopsideae** **40. *Maesopsis***
 – Ovary 2- or 4-locular; stigma not as above. **Rhamneae** 19
19. Fruit a fleshy, one-stoned drupe 20
 – Fruit with 2–4 free pyrenes 29
20. Leaves alternate 21
 – Leaves generally opposite 24
21. Armed shrubs or trees **36. *Condalia***
 – Unarmed shrubs or small tree 22
22. Leaves leathery, margin entire **30. *Berchemiella***
 – Leaves papery, margin serrate 23
23. Endosperm present **31. *Rhamnella***
 – Endosperm absent **32. *Dallachya***
24. Petals present 25
 – Petals absent 28
25. Leaves never gland-dotted **33. *Berchemia***
 – Leaves often distinctly gland-dotted, pinnately veined and with strongly parallel tertiaries 26
26. Ovules 2 per locule **35. *Karwinskia***
 – Ovule 1 per locule 27
27. Seeds albuminous; cotyledons flat **37. *Auerodendron***
 – Seeds exalbuminous; cotyledons convex **34. *Rhamnidium***
28. Leaves coriaceous **38. *Reynosia***
 – Leaves papery, margin undulate **39. *Krugiodendron***
29. Flowers and fruits almost sessile, inflorescences spike- or panicle-like, medifixed hairs present, disc cylindrical **29. *Sageretia***
 – Characters not as above 30
30. Petals emarginate or notched; flowers often unisexual; if plants armed, seeds furrowed **27. *Rhamnus***
 – Petals deeply obcordate or bilobed; flowers always bisexual; plants mostly armed, but then seeds not furrowed **28. *Scutia***
31. Ovary 3-locular 32
 – Ovary 2- or 4-locular 33
32. Climber; endosperm absent. **Ampelozizipheae** **43. *Ampeloziziphus***
 – Tree; endosperm present. **Bathiorhamneae** **45. *Bathiorhamnus***
33. Disc adnate to ovary only; endosperm present. **Doerpfeldieae** **44. *Doerpfeldia***
 – Disc adnate to ovary and floral tube; endosperm absent. **Paliureae p.p.** **2. *Ziziphus***
34. Plant stellate-hairy. **Pomaderreae** 35
 – Plant glabrous or with simple hairs 41
35. Flowers in pendent, terminal heads enclosed in an involucre of large brownish bracts; petals absent; leaves opposite **21. *Siegfriedia***
 – Flowers not as above; leaves alternate 36
36. Floral tube tubulate to campanulate 37

– Floral tube not as above 38
37. Disc pubescent **25. *Cryptandra***
– Disc glabrous **24. *Stenanthemum***
38. Flowers sessile, in heads enclosed by persistent brownish bracts **23. *Spyridium***
– Flowers pedicellate, solitary or in umbel-like cymes forming terminal racemes or panicles 39
39. Schizocarp often irregularly splitting; mericarps indehiscent; stamens incurved, enclosed by the hooded petals at early anthesis **22. *Trymalium***
– Schizocarp splitting regularly; mericarps dehiscent; stamens straight, never enclosed by the petals 40
40. Floral tube absent, sepals wide-spreading at anthesis, ovary (half-)inferior **20. *Pomaderris***
– Floral tube present, ovary superior **26. *Blackallia***
41. Roots with actinorhizal nodules; petals sometimes blue; North America **47. *Ceanothus***
– Roots without actinorhizal nodules; petals never blue; rarely in North America 42
42. Climber with tendrils. **Gouanieae p.p.** **17. *Alvimiantha***
– Tree or shrub without tendrils 43
43. Seed persisting on receptacle after dehiscence 44
– Seed not persisting on receptacle after dehiscence 45
44. Exocarp thin and leathery; ovary usually 2-locular **49. *Emmenosperma***
– Exocarp thick, spongy and crumbly at maturity; ovary usually 3-locular **46. *Alphitonia***
45. Leaves opposite 46
– Leaves usually alternate 47
46. Leaf margin not revolute; stipules interpetiolar **51. *Lasiodiscus***
– Leaf margin revolute; stipules not interpetiolar. **Phyliceae p.p.** **12. *Nesiota***
47. Branches clustered; stipules nearly always absent. **Phyliceae p.p.** 48
– Branches not clustered; stipules present 49
48. Stipules absent; disc glabrous **10. *Phylica***
– Stipules present; disc pubescent **11. *Trichocephalus***
49. Ovary superior **52. *Schistocarpaea***
– Ovary (half-)inferior 50
50. Plant glabrous; leaf margins always dentate. **Phyliceae p.p.** **13. *Noltea***
– Plant sometimes pubescent; leaf margins not always dentate 51
51. Pericarp with outer mealy layer; Western Australia **50. *Granitites***
– Pericarp without mealy layer; pantropical **48. *Colubrina***

Genera of Rhamnaceae

1. Tribe Paliureae Reiss. ex Endl. (1840).

Armed or unarmed trees or shrubs with alternate leaves. Disc sometimes hairy, adnate to ovary and hypanthium. Fruit dry with a membranous, transversal ring (*Paliurus*), a drupe (*Ziziphus*), or dry and indehiscent with fleshy pedicel (*Hovenia*). Three genera.

1. *Paliurus* Tourn. ex Mill.

Paliurus Tourn. ex Mill., Gard. Dict. abr. ed. 4 (1754); Schirarend & Olabi, Bot. Jahrb. Syst. 116: 333–359 (1994), rev.

Evergreen or deciduous, often thorny shrubs or small to medium-sized trees. Leaves distinctly triplinerved. Flowers in axillary cymes; hypanthium hemispherical to dish-shaped; disc fleshy, adnate to the intrastaminal surface of the hypanthium; ovary half-inferior, 2–3-locular. Fruit dry, indehiscent, with a disc- to cup-shaped or hemispherical wing around the apex. Four species in E Asia, one species (*P. spina-christi*) in the Mediterranean region.

2. *Ziziphus* Mill.

Ziziphus Mill., Gard. dict., abr. ed. 4 (1754); Johnston, Am. J. Bot. 51: 1113–1118 (1964), West Indian spp.
Condaliopsis (A. Weberb.) Suesseng. (1953).

Deciduous or evergreen, erect or straggling, often climbing shrubs or small to medium sized trees. Leaves distinctly triplinerved rarely pinnately veined; stipules often spinose. Flowers in axillary, corymb-like cymes or terminal or axillary thyrses; hypanthium shallow, patelliform to hemispherical; petals usually present; disc flat, fleshy, 5–10-lobed; ovary superior, 2–3(4)-locular. Fruit a single-stoned, (1)2–3-locular drupe. Almost pantropical, about 100 spp., centred in tropical America and SE Asia.

3. *Hovenia* Thunb.

Hovenia Thunb., Nov. Gen. 7 (1781).

Deciduous, unarmed trees or shrubs. Leaves long-petiolate. Flowers in axillary or terminal, 3- to several-flowered cymes; hypanthium obconical to dish-shaped; disc fleshy, plane, tomentose, filling the hypanthium; ovary half-inferior, 3-locular. Fruit a 3-seeded nut, peduncles and pedicels becoming fleshy and juicy at fruit maturity. Seven spp., restricted to E Asia.

2. Tribe Colletieae Reiss. ex Endl. (1840).

Usually armed, often nearly aphyllous, erect shrubs or small trees. Leaves decussate, each subtending 2 serial buds. Root nodules actinorhizal. Disc either a ring-like, laminar projection of the hypanthium or an inconspicuous nectary adnate to lower hypanthium. Fruit a papery nut, a barely fleshy drupe or an explosive capsule. Six genera (Aagesen 1999).

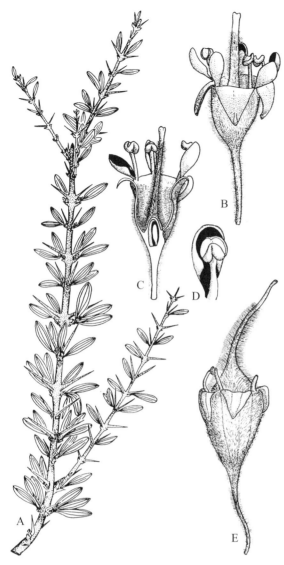

Fig. 103. Rhamnaceae–Colletieae. *Trevoa quinquenervia.* **A** Branch. **B** Flower. **C** Same, vertical section. **D** Petal and stamen seen from inside. **E** Fruit. (Tortosa 1992)

4. *Trevoa* Miers   Fig. 103

Trevoa Miers, Bot. Misc. 1: 158 (1829); Tortosa, Darwiniana 31: 223–252 (1992), rev.

Leafy shrubs up to 3 m tall. Flowers pubescent, perfect or staminate, solitary or in 1–7-flowered cymes; hypanthium slightly campanulate, persistent in fruit, pubescent inside; petals cucullate, each partly covering the pollen-presenting surface of the anthers; stamen filaments distally geniculate, anthers 4-locular; disc inconspicuous, adnate to the lower part of hypanthium; style persistent in fruit, stigma exserted. Ovary half-inferior,

3-locular. Fruit an indehiscent, thin-walled nut containing 1–2(3) seeds. One species in central Chile.

5. *Retanilla* (DC.) Brongn.

Retanilla Brongn., Ann. Sci. Nat. (Paris) I, 10: 364 (1827); Tortosa, Darwiniana 31: 223–252 (1992), rev.

Leafy to aphyllous, unarmed to spiny shrubs up to 4 m tall. Leaves subtending 1–2 buds. Flowers pubescent, 4–5-merous, terminal or in 3–5(7)-flowered cymes; hypanthium campanulate to urceolate, caducous to persistent in fruit, pubescent inside; petals cucullate, partly covering the pollen-presenting surface of the anthers; filaments distally geniculate, anthers 2-locular; disc indistinct, nectariferous surface lining the lower part of hypanthium; ovary half-superior, 2–3(4)-locular. Fruit somewhat fleshy, indehiscent or splitting into indehiscent endocarpids. Four species, S South America.

6. *Adolphia* Meisner

Adolphia Meisner, Pl. Vasc. Gen. tab. 70, comm. 50 (1837); Tortosa, Darwiniana 32: 185–189 (1993), rev.

Subaphyllous shrubs up to 3 m tall; the distal bud in each node forming a spine and the proximal one a vegetative shoot or a proliferating synflorescence. Flowers in 3–7-flowered cymes, pubescent; hypanthium campanulate, persistent in fruit; petals present; stamen filaments distally geniculate, anthers 4-locular; disc inconspicuous, adnate to the lower hypanthium; ovary half-inferior, 3-locular. Fruit an explosive capsule. One species in southern North America.

7. *Kentrothamnus* Suessenguth & Overkott

Kentrothamnus Suessenguth & Overkott, Repert. Spec. Nov. Regni Veg. 50: 326 (1941); Johnston, J. Arnold Arb. 54: 471–473 (1973), rev.

Subaphyllous shrubs up to 4 m tall; distal bud in each node producing a spine, the proximal one a vegetative shoot or a proliferating synflorescence. Flowers pubescent, 5-merous, in 1–3(4)-flowered cymes; hypanthium urceolate-campanulate, caducous in fruit, leaving a sinusoid scar that includes the filaments; petals present; stamen filaments distally geniculate, anthers 4-locular; disc inconspicuous, adnate to the lower part of hypanthium; ovary half-superior, 3-locular. Fruit

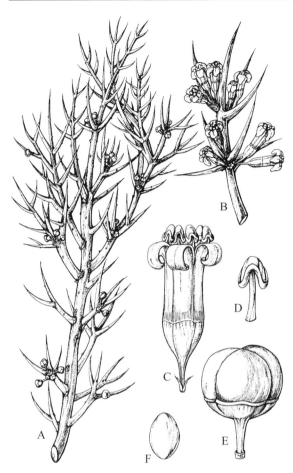

Fig. 104. Rhamnaceae–Colletieae. *Colletia hystrix.* **A** Fruiting branch. **B** Flowering branchlet. **C** Flower. **D** Stamen. **E** Fruit. **F** Seed. Drawn by Nora Mugaburu. (Dimitri 1972)

an explosive capsule. One species in Bolivia and northernmost Argentina.

8. *Colletia* Juss.   Fig. 104

Colletia Juss., Gen. Pl. 380 (1789), nom. cons.; Tortosa, Parodiana 5: 279–332 (1989), rev.

Virtually aphyllous shrubs up to 4 m tall. Leaves minute, early caducous; the distal bud in each node producing a spine or a spiniform twig ramified up to 3rd order, the proximal one producing a flowering short-shoot. Flowers 4–5-merous; hypanthium urceolate, caducous; petals present; anthers usually 2-locular; disc on a revolute laminar projection of the hypanthium; ovary half-inferior, 3(4)-locular. Fruit an explosive capsule. Five species, South America.

9. *Discaria* Hook.

Discaria Hook., Bot. Misc. 1: 156 (1829); Tortosa, Bol. Soc. Argent. Bot. 22: 301–335 (1983), rev.

Prostrate to erect, leafy or subaphyllous, often spiny shrubs or trees up to 8 m tall. Leaf bases united, forming a line. Flowers perfect, rarely unisexual, 4–5(6)-merous, in 1–3(6)-flowered cymes; hypanthium campanulate to urceolate, caducous; petals present or absent; anthers generally 4-locular; disc ring-like, more or less lobed, encircling the ovary at bottom of hypanthium; ovary half-inferior, 3-locular. Fruit an explosive capsule. Eight species, 5 in South America, 2 in Australia and 1 in New Zealand.

3. TRIBE PHYLICEAE Reiss. ex Endl. (1840).

Often ericoid shrubs or trees. Leaves alternate or opposite, usually densely tomentose beneath, leaf margins usually revolute. Stipules present or absent. Disc epigynous or adnate to the hypanthium. Fruit splitting into 3 one-seeded endocarpids. Four genera.

10. *Phylica* L.

Phylica L., Spec. Pl.: 195 (1753); Pillans, J. S. Afr. Bot. 8: 1–164 (1942), rev.

Evergreen ericoid shrubs or rarely small trees. Leaves alternate; stipules absent. Flowers bisexual (plants polygamous) in tight head-like thyrses with extremely pubescent bracts; hypanthium bell-shaped or cylindrical, surpassing the ovary; petals usually present; disc present or absent; ovary inferior, 3-locular. Fruit explosive; seeds arillate. About 150 spp. in Southern Africa to S Tanzania, Madagascar, southern Antarctic islands.

11. *Trichocephalus* Brongn.

Trichocephalus Brongniart, Ann. Sci. Nat. (Paris) I, 10: 374 (1827).

Evergreen ericoid shrubs. Leaves alternate, stipulate. Flowers bisexual in tight head-like thyrses with extremely pubescent bracts; hypanthium cylindrical, surpassing the ovary; petals present; disc adnate to hypanthium, pubescent; ovary inferior, pubescent, 3-locular. Fruit explosive; seeds arillate. One species, *T. stipularis* (L.) Brongn. from Southern Africa.

12. *Nesiota* Hook. f.

Nesiota Hook. f., in Bentham & Hook. f., Gen. 1: 380 (1862).

Evergreen small trees. Leaves opposite. Flowers in few-flowered lateral cymes; hypanthium shallow campanulate; petals deeply hooded; stamen filaments stout; disc thin, intrastaminal, densely pubescent around ovary at early anthesis; ovary inferior, 3–4-locular. One species, *N. elliptica* (Roxb.) Hook. f., St. Helena Island.

13. *Noltea* Reichb.

Noltea Reichb., Consp.: 145 (1828).

Evergreen shrubs or small trees up to 4 m high. Leaves alternate, discolorous. Flowers in axillary or terminal, few-flowered cymes in small panicles; hypanthium cup-shaped; disc very thin, adnate to the lower part of the hypanthium; ovary half-inferior, 3-locular. Endocarpids dehiscent along inner face. One species, *N. africana* (L.) Reichb. in Southern Africa.

4. TRIBE GOUANIEAE Reiss. ex Endl. (1840).

Unarmed climbers or herbs with tendrils. Leaves alternate, petiolate. Ovary inferior, usually 3-locular; disc fleshy, stellate or 5-angled. Fruit a longitudinally 3-winged schizocarp, a schizocarp releasing inflated, winged (hemi-) mericarps, a winged nut, or an explosive capsule. Six genera.

14. *Gouania* Jacq.

Gouania Jacq., Select. Stirp. Am. Hist.: 263 (1763).

Evergreen climbing shrubs, or up to 20 m high woody vines; tendrils circinnate. Flowers bisexual to polygamous, in axillary small cymes, usually aggregated into ample, leafless panicles; hypanthium shallow obconical to campanulate; petals rarely absent; disc broadly annular, with five more or less distinct, antesepalous appendages. Schizocarp dry, longitudinally 3-winged, separating septicidally into three 2-winged, indehiscent mericarps. A pantropical genus of about 50 species in need of revision.

15. *Helinus* E. Mey. ex Endl.

Helinus E. Mey. ex Endl., Gen.: 1102 (1840), nom. cons.

Evergreen woody climbers; tendrils coiled, axillary. Flowers in axillary, (1–)3–15-flowered umbelliform cymes; hypanthium broadly campanulate; disc massive, thickened around and adnate to ovary, never lobed. Fruit an explosive capsule. A tropical genus with five species, three in Eastern and Southern Africa, one in India, and one in Madagascar.

16. *Reissekia* Endl.

Reissekia Endl., Gen.: 1103 (1840).

Evergreen shrubs. Flowers in terminal and axillary cymes; hypanthium shallow, cup-shaped, sepals slightly winged laterally; petals distinctly hooded, enclosing the stamens; disc thin, adnate to the hypanthium, never 5-lobed or appendaged; ovary 3–4-locular. Schizocarp separating into 3–4 indehiscent, winged and inflated mericarps. One species, *R. smilacina* (Smith) Endl., Brazil.

17. *Alvimiantha* Grey-Wilson   Fig. 105

Alvimiantha Grey-Wilson, Bradea 43: 287–290 (1978).

Evergreen climbing shrubs. Leaves triplinerved. Flowers bisexual or functionally male, in axillary inflorescences. Hypanthium cup-shaped, glabrous; disc very thin. Schizocarp inflated, unwinged, subglobose, hemimericarps 3, winged, with a cordate seed each. One species, *A. tricamerata* Grey-Wilson, Brazil.

18. *Crumenaria* Mart.

Crumenaria Mart., Nov. Gen. Sp. Pl. 2: 68, pl. 160 (1826).

Annual or perennial herbs. Leaves often more or less reduced to completely absent. Flowers bisexual to polygamous, solitary or in axillary or terminal cymes; hypanthium obconical to campanulate; disc very thin or absent. Schizocarp dry, mericarps 3, winged, fruit axis persistent. Six species in tropical Brazil and NE Argentina, one in Guatemala.

19. *Pleuranthodes* A. Weberb.

Pleuranthodes A. Weberb. in Engler & Prantl, Nat. Pflanzenfam. III, 5: 424 (1896).

Evergreen shrubs. Flowers in axillary, long-pedunculate dichasia; hypanthium shallow, dish-shaped; disc distinctly 5-lobed, ovary (2)3–4-locular. Fruit

Fig. 105. Rhamnaceae–Gouanieae. *Alvimiantha tricamerata.* **A** Habit. **B** Functional male flower. **C** Same, vertical section. **D** Hermaphrodite flower, vertical section. **E** Fruit. **F** Dehisced fruit releasing seeds. **G** Half-fruit, showing position of seed with the pericarp valve and the central column in part. Drawn by C. Grey-Wilson. (Grey-Wilson 1978)

dry, longitudinally 2–4-winged, indehiscent. Two species, Hawaii.

5. Tribe Pomaderreae Reiss. ex Endl. (1840).

Unarmed or armed shrubs or small trees with stellate hairs. Leaves stipulate, opposite or alternate. Disc surrounding the ovary and adnate to the hypanthium. Ovary 3(2)-locular, superior to inferior. Fruit schizocarpic, the mericarps usually dehiscent ventrally. Seeds conspicuously arillate. Seven genera (but ongoing morphological research will most probably result in the addition of five new genera to the tribe; Kevin Thiele, pers. comm.).

20. *Pomaderris* Labill.

Pomaderris Labill., Nov. Holl. Pl. Sp. 1: 61, pl. 86–87 (1804).

Evergreen, unarmed shrubs or small, often multistemmed trees up to 10 m high. Leaves alternate, the margins distinctly revolute. Flowers in small, umbelliform cymes forming terminal racemes or panicles, sometimes head-like clusters, rarely solitary; hypanthium absent; petals present or absent; disc very thin and annular or absent; ovary half-inferior. Mericarps generally dehiscing ventrally through a basal valve. About 40 species, Australia and New Zealand.

21. *Siegfriedia* C.A. Gardner

Siegfriedia C.A. Gardner, Enum. Pl. Austral. Occ.: 76 (1931), nomen; J. Roy. Soc. W. Austral. 19: 85 (1934), descr.

Evergreen, unarmed shrubs. Leaves decussate, the margins distinctly revolute. Flowers congested in head-like, pendent clusters, covered by a whorl of 10–12 reddish-brown bracts; hypanthium 0; petals 0; stamens exserted, with about 6 mm long filaments; disc thin and delicate; ovary inferior. Mericaps dehiscing ventrally through a basal valve. One species, *S. darwinioides* C.A. Gardner, Western Australia.

22. *Trymalium* Fenzl

Trymalium Fenzl, Enum. Pl. Hueg.: 20 (1837).

Evergreen, usually unarmed shrubs. Leaves alternate, flat or with revolute margins. Flowers pedicellate, in raceme-like cymes or narrow panicles; hypanthium very short or 0; disc ring-like or distinctly lobed. Schizocarp often irregularly splitting in dehiscent or indehiscent mericarps. 14 spp., most in Western Australia.

23. *Spyridium* Fenzl

Spyridium Fenzl, Enum. Pl. Hueg.: 24 (1837).

Small, evergreen, unarmed, densely pubescent shrubs. Leaves alternate, comparatively small, coriaceous. Flowers in sessile heads or clusters, surrounded by persistent brown bracts, the heads often united into compound heads; hypanthium short or 0; disc more or less lobed; ovary inferior. Mericarps ventrally dehiscing or indehiscent. About 30 spp., Western and Southern Australia.

24. *Stenanthemum* Reissek

Stenanthemum Reissek, Linnaea 29: 295 (1858).

Usually unarmed, pubescent shrubs. Leaves entire or toothed, the margins revolute or not. Flowers aggregated in dense head-like clusters surrounded by bracts and leaves; hypanthium short to long, the glabrous disc adnate to it or inconspicuous. Mericarps enclosed in persistent hypanthium, apically and ventrally dehiscent; aril 3-lobed. At least 28 Australian species.

25. *Cryptandra* Sm.

Cryptandra Sm., Trans. Linn. Soc. 4: 217 (1798).

Evergreen, often spinescent shrubs. Leaves very small and narrow, often clustered, lower surface usually whitish-tomentose. Flowers solitary or in terminal, clustered spike- or head-like cymes; hypanthium campanulate to tubular; disc annular, often slightly lobed, stellate pubescent; ovary inferior. Mericarps dehiscing apically and ventrally, aril succulent, 3-lobed. About 40 spp., extratropical Australia.

26. *Blackallia* C.A. Gardner

Blackallia C.A. Gardner, J. Roy. Soc. W. Austral. 27: 183 (1942).

Evergreen, spinescent shrubs with small, alternate, bilobate leaves. Flowers solitary or in fascicles, long pedicellate; hypanthium tubular; sepals erect; petals cucullate; disc annular; ovary superior. Mericarps dehiscent. One species, *B. biloba* C.A. Gardner, Western Australia.

6. TRIBE RHAMNEAE Hook. f. (1862).

Usually unarmed trees, shrubs, or climbers. Leaves alternate or opposite, entire or serrate. Petals + or 0; disc free or adnate to hypanthium, style often persisting in fruit. Ovary superior or half-inferior. Fruit a 1–4-locular drupe, the pyrenes sometimes secondarily releasing the seeds. Thirteen genera.

Note that generic limits among *Reynosia*, *Rhamnidium*, *Auerodendron* and *Karwinskia* are problematic and these genera are in need of critical study.

27. *Rhamnus* L.

Rhamnus L., Sp. Pl. 1: 193 (1753).
Frangula Mill. (1754).
Oreorhamnus Ridl. (1920).
Oreoherzogia W. Vent (1962).

Deciduous or evergreen, unarmed or thorny trees or shrubs. Leaves alternate, rarely opposite. Flowers perfect or unisexual (then usually dioecious), mostly in sessile or peduncled, axillary, umbelliform cymes or fascicles; hypanthium campanulate to cup-like; petals usually +; disc conspicuous, adnate to the hypanthium, or apparently 0; ovary superior. Drupe 2–4-stoned, the stones indehiscent or ventrally deshiscing. Seeds unfurrowed or with dorsal or lateral furrows. About 100 spp., widespread, absent only from Madagascar, Australia and Polynesia.

28. *Scutia* Comm. ex Brongn.

Scutia Comm. ex Brongn., Ann. Sci. Nat. I, 10: 362 (1827), nom. cons.; Johnston, Bull. Torrey Bot. Club 101: 64–71 (1974), rev.

Evergreen, thorny, rarely scandent shrubs or small trees with opposite to subopposite leaves. Flowers solitary or in sessile, axillary, fasciculate cymes; hypanthium short, cup-like; disc fleshy, adnate to the hypanthium; ovary half-inferior. Drupe with 2–4 indehiscent pyrenes. Five species, one in the Old World Tropics and 4 in tropical South America.

29. *Sageretia* Brongn.

Sageretia Brongn., Ann. Sci. Nat. I, 10: 359, pl. 18, fig. 2 (1827).
Lamellisepalum Engl. (1897).

Evergreen or deciduous, spinescent or unarmed small trees or shrubs, rarely climbing by thorns representing modified inflorescence axes. Leaves alternate or subopposite. Flowers minute, sessile, in glomerules grouped in terminal or axillary spike-like thyrses; hypanthium short, campanulate to patelliform, or absent; disc thin to fleshy, annular to cupular, sometimes adnate to the hypanthium; ovary superior. Drupe ± fleshy with 2–3 indehiscent pyrenes. About 35 spp. mostly from Asia, but also in Africa and America.

30. *Berchemiella* Nakai

Berchemiella Nakai, Bot. Mag. Tokyo 37: 30 (1923).

Evergreen shrubs or small trees. Leaves alternate, glabrous. Flowers in axillary or terminal fascicles or cymes; hypanthium dish-shaped, persistent; disc comparatively thick, filling the hypanthium; ovary half-inferior. Drupe 1-locular. Two species, China and Japan.

31. *Rhamnella* Miq.

Rhamnella Miq., Ann. Mus. Bot. Lugd.-Batav. 3: 30 (1867).
Chaydaia Pit. in Lecomte (1912).

Evergreen shrubs or small trees. Leaves alternate. Flowers in axillary or terminal, contracted, umbelliform cymes; hypanthium cup- to dish-shaped; sepals adaxially keeled; disc conspicuous, filling, or adnate to, the hypanthium; ovary half-inferior. Drupe single-stoned; endocarp very hard, 1- or incompletely 2-locular. About 10 species, Central China, Japan and Korea, in the Himalayas up to 3000 m alt.

32. *Dallachya* F. Muell.

Dallachya F. Muell., Fragm. 9: 140, 1875.

Small tree. Leaves alternate. Flowers in sessile axillary umbel-like cymes; petals +; hypanthium short, with the disc adnate to it; ovary superior. Drupe single-stoned, the cartilaginous stone generally 1-locular. One species, *D. vitiensis* (Seem.) F. Muell., northern Australia, Papua New Guinea and the Pacific region.

33. *Berchemia* Necker ex DC.

Berchemia Necker ex DC., Prodr. 2: 22 (1825), nom. cons.
Phyllogeiton (A. Weberb.) Herzog (1903).

Evergreen to deciduous, climbing or erect, unarmed shrubs or small trees. Leaves mostly alternate. Flowers in thyrses composed of few-flowered, pedunceled to sessile, corymb-like cymes; hypanthium dish-shaped to hemispherical; disc fleshy, filling the hypanthium; ovary superior. Drupe single-stoned, 2-locular. About 20 species, most in SE Asia, one in New Caledonia and one in North America.

34. *Rhamnidium* Reissek

Rhamnidium Reissek in Mart., Fl. Brasil. 11(1): 94, pl. 31 (1861).

Evergreen shrubs or small trees. Leaves opposite, often ± distinctly gland-dotted, blackish. Flowers in small, long-pedunceled, umbelliform axillary cymes; hypanthium hemispherical to obconical, persistent in fruit; disc very thin, adnate to the hypanthium; ovary superior to half-inferior. Drupe single-stoned, incompletely 2-locular with 1–2 exalbuminous seeds. About 12 species, Cuba, Jamaica, Panama and tropical South America, in need of revision.

35. *Karwinskia* Zucc.

Karwinskia Zucc., Nov. Stirp. 1: 349, t. 16 (1832).

Evergreen shrubs or small trees. Leaves opposite, often gland-dotted. Flowers solitary or in pedunceled, axillary umbel-like cymes; hypanthium hemispherical to dish-shaped; disc thin, adnate to hypanthium; ovary superior, incompletely 2-locular, with 2 ovules per locule. Drupe single-stoned, 1–3-seeded. About 15 species, from southern North America to northern South America and the Caribbean.

36. *Condalia* Cav.

Condalia Cav., Anal. Hist. Nat. Madrid 1: 39 (1799), nom. cons.;
   Johnston, Brittonia 14: 332–368 (1962), rev.
Microrhamnus A. Gray (1852).

Evergreen shrubs or small trees often with numerous divaricate, thorn-tipped short shoots. Leaves alternate or appearing fasciculate on short shoots. Flowers solitary or fascicled on short shoots; hypanthium cup-shaped, persistent in fruit; disc thin, adnate to hypanthium; ovary superior. Drupe single-stoned, the stone 1–2-locular. About 20 spp., North and South America.

37. *Auerodendron* Urb.

Auerodendron Urb., Symb. Antill. 9: 221 (1924).

Evergreen shrubs or trees up to 20 m high. Leaves opposite, glabrous, ± conspicuously gland-dotted. Flowers in sessile or pedunceled cymes; hypanthium hemispherical; disc comparatively thin, adnate to the hypanthium; ovary superior. Drupe single-stoned, 2-locular. Seven species, Greater Antilles, in need of revision.

38. *Reynosia* Griseb.

Reynosia Griseb., Catal. Pl. Cub.: 33 (1866); Urban, Symbol. Antill. 9: 225–228 (1924).

Evergreen shrubs or small trees. Leaves opposite. Flowers solitary or in axillary, sessile, subumbellate, cymose fascicles; hypanthium short-campanulate to hemispherical; petals + or 0; disc fleshy, adnate to the hypantium; ovary superior. Drupe single-stoned, usually 1-locular by abortion of the second locule; seeds with copious, hard and ruminate endosperm. About 15 species, southern North America, Central America and the Caribbean, in need of revision.

39. *Krugiodendron* Urb.

Krugiodendron Urb., Symb. Antill. 3: 313 (1902).

Evergreen shrubs or small trees. Leaves opposite. Flowers in axillary, sessile or shortly peduncled, umbelliform cymes; hypanthium short, broadly obconical to dish-shaped; petals 0; disc broadly annular, crenate, fleshy, surrounding the base of the ovary; ovary superior. Drupe single-stoned, pericarp thin, fleshy, endocarp bony, 1(2)-locular; seeds without endosperm. One species, *K. ferreum* (Vahl) Urban, West Indies, Southern Florida, Mexico and parts of Central America, in need of revision.

7. TRIBE MAESOPSIDEAE A. Weberb. (1895).

Unarmed, large trees. Leaves opposite or alternate, strongly toothed. Flowers in peduncled, axillary cymes; hypanthium campanulate; petals sessile, deeply hooded; disc thin, adnate to the hypanthium; ovary superior, 1-locular, style thick, apically expanded into a mushroom-like stigma. Fruit a 1(2)-seeded drupe. One genus.

40. *Maesopsis* Engl.   Fig. 106

Maesopsis Engl., Pflanzenw. Ost-Afr. C: 255 (1895); Johnston, Fl. Tr. E. Afr. Rhamnaceae: 36 (1972).

Characters as for the tribe. One species, *M. eminii* Engl., tropical Africa.

8. TRIBE VENTILAGINEAE Hook. f. (1862).

Evergreen unarmed climbers or rarely small trees without tendrils. Leaves alternate. Petals present or absent, disc thick, filling the short hypanthium; ovary half-inferior. Fruit an apically-winged samara or a rostrate capsule; seeds exalbuminous. Two genera.

Fig. 106. Rhamnaceae–Maesopsideae. *Maesopsis eminii.* **A** Flowering branch. **B** Flower. **C** Petals, dorsal and ventral view. **D** Stamens, ventral and dorsal view. **E** Pistil. **F** Portion of fruiting branch. **G** Fruit, vertical section. (Johnston 1972)

41. *Smythea* Seem.

Smythea Seem., Bonplandia 9: 255 (1861).

Shrubs, often climbing. Flowers solitary or in axillary glomerules, sometimes in spike-like inflorescences; hypanthium dish-shaped; disc thin, adnate to the hypanthium. Capsule compressed, 1-locular, 1-seeded, dehiscing with 2 valves. About 10 spp., tropical SE Asia.

42. *Ventilago* Gaertn.

Ventilago Gaertn., Fruct. 1: 223, pl. 49, fig. 2 (1788).

Climbing shrubs or strong lianes, rarely small trees. Flowers solitary or in axillary glomerules, sometimes in spike-like inflorescences; hypanthium dish-shaped; disc thick, adnate to the hypanthium; petals usually present, cucullate.

Fruit a 1-seeded samara with an elongate terminal wing. About 40 spp., Old World tropics.

9. TRIBE AMPELOZIZYPHEAE J.E. Richardson (2000).

Unarmed climbers without tendrils. Leaves alternate, distichous. Flowers in axillary, clustered cymes, or composed of several cymes forming a panicle; hypanthium shortly turbinate; petals cucullate; disc thick, annular; ovary half-inferior. Fruit a 3-locular, 3-seeded, stipitate, explosive capsule; seeds exalbuminous. One genus.

43. *Ampelozizyphus* Ducke

Ampelozizyphus Ducke, Arch. Inst. Biol. Veg. Rio de Janeiro 2: 157 (1935).

Characters as for the tribe. One species, *A. amazonicus* Ducke, Amazonian Brazil.

10. TRIBE DOERPFELDIEAE J.E. Richardson (2000).

Evergreen, unarmed trees. Leaves alternate, entire or emarginate. Flowers solitary or in axillary fascicles; hypanthium very short and shallow; petals 0; disc thin, adnate to ovary; ovary superior, pseudobilocular, style bifid. Fruit a single-stoned, unequally 2-locular drupe with a single seed in the larger locule; seed albuminous. One genus.

44. *Doerpfeldia* Urban

Doerpfeldia Urban, Symb. Ant. 9: 218 (1924).

Characters as for the tribe. One species, *D. cubensis* Urban, Cuba.

11. TRIBE BATHIORHAMNEAE J.E. Richardson (2000).

Evergreen, unarmed trees. Leaves triplinerved, entire to toothed, alternate. Flowers in axillary fascicles; disc thick; ovary superior, 3-locular. Fruit splitting into 3 indehiscent mericarps; seeds exalbuminous. One genus.

45. *Bathiorhamnus* Capuron

Bathiorhamnus Capuron, Adansonia II, 6: 121 (1966).

Characters as for the tribe. Two species, Madagascar.

Genera *incertae sedis*

46. *Alphitonia* Reissek ex Endl.

Alphitonia Reissek ex Endl., Gen. Pl.: 1098 (1840); Braid, Kew Bull. 1925: 171–186 (1925), rev.

Evergreen, unarmed shrubs or trees, sometimes tall. Leaves entire, alternate. Flowers perfect and staminate, in dichasial or trichasial axillary or terminal cymes; hypanthium short; disc comparatively thick, filling the hypanthium; ovary inferior. Fruit a 2–3-seeded capsule. Seeds arillate, temporarily persistent on the receptacle after the fall of the pericarp, albuminous. About 15 species, Malaysia, West Pacific islands, New Caledonia and Australia.

47. *Ceanothus* L.

Ceanothus L., Sp. Pl.: 195 (1753); van Rensselaer & McMinn, *Ceanothus*. Sta. Barbara Bot. Gard. (1942), rev.

Deciduous or evergreen shrubs or rarely small trees, sometimes thorny, mostly with nitrogen-fixing actinorhizal root nodules. Leaves alternate or opposite. Flowers white, pink, or blue, borne in terminal or axillary, raceme- or corymb-like thyrses composed of few-flowered umbelliform cymes; hypanthium shallow, cupulate to hemispherical; petals +; disc annular, ovary half-inferior, 3–4-locular. Capsule 3–4-seeded, explosive; seeds albuminous, sometimes conspicuously arillate. 55 species, North and Central America.

48. *Colubrina* Rich. ex Brongn.   Fig. 107

Colubrina Rich. ex Brongn., Mem. Fam. Rham. 61: (1826), nom. cons.; Johnston, Brittonia 23: 2–53 (1971), rev.
Hybosperma Urban (1899).

Evergreen or deciduous, unarmed or spinescent, rarely scandent trees or shrubs. Leaves alternate, rarely opposite, often glandular. Flowers in sessile or shortly peduncled, axillary cymes or small thyrses or fascicles, rarely solitary; hypanthium cup- to dish-shaped; petals present; disc fleshy, almost filling the hypanthium; ovary half-inferior, 3(4)-locular. Capsule 3-seeded, explosively or slowly dehiscent with the outer pericarp irregularly breaking away. Seeds albuminous, sometimes

Fig. 107. Rhamnaceae. *Colubrina asiatica.* **A** Flowering and fruiting branch. **B** Flower. **C** Same, vertical section. **D** Transverse section of ovary. **E** Portion of fruiting branch. **F** Capsule. **G** Seed. (Johnston 1972)

conspicuously arillate. About 33 species, Old World and New World tropics.

49. *Emmenosperma* F. Muell.

Emmenosperma F. Muell., Fragm. 3: 62 (1862).

Evergreen, unarmed trees or shrubs. Leaves alternate or opposite. Flowers in trichotomous cymes arranged in axillary or terminal panicles; hypanthium cup-shaped; disc thin, adnate to the hypanthium; ovary half-inferior to superior, 2–3-locular. Fruit a capsule; seeds albuminous, persistent on the pedicel after the pericarp has fallen. Five species, Australia, New Caledonia, New Guinea and Fiji.

50. *Granitites* B.L. Rye

Granitites B.L. Rye, Nuytsia 10: 451 (1996).

Spinescent shrubs. Leaves alternate. Flowers long-pedicellate, solitary or condensed in subterminal,

leafy clusters; hypanthium short; disc prominent, adnate to the hypanthium; petals +; ovary inferior, 3-locular. Fruit beaked, apparently a 3-seeded capsule with a mealy layer in the outer pericarp. Seeds arillate. One species, *G. intangendus* (F. Muell. ex F. Muell.) Rye, Western Australia; related to *Alphitonia* (Fay et al. 2001).

51. *Lasiodiscus* Hook. f.

Lasiodiscus Hook. f. in Bentham & Hook. f., Gen. Pl. 1: 381 (1862); Figueiredo, Kew Bull. 50: 495–526 (1995), rev.

Evergreen, unarmed, often climbing shrubs or small trees. Leaves opposite; stipules comparatively large and interpetiolar. Flowers ± densely ferrugineous pubescent, congested in long-pedunculated, axillary, divaricate cymes or pseudo-umbels; hypanthium broadly campanulate; disc massive, fleshy, broadly annular; ovary inferior or half-inferior, 3-locular. Fruit a velvety, elastically dehiscent, 3-seeded capsule; seeds exalbuminous. 12 spp., tropical Africa and Madagascar.

52. *Schistocarpaea* F. Muell.

Schistocarpaea F. Muell., Victor. Nat. 7: 182 (1891).

Evergreen, unarmed trees with alternate leaves. Flowers cream or pinkish, in loose panicles of 3–many-flowered cymes in the upper axils; hypanthium cup-shaped, persistent in fruit; disc conspicuous, forming a narrow rim at the base of the ovary; ovary superior, 3-locular. Fruit a subglobose capsule; seeds exalbuminous. One species, *S. johnsonii* F. Muell., Australia.

Selected Bibliography

Aagesen, L. 1999. Phylogeny of the tribe *Colletieae*, Rhamnaceae. Bot. J. Linn. Soc. 131: 1–43.
Andersen, A.N., Ashton, D.H. 1985. Rates of seed removal by ants at heath and woodland sites in southeastern Australia. Austr. J. Ecol. 10: 381–390.
APG II (Angiosperm Phylogeny Group) 2003. See general references.
Aronne, G., Wilcock, C.C. 1994. First evidence of myrmecochory in fleshy-fruited shrubs of the Mediterranean region. New Phytol. 127: 781–788.
Aronne, G., Wilcock, C.C. 1995. Reproductive lability in predispersal biology of *Rhamnus alaternus* L. (Rhamnaceae). Protoplasma 187: 49–59.
Baker, D.D., Schwintzer, C.R. 1990. Introduction. In: Schwintzer, C.R., Tjepkema, J.D. (eds.) The biology of *Frankia* and actinorhizal plants. San Diego: Academic Press, pp. 1–13.
Basinger, J.F., Dilcher, D.L. 1984. Ancient bisexual flowers. Science 224: 511–513.

Behnke, H.-D. 1974. P- und S-Typ Siebelement-Plastiden bei Rhamnales. Beitr. Biol. Pflanzen 50: 457–464.

Bellingham, P.J. 1998. Shrub succession and invasibility in a New Zealand montane grassland. Austr. J. Ecol. 23: 562–573.

Bennek, C. 1958. Die morphologische Beurteilung der Staub- und Blumenblätter der Rhamnaceen. Bot. Jahrb. Syst. 77: 423–457.

Bouchet, P. 1974. Étude ultrastructurale des cellules constituant les poches "lysigènes" à mucilage de la bourdaine: *Rhamnus frangula* L. C.R. Acad. Sci. Paris D, 279: 1073.

Braid, K.W. 1925. Revision of the genus *Alphitonia*. Kew Bull. 1925: 171–186.

Brizicky, G.K. 1964. The genera of Rhamnaceae in the south-eastern United States. J. Arnold Arb. 45: 439–463.

Carlquist, S. 1966. The biota of long-distance dispersal. III. Loss of dispersibility in the Hawaiian flora. Brittonia 18: 310–355.

Cruz Cisneros, R., Valdes, M. 1991. Actinorhizhal root nodules on *Adolphia infesta* (H.B.K.) Meissner (Rhamnaceae). Nitrogen Fixing Tree Res. Rep. 9: 87–89.

D'Ambrogio, A.C., Medan, D. 1993. Comportamiento reproductivo de *Colletia paradoxa* (Rhamnaceae). Darwiniana 32: 1–14.

Davis, G.L. 1966. See general references.

Diem, H.G. 1996. Les mycorhizes des plantes actinorhiziennes. Acta Bot. Gallica 143: 581–592.

Dimitri, M.J. (ed.) 1972. La región de los bosques andino-patagónicos. Buenos Aires: INTA.

Evans, R.A., Biswell, H.H., Palmqvist, D.E. 1987. Seed dispersal in *Ceanothus cuneatus* and *C. leucodermis* in a sierran oak-woodland savanna. Madroño 34: 283–293.

Fay, M.F., Lledó, M.D., Richardson, J.E., Rye, B.L., Hopper, S.D. 2001. Molecular data confirm the affinities of the south-west Australian endemic *Granitites* with *Alphitonia* (Rhamnaceae). Kew Bull. 56: 669–675.

Figueiredo, E. 1995. A revision of *Lasiodiscus* (Rhamnaceae). Kew Bull. 50: 495–526.

Frett, J.J. 1989. Germination requirements of *Hovenia dulcis* seeds. Hort. Sci. 24: 152.

Galil, J., Zeroni, M. 1967. On the pollination of *Zizyphus spina-christi* (L.) Willd. in Israel. Israel J. Bot. 16: 71–77.

Gardner, I.C. 1986. Mycorrhizae of actinorhizal plants. Mircen J. 2: 147–160 (not seen, cited by Diem 1996).

Gemoll, K. 1902. Anatomisch-systematische Untersuchung des Blattes der *Rhamneen* aus der Triben: Rhamneen, Colletieen und Gouanieen. Beih. Bot. Centralbl. 12: 351–421.

Godt, M.J.W., Race, T., Hamrick, J.L. 1997. A population genetic analysis of *Ziziphus celata*, an endangered Florida shrub. J. Heredity 88: 531–533.

Grey-Wilson, C. 1978. *Alvimiantha*, a new genus of Rhamnaceae from Bahia, Brazil. Bradea 2: 287–290.

Grice, A.C. 1996. Seed production, dispersal and germination in *Cryptostegia grandiflora* and *Ziziphus mauritiana*, two invasive shrubs in tropical woodlands in northern Australia. Austr. J. Ecol. 21: 324–331.

Grice, A.C. 1998. Ecology in the management of Indian jujube (*Ziziphus mauritiana*). Weed Sci. 46: 467–474.

Guppy, H.B. 1906. Observations of a naturalist in the Pacific between 1896 and 1899. II. Plant dispersal. London: Macmillan.

Hall, K.F.M., Parsons, R.F. 1987. Ecology of *Discaria* (Rhamnaceae) in Victoria. Proc. Roy. Soc. Victoria 99: 99–108.

Hanácková, Z., Piñeyro López, A. 1999. The *Karwinskia parvifolia* flower. Biologia (Bratislava) 54: 85–90.

Harvey, C.F., Braggins, J.E. 1985. Reproduction of the New Zealand taxa of *Pomaderris* Labill. (Rhamnaceae). New Zeal. J. Bot. 23: 151–156.

Herzog, T. 1903. Anatomisch-systematische Untersuchung des Blattes der Rhamnaceen aus den Triben: Ventilagineen, Zizypheen und Rhamneen. Beih. Bot. Centralbl. 15: 95–207.

Hollick, A. 1930. The Upper Cretaceous floras of Alaska. Geol. Surv. Prof. Paper 159.

Huss-Danell, K. 1997. Actinorhizal symbioses and their N_2 fixation. New Phytol. 136: 375–405.

Jakovleva, O.V. 1988. Slime cells of the leaf epidermis in the dicotyledonous plants (electron microscope data). Bot. Zhurn. 73: 977–987.

Johnston, M.C. 1962. Revision of *Condalia* including *Microrhamnus* (Rhamnaceae). Brittonia 14: 332–368.

Johnston, M.C. 1964. The fourteen species of *Ziziphus* including *Sarcomphalus* (Rhamnaceae) indigenous to the West Indies. Am. J. Bot. 51: 1113–1118.

Johnston, M.C. 1971. Revision of *Colubrina* (Rhamnaceae). Brittonia 23: 2–53.

Johnston, M.C. 1972. Rhamnaceae. Flora of Tropical East Africa. London: Crown Agents.

Johnston, M.C. 1973. Revision of *Kentrothamnus* (Rhamnaceae). J. Arnold Arb. 54: 471–473.

Johnston, M.C. 1974. Revision of *Scutia* (Rhamnaceae). Bull. Torrey Bot. Club 101: 64–71.

Johnston, M.C. 1977. Rhamnales. The New Encyclopaedia Britannica (Macropedia) 15. Chicago: Hemingway Benton, pp. 794–796.

Jossang, A., Zahir, A., Diakite, D. 1996. Mauritine J, a cyclopeptide alkaloid from *Zizyphus mauritiana*. Phytochemistry 42: 565–567.

Keeley, J.E. 1992. Seed germination and life history syndromes in the California chaparral. Bot. Rev. 57: 81–116.

Keogh, J.A., Bannister, P. 1993. Transoceanic dispersal in the amphiantarctic genus *Discaria*: an evaluation. New Zeal. J. Bot. 31: 427–430.

Kilian, D., Cowling, R.M. 1992. Comparative seed biology and co-existence of two fynbos shrub species. J. Veg. Sci. 3: 637–646.

Lundström, A.N. 1887. Pflanzenbiologische Studien. II. Die Anpassungen der Pflanzen an Thiere. Nova Acta Reg. Soc. Sci. Upsal. III, 13: 1–88.

Lux, A., Lišková, D., Piñeyro-López, A., Ruiz-Ordóñez, J., Kákoniová, D. 1998. Micropropagation of *Karwinskia parvifolia* and the transfer of plants to *ex vitro* conditions. Biol. Plantarum 40: 143–147.

Malicky, H., Sobhian, R., Zwölfer, H. 1970. Investigations on the possibilities of a biological control of *Rhamnus cathartica* L. in Canada: host ranges, feeding sites, and phenology of insects associated with European Rhamnaceae. Z. Angew. Entomol. 65: 77–97.

Martin, A.C. 1946. The comparative internal anatomy of seeds. Am. Midl. Nat. 36: 513–660.

Medan, D. 1985. Fruit morphogenesis and seed dispersal in the Colletieae (Rhamnaceae). I. The genus *Discaria*. Bot. Jahrb. Syst. 105: 205–262.

Medan, D. 1986. Anatomía y arquitectura foliares de *Discaria* (Rhamnaceae). Kurtziana 18: 133–151.

Medan, D. 1988. Gynoecium ontogenesis in the Rhamnaceae – A comparative study. In: Leins, P., Tucker, S.C., Endress, P.K. (eds.) Aspects of floral development. Berlin: J. Cramer, pp. 133–141.

Medan, D. 1989. Diaspore diversity in the anemochorous Gouanieae (Rhamnaceae). Plant Syst. Evol. 168: 149–158.

Medan, D., Aagesen, L. 1995. Comparative flower and fruit structure in the Colletieae (Rhamnaceae). Bot. Jahrb. Syst. 117: 531–564.

Medan, D., Arce, M.E. 2000. Reproductive biology of the Andean-disjunct genus *Retanilla* (Rhamnaceae). Plant Syst. Evol. 218: 281–298.

Medan, D., D'Ambrogio, A.C. 1998. Reproductive biology of the andromonoecious shrub *Trevoa quinquenervia* (Rhamnaceae). Bot. J. Linn. Soc. 126: 191–206.

Medan, D., Hilger, H.H. 1992. Comparative flower and fruit morphogenesis in *Colubrina* (Rhamnaceae) with special reference to *C. asiatica*. Am. J. Bot. 79: 809–819.

Mills, J.N., Kummerow, J.K. 1989. Herbivores, seed predators and chaparral succession. In: Keeley, S.C. (ed.) The California chaparral. Paradigms reexamined. Los Angeles County Mus. Contrib. Sci. 34: 49–56.

Muller, J. 1981. See general references.

Pillans, N.S. 1942. The genus *Phylica* Linn. J. S. Afr. Bot. 8: 1–164.

Possessky, S.L., Williams, C.E., Moriarty, W.J. 2000. Glossy buckthorn, *Rhamnus frangula* L. A threat to riparian plant communities of the Northern Allegheny Plateau (USA). Nat. Areas J. 20: 290–292.

Reiche, K. 1907. Pflanzenverbreitung in Chile. Leipzig: W. Engelmann.

Rensselaer, M. van, McMinn, H.E. 1942. *Ceanothus*. Santa Bárbara Botanic Garden.

Richardson, J.E., Fay, M.F., Cronk, Q.C.B., Chase, M.W. 2000a. A revision of the tribal classification of Rhamnaceae. Kew Bull. 55: 311–340.

Richardson, J.E., Fay, M.F., Cronk, Q.C.B., Bowman, D., Chase, M.W. 2000b. A phylogenetic analysis of Rhamnaceae using *rbc*L and *trn*L-F plastid DNA sequences. Am. J. Bot. 87: 1309–1324.

Ridley, H.N. 1930. The dispersal of plants throughout the world. Ashford: Reeve.

Ronse Decraene, L.P., Clinckemaillie, D., Smets, E. 1993. Stamen–petal complexes in Magnoliatae. Bull. Jard. Bot. Natl. Belg. 62: 97–112.

Rye, B. 1996. *Granitites*, a new genus of Rhamnaceae from the south-west of Western Australia. Nuytsia 10: 451–457.

Sattler, R. 1973. Organogenesis of flowers. A photographic text–atlas. Toronto: University of Toronto Press.

Savolainen, V., Chase, M.W. et al. 2000. See general references.

Schirarend, C. 1984. Holzanatomische Untersuchungen als Beiträge zur Systematik der Familie Rhamnaceae Jussieu. Diss. Berlin: Humboldt-Universität.

Schirarend, C., Köhler, E. 1993. Rhamnaceae Juss. World Pollen Spore Flora 17/18: 1–53.

Schirarend, C., Olabi, M.N. 1994. Revision of the genus *Paliurus* Tourn. ex Mill. (Rhamnaceae). Bot. Jahrb. Syst. 116: 333–359.

Soltis, D.E., Soltis, P.S., Nickrent, D.L., Johnson, L.A., Hahn, W.J, Hoot, S.B., Sweere, J.A., Kuzoff, R.K., Kron, K.A., Chase, M.W., Swensen, S.M., Zimmer, A., Chaw, S.M., Gillespie, L.J., Kress, W.J., Sytsma, K.J. 1997. Angiosperm phylogeny inferred from 18 S ribosomal DNA sequences. Ann. Miss. Bot. Gard. 8: 1–49.

Soltis, D.E. et al. 2000. See general references.

Soriano, A. 1960. Germination of twenty dominant plants in Patagonia in relation to regeneration of the vegetation. In: 8th Int. Grassland Congr., session 6 A, pp. 2–6.

Suessenguth, K. 1953. Rhamnaceae. In: Engler & Prantl, Die natürlichen Pflanzenfamilien, ed. 2, 20d. Berlin: Duncker & Humblot.

Sun, D., Dickinson, G.R. 1996. The competition effect of *Brachiaria decumbens* on the early growth of direct-seeded trees of *Alphitonia petriei* in tropical north Australia. Biotropica 28: 272–276.

Swensen, S.M. 1996. The evolution of actinorhizal symbioses: evidence for multiple origins of the symbiotic association. Am. J. Bot. 83: 1503–1512.

Takhtajan, A. 1997. See general references.

Thorne, R.F. 1992. Classification and geography of the Flowering Plants. Bot. Rev. 58: 225–348.

Thulin, M., Bremer, B., Richardson, J.E., Niklasson, J., Fay, M.F., Chase, M.W. 1998. Family relationships of the enigmatic rosid genera *Barbeya* and *Dirachma* from the Horn of Africa region. Plant Syst. Evol. 213: 103–119.

Tieghem, P. van 1875. Recherches sur la structure du pistil et sur l'anatomie comparée de la fleur. Mém. Prés. Divers Savants Acad. Inst. Impérial France 21: 1–261, pl. 1–16.

Tortosa, R.D. 1983. El género *Discaria* (Rhamnaceae). Bol. Soc. Argent. Bot. 22: 301–335.

Tortosa, R.D. 1984. El gineceo de *Condalia* (Rhamnaceae) y su relación con el de otros géneros afines. Kurtziana 17: 49–54.

Tortosa, R.D. 1989. El género *Colletia* (Rhamnaceae). Parodiana 5: 279–332.

Tortosa, R.D. 1992. El complejo *Retanilla-Talguenea-Trevoa* (Rhamnaceae). Darwiniana 31: 223–252.

Tortosa, R.D. 1993. Revisión del género *Adolphia* (Rhamnaceae-Colletieae). Darwiniana 32: 185–189.

Tortosa, R.D., Medan, D. 1992. Rhamnaceae with multiple lateral buds: an architectural analysis. Bot. J. Linn. Soc. 108: 275–286.

Tortosa, R.D., Aagesen, L., Tourn, G.M. 1996. Morphological studies in the tribe Colletieae (Rhamnaceae): analysis of architecture and inflorescences. Bot. J. Linn. Soc. 122: 353–367.

Tourn, G.M., Bartoli, A., Tortosa, R.D. 1991. The morphology and growth of *Gouania ulmifolia* Triana et Planch. (Rhamnaceae): an architectural analysis. Naturalia Monspeliensia n.s.: 666–667.

Tourn, G.M., Tortosa, R.D., Medan, D. 1992. Rhamnaceae with multiple lateral buds: an architectural analysis. Bot. J. Linn. Soc. 108: 275–286.

Traveset, A. 1999. Ecology of plant reproduction: mating systems and pollination. In: Pugnaire, F.I., Valladares, F. (eds.) Handbook of functional plant ecology. New York: Marcel Dekker, pp. 545–588.

Vikhireva, V.V. 1952. Morfologo-anatomiceskoe issledovanie plodov krusinovikh. Trudy Botaniceskogo instituta imeni V. L. Komarova Akademii nauk SSSR. Ser. 7. Morf. Anat. Rastenij 3: 241–292.

Waksman, N., Martínez, L., Fernández, R. 1989. Chemical and toxicological screening in genus *Karwinskia* (Mexico). Rev. Latinoam. Quím. 20: 27–29.

World Conservation Monitoring Centre, 1996. Rhamnaceae. Regionally threatened taxa. Status Report 14 May 1996. Cambridge: World Conservation Monitoring Centre.

Yakovlev, M.S., Zhukova, G.Y. 1980. Chlorophyll in embryos of angiosperm seeds, a review. Bot. Notiser 133: 323–336.

Zedler, P.H. 1995. Plant life history and dynamic specialization in the chaparral/coastal sage shrub flora in southern California. In: Arroyo, M.T.K., Zedler, P.H., Fox, M.D. (eds.) Ecology and biogeography of Mediterranean ecosystems in Chile, California, and Australia. Berlin Heidelberg New York: Springer, pp. 89–115.

Zietsman, P.C. 1990. Pollination of *Ziziphus mucronata* subsp. mucronata (Rhamnaceae). S. Afr. J. Bot. 56: 350–355.

Zietsman, P.C., Botha, F.C. 1992. Flowering of *Ziziphus mucronata* subsp. *mucronata* (Rhamnaceae): anthesis, pollination and protein synthesis. Bot. Bull. Acad. Sinica 33: 33–42.

Roridulaceae

J.G. CONRAN

Roridulaceae Engl. & Gilg. in Engl., Syllabus ed. 9 & 10: 226 (1924), nom. cons.

Perennial insect-trapping evergreen woody shrubs from a taproot with limited lateral root development; stems perennial, erect and few-branched. Leaves alternate, crowded apically, linear to tapering, exstipulate, margins entire or with short lateral linear lobes, with insect-trapping stalked glands. Flowers in terminal, few-flowered botryoids, hermaphroditic, medium-sized and showy; calyx and corolla 5-merous, well developed; perianth actinomorphic, free, rotate, bright purple, reddish pink or white; petals imbricate in bud; stamens 5, opposite the sepals; anthers hypogynous; filaments free, filiform; anthers sub-basifixed, 2-thecate, 4-sporangiate, introrse, incurved in bud, subtended by a basal swelling containing a nectariferous cavity, irritable and swinging up when touched at anthesis to become erect, dehiscing by four short apical pores or slits; gynoecium of three united carpels; ovary superior, 3-locular with axile placentation; ovules anatropous, unitegmic, solitary or 2–4 per locule; style terminal, tapering with a small capitate stigma or expanding terminally with an obconical stigma, papillate, erect. Fruit a smooth loculicidal, 3-valved cartilaginous capsule. Seeds exotestal, small, ellipsoid and smoothly reticulate or angular-trilete and prominently warty or honeycomb-sculptured, dark reddish-brown; endosperm copious; embryo linear.

One genus, *Roridula*, with two species endemic to the Cape Province of South Africa. Both species are occasionally cultivated as ornamental curiosities.

VEGETATIVE MORPHOLOGY (Fig. 108). There is a well-developed taproot with relatively limited lateral root development (Bruce 1907). The few-branched woody stems show a more or less divaricate branching pattern with the leaves restricted to clusters near the branch apices. The leaves are linear and tapering. They are simple in *R. gorgonias* and finely laterally linear-lobed in *R. dentata*. Both species have densely glandular leaves.

VEGETATIVE ANATOMY. Calcium oxalate druses are present in the epidermal cells near the major veins (Bruce 1907). Tannin cells occur in the endosperm and seed integuments (Dahlgren and van Wyk 1988).

Primary stems have a ring of fibres connecting the vascular bundles (Solereder 1908). The wood anatomy of the family was described by Carlquist (1976) who observed that there was distinct secondary thickening in both species and clear growth rings in *R. dentata*. Vessels occur scattered and singly in the roots and stems, with spiral thickening, scalariform perforation plates and bordered bars in both species, as well as vessel plates with a meshwork of rhomboid perforations in *R. dentata*. Xylem parenchyma rays are uniseriate with uniquely upright ray cells. The tracheids have bordered pits (Carlquist 1976), not simple, as reported by Solereder (1908). Behnke (1991) reported Ss-type phloem sieve-tube plastids.

The leaves show brochidodromous venation with a clear marginal vein. The cuticle is thin and the anomocytic stomata are present on both surfaces and oriented along the leaf axis. The non-vascularised tentacles vary from uni- to 5-seriate stalks with six-celled to multicellular glandular heads. The larger hairs have a central head column three cells wide onto which the otherwise free resin-secreting cells are basally attached (Fenner 1904; Dahlgren and van Wyk 1988). The mesophyll is uniformly spongy, with prominent abaxial lacunae. The vascular bundles are surrounded by sclerenchyma (Solereder 1908).

INFLORESCENCE STRUCTURE. Although the inflorescence in Roridulaceae superficially resembles a raceme, it is determinate with a terminal flower, and is thus derived-paniculate (botryoid) (Dahlgren and van Wyk 1988).

FLORAL MORPHOLOGY. The rotate actinomorphic corolla consists of five broadly obovate, entire, glabrous and free petals which are imbricate in bud. The sepals have three non-anastomosing vascular traces (Chrtek et al. 1989). The anthers are strongly outwardly and ventrally

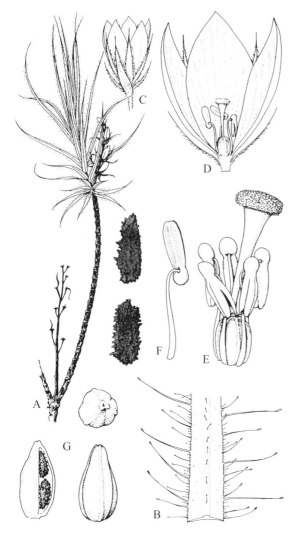

Fig. 108. Roridulaceae. *Roridula gorgonias*. **A** Floriferous branch. **B** Part of leaf. **C** Flower. **D** Same, vertical section. **E** Stamens and pistil. **F** Stamen. **G** Young fruits. **H** Seeds. (Dahlgren and van Wyk 1988)

recurved in bud but, when disturbed at anthesis, they swing upwards through 180° and dehisce by short, pore-like apical slits (Fig. 108E, F).

EMBRYOLOGY. Anther wall formation is of the basic type. There are no fibrous thickenings of the endothecium and the glandular tapetal nuclei exhibit both fusion and division. One of the middle layers if crushed at maturity and the other is tanniniferous. Microsporogenesis is simultaneous and the mature pollen grains are binucleate.

The anatropous ovule is unitegmic and there is a well-developed endothelium. The chalazal megaspore follows Polygonum-type development. The polar nuclei fuse prior to fertilisation and the antipodals persist until the first division of the endosperm nucleus. Fertilisation is porogamous, there are micropylar haustoria and endosperm formation is cellular. Embryogeny conforms to the Solanad type.

The mature endosperm is tanniniferous and the hypostase and testa walls become thickened. The outer epidermal cells of the testa are enlarged with strongly thickened inner tangential and radial walls (Vani-Hardev 1972).

POLLEN MORPHOLOGY. Pollen grains differ markedly between the two *Roridula* species. *R. dentata* has isopolar, actinomorphic tricolporate prolate-spheroidal pollen 36.5 µm (polar) by c. 36 µm (equatorial). The exine has a spinuliferous perforated tectum and infratectal baculae. Ora are indistinct and the colpi are marginally granular and centrally warty. *R. gorgonias* pollen is apolar, asymmetrical, spheroidal, c. 50 µm in diameter without obvious apertures. The exine is insular with bacula-supported polygonal to rounded segments (Erdtman 1952; Dahlgren and van Wyk 1988).

KARYOLOGY. Chromosomes in Roridulaceae were reported by Kress (1970) and Peng and Goldblatt (1983) as $2n = 12$ ($n = 6$).

POLLINATION. Hemiptera of the genus *Pameridia* (Miridae), one species for each *Roridula* species, pierce and feed from the anther connective nectaries. The piercing causes the anthers to spring up through 180°, dusting the insects with pollen (Marloth 1925). Plants in cultivation are also visited by hover flies and are self-fertile (Conran, pers. obs.)

FRUIT AND SEED. The fruit is a smooth loculicidal, 3-valved cartilaginous capsule. Seeds are small, ellipsoid and smoothly reticulate in *R. dentata*, and angular-trilete and prominently warty to honeycomb-sculptured in *R. gorgonias*, dark reddish-brown; endosperm is copious; the embryo is linear. The seeds of both species become sticky and mucilaginous when wet.

ECOLOGY. The plants grow in areas of permanent moisture, although *R. dentata* can apparently inhabit slightly drier areas than *R. gorgonias* (Carlquist 1976). The rootstock is fire-resistant and the roots are considered to be mycorrhizal (Marloth 1910), although this fungal association was questioned by Obermeyer (1970). The glandular tentacles secrete a balsam-like resin which, although it appears to trap insects, was not digestive in function (Marloth 1903, 1925); however, these trapped insects are preyed upon by Hemiptera of the genus *Pameridia* (Miridae: Heteroptera), *P. marlothii* on *R. dentata* and *P. roridulae* on *R. gorgonias*, and by a crab spider, *Synema*, all of which can move easily over the sticky leaves. Ellis and Midgley (1996) demonstrated that the excreta from these predators are absorbed by the leaves of *Roridula*, resulting in carnivory by mutualism. In addition, the foliage of *Roridula* is highly UV-adsorptive (Midgley and Stock 1998), a feature shared with other flypaper carnivorous plants such as the Droseraceae, and which is thought to make the plants attractive to potential insect prey (Joel et al. 1985).

PHYTOCHEMISTRY. Iridoids of unknown structure have been reported for Roridulaceae by Jensen et al. (1975). Naphthoquinones are absent (Zenk et al. 1969).

AFFINITIES. *Roridula* has been associated variously with the Droseraceae (e.g. Bentham and Hooker 1865), Ochnaceae (Planchon 1848), Clethraceae (Hallier 1912), mostly on differently emphasised features of gross morphology. More recently, the genus has been included in the Byblidaceae (e.g. Hutchinson 1959; Cronquist 1981) or treated as a monogeneric family (e.g. Takhtajan 1987; Dahlgren and van Wyk 1988; Juniper et al. 1989), generally with affinities near the Ericales. Molecular studies of its position, using chloroplast *rbc*L (Albert et al. 1992) and mitochondrial 18 S RNA sequences (Conran and Dowd 1994), confirmed that it was unrelated to the Byblidaceae. In particular, Albert et al. (1992) placed it in the Ericales as sister to the Northern Hemisphere carnivorous pitcher plant family Sarraceniaceae. In the recent classification of the flowering plants based on combined multigene studies, Roridulaceae are retained as a separate family in the Ericales (APG 1998).

DISTRIBUTION AND HABITATS. The family is endemic to the Cape Province region of South Africa in mountainous regions, with *R. gorgonias* found between Gordon's Bay and Hermanus to the Genadenal region. *R. dentata* occurs in the Tulbagh and Ceres districts through to the Cedarburg mountains. The plants grow on sandstone slopes in areas of seepage or near streams (Carlquist 1976).

ECONOMIC IMPORTANCE. Roridulaceae are of interest to specialist carnivorous plant collectors.

A single genus:

Roridula L.   Fig. 108

Roridula L., Gen. ed. 6: 567 (1764).

Description as for family.

Selected Bibliography

Albert, V.A., Williams, S.E., Chase, M.W. 1992. Carnivorous plants: phylogeny and structural evolution. Science 257: 1491–1495.

APG (Angiosperm Phylogeny Group) 1998. See general references.

Behnke, H.-D. 1991. See general references.

Bentham, G., Hooker, J.D. 1865. Genera Plantarum vol. I. London: L. Reeve.

Bruce, A.N. 1907. On the distribution of the tentacles of *Roridula*. Notes Roy. Bot. Gard. Edinburgh 17: 83–98.

Carlquist, S. 1976. Wood anatomy of Roridulaceae: ecological and phylogenetic implications. Am. J. Bot. 63: 1003–1008.

Chrtek, J., Slavíková, Z., Studnička, M. 1989. Beitrag zur Leitbündelanordnung in den Kronblättern von ausgewählten Arten der fleischfressenden Pflanzen. Preslia 61: 107–124.

Conran, J.G., Dowd, J.W. 1994. The phylogenetic relationships of the *Byblis-Roridula* (Byblidaceae-Roridulaceae) complex inferred from 18 S rRNA partial sequences. Plant Syst. Evol. 188: 73–86.

Cronquist, A. 1981. See general references.

Dahlgren, R.M.T., van Wyk, A.E. 1988. Structures and relationships of families endemic to, or centered in Southern Africa. Monogr. Syst. Bot. Missouri Bot. Gard. 25: 1–94.

Diels, L. 1930. Roridulaceae. In: Engler, A., Prantl, K. Die natürlichen Pflanzenfamilien ed. 2, 18a. Leipzig: W. Engelmann, pp. 346–348.

Ellis, A.G., Midgley, J.J. 1996. A new plant-animal mutualism involving a plant with sticky leaves and a resident Hemipteran insect. Oecologia 106: 478–481.

Erdtman, G. 1952. See general references.

Fenner, C.A. 1904. Beiträge zur Kenntnis der Anatomie, Entwicklungsgeschichte und Biologie der Laubblätter und Drüsen einiger Insectivoren. Flora 93: 335–434.

Hallier, H. 1912. L'origine et le système phylétique des angiospermes. Arch. Neerl. Sci. Exactes Nat., III. B 1: 146–234.

Hutchinson, J. 1959. The families of flowering plants. Vol. 2. Monocotyledons, 2nd edn. Oxford: Clarendon Press.

Jensen, S.R., Nielsen, B.J., Dahlgren, R. 1975. Iridoid compounds, their occurrence and systematic importance in the angiosperms. Bot. Not. 128: 148–180.

Joel, D.M., Juniper, B.E., Dafni, A. 1985. UV patterns in the traps of carnivorous plants. New Phytol. 101: 585–594.

Juniper, B.E., Robins, R.J., Joel, D.M. 1989. Carnivorous plants. London: Academic Press.

Kress, A. 1970. Zytotaxonomische Untersuchungen an einigen Insektenfängern (Droseraceae, Byblidaceae, Cephalotaceae, Roridulaceae, Sarraceniaceae). Ber. Deutsch Bot. Ges. 83: 55–62.

Lloyd, F.M. 1934. Is *Roridula* a carnivorous plant? Can. J. Res. 10: 780–786.

Lloyd, F.M. 1942. The carnivorous plants, 2nd edn. Waltham, Mass.: Chronica Botanica Co.

Marloth, R. 1903. Some recent observations on the biology of *Roridula*. Ann. Bot. 17: 151–157.

Marloth, R. 1910. Further observations on the biology of *Roridula*. Trans. Roy. Soc. S. Afr. 2: 59–61.

Marloth, R. 1925. Roridulaceae. In: The Flora of South Africa vol. 2, part I. Cape Town: Darter Bros., pp. 26–30.

Midgley, J.J., Stock, W.D. 1998. Natural abundance of Delta N-15 confirms insectivorous habit of *Roridula gorgonias*, despite it having no proteolytic enzymes. Ann. Bot. 82: 387–388.

Obermeyer, A.A. 1970. Roridulaceae. In: Codd, L.E., DeWinter, B., Killick, D.J.B., Rycroft, H.B. (eds.) Flora of Southern Africa, vol. 13. Pretoria: National Botanical Institute, pp. 201–204.

Peng, C.-I., Goldblatt, P. 1983. Confirmation of the chromosome number in Cephalotaceae and Roridulaceae. Ann. Missouri Bot. Gard. 70: 197–198.

Planchon, J.É. 1848. Sur la famille des Droséracées. Ann. Sci. Nat. Bot. III, 9: 79–99.

Solereder, H. 1908. Systematic anatomy of the Dicotyledons. Vol. 1, Introduction, Polypetalae and Gamopetalae. Oxford: Clarendon Press.

Takhtajan, A.L. 1987. Systema Magnoliofitorum. Leningrad: Nauka (in Russian).

Vani-Hardev, 1972. Systematic embryology of *Roridula gorgonias* Planch. Beitr. Biol. Pflanzen. 48: 339–351.

Zenk, M.H., Fürbringer, M., Steglich, W. 1969. Occurrence and distribution of 7-methyljuglone and plumbagin in the Droseraceae. Phytochemistry 8: 2199–2200.

Rosaceae

C. Kalkman[1]

Rosaceae Juss., Gen. Pl.: 334 (1789), nom. cons.
Spiraeaceae Maxim. (1879).
Amygdalaceae (Juss.) D. Don (1825), nom. cons.
Malaceae Small ex Britton & Small (1903), nom. cons.

Woody or herbaceous. Leaves usually alternate, sometimes distichous, rarely opposite, simple or compound; stipules on the twig or on the base of the petiole, free or adnate to the petiole, rarely 0. Inflorescences various, usually terminal, usually (compound) racemes. Flowers actinomorphic, mostly (4)5-merous, mostly bisexual, rarely unisexual and then the plants monoecious or dioecious; hypanthium usually well-developed (not evident in some staminate flowers), from saucer-shaped to tubular or campanulate, the epicalyx, sepals, petals, and stamens inserted on its rim, its inside usually lined by nectariferous tissue; disk sometimes distinct, intrastaminal; epicalyx + in some genera; sepals free; petals free, from large and showy to small and not or hardly distinct from sepals, rarely 0; stamens few to numerous, often their number distinctly related to the number of perianth parts; filaments free; anthers bilocular, dehiscing longitudinally; carpels 1–many, free or variously connate with each other and/or adnate to the hypanthium, forming 1 or more superior to inferior ovary(ies); stylodia (in monocarpellate ovaries styles) +, these sometimes (some Maleae) fused into a common, branched style; ovules 1–several (often 2) per carpel, anatropous, ascending or pendulous. Fruits various, fleshy or dry, dehiscent or not; seeds 1–several, testa usually firm, endosperm 0 or a thin layer, cotyledons fleshy or flat.

As recognized here, a moderately large, almost cosmopolitan family with 85 genera and c. 2000 sexual species. Apart from those, there are a large number of obligately or facultatively apomictic 'microspecies' in several genera.

Vegetative Morphology. Variation in vegetative characters is very great in this family and there is not a single trait, not even the presence of stipules, that is characteristic of the family as a whole.

Woody Rosaceae are mostly shrubs or small trees, not more than 12 m high. Small and low-growing shrubs with creeping woody branches are present in several genera, while large forest trees over 25–30 m high are rare and occur in few genera (*Eriobotrya*, *Sorbus*, *Prunus*). Woody species may be evergreen or leaf-shedding. Thorns as cauline metamorphoses are sometimes developed, and prickles (emergences) are common in some genera (*Rosa*, *Rubus*). Buds are normally protected by bud-scales in woody plants of climates with an unfavourable season. Woody genera not rarely show a differentiation in long and short shoots.

Herbaceous members of the family are mostly perennial and when growing in temperate climates they perennate by means of a subterranean vertical rhizome or horizontal rootstock. Annuals are extremely rare (*Alchemilla*, *Potentilla*). Herbaceousness may with some confidence be seen as an apomorphic state.

Many Rosaceae have simple leaves but in more than 30 genera compound leaves, mostly of a pinnate type, are the rule. That raises the question of what state is primitive in the family. Wolfe and Wehr (1988), in relation to their study of some fossil North American genera, made a distinction between true, pinnately compound leaves "formed by discrete laminar units that dehisce" (p. 178), and "paracompound" leaves that look pinnate but are in fact deeply dissected simple leaves, detaching (or, in Wolfe and Wehr's terminology, "dehiscing")

---

[1] In his contribution to this series the late Dr. Kalkman (†19 January 1998) had included *Quillaja* in, and excluded *Lyonothamnus* from Rosaceae, although he had been aware of the results of the *rbc*L analysis by Morgan et al. (1994) that were in favour of the contrary. I have found no reason for rejecting these molecular findings and consequently have changed Dr. Kalkman's manuscript and included *Lyonothamnus* in Rosaceae, although its position within the family remains unclear for the time being, whereas *Quillaja* has been removed from it and will be treated as a separate family in a further volume. After completion of Dr. Kalkman's manuscript, some important papers dealing with the morphology and molecular systematics of Rosaceae have appeared, which may indicate that the maintenance of tribe Exocordeae is probably not justified. I have inserted references to these papers into the bibliography but in general have left untouched Dr. Kalkman's argument, which to a large degree is supported by recent findings.

as a whole. Their contention is that in Rosaceae the trend is from simple and pinnatifid to pinnatisect leaves to paracompound leaves and probably (certainly, I would say) also to truly compound leaves. As a descriptive term, 'paracompound' may be useful and in some groups of the family the distinction may even have some morphological and phylogenetic value but at the moment there is too little evidence for making the choice between phylogenesis going in either of the two directions – from simple to compound, or vice versa (or both!). That the majority of the woody genera have simple leaves, whereas the herbaceous genera have (with few exceptions) compound leaves may probably be seen as an indication that the simple leaf is the plesiomorphic state in the family. Troll's typological approach of the compound leaf types in Rosaceae (Troll 1935) does not give the solution but stresses the diversity.

Leaves are normally alternate, only in three genera (*Coleogyne*, *Rhodotypos* and *Lyonothamnus*) are they opposite. In many woody genera there are erect orthotropic shoots with spirally arranged leaves, and horizontally growing plagiotropic shoots, with distichous leaves. The dorsiventral orientation of the plagiotropic shoots may already be visible in bud, as demonstrated by Charlton (1993) for *Prunus laurocerasus* (his "rotated lamina syndrome").

Especially in the maloids, the nervation type has some taxonomic value at the generic or subgeneric level. Nervation may there be craspedodromous (secondary nerves going straight to the margin and terminating there) or camptodromous (secondaries not reaching the margin). The lowermost secondaries are often stout and pedately divided.

Extrafloral glands are present on the leaves of some woody genera: in *Prunus* on the marginal teeth, on the petiole, or on the blades, in several maloids (*Aronia* and others) on the upper midrib. Water pores and/or guttation have been demonstrated in a number of common, Northern Hemisphere Rosaceae. Well-developed hydathodes with their anatomical specialization have been reported and described for only few species belonging to Sanguisorbeae (*Agrimonia*, *Sanguisorba*), *Alchemilla*, *Geum*, and *Physocarpus* (see Lersten and Curtis 1982).

Stipules are absent in one tribe, viz. Spiraeeae, and exceptionally in single species of normally stipulate genera. Often stipules are small and caducous but in several groups they are relatively large and persistent, more or less adnate with the petiole, and forming a sheath around the axillary bud (Potentilleae, *Rosa*, etc.).

WOOD ANATOMY. A comprehensive account was published by Zhang (1992), based on a survey of about 80% of the shrubby or arborescent genera. The following is a short summary of his family description, with the rare exceptions omitted.

Growth rings distinct except in tropical genera. Vessels vary substantially in number per sq. mm, in degree of grouping and pattern, as well as in diameter and length of the elements. In the temperate and shrubby taxa, vessel diameter is typically small and vessel frequency is low; the reverse holds true for the tropical trees. In most maloids, vessels are typically solitary (over 80%) while Pruneae have a high percentage (over 60%) of the vessels in radial multiples. The other groups show a variation range in between these two extremes. Perforations mostly exclusively simple, but rare scalariform or reticulate plates occur in some taxa. Intervessel pits alternate, vessel-ray and vessel-parenchyma pits usually half-bordered. Vessel walls with or without helical thickenings. Tyloses absent. Ground tissue mostly of fibre-tracheids, rarely mixed with libriform fibres in some Pruneae, non-septate. Parenchyma scanty to abundant, differently patterned but mainly diffuse and scanty paratracheal. Rays 1–16-seriate, very variable within and between taxonomic groups. Crystals abundant to absent, when present mostly prismatic but in the Pruneae druses may also be present.

Zhang (1992) recognized 12 phenetic groups, some of which clearly reflect taxonomic groupings based on morphological characters. His groups VI and VII are exclusively composed of maloid genera (with only *Cercocarpus* as an exception in group VII), and groups XI and XII consist only of (split-)genera that in the present treatment have been taken together as *Prunus* s.l. However, also the other, more mixed groups are not without merit. In Zhang's words (p. 79): "... with few exceptions all woody tribes, if not characterised by a single wood type, encompass few closely adjacent wood anatomical groups".

INFLORESCENCES. According to Troll (1964, and especially 1969) the basic synflorescence in the Rosaceae is monotelic (= with a terminal flower). Rich-flowered inflorescences may have the appearance of a panicle or a corymb. However, they are better described as thyrsoids since their partial inflorescences are cymose. Thyrsoids are found in many Maleae and Crataegeae but also, e.g., in *Sorbaria* (Gillenieae). The anthela of *Filipendula* is a special case. Its shape results from the fact that the side-branches are crowded and of unequal length, overtopping the terminal flower of the main axis.

Terminal flowers or even a larger terminal part of the synflorescence may be more or less reduced, either consistently or as one possibility within the variation in a species. *Holodiscus*, *Sibiraea* and some *Spiraea* species (all Spiraeeae) are good examples, but possibly also *Agrimonia* (Sanguisorbeae).

The axillary partial inflorescences of a thyrsoid may represent dichasia or monochasia. On some occasions, they are reduced to single flowers. In such cases, the inflorescence is a botryoid (= raceme with a terminal flower, e.g. in some *Rubus*; Troll 1969). These simple-looking inflorescences are not rare, occurring, e.g., in spiraeoid groups, Sanguisorbeae, maloid genera and in *Prunus*.

The occurrence of single terminal flowers can be interpreted as the result of a complete reduction of lateral partial inflorescences. Inflorescences reduced to one terminal flower occur in several groups: *Lindleya* (Exochordeae), *Kerria* and *Rhodotypos* (Kerrieae), *Cowania* and *Dryas* (Dryadeae), some species in *Rosa* and *Rubus*, and in *Cydonia*. In *Fallugia* (Geum group?) and *Mespilus* (Crataegeae), there are often or always one or few (up to 3 in *Mespilus*) lateral flowers or at least buds under the terminal flower.

It must not be imagined that the first Rosaceae had a very elaborate thyrsus, evolving only by reductions to other inflorescence types. It seems more plausible to envisage the first Rosaceae as having solitary terminal flowers, evolving by progressive branching to more elaborate types (see Parkin 1914). It must be recognized, however, that the presence of solitary flowers is not always a plesiomorphy: they certainly have also evolved secondarily as reductions.

FLOWERS. Although there are some genera with extreme reductions in the flowers, the generalized image of a rosaceous flower is that of a discoid structure, displaying the petals, the stamens and the nectar on the same level to passing insects.

The only character uniting the family is the presence of a hollowed hypanthium (also called flower tube, floral cup, or calyx tube). A recognizable hypanthium is wanting only in staminate flowers of some genera like *Bencomia* and *Sarcopoterium*.

In the majority of the Rosaceae the hypanthium is cupular, tubular, campanulate, etc., rarely saucer-shaped. The inside of an open hypanthium is normally lined with nectariferous tissue (see below, under disk). In several genera, however, the hypanthium is constricted at top and closed over the carpel(s), so that only the stylodia emerge through the central orifice. This is the case in *Alchemilla*, Sanguisorbeae, *Rosa*, and in a large part of the maloids, where hypanthium and ovary(ies) are fused so that the ovary becomes (semi-)inferior.

In few genera (*Aremonia*, *Spenceria*), under the hypanthium, that is under the flower, an involucrum is developed, consisting of bracts/prophylls (bracteoles). These organs must be distinguished well from an epicalyx which is, where present, always inserted at the apex of the hypanthium, with the sepals.

The phyllomatic organs forming the epicalyx are only present in genera of Potentilleae, the Geum group, Sanguisorbeae, Kerrieae (see below), and *Alchemilla* (all rosoids). The segments are normally present in the same number as the sepals. The general opinion was, and is that the epicalyx segments represent stipular appendages of the sepals. Isomery of sepals and epicalyx originates in this vision by connation of the stipular appendages of neighbouring, valvate sepals. Based on a teratological specimen of *Geum rivale*, Bolle (1935) proposed another explanation of the isomery: the outer 2 sepals have two stipules, the inner 2 none, the number-3 sepal one. Kania (1973) investigated the ontogeny of the flower in several species and saw that in *Aremonia* and *Potentilla* the epicalyx primordia develop after those of the sepals, in *Geum* and *Fragaria* after the petal primordia, in *Alchemilla* after the stamen primordia. In *Rhodotypos* (Kerrieae) the basal appendages of the sepals are also interpreted as an epicalyx (Schaeppi 1953), but according to Kania (1973) they are formed at the same time as the marginal teeth of the sepals. More study of the ontogenetical development of epicalyces is needed to elucidate their morphological nature. It is not at all certain that all these organs are homologous.

Sepals are usually rather small and often valvate, and protect the flowerbud only in a very juvenile stage. Sometimes, however, they are relatively large, as in *Rosa* and *Rubus*, with an imbricate vernation also expressed in the unequal development of the marginal lobes.

Petals are normally showy and the colour white predominates. Yellow petals are, however, most common in some groups: Sanguisorbeae (insofar as petaliferous), *Potentilla* and the Geum group. The petals may be clawed and they are caducous after anthesis, as is normal for petals. The absence of petals, as in some genera (*Alchemilla*, *Neviusia*, *Cercocarpus*, several Sanguisorbeae), is obviously a derived state, probably at least in part related to the evolution of wind pollination.

Unisexual flowers occur scattered through several of the tribes but hermaphroditic flowers are by far most common. Plants with unisexual flowers are rarely dioecious (some Spiraeeae, some Sanguisorbeae), more frequent is polygamy with hermaphroditic and staminate flowers in one plant or even in one inflorescence. In several genera the situation has not been observed in living populations and remains uncertain.

Stamens are free from each other and their number is often about 20. Depending on the dimensions of the flower, a disposition in one row or in several whorls may be evident. The ontogenetic studies of Lindenhofer and Weber (1999a, 1999b, 2000) suggest a preponderance of a 10 + 5 + 5 arrangement of the stamens, with antesepalous stamen pairs in the outer cycle. This pattern is interpreted as derived from a helical arrangement, whereas the possibility of intercalation or splitting of stamens is refuted. Any evidence that the primitive number of stamens should be 5 or 5 + 5 is lacking. In small, reduced flowers, the number of stamens may also be very low. In filled (double) flowers the extra petals are clearly petaloid stamens.

A disk is only mentioned as such in the generic descriptions when it is structurally visible as a separate organ, ring-shaped or divided into parts. It is probably always covered with nectariferous tissue, as may be the inside of the hypanthium. When a disk is present, it is intrastaminal on the upper inside rim of the hypanthium (but see *Alchemilla*). Where the hypanthium is closed, it may be the disk that surrounds the orifice.

Five carpels is possibly the plesiomorphic state in the family and, if that is the case, there have been evolutionary changes to larger numbers or reductions in number. When carpels are isomerous, they are either antesepalous or antepetalous. When there is a large number of carpels, a more or less highly elevated torus is usually developed (Potentilleae), the carpels either covering the whole torus or being absent in its basal part. Sometimes (in the Geum group, see Iltis 1913) the carpels are placed on gynophores, obviously homologous with the torus but shaped as a rather thin stalk with all (or at least most) carpels at the top. Connation of carpels (with one another) and adnation (to the hypanthium) define structure and shape of the fruit in the maloid groups. In most groups, however, the carpels are free from one another and from the hypanthium, each carpel forming a separate pistil and developing into a fruit. The anatomy of the carpels and their vascu-

lature has been studied by Sterling (1964–1969) over a wide range of genera. Stylodia are usually well developed and stigmas are generally of a simple shape, e.g. linear, capitate or bifid. In several genera of Sanguisorbeae the stigmas are fimbriate to penicellate, probably in connection with wind pollination.

Ovules are anatropous, rarely hemi-anatropous. In the groups usually considered to have maintained primitive flower characters (spiraeoids), there are several ovules per carpel. In many other groups the number is reduced to 2 (then usually only one developing to a seed) or 1. Where the number is 1 or 2, the ovules are often basal, ascending, and collateral, more rarely apically inserted and pendulous; horizontally oriented ovules also occur. Descending ovules are always epitropous, ascending ones apotropous. An obturator, guiding the pollen tube, may be present (especially in Pruneae, but also in *Exochorda* and *Adenostoma*); sometimes this function is performed by a swelling of the funiculus. The structure of the ovary, style/stylodium, stigma, and ovules in several genera all through the family are treated extensively by Juel (1918, 1927). Not easily explained is the large number of ovules in what is here called the Cydonia group. Is this a plesiomorphic character state ('left over' from the ancestral spiraeoids) or the result of a reversal from 2-ovulate (as in other maloids) to multiovulate?

FRUITS AND SEEDS. The follicle, originating from a monocarpellate, multi-ovulate pistil, occurs in spiraeoids and may confidently be regarded as the plesiomorphic fruit type. Derived from this type are the very common achene and the drupaceous achene with a slightly fleshy layer in the pericarp. The drupaceous achenes as in Kerrieae cannot be sharply separated from the true drupes with hard endocarp and fleshy mesocarp as in *Prunus* and unrelated *Rubus*. So-called pomes are morphologically diverse, and the term should be restricted to the fruits derived from a flower with the carpels adnate to the hypanthium forming a completely or incompletely inferior ovary, with a fleshy layer either entirely hypanthial or mixed hypanthial/carpellary, and with the endocarpal layers not woody. The carpels may be partly or entirely free from one another. When so defined, the pome is restricted to the Cydonia group and most of the genera in the tribe Maleae. When a flower as described above develops woody endocarpal layers in the fruiting stage (as in Crataegeae), such a fruit may be called a multipyrenous drupe but actually this is not a correct description (see note under tribe Crataegeae, p. 380).

Fleshy fruits may also be produced in other ways. In *Rosa* it is the closed hypanthium that becomes fleshy and the fruits are achenes with a hard pericarp. That reminds of the situation in quite unrelated genera with a fleshy hypanthium, like *Margyricarpus* and *Sarcopoterium*, and also of the situation in some genera of Maleae, viz. *Heteromeles* and *Dichotomanthes*. In *Fragaria* the fruits have a fleshy torus.

In the pericarp of the fruits of maloid genera, the fleshy layer shows anatomical differences the systematic value of which has been partially explored by Iketani and Ohashi (1991a). Heterogeneous flesh contains either large pigmented cells or large sclereids, or both; homogeneous flesh does not contain these specialized cells.

Seeds from follicles have a thick testa but, when the seeds remain enclosed in an achene or other fruit type, the testa is usually thin, although sometimes of a quite firm consistency. The embryo has flat to thick cotyledons. In full-grown seeds there is either nothing left of the endosperm or only a very thin layer.

EMBRYOLOGY. Pollen grains are 2-celled when shed. The ovules are anatropous or hemianatropous and bitegmic (the primitive state) or by connation unitegmic. In many genera the micropyle is (almost) absent by growth of the integuments over the nucellus. Meiosis usually occurs in only one megaspore mother cell, leading to a tetrad of which the chalazal cell develops into an 8-celled Polygonum-type embryo sac with one haploid egg cell. In some genera, however, the megaspore mother cell is replaced by one or more secondary megaspore mother cells undergoing meiosis.

The polar nuclei in the embryo sac generally fuse prior to fertilization. The secondary endosperm is first free-nuclear, later becoming cellular (Nuclear type). The embryo is mostly of the so-called Asterad type. For a full account of embryological data and literature, see Johri et al. (1992).

POLLEN MORPHOLOGY. No recent comprehensive account of pollen morphology in Rosaceae has been published, although some detailed regional accounts do exist. The following short account has been extracted from Van der Ham in Kalkman (1993: 231–232), where also the relevant references may be found.

Pollen grains of Rosaceae do not display large variation. They are monads, more or less spheroidal, and generally tricolporate. The colpi are usually relatively long. In some Rosaceae the ectoapertures are operculate. As far as is known, all Potentilleae display this character, also several Sanguisorbeae and some species of *Rosa*.

Transmission electron micrographs show a columellate infratectal layer. The outside pattern of ornamentation is variable but mostly striate.

As far as present knowledge goes, the characters of the pollen do not seem to be of much critical relevance for classification within the family.

KARYOLOGY. Summarizing the present knowledge of chromosome (base) numbers, the following picture emerges.

Spiraeoids

Exochordeae		Not sufficient evidence for establishing a base number. $2n = 16, 30, 34$
Spiraeeae	$x = 9$	Counts for all 8 genera
Neillieae	$x = 9$	1 of the 3 genera insufficiently known
Gillenieae	$x = 9$	Counts for all 4 genera

Rosoids?

Kerrieae	$x = 9$	1 out of 3 genera not counted
Dryadeae	$x = 9$	Counts for all 5 genera

Rosoids

Ulmarieae	$x = 7$	1 genus only
Sanguisorbeae	$x = 7$	6 out of 12 genera not counted
Potentilleae	$x = 7$	Counts for all 3 genera
Geum group	$x = 7$	2 out of 9 genera, incl. those of uncertain position, not counted
Rubeae	$x = 7$	1 genus only
Roseae	$x = 7$	1 genus only

Rosoids?

Alchemilla group	$x = 8$	1 genus only

Prunoids

Pruneae	$x = 8$	1 out of 4 genera not counted

Maloids

Cydonia group	$x = 17$	2 out of 4 genera not counted
Maleae	$x = 17$	3 out of 11 genera not counted
Crataegeae	$x = 17$	2 out of 7 genera not counted. See note under *Osteomeles*, p. 381.

A number of unresolved problems arise:

a) In few cases aberrant chromosome counts confuse the picture without sufficient evidence to consider the aberrant numbers as either error or aneuploidy.
b) In several of the tribes, establishing the base number rests on (too) little evidence, most notably in the Lindleya group but also in Neillieae and Sanguisorbeae. The natural character of these groups is, consequently, also more or less in doubt.
c) In some cases the chromosome base number is part of the evidence on which higher taxa are based (Kerrieae, Dryadeae, Alchemilleae). See also the paragraph on Classification, p. 349.

Much has been written on the origin of the high base number x = 17 for the maloid groups. Well-known is the postulate (going back to Sax 1932) that the number and the group are the result of one or more, old hybridization events between plants with $n = 9$ and with $n = 8$, followed by polyploidization. The $n = 9$ parent is usually referred to as spiraeoid and the other parent as prunoid or, more prudently, as proto-spiraeoid and proto-prunoid. Phipps et al. (1991) are strong supporters of this hypothesis. There are, of course, several other ways to arrive at $2n = 34$ and at the present stage there is, in my opinion, not sufficient evidence to make a responsible choice.[1] That polyploidization played a role in the history of the family is obvious, but successful hybridization between a member of the (primitive) spiraeoids and a member of the (much derived) prunoids is in my opinion not very likely. The counts of $2n = 34$ in *Lindleya* do not make things easier, although the taxonomic relationship of this genus with maloids is not at all a wild guess.

POLLINATION. In the terminology of Faegri and van der Pijl (1979), the blossom class most Rosaceae belong to is the dish- to bowl-shaped class and the individual flowers are the pollinator-attracting units. Some members of the family have brush-shaped blossoms. The individual flowers are then rather small, not individually conspicuous, and the visually attractive unit is the head- or spike-shaped inflorescence. Several Sanguisorbeae belong to this blossom class.

According to the classic work by Knuth (1898: 345–396; 1904: 334–347), pollinating insects are (in Europe, and there is no reason to suppose that it is different elsewhere) especially flies and short-tongued bees like *Anthrena* but, to a lesser degree, also long-tongued bees, beetles and even butterflies.

Cross-pollination is mostly ensured by the more or less distinct, protogynous development of the flowers, rarely by protandry, dioecy or by structural peculiarities (movement and place of stamens) that more or less impede self-pollination. The latter is, however, by no means rare.

Anemogamy may be supposed to be present in several genera belonging to the Sanguisorbeae. It can be deduced from the fimbriate or penicellate stigma and, in some cases, by the scabrate pollen. Experimental evidence is usually lacking.

Uncertainties are many-fold. For instance, the staminate panicles of *Aruncus dioicus* (Spiraeeae) are visited by hundreds of insects that take away the pollen and feed on it, but whether this contributes to pollination of the nectarless pistillate flowers is quite uncertain, the species being normally dioecious. However, there is not even unanimity in the literature about the pistillate flowers being nectariferous or not.

APOMIXIS. In a number of genera scattered over several tribes, apomixis has been established. Most common in Rosaceae is apospory where the embryo sac develops from a somatic cell (with unreduced nuclei) of the nucellus. Diplospory, where the embryo sac develops from a generative but unreduced cell derived from a megaspore mother cell, has also been established. A third type, nucellar (adventitious) embryony, where the embryos develop from somatic cells, seems to be unknown in Rosaceae (this type is not included into apomixis by all authors).

The egg cell with its unreduced nucleus of the aposporous or diplosporous embryo sac, stimulated in some way by pollination or not, will form the embryo without syngamy, i.e. without fusion with a microgamete from the pollen tube. In the words of Asker (1979: 233), "it may be questioned if there exists any 100% obligatory apomicts", although this has been claimed for a few Rosaceae (*Alchemilla*). In most cases apomixis is certainly facultative and mixed with sexual reproduction.

SEED DISPERSAL. Four main modes of seed/fruit dispersal are realized in Rosaceae, linked of course to the different fruit types (dry and dehiscent, dry and not dehiscent, fleshy).

---

[1] Evans and Campbell (2002) have now discarded the wide hybridisation hypothesis for Maloideae and instead suggest a polyploid origin involving only members of a lineage that contained the ancestors of *Gillenia*.

Endozoochory. The animals involved may be small ones (snails) in the case of low herbaceous plants like *Fragaria.* In the case of woody plants mostly birds, bats and other mammals are the active party. In most cases endozoochory is only surmised on the basis of morphology and often there are no observations or experiments. Myrmecochory may be present in *Aremonia,* where an extrafloral elaiosome seems to be present. The fleshy part is never an arilloid or sarcotesta.

Epizoochory. This method relies on accidentally passing animals. Specializations on the side of the plant include hooks on the hypanthium or a hooked style.

Ballistochory. The receptacle containing the diaspores is composed of hypanthium and sepals (and epicalyx), or it is the open follicle.

Anemochory is rare. *Hagenia* and *Leucosidea* have a flying apparatus composed of the epicalyx, *Cowania, Dryas* and *Geum* have feathery, persistent stylodia. A hypanthial wing is present on the fruits of some *Bencomia* spp. Whether the wings on the seeds of the Exocordeae are functional for dispersal by wind is unknown.

ECOLOGY. *Habitats.* Rosaceae range from semi-desert to lowland rainforest and open, alpine vegetation in a large number of different habitats. Nevertheless, a large number of Rosaceae may be found on wooded mountain slopes at medium altitudes and temperate latitude.

Mycorrhiza. Vesicular-arbuscular mycorrhiza seems to occur as common and erratic in Rosaceae as in Dicotyledons in general. Ectomycorrhiza has been demonstrated in only a small number of woody species belonging to Dryadeae (*Cercocarpus, Chamaebatia, Dryas*), maloids (*Crataegus, Malus, Pyrus, Sorbus*), *Prunus* and *Rosa.* This list was taken from Harley and Smith (1983) and it may be regarded as quite incomplete. Genera from areas outside Europe and North America have rarely or not been checked.

Nitrogen fixation. Root nodules with *Frankia* as bacterial component, in which nitrogen fixation has been demonstrated, also occur in Rosaceae, especially in Dryadeae (see Newcomb and Heisey 1984). Surprisingly, also a tropical Asian species of *Rubus* has to be included in the list (Becking 1979) and, just as is the case for ectomycor-

rhiza, it can be stated that knowledge is still incomplete.

DISTRIBUTION. Three genera, *Prunus, Alchemilla* and *Rubus,* are cosmopolitan. Of these only *Prunus* with certainty occurs naturally in Asia, Europe, the Americas, Africa, and the Australian region. The present-day distribution of the two other genera is grossly distorted by human influence, and specialist research would be needed to detect their original areas of distribution.

Sixteen genera are distributed over the entire Northern temperate zone with species in North America, Europe and Asia (*Agrimonia*, Amelanchier*, Aruncus, Crataegus*, Dryas, Filipendula*, Fragaria*, Geum*, Malus, Parageum, Potentilla*, Rosa*, Sanguisorba*, Sorbus*, Spiraea*, Waldsteinia*). Most of them (indicated by*) extend southward into regions with Mediterranean climate or even into tropical latitudes (often montane). *Geum* and *Agrimonia* are exceptional in being distributed also in the Southern Hemisphere.

Eight genera are Eurasian (*Aremonia, Cotoneaster*, Eriolobus, Mespilus, Orthurus, Pyracantha, Pyrus*, Sibiraea*). Their distribution ranges from rather small areas in Southeast Europe and West Asia to wider areas, some (indicated by *) extending to part(s) of Africa.

There are only two genera restricted to, but widely disjunct in, the Southern Hemisphere: *Acaena* and *Oncostylis.*

Endemicity at the level of genus is high in Asia (20 genera) and North America (17 genera), but less significant in South America, Africa (with 3 and 5 genera respectively) and Europe (1 genus); Australia has no endemic genus.

Croizat (1952), Goldblatt (1976), Thorne (1983) and Kalkman (1988) postulated a Southern (Gondwanan) origin for the family, or at least found it a hypothesis seriously to be considered. It must be admitted, however, that most authors favour a Laurasian origin for the family, and support for a Gondwanan origin is still weak.

CLASSIFICATION. *Subfamily level.* Four subfamilies are usually distinguished: Spiraeoideae, Rosoideae, Maloideae (or Pomoideae), and Prunoideae (or Amygdaloideae). This subdivision, which still is much used, obviously goes back to Focke (1888). At one time or another, these subfamilies have all been elevated to families: Spiraeaceae Maxim., Malaceae Small, Pomaceae Lois., Amygdalaceae (Juss.) D. Don, and Drupaceae DC. Several smaller groups have in the past also been elevated to subfamily or family rank, either formally or only as a suggestion (e.g. Kerriaceae, Sanguisorbaceae, *Dichotomanthes*).

While Prunoideae and Maloideae are probably monophyletic groups, this cannot be said for the two other subfamilies. Spiraeoideae contain the genera with dry, multi-seeded, dehiscent fruits (follicles), and thus are characterized by a plesiomorphic character. In other, derived groups, the fruits evolved to different states. Spiraeoideae are clearly linked with both Rosoideae and Maloideae, the delimitation of rosoids against spiraeoids being uncertain (see Table in the paragraph on Karyology). Prunoideae are most probably an offshoot from some part of primitive Rosoideae.

If one wants to intercalate a taxonomic rank between tribe and family, a logical and phylogenetically defendable distinction would be to recognize two subfamilies, one covering the classical Spiraeoideae and Maloideae, the other including the classical Rosoideae and Prunoideae. However, this classification is not formally proposed here, since the phylogeny within the family has not been worked out sufficiently. Some tribes such as Kerrieae are difficult to place. Nevertheless, for the sake of orientation, the conventional groups (spiraeoids, rosoids, maloids and prunoids) are used, although the delimitation of the former two is not undisputed.

Tribal level. Names for tribes were coined by Maximowicz (1879) (for his Spiraeaceae), by Focke (1888) for Rosoideae, and by Koehne (1890) for Pomaceae (Maloideae). The Prunoideae have never been subdivided into tribes. Hutchinson (1964) divided the family into some twenty tribes but did not distinguish subfamilies. The classification given in the taxonomic part of the present treatment is rather similar to Hutchinson's system and contains 18 tribes or (if a formal name did not yet exist) 'groups'. I have tried to make them homogeneous, but some problems remain and are discussed at the appropriate places.

Generic level. Generally, genera are taken in a moderately narrow sense, and some of them may be united when more data are available (e.g. *Purshia* and *Cowania*). In *Alchemilla* and *Potentilla* I have taken a wide view by lack of evidence in favour of splitting, and *Sorbus* has been recognized in a wide sense mainly for practical reasons (many cultivated species). Synonymy for genera is not exhaustive. However, all genus names cited under the family in Brummitt (1992) and all names recognized in Gunn et al. (1992) have been

mentioned, either as accepted or excluded names or as a synonym.

Intergeneric hybrids. Not listed in the taxonomic part are the intergeneric hybrids which sometimes have been named, especially in maloids. In the maloid tribes there are only few natural intergeneric hybrids, most having been made artificially or having been formed spontaneously when plants were brought together in cultivation (Robertson et al. 1991; Krügel 1992b). It is obvious that the (in)ability to hybridize cannot be used as an argument for splitting or lumping genera.

PHYLOGENY. A phylogenetic analysis of the family, considering only rather few morphological characters and tribes as operational units (Kalkman 1988), was not successful. Another attempt, based on 36 characters in 96 species representing almost all genera, was made by Phipps et al. (1991). A phylogenetic analysis on the basis of wood anatomical characters was conducted by Zhang (1992).

A molecular analysis of *rbc*L sequence data (Morgan et al. 1994), based on 40 species representing nearly all tribes and groups recognized in the present survey, established the family as a monophylum. The 'quillajoid' genera *Kageneckia*, *Lindleya* and *Vauquelinia* (but not *Quillaja* which appeared together with Surianaceae and Polygalaceae and consequently here is excluded from Rosaceae) were placed in one clade with the five maloid genera sampled (*Amelanchier*, *Crataegus*, *Eriobotrya*, '*Photinia*' and *Sorbus*). This stresses the relationship between this part of the spiraeoids and the maloids. A subdivision into subfamilies of the traditional circumscription was not supported by the resulting tree.

In another molecular study, Campbell et al. (1995) employed sequences from internally transcribed spacers ITS and part of the 5.8 S rDNA of 19 species of maloids, and included also morphological characters. Of the three subgenera of *Sorbus* that were sampled, *Aria* is not nested with *Cormus* and *Sorbus*, and the coherence of the Cydonia group is not unequivocally supported (*Chaenomeles* not nesting with *Cydonia* and *Pseudocydonia*). *Vauquelinia* was, as in the analysis by Morgan et al. (1994), included in the maloid clade in close proximity to *Eriobotrya* and *Rhaphiolepis*.

FAMILY RELATIONSHIPS. Rosaceae have always been placed in an order Rosales but the contents of the order varies much in different classifications. Families such as Calycanthaceae, Dichapetalaceae, Connaraceae, Cunoniaceae, Grossulariaceae, Saxifragaceae, Cunoniaceae, Leguminosae, Anisophylleaceae, Crossosomataceae, Neuradaceae, Rhabdodendraceae and Surianaceae have been considered as relatives of the Rosaceae. In contrast, recent molecular studies agree in placing Rosaceae in a clade together with Barbeyaceae, Rhamnaceae and Elaeagnaceae, and sister to Urticales families (see APG II 2003).

PHYTOCHEMISTRY. The extensive literature has been reviewed and evaluated by Hegnauer (1973, 1990), to which the reader is referred for details and references.

Phenolics are accumulated often in considerable amount and often appear species- or genus-characteristic. Tannins of the condensed type are almost ubiquitous, while hydrolysable tannins seem to be restricted to some groups of genera (Okuda et al. 1992).

Sorbitol, a sugar alcohol, replaces part of the saccharose in many of the Rosaceae. The absence of it seems to be characteristic of some rosoids, especially those with x = 7. Kerrieae and *Adenostoma* have sorbitol and in this respect are anomalous among rosoids.

Cyanogenic compounds (amygdalin, prunasin, dhurrin, and some unidentified glycosides) are well known from seeds and/or other parts in many Rosaceae but often show a patchy distribution. Of the spiraeoids only some genera have been proven to contain prunasin. The rosoid genera are normally devoid of cyanogenic compounds, with the exception of Kerrieae and some Dryadeae. In maloids many genera and species have been proven to be strongly cyanogenic, the seeds often containing amygdalin while other plant parts may have prunasin. Pruneae are more or less the same in this respect but the cyanogenic compounds in *Oemleria* and *Prinsepia* have not yet been identified. *Kageneckia* contains prunasin, which may be interpreted as a support for the intermediate place of the genus between spiraeoids and maloids. *Exochorda* leaves and twigs are weakly cyanogenic but the compound responsible is unidentified. *Adenostoma* and *Coleogyne* have small amounts of an unidentified cyanogenic compound, which does not help finding a good place for them.

Saponins, derivatives of pentacyclic triterpenes, are widespread in the family, normally as complex mixtures.

USES. Ethnobotanical value (usefulness to non-Western societies) is generally low, as evidenced by many published ethnobotanical lists. Sometimes this is easily understood when the region involved is poor in Rosaceae (see, e.g. Barr et al. (1993) for Australia, Morgan (1981) and Bhat et al. (1990) for parts of Africa, and Cambie and Ash (1994) for Fiji). In these four lists Rosaceae either are completely absent or represented by only one species. Less understandable is the absence of Rosaceae in many published ethnobotanical lists relating to parts of South and Central America, where Rosaceae are more diverse and more abundant. North American Indians, on the other hand, have found many uses (fruits or, in a single case, the taproot as food, several medicines, technical applications of wood or bark). For parts of Asia, ethnobotanical lists are usually also poor in Rosaceae, except as folk medicine. Perry (1980) lists species in 14 genera in her book on medicinal plants of East and Southeast Asia, the majority of the records being for China, a country that is very rich in Rosaceae. The difference between folk medicine and 'official' medicine becomes obvious when Perry's list is compared with the Chinese pharmacopoeia (Stöger 1989), where only seven species are treated.

In Western civilization, the economic importance of the family rests mainly on the edibility of the fleshy fruits of many of its members. The apple is the second most cultivated fruit in the world, after (wine-)grape. Also, pears, prunes, cherries, peaches, apricots, loquats, strawberries, blackberries, raspberries, etc., are all more or less widely grown, traded, and consumed, either in their fresh state, or preserved in tins or bottles, or in the form of juices, jellies, or jams.

Economically important is also the use of many species as ornamentals in gardens and parks or as street trees, mainly in temperate climates. The beautiful flowers and often decorative fruits make many species useful for such purposes. *Rosa*, the 'queen of flowers', may be economically the most important genus in this category.

As suppliers of timber the family is not important. Only few species reach timber size, and a minority of them may reach the world market. Some tree species, however, have a local importance for construction purposes.

Other useful products include pectin from apples, rose oil from the flowers of *Rosa* species, and tool handles from the hard wood of *Holodiscus* and *Cercocarpus*. Except for the first-mentioned one, these products are not of high market value. For pharmaceutical products, see the end of this paragraph.

The best-known medicine, originally extracted from a rosaceous species, is aspirin, the brand under which the German firm Bayer marketed acetylsalicylic acid in 1899. The brand name memorizes *Spiraea ulmaria* (now *Filipendula ulmaria*), source of old European folk medicines (Balick and Cox 1996: 32). From the flowerbuds of this plant, salicylic acid was isolated and a synthesized derivate became world-famous, aspirin now being a household-name for any over-the-counter febrifuge/sedative.

CONSERVATION. The database 'Globally threatened species' of the World Conservation Monitoring Centre WCMC at Cambridge (U.K.) contains (May 1997) 475 records pertaining to species or infraspecific taxa of Rosaceae. Five species are considered to be extinct: two *Crataegus* from Canada, two *Potentilla* (1 from U.S.A., 1 from Russia), and one *Rubus* from U.S.A. Under category E (endangered) or Ex/E (endangered or maybe already extinct), there are more then 100 entries listed in c. 25 genera (my classification). Cases are very diverse and I would not consider the loss of c. 30 North American microspecies of the *Rubus fruticosus* complex (see p. 370) or an apomictic hybrid species of *Sorbus* as alarming as the loss of one of the eight South American *Margyricarpus* species or the only *Chamaemeles* species endemic to Madeira.

Species not present in the database but, in view of their small range of distribution, probably at least of the 'vulnerable' category will almost certainly be added in the updating process. We may think of the Asian *Dichotomanthes tristaniifolia* and *Spenceria ramalana*, the North American *Chamaebatiaria millefolium*, and the Mexican *Lindleya mespiloides* and *Xerospiraea hartwegiana*, all species from monotypic genera, not (yet) in the WCMC listing.

KEY TO THE GENERA

1. Leaves compound   2
 - Leaves simple, lobed, sometimes deeply, or not (*Lyonothamnus* pinnate to pinnatifid)   42
2. Fruit a dehiscent follicle with more than 2 seeds   3
 - Fruit not dehiscent (achene, drupe or pome)   8
3. Plant herbaceous   4
 - Plant with woody branches, erect or prostrate   5
4. Leaves 2 or 3 times ternate to bi- or tripinnate, stipules 0. Flowers mostly unisexual; bisexual ones sometimes + in staminate inflorescences. Ovules inserted apically   **12. Aruncus**
 - Leaves trifoliolate, stipules +. Flowers bisexual. Ovules inserted near base   **19. Gillenia**
5. Prostrate shrublets. Leaves 2 or 3 times ternate   **9. Eriogynia**
 - Erect shrubs. Leaves (bi)pinnate   6
6. Ovules 2, collateral, inserted near base of cell. Leaves imparipinnate   **18. Spiraeanthus**

 - Ovules several, pendulous   7
7. Leaves imparipinnate, leaflets several cm long and wide. Seeds with small wings at both ends   **16. Sorbaria**
 - Leaves bipinnate or pinnate with deeply pinnatipartite leaflets, the ultimate segments only some mm long. Seeds without wings   **17. Chamaebatiaria**
8. Epicalyx 0 (hooked or barbed spines at the top of the hypanthium are not considered to represent an epicalyx)   9
 - Epicalyx +. Fruits achenes   27
9. Fruits dry achenes   10
 - Fruits of other types (drupes or pomes)   24
10. Herbs, stem base usually woody   11
 - Shrubs or trees   17
11. Hypanthium open above. Pistils 5–15   12
 - Hypanthium closed above or almost so, achenes remaining enclosed in it. Pistils 1–51   14
12. Leaves 1–3 times deeply tripartite. Hypanthium campanulate, more or less clearly differentiated into a lower part surrounding the ovaries and an upper tubular part. Stylodia basal. Ovule 1 per pistil   **41. Chamaerhodos**
 - Leaves interruptedly pinnate. Hypanthium cupular, ovaries and achenes not enclosed in it. Stylodia terminal   13
13. Pistils on bottom of hypanthium or on low torus, often shortly stalked. Ovules 2 per pistil   **28. Filipendula**
 - Pistils on torus on top of distinct gynophore. Ovule 1 per pistil   **44. Geum (*vernum*)**
14. Petals +, yellow   15
 - Petals 0   16
15. Hypanthium with hooked spines at top   **31. Agrimonia**
 - Hypanthium almost rudimentary, without spines   **31. Aremonia**
16. Hypanthium in fruit 4-angled to 4-winged, without spines   **34. Sanguisorba**
 - Hypanthium in fruit with 2–4 barbed spines at apex or with many scattered spines   **40. Acaena**
17. Flowers bisexual   18
 - Flowers unisexual. Petals 0   22
18. Petals 0   19
 - Petals +, 3–5   20
19. Flowers in inflorescences   **38. Polylepis**
 - Flowers solitary   **39. Margyricarpus**
20. Sepals and petals 3   **84. Potaninia**
 - Sepals and petals 5   21
21. Leaves very finely compound, 3 times pinnate with very small ultimate leaflets. Pistil 1   **27. Chamaebatia**
 - Leaves pinnate. Pistils many   **54. Rosa**
22. Flowers usually solitary. Often ericoid shrubs or shrublets, rarely small trees, unarmed, leaves not in rosettes   **37. Cliffortia**
 - Flowers in inflorescences   23
23. Small thorny shrubs with small leaves. Disk in flower not evident   **35. Sarcopoterium**
 - Candelabra shrubs, leaves in rosettes at end of branches. Flower with thick disk   **36. Bencomia**
24. Carpels several to many, superior, free from hypanthium, forming a collective fruit, composed of monopyrenous drupes   **53. Rubus**
 - Carpels up to 5, dorsally adnate to hypanthium and forming with it one spurious fruit (pome)   25
25. Stones (endocarps) hard and woody   **78. Osteomeles**
 - Endocarp not woody   26
26. Leaves pinnate   **64. Sorbus** (subg. ***Cormus*** and ***Sorbus***)
 - Leaves 3-foliolate   **72. Eriolobus**
27. Trees or (small) shrubs   28
 - Herbs with often stout, sometimes slightly woody caudex or rhizome   31

28. Stylodia subterminal to subbasal   **42. *Potentilla***
 – Stylodia termina   29
29. Flowers solitary. Small shrubs   **50. *Sieversia***
 – Flowers in inflorescences. Trees or shrubs   30
30. Flowers in large, axillary panicles, normally unisexual, trees monoecious   **29. *Hagenia***
 – Flowers in terminal spikes, bisexual   **30. *Leucosidea***
31. Involucre under flower +   32
 – No involucre under the flower   33
32. Stamens 5–10   **32. *Aremonia***
 – Stamens 35–40   **33. *Spenceria***
33. Stylodia subterminal, lateral or (sub)basal   34
 – Stylodia terminal   36
34. Torus not swollen after anthesis, or torus 0 and pistils on bottom of hypanthium   **42. *Potentilla***
 – Torus swollen after anthesis   35
35. Petals white   **43. *Fragaria***
 – Petals yellow or red   **42. *Potentilla***
36. Stylodium hooked   37
 – Stylodium not hooked   38
37. Upper part of stylodium caducous, lower part persistent, hooked   **44. *Geum***
 – Top of stylodium hooked in fruit   **47. *Oncostylus***
38. Stylodium articulate at base   39
 – Stylodium not articulate at base   40
39. Leaves trifoliolate   **52. *Waldsteinia***
 – Leaves pinnate to pinnatisect
    49. *Coluria* (see also **42. *Potentilla***, where the stylodia are ± always subbasal to subterminal. In very few species only the stylodia could be called terminal)
40. Pistils up to c. 15, most of them on top of stalk-like gynophore, some lower or on bottom of hypanthium   **46. *Orthurus***
 – Pistils many, not on a stalk-like gynophore   41
41. Terminal leaflet and uppermost lateral leaflets much larger than lower leaflets. Pistils on bottom of hypanthium or on low torus   **45. *Parageum***
 – Terminal leaflet not much larger than lateral ones. Pistils on bottom of hypanthium or on cylindrical torus   **48. *Acomastylis***
42. Fruit a 5-celled, woody, loculicidal capsule   **2. *Lindleya***
 – Fruit of a different type: follicle, achene, drupe or pome   43
43. Fruit follicular. Woody plants   44
 – Fruit of a different type, not dehiscent. Woody or herbaceous plants   55
44. Carpels 5, ventrally connate. Ovules 2 per carpel   45
 – Carpels 1–5, free from each other or at most basally shortly connate   46
45. Intrastaminal disk evident. Ovules pendulous   **4. *Exochorda***
 – Disk 0. Ovules ascending   **3. *Vauquelinia***
46. Stipules +. Ovules 4–many per carpel. Seeds winged or flat   47
 – Stipules 0. Ovules 2–several per carpel. Seeds not winged   48
47. Follicles 5, each with many winged seeds   **1. *Kageneckia***
 – Follicles 2, each with 4 flat seeds   **85. *Lyonothamnus***
48. Intrastaminal disk well developed   49
 – Disk inconspicuous or 0   51
49. Dioecious plants with unisexual flowers   **6. *Sibiraea***
 – Flowers bisexual   50
50. Low, prostrate, evergreen shrubs   **7. *Petrophytum***
 – Small or large, erect or prostrate shrubs, leaf shedding   **5. *Spiraea***
51. Pistil 1(2) per flower   52
 – Pistils 2–5   53

52. Seeds 1–2 per fruit   **15. *Stephanandra***
 – Seeds several in each fruit   **13. *Neillia***
53. Shrubs, the old twigs thorny   **10. *Xerospiraea***
 – Unarmed shrubs   54
54. Evergreen cushion plants. Hypanthium saucer-shaped   **8. *Kelseya***
 – Leaf-shedding shrubs. Hypanthium campanulate   **14. *Physocarpus***
55. Only 1 ovule per carpel   56
 – Two or more ovules per carpel   69
    (see ***Brachycaulos*** (p. 383), where the number of ovules is unknown)
56. Fruit a dry achene   57
 – Fruit drupaceous, i.e. with woody endocarp(s) and thin, slightly fleshy mesocarp   67
57. Petals 0   58
 – Petals +   61
58. Stipules adnate to petiole   **55. *Alchemilla***
 – Stipules free   59
59. Intrastaminal disk +, a tubular sheath around the ovary   **83. *Coleogyne***
 – Disk not evident   60
60. Sepals 5. Hypanthium narrowly tubular in lower part, the upper part widened and open   **26. *Cercocarpus***
 – Sepals 3(4). Hypanthium 0 in staminate flowers, ellipsoid and closed at mouth in pistillate flowers   **37. *Cliffortia***
61. Hypanthium closed at top, fleshy when fruiting   **54. *Rosa* (*persica*)**
 – Hypanthium open above, not fleshy when in fruit   62
62. Flowers (sepals and petals) 7–10-merous   **23. *Dryas***
 – Flowers 5-merous   63
63. Shrubs   64
 – Herbs   66
64. Epicalyx +, small   **51. *Fallugia***
 – Epicalyx 0   65
65. Pistils 1(–3)   **25. *Purshia***
 – Pistils 5–12   **24. *Cowania***
66. Stamens 5   **41. *Chamaerhodos***
 – Stamens many   **52. *Waldsteinia***
67. Flowers solitary. Petals large   **20. *Kerria***
 – Flowers in terminal racemes   68
68. Carpels free from hypanthium. Petals small or 0   **21. *Neviusia***
 – Carpels dorsally adnate to hypanthium. Petals large   **80. *Hesperomeles***
69. Fruit a drupe (the mesocarp fleshy or leathery) or at least drupaceous (the mesocarp thin and slightly fleshy). Fleshy layer only of carpellary origin   70
 – Fruit otherwise, the fleshy part, if present, not (or not only) of carpellary origin   74
70. Pistil normally 1   71
 – Pistils 2 or more   72
71. Style terminal
    57. *Prunus* (see also **58. *Maddenia***, which is hardly or not different from part of ***Prunus***)
 – Style becoming (sub)basal in fruit   **59. *Prinsepia***
72. Leaves opposite, rarely whorled in 3s   **22. *Rhodotypos***
 – Leaves alternate   73
73. Stipules 0 or rudimentary   **56. *Oemleria***
 – Stipules +   **53. *Rubus***
74. More than 2 ovules per carpel   75
 – Two ovules per carpel   78
75. Axillary thorns +, collateral with short shoots   **60. *Chaenomeles***
 – Plants unarmed   76
76. Flowers in terminal umbel-like inflorescences with up to five flowers   **62. *Docynia***

– Flowers solitary, terminal on leafy shoots 77

77. Flowers bisexual or staminate. Stylodia connate at very base, at base surrounded by a cup-shaped disk on top of the ovary **63. Pseudocydonia**
– Flowers bisexual. Stylodia free, not surrounded by a cup-shaped disk **61. Cydonia**

78. Fruits dry achenes, indehiscent 79
– Fruits with fleshy parts of carpellary and/or hypanthial origin 81

79. Pistil 1 **82. Adenostoma**
– Pistils 4 or more in each flower 80

80. Intrastaminal disk +, rim-like. Pistils 4–5 **11. Holodiscus**
– Disk not evident. Pistils 5–15, rarely fewer **28. Filipendula**

81. Carpel (pistil) always 1, free from the hypanthium or basally adnate to it 82
– Carpels 2 or more, exceptionally only 1 developed 83

82. Fruit a drupe with a woody stone **79. Chamaemeles**
– Fruit an achene, entirely free from, but almost fully surrounded by the enlarged and fleshy hypanthium **69. Dichotomanthes**

83. Carpel walls (or only the endocarps) forming woody pyrenes in the fruit, surrounded by fleshy tissue of hypanthial (and sometimes partly carpellary) origin 84
– Carpel wall not woody 87

84. Carpels completely connate with each other and adnate to the hypanthium **81. Mespilus**
– Carpels ventrally and laterally free from each other, dorsally adnate to hypanthium 85

85. Leaves lobed. Ovules superposed, one-stalked **77. Crataegus**
– Leaves not lobed. Ovules collateral 86

86. Thorns +. Carpels 5 **76. Pyracantha**
– Plants unarmed. Carpels 2–3 **75. Cotoneaster**

87. Inflorescence a simple raceme 88
– Inflorescence a compound raceme 92

88. Leaves deeply tripartite **72. Eriolobus**
– Leaves slightly or not lobed 89

89. Carpel apices free from hypanthium, exposed in flower 90
– Carpel apices adnate to hypanthium, not exposed 91

90. Flesh of fruit soft, core thin, sepals persistent. Carpels with incomplete septs **74. Amelanchier**
– Flesh of fruit hard, core cartilagnous, sepals ultimately caducous. No incomplete septs in carpels **70. Macromeles**

91. Stylodia free, emerging through a central hole in the disk that covers top of ovary **73. Pyrus**
– Stylodia connate at base **71. Malus**

92. Leaves deeply lobed **64. Sorbus** subg. **Torminaria**
– Leaves unlobed 93

93. Nerves going straight to the margin 94
– Nerves not going straight to the margin 96

94. Carpel apices not exposed. Sepals and part of hypanthium caducous **64. Sorbus** subg. **Micromeles**
– Carpel apices usually free from the hypanthium. Sepals persistent on the fruit 95

95. Flesh of fruit with large pigment cells **64. Sorbus** subg. **Aria**
– Flesh of fruit without large pigment cells **65. Eriobotrya**

96. Carpels completely adnate to hypanthium. Hypanthium and sepals caducous **67. Rhaphiolepis**
– Carpel apices free from hypanthium. Sepals (long) persistent on the fruit 97

97. Spurious fruit consisting of leathery achenes surrounded by a fleshy hypanthium **68. Heteromeles**
– Fruit a true pome 98

98. Petals upright. Carpels 2. Pigment cells in flesh of fruit **64. Sorbus** subg. **Chamaemespilus**
– Petals spreading. Carpels 2–5. Sclereids in flesh of fruit **66. Aronia**

TRIBES AND GENERA OF ROSACEAE

1. TRIBE EXOCHORDEAE Schulze-Menz (1964).

Woody, unarmed. Leaves alternate, simple, penninerved; stipules +. Hypanthium open at top. Epicalyx 0. Petals +. Carpels 5, free or ventrally connate; stylodia terminal; ovules numerous–2 per carpel. Fruits with woody pericarp. Seeds winged; endosperm thin or 0. Chromosome base number(s) unclear.

1. *Kageneckia* Ruiz & Pavon

Kageneckia Ruiz & Pavon, Prodr.: 145 (1794).

Shrubs or small trees. Leaves serrate. Dioecious. Flowers showy, male ones in poor racemes or corymbs, female ones solitary. Hypanthium campanulate to obconical. Petals white to yellowish. Disk rim-shaped. Stamens or staminodes c. 15, inserted on rim of disk. Pistils with obliquely dilated, 2-fid stigma; ovules many, 2-seriate, horizontal; pistillodes in male flowers small. Follicles spreading. Seeds many, winged at apex. $2n = 34$. About three species in South America (Bolivia, Peru, Chile, Argentine).

2. *Lindleya* H.B.K.

Lindleya Humb., Bonpl. & Kunth, Nova Gen. Sp., 6 ed. fol.: 188; ed. qu.: 239 (1824), nom. cons.

Trees, evergreen. Leaves with glandular-toothed margins. Flowers solitary, terminal on axillary short shoots towards the ends of branchlets, bisexual, 5-merous, showy. Hypanthium obconical. Sepals imbricate. Petals white. Stamens 15–20. Disk not evident. Carpels antesepalous, free from hypanthium, laterally connate, forming a 5-locular ovary; stigmas 2-lobed, oblique; ovules 2 per cell, ascending. Fruit a 5-celled, woody capsule, loculicidally 5-valved. Seeds 2 per cell, flat, with a thin wing. $n = 17$ (Goldblatt 1976). One species, *Lindleya mespiloides* H.B.K., in Mexico.

3. *Vauquelinia* Humb. & Bonpl. Fig. 109

Vauquelinia Correa ex Humb. & Bonpl., Pl. Aequin. 1: 140 (1807); Hess & Henrickson, Sida Contrib. Bot. 12: 101–163 (1987), rev.

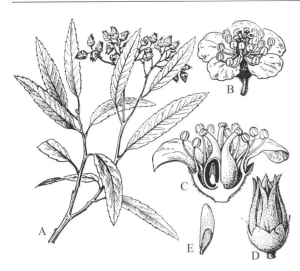

Fig. 109. Rosaceae–Exocordeae. *Vauquelinia californica*. **A** Fruiting branch. **B** Flower. **C** Flower, vertical section. **D** Fruit. **E** Seed. (Takhtajan 1981)

Large shrubs or small trees, evergreen. Leaves coriaceous, with serrate margins. Inflorescences terminal, corymb-shaped compound racemes. Flowers bisexual, 5-merous. Hypanthium hemispherical. Sepals valvate. Petals white. Stamens 15–20. Disk not evident. Carpels 5, free from hypanthium, ventrally connate; ovules 2 per cell, ascending, apotropous. Fruit separating into 5 follicles with woody pericarp, dehiscing ventrally and dorsally. Seeds 2 per follicle. $2n = 30$. Three species in Southern U.S.A. (Arizona, New Mexico, Texas) and Mexico, in chaparral and other arid to semiarid habitats, up to 2700 m altitude, often on limestone.

4. *Exochorda* Lindl.

Exochorda Lindl., Gard. Chron.: 925 (1858).

Shrubs, leaf-shedding. Leaves with entire or serrate margins. Inflorescence a terminal raceme. Flowers large, (4)5-merous, bisexual (unisexual?). Hypanthium campanulate to obconical. Sepals imbricate. Petals white. Stamens 15–30, 3–5 together before each petal, the groups separated by a larger space. Disk well developed as a rim. Carpels 5, free from hypanthium and free at top, but ventrally connate; stigma dilated; ovules 2, collateral, pendulous from apex. Fruit separating as 5 follicles with woody pericarp. Seeds 1 or 2 per follicle, flat. $2n = 16$ (one record of $2n = 18$). Some 4 species in Siberia, Korea, China, Japan.

2. Tribe SPIRAEEAE Maxim. (1879).

Woody (exception: *Aruncus*). Leaves alternate, simple (exceptions: *Aruncus, Eriogynia*), penninerved; stipules 0. Inflorescences terminal (compound) racemes (exception: *Kelseya*). Hypanthium open at the top, dry at maturity. Epicalyx 0. Sepals valvate. Petals +. Disk + (exception: *Xerospiraea*). Carpels 2–5, free; stylodia free; ovules inserted in apical part of suture. Follicles (exception: *Holodiscus*). $x = 9$.

5. *Spiraea* L.

Spiraea L., Sp. Pl.: 489 (1753); K.M. Purohit & G. Panigrahi, Rosaceae in India 1: 48–138 (1991), regional rev.
Pentactina Nakai (1917).

Shrubs, leaf-shedding. Leaves simple, rarely lobed, margins incised at least in apical part or sometimes entire. Inflorescences terminal racemes, umbels or compound racemes. Flowers 5(4)-merous, bisexual, rarely unisexual and then plants dioecious. Hypanthium campanulate to obconic. Disk at the rim of the hypanthium, collar-shaped or divided into lobes, intrastaminal. Petals white, rarely yellow, red. Stamens 15–many. Pistils usually 5, antepetalous, free; stylodia (sub)terminal; stigmas capitate; ovules 2–several, pendulous. Follicles with dry pericarp. Seeds few, small, with membranous testa, endosperm thin or 0. $2n = 18$, 27, 36, 54; aneuploids also recorded. About 80 species in the northern temperate zone, richest in Asia, with a centre in the Himalayas (c. 30 spp.), 10 species wild in North America, and 8 species wild in Europe. Usually in sunny places or forest margins, more rarely in forest, in low altitudes and montane.

6. *Sibiraea* Maxim.

Sibiraea Maxim., Acta Hort. Petropol. 6: 213 (1879).

Shrubs, more or less prostrate, unarmed, leaf-shedding. Leaves simple, entire. Inflorescence a terminal, compound raceme. Flowers 5-merous, unisexual, plants dioecious. Hypanthium hemispheric. Petals white or pinkish. Stamens resp. staminodes c. 20. Disk intrastaminal, lobed. Pistils/pistillodes mostly 5, very shortly connate at base; stylodia terminal; stigma capitate; ovules 4–8, pendulous from ventral placenta. Follicles with bony pericarp, opening ventrally, partly also dorsally. Seeds few, linear, with thin endosperm. $2n = 18$. One species, *S. laevigata* (L.) Maxim., in

Central and East Asia and south-eastern Europe. Rocky places in montane altitudes.

7. *Petrophytum* (Torr. & A. Gray) Rydb.

Petrophytum (Torr. & A. Gray) Rydb., Mem. New York Bot. Gard. 1: 206 (1900), 'Petrophyton'.
Spiraea § *Petrophytum* Nutt. ex Torr. & A. Gray (1840), rank not indicated.
Eriogynia sect. *Petrophytum* S. Wats. (1890).

Low, prostrate shrubs, unarmed, evergreen. Leaves crowded, simple, 1- or 3-nerved, entire, coriaceous. Inflorescence a terminal (compound) raceme. Flowers 5-merous, bisexual. Hypanthium saucer-shaped. Disk intrastaminal, entire. Petals white. Stamens c. 20. Pistils 3–5, free; stylodia terminal; stigmas minute; ovules 2–4, pendulous from top. Follicles opening dorsally and ventrally, coriaceous. Seeds linear. $2n = 18$. Three very similar species in the Western half of North America, possibly better combined into one. Montane, often on limestone.

The relationships of this genus and the following three with *Spiraea* were discussed by Henrickson (1986a). The reasons to keep them separate seem to be sufficiently compelling.

8. *Kelseya* (S. Wats.) Rydb.

Kelseya (S. Wats.) Rydb., Mem. New York Bot. Gard. 1: 207 (1900).
Eriogynia Hook. sect. *Kelseya* S. Wats. (1890).

Shrublets, growing in cushions, unarmed, evergreen. Leaves simple, entire, coriaceous. Flowers solitary, axillary, 5-merous, bisexual. Hypanthium saucer-shaped. Disk inconspicuous, entire. Petals purplish pink. Stamens c. 10. Carpels 3–5, free; stylodia terminal; ovules 3–7, pendulous. Follicles opening along ventral and dorsal sutures, pericarp coriaceous. Seeds fusiform. $2n = 18$. One species, *K. uniflora* (S. Wats.) Rydb. in some north-western states of U.S.A. Montane, on limestone.

9. *Eriogynia* Hook.

Eriogynia Hook., Fl. Bor. Am. 1: 255, pl. 88 (1832), pro parte, incl. type (*Spiraea pectinata* Pursh).
Luetkea Bong. (1833).

Evergreen shrublets, unarmed, with prostrate woody branches and erect herbaceous scape-like flowering stems. Leaves crowded at base of stems and smaller ones on the scapes, much divided, 2 or 3 times ternate. Inflorescence a terminal raceme, bracts leaf-like. Flowers 5-merous, bisexual. Hypanthium obconic. Disk a shallow, intrastaminal rim. Petals white. Stamens c. 20. Pistils 4–6, mostly 5, free, antepetalous; stylodia terminal; ovules several, pendulous from upper part of ventral suture. Follicles opening ventrally and dorsally. Seeds fusiform, small, endosperm 0. $2n = 18$. One species, *E. pectinata* (Pursh) Hook., in the north-western part of North America, from Bering Strait to Cascade Mts. Subalpine to alpine habitats.

10. *Xerospiraea* Henr.

Xerospiraea Henrickson, Aliso 11: 206 (1986).

Evergreen shrubs, with long and short shoots, old twigs often thorny. Leaves alternate, small, simple, with (sub)entire margins. Inflorescences terminal (compound) racemes, often leafy at base. Flowers 5-merous, bisexual. Hypanthium obconic. Petals white. Stamens 15–20. Disk not evident. Pistils 2–5, free; stylodia terminal; stigma punctiform; ovules 2, pendulous. Follicles thin-walled, opening ventrally and in the upper part also dorsally. Seeds (not observed) 1 or 2, testa thin, endosperm 0. $2n = 18$. One species, *X. hartwegiana* (Rydb.) Henr., in Mexico. Montane, on limestone.

11. *Holodiscus* (K. Koch) Maxim.

Holodiscus (K. Koch) Maxim., Acta Hort. Petropol. 6: 253 (1879), nom. cons., see also Taxon 45: 672 (1996).
Spiraea 'Gruppe' *Holodiscus* K. Koch, Dendrologie 1: 309 (1869), rank not indicated, basionym.

Shrubs, unarmed, leaf-shedding. Leaves simple, alternate, shallowly to deeply lobed, nervation pinnate, margins incised. Inflorescences terminal, panicle-shaped compound racemes. Flowers 5-merous, bisexual. Hypanthium saucer-shaped. Petals cream-coloured. Stamens c. 20. Disk intrastaminal, rim-like, not lobed. Pistils 4–5, free, antepetalous; stylodia terminal; ovules 2, collateral, pendulous from apex. Fruits achenes, follicle-like, with a bony pericarp, but obviously not dehiscing. Seed 1 (2?), endosperm thin. $2n = 18, 36$. One species, *H. discolor* (Pursh) Maxim., from NW North America through Central America to Northern South America. Montane habitats, in forest and on rocks.

12. *Aruncus* L.

Aruncus L., Opera varia: 259 (1758).
Pleiosepalum Hand.-Mazz. (1922); Symb. Sin. 7: 455, f. 13 (1933), reduction.

Large perennial herbs with thick rhizomes. Leaves in rosette and also cauline, 2–3-ternate to -pinnate, leaflets pinnatisect to toothed. Inflorescence a much branched terminal panicle-shaped compound raceme. Flowers 5-merous, usually unisexual and plants dioecious. Hypanthium saucer-shaped. Epicalyx 0. Petals white or yellowish-white. Stamens 15–30. Disk rim-shaped, not conspicuous. Pistils 3–5, free; stylodia terminal but often oblique; ovules few to several, pendulous. Follicles with coriaceous pericarp. Seeds few, very small, endosperm thin or 0. $2n = 18, 36$, but also $2n = 14, 16, 42$ have been recorded. Usually recognized as one polymorphic species, *A. dioicus* (Walter) Fernald, widespread over the northern temperate zone. Usually in shaded, damp to wet places, montane to subalpine.

3. TRIBE NEILLIEAE Maxim. (1879).

Unarmed, leaf-shedding shrubs. Leaves alternate, simple, palmatinerved; stipules free, on the twig. Flowers 5-merous, bisexual. Hypanthium open at top, dry at maturity. Epicalyx 0. Petals white. Disk not evident. Stylodia terminal. Follicles large. Seeds with abundant endosperm. Probably x = 9, see Note under *Neillia*.

13. *Neillia* D. Don   Fig. 110

Neillia D. Don, Prodr. Fl. Nepal.: 228 (1825); J. Vidal, Adansonia II, 3: 142–166 (1963), rev.; J. Cullen, J. Arnold Arbor. 52: 137–158 (1971), rev.

Leaves variously lobed; stipules rather large, persistent or caducous. Inflorescence a terminal raceme or leafy panicle. Hypanthium campanulate to tubular, in fruiting stage often with long-stalked glands outside and on pedicels. Sepals imbricate. Stamens 5–many. Pistil 1(2, free); ovules 2–12, biseriate on ventral placenta, pendulous. Follicle coriaceous, enclosed in to slightly protruding from the enlarged hypanthium, dehiscing ventrally. Seeds several, testa hard and shiny, raphe distinct. $n = 9$, but see Note below. About 12 species, in continental Asia from India to Korea, China, and Vietnam, one species also in Sumatra and Java. Montane thickets, grasslands and forest edges, up to 2500 m altitude.

Ratter and Milne (1973) reported $n = 9$ for *Neillia sinensis*; Sharma (1970, not seen) counted $n = 10$, a rather deviant and possibly aneuploid number.

Fig. 110. Rosaceae–Neillieae. *Neillia thyrsiflora.* **A** Flowering branch. **B** Flower. **C** Hypanthium with fruits. (Kalkman 1993)

14. *Physocarpus* (Cambess.) Raf.

Physocarpus (Cambess.) Raf., New Fl. 3: 73 (838), nom. et orth. cons.
Spiraea L. sect. *Physocarpus* Cambess. (1824).

Leaves more or less distinctly 3-lobed; stipules early caducous. Inflorescence a terminal raceme, ± globular to corymb-shaped. Hypanthium campanulate. Sepals valvate. Stamens many. Pistils 2–5, connate in basal part; ovules 2–4, in lower half of ventral suture, variously oriented. Follicles shortly stalked, inflated, protruding from hypanthium, opening along both sutures. Seeds few, testa hard, shiny. $2n = 18, 54$. Few species in North America, one in Northeast Asia. Mostly in moist places, along streambanks, etc. The bark scales off in narrow strips.

15. *Stephanandra* Siebold & Zucc.

Stephanandra Siebold & Zucc., Abh. Math.-Phys. Cl. Kön. Bay.
Akad. Wiss. 3: 739 (1843).

Leaves 3–more-lobed; stipules rather large, persistent. Inflorescences terminal, leafy racemes or panicles. Hypanthium obconic-campanulate. Sepals slightly imbricate. Stamens 10–20. Pistil 1; ovules 2, nearly apical, pendulous. Follicle protruding from hypanthium, not inflated, opening along ventral suture. Seeds 1–2, filling the follicle entirely, testa hard. $2n = 18$. About 3 species, Korea, Japan, and China. In thickets at low altitudes and montane.

4. TRIBE GILLENIEAE Maxim. (1879).

Unarmed. Leaves alternate, compound; stipules on junction twig-leaf. Flowers in terminal inflorescences, 5-merous, bisexual. Hypanthium open at top, dry at maturity. Epicalyx 0. Petals +. Disk not evident. Carpels up to 5, free. Follicles. Seeds with thin layer of endosperm. x = 9 (?).

16. *Sorbaria* (Ser.) A. Braun

Sorbaria (Ser. in DC.) A. Braun in Asch., Fl. Prov. Brandenb. 1:
177 (1860), nom. cons.; Rahn, Nord. J. Bot. 8: 557–563 (1989),
rev.
Spiraea L. sect. *Sorbaria* Ser. in DC. (1825), basionym.

Shrubs, leaf-shedding. Leaves imparipinnate, leaflets with incised margins. Inflorescence a terminal panicle-shaped compound raceme. Hypanthium hemispherical. Sepals valvate, persistent. Petals white or creamy. Stamens 20–50. Carpels usually 5, antesepalous, ventrally connate in lower half and at base also connate over a part of the lateral walls, otherwise free but closely adhering, forming a superior ovary with deep longitudinal grooves between the cells; stylodia (sub)terminal, recurving after anthesis; ovules several, pendulous, in 2 rows. Follicles much exserted from hypanthium, pericarp cartilaginous. Seeds few, fusiform, with small wings at both ends. $2n = 36$. According to the recent revision, 4 species in Central and East Asia. From low to montane and alpine altitudes.

17. *Chamaebatiaria* Maxim.

Chamaebatiaria (Brewer & Watson) Maxim., Acta Hort.
Petropol. 6: 225 (1879).
Spiraea § *Chamaebatiaria* Porter ex Brewer & Watson (1876).

Small shrubs, ± evergreen, aromatic-glandular. Leaves pinnate with pinnatipartite segments, or bipinnate. Inflorescence a terminal panicle-shaped compound raceme, with some leaves. Hypanthium hemispheric. Sepals valvate. Petals white. Stamens c. 60. Carpels 5, alternipetalous, ventrally connate at very base, otherwise free but closely adhering; stylodia terminal; stigmas capitate; ovules several, pendulous. Follicles dehiscent ventrally and at apex dorsally. Seeds few, fusiform. $2n = 18$ (16?). One species, *C. millefolium* (Torr.) Maxim., in the western part of the U.S.A., in montane altitudes.

18. *Spiraeanthus* (Fisch. & C.A. Mey.) Maxim.

Spiraeanthus (Fisch. & C.A. Mey.) Maxim., Acta Hort. Petropol.
6: 226 (1879).
Spiraea subg. *Spiraeanthus* Fisch. & C.A. Mey. (1842).

Shrubs, probably evergreen. Leaves imparipinnate; leaflets numerous, small, entire, coriaceous; stipules small. Inflorescence a terminal, panicle-shaped compound raceme. Hypanthium campanulate. Petals pink. Stamens 20–25. Carpels 2–5, alternipetalous, connate at base; stylodia terminal, thickened towards apex; stigmas capitate; ovules 2, collateral, inserted near the base. Follicles connate at base, ventrally dehiscent and dorsally at apex. Seeds 1–2, shortly winged at both ends. $2n = 18$. One species, *S. schrenckianus* (Fisch. & C.A. Mey) Maxim., in Kazakhstan: Karatau Range and Golodnaya Step.

19. *Gillenia* Moench

Gillenia Moench, Meth. Suppl.: 286 (1802).
Porteranthus Small (1894), nom. superfl., see Taxon 37: 139
(1988).

Perennial herbs, with rhizomes. Leaves trifoliolate, shortly petiolate, in inflorescence subsessile, leaflets with incised margins; stipules large or small. Inflorescence a terminal, corymb- or panicle-shaped compound raceme. Flowers on long pedicels, rather large. Hypanthium tubular to campanulate. Sepals imbricate. Petals white to pale pink. Stamens 10–20. Pistils 5, alternipetalous, free; stylodia terminal; ovules 2–4, inserted near the base, ascendent, apotropous. Follicles soon rupturing the hypanthium, pericarp bony. Seeds 1–4, rather large. $2n = 18$. Two species in North America: Canada (Ontario) and U.S.A. (Eastern part), in montane forests.

5. TRIBE KERRIEAE Focke (1888).

Unarmed, leaf-shedding shrubs. Leaves simple; stipules +, free. Flowers bisexual. Hypanthium open at top, dry at maturity. Sepals imbricate. Stamens many. Pistils several, free; stylodia terminal; ovules attached in the middle. Achenes drupaceous, i.e. with slightly fleshy mesocarp and woody endocarp. x = 9 (few data).

20. *Kerria* DC.

Kerria DC., Trans. Linn. Soc. Lond. 12: 156 (1818).

Leaves alternate, with incised margins. Flowers solitary, terminal on short lateral branches, large, 5-merous. Hypanthium shallowly cup-shaped. Epicalyx 0. Sepals imbricate, large. Petals large, (orange-)yellow (white in a cultivar). Stamens many. Disk not evident. Pistils 5–8; stigma punctiform; ovule 1. Achenes yellow, mesocarp thin, fleshy, endocarp woody. Seed 1, with thin testa, endosperm rather thick. $2n = 18$. One species, *K. japonica* (L.) DC., in China and Japan. In forest and open places.

21. *Neviusia* A. Gray

Neviusia A. Gray, Mem. Am. Acad. Arts II, 6: 374 (1858).

Leaves alternate, with incised margins; stipules small. Inflorescence a short, terminal raceme, flowers on long pedicels, 5-merous, showy although apetalous. Hypanthium saucer- to cup-shaped. Epicalyx 0. Sepals imbricate, large and leafy, green to greenish white, margins incised. Petals small or 0. Stamens many, white. Disk not evident. Pistils 2–5; stigma linear, on one side of stylodium; ovule 1. Achenes with thin fleshy mesocarp, endocarp crustaceous. Seed 1, endosperm very thin. x = ? Two spp., *N. alabamensis* A. Gray in south-eastern states of U.S.A., and *N. cliftonii* Shevock, Ertter & D.W. Taylor, California. On slopes and streambanks, often on limestone, shaded and more open places.

22. *Rhodotypos* Siebold & Zucc.

Rhodotypos Siebold & Zucc., Fl. Japon. 1: 185 (1841).

Leaves opposite (sometimes whorled in 3), with incised margins. Flowers solitary, terminal on short shoots, 4-merous, bisexual, showy. Hypanthium small, saucer-shaped. Sepals imbricate, leafy, rather large, margins incised, lower teeth

larger, directed more outwards and equivalent to or simulating an epicalyx (see below), persistent. Petals large, white. Stamens many, some of them adnate to the disk. Disk intrastaminal, large, covering the pistils, 4-lobed. Pistils usually 4 (2–6); stigmas capitate; ovules 2, collateral. Achenes black, mesocarp slightly fleshy, endocarp woody. Seed 1, with thin testa, endosperm a thin layer. $2n = 18$. One species, *R. scandens* (Thunb.) Makino in Japan and Central China.

The opposite leaves are an exceptional character in the family. A relationship with *Kerria* is assumed, mainly because of the similarity in fruits and leaves. The two outer sepals and sometimes also the inner ones have at one or both sides a large basal tooth, usually separated from other marginal teeth, which was interpreted by Schaeppi (1953) as stipular organs. The so-called disk was ontogenetically studied by Schaeppi (1977); it is not innervated and does not produce nectar. It cannot be a staminodial structure but seems to be homologous with nectariferous, annular, intrastaminal disks, as occur in other genera.

6. TRIBE DRYADEAE (Focke) Hutch. (1964).

Woody. Leaves alternate, simple (exception: *Chamaebatia*); stipules +. Flowers solitary or in fascicles, normally bisexual. Hypanthium open at the top, dry at maturity. Epicalyx 0. Petals + (exception: *Cercocarpus*). Disk 0. Pistils with long, terminal stylodia; ovule 1, basally attached, obturator 0. Achenes with persistent stylodia or beak. x = 9.

23. *Dryas* L.

Dryas L., Sp. Pl.: 501 (1753).

Dwarf shrubs with creeping branches, evergreen. Leaves close together, margin usually incised but entire in some forms. Flowers solitary, axillary, long-pedicelled, bisexual or one of the sexes reduced, showy. Hypanthium cup- to saucer-shaped. Sepals 7–10, imbricate or valvate. Petals usually isomerous with sepals, white, whitish, or yellow. Stamens many. Pistils many, free, on bottom of hypanthium or on low torus; stylodia terminal, long-hairy; stigmas minute, terminal, linear, longitudinally canaliculate. Fruits achenes terminated by persistent and elongate, hairy stylodia. Seed with membranous testa, endosperm a thin layer. $2n = 18$. Two species: *D. octopetala* L., circumpolar-alpine, and *D. drummondii* Rich. in

Northeast Asia and North America. On rocky places, up to high altitudes.

24. *Cowania* D. Don

Cowania D. Don, Trans. Linn. Soc. Lond. 14: 574 (1825).

Shrubs, evergreen, glandular. Leaves small, simple, pinnatipartite with 3–5 lobes to entire; stipules small, adnate to petiole. Flowers solitary, terminal on long and short shoots, sessile, 5-merous, normally bisexual, fragrant. Hypanthium funnel-shaped to obconical or ± campanulate. Sepals 5, slightly imbricate. Petals 5, white or pale yellowish. Stamens many. Pistils 5–12, free, on bottom of hypanthium, shortly stalked; stylodia terminal, long-hairy; stigma linear. Fruits achenes terminated by persistent, long-hairy stylodia. Seed with membranous testa, endosperm thin. $2n = 18$. One species, *C. mexicana* D. Don, in south-western U.S.A. and Mexico. Mountain slopes, often on limestone. Hybridizing with species of *Purshia* (Koehler and Smith 1981) and merged by Henrickson (1986b).

25. *Purshia* Poiret

Purshia DC. ex Poiret in Lam., Enc., Suppl. 4: 623 (1816).

Shrubs with erect or prostrate branches, leaf-shedding, glandular. Leaves small, simple, wedge-shaped, 3(5)-lobed from the top; stipules small, adnate to petiole. Flowers solitary, terminal on short shoots, (shortly) pedicelled, 5-merous, bisexual. Hypanthium obconical. Sepals imbricate. Petals yellowish. Stamens 20–50. Pistil 1 (2, 3, free), on bottom of hypanthium; stylodium terminal; stigma linear. Fruit achene, beaked by persistent stylodium, pericarp fibrous. Seed with dark testa, endosperm thin. $2n = 18$. Two closely related species (or varieties?) in western North America: British Columbia to California and New Mexico. Open mountain slopes and also in forest.

26. *Cercocarpus* Humb., Bonpl. & Kunth   Fig. 111

Cercocarpus Humb., Bonpl. & Kunth, Nova Gen. Sp. 6, ed. fol.: 183, ed. quart.: 232 (1824); F.L. Martin, Brittonia 7: 91–111 (1950), rev.

Small trees or shrubs, evergreen, with long and short shoots. Leaves usually leathery, with entire or incised margins; stipules free from petiole, caducous. Flowers solitary or fasciculate, terminal, 5-merous, bisexual. Hypanthium narrowly tubular

Fig. 111. Rosaceae–Dryadeae. *Cercocarpus traskiae*. **A** Branch with short-shoot. **B** Flowering branch. **C** Fruits. (Sudworth 1967)

(pedicel-like) in lower part, the upper part widened, caducous. Sepals continuous with upper part of hypanthium, valvate. Petals 0. Stamens 15–25(–40). Pistil 1(2), on bottom of hypanthium; stylodium terminal, long-hairy; stigma linear, on the ventral side of the stylodium; ovule with enlarged chalazal part. Fruit achene, stylodium after anthesis growing into a persistent, very long, twisted, plumose 'tail'. Seed 1, testa fibrous, stout, the apex an empty appendage, endosperm 0. $2n = 18$. About 5 species in western North America. In more or less arid habitats, especially montane.

27. *Chamaebatia* Benth.

Chamaebatia Benth., Pl. Hartweg.: 308 (1849).

Shrubs, aromatic-glandular, evergreen. Leaves alternate, 3 times pinnate, very finely compound and fern-like; stipules small, lower part adnate to the petiole, tips free. Inflorescence terminal on leafy twigs, cymose, few-flowered. Flowers 5-merous, bisexual. Hypanthium obconical. Epicalyx 0. Sepals imbricate. Petals white. Stamens many.

Disk not developed. Pistil 1, on bottom of hypanthium; stylodium rather short, constricted and articulate at its base; stigma linear; ovule 1, basal, ascending. Fruits achenes, with coriaceous pericarp. Seed with thin testa, endosperm thin. $2n = 18$. One species, *C. foliolosa* Benth., in California. Mountain slopes and open forest.

7. TRIBE ULMARIEAE Vent.

Herbs. Leaves pinnate; stipules +. Hypanthium open at top. Epicalyx 0. Petals +. Pistils several; stylodia terminal; ovules 2, pendulous. Fruits indehiscent, 1-seeded. x = 7.

28. *Filipendula* Mill.

Filipendula Mill., Gard. Dict., abr. ed. 4 (1754); Schanzer, J. Jap. Bot. 69: 290–319 (1994), rev., without descriptions, with key.

Perennial herbs. Leaves alternate, in rosette and cauline, interruptedly pinnate; stipules rather large, with incised margins. Inflorescence an anthela-shaped compound raceme. Flowers 5(-7)-merous, bisexual, rarely unisexual and plants polygamous. Hypanthium low cupular. Sepals valvate, reflexed and persistent after flowering. Petals white, cream-coloured, pink, or red. Stamens 20–45. Disk 0. Pistils 5–15, rarely fewer, on bottom of hypanthium or on low torus. Fruits achenes with the appearance of follicles but indehiscent, pericarp tough. Seed 1, testa thin, endosperm very scarce. $2n = 14, 28, 42$, aneuploid numbers also reported. According to the recent revision 15 species, in Europe, temperate Asia (also Himalayas and Yunnan) and Eastern North America. In forest and in open (grassy) vegetation, often in damp places.

8. TRIBE SANGUISORBEAE Focke (1888).

Leaves alternate, imparipinnate or 3-foliolate; stipules, at least those of rosette leaves, adnate to petiole. Hypanthium almost closed at the throat, enlarging in fruiting stage and persistent around the achene(s). Pistils few or only 1; stylodia terminal; ovule 1, pendulous. Achenes. Endosperm 0. x = 7, but chromosome number of several genera unknown.

29. *Hagenia* J.F. Gmel.   Fig. 112

Hagenia J.F. Gmelin, Syst. Nat. 2, 1: 613 (1791).
Brayera Kunth ex A. Richard (1822).

Trees up to 20 m, evergreen. Leaves interruptedly imparipinnate; stipules large, adnate to petiole, forming a wing over its entire length. Inflorescences panicles, axillary, pendulous. Flowers 4–5-merous, unisexual and trees monoecious, rarely bisexual; pedicel with 2 large prophylls. Hypanthium in male flowers small, saucer-shaped, in female flowers turbinate, constricted at throat and closed by a lobed disk. Epicalyx segments large, nerved. Sepals smaller than epicalyx, folded back. Petals small to minute, linear. Stamens c. 16–20, in female flowers staminodial and very small. Pistils 2–3, included; stigmas spathulate, fimbriate. Fruits enclosed in hardened hypanthium which is crowned by enlarged epicalyx. $2n = ?$ One species, *H. abyssinica* (Bruce) J.F. Gmel., in mountains of continental Africa, from Central African Republic, Sudan, and Ethiopia to Zimbabwe, up to 3500 m altitude.

30. *Leucosidea* Eckl. & Zeyh.

Leucosidea Ecklon & Zeyher, Enum.: 265 (1836).

Shrubs or small trees, probably evergreen. Leaves densely, rarely interruptedly imparipinnate, leaflets in few pairs, margins incised; stipules forming an amplexicaul sheath, tips free, intrapetiolar. Inflorescences terminal spikes. Flowers 5-merous, bisexual, bracteate. Hypanthium obconic, thick disk at the throat. Epicalyx shorter than sepals. Sepals valvate. Petals green, smaller than sepals, early caducous. Stamens 10. Pistils 2(-4); stylodia terminal, curved, rarely hooked at apex; stigmas linear on curved end. Fruits included in hardened hypanthium, crowned by persistent epicalyx and sepals. $2n = ?$ One species, *L. sericea* Eckl. & Zeyh., in Zimbabwe and South Africa. Montane grassy slopes, riverbanks.

31. *Agrimonia* L.

Agrimonia L., Sp. Pl.: 448 (1753).

Perennial rhizomatous herbs. Leaves interruptedly imparipinnate. Stipules large, free. Inflorescences terminal and axillary, simple, often interrupted racemes. Flowers 5-merous, bisexual; pedicel at base with two lobed prophylls. Hypanthium almost closed at the throat by a dome-shaped disk, the outside armed with hooked spines. Epicalyx 0. Sepals imbricate, after anthesis closing over the fruit. Petals yellow, rarely white. Stamens 5–10(-20). Pistils usually 2, free. Achene usually one per flower developing, enclosed in armed

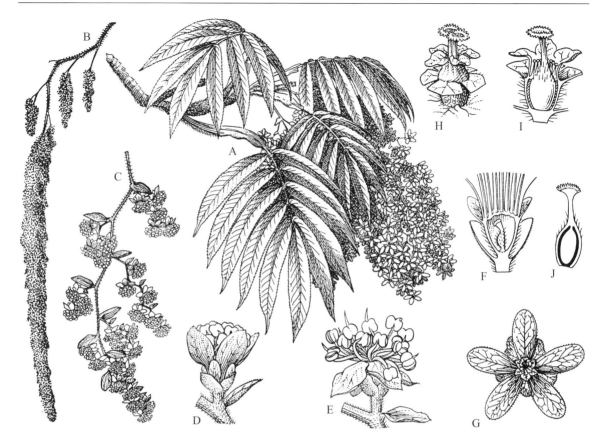

Fig. 112. Rosaceae–Sanguisorbeae. *Hagenia abessynica*. **A** Branch with female inflorescence. **B** Male inflorescence. **C** Part of the same. **D** Male flower, partly open. **E** Male flower. **F** Part of male flower halved to show pistillode. **G** Female flower, seen from above. **H** Female flower, lateral view, with outer calyx lobes removed. **I** Same, with hypanthium opened to show staminodes and pistil. **J** Pistil. (Graham 1960)

hypanthium, pericarp thin, coriaceous. Seed with membranous testa. $2n = 28, 42$ (in hybrid), 56. Between 10 and 15 species, widespread in Europe and Asia southward to Sri Lanka and Java, few species in North America and Mexico, 1 species in Haiti, one or few species in Brazil (Skalicky 1971; Fuks 1987). Some species introduced and sometimes naturalized outside the native area, e.g. in South Africa. Temperate, in tropical regions montane. Mostly in open vegetation, some species weedy.

32. *Aremonia* Nestl.

Aremonia Necker ex Nestler, Monogr. Potentilla: IV, 17 (1816), nom. cons.

Perennial, rosette-forming herbs. Leaves interruptedly imparipinnate; stipules adnate to the petiole but tips free, triangular. Flowers solitary or few fascicled in the axils of reduced leaves in apical part of flowering stems, 5-merous, bisexual. Hypanthium surrounded by, but free from a 6–12-lobed involucre, almost closed at the throat by the intrastaminal disk. Epicalyx much smaller than sepals, almost rudimentary. Sepals imbricate. Petals yellow. Stamens 5–10. Pistils 2; stigmas reniform. Fruits achenes with thin, hard pericarp, often only 1 per flower developed, enclosed in hardened but thin hypanthium, involucre enlarged in fruiting stage. $2n = 42$. One species, *A. agrimonoides* (L.) DC., in southeast Europe and west Asia. In forest, in montane and subalpine altitudes.

After flowering, the axis between the involucre and the fruit-bearing hypanthium develops into an elaiosome (Gams 1923) and the fruit (hypanthium plus achene) is dispersed by ants (Ridley 1930: 521).

33. *Spenceria* Trimen

Spenceria Trimen, J. Bot. 17: 97, t. 201 (1879).

Perennial rhizomatous herbs. Leaves (interruptedly) imparipinnate, most in basal rosette, few reduced ones cauline; stipules forming narrow wings. Flowering axis erect, inflorescence a narrow, simple or compound raceme. Flowers rather large, 5-merous, bisexual, bracteate. Involucre under the flower 7–9-lobed. Hypanthium free from involucre, obconic. Epicalyx leaves with rounded apex, much smaller than pointed sepals. Petals large, yellow. Stamens 35–40. Intrastaminal disk rim-like, with entire margin, closely surrounding the stylodia. Pistils 2, included; stigmas punctiform. Fruits achenes, usually one developed, included in the persistent, unarmed hypanthium. $2n = ?$ Possibly only one species, *S. ramalana* Trim., in China.

34. *Sanguisorba* L.

Sanguisorba L., Sp. Pl.: 116 (1753); Nordborg, Opera Bot. 11, 2: 1–103 (1966), partial rev.; Nordborg, Opera Bot. 16: 1–166 (1967), rev. of sect. *Poterium*.
Poterium L. (1753).
Poteridium Spach (1846).

Perennial, rhizomatous herbs. Leaves in radical rosettes and fewer cauline, imparipinnate; stipules of rosette leaves membranous, those of cauline leaves free. Inflorescences terminal on erect, branched stems, dense spikes or heads. Flowers bi- or unisexual, the plants normally monoecious. Hypanthium constricted and almost closed at apex, 4-angled, unarmed. Epicalyx 0. Sepals 4, petaloid, white, green, or reddish, caducous. Petals 0. Stamens 2–4 or up to 30 or more. Intrastaminal disk 0 but bases of sepals rather thickened. Pistils 1 or 2(–5), free; stigmas penicellate. Achene(s) remaining included in the hardened, 4-angled to 4-winged, dry hypanthium. $2n = 14, 28, 56$. About 15 species, if species limits are not drawn too narrowly. Most species in Eurasia and North Africa, 2 in North America. In grassy vegetation, shrubland, forest-edges, sometimes under marshy to damp circumstances.

Most of the species seem to be wind-pollinated, but insect pollination also occurs in some of the species (see Saïd 1979).

35. *Sarcopoterium* Spach

Sarcopoterium Spach, Ann. Sci. Nat. Bot. III, 5: 43 (1846); Nordborg, Opera Bot. 11, 2: 73–76 (1966).

Small shrubs, often cushionplants, branches terminating as thorns. Leaves small, imparipinnate, coriaceous; stipules small, tooth-like. Inflorescences spikes, opposite the leaves. Flowers unisexual, plants monoecious, female flowers at the end of the spike, greenish; floral bract and 2 prophylls

fimbriate. Hypanthium in male flowers minute, in female flowers large, constricted at apex, unarmed. Epicalyx 0. Sepals 4, caducous. Petals 0. Stamens many in male flowers, 0 in female ones. Intrastaminal disk not evident. Pistils 2 or 3, free; stigmas penicellate. Achene(s) remaining enclosed in smooth, slightly fleshy, red-coloured hypanthium. $2n = 28$. One species, *S. spinosum* (L.) Spach, East Mediterranean area. In xerophytic shrubland. Dispersal, however, is often hydrochorous, see Litav and Orshan (1971).

36. *Bencomia* Webb & Berthel.

Bencomia Webb & Berthelot, Hist. Nat. Iles Canaries 3: 10 (1842).
Marcetella Sventenius (1948).
Dendriopoterium Sventenius (1948).

Candelabra shrubs, leaf-shedding. Leaves in rosettes at the end of branches, imparipinnate; stipules in lower part adnate to petiole, upper part free. Inflorescences axillary spikes, or racemes of spikes. Flowers unisexual, plants dioecious, incompletely dioecious, or monoecious with both sexes in one inflorescence; floral bract and prohylls connate. Hypanthium in male flowers minute, in female flowers large, constricted and almost closed at apex, unarmed. Epicalyx 0. Sepals 3–5. Petals 0. Stamens many in male flowers, 0 in female flowers. Disk a thick collar around the orifice of the hypanthium. Pistils 2(–4), entirely enclosed in hypanthium; stigmas penicilliform. Achene(s) remaining enclosed, fruiting hypanthium globose or pear-shaped, ± 4-angled, or with 2 wings. $2n = 28$. About six species restricted to the Canary Islands, and one endemic to Madeira. In rocky places, montane.

Recognized as 3 separate genera by D. & Z.I. Bramwell (1974) and D. Bramwell (1978). In Vieira (1992) and Press and Short (1994), *Marcetella* is kept separate from *Bencomia*. Nordborg (1966) included *Dendriopoterium* and *Marcetella* in *Sanguisorba* and kept *Bencomia* separate.

37. *Cliffortia* L.

Cliffortia L., Sp. Pl.: 1038 (1753); Weimarck, Monogr. Genus *Cliffortia*: 1–229 (1934); Bot. Not. 1948: 167–203 (1948), survey.

Shrubs, erect or procumbent, or shrublets, rarely treelets, often ericoid, evergreen. Leaves trifoliolate, bifoliolate, or simple, usually strongly xeromorphic, blades of leaves/leaflets linear or wider; sheath amplexicaul, distinct or short, stipules on the sheath, rarely 0, usually scarious, rarely green, free. Flowers axillary, usually solitary, rarely in small fascicles, rarely female ones in cone-like inflorescences, inconspicuous, wind-pollinated, unisexual, exceptionally bisexual, plants dioecious or monoecious. Hypanthium 0 in male flowers, ellipsoid in female flowers, ribbed or smooth outside, closed at mouth. Epicalyx 0. Sepals 3(4), caducous or persistent, often coherent. Petals 0. Stamens 3–20(–30). Pistils 1 or 2, enclosed; stigma feathery. Fruits achenes, enclosed in bony to slightly fleshy, ribbed or winged hypanthium. $2n = ?$ At least 100 species in South Africa, richest in the West Cape Province. Only one species extending to Angola and Kenya. Often xerophytic and in dry places but some species preferring moist habitats.

The striking parallelism of the leaves with those of *Aspalathus* (Papilionaceae) in the same region was discussed by Dahlgren (1971).

38. *Polylepis* Ruiz & Pavon   Fig. 113

Polylepis Ruiz & Pavon, Fl. Peruv. Prodr.: 80 (1794); B.B.
   Simpson, Smithson. Contrib. Bot. 43: 1–62 (1979), monogr.;
   Kessler, Candollea 50: 131–171 (1995), rev. for Bolivia, 9 spp.;
   Romoleroux in Fl. Ecuador 56: 71–89 (1996), 7 spp.

Trees up to c. 25 m, or shrubs. Leaves imparipinnate or trifoliolate; stipules forming a closed sheath. Inflorescences racemes, pendent or erect. Flowers bisexual, bracteate. Hypanthium urceolate, constricted at throat and intrastaminal disk almost closing the orifice. Epicalyx 0. Sepals 3 or 4. Petals 0. Stamens 6 to many, anthers hairy. Pistil 1(–3), included; stylodium short or long; stigma peltate, fimbriate or penicellate. Achene(s) included in and closely accumbent to the hard and tough hypanthium with irregular spine- or knob-like protuberances or ridges, the latter sometimes expanded into wings. $2n = ?$ About 20 species in the Andean region of South America, from Venezuela to N Chile and N Argentina. In montane and subalpine habitats, often up to the tree line, to 5000 m altitude.

39. *Margyricarpus* Ruiz & Pavon

Margyricarpus Ruiz & Pavon, Fl. Peruv. Prodr.: 7, t. 33 (1794).
Tetraglochin Poepp., Fragm. Syn. Phan.: 26 (1833); Rothmaler,
   Darwiniana 3: 429–437 (1939), rev.

Shrubs, erect or creeping, possibly evergreen. Leaves imparipinnate, sometimes on short shoots 1-foliolate, petiole and rachis of long-shoot leaves in most species persistent as hard spines, leaflets

Fig. 113. Rosaceae–Sanguisorbae. *Polylepis quadrijuga.* **A** Flowering twig. **B** Flower with penicillate stigma. **C** Flower in the male phase. **D** Fruit. (Simpson 1979)

linear; stipules adnate to the petiole over their entire length, forming an amplexicaul sheath. Flowers solitary, axillary, bisexual. Hypanthium urceolate, constricted at apex by a disk, mostly 4-angled. Epicalyx 0. Sepals (3)4(5). Petals 0. Stamens (1)2(3). Pistil 1, included; stigma penicelliform. Fruit achene, enclosed in hypanthium, the latter in fruiting stage either dry, hard to leathery, with 3–5 distinct wings, or swollen into a fleshy pseudoberry without wings. $2n = ?$ About 8 species but badly in need of a critical revision. In the Andean region of South America from Ecuador to Chile and Argentina, and also in South Brazil and Uruguay.

Margyricarpus and *Tetraglochin* have often, e.g. by Rothmaler, been kept separate on the strength of the fleshiness of the fruiting hypanthium in probably only one species, *M. pinnatus* (Lam.) Kuntze. *Tetraglochin* was described as being dioecious and it is certainly not impossible that unisexual flowers are sometimes present.

The intergeneric hybrid × *Margyracaena skottsbergii* Bitter was described from the Juan Fernandez Archipelago. Using RAPD markers, Crawford et al. (1993) produced evidence that it is a hybrid between the indigenous *Margyricarpus digynus* (Bitter) Skottsb. and the introduced, weedy *Acaena argentea* Ruiz & Pavon.

40. *Acaena* L.   Fig. 114

Acaena Mutis ex L., Mant. 2: 145, 200 (1771); Bitter, Bibl. Bot. 74: 1–336, 37 pl. (1911), monogr.
Ancistrum J.R. & G. Forst. (1776).

Perennial herbs, usually woody at base, with creeping stems and erect flowering shoots. Leaves imparipinnate; stipules adnate to petiole, with small free tips. Inflorescences terminal or axillary, in heads or interrupted spikes. Flowers small, normally bisexual. Hypanthium narrowed at the throat. Epicalyx 0. Sepals (3)4(–7), valvate, caducous or persistent. Petals 0. Stamens often 4 (1–10). Disk 0. Pistil usually 1(2–4), free; stylodium terminal; stigma usually plumose. Achene(s) in hardened hypanthium, the latter with 2–4 barbed spines under the sepals, or with many spines scattered over the surface. $2n = 42$, c. 72, 84, 126. About 40 species when a not too narrow species concept is followed. Southern hemisphere, richest in extratropical South America, extending to Mexico and California (1 species), few in South Africa, several in Australia and New Zealand extending to Hawai'i (1 species) and New Guinea (1 species), some in subantarctic island groups. Bitter (l.c.) mentioned c. 110 species. In forests and open vegetation from the lowland up to 4500 m altitude.

Possibly two subgenera may be distinguished, on the basis of the number of spines on the fruiting hypanthium, 2–4 in one subgenus, many in the other. Bitter, however, was of the opinion that reductions in the number of spines have occurred repeatedly, in connection with greater compactness of the inflorescence.

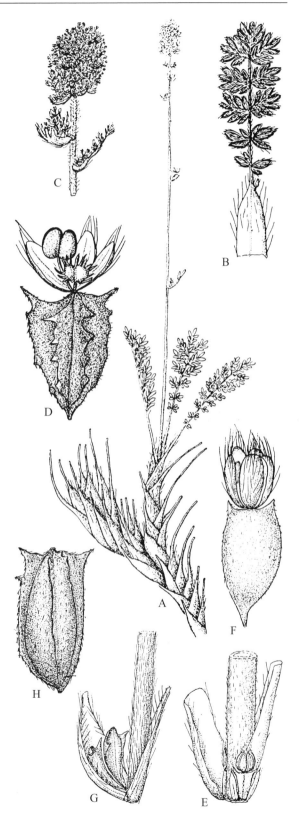

Fig. 114. Rosaceae–Sanguisorbeae. *Acaena poeppigiana.* **A** Flowering plant. **B** Leaf, the sheath with a rudimentary stipule. **C** Apex of inflorescence. **D** Chasmogamic flower from inflorescence. **E** Cleistogamic flowers hidden among the sheaths of rosette leaves. **F** Cleistogamic flower. **G** Leaf sheaths with fruits from cleistogamic flowers. **H** Fruit from cleistogamic flower. (Grandoso 1964)

9. TRIBE POTENTILLEAE Focke (1888).

Unarmed. Leaves alternate, compound. Stipules adnate to petiole. Flowers normally 5-merous, almost always bisexual. Hypanthium open at top, dry at maturity. Epicalyx + (exception: *Chamaerhodos*). Petals +. Carpels several to many, in some species in reduced number; stylodia not terminal; ovules 1 per carpel. Achenes. Endosperm 0. x = 7.

41. *Chamaerhodos* Bunge

Chamaerhodos Bunge in Ledeb., Fl. Alt. 1: 429 (1829).

Perennial herbs, more or less woody at base. Leaves in dense rosette-cushions and few on flowering stems, 1–3 times tripartite, finely dissected; stipules in rosette-leaves adnate to base of petiole as a membranous widened part, in the ± sessile cauline leaves practically 0. Inflorescences terminal, composed of cymes. Flowers 5-merous, bisexual. Hypanthium campanulate, its lower part surrounding the ovaries, the upper part ± tubular. Epicalyx 0. Sepals valvate. Petals white or purplish. Stamens 5, antepetalous. Disk not evident. Pistils 5–10 (+), on bottom of hypanthium; stylodia basal, articulated. Fruits dry achenes. Seed with firm testa. $2n = 14$. About five species in Central and East Asia, one also in the western part of North America (U.S.A. incl. Alaska, Canada). In open vegetation on mountain slopes, also in forest.

42. *Potentilla* L.

Potentilla L., Sp. Pl.: 495 (1753).
Comarum L. (1753).
Sibbaldia L. (1753).
Duchesnea Sm. (1811).
Horkelia Cham. & Schldl. (1827).
Ivesia Torr. & A. Gray (1858).
Comarella Rydb. (1896).
Stellariopsis (Baill.) Rydb. (1898).
Purpusia Brandegee (1899).
Sibbaldiopsis Rydb. (1901).
Drymocallis Fourr. ex Rydb. (1908).
Horkeliella (Rydb.) Rydb. (1908).

Perennial herbs with a rhizome or caudex, sometimes forming cushions, sometimes with stolons (shrubs, annual herbs). Leaves all radical or also alternate on longer stems, (interruptedly) pinnate, palmate, or tri(uni)-foliolate; stipules in their lower parts adnate to the petiole, membranous or the upper free parts leafy and green. Inflorescences cymose or thyrsoid, terminal or axillary, or flowers solitary and axillary or opposite the leaves. Flowers (4)5(6)-merous, bisexual. Hypanthium saucer-shaped, cup-shaped, hemispheric or campanulate, often with hairs on the inside. Epicalyx as large as or smaller than sepals. Sepals valvate, usually entire. Petals yellow or white, rarely red or purple. Stamens 10 to many. Disk mostly not evident, sometimes + as a rim on which the stamens are inserted. Pistils many–few, rarely 1, on the bottom of the hypanthium or on a low to elevated, usually hairy torus; stylodia subapical, ventral or subbasal, sometimes partly thickened, persistent or articulated and caducous. Achenes usually remaining enclosed by the persistent hypanthium with epicalyx and sepals closing above them, torus dry, rarely after anthesis enlarging and becoming spongy to somewhat fleshy. Seed with thin testa. $2n = 14$ or a multiple, recorded up to $2n = 112$, aneuploidy obviously rare. Probably more than 300 species in the northern temperate zone, and with few species extending to northern and eastern Africa and Malesia (New Guinea). Mostly in open habitats, many species montane to alpine. In need of a comprehensive, genus-wide study.

Molecular studies (Eriksson et al. 1998) raise doubt on the monophyly of the genus. Some Eurasian species complexes (*Potentilla arguta*, *P. argentea* a.o.) have a facultative apomictic reproduction mode, implying the usual taxonomic difficulties.

43. *Fragaria* L.

Fragaria L., Sp. Pl.: 494 (1753).

Perennial rhizomatous herbs, often stoloniferous. Leaves radical, trifoliolate, only in anomalous cases with 5 leaflets; stipules adnate to petiole. Inflorescences terminal on leafless erect stems, cymose, few-flowered (except in cultivars). Flowers 5-merous, bisexual or unisexual and then plants (polygamo-)dioecious. Hypanthium obconic to saucer-shaped. Epicalyx of ± same size as sepals. Sepals valvate. Petals white, rarely yellowish. Stamens c. 20–numerous. Disk sometimes visible as a rim on which the stamens are inserted. Pistils many, free, on cylindrical torus; stylodia lateral to subbasal, short; stigma capitellate; ovule inserted in the middle, descendent, epitropous. Achenes dry, on the much-enlarged, fleshy to juicy torus, forming an oblong, ovoid, or globose fruit. $2n = 14$ or a multiple, reported up to $2n = 56$. About 10 species, in temperate Eurasia and North America, one species, *F. chiloensis* (L.) Mill., also in South America (Chile). In open vegetation and in forest.

Kept separate from *Potentilla*, although the case for reduction is almost as strong as in *Duchesnea* and *Comarum*, which also have a torus enlarging after anthesis, although there becoming more spongy than fleshy.

Fragaria × *ananassa* (Duchesne) Guédès, the Garden strawberry, widely cultivated as table fruit and for preserves, originated (maybe more than once) in French gardens as an octaploid hybrid between the North American *F. virginiana* Mill. and *F. chiloensis*, both octaploid species.

10. GEUM GROUP

Perennial herbs with caudex, rarely rhizome, unarmed. Leaves imparipinnate, in rosette and few often reduced ones cauline, alternate. Stipules adnate to petiole, at least in rosette leaves. Flowers 5-merous, bisexual, in scapose, mostly few-flowered cymes. Hypanthium open at top, dry at maturity. Epicalyx + (one exception in *Geum*). Petals mostly yellow. Stamens many. Ovule 1, basal, ascending. Achenes. Endosperm 0 (not known in *Coluria*). x = 7 (but see *Orthurus*).

The characters of three genera (nos. 50–52), of uncertain position but probably related to the present group, have not been taken into account in the above summary of characters common to all genera. See p. 368.

Lack of material precludes a definite decision on the status of *Novosieversia*, *Taihangia*, and *Woronowia*.

44. *Geum* L.

Geum L., Sp. Pl.: 500 (1753); F. Bolle, Fedde Rep., Beih. 72: 1–119 (1933), monogr. of *Geum* in a slightly wider sense than accepted here.

Perennial. Rosette leaves (interruptedly) imparipinnate, end-lobe (and upper lateral lobes) larger, cauline leaves less divided. Inflorescence terminal on erect leafy stems in the axils of rosette leaves, cymose, mostly few-flowered. Hypanthium saucer-shaped to cup-shaped or obconical. Epicalyx + (lacking in *G. vernum* (Raf.) Torr. & A. Gray), lobes smaller than sepals. Sepals valvate or slightly imbricate. Petals mostly yellow, also white, pink, red, or purplish. Stamens 20–many. Intrastaminal disk sometimes + but never prominent. Pistils several to many, on the bottom of the hypanthium or on a conoid to cylindric torus, lower part of torus in some species developed as a gynophore; stylodia terminal, not articulate with ovary, lower

part elongating after anthesis, persistent and terminating in a hook, upper part caducous; stigmas punctiform. Achenes tipped by long persistent stylodia. Seeds with membranous testa. Most species hexaploid, 2n = 42, but 2n = 28, 56, 70 and 84 are also reported. About 30 species, widely distributed in Europe (including the Mediterranean), Asia, and North and South America, one species, *G. capense* Thunb., in South Africa. In open vegetation and forests, some weedy.

45. *Parageum* Nakai & H. Hara

Parageum Nakai & H. Hara, Bot. Mag. Tokyo 49: 124 (1935); Král, Preslia 38: 151 (1966), emend.
? *Taihangia* Te T. Yü & C.L. Li (1980), see Note below.
Sieversia auctt., pro parte, not incl. type.
Acomastylis auctt., pro parte, not incl. type.

Perennial, rarely stoloniferous. Rosette leaves imparipinnate, lyrate, with terminal leaflets much larger than lateral ones, cauline leaves few, reduced. Inflorescence terminal on erect or ascending leafy stems in axils of rosette leaves, cymose, few(–1)-flowered. Hypanthium obconoid. Epicalyx lobes smaller than sepals. Sepals valvate. Petals yellow. Stamens many. Intrastaminal disk low. Pistils many, on bottom of hypanthium or on low torus; stylodia terminal, not articulate with ovary, variously hairy, upper part not caducous; stigma linear to punctiform. Achenes crowned by elongated stylodia. Seeds with thin testa. 2n = 28, 42. About 6 species, Eurasia and North America, in the montane and subalpine belt.

The Chinese *Taihangia rupestris* Yü & Li has unisexual and bisexual flowers. It is possibly well accommodated in this genus, but no material was seen.

46. *Orthurus* Juz.

Orthurus Juz., Fl. USSR 10: 616 (1941, orig. ed.), 458 (1971, Engl. transl.).
Geum L. subg. *Orthostylus* (Fisch. & C.A. Mey.) F. Bolle, Fedde Rep., Beih. 72: 43–46 (1933), monogr.
? *Woronowia* Juz. (1941), see Note below.

Perennial. Rosette leaves interruptedly imparipinnate, lyrate, cauline leaves less divided but still large. Flowering stems in the axils of rosette leaves, laxly branched in upper part and forming a leafy inflorescence. Hypanthium narrowly obconical. Epicalyx lobes not very different from sepals. Sepals valvate. Petals yellowish. Stamens many. Intrastaminal disk not evident. Pistils rather few, up to c. 15, most in cluster on top of gynophore but always 1 or 2 lower on gynophore and/or on bottom of hypanthium; stylodia terminal, lower part persistent after anthesis, top with stiff retrorse hairs, upper part shorter, caducous; stigmas short-linear. Achenes tipped by persistent parts of stylodia, on lengthening gynophore. Seeds with firm testa. $2n = 28, 30$ (see below). One species, *O. heterocarpus* (Boiss.) Juz. in France, the Mediterranean, West and Central Asia. Montane, in shaded places.

Woronowia speciosa (Albov) Juz., based on *Geum speciosum* Albov (not seen) from the Caucasus Mts., also has a long gynophore with the pistils partly at its base but, according to the descriptions, the stylodia are not retrorsely barbate. Still, it may belong in this genus.

47. *Oncostylus* (Schltdl.) F. Bolle

Oncostylus (Schltdl.) F. Bolle, Fedde Rep., Beih. 72: 27–32 (1933), monogr.
Geum L. subg. *Oncostylus* Schltdl., Linnaea 28: 465 (1856), basionym.

Perennial. Rosette leaves imparipinnate, lyrate with terminal leaflet much larger than lateral ones, cauline leaves reduced. Flowering stems in axils of rosette leaves, stiffly branched in upper part and forming a bracteate, few-flowered inflorescence, or stems not branched and scapose with one flower. Hypanthium obconical. Epicalyx lobes narrower and usually shorter than sepals. Sepals valvate. Petals yellow or white. Stamens 10–many. Disk 0. Pistils many, on bottom of hypanthium or on cylindrical torus; stylodia terminal, more or less glabrous, after anthesis hooked at apex; stigmas punctiform. Achenes crowned by hooked stylodia. Seeds with thin testa. $2n = ?$ About 5 species in New Zealand incl. Auckland Islands, one in Tasmania, one or two in southern South America. In grassland and on rocky places, at low to montane altitudes.

48. *Acomastylis* Greene

Acomastylis Greene, Leaflets Bot. Obs. 1: 174 (1906); F. Bolle, Fedde Rep., Beih. 72: 78–90 (1933), monogr. of the genus in a wider sense than here.
Erythrocoma Greene (1906).
? *Novosieversia* F. Bolle (1933), see Note below.
Sieversia auctt., pro parte, not incl. type.

Perennial. Rosette leaves imparipinnate, often interruptedly, terminal leaflet not strikingly larger than lateral ones. Flowering stems in axils of rosette leaves. Inflorescence terminal, cymose, few- or 1-flowered. Hypanthium campanulate to obconical. Epicalyx lobes smaller than sepals, sometimes incised from apex. Sepals valvate. Petals yellow(ish). Stamens many. Intrastaminal disk evident or not. Pistils many, on bottom of hypanthium or on a cylindric torus; stylodia terminal, not articulate with ovary nor elsewhere, not elongating after anthesis, glabrous or hairy; stigmas punctiform. Seeds with firm testa. $2n = 28$, 70. In this restricted circumscription about twelve species, most of them in North America, one also in Northeast Asia. Two species are known from the Himalayas, but further research may reveal that these are better placed in *Parageum*.

Bolle (l.c.) recognized two subgenera, subg. *Megacomastylis*, part of which is here placed in *Parageum* (subg. *Parageum* sensu Král). The other part, together with the second subgenus, subg. *Micracomastylis*, here united with *Erythrocoma*, is placed here in *Acomastylis*.

The genus *Novosieversia*, with its only species *N. glacialis* (Adams) F. Bolle, occurs in Northeast Asia and is probably congeneric with the present one.

49. *Coluria* R. Br.

Coluria R. Br., J. Voy. N.-W. Pass., Bot.: 276 (1824); F. Bolle, Fedde Rep., Beih. 72: 90–93 (1933), monogr.

Perennial. Rosette leaves interruptedly imparipinnate to pinnatisect, end-lobe and upper lateral lobes distinctly larger than lower ones or all (primary) leaflets ± equal in size. Inflorescence terminal on axillary branches, cymose, mostly few-flowered. Hypanthium obconical. Epicalyx lobes smaller than sepals. Sepals valvate. Petals yellow (white in one species?). Stamens c. 15 to many, on often distinct rim (disk). Pistils (3–)12–25, on bottom of hypanthium or on short torus; stylodia (sub)terminal, articulate with ovary and caducous. Seeds not seen. $2n = 14$. About 4 species in Siberia and China. Mountain slopes at high altitudes.

Doubtfully belonging to the Geum group

The three genera placed here have in common that they seem to be related to the Geum group but differ from it in having distinctly to slightly woody stems. The differences between them are such that they do not form a separate group.

50. *Sieversia* Willd.

Sieversia Willd., Ges. Naturf. Fr. Berlin Mag. Neuesten Entd. 5: 397 (1811).

Small, evergreen shrubs, branches mostly prostrate, unarmed. Leaves in rosettes at the tops of the branches, pinnate; lower part of stipules adnate to the petiole, tips free, long and thin. Flowering stems scapose, not branched. Flowers solitary, 5-merous, bisexual. Hypanthium obconical. Epicalyx segments about as long as sepals but narrower, sometimes 2-topped. Sepals valvate. Petals white. Stamens very many. Intrastaminal disk 0. Pistils many (40 or more), free, on bottom of hypanthium, in fruit on a short gynophore; stylodia terminal, not articulate, straight, hairy except tip; stigmas punctiform; ovule 1, basal, ascending. Fruits achenes, crowned by persistent and elongate hairy stylodia. Seeds with membranous testa, endosperm 0. $2n = 14$. One species, *S. pentapetala* (L.) Greene, in Northeast continental Asia, Japan and the islands north of it. Often in damp situations.

Sieversia has in the past been considered as a much larger genus, including also herbaceous species now placed in *Parageum* and *Acomastylis*.

51. *Fallugia* Endl.

Fallugia Endl., Gen.: 1246 (1840).

Small shrubs, more or less evergreen, unarmed. Leaves alternate, deeply 3–7-partite, rarely simple, wedge-shaped in general outline, small; stipules at base shortly adnate to petiole, largest part free, basal part of petiole together with stipules remaining on the twig. Flowers solitary, terminal on the long shoots (growth sympodial), 5-merous, normally bisexual, under the terminal flower often two lateral ones. Hypanthium cup-shaped. Epicalyx segments small. Sepals imbricate. Petals white. Stamens many, inserted on a low ring. Intrastaminal disk not developed. Pistils many, on conoid torus, shortly stalked; stylodia terminal, not articulate, hairy; stigmas minute, terminal, longitudinally canaliculate; ovule 1 (see Note below), basally inserted. Fruits achenes, crowned by the persistent and elongated, hairy stylodia. Seed with membranous testa, endosperm 0. $2n = 28$. One species, *Fallugia paradoxa* (D. Don) Torr., in North America from Utah to Texas and Mexico. In montane shrubland.

In habit this genus reminds of *Cowania* in the Dryadeae, but it differs conspicuously by the presence of an epicalyx. According to Juel (1918: p. 61, figs. 115–119) there are two ovules, one long-stalked and mostly not fertile, one sessile and fertile. I have investigated several specimens but have not found this situation.

52. *Waldsteinia* Willd.

Waldsteinia Willd., Ges. Naturf. Fr. Berlin Neue Schr. 2: 105 (1799); F. Bolle, Fedde Rep., Beih. 72: 93–96 (1933), monogr.

Perennial herbs, unarmed, stems firm, mostly creeping. Leaves in rosette at the top of the stem, simple and 3–5(–7)-lobed or 3-foliolate; stipules of radical leaves completely adnate to the long petiole, forming a membranous wing. Inflorescence terminal on axillary stems, cymose, cauline leaves reduced, stipulate; bracts in inflorescence mostly small. Flowers 5-merous, bisexual. Hypanthium funnel-shaped to obconical. Epicalyx small or 0. Sepals valvate. Petals yellow. Stamens many. Intrastaminal disk +, thin. Pistils (1)2–6(–15), free, on bottom of hypanthium or on short torus; stylodia terminal, constricted and articulate at base, deciduous; stigmas punctiform; ovule 1, basally inserted. Fruits achenes. Seeds with membranous testa, closely adhering to pericarp, endosperm 0. $2n = 14, 21, 28, 35, 42$. About 5 species, 2 or 3 in North America, 1 in Southeast Europe, 1 (*W. ternata* (Steph.) Fritsch) disjunct from Alps to Korea. Montane, in heath- and shrubland and other sunny places, also in forest.

11. TRIBE RUBEAE (Focke) Hutch. (1964).

Stipules free. Epicalyx 0. Pistils many, free, on elevated torus; ovules 2, pendulous, one developing. Drupes. x = 7.

53. *Rubus* L.

Rubus L., Sp. Pl.: 492 (1753); Focke, Bibl. Bot. 72: 1–223 (1910–1911); ibid. 83: 1–274 (1914), monogr.

Shrubs (i.e. determinate flowering shoots developing on indeterminate woody stems of 1 or more years old), usually climbing, straggling or creeping, rarely erect, only few species herbaceous, perennial. Twigs and other parts nearly always with prickles. Leaves pinnately, palmately, or pedately compound or simple, when simple usually lobed; stipules free, on the base of the petiole or at the junction of twig and petiole, persistent or fugacious, rarely 0. Inflorescences terminal, determinate, elaborately branched, or little or not branched and in axillary bundles, rarely strongly reduced and flowers (sub)solitary. Flowers usually 5-merous, mostly bisexual, rarely unisexual and the plants dioecious. Hypanthium saucer- to cup-shaped. Epicalyx 0. Sepals imbricate, often unequal, exposed margins often lobed.

Petals white, cream, pink, purplish or red, in few species partly or entirely 0. Stamens many, rarely few. Disk 0. Pistils many, rarely few, free, on a mostly elevated torus; stylodia (sub)terminal; stigmas capitate or bifid; ovules 2, inserted almost at same height but situated one above the other, usually only the lower one developing, descendent, epitropous. Fruits drupes, exocarp black, red, yellow, orange or white, mesocarp juicy or fleshy, endocarp hard, drupelets cohering and falling as a collective fruit without or together with the dry torus, rarely separating individually. Seed with thin testa, endosperm scarce. $2n = 14$ and multiples, recorded up to $2n = 98$. Apart from several thousands of 'microspecies' in subg. *Rubus* (see below), possibly about 250 species. Distribution almost cosmopolitan in temperate climates. Several species have (intentionally or accidentally) been introduced to places outside their natural area; sometimes they have naturalized completely and some have become noxious weeds. Originally in natural open habitats but also in forest, now in many anthropogenic vegetations like hedges, shrubberies, etc.

The revision by Focke (l.c.) is now outdated and especially the infrageneric subdivision should be studied anew. There are three large subgenera: *Rubus*, *Malachobatus* (Focke) Focke, and *Idaeobatus* (Focke) Focke. Several monotypic to species-poor subgenera have also been recognized, e.g. *Anoplobatus* Focke, *Chamaebatus* (Focke) Focke, *Chamaemorus* Focke, *Cylactis* (Raf.) Focke, *Dalibarda* (L.) Focke, of which the taxonomic and phylogenetic status has to be reviewed. Sometimes these subgenera were recognized as separate genera.

Subgenus *Rubus* is taxonomically the most problematic one. It comprises a group, variously called 'Eubatus', the 'Moriferi', or the '*Rubus fruticosus/caesius* complex', that consists of sexually reproducing, self-sterile, diploid species and a larger number of facultatively apomictic polyploids of hybrid origin. The complex is represented in Europe and North America and it has been studied most extensively in Europe where c. 2000 species have been described. See Nybom (1988) for a recent discussion on apomixis versus sexuality in the subgenus. Apomixis seems to be absent or very rare in the other subgenera.

The fruits are edible. Blackberries, dewberries and loganberries (selections from species and crosses in subg. *Rubus*), and raspberries (cultivars of *Rubus idaeus* L. and crosses with other species in subg. *Idaeobatus*) are commercially cultivated in many places. Other species are only harvested from the wild (e.g. cloudberry, *Rubus chamaemorus* L. of subg. *Chamaemorus*) or grown on a small and often non-commercial scale (e.g. the wineberry, *Rubus phoenicolasius* Maxim. of subg. *Idaeobatus*).

Nitrogen fixation in root nodules has been established in *Rubus ellipticus* J.E. Sm. from Java, see Becking (1979), so *Frankia* symbiosis in Rosaceae seems not be to be restricted to the Dryadeae.

12. TRIBE ROSEAE Focke (1888).

Woody. Leaves once pinnate (exception: *Rosa persica*); stipules adnate to petiole (exception: *Rosa persica*). Epicalyx 0. Stamens many. Pistils many, entirely enclosed in hypanthium, fleshy after anthesis; ovule 1, pendulous. Achenes. x = 7.

54. *Rosa* L.

Rosa L., Sp. Pl.: 491 (1753).
Hulthemia Dumort. (1824).
× *Hulthemosa* Juz. (1941).

Shrubs, erect, climbing, straggling, or prostrate, evergreen or leaf-shedding, often glandular-hairy, branches and petioles nearly always armed with straight or curved prickles and often also with thinner, straight bristles, often paired prickles under the leaves. Leaves alternate, imparipinnate, leaflets penninerved, in subg. *Hulthemia* Dumort. leaves reduced to one leaflet and without stipules. Stipules in lower part adnate to the petiole, upper parts free. Inflorescence a terminal thyrsus or raceme, or flowers solitary and terminal. Flowers (4)5-merous, bisexual, large and showy, often fragrant, rarely functionally unisexual and plants cryptically dioecious (see Kevan et al., Biol. J. Linn. Soc. 40: 229–243, 1990). Hypanthium globular to urceolate, throat usually almost closed by a thickened annular disc. Epicalyx 0. Sepals imbricate, the exposed margins usually pinnately incised, persistent or caducous after flowering. Petals imbricate, in cultivars often more than 5, large, red, purple, white, or yellow. Stamens very many, on relatively short filaments. Pistils usually many, on hairy bottom of hypanthium and on the lower part of its wall, shortly stalked or subsessile; stylodia terminal, free or their upper parts coherent to connate, protruding through the orifice in the disk; stigmas capitate. Fruits achenes with bony, leathery or woody pericarp, remaining enclosed in the accrescent, ± fleshy hypanthium (hip). Seeds with thin testa, endosperm 0. $2n = 14$ or polyploids, recorded up to $2n = 56$. More than 100 wild species, in the temperate parts of the Northern Hemisphere, richest in Asia, in some places (Philippines, Ethiopia, Mexico) extending into the mountains of the tropical zone. In forest edges, shrubland, riverbanks, etc., usually not in closed forest.

Two species (*Rosa roxburghii* Tratt. from China and Japan, *R. minutifolia* Engelm. from North America) have been considered the types of two subgenera, resp. *Platyrhodon* Rehder (probably monotypic) and *Hesperhodos* (Cockerel) Rehder with maybe two species. They are rather ordinary roses but differ conspicuously from the other species by having the hypanthium armed with stout, straight prickles. The two species do not seem to be closely related. Recognition of these subgenera does not seem taxonomically useful. A third aberrant species is *Rosa persica* Juss. (*R. berberifolia* Pall.) from the Iran-Afghanistan region, Kazachstan and Uzbekistan. Apart from having armed hypanthia, it has peculiar and very strongly reduced leaves: unifoliolate and without stipules. It is the type species of subgenus *Hulthemia* (Dumort.) Focke and the genus *Hulthemia* Dumort., and there may be sufficient reason for a subgeneric status in this case. See also Zielinski (1980).

Subgenus *Rosa* is usually divided into some ten sections but its taxonomy is difficult for several reasons: many species cultivated, others very variable and split by some into too many species, polyploidy, apomixis, deviating reproduction methods (like in *Caninae* where the male gametes have $n = 7$, the female gametes $n = 21, 28,$ or 35).

Many of the garden roses have an untraceable breeding history with lots of undocumented interspecific hybridisation since times immemorial and still going strong. Most of the cultivars cannot be brought to one species and must bear only the genus name and the cultivar epithet. Cultivar groups have not yet been formalized.

13. ALCHEMILLA GROUP

Leaves usually palmately structured; stipules adnate to petiole. Hypanthium closed by disk. Epicalyx +, exceptionally 0. Petals 0. Stamens and pistils few to 1. Stylodium basal; ovule 1, ascending. x = 8.

Alchemilla has often been placed in tribe Potentilleae but the closed hypanthium of the former contrasts so much with the open hypanthium of

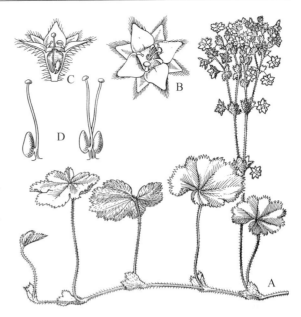

Fig. 115. Rosaceae. *Alchemilla rothii*. **A** Plant with stolon and inflorescence. **B** Flower, seen from above, the stylodia drawn to one side. **C** Flower, halved to show pistil. **D** Pistils. (Graham 1960)

the latter that I prefer to give *Alchemilla* s.l. its own group. The distinction is stressed by the difference in chromosome base number. In the tree generated by the molecular analysis of Morgan et al. (1994), the one species of *Alchemilla* is associated with *Potentilla* (1 species) and *Fragaria*. Hutchinson (1964) placed the genus in tribe Sanguisorbeae. A name on tribal level does not seem to exist.

55. *Alchemilla* L.   Fig. 115

Alchemilla L., Sp. Pl. 123 (1753); S. Froehner in Hegi, Ill. Fl. Mitteleuropa, ed. 2, IV, 2B: 13–248 (1990), regional rev.
Aphanes L. (1753).
Lachemilla (Focke) Lagerh. (1894); L.M. Perry, Contrib. Gray Herb. 84: 1–57 (1929), rev.; W. Rothmaler, Trab. Mus. Nac. Cienc. Nat. Madrid, Ser. Bot. 31: 52 pp. (1935), survey Colombian spp.; Romoleroux in Fl. Ecuador 56: 89–133 (1996), 24 spp.
Zygalchemilla Rydb. (1908).

Perennial herbs with woody caudex, annuals, or small shrubs, unarmed, evergreen or leaf-shedding. Leaves often partly in a radical rosette, palmately structured, entire or lobed, folded in bud; stipules adnate to the petiole, large, lower part membranous and circling the stem, upper part leaf-like. Inflorescences cymose, never truly terminal on the main stems, ranging from rich-flowered thyrses to reduced bundles in the axils of cauline leaves. Flowers small, 4(5)-merous, bisexual. Hypanthium ± urceolate, throat almost closed by the disk. Epicalyx segments usually smaller than sepals, rarely epicalyx 0. Sepals valvate. Petals 0. Stamens 1–4(5), ante- or alternisepalous, pollen in many species sterile and reproduction apogamous. Disk intra- or extrastaminal, nectariferous. Pistils 1–4(–12), free, more or less stalked. Fruits achenes, remaining enclosed in the dry, more or less hardened hypanthium. Seeds with thin testa, endosperm 0. $2n = 16$ or higher, mostly high-polyploid, reported up to $2n = $ c. 224, aneuploid numbers common. A genus of more than 1000 species, many of which in cold- to warm-temperate Eurasia including the Mediterranean region, few reaching Sri Lanka and Java, East Africa, South Africa and Madagascar; few in northern North America, many in South America, one indigenous in Australia. Usually in open, stony or grassy places, also (high) montane, some weedy.

The genus has not been monographed in the last century. Rothmaler published extensively on it and first recognized the genus in the widest sense, with three subgenera *Alchemilla*, *Aphanes*, and *Lachemilla*. Later (Rothmaler 1937) he elevated them to their original generic rank. They seem to be well recognizable:

Subg. *Alchemilla*. Perennial herbs or shrubs; stamens 4, alternisepalous; disk intrastaminal; up to 4 pistils. More than 1000 (micro)species (see Note below). Temperate Northern Hemisphere, few spp. in montane tropical Asia, East Africa, Madagascar, South Africa.

Subg. *Lachemilla*. Perennial herbs or shrubs; stamens 2, antesepalous; disk extrastaminal; pistils 1–12. About 80 species. Only in Mexico, Central and South America, in Andes up to 5000 m altitude.

Subg. *Aphanes*. Annual herbs; stamen 1, antesepalous; disk extrastaminal; pistil 1. About 20 species. Remarkably scattered area: Mediterranean area, adventitious in large part of Europe, Atlantic islands, North America (Southeast and West), South America (Pacific & Atlantic), Australia (1 species), Ethiopia (1 species).

Distinction as three genera seems at first sight indicated, but there remains considerable doubt about *Aphanes*. Its area suggests that this is an old residual group but its distinguishing characters are certainly apomorphic. So, it may be an unnatural grouping of species derived from different lineages within *Alchemilla* s.str. or *Lachemilla*. A fusion of *Aphanes* and *Lachemilla*, which have some characters in common, is not an obvious solution because of their distribution. Lacking a thorough modern phylogenetic analysis to lay the foundation for a definitive classification, it seems the best solution to keep the three groups together in one genus.

When the genus is thought to be related to the Potentilleae, the palmately structured leaf is considered to be homologous with the apical part of the pinnate leaves of genera in Potentilleae. One species, *Alchemilla pinnata* Rydb., has distinctly pinnate leaves and is the type species of monotypic *Zygalchemilla* Rydb. The latter genus is usually synonymized with either *Lachemilla* or *Alchemilla* s.l.

In *Alchemilla* s.str., and possibly also in part of *Lachemilla*, the species are apomictic and have sterile pollen. Recently Froehner (l.c.) studied the European species and recognized 137 species in Central Europe, stating that there would be certainly more than 1000 over the entire area. According to him, hybridisation between the (micro)species does never occur (with only one established exception!), although it played a major role in the history of the group and in establishing the multitude of forms. Other authors also see apomixis in this genus as obligatory, which would contrast with the facultative apomixis in other genera.

14. TRIBE PRUNEAE Hutch. (1964).

Woody plants. Leaves alternate, simple, penninerved. Hypanthium open at top. Epicalyx 0. Disk 0. Pistil(s) on bottom of hypanthium, free from it; ovules 2, descendent and epitropous, over a considerable length attached to the ventral wall, obturator distinct. Fruit drupaceous. Seed usually 1 per pistil developing, endosperm 0. x = 8. *Exochorda* is often considered to be closely related, see there.

56. *Oemleria* Rchb.

Oemleria Rchb., Deutsche Bot., Herb. Buch: Index 195, Emend. 236 (1841).
Nuttalia Torr. & A. Gray (1840), nom. illeg., non Raf. (1817).
Osmaronia Greene (1891).

Shrubs or small trees, leaf-shedding. Leaves with entire margins; stipules 0 or rudimentary. Inflorescence a short raceme, axillary(?) on the short shoots, with large bracts. Flowers fragrant, 5-merous, unisexual, plants dioecious. Hypanthium campanulate. Sepals valvate. Petals shortly

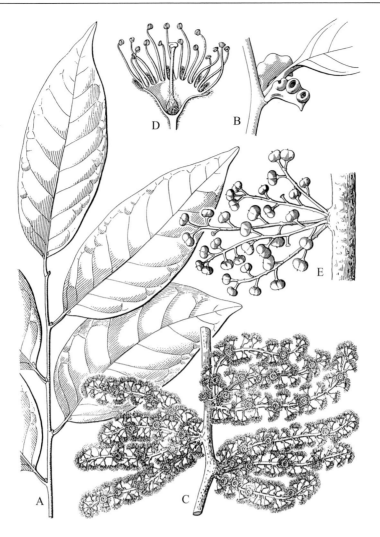

Fig. 116. Rosaceae–Pruneae. *Prunus arborea.* **A** Branch with leaves. **B** Stipules with glands. **C** Fascicled racemes. **D** Halved flower. **E** Infructescences. Drawn by R. van Crevel. (Kalkman 1993)

clawed, white. Stamens 15, inserted in 3 series, small and sterile in female flowers. Pistils 5 in female flowers, free, antesepalous, none in male flowers; stylodia (sub)terminal; stigmas dilated. Fruits drupes, usually only one or two developing in one flower, exocarp black-blue, mesocarp thin, fleshy, endocarp leathery, smooth. Seed with membranous testa. $2n = 16$ (only one record seen: Moffett 1931: 443, footnote). One species, *Oe. cerasiformis* (Hook. & Arn.) J.W. Landon, in the western part of North America from British Columbia to California. Thickets, etc., from sea level to montane altitudes, see Antos and Allen (1990).

57. *Prunus* L.   Fig. 116

Prunus L., Sp. Pl.: 473 (1753); Ghora & Panigrahi, Fam. Rosac. India 2: 5–171 (1995), regional rev.; Bortiri et al., Syst. Bot. 26: 797–807 (2001), molec. analysis.
Pygeum Gaertn. (1788); Kalkman, Blumea 13: 1–115 (1965), reduction.

Trees or shrubs, rarely with thorns, evergreen or leaf-shedding. Leaves with incised or entire margins, usually with glands in the margin or on the underside of the lamina or on the petiole; stipules on the twigs at the junction with the petioles, free, rarely connate. Inflorescence basically a raceme, rarely branched, often reduced to a few-flowered umbel or to only 1 or 2 flowers. Flowers normally 5-merous and bisexual. Hypanthium usually partly or entirely caducous after flowering. Sepals imbricate, usually caducous. Petals white or pink. Stamens c. 15–many. Pistil 1, sessile; stylodium terminal; stigma capitate. Fruit a drupe, mesocarp in wild species thin, fleshy or juicy to rather dry, sometimes dehiscent (almonds), endocarp forming a bony to woody, smooth or rugose pyrene. Seed with thin testa. $2n = 16, 24, 32, 48, 56, 64$, and aneuploid numbers. At least 200 spp., cosmopolitan (in Africa only very few species indigenous). In forests and open vegetation, often in montane or even subalpine altitudes.

Subdivision is controversial; a distinction of five subgenera seems the most useful one to me: *Prunus, Amygdalus* (L.) Focke, *Cerasus* (Mill.) Focke, *Padus* (Mill.) Focke, and *Laurocerasus* (Tourn. ex Duhamel) Rehd.

Many species and species hybrids were domesticated long ago and are cultivated for the edible fruits. Several species and hybrids are grown as ornamentals. Also used in folk-medicine and phytotherapy.

58. *Maddenia* Hook. f. & Thoms.

Maddenia Hook. f. & Thoms., Hook. J. Bot. Kew Gard. Misc. 6: 381 (1854).

Small trees, leaf-shedding, unarmed. Leaves with glandular-dentate margins; stipules large, on the junction of the petiole with the twig, caducous. Inflorescences terminal, short racemes, dense-flowered. Flowers bisexual. Hypanthium obconical, circumscissile above base after anthesis, upper part deciduous. Perianth segments c. 10, subequal or some more petal-like. Stamens 20–40. Pistil 1 (2, 0); stylodium terminal; stigma widened; ovules collateral, pendulous. Fruits drupes, mesocarp fleshy, endocarp woody. Seed with membranous testa. $2n = ?$ Possibly only one species. Eastern part of Himalayas, China, in montane and subalpine altitudes.

59. *Prinsepia* Royle

Prinsepia Royle, Ill. Bot. Himal. Mts. 1: 206 (1834), descr.; 2: t. 38,1 (1839); Baronov, Taiwania 11: 99–112 (1965), rev. *Plagiospermum* Oliv. (1886).

Shrubs, leaf-shedding, thorny with 'stem-spines' serially above the leaves and their axillary shoots. Leaves with distinctly incised margin to entire; stipules minute. Inflorescences racemes, sometimes only 1-flowered. Flowers 5-merous, bisexual. Hypanthium widely obconical to campanulate, persistent with sepals after flowering. Sepals imbricate or valvate. Petals narrowed at base, white or yellow, fugacious. Stamens 10–30. Pistil 1, sessile; stylodium by unequal development of ovary becoming lateral and ultimately subbasal;

stigma capitate. Exocarp purple, mesocarp pulpy to fleshy, leathery when dried, endocarp (slightly) woody. Seed subbasally attached, with membranous testa. $2n = 32$, once reported as $2n = 28$ (see Missouri Bot. Gard. Monogr. Syst. Bot. 30, 1990). Three (or maybe four) species, in China, Mongolia, Himalayas, Taiwan. Thickets on slopes and in valleys, montane, up to c. 3000 m altitude.

Note: As to generic delimitation in tribes/groups 15–17 ('maloids') I fully agree with Weber (1964) who had stated: "The genera of the Maloideae are thus delimited mostly for the sake of convenience, and there will always be disagreement among taxonomists as to how many genera and which ones should be recognized".

15. CYDONIA GROUP

Woody. Leaves alternate, simple, penninerved; stipules on junction of twig and petiole. Flowers 5-merous. Hypanthium with free rim above the pistil. Epicalyx 0. Carpels 5, dorsally adnate to hypanthium; stylodia terminal; ovules more than 2 per carpel. Fruit a pome, flesh (mesocarp) hypanthial and carpellary, core (endocarp) not woody. Seeds with thin layer of endosperm. $x = 17$, but chromosome number of two genera unknown.

This group of maloid genera is here distinguished as a separate unit characterized by the high number of ovules per carpel, although its monophyly is not certain (see the general part under Phylogeny).

60. *Chaenomeles* Lindl.

Chaenomeles Lindl., Trans. Linn. Soc. Lond. 13: 96, 97 (1821), 'Choenomeles', orth. cons.; C. Weber, J. Arnold Arbor. 45: 161–205, 302–345 (1964), rev.

Shrubs, leaf-shedding, thorns axillary on long shoots, collateral with the short shoots. Leaves with incised margins; stipules large on long shoots, very small on short shoots. Flowers few in axillary fascicles or solitary, bisexual or staminate. Hypanthium campanulate, free upper part lined with a nectary of which the rim is disk-like. Sepals unequal, slightly imbricate when very young, deciduous with the upper part of the hypanthium. Petals large, red (white in some cultivars), clawed. Stamens c. 20–60. Carpels 5, completely connate and adnate to the hypanthium, forming an inferior, 5-celled ovary; stylodia connate in basal part; stigmas thickened; ovules more than 20 per cell, axile; pistillode in staminate flowers minute. Pome with thick, hard flesh with sclereids, endocarp forming one core, rather thin, bony. Seeds many, dark brown, with firm testa. $2n = 34$. Three species in the mountains of East Asia. *Ch. japonica* Lindl. is much cultivated as ornamental garden plant, sometimes escaped and established.

61. *Cydonia* Mill.

Cydonia Mill., Gard. Dict., abr. ed.: 4 (1754).

Shrub or small tree, unarmed, leaf-shedding. Leaves entire; stipules large, long persistent. Flowers solitary, terminal on leafy shoots, showy, bisexual. Hypanthium narrowly campanulate, upper free part lined with a nectary. Sepals imbricate, glandular-dentate, soon reflexed, persistent. Petals white or pink. Stamens 15–25. Carpels 5, dorsally adnate to the hypanthium, laterally and ventrally free from each other, forming an inferior, 5-celled ovary with a central cavity; stylodia free; stigmas oblique; ovules many, in 2 rows, horizontal, pleurotropous. Pome large, flesh with sclereids, endocarps forming one core, coriaceous. Seeds several in each cell, with firm testa. $2n = 34$. One species, *C. oblonga* Mill., probably indigenous in the mountain areas of Caucasus and Kurdistan, but since long in many places cultivated as a fruit tree (quince), sometimes naturalized.

62. *Docynia* Decne

Docynia Decne, Nouv. Arch. Mus. Hist. Nat. 10: 125, 131 (1874).
Eriolobus sect. *Docynia* (Decne) C.K. Schneid. (1906).

Trees, unarmed, leaf-shedding (sometimes semi-evergreen?). Leaves sometimes lobed; stipules small, fugitive. Inflorescences umbel-like with (1)2–5 flowers, terminal on new shoots. Flowers bisexual. Hypanthium tubular. Sepals valvate, persistent. Petals large. Stamens 30–50. Carpels 5, completely connate with each other and adnate with basal part of hypanthium, forming an almost inferior, 5-celled ovary; stylodia connate at base; stigmas oblique; ovules 3–10 per cell. Pome with hard and gritty flesh, with many sclereids, endocarps bony, separated by flesh. Seed with firm testa. x = ? One species, *D. indica* (Wall.) Decne, in eastern Himalayas, eastwards to North Thailand and South China. In montane forest. See also under *Macromeles* in tribe Maleae, p. 378.

63. *Pseudocydonia* (C.K. Schneid.) C.K. Schneid.

Pseudocydonia (C.K. Schneid.) C.K. Schneid. in Fedde, Rep. 3: 180 (Nov. 1906).
Chaenomeles sect. *Pseudocydonia* C.K. Schneid., Ill. Handb. Laubholzk. 1: 729 (May 1906), basionym.

Shrub or small tree, unarmed, leaf-shedding. Leaves with glandular-serrate margins; stipules rather large. Flowers solitary on leafy short shoots, bisexual or staminate. Hypanthium in bisexual flowers narrowly campanulate. Sepals slightly imbricate, caducous. Petals large, pink. Stamens 20 or more. Carpels 5, completely connate and also adnate to the hypanthium, forming an inferior, 5-celled ovary; stylodia connate at the base and there surrounded by a cup-shaped 'disc' on top of the ovary; ovules many per cell. Pome not seen, according to descriptions very hard with many clusters of sclereids and with brown seeds. $2n = ?$ One species, *P. sinensis* (Thouin) C.K. Schneid., in China.

16. TRIBE MALEAE Schulze-Menz (1964).

Woody. Leaves alternate, penninerved. Inflorescence terminal. Flowers 5-merous, bisexual. Hypanthium adnate to carpels in lower part, upper part free (entirely free in *Dichotomanthes*). Epicalyx 0. Stylodia terminal (exception: *Dichotomanthes*), sometimes fused to form a branched style. Ovules 2 (exceptions occur, see *Malus, Rhaphiolepis, Sorbus*), normally basal and collateral. Fruit a pome (exception: *Dichotomanthes*, and see Note under *Heteromeles*), endocarp not woody. Seeds with thin layer of endosperm. x = 17.

64. *Sorbus* L.

Sorbus L., Sp. Pl.: 477 (1753); Aldasoro et al., Syst. Bot. 23: 189–212 (1998), rev. Europ. & N Afr. spp.
Chamaemespilus Medik. (1789).
Torminalis Medik. (1789).
Aria Host (1831).
Cormus Spach (1834).
Micromeles Decne (1874).

Trees or shrubs, mostly leaf-shedding, unarmed. Leaves simple or imparipinnate (in hybrids also intermediary), stipulate. Inflorescence a compound raceme, corymb- or panicle-shaped. Hypanthium obconical to cupular, free upper part lined with a ± obvious nectary. Sepals persistent or deciduous. Petals white or pink. Stamens c. 20. Carpels 2–5, dorsally adnate to hypanthium, apical parts exposed or not, ventrally and laterally at least in upper part usually free from each other, lower down completely connate; stylodia free or connate in lower part; stigmas obliquely truncate; ovules 2(–4) per cell. Pome variously coloured, flesh (hypanthial and carpellary) with or without sclereids, with or without large pigment cells; endocarp firm-membranous, leathery, bony or slightly woody. Seeds with firm, leathery to membranous testa, endosperm thin. $2n = 34, 51, 68$. About 150 species, in Europe, Asia, and North America. In forests and more open habitats, from low altitudes to high montane.

Sorbus presents a taxonomic problem of long standing, the two extreme possibilities being one genus or six, and several intermediary classifications also proposed by different authors. As many former authors had done, Hutchinson (1964) saw it as one genus. More recently, Robertson et al. (1991) took an opposite position and recognized five genera. After careful consideration I decided to recognize *Sorbus* as an inclusive genus. That so many species are of horticultural value and are in the pertinent literature, mostly called by their *Sorbus* name, has played a role in my decision.

Distinction of six subgenera seems indicated, but with misgivings about some of them. They are enumerated here, with the characters that have been used as differential.

Subg. *Sorbus*. Leaves imparipinnate, primary nerves not running straight to the margin. Sepals persistent. Carpel apices and stylodia free. Flesh of pome usually without sclereids and large pigment cells. The number of 'true' species is difficult to assess because of extensive hybridization and apomixis in some complexes. The c. 80 species names, enumerated in Phipps et al. (1990), may be considered a maximum. Temperate Eurasia and North America (south to Mexico). *S. aucuparia* L., grown as an ornamental.

Subg. *Cormus* (Spach) Duch. Differing from subg. *Sorbus* in: sepals deciduous but hypanthium persistent. Flesh of pome with sclereids, without large pigment cells. *S. domestica* L. from warmer parts of Europe, the fruits edible when very old.

Subg. *Aria* (Pers.) Beck. Leaves simple, primary nerves running straight to the margin. Sepals usually persistent. Carpels 2 or 3, carpel apices usually free from hypanthium, stylodia free. Flesh of pome without sclereids, with large pigment cells. About 50 species, in Europe, North Africa and temperate to tropical Asia.

Subg. *Chamaemespilus* (Medik.) K. Koch. Leaves simple, nerves not running straight to the margin. Sepals persistent. Carpels 2, carpel apices exposed, stylodia free or slightly connate. Flesh of pome without sclereids, with large pigment cells. Only species *S. chamaemespilus* (L.) Crantz in southern and eastern Europe.

Subg. *Torminaria* (DC.) K. Koch. Leaves simple, deeply lobed, primary nerves running straight to the margin. Sepals caducous, part of hypanthium persistent. Carpels 2 or 3, completely adnate to hypanthium, apices not exposed, stylodia basally connate. Flesh of pome with sclereids and large pigment cells. Only species *S. torminalis* (L.) Crantz in the Mediterranean region, a second species described from Iran deserves no more than varietal rank.

Subg. *Micromeles* (Decne) Phipps et al. Leaves simple, primary nerves running straight to margin. Sepals caducous, free part of hypanthium often also. Carpels 3 or 4, carpel apices not exposed, stylodia basally connate. Flesh of pome with sclereids and large pigment cells. 12–15 species, in Southeast and East Asia.

65. *Eriobotrya* Lindl.   Fig. 117B

Eriobotrya Lindl., Trans. Linn. Soc. Lond. 13: 102 (1821).

Trees or shrubs, unarmed, evergreen. Leaves simple, penninerved; stipules free, rarely intrapetiolarly connate. Inflorescence a compound raceme. Hypanthium obconoid, open at top. Sepals persistent. Petals white. Stamens 15–20(–40). Carpels 2–5, ventrally and laterally connate (in upper part ventrally free), dorsally adnate to the hypanthium, the hairy apex exposed; stylodia usually connate at base, hairy; stigmas capitate. Pome yellow, orange or red, flesh mostly of hypanthial origin, sclereids 0 or +, endocarps thin, membranous. Seeds 1–3, large, with a thin but firm testa; endosperm 0, cotyledons thick. $2n = 34$. About 15–20 species, from the Himalayas throughout continental southeast Asia to Japan and the islands of western Malesia. Montane or in tropical lowland forest. *E. japonica* (Thunb.) Lindl. widely cultivated (loquat).

66. *Aronia* Medik.   Fig. 117A

Aronia Medik., Philos. Bot. 1: 155 (1789), nom. cons.; Robertson et al., Syst. Bot. 16: 391 (1991), united with *Photinia*.
Photinia Lindl. (1821).
Stranvaesia Lindl. (1837).
Pourthiaea Decne (1874).

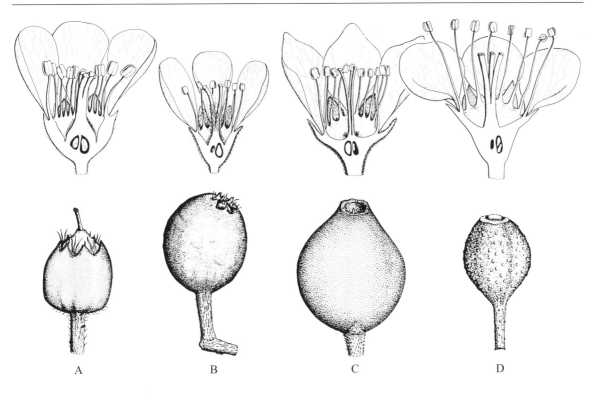

Fig. 117. Rosaceae–Cydonia group, flowers and fruits. **A** *Aronia* sp. (= *Photinia davidiana*). **B** *Eriobotrya bengalensis*. **C** *Rhaphiolepis philippinensis*. **D** *Macromeles corymbifera*. (Kalkman 1993)

Trees or shrubs, unarmed, evergreen or leaf-shedding. Leaves penninerved; stipules free, small and caducous. Inflorescence a panicle- or corymb-shaped compound raceme. Hypanthium obconoid to campanulate, open at top. Sepals long persistent, but ultimately often (partly) weathering off. Petals white to pinkish. Stamens 15–25, often 20. Carpels 2–5, ventrally and laterally completely connate, dorsally adnate to the hypanthium but the apical part free and protruding as a usually hairy dome; stylodia entirely free or connate at base; stigma slightly broadened, truncate. Pome red or black, flesh (almost) entirely of hypanthial origin, inner layer with sclereids, carpel wall rather thin, core bony. Seeds usually several per fruit, rather small, testa firm, endosperm thin or sometimes 0. $2n =$ 34, 68. Some forty species, in East Asia, and North & Central America. In forest and thickets, in low as well as montane and subalpine altitudes. In China 40 and more species have been recognized but not all of them will survive closer taxonomic scrutiny. Some species cultivated as ornamentals.

Photinia was described in 1821 from Asia, and subsequently enlarged by description of several Asian and North and Central American species. Robertson et al. (1991) merged the small American genus *Aronia* (indigenous in East Canada and the eastern half of U.S.A.) with *Photinia*, thereby creating a new example of the East-American/East Asian disjunction. They correctly state that flowers and fruits of the two genera are indistinguishable, but they kept the genus under the younger name *Photinia*. Their taxonomic decision is followed here but the generic name *Photinia* had to be relegated to synonymy. I am still uneasy about the monophyly of the enlarged genus *Aronia/Photinia* and a worldwide monograph is badly needed. The Asiatic species of *Photinia* will generally not have a synonym under *Aronia*.

67. *Rhaphiolepis* Ker Gawl.   Fig. 117C

Rhaphiolepis Lindley ex Ker Gawl., Bot. Reg. 6: t. 468 (1820), nom. et orth. cons.

Small trees or shrubs, unarmed, evergreen. Leaves simple, penninerved; stipules on extreme base of petiole, free. Inflorescence a compound, rarely simple raceme. Hypanthium obconoid, the free part at inside lined with an intrastaminal disk. Sepals caducous after flowering, together with the upper rim of the hypanthium. Petals white to pinkish. Stamens 15–20. Carpels 2, completely connate with each other and adnate to the hypanthium; stylodia free or basally connate; stigmas truncate; ovules normally 2 per carpel (rarely more, Sterling 1965c). Pome (sub)globular, with a distinct circular scar at top, purplish or bluish black, flesh rather thin, with many sclereids, carpel walls thin, membranous. Seeds 1 or 2 per fruit, large, globose; testa thin but firm, endosperm 0, cotyledons thick. $2n = 34$. Some 5 species, in Southeast and East Asia, see Ohashi (1988). In forest, low and montane altitudes.

68. *Heteromeles* M. Roem.

Heteromeles M. Roem., Fam. Nat. Syn. Monogr. 3: 100, 105 (1847); Phipps, Can. J. Bot. 70: 2138–2162 (1992), rev.

Small trees or shrubs, unarmed, evergreen. Leaves simple. Inflorescence a panicle-shaped compound raceme. Hypanthium obconical. Sepals persistent. Petals white. Stamens 10. Carpels 2, free from each other except at entire base, the lower half or less adnate to hypanthium, the free tops conical, long-hairy; stylodia free; stigmas capitate. 'Pome' (see Note below) red or yellow, crowned by apical rim of hypanthium and persistent sepals, carpel walls leathery, surrounded by and in the lower half adnate to the fleshy hypanthium, no sclereids in the flesh. Seeds 1 or 2 (maybe sometimes more), testa firm, endosperm 0. $2n = ?$ One species, *H. salicifolia* (C. Presl) Abrams (usually incorrectly called *H. arbutifolia* (Lindl.) M. Roem.), in western states of U.S.A. and Northwest Mexico. In sclerophyllous scrub and (pine) woods at montane altitudes up to c. 2000 m.

The inadequacy of the term pome is well illustrated in the fruits of this species: actually they are 'spurious' fruits consisting of a fleshy hypanthium and two practically free, indehiscent achenes with a leathery pericarp.

69. *Dichotomanthes* Kurz

Dichotomanthes Kurz, J. Bot. 11: 194 (1873); Gladkova, Bot. Zhurn. 54: 431–436 (1969), new subfamily.

Trees or shrubs, leaf-shedding, unarmed. Leaves simple, entire. Inflorescence a compact compound raceme, the last branches dichasial. Flowers scented. Hypanthium cup-shaped, open. Sepals persistent. Petals cream-coloured to white. Stamens 20, on rim of disc. Pistil 1, at base of and free from hypanthium; ovary 1-locular; stylodium slightly lateral, often sigmoid; stigma capitate,

lobed; ovules 2, basal, collateral, erect, apotropous. Fruit an achene, slightly exserted from enlarged and fleshy hypanthium, pericarp coriaceous to woody. Seed not observed. $2n = 34$ (Gladkova, l.c.). One species, *D. tristaniifolia* Kurz, in tropical parts of China. Montane on open slopes, forest and forest margins.

This, of course, is a deviating genus but the relationship with *Heteromeles* seems obvious to me. The latter has two carpels in the flower, basally adnate to the hypanthium, while *Dichotomanthes* has only one, entirely free carpel. In both cases the hypanthium becomes fleshy and the fruit an achene. A separate subfamily certainly does not seem justified, and a likeness to *Prunus* is only superficial. Juel (1927) placed the genus in a separate tribe near his Quillajeae and considered *Vauquelinia* (here in the Exochordeae) as the nearest relative.

70. *Macromeles* Koidz.   Fig. 117D

Macromeles Koidz., Fl. Symb. Or. As.: 53 (1930).
Malus sect. *Docyniopsis* C.K. Schneid. (1906), validly published?
Docyniopsis (C.K. Schneid.) Koidz. (1934).

Trees, leaf-shedding, sometimes thorny? (not observed). Leaves simple; stipules caducous. Inflorescence a few-flowered raceme. Hypanthium obconical. Sepals long persistent but ultimately deciduous. Petals white. Stamens 30–50. Carpels 5, dorsally adnate to hypanthium but apical part forming a small, free cone, laterally connate, ventrally with open sutures; stylodia connate only at very base, cohering by hairs in lower part; stigmas not broadened. Pome globose, yellowish (always?), flesh hard, with a layer full of sclereids round the core, the core (endocarps) cartilaginous, thin. Seeds 1 or 2 per cell, large, with thick and soft testa, endosperm very thin. $2n = ?$ Three species of which the Japanese *M. tschonoskii* (Maxim.) Koidz. is sometimes cultivated as an ornamental. Other species include one or two Chinese species and the East Asian *Pirus doumeri* Bois, which does not have a valid name in *Macromeles*. It was treated as *Malus doumeri* (Bois.) Chev. by Vidal in Fl. Camb., Laos & Vietn. 6: 99–101 (1968). Growing in forest at montane altitudes up to 2000 m.

71. *Malus* Mill.

Malus Mill., Gard. Dict., abr. ed.: 4 (1754).

Trees or shrubs, leaf-shedding, rarely evergreen and/or thorny. Leaves simple, lobed or not; stipules caducous. Inflorescences few-flowered racemes terminal on short shoots. Hypanthium obconical to campanulate, upper part persistent. Sepals persistent or deciduous. Petals white, pink or red. Stamens 15–20. Carpels 3–5, dorsally over their entire length adnate to hypanthium and also the apical part not exposed, laterally connate, ventrally with open sutures (see Note below); stylodia connate in lower part; stigmas slightly broadened to capitate; ovules 2 (4?) per carpel. Pome ± globose, green to red or yellow, rarely lenticellate, flesh soft and mealy to rather hard, without sclereids, the core (endocarps) cartilaginous to bony (even slightly woody). Seeds with thin but firm, dark coloured testa, endosperm a thin layer. Chromosome number $2n = 34, 51, 68, 85$. Between 30 and 50 species, northern temperate. In woods and thickets.

The circumscription of this genus is difficult and much disputed. See Notes under *Macromeles*, *Eriolobus*, and *Pyrus*, genera with which *Malus* has often been united. Langenfeld (1971) considered the *Macromeles* spp. (as sect. *Docyniopsis*) and *Eriolobus* spp. (as sect. *Eriolobus*) to be relicts of the original Tertiary *Malus* from which the other sections evolved. Merging *Malus* with *Sorbus*, as has also been advocated, does not seem very sensible or enlightening.

In the present circumscription two or three subgenera or sections may be distinguished: **subg. *Malus*** with unlobed leaves in Eurasia, and **subg. *Sorbomalus*** with often lobed leaves (at least on the long shoots) in Asia. A third subgenus or section, ***Chloromeles***, is North American. Its species have usually lobed leaves and, according to Robertson et al. (1991), "a dense layer of sclereids around the core" of the pome, but I did not have the opportunity to observe this myself in the limited material at my disposition. If confirmed, it would be a sufficient reason to separate the group as a genus, or it may lead to fusion of *Macromeles* and *Eriolobus* with *Malus*.

72. *Eriolobus* (DC.) Roem.

Eriolobus (DC.) Roem., Fam. Nat. Syn. Monogr. 3: 104, 216 (1847).
Pyrus sect. *Eriolobus* DC., Prodr. 2: 636 (1825), basionym; Browicz, Arbor. Kórn. 14: 5–23 (1969), taxonomic survey, distribution.

Trees, leaf-shedding, unarmed. Leaves simple, deeply 3-partite, the terminal and lateral lobes more or less deeply lobed again, rarely the leaf 3-foliolate; stipules caducous. Inflorescence a simple raceme, corymb-shaped, with few flowers. Hypan-

thium obconical. Sepals reflexed, persistent. Petals white. Stamens c. 20. Carpels 5, dorsally adnate to hypanthium and their apical part not exposed, laterally fully connate with each other, ventrally with open sutures; stylodia connate at base, the connate part persistent; stigmas capitate. Pome ± globose, red or yellow, rather hard and gritty, with a layer rich in sclereids around the cartilaginous core (endocarps). Seeds 1 or 2 per fertile cell, often only 1 cell fertile, testa thin, endosperm thin. $2n = ?$ Monotypic, containing only *E. trilobatus* (Poir.) Roem., in Southeast Europe and Southwest Asia. In forest and open places, at altitudes up to c. 1500 m.

This remarkable species has, apart from having been recognized as a genus, been moving around in *Crataegus* (basionym), *Pyrus*, *Sorbus*, *Cormus* and *Malus*. When not recognized as a genus, it will be best in place in *Malus* from which it differs chemically by not possessing the flavonoid phloridzin in the leaves but trilobatin (Williams 1982). The most important difference with *Macromeles* seems to be that in *Eriolobus* the apical part of the carpels is not free from the hypanthium.

73. *Pyrus* L.

Pyrus L., Sp. Pl.: 479 (1753); Browicz, Arbor. Korn. 38: 17–33, conspectus, distr.

Trees or shrubs, sometimes thorny, leaf-shedding. Leaves simple, rarely lobed; stipules caducous. Inflorescence a simple raceme, corymb-shaped, with few flowers. Hypanthium urceolate, free upper part with a cushion-like outgrowth (disc) covering the top of the carpels, leaving a central hole for the stylodia, ± persistent. Sepals persistent or caducous. Petals white, rarely pinkish. Stamens 15–30. Carpels 2–5, dorsally adnate to hypanthium, laterally and ventrally at least in the apical part often free from each other, the ventral sutures open; stylodia free; stigmas truncate to 2-partite. Pome pyriform to globose, brownish to yellow, sometimes lenticellate, flesh containing many sclereids, these sometimes compacted to a woody layer, core (endocarps) membranous to cartilaginous. Seeds with thin testa, endosperm a very thin layer. $2n = 34, 51, 68$, and aneuploid numbers also reported. Between 10 and 20 species, but an exact determination of their number is difficult because of extensive natural hybridization and establishment of escapes from cultivation. Eurasian, also in North Africa. In woods, thickets, hedges, also on sunny slopes. *Pyrus communis* L., *P. pyrifolia* (Burm.f.) Nakai and some other species are cultivated as fruit trees.

74. *Amelanchier* Medik.

Amelanchier Medik., Philos. Bot. 1: 155 (1789).
Peraphyllum Nutt. ex Torr. & Gray (1840).
Nägelia Lindl. ('Nagelia') (1842), nom. illeg. non Rabenh. (1844).
Cotoneaster § *Malacomeles* Decne, Nouv. Arch. Hist. Nat. 10: 177 (1874), rank not indicated.
Malacomeles (Decne) Engl. in Engl. & Prantl, Nat. Pflanzen-fam., Nachtr. 1: 186 (1897).

Shrubs or small trees, leaf-shedding, unarmed. Leaves simple; stipules small, caducous. Inflorescence a simple raceme. Hypanthium obconic to campanulate, free upper part saucer- to cup-shaped. Sepals persistent. Petals white, rarely pink. Stamens usually c. 20, sometimes fewer. Carpels 2–5, ventrally (partly) free from each other, laterally more or less connate, dorsally more than half adnate to hypanthium, top dome-shaped and glabrous or hairy, cells of the ovary usually divided by an incomplete sept between the ovules; stylodia free to highly connate; stigma(s) truncate to capitate. Fruit a small pome, red to black, rarely yellow, flesh thin, uniform of consistency, with sclereids. Seeds several, testa rather thick, endosperm a very thin layer. $2n = 34, 68$, and aneuploid numbers also reported. Maybe around 20 spp., northern temperate: North America, East Asia, Europe (one species), in America south to Mexico and Guatemala. Extensive hybridization between species and apomictic reproduction established. In rocky places of montane altitude, and open woods. Edible fruits.

17. TRIBE CRATAEGEAE Koehne (1890).

Woody. Buds perulate. Leaves alternate, simple (pinnate in *Osteomeles*); stipules on petiole-base. Flowers 5-merous, bisexual (sometimes male flowers in *Mespilus*). Hypanthium open, lower part adnate to carpels. Epicalyx 0. Carpels 2–5 (exceptions: 1 in *Chamaemeles*, rarely 1 in some *Cotoneaster* and some *Crataegus*), dorsally adnate to hypanthium; stylodia terminal. Fruit a fleshy hypanthium containing 5–1 pyrenes (stones); pyrenes 1-seeded; endosperm a thin layer. $x = 17$.

75. *Cotoneaster* Medik.

Cotoneaster Medik., Phil. Bot. 1: 154 (1789); Klotz, Beitr. Phytotax. Univ. Jena 10: 7–82 (1982), synopsis.

Shrubs, sometimes prostrate, rarely small trees, leaf-shedding or more or less evergreen, unarmed. Leaves simple, usually leathery; stipules rather persistent. Inflorescences cymes, rarely flowers solitary. Flowers rather small. Hypanthium obconical to campanulate. Sepals persistent, incurved after anthesis. Petals erect or spreading, white or pink. Stamens (6–)10–20(–30). Carpels (1)2 or 3(–5), ventrally and laterally entirely free from one another or very shortly connate at base, with exposed, hairy apex, dorsally adnate to hypanthium; ovules 2 per carpel. Fruit yellow, red or black, flesh rather dryish, without sclereids; stones mostly 2 or 3, their upper parts free from hypanthium. Seed with thin testa, endosperm 0. $2n = 34, 51, 68$, rarely 85, 102. Variously recorded as having c. 60 or up to c. 300 species, with widespread facultative apomixis and hybridization. Distributed over Asia, with few species in Europe and few in North Africa and Ethiopia. Tetraploids are far more numerous than diploids, the latter are mainly in the Himalayas and part of China, see Krügel (1992a). In open, often dry vegetation, and in thickets and forests, up to c. 4000 m altitude. Many species and hybrids are cultivated as ornamentals or as landscaping elements. For the difference with *Pyracantha*, see there.

76. *Pyracantha* M. Roem.

Pyracantha M. Roem., Fam. Nat. Syn. Monogr. 3: 104, 219 (1847).

Shrubs, evergreen, mostly with thorns. Leaf-margins incised or entire; stipules small, caducous, but on long shoots longer persistent and sometimes leafy. Inflorescence a terminal corymb-shaped thyrse. Hypanthium obconical. Sepals persistent and incurved after flowering. Petals white. Stamens c. 20. Carpels 5, ventrally and laterally free from each other, with exposed, hairy apex, dorsally adnate to hypanthium; ovules 2 per carpel. Fruit red, orange or yellow, flesh rather thin, without sclereids; stones 5, free, upper part exposed. Seed with thin testa. $2n = 34$. About three species (up to 10 reported in horticultural literature). Distributed from southern parts of Europe to East Asia. *Pyracantha coccinea* M. Roem. much cultivated with many cultivars.

The differences with *Cotoneaster* are slight and probably the number of carpels per flower and the presence/absence of thorns are the only decisive characters. Up to 5 carpels have been reported as occurring in some species of *Cotoneaster* but I myself have not observed that number.

77. *Crataegus* L.

Crataegus L., Sp. Pl.: 475 (1753); Phipps, Ann. Missouri Bot. Gard. 70: 667–700 ('1983' 1984), E As. and N Am. spp.

Shrubs or small trees, often with thorns, with short and long shoots, evergreen or leaf-shedding. Leaves pinnately lobed or pinnatifid or 3-lobed, rarely unlobed, those on flowering branches often unlike those on strictly vegetative branches; stipules usually small, mostly early caducous. Inflorescence terminal, a corymb-shaped thyrse or cyme. Hypanthium obconical to campanulate. Sepals persistent or caducous. Petals orbicular to obovate, spreading, white, pink or red. Stamens mostly c. 20, inserted on a disk-like rim, well visible on the fruit. Carpels 1–5, when more than 1 ventrally free from each other, laterally free, at least partly, dorsally adnate to hypanthium, apex free, hairy; ovules 2, basal. Fruit ± globose, red, black or yellow, crowned by the apical part of the hypanthium, flesh mostly thin and rather dry, without sclereids; stones 1–5, with thick, hard and woody walls, forming one more-celled stone. Seed compressed, with thin but firm testa. $2n = 34, 51, 68$. Between 100 and 200 species, although more than 1000 have been described, northern temperate zone and in Central America, richest in eastern North America. According to some sources also in the Andes but most probably in error. In woods, forest margins, thickets, in low and montane altitudes. Often planted as ornamental or in hedges. The taxonomy is complicated by hybridization and apomixis.

78. *Osteomeles* Lindl.

Osteomeles Lindl., Trans. Linn. Soc. Lond. 13: 96, 98 (1821).

Shrubs, often prostrate, rarely treelets, evergreen (?), unarmed. Leaves imparipinnate; leaflets in 5–15(–20) pairs; stipules small. Inflorescence a terminal corymb-shaped thyrse. Hypanthium obconical to campanulate, covering the carpels at apex and almost entirely adnate to them. Sepals pointed, persistent and erect on the fruit. Petals spreading, white. Stamens c. 20. Carpels 5, ventrally and laterally connate, dorsally adnate to hypanthium, forming an inferior, 5-celled ovary; stylodia mostly long-hairy at base; ovule 1 per carpel, basal, ascending. Fruit white, flesh thin, hard; stones 5, hard and woody, rather easily loosened from each other. Seed with membranous testa. $2n = 34$; according to Goldblatt in Missouri Monogr. Syst. Bot. 8 (1984), $2n = 32$ was reported for two Japanese species. Usually cited to have 3 to 5 species but rather a single variable species, *O. anthyllidifolia* Lindl., distributed in China (Sichuan to Guangdong), Taiwan, Ryu Kyu Islands, Bonin Islands, and Hawai'i. Often indicated from Myanmar but I have not seen any specimen from there and it is not present in Thailand or Indo-China. Forest margins, mountain slopes, also on dry places, up to 3000 m altitude but also near the coast. Sometimes cultivated as an ornamental.

79. *Chamaemeles* Lindl.

Chamaemeles Lindl., Trans. Linn. Soc. Lond. 13: 96, 104 (1821).

Evergreen shrubs, unarmed. Leaf margins entire to serrulate; stipules small, caducous. Inflorescence an axillary compound raceme. Flowers small. Hypanthium obconical. Sepals persistent. Petals clawed, small, pink to white. Stamens 10–18. Pistil 1, monocarpellate, lower half adnate to hypanthium; style terminal, rather thick, broadened at apex; stigma oblique; ovules 2, basal, collateral. Fruit with thin flesh. Seed 1, cotyledons convoluted. $2n = ?$ One species, *C. coriacea* Lindl., endemic to Madeira, on inaccessible places on cliffs and ravines.

80. *Hesperomeles* Lindl.

Hesperomeles Lindl., Bot. Reg. 23: t. 1956 (1837).

Shrubs or small trees, probably evergreen, with thorns or unarmed. Leaves only on long shoots; stipules small, caducous. Inflorescence a terminal, leafy compound raceme. Hypanthium obconical. Sepals persistent. Petals white or cream-coloured. Stamens c. 20. Carpels 4–6, dorsally adnate to hypanthium, ventrally free, laterally ± free from each other, apex exposed; stylodia terminal, free; stigma truncate; ovule 1, basal. Fruit red, flesh with sclereids; stones c. 5, walls thick, hard and woody, upper parts exposed and free from upper rim of hypanthium, covered by sepals. Seeds not observed. $2n = ?$ Probably about 10 species, in South America (Andes from Colombia to Bolivia) and Central America (Costa Rica, Panama). In cloud forest, up to 4000 m alt.

By its single ovules and other characters, connected with *Osteomeles* and formerly included in that genus. However, the leaves are quite different and recognition as a separate genus seems indicated.

81. *Mespilus* L.

Mespilus L., Sp. Pl.: 478 (1753).

Trees or shrubs, with thorns (in the wild) or unarmed (mostly in cultivation), leaf-shedding. Leaf-margins serrate; stipules caducous. Flowers solitary, terminal on short shoots, large, sometimes male, beneath them rarely with 1–3 later or not developing lateral flowers. Hypanthium obconical. Sepals large, persistent. Petals white. Stamens 30–40, on the rim of a disk covering the free part of the hypanthium. Carpels (4)5, completely connate and adnate to the hypanthium, forming an inferior, 5-celled ovary, apex hairy; stylodia free (to halfway connate: Gams 1922); ovules 2(3) per cell, only the basal one fertile. Fruit brownish, flesh mealy with many sclereids; stones with thick, hard, woody wall. Seed 1 per cell, testa thin but firm. $2n = 34$. One species, *M. germanica* L., from southeast Europe to Iran, now widely cultivated outside its homeland; the fruit edible when overripe. Wood very tough.

From Arkansas in U.S.A. a second species has been described by Phipps (1990). From a phytogeographical standpoint this is unlikely. I wonder whether this single population could have originated from a crossing between a cultivated medlar and some indigenous *Crataegus* species.

Rosaceous genera of unclear tribal position

82. *Adenostoma* Hook. & Arn.

Adenostoma Hook. & Arn., Bot. Beechey Voy. 3: 139 (1832).

Shrubs, evergreen, resinous. Leaves simple, rigid; stipules basally adnate to leafbase, minute on short shoot leaves. Inflorescence a panicle with short spike-like branches. Flowers many, very small, sessile, 5-merous, bisexual. Hypanthium funnel-shaped, open at the top, 10-ribbed. Epicalyx 0. Sepals mucronate, imbricate. Petals orbicular, white. Stamens 10–15. Intrastaminal disk lobed. Pistil 1, monocarpellate, on bottom of hypanthium; ovary oblique with 'cushion' on one side; style terminal but put aside by the dorsal cushion; stigma capitate; ovules 2, inserted below the apex. Fruit achene, covered by indurated hypanthium. $2n = ?$

Three species, in western North America. Dry mountain slopes, often in pure vegetation, see Jepson (1936).

Hutchinson (1964) and Schulze-Menz in Melchior (1964) placed the genus in a separate tribe, Adenostomateae (Adenostomeae). In Hutchinson's description the pistil is said to contain only one ovule, but Juel (1918) correctly described the genus as having two collateral ovules. The combination simple leaves/one pistil/two ovules is found in all Pruneae except *Oemleria*, in some maloids and in part of Neillieae. *Purshia* (Dryadeae) has some resemblance but differs in having a basal ovule and no intrastaminal disk. Zhang (1992) placed the genus in his wood-anatomical phenetic group V, together with *Hagenia* (Sanguisorbeae) and *Sorbaria* (Gillenieae), and also in his cladogram the three genera are closely associated. Morgan et al. (1994), in their molecular-phylogenetical analysis, found *Adenostoma* together with *Sorbaria* and *Chamaebatiaria* (Gillenieae), so this option must not be neglected. Juel (1918) placed the genus in a group derived from Quillajeae and consisting of *Exochorda*, *Lindleya* and the Pruneae. I could not find any reference to a published chromosome number.

83. *Coleogyne* Torr.

Coleogyne Torr., Smiths. Contrib. Knowl. 6: 8 (1853).

Shrubs, leaf-shedding, branches spinescent. Leaves opposite, crowded on the short shoots, simple, small, linear, entire, coriaceous and grooved below, sessile; stipules small, persistent on twig. Flowers solitary, terminal, bisexual, with 2–3-lobed prophylls close under the hypanthium. Hypanthium low obconical. Epicalyx 0. Sepals 4, large, persistent, yellow inside. Petals 0 (see below). Stamens 20–40. Intrastaminal disk forming a tubular 'sheath' around the ovary. Pistil 1, monocarpellate, on bottom of hypanthium; style lateral, twisted, exserted from disc, persistent; stigma linear, on ventral side of style; ovule 1, attached in the middle, descending, epitropous. Fruit achene, pericarp coriaceous. Seeds not seen, no data in literature. $2n = 16$, two counts, recorded by McArthur and Sanderson (1985). One species, *C. ramossissima* Torr. in North America (California, Nevada, Arizona, Colorado). Open and dry mountain slopes, sometimes in pure stands.

The tubular disk is obviously a much derived character and of no help in finding the closest relative. Kearney and Peebles c.s. (1960: 391) record a specimen with two pale yellow petals. Because of the opposite leaves and the absence of petals, Hutchinson (1964) placed the genus in Cercocarpeae but differences with *Cercocarpus* are large. Dryadeae and especially Kerrieae are often mentioned as possible places to accommodate the genus. Juel (1918), for example, saw *Rhodotypos* as the closest relative and detected a likeness with *Neviusia*. Zhang (1992) put the genus in his wood anatomical group I, together with *Kerria*, *Petrophytum* and *Stephanandra* (Neillieae), *Rosa*, *Rubus*, and *Sibiraea* (Spiraeeae). The chromosome number suggests a relationship with *Alchemilla*, also with sepals 4 and petals 0, but otherwise very different.

84. *Potaninia* Maxim.

Potaninia Maxim., Bull. Acad. Imp. Sci. St. Pétersb. 27: 465 (1881).

Much branched, small shrub, up to 40 cm. Leaves crowded in short shoots, alternate (opposite according to Hutchinson), imparipinnate with (1)3 or 5 small leaflets with entire margins; petiole persistent, spiny; stipules adnate to the petiole, relatively large. Flowers solitary, axillary, minute, 3-merous, bisexual. Hypanthium funnel-shaped. Epicalyx 0. Sepals acuminate. Petals white or pinkish. Stamens 3, inserted on a swollen disk. Pistil 1, monocarpellate, on the bottom of the hypanthium; style basal, stigma capitate; ovule 1, basal, ascending. Fruits achenes, pericarp light yellow. $2n = ?$ One species, *P. mongolica* Maxim., in Mongolia and China (Inner Mongolia). In sandy desert. Locally used as cattle feed (camels a.o.).

There are some mystifications about the flower structure. Hutchinson (1964) described the flower as having an epicalyx as well as sepals and petals, but Yü Te-tsun et al. in Fl. Reip. Pop. Sin. 37: 455–456 (1985) do not mention an epicalyx in their elaborate description (translated for me by Dr. Ding Hou) nor is it drawn in their plate. The place of the stamens also is reported in different ways: antesepalous according to Hutchinson, antepetalous according to Yü Te-tsun et al. I have seen only one very insufficient herbarium specimen myself, and my interpretation of the flower fragments available is in agreement with the description in Yü et al.: epicalyx 0, sepals 3, petals 3, stamens opposite the petals.

Hutchinson placed the genus together with *Cercocarpus* and *Coleogyne* in a separate tribe Cercocarpeae but in my opinion the three genera are not closely related. The best place for *Potaninia* may be tribe Potentilleae but more evidence is needed.

85. *Lyonothamnus* A. Gray

Lyonothamnus A. Gray, Proc. Am. Acad. Arts 20: 291 (1885).

Evergreen tree with exfoliating bark. Leaves opposite, petioled, simple and entire to pinnatifid or pinnately compound. Stipules deciduous. Flowers many, perfect, in large terminal corymbose compound panicles, with short pedicels. Flower tube campanulate, free from ovary, subtended by 1–3 bractlets. Sepals 5, persistent. Petals 5, clawless. Stamens c. 15, inserted on a wooly disk lining the flower tube. Pistils 2, distinct, 1-loculed. Stylodia stout, with subcapitate stigma. Ovules and seeds 4 in each ovary, the ovules bitegmic. Fruit a pair of small woody follicles. Seeds flat. $n = 27$. One sp., *L. floribundus* Gray, with 2 subspecies, on Santa Catalina Isl. and adjacent islands off the coast of California.

This genus is mostly recognized as spiraeoid, sometimes placed in Sorbarieae (Juel 1927), by others in Quillajeae (Hutchinson 1964). In the molecular-based tree published by Morgan et al. (1994), it associates with *Cercocarpus* and *Purshia* (Dryadeae), although the molecular support for this relationship is not strong, and morphologically it has little in common with these genera either. Morgan et al. (1994) therefore conclude that *Lyonothamnus* may be considered as an isolated descendant of the ancestral spiraeoid complex; it certainly would merit a more profound study.

Doubtfully Rosaceous genus

Brachycaulos Dikshit & Panigrahi

Brachycaulos Dikshit ('Dixit') & Panigrahi, Bull. Mus. Natl. Hist. Nat. Paris IV, 3, sect. B. Adansonia 1: 57–60 (1981).

Compact dwarf herbs, all parts glabrous (but see stipules). Leaves simple, entire, small; stipules membranous, fully adnate to petiole, with shortly ciliated margins, brown, clothing the stems. Flowers solitary, terminal, bisexual. Hypanthium flat. Epicalyx 0. Sepals 5, free, entire. Petals 5, clawed, slightly exceeding the sepals. Stamens 5, alternipetalous. Disk not evident. Pistils 2, free, on bottom of hypanthium; stylodia terminal; stigmas inconspicuous; ovules unknown. Fruit achene, smooth. $x = ?$ One species described: *B. simplicifolius* Dikshit & Panigrahi, from Sikkim, India, in the alpine belt (4575 m alt.).

This plant is only known from one collection, W.W. Smith 3993 (CAL). The authors ascribed it to tribe Potentilleae and compared it with *Chamaer-*

hodos. I am not at all sure that it is rosaceous and if so, it almost certainly does not belong to Potentilleae. Dr. G. Panigrahi (in litt.) kindly informed me that other collections of this species have not come to his knowledge and that duplicates do not seem to exist in other herbaria. Dr. K. Brummitt (K) suggested (pers. comm.) that it may belong to Saxifragaceae. For the time being I consider it as a genus incertae sedis.

Guamatela Donn. Sm. from Guatemala was placed in tribe Neillieae by Hutchinson (1964), but is probably not rosaceous. See Juel (1927) who thinks of a malvalean affinity (but the leaves are opposite), and Standley & Steyermark, Fieldiana, Bot. 24 (IV): 449 (1946). Its flowers are diplostemonous (Lindenhofer and Weber 1999a: 571). The only species is *G. tuerckheimii* Donn. Sm.

Excluded

Apopetalum Pax. Its only species, *A. pinnatum* Pax from Bolivia, belongs to *Brunellia* (Brunelliaceae). See Cuatrecasas in Flora Neotropica, Monogr. no. 2 (1970), and Suppl. (1985).

Quillaja Molina has first been excluded from Rosaceae on the basis of molecular data, and Lindenhofer and Weber (1999a: 572) list many structural and chemical characters that support this decision.

Selected Bibliography

Allen, G.A. 1986. Flowering pattern and fruit production in the dioecious shrub *Oemleria cerasiformis* (Rosaceae). Can. J. Bot. 64: 1216–1220.

Antos, J.A., Allen, G.A. 1990. Habitat relationships of the Pacific coast shrub *Oemleria cerasiformis* (Rosaceae). Madroño 37: 249–260.

APG II (Angiosperm Phylogeny Group) 2003. See general references.

Asker, S. 1979. Progress in apomixis research. Hereditas 91: 231–240.

Balick, M.J., Cox, P.A. 1996. Plants, people, and culture. The science of ethnobotany, ix + 228 pp. New York: Scientific American Library.

Barr, A. et al. 1993. Traditional aboriginal medicines in the Northern Territory of Australia, xxiv + 650 pp. Darwin: Conservation Commission.

Becking, J.H. 1979. Nitrogen fixation by *Rubus ellipticus* J.E. Smith. Plant Soil 53: 541–545.

Bhat, R.B., Etejere, E.O., Oladipo, V.T. 1990. Ethnobotanical studies from Central Nigeria. Econ. Bot. 44: 382–390.

Bolle, F. 1935. Ueber eine bemerkenswerte Missbildung bei *Geum*. Notizbl. Bot. Gart. Mus. Berlin-Dahlem 12, nr. 113: 349–354.

Bowden, W. 1945. A list of chromosome numbers in the higher plants II. Am. J. Bot. 32: 191–201.

Bramwell, D. 1978. The endemic genera of Rosaceae (Poterieae) in Macaronesia. Bot. Macar. 6: 67–73.

Bramwell, D., Bramwell, Z.I. 1974. Wild flowers of the Canary Islands, x + 261 pp. London: Thornes.

Brummitt, R.K. 1992. Vascular plant families and genera. Kew: Royal Botanic Gardens.

Cambie, R.C., Ash, J. 1994. Fijian medicinal plants. Canberra: CSIRO.

Campbell, C.S., Donoghue, M.J., Baldwin, B.G., Wojciechowski, M.F. 1995. Phylogenetic relationships in Maloideae (Rosaceae): evidence from sequences of the internal transcribed spacers of nuclear ribosomal DNA and its congruence with morphology. Am. J. Bot. 82: 903–918.

Charlton, W.A. 1993. The rotated-lamina syndrome. III. Cases in *Begonia*, *Corylus*, *Magnolia*, *Pellionia*, *Prunus*, and *Tilia*. Can. J. Bot. 71: 229–247.

Crawford, D.J., Brauner, S., Cosner, M.B., Stuessy, T.F. 1993. Use of RAPD markers to document the origin of the intergeneric hybrid x *Margyracaena skottsbergii* (Rosaceae) on the Juan Fernandez Islands. Am. J. Bot. 80: 89–92.

Croizat, L. 1952. Manual of phytogeography. The Hague: Junk.

Cronquist, A. 1981. See general references.

Cronquist, A. 1988. The evolution and classification of Flowering Plants, ed. 2. New York: New York Botanic Garden.

Dahlgren, R. 1971. Multiple similarity of leaf between two genera of Cape plants, *Cliffortia* L. (Rosaceae) and *Aspalathus* L. (Fabaceae). Bot. Not. 124: 292–304.

Dahlgren, R. 1983. General aspects of angiosperm evolution and macrosystematics. Nord. J. Bot. 3: 119–149.

Endress, P.K. 1994. Diversity and evolutionary biology of tropical flowers. Cambridge, U.K.: Cambridge University Press.

Eriksson, T., Donoghue, M.J., Hibbs, M.S. 1998. Phylogenetic analysis of *Potentilla* using DNA sequences of nuclear ribosomal internal transcribed spacers (ITS), and implications for the classification of *Rosoideae* (Rosaceae). Plant Syst. Evol. 211: 155–179.

Evans, R.C., Campbell, C.S. 2002. The origin of the apple subfamily (Maloideae; Rosaceae) is clarified by DNA sequence data from duplicated GBSSI genes. Am. J. Bot. 89: 1478–1484.

Evans, R.C., Dickinson, T.A. 1999a. Floral ontogeny and morphology in subfamily Amygdaloideae T. & G. (Rosaceae). Int. J. Plant Sci. 160: 955–979.

Evans, R.C., Dickinson, T.A. 1999b. Floral ontogeny and morphology in subfamily Spiraeoideae Endl. (Rosaceae). Int. J. Plant Sci. 160: 981–1012.

Faegri, K., van der Pijl, L. 1979. The principles of pollination ecology, ed. 3. Oxford: Pergamon Press.

Focke, W.O. 1888. Rosaceae. In: Engler & Prantl, Nat. Pflanzenfam. III, 3. Leipzig: Engelmann, pp. 1–61. (The pages 49–61 were published later than pp. 1–48, but there is no clarity about the publication dates, compare p. 767 and 849 in Stafleu & Cowan, Taxonomic Literature, ed. 2, vol. 1).

Fuks, R. 1987. O gênero Agrimonia L. (Rosaceae) no Brasil. Albertoa 1: 73–84.

Gams, H. 1922–1923. Rosaceae. In: Hegi, G. (ed.) Illustrierte Flora von Mittel-Europa, ed. 1, IV. 2, pp. 662–1112a. (The pages 662–908 were published in 1922, the remainder in 1923).

Gleason, H.A., Cronquist, A. 1991. Manual of vascular plants of Northeastern United States and adjacent Canada, ed. 2. New York: New York Botanic Garden.

Goldblatt, P. 1976. Cytotaxonomic studies in the tribe Quillajeae (Rosaceae). Ann. Missouri Bot. Gard. 63: 200–206.

Graham, R.A. 1960. Rosaceae. In: Fl. Trop. E. Africa, 61 pp. London: Crown Agents.

Grandoso, E. 1964. Las especies argentinas del género *Acaena* (Rosaceae). Darwiniana 13: 209–342.

Gunn, C.R., Wiersema, J.H., Ritchie, C.A., Kirkbride, H.J. 1992. Families and genera of Spermatophyta recognized by the Agricultural Research Service. USDA/ARS Tech. Bull. nr. 1796, 500 pp.

Harley, J.L., Smith, S.E. 1983. Mycorrhizal symbiosis. London: Academic Press.

Hegnauer, R. 1973, 1990. See general references.

Henrickson, J. 1986a. *Xerospiraea*, a generic segregate of *Spiraea* (Rosaceae) from Mexico. Aliso 11: 199–211.

Henrickson, J. 1986b. Notes on Rosaceae. Phytologia 60: 468.

Hutchinson, J. 1964. The genera of Flowering Plants, vol. 1. Oxford: Clarendon Press.

Hutchinson, J. 1973. The families of Flowering Plants, ed. 3. Oxford: Clarendon Press.

Iketani, H., Ohashi, H. 1991a. Anatomical structure of fruits and evolution of the tribe Sorbeae in the subfamily Maloideae (Rosaceae). J. Jap. Bot. 66: 319–351.

Iketani, H., Ohashi, H. 1991b. *Pourthiaea* (Rosaceae) distinct from *Photinia*. J. Jap. Bot. 66: 352–355.

Iltis, H. 1913. Ueber das Gynophor und die Fruchtausbildung bei der Gattung *Geum*. Sitz. Ber. Math.-Naturwiss. Kl. Kais. Akad. Wiss. 122: 1177–1212, 2 pl.

IUCN, 1994. IUCN Red list categories. Prepared by IUCN species survival commission, as approved at 40th Meet. IUCN Council, 21 pp. Gland: IUCN.

Jansen, P.C.M. 1981. Spices, condiments and medicinal plants in Ethiopia. . . . Ph. D. Wageningen, 327 pp. (also published in Belmontia n.s. 12).

Jepson, W.L. 1936. A flora of California, vol. 2. Berkeley: University of California Press.

Johri, B.M. et al. 1992. See general references.

Juel, H.O. 1918. Beiträge sur Blütenanatomie und zur Systematik der Rosaceen. Kungl. Svenska Vetensk. Akad. Handl. 58, 5, 81 pp.

Juel, H.O. 1927. Ueber die Blütenanatomie einiger Rosaceen. Nova Acta Reg. Soc. Sci. Upsal., vol. extr., 31 pp., 1 pl.

Kalkman, C. 1968. *Potentilla*, *Duchesnea*, and *Fragaria* in Malesia (Rosaceae). Blumea 16: 325–354.

Kalkman, C. 1988. The phylogeny of the Rosaceae. Bot. J. Linn. Soc. 98: 37–59.

Kalkman, C. 1993. Rosaceae. In: Flora Malesiana I, 11: 227–351.

Kania, W. 1973. Entwicklungsgeschichtliche Untersuchungen an Rosaceenblüten. Bot. Jahrb. 93: 175–246.

Kearney, T.H., Peebles, R.H. c.s. 1960. Arizona Flora, ed. 2. Berkeley: University of California Press.

Knuth, P. 1898, 1904. Handbuch der Blütenbiologie II,1 & III,1. Leipzig: Engelmann.

Koehler, D.L., Smith, D.M. 1981. Hybridization between *Cowania mexicana* var. *stansburiana* and *Purshia glandulosa* (Rosaceae). Madroño 28: 13–25.

Koehne, E. 1890. Die Gattungen der Pomaceen. Wiss. Beil. Progr. Falk-Realgymn. Berlin, 33 pp., 2 pl. Berlin: Gaertner.

Krügel, T. 1992a. Zur zytologischen Struktur der Gattung *Cotoneaster* (Rosaceae, Maloideae) III. Beitr. Phytotax. Univ. Jena 15: 69–86.

Krügel, T. 1992b. Zur zytologischen Struktur von x *Sorbocotoneaster pozdnjakovii* Pojark. Beitr. Phytotax. Univ. Jena 15: 87–92.

Langenfeld, W. 1971. Die Evolution der Gattung *Malus* Mill. Wiss. Zeitschr. Univ. Rostock 20, Math.-Naturwiss. Reihe 1: 49–51.

Lemordant, D. 1974 ('1972'). Histoire et ethnobotanique du Kosso. J. Agr. Trop. Bot. Appl. 19: 560–582.

Lersten, N.R., Curtis, J.D. 1982. Hydathodes in *Physocarpus* (Rosaceae: Spiraeoideae). Can. J. Bot. 60: 850–855.

Lindenhofer, A., Weber, A. 1999a. Polyandry in Rosaceae: evidence for a spiral origin of the androecium in Spiraeoideae. Bot. Jahrb. Syst. 121: 553–582.

Lindenhofer, A., Weber, A. 1999b. The spiraeoid androecium of Pyroideae and Amygdaloideae (Rosaceae). Bot. Jahrb. Syst. 121: 583–605.

Lindenhofer, A., Weber, A. 2000. Structural and developmental diversity of the androecium of Rosoideae (Rosaceae). Bot. Jahrb. Syst. 122: 63–91.

Litav, M., Orshan, G. 1971. Biological flora of Israel. 1. *Sarcopoterium spinosum* (L.) Sp. Israel J. Bot. 20: 48–64.

Long, A.A. 1989. Disjunct populations of the rare shrub, *Neviusia alabamensis* Gray (Rosaceae). Castanea 54: 29–39.

Maximowicz, C.J. 1879. Adnotationes de Spiraeaceis. Acta Hort. Petropol. 6, 1: i–xi, 105–261.

McArthur, E.D., Sanderson, S.C. 1985. A cytotaxonomic contribution to the Western North American Rosaceaous flora. Madroño 32: 24–28.

Melchior, H. 1964. A. Engler's Syllabus der Pflanzenfamilien, ed. 12, vol. 2. Berlin: Borntraeger. (Rosales by G.K. Schulze-Menz, pp. 193–242).

Mendes, E.J. 1978. Rosaceae. In: Launert, E. (ed) Flora Zambesiaca vol. 4. London: Fl. Zambes. Committee, pp. 7–33.

Moffett, A.A. 1931. The chromosome constitution of the Pomoideae. Proc. Roy. Soc. B 108: 423–446, 1 pl.

Morgan, W.T.W. 1981. Ethnobotany of the Turkana: use of plants by a pastoral people and their livestock in Kenya. Econ. Bot. 35: 96–130.

Morgan, D.R., Soltis, D.E., Robertson K.R. 1994. Systematic and evolutionary implications of *rbcL* sequence variation in Rosaceae. Am. J. Bot. 81: 890–903.

Newcomb, W., Heisey, R.M. 1984. Ultrastructure of actinorhizal root nodules of *Chamaebatia foliolosa* (Rosaceae). Can. J. Bot. 62: 1697–1707.

Nordborg, G. 1966. *Sanguisorba* L., *Sarcopoterium* Spach, and *Bencomia* Webb. et Berth. Delimitation and subdivision of the genera. Opera Bot. 11, 2: 1–103, pl. i–vi.

Nybom, H. 1988. Apomixix versus sexuality in blackberries (*Rubus* subg. *Rubus*, Rosaceae). Plant Syst. Evol. 160: 207–218.

Ohashi, H. 1988. *Rhaphiolepis* (Rosaceae) of Japan. J. Jap. Bot. 63: 1–7.

Okuda, T. et al. 1992. Hydrolysable tannins as chemotaxonomic markers in the Rosaceae. Phytochemistry 31: 3091–3096.

Parkin, J. 1914. The evolution of the inflorescence. J. Linn. Soc., Bot. 42: 511–562, 1 pl.

Perry, L.M. 1980. Medicinal plants of East and Southeast Asia. Attributed properties and uses (with the assistance of J. Metzger). Cambridge, Mass.: MIT Press.

Phipps, J.B. 1990. *Mespilus canescens*, a new Rosaceous endemic from Arkansas. Syst. Bot. 15: 26–32.

Phipps, J.B. 1992. *Heteromeles* and *Photinia* (Rosaceae, subfam. Maloideae) of Mexico and Central America. Can. J. Bot. 70: 2138–2162.

Phipps, J.B., Robertson, K.R., Smith, P.G., Rohrer, J.R. 1990. A checklist of the subfamily Maloideae (Rosaceae). Can. J. Bot. 68: 2209–2269.

Phipps, J.B., Robertson, K.R., Rohrer, J.R., Smith, P.G. 1991. Origins and evolution of subfam. Maloideae (Rosaceae). Syst. Bot. 16: 303–332.

Potter, D., Gao, F., Bortiri, P.E., Oh, S.-H., Baggett, S. 2002. Phylogenetic relationships in Rosaceae inferred from chloroplast *mat*K and *trn*L-*trn*F nucleotide sequence data. Plant Syst. Evol. 231: 77–89.

Press, J.R., Short, M.J. (eds.) 1994. Flora of Madeira. London: HSMO.

Ratter, J.A., Milne, C. 1973. Some Angiosperm chromosome numbers. Notes Roy. Bot. Gard. Edinburgh 32: 429–438.

Ridley, H.N. 1930. Dispersal of plants throughout the world. Ashford: Reeve.

Robertson, K.R. 1974. The genera of Rosaceae in the Southeastern United States. J. Arnold Arbor. 55: 303–332, 344–401, 611–662.

Robertson, K.R., Phipps, J.B., Rohrer, J.R., Smith, P.G. 1991. A synopsis of genera in Maloideae (Rosaceae). Syst. Bot. 16: 376–394.

Robertson, K.R., Weeden, N.F., Rohrer, J.R. 1995. The current status of *Chamaemeles* (Rosaceae: Maloideae), a Madeiran endemic. Bol. Mus. Mun. Funchal, suppl. 4: 621–636.

Romoleroux, K. 1996. Rosaceae. In: Flora of Ecuador nr. 56: 3–151.

Rothmaler, W. 1937. Systematische Vorarbeiten zu einer Monographie der Gattung *Alchemilla* (L.) Scop. Feddes Repert. 42: 164–173.

Saïd, C. 1979. Quelques aspects de l'écologie florale chez les Rosaceae: étude morphologique et histologique comparée chez *Sanguisorba officinalis* L. et *Poterium sanguisorba* L. Bull. Soc. Bot. Fr. 126, Lett. Bot.: 311–324.

Sax, K. 1932. The origin of the Pomoideae. Proc. Am. Hort. Soc. 30: 147–150.

Schaeppi, H. 1953. Kelch und Aussenkelch von *Rhodotypus kerrioides*. Viertelj.schr. Naturf. Ges. Zürich 98: 30–36.

Schaeppi, H. 1977. Ueber den "doppelten Fruchtknoten" von *Rhodotypos*. Beitr. Biol. Pflanzen 53: 165–175.

Schneider, C.K. 1905–1906a. Illustriertes Handbuch der Laubholzkunde I, iv + iv + 808 pp (Rosaceae on pp. 440–802). Jena: Fischer. (The pages 440–592 were published in 1905, the remainder in 1906).

Schneider, C.K. 1906b. Species varietatesque Pomacearum novae IV. In: Feddes Repert. 3: 177–183.

Sharma, A.K. 1970. Res. Bull. Univ. Calc. (Cytog. Lab.) 2 (cited in P. Goldblatt, ed. 1981. Index to plant chromosome numbers 1975–1978, Missouri Monogr. Syst. Bot. 5).

Simpson, B.B. 1979. A revision of the genus *Polylepis* (Rosaceae: Sanguisorbeae). Smithson. Contrib. Bot. 43, 62 pp. Washington, DC.: Smithsonian Institution.

Skalicky, V. 1971. Amerikanische Odermennige, *Agrimonia* L. ser. *Parviflorae* ser. n. Nov. Bot. Inst. Bot. Univ. Carol. Prag. 1970: 9–16.

Stebbins, G.L. 1974. Flowering plants. Evolution above the species level. Cambridge, Mass.: Harvard University Press.

Sterling, C. 1964a, b, c. Comparative morphology of the carpel in the Rosaceae. I–III. Am. J. Bot. 51: 36–44, 354–360, 705–712.

Idem. 1965a, b, c. IV–VI. Ibid. 52: 47–54, 418–426, 938–946.

Idem. 1966a, b, c. VII–IX. Ibid. 53: 225–231, 521–530, 951–960.

Idem. 1969. X (Evaluation and summary). Oest. Bot. Zeitschr. 116: 46–54.

Stöger, E.A. 1989. Arzneibuch der Chinesischen Medizin I, II. Wasserburg: DECA. (Loose-leaf edition of an edited translation from the Chinese Pharmacopoeia, 1985).

Sudworth, G.B. 1967. Forest trees of the Pacific slope. 2nd edn. New York: Dover.

Takhtajan, A. 1981. See general references.

Thorne, R.F. 1983. Proposed new realignments in the Angiosperms. Nord. J. Bot. 3: 85–117.

Thorne, R.F. 1992. Classification and geography of the Flowering Plants. Bot. Rev. 58: 225–348.

Troll, W. 1935. Vergleichende Morphologie der Fiederblätter. Nova Acta Leopoldina, N.F. 2: 315–455.

Troll, W. 1964, 1969. Die Infloreszenzen. Typologie und Stellung im Aufbau des Vegetationskörpers. I, II/1. Jena: Fischer.

Vieira, R. 1992. Flora da Madeira o interesse das plantas endémicas Macaronésicas, 155 pp. Lisboa?: Serv. Nac. de Parques.

Weber, C. 1964. The genus *Chaenomeles* (Rosaceae). J. Arnold Arbor. 45: 161–205, 302–345.

Wieffering, J. 1979. Het basis-chromosoomgetal en de taxonomische positie van de tribus Quillajeae binnen de Rosaceae. Danseria 16: 122–123.

Williams, A.H. 1982. Chemical evidence from the flavonoids relevant to the classification of *Malus* species. Bot. J. Linn. Soc. 84: 31–39.

Wolfe, J.A., Wehr, W. 1988. Rosaceous *Chamaebatiaria*-like foliage from the Paleogene of Western North America. Aliso 12: 177–200.

Zardini, E.M. 1971. Especies nuevas o criticas de la flora Jujeña, II. Bull. Soc. Argent. Bot. 14: 107–110.

Zhang, Shu-yin 1992. Systematic wood anatomy of the Rosaceae. Blumea 37: 81–158.

Zhang, Shu-yin, Baas, P. 1992. Wood anatomy of trees and shrubs from China III, Rosaceae. IAWA Bull., n.s. 13: 21–91.

Zielinski, J. 1980. Distribution of *Rosa persica* Michx ex Juss. and its hybrids. Arbor. Korn. 25: 41–51.

Samolaceae

B. Ståhl

Samolaceae Raf., Ann. Gén. Sci. Phys. Bruxelles 5: 349 (1820).

Perennial or sometimes annual herbs, rarely sub-shrubs, often woody at base, sometimes rhizomatous; secretory system usually appearing as a fine, purple or dark brown striation on inflorescence axes or short irregular lines or dots on calyx or bracts; indumentum mostly lacking, when present then consisting of stalked or sessile glandular hairs. Leaves alternate, often in a basal rosette, sometimes lacking on aerial parts, rarely reduced to scales, simple, petiolate but with petiole often poorly differentiated from blade; blades spathulate or oblong, sometimes linear or reniform, margins entire. Inflorescences racemes, corymbs or sparsely branched compound racemes; pedicels often long and slender, often subtended by a single bract inserted at base or midway on the pedicel, sometimes ebracteate; prophylls always lacking. Flowers 5-merous, hermaphrodite, the perianth perigynous, actinomorphic; calyx herbaceous, the tube ± fused to the ovary, the lobes narrowly to broadly ovate or triangular, persistent; corolla white, sometimes pink or purplish, campanulate or urceolate, lobes broadly ovate, imbricate in bud; stamens 5, opposite the corolla lobes, included or sometimes somewhat exserted; filaments largely fused with the corolla tube, free parts as long as the anthers or shorter; anthers narrowly sagittate to oblong, dithecal, opening introrsely by longitudinal slits, the connective sometimes apically produced; staminodes often present, linear to narrowly triangular, inserted at the mouth of the corolla tube, alternating with the lobes; ovary semi-inferior, globose to ovoid; style well developed, shorter than to somewhat longer than the ovary; stigma truncate to capitate; ovules numerous, hemitropous, inserted in several series on a free central placenta. Fruit a ± globose, loculicidally dehiscent capsule. Seeds small, angular; seed coat pale to dark brown, reticulate, two-layered, with rhomboid crystals; endosperm abundant, cell walls smooth; embryo short, straight, cotyledons narrow. x = 12, 13.

A monogeneric family of worldwide distribution, comprising about 12 species.

VEGETATIVE MORPHOLOGY. Some species are more or less woody at base and rhizomatous; the Mexican *S. cinerascens* is a dwarf shrub. The leaves are simple and have entire margins that usually are decurrent or sometimes expanded on the petiole, which thus is poorly demarcated from the blade. In many species, all or most of the leaves are condensed into a rosette at base of the plant. *Samolus junceus* and *S. porosus* have stems without leaves or with reduced, scale-like leaves.

VEGETATIVE ANATOMY. Although mostly glabrous, short protrusive or immersed glandular hairs occur on reproductive and vegetative parts in some species. *Samolus ebracteatus* is glandular-puberulous and glandular-punctate throughout a large part of its distribution (Henrickson 1983). The protrusive hairs have usually one or two stalk cells only, and a few-celled head. In this and a few other species, the corolla mouth and petal bases are densely puberulous from stalked glandular hairs with two to four stalk cells and a single-celled head.

Although not always evident, a secretory system seems to be present in all species.

INFLORESCENCE STRUCTURE. The inflorescence in Samolaceae is basically racemose, being developed as a raceme, corymb or irregularly and sparsely branched compound raceme, sometimes on a long peduncle. The pedicels are often long and slender. Prophylls are lacking. In some species, the bract is inserted near the middle of the pedicel.

FLOWER STRUCTURE. In several species, ligulate, linear or obtrullate staminodes are inserted between the lobes at the mouth of the corolla. These structures are provided with vascular tissue (Thenen 1911); vestigial anthers are lacking. Floral development in *S. valerandi* was studied by Dickson (1936).

POLLEN MORPHOLOGY. The pollen of one examined species, *S. valerandi*, is triocolporate, elliptic, relatively small (12–17 µm diam.), and has a psilate ornamentation (Punt et al. 1974).

FRUIT AND SEED. The fruit is a globose capsule opening loculicidally by five equal valves. The seeds are numerous, obconic or more or less angular, light to dark brown, and reticulate; the testa is two-layered and has rhomboid crystals. The seeds lack starch but contain amyloid and have a short embryo with narrow cotyledons.

AFFINITIES. The family consists of a single genus, *Samolus*, which usually has been referred to Primulaceae. Within that family, Pax and Knuth (1905) placed it in a separate tribe, Samoleae. Because of similarities in floral characters, Bartling (1830) and Mez (1902) suggested a close relationship between *Samolus* and *Maesa* (Maesaceae). However, phylogenetic analyses (Anderberg and Ståhl 1995; Anderberg et al. 1998) did not support a close relationship with Maesaceae, although its true position within primuloid taxa remained obscure. However, in a study combining morphological and DNA sequence data (Källersjö et al. 2000), *Samolus* is placed as sister to Theophrastaceae.

DISTRIBUTION AND HABITATS. Unlike other primuloid taxa, Samolaceae are well represented in the Southern Hemisphere, with three species in temperate South America, four in Australia and Tasmania, and one in South Africa. Four or five species are confined to southern North America (S USA, Cuba and Mexico). One species, *S. valerandi*, is cosmopolitan.

Most species grow in damp, temporarily inundated areas near lakes, ponds or rivers. Some species grow in more or less saline habitats such as small, brackish ponds near the seashore or in salt marshes.

ECONOMIC IMPORTANCE. At least one species, *S. valerandi*, is a popular aquarium and terrarium plant (Crusio 1981), but the family is otherwise of no economic importance.

A single genus:

Samolus L.   Fig. 118

Samolus L., Sp. Pl.: 171 (1753); Pax & Knuth in Engler, Pflanzenreich IV. 237: 336–344 (1905); Lourteig, Lilloa 8: 241–252 (1942), Argentinean spp.; Crusio, Mededel. W.A.P. 2: 13–25 (1982), 6: 13–16 (1984).

Description as for family. About 12 spp.

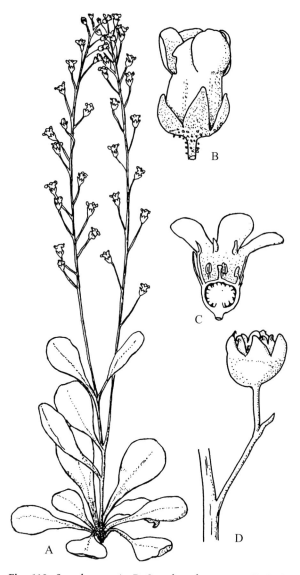

Fig. 118. Samolaceae. A, B *Samolus ebracteatus*. C, D *S. valerandi*. A Habit, flowering plant. B Flower. C Flower in vertical section showing ovary, stamens and staminodes. D Fruit with pedicel and bract. (Orig., A redrawn from Pax and Knuth 1905)

Selected Bibliography

Anderberg, A.A., Ståhl, B. 1995. Phylogenetic interrelationships in the order Primulales, with special emphasis on the family circumscriptions. Can. J. Bot. 73: 1699–1730.

Anderberg, A.A., Ståhl, B., Källersjö, M. 1998. Phylogenetic relationships in the Primulales inferred from *rbc*L sequence data. Plant Syst. Evol. 211: 93–102.

Bartling, F.G. 1830. Ordines naturales plantarum. Göttingen: Dietrich.

Crusio, W.E. 1981. Die Gattung *Samolus*. Inform. ZAG Wasserpfl. 11: 3–8.

Dickson, J. 1936. Studies in floral anatomy. III. An interpretation of the gynaecium in the Primualceae. Am. J. Bot. 23: 385–393.

Henrickson, J. 1983. A revision of *Samolus ebracteatus* (sensu lato) (Primulaceae). SouthWestern Nat. 28: 303–314.

Källersjö, M. et al. 2000. See general references.

Mez, C. 1902. Myrsinaceae. In: Engler, Pflanzenreich IV. 236. Leipzig: W. Engelmann, pp. 1–437.

Pax, F., Knuth, R. 1905. Primulaceae. In: Engler, Pflanzenreich IV. 237. Leipzig: W. Engelmann, pp. 1–386.

Punt, W., de Leeuw van Weenen, J.S., van Oostrum, A.P. 1974. Primulaceae. The Northwest European pollen flora, 3. Rev. Palaeobot. Palynol. 17: 31–70.

Thenen, S. 1911. Zur Phylogenie der Primulaceenblüte. Studien über den Gefässbündelverlauf in Blütenachse und Perianth. Jena: G. Fischer

Sapotaceae

T.D. Pennington

Sapotaceae Juss., Gen. Pl.: 151 (1789) ("Sapotae").
Sarcospermataceae H.J. Lam (1925).

Trees or shrubs, rarely geoxylic suffrutices or lianas, sometimes spiny; branching usually sympodial; latex nearly always present in trunk, branches and fruits, usually white, rarely yellow or blue; indumentum nearly always of malpighiaceous hairs (simple in *Delpydora*). Leaves alternate, spirally arranged or distichous, less frequently opposite or verticillate, simple, entire or very rarely spinous-toothed; petiole rarely bearing a pair of minute stipels; stipules + or 0. Inflorescence fasciculate or flowers occasionally solitary, axillary, ramiflorous or cauliflorous; fascicles occasionally arranged along short leafless axillary, panicle-like shoots; fascicle base sometimes developing into short, densely scaly brachyblasts. Flowers bisexual or unisexual (plants monoecious or dioecious), actinomorphic; calyx a single whorl of 4–6 free or partly fused, imbricate, sometimes quincuncial sepals, or 6–11 sepals in a closely imbricate spiral, or with 2 whorls of 2–4 sepals and then the outer whorl valvate or only slightly imbricate; corolla rotate, cyathiform or tubular, sympetalous; tube shorter than, equalling or exceeding the petals; petals 4–18, entire, lobed or partly divided or divided to the base into 3 segments and then median segment entire, 2 lateral or dorsal segments entire, laciniate or shallowly or deeply divided; stamens 4–35(–43), fixed in lower or upper half of corolla tube or at the base of the lobes, rarely free, in a single whorl opposite the corolla lobes, or, when more numerous than the corolla lobes, some opposite and some alternate with the corolla lobes, or sometimes several stamens clustered opposite each lobe, or arranged in 2–3 alternating whorls within the corolla tube, exserted or included; filaments often geniculate in bud, free, rarely fused into a staminal tube, or partially fused to the staminodes; anthers often extrorse; staminodes 0–8(–12) in a single whorl alternating with the stamens or fixed in the corolla lobe sinuses, simple or variously lobed, toothed or divided, sometimes petaloid; disk annular or patelliform, surrounding the ovary base and sometimes fused with it, or absent; ovary superior, 1–15(–30)-locular, loculi usually uniovulate, rarely 2–5-ovulate, placentation axile, basi-ventral or basal; style simple, included or exserted; style-head simple or minutely lobed. Fruit a berry or rarely a drupe, or tardily dehiscent by a single lateral valve; pericarp fleshy or less frequently leathery or woody; seeds 1–many, globose, ellipsoid, oblong, often strongly laterally compressed, testa usually smooth and shining, free from the pericarp, less frequently roughened, wrinkled or pitted and then often adherent to the pericarp; hilum adaxial, basi-ventral or basal, narrow or broad, sometimes extending to cover most or all of the seed; embryo vertical, oblique or horizontal, with thin foliaceous or thick flat or plano-convex, usually free, cotyledons; radicle included or exserted; endosperm + or 0. x = 10, 11, 12, 13, 14.

A pantropical family of 53 genera and about 1100 species, mostly in humid forest, but some genera (e.g. *Sideroxylon, Argania*) extending into semi-arid and arid regions.

MORPHOLOGY. All Sapotaceae are woody plants, the great majority trees or shrubs. They are predominantly plants of moist lowland forest below 1000 m altitude, and are commonly canopy trees or emergents. They are also well represented in semi-arid zones where they occur as small trees or dense multi-branched shrubs often equipped with sharp axillary spines derived from modified axillary shoots.

Bole section (fluted or cylindrical), presence and type of buttressing, bark type (fissured, scaling, smooth) are often specifically constant, but less useful at higher levels of classification. Nearly all species of Sapotaceae have fissured or scaling bark, and only a few examples are known of large Sapotaceous trees with smooth bark (e.g. *Micropholis guyanensis* and *Pouteria laevigata* in tropical America.)

The growth and branching pattern of the main axis and branches gives rise to the characteristics of the crown, and several architectural models recognized by Hallé and Oldeman (1970) occur in the Sapotaceae.

1. *Model of Aubréville.* Branched trees in which the main axis is monopodial with rhythmic growth producing subverticillate branches and a layered crown. The branches are plagiotropic and sympodial with a series of long, bare horizontal shoots each terminated by a short, vertically orientated section bearing densely clustered spirally arranged leaves. The vertically orientated part of a branch continues to grow slowly, and axillary inflorescences are produced indefinitely. The most frequently occurring architectural type in Sapotaceae, found in *Autranella, Baillonella, Vitellaria, Manilkara, Neolemonniera, Tieghemella, Palaquium, Madhuca, Aubregrinia, Englerophytum, Pouteria, Synsepalum, Omphalocarpum.*

2. *Model of Corner.* Unbranched treelets with lateral inflorescences and indeterminate growth of the shoot apex. *Delpydora, Niemeyera.*

3. *Model of Roux.* Branched trees with the trunk monopodial and orthotropous, with continuous growth producing spirally arranged branches. The branches are plagiotropic with distichous phyllotaxis, and they can be either monopodial or sympodial. This type follows taxonomic lines closely and is usually consistent for a genus or section of a genus. *Micropholis, Payena, Sarcaulus.*

4. *Model of Troll.* In this type all axes are plagiotropic from early in their development. Certain axes, however, especially in the juvenile state, may have a short orthotropic part. All branches have horizontal growth and distichous phyllotaxis, and growth in height is achieved by the straightening out of the basal part of certain shoots after leaf fall. Documented for *Chrysophyllum cainito* and probably in other species of this genus.

Phyllotaxis is usually spiral or distichous; opposite leaf arrangement is typical of *Sarcosperma, Pichonia* and *Leptostylis,* but occurs sporadically in many other genera, e.g. *Pouteria, Pradosia, Sideroxylon.*

The leaves of all Sapotaceae are simple. The leaf margin is entire in all species except *Chrysophyllum imperiale* and *C. subspinosum* (both South American), which are spinous-serrate. Leaf venation provides many characters of use at generic and sectional level. The principal venation types, which are defined in terms of the primary (midrib), secondary, tertiary and quaternary veins, are craspedodromous, eucamptodromous and brochidodromous (Hickey 1973). Stipules are commonly found in tribes Mimusopeae and Isonandreae. They are usually small and caducous, but less frequently large, conspicuous and sometimes persistent (e.g. *Madhuca* spp.). Elsewhere in the family stipules are generally absent, with a few notable exceptions, viz. *Sarcosperma, Chromoluma, Ecclinusa, Synsepalum, Englerophytum, Capurodendron* and *Tsebona.* In these latter cases the stipulate condition is well correlated with other generic characters. The petiole is generally unappendaged, but in a small number of species (some *Sarcosperma* and *Pradosia*) it bears a pair of small, undivided appendages near the base or apex. They have been called auricles or stipels. Their origin and function is not understood.

VEGETATIVE ANATOMY. The reader is referred to Metcalfe and Chalk (1950). A family characteristic is the presence of sticky, slow flowing latex in the cut bark, branches and fruit; its colour is usually white, but in a few species it may be yellow (*Pouteria congestifolia*) or blue-green (*Niemeyera acuminata*). The laticiferous ducts occur in the leaf, where they usually accompany the veins, but sometimes are also interspersed in the mesophyll. In the stem they can be found in the cortex, phloem, and pith. They are also present in flowers and fruits. An indumentum of malpighiaceous hairs is present in all genera except *Delpydora,* which has simple hairs. A common condition is a mixture of appressed 2-branched hairs with erect or spreading simple hairs. In a majority of species the malpighiaceous hairs are more or less sessile, with one long branch and one short branch, but in others both branches may be well-developed. Less frequently the 2-branched hairs have a long stalk. Some species of *Manilkara* and *Pouteria* sect. *Pouteria* have a minute, closely appressed indumentum which forms a pellicle on the lower leaf surface. The mesophyll often includes sclerenchymatous fibres; stomata are ranunculaceous and nearly always confined to the lower leaf surface. Cork in the axis arises superficially. Solitary crystals are always present and often abundant in the cortex. In the wood, vessels occur in loose, radial or oblique lines and often in multiples of 4 or more cells. The perforations are simple, and the intervascular pitting is alternate and usually small. The vessel elements are usually of medium length, but sometimes are moderately or very long. The axial parenchyma is apotracheal and tends to be banded. The rays are heterogeneous and 1–6 cells wide. The fibres have simple pits, but pits may occasionally be bordered. Vasicentric tracheids are present in some genera. Silica is often present in elements other than the vasicentric tracheids.

INFLORESCENCE MORPHOLOGY. The inflorescence is a simple fascicle of several to many pedicellate flowers arising directly from the leaf axil or rarely reduced to a solitary axillary flower (e.g. *Manilkara zapota*).

Scattered throughout the family are species in which the inflorescence is an axillary shoot of limited growth bearing several lateral fascicles, each subtended by a leaf or bract, and the fascicles may themselves be reduced to a few or only one flower. This type of compound inflorescence may be found together with the more usual simple axillary fascicle in the same species (e.g. *Pouteria ramiflora*) or it may be constant within a species (e.g. *Sideroxylon racemosum, Pouteria macahensis*), and in *Sarcosperma* it is found in every species of the genus. In *Sarcosperma* the inflorescence axis may itself be further branched.

Another modification of the simple axillary fascicle occurs sporadically in several genera, e.g. *Micropholis*. Here the fascicle produces flowers over a long period, during which time it develops a short stout and densely scaly axis (brachyblast terminated by a single, few-flowered fascicle.

FLORAL STRUCTURE. The calyx provides some of the best taxonomic characters at generic and tribal levels of classification. There are two major calyx types, the uniseriate and the biseriate.

The uniseriate calyx has a single whorl of (4)5(6) free or partially fused imbricate sepals (Fig. 123D), as in most *Pouteria* or *Chrysophyllum*, or 5–12 sepals in a closely imbricate spiral (e.g. *Pouteria* sect. *Aneulucuma*).

The biseriate calyx, which is characteristic of tribes Isonandreae and Mimusopeae, has two whorls of 2, 3 or 4 free sepals, the outer whorl being valvate (Fig. 121G). The number of sepals in each whorl is very consistent and closely follows generic lines. *Isonandra* and *Madhuca* have 2 whorls of 2 sepals, *Palaquium* and *Manilkara* have 2 whorls of 3 sepals, and *Mimusops* has 2 whorls of 4 sepals.

The corolla is always actinomorphic and gamopetalous. It may be rotate (e.g. *Pradosia, Niemeyera*), with the tube very short in relation to the lobes, or the tube may be shortly cyathiform, with spreading lobes (e.g. *Sideroxylon*), or it is short or long tubular usually with the tube exceeding the lobes, and with the lobes erect or only slightly spreading (Chrysophylleae). The number of corolla lobes relative to the number of sepals is a character of primary importance. In Mimusopeae, which have a biseriate calyx, the number of corolla lobes is usually the same as the number of sepals (Figs. 120C, 121D), but in Isonandreae, which also have a biseriate calyx, the number of corolla lobes is often 2–3 times that of the sepals (e.g. *Madhuca*) and occasionally more (e.g. *Burckella, Pouteria* sect. *Antholucuma*). The division of the corolla lobes into a median and two lateral segments is characteristic of many members of Mimusopeae and Sideroxyleae. In *Mimusops, Manilkara* and related genera, the two lateral segments equal the median segment and are themselves sometimes subdivided to the base. The arrangement of the various corolla lobe segments in the complex flowers of *Mimusops* and *Manilkara* is distinctive. The central segment remains erect and clasps the opposing stamen, but the lateral segments spread horizontally (Figs. 120C, 121D). As the anthers are generally extrorse, this arrangement ensures that they dehisce into the clasping corolla segment. In members of Sideroxyleae, the lateral corolla lobe segments are generally poorly developed.

All genera of Sapotaceae have the stamens opposite the corolla lobes inserted within the corolla tube or occasionally (e.g. some *Pradosia*) at the base of the corolla lobes, or rarely (e.g. *Pouteria bangii*) free. The position of the stamens relative to the corolla (exserted or included) follows tribal, generic or sectional lines. Exserted stamens are associated with the short-tubed, rotate or spreading corolla and are typical of genera in Sideroxyleae, and of such genera as *Elaeoluma, Pradosia* and *Niemeyera* in Chrysophylleae. The arrangement in Mimusopeae is essentially the same, except that the stamens are exserted relative only to the lateral corolla segments, while they are clasped by the median segment. Included stamens, found in many genera of Chrysophylleae, are associated with the tubular or cyathiform corolla with short erect lobes.

The number of stamens relative to the number of corolla lobes is diagnostic for the tribes. In Chrysophylleae, Mimusopeae, Sideroxyleae there is nearly always one stamen opposite each corolla lobe, whereas in Isonandreae and Omphalocarpeae there are 2 or more stamens opposite each lobe. The stamens may be grouped opposite the corolla lobes as in *Omphalocarpum*, or they may be in a single continuous whorl, or in 2–3 alternating whorls within the corolla tube (e.g. *Madhuca*).

The increase in number of stamens in Isonandreae is correlated with an increase in size of the anthers, which is possibly associated with bat pollination.

The stamens are generally free, but in a few genera become fused to the adjacent staminode (e.g. *Autranella*), or to the adjacent stamen, thus forming a partial or complete staminal tube (e.g. *Aulandra*, *Englerophytum*). *Magodendron* is unique in having locellate anthers.

Petaloid or scale-like structures alternating with the fertile stamens (opposite to them in *Gluema*) are widespread, and their presence or absence is highly correlated with other structures of generic importance. They are apparently derived from stamens, by loss of the anther, and then by further development or reduction of the filament. Staminodes are occasionally seen to bear apical anthers (Fig. 122D), and genera such as *Manilkara*, which typically have a whorl of stamens alternating with a whorl of sterile staminodes, occasionally (e.g. *M. valenzuelana*) present two whorls of fertile stamens. Sometimes the staminode whorl is incomplete by reduction in number and the staminodes present are variable in size. They reach their greatest development in *Mimusops* and related genera, where they are large, hairy, hooded structures alternating with the stamens, erect or incurved, forming a closed receptacle around the ovary. The structure so formed acts as a nectar receptacle in the newly opened flower. Elsewhere in the family they are smaller or absent, and in some of the simple-flowered 5-merous genera such as *Pouteria*, they are inserted in the corolla lobe sinuses, above the level of the fertile stamens.

Although the flowers secrete nectar, the nectar-secreting area is morphologically poorly differentiated, and represented only by a small ring-shaped disk around the ovary base.

The ovary is 1- to multilocular, with uniovulate loculi; exceptions are *Diploon* with 1 loculus and 2 basal ovules, and some *Sideroxylon* which have lost some or all of the ovary septa, resulting in up to 5 basal ovules. Placentation is axile or less frequently basal or basi-ventral.

Members of Mimusopeae, Isonandreae and Omphalocarpeae have ovary loculi equalling the number of sepals, twice as many or even more (up to 30 in *Omphalocarpum*) whereas in Sideroxyleae and Chrysophylleae the trend is generally towards reduction in the number of loculi towards the unilocular condition, which is reached in some *Pouteria* species (sect. *Franchetella*).

Five floral types are recognized in the family; they are based on the number and degree of complexity of the floral parts and the precise arrangement of the organs relative to each. They correlate well with the tribal divisions.

1. Mimusops type. Occurs in *Mimusops*, *Manilkara* and other genera of tribe Mimusopeae. This is the most complex floral structure of the family. The calyx is biseriate, with 2 whorls of 3 or 4 sepals. The corolla, which has a short tube exceded by the lobes, opens widely, exposing the erect staminodes and stamens. The corolla lobes are subdivided into 3 segments, the median segment is erect and clasping the opposing stamen, while the 2 lateral segments spread horizontally. The lateral corolla lobe segments are often subdivided to the base, so that each corolla lobe is subdivided into 5 equal-sized petaloid segments. The well-developed and variously shaped staminodes alternate with the stamens, and are held erect to form a closed sheath around the ovary. The majority of species in this group have equal numbers of sepals, corolla lobes, stamens and staminodes, but modification and reduction of corolla lobe appendages and staminodes occur in all the main geographical regions. Two genera, *Labourdonnaisia* and *Letestua*, have 2–3 times as many corolla lobes and stamens as sepals. Bat pollination has been recorded for this floral type (*Manilkara*).

2. Madhuca type. Characterized by a biseriate calyx of 2 whorls of 2 or 3 sepals, and a short-tubed open corolla with spreading lobes. The number of corolla lobes may equal the number of sepals or be 2–3 times as many, but the lobes are undivided. There are 2–3 times as many stamens as corolla lobes, usually with large exserted anthers. Staminodes are absent. The style is usually very long and exserted. This floral type is characteristic of most species of Isonandreae, and it also appears in the Omphalocarpeae (*Tsebona*). The latter differs, however, in the presence of well-developed staminodes. This floral type is also associated with bat pollination in *Madhuca*.

3. Sideroxylon type. Present in most species of tribe Sideroxyleae, and less complex than the 2 previous types, with 5-merous flowers, a uniseriate calyx of usually 5 imbricate sepals, and a short-tubed open corolla with wide-spreading lobes. The corolla lobes of many species are subdivided into 3 segments, but the two lateral segments are greatly reduced and much smaller than the median segment, and in many cases they have disappeared altogether. The stamens are attached at the top of the corolla tube and exserted and alternate with 5 variously developed staminodes. The style is generally short.

4. Pradosia type. This and the *Pouteria* type represent the simplest floral arrangement. It occurs only in several genera of Chrysophylleae, and is most commonly found in pentamerous flowers (e.g. *Pradosia, Pichonia, Elaeoluma, Synsepalum*). The flowers have a uniseriate calyx of 5 imbricate sepals, short-tubed rotate or spreading corolla with 5 simple corolla lobes and 5 exserted stamens fixed at the apex of the corolla tube. Simple staminodes alternating with the stamens are sometimes present. The style is short. No observations on the pollination of this type of flower have been recorded, but it is probably effected by short-tongued insects.

5. Pouteria type. Confined to genera in Chrysophylleae such as *Pouteria, Chrysophyllum, Micropholis* and *Ecclinusa*. The flowers have a uniseriate calyx and are usually pentamerous but, in some sections of *Pouteria*, the calyx is of only 4 imbricate sepals, and the corolla lobes may number 6–8 or occasionally more. The corolla lobes are simple, and staminodes if present are small and undivided. It differs from the *Pradosia* type in the shape of the corolla, which is tubular or cyathiform with small, more or less erect lobes, and included stamens. The stamens may be fixed at any level in the corolla tube, but if at the apex, then the filaments are poorly developed, so that the anthers remain within the erect corolla lobes. The corolla tube is often longer than the corolla lobes. The style is short and included, or sometimes slightly exserted beyond the apex of the corolla tube. Pollinators for this floral type are postulated to be short- or long-tongued insects.

Bisexual flowers predominate in tribes Mimusopeae, Isonandreae and Sideroxyleae. Only 1–2% of the species of these tribes have unisexual flowers, but among the latter are several species which exhibit the strongest dimorphism of any in the family. These are *Sideroxylon puberulum* (Mascarenes), *Neohemsleya usambarensis* (Tanzania), and *Nesoluma polynesicum* (Hawaiian Islands), all of which are dioecious. The functionally male plants bear well-developed rudiments of the opposite sex, the pistillode containing small ovules, but at least where field observations have been made, they never set fruit. Female flowers of these species are generally smaller than the male, and corolla, stamens and staminodes are strongly modified. The corolla lobes are reduced in size, almost free in *Neohemsleya*, and are represented by small, free vestigial petals in *Nesoluma*, or they are absent. The stamens are reduced to small sterile vestiges, or they are absent.

Unisexual flowers are more common in Chrysophylleae, especially in the large genus *Pouteria*, where they reach levels of around 50%. The true percentage of unisexual species may be much higher owing to the lack of observations, and to the fact that functionally male flowers have well-developed rudiments of the opposite sex and may be mistaken for hermaphrodite flowers. Within Chrysophylleae both monoecious and dioecious conditions are known, but dioecy seems to predominate.

The simplest form of sexual dimorphism seen in this group involves the loss of anthers in the female, the filaments remaining. In others the anther may be converted into a flattened appendage. Further reduction involves the complete loss of stamens in the female flower (e.g. *Pouteria nudipetala*), and loss of stamens may also be accompanied by reduction in size of the staminodes (e.g. *Pouteria rufotomentosa*). Finally, loss of stamens may be correlated with differences in flower size, with generally the female much smaller than the male. At the same time, differences in flower size may also be accompanied by a change in the relative length of corolla tube and corolla lobes. The female flowers of *Pouteria* have corolla lobes free almost to the base, whereas in the male the corolla tube is only slightly shorter than the lobes.

EMBRYOLOGY. The anther has a secretory tapetum with multinucleate cells. The pollen grains are 2-celled when shed. The ovule is erect, anatropous, unitegmic and tenuinucellate. Presence of a hypostase is contentious. The archesporial cell functions as megaspore mother cell and the embryo sac is of the Polygonum type. Endosperm development is of the nuclear type; the endosperm is thin-walled or with thick, often amyloid walls and oily (Corner 1976; Johri et al. 1992).

POLLEN MORPHOLOGY (by M.M. HARLEY). With few exceptions, Sapotaceae pollen is easily recognised based on its morphology. Exine stratification together with shape and apertures characterise the pollen of the family. Pollen grains are single, isopolar, angulaperturate or, less frequently, planaperturate (in 4-aperturate grains). The amb is more or less circular, or governed by the number of apertures, for example, 3-triangular (Fig. 119H), 4-square (Fig. 119I) or 5-pentagonal (Fig. 119J). The grains are predominantly prolate-spheroidal, subprolate or prolate, occasionally spheroidal or, rarely, oblate-spheroidal. Within single collections the colporate apertures may be 3 only, 3 and 4, 4 only, 4 and 5 or, rarely, 5 only in number. Very rarely some 6-colporate grains are recorded in samples which have mainly 4- or 5-colporate grains. Average polar length is between 16 and 93 μm. The colpi are usually narrow, occasionally broad, and range from vestigial (Fig. 119A) to c. 4/5th (Fig. 119B) or 5/6th of the polar length (Fig. 119C). They are provided with a tough, granular membrane (Fig. 119B) that does not rupture during acetolysis, except in the endoapertural region where there is no underlying endexine. The endoapertures are narrowly or broadly lalongate (Fig. 119F), less frequently circular or, extremely rare, broadly lolongate. The tectum in the central apertural region in many species is protrudent (Fig. 119A, K) or semi-protrudent. This is usually associated with narrow endoapertures and vestigial or short colpi. Wall thickness at the poles (Fig. 119G, K) is 1–3(–5) μm; at the equator 1–5(–8) μm. The ektexine is nearly always composed of a thick tectum, a very narrow infratectum ('columellar layer'), which is either granular or with very reduced columellae, that are sometimes interspersed with granulae, plus a foot layer of similar width, or thicker than, the tectum (Fig. 119G). The endexine is usually absent in the polar region but greatly thickened, especially in the area surrounding the endoaperture and underlying the colpus (Fig. 119D, E, K). In some species it is absent or very thin in the mesocolpial areas, while in others it may be slightly thinner, or as thick, as in the endoapertural areas, resulting in a continuously thickened band of exine in the equatorial region. There is, however, a gradual reduction in the thickness of this band towards the poles. The tectum is usually sparsely punctate or perforate (Fig. 119B). The perforations, which may occur in shallow pits, may be more dense and/or coarse, in either the mesocolpia or the apocolpia. Microfossulae are recorded for a few species. In some species the entire surface, seen with SEM, appears psilate, subpsilate (Fig. 119C), scabrate, finely or coarsely granular (Fig. 119L), anastomosed granular, finely striate-rugulate, striate (Fig. 119M), low-relief angular rugulate (Fig. 119N), or weakly to coarsely rugulate (Fig. 119O). In other species there is a clear differentiation between the more or less smooth apocolpia and a weakly or distinctly rugulate mesocolpia (Fig. 119A). Rarely, *Diploön cuspidata* and *Micropholis retusa*, the exine has supratectal spines (Harley 1990b). In *Sarcaulus* the exine is spinulose.

Two major pollen groups, A and B, were defined for Sapotaceae (Harley 1986a, 1990a), and 12 pollen types (Harley 1990b, 1991), based on presence or absence of continuous endexinous thickening in the equatorial region, number of apertures, surface patterning and colpus length. The pollen morphologies of *Chrysophyllum marginatum*, *C. inornatum* and *Englerophytum stelecantha* are more or less anomalous within the family (Harley 1990b).

KARYOLOGY. For a review of chromosome numbers in Sapotaceae, counts for 95 species have been evaluated (Johnson 1991), all based on x = 10, 11, 12, 13 or 14, nearly all at diploid level. The higher numbers (x = 13, 14) predominate in tribes Chrysophylleae and Omphalocarpeae, whereas numbers in Mimusopeae, Isonandreae and Sideroxyleae are, with one exception, based on x = 10 or 11. Johnson (1991) hypothesised that these numbers may reflect a series of descending dysploidy, but the reverse may also be a possibility, particularly in view of the correlation with floral structure.

POLLINATION (from notes kindly provided by S. VOGEL to the editor). Although the majority of Sapotaceae certainly is entomophilous, in the literature only instances of bat pollination seem to have been recorded. *Madhuca indica* (= *Bassia latifolia*) has been observed being visited and pollinated by flying foxes in India (Cleghorn 1922); the trees start blooming there in the leafless state and the pendent flowers produce a strong, unpleasant odour. The animals feed on the fleshy petals, get dusted with pollen when exploiting the protandrous flowers and, in turn, deposit the pollen on receptive stigmas. The same tree species was found to be bat-exploited in Mauritius and Java, where the Megachiroptera *Pteropus niger* and *Cynopterus sphinx*, respectively, were feeding on the petals (van der Pijl 1936; Cheke and Dahl 1981). *Petropus* bats were also observed as visitors to the flowers of *Palaquium gutta* and *P. quercifolium* and *Madhuca macrophylla* on Java, where they ate the petals of the bad-smelling flowers, those of *Madhuca macrophylla* being reportedly nectariferous in the evening (van der Pijl 1936).

FRUIT AND SEED. The fruit is typically a 1–many-seeded berry, with a fleshy, leathery or rarely woody outer pericarp. Differentiation of the endocarp to form a distinct cartilaginous layer occurs in *Pradosia*, the fruit of which is therefore

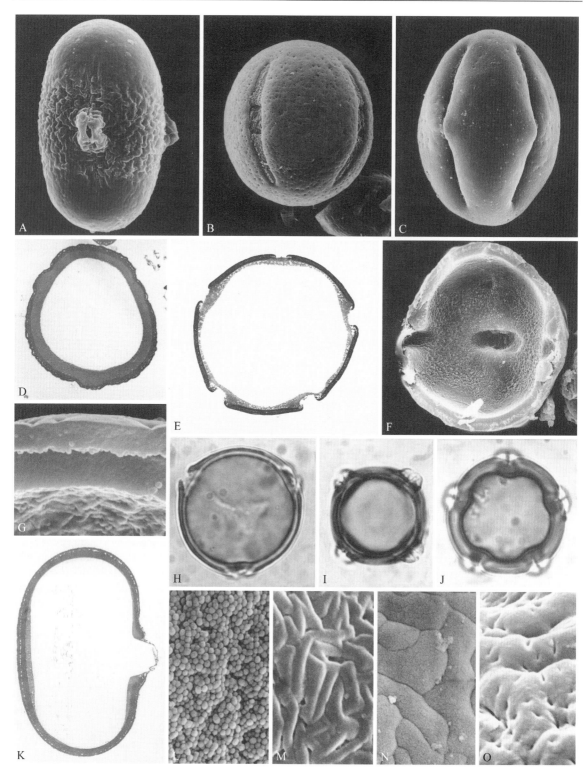

Fig. 119. Sapotaceae, pollen. **A** *Pouteria caimito*, equatorial view ×1200 SEM. **B** *Payena endertii*, equatorial view ×610 SEM. **C** *Vitellaria paradoxa*, equatorial view ×1310 SEM. **D** *Pouteria ucuqui*, equatorial plane ×2330 TEM. **E** *Mimusops schimperi*, equatorial plane ×1160 TEM. **F** *Pycnandra* aff. *vieillardii*, polar orientation, fracture to show interior of grain ×1085 SEM. **G** *Pouteria multiflora*, wall fracture, pole ×10000 SEM. **H** *Pouteria laurifolia*, equatorial plane ×1000 LM. **I** *Tieghemella heck-* *elii*, equatorial plane ×1000 LM. **J** *Palaquium quercifolium*, equatorial plane ×1000 LM. **K** *Sideroxylon glanduliferum*, polar plane ×1200 TEM. **L** *Madhuca aristulata*, tectum surface, mesocolpium, ×10000 SEM. **M** *Sideroxylon celastrinum*, tectum surface, mesocolpium, ×10000 SEM. **N** *Pouteria baillonii*, tectum surface, mesocolpium, ×10000 SEM. **O** *Madhuca pallida*, tectum surface, mesocolpium, ×10000 SEM. Photos M.M. Harley

considered to be a drupe. Four genera of the Mimusopeae (*Lecomtedoxa*, *Neolemonniera*, *Gluema*, *Eberhardtia*) have a dry fruit which dehisces tardily to form a 1–several-seeded loculicidal capsule.

The general appearance of the sapotaceous seed is so distinctive as to be one of the family characteristics. It has a smooth, hard, shining brown testa which strongly contrasts with the pale, rough "scar" area (hilum) by which the seed is attached to the fruit. The smooth shining testa is normally free from the endocarp, but several modifications of this arrangement are found. In some species, e.g. *Pouteria gongrijpii*, the testa, although still smooth, is adherent to the fruit over its whole surface. In this case the seed can still be removed from the fruit, but with difficulty. In others, such as in species of *Chrysophyllum* sect. *Prieurella*, the testa becomes roughened and is strongly adherent to the fruit. The fruit and seed are thus effectively fused and cannot be separated. In other species, such as *Pouteria glomerata*, fusion of the seed to the fruit is achieved by the extension of the hilun over almost all the seed surface. The seed shape, position of the hilum, and its extent relative to the smooth, shining area of testa all provide taxonomic characters useful at the generic and species level.

The position of the hilum is a most important character at the generic level and above, as it is highly correlated with other floral features. Sideroxyleae are typified by a basal hilum, derived from the basal or basi-ventral ovule position. A basal or basi-ventral hilum is also characteristic of *Mimusops* and some *Manilkara* species. In the majority of Chrysophylleae, Omphalocarpeae and Isonandreae the hilum is adaxial (derived from axile ovule placentation).

Within the two general hilum positions, there is considerable variability in the extent and shape of the hilar area. Thus, the basal hilum in members of the Sideroxyleae is generally as broad as long, or broader than long, whereas the basi-ventral hilum of *Manilkara* tends to be elongate. The lateral adaxial hilum may be short and narrow (confined to the basal half of the seed), as in some *Pouteria* species, or long and narrow (extending the full length of the seed), as in many *Chrysophyllum*. In other genera such as *Niemeyera*, the adaxial hilum extends to cover up to half or more of the seed surface, and in extreme cases covers the whole seed surface except for a thin abaxial strip which still retains the smooth, shining testa (e.g. *Pouteria speciosa*). In *Ecclinusa*, the narrow hilum extends down the adaxial face and around the

base. The width of the hilum is to some extent correlated with seed shape. The strongly laterally compressed seed of *Chrysophyllum* nearly always has a narrow adaxial hilum, and conversely, the broad ellipsoid seed of *Pouteria* usually has a broad adaxial hilum, but there are numerous exceptions to the latter.

The orientation of the embryo within the seed is generally vertical, with the cotyledons superior and the radicle inferior. The only variation is found in *Sideroxylon* where the arrangement varies from vertical, through oblique to fully horizontal.

There are two principal embryo types, strongly correlated with seed shape and with the presence or absence of endosperm:

a) Embryo with thin flat foliaceous cotyledons and exserted radicle, which is generally associated with a laterally compressed seed, and with the presence of endosperm. On germination the cotyledons emerge from the seed and are photosynthetic (phanerocotylar). This type is well represented in *Mimusops*, *Manilkara*, *Sideroxylon* and *Chrysophyllum*.
b) Embryo with thick plano-convex cotyledons with radicle included or only extending to the surface. This is generally associated with a broader seed which is not laterally compressed, and with the absence of endosperm. The cotyledons of this type are retained within the seed coat on germination and do not photosynthesise (cryptocotylar). Well represented in *Pouteria*.

All species of *Chrysophyllum* have an embryo with thin foliaceous cotyledons, with the presence of endosperm, correlated with the simple 5-merous flower lacking staminodes. Many of the smaller genera of Sapotaceae have only one embryo type, but the majority of the larger genera have a significant number of exceptions, e.g. *Palaquium* is predominantly non-endospermous, with thick plano-convex cotyledons, but a minority of species have thin foliaceous cotyledons with endosperm. Similar exceptions occur in *Sideroxylon* and *Madhuca*. *Pouteria* is predominantly non-endospermous with thick cotyledons, but about one third of the total species have the other embryo type, and between the two is an extensive series of intermediate conditions, in which the cotyledons are flattened but thick and fleshy, and the endosperm is reduced.

Corner (1976) found the testa multiplicative in most layers, including the outer epidermis; the outer part, 8–25 cells thick or more, forming a heavily lignified, sclerotic layer, the inner part thin-walled and eventually crushed.

DISPERSAL. There is great variation in the size and texture of the sapotaceous berry, correlated with the dispersal agent. Primates are the major dispersers, taking both small-fruited and large-fruited species. They are attracted to species with both soft fleshy pericarps or hard leathery pericarps, and the seeds are spat out, or pass unharmed through the gut. Some species such as *Manilkara zapota* are bat-dispersed, and these have large berries with a soft pericarp. Birds (especially parrots) are also seen to take Sapotaceae fruit, although they often destroy the seed.

SEEDLINGS. Bokdam (1977) studied the seedlings of African Sapotaceae and recognized two major groups within them. Type 1 has enlarged photosynthetic foliaceous cotyledons and develops from endospermous seeds; Type 2 has unenlarged and often non-photosynthetic fleshy cotyledons developing from seeds with scant or no endosperm. Bokdam subdivided these groups into phanerocotylar and cryptocotylar, and also on the thickness of the cotyledons, and on the development of cotyledonary nervation. He found correlations between seedling types and other morphological characters including the level of insertion of stamens in the corolla tube, number of ovary locules, number of seeds in ripe fruit and shilum position, shape and size, but the relationships are imperfect, even on the basis of African species alone. As the loss of endosperm has apparently occurred several times in all the major groups of the family, the associated seed, hilum and seedling characters have also changed, and therefore none of these characters is of use in defining them. They are, however, useful at the lower level of genus, section and species.

PHYTOCHEMISTRY. Data are from Hegnauer (1973, 1990), and Waterman and Mahmoud (1991). By far the most characteristic compound synthesised by Sapotaceae is the latex, a mixture of polyisoprenes and resins, from which the commercial coagulation products gutta percha, balata and chicle are derived. Sapotaceous latex is a mixture of *trans*-polyisoprenes and resins, in which the polyisoprene content, according to quality, ranges from 80 to about 20%, whereas ruber is a pure *cis*-isomer of isoprene. Gutta percha is a product of Malakka, Indonesia and the Philippines, mostly derived from *Palaquium* species; good qualities

have about 80% polyisomere content. Balata is a product of tropical America; it contains more resinous material and its source are the genera *Manilkara* and *Mimusops*. Chicle is a polymer containing both *trans*- and *cis*-isoprene in a ratio 2:1; the classical source is *Manilkara zapota*. The distribution of these polymers within the family is still unknown.

Further characteristic compounds are saponins and pentacyclic triterpenes. Saponins are frequently found in bark, wood and seed; in the latter they may act as defence and storage compounds. The pentacyclic triterpenes are based on widespread skeleta, to which unusual esterifying groups are attached at the C-3 alcohols.

Alkaloids are of limited occurrence in Sapotaceae and are represented by the pyrrolizidine group, which is widely distributed in other plant families. Phenolic compounds are represented by common flavonols including myricetin, proanthocyanidins (procyanidin and prodelphinidin) as constituents of condensed tannins, and gallic acid and the hydrolysable tannins based upon it.

AFFINITIES. Sapotaceae have usually been included in Ebenales but, as Corner (1976) remarked, in Ebenaceae the ovule is bitegmic and suspended, and seeds are exotestal in contrast with Sapotaceae, where the ovule is erect, unitegmic and the seeds are mesotestal. Moreover, molecular studies have revealed Ebenales as an artificial construct, and place Sapotaceae in a well-supported clade, the expanded Ericales. In a five gene-analysis of this clade (Anderberg et al. 2002), Sapotaceae appear sister to Lecythidaceae, although with very low support (52%) and backed by weak apo(rather plesio?)-morphies such as trilacunar nodes, stipules and nuclear endosperm, so that their affinity appears to remain obscure.

DISTRIBUTION AND HABITATS. The distribution of *Sapotaceae* extends across the humid tropics of Central and South America, Africa, Madagascar, and Asia east to the Pacific. Generic diversity is greatest in Africa and Asia, while the greatest concentration of species is in Amazonian South America and Malesia. In South America the family occupies a broad swathe from the Guianas to the eastern foothills of the Andes, with very high levels of species diversity and individual abundance, equalling or exceeding that of any other tree family. Madagascar and New Caledonia are also exceptionally rich in generic and species diversity and endemism.

The majority of species are confined to lowland rainforest below 1000 m altitude, and they vary in habit from small understorey treelets to large canopy emergents. Most occupy non-flooded forest, but a significant number are confined to periodically or permanently flooded forest. A few genera extend into savannah, as *Ecclinusa* in the Guianas, *Vitellaria* in West Africa, and *Sideroxylon* occurs widely in tropical dry forest in the West Indies and Central America, and also extends into North America as far north as Illinois.

POLLEN FOSSIL RECORD. The fossil record of Sapotaceae-like pollen is extensive. It has been summarised and discussed in Harley (1990b, 1991). It is clear from this record that by the lower Eocene Sapotaceae were represented in all continents and, by this time, pollen similar to all the major modern groups was in existence.

ECONOMIC IMPORTANCE. The family provides four important economic products: latex, oil, fruit and timber. The latex of *Palaquium* and *Manilkara* is used to make gutta percha and balata, which was formerly used on a large scale for insulating marine cables and for chewing gum, but these products are now largely replaced by synthetic materials. However, gutta percha is still important in dentistry, where it is used for root fillings.

The most important oil-producing species is *Vitellaria paradoxa*, the Shea Butter Nut, which replaces the oil palm as the source of cooking oil in N Nigeria (Hall 1996). *Argania spinosa* (Morocco) is also used for cooking oil, and several American species of *Pouteria* have similar oily seeds, which have yet to be investigated.

Virtually all species of *Sapotaceae* have edible fruit, and some of the better ones have been protected and improved by man over many centuries. Among the most well-known species are *Pouteria caimito* (western Amazonia), *Pouteria lucuma* (Andes) and *Manilkara zapota* (Mexico), and the latter is now grown commercially in Asia. Many others are sold locally throughout the range of the family.

Several Malesian genera, mainly *Palaquium*, *Madhuca* and *Payena*, provide good-quality medium-density timber, used for plywood and veneer (Soerianegara and Lemmens 1993). Other species of the same genera produce very dense, durable timber for heavy construction. Elsewhere, in Africa and tropical America, the timber-producing species are used principally for heavy construction and parquet.

CONSPECTUS OF SAPOTACEAE

I. Tribe Mimusopeae
  1. Subtribe Mimusopinae
    Genera 1–6
  2. Subtribe Manilkarinae
    Genera 7–12
  3. Subtribe Glueminae
    Genera 13–17
II. Tribe Isonandreae
  Genera 18–24
III. Tribe Sideroxyleae
  Genera 25–30
IV. Tribe Chrysophylleae
  Genera 31–49
V. Tribe Omphalocarpeae
  Genera 50–53

KEY TO THE TRIBES OF SAPOTACEAE

1. Sepals in 2 whorls, outer whorl valvate   2
– Sepals in 1 imbricate whorl   3
2. Stamens as many as corolla lobes, corolla lobes often divided into 3 or more segments   **I. Mimusopeae**
– Stamens 2–3 times as many as the corolla lobes, corolla lobes undivided   **II. Isonandrae** (p. 407)
3. Stamens 2–6 times as many as the corolla lobes   **V. Omphalocarpeae** (p. 418)
– Stamens as many as the corolla lobes   4
4. Corolla lobes often divided, stamens exserted, staminodes +, hilum basal or basiventral   **III. Sideroxyleae** (p. 409)
– Corolla lobes undivided, stamens exserted or included, staminodes + or 0, hilum usually adaxial   **IV. Chrysophylleae** (p. 411)

I. TRIBE MIMUSOPEAE Hartog (1878).

Unarmed trees or shrubs. Leaves spirally arranged, sometimes clustered. Inflorescences axillary. Flowers bisexual (unisexual in *Vitellariopsis*). Calyx usually 2 whorls of 3 or 4 sepals, the outer whorl valvate, less frequently (subtribe Glueminae) a single whorl of 5 imbricate, sometimes quincuncial sepals; corolla lobes, stamens and staminodes usually same number as sepals; corolla rotate or cyathiform; corolla lobes usually divided into 3 segments, a median one and 2 lateral or abaxial ones; lateral segments exceeding, equalling or smaller than the median segment, sometimes subdivided; stamens exserted, usually 8, inserted in a single whorl at the top of the corolla tube; anthers extrorse, usually glabrous; staminodes usually well-developed, usually 8, alternating with the stamens; ovary pubescent; hilum usually basal or basi-ventral, less frequently adaxial.

KEY TO THE SUBTRIBES OF MIMUSOPEAE

1a. Calyx of 2 whorls of 4 sepals, the outer whorl valvate; corolla lobes, stamens, staminodes, and ovary loculi usually 8; staminodes usually hairy; fruit indehiscent; hilum usually small and basal **1. Mimusopinae**

1b. Calyx of 2 whorls of 3 sepals, the outer whorl valvate, corolla lobes, stamens usually 6, less frequently 12–18; staminodes 6 or 0, glabrous; fruit indehiscent; hilum usually elongate, basi-ventral **2. Manilkarinae** (p. 402)

1c. Calyx a single whorl of 5 imbricate or quincuncial sepals; corolla lobes, stamens, staminodes and ovary loculi usually 5; staminodes hairy or glabrous; fruit dehiscent or not, hilum long, usually narrow, adaxial **3. Glueminae** (p. 405)

1. Subtribe Mimusopinae Aubréville (1964).

Calyx of 2 whorls of 4 sepals, the outer whorl valvate; corolla lobes, stamens, staminodes, and ovary loculi usually 8; staminodes usually hairy; fruit indehiscent; hilum usually small and basal.

Key to the Genera of Mimusopinae

1. Stipules large and conspicuous 2
- Stipules 0 or minute 3
2. Corolla tube much longer than lobes, filaments united with staminodes for most of their length; hilum basi-ventral, embryo with thin foliaceous cotyledons, endosperm copious **3. *Autranella***
- Corolla tube shorter than lobes, filaments free; hilum adaxial, embryo with plano-convex cotyledons, endosperm 0 **5. *Baillonella***
3. Corolla lobes entire; inflorescences clustered at shoot apex in axils of scale leaves; venation craspedodromous **6. *Vitellaria***
- Corolla lobes divided into 3 segments (entire in *Mimusops antsiranensis*); inflorescences axillary or at defoliated nodes, venation brochidodromous 4
4. Hilum small, basal or basi-ventral, embryo with thin foliaceous cotyledons and copious endosperm **1. *Mimusops***
- Hilum large, adaxial, embryo with thick plano-convex cotyledons, endosperm 0 5
5. Corolla tube well-developed, central corolla lobe segment narrowed or reduced to a filament; staminodes glabrous; seed coat thick **4. *Tieghemella***
- Corolla tube very short; corolla lobe segments more or less equal in size; staminodes hairy; seed coat thin, crustaceous or membranous **2. *Vitellariopsis***

1. *Mimusops* L.

Mimusosps L., Sp. Pl.: 349 (1753).

Trees or shrubs. Leaves clustered or not. Venation brochidodromous, tertiaries usually parallel to the secondaries and descending from the margin, or reticulate. Inflorescence axillary or in the axils of leaf scars. Stipules minute, caducous, sometimes 0. Flowers solitary or fasciculate. Corolla hairy or glabrous; tube much shorter than the lobes; lobes 8, nearly always divided to the base into 3 segments; median segment slender, usually erect and clasping the stamen, sometimes inflexed against the style; lateral segments widely spreading, often equalling or slightly exceeding the median segment, entire or deeply divided or laciniate. Stamens (7)8; filaments free or partially fused to the staminodes; anthers hairy or glabrous. Staminodes 8, inflexed and often forming an envelope round the gynoecium, lanceolate, hairy. Ovary (7)8-locular, placentation basi-ventral or basal; style exserted or not. Fruit 1–6-seeded, fleshy. Seed laterally compressed, usually with a smooth shining, rarely rugose woody testa; hilum small, often circular or elliptic, basal or basi-ventral. Embryo with foliaceous cotyledons and exserted radicle; endosperm copious. About 20 species in Africa, 15 in Madagascar, 4 in the Mascarenes, 1 in the Seychelles, and 1 in Asia and the Pacific.

2. *Vitellariopsis* Baillon Fig. 120

Vitellariopsis Baillon ex Dubard, Ann. Mus. Bot.-Géol. Colon. Marseille III, 3: 44 (1915); Aubréville, Adansonia II, 3: 41 (1963); Baehni, Boissiera 11: 112 (1965); Kupicha, Candollea 33: 29 (1978).
Mimusops sect. *Vitellariopsis* Baillon (1891).

Trees or shrubs. Leaves spirally arranged, clustered at the shoot apex. Venation brochidodromous; tertiaries forming a distinctive finely areolate reticulum. Stipules small, caducous. Flowers bisexual, rarely unisexual, solitary or fasciculate. Corolla glabrous; tube much shorter than the lobes; lobes (6–)8(–10), divided to the base into 3 segments; median segment erect, clasping a stamen, lateral segments widely spreading or reflexed, equalling or slightly exceeding the median segment, undivided. Stamens (6–)8(–10); filaments free; anthers glabrous. Staminodes (6–)8(–10), inflexed and forming an envelope around the gynoecium, lanceolate, hairy. Ovary 7–9-locular, placentation axile; style exserted or not. Fruit 1–several-seeded, fleshy or leathery. Seed broadly ellipsoid to oblong-ellipsoid, or plano-convex when several in a fruit, not laterally compressed, with a rather thin, dull, membranous or crustaceous testa; hilum large, broad, adaxial, covering up to half the seed surface. Embryo with thick fleshy plano-convex cotyledons; radicle extending to the surface, not exserted; endosperm 0. Six species in East Africa.

3. *Autranella* A. Chev.

Autranella A. Chev., Vég. Ut. Afr. Trop. Franc. 9: 271, fig. 29 (1917).

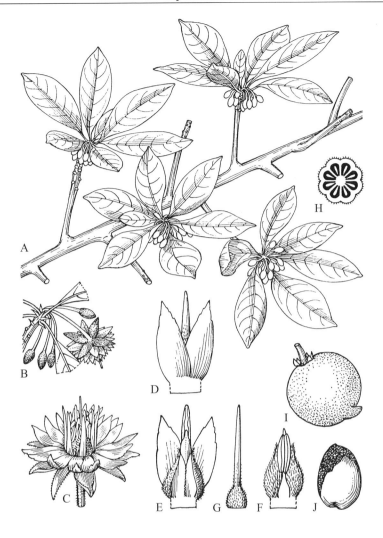

Fig. 120. Sapotaceae-Mimusupeae. *Vitellariopsis kirkii.* **A** Flowering branch. **B** Inflorescence. **C** Flower. **D** Trifid corolla lobe, viewed from outside. **E** Same, viewed from inside, with stamen and staminodes. **F** Stamen and staminodes, viewed from outside. **G** Pistil. **H** Ovary, schematic transverse section. **I** Fruit. **J** Seed. (Hemsley 1968)

Trees. Leaves clustered at the shoot tip. Venation brochidodromous; tertiaries parallel to the secondaries or reticulate. Stipules large, caducous. Corolla tube glabrous, lobes hairy, tube 2–3 times as long as the lobes; lobes 8, divided to base into 3 segments; median segment erect, clasping the stamen; lateral segments 1.5–2 times the length of the median segment, widely spreading or reflexed, undivided. Stamens 8; filaments united with the staminodes for most of their length, the free portion strongly reflexed; anthers glabrous. Staminodes 8, fused to the filaments for most of their length, free portion truncate or irregularly lobed, glabrous; stamen-staminode tube narrowed at the throat. Ovary large, filling the corolla tube, and gradually tapering from near the base, 8-locular, placentation basi-ventral; style short, included. Fruit 1(–2)-seeded, fleshy. Seed broadly obovoid, slightly laterally compressed, with a thick shining woody testa; hilum moderately large, rectangular, basi-ventral. Embryo with thin foliaceous cotyle-dons and exserted radicle; endosperm copious. A single species, *A. congolensis* (De Wild.) A. Chevalier, in West Africa.

4. *Tieghemella* Pierre

Tieghemella Pierre, Not. Bot. Sapot.: 18 (1890); Aubréville, Fl. Gabon 1, Sapotacées: 45 (1961).

Trees. Leaves weakly clustered. Venation brochido-dromous, tertiaries reticulate or parallel to the secondaries. Stipules 0. Corolla glabrous, tube well-developed, only slightly shorter than the lobes; lobes 8, partly or completely divided into 3 segments; median segment narrow or reduced to a small filamentous rudiment; lateral segments longer and broader than the median segment, undivided. Stamens 8; filaments free; anthers, glabrous. Staminodes 8, erect, carnose, narrowly lanceolate, glabrous. Ovary 8-locular, placentation basi-ventral; style included. Fruit 1–several-seeded, fleshy. Seed broadly ellipsoid, slightly laterally compressed, with a thick, woody, shining testa; hilum broad, adaxial, covering up to half the seed. Embryo with thick, fleshy, plano-convex cotyledons and exserted radicle; endosperm 0. Two species in West Africa.

5. *Baillonella* Pierre

Baillonella Pierre, Not. Bot. Sapot.: 13 (1890); Aubréville, Fl. Gabon 1, Sapotacées: 51 (1961).
Mimusops sect. *Baillonella* (Pierre) Engler (1897).

Unarmed trees. Leaves clustered at the shoot apex. Venation brochidodromous, tertiaries oblique. Stipules large, persistent. Inflorescences densely clustered at shoot apex. Corolla hairy (tube only), tube shorter than the lobes; lobes 8, divided to base into 3 segments; median segment erect; lateral segments spreading, exceeding the median segment, undivided. Stamens 8; filaments free; anthers, glabrous. Staminodes 8, erect, narrowly lanceolate, hairy. Ovary 8-locular, placentation basi-ventral; style slightly exserted. Fruit 1–2-seeded, fleshy. Seed broadly ellipsoid, with a thick, hard, shining testa; hilum broad, covering the adaxial surface of the seed. Embryo with thick fleshy, plano-convex cotyledons and exserted radicle; endosperm 0. A single species in West Africa (*B. toxisperma* Pierre).

6. *Vitellaria* Gaertner

Vitellaria Gaertner, Fruct. 3: 131, t. 205 (1807); Hepper, Taxon 11: 226 (1962).
Butyrospermum Kotschy (1865).

Trees or shrubs. Leaves in a dense terminal cluster. Venation craspedodromous with a prominent marginal vein; tertiaries parallel to the secondaries or reticulate; quaternaries finely areolate. Stipules small, caducous. Inflorescences densely clustered at the shoot apex. Corolla glabrous; tube much shorter than the lobes; lobes (6–)8, entire, contorted in bud, spreading. Stamens (6–)8; anthers glabrous. Staminodes (6–)8, erect or inflexed and forming an envelope round the gynoecium, lanceolate, margin erose, terminated

by a filiform point, subglabrous. Ovary (5)6-locular, placentation axile; style slightly exserted. Fruit 1–2-seeded, fleshy. Seed globose or broadly ellipsoid, not laterally compressed, with a rather thin shining testa; hilum broad, adaxial. Embryo with thick, fleshy, fused cotyledons; radicle not exserted; endosperm 0. Possibly two species in western tropical Africa and Cameroon.

2. SUBTRIBE MANILKARINAE H.J. Lam (1938).

Unarmed trees or shrubs. Leaves spirally arranged, clustered at the shoot apex, secondaries parallel. Inflorescence axillary or in the axils of fallen leaves. Flowers bisexual (few *Manilkara* dioecious). Calyx of 2 whorls of 3 sepals, the outer whorl valvate, corolla lobes, stamens usually 6, less frequently 12–18, inserted at the top of the corolla tube (within: some *Manilkara*), filaments free (partly fused with staminodes in some *Manilkara*), anthers extrorse, glabrous (exceptionally pubescent in *Manilkara* and *Faucherea*); staminodes 6 or 0, often small, glabrous, alternating with stamens; fruit indehiscent, fleshy, usually 1-seeded; hilum usually elongate, basi-ventral.

KEY TO THE GENERA OF MANILKARINAE

1. Corolla lobes, stamens and staminodes usually 6; staminodes well-developed, vestigial or rarely 0   2
 – Corolla lobes and stamens (10–)12–18, staminodes 0   5
2. Corolla lobes nearly always divided into 3 segments   3
 – Corolla lobes entire (vestigial lateral segments present in *Northia*)   4
3. Seed laterally compressed, hilum narrow, nearly always basi-ventral; staminodes usually well-developed   **7.** *Manilkara*
 – Seed not or only slightly laterally compressed, hilum broad, covering the adaxial surface; staminodes usually vestigial   **8.** *Labramia*
4. Venation craspedodromous; corolla tube equalling lobes; seed c. 7 cm long, with large adaxial hilum, endosperm 0   **10.** *Northia*
 – Venation brochidodromous, leaves striate; corolla tube much shorter than lobes; seed 1–2 cm long, with small basi-ventral hilum, endosperm +   **9.** *Faucherea*
5. Corolla lobes entire, ovary loculi 5–10; hilum basal or basi-ventral, hollowed   **11.** *Labourdonnaisia*
 – Corolla lobes divided into 3 segments; ovary loculi 16–18; hilum adaxial, narrow   **12.** *Letestua*

7. *Manilkara* Aubréville   Fig. 121

Manilkara Aubréville, Adansonia II, 11: 251–300 (1971).
Achras L. (1753).

Trees, rarely shrubs, nearly always with sympodial branching, rarely dioecious. Venation nearly al-

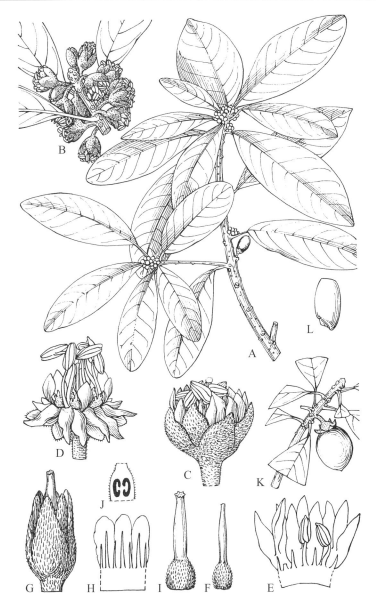

ways brochidodromous, secondaries often looping to form a submarginal vein (eucamptodromous and then with convergent secondaries); tertiaries often descending from the margin and parallel to the secondaries; higher order venation usually reticulate or areolate. Stipules small, caducous, or 0. Flowers solitary or fasciculate. Corolla nearly always glabrous, occasionally carnose; tube usually much shorter than the lobes, rarely equalling or exceeding them; lobes 6(–9), usually spreading, usually divided to the base into 3 segments; median segment usually erect, often clawed, clasping the stamen; two lateral segments spreading, shorter than, equalling or exceeding the median segment, entire or deeply divided or laciniate, or less frequently corolla lobe only par-

Fig. 121. Sapotaceae–Mimusupeae. *Manilkara discolor.* **A** Branch. **B** Inflorescences. **C** Young flower. **D** Flower at anthesis. **E** Corolla opened out, viewed from inside to show stamens and staminodes. **F** Pistil. **G** Female flower. **H** Part of corolla of female flower. **I** Pistil of female flower. **J** Section of ovary. **K** Young fruit, attached. **L** Seed. (Hemsley 1968)

tially divided or 2–3-lobed at the apex, or rarely entire. Stamens 6(–12), very rarely fixed within the tube; filaments sometimes fused with staminodes. Staminodes (0–)6(–12), bifid, laciniate, truncate or irregularly divided or vestigial, erect or rarely incurved, but not forming an envelope round the gynoecium, nearly always glabrous. Small annular disk occasionally +. Ovary 6–14-locular, hairy or glabrous, placentation axile or basi-ventral; style exserted. Fruit 1–several-seeded. Seed ellipsoid to obovoid, laterally compressed, with a hard shining woody testa; hilum nearly always narrowly elongate, basi-ventral or less frequently extending along most of the adaxial face, very rarely broad. Embryo with foliaceous cotyledons and exserted radicle; endosperm copious. Pantropical, 30 species in America, c. 20 in Africa and Madagascar and c. 15 in Asia and the Pacific.

8. *Labramia* A. de Candolle

Labramia A. de Candolle, Prodr. 8: 672 (1844).

Trees or shrubs. Venation brochidodromous, higher order venation obscure. Stipules small, caducous or 0. Flowers fasciculate. Corolla glabrous; tube shorter than the lobes; lobes 6(–8), divided to the base into 3 segments; median segment erect, clasping the stamen; two lateral segments erect or spreading; almost equalling or much shorter than the median segment, often deeply divided or laciniate. Stamens 6(–8). Staminodes 6(–8), usually reduced to a small fleshy vestige. Ovary 8–12-locular, glabrous, placentation axile; style exserted. Fruit 1-seeded. Seed ellipsoid, not or only slightly laterally compressed, with a hard shining woody testa; hilum broad, covering the adaxial surface. Embryo with foliaceous cotyledons and exserted radicle; endosperm copious. Eight species in Madagascar.

9. *Faucherea* Lecomte

Faucherea Lecomte, Bull. Mus. Hist. Nat. (Paris) 26: 245 (1920).

Trees. Venation brochidodromous; secondaries numerous, higher order venation parallel to the secondaries, obscure, leaves appearing striate. Stipules 0. Flowers fasciculate. Corolla glabrous; tube much shorter than lobes; lobes 6(–11), entire. Stamens 6(–11), anthers hairy or glabrous. Staminodes 6(–11), usually very short, irregular or dentate. Small annular disk sometimes + around base of ovary. Ovary (5–)6(–10)-locular, hairy; placentation axile; style exserted. Fruit 1(–4)-seeded. Seed slightly laterally compressed, with a hard shining testa; hilum basi-ventral, less than half as long as the seed. Embryo with foliaceous cotyledons and exserted radicle; endosperm copious. Eleven species in Madagascar.

10. *Northia* J.D. Hooker

Northia J.D. Hooker in Hooker's Icon. Pl. t. 1473 (1884).

Trees. Venation craspedodromous with a prominent marginal vein; tertiaries obscurely reticulate. Stipules 0. Flowers in small fascicles. Corolla hairy; tube equalling the lobes, carnose; lobes 6, contorted in bud, median segment erect or slightly spreading, abruptly narrowed at base, two lateral segments reduced to small irregularly shaped vestiges at the base of the median segment or 0. Stamens 6. Staminodes 0 or minute. Ovary 6-locular, hairy, placentation axile; style slightly exserted. Fruit 1-seeded. Seed broadly ellipsoid, not laterally compressed, with a woody testa; hilum broad, adaxial, covering about a third of the seed surface. Embryo with thick plano-convex, partially fused cotyledons; radicle extending to the surface; endosperm 0. A single species, *N. seychellana* J.D. Hooker, in the Seychelles.

11. *Labourdonnaisia* Bojer

Labourdonnaisia Bojer, Mém. Soc. Phys. Genève 9: 295 (1841).

Trees with sympodial branching. Venation brochidodromous, secondaries often joining to form a submarginal vein; tertiaries parallel to the secondaries. Stipules 0. Flowers solitary or in small fascicles. Corolla glabrous; tube much shorter than the lobes; lobes (10–)12–18, imbricate in bud, spreading or reflexed, entire or with a few small irregular lateral teeth. Stamens 11–18(–21). Staminodes 0 or a few small irregular vestiges. Ovary 5–10-locular, hairy, placentation basi-ventral; style included or slightly exserted. Fruit 1(–2)-seeded. Seed narrowly ellipsoid, not or only slightly laterally compressed, with a shiny woody testa; hilum basal or basi-ventral, broad, often strongly concave, embryo with foliaceous cotyledons and long exserted radicle; endosperm copious. Three imperfectly known species in Madagascar.

12. *Letestua* Lecomte

Letestua Lecomte, Notul. Syst. (Paris) 4: 4 (1920).

Trees. Venation eucamptodromous with widely spaced secondaries; tertiaries horizontal; higher order venation reticulate. Stipules 0. Flowers fasciculate. Corolla glabrous; tube much shorter than lobes; lobes 12–18, erect or slightly spreading, divided to the base into 3 segments; median segment slightly exceeding the lateral segments, all petaloid. Stamens 12–18. Staminodes 0 (occasionally a few stamens lack anthers). Ovary 16–18-locular, hairy, placentation basi-ventral; style included. Fruit 1-seeded. Seed ellipsoid, laterally compressed, with a shining woody testa; hilum long, narrow, adaxial. Embryo with foliaceous cotyledons and exserted radicle; endosperm copious. A single species in West Africa, *L. durissima* (A. Chevalier) Lecomte.

3. Subtribe Glueminae Baehni ex T.D. Pennington (1991).

Unarmed trees. Leaves spirally arranged. Inflorescences axillary or in the axils of fallen leaves; flowers fasciculate, bisexual. Corolla tube shorter than the lobes (± equal in some *Eberhardtia*). Calyx a single whorl of 5 imbricate or quincuncial sepals; corolla lobes, stamens, staminodes and ovary loculi usually 5; stamens in a single whorl inserted at the corolla tube, filaments free; staminodes hairy or glabrous; ovary 5-locular, placentation axile; fruit dehiscent or not, seed ellipsoid, laterally compressed, radicle exserted; hilum long, usually narrow, adaxial.

Key to the Genera of Glueminae

1. Fruit indehiscent   **13.** *Inhambanella*
- Fruit dehiscent   2
2. Stipules +   3
- Stipules 0   4
3. Leaf without fine parallel striations, corolla & staminodes glabrous, staminodes with terminal sagittate appendage; fruit 3–5-seeded   **17.** *Eberhardtia*
- Leaf with numerous fine parallel striations perpendicular to midrib; corolla and staminodes hairy, staminodes without terminal appendage; fruit 1-seeded   **14.** *Neolemonniera*
4. Staminodes alternating with stamens, free   **15.** *Lecomtedoxa*
- Staminodes opposite the stamens, partially fused   **16.** *Gluema*

13. *Inhambanella* (Engler) Dubard   Fig. 122

Inhambanella (Engler) Dubard, Ann. Inst. Bot.-Géol. Colon. Marseille II, 3: 42 (1915).
Mimusops sect. *Inhambanella* Engler (1904).
Kantou Aubréville & Pellegrin (1957).

Leaves spaced or clustered at the shoot apex. Venation eucamptodromous with slightly convergent secondary veins; tertiaries variable. Stipules small, caducous. Corolla glabrous, lobes (4–)5(–6), erect or slightly spreading, subdivided to halfway or more into 3 segments; median segment generally longer and broader than the lateral segments. Stamens (4–)5(–6); anthers extrorse, glabrous. Staminodes (4–)5(–6), alternating with the stamens,

well-developed, erect, lanceolate, glabrous. Ovary hairy; style included. Fruit 1-seeded, indehiscent, fleshy. Testa smooth or rugose; hilum long and rather broad, adaxial. Embryo with thinly planoconvex cotyledons; endosperm 0. Two species, one in East Africa, one in West Africa.

14. *Neolemonniera* Heine

Neolemonniera Heine, Kew Bull. 16: 301 (1960).

Leaves densely clustered at shoot tip. Venation eucamptodromous-brochidodromous with slightly convergent secondary veins, the tertiaries forming a lax reticulum; numerous fine minute striations perpendicular to the mibrib. Stipules +. Corolla hairy; lobes 5, suberect, divided to near the base into 3 segments, median segment exceeding the 2 lateral ones. Stamens 5; anthers extrorse, pubescent or glabrous. Staminodes 5, alternating with the stamens, lanceolate, hairy, inflexed and forming an envelope around the gynoecium. Ovary hairy; style slightly exserted. Fruit a 1-seeded, leathery capsule dehiscent by a single lateral valve. Seed somewhat asymmetric; testa smooth, shining; hilum long, narrow, adaxial. Embryo with thick flat cotyledons; endosperm a thin layer. Five species in West Africa.

15. *Lecomtedoxa* (Pierre ex Engler) Dubard

Lecomtedoxa (Pierre ex Engler), Notul. Syst. (Paris) 3: 46 (1914).
Mimusops subgen. *Lecomtedoxa* Pierre ex Engler (1904).

Leaves clustered at shoot apex. Venation eucamptodromous with convergent secondary veins, and a lax tertiary reticulum, or brochidodromous and then secondaries parallel and tertiaries parallel to the secondaries and descending from the margin. Stipules 0. Inflorescence axillary or ramiflorous. Corolla hairy or glabrous; lobes 5, erect or slightly spreading, deeply divided into 3 segments; median segment exceeding the lateral segments. Stamens 5; anthers extrorse, hairy. Staminodes 5, free, alternating with the stamens, lanceolate, often terminated by a fine point, erect, glabrous or occasionally hairy. Ovary hairy; style exserted. Fruit 1-seeded, dehiscent by a single lateral valve, leathery. Seed asymmetric, strongly laterally compressed; testas smooth, shining, woody; hilum long, narrow, adaxial. Embryo with thick flat cotyledons; endosperm a thin layer. Five species in Gabon.

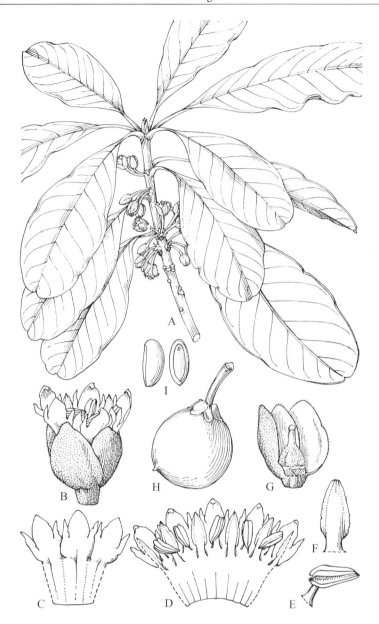

Fig. 122. Sapotaceae–Mimusupeae. *Inhambanella henriquesii.* A Flowering branch. B Flower. C Part of corolla, viewed from outside. D Corolla, opened out, viewed from inside, showing stamens and staminodes. E Stamen. F Staminode. G Flower with two sepals and corolla removed to show pistil. H Fruit. I Seeds. (Hemsley 1968)

16. *Gluema* Aubréville & Pellegrin

Gluema Aubréville & Pellegrin, Bull. Soc. Bot. France 81: 797 (1935).

Leaves clustered at the shoot apex. Venation eucamptodromous-brochidodromous with slightly convergent or parallel secondaries; tertiaries finely areolate. Stipules 0. Corolla hairy; lobes 5, divided to the base into 3 segments; median segment erect, clasping the stamen; lateral segments equalling the median segment, widely spreading. Stamens 5; anthers extrorse, hairy. Staminodes 5, fixed inside and opposite the stamens, fused at the base, lanceolate, densely hairy, inflexed and forming an envelope around the gynoecium. Ovary hairy; style exserted. Fruit 1-seeded, dehiscent by a single lateral valve, leathery. Testa smooth, shining, woody; hilum long, narrow, adaxial. Embryo, with thinly plano-convex cotyledons; endosperm 0. A single species in West Africa (*G. ivorensis* Aubréville & Pellegrin).

17. *Eberhardtia* Lecomte

Eberhardtia Lecomte, Bull. Mus. Hist. Nat. (Paris) 26: 345 (1920).

Venation eucamptodromous, with parallel second-aries, tertiaries oblique, higher order venation finely reticulate-areolate. Stipules well-developed. Corolla glabrous, tube equalling or slightly shorter than the lobes; lobes 5, divided for c. 2/3 of their length into 3 segments; median segment linear or narrowly subulate, lateral segments slightly longer and much broader than the median segment. Stamens 5, included; anthers dehiscing laterally, glabrous. Staminodes 5, alternating with the stamens, exceeding the stamens, narrowly lanceo-late, glabrous, bearing at the apex a large sagittate, versatile, caducous appendage. Ovary hairy or glabrous; style included. Fruit a 3–5-seeded loculi-cidal capsule, leathery-fleshy, slightly constricted between the seeds. Testa smooth, shining; hilum narrow or broad, adaxial, extending the length of the seed. Embryo with thin foliaceous cotyledons; endosperm copious. Three poorly defined species of montane forest in South China, Vietnam and Laos; recently collected in Sabah.

II. Tribe Isonandreae Hartog (1878).

Calyx usually 2 whorls of 2 or 3 sepals, the outer whorl valvate, less frequently a single whorl of 4–5 imbricate or quincuncial sepals; corolla cyathi-form, lobes as many as or 2–3 times as many as sepals, undivided; stamens 2–3(–5) times as many as corolla lobes, in 1–3 whorls; staminodes 0; hilum adaxial.

Key to the Genera of Isonandreae

1. Calyx biseriate, 2 whorls of 3 sepals 2
- Calyx uniseriate or if biseriate then with 2 whorls of 2 sepals 3
2. Stamens free; inflorescences axillary or just below the leaves, fascicles sessile **18. Palaquium**
- Filaments partially united in a staminal tube; plant cauliflorous or ramiflorous, fascicles produced on scaly brachyblasts **19. Aulandra**
3. Calyx biseriate with 2 whorls of 2 sepals 4
- Calyx uniseriate with 4 or 5 imbricate sepals 6
4. Flowers subsessile; corolla lobes 4(–5); stamens 8(–10); ovary loculi 4(–5); seed with copious endosperm **20. Isonandra**
- Flowers distinctly pedicellate; corolla lobes (6)7–12(–17); stamens (12)13–36(–43); ovary loculi (4–)6–9(–15); endosperm + or 0 5
5. Leaves nearly always spirally arranged; tertiary venation usually oblique or reticulate; embryo usually with plano-convex or thick flat cotyledons, endosperm thin or 0 **21. Madhuca**

- Leaves alternate and distichous; tertiary venation parallel to secondaries and descending from the margin; embryo with thin foliaceous cotyledons, endosperm copious **22. Payena**
6. Fascicles densely clustered at shoot apex forming a pseudo-terminal inflorescence; sepals 4; corolla lobes 8(9), imbricate; ovary (3)4(5)-locular **23. Burckella**
- Fascicles axillary or ramiflorous; sepals usually 4–5 or 5; corolla lobes 8–16, contorted; ovary (5)6–15-locular **24. Diploknema**

18. *Palaquium* Blanco

Palaquium Blanco, Fl. Filip.: 403 (1837); ed. 2: 282 (1845).

Leaves spirally arranged; venation usually eucamptodromous with oblique to horizontal ter-tiaries or rarely brochidodromous and then with tertiaries descending from the margin and paral-lel to the secondaries. Stipules +, usually small and caducous, sometimes 0. Flowers usually bisexual, rarely unisexual and plant dioecious, in 1–many-flowered fascicles, axillary or in axils of fallen leaves. Sepals 3 + 3, free or slightly united, the outer ± valvate. Corolla lobes (5)6, usually contorted, less frequently imbricate, usually spreading or reflexed, nearly always exceeding the tube, rarely only equalling it; corolla tube usually glabrous inside. Stamens usually 12, less frequently 10, 13, 18, 24 or c. 30, exserted, in a single whorl or rarely in 2 whorls, inserted near top of corolla tube or at base of corolla lobes; filaments free or occasionally a few partly fused, tapering to an acute apex, hairy or glabrous, 0 in female flowers. Nectary 0 (repre-sented by a small annular swelling fused to ovary). Ovary hairy, rarely glabrous, (5)6(–10)-locular; style long exserted. Fruit 1–2(–several)-seeded. Seed usually broadly oblong or ellipsoid, with broad adaxial hilum covering up to 2/3 of the surface and then without endosperm and embryo with plano-convex cotyledons, or less frequently seed laterally compressed with a narrow adaxial hilum and then usually with endosperm and embryo with foliaceous cotyledons, radicle usually extending to the surface, less frequently exserted. About 110 species from India, through SE Asia to the Pacific Islands.

19. *Aulandra* H.J. Lam

Aulandra H.J. Lam, Bull. Jard. Bot. Buitenzorg III, 8: 415 (1927).

Leaves spirally arranged, venation eucamptodro-mous with oblique tertiaries. Stipules small, caducous. Flowers bisexual, cauliflorous or rami-florous, in few-flowered fascicles borne terminally on densely scaly, simple or divaricately branched brachyblasts up to 3 cm long. Sepals 3 + 3, free or united at base, the outer ± valvate. Corolla lobes 6, imbricate or contorted, spreading or reflexed, exceeding tube; corolla glabrous. Stamens 18–19, exserted, in a single whorl inserted at the top of the corolla tube; filaments united for more than half their length, free portion usually geniculate in bud; anthers extrorse, hairy. Nectary 0. Ovary hairy, 6(–7)-locular; style exserted. Berry 1-seeded. Seed broadly ovoid, with a broad adaxial hilum covering of seed surface. Embryo with plano-convex cotyledons, radicle extending to the surface; endosperm 0. Three species in Borneo.

20. *Isonandra* Wight

Isonandra Wight, Ic. Pl. 2: 4, t. 359, 360 (1840); 4: 9, t. 1219, 1220 (1848).

Leaves spirally arranged. Venation eucamptodromous with oblique or horizontal tertiaries. Stipules often small and caducous. Flowers bisexual, in small dense-flowered subsessile fascicles, sometimes ramiflorous. Sepals ± free, 2 + 2 (2 + 3), the outer pair open or imbricate. Corolla lobes 4(5), imbricate, rarely contorted, erect or slightly spreading, equalling or exceeding the tube; corolla tube glabrous within. Stamens 8(–10), inserted in a single whorl at the top of the tube; filaments free or only slightly fused at base, strongly geniculate in bud; anthers extrorse, usually with an apical tuft of hairs. Nectary 0. Ovary hairy, 4(–5)-locular; style slightly exserted. Berry 1-seeded. Seed laterally compressed, with a long narrow adaxial hilum. Embryo with foliaceous cotyledons and exserted radicle; endosperm copious. About ten species in South India, Sri Lanka, Malay Peninsula and Borneo.

21. *Madhuca* Hamilton ex Gmelin

Madhuca Hamilton ex Gmelin, Syst. 2: 773, 799 (1791).
Ganua Pierre ex Dubard (1908).

Leaves spirally arranged, very rarely distichous on horizontal shoots; venation eucamptodromous or brochidodromous, with oblique, horizontal or reticulate tertiaries, rarely intersecondaries and tertiaries descending from the margin and parallel to the secondaries. Stipules +, sometimes large and persistent. Flowers bisexual in 2–many-flowered fascicles, usually on young branches in the axils of fallen leaves or axillary. Sepals 2 + 2, free, rarely a mixture of 2 + 2 and 5 imbricate sepals

in a single whorl on the same individual, aestivation of outer pair valvate, imbricate or open. Corolla lobes (6–)8–12(–17), imbricate, usually spreading or reflexed, nearly always equalling or exceeding the tube; corolla tube usually hirsute or barbate at the throat. Stamens (12–)14–36(–43), exserted (uppermost whorls only), in 1–3 alternating whorls inserted in the throat of the corolla tube, or occasionally the lowermost whorl inserted near the base of the tube; filaments usually free, rarely partially or completely fused in pairs or into a short tube, not geniculate, or anthers sessile. Anthers extrorse or laterally dehiscent, usually large and tapering gradually to the acute apex, hairy or glabrous. Nectary sometimes +, poorly developed, annular. Ovary usually glabrous, less frequently hairy, (5–)8–9(–15)-locular, style exserted, often long, tapering gradually to the apex. Fruit 1-seeded. Seed broadly ellipsoid, oblong, plano-convex or laterally compressed; hilum usually long narrow adaxial, less frequently, broader and covering up to of the seed surface. Embryo with plano-convex, thick but flat, rarely foliaceous cotyledons; radicle exserted. Endosperm + or 0. About 100 species from India through Malesia and South China to New Guinea.

22. *Payena* A. de Candolle

Payena A. de Candolle, Prodr. 8: 196 (1844).

Leaves distichous, sometimes spirally arranged on vertical shoots. Venation brochidodromous, intersecondary and tertiary venation descending from the margin and more or less parallel to the secondaries. Stipules usually caducous. Flowers bisexual, in axillary fascicles. Sepals 2 + 2, ± free, the outer valvate or open. Corolla lobes 7–9, imbricate, erect or slightly spreading, exceeding tube; tube usually glabrous inside. Stamens 13–20(–30), inserted in 1(2) whorls at the top of the tube; filaments free or fused in pairs or bundles, often geniculate in bud. Anthers extrorse or laterally dehiscent, large, tapering gradually to an acute apex. Nectary 0, or a poorly developed annulus around the base of the ovary. Ovary hairy or glabrous, (4–)6–8(–9)-locular; style exserted, often long, tapering gradually to the apex. Fruit 1–2-seeded. Seed slightly to strongly laterally compressed, with a long narrow adaxial hilum. Embryo with foliaceous cotyledons and exserted radicle; copious endosperm. About 15 species in western Malesia from the Andaman Islands and Burma to the Malay Peninsula, Sumatra, Java to Borneo and Mindanao.

23. *Burckella* Pierre

Burckella Pierre, Not. Bot. Sapot.: 3 (1890).

Leaves spirally arranged. Venation brochidodromous, tertiaries often descending from the margin
and parallel with the secondaries, or reticulate or
oblique. Stipules + or 0. Flowers bisexual, the fascicles clustered densely in the axils of scale leaves
to form a pseudo-terminal inflorescence. Sepals 2
+ 2, strongly imbricate. Corolla lobes 8(9), imbricate, erect or slightly spreading, exceeding the
tube; corolla tube nearly always barbate in
the throat. Stamens 16–18(–30) in 1–2 whorls at
the top of the corolla tube; filaments free, not
geniculate, or anthers sessile. Anthers extrorse.
Nectary annular or patelliform, often enveloping
the ovary, rarely 0. Ovary hairy or glabrous,
(3)4(5)-locular, style long, exserted. Fruit 1-
seeded. Seed broadly ellipsoid or dorso-ventrally
compressed, hilum covering at least 1/2 the surface
of the seed and frequently nearly all of it and then
leaving only a narrow adaxial strip; hilum area
often rugose or ruminate, sometimes with prominent protuberances. Embryo with plano-convex
cotyledons, radicle extending to the surface, not
exserted, endosperm 0. About 14 species in the
western Pacific from the Moluccas and New
Guinea to Fiji, Samoa and Tonga.

24. *Diploknema* Pierre

Diploknema Pierre, Arch. Néerl. Sci. Exact. Nat. 19: 103 (1884).

Leaves spirally arranged. Venation eucamptodromous, rarely brochidodromous, tertiaries oblique
or reticulate. Stipules often small and caducous.
Flowers bisexual or unisexual (plant dioecious), in
axillary fascicles or ramiflorous. Sepals a single
whorl of (4)5(6) free or partly united, imbricate or
quincuncial sepals. Corolla lobes 8–16, contorted
(not known in some species), usually spreading or
reflexed, exceeding tube; tube hairy or glabrous.
Stamens (10–)16–30(–80), inserted in 1 or 2, often
dense whorls at the top of the corolla tube; filaments free or partially united in pairs or small
bundles, usually geniculate in bud. Anthers
extrorse, large, tapering to an acute apex, hairy or
glabrous; stamens in female flowers converted into
flat narrow staminodes. Nectary 0 or a small
annulus at base of ovary. Ovary hairy or glabrous,
(5)6–15-locular; style often exserted, apex tapering and simple or truncate and then with minute
stigmatic lobes. Fruit 1–4-seeded. Seed ellipsoid or
laterally compressed (known in only 4 spp.), with

a broad or narrow adaxial hilum. Embryo with
plano-convex or thin foliaceous cotyledons,
radicle extending to the surface or exserted;
endosperm + or 0. About 10 species from northern India, Nepal and Bhutan to Burma, Thailand,
Cambodia and Vietnam; also in Borneo,
Philippines and Amboina.

III. TRIBE SIDEROXYLEAE (Engler) H.J. Lam (1938).

Unarmed or spinose trees or shrubs. Sepals 5, in a
single whorl, imbricate, sometimes quincuncial;
corolla lobes, stamens and staminodes as many as
sepals; corolla rotate or cyathiform; corolla lobes
often divided into 3 segments, a larger median one
and 2 smaller lateral ones; stamens in a single
whorl, exserted, fixed at top of corolla tube; staminodes usually well-developed, rarely 0; disk 0;
seed hilum basal or basi-ventral.

KEY TO THE GENERA OF SIDEROXYLEAE

1. Leaves usually opposite; stipules +; inflorescence a raceme
   or panicle **29. Sarcosperma**
 - Leaves usually spirally arranged or fascicled on short
   shoots; stipules 0; inflorescence nearly always a fascicle 2
2. Corolla lobes and stamens usually 7–10, staminodes 1–2
   27. Nesoluma
 - Corolla lobes and stamens usually 4–5, staminodes
   equalling the number of corolla lobes or 0 3
3. Staminodes 0 **30. Diploon**
 - Staminodes + 4
4. Ovary 2–3-locular; fruit containing several seeds completely fused to form a single pyrene, hilum 0 **28. Argania**
 - Ovary usually 5-locular; fruit containing 1 or more free
   seeds with usually basal hilum 5
5. Flowers unisexual and plant dioecious, female flowers with
   vestigial corolla, and without stamens and staminodes; filaments in male flowers short, anthers included or only
   slightly exserted; hilum almost full-length, adaxial
   26. Neohemsleya
 - Flowers nearly always bisexual, rarely unisexual but then
   female flower lacking only anthers; filaments in bisexual or
   male flowers well-developed, anthers clearly exserted;
   hilum nearly always basal or basi-ventral **25. Sideroxylon**

25. *Sideroxylon* L.

Sideroxylon L., Gen. Pl. ed. 5: 89 (1754).
Bumelia Swartz, Prod. Veg. Ind. Occ.: 49 (1788), nom. cons.
Monotheca A. de Candolle (1844).
Dipholis A. de Candolle (1844), nom. cons.
Mastichodendron (Engler) H.J. Lam (1939).
Sinosideroxylon (Engler) Aubréville (1963).
Spiniluma Baillon ex Aubréville (1963).

Spinous or unarmed trees or shrubs. Leaves spirally arranged, rarely opposite, often fascicled on

short lateral shoots. Venation variable. Stipules 0. Inflorescence axillary or in the axils of fallen leaves, sessile or very rarely pedunculate. Flowers solitary or fasciculate, bisexual or rarely unisexual (?dioecious). Sepals one whorl of 5(–8), quincuncial, free. Corolla cyathiform, usually glabrous; tube nearly always shorter than the lobes, rarely equalling or exceeding it; lobes (4)5(–8), imbricate or quincuncial, spreading, entire or divided into a larger median segment and two smaller lateral segments. Stamens (4)5(–8), exserted; filaments well-developed; anthers extrorse, usually glabrous; stamens sometimes converted into sterile staminodes in male flowers. Staminodes (4)5(–8), usually well-developed, often lanceolate, erose, infolded and incurved against the style, usually glabrous, less frequently hairy. Ovary (1–)5(–8)-locular, hairy or glabrous; placentation basi-ventral or basal; style exserted or included. Fruit 1(–2)-seeded, fleshy, usually glabrous. Seed globose, ovoid, oblong or ellipsoid, not laterally compressed, rarely plano-convex when 2 seeds in a fruit; testa smooth, shining, free from the pericarp, often thick and woody, often sculptured on the adaxial surface with several prominent thickened plates; hilum nearly always basal or basi-ventral, small, circular, lanceolate or elliptic, rarely adaxial and then broad. Embryo vertical, oblique or horizontal, with thin foliaceous cotyledons, and copious endosperm, or with plano-convex cotyledons and then with a thin sheath of endosperm, or endosperm 0; radicle exserted. Forty nine species in the Neotropics, about 25 elsewhere (6 Africa, 6 Madagascar, 8 Mascarenes, 4 Asia, 1 in NW Pakistan, Afghanistan, Oman, Somalia, Ethiopia, Djibouti).

26. *Neohemsleya* T.D. Pennington

Neohemsleya T.D. Pennington, The genera of Sapotaceae: 175 (1991).

Unarmed dioecious trees. Leaves laxly spirally arranged, not clustered or fascicled, simple. Venation eucamptodromous-brochidodromous with convergent secondaries, tertiaries few, reticulate to horizontal. Stipules 0. Flowers axillary, solitary or paired, unisexual. Sepals 5, free or slightly fused, quincuncial. Corolla (male) tube much shorter than lobes; lobes (4)5, imbricate, entire, spreading, (female) reduced to 5 minute, almost free petals, persisting in fruit as slightly accrescent membranous structures. Stamens (4)5, included or slightly exserted; filaments short; anthers extrorse or laterally dehiscent; stamens 0 in female flower. Sta-

minodes (4)5, much shorter than corolla lobes, subulate, 0 in female flower. Ovary 5-locular, placentation axile; style short, stout, included; stigma minutely 5-lobed, apex more or less truncate. Berry 1–2-seeded. Seed ellipsoid, not or slightly laterally compressed, or plano-convex (when fruit 2-seeded); testa smooth, shining; hilum almost full length, adaxial, almost as wide as seed. Embryo vertical, with thin foliaceous cotyledons and long-exserted radicle; endosperm copious. A single species in Tanzania (*N. usambarensis* Pennington).

27. *Nesoluma* Baillon

Nesoluma Baillon, Bull. Mens. Soc. Linn. Paris 2: 964 (1891).

Unarmed trees or shrubs. Leaves spirally arranged. Venation brochidodromous, secondaries parallel, higher order venation finely areolate. Stipules 0. Inflorescence axillary and in the axils of leaf scars, fasciculate. Flowers bisexual or unisexual (plant dioecious). Sepals 4–5, strongly imbricate or quincuncial. Corolla glabrous, shortly tubular, tube shorter than lobes; lobes (5–)7–10(–12), erect or spreading, entire or sometimes with a small lateral segment, unequal, in female flower reduced to free vestigial scales or 0. Stamens 7–10(–12), often more than 1 opposite each corolla lobe, exserted on well-developed filaments, with prominent traces to the base of the tube; sometimes a few stamens converted into sterile petaloid or staminode-like structures; anthers extrorse, glabrous; stamens 0 in female flower. Staminodes irregular, usually only 1–2, well-developed, sometimes petaloid, glabrous; 0 in female flower. Ovary 3–5(6)-locular, slightly hairy, placentation axile; style slightly exserted. Berry usually 1-seeded. Seed obovoid or ellipsoid, not laterally compressed; testa smooth, shining, thick, woody; hilum large, rounded, basi-ventral or basal. Embryo vertical, oblique or horizontal, with thin foliaceous cotyledons and exserted radicle; copious endosperm. Three poorly defined species, Hawai'i, Henderson Island, Rapa and Tahiti.

28. *Argania* Roemer & Schultes

Argania Roemer & Schultes, Syst. Veg. 4: XLVI, 502 (1819), nom. cons.

Spinous shrubs or small trees. Leaves spirally arranged, becoming fascicled on short lateral shoots. Venation eucamptodromous to brochidodromous, with convergent secondaries, higher order veins forming an open reticulum. Stipules 0. Inflorescence axillary or in the axils of leaf scars, fasciculate. Flowers bisexual. Sepals 5, quincuncial. Corolla glabrous; tube shorter than lobes; lobes 5, imbricate, entire, spreading. Stamens 5, exserted on well-developed filaments, anthers extrorse, glabrous. Staminodes 5, subulate or toothed at base, glabrous. Ovary 2–3-locular, hairy; placentation basi-ventral; style exserted. Fruit 1–3-seeded, fleshy. Seeds completely fused by their adaxial surfaces, the resultant woody pyrene ellipsoid to ovoid; testa smooth, shining, thick, woody, attached to the pericarp along the lines of fusion of the seeds. Embryo often solitary by abortion, vertical, with thin foliaceous cotyledons and exserted radicle; endosperm abundant. A single species in Morocco, introduced into Libya; naturalized in Southern Spain.

29. *Sarcosperma* J.D. Hooker

Sarcosperma J.D. Hooker in Bentham & J.D. Hooker, Gen. Pl. 2: 655 (1876).

Unarmed trees. Leaves usually opposite, less frequently spirally arranged. Venation eucamptodromous, with convergent or parallel secondaries, tertiaries horizontal. Hollow pits sometimes present on lower surface in axils of secondary veins. Petiole sometimes bearing 2 small scales (stipels). Stipules +. Inflorescence axillary or in axils of leaf scars, a small raceme or panicle. Flowers bisexual. Sepals 5, quincuncial. Corolla glabrous; tube shorter than lobes; lobes 5, imbricate or quincuncial, entire, sometimes auriculate at base, widely spreading. Stamens 5, slightly exserted, with very short filaments; anthers introrse or laterally dehiscent, glabrous. Staminodes 5, small, glabrous. Ovary (1)2-locular, glabrous, placentation basal; style short, included; stigma minutely 2–4-lobed. Berry 1(–2)-seeded. Seed ellipsoid or oblong, not laterally compressed (plano-convex in 2-seeded fruit); testa smooth, thin; hilum small, round, basal or basi-ventral. Embryo vertical, with plano-convex, usually fused, cotyledons, radicle included or extending to the surface, endosperm 0. About 8 species from India to South China and Malesia.

30. *Diploon* Cronquist

Diploon Cronquist, Bull. Torrey Bot. Club 73: 466 (1946).

Leaves spaced, distichous or weakly spirally arranged, venation brochidodromous, the second-

aries joining below the margin to form a submarginal vein, intersecondaries long, usually extending to near the margin, giving the leaves a slightly striate appearance. Stipules 0. Flowers bisexual, in axillary fascicles. Sepals 4–5, free. Corolla rotate, tube very short, greatly exceeded by lobes; lobes 4–5, widely spreading, simple. Stamens 4–5, exserted; filaments thickened basally. Staminodes 0. Ovary glabrous, 1-locular with 2 basal ovules. Fruit a 1-seeded berry. Seed with small broad basal or basi-ventral hilum; embryo with plano-convex, free cotyledons, radicle extending to the surface; endosperm 0. One species, *D. cuspidatum* (Hoehne) Cronquist, South America.

IV. TRIBE CHRYSOPHYLLEAE Hartog (1878).

Calyx a single whorl of 4–5(–11) imbricate or quincuncial sepals; corolla lobes and stamens usually same number as sepals; corolla tubular, cyathiform or rotate; corolla lobes undivided; stamens exserted or included; staminodes small, in a single whorl alternating with stamens or 0; hilum adaxial or rarely basi-ventral.

KEY TO THE GENERA OF CHRYSOPHYLLEAE

1. Corolla rotate or tubular with spreading lobes; stamens fixed at the top of the corolla tube and exserted — 2
- Corolla tubular or cyathiform with more or less erect lobes; stamens fixed within or at the top of the corolla tube, included — 18
2. Staminodes +, seed endospermous or not — 3
- Staminodes 0; seed without endosperm or rarely with a thin sheath of endosperm — 12
3. Seed nearly always endospermous; embryo with thin foliaceous cotyledons and exserted radicle — 4
- Seed without endosperm; embryo with plano-convex cotyledons, radicle usually included — 6
4. Leaves alternate and distichous; secondary venation closely parallel, secondary and higher order venation indistinguishable, the leaf appearing finely striate — **34. *Micropholis*** (sect. *Exsertistamen*)
- Leaves spirally arranged; venation not closely parallel, leaves not finely striate — 5
5. Plant often spiny; flower buds slender, acute; large anthers with connective produced at apex, usually closely pressed against corolla lobes; staminodes often aristate and hairy; disk 0; style long-exserted — **48. *Xantolis***
- Plant not spiny; flower buds short and rounded; anther connective not produced, anthers not pressed against corolla lobes; staminodes not aristate, glabrous; annular disk often +; style included or slightly exserted — **31. *Pouteria*** (sect. *Pierrisideroxylon*)
6. Leaves usually alternate and distichous; corolla tube, filaments and staminodes carnose; seed laterally compressed with narrow adaxial hilum — **40. *Sarcaulus***
- Leaves spirally arranged, opposite or verticillate; corolla tube, filaments and staminodes not carnose; seed not laterally compressed, hilum nearly always broad — 7

7. Usually stipulate; secondary veins close, parallel; higher order venation parallel to the secondaries, the leaf appearing coarsely striate; filaments often partially or completely fused into a staminal tube   **47. *Englerophytum***
 - Stipulate or not; secondary veins not closely parallel, leaves not striate; filaments free   8
8. Small caducous stipules +; corolla lobes with contorted aestivation; staminodes densely woolly, incurved, forming a cap above the ovary; anthers closely applied to corolla lobes   **49. *Capurodendron***
 - Stipulate or not; corolla lobes with imbricate or valvate aestivation; staminodes not densely woolly, not incurved over the ovary; anthers not closely applied to the corolla lobes   9
9. Stipules +; leaves spirally arranged   **46. *Synsepalum***
 - Stipules 0; leaves often opposite or verticillate   10
10. Higher order venation nearly always finely areolate; annular disk often +   **39. *Pichonia***
 - Higher order venation not finely areolate; disk 0   11
11. Leaves spirally arranged; hilum covering at least half the seed surface   **46. *Synsepalum***
 - Leaves often opposite or verticillate; hilum covering less than half the seed surface   **31. *Pouteria*** (sect. *Gayella*)
12. Leaves opposite; calyx of 4 sepals; long filiform exserted style   **44. *Leptostylis***
 - Leaves spirally arranged; calyx of 5(–6) sepals; style exserted or not   13
13. 2–4 stamens opposite each corolla lobe   **45. *Pycnandra***
 - 1 stamen opposite each corolla lobe   14
14. Stipules +   15
 - Stipules 0   16
15. Secondary veins close, parallel; higher order venation parallel to the secondaries, the leaf appearing coarsely striate; filaments often partially or completely fused into a staminal tube   **47. *Englerophytum***
 - Secondary veins not closely parallel, leaves not striate; filaments free   **46. *Synsepalum***
16. Lower leaf surface usually minutely punctate; higher order venation obscure; ovary 2–3(4)-locular, hilum covering less than half the seed surface; embryo with exserted radicle; thin sheath of endosperm +   **41. *Elaeoluma***
 - Lower leaf surface not punctate; higher order venation not obscure; ovary (2)3-locular; hilum often covering more than half the seed surface; embryo with exserted or included radicle; endosperm usually 0   17
17. Corolla lobes and stamens (4)5–10; fruit a berry; hilum often covering or more of seed surface; radicle included   **42. *Niemeyera***
 - Corolla lobes and stamens 5; fruit a drupe; hilum not covering more than of the seed surface; radicle often exserted   **43. *Pradosia***
18. Staminodes +   19
 - Staminodes 0   24
19. Seed without endosperm   20
 - Seed endospermous   21
20. Plant with large stipules   **35. *Chromolucuma***
 - Stipules 0 (+ in *Pouteria congestifolia*)   **31. *Pouteria***
    (sects. ***Rivicoa, Aneulucuma, Antholucuma, Pouteria, Oxythece, Franchetella***)
21. Leaves alternate and distichous; secondary venation closely parallel; leaf appearing striate   **34. *Micropholis*** (sect. *Micropholis*)
 - Leaves spirally arranged; secondary venation not closely parallel; leaf not striate   22
22. Ovary 7–9-locular   23
 - Ovary 5-locular   **31. *Pouteria*** (sect. *Oligotheca*)

23. Flowers bisexual; stamens inserted near the top of the corolla tube; hilum adaxial and extending around the base   **33. *Breviea***
 - Flowers unisexual; stamens inserted halfway up the corolla tube; hilum adaxial   **32. *Aubregrinia***
24. Stipules +   **37. *Ecclinusa***
 - Stipules 0   25
25. Seed without endosperm   26
 - Seed endospermous   **36. *Chrysophyllum***
26. Indumentum of long, stiff, simple hairs; leaves glandular-striate   **38. *Delpydora***
 - Indumentum of short malpighiaceous hairs; leaves not glandular-striate   **31. *Pouteria*** (sect. *Oxythece*)

31. *Pouteria* Aublet

Pouteria Aublet, Hist. Pl. Guiane 1: 85, pl. 33 (excl. fruit) (1775).

For generic synonymy see Pennington (1991).

Trees or shrubs, rarely geoxylic suffrutices. Leaves spirally arranged, rarely opposite. Venation eucamptodromous or brochidodromous, usually without a submarginal vein, never finely striate (except *P. keyensis*). Stipules 0 (+ in *P. congestifolia*). Inflorescence axillary or ramiflorous, fasciculate, fascicles single or occasionally arranged along short leafless shoots. Flowers often unisexual (plant dioecious). Sepals 4–6, free, imbricate or quincuncial, or 6–11 in a closely imbricate spiral. Corolla cyathiform to tubular, rarely rotate, tube shorter to longer than the lobes, lobes 4–6(–9), usually erect, rarely spreading, simple, sometimes fringed-ciliate or papillose. Stamens 4–6(–9), fixed inside corolla tube, or rarely at base of lobes, rarely free, usually included, less frequently exserted; filaments generally short; anthers usually extrorse or laterally dehiscent, usually glabrous. Staminodes usually same number as corolla lobes, less frequently lacking, inserted in the corolla sinus or inside the tube, sometimes fringed-ciliate or papillose. Disk + or 0. Ovary 1–6(–15)-locular, placentation axile; style included or exserted. Fruit a 1–several-seeded berry. Seed broadly ellipsoid, plano-convex, shaped like the segment of an orange or laterally compressed, testa smooth, wrinkled or pitted; hilum adaxial, usually full-length, narrow, broad or sometimes covering almost all the seed surface. Embryo vertical, with plano-convex or thin foliaceous cotyledons, radicle exserted or included; endosperm + or 0. About 200 species in the Neotropics, c. 120 species in Asia, Malesia, Australia and the Pacific, c. 5 species in Africa. Nine sections distinguished by Pennington (1991); see there for sectional descriptions and synonymy.

32. *Aubregrinia* Heine

Aubregrinia Heine, Kew Bull. 14: 301 (1960).

Dioecious trees. Leaves spirally arranged. Venation eucamptodromous. Inflorescence axillary, fasciculate. Stipules 0. Flowers unisexual. Sepals 5, free, quincuncial. Corolla broadly tubular, tube exceeding the lobes, lobes 5, erect, simple. Stamens 5, fixed about halfway up the corolla tube, included; anthers extrorse, glabrous, 0 in female flower. Staminodes 5, inserted in the corolla sinuses. Disk +? (according to Aubréville & Pellegrin present and fused to ovary). Ovary 7–8-locular, placentation axile; style included. Berry large, several-seeded. Seed laterally compressed, testa smooth; hilum adaxial, full-length, narrow. Embryo vertical, with thin foliaceous cotyledons, radicle exserted; endosperm copious. A single species, *A. taiensis* (Aubréville & Pellegrin) Heine, in West Africa.

33. *Breviea* Aubréville & Pellegrin

Breviea Aubréville & Pellegrin, Bull. Soc. Bot. France 81: 792 (1934); Aubréville, Fl. For. Cote d'Ivoire, ed. 2, 3: 130 (1959); Heine, Kew Bull. 14: 302 (1960).

Trees. Leaves distichous. Venation eucamptodromous-brochidodromous. Stipules 0. Inflorescence axillary, fasciculate. Flowers bisexual. Sepals 5, free, quincuncial. Corolla tubular, tube greatly exceeding the lobes; lobes 5, erect, simple. Stamens 5, fixed near the top of the corolla tube, included; anthers laterally dehiscent, glabrous. Staminodes 5, fixed in the corolla sinuses. Disk annular and slightly lobed, fused to the ovary. Ovary 8–9-locular, placentation axile; style included. Berry large, several-seeded. Seed strongly laterally compressed, testa smooth; hilum adaxial and extending around the base, narrow. Embryo vertical, with thin foliaceous cotyledons and exserted radicle; endosperm copious. A single species, *B. sericea* Aubréville & Pellegrin, in West Africa.

34. *Micropholis* (Grisebach) Pierre

Micropholis (Grisebach) Pierre, Not. Bot. Sapot.: 37 (1891); Pierre & Urban, Symb. Ant. 5: 111 (1904); Pennington, Fl. Neotrop. 52: 172 (1990).
Sapota sect. *Micropholis* Grisebach (1861).

Trees or shrubs. Leaves spaced, distichous or spirally arranged. Venation brochidodromous with a submarginal vein, or craspedodromous, secondary veins closely parallel, often not differentiated from the higher order venation, and then the leaf appearing finely striate. Stipules 0. Inflorescence axillary, ramiflorous or cauliflorous, fasciculate. Flowers often unisexual. Sepals (4)5, free, imbricate or quincuncial. Corolla campanulate to short- or long-cylindrical, the tube nearly always exceeding the lobes, rarely equalling them, lobes (4)5, erect to reflexed, simple. Stamens (4)5, fixed near the top of the corolla tube, included or exserted; filaments short and straight or long and geniculate (at least in bud); anthers extrorse in bud, glabrous. Staminodes (4)5, in the corolla sinuses, alternating with the stamens, usually lanceolate or subulate, or rarely petaloid. Disk + or 0. Ovary (4)5-locular, placentation axile; style included or exserted. Fruit 1–several-seeded. Seed laterally compressed, testa smooth or often minutely transversely wrinkled, shining or dull; hilum adaxial, extending the length of the seed, usually narrow. Embryo vertical, with thin foliaceous cotyledons and exserted radicle, surrounded by thick endosperm. Thirty eight species in Central and South America and the West Indies.

Two sections, 1. **sect. *Micropholis***, corolla less than 10 mm long, lobes erect or only slightly spreading; stamens included stamens; thirty species throughout tropical America; 2. **sect. *Exsertistamen*** T. Pennington, corolla usually more than 10 mm long; corolla lobes spreading or reflexed, stamens exserted; nine spp. centred on the Guianas and extending across Brazilian Amazonia to Peru.

35. *Chromolucuma* Ducke

Chromolucuma Ducke, Arch. Jard. Bot. Rio de Janeiro 4: 160 (1925); Trop. Woods 71: 20 (1942); Pennington, Fl. Neotrop. 52: 229–232 (1990).
Pouteria sect. *Chromolucuma* (Ducke) Baehni (1942).

Latex yellow. Leaves spirally arranged, venation eucamptodromous. Stipules large. Flowers unisexual (plants monoecious or dioecious), pedicellate. Sepals 5, imbricate. Corolla cyathiform or shortly tubular, tube usually equalling or slightly longer than lobes (slightly shorter in male flowers of *C. rubriflora*). Stamens fixed in the upper half or at the top of the corolla tube, included; filaments short. Staminodes +, vestigial in female flower of *C. baehniana*. Disk 0. Ovary broadly truncate to ovoid, 2–5-locular. Seed with dull rough testa, and broad adaxial hilum covering up to two thirds of seed; embryo with plano-convex, free cotyledons, radicle slightly exserted; endosperm 0. Two species in the Guianas, southern Venezuela and central Amazonian Brazil.

36. *Chrysophyllum* L.

Chrysophyllum L., Sp. Pl.: 192 (1753).

Trees or shrubs, very rarely lianas. Leaves distichous or spirally arranged. Venation brochidodromous or eucamptodromous, tertiary veins often parallel to the secondaries and descending from the margin, or oblique and closely parallel, or reticulate. Stipules 0. Inflorescence axillary, ramiflorous or cauliflorous. Flowers unisexual or bisexual, fasciculate or rarely solitary. Sepals (4)5(6), imbricate or quincuncial, sometimes accrescent in fruit, frequently ciliate. Corolla globose, campanulate or cylindrical, tube shorter than, equalling or exceeding the lobes, lobes (4)5(-8), simple. Stamens (4)5(-8), fixed in the lower or upper part of the corolla tube, included; anthers extrorse in bud, hairy or glabrous. Staminodes rarely + as small lanceolate or subulate structures in the corolla lobe sinuses, alternating with the stamens. Disk 0. Ovary (4)5(-12)-locular, placentation axile, style included. Fruit 1-many-seeded. Seed laterally compressed, with a narrow adaxial hilum, sometimes extending around base of seed, or not laterally compressed and then the hilum broader, basi-ventral or adaxial; testa smooth and shining, or rough and then adherent to the pericarp. Embryo vertical, with thin foliaceous or thick flat cotyledons and exserted radicle, endosperm abundant or about equalling the thickness of the cotyledons. Forty three species in the Neotropics, c. 15 in Africa, c. 10 in Madagascar, and 2-3 extending from India to Malesia and Australia.

Six sections; for descriptions, keys and synonymy see Pennington (1991).

37. *Ecclinusa* Martius

Ecclinusa Martius, Flora 22, Beibl. 1: 2 (1839); Pennington, Fl. Neotrop. 52: 622-639 (1990).

Trees or rarely shrubs. Leaves spirally arranged, usually loosely clustered at the shoot apex; venation usually eucamptodromous or rarely brochidodromous, intersecondaries usually 0; tertiaries usually oblique, numerous, close, parallel, rarely reticulate or areolate. Stipules caducous, leaving a conspicuous hilum. Inflorescence axillary or in the axils of fallen leaves. Flowers sessile, subtended by small persistent bracts, usually unisexual (monoecious or dioecious). Sepals (4)5, free, quincuncial. Corolla small (usually less than 5 mm long), campanulate or shortly tubular, the lobes usually exceeding the tube, rarely equalling it; lobes 5(-7),

simple. Stamens 5(-7) included, usually fixed near halfway or in the upper half of the corolla tube, rarely in the lower half; filaments well-developed, free; anthers extrorse, glabrous. Staminodes 0. Disk 0. Ovary (3-)5(-9)-locular, placentation axile or basi-ventral; style included. Fruit 1-several-seeded, often thin-walled and constricted between the seeds. Seed globose, ellipsoid, sometimes slightly laterally compressed or shaped like the segment of an orange, testa smooth, thin, shining; hilum adaxial and nearly always extending around the base of the seed, usually narrow. Embryo with thick plano-convex cotyledons, radicle not exserted, extending to the surface; endosperm 0. Eleven species, from Panama throughout tropical South America.

38. *Delpydora* Pierre

Delpydora Pierre, Bull. Mens. Soc. Linn. Paris 2: 1275 (1897).

Small trees. Indumentum of long stiff simple hairs. Leaves spirally arranged, pellucid-striate, venation eucamptodromous or brochidodromous, tertiaries oblique. Stipules 0. Inflorescence axillary or ramiflorous, fasciculate. Flowers bisexual. Sepals 5, free, quincuncial. Corolla shortly tubular, tube much longer than lobes; lobes 5, erect or slightly spreading, simple. Stamens 5, fixed in the lower half of the corolla tube, included; filaments well-developed, free; anthers extrorse, closely applied to the style and sometimes laterally connivent, glabrous. Staminodes 0. Disk 0. Ovary 5-locular, placentation axile; style included. Berry several-seeded. Seed broadly oblong or shaped like the segment of an orange, not laterally compressed, testa thin, smooth; hilum adaxial and extending around the base, very narrow. Embryo with plano-convex cotyledons and included radicle; endosperm 0. Two species in West Africa.

39. *Pichonia* Pierre

Pichonia Pierre, Not. Bot. Sapot.: 22 (1890).

Trees. Leaves opposite, rarely spirally arranged. Venation eucamptodromous-brochidodromous, higher order venation finely reticulate-areolate. Stipules 0. Inflorescence axillary, fasciculate, rarely on leafless short shoots. Flowers bisexual. Sepals (4)5, free, quincuncial. Corolla cyathiform to rotate, tube shorter than the lobes; lobes 5, widely spreading, simple. Stamens 5, inserted at the top of the corolla tube, exserted, with long filaments; anthers laterally dehiscent or extrorse, with apical tuft of hairs or glabrous. Staminodes 5, well-developed, fixed in the corolla sinuses. Disk annular, stipitate or 0. Ovary (4)5(6)-locular, placentation axile; style exserted. Berry 1-seeded. Seed narrowly to broadly ellipsoid, not laterally compressed, testa smooth; hilum adaxial, full-length, broad, covering from to almost all the seed surface, and then only a narrow abaxial strip remaining free. Embryo with thick plano-convex cotyledons and included radicle; endosperm 0. About 5 species in New Caledonia, Papua New Guinea and the Solomon Islands.

40. *Sarcaulus* Radlkofer

Sarcaulus Radlkofer, Sitzungsber. Math.-Phys. Cl. Königl. Bayer. Akad. Wiss. München 12: 310 (1882); Pennington, Fl. Neotrop. 52: 232–239 (1990).

Leaves spaced, distichous or less frequently weakly spirally arranged, venation eucamptodromous or brochidodromous. Stipules 0. Flowers unisexual (plant dioecious). Sepals usually 5. Corolla globose, broadly cyathiform or subrotate, tube weakly to strongly carnose, equalling or slightly exceeding the lobes; lobes usually 5, slightly imbricate or subvalvate. Stamens usually 5, fixed at the top of the corolla tube, exserted; filaments short, swollen; anthers strongly inflexed. Staminodes usually 5, thick, carnose, erect or incurved against the style. Disk 0. Ovary 2–5-locular. Berry 1–several-seeded. Seeds laterally compressed, testa smooth or wrinkled, shining; hilum adaxial, full-length, rather narrow; embryo with plano-convex, free cotyledons, radicle extending to the surface; endosperm 0. Five species in tropical South America.

41. *Elaeoluma* Baillon

Elaeoluma Baillon, Hist. Pl. 11: 293 (1891); Pennington, Fl. Neotrop. 52: 240–247 (1990).

Trees or shrubs. Leaves spirally arranged, usually minutely punctate on the lower surface. Venation eucamptodromous or brochidodromous, higher order veins often obscure, forming a lax reticulum. Stipules 0. Inflorescence mostly axillary. Flowers unisexual. Sepals (4)5, imbricate or quincuncial. Corolla broadly cyathiform to rotate, tube shorter than the lobes or rarely equalling the lobes in female, lobes 5(6), spreading, simple. Stamens 5(6), fixed at the top of the corolla tube and exserted; filaments well-developed; anthers glabrous. Staminodes 0 (1–2). Disk 0. Ovary 2–3(–4)-locular,

placentation axile; style short. Berry 1-seeded. Seed broadly ellipsoid, not or sometimes laterally compressed, testa smooth to slightly wrinkled, shining; hilum adaxial, full-length, narrow or broad. Embryo cotyledons plano-convex, radicle slightly exserted; endosperm thin. Four species in southern Venezuela, the Guianas, Brazilian Amazonia and Panama.

42. *Niemeyera* F. Muell.

Niemeyera F. Muell., Fragm. 7: 114 (1870).

Trees or treelets, sometimes pachycaulous. Leaves spirally arranged; venation eucamptodromous or brochidodromous, higher order venation oblique, or descending from the margin and parallel to secondaries; stipules 0. Inflorescence fasciculate, axillary or ramiflorous. Flowers bisexual. Sepals 5, free, imbricate or quincuncial. Corolla rotate or cyathiform with spreading lobes, tube nearly always shorter than the lobes or equalling them; lobes (4)5–10, simple. Stamens (4)5–10, exserted, usually inserted at or near the top of the corolla tube, rarely only halfway up the tube; filaments well-developed, free, geniculate in bud; anthers versatile in open flower, hairy or glabrous. Staminodes 0. Disk usually 0. Ovary (2)3–5-locular; placentation axile; style exserted or not. Berry 1–2-seeded. Seed narrowly ellipsoid to globose, not laterally compressed; hilum adaxial, broad, often covering up to or more of the seed surface. Embryo with plano-convex cotyledons; radicle included; endosperm 0. About 20 species in New Caledonia and Australia.

43. *Pradosia* Liais

Pradosia Liais, Climat., Geol., Faune Brésil: 614 (1872); Pennington, Fl. Neotrop. 52: 639–668 (1990).

Leaves opposite or verticillate, less frequently spirally arranged; venation usually eucamptodromous, less frequenly brochidodromous, midrib sunken on the upper surface, rarely flat or raised, secondaries often impressed on the upper surface; intersecondaries usually 0, rarely well-developed; tertiaries usually oblique or horizontal; minute paired scales (stipels) sometimes present on petiole. Stipules 0. Inflorescences cauliflorous or ramiflorous, less frequently axillary. Flowers bisexual. Sepals usually 5. Corolla rotate, tube nearly always shorter than the widely spreading lobes, corolla lobes 5. Stamens 5, fixed at top of corolla tube or on base of lobes, exserted; filaments long, geniculate below the apex, and strongly narrowed below insertion of anthers. Staminodes 0. Disk 0. Ovary (4)5(6)-locular, style short. Fruit a drupe with thinly cartilaginous endocarp, often slightly asymmetric. Seed solitary, with smooth, shining testa and full-length adaxial hilum covering up to one third of the seed surface; embryo with thinly plano-convex cotyledons, radicle often exserted or only extending to the surface; endosperm a thin sheath or 0. Twenty three species in South America, with one extending into Panama and Costa Rica.

44. *Leptostylis* Bentham

Leptostylis Bentham in Bentham & Hooker, Gen. Pl. 2: 659 (1876).

Shrubs or small trees. Leaves (sub)opposite; venation usually brochidodromous, less frequently eucamptodromous; higher order venation reticulate. Stipules 0. Inflorescence fasciculate, axillary or ramiflorous. Flowers bisexual. Sepals 4, free or slightly united, strongly imbricate. Corolla short to long-tubular, infundibuliform or rotate, tube greatly exceeding, equalling or shorter than the lobes; lobes 4–10, widely spreading or reflexed, simple. Stamens 4–10, exserted, inserted in the upper half or at the apex of the corolla tube; filaments well-developed, free, geniculate in bud; anthers extrorse, usually glabrous, sometimes with a tuft of hairs at the apex. Staminodes 0. Disk 0. Ovary 3–4(5)-locular, placentation axile; style exserted, often very long and filiform. Fruit (known in only *L. filipes* and *L. petiolata*) a 1-seeded berry. Seed ellipsoid to narrowly ovate, not laterally compressed; hilum adaxial, full-length, covering about to the seed surface. Embryo with plano-convex cotyledons, radicle included or slightly exserted; endosperm 0. Eight species in New Caledonia.

45. *Pycnandra* Bentham

Pycnandra Bentham in Bentham & Hooker, Gen. Pl. 2: 658 (1876).

Trees or treelets, often pachycaulous. Leaves spirally arranged or occasionally opposite; venation usually eucamptodromous, less frequently brochidodromous; higher order venation oblique or reticulate. Stipules 0. Inflorescence axillary or ramiflorous. Flowers bisexual. Sepals 5(6), free, quincuncial. Corolla tube cyathiform, usually shorter than or equalling the lobes, rarely slightly exceed-

ing the lobes; lobes 5–10, widely spreading or reflexed, simple. Stamens (7–)10, 1–2 opposite each corolla lobe, less frequently up to 25 and then 3–4 opposite each corolla lobe, exserted, fixed in a single whorl in the upper half or at the top of the corolla tube; filaments well-developed, free or sometimes partially united in pairs, geniculate in bud; anthers extrorse, glabrous or rarely hairy. Staminodes 0. Disk 0 (small annulus at ovary base in *P. vieillardii*). Ovary 5–8(–10)-locular, placentation axile; style usually included. Fruit a 1-seeded berry. Seed broadly ellipsoid, not laterally compressed, hilum adaxial, full-length, covering about 1/3 to 1/2 the surface, or almost all the surface, hilum area sometimes strongly rugose or ruminate. Embryo with plano-convex, smooth or ruminate cotyledons, radicle extending to the surface; endosperm 0. About 12 species confined to New Caledonia.

46. *Synsepalum* (A. de Candolle) Daniell Fig. 123

Synsepalum (A. de Candolle) Daniell, Pharm. J. Trans. 11: 445 (1852).
Sideroxylon sect. *Synsepalum* A. de Candolle (1844).
Vincentella Pierre (1891).
Pachystela Pierre ex Baillon (1891).
Afrosersalisia A. Chevalier (1943).

Shrubs or trees. Leaves spirally arranged. Venation nearly always eucamptodromous, rarely brochidodromous, higher order venation oblique or reticulate; leaves not striate. Stipules + and often conspicuous, or 0. Inflorescence axillary or ramiflorous, rarely cauliflorous. Flowers in fascicles, bisexual. Sepals 5(6), free or partly united, quincuncial or rarely open. Corolla usually with short tube exceeded by the widely spreading lobes, rarely exceeding the lobes and then cyathiform or tubular, lobes 5(6), usually widely spreading, simple; aestivation imbricate or less frequently induplicate valvate. Stamens 5(6), fixed at the top of the corolla tube or rarely slightly inside the throat, usually strongly exserted; filaments usually well-developed, free; anthers usually extrorse, glabrous or less frequently hairy. Staminodes variable in size, glabrous, or 0. Disk 0. Ovary 5(–7)-locular, placentation axile; style usually exserted. Berry 1-seeded. Seed ellipsoid, not laterally compressed; testa smooth; hilum adaxial, broad, usually covering at least one third of the surface and sometimes nearly all of the surface. Embryo with plano-convex cotyledons, radicle extending to the surface or slightly exserted; endosperm 0. Possibly 20 species in tropical Africa.

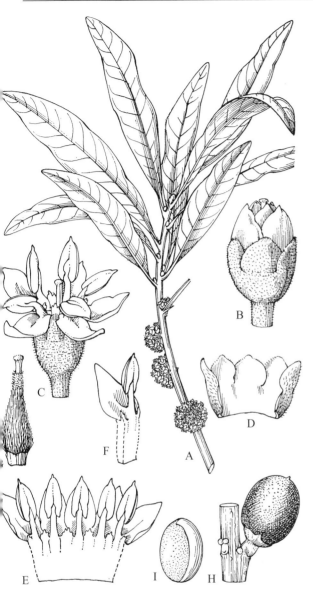

Fig. 123. Sapotaceae–Chrysophylleae. *Synsepalum cerasiferum.* **A** Flowering branch. **B** Young flower. **C** Anthetic flower. **D** Dissected calyx. **E** Dissected corolla showing stamens and small staminodes. **F** Corolla segment with stamens. **G** Ovary. **H** Fruit. **I** Seed. (Hemsley 1968)

47. *Englerophytum* Krause

Englerophytum Krause, Bot. Jahrb. Syst. 50, Suppl.: 343 (1914). *Bequaertiodendron* De Wildeman (1919).

Shrubs or trees. Leaves spirally arranged. Venation brochidodromous, usually with a submarginal vein, secondaries usually close and parallel, higher order venation parallel to the secondaries, the leaves appearing coarsely striate. Stipules usually +, often large and persistent. Inflorescence axillary, ramiflorous or cauliflorous. Flowers in fascicles, bisexual or unisexual. Sepals 5(6), free or slightly united, quincuncial. Corolla shortly tubular or cyathiform, the tube slightly shorter than, equalling or slightly exceeding the lobes; lobes 5(–10), often widely spreading, simple, imbricate or quicuncial. Stamens 5(–10), fixed at the top of the corolla tube, usually exserted; filaments usually well-developed, and often partially or completely fused into a short tube or thick fleshy collar; anthers usually extrorse, glabrous; stamens sometimes converted into petaloid staminodes in female. Staminodes 0 or occasionally a single whorl of small subulate structures. Disk 0. Ovary (4)5(–10)-locular, placentation axile; style included or slightly exserted. Berry 1-seeded. Seed ellipsoid, not or only slightly laterally compressed; testa smooth, often thin; hilum adaxial; variable in width, from narrow to broad, and sometimes extending to cover most of the seed surface. Embryo with plano-convex cotyledons, radicle extending to the surface; endosperm 0. Five to ten poorly defined species in tropical Africa.

48. *Xantolis* Rafinesque

Xantolis Rafinesque, Sylva tellur. 36 (1838); van Royen, Blumea 8: 207 (1957); Aubréville, Fl. Cambodge, Laos et Vietnam 3. Sapotacées: 74 (1963).

Trees or shrubs, often spiny. Stipules 0. Leaves spirally arranged, sometimes fascicled on short shoots. Venation eucamptodromous or brochidodromous. Inflorescence axillary or ramiflorous, fasciculate, sometimes clustered along short leafless shoots. Flowers bisexual. Sepals 5, free, quincuncial. Corolla with short tube nearly always exceeded by the spreading lobes; lobes 5, simple, imbricate; corolla tube nearly always hairy around the base of the filaments. Stamens 5, fixed at the top of the corolla tube, exserted or closely applied to the corolla lobes; anthers large, connective produced at the apex, extrorse, glabrous. Staminodes 5, well-developed, often equalling the corolla lobes, often fimbriate and aristate, often hairy. Disk 0. Ovary (4)5-locular, placentation axile; style long-exserted. Berry 1(2)-seeded. Seed slightly to strongly laterally compressed; hilum adaxial, narrow, equalling the length of the seed to only half as long. Embryo with thin, flat, foliaceous cotyledons and long exserted radicle; copious endosperm. About 14 species from southern India to Vietnam and south China, one species in the Philippine Islands.

49. *Capurodendron* Aubréville

Capurodendron Aubréville, Adansonia II, 2: 921 (1962); Fl. Madag. 164. Sapotacées: 68 (1974).

Trees or shrubs. Leaves spirally arranged. Venation eucamptodromous or brochidodromous. Stipules usually +, caducous. Inflorescence axillary, fasciculate. Flowers bisexual. Sepals 5, free or slightly fused, quincuncial. Corolla tube often slightly constricted at the apex, slightly shorter than, equalling or slightly longer than the lobes; lobes 5, spreading, simple, contorted. Stamens 5, fixed at the top of the corolla tube; anthers extrorse, usually closely applied to the corolla lobes, hairy or glabrous. Staminodes 5, well-developed, nearly always narrowly triangular and densely woolly, closely applied to the style. Disk 0. Ovary 5(6)-locular, placentation axile; style exserted. Berry 1-seeded. Seed globose to narrowly ellipsoid, not laterally compressed; testa smooth; hilum adaxial (basi-ventral), occupying about half the length of the seed or full-length, broad, sometimes covering about one third of the seed surface. Embryo with plano-convex cotyledons, radicle extending to the surface; endosperm 0. Twenty three species confined to Madagascar.

V. Tribe Omphalocarpeae Dubard ex Aubréville (1964).

Calyx a single whorl of 5 valvate or quincuncial sepals; corolla lobes same number as sepals, simple; corolla cyathiform or tubular, tube shorter than or exceeding the lobes; stamens in a single whorl in groups of 2–6 opposite each corolla lobe; staminodes +; hilum adaxial.

Key to the Genera of Omphalocarpeae

1. Stipules +; anthers c. length of corolla lobes   **50. *Tsebona***
– Stipules 0; staminodes much shorter than corolla lobes 2
2. Anthers locellate; disk patelliform; seed not laterally compressed, without endosperm   **51. *Magodendron***
– Anthers not locellate; disk 0; seed strongly laterally compressed, endospermous   3
3. Corolla lobes slightly longer than tube; stamens fixed near top of corolla tube; filaments mostly free, glabrous; anthers glabrous   **52. *Omphalocarpum***
– Corolla lobes free almost to base, stamens fixed in lower half of corolla tube, almost completely fused in pairs or 3s; filaments hairy, anthers hairy or glabrous   **53. *Tridesmostemon***

50. *Tsebona* Capuron

Tsebona Capuron, Adansonia II, 2: 122 (1962); Baehni, Boissiera 1: 105 (1965); Aubréville, Fl. Madag. 164, Sapotacées: 123 (1974).

Tree. Leaves spirally arranged. Venation eucamptodromous. Stipules +, well-developed, caducous. Flowers solitary, axillary, bisexual. Sepals 5, free, quincuncial, both margins of the outer 2 sepals and 1 margin of the third sepal induplicate valvate, and completely enclosing the remaining sepals in bud. Corolla tube much shorter than the lobes, lobes 5, slightly spreading, simple, contorted. Stamens in groups of 2–4 opposite each corolla lobe, fixed at the top of the corolla tube; filaments free, short, hairy; anthers about the length of the corolla lobes, extrorse, hairy. Staminodes 5, much shorter than the corolla lobes, doubly geniculate, and pressed against the style, densely hairy. Disk minute, free, annular, surrounding the base of the ovary. Ovary 5-locular, placentation axile; style exserted. Berry 1–several-seeded. Seed broadly ovoid to globose, not laterally compressed; testa smooth; hilum covering about 3/4 to 7/8 of the seed surface. Embryo with plano-convex cotyledons and included radicle; endosperm 0. A single species in Madagascar (*T. macrantha* Capuron).

51. *Magodendron* Vink

Magodendron Vink, Nova Guinea II 8: 124 (1957); Baehni, Boissiera 11: 116 1965.

Tree. Leaves spirally arranged. Venation eucamptodromous. Stipules 0. Inflorescence fasciculate, cauliflorous. Flowers bisexual. Sepals 5, free, quincuncial. Corolla broadly tubular, expanded above, tube exceeding the lobes; lobes 5, spreading, simple, quincuncial. Stamens 10 with 2 opposite each corolla lobe and fixed at the top of the corolla tube, long-exserted; filaments free, well-developed; anthers locellate, glabrous. Staminodes 5, alternating with the corolla lobes and fixed at the level of the stamens, strongly inflexed, margin fringed, glabrous. Disk patelliform, free, surrounding the base of the ovary, sparsely hairy. Ovary 5-locular, placentation axile; style included. Berry 1-seeded. Seed broadly ellipsoid, testa smooth, shining; hilum adaxial, full-length, covering about half the seed surface. Embryo with plano-convex cotyledons; radicle slightly exserted; endosperm 0. A single species in Papua New Guinea.

52. *Omphalocarpum* Palisot de Beauvois

Omphalocarpum Palisot de Beauvois ex Ventenat, Bull. Sci. Soc.
Philom. Paris 2: 146 (1800); Aubréville, Fl. Forest. Cote
d'Ivoire ed. 2, 3: 109 (1959), Fl. Gabon 1, Sapotacées: 75
(1961), Fl. Cameroun 2, Sapotacées: 57 (1964).
Ituridendron De Wildeman (1926).

Trees. Leaves spirally arranged. Venation eucamptodromous. Stipules 0. Cauliflorous or ramiflorous. Flowers in fascicles, bisexual or unisexual
(dioecious), often subtended by spirally arranged
bracts. Sepals 5, free, quincuncial. Corolla broadly
tubular or cyathiform, tube shorter than the lobes;
lobes 5–11, erect or slightly spreading, simple,
imbricate or quincuncial. Stamens in groups of
3–6 opposite each corolla lobe and inserted at the
top of the corolla tube, exserted or not; filaments
well-developed, exceeding the anthers, free or a
few partially united near the base (rarely irregularly united to halfway or above), glabrous; anthers
extrorse, glabrous, 0 in female flowers. Staminodes
5–11, much shorter than the corolla lobes, usually
petaloid, fimbriate, glabrous, often inflexed and
covering the ovary. Disk 0. Ovary 5–30-locular
(number of loculi reduced in male flowers), placentation axile; style usually included. Fruit
usually depressed globose, the outer pericarp
often woody, several to many-seeded. Seed
strongly laterally compressed, testa smooth,
shining; hilum nearly full-length, very narrow.
Embryo with thin foliaceous cotyledons and
exserted radicle; endosperm copious. About six
species in West and Central Africa.

53. *Tridesmostemon* Engler

Tridesmostemon Engler, Bot. Jahrb. Syst. 38: 99 (1905); De
Wildeman, Pl. Bequaert. 4: 144 (1926); Pellegrin, Bull. Soc.
Bot. France 85: 179 (1938); Aubréville, Fl. Gabon 1.
Sapotacées: 82 (1961); Fl. Cameroun 2. Sapotacées: 66
(1964).

Trees. Stipules 0 (or caducous?). Leaves spirally
arranged. Venation eucamptodromous. Flowers
bisexual or rarely unisexual?, solitary or in few-
flowered fascicles, axillary or in the axils of fallen
leaves. Sepals 5, free, imbricate. Corolla cyathiform, corolla lobes 5, simple, imbricate, spreading
or reflexed at the tip, free almost to the base.
Stamens 10–15, 2–3 opposite each corolla lobe,
fixed to way up the corolla lobes, with prominent
traces to the base; filaments of each pair or group
of 3 fused almost to the apex, hairy; anthers
extrorse, hairy or glabrous. Staminodes 5, usually
3-dentate at apex, hairy. Disk 0. Ovary 10-locular,
placentation axile; style slightly exserted. Fruit

subglobose, several seeded. Seed strongly laterally
compressed, testa smooth, shining; hilum adaxial,
almost full-length, very narrow. Embryo with thin
foliaceous cotyledons and exserted radicle;
endosperm +. Two or three species in Central
Africa (Cameroon, Gabon, Zaire).

Doubtful Genus

Boerlagella Cogniaux

Boerlagella Cogniaux in A. de Candolle & C. de Candolle,
Monogr. Phan. 7: 1173. (1891); H.J. Lam, Bull. Jard. Bot.
Buitenzorg III, 7: 250 (1925).
Boerlagia Pierre (1890).

Known only from incomplete material of leaf, fruit
and seed. It was placed in *Planchonella* by Dubard
(1912: 61) on account of the narrow hilum and
exendospermous seed, and subsequent authors,
except Lam, have regarded it as doubtfully Sapotaceous (Aubréville 1964a; Baehni 1938, 1965).

Lam (1925: 250) placed it (with *Dubardella*) in
the Boerlagellaceae, for reasons which are not
clear, although possibly on account of the cotyledons which are described as involute or inrolled.

As no new material has been forthcoming, it is
still not possible to place this plant more accurately. The leaf and seed structure appear to be
perfectly consistent with the Sapotaceae, as are the
malpighiaceous hairs at the base of the fruit. The
cotyledon condition, described by Lam, is anomalous and not recorded elsewhere in the family, but
I have not been able to confirm it, as the only available seed seen by me is immature.

In the absence of any further material, I agree
with Dubard and place it provisionally in *Pouteria*
sect. *Oligotheca*.

Selected Bibliography

Anderberg, A.A. et al. 2002. See general references.
Arends, J.C. 1976. Somatic chromosome numbers of some
African Sapotaceae. Acta Bot. Neerl. 25: 449–457.
Assem, J. van den. 1953. Revision of the Sapotaceae of the
Malaysian area in a wider sense. 4. *Ganua* Pierre ex Dubard.
Blumea 7: 364–400.
Aubréville, A. 1961a. Notes sur les Sapotacées de l'Afrique
equatoriale. Notul. Syst. (Paris) 16: 223–279.
Aubréville, A. 1961b. Notes sur les Pouteriées africaines et sud
américaines. Adansonia II, 1: 6–38.
Aubréville, A. 1962a. Notes sur des Pouteriées américaines.
Adansonia II, 1: 150–191.
Aubréville, A. 1962b. Notes sur les Sapotacées de la Nouvelle
Calédonie. Adansonia II, 2: 172–199.
Aubréville, A. 1963a. Notes sur les Sapotacées. Adansonia II, 3:
19–42.

Aubréville, A. 1963b. Flore du Cambodge, du Laos et du Vietnam, Sapotacées, 1–105.

Aubréville, A. 1964a. Sapotacées. Adansonia, Mémoire no. 1: 1–157.

Aubréville, A. 1964b. Notes sur des Sapotacées 3. 1. Réhabilitation des genres américains *Ragala* Pierre et *Prieurella* Pierre. Adansonia II, 4: 367–391.

Aubréville, A. 1964c. Système de classification des Sapotacées. Adansonia II, 4: 38–42.

Aubréville, A. 1965. Notes sur des Sapotacées australiennes. Adansonia II, 5: 21–26.

Aubréville, A. 1967. Flore de la Nouvelle Calédonie et Dépendences. 1. Sapotaceae 1–168.

Aubréville, A. 1971. Essais de géophylétique des Sapotacées 2. Adansonia II, 11: 425–436.

Aubréville, A., Pellegrin, F. 1934. De quelques Sapotacées de la Cote d'Ivoire. Bull. Soc. Bot. France 81: 792–800.

Baehni, C. 1938. Mémoires sur les Sapotacées. 1. Système de classification. Candollea 7: 394–507.

Baehni, C. 1942. Mémoires sur les Sapotacées. 2. Le genre *Pouteria*. Candollea 9: 147–475.

Baehni, C. 1964. Genres nouveaux de Sapotacées. Arch. Sci. 17: 77–79.

Baehni, C. 1965. Mémoire sur les Sapotacées. 3. Inventaire des genres. Boissiera 11: 1–261.

Bokdam, J. 1977. Seedling morphology of some African Sapotaceae and its taxonomical significance. Meded. Landbouwhogesch. Wageningen 77/20: 1–84.

Cheke, A.S., Dahl, J.F. 1981. The status of bats in western Indian Ocean islands, with special reference to *Pteropus*. Mammalia 45: 205–238.

Cleghorn, M.L. 1922. Observations on the bat flowers of the Mohwa (*Bassia latifolia* [= *Illipe l.*]). J. Proc. Asiat. Soc. Beng. II, 18: 571–576.

Corner, E.J.H. 1976. See general references.

Cronquist, A. 1945a. Studies in the Sapotaceae 3. *Dipholis* and *Bumelia*. J. Arnold Arbor. 26: 435–471.

Cronquist, A. 1945b. Studies in the Sapotaceae 4. The North American species of *Manilkara*. Bull. Torrey Bot. Club 72: 550–562.

Cronquist, A. 1946. Studies in the Sapotaceae 2. Survey of the North American Genera. Lloydia 9: 241–292.

Cronquist, A. 1946a. Studies in the Sapotaceae 5. The South American species of *Chrysophyllum*. Bull. Torrey Bot. Club 73: 286–311.

Cronquist, A. 1946b. Studies in Sapotaceae 6. Misc. Notes Bull. Torrey Bot. Club. 73: 465–471.

Dubard, M. 1908. Les Sapotacées du groupe des Illipées. Rev. Gén. Bot. 20: 193–206.

Dubard, M. 1909. Les Sapotacées de groupe des Isonandrées. Rev. Gén. Bot. 21: 392–398.

Dubard, M. 1912. Les Sapotacées de groupe des Sideroxylinées. Ann. Inst. Bot.-Géol. Colon. Marseille II, 10: 1–98.

Dubard, M. 1915. Les Sapotacées du groupe des Sideroxylinées-Mimusopées. Ann. Inst. Bot. Géol. Colon. Marseille III, 3: 1–62.

Engler, A. 1904. Monographien afrikanischer Pflanzen-Familien und -Gattungen 8. Sapotaceae. Leipzig: W. Engelmann.

Eyma, P.J. 1936. Notes on Guiana Sapotaceae. Recueil Trav. Bot. Néerl. 33: 156–210.

Friedmann, F. 1981. Flore des Mascareignes 116. Sapotacées. 1–27.

Friis, I. 1978. A reconsideration of the genera *Monotheca* and *Spiniluma* (Sapotaceae). Kew Bull. 33: 91–98.

Gilly, C.L. 1943. Studies in the Sapotaceae 2. The *Sapodilla-Nispero* complex. Trop. Woods 73: 1–22.

Govaerts, R., Frodin, D.G., Pennington, T.D. 2001. World checklist and bibliography of Sapotaceae. Kew: Royal Botanic Gardens, Kew.

Hall, J.B. 1996. *Vitellaria paradoxa*: a monograph. Bangor: University of Wales, 125 p.

Hallé, F., Oldeman, R.A.A. 1970. Essai sur l'architecture et la dynamique de croissance des arbres tropicaux. Paris: Masson.

Harley, M.M. 1986a. Distinguishing pollen characters for the Sapotaceae. Can. J. Bot. 64: 3091–3100.

Harley, M.M. 1986b. The nature of the endoaperture in the pollen of the Sapotaceae. In: Blackmore, S., Ferguson, I.K. (eds.) Pollen and spores: form and function. London: Academic Press. Linn. Soc. Symp. Ser. 12: 417–419.

Harley, M.M. 1990a. Pollen morphology of neotropical Sapotaceae. In: Pennington, T.D. 1990: A monograph of neotropical Sapotaceae. Flora Neotrop. 52: 11–29, 710–741.

Harley, M.M. 1990b. The pollen morphology of the Sapotaceae. Kew Bull. 46: 379–491.

Harley, M.M. 1991. Pollen morphology of the Sapotaceae. In: Pennington, T.D. 1990: The genera of Sapotaceae. Kew: Royal Botanic Gardens Kew, New York: New York Botanical Garden, pp. 23–50.

Harley, M.M., Kurmann, M.H., Ferguson, I.K. 1991. Systematic implications of comparative morphology in selected fossil and extant pollen from the Palmae and the Sapotaceae. In: Blackmore, S., Barnes, S.H. (eds.) Pollen and spores: patterns of diversification. Oxford: Clarendon Press, pp. 225–238.

Hartog, M.M. 1878. On the floral structure and affinities of Sapotaceae. J. Bot. 16: 65–72, 145.

Hartog, M.M. 1879. Notes on Sapotaceae II. J. Bot. 17: 356–359.

Hegnauer, R. 1973, 1990. See general references.

Heine, H. 1963. Sapotaceae in Flora of West Tropical Africa, 2nd ed., vol. II: 16–30.

Heine, H., Hemsley, J.H. 1960. Notes on African Sapotaceae II. The genus *Bequaertiodendron* De Wild. Kew Bull. 14: 304–309.

Hemsley, J.H. 1966. Notes on African Sapotaceae. Kew Bull. 20: 461–510.

Hemsley, J.H. 1968. Sapotaceae. In: Flora of tropical East Africa. London: Crown Agents.

Hermann-Erlee, M.P.M., Royen, P. van 1957. Revision of the Sapotaceae of the Malaysian area in a wider sense. 9. *Pouteria* Aublet. Blumea 8: 452–509.

Hickey, L.J. 1973. Classification of the architecture of dicotyledonous leaves. Am. J. Bot. 60: 17–33.

Johnson, M.A.T. 1991. Cytology. In: Pennington, T.D. The genera of Sapotaceae, chap. 2. Kew: Royal Botanic Gardens Kew, New York: New York Botanical Garden, pp. 15–22.

Johri, B.D. et al. 1992. See general references.

Kukachka, B.F. 1978. Wood anatomy of the neotropical Sapotaceae. 3. *Dipholis*. Forest Products Laboratory, U.S.D.A. Madison. Res. Pap. 327: 1–7.

Kukachka, B.F. 1979. Wood Anatomy of the Neotropical Sapotaceae. 8. *Diploon*. Forest Products Research Laboratory, U.S.D.A. Madison. Res. Pap. 349: 1–4.

Kukachka, B.F. 1981. Wood Anatomy of the Neotropical Sapotaceae. 23. *Gayella*. Forest Products Research Laboratory, U.S.D.A. Madison. Res. Pap. 374: 1–3.

Kupicha, F.K. 1978. Notes on East African Sapotaceae. Candollea 33: 29–41.

Lam, H.J. 1925. The Sapotaceae, Sarcospermaceae and Boerlagellaceae of the Dutch East Indies and surrounding countries. Bull. Jard. Bot. Buitenzorg III, 7: 1–289.

Lam, H.J. 1927. Further studies on Malayan Sapotaceae. 1. Bull. Jard. Bot. Buitenzorg III, 8: 381–493.

Lam, H.J. 1938. Monograph of the genus *Nesoluma* (Sapotaceae). Occas. Pap. Bernice Pauahi Bishop Mus. 14: 127–165.

Lam, H.J. 1939. On the system of the Sapotaceae, with some remarks on taxonomical methods. Recueil Trav. Bot. Néerl. 36: 509–525.

Lam, H.J. 1941. Note on the Sapotaceae-Mimusopoideae in general and on the far-eastern *Manilkara* allies in particular. Blumea 4: 323–358.

Lam, H.J., van Royen, P. 1952. Concise revision of the Sarcospermataceae. Blumea 7: 148–153.

Lam, H.J., Varossieau, W.W. 1938. Revision of the Sarcospermataceae. Blumea 3: 183–200.

Liben, L. 1989. La veritable identité des genres et espèces confondus sous le nom de *Bequaertiodendron magalismontanum* (Sond.) Heine & Hemsley (Sapotaceae) en Afrique centrale et occidentale. Bull. Jard. Bot. Natl. Belg. 59: 151–169.

Meeuse, A.D.J. 1960. Notes on the Sapotaceae of Southern Africa. Bothalia 7: 317–379.

Melcalfe, C.R., Chalk, L. 1950. See general references.

Miquel, F.A.W. 1863. Sapotaceae. In: Martius, Fl. Bras. 7: 38–118.

Pellegrin, F. 1938. Sur un genre africain peu connu: *Tridesmostemon* Engl. (Sapotacées). Bull. Soc. Bot. France 85: 179–181.

Pennington, T.D. 1990. Flora Neotropica, 52. Sapotaceae. New York: New York Botanical Garden.

Pennington, T.D. 1991. The genera of Sapotaceae. Kew: Royal Botanic Gardens Kew, New York: New York Botanical Garden.

Pierre, J.B.L. 1890–1891. Notes Botaniques Sapotacees. Paris, 1–83.

Pijl, L. van der 1936. Fledermäuse und Blumen. Flora II, 31: 1–40.

Rivera Nunez, D., Ruiz Liminana, J.B. 1987. *Argania spinosa* (L.) Skeels (Sapotaceae) subespontanea en la peninsula Iberica. An. Jard. Bot. Madrid 44(12): 173.

Royen, P. van. 1957. Revision of the Sapotaceae of the Malaysian area in a wider sense. 7. *Planchonella*. Blumea 8: 235–445.

Royen, P. van 1958. Revision of the Sapotaceae of the Malaysian area in a wider sense. 14. *Diploknema* Pierre. Blumea 9: 75–88.

Royen, P. van 1959. Revision of the Sapotaceae of the Malaysian area in a wider sense. 19. *Chelonespermum* Hemsley. Nova Guinea II, 10: 137–142.

Royen, P. van. 1960. Revision of the Sapotaceae of the Malaysian area in a wider sense. 22. *Mastichodendron*. Blumea 10: 122–125.

Soerianegara, I., Lemmens, R.H.M.J. (eds.) 1993. Plant resources of South East Asia 5(1). Timber trees: major commercial timbers. Bogor, 610 pp.

Stearn, W.T. 1968. Jamaican and other species of *Bumelia* (Sapotaceae). J. Arnold Arbor. 49: 280–289.

Vink, W. 1958. Revision of the Sapotaceae of the Malaysian area in a wider sense. 13. *Chrysophyllum*. Blumea 9: 21–74.

Waterman, P.G., Mahmoud, E.N. 1991. Chemical taxonomy of the Sapotaceae: patterns in the distribution of some simple phenolic compounds. In: Pennington, T.D.: The genera of Sapotaceae, chap. 4. Kew: Royal Botanic Gardens Kew, New York: New York Botanical Garden, pp. 51–74.

Sarraceniaceae

K. KUBITZKI

Sarraceniaceae Dumort., Anal. Fam. Pl.: 53 (1829), nom. cons.

Perennial, rhizomatous, mostly acaulescent, car-nivorous herbs. Leaves alternate, shortly petio-late, exstipulate, borne in a rosette or (some *Heliamphora*) on an upright, sometimes branching stem, all or some highly modified and transformed into more or less elongate, ascidiate, often pitcher-like traps ("amphores") partly filled with digestive liquid; petiole short and passing into the ascidiate portion of the leaf which, on the ventral side, bears a double or simple ridge or laminar wing and, on the apex of the dorsal side, a flattened, small or large, often hood-like appendix; the opening of the trap provided with a more or less distinctive collar; the outer and inner side of the pitcher around the opening provided with specialized glands and retrorse hairs that serve to entrap insects and other small animals; scale-like or sword-like leaves sometimes produced late in the season. Flowers perfect, actinomorphic, nodding, solitary on a scape or (*Heliamphora*) in few-flowered, some-times axillary racemoids; prophylls present (*Heliamphora*); perianth (3)4–6-merous, uniseri-ate, petaloid (*Heliamphora*), or biseriate of persist-ent sepals and showy, caducous petals; all perianth members distinct, imbricate; stamens 10–20 or numerous, in *Sarracenia* usually arising from 10 primordia; filaments short; anthers basifixed or (*Sarracenia*) versatile, introrse, dehiscing with longitudinal slits or (*Heliamphora*) short slits on caudal anther appendages; gynoecium of 5 or (*Heliamphora*) 3 united carpels; ovary superior, 5(3)-locular, but the partition of the upper part often incomplete and placentation therefore axile below and parietal above; style slightly 3-lobed and more or less truncate at the apex (*Heliamphora*), with 5 short style branches (*Darlingtonia*), or apically expanded and peltate, umbrella-shaped, with small stigmas under the tip of each of the 5 lobes (*Sarracenia*); ovules numerous, anatropous, unitegmic or (*Darlingtonia*?) bitegmic, tenuinucel-late. Fruit a loculicidal capsule with numerous seeds; seeds small, often winged, with exotestal seed coat, copious endosperm rich in oil and protein, and small linear embryo.

A distinctive family of three genera and ca. 15 species from North and northern South America, all occurring in nutrient-poor habitats.

VEGETATIVE STRUCTURES. Sarraceniaceae are predominantly rhizomatous herbs with a rosette of pitcher-like leaves. Only some species of *Heliamphora*, such as *H. nutans* and *H. minor*, have erect stems and can attain a dendroid habit.

The comparative morphology of the highly modified ascidiate leaves has been dealt with by numerous authors, among them Troll (1939), Arber (1941) and Markgraf (1955). The pitchers are best interpreted in terms of Troll's (1932) peltation theory, which implies unifaciality and peltation including a "Querzone", although some difficulties are inherent in this explanation (see Juniper et al. 1989: 57 seq.).

The pitcher rosettes of *Heliamphora* and *Sar-racenia* form ± symmetrical circles, facing inwards with their openings, whereas the openings of *Dar-lingtonia* face outwards as a result of a tubular twist of about 180°.

As in other carnivorous families, the pitcher traps of Sarraceniaceae are covered with numer-ous nectarial glands that attract insects; these glands are densely packed around the entrance of the pitchers. Downwards-directed trichomes of varying length are another important part of the trap mechanism, and *Darlingtonia* shares with the unrelated *Nepenthes* and *Brocchinia* the use of detachable wax platelets to prevent arthropod escape.

Heliamphora has the simplest pitcher structure of all Sarraceniaceae (Arber 1941), possessing a poorly developed lid and lacking a rolled collar on the rim, whereas the ventral keel (interpreted as the leaf margins) is a double structure (Fig. 124A). In *Darlingtonia*, the pitcher consists of a broadly winged, bifacial sheathing base, a tubular region provided with a simple keel, and a strongly curved hood, which closes the top of the pitcher. The tissue of the hood becomes transparent and forms a "window". Access for insects is possible only from underneath and leads over a peristome rim. Additionally, a fishtail-shaped appendage covered with nectaries hangs down from the entrance and is obviously homologous with the lid or flap of the other genera. *Sarracenia* also has a single keel, a peristome rim and a lid that usually does not occlude the mouth of the pitcher.

One could speculate that the flaps or hoods of the pitchers have the function of preventing rain-water from entering into the pitcher and diluting the digestive fluid. However, the *Heliamphora* species, which are most ineffective in this respect, grow in regions where heavy rains fall all over the year. Various insects dwell in the pitchers of *Sarracenia* (Uphof 1936), and the larvae of the tropical mosquito *Wyeomyia* even overwinter in them (Wood 1960).

Since the axial cambium in Sarraceniaceae is active only for a short period of time, a limited amount of secondary xylem is present in the rhizomes. It has a number of primitive features including solitary vessels with scalariform perforation plates and long and oblique end walls, tracheids with scalariform pitting, and diffuse axial parenchyma. Vessel elements in *Heliamphora* have the greatest number of allegedly primitive traits (DeBuhr 1977).

EMBRYOLOGY. Pollen grains are 2-celled when shed. Ovules are anatropus, unitegmic (bitegmic in *Darlingtonia*?), and tenuinucellar. The archespore functions directly as the megaspore mother cell; the chalazal megaspore of the tetrad develops into a Polygonum-type embryo sac. An endothelium encloses the embryo sac. Endosperm formation is cellular (Davis 1966).

KARYOLOGY AND HYBRIDISATION. Chromosome numbers are $n = 13$ (*Sarracenia*), 15 (*Darlingtonia*), and 21 (*Heliamphora*); polyploidy has not been observed. All *Sarracenia* species are interfertil and many natural hybrids are known from the wild, yet the species remain morphologically distinct (Bell 1952).

POLLEN MORPHOLOGY. Pollen is oblate to subspherical and 3–9-colporate, in *Heliamphora* (3)4–5-colporate and verruculate, in *Darlingtonia* 4–6-colporate and minutely scrobiculate, and in *Sarracenia* 6–9-colporate and minutely scrobiculate (Thanikaimoni and Vasanthy 1972).

POLLINATION. In *Sarracenia*, nectar is produced at the base of the style; pollen is shed from the versatile anthers and falls onto the stylar disk of the pendent flowers; in *S. rubriflora* it sticks together in little pellets (Burr 1979); pollination takes place by bumble-bees (Renner 1989 and literature cited therein). *Darlingtonia* does not seem to produce nectar, and pollinators are unknown. *Heliamphora* flowers are scentless and produce no nectar; in the pendent flowers the short slits of the caudal appendages of the reflexed anthers face downward and make the anthers technically poricidal; *H. tatei* was observed to be buzz-pollinated by bumble-bees, *Xylocopa* and *Eulaema* (Renner 1989). Since buzz-pollinated flowers are frequent in the montane scrub communities of the Roraima formation, including several genera of Melastomataceae and Rapateaceae, *Heliamphora* obviously shares the common pollinator-pool of these communities.

SEED. The seed coat is formed by the outer epidermis of the integument, which has thickened radial and inner walls; the rest of the tissue is crushed. The endosperm is cellular and oily, the embryo is usually minute (Netolitzky 1926).

PHYTOCHEMISTRY. Chemosystematically significant is the occurrence of secoiridoids of route I of Jensen (1991). Sarracenin, an enol diacetal monterpene, seems to be a compound unique to Sarraceniaceae and has been recorded from five species of *Sarracenia* and two of *Heliamphora*. For further compounds see Hegnauer (1990).

AFFINITIES. On the basis of a broad comparison of the floral morphology and embryology, DeBuhr (1975) refuted the former hypothesis of relationship between Sarraceniaceae and other insectivorous families such as Droseraceae and Nepenthaceae. Instead, he noted a significant agreement in these characters with Actinidiaceae, Marcgraviaceae and Theaceae. Analysis of nuclear and plastid DNA sequences (Bayer et al. 1996) revealed Sarraceniaceae as a monophylum sister to *Roridula*; *Darlingtonia* appears sister to *Sarracenia* + *Heliamphora*. Other analyses (Albach et al. 2001) support the association of Sarraceniaceae and Roridulaceae, with Actinidiaceae the next closest family, and Cyrillaceae and Clethraceae more distantly related. These families share many morphological characters.

Up to the present, on account of its relatively unspecialised pitchers and the racemoid inflorescence, *Heliamphora* had usually been considered the most basal genus of Sarraceniaceae. A comparison with the sister family Roridulaceae shows that the racemoids but not the uniseriate perianth of *Heliamphora* may be plesiomorphic.

DISTRIBUTION AND HABITATS. The six species of *Heliamphora* are restricted to the Roraima sandstone formation of the Guayana Highlands in northern South America, where they grow in open, marshy or wet savannahs and altitudinal heath formations. The monotypic *Darlingtonia* grows in bogs and wet meadows in the western United States (California and Oregon). Eight of the nine species of *Sarracenia* grow in flatwoods, open bogs and savannahs of the south-eastern United States; *S. purpurea* extends north to Canada and west to British Columbia.

ECONOMIC IMPORTANCE AND CONSERVATION. Because of their beauty and amazing insectivorous lifestyle, all members of the family have attracted the interest of gardeners, traders and the public (see Simpson 1994). Whereas the habitats of *Heliamphora* are of difficult access, *Sarracenia* and *Darlingtonia* are considered under such threat in the wild that all taxa of these genera are included in CITES. Apart from efforts of in-situ conservation of the nutrient-deficient wetlands, the creation of collections in Botanic Gardens and public collections is an additional safeguard against loss in the wild.

KEY TO THE GENERA

1. Flowers in a racemoid; perianth simple; ovary 3(4)-locular
   1. *Heliamphora*
 - Flowers solitary on a scape; perianth of calyx and corolla; ovary 5-locular 2
2. Scape aphyllous; stamens numerous; style expanded into broad umbraculiform disk **3. *Sarracenia***
 - Scape bracteose; stamens 15; style with 5 horizontally patent style branches **2. *Darlingtonia***

1. *Heliamphora* Benth. Fig. 124

Heliamphora Benth., Trans. Linn. Soc. 18: 432 (1840); Steyermark, Ann. Missouri Bot. Gard. 71: 302–312 (1984), rev.

Pitcher with a double keel. Tepals petaloid; stamens 10–20; anthers basifixed, at anthesis reflexed, dehiscing through caudal appendages; ovary 3(4)-locular; style short, stigma truncately 3(4)-lobed; seed compressed, irregularly scariofimbriately winged; embryo small, embedded in copious endosperm. $2n = 42$. Five spp., partly polymorphic, at middle and upper altitudes of the Guayana Highland in Venezuela and northern Brazil.

2. *Darlingtonia* Torrey

Darlingtonia Torrey, Smithson. Contrib. 6: 4, t. 12 (1854).

Fig. 124. Sarraceniaceae. *Heliamphora tatei* var. *neblinae*. **A** Two amphores. **B** Trichomes above the "waist". **C** Distal part of racemoid. **D** Flower, one tepal removed. **E** Flower with perianth removed, anthers still erect. **F** Same, later stage with anthers inverted. **G** Young stamens. **H** Inverted stamen. **I** Inverted stamens and pistil, with stigma enlarged. **J** Pistil. **K** Same, transverse section. **L** Placenta with ovules. (Maguire 1978)

Pitchers with simple keel, twisted for about 180°. Anthers basifixed, erect, dehiscing longitudinally; ovary 5-locular; style with 5 stylodia. Seeds oval, basally tailed; embryo relatively large, embedded in endosperm. $2n = 30$. One sp., *D. californica* Torrey, N California, Oregon.

3. *Sarracenia* L.

Sarracenia L., Sp. Pl. 1: 510 (1753); McDaniel, Bull. Timbers Res. Stat. 9: 1–36 (1971), rev.

Pitchers with simple keel. Sepals 5, with 3 appressed bracts; petals 5; stamens 70–80, arising in 10 groups; anthers dorsifixed, not reflexed; style expanded into broad disk, with a small stigma under each of the notched lobes. Seeds clavate to obovate, unilaterally inconspicuously winged or ridged; embryo small, embedded in copious endosperm. $2n = 26$. Two sections: sect. *Erectae* Uphof with upright, tubular or trumpet-shaped leaves, seven or eight spp. in the SE United States; and sect. *Sarracenia*, with highly modified, decumbent pitchers and maroon petals, two spp., North America from Labrador to Florida.

Selected Bibliography

Albach, D., Soltis, P.S., Soltis, D.E., Olmstead, R.G. 2001. Phylogenetic analysis of asterids based on sequences of four genes. Ann. Missouri Bot. Gard. 88: 163–212.

Arber, A. 1941. On the morphology of the pitcher-leaves in *Heliamphora, Sarracenia, Darlingtonia, Cephalotus*, and *Nepenthes*. Ann. Bot. II, 5: 563–578.

Bayer, R.J., Hufford, L., Soltis, D.E. 1996. Phylogenetic relationships in Sarraceniaceae based on *rbc*L and ITS sequences. Syst. Bot. 21: 121–134.

Bell, C.R. 1949. A cytotaxonomic study of the Sarraceniaceae of North America. J. Elisha Mitchell Sci. Soc. 65: 137–166.

Bell, C.R. 1952. Natural hybrids in the genus *Sarracenia*. I. History, distribution, and taxonomy. J. Elisha Mitchell Sci. Soc. 68: 55–80.

Davis, G.L. 1966. Systematic embryology of the angiosperms. New York: Wiley.

DeBuhr, L.E. 1975. Phylogenetic relationships of the Sarraceniaceae. Taxon 24: 297–306.

DeBuhr, L.E. 1977. Wood anatomy of the Sarraceniaceae: ecological and evolutionary implications. Plant Syst. Evol. 128: 159–169.

Hegnauer, R. 1990. Chemotaxonomie der Pflanzen, vol. 9. Basel: Birkhaeuser.

Jensen, S.R., Nielsen, B.J., Dahlgren, R. 1975. Iridoid compounds, their occurrence and systematic importance in the angiosperms. Bot. Notiser 128: 148–180.

Maguire, B. 1978. Sarraceniaceae. In: The Botany of the Guayana Highland – Part X. Mem. New York Bot. Gard. 29: 36–62.

Markgraf, F. 1955. Über Laubblatt-Homologien und verwandtschaftliche Zusammenhänge bei Sarraceniales. Planta 46: 414–446.

Newman, T., Ibrahim, S., Wheeler, J.W., McLaughlin, W.B., Petersen, R.L., Duffield, R.M. 2000. Identification of sarracenin in four species of *Sarracenia* (Sarraceniaceae). Biochem. Syst. Ecol. 28: 193–195.

Renner, S.S. 1989. Floral biological observations on *Heliamphora tatei* (Sarraceniaceae) and other plants from Cerro de la Neblina in Venezuela. Plant Syst. Evol. 163: 21–29.

Shreve, F. 1906. The development and anatomy of *Sarracenia purpurea*. Bot. Gaz. 42: 107–126.

Simpson, R.B. 1994. Pitchers in trade: a conservation review of the carnivorous genera of *Sarracenia, Darlingtonia*, and *Heliamphora*. Kew: Royal Botanic Gardens Kew.

Thanikaimoni, G., Vasanthy, G. 1972. Sarraceniaceae: Palynology and systematics. Pollen Spores 14: 143–155.

Uphof, J.C.Th. 1936. Sarraceniaceae. In: Engler, A., Harms, H. (eds.) Die natürlichen Pflanzenfamilien, 2nd edn., 17b. Leipzig: W. Engelmann, pp. 704–727.

Wood, C.R. 1960. The genera of Sarraceniaceae and Droseraceae in the southeastern United States. J. Arnold Arbor. 41: 152–163.

Scytopetalaceae

O. Appel

Scytopetalaceae Engl. in Engler & Prantl, Nat. Pflanzenfam. Nachtr. I: 242–245 (1897); nom. cons.

Evergreen trees or shrubs; young stems with cortical bundles; indumentum, if present, of uniseriate hairs. Leaves alternate, simple, entire or serrate, papery to coriaceous, minutely stipulate. Inflorescences axillary, often ramiflorous or cauliflorous, panicles, thyrsoids, botryoids or racemes, rarely reduced to single flowers (*Asteranthos*). Flowers regular, hypogynous or rarely perigynous; sepals united; calyx cupular, thick, leathery, persistent on ripe fruit, occasionally accrescent; petals lacking; staminodes 6–28, completely fused, forming a showy corolla-like structure; pseudocorolla glabrous or rarely pubescent, usually very thick and leathery (but thin in *Asteranthos*); stamens numerous; filament bases united, adnate to pseudocorolla; anthers basifixed, tetrasporangiate; anther dehiscence lengthwise, complete or apical and pore-like; ovary syncarpous, 3–8-locular; placentation axile; style solitary; stigma undifferentiated or lobed; ovules bitegmic, anatropous, pendulous, 2, 4, or many in each locule. Fruits 1(2)- or many-seeded, drupes, loculicidally dehiscent capsules, or leathery and berry-like. Seeds large, ca. 1 cm long; testa poorly differentiated; endosperm horny, often markedly ruminate; embryo well differentiated; cotyledons large, flat, cordate, rarely reduced (*Asteranthos*).

Six genera and about 21 species in humid tropical West Africa and South America.

VEGETATIVE STRUCTURES. The shoot system of Scytopetalaceae proliferates by sympodial branching. At the base of each module usually the scars of several scale leaves can be recognized, which in Rhaptopetaloideae may subtend inflorescences. Young shoots are often angular to broadly winged. The position of the wings is correlated with the path of a peculiar system of cortical bundles, which are borne at one node and usually enter the leaf of the node after next (van Tieghem 1905). The primary cortex, phloem and medullary rays are remarkable for the abundance of octaedric calcium oxalate crystals.

The leaves of the Scytopetalaceae are simple, estipulate or rarely stipulate (*Asteranthos*) and on side branches usually arranged distichously. The venation is generally brochidodromous. Usually, the leaf-margin is entire and often undulate, but dentate in *Brazzeia soyauxii* and *Pierrina zenkeri*. Some species are characterized by slanting leaf blades. Stomata are anisocytic. They are present on both leaf surfaces in *Oubanguia* and *Scytopetalum*, but restricted to the lower surface in the remaining genera. In species with thin leaves, the palisade layer is lacking. The leaves of all Scytopetalaceae are distinguished by the occurrence of spicular cells. In *Oubanguia* and *Scytopetalum* calcium oxalate crystals were observed in tissues sheathing the veins. Van Tieghem (1905), Metcalfe and Chalk (1950), and Amaral (1991) described these crystalliferous cells as "cristarque cells".

Metcalfe and Chalk (1950), Carlquist (1988), and de Zeeuw (1990) studied the wood anatomy of several members of the family. Important features are scalariform perforation plates, alternate medium-sized intervascular pits, multi- and uniseriate rays, and apotracheal parenchyma in numerous uniseriate bands. The phloem is stratified in lignified and unlignified portions. Cork develops surficially.

INFLORESCENCE MORPHOLOGY. The inflorescences of the Scytopetalaceae are always axillary in position. The basic inflorescence type is a panicle. All other types found in the family can be interpreted as derived from panicles through reductions and structural changes (Appel 1996). Thyrsoids are most common within the family (Fig. 125A) but botryoids, racemes and single flowers also occur. In Scytopetaloideae the inflorescences appear in the leaf axils of the youngest branches, whereas in Rhaptopetaloideae they have shifted to a ramiflorous or cauliflorous position.

FLOWER STRUCTURE. In all members of Scytopetalaceae the sepals are thick, leathery, and completely fused. Early in ontogeny the cupular calyx opens rhexigenously, thus often having irregular incisions or dentations that do not correspond to the number of sepals. In *Asteranthos* the calyx is strongly accrescent and serves as a floating device for its hydrochorous fruits (Ducke 1948). In *Oubanguia* the abaxial surface of the calyx is densely covered with uniseriate hairs.

The flowers are here interpreted as apetalous, but possessing a showy corona or corolla-like structure of presumably staminodial origin. The number of staminodes varies widely from 6 (*Oubanguia*) to 16 (*Scytopetalum*) or even 24–28 (*Asteranthos*). They are always completely fused, forming a pseudocorolla that bears several structural differences: usually the pseudocorolla has a thick leathery texture and opens rhexigenously at anthesis, whereas in *Asteranthos* it is thin and pleated in bud (Fig. 125B), unfolding like an umbrella at anthesis. In *Oubanguia* and *Scytopetalum* the pseudocorolla splits into lobes that presumably correspond to single staminodes, whereas in *Rhaptopetalum, Pierrina*, and *Brazzeia* the number of pseudocorolla lobes in opened flowers does not reflect the number of component staminodes. In *Scytopetalum* and *Oubanguia* the pseudocorolla bears apical outgrowths that project into the bud, filling its apex like a plug. In *Asteranthos* the margin of the pseudocorolla is ciliate and curved inward in bud. The stamens are numerous and arise in centrifugal succession from a ring-shaped meristematic wall. The gynoecium is syncarpous with axile placentation. Usually, the ovaries are superior, but semi-inferior in some species of *Rhaptopetalum* and in *Asteranthos*. The stigma is entire and somewhat papillose in *Rhaptopetalum*, but lobed in the other genera.

EMBRYOLOGY. Vijayaraghavan and Dhar (1976) and Tsou (1994) studied the embryology of *Scytopetalum tieghemii* and *Asteranthos brasiliensis*. Pollen grains are shed at the 2-celled stage. Ovules are anatropous, bitegmic and tenuinucellate. The inner integument is longer than the outer and forms the mycropyle. The development of the embryo sac conforms to the Polygonum type.

POLLEN MORPHOLOGY. The pollen of several members of the family was studied by Erdtman (1952), Muller (1972) and Tsou (1994). The grains are monads of medium size (20–30 µm), suboblate to spheroidal, tricolpate (Scytopetaloideae) or tricolporoidate (Rhaptopetaloideae) and circumaperturate to angulaperturate. The sexine sculpture is microreticulate with duplibaculate muri and granulate lumina. The colpus membranes are granulate.

FRUIT AND SEED. In all genera except *Oubanguia* the fruits appear to be indehiscent. The mesocarp is always woody, endocarp and exocarp are usually leathery but in *Scytopetalum* the exocarp is fleshy.

The seeds are glabrous (Scytopetaloideae) or embedded into a shiny, white haircoat of long unicellular hairs formed by epidermal cells along the testal bundles (Rhaptopetaloideae). The testa is poorly differentiated and lacks any sclerotic tissue. The abundant endosperm has strongly thickened cell walls that contain amyloid, a hemicellulose, as main storage compound. Starch and oil are of minor importance. In all genera except *Oubanguia* the endosperm is ruminate (Fig. 125E, F). Both subfamilies are characterized by distinct rumination forms, which apparently evolved independently (Appel 1996). In Rhaptopetaloideae the position of the endosperm foldings is correlated with the path of the testal bundles, whereas in Scytopetaloideae there is no such correlation (Appel 1996).

RELATIONSHIPS WITHIN THE FAMILY. Within the Scytopetalaceae two clearly delimitated subfamilies can be distinguished (for distinguishing characters see key). The neotropical *Asteranthos* is still extremely similar to *Scytopetalum* and *Oubanguia*, although it has apparently been separated from its African stock long ago.

AFFINITIES. Traditionally the Scytopetalaceae were placed in the Malvales. This position was supported mainly by the stratified phloem common to both groups, and by the embryological characters provided by Vijayaraghavan and Dhar (1976).

In more recent publications, generally a position near Ochnaceae and Quiinaceae in a broadly defined order Theales or in Ochnales was suggested (see Amaral 1991). Carlquist (1988), however, revealed striking similarities in wood characters between Scytopetalaceae and Lecythidaceae. *Asteranthos*, which is now included in the Scytopetalaceae (see Tsou 1994), formerly has often been placed in Lecythidaceae or (together with Napoleonaea) at least close to this family. Affinities between Scytopetalaceae, Lecythidaceae, Foetidiaceae and Napoleonaeaceae are supported not only by wood anatomical features, such as the occurrence of cortical bundles, stratified phloem, and crystalliferous strands in secondary xylem and primary cortex (Lignier 1890; de Zeeuw 1990), but also with regard to embryology and palynology (Erdtman 1952; Muller 1972; Vijayaraghavan and Dhar 1976; Tsou 1989, 1994). Moreover, a systematic position of Scytopetalaceae near

Lecythidaceae, Foetidiaceae and Napoleonaeaceae is further substantiated by *rbc*L sequence data. According to Morton et al. (1997), Lecythidaceae s.l. would form a monophylum if Scytopetalaceae were included. This inclusion was formally substantiated by Morton et al. (1998). However, since Scytopetalaceae form a readily separated and well-established family, their inclusion into the large and highly diversified Lecythidaceae is not followed here.

Key to the Genera

1. Inflorescences (or single flowers) inserted in leaf axils of youngest branches; number of staminodes recognizable in bud, their apices curved inwards in bud; anther dehiscence complete; ovules 2 or 4 in each locule; fruits 1(2)-seeded; seeds glabrous. **Scytopetaloideae**   2
 – Inflorescences in ramiflorous or cauliflorous position; number of staminodes not recognizable in bud, their apices not curved inwards; anther dehiscence apical, pore-like; ovules usually many in each locule; fruits many-seeded; seeds pubescent. **Rhaptopetaloideae**   4
2. Flowers solitary; calyx accrescent; pseudocorolla more than 20-partite, thin, pleated in bud, unfolding like an umbrella at anthesis; ovules usually 4 in each locule
                                                   1. *Asteranthos*
 – Flowers in few- to many-flowered inflorescences; pseudocorolla up to 16-partite, thick, leathery, opening rhexigenously; ovules usually 2 in each locule   3
3. Inflorescences determinate, with pubescent axes; carpels 3–5; fruits capsular; endosperm not ruminate
                                                   2. *Oubanguia*
 – Inflorescences indeterminate, with glabrous axes; carpels (6)7–8; fruits drupaceous; endosperm ruminate
                                                   3. *Scytopetalum*
4. Flower stalks with distinct abscission zones; filaments much shorter than anthers; stigma simple; seeds not flattened, without a circular notch   **4. *Rhaptopetalum***
 – Flower stalks without distinct abscission zones; filaments longer than anthers; stigma with as many lobes as carpels; seeds flattened, with a circular notch   5
5. Older inflorescences forming dense clusters; fruits globose, rounded at apex   **5. *Brazzeia***
 – Inflorescences never forming dense clusters; fruits fusiform, acute at apex   **6. *Pierrina***

I. Subfam. Scytopetaloideae O. Appel (1996).

Inflorescences or single flowers on youngest branches. Staminodes 6–28, their number distinguishable in bud, apically curved inwards. Stamens 100–240; filaments more than 3 times longer than anthers; anther dehiscence complete, by longitudinal slits. Fruits usually 1-seeded. Seeds glabrous. Pollen grains tricolpate.

1. *Asteranthos* Desf.   Fig. 125B, C

Asteranthos Desf., Mém. Mus. Hist. Nat. 6: 9 (1820).

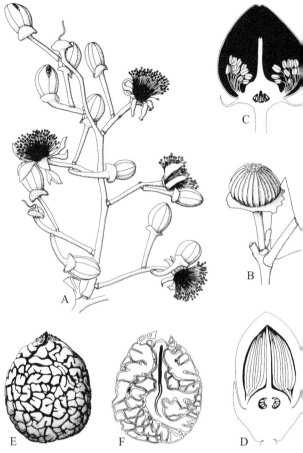

Fig. 125. Scytopetalaceae. **A** *Oubangia africana*, inflorescence; note basal articulations of pedicels. **B, C** *Asteranthos brasiliensis*. **B** Single axillary flower. **C** Vertical section of flower. **D** *Rhaptopetalum* sp., vertical section of flower. **E, F** *Scytopetalum tieghemii*, seed. **E** Surface view. **F** Vertical section. (Appel 1996)

Shrubs or small trees; leaf-margin entire; flowers solitary in leaf axils; calyx accrescent; staminodes 24–28, glabrous; pseudocorolla thin, pleated in bud, at anthesis unfolding like an umbrella; ovary semi-inferior; carpels 5–8; ovules 4 in each locule; endosperm ruminate. $n = 21$. One sp., *A. brasiliensis* Desf., N tropical South America.

2. *Oubanguia* Baill.   Fig. 125A

Oubanguia Baill., Bull. Soc. Linn. Paris 2: 869 (1890).

Trees or shrubs; leaf-margin entire or serrate; inflorescences determinate, with pubescent axes; calyx not accrescent; staminodes 6–12, pubescent; pseudocorolla thick leathery, rhexigenously opening; ovary superior; carpels 3–5; ovules 2 in each locule; fruits loculizidally dehiscent capsules.

Endosperm non-ruminate. About 3 spp., tropical West Africa.

3. *Scytopetalum* Pierre ex Engl.   Fig. 125E, F

Scytopetalum Pierre ex Engl. in Engler & Prantl, Nat. Pflanzenfam. Nachtr. I: 244 (1897).

Trees or shrubs; leaf-margin entire; inflorescences indeterminate, with glabrous axes; petioles with a basal abscission zone; calyx not accrescent; staminodes 12–16, glabrous; pseudocorolla thick, leathery, opening rhexigenously; ovary superior; carpels (6)7–8, ovules 2 in each locule; fruits drupaceous; endosperm ruminate. About 3 spp., tropical West Africa.

II. Subfam. Rhaptopetaloideae
O. Appel (1996).

Inflorescences ramiflorous or cauliflorous; staminodes completely united into a thick, leathery pseudocorolla, their number not distinguishable in bud, apically not curved inwards; stamens 60–100; anthers dehiscent by short, pore-like longitudinal slits; ovules many in each locule; fruits many-seeded; seeds pubescent along testal bundles; pollen grains tricolporoidate.

4. *Rhaptopetalum* Oliv.   Fig. 125D

Rhaptopetalum Oliv., J. Linn. Soc. 8: 159 (1865).

Trees or shrubs; leaf-margin entire; inflorescences not in dense clusters; petioles with distinct abscission zones; filaments much shorter than anthers; carpels 3–5; stigma simple, papillose; fruits globose; testal bundle branched. About 10 spp., tropical West Africa.

5. *Pierrina* Engl.

Pierrina Engl., Bot. Jahrb. Syst. 43: 374 (1909).

Shrubs; leaf-margin entire or serrate; inflorescences not in dense clusters; petioles without abscission zones; filaments longer than anthers; carpels (3)4(5); stigma lobed; fruits fusiform, acute; testal bundle unbranched. One sp., *P. zenkeri* Engl., humid tropical West Africa.

6. *Brazzeia* Baill.

Brazzeia Baill., Bull. Soc. Linn. Paris 1: 609 (1886).

Trees or shrubs; leaf-margin entire or serrate; inflorescences usually in dense clusters. Petioles without abscission zones; filaments longer than anthers; carpels (5)6(7); stigma lobed; fruits globose; testal bundle unbranched. About 3 spp., tropical West Africa.

Selected Bibliography

Amaral, M.C.E. 1991. Phylogenetische Systematik der Ochnaceae. Bot. Jahrb. Syst. 113: 105–196.

Appel, O. 1996. Morphology and systematics of the Scytopetalaceae. Bot. J. Linn. Soc. 121: 207–227.

Carlquist, S. 1988. Wood anatomy of Scytopetalaceae. Aliso 12: 63–76.

Dehay, C. 1955. Caractères de la feuille chez les Scytopétalacées. Bull. Soc. Bot. Nord France 8: 76–81.

Ducke, J.A. 1948. Arvores Amazônicas e sua propagacâo. Bol. Mus. Paraense Hist. Nat. 10: 81–92.

Engler, A. 1897. Scytopetalaceae. In: Engler & Prantl, Nat. Pflanzenfam. Nachtr. 1. Leipzig: Engelmann, pp. 242–245.

Engler, A. 1909. Scytopetalaceae africanae II. Bot. Jahrb. Syst. 43: 373–377.

Erdtman, G. 1952. See general references.

Knuth, R. 1934. Über die Gattung *Asteranthos*. Notizbl. Bot. Gart. Berlin-Dahlem 11: 1034–1036.

Knuth, R. 1939. Asteranthaceae. In: Engler, Pflanzenreich IV, 219b. Leipzig: W. Engelmann.

Kowal, R.R. 1989. Chromosome numbers of *Asteranthos* and the putatively related Lecythidaceae. Brittonia 41: 131–135.

Kravcoca, T.I. 1991. Scytopetalaceae. In: Takhtajan, A. (ed.) Anatomia seminum comparativa. Vol. 3. Caryophyllidae – Dilleniidae. Leningrad: Nauka, pp. 224–227.

Letouzey, R. 1961. Notes sur les Scytopétalacées (Révision des Scytopétalacées de l'herbier de Paris). Adansonia II, 1: 106–142.

Letouzey, R. 1977. Nouvelle espèces de *Rhaptopetalum* Oliv. (Scytopétalacées) du Cameroun et du Gabon. Adansonia II, 17: 129–138.

Letouzey, R. 1978a. Scytopétalacées. In: Aubréville, A., Leroy, J.-F. (eds.) Flore du Cameroun, vol. 20, pp. 139–193.

Letouzey, R. 1978b. Scytopétalacées. In: Aubréville, A., Leroy, J.-F. (eds.) Flore du Gabon 24, pp. 139–193.

Lignier, O. 1890. Recherches sur l'anatomie des organes végétatifs des Lecythidées, des Napoléonées et des Barringtoniées. Bull. Sci. France Belgique 21: 291–420.

Metcalfe, C.R., Chalk, L. 1950. See general references.

Morton, C.M., Mori, S.A., Prance, G.T., Karol, K.G., Chase, M.C. 1997. Phylogenetic relationships of Lecythidaceae: a cladistic analysis using *rbcL* sequence and morphological data. Am. J. Bot. 84: 530–540.

Morton, C.M., Prance, G.T., Mori, S.A., Thorburn, L.G. 1998. Recircumscription of Lecythidaceae. Taxon 47: 817–827.

Muller, J. 1972. Pollen morphological evidence for subdivision and affinities of Lecythidaceae. Blumea 20: 350–355.

Tsou, C.-H. 1989. The floral morphology and embryology of *Asteranthos* and its systematic consideration. Am. J. Bot. 76 suppl.: 275–276.

Tsou, C.-H. 1994. The embryology, reproductive morphology and systematics of Lecythidaceae. Mem. New York Bot. Gard. 71: 1–112.

van Tieghem, P. 1905. Sur les Rhaptopetalacées. Ann. Sci. Nat., Bot. IX, 1: 321–388.

Vijayaraghavan, M.R., Dhar, U. 1976. *Scytopetalum tieghemii* – embryologically unexplored taxon and affinities of the family Scytopetalaceae. Phytomorphology 26: 16–22.

Zeeuw, C.H. de 1990. Secondary xylem of Neotropical Lecythidaceae. In: Mori, S.A., Prance, G.T., Lecythidaceae – Part II, chap. II. The zygomorphic-flowered New World genera. Fl. Neotrop. Monogr. 21 (II), pp. 4–59.

Sladeniaceae

P.F. STEVENS and A.L. WEITZMAN

Sladeniaceae (Gilg & Werdermann) Airy Shaw, Kew Bull. 18: 267 (1964).

Evergreen trees; hairs unicellular. Leaves spiral or distichous, margins toothed or entire, secondary veins pinnate, stipules none. Inflorescences axillary, dichasial. Flowers small, 5(6)-merous; sepals and petals imbricate; sepals free; petals free or basally connate; stamens (8–)10–15, free or adnate to corolla, anthers basifixed, dehiscing by apical pores or slits; nectary none; gynoecium 3- or 5-carpellate; ovules axile, 2 or many per carpel; style short, with sometimes very short style branches; fruit schizocarpic, endocarp crustaceous, or a loculicidal capsule with persistent columella, calyx persistent; seeds winged or not, testa crustose; embryo straight, endosperm copious; $n = 24$.

Two monotypic genera, one from East Africa, the other from SE Asia.

VEGETATIVE STRUCTURES. All hairs are unicellular. Those of *Sladenia* are very short, and the plant may even be glabrous, while those of *Ficalhoa* may be over 1 mm long. Petiole bundles are arcuate, but associated fibers are lacking in *Ficalhoa*; there are also wing bundles in *Sladenia*. *Ficalhoa* contains copious white "latex" (e.g., Verdcourt 1962); mucilaginous substances in the epidermis are reported from *Sladenia*; there is no hypodermis. Stomata are anomocytic.

Growth rings are faint. The vessels are diffuse-porous, and are in radial multiples (*Sladenia*) or not, with scalariform perforation plates that have (11–)16–17(–30) bars. Intervessel pits are scalariform (*Ficalhoa* only) and opposite to alternate. Fiber tracheids are thick- to thin-walled and have distinctly bordered pits. Parenchyma is diffuse-in-aggregates, vasicentric, or scanty paratracheal. Rays are 1–3(4) cells across (*Sladenia*) or of two distinct sizes, to 1.4 mm high, 1–2-seriate and (*Ficalhoa*) (3–)5–7-seriate, heterogeneous, with 1–6(–10) rows of square to upright marginal cells. There are chambered crystals in strands or axial parenchyma cells (*Sladenia*), or crystals are absent (*Ficalhoa*) (Liang and Baas 1990, 1991). Sclereids are absent; druses are widespread. The nodes are unilacunar and the phellogen is deep-seated.

INFLORESCENCE AND FLOWER STRUCTURE. The inflorescences are dichasia, simple in the case of *Sladenia* (Fig. 126A). Those of *Ficalhoa* are often paired, originating in the axils of the prophylls of the axillary bud. The flowers are sessile (apart from the ultimate flowers of *S. integrifolia*), with 2 prophylls, small, usually 5-merous and apparently perfect. The calyx is quincuncial and the corolla is imbricate (*Sladenia*) or quincuncial (*Ficalhoa*); the sepals are free. The corolla of *Sladenia* is connate basally, that of *Ficalhoa* may be very shortly connate (Shui et al. 2002). The stamens of *Sladenia* are borne in a single whorl, and the swollen filaments are tightly pressed to one another, but apparently free. The thin filaments of *Ficalhoa* are adnate to the corolla. *Sladenia* is tricarpellate and there are two apical epitropous ovules per carpel. *Ficalhoa* has five carpels opposite the petals; the placentas are axile and swollen and bear numerous ovules. *Sladenia* has a rather stout or almost non-existent style continuous with the ovary (Fig. 126E); the short style of *Ficalhoa* is impressed.

POLLEN MORPHOLOGY. The pollen grains of *Sladenia* are tricolpate, broadly ellipsoid and ca. 15 μm across. The colpi lack internal thickenings and a well-defined pore, although it more or less bulges in the center. The exine is moderately thick, smooth or faintly and finely granular; the intine is thin, but slightly thickened beneath the colpi (Wodehouse in Kobuski 1951).

FRUIT AND SEEDS. It seems that no mature fruits of *Sladenia* have been studied so far (Gilg and Werdermann 1925; Kobuski 1951; Airy Shaw 1964; pers. obs. S. Dressler). They were described as indehiscent but, since the carpels split easily along the septae, they are most probably schizocarpic and break into 3 mericarps. The exocarp is papery and the endocarp crustaceous. The small long-ellipsoid seeds have a ventral raphe and a thin, fine reticulate testa. They have been described as winged (Keng 1962; Shui et al. 2002) but this seems to be a drying artifact. When fresh they are swollen and apparently triangular (Kobuski 1951). Testal cells are polygonal in surface view and probably little thickened. *Ficalhoa* has small loculicidal capsules with several seeds on swollen, spongy placentae. The seed is tiny, compressed ovoid with an obscure circumferential wing which suggests anemochory (pers. obs. S. Dressler). The walls of the testa are unthickened, and the embryo is terete, with short cotyledons, and surrounded by copious endosperm.

AFFINITIES. Sladeniaceae are a very poorly known family. *Sladenia* has often been included in Theaceae, either as a separate subfamily (Takhtajan 1997) or as a tribe of Ternstroemioideae (Keng 1962); Airy Shaw (1964) described it as a family distinct from Actinidiaceae, in which it had been included; others have placed it in Dilleniaceae or Linaceae (see Liang and Baas 1990 for a summary). *Ficalhoa* was originally placed in Ericaceae but it, too, has been placed in Actinidiaceae (and also Sapotaceae) and is now often included in Theaceae (Liang and Baas 1991; these authors favored an association with Ternstroemioideae). *Sladenia* was sister to Ternstroemiaceae in *rbc*L studies (Savolainen, Fay et al. 2000), although its DNA was rather degraded. A more comprehensive analysis has confirmed this position (support moderate), in addition placing *Ficalhoa* as its sister group (support also moderate: Anderberg et al. 2002). The two share a number of similarities, most conspicuously their deep-seated, non-surficial phellogen (polarity uncertain), their cymose inflorescences with small flowers (possible apomorphies), and their straight embryos (a probable plesiomorphy). Their recognition as a family separate from Ternstroemiaceae is justified.

Despite their evident similarities, *Sladenia*, from the South East Asian mainland, and *Ficalhoa*, from tropical East Africa, show substantial differences in flower, fruit and wood anatomy. Fossil wood named *Sladenioxylum africanum* has been found in Albian-Cenomanian deposits from the northern Sudan (Giraud et al. 1992); it is remarkably like the wood of extant *Sladenia*.

KEY TO THE GENERA

1. Plant glabrous or with short hairs; petals free or almost so; stamens forming a whorl, filaments swollen, anthers sagittate, porose; fruit few-seeded **1. *Sladenia***
- Plant often with hairs to 1 mm long; petals basally connate; stamens in groups of three, filaments linear, anthers linear, dehiscing by terminal transverse slits; fruit many-seeded **2. *Ficalhoa***

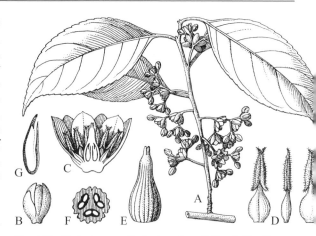

Fig. 126. Sladeniaceae. *Sladenia celastrifolia.* **A** Floriferous branch. **B** Flower bud. **C** Flower, vertical section. **D** Stamens. **E** Ovary. **F** Same in transverse section. **G** Ovule. (Gilg and Werdermann 1925)

1. *Sladenia* Kurz   Fig. 126

Sladenia Kurz, J. Bot. 11: 194 (1873); Sprague, Hook. Icon. 51: t. 3026 (1915); Shui et al., Novon 12: 539–542 (2002).

Plant glabrous or with short hairs; leaves spiral, serrate or entire; inflorescences simply cymose; calyx and corolla free, similar in size, different in texture; stamens 8–13, filaments swollen, anthers sagittate, pilose, porose; gynoecium tricarpellate, placentation apical-axile, ovules pendulous, 2 per carpel, style with short branches; fruit schizocarpic, sepals and style persistent, endocarp crustaceous; testa crustose, endosperm none; $2n = 48$. Two species, SE Asia.

2. *Ficalhoa* Hiern

Ficalhoa Hiern, J. Bot. 36: 329 (1898); Robson, Fl. Zambesiaca 1: 405–406 (1961), tab. 79.

Plant often with long, unicellular hairs; leaves distichous, serrate; inflorescences cymose, with two main spreading branches at base; corolla connate; stamens 15, in antesepalous groups of 3 and adnate to corolla, connective much swollen at base of anthers, anthers transversely dehiscent apically; gynoecium 5-carpellate, placentation axile, ovules many per carpel, style impressed, style branches spreading, recurved, stigmas punctate; fruit a loculicidal capsule with a central columella, calyx persistent; seeds minute. One species, *F. laurifolia* Hiern, a prominent tree of montane rain forest on the east African mountains.

Selected Bibliography

Airy-Shaw, H.K. 1964. Diagnoses of new families, new names, etc. for the seventh edition of Willis's dictionary. Kew Bull. 18: 249–273.

Anderberg, A.A. et al. 2002. See general references.

Gilg, E., Werdermann, E. 1925. Actinidiaceae. In: Engler & Prantl, Nat. Pflanzenfam. ed. 2, 21. Leipzig: W. Engelmann, pp. 36–47.

Giraud, B., Bussert, R., Schrank, E. 1992. A new Theacean wood from the Cretaceous of northern Sudan. Rev. Palaeobot. Palynol. 75: 289–299.

Hiern, W.P. 1898. A new genus of Ericaceae from Angola. J. Bot. 36: 329–330, pl.

Keng, H. 1962. Comparative morphological studies in the Theaceae. Univ. Calif. Publ. Bot. 33: 269–384.

Kobuski, C.E. 1951. Studies in the Theaceae. XXIV. The genus *Sladenia*. J. Arnold Arbor. 32: 402–408, pl. 1.

Li, L. 2001. Chromosome number of *Sladenia celastrifolia*. Acta Bot. Yunn. 23: 223–224.

Li, L., Liang, H.X., Peng, H. 2003. Karyotype of *Sladenia* and its systematic insights. Acta Bot. Yunn. 25: 321–326. (In Chinese, with English summary). *Sladenia* for from Theaceae, not close to Ternstroemiaceae.

Liang, D., Baas, P. 1990. Wood anatomy of trees and shrubs from China II. Theaceae. I.A.W.A. Bull. II, 11: 337–378.

Liang, D., Baas, P. 1991. The wood anatomy of the Theaceae. I.A.W.A. Bull. II, 12: 333–353.

Savolainen, V., Fay, M.F. et al. 2000. See general references.

Shui, Y.-M., Zhang, G.-J., Zhou, Z.-K., Mo, M.-Z. 2002. *Sladenia integrifolia* (Sladeniaceae), a new species from China. Novon 12: 539–542.

Takhtajan, A.L. 1997. See general references.

Verdcourt, B. 1962. Flora of Tropical East Africa. Theaceae. London: Crown Agents.

Wei, Z.-X., Li, D.-Z., Fan, X.-K., Zhang, X.-L. 1999. Pollen ultrastructure of Pentaphylacaceae and their relationships to the family Theaceae. Acta Bot. Yunn. 21: 202–206, 3 pl. (in Chinese, with English summary).

Styracaceae

P.W. FRITSCH

Styracaceae Dumort., Anal. Fam. Pl.: 28, 29 (1829), nom. cons.

Evergreen or deciduous trees or shrubs. Leaves alternate, simple, pinnately nerved, petiolate, exstipulate. Plants with stellate or peltate trichomes. Inflorescences terminal and/or axillary cymes, racemes, or panicles, or sometimes 1–2-flowered, usually bracteolate. Flowers bisexual or rarely female in gynodioecious species, actinomorphic, with hypanthium adnate to ovary wall at various levels, usually pendent; calyx synsepalous, truncate or 4–5(–9)-toothed, teeth valvate or open in bud; corolla sympetalous, (4)5(–8)-lobed or -parted, imbricate or subinduplicate-valvate in bud, generally campanulate to open with the lobes spreading to reflexed; stamens usually twice, rarely up to four times or equal the number of corolla lobes, uniseriate, replaced by 5 staminodes in female flowers; filaments usually flattened at least at the base, adnate to the corolla, sometimes forming a tube distally; anthers basifixed, introrse, tetrasporangiate, longitudinally dehiscent, oblong to linear, the connective roughly equal to or surpassing the thecae; ovary 2–4(5)-carpellate, 2–4(5)-septate at the base but usually 1-locular through the distal attenuation of the septa, partly to completely inferior, with essentially axile placentation; style filiform, usually hollow; stigma terminal, truncate or minutely lobed; ovules (1–)4–9(–ca. 30) per carpel, anatropous, unitegmic or bitegmic, tenuinucellate. Fruits mostly dry and usually capsular with loculicidal dehiscence, or indehiscent and sometimes samaroid, or drupaceous, with persistent calyx. Seeds 1–4(–ca. 50), ± globose to fusiform, rarely winged; testa brown, thin to indurate; cotyledons flattened or nearly terete; endosperm copious, oily.

A family of 11 genera and about 160 spp., tropical and warm-temperate regions of North and South America, S Europe, E and SE Asia, and Malesia.

VEGETATIVE MORPHOLOGY. Members of Styracaceae are typically medium-sized trees up to 20 m tall; occasionally they reach 30 m or are shrubs 1–4 m tall. *Styrax grandifolius* produces underground runners, forming open thickets. In *Alniphyllum, Bruinsmia, Huodendron,* and *Styrax* the buds are naked, i.e., the outer leaves that constitute the buds on twigs of the dormant season typically expand fully during the expansion of the new shoot. The remaining genera possess at least two bud scales that abscise quickly after shoot expansion. In several of the latter genera, a ± continuous series exists from indurate, non-expansive bud scales proximally through green, leafy bracts and finally fully formed leaves distally. Thorns are present in the species of *Sinojackia.* Several species of *Styrax* in Brazil develop xylopodia, which have a strong capacity to form new sprouts after fire. Phyllotaxis is spiral. In *Bruinsmia* and most tropical species of *Styrax,* the leaves are persistent, whereas in the rest of the genera they are deciduous (the condition of *Parastyrax* is unknown). Leaves are usually chartaceous (Fig. 127A), but are membranous or coriaceous in some species of *Styrax.* They can be entire or serrate; serrations are capped by a gland-like structure. Stellate hairs of various shapes or sometimes radiate to peltate scales are always present (Fig. 127B, C); rust-colored to black glands are scattered over the petioles, inflorescence branches, pedicels, and calyces in many species of *Styrax.*

VEGETATIVE ANATOMY. Wood anatomical characters occurring in most or all genera include growth rings, diffuse porosity, combinations of both solitaries and pore multiples, scalariform perforation plates, opposite to alternate intervessel pitting, imperforate tracheary elements with indistinctly bordered pits, both uniseriate and multiseriate heterocellular rays, and scanty axial parenchyma distributed as a combination of diffuse and diffuse-in-aggregates. Prismatic crystals are present in some genera, and silica is present in many species of *Styrax* series *Valvatae.* Pore outline is generally a combination of angular and circular. Imperforate tracheary elements range from fiber-tracheids with distinctly bordered pits to elements having small, indistinctly bordered pits that approach the simple condition. The species of *Styrax* series *Styrax* have a number of characters often associated with dry climates, e.g., simple perforation plates in the wider, early-wood vessel elements or plates with a reduced number of bars, increased pore frequency, and decreased vessel element length (Metcalfe and Chalk 1950; Dickison and Phend 1985).

Anatomical characters that are found in leaves of nearly all genera include bifacial mesophyll; anomocytic stomata confined to the abaxial surface; crystals in the form of druses or prisms or both; stellate, peltate, or simple-cylindrical trichomes; sheathing and supporting elements associated with the venation; and pinnate venation accompanied by camptodromous secondary venation. Almost all variation observed in the leaf structure of the ten smaller genera of Styracaceae is found within *Styrax*. Petiolar vasculature, except for *Parastyrax*, consists of an arc, an arc with invaginated ends, or a medullated cylinder (Schadel and Dickison 1979).

FLORAL MORPHOLOGY AND ANATOMY. Flowers of Styracaceae are perfect except in females of the gynodioecious species. Female flowers have staminodes with nonfunctional pollen sacs. At least some parts of the flower usually have stellate or peltate trichomes (Fig. 127E, L, R). The pedicel is articulated apically in all genera except *Huodendron* and *Styrax*, in which pedicel articulation is absent.

The degree of connation and adnation of parts is highly variable. All Styracaceae have a hypanthium that is adnate to the ovary wall from just above the base of the ovary to nearly the complete length of the ovary. The hypanthium and free portion of the calyx appear as a continuous, generally cupuliform or funnelform structure that is thickened below and typically gradually diminishes in thickness distally. Above the hypanthium, sepals range from distinct to completely connate, petals and stamens range from nearly distinct (coherent at base) to connate most of their length, and carpels are nearly always completely connate; stylodia are usually distally distinct in *Huodendron*). Distal to the hypanthium, the calyx is always free from the corolla and androecium (Fig. 127E, R), whereas the stamens are slightly to nearly completely adnate to the petals (Fig. 127N). Connation of either the petals or the stamens or both may continue for a variable distance distal to the point of corolla and androecium divergence.

The perianth is always differentiated into a calyx and a usually white, sometimes rose-pink or yellow, corolla. Corolla thickness ranges from membranous to ± fleshy. Sepals are either three-trace or one-trace structures, whereas the petals are always one-trace.

The androecium is seemingly 1-whorled (Fig. 127N), but in some genera the stamens that alternate with the petals are longer, with their anthers nearer the style. This arrangement led van Steenis (1932) to hypothesize an obdiplostemonous ancestor for the family. This hypothesis is supported by the lower level of origin of the vascular traces to the stamens opposite the petals relative to those that alternate with the petals in some flowers, as well as the fact that petal and stamen traces are more commonly fused than are sepal and stamen traces (Dickison 1993). The stamens that alternate with the petals develop first in *Styrax* and *Halesia* (Payer 1966), from which it can be deduced – if the hypothesis of an obdiplostemonous ancestor for the family is correct – that stamen development in Styracaceae is centrifugal.

In many Neotropical species of *Styrax*, a mass of trichomes occurs on the ventral portion of the stamen filament (Fig. 127P). The connective is ± linear in all genera (Fig. 127O). The anther connectives are characterized by conspicuously enlarged, thin-walled epidermal cells on both the ventral and dorsal surfaces that may aid in anther dehiscence through changes in turgor. At maturity the microsporangium wall has an endothecium (consisting of a single layer), except in *Huodendron* and some species of *Styrax*. The stamens are always one-trace structures.

Ovaries are incompletely septate in all genera except *Parastyrax*, which has completely septate ovaries. In most of the incompletely septate members, the septa are united along a central axis until a point ± midway distally, where the axis ends and the septa begin to diverge. The edges of the septa recede distally toward the ovary wall and continue uninterrupted into the hollow style in the form of ridges. This results in a single, continuous chamber within the ovary, partially compartmentalized proximally, that is also continuous with the ridged and angled stylar canal (Fig. 127E–G, R). Placentation is axile in all genera except some of the gynodioecious species of *Styrax*, in which placentation is near-basal. In all genera except *Styrax* placentas are divided distally into two lobes by ventral carpellary sutures that meet at the apex of the central axis. In all species of *Styrax* except some of the gynodioecious species, placental outgrowths (obturators) occur between the ovules (Fig. 127E). The style always appears as an attenuation of the ovary (Fig. 127D, E, R). The stigma may be flat and not wider than the style, or lobed to

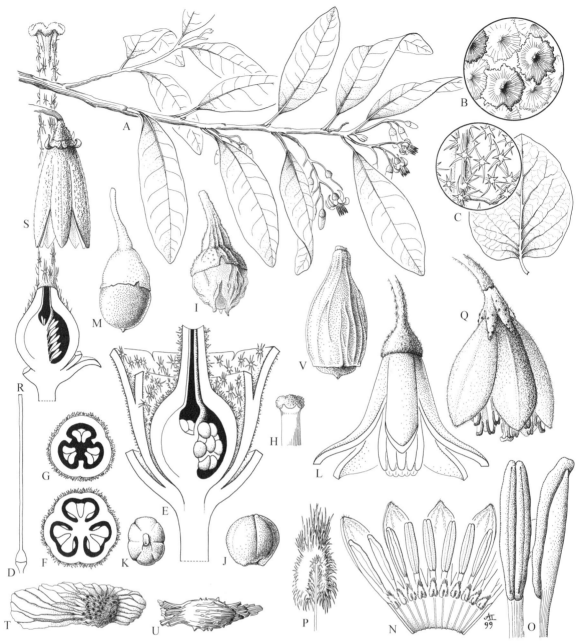

Fig. 127. Styracaceae. **A, B** *Styrax leprosus*. **A** Flowering shoot. **B** Close-up of abaxial leaf surface. **C–K** *Styrax redivivus*. **C** Leaf, abaxial view and close-up of abaxial leaf surface. **D** Pistil. **E** Ovary with perianth and androecium, vertical section. **F** Ovary in transverse section, lower part. **G** Ovary in transverse section, upper part. **H** Stigma. **I** Fruit. **J** Seed in lateral view. **K** Seed in polar (hilum end) view. **L–M** *Styrax ramirezii*. **L** Flower. **M** Fruit. **N–P** *Styrax pallidus*. **N** Corolla and androecium opened out. **O** Upper portion of stamen in ventral and lateral view. **P** Lower free portion of stamen filament showing ventral mass of trichomes. **Q–T** *Alniphyllum fortunei*. **Q** Flower. **R** Ovary with perianth and androecium, vertical section, and style in lateral view. **S** Capsular fruit. **T** Seed. **U** *Huodendron biaristatum* seed. **V** *Melliodendron xylocarpum* fruit (distal end at bottom). (Orig.)

nearly capitate and wider than the style (Fig. 127H, R). The orientation of the ovules is usually apotropous, but is modified in several genera such that the upper ovules are apotropous and the lower are epitropous. Several genera possess intra-ovarian trichomes. Gynoecial vascular features include three-trace carpels; numerous carpellary wall bundles most commonly arising on the inner and lateral edges of the hypanthial wall bundles; a single dorsal bundle per carpel; and ovules vascularized by a single vein that branches from a placental bundle derived from the ventral ovarian venation (Dickison 1993).

EMBRYOLOGY. Embryological characters have been studied in *Alniphyllum*, *Halesia*, *Pterostyrax*, and *Styrax* (Manshard 1936; Copeland 1938; Veillet-Bartoszewska 1960; Yamazaki 1970; Yakovlev 1983; Johri et al. 1992). The microspore mother cells undergo meiotic divisions, cytokinesis is simultaneous, and tetrads are tetrahedral and isobilateral. Ovules of *Styrax* are bitegmic, whereas those of the other genera are unitegmic. An integumentary tapetum is missing in most Styracaceae. The archesporial cell functions as the megaspore mother cell, undergoes meiotic divisions, produces a linear or T-shaped tetrad, and the chalazal megaspore develops into a Polygonum-type embryo sac. The polar nuclei fuse upon fertilization, and antipodals are ephemeral. Numerous large, unstained starch grains occur in the megagametophyte of some genera (Dickison 1993). Endosperm development is cellular. Embryo development is of the Polygonad type–Solanad variation and the embryo is straight or slightly curved (Johri et al. 1992).

POLLEN MORPHOLOGY. Pollen morphology of the family is fairly uniform. Grains of nearly all genera are solitary, radially symmetrical, tricolporate with lalongate endoapertures each with an equatorial bridge, and a tectate-perforate-columellate exine. Mean grain size ranges from $15.3 \times 21.4\,\mu m$ to $43.3 \times 37.7\,\mu m$, with an overall mean of $32.7 \times 37.7\,\mu m$. The exine is stratified into ektexine and endexine, with the ektexine composed of a distinct tectum, columellae, and foot layer. Outer tectal surfaces are smooth, granulose, microspinulose, beaded, or banded (Morton and Dickison 1992).

KARYOLOGY. The basic chromosome number of *Styrax* is $x = 8$. In *Styrax* two polyploid species are known (Gonsoulin 1974). *Halesia*, *Sino-jackia* and *Pterostyrax*, the only other genera of Styracaceae sampled for chromosome number, have $n = 12$ (Manshard 1936; Lewis et al. 1962, Hsu et al. 1994).

POLLINATION. Species of the family are primarily bee-pollinated. *Halesia* and *Styrax* species are generally most frequently pollinated by bumble bees and honey bees (Chester 1966; Gonsoulin 1974; Sugden 1986). Other pollinators reported for *Styrax* species are papilionoid butterflies, syrphid flies, sphingid moths, wasps of various families, and a wide variety of bees (e.g., carpenter bees, halictids, anthophorids; Copeland 1938; Gonsoulin 1974; Sugden 1986; Saraiva et al. 1988; Tamura and Hiura 1998). *Styrax redivivus* has

been described as having a mix of floral features characteristic of bee, butterfly, moth, and perhaps hummingbird syndromes (Sugden 1986). Both nectar and pollen serve as floral rewards for pollinators. The stellate hairs present on the exterior surface of the corolla in most species of Styracaceae have been suggested as an adaptation for supporting large pollinators, which use them as "toe holds" to gather nectar and pollen (Sugden 1986). Both *Halesia* and *Styrax* flowers are sweetly fragrant (Perkins 1907; Copeland 1938).

REPRODUCTIVE SYSTEMS. Nearly all species have exclusively hermaphroditic flowers. Six species of *Styrax* are known to have hermaphroditic and female flowers on separate plants (Wallnöfer 1997; Fritsch 1999). These species are assumed to be functionally gynodioecious, although this has not been confirmed with breeding experiments (Fritsch 1999). Reports of polygamo-dioecy in *Bruinsmia* are unfounded (van Steenis 1932). Observations of andromonoecy in *Halesia* are suspected to have been based on immature hermaphroditic flowers (Wood and Channell 1960). Predominant to obligate xenogamy is documented for the several species of *Styrax* examined for breeding system type (Sugden 1986; Saraiva et al. 1988; Tamura and Hiura 1998).

FRUIT AND SEED. Styracaceae possess a wide array of fruit types, including loculicidal capsules (*Alniphyllum*, *Huodendron*, *Styrax*; Fig. 127I, S), drupes (*Parastyrax*, *Styrax*; Fig. 127M), samaras (*Halesia*, *Pterostyrax*), baccoid fruits (*Bruinsmia*), and nut-like (i.e., dry, indehiscent, one- to several-seeded) fruits (*Changiostyrax*, *Pterostyrax*, *Melliodendron*, *Rehderodendron*, *Sinojackia*, *Styrax*; Fig. 127V). The fruits of *Styrax* series *Valvatae* possess a dark purple, often sweet-tasting drupe (Fig. 127M), whereas those of the other three series of the genus are almost always dry (fleshy in one species) and either indehiscent or 3-valvate-dehiscent (Fig. 127I). In all genera the calyx and hypanthium are persistent in fruit (Fig. 127M, S, V), and in seven genera the hypanthial tissue greatly elongates during fruit development, eventually constituting most of the fruit wall (Fig. 127V). The style base also elongates and thickens in these genera to form a narrow to broad rostrum in the mature fruit (Fig. 127W). The mesocarp can be fleshy (*Parastyrax*, some *Styrax*), spongy (*Sinojackia*), woody (*Melliodendron*), or not differentiated from the remaining pericarp (*Huodendron*, *Pterostyrax*). The endocarp is thick, indurate, and ribbed in many genera.

There are usually one to four seeds per fruit; in *Alniphyllum*, *Bruinsmia*, and *Huodendron* there are many. The seed coat is indurate and thick in *Styrax* (Fig. 127J, K), thin in all others (Fig. 127T, U). Papillae or hairs cover the seeds of some species of *Styrax*. Seed shape can be ± globose or ellipsoid (*Styrax*; Fig. 127J, K), irregularly angular (*Bruinsmia*), flattened-alate or -auriculate (*Alniphyllum*, *Huodendron*; Fig. 127U, V), or fusiform to narrowly cylindrical (other genera). In *Halesia* and *Styrax*, the testa is at first thick, with postchalazal vascular bundles, then crushed. The endosperm is cellular, oily, and copious (Netolitzky 1926; Corner 1976).

DISPERSAL. Little data exist on the dispersal mechanisms of Styracaceae. Fruits of *Halesia* are dispersed by wind, water, and possibly wildlife (Ridley 1930; U.S.D.A. Forest Service 1948). Fruits of *Styrax obassia* are dispersed by ground mammals and birds (Kato and Hiura 1999). After the fruit wall has become detached, the seeds of the riparian species *S. faberi* remain attached to the infructescence. The seeds, which would otherwise sink, can thus be transported down-river by the floating infructescence. The seeds of *S. americanus* reportedly have been found attached to the feet of waterfowl (Ridley 1930), but this is probably not a primary means of dispersal of the species because the seed is smooth and ellipsoid.

PHYTOCHEMISTRY. Leaf extracts of *Halesia* and *Pterostyrax* contain quercetin, kaempferol, tannins, a small amount of leucoanthocyanin, saponins, and caffeic, p-coumaric and ferulic acids. The benzofuran egonol and its glycosides occur in the seed oil of a number of species of *Styrax*. Fruits of *Styrax* contain significant amounts of jegosaponin, a potent defense chemical. Various species of *Styrax* also contain styracitol, β-phenyl ethyl alcohol, and coniferin (Hegnauer 1962; Gibbs 1974).

In many species of *Styrax* a balsamic resin (benzoin, gum benjamin) exudes from the bark and wood tissues following injury to the cambium. This resin consists chiefly of coniferyl cinnamate, cinnamyl cinnamate (styracin), and coniferyl benzoate associated with cinnamic and benzoic acids; minor components are fragrant benzaldehyde, vanillin, and styrene (Hegnauer 1962; Langenheim 2003).

AFFINITIES. Traditionally Styracaceae have been placed in Ebenales (Dahlgren 1980; Cronquist 1981; Thorne 1992; Takhtajan 1997). Pollen morphology suggests a link to Theaceae (Morton and Dickison 1992), whereas floral morphology suggests a close relationship with members of Ericales sensu Cronquist (1981). Data from *rbc*L and *atp*B chloroplast DNA sequences, as well as 18 S ribosomal DNA sequences (Soltis et al. 1997), strongly support the placement of the family within a clade comprising the families of Ericales s. l. (APG 1998), which includes the Ebenales and many members of Cronquist's Theales (Morton et al. 1996; Savolainen, Fay et al. 2000), although unequivocal morphological synapomorphies for this clade have not been identified. Close relationships between Styracaceae, Diapensiaceae, Actinidiaceae, Clethraceae, Ericaceae, and Cyrillaceae have been suggested by different phylogenetic studies based on DNA sequences, but statistical support for these relationships was weak and contradictory. The five-gene analysis of Anderberg et al. (2002) has now revealed a strongly supported relationship between Styracaceae and Diapensiaceae, which is also reflected in embryological traits.

RELATIONSHIPS WITHIN THE FAMILY. Infrafamilial relationships have recently been investigated with morphological and molecular data (Fritsch et al. 2001). A combined analysis of three DNA sequence datasets and a morphological dataset supports Styracaceae as monophyletic with *Styrax* and *Huodendron* in one clade, and the rest of the genera in another. *Alniphyllum* and *Bruinsmia* form a clade that is sister to the remaining genera. *Halesia macgregori* (China) groups with *Rehderodendron* (China) rather than to the other species of *Halesia* (United States). Unambiguously placed synapomorphies supporting the monophyly of the Styracaceae are stellate trichomes, linear anther connectives of equal width throughout, and apotropous ovules (modified within the family to mixed apotropous/epitropous). The genus *Lissocarpa* was originally included in Styracaceae (Bentham and Hooker 1873) but is now commonly placed within the monogeneric family Lissocarpaceae. *Lissocarpa* lacks the stellate or peltate trichomes that characterize the Styracaceae.

DISTRIBUTION AND HABITATS. In the Americas, the family occurs from the United States to northwestern Argentina and Uruguay, including the Antilles. In Asia, the family ranges from Japan and Korea, south to Nepal, eastern India, Myanmar, southeast Asia, Indonesia, and Papua New Guinea. One species of *Styrax* occurs in the Mediterranean region, and another extends into several island chains of the South Pacific. The distribution of

Styrax essentially mirrors that of the family, and it is the only genus occurring in the Neotropics. *Halesia* is disjunct between the eastern United States and China (but see above). The rest of the genera are distributed among Asia and Malesia. *Changiostyrax*, *Melliodendron*, and *Sinojackia* are endemic to mainland China. Most species are understorey trees of humid warm-temperate and tropical lowland or montane forests from 100 to 3000 m elevation. Two species of *Styrax* occur under summer-dry climates in California and the Mediterranean region; several others occur in the cerrado of southern Brazil and Andean subparamo vegetation. The several intercontinental disjunctions among species of *Styrax* (Fritsch 1999) and *Halesia*, as well as the occurrence of Tertiary fossils in areas of the northern hemisphere in which the family is not known to occur today, suggest that this family once had a more continuous and wider distribution in North America and Eurasia.

PARASITES. Some species of *Styrax* (series *Cyrta* in part and series *Benzoin*) serve as the primary host for species of gall-forming aphids constituting the family Hormaphididae (tribe Cerataphidini). Each cerataphidine species produces galls and reproduces sexually on only one species of *Styrax*. The galls are formed on either vegetative or floral meristems and come in a wide variety of shapes, including globose, spiral, coralliniform, alcicorniform, and tubular (Docters van Leeuwen 1922; Docters van Leeuwen-Reijnvaan and Docters van Leeuwen 1926). Although cerataphidines are distributed pantropically, the galls are found only on Asian species of *Styrax*. Phylogenetic patterns provide evidence for host-switching between members of series *Benzoin* and series *Cyrta* (Stern 1994; Fritsch 1999).

PALEOBOTANY AND DISTRIBUTIONAL HISTORY. The fossil record of Styracaceae has been reviewed in Fritsch et al. (2001). Eocene and Miocene fossil fruits of *Rehderodendron* are known from France (Mai 1970; Vaudois-Miéja 1983). Fruits of *Halesia* are well represented by fossil endocarps in the European Tertiary (Kirchheimer 1957; Manchester 1999). Fossil fruits of *Melliodendron* and *Pterostyrax* are known from the late Tertiary of Japan (Miki 1941, 1963, 1968). *Styrax* fossil seeds are reported for various European localities ranging from late Eocene to upper Pliocene (Kirchheimer 1957) and for Japan from late Miocene through late Pliocene (Miki 1941). Reports of various genera of Styracaceae in the fossil record based on leaf impressions cannot be considered reliable unless detailed anatomical features, such as stellate or scale-like trichomes, can be observed; hence, the fossil occurrence of the family in the New World is unproven.

The results of phylogenetic analysis of the family (Fritsch 2001; Fritsch et al. 2001) are consistent with the fossil record in suggesting an Eurasian origin of the family. During the Eocene, Styracaeae must have crossed the North Atlantic land bridge and later.

ECONOMIC IMPORTANCE. Several Asian species of *Styrax* (*S. benzoin*, *S. paralleloneurus*, *S. tonkinensis*) are commercial sources of the balsamic resin benzoin, used in the pharmaceutical, confection, and fragrance industries (Duke 1985). The resin of several Neotropical species of *Styrax* is used medicinally and for incense on a local scale (Standley 1924; Schultes and Raffauf 1990). Resin production in the Mediterranean species *S. officinalis* has been widely reported anecdotally, but bark-incision experiments have failed to induce resin formation in this species (Zeybek 1970). This discrepancy may be a result of nomenclatural confusion with the resin produced by *Liquidambar*. Alternatively, genetically based or environmentally controlled variation in resin production may exist among individuals of *S. officinalis* (Langenheim 2003).

KEY TO THE GENERA

1. Fertile shoots produced both laterally and (on at least some twigs) terminally on twig of the previous growth period; ovules not arranged in 2 longitudinal rows in each carpel   2
– Fertile shoots produced only laterally on twig of the previous growth period; ovules arranged in 2 longitudinal rows in each carpel   4
2. Stamen connectives 2–3-toothed at apex; stylodia usually distinct distally; fruiting calyx epicarpous 1/3 to 2/3 the length of the fruit   **2. Huodendron**
– Stamen connectives not toothed at apex; style completely unbranched; fruiting calyx epicarpous only slightly beyond the fruit base   3
3. Pedicels articulated; seed to carpel ratio >2; seed coat thin, fragile   **3. Bruinsmia**
– Pedicels not articulated; seed to carpel ratio ≤1; seed coat thick, indurate   **1. Styrax**
4. Fruit capsular; fruiting calyx epicarpous only slightly beyond the fruit base; seed to carpel ratio >2; seeds winged   **4. Alniphyllum**
– Fruit indehiscent; fruiting calyx epicarpous the complete length of the fruit (excluding rostrum); fruit indehiscent; seed to carpel ratio ≤2; seeds unwinged   5
5. Corolla 4-lobed   6
– Corolla 5-lobed   7
6. Calyx 4-toothed; fruit winged, rostrum <1.5 cm long   **9. Halesia**
– Calyx truncate; fruit not winged, rostrum 4–5.5 cm long   **5. Changiostyrax**
7. Thorns present on trunk; stamen filament trichomes glossy, arms cylindrical; ovules all apotropous   **6. Sinojackia**

– Thorns absent; stamen filament trichomes dull, arms flattened; upper 2 ovules of each carpel apotropous, lower 2 ovules epitropous 8

8. Ovary 3-locular; fruit drupaceous; exocarp not ribbed
7. Parastyrax

– Ovary 1-locular through distal attenuation of the septa; fruit dry; exocarp ribbed (and sometimes winged) 9

9. Inflorescences ca. 30–70-flowered, branches one-sided; corolla lobes connate (coherent) at the base only
8. Pterostyrax

– Inflorescences 1–10-flowered, branches not one-sided; corolla lobes connate distinctly beyond the base 10

10. Pedicels borne on inflorescence axis; calyx toothed; mesocarp indistinct from endocarp, internally lacunate
10. Rehderodendron

– Pedicels borne directly on twig; calyx truncate; mesocarp distinct from endocarp, both solid **11. Melliodendron**

Genera of Styracaceae

1. *Styrax* L. Fig. 127A–P

Styrax L., Sp. Pl. 1: 444 (1753); Perkins, Pflanzenreich IV, 241: 1–111 (1907), rev.; van Steenis, Bull. Jard. Bot. Buitenzorg III, 12: 212–272 (1932), Malesian spp.; Hwang, Fl. Rei. Pop. Sin. 60(2): 77–150 (1987), Chinese spp.; Svengsuksa & Vidal, Fl. Cam. Laos Viêt. 26: 145–195 (1992), Indochinese spp.; Fritsch, Ann. Missouri Bot. Gard. 84: 705–761 (1997), Mesoamerican spp.; Wallnöfer, Ann. Naturhist. Mus. Wien 99B: 681–720 (1997), 5-anthered spp..
Pamphilia Mart. ex A.DC. (1844). *Styrax* sectio *Pamphilia* (Mart. ex A.DC.) B. Walln.

Bud scales absent; fertile shoots produced both laterally and (on at least some twigs) terminally on twig of the previous growth period; pedicels not articulated; stamens attached high on corolla tube; stamen filament trichomes glossy, arms cylindrical; ovules 1–many per carpel; placental obturators usually present; mesocarp present; seeds 1–3; seed coat thick, indurate. $n = 8, 16, >20$. About 130 spp., E and SE Asia, Malesia, North and South America, Mediterranean. Fritsch (1999) has divided the genus into sect. *Styrax*, subdivided into series *Styrax* and *Cyrta* (Lour.) P.W. Fritsch; and sect. *Valvatae* Gürke, subdivided into series *Valvatae* Perkins and *Benzoin* P.W. Fritsch.

2. *Huodendron* Rehder Fig. 127U

Huodendron Rehder, J. Arnold Arbor. 16: 341 (1935).

Bud scales absent; fertile shoots produced both laterally and (on at least some twigs) terminally on twig of the previous growth period; pedicels not articulated; calyx 5-toothed; stamen filament trichomes glossy, arms cylindrical; stamen connectives 2–3-toothed at apex; ovules many per carpel, irregularly arranged; fruit a 3-locular capsule; mesocarp absent; seeds numerous. Four spp., China, Myanmar, Thailand, Vietnam.

3. *Bruinsmia* Boerl. & Koord.

Bruinsmia Boerl. & Koord., Natuurk. Tijdschr. Ned.-Indië 53(1): 68 (1893).

Bud scales absent; fertile shoots produced both laterally and (on at least some twigs) terminally on twig of the previous growth period; pedicels articulated; calyx truncate or slightly 5-toothed; stamen filament trichomes glossy, arms cylindrical; ovules many per carpel, irregularly arranged; fruit baccoid; fruiting calyx epicarpous only slightly beyond the fruit base; seeds numerous. Two spp., India, China, Malaysia, Papua New Guinea.

4. *Alniphyllum* Matsum. Fig. 127Q–T

Alniphyllum Matsum., Bot. Mag. (Tokyo) 15: 67 (1901).

Bud scales absent; fertile shoots produced only laterally on twig of the previous growth period; pedicels articulated; calyx 5-toothed; stamen filament trichomes glossy, arms flattened; ovules 8–10 per carpel, in 2 longitudinal rows, all apotropous; fruit a 5-locular capsule; fruiting calyx epicarpous only slightly beyond the fruit base; seeds numerous, winged. Three spp., India, China, Laos, Myanmar, Vietnam.

5. *Changiostyrax* C.T. Chen

Changiostyrax C.T. Chen, Guihaia 15: 291 (1995).

Bud scales present; pedicels articulated; calyx truncate; corolla 4-lobed; ovules 8 per carpel, in 2 longitudinal rows, all apotropous; fruit dry, indehiscent; ectocarp 8-ribbed; mesocarp corky; fruiting calyx epicarpous the complete length of the fruit (excluding the 4–5.5-cm-long rostrum). A single species, *C. dolichocarpus* (C.J. Qi) C.T. Chen, SE China.

6. *Sinojackia* Hu

Sinojackia Hu, Contrib. Biol. Lab. Chin. Assoc. Adv. Sci., Sect. Bot. 4(1): 1 (1928).

Thorns present on trunk; bud scales present; calyx 5-toothed; pedicels articulated; stamen filament trichomes glossy, arms cylindrical; ovules 6–8 per locule, in 2 longitudinal rows, all apotropous; fruit indehiscent; exocarp distinctly lenticellate, not ribbed; mesocarp spongy; fruiting calyx epicarpous the complete length of the fruit (excluding the broad rostrum). $n = 12$. Five spp., China.

7. *Parastyrax* W.W. Sm.

Parastyrax W.W. Sm., Notes Roy. Bot. Gard. Edinburgh 12: 231 (1920).

Pedicels articulated; calyx truncate; stamen filament trichomes dull, arms flattened; ovary 3-locular throughout; style solid; ovules 4 per carpel, the upper 2 of each carpel apotropous, lower 2 epitropous; fruit drupaceous; exocarp distinctly lenticellate, not ribbed; fruiting calyx epicarpous the complete length of the fruit (excluding the short rostrum). Two spp., China, Myanmar.

8. *Pterostyrax* Sieb. & Zucc.

Pterostyrax Sieb. & Zucc., Fl. Jap. 1: 94 (1839).

Bud scales present; inflorescences ca. 30–70-flowered, branches one-sided; pedicels articulated; stamen filament trichomes dull, arms flattened; ovules 4 per carpel, the upper 2 of each carpel apotropous, lower 2 epitropous; fruit dry, indehiscent; exocarp 10-ribbed or 5-winged; mesocarp absent; fruiting calyx epicarpous the complete length of the fruit (excluding the rostrum). $n = 12$. Four spp., China, Japan, Myanmar.

9. *Halesia* J. Ellis ex L.

Halesia J. Ellis ex L., Syst. Nat. ed. 10, 1044 (1759); Fritsch & Lucas, Syst. Bot. 25: 197–210 (2000), part rev. of North American spp.

Bud scales present; calyx 4-toothed; pedicels articulated; corolla 4-lobed; stamen filament trichomes dull, arms flattened; ovules 4 per carpel, the upper 2 of each carpel apotropous, lower 2 epitropous; fruit dry, indehiscent, 2–4-winged; fruiting calyx epicarpous the complete length of the fruit (excluding the <1.5-cm-long rostrum). $n = 12$. Three spp., eastern U.S.A., China.

10. *Rehderodendron* Hu

Rehderodendron Hu, Sinensia 2: 109 (1932).

Bud scales present; inflorescences 3–10-flowered; calyx 5-toothed; pedicels articulated; ovules 4 per carpel, the upper 2 of each carpel apotropous, lower 2 epitropous; fruit dry, indehiscent, broadly cylindrical; exocarp ca. 10-ribbed; mesocarp indistinct from endocarp, internally lacunate; fruiting calyx epicarpous the complete length of the fruit (excluding the short rostrum). Five spp., China, Myanmar, Vietnam.

11. *Melliodendron* Hand.-Mazz.   Fig. 127V

Melliodendron Hand.-Mazz., Anz. Akad. Wiss. Wien., Math.-Naturwiss. Kl. 59: 109 (1922).

Bud scales present; flowers 1–2 per leaf axil; calyx truncate; pedicels articulated; ovules 4 per carpel, the upper 2 of each carpel apotropous, lower 2 epitropous; fruit dry, indehiscent, turbinate; mesocarp distinct from endocarp, both solid; fruiting calyx epicarpous the complete length of the fruit (excluding the broad rostrum). A single species, S China.

Selected Bibliography

Anderberg, A.A. et al. 2002. See general references.

APG (Angiosperm Phylogeny Group) 1998. See general references.

Bentham, G., Hooker, J.D. 1873. Genera plantarum. Vol. 2. London: Reeve.

Chester, E. 1966. A biosystematic study of the genus *Halesia*, Ellis (Styracaceae). Ph.D. dissertation. Ann Arbor, MI: The University of Tennessee, University Microfilms.

Copeland, H.F. 1938. The *Styrax* of northern California and the relationships of the Styracaceae. Am. J. Bot. 25: 771–780.

Corner, E.J.H. 1976. See general references.

Cronquist, A. 1981. See general references.

Dahlgren, R.T.M. 1980. A revised system of classification of the angiosperms. Bot. J. Linn. Soc. 80: 91–124.

De Candolle, A.L.P.P. 1844. Styraceae. In: Prodromus Systematis Naturalis Regni Vegetabilis. Vol. 8. Paris: Treuttel & Wurtz, pp. 244–272.

Dickison, W.C. 1993. Floral anatomy of the Styracaceae, including observations on intra-ovarian trichomes. Bot. J. Linn. Soc. 112: 223–255.

Dickison, W.C., Phend, K.D. 1985. Wood anatomy of the Styracaceae: evolutionary and ecological considerations. I.A.W.A. Bull. N.S. 6: 3–22.

Docters van Leeuwen, W.M. 1922. Über einige von Aphiden an *Styrax*-Arten gebildete Gallen. Bull. Jard. Bot. Buitenzorg, sér. 3 4: 147–162.

Docters van Leeuwen-Reijnvaan, J., Docters van Leeuwen, W.M. 1926. The zoocecidia of the Netherlands East Indies. Batavia: Drukkerij de Unie.

Duke, J.A. 1985. CRC handbook of medicinal herbs. Boca Raton, FL: CRC Press.

Fritsch, P.W. 1996a. Population structuring and patterns of morphological variation in California *Styrax* (Styracaceae). Aliso 14: 205–218.

Fritsch, P.W. 1996b. Isozyme analysis of intercontinental disjuncts within *Styrax* (Styracaceae): implications for the Madrean-Tethyan hypothesis. Am. J. Bot. 83: 342–355.

Fritsch, P.W. 1997. A revision of *Styrax* (Styracaceae) for western Texas, Mexico, and Mesoamerica. Ann. Missouri Bot. Gard. 84: 705–761.

Fritsch, P.W. 1999. Phylogeny of *Styrax* based on morphological characters, with implications for biogeography and infrageneric classification. Syst. Bot. 24: 356–378.

Fritsch, P.W. 2001. Phylogeny and biogeography of the flowering plant genus *Styrax* (Styracaceae) based on chloroplast DNA restriction sites and DNA sequences of the internal transcribed spacer region. Molec. Phylog. Evol. 19: 387–408.

Fritsch, P.W. In press. Styracaceae. In: Flora of North America. New York: Oxford University Press.

Fritsch, P.W., Lucas, S.D. 2000. Clinal variation in the *Halesia carolina* complex (Styracaceae). Syst. Bot. 25: 197–210.

Fritsch, P.W., Morton, C.M., Chen, T., Meldrum, C. 2001. Phylogeny and biogeography of the Styracaceae. Int. J. Pl. Sci. 162 (6 Suppl.): S95–S116.

Gibbs, R.D. 1974. Chemotaxonomy of the flowering plants, 4 vols. Montreal: McGill-Queen's University Press.

Gonsoulin, G.J. 1974. A revision of *Styrax* (Styracaceae) in North America, Central America, and the Caribbean. Sida 5: 191–258.

Gürke, M. 1891. Styracaceae. In: Engler, A., Prantl, K. Die natürlichen Pflanzenfamilien, IV, 1. Leipzig: W. Engelmann, pp. 172–180.

Hegnauer, R. 1962. See general references.

Hsu, P.S., Weng, R.F., Kurita, S. 1994. New chromosome counts of some dicots in the Sino-Japanese region and their systematic and evolutionary significance. Acta Phytotax. Sin. 32: 411–418.

Hwang, S.-M. 1987. Styracaceae. In: Flora Reipublicae Popularis Sinicae. Vol. 60(2). Beijing: Science Press, pp. 77–150.

Hwang, S.-M., Grimes, J. 1996. Styracaceae. In: Wu, Z.-Y., Raven, P.H. (eds.) Flora of China. Vol. 15. Beijing: Science Press, pp. 253–271.

Johri, B.M. et al. 1992. See general references.

Kato, E., Hiura, T. 1999. Fruit set in *Styrax obassia* (Styracaceae): the effect of light availability, display size, and local floral density. Am. J. Bot. 86: 495–501.

Kirchheimer, F. 1957. Die Laubgewächse der Braunkohlenzeit. Halle (Saale): VEB Wilhelm Knapp.

Langenheim, J. 2003. Plant resins: chemistry, evolution, ecology, and ethnobotany. Portland: Timber Press.

Lewis, W.H., Stripling, H.L., Ross, R.G. 1962. Chromosome number reports LXIX. Taxon 29: 728.

Magellón, S., Crane, P.R., Herendeen, P.S. 1999. Phylogenetic pattern, diversity, and diversification of eudicots. Ann. Missouri Bot. Gard. 86: 297–372.

Mai, D.H. 1970. Subtropische Elemente im europäischen Tertiär I. Paläont. Abh. B 3: 441–503.

Mai, D.H. 1988. Über antillanische Styracaceae. Feddes Repert. 99: 173–181.

Manchester, S.R. 1999. Biogeographical relationships of North American Tertiary floras. Ann. Missouri Bot. Gard. 86: 472–522.

Manshard, E. 1936. Embryologische Untersuchungen an *Styrax obassia* Sieb. et Zucc. Planta 25: 264–383.

Metcalfe, R.C., Chalk, L. 1950. Anatomy of dicotyledons. Oxford: Clarendon Press.

Miers, J. 1859. On the natural order Styraceae, as distinguished from the Symplocaceae. Ann. Mag. Nat. Hist. III, 3: 394–404.

Miki, S. 1941. On the change of flora in Eastern Asia since Tertiary period. I. The clay of lignite beds flora in Japan with special reference to the *Pinus trifolia* beds in Central Hondo. Japan. J. Bot. 11: 237–304.

Miki, S. 1963. Further study on plant remains in *Pinus trifoliata* beds, Central Hondo, Japan. Special issue, Chigakukenkyu 80–93.

Miki, S. 1968. *Paleodavidia*, synonym of *Melliodendron* and fossil remains in Japan. Bull Mukogawa Women Univ. 16: 287–291.

Morton, C.M., Dickison, W.C. 1992. Comparative pollen morphology of the Styracaceae. Grana 31: 1–15.

Morton, C.M., Chase, M.W., Kron, K.A., Swensen, S.M. 1996. A molecular evaluation of the monophyly of the order Ebenales based upon *rbc*L sequence data. Syst. Bot. 21: 567–586.

Netolitzky, F. 1926. Anatomie der Angiospermen-Samen. In: Linsbauer, K. (ed.) Handbuch der Pflanzenanatomie. Vol. 10, 4. Berlin: Bornträger.

Payer, J.-B. 1966. Traité d'organogénie comparée de la fleur. New York: J. Cramer.

Perkins, J. 1907. Styracaceae. In: Engler, A. (ed.) Pflanzenreich IV. Vol. 241 (Heft 30). Leipzig: W. Engelmann, pp. 1–111.

Perkins, J. 1928. Übersicht über die Gattungen der Styracaceae. Leipzig: W. Engelmann.

Ridley, H.N. 1930. The dispersal of plants throughout the world. Ashford, Kent: L. Reeve.

Saraiva, L., Cesar, O., Monteiro, R. 1988. Biologia da polinização e sistema de reprodução de *Styrax camporum* Pohl e *S. ferrugineus* Nees et Mart. (Styracaceae). Revista Brasil. Bot. 11: 71–80.

Savolainen, V., Fay, M.F. et al. 2000. See general references.

Schadel, W.E., Dickison, W.C. 1979. Leaf anatomy and venation patterns of the Styracaceae. J. Arnold Arbor. 60: 8–37.

Schultes, R.E., Raffauf, R.F. 1990. The healing forest. Portland, OR: Dioscorides Press.

Soltis, D.E. et al. 1997. See general references.

Spongberg, S.A. 1976. Styracaceae hardy in temperate North America. J. Arnold. Arbor. 57: 54–73.

Standley, P.C. 1924. Trees and shrubs of Mexico. Contrib. U.S. Natl. Herb. 23: 1–1721.

Stern, D.L. 1994. Phylogenetic evidence that aphids, rather than plants, determine gall morphology. Proc. Roy. Soc. Lond. Ser. B, Biol. Sci. 256: 203–209.

Sugden, E.A. 1986. Anthecology and pollinator efficacy of *Styrax officinale* subsp. *redivivum* (Styracaceae). Am. J. Bot. 73: 919–930.

Svengsuksa, B.K., Vidal, J.E. 1992. Styracacées. In: Morat, P. (ed.) Flore du Cambodge du Laos et du Viêtnam. Vol. 26. Paris: Muséum National d'Histoire Naturelle, pp. 145–195.

Takhtajan, A. 1997. See general references.

Tamura, S., Hiura, T. 1998. Proximate factors affecting fruit set and seed mass of *Styrax obassia* in a masting year. Ecoscience 5: 100–107.

Thorne, R.F. 1992. Classification and geography of the flowering plants. Bot. Rev. 58: 225–348.

U.S.D.A. Forest Service. 1948. Woody-plant seed manual. Washinton, D.C.: U.S. Government Printing Office.

van Steenis, C.G.G.J. 1932. The Styracaceae of Netherlands India. Bull. Jard. Bot. Buitenzorg III, 12: 212–272.

Vaudois-Mieja, N. 1983. Extention palaeogéographique en Europe de l'actuel genre *Rehderodendron* Hu (Styracaceae). C. R. Acad. Sci. Paris II, 296: 125–130.

Veillet-Bartoszewska, M. 1960. Embryogénie des Styracacées. Développement de l'embryon chez le *Styrax officinalis* L. C. R. Acad. Sci. Paris II, 250: 905–907.

Wallnöfer, B. 1997. A revision of *Styrax* L. section *Pamphilia* (Mart. ex A. DC.) B. Walln. (Styracaceae). Ann. Naturhist. Mus. Wien 99B: 681–720.

Wood, C.E., Channell, R.B. 1960. The genera of the Ebenales in the southeastern United States. J. Arnold Arbor. 41: 1–35.

Yakovlev, M.S. (ed.) 1983. Comparative embryology of flowering plants, Phytolaccaceae-Thymelaeaceae. Leningrad: Nauka (in Russian).

Yamazaki, T. 1970. Embryological studies in Ebenales 1. Styracaceae. J. Japan. Bot. 45: 267–273.

Zeybek, N. 1970. Liefert *Styrax officinalis* L. ein Harz? Ber. Schweiz. Bot. Ges. 80: 189–193.

Symplocaceae

H.P. Nooteboom

Symplocaceae Desf., Mém. Mus. Hist. Nat. Paris 6: 9 (1820), nom. cons.

Evergreen shrubs or trees (but *S. paniculata* is deciduous). Leaves spirally or distichously arranged, exstipulate, simple, petiolate; leaf margin blade dentate, glandular dentate, or entire, midvein adaxially impressed (rarely flat or prominent). Inflorescence a thyrse or more often reduced or compound to a spike, raceme, panicle, glomerule, or a solitary flower. Flowers actinomorphic, bisexual, rarely (unisexual), in spikes, racemes, panicles, or glomerules (solitary), supported by 1 bract and 2 prophylls (exceptionally bractless or with several bracts in the axil of a leaf); calyx lobes (3–)5, valvate or imbricate, persistent; corolla of (3–)5(–11) imbricate lobes, white or rarely yellow, gamopetalous but divided nearly to base (or sometimes to middle in subg. *Symplocos*); stamens many (4 or 5), rarely few, adnate to base of the corolla tube, monadelphous filaments all united in subg. *Symplocos*, and all united or monadelphous or pentadelphous, the same or in 5 antesepalous bundles in subg. *Hopea*; anthers subglobose, 2-locular, opening with longitudinal slits; ovary inferior or half-inferior, (2)3(–5)-celled, usually with an apical 5-glandular, annular, cylindrical, or 5-lobed disk; style filiform to thick; stigma small, capitate or 2–5-lobed; ovules 2–4 per cell. Fruit a drupe; seeds with copious endosperm; embryo straight or curved; cotyledons very short.

A monogeneric family of ca. 3200 species, widely distributed in tropical and subtropical Asia, Australia, and America, with few species extending to temperate Asia and North America.

VEGETATIVE MORPHOLOGY. Symplocaceae are shrubs or trees, varying, in some species, from small shrubs to moderately large trees. The twigs are slender to rather thick and in some species are covered with pulvinate leaf scars. When the leaves are distichously arranged, the twigs are often zigzag. In most species growth is discontinuous and the twigs are terminated by a bud protected by large, leathery scales. In the species with continuous growth, the buds are naked or covered by small scales. The hairs are simple and 1-celled but have transverse septa; rarely they are bulbous-based. The indumentum varies greatly and sometimes is important for the discrimination of the species but often it is of no systematic value at all. Sometimes the absence of hairs is taxonomically important.

The leaves are mostly spirally or distichously arranged, rarely they are pseudoverticillate. In general, phyllotaxis is constant for a species but, in some species with distichous leaves, the 'leaders' often have spirally arranged leaves. The shape, size and venation of the leaves often yield good specific characters but in some species vary so widely that species cannot be distinguished vegetatively.

With the exception of a few species, in the dry state the midrib is sulcate. The number of lateral nerves varies greatly. The reticulation of the veins is often an important character. In some species the secondary veins are transverse to the nerves. The tertiary and the quaternary veins can be prominent or obscure, the reticulation dense or lax, and sometimes a fine reticulation is visible only in translucent light. Usually details of the venation are only distinct on the undersurface. The margin of the leaves is often provided with vesicular or tooth-like glands. The latter are usually lost during the expansion of the leaves. The length of the petiole is very variable in some species, but rather constant and providing a good character in others. In cross-section the petiole is usually convex beneath and grooved or flat above. In some species it has decurrent edges and may even appear winged by the decurrent leafblade.

VEGETATIVE ANATOMY. The leaves of *Symplocos* are bifacial. The cuticle is smooth, striated or ridged. The lower epidermis is generally smooth but sometimes papillate. The thickness of the leaves is very variable for the genus as a whole, but relatively constant for the species. The size of the epidermal cells varies considerably from species to species; their radial walls are straight or undulate. A hypodermal layer is present in several species, but is sometimes not continuous. In one and the same species, the number of layers of palisade parenchyma can vary from 1 to 3, but mostly this character is rather constant. The spongy parenchyma is very compact to very loose; in some species it contains sclereids. The midrib encloses one large vascular bundle, usually enveloped by a sclerenchymatic sheath which can be continuous or not. Above the vascular bundle, usually several layers of parenchymatic or collenchymatic tissue can be found. Beneath the vascular bundle, mostly a thick sclerenchymatic cap is present. In all but three species, the midrib is grooved in the dried state.

The leaf surface is totally glabrous in some species, in other species it possesses a more or less dense indumentum, especially underneath on the nerves. The hairs are nearly always septate; they are appressed to erect and sometimes lignified and heavily cutinised, especially in the basal part.

The stomata are of the paracytic type. In most (or all?) species there are two kinds of stomata: besides the "normal" ones, a few, very large, so-called water stomata occur. Within certain limits, the size of the stomata is constant for a species.

Clustered calcium Ca-oxalate crystals probably occur in all species and can be found in most parts of the leaves. Tannin-like compounds, probably condensed proleucoanthocyanins and catechins (so-called Inklusen) are common in the epidermal cells, sometimes giving them a very characteristic, brown colour. The leaves of many species contain a high amount of aluminium compounds; these are often visible as dark bodies in the leaf tissues (cf. fat-like bodies of Metcalfe and Chalk 1950).

Leaf anatomy is rather constant for some species and highly variable for others, probably often due to altitudinal variation. For further information on leaf anatomy, see Loesener (1896), Cador (1900), Wehnert (1906), Neger (1923), and Metcalfe and Chalk (1950).

The wood is light and soft, white to yellowish, sometimes with a pinkish tinge. Growth rings, when present, are usually visible to the naked eye, as are the vessels and broad rays. Microscopically, the wood is diffuse-porous, very rarely tending to semi-ring-porous. The frequency of vessels is 12–271/mm^2; typically they are solitary but sometimes are arranged in radial pairs or more rarely in multiples of up to 4. Vessels are angular to oval in transverse section and the average vessel member length is 772–1836 µm. Inter-vessel pits are scalariform, transitional or opposite with a tangential diameter of 4–15 µm, or in scalariform pits even wider, with slit-like, included apertures. Vessel-ray and vessel-parenchyma pits are half-bordered to almost simple, scalariform, transitional, opposite or even tending to alternate. In very oblique end walls, perforations are scalariform with 9–59 bars. Spiral thickenings are absent or present in tails only, or are well developed. Tyloses or warty layers are noted in few specimens only. The ground tissue is composed of long fibre-tracheids (1170–2920 µm). The pits are conspicuously bordered (4–8 µm in diameter) on both radial and tangential walls, with slit-like, included or extending apertures. Spiral or annular thickenings are absent, restricted to fibre tips, or well developed. Axial parenchyma is scanty paratracheal and apotracheally diffuse, very sparse to fairly abundant, in strands of 4–8 cells. Rays are of the Kribs' type heterogeneous I, sometimes tending to heterogeneous II. Crystals are mostly absent; if present they are irregularly rhomboidal or fragmented and form crystalline masses mostly deposited in ordinary ray cells, rarely in chambered ray cells or axial parenchyma. Silica bodies are absent. Pith flecks are present in some specimens only (Metcalfe and Chalk 1950; van den Oever et al. 1981).

A similar wood structure anatomy is found in quite a few other families, such as Theaceae, Cornaceae, Nyssaceae, Saxifragaceae, Clethraceae, Cyrillaceae, and Hamamelidaceae (e.g. Janssonius 1920), representing the Cornus type of Huber (1963).

The results of a fairly large sample of *Symplocos* species and specimens (van den Oever et al. 1981) confirm the general validity of most of the altitudinal and latitudinal trends in wood anatomical characters reported by Baas (1973) for the genus *Ilex*.

Inflorescence and Flower Structure. The inflorescences are inserted in the axils of the upper leaves, or on the wood beneath them; rarely, they are terminal. Racemose in appearance, they are often true racemes or spikes, but with a tendency to branch. Sometimes the inflorescences are true panicles, or condensed to fascicles or very short clusters of flowers, or even reduced to one flower. Morphologically, they are probably thyrses or derived there from.

Principally one bract and two prophylls are present below each flower, the latter directly beneath the flower (Fig. 128B). Occasionally, a pedicel is present between the bract and the prophylls. The bracts and/or prophylls are persistent or caducous. Sometimes the presence of additional bracts (metaxyphylls) are indicatives of an origin from a more elaborate inflorescence. Sometimes the bracts and prophylls bear vesicular glands on the margin.

The flowers are usually bisexual, but rarely functionally unisexual flowers occur in taxa that are either dioecious or polygamous. In male flowers the style is small without a stigma; in female flowers the number of stamens, the anthers of which are sterile, is reduced, sometimes even to less than ten.

The ovary is inferior and the calyx tube is adnate to it; the free part of the calyx is called the calyx limb, consisting of (3–)5 free or basely connate lobes. The calyx lobes are normally ± equal, but in several species the calyx splits after anthesis, thus becoming symmetrically cleft. The incision between the calyx lobes sometimes becomes longer with age. The presence or absence of an indumentum on the calyx, and the shape and the length of the calyx lobes sometimes offer useful characters.

In subg. *Symplocos* the petals are connate into a tube which often can be seen only from the inside because the overlapping margins are free; in subg. *Hopea* the petals are connate only at their very base. Mostly the petals are glabrous or provided with some minute hairs towards the outer base. Rarely they are covered with a dense indumentum, at least in bud. For most species the corolla does not offer good specific characters.

In subg. *Hopea* the stamen filaments are connate only towards the base for at most 2 mm, and form 5 antepetalous-alternipetalous bundles or are evenly arranged around the gynoecium. In subg. *Symplocos* the stamen filaments are fused to form a long tube. The number of the filaments and the relative length of their free part provide good characters. In some species the stamens are hairy.

The upper part of the ovary is covered by an often stellate disk which bears 5 glands. In some species the shape of this part of the ovary is variable. The disk can be globose, low-cylindrical, or flat and then inconspicuous. It is often covered by an indumentum.

The style is of the same shape in the whole genus; its indumentum is often coincident with that of the disk. In male flowers of dioecious or polygamous species, the peltate or punctiform stigma is absent.

The ovary is mostly rather constantly 2-, 3-, 4-, or 5-celled; the number of ovules being 2–4, but usually 4 in each locule. Placentation is mostly considered axile but, according to Chirtoiü (1918) and Nakai (1924), for Asiatic species placentation is parietal. This was denied by Handel-Mazzetti and Peter-Stibal (1943). Re-examination of this controversial character by Nooteboom (1975) showed that the ovules are attached on the induplicate part of the carpels, close to the centre. The septa are not connate to each other at the centre in the upper part of the ovary (see also the developmental study of *Symplocos paniculata* flowers by Caris et al. 2001).

EMBRYOLOGY. The ovule is anatropous, unitegmic, and tenuinucellar; the archesporial cell functions as the megaspore mother cell; the chalazal or subchalazal megaspore develops into a Polygonum-type embryo sac (Davis 1966; Corner 1976).

POLLEN MORPHOLOGY (by R.W.E.J.M. van der Ham). Pollen grains of *Symplocos* are 3(2 or 4)-aperturate monads of 20–70 (av. 30) μm equatorial diameter. In polar view the grains are circular, with more or less protruding apertures, to angular; in equatorial view they are oblate to oblate spheroidal. The colpate ectoapertures are usually (very) short and often inconspicuous in light microscopy, unless a thickened margo is present. The endoapertures are mostly circular to lalongate pores, sometimes lalongate colpi, provided with an irregularly delimited and mostly fragmented endoannulus or polar costae. The total thickness of the exine is 1–4 μm. The nexine is usually thicker than the sexine, and sometimes shows endocracks, especially around the apertures. A very thin, fragmented endexine is probably present (Barth 1982; Nagamasu 1989). The infratectum is distinct (columellate) to thin and indistinct. The tectum is finely punctate, perforate to ± reticulate, and psilate or provided with supratectal elements (scabrate, rugulate, areolate, verrucate, echinate). Irregularly distributed globules of 2–6 μm have been interpreted as Ubisch bodies.

On the basis of the presence/absence of grooves around the endoannulus (or along the equatorial sides), Van der Meijden (1970) distinguished two main pollen types, which correspond with the two subgenera *Symplocos* and *Hopea* delimited by Nooteboom (1975). Other attempts at relating pollen morphology and infrageneric classification of *Symplocos* have been undertaken by Gupta and Sharma (1977) and Barth (1979).

KARYOLOGY. The base number seems to be x = 11; in subgen. *Hopea* n = 11 is most frequent (see tabulation by Nooteboom 1975); the very few records of 2n = 23 and 24 may be due to B-chromosomes (Mehra and Gill 1968). The chromosome number of *Symplocos pendula* (the only species of subg.

Symplocos studied so far) was determined as approximately $2n = 90$ and is probably octoploid, based on x = 11 (Nooteboom 1975).

Karyotypes of all species of subg. *Hopea* are very similar. This is consistent with the assumption that hybridisation between them is possible. This is also indicated by the frequency of intermediate or transitional specimens found in the herbarium. There are also certain hermaphroditic taxa which always show a high proportion of sterile pollen, which may be due to their hybrid origin.

FRUIT AND SEED. The fruit is a drupe with a fleshy, corky or woody mesocarp and a very hard stone (endocarp). Shape and size of the fruit provide good characters, particularly for groups of related species. The shape can be globose, ellipsoid, ovoid, obovoid, ampulliform, spindle-shaped, and cylindrical. The shape of the stone may differ from the shape of the fruit, and ovoid fruits sometimes contain ampulliform stones.

The seeds are exotegmic-ericoid (Fig. 2D). The seed coat is very thin. The embryo can be straight or curved to U-shaped, or curved twice in two planes. Almost always the shape of the embryo is similar to that of the seed. Globose and ampulliform fruits are mostly correlated with curved embryos; cylindrical fruits always have straight embryos. Ovoid and ellipsoid fruits may contain straight or curved seeds and embryos; in some species there are even intergradations between them. The endosperm often contains starch as well as fatty oil and probably aleuron (pers. obs.). Seeds of *S. paniculata* are reported to be starch-free and to contain 21.2% proteins and 51.7% fatty oil. Palmitinic, oleic and linolic acids were found to be the main fatty acids in samples of *Symplocos* seeds investigated to date.

Germination of *S. paniculata* is phanerocotylar, and has been described by Lubbock (1892).

DISPERSAL AND REPRODUCTION. Birds are probably the main dispersal agents. Seedlings are, however, rarely encountered in the forest (pers. obs.). In *S. cochinchinensis* var. *laurina* on the Doi Inthanon (Thailand), it was observed that most, if not all, young trees sprouted from fallen branches. One fallen treelet gave rise to more than ten new trees. This probably reflects a high degree of seed sterility (Nooteboom 1975).

PHYTOCHEMISTRY. The reader is referred to Hegnauer (1973, 1990). Phenolic compounds, including gallic and ellagic acids and the hydrolysable tannins based on them, and leucoan-thocyanins seem to be rather common. The record of the iridoid compound cornin is remarkable and agrees with a position within the Ericales. The leaves of subg. *Hopea* contain much aluminium and turn yellow at drying when the aluminium forms complexes with the leaf flavonoids (Chenery 1948a; Nooteboom 1975). In subg. *Symplocos* no accumulation of aluminium has been observed, and their leaves never turn yellow.

Already Radlkofer (1904) reported that the ash of the leaves contain ca. 50% aluminium oxide. He also described the so-called Tonerdekörper in the leaves of *Symplocos*. These are masses of colourless material filling often large parts of the cells, predominantly in the palisade parenchyma. Radlkofer stated that most of the aluminium of the leaves is accumulated in these masses. Neger (1923) cultivated individuals of *S. lucida* S. & Z. (= *S. japonica* DC) in solutions containing different amounts of aluminium salts. He observed that plants grew best on a solution of 1% aluminium, an amount that is toxic for other xerophyllous plants.

AFFINITIES. The taxonomic history of the family has been dealt with by Nooteboom (1975). Following Jussieu (1789), most later authors placed Symplocaceae in the same alliance with Ebenaceae, Sapotaceae, and Styracaceae, a position that was maintained. All these taxa were part of the Ebenales of most later authors, up to Takhtajan (1997).

Formerly *Symplocos* was frequently included in Styracaceae but, besides the well-known morphological and anatomical differences exhibited with this family, Kirchheimer (1949) noted that in *Symplocos* fruits each locule possesses an apical pore which is absent in the fruits of Styracaceae; according to him this difference goes back to Eocene times at least.

In the ordinal classification of APG (1998), Symplocaceae were assigned to a broadly construed order Ericales, comprising some 24 families including those mentioned above. This affiliation has been confirmed by several recent molecular studies, but none of them, including the five-gene analysis by Anderberg et al. (2002), has been able to determine a precise position of the family within this clade. When comparing the broader based, recent molecular studies (as available at the time of writing, Nov. 2000), Symplocaceae invariably form part of this Ericales clade or grade, but the closest relatives vary considerably and include families such as Lecythidaceae, Sapotaceae,

Styracaceae, Theaceae, Ternstroemiaceae, and Diapensiaceae.

DISTRIBUTION AND HABITATS. The large genus of nearly 300 species is distributed in the eastern parts of the Old World from the Deccan Peninsula to northern China/Japan, extending beyond the Tropic of Cancer, in Australia reaching as far as New South Wales and Lord Howe Island, and in the Pacific as far as Fiji. In the New World *Symplocos* is found from the south-eastern United States to southern Brazil.

Subg. *Symplocos* is represented in Asia only by two species, while the New World harbours approximately 130 species. The range of this subgenus is almost entirely tropical. Subg. *Hopea* has ca. 110 species in Asia, one species extending north to Korea, and ca. 50 in the New World.

Symplocos grows in warm-temperate and tropical zones, preferably in moist to wet mixed, mostly evergreen rainforest. The genus is most abundant in mountain forests. On Mt. Kinabalu and the New Guinean mountains, it ascends up to an altitude of 4000 m, where it is represented by microphyllous dwarf shrubs. In the Andes it extends from 500 to 4000 m altitude but is most frequent from 2500 to 3500 m (Ståhl 1995). *Symplocos* avoids arid regions.

DISTRIBUTIONAL HISTORY AND PALAEOBOTANY. The oldest record of fossil *Symplocos* consists of pollen grains from the Maastrichtian of California (Muller 1981). According to Krutzsch (1989), who has dealt with the distributional history of *Symplocos*, by the Upper Eocene the genus had acquired a coherent north-Tethyan distribution area from western N America to E Asia. This belt disintegrated in the younger Tertiary, and in Europe the genus became extinct at the beginning of the Pleistocene. The expansion southward in America and south-eastward in SE Asia occurred from the Pliocene onward, concomitantly with the formation of secondary centres in America and SE Asia.

The oldest fossil fruits are from the Lower Eocene of C and W Europe. In some instances the fruit stones of up to as much as four species were found together in European brown-coal beds. This makes it probable that in the Lower Tertiary *Symplocos* was already as diversified as it is today. About 25 species have been described from fossil fruits. About 22 of them, all occurring in Europe, possessed straight seeds; they belonged to subg. *Hopea*. This is also the condition in all species of subg. *Symplocos*, both in the Old and New World.

Only the Japanese Pliocene fossils of *S. paniculata*, *S. lancifolia* and *S. sumuntia* have curved seeds. Of the extant Old World species, ca. 20% possess 1- or 2-curved seeds, among them *S. cochinchinensis*. This species not only occupies nearly the whole range of *Symplocos* in the Old World, but also has many infraspecific taxa and in some places dominates the vegetation. Taken together, the species with curved embryos represent the vast majority of individual plants of the genus in SE Asia. As far as is known from the fossil record, no relatives of this species with curved seeds ever reached Europe.

Several fossil *Symplocos* endocarps are so similar to endocarps of Recent species that it is almost certain that, from the Lower Tertiary up to Recent times, no major evolutionary changes have occurred. Some fossil species, particularly from Japan, such as *S. paniculata*, *S. lancifolia*, and *S. sumuntia*, are hardly distinguishable from Recent species.

Hardly any other genus of dicotyledons is known to have occurred in such local abundance and diversity as *Symplocos*. Kirchheimer (1949) calculated the total mass of carpolithic coal in the Vogelsberg mountains to be 3500 m³. In 125 cm³ of coal he found 99 *Symplocos* endocarps and 19 fruits or seeds of other plants. The total number of *Symplocos* endocarps was estimated to be more than 2.5 milliards, mainly belonging to *S. minutula* (v. Sternb.) Kirchheimer, but also to *S. lignitarum* (Quensted) Kirchheimer. According to Kirchheimer, this enormous mass of fruits must have been produced in situ because the coal layers provided no evidence of the presence of running water which could have transported the endocarps. Kirchheimer estimated that the coal layer was deposited during a period of about 100 years, and that the surrounding plants would have furnished at least 25 million endocarps a year. The vegetation certainly was dominated by *Symplocos* trees, although fruits/seeds of *Mastixia*, *Brasenia*, *Magnolia*, and *Vitis* were found in the same beds.

ECONOMIC IMPORTANCE. The reader is referred to Nooteboom (1991). The inner bark of *S. cochinchinensis* ssp. *cochinchinensis* var. *cochinchinensis* and *S. fasciculata* was often used as a mordant in the batik industry and, mixed with other plants, as a dye. It gives a yellow colour by itself but is more frequently used in the preparation of reds derived from *Morinda* spp., *Caesalpinia sappan*, *Butea* spp., and other dye plants. Also the leaves are used as a yellow dye or mordant, as in *S. cochinchinensis* ssp. *laurina* var. *laurina* and *S. lucida*. In

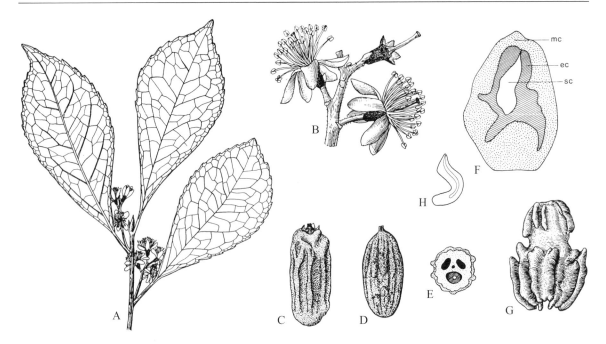

Fig. 128. Symplocaceae. *Symplocos ophirensis* var. *cumingiana.* **A** Flowering twig. **B** Flowers. **C** Fruit. **D** Fruit stone. **E** Same in transverse section. **F** Fruit, vertical section. mc = mesocarp; ec = endocarp; sc = seed cavity. **G** Seed. **H** Embryo. (Nooteboom 1975)

western Java the inner bark of *S. odoratissima* var. *odoratissima,* known as "kulit seriawan", is exhibited in every drugstore. It is pulped and rubbed on the gums to cure thrush. An infusion of the leaves of this variety is used for the same purpose. All these uses seem to be based on the astringent properties of the plant tissue.

Most species do not reach a volume adequate for timber; usually the wood is reported as soft and light and used for light construction, but in Vietnam the wood of *S. adenophylla* var. *adenophylia* is reportedly very hard and good for columns. In S America the leaves of several species are used for making mate (Cador 1900).

A single genus:

Symplocos Jacq.   Fig. 128

Symplocos Jacq., Enum. Pl. Carib.: 5, 24 (1760); Nooteboom, Leiden Bot. Ser. 1 (1975), rev. Old World; Ståhl, Fl. Ecuador 43: 1–4 (1991), Candollea 48: 351–382 (1993), Peruvian spp., 49: 369–388 (1994), Bolivian spp., 51: 323–364 (1996), Colombian spp.

Description as for family. About 300 spp., subtropical and tropical regions except Africa. Two subgenera are distinguished by Nooteboom (1975): **subg.** *Symplocos,* with petals connate at least halfway up, leaves usually not becoming yellow when drying, and straight embryo; about 130 spp., mostly New World; **subg.** *Hopea,* with petals connate only at the base, leaves usually becoming yellow when drying, and embryo straight or curved; ca. 160 spp., Old and New World.

Selected Bibliography

Airy Shaw, H.K. 1966. A dictionary of the flowering plants and ferns, by J.C. Willis, ed. 7, 1214 pp.

Anderberg, A.A. et al. 2002. See general references.

APG (Angiosperm Phylogeny Group) 1998. See general references.

Baas, P. 1973. The wood anatomical range in Ilex (Aquifoliaccae) and its ecological and phylogenetical significance. Blumea 21: 193–258.

Barth, O.M. 1979. Pollen morphology of Brazilian *Symplocos.* Grana 18: 99–107.

Barth, O.M. 1982. The sporoderm of Brazilian *Symplocos* pollen types (Symplocaceae). Grana 21: 65–69.

Borgmann, E. 1964. Anteil der Polyploiden in der Flora des Bismarckgebirges von Ostneuguinea. Zeitschr. Bot. 52: 134.

Brand, A. 1901. Symplocaceae. In: Engler, A. (ed.) Das Pflanzenreich IV. 242. Leipzig: W. Engelmann.

Cador, L. 1900. Blattstruktur der matelieferenden *Symplocos-*Arten. Bot. Centralbl. 2184: 248–251, 345, 369–371.

Candolle, A. de. 1844. Prodromus Systematis Naturalis Regni Vegetabilis 8. Paris: Treutel et Würtz, 684 pp.

Caris, P., Ronse Decraene, L.P., Smets, E., Clinckemaillie, D. 2001. The uncertain systematic position of *Symplocos* (Symplocaceae): evidence from a floral ontogenetic study. Int. J. Plant Sci. 163: 67–74.

Chenery, E.M. 1948a. Aluminium in the plant world. Kew Bull.: 173–183.

Chenery, E.M. 1948b. Thioglycolic acid as an inhibitor for iron. Analyst: 501.

Chirtoiü, M. 1918. Lacistemacées et Symplocacées. Bull. Soc. Bot. Genève XI, 10: 350–361.

Corner, E.J.H. 1976. See general references.

Davis, G.L. 1966. See general references.

Endlicher, S.L. 1839. Genera plantarum: 744. Vienna: Fr. Beck.

Erdtman, G. 1952. See general references.

Gupta, H.P., Sharma, C. 1977. Palynotaxonomy and phylogeny of Indian Symplocaceae and Sapotaceae. Geophytology 7: 147–159.

Hallier, H. 1923. Beiträge zur Kenntnis der Linaceae. Beih. Bot. Centralbl. 39B b: 1–178.

Handel-Mazzetti, H., Peter-Stibal, E. 1943. Eine Revision der chinesischen Arten der Gattung Symplocos Jacq. Beih. Bot. Centralbl. 62B b: 1–42.

Hardin, J.W. 1966. An analysis of variation in Symplocos tinctoria. J. Elisha Mitchell Sci. Soc. 82: 6–12.

Hegnauer, R. 1973, 1990. See general references.

Huang, T.C. 1967. Pollen grains of Formosan plants 2. Taiwania 13: 15–110.

Huang, T.C. 1972. Pollen flora of Taiwan. Taipei: National Taiwan University.

Huber, H. 1963. Die Verwandtschaftsverhältnisse der Rosifloren. Mitt. Bot. Staatssamml. München 5: 1–48.

Hutchinson, J. 1959. The families of flowering plants, ed. 2, 1. Oxford: Clarendnon Press, 510 pp.

Janssonius, H.H. 1920. Mikrographie des Holzes. 4. Gamopetalac, pp. 471–499.

Jussieu, A. de. 1789. Genera plantarum. Paris: Viduam Héerissant, 498 pp.

Keng, H. 1962. Comparative morphological studies in Theaceae. Univ. Calif. Publ. Bot. 33: 269–384.

Kircheimer, F. 1949. Die Symplocaceae der erdgeschichtlichen Vergangenheit. Palaeontographica 90B b: 1–52, t. 1–2.

Kratzmann, E. 1913. Der mikrochemische Nachweis und die Verbreitung des Aluminiums im Pflanzenreich. Sitz. Ber. Akad. Wiss. Wien: 311–336.

Krutzsch, W. 1989. Paleogeography and historical plant geography (paleochorology) in the Neophyticum. Plant Syst. Evol. 162: 5–61.

Liang, Y.H., Yu, C.H. 1985. Pollen morphology of Styracaceae and its taxonomic significance. Acta Phytotax. Sin. 23(2): 81–90.

Lieux, M.H. 1982. An atlas of pollen of trees, shrubs, and woody vines of Louisiana and other southeastern states, part 4, Sapotaceae to Fabaceae. Pollen Spores 24: 331–368.

Loesener, Th. 1896. Beiträge zur Kenntnis der Mateépflanzen. Ber. Deutsch. Pharm. Ges. 6: 203.

Lubbock, J. 1892. A contribution to our knowledge of seedlings. London: 206–208.

Maguire, B., Huang, Y.C. 1978. Symplocaceae. In: Maguire, B. et al., The botany of the Guayana Highlands 10. Mem. New York Bot. Gard. 29: 223–230.

Mehra, P.N., Bawa, K.S. 1969. Chromosomal evolution in tropical hardwoods. Evolution 23: 466–481.

Mehra, P.N., Gill, B.S. 1968. IOPB chromosome number reports XIX. Taxon 17: 575.

Meijden, R. Van der, 1970. A survey of the pollen morphology of the Indo-Pacific species of Symplocos (Symplocaceae). Pollen Spores 12: 513–551.

Metcalfe, C.R., Chalk, L. 1950. See general references.

Muller, J. 1981. Fossil pollen records of extant angiosperms. Bot. Rev. 47: 1–142.

Nagamasu, H. 1989. Pollen morphology of Japanese Symplocos (Symplocaceae). Bot. Mag. Tokyo 102: 149–164.

Nakai, T. 1924. Abstract from T. Nakai: trees and shrubs indigenous in Japan proper, vol. 1 (1922), with additional remarks on some species. Bot. Mag. Tokyo 38: 23–48.

Neger, F.W. 1923. Neue Methoden und Ergebnisse der Mikrochemie der Pflanzen. Flora N.F. 16: 26–330.

Nevling, L.I. Jr. 1969. The ecology of an elfin forest in Puerto Rico. 5. Chromosome numbers of some flowering plants. J. Arnold Arbor. 50: 99–103.

Nooteboom, H.P. 1975. Revision of the Symplocaceae of the Old World, New Caledonia excepted. Leiden Bot. Ser. 1. Leiden: Universitaire Pers.

Nooteboom, H.P. 1991. Symplocos. In: Prosea. Plant resources of South-East Asia 3. Dye and tannin producing plants. Wageningen: Pudoc, pp. 115–118.

Oever, L. van den, Baas, P., Zandee, M. et al. 1981. Comparative wood anatomy of Symplocos and latitude and altitude. IAWA Bull. n.s. 2: 3–24.

Radlkofer. L. 1904. Über Tonerdekörper in Pflanzenzellen. Ber. Deutsch. Bot. Ges. 22: 216–224.

Ståhl, B. 1995. Diversity and distribution of Andean Symplocaceae. In: Churchill, S.P., Balslev, H., Forero, H., Luteyn, J.L. (eds.) Biodiversity and conservation of neotropical montane forests. Bronx, New York: The New York Botanical Garden, pp. 397–405.

Takhtajan, A.L. 1959. Die Evolution der Angiospermae. Jena: G. Fischer.

Takhtajan, A.L. 1973. Evolution und Ausbreitung der Blütenpflanzen. Jena: G. Fischer.

Takhtajan, A.L. 1997. See general references.

Webb, L.J. 1954. Aluminium accumulation in the Australian New Guinea flora. Austr. J. Bot. 2. 176–196.

Wehnert, A. 1906. Anatomisch-systematische Untersuchungen der Gattung Symplocos. Thesis. München.

Ternstroemiaceae

A.L. WEITZMAN, S. DRESSLER and P.F. STEVENS

Ternstroemiaceae Mirb. ex DC., Essai prop. méd. pl., ed. 2: 203 (1816).
Pentaphylacaceae Engler in Engler & Prantl, Nat. Pflanzenfam., Nachtr. 1: 214 (1897), nom. cons.[1]

Evergreen trees or shrubs, accumulating aluminium; indumentum of unicellular hairs. Leaves distichous or spiral, conduplicate-involute or supervolute, usually evergreen and coriaceous, margins entire or toothed, a small, deciduous seta near the apex of each tooth; stipules absent. Plant dioecious or flowers hermaphroditic. Inflorescences axillary, rarely pseudo-terminal, flowers solitary or fasciculate, rarely in short racemes of up to 15 flowers, pedicellate, regular, 5-merous, usually <2 cm in diameter, pedicel with more or less apical prophylls; sepals quincuncial, connate or distinct, persistent in fruit, usually thick, concave, subequal, rarely unequal; petals distinct or connate basally, quincuncial, thick or membranous; stamens 5–many, free, connate and/or adnate to corolla basally, filaments equal or unequal, wide and flat or thickened, often narrowing and incurved apically, shorter than to 4× as long as the anthers; anthers dithecal, basifixed, introrse, oblong, opening by longitudinal slits, rarely by flaps or pore-like slits, connective often elongated; carpels (2)3–5(6), ovary superior, rarely inferior, placentation axile, rarely apical or parietal; ovules anatropous to campylotropous, (1, 2)–many per carpel; style simple or with apical branches, or stylodia separate; stigmas usually separate. Fruit a berry, or rarely irregularly dehiscent, or a loculicidal capsule; seeds 1–numerous, reniform, small, reticulate or areolate, or larger, ellipsoid, ± smooth; endosperm slight to copious; embryo hippocrepiform to ± straight.

A family of about 12 genera and about 340 species; worldwide in tropical and subtropical regions, mostly in Asia and the neotropics, few in Africa.

VEGETATIVE MORPHOLOGY. Ternstroemiaceae are evergreen shrubs to usually rather small trees.

The hairs are unicellular, although sometimes aggregated into fascicles, as in some *Freziera*. In leaves of *Adinandra* the hair base may be large and reddish, making the lower surface, at least, distinctly punctate. However, not all hairs on the leaf have such bases (e.g. *A. clemensiae*), and it is unclear if the punctations on the leaves of all Ternstroemieae are comparable. *Pentaphylax* has perulate buds. Growth elsewhere in the family, as in *Ternstroemia* and *Cleyera*, is also clearly rhythmic, and the first leaves of the flush are small and fall off early. In many other genera, however, it is not easy to distinguish flushes of growth, at least on herbarium material. The leaves are distichous to spiral, this varying infragenerically in *Eurya* (Barker 1982). In *Ternstroemia* and *Anneslea* the expanded leaves are congested at the end of each innovation. The lamina margin is toothed to serrate, but this varies within genera, and perhaps within a species (*Cleyera japonica*). The teeth are glandular, black to red in colour and tend to be deciduous; they are sometimes described as setae (Hickey 1979). They are borne near the apex of the teeth. Interestingly, although the leaf blades of *Ternstroemia* are entire to crenulate, seedlings of *T. elongata* are drawn with strongly serrate leaves (de Vogel 1980). Venation is commonly brochidodromous, sometimes reticulodromous. The leaves are usually petiolate, but frequently only very shortly so or even sessile in genera such as *Freziera* and *Archboldiodendron*. Colleters are known in some genera, e.g. *Freziera*.

Leaf sclerification can be extensive (Keng 1962). Single cells or clusters of palisade cells may be sclerified, and may in turn be associated with sclerified hypodermal cells. In addition, single or clustered, more or less isodiametric, angular to shortly-branched sclereids are usually present in the spongy parenchyma; they are absent in *Balthasaria* and *Pentaphylax*. Both palisade sclereids and isodiametric sclereids may be densely clustered around veins or scattered evenly throughout the leaf and, in the dried leaf, are often visible as minute "papillae". Ternstroemieae have much-branched sclereids, whilst those of Freziereae are less branched to rounded (excep-

---

[1] See nomenclatural note below under "Affinities".

tions occur in *Adinandra* and *Cleyera*). On the upper surface of the lamina there may be one or two hypodermal layers of almost isodiametric cells. Mucilage has been reported in both upper and lower epidermes (Keng 1962). The palisade tissue consists of a variable number of layers. Stomata are present only on the lower surface and are usually anomocytic (Keng 1962). In some genera they tend to be cyclocytic (Kva ek and Walther 1984), and in *Freziera* there may be a considerable difference in size between subsidiary cells and epidermal cells (Weitzman 1987b); *Pentaphylax* has paracytic stomata. In general, Ternstroemieae and Freziereae show little epidermal variation (Kva ek and Walther 1984). Calcium oxalate druses are found throughout the plant, but are scanty in *Pentaphylax*.

Nodes in most Ternstroemiaceae are unilacunar with one trace (Beauvisage 1920; Keng 1962; Schofield 1968; Weitzman 1987b). The leaf trace leaves the stele as a flattened arc; lateral leaf bundles diverge from the ends of the trace, and there are from one to five bundles in the petiole. However, in *Archboldiodendron* and a few species of *Freziera*, the ends of the arc curve inwards adaxially; lateral leaf bundles diverge from the shoulders of the main bundle. *Freziera*, the only genus studied in detail, shows further variation in petiole anatomy, and trilacunar nodes also occur in the genus (Weitzman 1987b).

WOOD ANATOMY. Growth rings are usually indistinct. Vessels are usually solitary and vessel elements are 910–2340 µm long (measurements are species means), 30–95(–62) µm in tangential diameter, sometimes with spiral thickening, and with scalariform to opposite intervessel pits. *Balthasaria* and some species of *Eurya* have particularly wide vessels. Perforation plates are scalariform, with (10–)20–55(–108) bars. Vessel/ray pits are half to fully bordered (note that the pits in Theaceae lack a border or have only reduced borders, Liang and Baas 1990). Fibre tracheids are 1440–2930 µm long. Parenchyma is apotracheal diffuse to paratracheal. Prismatic crystals are occasional. Rays are 1–10 cells across, and are often of two different size classes; they are Kribs types heterogeneous I–III. The rays in *Adinandra* and especially *Cleyera* are narrow, usually one cell across (information is taken from Keng 1962; Baretta-Kuipers 1976; Carlquist 1984; Liang and Baas 1990, 1991).

INFLORESCENCE AND FLOWER STRUCTURE. Inflorescences are always axillary. In *Pentaphylax* and the Ternstroemieae there is only a single flower per axil of a reduced, non-photosynthetic leaf (but *Anneslea donnaiensis* may have flowers in the axils of photosynthetic leaves). In *Pentaphylax* such flowers are borne at the beginning of the innovation (Fig. 129A), and ordinary expanded leaves may never be produced later, so the inflorescence appears to be a raceme. In Freziereae flowers are either single or fasciculate (reduced axillary shoots) in the axils of expanded and sometimes also reduced leaves (Fig. 131A). The flowers may also be single in the axils of leaves of more or less developed shoots arising from the current innovation; almost all this variation can be seen within a single individual of *Cleyera japonica* (see also Keng 1962). The pedicels have two more or less apical prophylls; these may be persistent or caducous.

Flowers are either hermaphroditic or unisexual, although this latter condition may be difficult to observe, as in *Freziera* (see below). The flowers are pentamerous and radially symmetrical. Sepals and petals are opposite each other in Ternstroemieae. The sepals are quincuncial, persistent, and usually free, although in *Balthasaria* Verdcourt (1962) draws three sepals as being completely outside the others; there is no overlap of any of the sepals in what are depicted as two whorls. In most taxa there is a small apical seta on the sepals similar to those found on the leaf margins (see above), and these have been observed in *Pentaphylax* as well; setae may also occur on the margins of the sepals. The corolla is quincuncial, and is more or less campanulate and fleshy; there may be some sclereids. In *Freziera* (and some other taxa, as in *Eurya sandwicensis*) it is urceolate and very highly sclerified (Weitzman 1987b). The petals vary from free to more or less fused. *Archboldiodendron* alone has a biseriate corolla usually with ten petals (Fig. 130), although there may be as few as six, members of the inner series being absent (Kobuski 1940).

Stamens are usually indefinite in number, although in *Pentaphylax* and some *Eurya* there are only five, alternating with the petals. The filaments are commonly slightly connate at the base and slightly adnate to the petals; those of *Pentaphylax* are very broad and closely adpressed to the corolla, forming a tube. The stamens are whorled, although this can be difficult to see when they are connate, as in *Adinandra* (Barker 1980). In *Visnea* there are antepetalous stamen pairs (Payer 1857; Corner 1946; Ronse Decraene and Smets 1996). Although the family is sometimes characterised as having filaments at most only slightly longer than the anthers (Judd in Schoenenberger and Friis

2001), in Freziereae such as *Cleyera* they are appreciably longer (see also de Wit 1947). The anthers are quite often hairy, and the connective is often more or less produced, and it is especially conspicuous in *Anneslea*. Anthers of *Pentaphylax* open by flaps near the apex of the thecae, not simply by pores (Cronquist 1981; Takhtajan 1997; cf., amongst others, Gardner 1849; van Steenis 1955).

The gynoecium is generally superior, although inferior in *Anneslea* and *Symplococarpon*, and apparently inferior in *Visnea*, partly because the connate sepals closely invest the ovary in fruit. Carpel number is variable, although commonly three (De Wit 1947), and placentation varies from axile (the usual condition) to parietal, as in *Cleyera*. Ovule number varies considerably; when there are only a few, they are apical. There is either a style with a stigma with as many radii as carpels, as in *Pentaphylax*, or there may be as many style branches or stylodia as carpels. Infrageneric variation in this feature is common. The stigma is papillate.

FLORAL ANATOMY. There are sclereids in the sepals, petals and ovary. Crystal druses in anthers are reported (Tsou 1995). In *Cleyera*, sepals, petals and stamens all have single traces. A single dorsal trace and two ventral traces supply each carpel, the two ventral traces proceeding up the hollow style (the style is also hollow in *Adinandra* and *Pentaphylax*). *Symplocarpon* and *Anneslea*, with inferior ovaries, as well as *Visnea*, have recurrent vascular bundles from which the individual carpel traces diverge. This suggests that the ovary is immersed in receptacular or axial tissue (Keng 1962). Development of the androecium appears to be centrifugal in *Adinandra* (Corner 1976).

EMBRYOLOGY. Detailed anther and ovule development is known from only eight species of *Adinandra*, *Cleyera* and *Eurya* (Tsou 1995), although Corner (1976) includes some information on ovule morphology. Anther wall development is dicotyledonous, and the anther epidermis is heavily tanniniferous. Tapetal cells are 2-(*Eurya*) or 4-nucleate (*Cleyera*, *Adinandra*). Microspore initiation is simultaneous, and tetrahedral tetrads are usually produced. Ovules are amphitropous or campylotropous, and the ovule epidermis is more or less tanniniferous. Ovules are tenuinucellate and bitegmic, the integuments being three cells thick. However, there is a suggestion that *Pentaphylax* is crassinucellate (Mauritzon 1936). The apical ovules of *Visnea*, the Ternstroemieae

(Corner 1976) and *Pentaphylax* are apotropous, those of *Symplocarpon* epitropous. Orientation in *Eurya*, *Freziera* and *Cleyera* is variable. The micropyle is endostomal and the embryo sac is eight-celled, its formation being of the Polygonum type.

POLLEN MORPHOLOGY. Pollen grains are spheroidal to prolate, tricolporate and tectate; exine sculpturing is inconspicuous, varying from psilate to scabrate, rugulate, foveolate or rugose (Erdtman 1952; Keng 1962). The exine of *Euryodendron* is indistinctly rugose and sparsely and minutely perforate at the polar regions (Ying et al. 1993). Weitzman (1987b) found *Freziera* pollen to be very small (9–20 μm, usually about 10 μm long); that of *Balthasaria* is rather large (up to 28.5 μm long). The maximum reported is an equatorial diameter of 37.8 μm in *Ternstroemia mokof* (Lee 1987). *Pentaphylax* pollen is 14–16.5 μm and smooth, with thin tectum, poorly developed columellae and a thick endexine (Wei et al. 1999).

KARYOLOGY. Haploid numbers in *Ternstroemia* are 20 or 25, in Freziereae they are 12–13?, 15, 18, 21 (the commonest number), 22, 23, etc., suggesting polyploidy. However, the family is poorly known cytologically.

FRUIT AND SEED. *Pentaphylax* has loculicidal capsules and apparently wind-dispersed seeds. *Ternstroemia*, and probably *Anneslea*, have irregularly dehiscent fruits whose rupture may be caused by the expansion of the ripening seeds; when open, the seeds dangle and "the brilliant sarcotesta [is displayed] to birds" (Barker 1980: 23). The fruit of Freziereae is usually a berry. A few species of *Freziera* have drupes, with as many few-seeded pyrenes as there are carpels, while in *Visnea* the dry, indehiscent fruit is enclosed by the accrescent, more or less fleshy calyx. The fruits of Freziereae are probably eaten by birds or sometimes bats, as in *Adinandra* (Ridley 1930, cit. in Barker 1980). Seed size and number varies considerably. The fruits of *Symplococarpon* have a single seed ca. 5 mm long, while fruits of *Balthasaria* have hundreds of seeds ca. 1 mm long. *Ternstroemia* and *Anneslea* have fruits with rather few and large seeds (2–)5–10 mm long.

Testa structure is variable. The seeds of Ternstroemieae dry reddish and are reported to be arillate (e.g. Keng 1962). However, in both *Anneslea* and *Ternstroemia* there is a sarcotesta, and both have stout, multicellular, unthickened hairs. *Anneslea* has a sarcotesta, then a layer of thin-walled crystalliferous cells, and then a sclerotic mesotesta to 10 cells thick; other testal and tegmic cells are thin-walled. In *T. lowii*, the only species for which there is detailed information, there are pockets of "watery tissue with large cells" (Corner 1976: 506) on either side of the seed. The epidermal cells are small, the crystalline hypodermal cells are thickened, and again there is a well-developed sclerotic mesotesta. The testa of Freziereae and Pentaphylaceae is reticulate and black to brown in colour. In at least some Freziereae, the seeds are embedded in a fleshy placenta. *Cleyera japonica* has much enlarged exotestal cells, but *C. theoides* has very enlarged and thickened mesotestal cells with mucilaginous walls (Keng 1962). Other Freziereae have a thick-walled mesotesta which is one (*Adinandra*), two (*Eurya*) or 4–5 (*Visnea*) cells thick, the cells containing crystals; the last two genera have enlarged exotestal cells with more or less massive, U-shaped thickenings, while the exotesta of *Adinandra* is not notably thickened (Corner 1976). The inner walls of the exotestal cells of *Pentaphylax* are somewhat thickened, but underlying cells are thin-walled.

The embryo occupies the full length of the seed cavity, or almost so; although Takhtajan (1997) described the embryo of *Visnea* as minute, this is belied by the illustration in Corner (1976). The embryo is curved back on itself and hippocrepiform in Ternstroemieae and Pentaphylaceae. In Freziereae like *Freziera*, it is slightly curved, although it is hippocrepiform in *Freziera dudleyi*. Although the seed is clearly U-shaped in *Archboldiodendron* and *Eurya*, the embryo itself develops in only one arm of the "U", and hence is only slightly curved (Barker 1980), while in *Balthasaria* the seed and embryo are strongly curved, but are not as narrowly U-shaped as is common elsewhere.

Germination of *Cleyera* is epigeal, but the radicle soon stops growing; *Adinandra* also has epigeal germination, but there the behaviour of the radicle was not mentioned (Keng 1962). *Ternstroemia* has epigeal germination and a well-developed radicle (de Vogel 1980); in *T. bancana* the cotyledons seem to be short-lived (Corner 1976).

REPRODUCTIVE SYSTEMS AND POLLINATION. *Ternstroemia* is reported to be basically dioecious. However, the situation needs more study. Barker (1980: 14) qualified his description of its breeding system "?rarely monoecious, sometimes andro-dioecious or andro-monoecious . . . ," while Kobuski (1961: 263) noted of the Philippine species "the genus in the Philippines is, as far as I

know, dioecious, androdioecious, or perhaps, hermaphroditic. In most instances, three different specimens are necessary for a complete understanding of a single taxon" (see also Kobuski e.g. 1942a, 1942b, 1943, 1963). *Eurya* is dioecious. *Freziera* appears to be gynodioecious; however, all species for which there are sufficient data are functionally dioecious (Weitzman 1987a, 1987b).

The gynoecium in staminate flowers of *Ternstroemia* may be absent or represented by a withered and clearly non-functional ovary with styles (and rarely stigmas). Although the gynoecium in staminate flowers of *Freziera* appears to be functional, such flowers nearly always fall off soon after anthesis. The anthers of staminate flowers may open in bud (Weitzman 1987b). In both genera, staminodia in carpellate flowers look very much like filaments without anthers. Staminate flowers of *Eurya* lack or have only a very rudimentary gynoecium, while carpellate flowers usually lack staminodes.

The corolla is greenish to yellowish and the flowers are fairly small, usually less than 1.5 cm across, but substantially larger in Ternstroemieae and especially in *Balthasaria*, where they are up to 5 cm long, yellow-orange in colour, and perhaps even pollinated by birds. Generally, pollination is probably by bees or other insects, but little is known in detail. Vibrational ("buzz") pollination has been reported in two species of *Ternstroemia* (Bittrich et al. 1993). Although the family is supposed to lack nectaries, there is a presumably nectariferous ring around the base of the ovary in *Pentaphylax*, flowers of *Cleyera japonica* secrete substantial amounts of nectar within the staminal ring, and other species apparently have a nectary at the base of the ovary, while *Symplococarpon* is described as having a disc on top of the inferior ovary. In genera like *Ternstroemia* and *Adinandra*, the stamens closely surround the ovary, and one may suppose that any nectar would be hard to obtain.

PHYTOCHEMISTRY. Common proanthocyanidins, flavonols and ellagic acid are widespread, myricetin is rare; saponins are present, and large amounts of triterpenic acids (betulinic acid) are accumulated in the bark of *Eurya* and *Ternstroemia* spp. (the latter compounds are not known from Theaceae). Accumulation of aluminium and fluoride is general (Hegnauer 1973, 1990).

SUBDIVISION AND RELATIONSHIPS WITHIN THE FAMILY. Ternstroemiaceae as circumscribed here are divided into three tribes, Pentaphylaceae, Ternstroemieae and Freziereae.

Pentaphylaceae are the only Ternstroemiaceae to have regularly capsular fruits; these are loculicidal and probably plesiomorphic, as is the apparent absence of sclereids. The inflorescence of *Pentaphylax* is basically the same as that of Ternstroemieae; the cymose inflorescence of Sladeniaceae has little in common with either. The curved embryo of *Pentaphylax* also agrees with its position in Ternstroemiaceae.

Ternstroemieae and Freziereae are linked by their foliar sclereids. Ternstroemieae are morphologically very distinct in their pseudoverticillate leaves, short filaments, sepals opposite the petals, fruit type, etc. *Anneslea* is very like *Ternstroemia*, and it can be placed in a separate tribe only if the significance of its inferior ovary is overemphasized.

Freziereae are distinct enough because of their often short sclereids and baccate fruit, but generic limits in the tribe are very difficult. Bentham (1861: 55) noted that the then recognized genera of Ternstroemioideae were "so closely connected with each other that their distinct separation by positive characters is very difficult." Several genera have usually been considered to be separate for 150 years, yet as genera they are often distinguishable only by those who are very familiar with the individual species. The tribe has often been divided into two groups centred on *Adinandra* and *Eurya*. The character most frequently used to separate them has been stylar fusion, although from the descriptions it can be seen that it commonly varies even within genera. *Archboldiodendron*, which has separate stylodia, was segregated from *Adinandra* which has a single style (Kobuski 1940), yet no mention was made of other members of the tribe like *Eurya* which have separate stylodia. On the other hand, *Freziera* and *Cleyera* were segregated from *Eurya* largely on the basis of their single style! *Symplococarpon* has free stylodia and was segregated from *Cleyera*, which has a single style – but the former also has an inferior ovary.

Some genera of the Freziereae may not be monophyletic, but our understanding of the basic morphological variation of the group is poor, *Balthasaria* and *Euryodendron* being particularly little known. *Archboldiodendron*, *Visnea* and *Symplococarpon* are small genera, but are highly derived and consequently difficult to place. There may be relationships between *Adinandra* and Asian species of *Cleyera*, between Neotropical species of *Cleyera* and *Euryodendron*, and between *Freziera* and *Ternstroemiopsis* (here treated as a Hawaiian species of *Eurya*).

AFFINITIES. For details of the relationships of Theaceae s.l., in which Ternstroemiaceae have often been included, see the account of the former. Molecular data now place Theaceae s.l. in the asterids–Ericales (e.g. Morton et al. 1996). What then are the proper limits of Theaceae?

Within the old Theaceae, *Ternstroemia* and its relatives, whether or not including *Sladenia*, have long been considered separable from *Camellia* and its relatives, at least at some level. *Pentaphylax* has often not been considered part of Theaceae s.l. (Mattfeld 1942; Keng 1962; Liang and Baas 1990, 1991; Takhtajan 1997; see van Steenis 1955 for a summary). Its anthers are superficially like those of Diapensiaceae, placed in Theales (Takhtajan 1997); *Pentaphylax* and Theaceae are superficially similar, and both Theaceae and Diapensiaceae are included in Ericales here. The seed is Ericalean (Huber 1991), although the elaborated mesotesta is unlike that of Ericaceae, etc. There have been suggestions that Pentaphylacaceae go with Balsaminaceae, etc., also in Ericales (Nandi et al. 1998). Wei et al. (1999) compared the pollen of *Pentaphylax* with that of *Clematoclethra* (Actinidiaceae) – again, Ericales. Although *Pentaphylax* was associated with Cardiopteridaceae and *Gonocaryum* in Savolainen, Fay et al. (2000), the latter two are strongly associated with Aquifoliales in other analyses (Soltis et al. 2000; Kårehed 2001).

Recent studies confirm that Theaceae s.l. can easily be separated into two morphologically quite distinct groups, Theaceae s.str. and Ternstroemiaceae plus *Pentaphylax* and Sladeniaceae. There is no evidence that Theaceae s.l. are monophyletic, but both of these groups are monophyletic (Prince and Parks 2001; Anderberg et al. 2002), forming two smallish clades in a major polytomy in Ericales.

What are the limits of Ternstroemiaceae? Sladeniaceae (*Sladenia* and *Ficalhoa*) are a monophyletic group that differs from Ternstroemiaceae + *Pentaphylax* in such features as stem anatomy (deep-seated versus surficial phellogen), inflorescence type (cymose versus racemose inflorescence or single axillary flowers; see above) and embryo type (straight versus U-shaped). These and other differences allow the two to be readily distinguished. *Pentaphylax* and the other Ternstroemiaceae have similar anatomy, inflorescence and seeds. The only problem in combining the two is nomenclatural; the name Pentaphylacaceae is conserved. Conservation of Ternstroemiaceae will be proposed. The name Pentaphylaceae used by Hallier (1923: pp. 133, 176) is invalid, lacking a description or reference to one.

DISTRIBUTION AND HABITATS. Ternstroemiaceae prefer warm temperate to subtropical conditions, being commoner in tropical colline and montane habitats than in the lowlands. Some species of *Eurya* at times almost dominate subalpine habitats in New Guinea, although other species grow in the lowlands. Species of *Eurya* and *Freziera* in particular do well in disturbed habitats.

PALAEOBOTANY. *Paradinandra* was described (Schönenberger and Friis 2001) from flowers from the Late Cretaceous of southern Sweden. Although they did not place it in a family, they compared it with Ternstroemiaceae, specifically *Cleyera*, but their comparison was hindered by the prevailing belief that Ternstroemiaceae lack nectar and have very short filaments, while *Paradinandra* had relatively longer filaments and an intrastaminal nectary (see above, both are features of *Cleyera*). A perhaps more distinctive feature is its tricolpate pollen, an unusual feature for Ternstroemiaceae. A fossil flower and leaf fragments of *Pentaphylax* were described from Baltic amber (Conwentz 1886). Mai (1971) and Knobloch and Mai (1986) described fossil fruits and seeds from the Late Cretaceous of central Europe as *Allericarpus* and *Pentaphylax* resp. *Eurya*, *Visnea* and *Protovisnea*, although confirmation is needed; both Theaceae and Ternstroemiaceae appear to have been common in the lauraceous forests that were then widespread (summary in Grote and Dilcher 1989). *Cleyera* seed is known from the Middle Tertiary of Vermont (Tiffney 1994). Wijninga and Kuhry (1990) and Wijninga (1996) report seeds perhaps of *Freziera* from the Pliocene of Colombia near Bogotá. *Ternstroemia*-like leaves are widespread in the northern hemisphere, but such leaves (in the form genus *Ternstroemites*; also the form genera *Ternstroemiacinium* and *Ternstroemioxylon*) resemble genera in both Theaceae and Ternstroemiaceae (Grote and Dilcher 1989).

ECONOMIC IMPORTANCE. Species like *Eurya japonica*, *Eurya chinensis*, *Ternstroemia japonica* and *Cleyera japonica* are used as ornamentals.

CONSPECTUS OF TERNSTROEMIACEAE

I. Tribe Pentaphylaceae
 Genus 1
II. Tribe Ternstroemieae
 Genera 2–3
III. Tribe Freziereae
 Genera 4–12

KEY TO THE GENERA

1. Stamens 5; fruit a loculicidally dehiscent capsule
   1. *Pentaphylax*
 – Stamens (5–)15–many; fruit a berry or irregularly dehiscent, rarely leathery capsule   2
2. Leaves pseudoverticillate; flowers solitary; seeds often 4 mm long, with reddish sarcotesta   3
 – Leaves scattered; flowers usually in axillary fascicles of 2 or more; seeds usually <4 mm long, with brown or black reticulate testa   4
3. Flowers borne below the leaves; ovary superior; petals nearly free   **2. *Ternstroemia***
 – Flowers borne above the leaves; ovary inferior ; petals often connate in their basal half   **3. *Anneslea***
4. Petals (6–9)10   **6. *Archboldiodendron***
 – Petals 5   5
5. Corolla tubular, at least 3 cm long; style longer than 2 cm
   12. *Balthasaria*
 – Corolla urceolate, campanulate, rarely tubular, up to 1.5 cm long; style shorter than 2 cm long   6
6. Ovary 2-locular, inferior; calyx on top of fruit
   9. *Symplococarpon*
 – Ovary (2)3–5-locular, superior; fruit base ± surrounded by the calyx   7
7. Sepals ca. 1 cm long, half connate (ovary appears to be inferior), enclosing the woody indehiscent fruit   **11. *Visnea***
 – Sepals often shorter, free, persisting at the base of the baccate fruit   8
8. Ovaries 2–3-locular; ovules 8–16; filaments more than twice as long as the anthers   **5. *Cleyera***
 – Ovaries 3–5-locular, ovules 20–100; filaments usually up to twice as long as anthers   9
9. Staminate flowers with clearly non-functional gynoecium; carpellate flowers often lacking staminodia; styles separate   **7. *Eurya***
 – All flowers with apparently functional gynoecia; carpellate flowers (when recognisable) with staminodia; style single, divided or not   10
10. Leaves spiral; flowers with style about the length of the ovary, ca 3 mm long   **10. *Euryodendron***
 – Leaves distichous, if spiral, styles apically free   11
11. Corolla campanulate; style longer than ovary, to 1.5 cm long; leaves entire or serrate   **4. *Adinandra***
 – Corolla urceolate; style shorter than ovary, ca. 2 mm long; leaves serrate   **8. *Freziera***

Basal condition for the family

Growth rhythmic; hairs unicellular; leaves exstipulate; flowers 5-merous, regular, ± campanulate; sepals 5, free, quincuncial; petals 5, connate basally, quincuncial; placentation axile; embryo long, U-shaped, endosperm +.

1. TRIBE PENTAPHYLACEAE P.F. Stevens & A.L. Weitzman, trib. nov.

Pentaphylacaceae Engler in Engler & Prantl, Nat. Pflanzenfam., Nachtr. 1: 214 (1897).

Leaves scattered; flowers single, in the axils of reduced leaves; stamens 5, opposite the sepals.

Fig. 129. Ternstroemiaceae. *Pentaphylax euryoides*. **A** Flowering branch. **B** Flower. **C** Stamens. **D** Dehisced fruit. **E** Young fruit. **F** Pistil. **G** Seed. Drawn by R. van Crevel. (van Steenis 1955)

Fruit a loculicidal capsule, the midribs of the carpels forming long "teeth" between the valves, columella persistent; cotyledons longer than radicle.

1. *Pentaphylax* Gardner & Champ.　　　Fig. 129

Pentaphylax Gardner & Champ., Hooker's J. Bot. Kew Gard. Misc. 1: 244 (1849); Mattfeld in Engler & Prantl, Nat. Pflanzenfam. ed. 2, 20b: 13–21 (1942); van Steenis, Fl. Males. I, 5: 121–124 (1955).

Trees. Leaves spiral, entire. Flowers subsessile; prophylls persistent; petals emarginate; stamens 5, filaments very broad, anthers incurved, dehiscing by two valves; ovary 5-locular, nectary basal, 2 apical pendulous ovules per loculus, style +. Seed reticulate. One sp., *P. euryoides* Gardner & Champ., China (Kwangtung and Hainan) to Sumatra, scattered.

Ternstroemieae + Freziereae

Stamens (5–)<8; fruit a berry; radicle longer than cotyledons, cotyledons incumbent.

2. TRIBE TERNSTROEMIEAE DC. (1822).

Leaves pseudoverticillate, often with black dots on the lower surface; sclereids usually much branched. Flowers single, in the axils of non-expanded leaves, rather large (usually >1.5 cm across); sepals opposite petals, filaments shorter than the anthers; ovules pendulous. Fruit irregularly dehiscent; seeds usually >4 mm long, reddish-brown; testa fleshy.

2. *Ternstroemia* Mutis ex L. f.

Ternstroemia Mutis ex L. f., Suppl. Pl.: 39, 264 (1782), nom. cons.; Kobuski, J. Arnold Arbor. 23: 298–343, 464–478 (1942), 24: 60–76 (1943), 42: 81–86, 263–275, 426–429 (1961), 44: 421–435 (1963); Barker, Brunonia 1: 14–42 (1980); Hung T. Chang & Kuan, Higher plants of China 4: 620–624 (2000).

Shrubs or trees. Leaf margin entire, rarely crenulate, dots on lower surface often lacking. Flowers borne below expanded leaves, prophylls deciduous or persistent; petals at most connate basally; stamens many, 3-seriate; anthers retuse to apiculate, base hardly raised; ovary 1–3-locular, or 4–6-locular with false septae, 1–20 ovules per loculus, stylodia or style +. Endosperm sometimes sparse. Ca. 100 spp., Sri Lanka to SE and E Asia, tropical and subtropical Americas (most species), Africa (ca. 2 species).

3. *Anneslea* Wall.

Anneslea Wall., Pl. Asiat. Rar. 1: 5, pl. 5 (1829), nom. cons.; Kobuski, J. Arnold Arbor. 33: 79–90 (1952).
Paranneslea Gagnep.

Trees. Leaf margin crenulate-serrulate. Flowers borne above expanded leaves; prophylls (sub)persistent; petals connate half way; stamens many, 1–2-seriate; anther base cordate, apiculus very long; ovary 2–3-locular, inferior, 2–10 ovules per loculus, style 2–5-fid at apex. Three spp., China, SE Asia to Sumatra.

3. TRIBE FREZIEREAE DC. (1822).

Growth rarely obviously rhythmic; leaves scattered, sclereids usually only slightly branched; flowers 1 or more together, at least sometimes in the axils of expanded leaves; placentae median (apical); seeds <4(-6) mm long, brown or black.

4. *Adinandra* Jack

Adinandra Jack, Malayan Misc. 2(7): 49 (1822); Kobuski, J. Arnold Arbor. 28: 1–98 (1947); Hung T. Chang & Kuan, Higher plants of China 4: 624–629 (2000).

Trees or shrubs. Leaves distichous, margins serrate. Flowers 1–2 together; prophylls deciduous to persistent; petals basally connate; stamens 15–many, 1–5-seriate (in bundles), more or less connate, anthers apiculate, base ± rounded; ovary (2)3(4)5-locular, many ovules per carpel; style (stylodia) up to 15 mm long. Embryo also J-shaped. Ca. 80 spp.; E and SE Asia to Malesia.

5. *Cleyera* Thunb.

Cleyera Thunb., Nov. Gen. Pl. 3: 68 (1783), nom. cons., p.p. emend. Sieb. & Zucc., Fl. Jap. 153, pl. 81 (1841); Kobuski, J. Arnold Arbor. 18: 118–129 (1937), 22: 395–416 (1941), Hung T. Chang & Kuan, Higher plants of China 4: 629–633 (2000).

Trees. Leaves distichous, margins entire to serrate. Flowers hermaphroditic, 1–4 together; prophylls deciduous; petals connate basally; stamens 25, uniseriate; anthers opening by short gaping slits, apiculate, base cuneate; ovary 2–3-locular, placentation axile, but sometimes seemingly parietal by long proliferating placentar strands, 5–10 ovules per loculus; style +, divided half way or not. Eight spp., India to Japan, Mexico, Central America, Caribbean.

6. *Archboldiodendron* Kobuski   Fig. 130

Archboldiodendron Kobuski, J. Arnold Arbor. 21: 140 (1940); Barker, Brunonia 3: 47–54 (1980).

Trees. Leaves distichous, margins serrate, petioles short. Flowers hermaphroditic, single; prophylls deciduous; petals (6–)10, biseriate, basally connate; stamens 30, uniseriate, anther apex acute, base rounded; ovary 5(–7)-locular, placentae T-shaped, many ovules per loculus, styles free, short. Embryo curved. One sp., *A. calosericeum* Kobuski, montane New Guinea.

7. *Eurya* Thunb.

Eurya Thunb., Nov. Gen. Pl. 67 (1783); Kobuski, J. Arnold Arbor. 16: 347–352 (1935), 25: 299–359 (1938), 20: 361–374 (1939); De Wit, Bull. Jard. Bot. Buit. III, 17: 329–375 (1947);

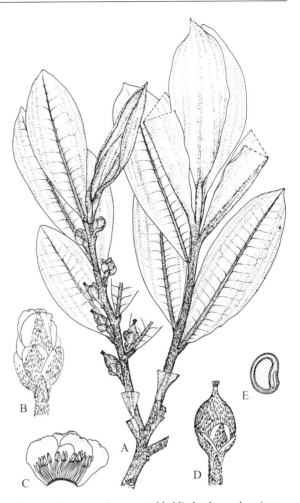

Fig. 130. Ternstroemiaceae. *Archboldiodendron calosericeum.* **A** Fruiting branch. **B** Flower. **C** Corolla spread open, with stamens. **D** Immature fruit with persistent style and style branches. **E** Seed, vertical section. Drawn by T. Iwagu. (Barker 1980)

Barker in van Royen, Alpine Fl. New Guinea 2: 1402–1454 (1982); Hung T. Chang & Kuan, Higher plants of China 4: 633–655 (2000).

Ternstroemiopsis Urban

Trees or shrubs; leaves distichous, rarely spirally arranged, margins serrate. Plant dioecious; flowers 1–8 together; prophylls persistent; petals connate ca. 1/3–1/2, staminate flowers: stamens 5–25, uniseriate, anthers often locellate, apex apiculate to obtuse, base rounded, rarely pistillode vestigial; pistillate flowers: staminodes rarely +, ovary 3(–5)-locular, 4–many ovules per loculus, style branched. Embryo curved. 50–100 spp., S China, SE Asia and nearby Pacific islands. The genus is badly in need of revision.

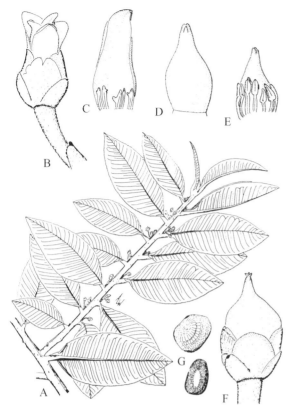

Fig. 131. Ternstroemiaceae. *Freziera carinata.* **A** Flowering branch. **B** Flower. **C** Petal of female flower, stamens adnate. **D** Pistil of female flower. **E** Pistil and stamens of male flower. **F** Fruit. **G** Seeds, side view and chalazal view. Drawn by A.L. Weitzman. (Weitzman 1987a)

8. *Freziera* Willd.   Fig. 131

Freziera Willd., Sp. Pl. 2(2): 1179 (1799) nom. cons.; Weitzman, Systematics of *Freziera* Willd. (Theaceae), Ph.D. thesis, Harvard University (1987).
Patascoya Urb.
Killipiodendron Kobuski

Small trees or shrubs. Leaves distichous, margins serrate, petioles short. Plant functionally dioecious; flowers 1–14 together; prophylls usually persistent. Corolla urceolate, petals free or slightly connate basally; staminate flowers: stamens (8–)15–35(–48), uniseriate, anthers apiculate, ± rounded basally; pistillode +; carpellate flowers: staminodes +; ovary (2–)3–5(–6)-locular, (6–)many ovules per loculus, style short, with short lobes. Fruit a drupe. Embryo curved, sometimes slightly, sometimes straight. 57 spp., S Mexico and the West Indies through Central and South America to S Bolivia, and E to the Guayana Highland; primarily montane. *F. dudleyi* Gentry has many characters anomalous in the genus.

9. *Symplococarpon* Airy-Shaw

Symplococarpon Airy-Shaw in Hooker's Icon. Pl. 34: pl. 3342 (1937); Kobuski, J. Arnold Arbor. 22: 188–196 (1941).

Trees. Leaves distichous, margins entire to slightly serrate. Plant hermaphroditic; flowers 1–5 together; prophylls persistent; petals ± free; stamens 25–40, uniseriate, filaments long, anthers caudate apically, base cuneate; ovary 2-locular, inferior, disk on top, 1(–2) pendulous ovules per loculus, style free or stylodia +. Nine spp., Mexico, Central America and NW South America.

10. *Euryodendron* Hung T. Chang

Euryodendron Hung T. Chang, Acta Sci. Nat. Univ. Sunyatseni 1963(4): 129 (1963); Ying et al., The endemic seed plants of China: 693–695 (1993); Hung T. Chang & Kuan, Higher plants of China 4: 633 (2000).

Trees. Leaves spiral, margins serrate. Flowers hermaphroditic, 1–2 together, prophylls persistent; petals ?free; stamens ca. 25, uniseriate, anthers hairy, apex acute, base rounded; ovary 3-locular, many ovules per loculus; style +. Embryo unknown. One sp., *E. excelsum* Hung T. Chang, Guangxi, China, low alt. The margins of the sepals are fimbriate.

11. *Visnea* L. f.

Visnea L. f., Suppl.: 36 (1782); Kobuski, J. Arnold Arbor. 33: 188–191 (1952).

Small trees or shrubs. Leaves distichous, margins serrate. Flowers hermaphroditic?, 1–3 together; prophylls persistent; sepals half connate; petals connate basally; stamens 12(–21), uniseriate, anthers apiculate, base rounded; ovary (2–)3-locular, 2–3 pendent ovules per loculus, stylodia free. Embryo J-shaped. One sp., *V. mocanera* L. f., from cloud/laurel forests of the Canary Islands and Madeira.

12. *Balthasaria* Verdc.

Balthasaria Verdc., Kew Bull. 23: 469 (1969).
Melchiora Kobuski, J. Arnold Arbor. 37: 154 (1956); Verdc., Fl. Trop. E. Afr. Theaceae: 3–6 (1962), non *Melchioria* Penz. & Sacc.

Large trees. Leaves spiral, margins serrate, petioles short. Flowers hermaphroditic, single; prophylls persistent; petals 3 cm long, tubular, free; stamens 15–35, uniseriate, anthers long-apiculate, base rounded; ovary 4–5 locular, many ovules per loculus, style >2 cm long. Perhaps one sp., *M. schliebenii* (Melchior) Verdc., montane E Africa.

Selected Bibliography

Anderberg A.A. et al. 2002. See general references.

Baretta-Kuipers, T. 1976. Comparative wood anatomy of Bonnetiaceae, Theaceae and Guttiferae. Leiden Bot. Ser. 3: 76–101.

Barker, W.R. 1980. Taxonomic revisions in Theaceae in Papuasia. I. *Gordonia, Ternstroemia, Adinandra* and *Archboldiodendron*. Brunonia 3: 1–60.

Barker, W.R. 1982. Theaceae. In: van Royen, P., The Alpine Flora of New Guinea. Vol. 3: Taxonomic part. Winteraceae to Polygonaceae. Vaduz: J. Cramer, pp. 1397–1454.

Beauvisage, L. 1920. Contribution à l'étude anatomique de la famille des Ternstroemiacées. Tours: Arrault.

Bentham, G. 1861. Notes on Ternstroemiaceae. J. Proc. Linn. Soc. Bot. 5: 53–65.

Bittrich, V., Amaral, M.C.E., Melo, G.A.R. 1993. Pollination biology of *Ternstroemia laevigata* and *T. dentata* (Theaceae). Pl. Syst. Evol. 185: 1–6.

Carlquist, S. 1984 (1985). Wood anatomy and relationships of Pentaphylacaceae: significance of vessel features. Phytomorphology 34: 84–90.

Conwentz, H. 1886. Die Angiospermen des Bernsteins. Danzig: W. Engelmann.

Corner, E.J.G.H. 1946. Centifugal stamens. J. Arnold Arbor. 27: 423–437.

Corner, E.J.G.H. 1976. See general references.

Cronquist, A. 1981. See general references.

De Vogel, E.F. 1980. Seedlings of dicotyledons. Wageningen: Pudoc.

De Wit, H.C.D. 1947. A revision of the genus *Eurya* Thunb. (Theac.) in the Malay Archipelago (including New Guinea and south of the Philippines). Bull. Jard. Bot. Buitenzorg, ser. III, 17: 329–375.

Erdtman, G. 1952. See general references.

Gardner, G. 1849. Descriptions of some new genera and species of plants, collected in the island of Hong Kong by Capt. J. G. Champion, 95th Regt. Hooker's J. Bot. Kew Gard. Misc. 1: 240–246.

Grote, P.J., Dilcher, D.L. 1989. Investigations of angiosperms from the Eocene of North America: a new genus of Theaceae based on fruit and seed remains. Bot. Gaz. 150: 190–206.

Hallier, H. 1923. Beiträge zur Kenntnis der Linaceae (DC. 1819) Dumort. Beih. Bot. Centralbl. 39(ii): 1–178.

Hegnauer, R. 1973, 1990. See general references.

Hickey, L.J. 1979. A revised classification of the architecture of dicotyledonous leaves. In: Metcalfe, C.R., Chalk, L. (eds.) Anatomy of the dicotyledons, ed. 2, vol. 1. Oxford: Oxford University Press, pp. 25–39.

Hitzemann, C. 1886. Beiträge zur vergleichende Anatomie der Ternstroemiaceen, Dilleniaceen, Dipterocarpaceen und Chlaenaceen. Osterode: von Giebel & Oehlschlägel.

Huber, H. 1991. Angiospermen. Leitfaden durch die Ordnungen und Familien der Bedecktsamer. Stuttgart: Gustav Fischer.

Hung T. Chang, Ling Lai Kuan, 2000. Theaceae. In: Fu, L., Chen, T., Lang, K., Hong, T., Lin, Q. (eds.) Higher plants of China, vol. 4. Quingdao: Quingdao Publishing House, pp. 572–656 (in Chinese).

Kårehed, J. 2001. Multiple origin of the tropical forest tree family Icacinaceae. Am. J. Bot. 88: 2259–2274.

Keng, H. 1962. Comparative morphological studies in the Theaceae. Univ. Calif. Publ. Bot. 33: 269–384.

Knobloch, E., Mai, D.H. 1986. Monographie der Früchte und Samen in der Kreide von Mitteleuropa. Rozpr. Ustr. Ust. Geol. 47: 1–219.

Kobuski, C.E. 1940. Studies in the Theaceae, V. The Theaceae of New Guinea. J. Arnold Arbor. 21: 134–162.

Kobuski, C.E. 1942a. Studies in Theaceae, XII. Notes on the South American species of *Ternstroemia*. J. Arnold Arbor. 23: 298–343.

Kobuski, C.E. 1942b. Studies in Theaceae, XIII. Notes on the Mexican and Central American species of *Ternstroemia*. J. Arnold Arbor. 23: 464–478.

Kobuski, C.E. 1943. Studies in Theaceae, XIV. Notes on the West Indian species of *Ternstroemia*. J. Arnold Arbor. 24: 60–76.

Kobuski, C.E. 1961. Studies in Theaceae, XXXII. A review of the genus *Ternstroemia* in the Philippine Islands. J. Arnold Arb. 42: 263–275.

Kobuski, C.E. 1963. Studies in Theaceae, XXXV. Two new species of *Ternstroemia* from the Lesser Antilles. J. Arnold Arb. 44: 434–435.

Kva ek, Z., Walther, H. 1984. Nachweis tertiärer Theaceen Mitteleuropas nach blatt-epidermalen Untersuchungen. I. Teil - Epidermale Merkmalskomplexe rezenter Theaceae. Feddes Repert. 95: 209–227.

Lee, S. 1987. A palynotaxonomic study on the Korean Theaceae. Korean J. Bot. 30: 215–223.

Liang, D., Baas, P. 1990. Wood anatomy of trees and shrubs from China II. Theaceae. I.A.W.A. Bull. II, 11: 337–378.

Liang, D., Baas, P. 1991. The wood anatomy of the Theaceae. I.A.W.A. Bull. II, 12: 333–353.

Mai, D.H. 1971. Über fossile Lauraceae und Theaceae in Mitteleuropa. Feddes Repert. 82: 313–341.

Mattfeld, J. 1942. Pentaphylacaceae. In: Engler, A., Prantl, K. Nat. Pflanzenfam., ed. 2, 20b. Leipzig: W. Engelmann, pp. 13–21.

Mauritzon, J. 1936. Zur Embryologie und systematischen Abgrenzung der Reihen Terebinthales und Celastrales. Bot. Not. (Lund) 1936: 161–212.

Melchior, H. 1925. Theaceae. In: Engler, A., Prantl, K. Nat. Pflanzenfam., ed. 2, 21. Leipzig: W. Engelmann, pp. 109–154.

Metcalfe, C.R., Chalk, L. 1950. See general references.

Morton, C.M. et al. 1996. See general references.

Nandi, O.I., Chase, M.W., Endress, P.K. 1998. A combined cladistic analysis of angiosperms using *rbc*L and non-molecular data sets. Ann. Missouri Bot. Gard. 85: 137–212.

Payer, J.-B. 1857. Traité d'organogénie comparée de la fleur. Paris: Victor Masson.

Prince, L.M., Parks, C.M. 2001. Phylogenetic relationships of Theaceae inferred from chloroplast DNA data. Am. J. Bot. 88: 2309–2320.

Ridley, H.N. 1930. The dispersal of plants throughout the world. Ashford, Kent.

Ronse Decraene, L.P., Smets, E.F. 1996. The morphological variation and systematic value of stamen pairs in the Magnoliatae. Feddes Repert. 107: 1–17.

Savolainen, V., Fay, M.F. et al. 2000. See general references.

Schofield, E.K. 1968. Petiole anatomy of Guttiferae and related families. Mem. New York Bot. Gard. 18: 1–55.

Schönenberger, J., Friis, E.M. 2001. Fossil flowers of Ericalean affinity from the Late Cretaceous of Southern Sweden. Am. J. Bot. 88: 467–480.

Soltis, D.E. et al. 2000. See general references.

Takhtajan, A.L. 1997. See general references.

Tiffney, B.H. 1994. Re-evaluation of the age of the Brandon lignite (Vermont, USA) based on plant megafossils. Rev. Palaeobot. Palynol. 82: 299–315.

Tsou, C.-H. 1995. Embryology of Theaceae – anther and ovule development of *Adinandra*, *Cleyera* and *Eurya*. J. Plant Res. 108: 77–86.

Van Steenis, G.G.G.J. 1955. Pentaphylacaceae. In: van Steenis, G.G.G.J. (ed.) Flora males. I, 5: 121–124. Leyden: Noordhoff.

Verdcourt, B. 1962. Theaceae. In: Hubbard, C.E., Milne-Redhead, E. (eds.) Flora of Tropical East Africa. Theaceae. London: Crown Agents for Overseas Governments, pp. 1–8.

Wawra von Fernsee, H.R. 1886. Ternstroemiaceae. In: Martius, C.F.P. (ed.) Flora brasiliensis, vol. 12(1). München: F. Fleischer, pp. 261–334, pl. 52–68.

Wei, Z.-X., Li, D.-Z., Fan, X.-K., Zhang, X.-L. 1999. Pollen ultrastructure of Pentaphylacaceae and Sladeniaceae and their relationships to the family Theaceae. Acta Bot. Yunn. 21: 202–206, pl. 1–3 (in Chinese).

Weitzman, A.L., 1987a. Taxonomic studies in *Freziera* (Theaceae), with notes on reproductive biology. J. Arnold Arbor. 68: 323–334.

Weitzman, A.L. 1987b. Systematics of *Freziera* Willd. (Theaceae). Ph.D. thesis, Department of Organismic and Evolutionary Biology, Harvard University.

Wijninga, V.M. 1996. Paleobotany and palynology of Neogene sediments from the High Plain of Bogotá (Colombia). Wageningen: Ponsen & Looijen BV.

Wijninga, V.M., Kuhry, P. 1990. A Pliocene flora from the Subachoque Valley (Cordillera Oriental, Colombia). Rev. Palaeobot. Palynol. 62: 249–290.

Ying, T.-S., Zhang, Y.-L., Boufford, D.E. 1993. The endemic genera of seed plants of China. Peking: Science Press.

Tetrameristaceae

K. Kubitzki

Tetrameristaceae Hutch., Fam. Fl. Pl. ed. 2: 277, fig. 140 (1959).

Trees or shrubs; nodes trilacunar. Leaves alternate, simple, entire, pinnately veined, petiolate or sessile; leaf bases decurrent; stipules wanting. Inflorescences axillary, pedunculate, umbelliform or compactly corymbiform racemes; flowers 4- or 5-merous throughout, small, with 2 persistent or caducous prophylls inserted directly below the sepals; sepals distinct, imbricate, with numerous scattered pits near the middle of the upper surface; petals distinct, imbricate, not much if at all longer than the sepals; filaments applanate, shortly connate at base; anthers tetrasporangiate, introrse, dehiscing by longitudinal slits; gynoecium syncarpous, 4–5-carpellate; ovary 4–5-locular, capped by a terminal style with a punctate or minutely lobed stigma; ovules solitary in each locule, axile-basal, anatropous, bitegmic, tenuinucellate. Fruit a 4–5-seeded coriaceous berry; seeds relatively large, with copious endosperm surrounding a straight, basal embryo; hypocotyl much longer than cotyledons.

Two genera, the monotypic *Pentamerista* in northern South America and *Tetramerista* (1 [3?] spp.) in SE Asia-Malesia.

MORPHOLOGY AND ANATOMY. Young leaves are provided with marginal glands. In *Tetramerista* the leaves are puncate with black glands below (Hutchinson 1959), and *Pentamerista* has pocket domatia along the median and lateral veins. The cortex of young stems contains numerous prominent isolated masses of stone cells. Raphide cells are commonly found in the parenchyma and especially frequent in the palisade leaf tissue. Stomata are mostly anomocytic. Sclereids are wanting. In the wood, vessel elements have predominantly simple perforations; rays are 2–3 cells wide, heterocellular Kribs type I; axial parenchyma is apotracheal diffuse to diffuse-in-aggregates (Maguire et al. 1972). Both genera agree in all essential features, which is confirmed by H.-G. Richter (pers. comm.) who compared wood specimens of the two genera available in the Bundesforschungsanstalt für Forst- und Holzwirtschaft, Hamburg.

Both the sepals and petals of *Pentamerista* are quincuncially arranged. Maguire et al. (1972) suggest that in *Tetramerista* it is the fifth sepal and petal that have been reduced. A derived state for *Tetramerista* would correlate with the lack of scalariform perforations of vessel elements, in contrast to the condition in *Pentamerista*.

Both genera have flask-shaped glands at the basis of the sepals, which appear as pores or slits on the inner surface and obviously function as nectaries; similar structures are known from *Pelliciera*.

EMBRYOLOGY, CHROMOSOMES, POLLINATION. Unknown.

POLLEN MORPHOLOGY. Pollen is tricolporate, suboblate, and reticulate (Erdtman 1952; Maguire et al. 1972).

AFFINITIES. *Tetramerista* has variously been included in families such as Ochnaceae, Marcgraviaceae or Theaceae and, after elevating *Tetramerista* to family rank, an affinity with Pellicieraceae (Maguire et al. 1972) and Theaceae (Takhtajan 1997) has been proposed. Both plastid and nuclear gene sequences provide strong support for a clade comprising Balsaminaceae, Marcgraviaceae, Pellicieraceae and Tetrameristaceae (Savolainen, Fay et al. 2000; Soltis et al. 2000; Anderberg 2002), with Pellicieraceae being closest to Tetrameristaceae.

KEY TO THE GENERA

1. Flowers 5-merous; prophylls and sepals early caducous; seed obtusely triangular in cross section. N South America   **1. *Pentamerista***
- Flowers 4-merous; prophylls and sepals persistent; seed compressed, broadly oblong in cross section. SE Asia/Malesia   **2. *Tetramerista***

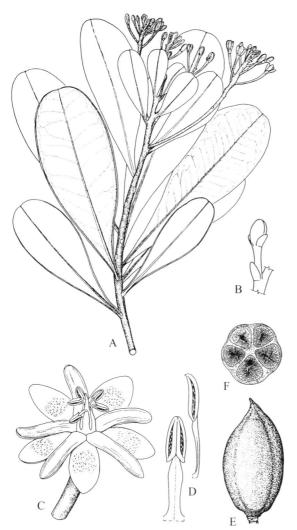

Fig. 132. Tetrameristaceae. *Pentamerista neotropica*. **A** Florif-
erous branch. **B** Flower with bract and prophylls. **C** Flower.
D Stamens with subsurface raphide bundles visible. **E** Fruit.
F Same, transverse section. (Maguire et al. 1972)

1. *Pentamerista* Maguire   Fig. 132

Pentamerista Maguire, Mem. New York Bot. Gard. 23: 187 (1972).

Shrub or small tree rarely up to 15 m high. One species, *P. neotropica* Maguire, lowland savannah edges of Upper Orinoco-Casiquiare-drainage, Venezuela, Colombia.

2. *Tetramerista* Miq.

Tetramerista Miq., Fl. Ind. Bat.: 534 (1860); Ic. Bog. 252 (4): tab. 83 (1901); Maguire, op.c.: 187; Sugan in Tree Fl. Sabal & Sarawak 2: 379–382 (1996).

Tree up to 30 m high; seeds ruminate (van Balgooy 1997). Malaya, Sumatra, Borneo. Probably only one species, *P. glabra* Miq., in lowland marshes and forests up to 1000 m altitude.

Selected Bibliography

Anderberg, A.A. et al. 2002. See general references.
Balgooy, M. van 1997. Malesian seed plants, vol. 1: spot characters. Leyden: Rijksherbarium.
Erdtman, G. 1952. See general references.
Hutchinson, J. 1959. The families of flowering plants, 2nd edn., vol. 1. Oxford: Clarendon Press.
Maguire, B., de Zeeuw, C., Huang, Y.-c. [sic], Clare, C.C. Jr. 1972. Tetrameristaceae. In: The Botany of the Guayana Highland, part IX. Mem. New York Bot. Gard. 23: 165–192.
Savolainen, V., Fay, M.F. et al. 2000. See general references.
Soltis, D.E. et al. 2000. See general references.
Takhtajan, A. 1997. See general references.

Theaceae

P.F. Stevens, S. Dressler and A.L. Weitzman

Theaceae Mirb., Bull. Soc. Philom. 3: 381 (1813).

Trees or shrubs, accumulating aluminium; indumentum of unicellular hairs or 0. Leaves spiral or distichous, simple, usually evergreen, usually coriaceous, margins toothed, rarely entire, with a small, deciduous gland associated with each tooth; stipules absent. Flowers hermaphroditic, solitary, axillary, large and showy, prophylls two, or several bracteoles intergrading with calyx and corolla; calyx of five or more sepals, imbricate, connate or distinct basally, often persistent in fruit, usually thick, concave, equal to unequal; corolla of five, rarely numerous petals, distinct or connate basally, imbricate; stamens 20+, free, rarely connate, often adnate to base of corolla, anthers versatile, rarely basifixed, opening by longitudinal slits; pollen tricolporate, pseudopollen present; gynoecium syncarpous, (3–)5(–10)-carpellate, ovary superior, placentation mostly axile; 2–few ovules/carpel, ovules bitegmic, tenuinucellate, styles simple, branched, or stylodia, stigmas usually lobed. Fruit a loculicidal capsule, rarely irregularly dehiscent or a drupe, columella persistent, rarely none; seeds few, sometimes winged; testa vascularised, more or less lignified; endosperm nuclear, usually slight; embryo straight; germination epigeal, rarely hypogeal; $n = 15, 18$. About 7 genera and 195–460 species; most speciose in Southeast Asia, but also Indo-Malesia, SE U.S.A., and the Caribbean and tropical America.

CHARACTERS OF RARE OCCURRENCE. *Gordonia imbricata* and a few other species have the connective produced at the anther apex. Hairy seeds are known in *Camellia* sections *Tuberculata* and *Pseudocamellia*. *Apterosperma* has reniform seeds.

VEGETATIVE MORPHOLOGY. The leaves are spiral or distichous, involute or supervolute in bud, and evergreen, or sometimes deciduous, and the buds are usually perulate (but in *Stewartia* leaves may be deciduous or evergreen and buds perulate or not). Venation is pinnate and more or less craspedromous. The leaf teeth are crowned by glandular, deciduous tips, the Theoid leaf tooth.

VEGETATIVE ANATOMY. Druses or sometimes single crystals of calcium oxalate may be found throughout the plant. The cork is usually deep-seated, being initiated near the pericycle, but it is subepidermal in *Franklinia*, *Schima*, and *Gordonia* s. str. The pericyclic sheath is either collenchymatous, lignified, or with groups of fibres. Nodes are unilacunar and with a single trace, although cotyledons may have three or more traces. The petiole bundle is more or less U- or V-shaped, but there may be additional smaller lateral bundles. Much-thickened sclereids are common in stem cortex, petiole and lamina, and they are usually branched and with pointed, rarely rounded (*Gordonia*) ends; in some genera they are much-elongated spicular cells which traverse the whole of the mesophyll. They are absent in *Stewartia*, except for the pedicels, and they are never found in seedlings (Beauvisage 1920; Keng 1962).

There may be large and much-thickened fibres in the phloem. The wood is diffuse porous or more or less ring porous. Vessels are small, usually 25–100 µm tangential diameter, usually solitary and 16–400 per mm². Spiral thickening is common, and the perforation plates are very oblique and with 7–42(–57) bars. Intervascular pitting is scalariform to opposite. Vessel-ray pits have at most narrow borders. Parenchyma is often scanty apotracheal, usually diffuse but sometimes in short tangential lines. Strands of crystalliferous cells occur in *Schima* and elsewhere. Wood fibres are about 800–2000 µm long and have thin to thick walls. Rays are all about the same size and are 1–3(–8) cells wide and up to 2 mm or more high. Uniseriate rays are made up of tall to low upright cells only. Multicellular rays are of Kribs types I and II, with narrow, procumbent cells in the body of the ray and uniseriate margins of square to upright cells (Metcalfe and Chalk 1952; Liang and Baas 1990, 1991).

Hairs are unicellular, rarely fasciculate. Epidermal cells are mucilaginous or not, and there is sometimes a single-layered hypodermis. The stomata are usually gordoniaceous, the guard cells being surrounded by 2–4 narrow subsidiary cells, but *Franklinia* and many species of *Pyrenaria* have

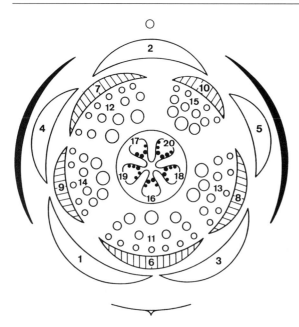

Fig. 133. Theaceae. Floral diagram of *Stewartia pseudocamellia*, with indication of sequence of inception of floral organs all of which originate in a consecutive sequence, following an angle of 144°, except for the smaller angle between organs 10/11 (the transition from corolla to androecium); were this divergence also 2/5, a superposed position of sepals and petals, as in Ternstroemiaceae–Ternstroemieae, would result. (Erbar 1986)

anomocytic stomata and *Stewartia* has paracytic stomata. There is also extensive variation in cuticle surface and details of stomatal morphology (Kvaček and Walther 1984a). There are 1–3 layers of palisade tissue and the fine venation is embedded. In some species of *Camellia* there are "cork-warts", rings of corky material surrounding shallow depressions, on the lower surface of the lamina (Beauvisage 1920; Metcalfe and Chalk 1952; Keng 1962).

INFLORESCENCE AND FLOWER STRUCTURE. Flowers are large, solitary and axillary, although sometimes appearing terminal. In *Camellia piquetiana* several flowers occur in the axils of scale-like bracts on a short axillary shoot with a terminal bud. There has been much discussion over the distinction between sepals, petals, and prophylls in Theaceae. In taxa such as species of *Camellia*, all three grade insensibly into one another whereas in others, like *Schima*, prophylls, calyx and corolla are all clearly distinguished. There are 20 to well over 100 stamens. In *Stewartia* there are five antepetalous primary primordia on which individual stamens are initiated centrifugally (Fig. 133; Erbar 1986); the same may be true of *Franklinia* and

Schima (Tsou 1998). In *Camellia*, by contrast, and probably other genera as well, the stamens are initiated centrifugally from a ring wall (a circular primordium) (Ronse Decraene 1989), certainly a derived state. The stamens are often shortly adnate to the base of the corolla, and in some species of *Camellia* the filaments are connate and form a prominent tube in the centre of the flower. The anthers are usually dorsifixed or versatile, the filaments being relatively long and the anthers short and usually without any prolongation of the connective; basifixed anthers are known in *Camellia*, etc. When the anthers are basifixed, the filament may be narrowed and effectively articulated with the anther, unlike the basifixed anthers (with short filaments) of Ternstroemiaceae. Although the ovary is superior, in *Franklinia* and *Schima* it is partly enclosed by the fused bases of the stamens plus petals, and so has been described as being semi-inferior (Tsou 1997). The ovary is often hairy and narrows more or less gradually into the style, which sometimes is branched (Figs. 134C, 136D) or cleft into stylodia; there is commonly infrageneric variation in this feature. The stigmas are more or less expanded or lobed. The carpels are usually opposite the petals, although they are apparently opposite the sepals in *Camellia*. This may be connected with the fact that sepals and petals intergrade and so the arrangement of the petals differs from that of other taxa. The basic orientation in *Camellia* is the same as that in *Gordonia*, where the calyx is clearly opposite the corolla (Eichler 1875–1878). The young gynoecium shows strongly ascidiate development and the degree of congenital fusion is usually low, although high in *Franklinia* and *Schima*.

EMBRYOLOGY. Anther cells can have crystals and/or tannin, or these may be absent. The connective usually has stomata, but not in *Camellia*. Anther wall formation is of the basic type. The tapetum is glandular and microsporogenesis is simultaneous. The pollen is 2-celled. Pseudopollen is produced from connective cells. It has large vacuoles and a marginally-situated nucleus when mature; the wall itself has either rib-like thickenings (e.g. *Camellia*) or circular thickenings, and pseudopollen with the latter thickenings can look like pantoporate pollen grains (e.g. *Schima*: see Tsou 1997, for details).

There are only a few apotropous or pleurotropous ovules in each carpel and they are spreading or more or less pendulous, except in *Stewartia* where they are more or less basal. The ovules are anatropous (campylotropous in *Schima*), bitegmic, and tenuinucellate. The micropyle is endostomal. The outer integument is 4–8 cells thick and the inner integument is 4–7 cells thick; in *Schima* the outer epidermis of the outer integument proliferates in the antiraphal region, forming a ridge of cells up to 12 cells high. In *Franklinia* the raphe can be massive, to 30 cells thick. There are vascular bundles in either the outer or the inner integument, or sometimes only at the base of the inner integument. There is an integumentary tapetum and a hypostase. The archesporium is 1-celled and the embryo sac is usually the Polygonum type, but it is of the Allium type in *Camellia*. (General information is taken mainly from Yang and Ming [1995a: *Pyrenaria, Tutcheria*] and Tsou [1997: *Camellia, Franklinia, Schima*]). The endosperm is nuclear.

POLLEN MORPHOLOGY. The pollen is tricolporate, suboblate or oblate, and over 28 µm and sometimes over 50 µm long. The exine is finely granulate or finely reticulate; sexine and nexine are indistinguishable. The endoaperture is rather large, lalongate, and indistinct (Keng 1962; Ying et al. 1993).

KARYOLOGY. Chromosome numbers in *Camellia* species are most frequently $2n = 30$, although some species are polyploids; triploids are common in cultivars of *Camellia japonica* (Kondo 1977). The other main number in the family is $2n = 36$ (e.g. *Franklinia, Stewartia*), again with possibly polyploid series based on this. Information on chromosome numbers is summarized by Tsou (1998).

FRUIT AND SEED. The fruit is usually a loculicidally dehiscent capsule with a persistent columella (Fig. 135D), although the latter is at most poorly developed in *Stewartia*. The fruit of *Franklinia* also dehisces septicidally from the base up. The valves may soon fall off, or they may remain attached to the fruit. Some species of *Camellia* have irregularly-dehiscent capsules, while many species of *Pyrenaria* have a drupaceous fruit. The calyx usually persists at the base of the fruit (Fig. 136B).

Schima, Gordonia and *Stewartia* have winged seeds, although the development of the wing differs in at least *Schima* and *Polyspora* (*Gordonia* s.l.) (Tsou 1997, 1998). The shiny seeds sometimes (e.g. *Pyrenaria*) have a large hilum and so look remarkably like those of Sapotaceae. The seed coat is very thick for an Asterid, often being more than 30 cells across. The testa is vascularised. The exotesta is lignified or not, and the cells are polyg-

onal, rarely elongated. The mesotesta is lignified or contains sclereids; sometimes the cells are fibrous. The endotesta is lignified or not; the cells are polygonal (Grote and Dilcher 1989, 1992). Mature seeds usually contain at most a little oily endosperm (described as being abundant in *Camellia*: Sealy 1958) and a large, straight or rarely curved embryo. The cotyledons are accumbent and are very large, being much longer than the radicle. They are either flat or longitudinally folded (*Pyrenaria*). Germination is hypogeal (*Camellia*) or epigeal, the latter being commonest, but in the former germination type the cotyledons generally separate and may even be photosynthetic.

REPRODUCTIVE SYSTEMS, POLLINATION AND DISPERSAL. Although most Theaceae have hermaphroditic flowers, there are reports that *Gordonia* (inc. *Laplacea*) may be dioecious or monoecious (e.g. Melchior 1925, but cf. Kobuski 1949, 1950). Theaceae lack nectar, and pollen is presumably the reward for pollinators. The role of the pseudopollen in pollination is not well understood. Tsou (1997) found both pollen and pseudopollen on three honey bees foraging in a tea garden, although the bees had many fewer grains of pseudopollen than of pollen. *Franklinia* is highly unusual in that the showy and strongly scented flowers appear in late summer, but the capsules do not dehisce and the seeds are not ripe until the following summer. Dispersal of the winged seeds is presumably largely by the wind. The wingless seeds of *Franklinia* and especially *Camellia* are not likely to travel far. It is not known what eats the drupes of *Pyrenaria*.

PHYTOCHEMISTRY. The reader is referred to Hegnauer (1973, 1990). Common polyphenols such as flavonols, flavones, and proanthocyanins are widely distributed in Theaceae; the same is true of ellagic acid; myricetin is rare in large amounts, and gallic acid and catechins (flavan-3-ols and flavan-3,4-diols) are restricted to *Camellia* sect. *Thea* (*C. sinensis, C. taliensis* and *C. irrawadiensis*). In young shoots of the tea bush, catechins amount to more than 30% of dry matter weight but later decrease in quantity. Caffeine and its precursors theobromine and theophylline are restricted to sect. *Thea*; they have not been found in wild species of *Camellia* or other Theaceae. Caffeine occurs in the tea bush in quantities of 2.5–4%, and the high content in catechins and caffeine in the tea bush results from selection by man for these substances. Triterpenes and their glycosides (saponins) are widely distributed in the family and found in seeds, bark, wood, and leaves. The plants often accumulate aluminium and fluoride.

SUBDIVISION AND RELATIONSHIPS WITHIN THE FAMILY. Relationships within Theaceae are still rather unclear. In earlier classifications (e.g. Melchior 1925, 1964; Airy Shaw 1936; Sealy 1958; Keng 1962), various combinations of simple and/or continuously varying characters like the winging of the seed and the relative size, distinctness and connation of bracts, sepals and petals were emphasized. More recently, Ye (1990a, 1990b) and Tsou (1998), although adding new information (especially the latter) and thinking about the characters in rather different ways from these earlier authors, still ended up with groupings relying heavily on a few characters. Thus, a variety of arrangements in the family have been proposed, Tsou (1998) suggesting there were two tribes, Camellieae and Gordonieae, the latter including Stewartiinae and Gordoniinae. Prince and Parks (2001) propose a three-tribe classification, the Stewartieae being sister to Gordonieae and Theaeae. Their work is based on an analysis of variation in two chloroplast genes and flanking spacer regions and, although apparently not that dissimilar from Tsou's classification, the tree suggests a major change of taxon limits in the family. In particular, major realignments within *Gordonia* s.l. and both Gordonieae and Theaeae are implied. *Gordonia lasianthus* and *G. brandegeei* remain in Gordonieae, but *Polyspora* (Old World) and *Laplacea* (New World), both commonly included in *Gordonia*, are separate lineages within the Theaeae. We have not adopted any tribal classification below, pending more detailed molecular work and a re-evaluation of the morphological data. Nevertheless, it is almost certain that *Gordonia* will not retain the circumscription it is given below.

Generic limits elsewhere in the family are a matter of some dispute. The limited sampling of Prince and Parks (2001) supports the broad circumscription of *Camellia* adopted below, while that of *Pyrenaria* is supported by the largely unpublished work of Yang (Yang 1988; Yang and Ming 1995a, 1995b). *Hartia*, with characters such as evergreen leaves, narrowly winged seeds, and buds without scales, has often been separated from *Stewartia*, with deciduous or semi-evergreen leaves, winged seeds, and buds with scales. Recognition of *Hartia* would make *Stewartia* paraphyletic (Prince and Parks 2001; Li et al. 2002).

Prince and Parks's (2001) classification of the family is a suggestion of how the tribal classification may look in the future:

- Theaeae: *Camellia, Pyrenaria, Polyspora, Laplacea, Apterosperma.*
- Gordonieae: *Gordonia, Franklinia, Schima.*
- Stewartieae: *Stewartia.*

There are perhaps even more problems with species delimitation than with generic boundaries, as comments after the descriptions below make clear.

AFFINITIES. In the past, Theaceae s.l. (including Ternstroemiaceae and *Sladenia*) seemed to be a pivotal family showing relationships with a number of other taxa. The separate stylodia or style branches of some Theaceae, as well as the intergradation between prophylls, calyx, and corolla, suggested a rather basal position in the system for some. Although Theaceae are largely polypetalous, the realization that their stamen development was centrifugal removed them from direct links with basal polystaminate angiosperms. Cronquist (1981) included 18 families in his Theales, and links between Theaceae and families like Clusiaceae, also multistaminate, seemed to be established by the "intermediate" Bonnetioideae, including genera like *Bonnetia, Kielmeyera*, both with spiral leaves and the former with minutely serrate leaves. Wood anatomy also seemed to support such a link (Baretta-Kuipers 1976), but DNA sequence analyses have demonstrated that Cronquist's Theales form three major groups that are now included in Malpighiales, Malvales, and Ericales, with three other families placed in the Caryophyllales and the Asterid I and II groups respectively. Nevertheless, even after the exclusion of *Bonnetia* and its allies (now in Malpighiales), of *Pelliciera, Tetramerista* (both in Ericales, see this volume) and *Asteropeia* (see vol. V of this series), all at one time or another associated with Theaceae, there is no evidence that the rest of Theaceae is monophyletic, although it is unambiguously ericalean (Morton et al. 1996; APG 2003). The molecular evidence (Prince and Parks 2001; and particularly the five-gene analysis of Anderberg et al. 2002) is consistent with its separation into three morphologically quite distinct, monophyletic groups, Ternstroemiaceae, Sladeniaceae, and Theaceae s.str.

DISTRIBUTION AND HABITATS. Theaceae are scattered in the north temperate region and in the tropics, especially in South East Asia, where there are numerous species in evergreen rainforest. There are clear links between Theaceae of SE U.S.A. and those of E Asia. Thus, the two species of *Stewartia* from the SE U.S.A. are sister to much of the rest of the genus and had diverged from them by the late Miocene (Li et al. 2002). However, understanding details of the biogeographical relationships within the family awaits further findings on its phylogeny (see above).

Franklinia alatamaha is one of the better-known North American plants, largely because it has become extinct since it was first discovered. It was first found by the Bartrams in 1765 along the Alatamaha River at Fort Barrington, Georgia, but it had become extinct by 1790, quite possibly because of over-collection (Wood 1959).

PALAEOBOTANY. Theaceae are known from the late Cretaceous and are common in the Tertiary. Leaves of the form genus *Ternstroemites*, common in North America, as well as the form genera *Ternstroemiacinium* and *Ternstroemioxylon*, resemble genera in both Theaceae and Ternstroemiaceae (Grote and Dilcher 1989), and so are of little value in understanding the history of the group. Seeds of Theaceae are known from the Cretaceous (Knobloch and Mai 1986). Grote and Dilcher (1989, 1992, for a literature summary) show that in the Tertiary a variety of extant and extinct genera (the latter including two genera of Theaceae from the Middle Eocene Claiborne flora) were found throughout the northern Hemisphere. A rather generalised theaceous flower has been found in Baltic amber, and a rather *Schima*-like wood is reported from Borneo (Kramer 1974). *Gordonia* was common in the early Miocene Brandon lignite from Vermont, U.S.A. (Tiffney 1994). Current evidence suggests that *Camellia* once grew in North America, two genera not now known from Japan once occurred there, while perhaps four genera are known from Europe, remains of both vegetative (e.g. Kvaček and Walther 1984b) and reproductive parts being quite common; no member of the family is currently found there. However, as generic limits of extant taxa are re-evaluated (see above), the fossil record will need similar treatment; detailed studies of leaf surface may be valuable.

ECONOMIC IMPORTANCE. The dried young shoots of *Camellia sinensis* constitute perhaps the most important caffeine drink (Ukers 1935 for general, if somewhat dated information). Some 2,600,000 tons are produced annually (1996 values), of which

about half comes from India and China. Green teas are not fermented during the drying process, other kinds of teas are; alkaloids in tea include caffeine and theobromine. Teas have been used as a beverage in China since at least 350 B.C.E., and tea and tea drinking were introduced to Japan in about 700 C.E., the tea ceremony taking its modern form in the fifteenth century. Tea drinking became common in England as from the middle of the seventeenth century, and the import of tea was a monopoly of the British East India Company until the nineteenth century.

Camellia japonica, together with thousands of cultivars derived from it and species like *C. saluensis*, *C. reticulata* and *C. sasanqua* and their hybrids, is a much-grown horticultural shrub with single or double white to red but often odourless flowers; less frequently-cultivated species have yellow flowers. The seeds of *Camellia oleifera* and some other species yield tea oil used mainly as hair oil, in cooking, and for lubrication. Other species used as ornamentals include *Franklinia altamaha*, which grows quite readily in cultivation, *Stewartia* spp., some with very beautiful bark, and *Gordonia lasianthus*. The timber of *Schima* is locally of some importance, while other genera such as *Gordonia* may yield woods of minor value.

KEY TO THE GENERA

1. Fruits drupes or leathery indehiscent capsules
   2. *Pyrenaria*
 - Fruits capsular 2
2. Anthers often basifixed; carpels mostly 3; seeds large (usually >1 cm), wingless **1. *Camellia***
 - Anthers often versatile; carpels mostly 5; if seeds large and wingless, then capsule valves deciduous and seeds angular (*Pyrenaria*) 3
3. Capsule valves deciduous 4
 - Capsules valves persistent, and/or capsules lacking clearly-defined valves (*Franklinia*) 5
4. Capsule <1 cm across; seeds <1 cm long, reniform; calyx eventually caducous **3. *Apterosperma***
 - Capsule >1 cm across; seeds >1 cm long, ovoid to angular; calyx persistent or caducous **2. *Pyrenaria.***
5. Capsules (conical-)ovoid to cylindrical 6
 - Capsules spherical to more or less depressed-globose 7
6. Calyx at the base of the fruit; capsule with columella; seeds with wing at one end **4. *Gordonia***
 - Calyx more or less investing the fruit; capsule lacking columella, seeds wingless or with a wing all round
   7. *Stewartia*
7. Plant deciduous; pedicels short; capsule loculicidal, also septicidal from the base; seeds wingless, not reniform **6. *Franklinia***
 - Plant evergreen; pedicels medium to long; capsule loculicidal; seeds surrounded by wing, reniform **5. *Schima***

Fig. 134. Theaceae. *Camellia vietnamensis.* **A** Habit. **B** Stamens. **C** Pistil. **D** Fruit. **E** Dehiscing fruit. **F** Seeds. (Hu 1965)

1. *Camellia* L. Fig. 134

Camellia L., Sp. Pl.: 698 (1753); Sealy, A revision of the genus *Camellia* (1958); H.T. Chang, Acta Sci. Nat. Univ. Sunyatseni, Monogr. Ser. 1: 1–180 (1981); H.T. Chang & Bartholomew, Camellias (1984).
Thea L. (1753).
Piquetia (Pierre) H. Hallier (1921).
Yunnanea Hu (1956).

Shrubs or trees; leaves spiral or distichous, evergreen, coriaceous, serrate. Pedicel none–medium; bracteoles 2 or more, not sharply differentiated from calyx, persistent or deciduous. Sepals 5–21; petals 5–12, shortly connate basally; stamens with outer whorl adnate to petals in basal half, connate into a tube or not, anthers more or less basifixed; ovary 3(–5)-locular, with 1–2(–10) ovules per carpel; style simple, branched, or free stylodia. Fruit a capsule, columella +; calyx persistent, at base of fruit, rarely deciduous. Seeds more or less globose, wingless; embryo straight, cotyledons hemispherical, endosperm +. 100 (B.

Bartholomew, pers. comm.)–300 spp., E Asia to Indomalesia, mainly S China. Species limits are very uncertain. The genus has been divided into four subgenera, *Protocamellia* Hung T. Chang (3 sections), *Camellia* (7 sections), *Thea* (L.) Hung T. Chang (8 sections), *Metacamellia* Hung T. Chang (2 sections: H.T. Chang 1981, see also H.T. Chang and Bartholomew 1984, Ye 1988).

2. *Pyrenaria* Blume

Pyrenaria Blume, Bijdr.: 1119 (1827); Keng, Gard. Bull. Singapore 33: 264–289 (1980); Hung T. Chang & Kuan, Higher plants of China 4: 619–620 (2000).
Tutcheria Dunn (1908); Hung T. Chang & Kuan, Higher plants of China 4: 603–607 (2000).
Sinopyrenaria Hu (1956).
Parapyrenaria H.T. Chang (1963); Ying et al., Endemic genera seed plants China: 696–698 (1993); Hung T. Chang & Kuan, Higher plants of China 4: 618–619 (2000).
Glyptocarpa Hu (1965).

Tree; leaves spiral, evergreen, coriaceous, serrate. Pedicels short, bracteoles 2, rarely many, persistent, rarely deciduous. Sepals 5–6, unequal; petals 5–6(–12), slightly connate basally; stamens basally connate, basally adnate to petals, anthers versatile; ovary (3–)5–6(–10)-locular, with 2–3(–5) ovules per carpel; style simple, branched, or free stylodia. Fruit a drupe, rarely a capsule, columella +; calyx persistent, at base of fruit, or deciduous. Seeds ellipsoid, faces flattened; embryo straight, cotyledons thin, tightly folded and twisted, endosperm 0. 42 spp., in evergreen forests, SE Asia, China (many species) southwards to W Malesia.

3. *Apterosperma* Hung T. Chang

Apterosperma Hung T. Chang, Acta Sci. Nat. Univ. Sunyatseni 1976: 90 (1976); Ying et al., Endemic genera seed plants China: 691–692 (1993); Hung T. Chang & Kuan, Higher plants of China 4: 613 (2000).

Small tree; leaves ?spiral, evergreen, coriaceous, serrate. Pedicels medium, bracts 2, deciduous. Sepals 5, equal; petals 5, slightly connate basally; stamens 22–24, anthers basifixed, ovary 5-locular, with 3–4 ovules per carpel; style short. Capsule depressed-globose, columella +; calyx and valves caducous. Seeds reniform, wingless. One sp., *A. oblatum* Hung T. Chang, S China, in open mixed forests, 600 m.

4. *Gordonia* Ellis

Gordonia Ellis, Phil. Trans. Lond. 60: 518, t. 11 (1771), nom. cons.; Keng, Gard. Bull. Singapore 37:1–47 (1984);

Hung T. Chang & Kuan, Higher plants of China 4: 607–609 (2000).
Laplacea Kunth (1822); Kobuski, J. Arnold Arbor. 30: 166–186 (1949), 31: 405–429 (1950).
Polyspora Sweet ex G. Don (1831).

Trees; leaves distichous or spiral, evergreen, coriaceous, entire or serrate. Bracteoles 2–7, at the apex of the pedicel, deciduous. Sepals 5, unequal; petals 5(–10), slightly connate basally; stamens adnate to the base of the corolla, base forming fleshy pads, anthers versatile; ovary (3–)5(–10)-locular, with (2–)3–5(–10) ovules per carpel; style simple, or style branches more or less developed, short to long; stigmas capitate. Capsule cylindrical, often longitudinally angled, columella +; calyx usually persistent, at the base. Seeds flat, with an apical wing; embryo slightly curved, cotyledons thin, flat, endosperm thin. 20–65 spp., perhaps 6 spp. in China, 7–35 spp. elsewhere in Southeast Asia, 3–25 spp. New World tropics, 1 sp. U.S.A. In *Polyspora* there is a gradual transition between the calyx and corolla.

5. *Schima* Blume   Fig. 135

Schima Reinw. ex Blume, Cat. Buitenz. 80 (1823); Bloembergen, Reinwardtia 2: 133–183 (1952); Keng, Gard. Bull. Singapore 46: 77–87 (1994); Hung T. Chang & Kuan, Higher plants of China 4: 609–613 (2000).

Trees; leaves spiral, evergreen, coriaceous, entire or crenate. Flowers often crowded near branch apices; pedicels long, prophylls 2, deciduous, in upper 1/3 of pedicel. Sepals 5, nearly equal; petals 5, slightly connate basally, the outermost one enveloping the others in bud; stamens adnate to the base of the corolla, anthers versatile; ovary 5(–7)-locular, with 2–6 ovules per carpel; style simple. Capsule subglobose, columella at most short; calyx persistent, at the base. Seeds flat, reniform, surrounded by a narrow wing; embryo curved, cotyledons thin, unequal, flat or somewhat longitudinally folded, endosperm thin. 1–30 spp., India, Southeast Asia to New Guinea, esp. variable in China.

6. *Franklinia* Marshall

Franklinia Marshall, Arbust. Am.: 48 (1785).

Suckering shrub; leaves spiral, deciduous, membranous, serrate. Pedicel short, prophylls 2, deciduous; sepals 5; petals 5, free, stamens many, adnate to the base of the corolla, anthers versatile; ovary 5-locular, with 6–12 ovules per carpel; style

Fig. 135. Theaceae. *Schima wallichii.* **A** Flowering branch. **B** Stamen. **C** Calyx with pistil. **D** Calyx and central columella of fruit. **E** Open fruit from above. **F** Seed. **G** Fruit valve, showing septum loosening from lower half of fruit-wall. Drawn by S. Kirno. (Bloembergen 1952)

simple. Capsule globose, also dehiscing septicidally from the base to half way up, columella +; calyx deciduous. Seeds angular; embryo straight, cotyledons thin, folded, endosperm thin. One species, *F. alatamaha* Marshall, SE U.S.A. (Georgia), now known only in cultivation.

7. *Stewartia* L.   Fig. 136

Stewartia L., Sp. Pl.: 698 (1753); Spongberg, J. Arnold Arbor. 55: 182–214 (1974); Ye, Acta Sci. Nat. Univ. Sunyatseni 1982(4): 108–116 (1982); Li, Acta Phytotax. Sinica 34: 48–67 (1996); Prince, Castanea 67: 290–301 (2002), molec. syst.
Hartia Dunn (1902); Hung T. Chang & Kuan, Higher plants of China 4: 613–616 (2000).

Shrubs or trees; leaves distichous, evergreen or deciduous, often membranous, serrate. Pedicels short, prophylls 2, persistent; sepals 5, equal; petals 5, connate basally; stamens many, filaments connate in lower half, anthers versatile; ovary 5-locular, with 1–7 basal ovules per carpel, style simple or branched. Capsule ovoid, usually

Fig. 136. Theaceae. **A–C** *Stewartia ovata*. **A** Flowering branch-let. **B** Apically dehiscent capsule with persistent style branches and persistent calyx. **C** Seed. **D, E** *Stewartia sinensis*. **D** Pistil. **E** Dormant winter buds. (Spongberg 1974)

enveloped by persistent calyx, columella 0. Seeds subovoid or flattened, winged or not; cotyledons thin and flat, endosperm at most slight. 30 spp., most in China, 3 spp. in Japan and Korea, 2 spp. in SE U.S.A.

Selected Bibliography

Airy-Shaw, H.K. 1936. Notes on the genus *Schima* and on the classification of the Theaceae-Camellioïdeae. Kew Bull. Misc. Inf., pp. 496–500.

Anderberg A.A. et al. 2002. See general references.

APG (Angiosperm Phylogeny Group) 2003. See general references.

Baretta-Kuipers, T. 1976. Comparative wood anatomy of Bonnetiaceae, Theaceae and Guttiferae. In: Baas, P., Bolton, A.J., Catling, D.M. (eds.) Wood structure in biological and technological research. Leiden Bot. Ser. 3: 76–101.

Beauvisage, L. 1920. Contribution à l'étude anatomique de la famille des Ternstroemiacées. Tours.

Bloembergen, S. 1952. A critical study in the complex-polymorphous genus *Schima* (Theaceae). Reinwardtia 2: 133–183.

Chang, H.T. 1981. A taxonomy of the genus *Camellia*. Acta Sci. Nat. Univ. Sunyatseni, Monogr. Ser. 1: 1–180 (in Chinese).

Chang, H.T., Bartholomew, B. 1984. Camellias. Portland: Timber Press, London: Batsford.

Cronquist, A. 1981. See general references.

Cuénoud, P., Savolainen, V., Chatrou, L.W., Powell, M., Grayer, R.J., Chase, M.W. 2002. Molecular phylogenetics of Caryophyllales based on nuclear 18s rDNA and plastid *rbc*L, *atp*B and *mat*K DNA sequences. Am. J. Bot. 89: 132–144.

Eichler, A.W. 1875–1878. See general references.

Erbar, C. 1986. Untersuchungen zur Entwicklung der spiraligen Blüten von *Stewartia pseudocamellia* (Theaceae). Bot. Jahrb. Syst. 106: 391–407.

Field, B.S. 1993. Theaceae. In: Heywood, V. (ed.) Flowering plants of the world. New York: Oxford University Press, pp. 82–83.

Grote, P.J., Dilcher, D.L. 1989. Investigations of angiosperms from the Eocene of North America: a new genus of Theaceae based on fruit and seed remains. Bot. Gaz. 150: 190–206.

Grote, P.J., Dilcher, D.L. 1992. Fruits and seeds of tribe Gordonieae (Theaceae) from the Eocene of North America. Am. J. Bot. 79: 744–753.

Hegnauer, R. 1973, 1990. See general references.

Hu, H.-H. 1965. New species and varietie[s] of *Camellia* and *Theopsis* of China (1). Acta Phytotax. Sinica 10: 131–142, pl. 23–28.

Keng, H. 1962. Comparative morphological studies in the Theaceae. Univ. Calif. Publ. Bot. 33: 269–384.

Knobloch, E., Mai, D.H. 1986. Monographie der Früchte und Samen in der Kreide von Mitteleuropa. Rozpr. Ustr. Ust. Geol. 47: 1–219.

Kobuski, C.E. 1949. Studies in the Theaceae. XVIII. The West Indian species of *Laplacea*. J. Arnold Arbor. 30: 166–186.

Kobuski, C.E. 1950. Studies in the Theaceae. XX. Notes on the South and Central American species of *Laplacea*. J. Arnold Arbor. 31: 405–429.

Kondo, K. 1977. Chromosome numbers in the genus *Camellia*. Biotropica 9: 86–94.

Kramer, K. 1974. Die tertiären Hölzer Südost-Asiens (unter Ausschluss der Dipterocarpaceae), 2. Teil. Palaeontographica B, 145: 1–150.

Kvaček, Z., Walther, H. 1984a. Nachweis tertiärer Theaceen Mitteleuropas nach blatt-epidermalen Untersuchungen. I. Teil – Epidermale Merkmalskomplexe rezenter Theaceae. Feddes Repert. 95: 209–227.

Kvaček, Z., Walther, H. 1984b. Nachweis tertiärer Theaceen Mitteleuropas nach blatt-epidermalen Untersuchungen. II. Teil – Bestimmung fossiler Theaceen-Sippen. Feddes Repert. 95: 331–346.

Li, J., del Tredici, P., Yang, S., Donoghue, M.J. 2002. Phylogenetic relationships and biogeography of *Stewartia* (Camellioïdeae, Theaceae) inferred from nuclear ribosomal DNS ITS sequences. Rhodora 104: 117–133.

Liang, D., Baas, P. 1990. Wood anatomy of trees and shrubs from China II. Theaceae. I.A.W.A. Bull. II, 11: 337–378.

Liang, D., Baas, P. 1991. The wood anatomy of the Theaceae. I.A.W.A. Bull. II, 12: 333–353.

Melchior, H. 1925. Theaceae. In: Engler, A., Prantl, K. Die natürlichen Pflanzenfamilien, ed. 2, 21. Leipzig: W. Engelmann, pp. 109–154.

Melchior, H. 1964. Theaceae. In: Melchior, H. (ed.) A. Engler's Syllabus der Pflanzenfamilien, ed. 12, 2. Berlin: Bornträger, pp. 109–154.

Metcalfe, C.R., Chalk, L. 1950. See general references.

Morton, C.M., Chase, M.W., Kron, K.A., Swensen, S.M. 1996. A molecular evaluation of the monophyly of the order Ebenales based on *rbc*L sequence data. Syst. Bot. 21: 567–586.

Prince, L.M. 1998. Systematics of the Theoideae (Theaceae) and the resolution of morphological, anatomical and molecular data. Ph.D. thesis. Chapel Hill: Department of Biology, University of North Carolina at Chapel Hill.

Prince, L.M., Parks, C.M. 2001. Phylogenetic relationships of Theaceae inferred from chloroplast DNA data. Am. J. Bot. 88: 2309–2320.

Ronse Decraene, L.P. 1989. Floral development of *Cochlospermum tinctorium* and *Bixa orellana* with special emphasis on the androecium. Am. J. Bot. 76: 1344–1359.

Sealy, J.R. 1958. A revision of the genus *Camellia*. London: Royal Horticultural Society.

Spongberg, S.A. 1974. A review of deciduous-leaved species of *Stewartia* (Theaceae). J. Arnold Arbor. 55: 182–214.

Tiffney, B.H. 1994. Re-evaluation of the age of the Brandon lignite (Vermont, USA) based on plant megafossils. Rev. Palaeobot. Palynol. 82: 299–315.

Tsou, C.-H. 1997. Embryology of the Theaceae – anther and ovule development of *Camellia*, *Franklinia*, and *Schima*. Am. J. Bot. 84: 369–481.

Tsou, C.-H. 1998. Early floral development of Camellioideae (Theaceae). Am. J. Bot. 85: 1531–1547.

Ukers, W.H. 1935. All about tea, 2 vols. New York: The Tea and Coffee Trade Journal Company.

Wood, C.E. 1959. The genera of Theaceae of the southeastern United States. J. Arnold Arbor. 40: 413–419.

Yang, S.-X. 1988. Systematics, diversification and geographical distribution of *Pyrenaria sensu lato* (Theaceae). Ph.D. dissertation. Yunnan: Kunming Institute of Botany.

Yang, S.-X., Ming, T.-L. 1995a. Embryological studies on genera *Pyrenaria* and *Tutcheria* of family Theaceae. Acta Bot. Yunn. 17: 67–71, pl. 1–4 (in Chinese).

Yang, S.-X., Ming, T.-L. 1995b. Studies on the systematic position of genera *Pyrenaria*, *Tutcheria* and *Parapyrenaria* of family Theaceae. Acta Bot. Yunn. 17: 192–196, pl. 1–3 (in Chinese).

Ye, C.-X. 1988. The subdivisions of genus *Camellia* with a discussion on their phylogenetic relationship. Acta Bot. Yunn. 10: 61–67 (in Chinese).

Ye, C.-X. 1990a. A discussion on relationship among the genera in Theoideae (Theaceae). Acta Sci. Nat. Univ. Sunyatsenia 29: 74–81 (in Chinese).

Ye, C.-X. 1990b. The range of Gordonieae (Theaceae) and limitation of genera in the tribe. Guihaia 10: 99–103 (in Chinese).

Ying, T.-S., Zhang, Y.-L., Boufford, D.E. 1993. The endemic genera of seed plants of China. Peking: Science Press.

Zomlefer, W.B. 1994. Guide to Flowering Plant Families. Chapel Hill: University of North Carolina Press.

Theophrastaceae

B. Ståhl

Theophrastaceae Link, Handbuch 1: 440 (1829), nom. cons.

Shrubs or small trees, usually evergreen. Leaves alternate, mostly arranged in one or several pseudowhorls at branch tips, petiolate, exstipulate; blades simple, glandular-punctate, often faintly striate. Inflorescences terminal or lateral, racemose, each flower subtended by a single bract. Flowers hypogynous, actinomorphic, 5(4)-merous, hermaphrodite or unisexual, the aestivation imbricate; calyx persistent, lobes free to base, the margins usually erose-ciliate; corolla sympetalous, usually firm and waxy; tube well developed, shorter to somewhat longer than the lobes; lobes oblong to suborbicular, usually somewhat unequal in size; stamens homomerous, antepetalous; staminodes alternipetalous, fused with the corolla tube; filaments flattened, fused to the lower part of the corolla, basally connate or (often in *Clavija*) united for their entire length; anthers dithecal, basifixed, extrorsely dehiscent through longitudinal slits, the upper and lower parts of thecae filled up with a white meal of calcium oxalate crystals; ovary superior to semi-inferior, ovoid to subglobose, unilocular; placenta a basal column; ovules few to numerous, anatropous, bitegmic, tenuinucellate, spirally inserted on but not immersed into a basal column; style shorter to somewhat longer than the ovary; stigma capitate or truncate, entire or vaguely lobed. Fruit a berry with a dry and sometimes woody pericarp, indehiscent, subglobose, oblong or ovoid, yellow, orange or red; pulp orange or yellow, juicy and sweet. Seeds 1 to many, brown or pale brown; endosperm abundant, hard; embryo straight, the cotyledons foliaceous or poorly differentiated.

A neotropical family of six genera and about 90 species.

CHARACTERS OCCURRING IN RELATIVELY FEW GENERA AND SPECIES. Stem spiny in *Theophrasta*; corolla green in *Deherainia*; anthers distinctly produced at apex in *Theophrasta* and *Neomezia*; fruits puberulous in *Neomezia* and *Clavija glandulifera*; flowers in *Clavija* mostly unisexual or functionally unisexual and then with the filaments fused throughout their length. *Jacquinia nervosa* is deciduous and leafless in the rainy season.

VEGETATIVE MORPHOLOGY. All Theophrastaceae are woody, varying in habit from dwarf-shrubs (*Neomezia* and a few species of *Clavija*) to shrubs or small trees to about 15 m high. Plants of *Jacquinia*, *Deherainia*, and *Votschia* are more or less richly branched, whereas those of *Clavija*, *Neomezia*, and *Theophrasta* are sparsely branched or unbranched. Among the latter genera, *Theophrasta* is always unbranched (Ståhl 1987), whereas plants of *Clavija* and *Neomezia* usually produce one or a few branches (Ståhl 1991).

The leaves are alternate but are usually condensed at stem apices into one or several pseudowhorls (Fig. 137A, B). A strictly alternate condition is met with in several species of *Jacquinia*. The rhythmic production of leaves is clearly expressed also in species growing under more or less aseasonal conditions and is also retained in plants cultivated in the greenhouse.

During shoot dormancy the shoot apex is covered by minute scale leaves, which are left on the stem as the shoot prolongates. These scale leaves are soft and ephemeral except in *Theophrasta* where they become lignified and persist on the trunk as spines.

The leaves lack stipules and in most taxa have a short, but well-demarcated petiole, the lower part of which is more or less swollen. Long petioles, to about half the length of the lamina, occur in a few species of *Clavija*. The family exhibits a very large variation in leaf size, from small and sometimes needle-shaped leaves in *Jacquinia* (Ståhl 1995), via medium-sized leaves in *Deherainia*, *Neomezia*, *Votschia* and some species of *Clavija* to large or very large leaves in *Theophrasta* and many species of *Clavija* (to 130 cm long; Ståhl 1991). The lamina is simple, usually oblanceolate or elliptic, and mostly coriaceous or chartaceous. The leaf margins are entire in *Deherainia*, *Jacquinia*, and *Votschia*, whereas they are spinose-serrate in *Neomezia* and *Theophrasta*; in *Clavija*, species with entire, serrulate, serrate, and spinose-serrate leaf margins occur. All species of *Jacquinia* have leaves with an apical spine (Fig. 137M), although in some species this is only visible in young, developing leaves (Ståhl 1992).

VEGETATIVE ANATOMY. All Theophrastaceae have immersed glandular trichomes consisting of a short stalk-cell and a multicellular, circular and somewhat flattened head. They occur on leaves and floral parts but are lacking on the corolla in *Clavija*. Long, unbranched, glandular-tipped trichomes occur on the leaves and floral parts of *Deherainia smaragdina* and the lower leaf surface of many species of *Clavija*, and branched, eglandular hairs are present on the leaves of some species of *Jacquinia*. The young shoots of *Theophrasta*, *Neomezia*, and many species of *Clavija* have a very dense and often light brown pubescence of minute, eglandular, and largely unbranched trichomes. Similar trichomes occur also on branchlets of several species of *Jacquinia*. A curious type of branched trichomes with thick cell walls provides a dense, white or rust-coloured pubescence on branchlets of some Antillean species of *Jacquinia*.

Transsections of the vascular strand in petioles of *Deherainia* and *Jacquinia* are open-crescentic and composed of 3–5 bundles. In the other genera the petiolar vascular strand is circular showing various degrees of complexity.

With the exceptions of a few species of *Clavija*, extraxylary sclerenchyma is present in the leaves of all Theophrastaceae (Votsch 1904; Ståhl 1989, 1991). It consists of short, unbranched fibre cells (Ståhl 1987), which are arranged in subepidermal bundles or layers. In most species this sclerenchyma occurs near the epidermis on both sides of the leaf, but in some species of *Jacquinia* and a few species of *Clavija* the foliar sclerenchyma is more or less sunken into the subjacent mesophyll. In addition, the adaxial foliar sclerenchyma is sometimes separated from the epidermis by one or sometimes two layers of hypodermal cells. In dried leaves the extraxylary foliar sclerenchyma appears as a fine striation running more or less parallel to the lateral veins.

Sclereids in the form of irregularly branched cells are present in the leaves of a few species of *Jacquinia*. Stone cells occur throughout the family in the pith of young stems and in tissues of reproductive organs.

According to Metcalfe and Chalk (1950), the wood of Theophrastaceae has small vessels (usually less than 50 μm diam.) with simple perforation plates. Parenchyma is extremely sparse in *Jacquinia* and absent in *Clavija*, *Deherainia*, and *Theophrasta*, and the strands are usually composed of 2–4 or 4 cells. Wood rays are typically multiseriate, being 3–6 cells wide in *Deherainia* and more than 10 cells wide in *Clavija*, *Jacquinia*, and *Theophrasta*. Silica grains in the wood have been reported in some species of *Clavija* (ter Welle 1976).

INFLORESCENCE STRUCTURE. In *Deherainia*, *Jacquinia*, and *Votschia*, the inflorescences are terminal, whereas in *Clavija* and *Neomezia* they are produced laterally, mainly among and just beneath the foliage. In *Theophrasta* the inflorescences appear above the uppermost leaf whorl, but originate just beneath the shoot tip.

The inflorescence is basically racemose with few to many flowers that mature successively during a period of one to several weeks. Some *Jacquinia* have corymbose or umbellate inflorescences. In *Deherainia* the inflorescence is usually reduced, often consisting of a single flower (Lepper 1989). The bracts that subtend the flowers are often shifted upward on the pedicel, particularly in species with long pedicels.

FLOWER STRUCTURE AND ANATOMY. The flowers of Theophrastaceae are hypogynous, actinomorphic, and 5- or (sometimes in *Clavija*) 4-merous. In size they vary from 3–4 mm diam., as in some species of *Clavija* and *Jacquinia*, to 35–45 mm diam. in *Deherainia*.

The calyx is persistent and usually more or less greenish; the lobes are well developed and mostly erose-margined. The corolla is thick and waxy and exhibits a large variation with regard to shape and colour. It is subrotate in *Clavija*, broadly campanulate in *Deherainia*, *Votschia*, and *Neomezia*, campanulate in *Theophrasta* and most species of *Jacquinia*, and urceolate in some *Jacquinia* species. The lobes, which are imbricate in bud, are broadly oblong or subrotate and usually somewhat unequal in size. Orange is the most widespread colour of the corolla and is found in all species of *Clavija* and any species of *Jacquinia*. Other corolla colours are green (*Deherainia*), white (*Jacquinia* spp.), and pale yellow (*Jacquinia* spp., *Votschia*). In *Neomezia*, the corolla is basically orange but has a distinct brownish tinge, which is also observed in one species of *Clavija*. The corolla of some species of *Jacquinia* are pale orange in bud and at the beginning of anthesis become orange-red. In *Theophrasta* the corolla is dull yellow or buff with a distinct orange-brown tinge in bud. The orange corolla colour is retained in properly dried herbarium specimens. In *Theophrasta* and *Jacquinia berterii* the corolla becomes dark brown and finally black when aged or upon drying.

All genera have a series of staminodial appendages alternating with the corolla lobes. These appendages are completely fused with the corolla, being inserted at the mouth of the tube, except in *Theophrasta*, where they are situated inside the tube. In *Jacquinia* and *Votschia* the appendages are large and flattened and resemble an extra whorl of corolla lobes; in *Clavija* they are gibbous-ellipsoid, in *Theophrasta* gibbous and transversely oblong, in *Deherainia* lanceolate to narrowly obovate, and in *Neomezia* small and triangular.

The stamens are of the same number as the corolla lobes and stand in front of them. The filaments are somewhat flattened and fused at the base, where they are adnate to the corolla tube. The flowers are protandrous and dehisce extrorsely with longitudinal slits. In all genera except *Clavija*, the stamens are loosely coherent during the male phase and spread during the female phase. In *Clavija*, the stamens of male, most hermaphrodite and functionally male flowers have the filaments fused into a permanent tube, whereas in female flowers the non-functional stamens are distinct throughout anthesis. In *Neomezia* and *Theophrasta*, the anthers are produced at the apex. The upper and lower parts of the thecae are filled up with a white meal of calcium oxalate druses and crystals, which at anthesis is exposed at the tip and base of the anther and forms a distinct visual contrast to the yellow pollen. First described for *Deherainia* by Chandler (1911) and Pohl (1931), these oxalate accumulations occur throughout the family.

The gynoecium is unilocular and has a globular central placenta. Although certainly syncarpous, the number of carpels forming the gynoecium is unknown. In all genera the ovules are densely set on the surface of the placenta and their number varies from 40 to 250 (Fig. 137L); only in *Clavija* are there usually less than 30 ovules, and these are often rather sparsely distributed on the placenta (Fig. 137F).

EMBRYOLOGY. The anthers are reported to have fibrous endothecium and a secretory tapetum with uninucleate cells (Johri et al. 1992). Like in most other members of the Primulales, the ovules in Theophrastaceae are anatropous, tenuinucellate and bitegmic (Mauritzon 1936). The micropyle is formed by both integuments, which are clearly distinguishable only during the early stages of ovule formation. The embryo sac is probably of the Polygonum type with ephemeral antipodes; endosperm formation is nuclear. Some, but not all Theophrastaceae (and Myrsinaceae) have an endosperm with thickened, abundantly pitted cells walls (see Anderberg and Ståhl 1995, fig. 7, and Takhtajan 1992, fig. p. 56). In *Deherainia*, Mauritzon (1936) classified this tissue as perisperm developed from the inner integument, but this interpretation still needs to be verified.

POLLEN MORPHOLOGY. Pollen grains in the family are spheroidal and vary in size from 17 to 42 μm in diam. (Erdtman 1952; Carrasquel 1970). They are 3-colporate, except in several species of *Clavija*, which have 4-colporate grains. The exine is foveolate in *Jacquinia*, fossulate in *Deherainia*, rugulate in *Neomezia*, *Theophrasta*, and *Votschia*, and reticulate in *Clavija*. Extratectal structures in the form of spinules are restricted to *Clavija*, although they are lacking there in at least one species.

KARYOLOGY. Various studies indicate a basic chromosome number of x = 18 for Theophrastaceae (Faure 1968), although $2n = 40$ and $2n = 28$ have been reported in one species of *Clavija* and *Jacquinia* respectively (Gadella et al. 1969; Hanson et al. 2001).

POLLINATION AND REPRODUCTIVE SYSTEMS. All genera of Theophrastaceae seem to be insect-pollinated. *Jacquinia* and *Clavija*, the flowers of which have a perfumed or fruity scent respectively, appear to be bee-pollinated, whereas *Neomezia*, *Theophrasta*, and *Deherainia*, which have flowers with a foetid scent, most likely are pollinated by Diptera. However, a species of gall midges was observed visiting flowers of *Clavija* in Ecuador (Gagné et al. 1997). The glandular trichomes on floral parts may produce small quantities of nectar, as has been observed in *J. macrocarpa* (Vogel 1986), but nectar as free liquid inside the corolla is lacking in all genera. Even though pollen is produced in small quantities, pollen-gathering bees may well be important as pollinators in *Jacquinia*, *Clavija*, and *Votschia*. In *Deherainia*, *Neomezia*, and *Theophrasta* pollination is clearly based on deceit. In *Theophrasta* the blossoms mimic fruit-bodies of mushrooms through colour, scent, and in being covered by plant litter, and are probably pollinated by Diptera that visit the flowers for oviposition (Ståhl 1987). The role of the calcium oxalate crystals in pollination, if any, is unknown.

All genera except *Clavija* have hermaphrodite, protandrous flowers. In *Clavija*, most species have flowers that are morphologically intermediate between the hermaphrodite and dioecious condition and usually appear as gynodioecious, although the female function is probably strongly suppressed (Ståhl 1991). Studies of herbarium material suggest that a few species are functionally androdioecious. *Deherainia* grown under greenhouse condition has been reported to be self-incompatible (Chandler 1911). A poor fruit set has also been observed in cultivated plants of *Theophrasta*.

FRUIT AND SEED. Theophrastaceae form indehiscent berries with dry, coriaceous or woody pericarps. The fruits are yellow or orange, globose or ovoid, glabrous or rarely brownish-puberulous, and vary in size from 0.5 to 5 cm diam. Each fruit has usually less than 5 and rarely more than 10 seeds, which are entirely or partly imbedded in an orange and sweet-tasting placental pulp. The seeds are 0.3–10 mm wide, globose-compressed, irregularly obtuse-angled, or flattened and oblong. They are light to dark brown and have a thin, smooth or faintly reticulate testa. The endosperm is abundant and hard and has smooth or irregularly thickened cell walls. The embryo is straight and has a well-developed hypocotyl and foliaceous or small and poorly differentiated cotyledons.

DISPERSAL. No reports on seed dispersal in the family have been published; however, the juicy and coloured fruits strongly suggest zoochory, probably involving birds as the most important vectors. Collections of *Clavija* and *Theophrasta* with fruits that evidently were pecked open by birds suggest that the seeds often are deposited near the mother-plant as the soft pulp is consumed. Other dispersal vectors that could be involved are forest rodents and monkeys. The brownish colour of the fruits of *Neomezia* and their position near to the ground suggest rodents or reptiles as dispersal agents of this genus.

PHYTOCHEMISTRY. Green parts are rich in saponins and the seeds have an oily endosperm lacking starch (Hegnauer 1973). The orange pigment in flowers of *Jacquinia nervosa* has been identified as *trans*-crocetindialdehyd and *trans*-crocetinhalbaldehyd (Eugster et al. 1969), and it seems likely that the same carotenoids are present in the other orange-flowered species of the family. The chemical composition of the floral scent differs to a large extent between the genera (Knudsen and Ståhl 1994). In *Clavija* the scent is characterised mainly by sesquiterpene hydrocarbons, in *Jacquinia* by benzenoids, phenyl propanoids, and (in orange-flowered species) trimethylcyclohexanes derivatives of carotenoids, and in *Deherainia* and *Theophrasta* by fatty-acid derivatives.

SUBDIVISION OF AND RELATIONSHIPS WITHIN THE FAMILY. Theophrastaceae are divided into two groups, one including the pachycaul genera *Clavija*, *Neomezia*, and *Theophrasta*, and one comprising the richly branched genera *Deherainia*, *Jacquinia*, and *Votschia*. This subdivision is in agreement with the classifications proposed by de Candolle (1844) and Votsch (1904), who referred the two groups to the tribes Clavijeae and Jacquinieae respectively. It is also consistent with a preliminary cladistic analysis (Ståhl 1991).

AFFINITIES. Theophrastaceae are firmly placed in Primulales because of their 5(4-)-merous flowers, sympetalous corolla, single whorl of antepetalous stamens, and unilocular ovary with a free, central placenta. Being woody and tropical, the family has usually been related to the Myrsinaceae. Phylogenetic studies based on morphological and *rbc*L sequence data (Anderberg and Ståhl 1995; Anderberg et al. 1998) placed Theophrastaceae as the sister group to the rest of primuloid families. However, in more recent work (Källersjö et al. 2000), *Maesa* (Maesaceae) replaces Theophrastaceae as the basal family in the Primulales, and Theophrastaceae with *Samolus* (Samolaceae) as the sister group forms the second branch of the primuloids, which is basal to the rest of the primuloids (Myrsinaceae and Primulaceae s.str.). The monophyly of the Theophrastaceae is well supported by features such as pseudowhorled leaves, extraxylary foliar sclerenchyma, and anthers with calcium oxalate crystals (Anderberg and Ståhl 1995).

DISTRIBUTION AND HABITATS. Theophrastaceae are strictly neotropical, being distributed from northern Mexico and southern Florida to southern Brazil and northern Paraguay. At the generic level the family is most diverse in and around the Caribbean. *Neomezia* and *Theophrasta* are endemic to the islands of Cuba and Hispaniola respectively; *Votschia* is restricted to a narrow strip of land along the Caribbean side of eastern Panama; *Deherainia* is distributed from eastern Mexico to Honduras; and *Jacquinia* has 12 species in the Mesoamerican region, about 18 in the Antilles, and a few in northern South America. *Clavija* is the only genus in the family centred on the South American mainland, where it is particularly diverse in north-western South America and the western Amazon.

The family as a whole has mainly a lowland distribution but species of *Jacquinia* and *Clavija* may reach an elevation of 2000–2500 m. Species occurring above 1500 m altitude usually have a restricted distribution.

Most Theophrastaceae grow in seasonally dry areas from semideciduous and deciduous forests to thorn scrubs. The last-mentioned vegetation type is a common habitat for *Jacquinia*, several species of which inhabit dry coastal vegetation and some may even be found in the upper edge of mangrove swamps. Although several species of *Clavija* grow in deciduous or semideciduous forests, many inhabit wet, evergreen forests. Some Amazonian species of *Clavija* are restricted to habitats with temporarily inundated soils.

ECONOMIC IMPORTANCE. Theophrastaceae are of little economic importance. Crushed roots and immature fruits of some species of *Jacquinia* have been used locally as a fish poison ("barbasco") and, in coastal Ecuador, crushed, immature fruits of *J. sprucei* are used in shrimp farms to clear baskets from unwanted fish. In the northern Caribbean the soft cortex of some *Jacquinia* species has been used as a substitute for soap. Flowers of *Jacquinia* were once used to decorate the Maya temples, and flowers of *Clavija* and *Jacquinia* are strung into necklaces and garlands by indigenous people in the western Amazon and Mesoamerica respectively (Lundell 1960; Standley and Williams 1966). Fruits of *Clavija* are eaten as a "forest-snack" throughout Central and South America.

KEY TO THE GENERA

1. Usually richly branched shrubs and trees; leaves small (<15 cm long) or very small, margins entire 2
- Sparsely branched or unbranched treelets, shrubs or dwarf-shrubs; leaves of medium size (rarely <15 cm long) to very large (to 130 cm long), margins serrate, serrulate, spinulose, or entire 4
2. Corolla campanulate or urceolate, orange, white or yellowish, 0.5–1.5 cm diam. at anthesis; leaves generally <5 cm long, mostly spine-tipped 1. *Jacquinia*
- Corolla broadly campanulate, green or yellow, 2.5–4 cm diam. at anthesis; leaves usually >5 cm long, not spine-tipped 3
3. Corolla green; fruit ovoid, tapering towards apex 2. *Deherainia*
- Corolla yellow; fruit subspherical 3. *Votschia*

4. Leaf margins entire to serrate, rarely spinose-serrate; flowers uni- or bisexual, sweet-scented; corolla crateriform 6. *Clavija*
- Leaf margins usually spinose-serrate; flowers bisexual, with foetid odour; corolla ± campanulate 5
5. Erect, unbranched treelets or shrubs; inflorescences many-flowered, appearing above the foliage; corolla appendages transversely oblong 4. *Theophrasta*
- More or less decumbent and sparsely branched dwarf-shrubs; flowers solitary or inflorescences few-flowered, appearing among and beneath the leaves; corolla appendages triangular 5. *Neomezia*

1. *Jacquinia* L. Fig. 137M–O

Jacquinia L., Fl. Jamaica: 27 (1759); Mez in Pflanzenreich IV. 236a: 28–44 (1903); Ståhl, Nord. J. Bot. 9: 15–30 (1989), 15: 493–511 (1995).

Richly branched shrubs and small trees. Leaves small, sometimes needle-like, mostly spine-tipped, margins entire. Racemes terminal, few- to many-flowered. Flowers 5-merous, hermaphrodite; corolla campanulate or urceolate, orange, white, or yellowish; corolla appendages flattened and petaloid; stamens filaments united at base; anthers obtuse-triangular. Fruit subglobose or oblong, pericarp thick and woody or thin and brittle. About 32 spp., West Indies, Mexico to northern South America.

2. *Deherainia* Decaisne

Deherainia Decaisne, Ann. Sci. Nat. Bot. VI, 3: 138 (1876); Ståhl, Nord. J. Bot. 9: 20–21 (1989).

Shrubs or small trees. Leaves medium-sized, margins entire. Racemes terminal, few-flowered, or flowers solitary. Flowers 5-merous, hermaphrodite; corolla broadly campanulate, green; corolla appendages lanceolate or narrowly obovate; stamens filaments united at base; anthers obtuse-triangular. Fruit ovoid, tapering towards apex, pericarp thin and brittle. Two spp., eastern Mexico to Honduras.

3. *Votschia* Ståhl

Votschia Ståhl, Brittonia 45: 204–207 (1993).

Branched shrubs. Leaves medium-sized, margins entire. Racemes terminal, few-flowered. Flowers 5-merous, hermaphrodite; corolla broadly campanulate, pale yellow; corolla appendages flattened, petaloid; stamens filaments united at base; anthers obtuse-oblong. Fruits subglobose, pericarp somewhat thick and woody. One sp., *V. nemophila* (Pittier) Ståhl, north-eastern Panama.

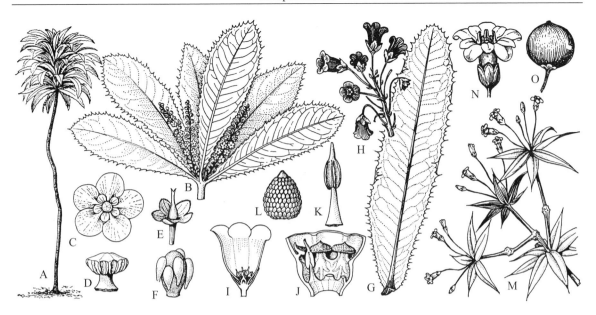

4. *Theophrasta* L. Fig. 137G–L

Theophrasta L., Sp. Pl. 1: 149 (1753); Ståhl, Nord. J. Bot. 7: 529–538 (1987).

Unbranched shrubs or treelets, the stem spiny. Leaves large, margins spinose-serrate or serrulate. Racemes subterminal, many-flowered. Flowers 5-merous, hermaphrodite; corolla campanulate, orange-brown to pale buff, turning brownish black with age and upon drying; corolla appendages gibbous, transversely oblong, inserted within the tube; stamens filaments united at base; anthers narrowly oblong, produced at apex. Fruit subglobose, pericarp rather thin and brittle. Two spp., Hispaniola.

5. *Neomezia* Votsch

Neomezia Votsch, Bot. Jahrb. Syst. 33: 541 (1904).

Unbranched or sparsely branched dwarf-shrubs. Leaves medium-sized to large, margins spinose-serrate. Racemes lateral, few-flowered, or flowers solitary. Flowers 5-merous, hermaphrodite; corolla brownish orange, broadly campanulate; corolla appendages triangular; stamen filaments united at base; anthers produced at apex. Fruits subglobose, puberulous, pericarp thin and brittle. One sp., *N. cubensis* (Radlk.) Votsch, Cuba.

6. *Clavija* Ruiz & Pav. Fig. 137A–F

Clavija Ruiz & Pav., Fl. Peruv. Prodr.: 142, tab. 30 (1794); Ståhl, Opera Bot. 107: 1–77 (1991), rev.

Fig. 137. Theophrastaceae. **A** *Clavija ornata*, habit. **B–F** *C. pungens*. **B** Flowering shoot. **C** Flower, seen from above. **D** Staminal tube. **E** Calyx and gynoecium of bisexual or morphologically male flower. **F** Placenta and ovules. **G–L** *Theophrasta jussieui*. **G** Leaf. **H** Inflorescence. **I** Flower, vertical section. **J** Part of floral tube with one stamen and staminodes. **K** Stamen, seen from outside. **L** Placenta with ovules. **M–O** *Jacquinia aculeata*. **M** Flowering shoot. **N** Flower. **O** Fruit. (Takhtajan 1981)

Unbranched or sparsely branched shrubs or small trees. Leaves large to very large, margins entire, serrulate, serrate, or rarely spinose-serrate. Racemes lateral, many- or few-flowered. Flowers 5- or 4-merous, often unisexual; corolla crateriform, orange; corolla appendages gibbous-oblong; stamens often fused into a tube; anthers obtuse-triangular, rarely with short apical production. Fruit subglobose, usually glabrous, pericarp mostly thin and brittle. About 55 spp., Central and South America, Haiti.

Selected Bibliography

Anderberg, A.A., Ståhl, B. 1995. Phylogenetic interrelationships in the order Primulales, with special emphasis on the family circumscriptions. Can. J. Bot. 73: 1699–1730.

Anderberg, A.A., Ståhl, B., Källersjö, M. 1998. Phylogenetic relationships in the Primulales inferred from *rbc*L sequence data. Plant Syst. Evol. 211: 93–102.

Candolle, A.L.P.P. de 1844. Ordo. CXXIII. Theophrastaceae. In: Candolle, A.P. de (ed.) Prodromus systematis naturalis regni vegetabilis 8: 144–153.

Carrasquel, N. 1970. Estudios anátomo-morfológicos de las especies del género *Jacquinia* en Venezuela para su interpretación taxonómica. Acta Bot. Venezuel. 4: 303–357.

Chandler, B. 1911. *Deherainia smaragdina* Dcne. Notes Roy. Bot. Gard. Edinburgh 5: 49–56.

Erdtman, G. 1952. See general references.

Eugster, C.H., Hürlimann, H., Leuenberger, H.J. 1969. 90. Crocetindialdehyd und Crocetinhalbaldehyd als Blütenfarbstoffe von *Jacquinia angustifolia*. Helvetica Chem. Acta 52: 806–807.

Faure, P. 1968. Contribution à l'étude caryo-taxonomique des Myrsinacées et des Théophrastacées. Mém. Mus. Natl. Hist. Nat. Sér. B. 18: 37–57.

Gadella, Th.W.J., Kliphuis, E., Lindeman, J.C., Mennega, E.A. 1969. Chromosome numbers and seedling morphology of some angiospermae collected in Brazil. Acta Bot. Neerl. 18: 74–83.

Gagné, R.J., Ott, C., Renner, S.S. 1997. A new species of gall midge (Diptera: Cecidomyiidae) from Ecuador associated with flowers of *Clavija* (Theophrastaceae). Proc. Entomol. Soc. Wash. 99: 110–114.

Hanson, L., McMahon, K.A., Johnson, M.A.T., Bennett, M.D. 2001. First nuclear DNA e-values for 25 angiosperm families. Ann. Bot. 87: 251–258.

Hegnauer, R. 1973. See general references.

Johri, B.M. et al. 1992. See general references.

Källersjö, M. et al. 2000. See general references.

Knudsen, J.T., Ståhl, B. 1994. Floral odours in the Theophrastaceae. Biochem. Ecol. Syst. 22: 259–268.

Lepper, L. 1989. Infloreszenzstruktur, Blütenbau und Blütenökologie von *Deherainia smaragdina* Decne. (Theophrastaceae). Wiss. Zeitschr. Friedrich-Schiller-Univ. Jena, Naturwiss. R. 38: 337–345.

Lundell, C.L. 1960. Plantae mayanae – I. Notes on collections from the lowland of Guatemala. Wrightia 2: 49–63.

Mauritzon, J. 1936. Embryologische Angaben über Theophrastaceen. Ark. Bot. 28B, No. 1: 1–4.

Metcalfe, C.R., Chalk, L. 1950. See general references.

Mez, C. 1903. Theophrastaceae. In: Engler, Pflanzenreich IV. 236a. Leipzig: W. Engelmann.

Pohl, F. 1931. Über sich öffnende Kristallräume in den Antheren von *Deherainia smaragdina*. Jahrb. wiss. Bot. 75: 481–493.

Ståhl, B. 1987. The genus *Theophrasta* (Theophrastaceae). Foliar structures, floral biology and taxonomy. Nord. J. Bot. 7: 529–538.

Ståhl, B. 1989. A synopsis of Central American Theophrastaceae. Nord. J. Bot. 9: 15–30.

Ståhl, B. 1991. A revision of *Clavija* (Theophrastaceae). Opera Bot. 107: 1–77.

Ståhl, B. 1992. On the identity of *Jacquinia armillaris* (Theophrastaceae) and related species. Brittonia 44: 54–60.

Ståhl, B. 1995. A synopsis of *Jacquinia* (Theophrastaceae) in the Antilles and South America. Nord. J. Bot. 15: 493–511.

Standley, P.C., Williams, L.O. 1966. Flora of Guatemala. Fieldiana Bot. 24 (8): 127–133.

Takhtajan, A. (ed.) 1981. See general references.

Takhtajan, A. (ed.) 1992. Anatomia seminum comparativa, Tomus 4. St. Petersburg: Nauka.

Vogel, S. 1986. Ölblumen und ölsammelnde Bienen. Zweite Folge. *Lysimachia* und *Macropis*. Trop. Subtrop. Pflanzenwelt 54: 1–168.

Votsch, W. 1904. Neue systematische-anatomische Untersuchungen von Blatt und Achse der Theophrastaceen. Bot. Jahrb. Syst. 33: 502–546.

Welle, B.J.H. ter 1976. Silica grains in woody plants of the Neotropics, especially Suriname. In: Baas, P. et al. (eds.) Wood structure in biological and technological research. Leiden Bot. Res. 3: 107–142. Leiden: Leiden University Press.

Index to Scientific Names

References to main entries in **bold-faced** print, to illustrations in *italics*.

Printing and Binding: Stürtz AG, Würzburg